REAL ANALYSIS
Theory of Measure and Integration

2nd Edition

2nd Edition

REAL ANALYSIS

Theory of Measure and Integration

J Yeh

University of California, Irvine

World Scientific

NEW JERSEY • LONDON • SINGAPORE • BEIJING • SHANGHAI • HONG KONG • TAIPEI • CHENNAI

Published by

World Scientific Publishing Co. Pte. Ltd.
5 Toh Tuck Link, Singapore 596224
USA office: 27 Warren Street, Suite 401-402, Hackensack, NJ 07601
UK office: 57 Shelton Street, Covent Garden, London WC2H 9HE

Library of Congress Cataloging-in-Publication Data
Yeh, J.
Real analysis : theory of measure and integration / J. Yeh. -- 2nd ed.
p. cm.
Rev. ed. of: Lectures on real analysis. 1st ed. c2000.
ISBN-13 978-981-256-653-9 -- ISBN-13 978-981-256-654-6 (pbk)
ISBN-10 981-256-653-8 -- ISBN-10 981-256-654-6 (pbk)
1. Measure theory. 2. Lebesgue integral. 3. Integrals, Generalized. 4. Mathematical analysis. 5. Lp spaces. I. Yeh, J. Lectures on real analysis. II. Title.

QA312.Y44 2006
515'.42--dc22

2006045274

British Library Cataloguing-in-Publication Data
A catalogue record for this book is available from the British Library.

First published 2006
Reprinted 2008 (paperback)

Printed in Singapore

To

Betty

Contents

Preface to the First Edition

This monograph evolved from a set of lecture notes for a course entitled Real Analysis that I taught at the University of California, Irvine. The subject of this course is the theory of measure and integration. Its prerequisite is advanced calculus. All of the necessary background material can be found, for example, in R. C. Buck's *Advanced Calculus*. The course is primarily for beginning graduate students in mathematics but the audience usually includes students from other disciplines too. The first five chapters of this book contain enough material for a one-year course. The remaining two chapters take an academic quarter to cover.

Measure is a fundamental concept in mathematics. Measures are introduced to estimate sizes of sets. Then measures are used to define integrals. Here is an outline of the book.

Chapter 1 introduces the concepts of measure and measurable function. §1 defines measure as a nonnegative countably additive set function on a σ-algebra of subsets of an arbitrary set. Measurable mapping from a measure space into another is then defined. §2 presents construction of a measure space by means of an outer measure. To have a concrete example of a measure space early on, the Lebesgue measure space on the real line $\mathbb{R}$ is introduced in §3. Subsequent developments in the rest of Chapter 1 and Chapter 2 are in the setting of a general measure space. (This is from the consideration that in the definition of a measure and an integral with respect to a measure the algebraic and topological structure of the underlying space is irrelevant and indeed unnecessary. Topology of the space on which a measure is defined becomes relevant when one considers the regularity of the measure, that is, approximation of measurable sets by Borel sets.) §4 treats measurable functions, in particular algebraic operations on measurable functions and pointwise limits of sequences of measurable functions. §5 shows that every measure space can be completed. §6 compares two modes of convergence of a sequence of measurable functions: convergence almost everywhere and convergence in measure. The Borel-Cantelli Lemma and its applications are presented. A unifying theorem (Theorem 6.5) is introduced from which many other convergence theorems relating the two modes of convergence are derived subsequently. These include Egoroff's theorem on almost uniform convergence, Lebesgue's and Riesz's theorems.

Chapter 2 treats integration of functions on an arbitrary measure space. In §7 the Lebesgue integral, that is, an integral with respect to a measure, is defined for a bounded real-valued measurable function on a set of finite measure. The Bounded Convergence Theorem on the commutation of integration and limiting process for a uniformly bounded sequence of measurable functions which converges almost everywhere on a set of finite measure is proved here. The proof is based on Egoroff's theorem. On the Lebesgue measure space on $\mathbb{R}$, comparison of the Lebesgue integral and the Riemann integral is made. §8 contains the fundamental idea of integration with respect to a measure. It is shown here that for every nonnegative extended real-valued measurable function on a measurable set the integral with respect to the measure always exists even though it may not be finite. The Monotone Convergence Theorem for an increasing sequence of nonnegative measurable functions, the most fundamental of all convergence theorems regarding commutation of integration and convergence of the sequence of integrands, is proved here. Fatou's Lemma concerning the limit inferior of a sequence of nonnegative measurable functions is derived from the Monotone Convergence Theorem. In §9 the integral of an extended real-valued measurable function on a measurable set is then defined as the difference of the integrals of the positive and negative parts of the function provided the difference exists in the extended real number system. The generalized monotone convergence theorem for a monotone sequence of extended real-valued measurable functions, generalized Fatou's lemma for the limit inferior and the limit superior of a sequence of extended real-valued measurable functions, and Lebesgue's Dominated Convergence Theorem are proved here. Fatou's Lemma and Lebesgue's Dominated Convergence Theorem under convergence in measure are included. In §10 a signed measure is defined as an extended real-valued countably additive set function on a σ-algebra and then shown to be the difference of two positive measures. In §11 the Radon-Nikodym derivative of a signed measure with respect to a positive measure is defined as a function which we integrate with respect to the latter to obtain the former. The existence of the Radon-Nikodym derivative is then proved under the assumption that the former is absolutely continuous with respect to the latter and that both are σ-finite. (The fact that the Radon-Nikodym derivative is a derivative not only in name but in fact it is the derivative of a measure with respect to another is shown for Borel measures on the Euclidean space in §25.)

Chapter 3 treats the interplay between integration and differentiation on the Lebesgue measure space on $\mathbb{R}$. §12 presents Lebesgue's theorem that every real-valued increasing function on $\mathbb{R}$ is differentiable almost everywhere on $\mathbb{R}$. The proof is based on a Vitali covering theorem. This is followed by Lebesgue's theorem on the integral of the derivative of a real-valued increasing function on a finite closed interval in $\mathbb{R}$. Functions of bounded variation are included here. §13 defines absolute continuity of a real-valued function on a finite closed interval in $\mathbb{R}$ and then shows that a function is absolutely continuous if and only if it is an indefinite integral of a Lebesgue integrable function. This is followed by Lebesgue's decomposition of a real-valued increasing function as the sum of an absolutely continuous function and a singular function. Such methods of calculating a Riemann integral in calculus as the Fundamental Theorem of Calculus, integration by parts, and change of variable of integration find their counterparts in the Lebesgue integral here. §14 treats convex functions and in particular their differentiability and absolute continuity property. Jensen's inequality is included here.

Chapter 4 treats the L^p spaces of measurable functions f with integrable $|f|^p$ for $p \in (0, \infty)$ and the space L^∞ of essentially bounded measurable functions on a general measure space. Here Hölder's inequality and Minkowski's inequality are proved for $p \in (0, \infty]$. §15 introduces the Banach space and its dual. §16 treats L^p spaces for $p \in [1, \infty]$ as well as for $p \in (0, 1)$. §17 treats relation among the L^p spaces for different values of p. The ℓ^p spaces of sequences of numbers $(\alpha_n : n \in \mathbb{N})$ with $\sum_{n \in \mathbb{N}} |\alpha_n|^p < \infty$ is treated as a particular case of L^p spaces in which the underlying measure space is the counting measure space on the set $\mathbb{N}$ of natural numbers. The Riesz representation theorem on the L^p spaces is proved in §18. §19 treats integration on a locally compact Hausdorff space. Urysohn's Lemma on the existence of a continuous function with compact support and partition of unity, Borel and Radon measures, the Riesz representation theorem on the space of continuous functions with compact support as well as Lusin's theorem on approximation of a measurable function by continuous functions are included here. (The placement of §19 in Chapter 4 is somewhat arbitrary.)

Chapter 5 treats extension of additive set functions to measures. It starts with extension of an additive set function on an algebra to a measure in §20 and completes the theory with extension of an additive set function on a semialgebra to a measure in §21. (Semialgebra of sets is an abstraction of the aggregate of left-open and right-closed boxes in the Euclidean space $\mathbb{R}^n$. Its importance lies in the fact that the Cartesian product of finitely many algebras and in particular σ-algebras is in general not an algebra, but only a semialgebra.) As an example of extending an additive set function on a semialgebra to a measure, the Lebesgue-Stieltjes measure determined by a real-valued increasing function on $\mathbb{R}$ is treated in §22. Theorems establishing the equivalence of the absolute continuity and singularity of a Lebesgue-Stieltjes measure with respect to the Lebesgue measure with the absolute continuity and singularity of the increasing function that determines the Lebesgue-Stieltjes measure are proved. As a second example of extending an additive set function on a semialgebra to a measure, the product measure on the product of finitely many measure spaces is included in §23. Tonelli's theorem and Fubini's theorem on the reduction of a multiple integral to iterated integrals are found here.

Chapter 6 specializes in integration in the Lebesgue measure space on $\mathbb{R}^n$. In §24 the Lebesgue measure on $\mathbb{R}^n$ is constructed as an extension of the notion of volumes of boxes in $\mathbb{R}^n$ to Lebesgue measurable subsets of $\mathbb{R}^n$. Then it is shown that the Lebesgue measure space on $\mathbb{R}^n$ is the completion of the n-fold product of the Lebesgue measure space on $\mathbb{R}$. Regularity of the Lebesgue measure and in particular approximation of Lebesgue measurable sets by open sets leads to approximation of the integral of a measurable function by that of a continuous function. Translation invariance of the Lebesgue measure and integral and linear transformation of the Lebesgue measure and integral are treated. §25 begins with the study of the average function of a locally integrable function. Hardy-Littlewood maximal theorem and Lebesgue differentiation theorem are presented. These are followed by differentiation of a set function with respect to the Lebesgue measure, in particular differentiation of a signed Borel measure with respect to the Lebesgue measure, and density of a Lebesgue measurable set with respect to the Lebesgue measure. §26 treats change of variable of integration by differentiable transformations.

Chapter 7 is an introduction to Hausdorff measures on $\mathbb{R}^n$. §27 defines s-dimensional Hausdorff measures on $\mathbb{R}^n$ for $s \in [0, \infty)$ and the Hausdorff dimension of a subset of $\mathbb{R}^n$.

§28 studies transformations of Hausdorff measures. §29 shows that a Hausdorff measure of integral dimension is a constant multiple of the Lebesgue measure of the same dimension.

Every concept is defined precisely and every theorem is presented with a detailed and complete proof. I endeavored to present proofs that are natural and inevitable. Counter-examples are presented to show that certain conditions in the hypothesis of a theorem can not be simply dropped. References to earlier results within the text are made extensively so that the relation among the theorems as well as the line of development of the theory can be traced easily. On these grounds this book is suitable for self-study for anyone who has a good background in advanced calculus.

In writing this book I am indebted to the works that I consulted. These are listed in the Bibliography. I made no attempt to give the origin of the theory and the theorems. To be consistent, I make no mention of the improvements that I made on some of the theorems. I take this opportunity to thank all the readers who found errors and suggested improvements in the various versions of the lecture notes on which this book is based.

J. Yeh

Corona del Mar, California
January, 2000

Preface to the Second Edition

In this new edition all chapters have been revised and additional material have been incorporated although the framework and organization of the book are unchanged. Specifically the following sections have been added:

§13 [VI] Length of Rectifiable Curves
§15 [VII] Baire Category Theorem
[VIII] Uniform Boundedness Theorem
[IX] Open Mapping Theorem
[X] Hahn-Banach Extension Theorems
§16 weak convergence in L^p spaces in [III] and [IV] of §16
the complete metric spaces L^p for $p \in (0, 1)$ in [V] of §16
§19 [V] Signed Radon Measures
[VI] Dual Space of $C(X)$
§23 [IV.2] Convolution of L^p Functions
[IV.3] Approximate Identity in Convolution Product
[IV.4] Approximate Identity Relative to Pointwise Convergence

Besides these topics there are additional theorems in sections: §1, §4, §5, §8, §10, §11, §13, §15, §16, §17, §19, §20, §21, §23, §24, §25, and §27. Also 64 problems have been added.

To use this book as a textbook, selection of the following sections for instance makes a possible one-year course at the graduate level:

§1 to §13, §15([I] to [VI]), §16 to §21, §23([I] to [III])

It is my pleasure to thank Abel Klein for his helpful comments on the first edition of this book.

J. Yeh

Corona del Mar, California
March, 2006

List of Notations

$\mathbb{N}$	the natural numbers	p.3
$\mathbb{Z}$	the integers	p.3
$\mathbb{Z}_+$	the nonnegative integers	p.3
$\mathbb{R}$	the real numbers	p.1
$\mathbb{C}$	the complex numbers	p.323
$\mathbb{K}$	$\mathbb{R}$ or $\mathbb{C}$	p.323
$\overline{\mathbb{R}}$	the extended real number system $\{-\infty\} \cup \mathbb{R} \cup \{\infty\}$	p.11
$\overline{\mathbb{C}}$	the set of $\zeta = \xi + i\eta$ where $\xi, \eta \in \overline{\mathbb{R}}$	p.376
$\mathbb{R}^n$	the n-dimensional Euclidean space	p.597
$\mathfrak{P}(X)$	the collection of all subsets of a set X	p.4
$\mathfrak{B}_X$	the σ-algebra of Borel sets in a topological space X	D1.16, p.10
$\mathfrak{B}_{\mathbb{R}}$	the σ-algebra of Borel sets in $\mathbb{R}$	D1.16, p.10
$\mathfrak{B}_{\mathbb{R}^n}$	the σ-algebra of Borel sets in $\mathbb{R}^n$	D1.16, p.10
$(\mathfrak{B}_{\mathbb{R}^n})_b$	the collection of bounded Borel sets in $\mathbb{R}^n$	D25.19, p.632
$\mathfrak{M}_L$	the σ-algebra of Lebesgue measurable sets in $\mathbb{R}$	D3.1, p.41
$\mathfrak{M}_L^n$	the σ-algebra of Lebesgue measurable sets in $\mathbb{R}^n$	D24.8, p.600
$\alpha(\mathfrak{E})$	algebra generated by $\mathfrak{E}$	D1.12, p.7
$\sigma(\mathfrak{E})$	σ-algebra generated by $\mathfrak{E}$	D1.12, p.7
$\mathfrak{M}(\mu^*)$	σ-algebra of μ^*-measurable sets	D2.2, p.28
$\mathfrak{A} \otimes \mathfrak{B}$	product σ-algebra $\sigma(\mathfrak{A} \times \mathfrak{B})$	D23.3, p.528
$\mathfrak{I}_o, \mathfrak{I}_{oc}, \mathfrak{I}_{co}, \mathfrak{I}_c$	classes of intervals in $\mathbb{R}$	D3.1, p.41
$\mathfrak{I}_o^n, \mathfrak{I}_{oc}^n, \mathfrak{I}_{co}^n, \mathfrak{I}_c^n$	classes of intervals in $\mathbb{R}^n$	D24.2, p.597
$\mathfrak{D}(f)$	domain of definition of a function f	p.19
$\mathfrak{R}(f)$	range of a function f	p.19
$\{D : f \le \alpha\}$	abbreviation of $\{x \in D : f(x) \le \alpha\}$	p.70
μ_L^*	Lebesgue outer measure on $\mathbb{R}$	D3.1, p.41
μ_L	Lebesgue measure on $\mathbb{R}$	D3.1, p.41
$(\mu_L^n)^*$	Lebesgue outer measure on $\mathbb{R}^n$	D24.5, p.598
μ_L^n	Lebesgue measure on $\mathbb{R}^n$	D24.8, p.600
$(\mathbb{R}, \mathfrak{M}_L, \mu_L)$	Lebesgue measure space on $\mathbb{R}$	D3.1, p.41
$(\mathbb{R}^n, \mathfrak{M}_L^n, \mu_L^n)$	Lebesgue measure space on $\mathbb{R}^n$	D24.8, p.600
$(\mathbb{R}, \mathfrak{M}(\mu_g^*), \mu_g)$	Lebesgue-Stieltjes measure space	D22.5, p.507
$\mathcal{L}^p(X, \mathfrak{A}, \mu)$		D16.8, p.378
$L^p(X, \mathfrak{A}, \mu)$		D16.22, p.385
$\mathcal{L}^\infty(X, \mathfrak{A}, \mu)$		D16.38, p.394
$L^\infty(X, \mathfrak{A}, \mu)$		D16.42, p.395
$\mathcal{L}_{loc}^1(\mathbb{R}^n, \mathfrak{M}_L^n, \mu_L^n)$		D25.1, p.620

$L_r^p(\mathbb{R}, \mathfrak{M}_L, \mu_L)$		D23.29, p.547
$\mathcal{S}(X, \mathfrak{A}, \mu)$		L18.2, p.432
$\mathcal{S}_0(X, \mathfrak{A}, \mu)$		L18.2, p.432
$S_0(X, \mathfrak{A}, \mu)$		T18.3, p.433
$C(X)$		D19.48, p.471
$C^+(X)$		D19.48, p.471
$C_c(X)$		D19.9, p.447
$C_b(X)$		D23.45, p.564
$C_0(X)$		D23.45, p.564
$C(\mathbb{R})$, $C_c(\mathbb{R})$		D23.38, p. 555
$C^n(\mathbb{R})$, $C_c^n(\mathbb{R})$		D23.38, p. 555
$C^\infty(\mathbb{R})$, $C_c^\infty(\mathbb{R})$		D23.38, p. 555
$C_b(\mathbb{R})$, $C_0(\mathbb{R})$		D23.45, p. 564
$:=$	equal by definition	
$E^c, E^\circ, \overline{E}, \partial E$	complement, interior, closure, boundary of E	
$\lvert E\rvert$	diameter of E	D27.1, p.675
$B_r(x)$, $B(x,r)$	open ball with center x and radius r	D15.33, p.338
$\overline{B}_r(x)$, $\overline{B}(x,r)$	closed ball with center x and radius r	D15.33, p.338
$S_r(x)$, $S(x,r)$	spherical (hyper)surface with center x and radius r	D15.33, p.338
f^+, f^-	positive and negative parts of a function f	D4.25, p.83
$\Re f$	real part of a function f	p.376
$\Im f$	imaginary part of a function f	p.376
$f(a+)$, $f(a-)$	$\lim_{x\downarrow a} f(x)$, $\lim_{x\uparrow a} f(x)$	p.251
$\text{supp}\{f\}$	support of f	D19.7, p.447
$dT(p;\cdot)$	differential of a mapping T at a point p	p.649
$J_T(p)$	Jacobian matrix of a mapping T at a point p	p.650
λ^+, λ^-	positive and negative parts of a signed measure λ	D10.22, p.213
$\lVert\lambda\rVert$	total variation of a signed measure λ	D10.22, p.213
$\lambda \ll \mu$	absolute continuity of λ with respect to μ	D11.4 , p.225
$\lambda \perp \mu$	singularity of λ and μ	D10.16, p.211
$\lVert\cdot\rVert_u$	uniform norm	T19.50, p.471
$\lVert\cdot\rVert_p$	L^p norm	D16.8, p.378
$\lVert\cdot\rVert_\infty$	essential supremum	D16.36, p.393
$\lVert L\rVert_{V,W}$	norm of a bounded linear mapping L of V into W	D15.29, p.337
$\lVert f\rVert_*$	nrom of a bounded linear functional f	D15.39, p344
$\mathcal{H}_\delta^s$		D27.3, p.675
$\mathcal{H}^s$	Hausdorff measure	D27.7, p.677
$\mathcal{H}^\bullet(E)$	$\mathcal{H}^s(E)$ as a function of $s \in [0,\infty)$	p.688
$\mathcal{F}_\delta^s, \mathcal{F}^s, \mathcal{G}_\delta^s, \mathcal{G}^s$		T27.13, p.680
$\mathcal{S}_\delta^s, \mathcal{S}^s, \mathcal{B}_\delta^s, \mathcal{B}^s$		P27.14, p.681
$\mathcal{C}_\delta^s, \mathcal{C}^s$		T27.24, p.684
$\mathcal{K}_\delta^s, \mathcal{K}^s$		C27.25, p.685
$\dim_H$	Hausdorff dimension	D27.34, p.691

Euler Fraktur and Script

Here is a list of the capital Roman letters, each followed by its corresponding Euler Fraktur and Script letters:

$A, \mathfrak{A}, \mathcal{A}$	$B, \mathfrak{B}, \mathcal{B}$	$C, \mathfrak{C}, \mathcal{C}$	$D, \mathfrak{D}, \mathcal{D}$	$E, \mathfrak{E}, \mathcal{E}$	$F, \mathfrak{F}, \mathcal{F}$	$G, \mathfrak{G}, \mathcal{G}$	$H, \mathfrak{H}, \mathcal{H}$
$I, \mathfrak{I}, \mathcal{I}$	$J, \mathfrak{J}, \mathcal{J}$	$K, \mathfrak{K}, \mathcal{K}$	$L, \mathfrak{L}, \mathcal{L}$	$M, \mathfrak{M}, \mathcal{M}$	$N, \mathfrak{N}, \mathcal{N}$	$O, \mathfrak{O}, \mathcal{O}$	$P, \mathfrak{P}, \mathcal{P}$
$Q, \mathfrak{Q}, \mathcal{Q}$	$R, \mathfrak{R}, \mathcal{R}$	$S, \mathfrak{S}, \mathcal{S}$	$T, \mathfrak{T}, \mathcal{T}$	$U, \mathfrak{U}, \mathcal{U}$	$V, \mathfrak{V}, \mathcal{V}$	$W, \mathfrak{W}, \mathcal{W}$	$X, \mathfrak{X}, \mathcal{X}$
$Y, \mathfrak{Y}, \mathcal{Y}$	$Z, \mathfrak{Z}, \mathcal{Z}$						

Chapter 1

Measure Spaces

§0 Introduction

Let us consider the problem of measuring an arbitrary subset of the real line $\mathbb{R}$. For a finite open interval $I = (a, b)$ in $\mathbb{R}$, we define the length of I by setting $\ell(I) = b - a$, and for an infinite open interval I in $\mathbb{R}$, we set $\ell(I) = \infty$. Let $\mathfrak{P}(\mathbb{R})$ be the collection of all subsets of $\mathbb{R}$. To extend the notion of length to an arbitrary $E \in \mathfrak{P}(\mathbb{R})$, let $(I_n : n \in \mathbb{N})$ be an arbitrary sequence of open intervals in $\mathbb{R}$ such that $\bigcup_{n\in\mathbb{N}} I_n \supset E$, take the sum of the lengths of the intervals $\sum_{n\in\mathbb{N}} \ell(I_n)$, and define $\mu^*(E)$ as the infimum of all such sums, that is,

$$(1) \qquad \mu^*(E) = \inf\left\{\textstyle\sum_{n\in\mathbb{N}} \ell(I_n) : (I_n : n \in \mathbb{N}) \text{ such that } \bigcup_{n\in\mathbb{N}} I_n \supset E\right\}.$$

The set function μ^* thus defined on $\mathfrak{P}(\mathbb{R})$ is nonnegative extended real-valued, that is, $0 \le \mu^*(E) \le \infty$ for every $E \in \mathfrak{P}(\mathbb{R})$, with $\mu^*(\emptyset) = 0$; monotone in the sense that $\mu^*(E) \le \mu^*(F)$ for any $E, F \in \mathfrak{P}(\mathbb{R})$ such that $E \subset F$; and $\mu^*(I) = \ell(I)$ for every open interval I in $\mathbb{R}$ so that μ^* is an extension of the notion of length to an arbitrary subset of $\mathbb{R}$. The set function μ^* also has the property that

$$(2) \qquad \mu^*(E_1 \cup E_2) \le \mu^*(E_1) + \mu^*(E_2)$$

for any two sets $E_1, E_2 \in \mathfrak{P}(\mathbb{R})$. We call this property the subadditivity of μ^* on $\mathfrak{P}(\mathbb{R})$. We say that a set function ν on a collection $\mathfrak{C}$ of subsets of $\mathbb{R}$ is additive on $\mathfrak{C}$ if we have $\nu(E_1 \cup E_2) = \nu(E_1) + \nu(E_2)$ whenever $E_1, E_2 \in \mathfrak{C}$, $E_1 \cap E_2 = \emptyset$ and $E_1 \cup E_2 \in \mathfrak{C}$. Our set function μ^* is not additive on $\mathfrak{P}(\mathbb{R})$, that is, there exist subsets E_1 and E_2 of $\mathbb{R}$ which are disjoint but not separated enough, as far as μ^* is concerned, to have $\mu^*(E_1 \cup E_2) = \mu^*(E_1) + \mu^*(E_2)$. Examples of such sets are constructed in §3 and §4.

Let us show that it is possible to restrict μ^* to a subcollection of $\mathfrak{P}(\mathbb{R})$ so that μ^* is additive on the subcollection. Let $E \in \mathfrak{P}(\mathbb{R})$ be arbitrarily chosen. Then for every $A \in \mathfrak{P}(\mathbb{R})$, $A \cap E$ and $A \cap E^c$, where E^c is the complement of E, are two disjoint members of $\mathfrak{P}(\mathbb{R})$ whose union is A. We say that the set E satisfies the μ^*-measurability condition and E is a μ^*-measurable set if

$$(3) \qquad \mu^*(A) = \mu^*(A \cap E) + \mu^*(A \cap E^c) \quad \text{for every } A \in \mathfrak{P}(\mathbb{R}).$$

It is clear that if E satisfies condition (3), then so does E^c. Note also that $\emptyset$ and $\mathbb{R}$ are two examples of members of $\mathfrak{P}(\mathbb{R})$ satisfying condition (3). Now let $\mathfrak{M}(\mu^*)$ be the subcollection of $\mathfrak{P}(\mathbb{R})$ consisting of all μ^*-measurable sets in $\mathfrak{P}(\mathbb{R})$. Let us show that $\mathfrak{M}(\mu^*)$ is closed under unions. Let $E_1, E_2 \in \mathfrak{M}(\mu^*)$. Then we have $\mu^*(A) = \mu^*(A \cap E_1) + \mu^*(A \cap E_1^c)$ for an arbitrary $A \in \mathfrak{P}(\mathbb{R})$ by (3). With $A \cap E_1^c$ as a testing set for our $E_2 \in \mathfrak{M}(\mu^*)$ replacing A in condition (3), we have $\mu^*(A \cap E_1^c) = \mu^*(A \cap E_1^c \cap E_2) + \mu^*(A \cap E_1^c \cap E_2^c)$. Thus

$$\mu^*(A) = \mu^*(A \cap E_1) + \mu^*(A \cap E_1^c \cap E_2) + \mu^*(A \cap E_1^c \cap E_2^c).$$

For the union of the first two sets on the right side of the last equality, we have

$$(A \cap E_1) \cup \big(A \cap (E_1^c \cap E_2)\big) = A \cap \big(E_1 \cup (E_1^c \cap E_2)\big) = A \cap \big(E_1 \cup (E_2 \setminus E_1)\big) = A \cap (E_1 \cup E_2).$$

Then $\mu^*(A \cap E_1) + \mu^*\big(A \cap (E_1^c \cap E_2)\big) \geq \mu^*\big(A \cap (E_1 \cup E_2)\big)$ by the subadditivity of μ^* on $\mathfrak{P}(\mathbb{R})$. Thus we have $\mu^*(A) \geq \mu^*\big(A \cap (E_1 \cup E_2)\big) + \mu^*\big(A \cap (E_1 \cup E_2)^c\big)$. On the other hand by the subadditivity of μ^* on $\mathfrak{P}(\mathbb{R})$, the reverse of this inequality holds. Thus $E_1 \cup E_2$ satisfies condition (3) and is therefore a member of $\mathfrak{M}(\mu^*)$. This shows that $\mathfrak{M}(\mu^*)$ is closed under unions. We show in §2 that $\mathfrak{M}(\mu^*)$ is closed under countable unions. A collection of subsets of a nonempty set X is called a σ-algebra of subsets of X if it includes X as a member, is closed under complementations and countable unions. Thus our $\mathfrak{M}(\mu^*)$ is a σ-algebra of subsets of $\mathbb{R}$. Let us show next that μ^* is additive on the σ-algebra $\mathfrak{M}(\mu^*)$ of subsets of $\mathbb{R}$. Thus let $E_1, E_2 \in \mathfrak{M}(\mu^*)$ and assume that $E_1 \cap E_2 = \emptyset$. Now with $E_1 \cup E_2$ as the testing set A in the μ^*-measurability condition (3) which is satisfied by E_1, we have

$$\mu^*(E_1 \cup E_2) = \mu^*\big((E_1 \cup E_2) \cap E_1\big) + \mu^*\big((E_1 \cup E_2) \cap E_1^c\big).$$

But $(E_1 \cup E_2) \cap E_1 = E_1$ and $(E_1 \cup E_2) \cap E_1^c = E_2$. Thus the last equality reduces to

$$\mu^*(E_1 \cup E_2) = \mu^*(E_1) + \mu^*(E_2). \tag{4}$$

This shows that μ^*, though not additive on $\mathfrak{P}(\mathbb{R})$, is additive on the subcollection $\mathfrak{M}(\mu^*)$ of $\mathfrak{P}(\mathbb{R})$. Now μ^* is additive on $\mathfrak{M}(\mu^*)$ so that we may regard it as the extension of the notion of length to sets which are members of $\mathfrak{M}(\mu^*)$. For this extension μ^* to be interesting, the collection $\mathfrak{M}(\mu^*)$ must be large enough to include subsets of $\mathbb{R}$ that occur regularly in analysis. In §3, we show that $\mathfrak{M}(\mu^*)$ includes all open sets in $\mathbb{R}$ and all subsets of $\mathbb{R}$ that are the results of a sequence of such set theoretic operations as union, intersection, and complementation, on the open sets.

§1 Measure on a σ-algebra of Sets

[I] σ-algebra of Sets

Notations. We write $\mathbb{N}$ for both the sequence $(1, 2, 3, \ldots)$ and the set $\{1, 2, 3, \ldots\}$. Whether a sequence or a set is meant by $\mathbb{N}$ should be clear from the context. Similarly we write $\mathbb{Z}$ for both $(0, 1, -1, 2, -2, \ldots)$ and $\{0, 1, -1, 2, -2, \ldots\}$ and $\mathbb{Z}_+$ for both $(0, 1, 2, \ldots)$ and $\{0, 1, 2, \ldots\}$.

Definition 1.1. *Let X be an arbitrary set. A collection $\mathfrak{A}$ of subsets of X is called an algebra (or a field) of subsets of X if it satisfies the following conditions:*

1° $X \in \mathfrak{A}$,

2° $A \in \mathfrak{A} \Rightarrow A^c \in \mathfrak{A}$,

3° $A, B \in \mathfrak{A} \Rightarrow A \cup B \in \mathfrak{A}$.

Lemma 1.2. *If $\mathfrak{A}$ is an algebra of subsets of a set X, then*

(1) $\emptyset \in \mathfrak{A}$,

(2) $A_1, \ldots, A_n \in \mathfrak{A} \Rightarrow \bigcup_{k=1}^n A_k \in \mathfrak{A}$,

(3) $A, B \in \mathfrak{A} \Rightarrow A \cap B \in \mathfrak{A}$,

(4) $A_1, \ldots, A_n \in \mathfrak{A} \Rightarrow \bigcap_{k=1}^n A_k \in \mathfrak{A}$,

(5) $A, B \in \mathfrak{A} \Rightarrow A \setminus B \in \mathfrak{A}$.

Proof. (1) follows from 1° and 2° of Definition 1.1. (2) is by repeated application of 3°. Since $A \cap B = (A^c \cup B^c)^c$, (3) follows from 2° and 3°. (4) is by repeated application of (3). For (5) note that $A \setminus B = A \cap B^c \in \mathfrak{A}$ by 2° and (3). ∎

Definition 1.3. *An algebra $\mathfrak{A}$ of subsets of a set X is called a σ-algebra (or a σ-field) if it satisfies the additional condition:*

4° $(A_n : n \in \mathbb{N}) \subset \mathfrak{A} \Rightarrow \bigcup_{n \in \mathbb{N}} A_n \in \mathfrak{A}$.

Note that applying condition 4° to the sequence $(A, B, \emptyset, \emptyset, \ldots)$, we obtain condition 3° in Definition 1.1. Thus 3° is implied by 4°. Observe also that if an algebra $\mathfrak{A}$ is a finite collection, then it is a σ-algebra. This follows from the fact that when $\mathfrak{A}$ is a finite collection then a countable union of members of $\mathfrak{A}$ is actually a finite union of members of $\mathfrak{A}$ and this finite union is a member of $\mathfrak{A}$ by (2) of Lemma 1.2.

Lemma 1.4. *If $\mathfrak{A}$ is a σ-algebra of subsets of a set X, then*

(6) $(A_n : n \in \mathbb{N}) \subset \mathfrak{A} \Rightarrow \bigcap_{n \in \mathbb{N}} A_n \in \mathfrak{A}$.

Proof. Note that $\bigcap_{n \in \mathbb{N}} A_n = \left(\bigcup_{n \in \mathbb{N}} A_n^c\right)^c$. By 2°, $A_n^c \in \mathfrak{A}$ and by 4°, $\bigcup_{n \in \mathbb{N}} A_n^c \in \mathfrak{A}$. Thus by 2°, we have $\left(\bigcup_{n \in \mathbb{N}} A_n^c\right)^c \in \mathfrak{A}$. ∎

Notations. For an arbitrary set X, let $\mathfrak{P}(X)$ be the collection of all subsets of X. Thus $A \in \mathfrak{P}(X)$ is equivalent to $A \subset X$.

Example 1. For an arbitrary set X, $\mathfrak{P}(X)$ satisfies conditions 1° - 3° of Definition 1.1 and condition 4° of Definition 1.3 and therefore a σ-algebra of subsets of X. It is the greatest σ-algebra of subsets of X in the sense that if $\mathfrak{A}$ is a σ-algebra of subsets of X and if $\mathfrak{P}(X) \subset \mathfrak{A}$ then $\mathfrak{A} = \mathfrak{P}(X)$.

Example 2. For an arbitrary set X, $\{\emptyset, X\}$ is a σ-algebra of subsets of X. It is the smallest σ-algebra of subsets of X in the sense that if $\mathfrak{A}$ is a σ-algebra of subsets of X and if $\mathfrak{A} \subset \{\emptyset, X\}$ then $\mathfrak{A} = \{\emptyset, X\}$.

Example 3. In $\mathbb{R}^2$, let $\mathfrak{R}$ be the collection of all rectangles of the type $(a_1, b_1] \times (a_2, b_2]$ where $-\infty \leq a_i < b_i \leq \infty$ for $i = 1, 2$ with the understanding that $(a_i, \infty] = (a_i, \infty)$. Let $\mathfrak{A}$ be the collection of all finite unions of members of $\mathfrak{R}$. We have $\mathfrak{R} \subset \mathfrak{A}$ since every $A \in \mathfrak{R}$ is the union of finitely many, actually one, members of $\mathfrak{R}$ so that $A \in \mathfrak{A}$. We regard $\emptyset$ as the union of 0 members of $\mathfrak{R}$ so that $\emptyset \in \mathfrak{A}$. It is easily verified that $\mathfrak{A}$ is an algebra of subsets of $\mathbb{R}^2$. However $\mathfrak{A}$ is not a σ-algebra. Consider for instance, $A_n = \left(n - \frac{1}{2}, n\right] \times (0, 1] \in \mathfrak{R} \subset \mathfrak{A}$ for $n \in \mathbb{N}$. Then $\bigcup_{n \in \mathbb{N}} A_n$ is not a finite union of members of $\mathfrak{R}$ and is thus not a member of $\mathfrak{A}$.

[II] Limits of Sequences of Sets

Definition 1.5. *Let $(A_n : n \in \mathbb{N})$ be a sequence of subsets of a set X. We say that $(A_n : n \in \mathbb{N})$ is an increasing sequence and write $A_n \uparrow$ if $A_n \subset A_{n+1}$ for $n \in \mathbb{N}$. We say that $(A_n : n \in \mathbb{N})$ is a decreasing sequence and write $A_n \downarrow$ if $A_n \supset A_{n+1}$ for $n \in \mathbb{N}$. A sequence $(A_n : n \in \mathbb{N})$ is called a monotone sequence if it is either an increasing sequence or a decreasing sequence. For an increasing sequence $(A_n : n \in \mathbb{N})$, we define*

(1) $\lim_{n\to\infty} A_n = \bigcup_{n \in \mathbb{N}} A_n = \{x \in X : x \in A_n \text{ for some } n \in \mathbb{N}\}.$

For a decreasing sequence $(A_n : n \in \mathbb{N})$, we define

(2) $\lim_{n\to\infty} A_n = \bigcap_{n \in \mathbb{N}} A_n = \{x \in X : x \in A_n \text{ for every } n \in \mathbb{N}\}.$

For a monotone sequence $(A_n : n \in \mathbb{N})$, $\lim_{n\to\infty} A_n$ always exists although it may be $\emptyset$. If $A_n \uparrow$, then $\lim_{n\to\infty} A_n = \emptyset$ if and only if $A_n = \emptyset$ for every $n \in \mathbb{N}$. If $A_n \downarrow$, we may have $\lim_{n\to\infty} A_n = \emptyset$ even if $A_n \neq \emptyset$ for every $n \in \mathbb{N}$. Consider for example $X = \mathbb{R}$ and $A_n = \left(0, \frac{1}{n}\right)$ for $n \in \mathbb{N}$. Then $\lim_{n\to\infty} A_n = \emptyset$. On the other hand if $A_n = \left[0, \frac{1}{n}\right)$ for $n \in \mathbb{N}$ then $A_n \downarrow$ and $\lim_{n\to\infty} A_n = \{0\}$.

In order to define a limit for an arbitrary sequence $(A_n : n \in \mathbb{N})$ of subsets of a set X we define first the limit inferior and the limit superior of a sequence.

Definition 1.6. *We define the limit inferior and the limit superior of a sequence* $(A_n : n \in \mathbb{N})$ *of subsets of a set* X *by setting*

(1) $\liminf_{n\to\infty} A_n = \bigcup_{n\in\mathbb{N}} \bigcap_{k\geq n} A_k,$

(2) $\limsup_{n\to\infty} A_n = \bigcap_{n\in\mathbb{N}} \bigcup_{k\geq n} A_k.$

Note that $\left(\bigcap_{k\geq n} A_k : n \in \mathbb{N}\right)$ is a increasing sequence of subsets of X and this implies that $\lim_{n\to\infty} \bigcap_{k\geq n} A_k = \bigcup_{n\in\mathbb{N}} \bigcap_{k\geq n} A_k$ exists. Similarly $\left(\bigcup_{k\geq n} A_k : n \in \mathbb{N}\right)$ is a decreasing sequence of subsets of X and thus $\lim_{n\to\infty} \bigcup_{k\geq n} A_k = \bigcap_{n\in\mathbb{N}} \bigcup_{k\geq n} A_k$ exists. Thus $\liminf_{n\to\infty} A_n$ and $\limsup_{n\to\infty} A_n$ always exist although they may be $\emptyset$.

Lemma 1.7. *Let* $(A_n : n \in \mathbb{N})$ *be a sequence of subsets of a set* X. *Then*

(1) $\liminf_{n\to\infty} A_n = \{x \in X : x \in A_n \text{ for all but finitely many } n \in \mathbb{N}\}.$

(2) $\limsup_{n\to\infty} A_n = \{x \in X : x \in A_n \text{ for infinitely many } n \in \mathbb{N}\},$

(3) $\liminf_{n\to\infty} A_n \subset \limsup_{n\to\infty} A_n.$

Proof. 1. Let $x \in X$. If $x \in A_n$ for all but finitely many $n \in \mathbb{N}$, then there exists $n_0 \in \mathbb{N}$ such that $x \in A_k$ for all $k \geq n_0$. Then $x \in \bigcap_{k\geq n_0} A_k \subset \bigcup_{n\in\mathbb{N}} \bigcap_{k\geq n} A_k = \liminf_{n\to\infty} A_n$. Conversely if $x \in \liminf_{n\to\infty} A_n = \bigcup_{n\in\mathbb{N}} \bigcap_{k\geq n} A_k$, then $x \in \bigcap_{k\geq n_0} A_k$ for some $n_0 \in \mathbb{N}$ and thus $x \in A_k$ for all $k \geq n_0$, that is, $x \in A_n$ for all but finitely many $n \in \mathbb{N}$. This proves (1).

2. If $x \in A_n$ for infinitely many $n \in \mathbb{N}$, then for every $n \in \mathbb{N}$ we have $x \in \bigcup_{k\geq n} A_k$ and thus $x \in \bigcap_{n\in\mathbb{N}} \bigcup_{k\geq n} A_k = \limsup_{n\to\infty} A_n$. Conversely if $x \in \limsup_{n\to\infty} A_n = \bigcap_{n\in\mathbb{N}} \bigcup_{k\geq n} A_k$, then $x \in \bigcup_{k\geq n} A_k$ for every $n \in \mathbb{N}$. Thus for every $n \in \mathbb{N}$, $x \in A_k$ for some $k \geq n$. This shows that $x \in A_n$ for infinitely many $n \in \mathbb{N}$. This proves (2).

3. (1) and (2) imply (3). ∎

Definition 1.8. *Let* $(A_n : n \in \mathbb{N})$ *be an arbitrary sequence of subsets of a set* X. *If* $\liminf_{n\to\infty} A_n = \limsup_{n\to\infty} A_n$, *then we say that the sequence converges and define* $\lim_{n\to\infty} A_n$ *by setting* $\lim_{n\to\infty} A_n = \liminf_{n\to\infty} A_n = \limsup_{n\to\infty} A_n$. *If* $\liminf_{n\to\infty} A_n \neq \limsup_{n\to\infty} A_n$, *then* $\lim_{n\to\infty} A_n$ *does not exist.*

Note that this definition of $\lim_{n\to\infty} A_n$ contains the definition of $\lim_{n\to\infty} A_n$ for monotone sequences in Definition 1.5 as particular cases and thus the two definitions are consistent. Indeed if $A_n \uparrow$ then $\bigcap_{k\geq n} A_k = A_n$ for every $n \in \mathbb{N}$ and $\bigcup_{n\in\mathbb{N}} \bigcap_{k\geq n} A_k = \bigcup_{n\in\mathbb{N}} A_n$ and therefore $\liminf_{n\to\infty} A_n = \bigcup_{n\in\mathbb{N}} A_n$. On the other hand, $\bigcup_{k\geq n} A_k = \bigcup_{k\in\mathbb{N}} A_k$ for every $n \in \mathbb{N}$ and $\bigcap_{n\in\mathbb{N}} \bigcup_{k\geq n} A_k = \bigcup_{k\in\mathbb{N}} A_k$ and thus $\limsup_{n\to\infty} A_n = \bigcup_{n\in\mathbb{N}} A_n$. Similarly for $A_n \downarrow$. Note also that if $(A_n : n \in \mathbb{N})$ is such that $\liminf_{n\to\infty} A_n = \emptyset$ and $\limsup_{n\to\infty} A_n = \emptyset$ also then $\lim_{n\to\infty} A_n = \emptyset$.

Example. Let $X = \mathbb{R}$ and let a sequence $(A_n : n \in \mathbb{N})$ of subsets of $\mathbb{R}$ be defined by $A_1 = [0,1]$, $A_3 = \left[0, \frac{1}{3}\right]$, $A_5 = \left[0, \frac{1}{5}\right], \ldots$, and $A_2 = [0,2]$, $A_4 = [0,4]$, $A_6 = [0,6], \ldots$. Then $\liminf_{n\to\infty} A_n = \{x \in X : x \in A_n \text{ for all but finitely many } n \in \mathbb{N}\} = \{0\}$ and $\limsup_{n\to\infty} A_n = \{x \in X : x \in A_n \text{ for infinitely many } n \in \mathbb{N}\} = [0,\infty)$. Thus $\lim_{n\to\infty} A_n$ does not exist. The subsequence $(A_{n_k} : k \in \mathbb{N}) = (A_1, A_3, A_5, \ldots)$ is a decreasing sequence with $\lim_{k\to\infty} A_{n_k} = \{0\}$ and the subsequence $(A_{n_k} : k \in \mathbb{N}) = (A_2, A_4, A_6, \ldots)$ is an increasing sequence with $\lim_{k\to\infty} A_{n_k} = [0,\infty)$.

Theorem 1.9. *Let $\mathfrak{A}$ be a σ-algebra of subsets of a set X. For every sequence $(A_n : n \in \mathbb{N})$ in $\mathfrak{A}$, the two sets $\liminf_{n\to\infty} A_n$ and $\limsup_{n\to\infty} A_n$ are in $\mathfrak{A}$. So is $\lim_{n\to\infty} A_n$ if it exists.*

Proof. For every $n \in \mathbb{N}$, $\bigcap_{k\ge n} A_k \in \mathfrak{A}$ by Lemma 1.4. Then $\bigcup_{n\in\mathbb{N}} \bigcap_{k\ge n} A_k \in \mathfrak{A}$ by 4° of Definition 1.3. This shows that $\liminf_{n\to\infty} A_n \in \mathfrak{A}$. Similarly $\bigcup_{k\ge n} A_k \in \mathfrak{A}$ by 4° of Definition 1.3. Then $\bigcap_{n\in\mathbb{N}} \bigcup_{k\ge n} A_k \in \mathfrak{A}$ by Lemma 1.4. Thus $\limsup_{n\to\infty} A_n \in \mathfrak{A}$. If $\lim_{n\to\infty} A_n$ exists, then $\lim_{n\to\infty} A_n = \liminf_{n\to\infty} A_n \in \mathfrak{A}$. ∎

[III] Generation of σ-algebras

Let A be an arbitrary set. If we select a set E_α corresponding to each $\alpha \in A$, then we call $\{E_\alpha : \alpha \in A\}$ a collection of sets indexed by A. Usual examples of indexing set A are for instance $\mathbb{N} = \{1, 2, 3, \ldots\}$, $\mathbb{Z} = \{0, 1, -1, 2, -2, \ldots\}$, and $\mathbb{Z}_+ = \{0, 1, 2, \ldots\}$. An arbitrary set A can serve as an indexing set.

Lemma 1.10. *Let $\{\mathfrak{A}_\alpha : \alpha \in A\}$ be a collection of σ-algebras of subsets of a set X where A is an arbitrary indexing set. Then $\bigcap_{\alpha\in A} \mathfrak{A}_\alpha$ is a σ-algebra of subsets of X. Similarly if $\{\mathfrak{A}_\alpha : \alpha \in A\}$ is an arbitrary collection of algebras of subsets of X, then $\bigcap_{\alpha\in A} \mathfrak{A}_\alpha$ is an algebra of subsets of X.*

Proof. Let $\{\mathfrak{A}_\alpha : \alpha \in A\}$ be an arbitrary collection of σ-algebras of subsets of X. Then $\bigcap_{\alpha\in A} \mathfrak{A}_\alpha$ is a collection of subsets of X. To show that it is a σ-algebra we verify 1°, 2°, and 3° in Definition 1.1 and 4° in Definition 1.3. Now $X \in \mathfrak{A}_\alpha$ for every $\alpha \in A$ so that $X \in \bigcap_{\alpha\in A} \mathfrak{A}_\alpha$ verifying 1°. To verify 2°, note that if $E \in \bigcap_{\alpha\in A} \mathfrak{A}_\alpha$, then $E \in \mathfrak{A}_\alpha$ so that $E^c \in \mathfrak{A}_\alpha$ for every $\alpha \in A$ and then $E^c \in \bigcap_{\alpha\in A} \mathfrak{A}_\alpha$. 3° is implied by 4°. To verify 4°, let $(E_n : n \in \mathbb{N}) \subset \bigcap_{\alpha\in A} \mathfrak{A}_\alpha$. Then for every $\alpha \in A$, we have $(E_n : n \in \mathbb{N}) \subset \mathfrak{A}_\alpha$ so that $\bigcup_{n\in\mathbb{N}} E_n \in \mathfrak{A}_\alpha$. Then $\bigcup_{n\in\mathbb{N}} E_n \in \bigcap_{\alpha\in A} \mathfrak{A}_\alpha$. ∎

Theorem 1.11. *Let $\mathfrak{E}$ be an arbitrary collection of subsets of a set X. There exists the smallest σ-algebra $\mathfrak{A}_0$ of subsets of X containing $\mathfrak{E}$, smallest in the sense that if $\mathfrak{A}$ is a σ-algebra of subsets of X containing $\mathfrak{E}$ then $\mathfrak{A}_0 \subset \mathfrak{A}$. Similarly there exists the smallest algebra of subsets of X containing $\mathfrak{E}$.*

Proof. There exists at least one σ-algebra of subsets of X containing $\mathfrak{E}$, namely $\mathfrak{P}(X)$. Let $\{\mathfrak{A}_\alpha : \alpha \in A\}$ be the collection of all σ-algebras of subsets of X containing $\mathfrak{E}$. Then $\bigcap_{\alpha \in A} \mathfrak{A}_\alpha$ contains $\mathfrak{E}$ and it is a σ-algebra according to Lemma 1.10. It is indeed the smallest σ-algebra containing $\mathfrak{E}$ since any σ-algebra $\mathfrak{A}$ containing $\mathfrak{E}$ is a member of $\{\mathfrak{A}_\alpha : \alpha \in A\}$ so that $\mathfrak{A} \supset \bigcap_{\alpha \in A} \mathfrak{A}_\alpha$. ∎

Definition 1.12. *For an arbitrary collection $\mathfrak{E}$ of subsets of a set X, we write $\sigma(\mathfrak{E})$ for the smallest σ-algebra of subsets of X containing $\mathfrak{E}$ and call it the σ-algebra generated by $\mathfrak{E}$. Similarly we write $\alpha(\mathfrak{E})$ for the smallest algebra of subsets of X containing $\mathfrak{E}$ and call it the algebra generated by $\mathfrak{E}$.*

It follows immediately from the definition above that if $\mathfrak{E}_1$ and $\mathfrak{E}_2$ are two collections of subsets of a set X and $\mathfrak{E}_1 \subset \mathfrak{E}_2$, then $\sigma(\mathfrak{E}_1) \subset \sigma(\mathfrak{E}_2)$. If $\mathfrak{A}$ is a σ-algebra of subsets of X, then $\sigma(\mathfrak{A}) = \mathfrak{A}$. In particular for an arbitrary collection $\mathfrak{E}$ of subsets of X, we have $\sigma(\sigma(\mathfrak{E})) = \sigma(\mathfrak{E})$.

Let f be a mapping of a set X into a set Y, that is, f is a Y-valued function defined on X. The image of X by f, $f(X)$, is a subset of Y. Let E be an arbitrary subset of Y. E need not be a subset of $f(X)$ and indeed E may be disjoint from $f(X)$. The preimage of E under the mapping f is a subset of X defined by $f^{-1}(E) = \{x \in X : f(x) \in E\}$, that is, the collection of every $x \in X$ such that $f(x) \in E$. Thus if $E \cap f(X) = \emptyset$ then $f^{-1}(E) = \emptyset$. For an arbitrary subset E of Y we have $f\big(f^{-1}(E)\big) \subset E$. Note also that

$$
\begin{aligned}
&f^{-1}(Y) = X, \\
&f^{-1}(E^c) = f^{-1}(Y \setminus E) = f^{-1}(Y) \setminus f^{-1}(E) = X \setminus f^{-1}(E) = \big(f^{-1}(E)\big)^c, \\
&f^{-1}\big(\textstyle\bigcup_{\alpha \in A} E_\alpha\big) = \textstyle\bigcup_{\alpha \in A} f^{-1}(E_\alpha), \\
&f^{-1}\big(\textstyle\bigcap_{\alpha \in A} E_\alpha\big) = \textstyle\bigcap_{\alpha \in A} f^{-1}(E_\alpha).
\end{aligned}
$$

For an arbitrary collection $\mathfrak{E}$ of subsets of Y, let $f^{-1}(\mathfrak{E}) := \{f^{-1}(E) : E \in \mathfrak{E}\}$.

Proposition 1.13. *Let f be a mapping of a set X into a set Y. If $\mathfrak{B}$ is a σ-algebra of subsets of Y then $f^{-1}(\mathfrak{B})$ is a σ-algebra of subsets of X.*

Proof. Let us show that $\mathfrak{A}$ is a σ-algebra of subsets of X by showing that $X \in f^{-1}(\mathfrak{B})$, $f^{-1}(\mathfrak{B})$ is closed under complementations in X, and $f^{-1}(\mathfrak{B})$ is closed under countable unions.

1. We have $X = f^{-1}(Y) \in f^{-1}(\mathfrak{B})$ since $Y \in \mathfrak{B}$.

2. Let $A \in f^{-1}(\mathfrak{B})$. Then $A = f^{-1}(B)$ for some $B \in \mathfrak{B}$. Since $B^c \in \mathfrak{B}$, we have $f^{-1}(B^c) \in f^{-1}(\mathfrak{B})$. On the other hand, $f^{-1}(B^c) = \big(f^{-1}(B)\big)^c = A^c$. Thus we have $A^c \in f^{-1}(\mathfrak{B})$.

3. Let $(A_n : n \in \mathbb{N})$ be an arbitrary sequence in $f^{-1}(\mathfrak{B})$. Then $A_n = f^{-1}(B_n)$ for some $B_n \in \mathfrak{B}$ for each $n \in \mathbb{N}$. Thus we have

$$\bigcup_{n\in\mathbb{N}} A_n = \bigcup_{n\in\mathbb{N}} f^{-1}(B_n) = f^{-1}\Big(\bigcup_{n\in\mathbb{N}} B_n\Big) \in f^{-1}(\mathfrak{B})$$

since $\bigcup_{n\in\mathbb{N}} B_n \in \mathfrak{B}$. This verifies that $f^{-1}(\mathfrak{B})$ is a σ-algebra of subsets of X. ∎

Theorem 1.14. *Let f be a mapping of a set X into a set Y. Then for an arbitrary collection $\mathfrak{E}$ of subsets of Y, we have $\sigma\big(f^{-1}(\mathfrak{E})\big) = f^{-1}\big(\sigma(\mathfrak{E})\big)$.*

Proof. Since $\mathfrak{E} \subset \sigma(\mathfrak{E})$, we have $f^{-1}(\mathfrak{E}) \subset f^{-1}\big(\sigma(\mathfrak{E})\big)$ and consequently $\sigma\big(f^{-1}(\mathfrak{E})\big) \subset \sigma\big(f^{-1}\big(\sigma(\mathfrak{E})\big)\big)$. Since $\sigma(\mathfrak{E})$ is a σ-algebra of subsets of Y, $f^{-1}\big(\sigma(\mathfrak{E})\big)$ is a σ-algebra of subsets of X by Proposition 1.13 so that $\sigma\big(f^{-1}\big(\sigma(\mathfrak{E})\big)\big) = f^{-1}\big(\sigma(\mathfrak{E})\big)$. Thus we have $\sigma\big(f^{-1}(\mathfrak{E})\big) \subset f^{-1}\big(\sigma(\mathfrak{E})\big)$.

To prove the reverse inclusion, let $\mathfrak{A}_1$ be an arbitrary σ-algebra of subsets of X and let

$$\mathfrak{A}_2 = \big\{A \subset Y : f^{-1}(A) \in \mathfrak{A}_1\big\}.$$

To show that $\mathfrak{A}_2$ is a σ-algebra of subsets of Y, note first of all that $f^{-1}(Y) = X \in \mathfrak{A}_1$ so that $Y \in \mathfrak{A}_2$. Secondly, for every $A \in \mathfrak{A}_2$ we have $f^{-1}(A^c) = \big(f^{-1}(A)\big)^c \in \mathfrak{A}_1$ so that $A^c \in \mathfrak{A}_2$. Finally for any $(A_n : n \in \mathbb{N}) \subset \mathfrak{A}_2$, we have $f^{-1}\big(\bigcup_{n\in\mathbb{N}} A_n\big) = \bigcup_{n\in\mathbb{N}} f^{-1}(A_n) \in \mathfrak{A}_1$ so that $\bigcup_{n\in\mathbb{N}} A_n \in \mathfrak{A}_2$ and thus $\mathfrak{A}_2$ is a σ-algebra of subsets of Y. In particular, if we let

$$\mathfrak{A} = \big\{A \subset Y : f^{-1}(A) \in \sigma\big(f^{-1}(\mathfrak{E})\big)\big\},$$

then $\mathfrak{A}$ is a σ-algebra of subsets of Y. Clearly $\mathfrak{A} \supset \mathfrak{E}$ and thus $\mathfrak{A} \supset \sigma(\mathfrak{E})$ and then $f^{-1}(\mathfrak{A}) \supset f^{-1}\big(\sigma(\mathfrak{E})\big)$. But $f^{-1}(\mathfrak{A}) \subset \sigma\big(f^{-1}(\mathfrak{E})\big)$ by the definition of $\mathfrak{A}$. Thus we have $\sigma\big(f^{-1}(\mathfrak{E})\big) \supset f^{-1}\big(\sigma(\mathfrak{E})\big)$. Therefore $\sigma\big(f^{-1}(\mathfrak{E})\big) = f^{-1}\big(\sigma(\mathfrak{E})\big)$. ∎

Notations. For an arbitrary collection $\mathfrak{E}$ of subsets of a set X and an arbitrary subset A of X, let us write $\mathfrak{E} \cap A = \{E \cap A : E \in \mathfrak{E}\}$. We write $\sigma_A(\mathfrak{E} \cap A)$ for the σ-algebra of subsets of A generated by the collection $\mathfrak{E} \cap A$ of subsets of A. Note that the subscript A in σ_A indicates that it is a σ-algebra of subsets of A, not a σ-algebra of subsets of X.

Theorem 1.15. *Let $\mathfrak{E}$ be an arbitrary collection of subsets of a set X and let $A \subset X$. Then $\sigma_A(\mathfrak{E} \cap A) = \sigma(\mathfrak{E}) \cap A$.*

Proof. Since $\mathfrak{E} \subset \sigma(\mathfrak{E})$ we have $\mathfrak{E} \cap A \subset \sigma(\mathfrak{E}) \cap A$. From the fact that $\sigma(\mathfrak{E})$ is a σ-algebra of subsets of X and $A \subset X$ it follows that $\sigma(\mathfrak{E}) \cap A$ is a σ-algebra of subsets of A. Thus

(1) $$\sigma_A(\mathfrak{E} \cap A) \subset \sigma(\mathfrak{E}) \cap A.$$

Therefore, to prove the theorem it remains to show

(2) $$\sigma(\mathfrak{E}) \cap A \subset \sigma_A(\mathfrak{E} \cap A).$$

Let $\mathfrak{K}$ be the collection of subsets K of X of the type

$$K = (C \cap A^c) \cup B, \tag{3}$$

where $C \in \sigma(\mathfrak{E})$ and $B \in \sigma_A(\mathfrak{E} \cap A)$. Observe that since $B \subset A$, the union in (3) is a disjoint union. By (3), $X \in \mathfrak{K}$ and $\mathfrak{K}$ is closed under countable unions. To show that $\mathfrak{K}$ is also closed under complementations, let $K \in \mathfrak{K}$ be as given by (3). Then

$$K^c = X \setminus K = \left[(X \cap A^c) \cup A\right] \setminus \left[(C \cap A^c) \cup B\right]$$
$$= \left[(X \cap A^c) \setminus (C \cap A^c)\right] \cup (A \setminus B)$$

since $X \cap A^c \supset C \cap A^c$ and $A \supset B$. But $(X \cap A^c) \setminus (C \cap A^c) = C^c \cap A^c$. Therefore

$$K^c = (C^c \cap A^c) \cup (A \setminus B) \in \mathfrak{K}.$$

Thus $\mathfrak{K}$ is closed under complementations and is therefore a σ-algebra of subsets of X. Next, observe that for any $K \in \mathfrak{K}$ as given by (3) we have $K \cap A = B \in \sigma_A(\mathfrak{E} \cap A)$ so that $\mathfrak{K} \cap A \subset \sigma_A(\mathfrak{E} \cap A)$. Thus to show (2) it suffices to show that $\sigma(\mathfrak{E}) \cap A \subset \mathfrak{K} \cap A$. Since $\mathfrak{K}$ is a σ-algebra of subsets of X, it remains only to show that $\mathfrak{E} \subset \mathfrak{K}$. Let $E \in \mathfrak{E}$ and write $E = (E \cap A^c) \cup (E \cap A)$. Since $E \cap A \in \sigma_A(\mathfrak{E} \cap A)$, E is a subset of X of the type (3). Thus $E \in \mathfrak{K}$ and therefore $\mathfrak{E} \subset \mathfrak{K}$. This completes the proof. ∎

[IV] Borel σ-algebras

To fix our terminology let us review definitions of some topological concepts. Let X be a set. A collection $\mathfrak{O}$ of subsets of X is called a topology on X if it satisfies the following axioms:

I $\emptyset \in \mathfrak{O}$,

II $X \in \mathfrak{O}$,

III $\{E_\alpha : \alpha \in A\} \subset \mathfrak{O} \Rightarrow \bigcup_{\alpha \in A} E_\alpha \in \mathfrak{O}$,

IV $E_1, E_2 \in \mathfrak{O} \Rightarrow E_1 \bigcap E_2 \in \mathfrak{O}$.

The pair $(X, \mathfrak{O})$ is called a topological space. The members of $\mathfrak{O}$ are called the open sets of the topological space.

A subset E of X is called a closed set if its complement E^c is an open set. Thus X is both an open set and a closed and so is $\emptyset$.

An arbitrary union of open sets is an open set and a finite intersection of open sets is an open set. An arbitrary intersection of closed sets is a closed set and a finite union of closed sets is a closed set.

The interior E° of a subset E of X is defined as the union of all open sets contained in E. Thus it is the greatest open set contained in E.

The closure $\overline{E}$ of E is defined as the intersection of all closed sets containing E. It is the smallest closed set containing E.

The boundary ∂E of E is defined by $\partial E = \left(E^\circ \cup (E^c)^\circ\right)^c$.

A subset E of X is called a compact set if for every collection $\mathfrak{V}$ of open sets such that $E \subset \bigcup_{V \in \mathfrak{V}} V$ there exists a finite subcollection $\{V_1, \ldots, V_N\}$ such that $E \subset \bigcup_{n=1}^N V_n$.

Let X be an arbitrary set. A function ρ on $X \times X$ is called a metric on X if it satisfies the following conditions:

1° $\rho(x, y) \in [0, \infty)$ for $x, y \in X$,
2° $\rho(x, y) = 0 \Leftrightarrow x = y$,
3° $\rho(x, y) = \rho(y, x)$ for $x, y \in X$,
4° triangle inequality: $\rho(x, y) \leq \rho(x, z) + \rho(z, y)$ for $x, y, z \in X$.

The pair (X, ρ) is called a metric space.

In $\mathbb{R}^n$, if we define $\rho(x, y) = |x - y| = \left\{\sum_{k=1}^n (x_k - y_k)^2\right\}^{1/2}$ for $x = (x_1, \dots, x_n)$ and $y = (y_1, \dots, y_n)$ in $\mathbb{R}^n$, then ρ satisfies conditions 1°, 2°, 3°, and 4° above and is thus a metric. This metric on $\mathbb{R}^n$ is called the Euclidean metric.

In a metric space (X, ρ), if $x_0 \in X$ and $r > 0$ the set $B(x_0, r) = \{x \in X : \rho(x, x_0) < r\}$ is called an open ball with center x_0 and radius r. A subset E of X is called an open set if for each $x \in E$ there exists $r > 0$ such that $B(x, r) \subset E$. An open ball is indeed an open set in the sense defined above. The collection of all open sets in a metric space satisfies the axioms I, II, III, and IV and is thus a topology. We call this topology the metric topology of X by the metric ρ.

A set E in a metric space (X, ρ) is said to be bounded if there exist $x_0 \in X$ and $r > 0$ such that $E \subset B(x_0, r)$. A set E in $\mathbb{R}^n$ is a compact set if and only if E is a bounded and closed set.

Definition 1.16. *Let $\mathfrak{O}$ be the collection of all open sets in a topological space X. We call the σ-algebra $\sigma(\mathfrak{O})$ the Borel σ-algebra of subsets of the topological space X and we write $\mathfrak{B}_X$ or $\mathfrak{B}(X)$ for it. We call its members the Borel sets of the topological space.*

Lemma 1.17. *Let $\mathfrak{C}$ be the collection of all closed sets in a topological space $(X, \mathfrak{O})$. Then $\sigma(\mathfrak{C}) = \sigma(\mathfrak{O})$.*

Proof. Let $E \in \mathfrak{C}$. Then $E^c \in \mathfrak{O} \subset \sigma(\mathfrak{O})$. Now since $\sigma(\mathfrak{O})$ is a σ-algebra, we have $E = (E^c)^c \in \sigma(\mathfrak{O})$. Thus $\mathfrak{C} \subset \sigma(\mathfrak{O})$ and consequently $\sigma(\mathfrak{C}) \subset \sigma(\sigma(\mathfrak{O})) = \sigma(\mathfrak{O})$. By the same sort of argument as above we have $\sigma(\mathfrak{O}) \subset \sigma(\mathfrak{C})$. Therefore $\sigma(\mathfrak{C}) = \sigma(\mathfrak{O})$. ∎

Definition 1.18. *Let $(X, \mathfrak{O})$ be a topological space. A subset E of X is called a G_δ-set if it is the intersection of countably many open sets. A subset E of X is called an F_σ-set if it is the union of countably many closed sets.*

Thus, if E is a G_δ-set, then E^c is an F_σ-set, and if E is an F_σ-set then E^c is a G_δ-set. Note that every G_δ-set is a member of $\mathfrak{B}_X$. So is every F_σ-set. Indeed if E is a G_δ-set, then $E = \bigcap_{n \in \mathbb{N}} O_n$ where $O_n \in \mathfrak{O}$ for $n \in \mathbb{N}$. Now $O_n \in \mathfrak{O} \subset \sigma(\mathfrak{O}) = \mathfrak{B}_X$ for every $n \in \mathbb{N}$. Since $\mathfrak{B}_X$ is a σ-algebra, we have $E = \bigcap_{n \in \mathbb{N}} O_n \in \mathfrak{B}_X$.

Let us note also that if E is a G_δ-set, then there exists a sequence $(O_n : n \in \mathbb{N})$ of open sets such that $E = \bigcap_{n \in \mathbb{N}} O_n$. If we let $G_n = \bigcap_{k=1}^n O_k$, then $(G_n : n \in \mathbb{N})$ is a decreasing sequence of open sets and $\bigcap_{n \in \mathbb{N}} G_n = \bigcap_{n \in \mathbb{N}} O_n = E$. Thus a G_δ-set is always the limit of a decreasing sequence of open sets. Similarly if E is an F_σ-set, then there exists a sequence $(C_n : n \in \mathbb{N})$ of closed sets such that $E = \bigcup_{n\in\mathbb{N}} C_n$. If we let $F_n = \bigcup_{k=1}^n C_k$, then $(F_n : n \in \mathbb{N})$ is an increasing sequence of closed sets and $\bigcup_{n\in\mathbb{N}} F_n = \bigcup_{n\in\mathbb{N}} C_n = E$. Thus an F_σ-set is always the limit of an increasing sequence of closed sets.

[V] Measure on a σ-algebra

Notations. Let $\overline{\mathbb{R}} = \{-\infty\} \cup \mathbb{R} \cup \{\infty\}$ and call it the extended real number system. We use the alternate notation $[-\infty, \infty]$ for $\overline{\mathbb{R}}$ also.

Definition 1.19. *Let $\mathfrak{E}$ be a collection of subsets of a set X. Let γ be a nonnegative extended real-valued set function on $\mathfrak{E}$. We say that*
(a) *γ is monotone on $\mathfrak{E}$ if $\gamma(E_1) \le \gamma(E_2)$ for $E_1, E_2 \in \mathfrak{E}$ such that $E_1 \subset E_2$,*
(b) *γ is additive on $\mathfrak{E}$ if $\gamma(E_1 \cup E_2) = \gamma(E_1) + \gamma(E_2)$ for $E_1, E_2 \in \mathfrak{E}$ such that $E_1 \cap E_2 = \emptyset$ and $E_1 \cup E_2 \in \mathfrak{E}$,*
(c) *γ is finitely additive on $\mathfrak{E}$ if $\gamma\left(\bigcup_{k=1}^n E_k\right) = \sum_{k=1}^n \gamma(E_k)$ for every disjoint finite sequence $(E_k : k = 1, \dots, n)$ in $\mathfrak{E}$ such that $\bigcup_{k=1}^n E_k \in \mathfrak{E}$,*
(d) *γ is countably additive on $\mathfrak{E}$ if $\gamma\left(\bigcup_{n\in\mathbb{N}} E_n\right) = \sum_{n\in\mathbb{N}} \gamma(E_n)$ for every disjoint sequence $(E_n : n \in \mathbb{N})$ in $\mathfrak{E}$ such that $\bigcup_{n\in\mathbb{N}} E_n \in \mathfrak{E}$,*
(e) *γ is subadditive on $\mathfrak{E}$ if $\gamma(E_1 \cup E_2) \le \gamma(E_1) + \gamma(E_2)$ for $E_1, E_2 \in \mathfrak{E}$ such that $E_1 \cup E_2 \in \mathfrak{E}$,*
(f) *γ is finitely subadditive on $\mathfrak{E}$ if $\gamma\left(\bigcup_{k=1}^n E_k\right) \le \sum_{k=1}^n \gamma(E_k)$ for every finite sequence $(E_k : k = 1, \dots, n)$ in $\mathfrak{E}$ such that $\bigcup_{k=1}^n E_k \in \mathfrak{E}$,*
(g) *γ is countably subadditive on $\mathfrak{E}$ if $\gamma\left(\bigcup_{n\in\mathbb{N}} E_n\right) \le \sum_{n\in\mathbb{N}} \gamma(E_n)$ for every sequence $(E_n : n \in \mathbb{N})$ in $\mathfrak{E}$ such that $\bigcup_{n\in\mathbb{N}} E_n \in \mathfrak{E}$.*

Note that in (c) while $\bigcup_{k=1}^n E_k \in \mathfrak{E}$ is required, it is not required that any of $\bigcup_{k=1}^2 E_k$, $\bigcup_{k=1}^3 E_k, \dots, \bigcup_{k=1}^{n-1} E_k$ be in $\mathfrak{E}$. Note also that (c) implies (b) and (f) implies (e).

Observation 1.20. Let γ be a nonnegative extended real-valued set function on a collection $\mathfrak{E}$ of subsets of a set X. Assume that $\emptyset \in \mathfrak{E}$ and $\gamma(\emptyset) = 0$.
(a) If γ is countably additive on $\mathfrak{E}$, then it is finitely additive on $\mathfrak{E}$.
(b) If γ is countably subadditive on $\mathfrak{E}$, then it is finitely subadditive on $\mathfrak{E}$.

Proof. Suppose γ is countably additive on $\mathfrak{E}$. To show that it is finitely additive on $\mathfrak{E}$, let $(E_k : k = 1, \dots, n)$ be a disjoint finite sequence in $\mathfrak{E}$ such that $\bigcup_{k=1}^n E_k \in \mathfrak{E}$. Consider the infinite sequence $(F_k : k \in \mathbb{N})$ in $\mathfrak{E}$ defined by $F_k = E_k$ for $k = 1, \dots, n$ and $F_k = \emptyset$ for $k \ge n+1$. Since $\emptyset \in \mathfrak{E}$, $(F_k : k \in \mathbb{N})$ is a disjoint sequence in $\mathfrak{E}$ with $\bigcup_{k\in\mathbb{N}} F_k = \bigcup_{k=1}^n E_k \in \mathfrak{E}$. Thus by the countable additivity of γ on $\mathfrak{E}$ and by the fact that $\gamma(\emptyset) = 0$, we have $\gamma\left(\bigcup_{k=1}^n E_k\right) = \gamma\left(\bigcup_{k\in\mathbb{N}} F_k\right) = \sum_{k\in\mathbb{N}} \gamma(F_k) = \sum_{k=1}^n \gamma(E_k)$. This proves the finite additivity of γ on $\mathfrak{E}$. We show similarly that if γ is countably subadditive on $\mathfrak{E}$, then it is finitely subadditive on $\mathfrak{E}$. ∎

Lemma 1.21. *Let $(E_n : n \in \mathbb{N})$ be an arbitrary sequence in an algebra $\mathfrak{A}$ of subsets of a set X. Then there exists a disjoint sequence $(F_n : n \in \mathbb{N})$ in $\mathfrak{A}$ such that*

(1) $\bigcup_{n=1}^{N} E_n = \bigcup_{n=1}^{N} F_n$ *for every* $N \in \mathbb{N}$,

and

(2) $\bigcup_{n\in\mathbb{N}} E_n = \bigcup_{n\in\mathbb{N}} F_n$.

In particular, if $\mathfrak{A}$ *is a* σ*-algebra, then* $\bigcup_{n\in\mathbb{N}} F_n = \bigcup_{n\in\mathbb{N}} E_n \in \mathfrak{A}$.

Proof. Let $F_1 = E_1$ and $F_n = E_n \setminus (E_1 \cup \ldots \cup E_{n-1})$ for $n \geq 2$. Since $\mathfrak{A}$ is an algebra, $F_n \in \mathfrak{A}$ for $n \in \mathbb{N}$. Let us prove (1) and (2) and then the disjointness of $(F_n : n \in \mathbb{N})$.

Let us prove (1) by induction. To start with, (1) is valid when $N = 1$ since $F_1 = E_1$. Next, assume that (1) is valid for some $N \in \mathbb{N}$, that is, $\bigcup_{n=1}^{N} E_n = \bigcup_{n=1}^{N} F_n$. Then we have

$$\bigcup_{n=1}^{N+1} F_n = \Big(\bigcup_{n=1}^{N} F_n\Big) \cup F_{N+1} = \Big(\bigcup_{n=1}^{N} E_n\Big) \cup \Big(E_{N+1} \setminus \bigcup_{n=1}^{N} E_n\Big) = \bigcup_{n=1}^{N+1} E_n,$$

that is, (1) holds for $N + 1$. Thus by induction, (1) holds for every $N \in \mathbb{N}$.

To prove (2), let $x \in \bigcup_{n\in\mathbb{N}} E_n$. Then $x \in E_n$ for some $n \in \mathbb{N}$ and thus we have $x \in \bigcup_{k=1}^{n} E_k = \bigcup_{k=1}^{n} F_k \subset \bigcup_{n\in\mathbb{N}} F_n$ by (1). We show similarly that if $x \in \bigcup_{n\in\mathbb{N}} F_n$ then $x \in \bigcup_{n\in\mathbb{N}} E_n$. Thus we have $\bigcup_{n\in\mathbb{N}} E_n = \bigcup_{n\in\mathbb{N}} F_n$. This proves (2).

Finally let us show that $(F_n : n \in \mathbb{N})$ is a disjoint sequence. Consider F_n and F_m where $n \neq m$, say $n < m$. We have $F_m = E_m \setminus (E_1 \cup \cdots \cup E_{m-1})$. By (1) and by the fact that $n < m$, we have $E_1 \cup \cdots \cup E_{m-1} = F_1 \cup \cdots \cup F_{m-1} \supset F_n$. Thus we have $F_n \bigcap F_m = \emptyset$. This prove the disjointness of $(F_n : n \in \mathbb{N})$. ∎

Lemma 1.22. *Let* γ *be a nonnegative extended real-valued set function on an algebra* $\mathfrak{A}$ *of subsets of a set* X.

(a) *If* γ *is additive on* $\mathfrak{A}$*, it is finitely additive, monotone, and finitely subadditive on* $\mathfrak{A}$.

(b) *If* γ *is countably additive on* $\mathfrak{A}$*, then it is countably subadditive on* $\mathfrak{A}$.

Proof. 1. Suppose γ is additive on $\mathfrak{A}$. Let $(E_k : k = 1, \ldots, n)$ be a disjoint finite sequence in $\mathfrak{A}$. Since $\mathfrak{A}$ is an algebra, we have $\bigcup_{i=1}^{k} E_i \in \mathfrak{A}$ for $k = 1, \ldots, n$. By the disjointness of $\bigcup_{k=1}^{n-1} E_k$ and E_n and by the additivity of γ on $\mathfrak{A}$, we have

$$\gamma\Big(\bigcup_{k=1}^{n} E_k\Big) = \gamma\Big(\bigcup_{k=1}^{n-1} E_k\Big) + \gamma(E_n).$$

Repeating the argument, we have $\gamma\left(\bigcup_{k=1}^{n} E_k\right) = \sum_{k=1}^{n} \gamma(E_k)$. This proves the finite additivity of γ on $\mathfrak{A}$. To prove the monotonicity of γ on $\mathfrak{A}$, let $E_1, E_2 \in \mathfrak{A}$ and $E_1 \subset E_2$. Then $E_1, E_2 \setminus E_1 \in \mathfrak{A}$, $E_1 \cap (E_2 \setminus E_1) = \emptyset$, and $E_1 \cup (E_2 \setminus E_1) = E_2 \in \mathfrak{A}$ so that by the additivity of γ on $\mathfrak{A}$, we have $\gamma(E_1) + \gamma(E_2 \setminus E_1) = \gamma(E_2)$. Then since $\gamma(E_2 \setminus E_1) \geq 0$, we have $\gamma(E_1) \leq \gamma(E_2)$. This proves the monotonicity of γ on $\mathfrak{A}$.

To show the finite subadditivity of γ on $\mathfrak{A}$, let $(E_k : k = 1, \ldots, n)$ be a finite sequence in $\mathfrak{A}$. If we let $F_1 = E_1$ and $F_k = E_k \setminus (E_1 \cup \cdots \cup E_{k-1})$ for $k = 2, \ldots, n$, then as we showed in the Proof of Lemma 1.21, $(F_k : k = 1, \dots, n)$ is a disjoint finite sequence in $\mathfrak{A}$ with $\bigcup_{k=1}^n F_k = \bigcup_{k=1}^n E_k$ so that by the finite additivity and the monotonicity of γ on $\mathfrak{A}$, we have $\gamma\left(\bigcup_{k=1}^n E_k\right) = \gamma\left(\bigcup_{k=1}^n F_k\right) = \sum_{k=1}^n \gamma(F_k) \le \sum_{k=1}^n \gamma(E_k)$. This proves the finite subadditivity of γ on $\mathfrak{A}$.

2. Suppose γ is countably additive on $\mathfrak{A}$. To show that it is countable subadditive on $\mathfrak{A}$, let $(E_n : n \in \mathbb{N})$ be a sequence in $\mathfrak{A}$ such that $\bigcup_{n\in\mathbb{N}} E_n \in \mathfrak{A}$. Let $F_1 = E_1$ and $F_n = E_n \setminus (E_1 \cup \dots E_{n-1})$ for $n \ge 2$. Then by Lemma 1.21, $(F_n : n \in \mathbb{N})$ is a disjoint sequence in $\mathfrak{A}$ and $\bigcup_{n\in\mathbb{N}} F_n = \bigcup_{n\in\mathbb{N}} E_n$. Thus by the countable additivity and the monotonicity of γ on $\mathfrak{A}$, we have $\gamma\left(\bigcup_{n\in\mathbb{N}} E_n\right) = \gamma\left(\bigcup_{n\in\mathbb{N}} F_n\right) = \sum_{n\in\mathbb{N}} \gamma(F_n) \le \sum_{n\in\mathbb{N}} \gamma(E_n)$. This proves the countable subadditivity of γ on $\mathfrak{A}$. ■

Proposition 1.23. *Let γ be a nonnegative extended real-valued set function on an algebra $\mathfrak{A}$ of subsets of a set X. If γ is additive and countably subadditive on $\mathfrak{A}$ then γ is countably additive on $\mathfrak{A}$.*

Proof. Suppose γ is additive and countably subadditive on $\mathfrak{A}$. To show that γ is countably additive on $\mathfrak{A}$, let $(E_n : n \in \mathbb{N})$ be a disjoint sequence in $\mathfrak{A}$ such that $\bigcup_{n\in\mathbb{N}} E_n \in \mathfrak{A}$. The additivity of γ on $\mathfrak{A}$ implies its monotonicity and finite additivity on $\mathfrak{A}$ by (a) of Lemma 1.22. Thus for every $N \in \mathbb{N}$, we have $\gamma\left(\bigcup_{n\in\mathbb{N}} E_n\right) \ge \gamma\left(\bigcup_{n=1}^N E_n\right) = \sum_{n=1}^N \gamma(E_n)$. Since this holds for every $N \in \mathbb{N}$, we have $\gamma\left(\bigcup_{n\in\mathbb{N}} E_n\right) \ge \sum_{n\in\mathbb{N}} \gamma(E_n)$. On the other hand, by the countable subadditivity of γ on $\mathfrak{A}$, we have $\gamma\left(\bigcup_{n\in\mathbb{N}} E_n\right) \le \sum_{n\in\mathbb{N}} \gamma(E_n)$. Thus $\gamma\left(\bigcup_{n\in\mathbb{N}} E_n\right) = \sum_{n\in\mathbb{N}} \gamma(E_n)$. This proves the countable additivity of γ on $\mathfrak{A}$. ■

Definition 1.24. *Let $\mathfrak{A}$ be a σ-algebra of subsets of a set X. A set function μ defined on $\mathfrak{A}$ is called a measure if it satisfies the following conditions:*

1° *nonnegative extended real-valued :* $\mu(E) \in [0, \infty]$ *for every* $E \in \mathfrak{A}$,

2° $\mu(\emptyset) = 0$,

3° *countable additivity:* $(E_n : n \in \mathbb{N}) \subset \mathfrak{A}$, *disjoint* $\Rightarrow \mu\left(\bigcup_{n\in\mathbb{N}} E_n\right) = \sum_{n\in\mathbb{N}} \mu(E_n)$.

Lemma 1.25. *A measure μ on a σ-algebra $\mathfrak{A}$ of subsets of a set X has the following properties:*

(1) *finite additivity:* $(E_1, \dots, E_n) \subset \mathfrak{A}$, *disjoint* $\Rightarrow \mu\left(\bigcup_{k=1}^n E_k\right) = \sum_{k=1}^n \mu(E_k)$,

(2) *monotonicity:* $E_1, E_2 \in \mathfrak{A}, E_1 \subset E_2 \Rightarrow \mu(E_1) \le \mu(E_2)$,

(3) $E_1, E_2 \in \mathfrak{A}, E_1 \subset E_2, \mu(E_1) < \infty \Rightarrow \mu(E_2 \setminus E_1) = \mu(E_2) - \mu(E_1)$,

(4) *countable subadditivity:* $(E_n : n \in \mathbb{N}) \subset \mathfrak{A} \Rightarrow \mu\left(\bigcup_{n\in\mathbb{N}} E_n\right) \le \sum_{n\in\mathbb{N}} \mu(E_n)$,

and in particular

(5) *finite subadditivity:* $(E_1, \dots, E_n) \subset \mathfrak{A} \Rightarrow \mu\left(\bigcup_{k=1}^n E_k\right) \le \sum_{k=1}^n \mu(E_k)$.

Proof. The countable additivity of μ on $\mathfrak{A}$ implies its finite additivity on $\mathfrak{A}$ by (a) of Observation 1.20. The finite additivity of μ on $\mathfrak{A}$ implies its additivity on $\mathfrak{A}$ and then its monotonicity on $\mathfrak{A}$ by (a) of Lemma 1.22.

To prove (3), let $E_1, E_2 \in \mathfrak{A}$ and $E_1 \subset E_2$. Then E_1 and $E_2 \setminus E_1$ are two disjoint members of $\mathfrak{A}$ whose union is equal to E_2. Thus by the additivity of μ on $\mathfrak{A}$, we have $\mu(E_2) = \mu(E_1) + \mu(E_2 \setminus E_1)$. If $\mu(E_1) < \infty$, then subtracting $\mu(E_1)$ for both sides of the last equality, we have $\mu(E_2) - \mu(E_1) = \mu(E_2 \setminus E_1)$. This proves (3).

The countable additivity of μ on $\mathfrak{A}$ implies its countable subadditivity on $\mathfrak{A}$ by (b) of Lemma 1.22. This then implies the finite subadditivity of μ on $\mathfrak{A}$ by (b) of Observation 1.20. ∎

Regarding (3) of Lemma 1.25, let us note that if $\mu(E_1) = \infty$ then by the monotonicity of μ we have $\mu(E_2) = \infty$ also so that $\mu(E_2) - \mu(E_1)$ is not defined.

[VI] Measures of a Sequence of Sets

Let μ be a measure on a σ-algebra $\mathfrak{A}$ of subsets of a set X. Let $(E_n : n \in \mathbb{N})$ be a sequence in $\mathfrak{A}$. If $\lim_{n\to\infty} E_n$ exists, does $\lim_{n\to\infty} \mu(E_n)$ exist? If it does, do we have $\mu\big(\lim_{n\to\infty} E_n\big) = \lim_{n\to\infty} \mu(E_n)$? The next theorem addresses this question for monotone sequences of measurable sets. It is based on the countable additivity of a measure. It is a fundamental theorem in that a subsequent theorem regarding the limit inferior and the limit superior of the measures of an arbitrary sequence of measurable sets as well as the monotone convergence theorem for the Lebesgue integral, Fatou's lemma, and Lebesgue's dominated convergence theorem are ultimately based on this theorem.

Theorem 1.26. (Monotone Convergence Theorem for Sequences of Measurable Sets) *Let μ be a measure on a σ-algebra $\mathfrak{A}$ of subsets of a set X and let $(E_n : n \in \mathbb{N})$ be a monotone sequence in $\mathfrak{A}$.*
(a) *If $E_n \uparrow$, then $\lim_{n\to\infty} \mu(E_n) = \mu\big(\lim_{n\to\infty} E_n\big)$.*
(b) *If $E_n \downarrow$, then $\lim_{n\to\infty} \mu(E_n) = \mu\big(\lim_{n\to\infty} E_n\big)$, provided that there exists a set $A \in \mathfrak{A}$ with $\mu(A) < \infty$ such that $E_1 \subset A$.*

Proof. If $E_n \uparrow$, then $\lim_{n\to\infty} E_n = \bigcup_{n\in\mathbb{N}} E_n \in \mathfrak{A}$. If $E_n \downarrow$, then $\lim_{n\to\infty} E_n = \bigcap_{n\in\mathbb{N}} E_n \in \mathfrak{A}$. Note also that if $(E_n : n \in \mathbb{N})$ is a monotone sequence in $\mathfrak{A}$, then $(\mu(E_n) : n \in \mathbb{N})$ is a monotone sequence in $[0, \infty]$ by the monotonicity of μ so that $\lim_{n\to\infty} \mu(E_n)$ exists in $[0, \infty]$.

1. Suppose $E_n \uparrow$. Then we have $\mu(E_n) \uparrow$. Consider first the case where $\mu(E_{n_0}) = \infty$ for some $n_0 \in \mathbb{N}$. In this case we have $\lim_{n\to\infty} \mu(E_n) = \infty$. Since $E_{n_0} \subset \bigcup_{n\in\mathbb{N}} E_n = \lim_{n\to\infty} E_n$, we have $\mu\big(\lim_{n\to\infty} E_n\big) \geq \mu(E_{n_0}) = \infty$. Thus $\mu\big(\lim_{n\to\infty} E_n\big) = \infty = \lim_{n\to\infty} \mu(E_n)$.

Consider next the case where $\mu(E_n) < \infty$ for every $n \in \mathbb{N}$. Let $E_0 = \emptyset$ and consider a disjoint sequence $(F_n : n \in \mathbb{N})$ in $\mathfrak{A}$ defined by $F_n = E_n \setminus E_{n-1}$ for $n \in \mathbb{N}$. We have

$\bigcup_{n=1}^{N} E_n = \bigcup_{n=1}^{N} F_n$ for every $N \in \mathbb{N}$ and hence $\bigcup_{n\in\mathbb{N}} E_n = \bigcup_{n\in\mathbb{N}} F_n$. Then we have

$$\mu\big(\lim_{n\to\infty} E_n\big) = \mu\Big(\bigcup_{n\in\mathbb{N}} E_n\Big) = \mu\Big(\bigcup_{n\in\mathbb{N}} F_n\Big) = \sum_{n\in\mathbb{N}} \mu(F_n)$$

$$= \sum_{n\in\mathbb{N}} \mu(E_n \setminus E_{n-1}) = \sum_{n\in\mathbb{N}} \{\mu(E_n) - \mu(E_{n-1})\},$$

where the third equality is by the countable additivity of μ and the fifth equality is by (3) of Lemma 1.25. Since the sum of a series is the limit of the sequence of partial sums we have

$$\sum_{n\in\mathbb{N}} \{\mu(E_n) - \mu(E_{n-1})\} = \lim_{n\to\infty} \sum_{k=1}^{n} \{\mu(E_k) - \mu(E_{k-1})\}$$

$$= \lim_{n\to\infty} \{\mu(E_n) - \mu(E_0)\} = \lim_{n\to\infty} \mu(E_n).$$

Thus we have $\mu\big(\lim_{n\to\infty} E_n\big) = \lim_{n\to\infty} \mu(E_n)$.

2. Suppose $E_n \downarrow$ and assume the existence of a containing set A with finite measure. Define a disjoint sequence $(F_n : n \in \mathbb{N})$ in $\mathfrak{A}$ by setting $F_n = E_n \setminus E_{n+1}$ for $n \in \mathbb{N}$. Then

(1) $$E_1 \setminus \bigcap_{n\in\mathbb{N}} E_n = \bigcup_{n\in\mathbb{N}} F_n.$$

To show this, let $x \in E_1 \setminus \bigcap_{n\in\mathbb{N}} E_n$. Then $x \in E_1$ and x is not in every E_n. Since $E_n \downarrow$, there exists the first set E_{n_0+1} in the sequence not containing x. Then $x \in E_{n_0} \setminus E_{n_0+1} = F_{n_0} \subset \bigcup_{n\in\mathbb{N}} F_n$. This shows that $E_1 \setminus \bigcap_{n\in\mathbb{N}} E_n \subset \bigcup_{n\in\mathbb{N}} F_n$. Conversely if $x \in \bigcup_{n\in\mathbb{N}} F_n$, then $x \in F_{n_0} = E_{n_0} \setminus E_{n_0+1}$ for some $n_0 \in \mathbb{N}$. Now $x \in E_{n_0} \subset E_1$. Since $x \notin E_{n_0+1}$, we have $x \notin \bigcap_{n\in\mathbb{N}} E_n$. Thus $x \in E_1 \setminus \bigcap_{n\in\mathbb{N}} E_n$. This shows that $\bigcup_{n\in\mathbb{N}} F_n \subset E_1 \setminus \bigcap_{n\in\mathbb{N}} E_n$. Therefore (1) holds. Now by (1), we have

(2) $$\mu\Big(E_1 \setminus \bigcap_{n\in\mathbb{N}} E_n\Big) = \mu\Big(\bigcup_{n\in\mathbb{N}} F_n\Big).$$

Since $\mu\left(\bigcap_{n\in\mathbb{N}} E_n\right) \le \mu(E_1) \le \mu(A) < \infty$, we have by (3) of Lemma 1.25

(3) $$\mu\Big(E_1 \setminus \bigcap_{n\in\mathbb{N}} E_n\Big) = \mu(E_1) - \mu\Big(\bigcap_{n\in\mathbb{N}} E_n\Big) = \mu(E_1) - \mu\big(\lim_{n\to\infty} E_n\big).$$

By the countable additivity of μ, we have

(4) $$\mu\Big(\bigcup_{n\in\mathbb{N}} F_n\Big) = \sum_{n\in\mathbb{N}} \mu(F_n) = \sum_{n\in\mathbb{N}} \mu(E_n \setminus E_{n+1})$$

$$= \sum_{n\in\mathbb{N}} \{\mu(E_n) - \mu(E_{n+1})\} = \lim_{n\to\infty} \sum_{k=1}^{n} \{\mu(E_k) - \mu(E_{k+1})\}$$

$$= \lim_{n\to\infty} \{\mu(E_1) - \mu(E_{n+1})\} = \mu(E_1) - \lim_{n\to\infty} \mu(E_{n+1}).$$

Substituting (3) and (4) in (2), we have

$$\mu(E_1) - \mu\big(\lim_{n\to\infty} E_1\big) = \mu(E_1) - \lim_{n\to\infty} \mu(E_{n+1}) = \mu(E_1) - \lim_{n\to\infty} \mu(E_n).$$

Subtracting $\mu(E_1) \in \mathbb{R}$ from both sides we have $\mu\big(\lim_{n\to\infty} E_n\big) = \lim_{n\to\infty} \mu(E_n)$. ∎

Remark 1.27. (b) of Theorem 1.26 has the following particular cases. Let $(E_n : n \in \mathbb{N})$ be a decreasing sequence in $\mathfrak{A}$. Then $\lim_{n\to\infty} \mu(E_n) = \mu\big(\lim_{n\to\infty} E_n\big)$ if any one of the following conditions is satisfied:
(a) $\mu(X) < \infty$,
(b) $\mu(E_1) < \infty$,
(c) $\mu(E_{n_0}) < \infty$ for some $n_0 \in \mathbb{N}$.

Proof. (a) and (b) are particular cases of (b) of Theorem 1.26 in which X and E_1 respectively are the containing set $A \in \mathfrak{A}$ with $\mu(A) < \infty$.

To prove (c), suppose $\mu(E_{n_0}) < \infty$ for some $n_0 \in \mathbb{N}$. Let $(F_n : n \in \mathbb{N})$ be a decreasing sequence in $\mathfrak{A}$ obtained by dropping the first n_0 terms from $(E_n : n \in \mathbb{N})$, that is, we set $F_n = E_{n_0+n}$ for $n \in \mathbb{N}$. Lemma 1.7 implies that $\liminf_{n\to\infty} F_n = \liminf_{n\to\infty} E_n$ and $\limsup_{n\to\infty} F_n = \limsup_{n\to\infty} E_n$ and thus $\lim_{n\to\infty} F_n = \lim_{n\to\infty} E_n$. Now since $(F_n : n \in \mathbb{N})$ is a decreasing sequence and $F_n \subset E_{n_0}$ for $n \in \mathbb{N}$ and since $\mu(E_{n_0}) < \infty$, (b) of Theorem 1.26 applies so that $\lim_{n\to\infty} \mu(F_n) = \mu\big(\lim_{n\to\infty} F_n\big) = \mu\big(\lim_{n\to\infty} E_n\big)$. Since $(\mu(F_n) : n \in \mathbb{N})$ is a sequence obtained by dropping the first n_0 terms of $(\mu(E_n) : n \in \mathbb{N})$, we have $\lim_{n\to\infty} \mu(F_n) = \lim_{n\to\infty} \mu(E_n)$. Therefore we have $\lim_{n\to\infty} \mu(E_n) = \mu\big(\lim_{n\to\infty} E_n\big)$. ∎

Let μ be a measure on a σ-algebra $\mathfrak{A}$ of subsets of a set X. Then for an arbitrary sequence $(E_n : n \in \mathbb{N})$ in $\mathfrak{A}$, $\liminf_{n\to\infty} E_n$ and $\limsup_{n\to\infty} E_n$ exist in $\mathfrak{A}$ by Theorem 1.9 and thus $\mu\big(\liminf_{n\to\infty} E_n\big)$ and $\mu\big(\limsup_{n\to\infty} E_n\big)$ are defined. Now $(\mu(E_n) : n \in \mathbb{N})$ is a sequence in $[0, \infty]$ and thus $\liminf_{n\to\infty} \mu(E_n) = \lim_{n\to\infty} \inf_{k\ge n} \mu(E_k)$ and $\limsup_{n\to\infty} \mu(E_n) = \lim_{n\to\infty} \sup_{k\ge n} \mu(E_k)$ exist in $[0, \infty]$. How are $\mu\big(\liminf_{n\to\infty} E_n\big)$ and $\mu\big(\limsup_{n\to\infty} E_n\big)$ related respectively to $\liminf_{n\to\infty} \mu(E_n)$ and $\limsup_{n\to\infty} \mu(E_n)$? The next theorem addresses this question.

Theorem 1.28. *Let μ be a measure on a σ-algebra $\mathfrak{A}$ of subsets of a set X.*
(a) *For an arbitrary sequence $(E_n : n \in \mathbb{N})$ in $\mathfrak{A}$, we have*

$$\mu\big(\liminf_{n\to\infty} E_n\big) \le \liminf_{n\to\infty} \mu(E_n). \tag{1}$$

(b) *If there exists $A \in \mathfrak{A}$ with $\mu(A) < \infty$ such that $E_n \subset A$ for $n \in \mathbb{N}$, then*

$$\mu\big(\limsup_{n\to\infty} E_n\big) \ge \limsup_{n\to\infty} \mu(E_n). \tag{2}$$

(c) *If both* $\lim_{n\to\infty} E_n$ *and* $\lim_{n\to\infty} \mu(E_n)$ *exist, then*

$$\mu\big(\lim_{n\to\infty} E_n\big) \le \lim_{n\to\infty} \mu(E_n). \tag{3}$$

(d) *If* $\lim_{n\to\infty} E_n$ *exist and if there exists* $A \in \mathfrak{A}$ *with* $\mu(A) < \infty$ *such that* $E_n \subset A$ *for* $n \in \mathbb{N}$, *then* $\lim_{n\to\infty} \mu(E_n)$ *exists and*

$$\mu\big(\lim_{n\to\infty} E_n\big) = \lim_{n\to\infty} \mu(E_n). \tag{4}$$

Proof. 1. Recall that $\liminf_{n\to\infty} E_n = \bigcup_{n\in\mathbb{N}} \bigcap_{k\ge n} E_k = \lim_{n\to\infty} \bigcap_{k\ge n} E_k$ by the fact that $\big(\bigcap_{k\ge n} E_k : n \in \mathbb{N}\big)$ is an increasing sequence in $\mathfrak{A}$. Then by (a) of Theorem 1.26, we have $\mu\big(\liminf_{n\to\infty} E_n\big) = \lim_{n\to\infty} \mu\left(\bigcap_{k\ge n} E_k\right) = \liminf_{n\to\infty} \mu\left(\bigcap_{k\ge n} E_k\right)$ since the limit of a sequence, if it exists, is equal to the limit inferior of the sequence. Since $\bigcap_{k\ge n} E_k \subset E_n$, we have $\mu\left(\bigcap_{k\ge n} E_k\right) \le \mu(E_n)$ for $n \in \mathbb{N}$ by the monotonicity of μ. This then implies $\liminf_{n\to\infty} \mu\left(\bigcap_{k\ge n} E_k\right) \le \liminf_{n\to\infty} \mu(E_n)$. Continuing the chain of equalities above with this inequality, we have (1).

2. Assume that there exists $A \in \mathfrak{A}$ with $\mu(A) < \infty$ such that $E_n \subset A$ for $n \in \mathbb{N}$. Now $\limsup_{n\to\infty} E_n = \bigcap_{n\in\mathbb{N}} \bigcup_{k\ge n} E_k = \lim_{n\to\infty} \bigcup_{k\ge n} E_k$ by the fact that $\big(\bigcup_{k\ge n} E_k : n \in \mathbb{N}\big)$ is a decreasing sequence in $\mathfrak{A}$. Since $E_n \subset A$ for all $n \in \mathbb{N}$, we have $\bigcup_{k\ge n} E_k \subset A$ for all $n \in \mathbb{N}$. Thus we have $\mu\big(\limsup_{n\to\infty} E_n\big) = \mu\big(\lim_{n\to\infty} \bigcup_{k\ge n} E_k\big) = \lim_{n\to\infty} \mu\left(\bigcup_{k\ge n} E_k\right)$ by (b) of Theorem 1.26. Now $\lim_{n\to\infty} \mu\left(\bigcup_{k\ge n} E_k\right) = \limsup_{n\to\infty} \mu\big(\bigcup_{k\ge n} E_k\big)$ since the limit of a sequence, if it exists, is equal to the limit superior of the sequence. Then by $\bigcup_{k\ge n} E_k \supset E_n$, we have $\mu\left(\bigcup_{k\ge n} E_k\right) \ge \mu(E_n)$. Thus $\limsup_{n\to\infty} \mu\left(\bigcup_{k\ge n} E_k\right) \ge \limsup_{n\to\infty} \mu(E_n)$. Continuing the chain of equalities above with this inequality, we have (2).

3. If $\lim_{n\to\infty} E_n$ and $\lim_{n\to\infty} \mu(E_n)$ exist, then $\lim_{n\to\infty} E_n = \liminf_{n\to\infty} E_n$ and $\lim_{n\to\infty} \mu(E_n) = \liminf_{n\to\infty} \mu(E_n)$ so that (1) reduces to (3).

4. If $\lim_{n\to\infty} E_n$ exists, then $\limsup_{n\to\infty} E_n = \lim_{n\to\infty} E_n = \liminf_{n\to\infty} E_n$. If there exists $A \in \mathfrak{A}$ with $\mu(A) < \infty$ such that $E_n \subset A$ for $n \in \mathbb{N}$, then by (2) and (1) we have

$$\begin{aligned} \limsup_{n\to\infty} \mu(E_n) &\le \mu\big(\limsup_{n\to\infty} E_n\big) = \mu\big(\lim_{n\to\infty} E_n\big) \\ &= \mu\big(\liminf_{n\to\infty} E_n\big) \le \liminf_{n\to\infty} \mu(E_n). \end{aligned} \tag{5}$$

Since $\liminf_{n\to\infty} \mu(E_n) \le \limsup_{n\to\infty} \mu(E_n)$ the inequalities (5) imply

$$\liminf_{n\to\infty} \mu(E_n) = \mu\big(\lim_{n\to\infty} E_n\big) = \limsup_{n\to\infty} \mu(E_n).$$

Thus $\lim_{n\to\infty} \mu(E_n)$ exists and then by (5) we have $\mu\big(\lim_{n\to\infty} E_n\big) = \lim_{n\to\infty} \mu(E_n)$. This proves (4). ∎

[VII] Measurable Space and Measure Space

Definition 1.29. *Let $\mathfrak{A}$ be a σ-algebra of subsets of a set X. The pair $(X, \mathfrak{A})$ is called a measurable space. A subset E of X is said to be $\mathfrak{A}$-measurable if $E \in \mathfrak{A}$.*

Definition 1.30. (a) *If μ is a measure on a σ-algebra $\mathfrak{A}$ of subsets of a set X, we call the triple $(X, \mathfrak{A}, \mu)$ a measure space.*
(b) *A measure μ on a σ-algebra $\mathfrak{A}$ of subsets of a set X is called a finite measure if $\mu(X) < \infty$. In this case, $(X, \mathfrak{A}, \mu)$ is called a finite measure space.*
(c) *A measure μ on a σ-algebra $\mathfrak{A}$ of subsets of a set X is called a σ-finite measure if there exists a sequence $(E_n : n \in \mathbb{N})$ in $\mathfrak{A}$ such that $\bigcup_{n\in\mathbb{N}} E_n = X$ and $\mu(E_n) < \infty$ for every $n \in \mathbb{N}$. In this case $(X, \mathfrak{A}, \mu)$ is called a σ-finite measure space.*
(d) *A set $D \in \mathfrak{A}$ in an arbitrary measure space $(X, \mathfrak{A}, \mu)$ is called a σ-finite set if there exists a sequence $(D_n : n \in \mathbb{N})$ in $\mathfrak{A}$ such that $\bigcup_{n\in\mathbb{N}} D_n = D$ and $\mu(D_n) < \infty$ for every $n \in \mathbb{N}$.*

Lemma 1.31. (a) *Let $(X, \mathfrak{A}, \mu)$ be a measure space. If $D \in \mathfrak{A}$ is a σ-finite set, then there exists an increasing sequence $(F_n : n \in \mathbb{N})$ in $\mathfrak{A}$ such that $\lim_{n\to\infty} F_n = D$ and $\mu(F_n) < \infty$ for every $n \in \mathbb{N}$ and there exists a disjoint sequence $(G_n : n \in \mathbb{N})$ in $\mathfrak{A}$ such that $\bigcup_{n\in\mathbb{N}} G_n = D$ and $\mu(G_n) < \infty$ for every $n \in \mathbb{N}$.*
(b) *If $(X, \mathfrak{A}, \mu)$ is a σ-finite measure space then every $D \in \mathfrak{A}$ is a σ-finite set.*

Proof. 1. Let $(X, \mathfrak{A}, \mu)$ be a measure space. Suppose $D \in \mathfrak{A}$ is a σ-finite set. Then there exists a sequence $(D_n : n \in \mathbb{N})$ in $\mathfrak{A}$ such that $\bigcup_{n\in\mathbb{N}} D_n = D$ and $\mu(D_n) < \infty$ for every $n \in \mathbb{N}$. For each $n \in \mathbb{N}$, let $F_n = \bigcup_{k=1}^{n} D_k$. Then $(F_n : n \in \mathbb{N})$ is an increasing sequence in $\mathfrak{A}$ such that $\lim_{n\to\infty} F_n = \bigcup_{n\in\mathbb{N}} F_n = \bigcup_{n\in\mathbb{N}} D_n = D$ and $\mu(F_n) = \mu\big(\bigcup_{k=1}^{n} D_k\big) \le \sum_{k=1}^{n} \mu(D_k) < \infty$ for every $n \in \mathbb{N}$.

Let $G_1 = F_1$ and $G_n = F_n \setminus \bigcup_{k=1}^{n-1} F_k$ for $n \ge 2$. Then $(G_n : n \in \mathbb{N})$ is a disjoint sequence in $\mathfrak{A}$ such that $\bigcup_{n\in\mathbb{N}} G_n = \bigcup_{n\in\mathbb{N}} F_n = D$ as in the Proof of Lemma 1.21, $\mu(G_1) = \mu(F_1) < \infty$ and $\mu(G_n) = \mu\big(F_n \setminus \bigcup_{k=1}^{n-1} F_k\big) \le \mu(F_n) < \infty$ for $n \ge 2$. This proves (a).

2. Let $(X, \mathfrak{A}, \mu)$ be a σ-finite measure space. Then there exists a sequence $(E_n : n \in \mathbb{N})$ in $\mathfrak{A}$ such that $\bigcup_{n\in\mathbb{N}} E_n = X$ and $\mu(E_n) < \infty$ for every $n \in \mathbb{N}$. Let $D \in \mathfrak{A}$. For each $n \in \mathbb{N}$, let $D_n = D \cap E_n$. Then $(D_n : n \in \mathbb{N})$ is a sequence in $\mathfrak{A}$ such that $\bigcup_{n\in\mathbb{N}} D_n = D$ and $\mu(D_n) \le \mu(E_n) < \infty$ for every $n \in \mathbb{N}$. Thus D is a σ-finite set. This proves (b). ∎

Definition 1.32. *Given a measure μ on a σ-algebra $\mathfrak{A}$ of subsets of a set X. A subset E of X is called a null set with respect to the measure μ if $E \in \mathfrak{A}$ and $\mu(E) = 0$. In this case we say also that E is a null set in the measure space* $(X, \mathfrak{A}, \mu)$. (Note that $\emptyset$ is a null set in any measure space but a null set in a measure space need not be $\emptyset$.)

Observation 1.33. A countable union of null sets in a measure space is a null set of the measure space.

Proof. Let $(E_n : n \in \mathbb{N})$ be a sequence of null sets in a measure space $(X, \mathfrak{A}, \mu)$. Let $E = \bigcup_{n\in\mathbb{N}} E_n$. Since $\mathfrak{A}$ is closed under countable unions, we have $E \in \mathfrak{A}$. By the countable subadditivity of μ on $\mathfrak{A}$, we have $\mu(E) \le \sum_{n\in\mathbb{N}} \mu(E_n) = 0$. Thus $\mu(E) = 0$. This shows that E is a null set in $(X, \mathfrak{A}, \mu)$. ■

Definition 1.34. *Given a measure μ on a σ-algebra $\mathfrak{A}$ of subsets of a set X. We say that the σ-algebra $\mathfrak{A}$ is complete with respect to the measure μ if an arbitrary subset E_0 of a null set E with respect to μ is a member of $\mathfrak{A}$ (and consequently has $\mu(E_0) = 0$ by the monotonicity of μ). When $\mathfrak{A}$ is complete with respect to μ, we say that $(X, \mathfrak{A}, \mu)$ is a complete measure space.*

Example. Let $X = \{a, b, c\}$. Then $\mathfrak{A} = \big\{\emptyset, \{a\}, \{b, c\}, X\big\}$ is a σ-algebra of subsets of X. If we define a set function μ on $\mathfrak{A}$ by setting $\mu(\emptyset) = 0$, $\mu\big(\{a\}\big) = 1$, $\mu\big(\{b, c\}\big) = 0$, and $\mu(X) = 1$, then μ is a measure on $\mathfrak{A}$. The set $\{b, c\}$ is a null set in the measure space $(X, \mathfrak{A}, \mu)$, but its subset $\{b\}$ is not a member of $\mathfrak{A}$. Therefore $(X, \mathfrak{A}, \mu)$ is not a complete measure space.

Definition 1.35. (a) *Given a measurable space $(X, \mathfrak{A})$. An $\mathfrak{A}$-measurable set E is called an atom of the measurable space if $\emptyset$ and E are the only $\mathfrak{A}$-measurable subsets of E.*
(b) *Given a measure space $(X, \mathfrak{A}, \mu)$. An $\mathfrak{A}$-measurable set E is called an atom of the measure space if it satisfies the following conditions :*

1° $\mu(E) > 0$,

2° $E_0 \subset E,\ E_0 \in \mathfrak{A} \Rightarrow \mu(E_0) = 0$ *or* $\mu(E_0) = \mu(E)$.

Observe that if E is an atom of $(X, \mathfrak{A})$ and $\mu(E) > 0$, then E is an atom of $(X, \mathfrak{A}, \mu)$.

Example. In a measurable space $(X, \mathfrak{A})$ where $X = \{a, b, c\}$ and $\mathfrak{A} = \{\emptyset, \{a\}, \{b, c\}, X\}$, if we define a set function μ on $\mathfrak{A}$ by setting $\mu(\emptyset) = 0$, $\mu\big(\{a\}\big) = 1$, $\mu\big(\{b, c\}\big) = 2$, and $\mu(X) = 3$, then μ is a measure on $\mathfrak{A}$. The set $\{b, c\}$ is an atom of the measure space $(X, \mathfrak{A}, \mu)$.

[VIII] Measurable Mapping

Let f be a mapping of a subset D of a set X into a set Y. We write $\mathfrak{D}(f)$ and $\mathfrak{R}(f)$ for the domain of definition and the range of f respectively. Thus

$$\mathfrak{D}(f) = D \subset X,$$

$$\mathfrak{R}(f) = \big\{y \in Y : y = f(x) \text{ for some } x \in \mathfrak{D}(f)\big\} \subset Y.$$

For the image of $\mathfrak{D}(f)$ by f we have $f\big(\mathfrak{D}(f)\big) = \mathfrak{R}(f)$.

For an arbitrary subset E of Y we define the preimage of E under the mapping f by

$$f^{-1}(E) := \big\{x \in X : f(x) \in E\big\} = \big\{x \in \mathfrak{D}(f) : f(x) \in E\big\}.$$

Note that E is an arbitrary subset of Y and need not be a subset of $\mathfrak{R}(f)$. Indeed E may be disjoint from $\mathfrak{R}(f)$, in which case $f^{-1}(E) = \emptyset$. In general we have $f\big(f^{-1}(E)\big) \subset E$. For an arbitrary collection $\mathfrak{E}$ of subsets of Y, we let $f^{-1}(\mathfrak{E}) := \{f^{-1}(E) : E \in \mathfrak{E}\}$.

Observation 1.36. Given sets X and Y. Let f be a mapping with $\mathfrak{D}(f) \subset X$ and $\mathfrak{R}(f) \subset Y$. Let E and E_α be arbitrary subsets of Y. Then

(1) $f^{-1}(Y) = \mathfrak{D}(f)$,

(2) $f^{-1}(E^c) = f^{-1}(Y \setminus E) = f^{-1}(Y) \setminus f^{-1}(E) = \mathfrak{D}(f) \setminus f^{-1}(E)$,

(3) $f^{-1}\big(\bigcup_{\alpha \in A} E_\alpha\big) = \bigcup_{\alpha \in A} f^{-1}(E_\alpha)$,

(4) $f^{-1}\big(\bigcap_{\alpha \in A} E_\alpha\big) = \bigcap_{\alpha \in A} f^{-1}(E_\alpha)$.

Proposition 1.37. *Given sets X and Y. Let f be a mapping with $\mathfrak{D}(f) \subset X$ and $\mathfrak{R}(f) \subset Y$. If $\mathfrak{B}$ is a σ-algebra of subsets of Y then $f^{-1}(\mathfrak{B})$ is a σ-algebra of subsets of the set $\mathfrak{D}(f)$. In particular, if $\mathfrak{D}(f) = X$ then $f^{-1}(\mathfrak{B})$ is a σ-algebra of subsets of the set X.*

Proof. Let $\mathfrak{B}$ be a σ-algebra of subsets of the set Y. To show that $f^{-1}(\mathfrak{B})$ is a σ-algebra of subsets of the set $\mathfrak{D}(f)$ we show that $\mathfrak{D}(f) \in f^{-1}(\mathfrak{B})$; if $A \in f^{-1}(\mathfrak{B})$ then $\mathfrak{D}(f) \setminus A \in f^{-1}(\mathfrak{B})$; and for any sequence $(A_n : n \in \mathbb{N})$ in $f^{-1}(\mathfrak{B})$ we have $\bigcup_{n \in \mathbb{N}} A_n \in f^{-1}(\mathfrak{B})$.

1. By (1) of Observation 1.36, we have $\mathfrak{D}(f) = f^{-1}(Y) \in f^{-1}(\mathfrak{B})$ since $Y \in \mathfrak{B}$.

2. Let $A \in f^{-1}(\mathfrak{B})$. Then $A = f^{-1}(B)$ for some $B \in \mathfrak{B}$. Since $B^c \in \mathfrak{B}$ we have $f^{-1}(B^c) \in f^{-1}(\mathfrak{B})$. On the other hand by (2) of Observation 1.36, we have $f^{-1}(B^c) = \mathfrak{D}(f) \setminus f^{-1}(B) = \mathfrak{D}(f) \setminus A$. Thus $\mathfrak{D}(f) \setminus A \in f^{-1}(\mathfrak{B})$.

3. Let $(A_n : n \in \mathbb{N})$ be a sequence in $f^{-1}(\mathfrak{B})$. Then $A_n = f^{-1}(B_n)$ for some $B_n \in \mathfrak{B}$ for each $n \in \mathbb{N}$. Then by (3) of Observation 1.36, we have

$$\bigcup_{n \in \mathbb{N}} A_n = \bigcup_{n \in \mathbb{N}} f^{-1}(B_n) = f^{-1}\Big(\bigcup_{n \in \mathbb{N}} B_n\Big) \in f^{-1}(\mathfrak{B}),$$

since $\bigcup_{n \in \mathbb{N}} B_n \in \mathfrak{B}$. ∎

Definition 1.38. *Given two measurable spaces $(X, \mathfrak{A})$ and $(Y, \mathfrak{B})$. Let f be a mapping with $\mathfrak{D}(f) \subset X$ and $\mathfrak{R}(f) \subset Y$. We say that f is a $\mathfrak{A}/\mathfrak{B}$-measurable mapping if $f^{-1}(B) \in \mathfrak{A}$ for every $B \in \mathfrak{B}$, that is, $f^{-1}(\mathfrak{B}) \subset \mathfrak{A}$.*

According to Proposition 1.37 for an arbitrary mapping f of $\mathfrak{D}(f) \subset X$ into Y, $f^{-1}(\mathfrak{B})$ is a σ-algebra of subsets of the set $\mathfrak{D}(f)$. $\mathfrak{A}/\mathfrak{B}$-measurability of the mapping f requires that the σ-algebra $f^{-1}(\mathfrak{B})$ of subsets of $\mathfrak{D}(f)$ be a subcollection of the σ-algebra $\mathfrak{A}$ of subsets of X. Note also that the $\mathfrak{A}/\mathfrak{B}$-measurability of f implies that $\mathfrak{D}(f) = f^{-1}(Y) \in \mathfrak{A}$.

Observation 1.39. Given two measurable spaces $(X, \mathfrak{A})$ and $(Y, \mathfrak{B})$. Let f be a $\mathfrak{A}/\mathfrak{B}$-measurable mapping.

(a) If $\mathfrak{A}_1$ is a σ-algebra of subsets of X such that $\mathfrak{A}_1 \supset \mathfrak{A}$, then f is $\mathfrak{A}_1/\mathfrak{B}$-measurable.
(b) If $\mathfrak{B}_0$ is a σ-algebra of subsets of Y such that $\mathfrak{B}_0 \subset \mathfrak{B}$, then f is $\mathfrak{A}/\mathfrak{B}_0$-measurable.

Proof. (a) follows from $f^{-1}(\mathfrak{B}) \subset \mathfrak{A} \subset \mathfrak{A}_1$ and (b) from $f^{-1}(\mathfrak{B}_0) \subset f^{-1}(\mathfrak{B}) \subset \mathfrak{A}$. ∎

Composition of two measurable mappings is a measurable mapping provided that the two measurable mappings form a chain. To be precise, we have the following:

Theorem 1.40. (Chain Rule for Measurable Mappings) *Given measurable spaces $(X, \mathfrak{A})$, $(Y, \mathfrak{B})$, and $(Z, \mathfrak{C})$. Let f be a mapping with $\mathfrak{D}(f) \subset X$, $\mathfrak{R}(f) \subset Y$, g be a mapping with $\mathfrak{D}(g) \subset Y$, $\mathfrak{R}(g) \subset Z$ such that $\mathfrak{R}(f) \subset \mathfrak{D}(g)$ so that the composite mapping $g \circ f$ is defined with $\mathfrak{D}(g \circ f) \subset X$ and $\mathfrak{R}(g \circ f) \subset Z$. If f is $\mathfrak{A}/\mathfrak{B}$-measurable and g is $\mathfrak{B}/\mathfrak{C}$-measurable, then $g \circ f$ is $\mathfrak{A}/\mathfrak{C}$-measurable.*

Proof. By the $\mathfrak{A}/\mathfrak{B}$-measurability of f, we have $f^{-1}(\mathfrak{B}) \subset \mathfrak{A}$, and by the $\mathfrak{B}/\mathfrak{C}$-measurability of g, we have $g^{-1}(\mathfrak{C}) \subset \mathfrak{B}$. Thus $(g \circ f)^{-1}(\mathfrak{C}) = f^{-1}\big(g^{-1}(\mathfrak{C})\big) \subset f^{-1}(\mathfrak{B}) \subset \mathfrak{A}$. ∎

The $\mathfrak{A}/\mathfrak{B}$-measurability condition can be reduced when $\mathfrak{B}$ is the σ-algebra generated by a collection $\mathfrak{E}$ of subsets of Y. Thus we have the following:

Theorem 1.41. *Given two measurable spaces $(X, \mathfrak{A})$ and $(Y, \mathfrak{B})$, where $\mathfrak{B} = \sigma(\mathfrak{E})$ and $\mathfrak{E}$ is an arbitrary collection of subsets of Y. Let f be a mapping with $\mathfrak{D}(f) \in \mathfrak{A}$ and $\mathfrak{R}(f) \subset Y$. Then f is a $\mathfrak{A}/\mathfrak{B}$-measurable mapping of $\mathfrak{D}(f)$ into Y if and only if $f^{-1}(\mathfrak{E}) \subset \mathfrak{A}$.*

Proof. If f is a $\mathfrak{A}/\mathfrak{B}$-measurable mapping of $\mathfrak{D}(f)$ into Y, then $f^{-1}(\mathfrak{B}) \subset \mathfrak{A}$ so that $f^{-1}(\mathfrak{E}) \subset \mathfrak{A}$. Conversely if $f^{-1}(\mathfrak{E}) \subset \mathfrak{A}$, then $\sigma\big(f^{-1}(\mathfrak{E})\big) \subset \sigma(\mathfrak{A}) = \mathfrak{A}$. Now by Theorem 1.14, $\sigma\big(f^{-1}(\mathfrak{E})\big) = f^{-1}\big(\sigma(\mathfrak{E})\big) = f^{-1}(\mathfrak{B})$. Thus $f^{-1}(\mathfrak{B}) \subset \mathfrak{A}$ and f is a $\mathfrak{A}/\mathfrak{B}$-measurable mapping of $\mathfrak{D}(f)$. ∎

Proposition 1.42. *Given two measurable spaces $(X, \mathfrak{A})$ and $(Y, \mathfrak{B}_Y)$, where Y is a topological space and $\mathfrak{B}_Y$ is the Borel σ-algebra of subsets of Y. Let f be a mapping with $\mathfrak{D}(f) \in \mathfrak{A}$ and $\mathfrak{R}(f) \subset Y$. Let $\mathfrak{O}_Y$ and $\mathfrak{C}_Y$ be respectively the collection of all open sets and the collection of all closed sets in Y.*
(a) *f is a $\mathfrak{A}/\mathfrak{B}_Y$-measurable mapping of $\mathfrak{D}(f)$ into Y if and only if $f^{-1}(\mathfrak{O}_Y) \subset \mathfrak{A}$.*
(b) *f is a $\mathfrak{A}/\mathfrak{B}_Y$-measurable mapping of $\mathfrak{D}(f)$ into Y if and only if $f^{-1}(\mathfrak{C}_Y) \subset \mathfrak{A}$.*

Proof. Since $\mathfrak{B}_Y = \sigma(\mathfrak{O}_Y) = \sigma(\mathfrak{C}_Y)$, the Proposition is a particular case of Theorem 1.41. ∎

Theorem 1.43. *Given two measurable spaces $(X, \mathfrak{B}_X)$ and $(Y, \mathfrak{B}_Y)$ where X and Y are topological spaces and $\mathfrak{B}_X$ and $\mathfrak{B}_Y$ are the Borel σ-algebras of subsets of X and Y respectively. If f is a continuous mapping defined on a set $D \in \mathfrak{B}_X$, then f is a $\mathfrak{B}_X/\mathfrak{B}_Y$-measurable mapping of D into Y.*

Proof. Let V be an open set in Y. The continuity of f on D implies that $f^{-1}(V) = U \cap D$ where U is an open set in X so that $f^{-1}(V) \in \mathfrak{B}_X$. Since this holds for every open set V in Y, f is a $\mathfrak{B}_X/\mathfrak{B}_Y$-measurable mapping of D into Y by (a) of Proposition 1.42. ∎

A particular case of Theorem 1.43 is when we have a real-valued continuous function f defined on a set $D \in \mathfrak{B}_X$ where $\mathfrak{B}_X$ is the Borel σ-algebra of subsets of a topological space X. In this case we have $(Y, \mathfrak{B}_Y) = (\mathbb{R}, \mathfrak{B}_{\mathbb{R}})$. By Theorem 1.43, f is a $\mathfrak{B}_X/\mathfrak{B}_{\mathbb{R}}$-measurable mapping of D into $\mathbb{R}$.

[IX] Induction of Measure by Measurable Mapping

Let μ be a measure on a σ-algebra $\mathfrak{A}$ of subsets of a set X. We show next that a measurable mapping of the measurable space $(X, \mathfrak{A})$ into another measurable space $(Y, \mathfrak{B})$ induces a measure on the σ-algebra $\mathfrak{B}$. The induced measure on $\mathfrak{B}$ is called the image measure induced by the measurable mapping.

Theorem 1.44. (Image Measure) *Given two measurable spaces $(X, \mathfrak{A})$ and $(Y, \mathfrak{B})$. Let f be a $\mathfrak{A}/\mathfrak{B}$-measurable mapping of X into Y. Let μ be a measure on $\mathfrak{A}$. The set function defined by $\nu = \mu \circ f^{-1}$ on $\mathfrak{B}$, that is, $\nu(B) = \mu\left(f^{-1}(B)\right)$ for $B \in \mathfrak{B}$, is a measure on $\mathfrak{B}$.*

Proof. Since f is a $\mathfrak{A}/\mathfrak{B}$-measurable mapping of X into Y, we have $f^{-1}(B) \in \mathfrak{A}$ for every $B \in \mathfrak{B}$ and then $\nu(B) = \mu\left(f^{-1}(B)\right) \in [0, \infty]$. Also $\nu(\emptyset) = \mu\left(f^{-1}(\emptyset)\right) = \mu(\emptyset) = 0$. Let $(B_n : n \in \mathbb{N})$ be a disjoint sequence in $\mathfrak{B}$. Then $(f^{-1}(B_n) : n \in \mathbb{N})$ is a disjoint sequence in $\mathfrak{A}$ and $f^{-1}\left(\bigcup_{n\in\mathbb{N}} B_n\right) = \bigcup_{n\in\mathbb{N}} f^{-1}(B_n) \in \mathfrak{A}$. Thus we have the equality $\nu\left(\bigcup_{n\in\mathbb{N}} B_n\right) = \mu\left(f^{-1}\left(\bigcup_{n\in\mathbb{N}} B_n\right)\right) = \sum_{n\in\mathbb{N}} \mu\left(f^{-1}(B_n)\right) = \sum_{n\in\mathbb{N}} \nu(B_n)$. This shows that ν is countably additive on $\mathfrak{A}$. Therefore ν is a measure on $\mathfrak{B}$. ∎

Problems

Prob. 1.1. Given two sequences of subsets $(E_n : n \in \mathbb{N})$ and $(F_n : n \in \mathbb{N})$ of a set X.
(a) Show that

$$\begin{aligned}\liminf_{n\to\infty} E_n \cup \liminf_{n\to\infty} F_n &\subset \liminf_{n\to\infty}(E_n \cup F_n) \subset \liminf_{n\to\infty} E_n \cup \limsup_{n\to\infty} F_n \\ &\subset \limsup_{n\to\infty}(E_n \cup F_n) \subset \limsup_{n\to\infty} E_n \cup \limsup_{n\to\infty} F_n.\end{aligned}$$

(b) State and prove a similar chain of inclusions for intersections.
(c) Show that if $\lim_{n\to\infty} E_n$ and $\lim_{n\to\infty} F_n$ exist, then $\lim_{n\to\infty}\left(E_n \cup F_n\right)$ and $\lim_{n\to\infty}\left(E_n \cap F_n\right)$ exist.

Prob. 1.2. (a) Let $(A_n : n \in \mathbb{N})$ be a sequence of subsets of a set X. Let $(B_n : n \in \mathbb{N})$ be a sequence obtained by dropping finitely many entries in the sequence $(A_n : n \in \mathbb{N})$. Show that $\liminf_{n\to\infty} B_n = \liminf_{n\to\infty} A_n$ and $\limsup_{n\to\infty} B_n = \limsup_{n\to\infty} A_n$. Show that $\lim_{n\to\infty} B_n$ exists if and only if $\lim_{n\to\infty} A_n$ exists and when they exist they are equal.

(b) Let $(A_n : n \in \mathbb{N})$ and $(B_n : n \in \mathbb{N})$ be two sequences of subsets of a set X such that $A_n = B_n$ for all but finitely many $n \in \mathbb{N}$. Show that $\liminf_{n\to\infty} B_n = \liminf_{n\to\infty} A_n$ and $\limsup_{n\to\infty} B_n = \limsup_{n\to\infty} A_n$. Show that $\lim_{n\to\infty} B_n$ exists if and only if $\lim_{n\to\infty} A_n$ exists and when they exist they are equal.

Prob. 1.3. (a) Let $a \in \mathbb{R}$ and let $(x_n : n \in \mathbb{N})$ be a sequence of points in $\mathbb{R}$, all distinct from a, such that $\lim_{n\to\infty} x_n = a$. Show that $\lim_{n\to\infty} \{x_n\} \neq \{a\}$.
(b) Let $E \subset \mathbb{R}$. For $t \in \mathbb{R}$ let $E + t = \{x + t : x \in E\}$, that is, the translate of E by t. Let $(t_n : n \in \mathbb{N})$ be a sequence in $\mathbb{R}$ such that $\lim_{n\to\infty} t_n = 0$ and let $E_n = E + t_n$ for $n \in \mathbb{N}$. Construct a set $E \subset \mathbb{R}$ such that $\lim_{n\to\infty} E_n \neq E$.

Prob. 1.4. The characteristic function $\mathbf{1}_A$ of a subset A of a set X is a function on X defined by

$$\mathbf{1}_A(x) = \begin{cases} 1 & \text{for } x \in A, \\ 0 & \text{for } x \in A^c. \end{cases}$$

Let $(A_n : n \in \mathbb{N})$ be a sequence of subsets of X and A be a subset of X.
(a) Show that if $\lim_{n\to\infty} A_n = A$ then $\lim_{n\to\infty} \mathbf{1}_{A_n} = \mathbf{1}_A$ on X.
(b) Show that if $\lim_{n\to\infty} \mathbf{1}_{A_n} = \mathbf{1}_A$ on X then $\lim_{n\to\infty} A_n = A$.

Prob. 1.5. Let $\mathfrak{A}$ be a σ-algebra of subsets of a set X and let Y be an arbitrary subset of X. Let $\mathfrak{B} = \{A \cap Y : A \in \mathfrak{A}\}$. Show that $\mathfrak{B}$ is a σ-algebra of subsets of Y.

Prob. 1.6. Let $\mathfrak{A}$ be a collection of subsets of a set X with the following properties:
1°. $X \in \mathfrak{A}$,
2°. $A, B \in \mathfrak{A} \Rightarrow A \setminus B = A \cap B^c \in \mathfrak{A}$.
Show that $\mathfrak{A}$ is an algebra of subsets of the set X.

Prob. 1.7. Let $\mathfrak{A}$ be an algebra of subsets of a set X. Suppose $\mathfrak{A}$ has the property that for every increasing sequence $(A_n : n \in \mathbb{N})$ in $\mathfrak{A}$, we have $\bigcup_{n\in\mathbb{N}} A_n \in \mathfrak{A}$. Show that $\mathfrak{A}$ is a σ-algebra of subsets of the set X.
(Hint: For an arbitrary sequence $(B_n : n \in \mathbb{N})$ in $\mathfrak{A}$, we have $\bigcup_{n\in\mathbb{N}} B_n = \bigcup_{n\in\mathbb{N}} A_n$ where $(A_n : n \in \mathbb{N})$ is an increasing sequence in $\mathfrak{A}$ defined by $A_n = \bigcup_{k=1}^n B_k$ for $n \in \mathbb{N}$.)

Prob. 1.8. (a) Show that if $(\mathfrak{A}_n : n \in \mathbb{N})$ is an increasing sequence of algebras of subsets of a set X, then $\bigcup_{n\in\mathbb{N}} \mathfrak{A}_n$ is an algebra of subsets of X.
(b) Show by example that even if $\mathfrak{A}_n$ in (a) is a σ-algebra for every $n \in \mathbb{N}$, the union still may not be a σ-algebra.

Prob. 1.9. Let $(X, \mathfrak{A})$ be a measurable space and let $(E_n : n \in \mathbb{N})$ be an increasing sequence in $\mathfrak{A}$ such that $\bigcup_{n\in\mathbb{N}} E_n = X$.
(a) Let $\mathfrak{A}_n = \mathfrak{A} \cap E_n$, that is, $\mathfrak{A}_n = \{A \cap E_n : A \in \mathfrak{A}\}$. Show that $\mathfrak{A}_n$ is a σ-algebra of subsets of E_n for each $n \in \mathbb{N}$.
(b) Does $\bigcup_{n\in\mathbb{N}} \mathfrak{A}_n = \mathfrak{A}$ hold?

Prob. 1.10. Let X be an arbitrary infinite set. We say that a subset A of X is co-finite if A^c

is a finite set. Let $\mathfrak{A}$ consist of all the finite and the co-finite subsets of a set X.
(a) Show that $\mathfrak{A}$ is an algebra of subsets of X.
(b) Show that $\mathfrak{A}$ is a σ-algebra if and only if X is a finite set.

Prob. 1.11. Let X be an arbitrary uncountable set. We say that a subset A of X is co-countable if A^c is a countable set. Let $\mathfrak{A}$ consist of all the countable and the co-countable subsets of a set X. Show that $\mathfrak{A}$ is a σ-algebra.
(This offers an example where an uncountable union of members of a σ-algebra is not a member of the σ-algebra. Indeed, let X be an uncountable set and let A be a subset of X such that neither A nor A^c is a countable set so that $A, A^c \notin \mathfrak{A}$. Let A be given as $A = \{x_\gamma \in X : \gamma \in \Gamma\}$ where Γ in an uncountable set. Then $\{x_\gamma\} \in \mathfrak{A}$ for every $\gamma \in \Gamma$, but $\bigcup_{\gamma \in \Gamma}\{x_\gamma\} = A \notin \mathfrak{A}$.)

Prob. 1.12. For an arbitrary collection $\mathfrak{E}$ of subsets of a set X, let $\alpha(\mathfrak{E})$ be the algebra generated by $\mathfrak{E}$, that is, the smallest algebra of subsets of X containing $\mathfrak{E}$, and let $\sigma(\mathfrak{E})$ be the σ-algebra generated by $\mathfrak{E}$. Prove the following statements:
(a) $\alpha(\alpha(\mathfrak{E})) = \alpha(\mathfrak{E})$,
(b) $\sigma(\sigma(\mathfrak{E})) = \sigma(\mathfrak{E})$,
(c) $\alpha(\mathfrak{E}) \subset \sigma(\mathfrak{E})$,
(d) if $\mathfrak{E}$ is a finite collection, then $\alpha(\mathfrak{E}) = \sigma(\mathfrak{E})$,
(e) $\sigma(\alpha(\mathfrak{E})) = \sigma(\mathfrak{E})$.
(Hint for (d): Use Prob. 1.13 below.)

Prob. 1.13. Let $\mathfrak{E} = \{E_1, \cdots, E_n\}$ be a finite collection of distinct, but not necessarily disjoint, subsets of a set X. Let $\mathfrak{D}$ be the collection of all subsets of X which have expressions of the type $F_1 \cap \cdots \cap F_n$ where F_i is either E_i or E_i^c for each $i = 1, \cdots, n$. Let $\mathfrak{A}$ be the collection of all finite unions of members of $\mathfrak{D}$.
(a) Show that $\mathfrak{D}$ has at most 2^n distinct members.
(b) Show that $\mathfrak{A} = \alpha(\mathfrak{E})$.
(c) Show that $\alpha(\mathfrak{E})$ has at most 2^{2^n} distinct members.
(d) Show that $\sigma(\mathfrak{E}) = \alpha(\mathfrak{E})$.
Remark. For an arbitrary collection $\mathfrak{E}$ of subsets of a set X, the smallest σ-algebra of subsets of X containing $\mathfrak{E}$, $\sigma(\mathfrak{E})$, always exists according to Theorem 1.11. Prob. 1.13 presents a method of constructing $\sigma(\mathfrak{E})$ for the case that $\mathfrak{E}$ is a finite collection.

Prob. 1.14. Let $(\mathfrak{A}_n : n \in \mathbb{N})$ be a monotone sequence of algebras of subsets of a set X and let $\mathfrak{A} = \lim_{n\to\infty} \mathfrak{A}_n$. Show that $\mathfrak{A}$ is an algebra.

Prob. 1.15. Let $(\mathfrak{A}_n : n \in \mathbb{N})$ be a monotone sequence of σ-algebras of subsets of a set X and let $\mathfrak{A} = \lim_{n\to\infty} \mathfrak{A}_n$.
(a) Show that if $(\mathfrak{A}_n : n \in \mathbb{N})$ is a decreasing sequence then $\mathfrak{A}$ is a σ-algebra.
(b) Show that if $(\mathfrak{A}_n : n \in \mathbb{N})$ is an increasing sequence then $\mathfrak{A}$ need not be a σ-algebra by constructing an example.

Prob. 1.16. Let $\{A_i : i = 1, \ldots, n\}$ be a disjoint collection of nonempty subsets of a set X such that $\bigcup_{i=1}^n A_i = X$.

(a) Show that the collection of all finite unions of members of $\{A_i : i = 1, \ldots, n\}$ is equal to $\sigma(\{A_i : i = 1, \ldots, n\})$.
(b) Show that $\sigma(\{A_i : i = 1, \ldots, n\})$ is a finite set.

Prob. 1.17. Let $\{A_n : n \in \mathbb{N}\}$ be a disjoint collection of nonempty subsets of a set X such that $\bigcup_{n \in \mathbb{N}} A_n = X$.
(a) Show that the collection of all countable unions of members of $\{A_n : n \in \mathbb{N}\}$ is equal to $\sigma(\{A_n : n \in \mathbb{N}\})$.
(b) Show that $\sigma(\{A_n : n \in \mathbb{N}\})$ is an uncountable set.

Prob. 1.18. Show that a σ-algebra of subsets of a set cannot be a countably infinite collection, that is, its cardinality is either finite or else it is at least that of the continuum.

Prob. 1.19. Let $\mathfrak{E}$ be an arbitrary collection of subsets of a set X. Show that for every $A \in \alpha(\mathfrak{E})$ there exists a finite subcollection $\{E_1, \cdots, E_n\}$ of $\mathfrak{E}$ depending on A such that $A \in \alpha(\{E_1, \cdots, E_n\})$.
(Hint: Show that the collection of all sets A with this property is an algebra and then show that this algebra contains $\mathfrak{E}$.)

Prob. 1.20. Let $\mathfrak{E}$ be an arbitrary collection of subsets of a set X. Show that for every $A \in \sigma(\mathfrak{E})$ there exists a countable subcollection $\mathfrak{E}_A$ of $\mathfrak{E}$ depending on A such that $A \in \sigma(\mathfrak{E}_A)$. (We say that every member of $\sigma(\mathfrak{E})$ is countably generated.)

Prob. 1.21. Let X be a countably infinite set and let $\mathfrak{A}$ be the σ-algebra of all subsets of X. Define a set function μ on $\mathfrak{A}$ by defining for every $E \in \mathfrak{A}$

$$\mu(E) = \begin{cases} 0 & \text{if } E \text{ is a finite set,} \\ \infty & \text{otherwise.} \end{cases}$$

(a) Show that μ is additive but not countably additive on $\mathfrak{A}$.
(b) Show that X is the limit of an increasing sequence $(E_n : n \in \mathbb{N})$ in $\mathfrak{A}$ with $\mu(E_n) = 0$ for all n, but $\mu(X) = \infty$.

Prob. 1.22. Let X be an infinite set and let $\mathfrak{A}$ be the algebra consisting of the finite and the co-finite subsets of X (cf. Prob. 1.10). Define a set function μ on $\mathfrak{A}$ by setting for every $A \in \mathfrak{A}$:

$$\mu(A) = \begin{cases} 0 & \text{if } A \text{ is finite,} \\ 1 & \text{if } A \text{ is co-finite.} \end{cases}$$

(Note that since X is an infinite set, no subset A of X can be both finite and co-finite although it can be neither.)
(a) Show that μ is additive on the algebra $\mathfrak{A}$.
(b) Show that when X is countably infinite, μ is not countably additive on the algebra $\mathfrak{A}$.
(c) Show that when X is countably infinite, then X is the limit of an increasing sequence $\{A_n : n \in \mathbb{N}\}$ in $\mathfrak{A}$ with $\mu(A_n) = 0$ for every $n \in \mathbb{N}$, but $\mu(X) = 1$.
(d) Show that when X is uncountable, then μ is countably additive on the algebra $\mathfrak{A}$.

Prob. 1.23. Let X be an uncountable set and let $\mathfrak{A}$ be the σ-algebra consisting of the countable and the co-countable subsets of X (cf. Prob. 1.11). Define a set function μ on $\mathfrak{A}$ by setting for every $A \in \mathfrak{A}$:

$$\mu(A) = \begin{cases} 0 & \text{if } A \text{ is countable,} \\ 1 & \text{if } A \text{ is co-countable.} \end{cases}$$

(Note that since X is an uncountable set, no subset A of X can be both countable and co-countable although it can be neither.) Show that μ is countably additive on $\mathfrak{A}$.

Prob. 1.24. Let $X = (0, \infty)$ and let $\mathfrak{J}$ be the collection of intervals of the type $(n-1, n]$ for $n \in \mathbb{N}$. Let $\mathfrak{A}$ be the collection of all arbitrary unions of members of $\mathfrak{J}$. For every $A \in \mathfrak{A}$ let us define $\mu(A)$ to be the number of elements of $\mathfrak{J}$ that constitute A.
(a) Show that $\mathfrak{A}$ is a σ-algebra of subsets of X.
(b) Show that μ is a measure on the σ-algebra $\mathfrak{A}$.
(c) Let $(A_n : n \in \mathbb{N}) \subset \mathfrak{A}$ where $A_n = (n, \infty)$ for $n \in \mathbb{N}$. Show that for the decreasing sequence $(A_n : n \in \mathbb{N})$ we have $\lim_{n\to\infty} \mu(A_n) \neq \mu\big(\lim_{n\to\infty} A_n\big)$.

Prob. 1.25. Let $(X, \mathfrak{A}, \mu)$ be a finite measure space. Let $\mathfrak{E} = \{E_\lambda : \lambda \in \Lambda\}$ be a disjoint collection of members of $\mathfrak{A}$ such that $\mu(E_\lambda) > 0$ for every $\lambda \in \Lambda$. Show that $\mathfrak{E}$ is at most a countable collection.

Prob. 1.26. Given a measure space $(X, \mathfrak{A}, \mu)$. We say that a sequence $(A_n : n \in \mathbb{N})$ in $\mathfrak{A}$ is almost disjoint if $\mu(A_j \cap A_j) = 0$ for $j \neq k$.
(a) Show that if $(A_n : n \in \mathbb{N}) \subset \mathfrak{A}$ is almost disjoint, then $\mu\left(\bigcup_{n\in\mathbb{N}} A_n\right) = \sum_{n\in\mathbb{N}} \mu(A_n)$.
(b) Show that if $(A_n : n \in \mathbb{N}) \subset \mathfrak{A}$ is such that $\mu\left(\bigcup_{n\in\mathbb{N}} A_n\right) = \sum_{n\in\mathbb{N}} \mu(A_n) < \infty$, then $(A_n : n \in \mathbb{N})$ is almost disjoint.
(c) If we remove the condition $\sum_{n\in\mathbb{N}} \mu(A_n) < \infty$ in (b), then the conclusion is not valid. Show this by constructing an example.

Prob. 1.27. Let μ be a measure on a σ-algebra $\mathfrak{A}$ of subsets of a set X and let $\mathfrak{A}_0$ be a sub-σ-algebra of $\mathfrak{A}$, that is, $\mathfrak{A}_0$ is a σ-algebra of subsets of X and $\mathfrak{A}_0 \subset \mathfrak{A}$. Show that the restriction of μ to $\mathfrak{A}_0$ is a measure on $\mathfrak{A}_0$.

Prob. 1.28. Let μ_1 and μ_2 be measures on a σ-algebra $\mathfrak{A}$ of subsets of a set X and let $\alpha_1, \alpha_2 \geq 0$. Show that the set function $\alpha_1\mu_1 + \alpha_2\mu_2$ on $\mathfrak{A}$ defined for $E \in \mathfrak{A}$ by setting $(\alpha_1\mu_1 + \alpha_2\mu_2)(E) = \alpha_1\mu_1(E) + \alpha_2\mu_2(E)$ is a measure on $\mathfrak{A}$.

Prob. 1.29. Let $(X, \mathfrak{A}, \mu)$ be a σ-finite measure space so that there exists a sequence $(E_n : n \in \mathbb{N})$ in $\mathfrak{A}$ such that $\bigcup_{n\in\mathbb{N}} E_n = X$ and $\mu(E_n) < \infty$ for every $n \in \mathbb{N}$. Show that there exists a disjoint sequence $(F_n : n \in \mathbb{N})$ in $\mathfrak{A}$ such that $\bigcup_{n\in\mathbb{N}} F_n = X$ and $\mu(F_n) < \infty$ for every $n \in \mathbb{N}$.

Prob. 1.30. Let $(X, \mathfrak{A}, \mu)$ be a measure space. Show that for any $E_1, E_2 \in \mathfrak{A}$ we have the equality: $\mu(E_1 \cup E_2) + \mu(E_1 \cap E_2) = \mu(E_1) + \mu(E_2)$.

Prob. 1.31. The symmetric difference of two subsets A and B of a set X is defined by $\mu(A \triangle B) = (A \setminus B) \cup (B \setminus A)$.
(a) Prove the triangle inequality for the symmetric difference of sets, that is, for any three subsets A, B, C of a set X we have $A \triangle B \subset (A \triangle C) \cup (C \triangle B)$.
(b) Let $(X, \mathfrak{A}, \mu)$ be a measure space. Show that $\mu(A \triangle B) \leq \mu(A \triangle C) + \mu(C \triangle B)$ for any $A, B, C \in \mathfrak{A}$.

Remark. Let $(X, \mathfrak{A}, \mu)$ be a finite measure space. Then a function ρ on $\mathfrak{A} \times \mathfrak{A}$ defined by $\rho(A, B) = \mu(A \triangle B)$ for $A, B \in \mathfrak{A}$ has the following properties:
1° $\rho(A, B) \in [0, \mu(X)]$,
2° $\rho(A, B) = \rho(B, A)$,
3° $\rho(A, B) \leq \rho(A, C) + \rho(C, B)$.
However ρ need not be a metric on the set $\mathfrak{A}$ since $\rho(A, B) = 0$ does not imply $A = B$.

Prob. 1.32. Let $(X, \mathfrak{A}, \mu)$ be a finite measure space. Let a relation $\sim$ among the members of $\mathfrak{A}$ be defined by writing $A \sim B$ when $\mu(A \triangle B) = 0$.
(a) Show that $\sim$ is an equivalence relation, that is,
1° $A \sim A$,
2° $A \sim B \Rightarrow B \sim A$,
3° $A \sim B, B \sim C \Rightarrow A \sim C$.
(b) Let $[A]$ be the equivalence class to which A belongs and let $[\mathfrak{A}]$ be the collection of all the equivalence classes with respect to the equivalence relation $\sim$. Define a function ρ^* on $[\mathfrak{A}] \times [\mathfrak{A}]$ by setting $\rho^*([A], [B]) = \mu(A \triangle B)$ for $[A], [B] \in [\mathfrak{A}]$.
(c) Show that ρ^* is well defined in the sense that its definition as given above does not depend on the particular representative A and B of the equivalence classes $[A]$ and $[B]$; in other words,

$$A' \in [A], B' \in [B] \Rightarrow \mu(A' \triangle B') = \mu(A \triangle B).$$

(d) Show that ρ^* is a metric on the set $[\mathfrak{A}]$.

§2 Outer Measures

[I] Construction of Measure by Means of Outer Measure

Definition 2.1. *Let X be an arbitrary set. A set function μ^* defined on the σ-algebra $\mathfrak{P}(X)$ of all subsets of X is called an outer measure on X if it satisfies the following conditions :*

1° *nonnegative extended real-valued : $\mu^*(E) \in [0, \infty]$ for every $E \in \mathfrak{P}(X)$,*

2° $\mu^*(\emptyset) = 0$,

3° *monotonicity : $E_1, E_2 \in \mathfrak{P}(X), E_1 \subset E_2 \Rightarrow \mu^*(E_1) \leq \mu^*(E_2)$,*

4° *countable subadditivity: $(E_n : n \in \mathbb{N}) \subset \mathfrak{P}(X) \Rightarrow \mu^*\big(\bigcup_{n\in\mathbb{N}} E_n\big) \leq \sum_{n\in\mathbb{N}} \mu^*(E_n)$.*

By definition an outer measure μ^* is a countably subadditive set function on the σ-algebra $\mathfrak{P}(X)$. If it is also additive on $\mathfrak{P}(X)$, that is, if it satisfies the condition that $\mu^*(E_1 \cup E_2) = \mu^*(E_1) + \mu^*(E_2)$ whenever $E_1, E_2 \in \mathfrak{P}(X)$ and $E_1 \cap E_2 = \emptyset$, then according to Proposition 1.23, μ^* is countably additive on $\mathfrak{P}(X)$ so that it is a measure on $\mathfrak{P}(X)$. In general an outer measure μ^* does not satisfy the additivity condition on $\mathfrak{P}(X)$. We shall show that there exists a σ-algebra $\mathfrak{A}$ of subsets of X, $\mathfrak{A} \subset \mathfrak{P}(X)$, such that when μ^* is restricted to $\mathfrak{A}$ it is additive on $\mathfrak{A}$. Then μ^* is countably additive on $\mathfrak{A}$ by Proposition 1.23 and is thus a measure on $\mathfrak{A}$.

Let $E \in \mathfrak{P}(X)$. For an arbitrary $A \in \mathfrak{P}(X)$, we have $(A \cap E) \cap (A \cap E^c) = \emptyset$ and $(A \cap E) \cup (A \cap E^c) = A$.

Definition 2.2. *Let μ^* be an outer measure on a set X. We say that a set $E \in \mathfrak{P}(X)$ is measurable with respect to μ^* (or μ^*-measurable) if it satisfies the following Carathéodory condition :*

$$\mu^*(A) = \mu^*(A \cap E) + \mu^*(A \cap E^c) \quad \text{for every } A \in \mathfrak{P}(X).$$

The set A is called a testing set in the Carathéodory condition. We write $\mathfrak{M}(\mu^*)$ for the collection of all μ^*-measurable $E \in \mathfrak{P}(X)$.

Observation 2.3. The countable subadditivity of μ^* implies its finite subadditivity on $\mathfrak{P}(X)$ by Observation 1.20. Thus $\mu^*(A) \leq \mu^*(A \cap E) + \mu^*(A \cap E^c)$ for any $E, A \in \mathfrak{P}(X)$. Therefore to verify the Carathéodory condition for $E \in \mathfrak{P}(X)$, it suffices to verify that $\mu^*(A) \geq \mu^*(A \cap E) + \mu^*(A \cap E^c)$ for every $A \in \mathfrak{P}(X)$.

Lemma 2.4. *Let μ^* be an outer measure on a set X. Consider the collection $\mathfrak{M}(\mu^*)$ of all μ^*-measurable $E \in \mathfrak{P}(X)$.*

(a) *If $E_1, E_2 \in \mathfrak{M}(\mu^*)$, then $E_1 \cup E_2 \in \mathfrak{M}(\mu^*)$.*

(b) *The set function μ^* is additive on $\mathfrak{M}(\mu^*)$, that is, $\mu^*(E_1 \cup E_2) = \mu^*(E_1) + \mu^*(E_2)$ for $E_1, E_2 \in \mathfrak{M}(\mu^*)$ such that $E_1 \cap E_2 = \emptyset$.*

Proof. 1. Suppose $E_1, E_2 \in \mathfrak{M}(\mu^*)$. Let $A \in \mathfrak{P}(X)$. Since $E_1 \in \mathfrak{M}(\mu^*)$ we have

$$(1) \qquad \mu^*(A) = \mu^*(A \cap E_1) + \mu^*(A \cap E_1^c).$$

With $A \cap E_1^c$ as a testing set for $E_2 \in \mathfrak{M}(\mu^*)$, we have

$$(2) \qquad \mu^*(A \cap E_1^c) = \mu^*(A \cap E_1^c \cap E_2) + \mu^*(A \cap E_1^c \cap E_2^c).$$

Substituting (2) into (1) we have

$$(3) \qquad \mu^*(A) = \mu^*(A \cap E_1) + \mu^*(A \cap E_1^c \cap E_2) + \mu^*(A \cap E_1^c \cap E_2^c).$$

Regarding the first two terms on the right-hand side of (3) note that

$$\begin{aligned}(A \cap E_1) \cup (A \cap E_1^c \cap E_2) &= A \cap \big(E_1 \cup (E_1^c \cap E_2)\big) \\ &= A \cap \big(E_1 \cup (E_2 \setminus E_1)\big) = A \cap (E_1 \cup E_2).\end{aligned}$$

Then by the subadditivity of μ^* we have

$$\begin{aligned}(4) \qquad \mu^*(A \cap E_1) + \mu^*(A \cap E_1^c \cap E_2) &\geq \mu^*\big((A \cap E_1) \cup (A \cap E_1^c \cap E_2)\big) \\ (5) \qquad &= \mu^*\big(A \cap (E_1 \cup E_2)\big).\end{aligned}$$

We have also $\mu^*(A \cap E_1^c \cap E_2^c) = \mu^*\big(A \cap (E_1 \cup E_2)^c\big)$. Substituting this and (4) in (3), we have

$$\mu^*(A) \geq \mu^*\big(A \cap (E_1 \cup E_2)\big) + \mu^*\big(A \cap (E_1 \cup E_2)^c\big).$$

By Observation 2.3 this shows that $E_1 \cup E_2$ satisfies the Carathéodory condition. Hence $E_1 \cup E_2 \in \mathfrak{M}(\mu^*)$.

2. To prove the additivity of μ^* on $\mathfrak{M}(\mu^*)$, let $E_1, E_2 \in \mathfrak{M}(\mu^*)$ and $E_1 \cap E_2 = \emptyset$. Since $E_1 \in \mathfrak{M}(\mu^*)$, we have $\mu^*(A) = \mu^*(A \cap E_1) + \mu^*(A \cap E_1^c)$ for every $A \in \mathfrak{P}(X)$. In particular, with $A = E_1 \cup E_2$ we have

$$\mu^*(E_1 \cup E_2) = \mu^*\big((E_1 \cup E_2) \cap E_1\big) + \mu^*\big((E_1 \cup E_2) \cap E_1^c\big).$$

Now the disjointness of E_1 and E_2 implies that $(E_1 \cup E_2) \cap E_1 = E_1$ and $(E_1 \cup E_2) \cap E_1^c = E_2$. Thus $\mu^*(E_1 \cup E_2) = \mu^*(E_1) + \mu^*(E_2)$. ■

The next theorem shows that for an arbitrary outer measure μ^* on a set X, there exists a non μ^*-measurable subset of X if and only if μ^* is not additive on $\mathfrak{P}(X)$.

Theorem 2.5. *Let μ^* be an outer measure on a set X. Then the following two conditions are equivalent:*

(i) *μ^* is additive on $\mathfrak{P}(X)$.*

(ii) *Every member of $\mathfrak{P}(X)$ is μ^*-measurable, that is, $\mathfrak{M}(\mu^*) = \mathfrak{P}(X)$.*

Thus there exist non μ^*-measurable sets in X if and only if μ^* is not additive on $\mathfrak{P}(X)$.

Proof. 1. Suppose μ^* is additive on $\mathfrak{P}(X)$. Let $E \in \mathfrak{P}(X)$. Then for an arbitrary $A \in \mathfrak{P}(X)$, the two sets $A \cap E$ and $A \cap E^c$ are disjoint members of $\mathfrak{P}(X)$ whose union is equal to A so that by the additivity of μ^* on $\mathfrak{P}(X)$ we have $\mu^*(A) = \mu^*(A \cap E) + \mu^*(A \cap E^c)$. This shows that every $E \in \mathfrak{P}(X)$ satisfies the Carathéodory condition. Thus $E \in \mathfrak{M}(\mu^*)$ and then $\mathfrak{P}(X) \subset \mathfrak{M}(\mu^*)$. On the other hand, since $\mathfrak{P}(X)$ is the collection of all subsets of X we have $\mathfrak{M}(\mu^*) \subset \mathfrak{P}(X)$. Therefore we have $\mathfrak{M}(\mu^*) = \mathfrak{P}(X)$.

2. Conversely suppose $\mathfrak{P}(X) = \mathfrak{M}(\mu^*)$. Then since μ^* is additive on $\mathfrak{M}(\mu^*)$ by Lemma 2.4, μ^* is additive on $\mathfrak{P}(X)$. ∎

Example. Let X be an arbitrary set and for every $E \in \mathfrak{P}(X)$ let $\mu^*(E)$ be equal to the number of elements in E. Then μ^* satisfies conditions 1° - 4° in Definition 2.1 and is thus an outer measure on X. Moreover μ^* is additive on $\mathfrak{P}(X)$.

Lemma 2.6. *Let μ^* be an outer measure on a set X. If $E \in \mathfrak{P}(X)$ and $\mu^*(E) = 0$, then every subset E_0 of E, and in particular E itself, is a member of $\mathfrak{M}(\mu^*)$.*

Proof. If $\mu^*(E) = 0$, then for any subset E_0 of E, we have $\mu^*(E_0) = 0$ by the monotonicity of μ^*. Then for every $A \in \mathfrak{P}(X)$, we have $\mu^*(A \cap E_0) + \mu^*(A \cap E_0^c) \leq \mu^*(E_0) + \mu^*(A) = \mu^*(A)$ by the monotonicity of μ^*. This shows that E_0 satisfies the Carathéodory condition by Observation 2.3. Hence $E_0 \in \mathfrak{M}(\mu^*)$. ∎

Lemma 2.7. *Let μ^* be an outer measure on a set X. If $E, F \in \mathfrak{P}(X)$ and $\mu^*(F) = 0$, then $\mu^*(E \cup F) = \mu^*(E)$.*

Proof. By the subadditivity of μ^*, we have $\mu^*(E \cup F) \leq \mu^*(E) + \mu^*(F) = \mu^*(E)$. On the other hand by the monotonicity of μ^*, we have $\mu^*(E) \leq \mu^*(E \cup F)$. Therefore we have $\mu^*(E \cup F) = \mu^*(E)$. ∎

Theorem 2.8. *Let μ^* be an outer measure on a set X. Then the collection $\mathfrak{M}(\mu^*)$ of all μ^*-measurable subsets of X is a σ-algebra of subsets of X.*

Proof. 1. For any $A \in \mathfrak{P}(X)$, we have $\mu^*(A \cap X) + \mu^*(A \cap X^c) = \mu^*(A) + \mu^*(\emptyset) = \mu^*(A)$. This shows that X satisfies the Carathéodory condition so that $X \in \mathfrak{M}(\mu^*)$.

2. Let $E \in \mathfrak{M}(\mu^*)$. Then for every $A \in \mathfrak{P}(X)$ we have

$$\mu^*(A) = \mu^*(A \cap E) + \mu^*(A \cap E^c) = \mu^*\big(A \cap (E^c)^c\big) + \mu^*(A \cap E^c).$$

This shows that E^c satisfies the Carathéodory condition so that $E^c \in \mathfrak{M}(\mu^*)$.

3. Let $(E_n : n \in \mathbb{N}) \subset \mathfrak{M}(\mu^*)$. By Observation 2.3, to show that $\bigcup_{n \in \mathbb{N}} E_n \in \mathfrak{M}(\mu^*)$, it suffices to show that for every $A \in \mathfrak{P}(X)$, we have

$$\mu^*(A) \geq \mu^*\Big(A \cap \Big[\bigcup_{n \in \mathbb{N}} E_n\Big]\Big) + \mu^*\Big(A \cap \Big[\bigcup_{n \in \mathbb{N}} E_n\Big]^c\Big).$$

Let us show first that for every $k \in \mathbb{N}$, we have

$$(1) \qquad \mu^*(A) = \sum_{j=1}^{k} \mu^*\Big(A \cap \Big[\bigcup_{i=1}^{j-1} E_i\Big]^c \cap E_j\Big) + \mu^*\Big(A \cap \Big[\bigcup_{j=1}^{k} E_j\Big]^c\Big)$$

with the understanding that $\bigcup_{i=1}^{0} E_i = \emptyset$. Let us prove (1) by induction on $k \in \mathbb{N}$. Now for $k = 1$, since $E_1 \in \mathfrak{M}(\mu^*)$, we have $\mu^*(A) = \mu^*(A \cap E_1) + \mu^*(A \cap E_1^c)$. Thus (1) is valid for $k = 1$. Next assume that (1) is valid for some $k \in \mathbb{N}$. Let us show that (1) is valid for $k+1$. Now with $A \cap \Big[\bigcup_{j=1}^{k} E_j\Big]^c$ as a testing set for $E_{k+1} \in \mathfrak{M}(\mu^*)$, we have

$$\begin{aligned}\mu^*\Big(A \cap \Big[\bigcup_{j=1}^{k} E_j\Big]^c\Big) &= \mu^*\Big(A \cap \Big[\bigcup_{j=1}^{k} E_j\Big]^c \cap E_{k+1}\Big) + \mu^*\Big(A \cap \Big[\bigcup_{j=1}^{k} E_j\Big]^c \cap E_{k+1}^c\Big)\\ &= \mu^*\Big(A \cap \Big[\bigcup_{j=1}^{k} E_j\Big]^c \cap E_{k+1}\Big) + \mu^*\Big(A \cap \Big[\bigcup_{j=1}^{k+1} E_j\Big]^c\Big).\end{aligned}$$

Substituting this equality into (1) which is valid for k by our assumption, we have

$$\begin{aligned}\mu^*(A) &= \sum_{j=1}^{k} \mu^*\Big(A \cap \Big[\bigcup_{i=1}^{j-1} E_i\Big]^c \cap E_j\Big) + \mu^*\Big(A \cap \Big[\bigcup_{j=1}^{k} E_j\Big]^c \cap E_{k+1}\Big) + \mu^*\Big(A \cap \Big[\bigcup_{j=1}^{k+1} E_j\Big]^c\Big)\\ &= \sum_{j=1}^{k+1} \mu^*\Big(A \cap \Big[\bigcup_{i=1}^{j-1} E_i\Big]^c \cap E_j\Big) + \mu^*\Big(A \cap \Big[\bigcup_{j=1}^{k+1} E_j\Big]^c\Big).\end{aligned}$$

This shows that (1) is valid for $k+1$ under the assumption that it is valid for k. Thus by induction, (1) is valid for every $k \in \mathbb{N}$.

Since $\bigcup_{j=1}^{k} E_i \subset \bigcup_{j\in\mathbb{N}} E_j$, we have $\big[\bigcup_{j=1}^{k} E_i\big]^c \supset \big[\bigcup_{j\in\mathbb{N}} E_j\big]^c$, and then by the monotonicity of μ^*, we have $\mu^*\big(A \cap \big[\bigcup_{j=1}^{k} E_i\big]^c\big) \geq \mu^*\big(A \cap \big[\bigcup_{j\in\mathbb{N}} E_j\big]^c\big)$. Using this in (1), we have

$$\mu^*(A) \geq \sum_{j=1}^{k} \mu^*\Big(A \cap \Big[\bigcup_{i=1}^{j-1} E_i\Big]^c \cap E_j\Big) + \mu^*\Big(A \cap \Big[\bigcup_{j\in\mathbb{N}} E_j\Big]^c\Big).$$

Since this holds for every $k \in \mathbb{N}$, we have

$$\begin{aligned}\mu^*(A) &\geq \sum_{j\in\mathbb{N}} \mu^*\Big(A \cap \Big[\bigcup_{i=1}^{j-1} E_i\Big]^c \cap E_j\Big) + \mu^*\Big(A \cap \Big[\bigcup_{j\in\mathbb{N}} E_j\Big]^c\Big)\\ &\geq \mu^*\Big(\bigcup_{j\in\mathbb{N}} \Big(A \cap \Big[\bigcup_{i=1}^{j-1} E_i\Big]^c \cap E_j\Big)\Big) + \mu^*\Big(A \cap \Big[\bigcup_{j\in\mathbb{N}} E_j\Big]^c\Big)\\ &= \mu^*\Big(A \cap \bigcup_{j\in\mathbb{N}} \Big(\Big[\bigcup_{i=1}^{j-1} E_i\Big]^c \cap E_j\Big)\Big) + \mu^*\Big(A \cap \Big[\bigcup_{j\in\mathbb{N}} E_j\Big]^c\Big),\end{aligned}$$

where the second inequality is by the countable subadditivity of μ^*. Now by Lemma 1.21

$$\bigcup_{j\in\mathbb{N}}\Big(\Big[\bigcup_{i=1}^{j-1}E_i\Big]^c\cap E_j\Big)=\bigcup_{j\in\mathbb{N}}\Big(E_j\setminus\Big[\bigcup_{i=1}^{j-1}E_i\Big]\Big)=\bigcup_{j\in\mathbb{N}}E_j.$$

Therefore we have

$$\mu^*(A)\geq\mu^*\Big(A\cap\Big[\bigcup_{j\in\mathbb{N}}E_j\Big]\Big)+\mu^*\Big(A\cap\Big[\bigcup_{j\in\mathbb{N}}E_j\Big]^c\Big).$$

Thus $\bigcup_{j\in\mathbb{N}}E_j$ satisfies the Carathéodory condition. Then we have $\bigcup_{j\in\mathbb{N}}E_j\in\mathfrak{M}(\mu^*)$. ∎

We have just shown that if μ^* is an outer measure on a set X, then the collection $\mathfrak{M}(\mu^*)$ of all μ^*-measurable subsets of X is a σ-algebra of subsets of X. For an arbitrary measurable space $(X,\mathfrak{A})$, by Definition 1.29 we say that a subset of X is $\mathfrak{A}$-measurable if it is a member of $\mathfrak{A}$. Since $\mathfrak{M}(\mu^*)$ is the collection of all μ^*-measurable subsets of X by Definition 2.2, μ^*-measurability and $\mathfrak{M}(\mu^*)$-measurability of a subset of X are equivalent.

Theorem 2.9. *Let μ^* be an outer measure on a set X. If we let μ be the restriction of μ^* to the σ-algebra $\mathfrak{M}(\mu^*)$, then μ is a measure on $\mathfrak{M}(\mu^*)$ and furthermore $(X,\mathfrak{M}(\mu^*),\mu)$ is a complete measure space.*

Proof. Since μ^* is countably subadditive on $\mathfrak{P}(X)$, its restriction on $\mathfrak{M}(\mu^*)$ is countably subadditive on $\mathfrak{M}(\mu^*)$. By Lemma 2.4, μ^* is additive on $\mathfrak{M}(\mu^*)$. Therefore by Proposition 1.23, μ^* is countably additive on the σ-algebra $\mathfrak{M}(\mu^*)$ and is therefore a measure on $\mathfrak{M}(\mu^*)$. If we write μ for the restriction of μ^* to $\mathfrak{M}(\mu^*)$, then we have a measure space $(X,\mathfrak{M}(\mu^*),\mu)$. According to Lemma 2.6, if $E\in\mathfrak{M}(\mu^*)$ and $\mu(E)=0$, then every subset of E is a member of $\mathfrak{M}(\mu^*)$. Thus $(X,\mathfrak{M}(\mu^*),\mu)$ is a complete measure space. ∎

[II] Regular Outer Measures

Definition 2.10. *An outer measure μ^* on a set X is called a regular outer measure if every subset of X is contained in a μ^*-measurable set with equal outer measure, that is, for every $E\in\mathfrak{P}(X)$, there exists $F\in\mathfrak{M}(\mu^*)$ such that $F\supset E$ and $\mu^*(F)=\mu^*(E)$.*

In a measure space $(X,\mathfrak{A},\mu)$, if $(E_n : n\in\mathbb{N})$ is an increasing sequence in $\mathfrak{A}$, then $\lim_{n\to\infty}\mu(E_n)=\mu\big(\lim_{n\to\infty}E_n\big)$. This is a consequence of the countable additivity of μ on $\mathfrak{A}$. (See the Proof of Theorem 1.26.) If μ^* is an outer measure on a set X, then for an increasing sequence $(E_n : n\in\mathbb{N})$ in $\mathfrak{P}(X)$ we may not have $\lim_{n\to\infty}\mu^*(E_n)=\mu^*\big(\lim_{n\to\infty}E_n\big)$. We show next that if μ^* is a regular outer measure then the equality holds.

Theorem 2.11. *Let μ^* be an outer measure on a set X and let $(E_n : n\in\mathbb{N})$ be an increasing sequence of subsets of X. Then*

(1) $$\lim_{n\to\infty}\mu^*(E_n)\leq\mu^*\big(\lim_{n\to\infty}E_n\big).$$

If μ^* is a regular outer measure then

(2) $$\lim_{n\to\infty} \mu^*(E_n) = \mu^*\big(\lim_{n\to\infty} E_n\big).$$

Proof. 1. Let $(E_n : n \in \mathbb{N})$ be an increasing sequence of subsets of X. By the monotonicity of μ^*, $(\mu^*(E_n) : n \in \mathbb{N})$ is an increasing sequence of extended real numbers so that $\lim_{n\to\infty} \mu^*(E_n)$ exists. Also by the monotonicity of μ^* we have $\mu^*(E_n) \le \mu^*\left(\bigcup_{n\in\mathbb{N}} E_n\right)$ for every $n \in \mathbb{N}$ so that $\lim_{n\to\infty} \mu^*(E_n) \le \mu^*\left(\bigcup_{n\in\mathbb{N}} E_n\right) = \mu^*\big(\lim_{n\to\infty} E_n\big)$. This proves (1).

2. Assume that μ^* is a regular outer measure. Then for every $n \in \mathbb{N}$, there exists $F_n \in \mathfrak{M}(\mu^*)$ such that $E_n \subset F_n$ and $\mu^*(E_n) = \mu^*(F_n) = \mu(F_n)$, where μ is the restriction of μ^* to $\mathfrak{M}(\mu^*)$. Now $E_n \subset F_n$ for $n \in \mathbb{N}$ implies that $\liminf_{n\to\infty} E_n \subset \liminf_{n\to\infty} F_n$. Then

$$\begin{aligned}
\mu^*\big(\lim_{n\to\infty} E_n\big) &= \mu^*\big(\liminf_{n\to\infty} E_n\big) \le \mu^*\big(\liminf_{n\to\infty} F_n\big) \\
&= \mu\big(\liminf_{n\to\infty} F_n\big) \le \liminf_{n\to\infty} \mu(F_n) = \liminf_{n\to\infty} \mu^*(F_n) \\
&= \liminf_{n\to\infty} \mu^*(E_n) = \lim_{n\to\infty} \mu^*(E_n),
\end{aligned}$$

where the second equality is by the fact that $\liminf_{n\to\infty} F_n \in \mathfrak{M}(\mu^*)$, the second inequality is by (a) of Theorem 1.28, and the last equality is by the existence of $\lim_{n\to\infty} \mu^*(E_n)$. This and (1) imply (2). ∎

Definition 2.12. *Let μ^* be an outer measure on a set X. We say that μ^* is σ-finite on $\mathfrak{P}(X)$ if there exists a sequence $(A_n : n \in \mathbb{N})$ in $\mathfrak{P}(X)$ such that $\bigcup_{n\in\mathbb{N}} A_n = X$ and $\mu^*(A_n) < \infty$ for every $n \in \mathbb{N}$.*

Let us note that the sequence $(A_n : n \in \mathbb{N})$ in Definition 2.12 can be chosen to be a disjoint sequence. In fact if we let $B_1 = A_1$ and $B_n = A_n \setminus \bigcup_{k=1}^{n-1} A_k$ for $n \ge 2$, then $(B_n : n \in \mathbb{N})$ is a disjoint sequence in $\mathfrak{P}(X)$, $\bigcup_{n\in\mathbb{N}} B_n = \bigcup_{n\in\mathbb{N}} A_n = X$, and $\mu^*(B_n) \le \mu^*(A_n) < \infty$ for every $n \in \mathbb{N}$ by the monotonicity of μ^* on $\mathfrak{P}(X)$.

In a measure space $(X, \mathfrak{A}, \mu)$, if $E, F \in \mathfrak{A}$, $E \subset F$, and $\mu(E) = \mu(F) < \infty$, then $\mu(F \setminus E) = \mu(F) - \mu(E) = 0$. This follows from the additivity of μ on $\mathfrak{A}$. The assumption $\mu(E) < \infty$ is to ensure that the difference $\mu(F) - \mu(E)$ is defined. If μ^* is an outer measure on a set X, and if $E \in \mathfrak{P}(X)$, $F \in \mathfrak{M}(\mu^*)$, $E \subset F$, and $\mu^*(E) = \mu^*(F) < \infty$, do we have $\mu^*(F \setminus E) = 0$? In the next theorem, we show that for an outer measure μ^* which is regular and σ-finite, this question is equivalent to the question as to whether μ^* is additive on $\mathfrak{P}(X)$, or equivalently according to Theorem 2.5, the question of non-existence of non μ^*-measurable sets in X.

Theorem 2.13. *Let μ^* be a regular and σ-finite outer measure on a set X. Then the following two conditions are equivalent:*

(i) $\mathfrak{M}(\mu^*) = \mathfrak{P}(X)$.

(ii) $E \in \mathfrak{P}(X), F \in \mathfrak{M}(\mu^*), E \subset F, \mu^*(E) = \mu^*(F) < \infty \Rightarrow \mu^*(F \setminus E) = 0$.

Proof. 1. To show that (i) implies (ii), assume $\mathfrak{M}(\mu^*) = \mathfrak{P}(X)$. Suppose E and F satisfy the hypothesis of (ii). Then E and F are members of $\mathfrak{M}(\mu^*)$. Since μ^* is a measure on the σ-algebra $\mathfrak{M}(\mu^*)$, it is additive on $\mathfrak{M}(\mu^*)$. This implies that $\mu^*(F) = \mu^*(F \setminus E) + \mu^*(E)$. Subtracting $\mu^*(E) \in \mathbb{R}$ from both sides, we have $\mu^*(F \setminus E) = \mu^*(F) - \mu^*(E) = 0$. This shows that (i) implies (ii).

2. To show that (ii) implies (i), let us assume (ii). Let E be an arbitrary member of $\mathfrak{P}(X)$. Since μ^* is σ-finite on $\mathfrak{P}(X)$, there exists a sequence $(A_n : n \in \mathbb{N})$ in $\mathfrak{P}(X)$ such that $\bigcup_{n\in\mathbb{N}} A_n = X$ and $\mu^*(A_n) < \infty$ for every $n \in \mathbb{N}$. Let $E_n = E \cap A_n$ for $n \in \mathbb{N}$. Then $(E_n : n \in \mathbb{N})$ is a sequence in $\mathfrak{P}(X)$ with $\bigcup_{n\in\mathbb{N}} E_n = E$ and $\mu^*(E_n) \leq \mu^*(A_n) < \infty$ for every $n \in \mathbb{N}$ by the monotonicity of μ^* on $\mathfrak{P}(X)$. Now since μ^* is a regular outer measure, there exists $F_n \in \mathfrak{M}(\mu^*)$ such that $F_n \supset E_n$ and $\mu^*(F_n) = \mu^*(E_n)$ for every $n \in \mathbb{N}$. By (ii), we have $\mu^*(F_n \setminus E_n) = 0$. This implies that $F_n \setminus E_n \in \mathfrak{M}(\mu^*)$ by Lemma 2.6. Since $E_n = F_n \setminus (F_n \setminus E_n)$ and F_n and $F_n \setminus E_n$ are members of the σ-algebra $\mathfrak{M}(\mu^*)$, E_n is a member of $\mathfrak{M}(\mu^*)$ for every $n \in \mathbb{N}$. Then $E = \bigcup_{n\in\mathbb{N}} E_n \in \mathfrak{M}(\mu^*)$. This shows that every member of $\mathfrak{P}(X)$ is a member of $\mathfrak{M}(\mu^*)$. Thus (ii) implies (i). ∎

Definition 2.14. *An outer measure μ^* on a topological space X is called a Borel outer measure if $\mathfrak{B}_X \subset \mathfrak{M}(\mu^*)$.*

Definition 2.15. *An outer measure μ^* on a topological space X is called a Borel regular outer measure if it is a Borel outer measure on X and if for every $E \in \mathfrak{P}(X)$ there exists $F \in \mathfrak{B}_X$ such that $F \supset E$ and $\mu^*(F) = \mu^*(E)$.*

Remark 2.16. A Borel regular outer measure μ^* on a topological space X is a regular outer measure on X in the sense of Definition 2.10. This follows from the fact that $\mathfrak{B}_X \subset \mathfrak{M}(\mu^*)$.

[III] Metric Outer Measures

Given a metric space (X, d). Let the topology on X be the metric topology by the metric d. The distance between a point $x \in X$ and a set $E \in \mathfrak{P}(X)$ is defined by $d(x, E) = \inf_{y\in E} d(x, y)$. If $x \in E$ then $d(x, E) = 0$ but the converse is false. If E is a closed set then $d(x, E) = 0$ if and only if $x \in E$. The distance between two sets $E, F \in \mathfrak{P}(X)$ is defined by $d(E, F) = \inf_{x\in E, y\in F} d(x, y)$. If $E \cap F \neq \emptyset$ then $d(E, F) = 0$. If $E \cap F = \emptyset$, $d(E, F)$ may still be equal to 0 even if E and F are closed sets. If E is a closed set and F is a compact set, then $d(E, F) = 0$ if and only if $E \cap F \neq \emptyset$.

Definition 2.17. *Given a metric space (X, d).*

(a) *Two sets $E_1, E_2 \in \mathfrak{P}(X)$ are said to be positively separated if $d(E_1, E_2) > 0$.*

(b) *An outer measure μ^* on X is called a metric outer measure if for every pair of positively separated sets $E_1, E_2 \in \mathfrak{P}(X)$, we have $\mu^*(E_1 \cup E_2) = \mu^*(E_1) + \mu^*(E_2)$.*

Lemma 2.18. *Let μ^* be a metric outer measure on a metric space (X, d). Let $(A_n : n \in \mathbb{N})$ be an increasing sequence in $\mathfrak{P}(X)$ and let $A = \lim_{n\to\infty} A_n$. If A_n and $A \setminus A_{n+1}$ are positively separated for every $n \in \mathbb{N}$, then $\mu^*(A) = \lim_{n\to\infty} \mu^*(A_n)$.*

Proof. Since $A_n \uparrow$, we have $\mu^*(A_n) \uparrow$ as $n \to \infty$ by the monotonicity of the outer measure μ^*. Since $A_n \subset A$ we have $\mu^*(A_n) \le \mu^*(A)$ for every $n \in \mathbb{N}$. Thus we have $\lim_{n\to\infty} \mu^*(A_n) \le \mu^*(A)$. It remains to show

$$\mu^*(A) \le \lim_{n\to\infty} \mu^*(A_n). \tag{1}$$

Let $A_0 = \emptyset$ and define a sequence $(B_n : n \in \mathbb{N})$ in $\mathfrak{P}(X)$ by setting $B_n = A_n \setminus A_{n-1}$ for $n \in \mathbb{N}$. Let us show

$$\sum_{k\in\mathbb{N}} \mu^*(B_{2k-1}) \le \lim_{n\to\infty} \mu^*(A_n), \tag{2}$$

$$\sum_{k\in\mathbb{N}} \mu^*(B_{2k}) \le \lim_{n\to\infty} \mu^*(A_n). \tag{3}$$

To prove (2), let us consider $\mu^*\big(\bigcup_{k=1}^{N+1} B_{2k-1}\big)$ for an arbitrary $N \in \mathbb{N}$. Let us write $\bigcup_{k=1}^{N+1} B_{2k-1} = \big(\bigcup_{k=1}^{N} B_{2k-1}\big) \cup B_{2N+1}$. Now we have $\bigcup_{k=1}^{N} B_{2k-1} \subset A_{2N-1}$ and $B_{2N+1} = A_{2N+1} \setminus A_{2N} \subset A \setminus A_{2N}$. This implies

$$d\Big(\bigcup_{k=1}^{N} B_{2k-1}, B_{2N+1}\Big) \ge d\,(A_{2N-1}, A \setminus A_{2N}) > 0$$

by our assumption on the sequence $(A_n : n \in \mathbb{N})$. Then since μ^* is a metric outer measure, we have

$$\begin{aligned}\mu^*\Big(\bigcup_{k=1}^{N+1} B_{2k-1}\Big) &= \mu^*\Big(\Big[\bigcup_{k=1}^{N} B_{2k-1}\Big] \cup B_{2N+1}\Big)\\ &= \mu^*\Big(\bigcup_{k=1}^{N} B_{2k-1}\Big) + \mu^*(B_{2N+1}).\end{aligned}$$

Repeating the argument to $\mu^*\big(\bigcup_{k=1}^{N} B_{2k-1}\big)$ and iterating the process N times, we obtain

$$\mu^*\Big(\bigcup_{k=1}^{N+1} B_{2k-1}\Big) = \sum_{k=1}^{N+1} \mu^*(B_{2k-1}).$$

Since $\bigcup_{k=1}^{N+1} B_{2k-1} \subset A_{2N+1}$, we have $\sum_{k=1}^{N+1} \mu^*(B_{2k-1}) \le \mu^*(A_{2N+1})$. Thus

$$\sum_{k\in\mathbb{N}} \mu^*(B_{2k-1}) = \lim_{N\to\infty} \sum_{k=1}^{N+1} \mu^*(B_{2k-1}) \le \lim_{N\to\infty} \mu^*(A_{2N+1}) = \lim_{n\to\infty} \mu^*(A_n).$$

This proves (2). We prove (3) likewise by starting with $\bigcup_{k=1}^{N+1} B_{2k}$ for $N \in \mathbb{N}$.

If at least one of the two series in (2) and (3) diverges, then $\lim_{n\to\infty} \mu^*(A_n) = \infty$ and (1) holds. Suppose that both series converge. Then we have

(4) $$\sum_{n\in\mathbb{N}} \mu^*(B_n) = \sum_{k\in\mathbb{N}} \mu^*(B_{2k-1}) + \sum_{k\in\mathbb{N}} \mu^*(B_{2k}) < \infty.$$

Now $A = \lim_{n\to\infty} A_n = \bigcup_{n\in\mathbb{N}} A_n = A_{n_0} \cup \left(\bigcup_{n\ge n_0+1} B_n\right)$ for an arbitrary $n_0 \in \mathbb{N}$. Thus by the countable subadditivity of the outer measure μ^*, we have

(5) $$\mu^*(A) \le \mu^*(A_{n_0}) + \sum_{n\ge n_0+1} \mu^*(B_n).$$

By (4) we have $\lim_{n_0\to\infty} \sum_{n\ge n_0+1} \mu^*(B_n) = 0$. Then $\mu^*(A) \le \lim_{n_0\to\infty} \mu^*(A_{n_0})$ by letting $n_0 \to \infty$ in (5). This proves (1). ∎

Theorem 2.19. *An outer measure μ^* on a metric space (X, d) is a Borel outer measure, that is, $\mathfrak{B}_X \subset \mathfrak{M}(\mu^*)$, if and only if μ^* is a metric outer measure.*

Proof. 1. Suppose μ^* is a metric outer measure. The Borel σ-algebra $\mathfrak{B}_X$ of subsets of X is the smallest σ-algebra of subsets of X containing all open sets in X and is thus also the smallest σ-algebra of subsets of X containing all closed sets in X by Lemma 1.17. Therefore to show that $\mathfrak{B}_X \subset \mathfrak{M}(\mu^*)$ it suffices to show that every closed set in X is a member of $\mathfrak{M}(\mu^*)$. Let E be a closed set in X. To show that $E \in \mathfrak{M}(\mu^*)$, according to Observation 2.3 it suffices to show that for every $A \in \mathfrak{P}(X)$

(1) $$\mu^*(A) \ge \mu^*(A \cap E) + \mu^*(A \cap E^c).$$

Let $B = A \cap E^c$. Let $(B_n : n \in \mathbb{N})$ be an increasing sequence of subsets of B defined by

(2) $$B_n = \left\{x \in B : d(x, E) \ge \tfrac{1}{n}\right\}.$$

Let us show that the sequence satisfies the conditions

(3) $$\lim_{n\to\infty} B_n = \bigcup_{n\in\mathbb{N}} B_n = B,$$

(4) $$d\big(B_n, B \setminus B_{n+1}\big) \ge \frac{1}{n} - \frac{1}{n+1} > 0.$$

Now since $B_n \subset B$ for every $n \in \mathbb{N}$, we have $\bigcup_{n\in\mathbb{N}} B_n \subset B$. To prove the reverse inclusion, let $x \in B$. Then $x \notin E$. Since E is a closed set, we have $d(x, E) > 0$. Thus there exists $n \in \mathbb{N}$ such that $d(x, E) \ge \frac{1}{n}$. Then by (2) we have $x \in B_n$. This shows that $B \subset \bigcup_{n\in\mathbb{N}} B_n$ and proves (3). To prove (4), let $x \in B_n$ and $y \in B \setminus B_{n+1}$. Since $y \in B$ and $y \notin B_{n+1}$, we have $d(y, E) < \frac{1}{n+1}$ by (2). Then there exists $z \in E$ such that $d(y, z) < \frac{1}{n+1}$. Since $x \in B_n$ and $z \in E$, we have $d(x, z) \ge \frac{1}{n}$ by (2). The triangle inequality of the metric

$d(x,z) \le d(x,y) + d(y,z)$ implies $d(x,y) \ge d(x,z) - d(y,z) > \frac{1}{n} - \frac{1}{n+1}$. Since this holds for arbitrary $x \in B_n$ and $y \in B \setminus B_{n+1}$, (4) holds.

Now (3) and (4) imply by Lemma 2.18 that

$$(5) \qquad \lim_{n\to\infty} \mu^*(B_n) = \mu^*(B) = \mu^*(A \cap E^c).$$

Since $A \cap E \subset E$, we have $d(A \cap E, B_n) \ge d(E, B_n) \ge \frac{1}{n}$ by (2). Thus $A \cap E$ and B_n are positively separated. Then the fact that μ^* is a metric outer measure implies that we have $\mu^*(A \cap E) + \mu^*(B_n) = \mu^*\big((A \cap E) \cup B_n\big) \le \mu^*(A)$ by the fact that $B_n \subset B \subset A$. Letting $n \to \infty$ and applying (5), we have (1).

2. Conversely suppose μ^* is a Borel outer measure on X, that is, $\mathfrak{B}_X \subset \mathfrak{M}(\mu^*)$. To show that μ^* is a metric outer measure, let $E_1, E_2 \in \mathfrak{P}(X)$ be such that $c := d(E_1, E_2) > 0$. For each $x \in E_1$, let $G_x = \{y \in X : d(x,y) < \frac{c}{2}\}$ and let $G = \bigcup_{x \in E_1} G_x$. Then G is an open set, $E_1 \subset G$, and $E_2 \cap G = \emptyset$. Since $G \in \mathfrak{B}_X \subset \mathfrak{M}(\mu^*)$, we have $\mu^*(E_1 \cup E_2) = \mu^*\big((E_1 \cup E_2) \cap G\big) + \mu^*\big((E_1 \cup E_2) \cap G^c\big)$. Since $E_1 \subset G$ and $E_2 \cap G = \emptyset$, we have $(E_1 \cup E_2) \cap G = E_1$ and $(E_1 \cup E_2) \cap G^c = E_2$. Thus $\mu^*(E_1 \cup E_2) = \mu^*(E_1) + \mu^*(E_2)$. This shows that μ^* is a metric outer measure. ∎

[IV] Construction of Outer Measures

Definition 2.20. *A collection $\mathfrak{V}$ of subsets of a set X is called a covering class if it satisfies the following conditions :*

1° *there exists $(V_n : n \in \mathbb{N}) \subset \mathfrak{V}$ such that $\bigcup_{n\in\mathbb{N}} V_n = X$,*

2° $\emptyset \in \mathfrak{V}$.

For $E \in \mathfrak{P}(X)$, a sequence $(V_n : n \in \mathbb{N}) \subset \mathfrak{V}$ such that $\bigcup_{n\in\mathbb{N}} V_n \supset E$ is called a covering sequence for E.

Theorem 2.21. *Let $\mathfrak{V}$ be a covering class of subsets of a set X. Let γ be an arbitrary set function on $\mathfrak{V}$ such that*

1° *nonnegative extended real-valued : $\gamma(V) \in [0, \infty]$ for every $V \in \mathfrak{V}$,*

2° $\gamma(\emptyset) = 0$.

Let us define a set function μ^* on $\mathfrak{P}(X)$ by setting for every $E \in \mathfrak{P}(X)$

$$\mu^*(E) = \inf\Big\{ \textstyle\sum_{n\in\mathbb{N}} \gamma(V_n) : (V_n : n \in \mathbb{N}) \subset \mathfrak{V}, \bigcup_{n\in\mathbb{N}} V_n \supset E \Big\}.$$

Then μ^* is an outer measure on X. We call μ^* the outer measure based on γ.

Proof. Let us show that our μ^* satisfies conditions 1°, 2°, 3°, and 4° of Definition 2.1. Clearly $\mu^*(E) \in [0, \infty]$ for every $E \in \mathfrak{P}(X)$. Since $\emptyset \in \mathfrak{V}$, $(\emptyset)$ is a one term covering sequence in $\mathfrak{V}$ for $\emptyset$ and thus $\mu^*(\emptyset) \le \gamma(\emptyset) = 0$.

To show the monotonicity of μ^* on $\mathfrak{P}(X)$, let $E_1, E_2 \in \mathfrak{P}(X)$ and $E_1 \subset E_2$. Then every covering sequence in $\mathfrak{V}$ for E_2 is also a covering sequence for E_1. This implies that the collection of extended nonnegative numbers on which we take the infimum to obtain

$\mu^*(E_1)$ is greater than the collection on which we take infimum to obtain $\mu^*(E_2)$. Now if A and B are two collections of extended real numbers and $A \supset B$ then $\inf A \le \inf B$. Thus $\mu^*(E_1) \le \mu^*(E_2)$. This proves the monotonicity of μ^*.

To show the countable subadditivity of μ^* on $\mathfrak{P}(X)$, let $(E_n : n \in \mathbb{N}) \subset \mathfrak{P}(X)$. Let $\varepsilon > 0$. For each $n \in \mathbb{N}$, by the definition of $\mu^*(E_n)$ as an infimum there exists a sequence $(V_{n,k} : k \in \mathbb{N}) \subset \mathfrak{V}$ such that $\bigcup_{k\in\mathbb{N}} V_{n,k} \supset E_n$ and $\sum_{k\in\mathbb{N}} \gamma(V_{n,k}) \le \mu^*(E_n) + \frac{\varepsilon}{2^n}$. Then $\bigcup_{n\in\mathbb{N}} \left(\bigcup_{k\in\mathbb{N}} V_{n,k}\right) \supset \bigcup_{n\in\mathbb{N}} E_n$. This implies

$$\mu^*\left(\bigcup_{n\in\mathbb{N}} E_n\right) \le \sum_{n\in\mathbb{N}} \left[\sum_{k\in\mathbb{N}} \gamma(V_{n,k})\right] \le \sum_{n\in\mathbb{N}} \{\mu^*(E_n) + \tfrac{\varepsilon}{2^n}\} = \sum_{n\in\mathbb{N}} \mu^*(E_n) + \varepsilon.$$

By the arbitrariness of $\varepsilon > 0$, we have $\mu^*\left(\bigcup_{n\in\mathbb{N}} E_n\right) \le \sum_{n\in\mathbb{N}} \mu^*(E_n)$. This proves the countable subadditivity of μ^*. ∎

Remark 2.22. The nonnegative extended real-valued set function γ on the covering class $\mathfrak{V}$ on which the outer measure μ^* is based in Theorem 2.21 satisfies no conditions other than that $\gamma(\emptyset) = 0$. In particular, γ is not required to be monotone on $\mathfrak{V}$. Thus it is possible that we have $V, W \in \mathfrak{V}$ such that $W \subset V$ but $\gamma(W) > \gamma(V)$. The existence of such a set W in $\mathfrak{V}$ has no effect on the definition of μ^*. If such a set W is a member of a covering sequence $(V_n : n \in \mathbb{N})$ for a set $E \in \mathfrak{P}(X)$, then by replacing W with V we have a covering sequence of E with a possibly smaller, but never greater, sum $\sum_{n\in\mathbb{N}} \gamma(V_n)$. Thus in taking infimum of $\sum_{n\in\mathbb{N}} \gamma(V_n)$ on the collection of all covering sequences of E, a sequence with W as a member has no effect.

Remark 2.23. Further regarding the set functions γ and μ^* in Theorem 2.21, we have:
(a) For $V \in \mathfrak{V}$, (V) is a one term covering sequence in $\mathfrak{V}$ for V and thus

$$\mu^*(V) = \inf\left\{\sum_{n\in\mathbb{N}} \gamma(V_n) : (V_n : n \in \mathbb{N}) \subset \mathfrak{V}, \bigcup_{n\in\mathbb{N}} V_n \supset V\right\} \le \gamma(V).$$

However $\mu^*(V) = \gamma(V)$ may not hold. In fact if there exist $W, V \in \mathfrak{V}$ such that $W \subset V$ and $\gamma(W) > \gamma(V)$, then since (V) is a one term covering sequence in $\mathfrak{V}$ for W, we have $\mu^*(W) \le \gamma(V) < \gamma(W)$.
(b) In general $\mathfrak{V} \subset \mathfrak{M}(\mu^*)$ does not hold. See §20 for an example. In §20 we show that if $\mathfrak{V}$ is an algebra of subsets of X and γ is additive on $\mathfrak{V}$, then $\mathfrak{V} \subset \mathfrak{M}(\mu^*)$, and if γ is countably additive on $\mathfrak{V}$, then $\gamma = \mu^*$ on $\mathfrak{V}$. These results are extended in §21 to the case where $\mathfrak{V}$ is a semialgebra (see Definition 21.1) of subsets of X. In §20 we show also that an outer measure based on a countably additive set function on an algebra is a regular outer measure.

For an outer measure μ^* based on an extended nonnegative valued function γ with $\gamma(\emptyset) = 0$ on a covering class $\mathfrak{V}$ of a set X as in Theorem 2.21, the scope of the testing sets in the Carathéodory condition for μ^*-measurability is reduced as we show in the next theorem.

Theorem 2.24. *Let $\mathfrak{V}$ be a covering class for a set X and let γ be a set function on $\mathfrak{V}$ such that $\gamma(V) \in [0, \infty]$ for every $V \in \mathfrak{V}$ and $\gamma(\emptyset) = 0$. Define an outer measure μ^* on*

X by setting for every $E \in \mathfrak{P}(X)$

$$\mu^*(E) = \inf\Big\{ \textstyle\sum_{n\in\mathbb{N}} \gamma(V_n) : (V_n : n \in \mathbb{N}) \subset \mathfrak{V},\ \bigcup_{n\in\mathbb{N}} V_n \supset E \Big\}.$$

Then the following two conditions, the first of which is the Carathéodory condition for the μ^*- measurability of E, are equivalent:

(i) $\mu^*(A) = \mu^*(A \cap E) + \mu^*(A \cap E^c)$ *for every* $A \in \mathfrak{P}(X)$

(ii) $\mu^*(V) = \mu^*(V \cap E) + \mu^*(V \cap E^c)$ *for every* $V \in \mathfrak{V}$.

Proof. Since (i) implies (ii), it remains to show that (ii) implies (i). Let us assume (ii). Let $E \in \mathfrak{P}(X)$. For an arbitrary $A \in \mathfrak{P}(X)$, let $(V_n : n \in \mathbb{N})$ be an arbitrary sequence in $\mathfrak{V}$ such that $\bigcup_{n\in\mathbb{N}} V_n \supset A$. By (ii) we have $\mu^*(V_n) = \mu^*(V_n \cap E) + \mu^*(V_n \cap E^c)$ for every $n \in \mathbb{N}$. Summing over $n \in \mathbb{N}$, we have

$$\begin{aligned}\sum_{n\in\mathbb{N}} \mu^*(V_n) &= \sum_{n\in\mathbb{N}} \mu^*(V_n \cap E) + \sum_{n\in\mathbb{N}} \mu^*(V_n \cap E^c) \\ &\geq \mu^*\Big(\bigcup_{n\in\mathbb{N}} (V_n \cap E)\Big) + \mu^*\Big(\bigcup_{n\in\mathbb{N}} (V_n \cap E^c)\Big) \\ &= \mu^*\Big(\Big(\bigcup_{n\in\mathbb{N}} V_n\Big) \cap E\Big) + \mu^*\Big(\Big(\bigcup_{n\in\mathbb{N}} V_n\Big) \cap E^c\Big) \\ &\geq \mu^*(A \cap E) + \mu^*(A \cap E^c),\end{aligned}$$

where the first inequality is by the countable subadditivity of the outer measure μ^* and the second inequality is by the monotonicity of μ^*. According to (a) of Remark 2.23, $\mu^*(V_n) \leq \gamma(V_n)$. Thus $\sum_{n\in\mathbb{N}} \gamma(V_n) \geq \mu^*(A \cap E) + \mu^*(A \cap E^c)$. This shows that $\mu^*(A \cap E) + \mu^*(A \cap E^c)$ is a lower bound for the collection of nonnegative extended real numbers $\big\{\sum_{n\in\mathbb{N}} \gamma(V_n) : (V_n : n \in \mathbb{N}) \subset \mathfrak{V}, \bigcup_{n\in\mathbb{N}} V_n \supset A\big\}$. Since $\mu^*(A)$ is the infimum, that is, the greatest lower bound, of this collection, we have $\mu^*(A) \geq \mu^*(A \cap E) + \mu^*(A \cap E^c)$. The reverse inequality holds by the monotonicity of μ^*. Thus (i) holds. ∎

Problems

Prob. 2.1. For an arbitrary set X let us define a set function μ^* on $\mathfrak{P}(X)$ by

$$\mu^*(E) = \begin{cases} \text{number of elements of } E \text{ if } E \text{ is a finite set,} \\ \infty \text{ if } E \text{ is an infinite set.} \end{cases}$$

(a) Show that μ^* is an outer measure on X.
(b) Show that μ^* is additive on $\mathfrak{P}(X)$, that is, $\mu^*(E_1 \cup E_2) = \mu^*(E_1) + \mu^*(E_2)$ for any $E_1, E_2 \in \mathfrak{P}(X)$ such that $E_1 \cap E_2 = \emptyset$.
(c) Show that μ^* is a measure on the σ-algebra $\mathfrak{P}(X)$. (This measure is called the counting measure.)
(d) Show that $\mathfrak{M}(\mu^*) = \mathfrak{P}(X)$, that is, every $E \in \mathfrak{P}(X)$ is μ^*-measurable.

Prob. 2.2. Let X be an infinite set and let μ be the counting measure on the σ-algebra $\mathfrak{A}$ of all subsets of X. Show that there exists a decreasing sequence $(E_n : n \in \mathbb{N})$ in $\mathfrak{A}$ such that $E_n \downarrow \emptyset$, that is, $\lim_{n\to\infty} E_n = \emptyset$, with $\lim_{n\to\infty} \mu(E_n) \neq 0$.

Prob. 2.3. Let μ^* be an outer measure on a set X. Show that if μ^* is additive on $\mathfrak{P}(X)$, then it is countably additive on $\mathfrak{P}(X)$.

Prob. 2.4. Let μ^* be an outer measure on a set X. Show that a non μ^*-measurable subset of X exists if and only if μ^* is not countably additive on $\mathfrak{P}(X)$.

§3 Lebesgue Measure on $\mathbb{R}$

[I] Lebesgue Outer Measure on $\mathbb{R}$

Definition 3.1. *Let $\mathfrak{I}_o$ be the collection of $\emptyset$ and all open intervals in $\mathbb{R}$, $\mathfrak{I}_{oc}$ be the collection of $\emptyset$ and all intervals of the type $(a, b]$ in $\mathbb{R}$, $\mathfrak{I}_{co}$ be the collection of $\emptyset$ and all intervals of the type $[a, b)$ in $\mathbb{R}$, and $\mathfrak{I}_c$ be the collection of $\emptyset$ and all closed intervals in $\mathbb{R}$, with the understanding that $(a, \infty] = (a, \infty)$ and $[-\infty, b) = (-\infty, b)$. Let $\mathfrak{I} = \mathfrak{I}_o \cup \mathfrak{I}_{oc} \cup \mathfrak{I}_{co} \cup \mathfrak{I}_c$, that is, the collection of $\emptyset$ and all intervals in $\mathbb{R}$. For an interval I in $\mathbb{R}$ with endpoints $a, b \in \mathbb{R}$, $a < b$, we define $\ell(I) = b - a$. For an infinite interval I in $\mathbb{R}$ we define $\ell(I) = \infty$. We set $\ell(\emptyset) = 0$.*

For a countable disjoint collection $\{I_n : n \in \mathbb{N}\}$ in $\mathfrak{I}$, we define $\ell\big(\bigcup_{n\in\mathbb{N}} I_n\big) = \sum_{n\in\mathbb{N}} \ell(I_n)$.

With the nonnegative extended real-valued set function ℓ on the covering class $\mathfrak{I}_o$ of $\mathbb{R}$, let us define a set function on $\mathfrak{P}(\mathbb{R})$ by setting for every $E \in \mathfrak{P}(\mathbb{R})$

$$\mu_L^*(E) = \inf\Big\{ \textstyle\sum_{n\in\mathbb{N}} \ell(I_n) : (I_n : n \in \mathbb{N}) \subset \mathfrak{I}_o,\ \bigcup_{n\in\mathbb{N}} I_n \supset E\Big\}.$$

By Theorem 2.21, μ_L^* is an outer measure on $\mathbb{R}$. We call μ_L^* the Lebesgue outer measure on $\mathbb{R}$. We write $\mathfrak{M}_L$ for the σ-algebra $\mathfrak{M}(\mu_L^*)$ of μ_L^*-measurable sets $E \in \mathfrak{P}(\mathbb{R})$ and call it the Lebesgue σ-algebra of subsets of $\mathbb{R}$. Members of the σ-algebra $\mathfrak{M}_L$ are called $\mathfrak{M}_L$-measurable or Lebesgue measurable sets. We call $(\mathbb{R}, \mathfrak{M}_L)$ the Lebesgue measurable space. We write μ_L for the restriction of μ_L^* to $\mathfrak{M}_L$ and call it the Lebesgue measure on $\mathbb{R}$. We call $\big(\mathbb{R}, \mathfrak{M}_L, \mu_L\big)$ the Lebesgue measure space on $\mathbb{R}$.

For some $E \in \mathfrak{P}(\mathbb{R})$, we have $\sum_{n\in\mathbb{N}} \ell(I_n) = \infty$ for any $(I_n : n \in \mathbb{N}) \subset \mathfrak{I}_o$, such that $\bigcup_{n\in\mathbb{N}} I_n \supset E$ so that we have $\inf\Big\{ \sum_{n\in\mathbb{N}} \ell(I_n) : (I_n : n \in \mathbb{N}) \subset \mathfrak{I}_o,\ \bigcup_{n\in\mathbb{N}} I_n \supset E\Big\} = \infty$.

Observation 3.2. Each of the four collections $\mathfrak{I}_o$, $\mathfrak{I}_{oc}$, $\mathfrak{I}_{co}$, $\mathfrak{I}_c$, is a covering class of $\mathbb{R}$ in the sense of Definition 2.20, and ℓ is an extended nonnegative valued set function with $\ell(\emptyset) = 0$ on each of these four collections. Thus if we define four set functions m_o^*, m_{oc}^*, m_{co}^*, and m_c^* on $\mathfrak{P}(\mathbb{R})$ by setting for every $E \in \mathfrak{P}(\mathbb{R})$

$$\begin{aligned}
\mu_o^*(E) &= \inf\{\textstyle\sum_{n\in\mathbb{N}} \ell(I_n) : (I_n : n \in \mathbb{N}) \subset \mathfrak{I}_o, \bigcup_{n\in\mathbb{N}} I_n \supset E\},\\
\mu_{oc}^*(E) &= \inf\{\textstyle\sum_{n\in\mathbb{N}} \ell(I_n) : (I_n : n \in \mathbb{N}) \subset \mathfrak{I}_{oc}, \bigcup_{n\in\mathbb{N}} I_n \supset E\},\\
\mu_{co}^*(E) &= \inf\{\textstyle\sum_{n\in\mathbb{N}} \ell(I_n) : (I_n : n \in \mathbb{N}) \subset \mathfrak{I}_{co}, \bigcup_{n\in\mathbb{N}} I_n \supset E\},\\
\mu_c^*(E) &= \inf\{\textstyle\sum_{n\in\mathbb{N}} \ell(I_n) : (I_n : n \in \mathbb{N}) \subset \mathfrak{I}_c, \bigcup_{n\in\mathbb{N}} I_n \supset E\},
\end{aligned}$$

then each of these set functions is an outer measure on $\mathbb{R}$ by Theorem 2.21. Among these four, μ_o^* is the Lebesgue outer measure μ_L^*. Actually for every $E \in \mathfrak{P}(\mathbb{R})$ we have the equalities $\mu_o^*(E) = \mu_{oc}^*(E) = \mu_{co}^*(E) = \mu_c^*(E)$.

Proof. Let us show that $\mu_o^*(E) = \mu_c^*(E)$ for every $E \in \mathfrak{P}(\mathbb{R})$. The same kind of argument can be used to show that $\mu_o^*(E) = \mu_{oc}^*(E)$ and $\mu_o^*(E) = \mu_{co}^*(E)$.

Let $\varepsilon > 0$ be arbitrarily given. Let $(I_n : n \in \mathbb{N})$ be a sequence in $\mathfrak{I}_o$ such that $\bigcup_{n\in\mathbb{N}} I_n \supset E$. Let $I_n = (a_n, b_n)$ and $J_n = \left[a_n - \varepsilon/2^{n+1}, b_n + \varepsilon/2^{n+1}\right]$ for $n \in \mathbb{N}$. Then $(J_n : n \in \mathbb{N})$ is a sequence in $\mathfrak{I}_c$, $\bigcup_{n\in\mathbb{N}} J_n \supset E$, and $\sum_{n\in\mathbb{N}} \ell(J_n) = \sum_{n\in\mathbb{N}} \ell(I_n) + \varepsilon$. The last equality implies by the definition of $\mu_c^*(E)$ as an infimum that $\mu_c^*(E) \leq \sum_{n\in\mathbb{N}} \ell(I_n) + \varepsilon$. Since this holds for an arbitrary sequence $(I_n : n \in \mathbb{N})$ in $\mathfrak{I}_o$ such that $\bigcup_{n\in\mathbb{N}} I_n \supset E$, we have $\mu_c^*(E) \leq \mu_o^*(E) + \varepsilon$. Then by the arbitrariness of $\varepsilon > 0$, we have $\mu_c^*(E) \leq \mu_o^*(E)$. Conversely starting with an arbitrary sequence $(I_n : n \in \mathbb{N})$ in $\mathfrak{I}_c$ given by $I_n = [a_n, b_n]$ for $n \in \mathbb{N}$ such that $\bigcup_{n\in\mathbb{N}} I_n \supset E$, and defining a sequence $(J_n : n \in \mathbb{N})$ in $\mathfrak{I}_o$ by setting $J_n = \left(a_n - \varepsilon/2^{n+1}, b_n + \varepsilon/2^{n+1}\right)$ for $n \in \mathbb{N}$, we show by the same argument as above that $\mu_o^*(E) \leq \mu_c^*(E)$. Thus $\mu_o^*(E) = \mu_c^*(E)$. ∎

Lemma 3.3. (a) *For every $x \in \mathbb{R}$ we have $\{x\} \in \mathfrak{M}_L$ and $\mu_L^*(\{x\}) = 0$.*
(b) *Every countable subset of $\mathbb{R}$ is a null set in $(\mathbb{R}, \mathfrak{M}_L, \mu_L)$.*

Proof. 1. Let $x \in \mathbb{R}$. For every $\varepsilon > 0$, we have $x \in (x - \varepsilon, x + \varepsilon) \in \mathfrak{I}_o$ and thus the sequence $(I_n : n \in \mathbb{N}) = \big((x - \varepsilon, x + \varepsilon), \emptyset, \emptyset, \dots\big)$ in $\mathfrak{I}_o$ is a covering sequence for $\{x\}$. This implies that $\mu_L^*(\{x\}) \leq \sum_{n\in\mathbb{N}} \ell(I_n) = 2\varepsilon$. By the arbitrariness of $\varepsilon > 0$, we have $\mu_L^*(\{x\}) = 0$. By Lemma 2.6, this implies $\{x\} \in \mathfrak{M}(\mu_L^*) = \mathfrak{M}_L$.

2. A singleton in $\mathbb{R}$ is a null set in $(\mathbb{R}, \mathfrak{M}_L, \mu_L)$ by (a). A countable subset of $\mathbb{R}$ is then a countable union of null sets and is therefore a null set by Observation 1.33. ∎

Lemma 3.4. *$\mu_L^* = \ell$ on $\mathfrak{I}$, that is, $\mu_L^*(I) = \ell(I)$ for every interval I in $\mathbb{R}$.*

Proof. 1. Consider the case where I is a finite closed interval given by $I = [a, b]$ where $a, b \in \mathbb{R}$, $a < b$. For an arbitrary $\varepsilon > 0$, we have $I = [a, b] \subset (a - \varepsilon, b + \varepsilon) \in \mathfrak{I}_o$. Thus $\big((a - \varepsilon, b + \varepsilon), \emptyset, \emptyset, \dots\big)$ is a covering sequence in $\mathfrak{I}_o$ for I. This implies that we have $\mu_L^*(I) \leq \ell\big((a - \varepsilon, b + \varepsilon)\big) + \ell(\emptyset) + \ell(\emptyset) + \cdots = \ell(I) + 2\varepsilon$. Since this holds for every $\varepsilon > 0$, we have

$$(1) \qquad \mu_L^*(I) \leq \ell(I).$$

Next let we show that for every covering sequence $(I_n : n \in \mathbb{N})$ in $\mathfrak{I}_o$ for I, we have

$$(2) \qquad \sum_{n\in\mathbb{N}} \ell(I_n) \geq \ell(I).$$

Now if a member I_n in a covering sequence is an infinite interval, then $\ell(I_n) = \infty$ so that $\sum_{n\in\mathbb{N}} \ell(I_n) = \infty \geq \ell(I)$ and (2) holds. Thus consider the case where every member in the covering sequence is a finite interval. Let us drop any member in the covering sequence that is disjoint from I and drop also any member that is contained in another member of the sequence. The resulting sequence $(J_n : n \in \mathbb{N})$ is a covering sequence for I with $\sum_{n\in\mathbb{N}} \ell(J_n) \leq \sum_{n\in\mathbb{N}} \ell(I_n)$. Since $(J_n : n \in \mathbb{N})$ is an open cover of the compact set I, it has a finite subcover. Renumber the members of the sequence if necessary so that $I \subset \bigcup_{k=1}^N J_k$ where $J_k = (a_k, b_k)$ for $k = 1, \dots, N$ and $a_1 \leq a_2 \leq \cdots \leq a_N$. Now if $a_i = a_j$ for some $i \neq j$, then either $J_i \subset J_j$ or $J_j \subset J_i$ which contradicts the fact that none in the sequence $(J_k : k = 1, \dots, N)$ is contained in another in the sequence. Thus we have

$a_1 < a_2 < \cdots < a_N$. Let us show that $a_2 < b_1$. Assume the contrary, that is, $b_1 \leq a_2$. Since J_1 and J_2 are not disjoint from I, there exist $x_1 \in (a_1, b_1) \cap I$ and $x_2 \in (a_2, b_2) \cap I$. Then $a_1 < x_1 < b_1 \leq a_2 < x_2 < b_2$. Since $x_1, x_2 \in I$ and I is an interval we have $[x_1, x_2] \subset I$. Since $b_1 \leq a_2$, there is at least one point in $[x_1, x_2]$ that is not covered by $(J_1, \ldots, J_N)$. Indeed if $b_1 = a_2$ then this point is not covered by $(J_k : k = 1, \ldots, N)$ and if $b_1 < a_2$ then (b_1, a_2) is not covered by $(J_k : k = 1, \ldots, N)$. This is a contradiction. Thus we have $a_2 < b_1$. Similarly we show that

$$\begin{aligned} a_1 &< a_2 < b_1, \\ a_2 &< a_3 < b_2, \\ &\vdots \\ a_{N-1} &< a_N < b_{N-1}, \\ a_N &< b_N. \end{aligned}$$

Now $\sum_{n \in \mathbb{N}} \ell(I_n) \geq \sum_{k=1}^{N} \ell(J_k)$ and

$$\begin{aligned} \sum_{k=1}^{N} \ell(J_k) &= (b_1 - a_1) + (b_2 - a_2) + \cdots + (b_{N-1} - a_{N-1}) + (b_N - a_N) \\ &\geq (a_2 - a_1) + (a_3 - a_2) + \cdots + (a_N - a_{N-1}) + (b_N - a_N) \\ &= b_N - a_1 \geq b - a = \ell(I). \end{aligned}$$

Thus $\sum_{n \in \mathbb{N}} \ell(I_n) \geq \ell(I)$. This proves (2). According to (2), $\ell(I)$ is a lower bound for $\left\{\sum_{n \in \mathbb{N}} \ell(I_n) : (I_n : n \in \mathbb{N}) \subset \mathfrak{I}_o, \bigcup_{n \in \mathbb{N}} I_n \supset I\right\}$. Since $\mu_L^*(I)$ is the infimum of this collection of nonnegative extended real numbers, we have $\mu_L^*(I) \geq \ell(I)$. Therefore $\mu_L^*(I) = \ell(I)$.

2. If I is a finite open interval given by $I = (a, b)$, then by the monotonicity and finite subadditivity of μ_L^* and by (a) of Lemma 3.3, we have

$$\mu_L^*((a, b)) \leq \mu_L^*([a, b]) \leq \mu_L^*(\{a\}) + \mu_L^*((a, b)) + \mu_L^*(\{b\}) = \mu_L^*((a, b)).$$

Thus $\mu_L^*\big((a, b)\big) = \mu_L^*\big([a, b]\big) = \ell\big([a, b]\big) = \ell\big((a, b)\big)$, where the second equality is by our result in **1**.

3. If I is a finite interval of the type $I = (a, b]$, then since $(a, b] = (a, b) \cup \{b\}$ and $\mu_L^*(\{b\}) = 0$ by (a) of Lemma 3.3, we have $\mu_L^*\big((a, b]\big) = \mu_L^*\big((a, b)\big)$ by Lemma 2.7. Thus $\mu_L^*\big((a, b]\big) = \ell\big((a, b)\big) = \ell\big((a, b]\big)$. Similarly for a finite interval I of the type $I = [a, b)$.

4. Let I be an infinite interval. Consider for instance $I = (a, \infty)$ where $a \in \mathbb{R}$. For any $n \in \mathbb{N}$ such that $n > a$, we have $(a, \infty) \supset (a, n)$ and $\mu_L^*\big((a, \infty)\big) \geq \mu_L^*\big((a, n)\big) = n - a$, by the monotonicity of μ_L^* and by our result in **2**. Since this holds for every $n > a$, we have $\mu_L^*\big((a, \infty)\big) = \infty = \ell\big((a, \infty)\big)$. Similar arguments apply to infinite intervals of other types. ∎

Proposition 3.5. *The Carathéodory condition for the μ_L^*-measurability for $E \in \mathfrak{P}(\mathbb{R})$*

(i) $\mu_L^*(A) = \mu_L^*(A \cap E) + \mu_L^*(A \cap E^c)$ *for every* $A \in \mathfrak{P}(\mathbb{R})$

is equivalent to the condition

(ii) $\mu_L^*(I) = \mu_L^*(I \cap E) + \mu_L^*(I \cap E^c)$ *for every* $I \in \mathfrak{I}_o$.

Proof. Since μ_L^* is an outer measure based on the set function ℓ on the covering class $\mathfrak{I}_o$ of subsets of $\mathbb{R}$, the equivalence of (i) and (ii) is a particular case of Theorem 2.24. ∎

Lemma 3.6. $\mathfrak{I} \subset \mathfrak{M}_L$, *that is, every interval in* $\mathbb{R}$ *is a* $\mathfrak{M}_L$*-measurable set.*

Proof. 1. Let us show first that every open interval in $\mathbb{R}$ is $\mathfrak{M}_L$-measurable. By Definition 3.1, $\mathfrak{M}_L$-measurability and $\mathfrak{M}_L$-measurability of a subset E of $\mathbb{R}$ are synonymous. Thus according to Proposition 3.5, a subset E of $\mathbb{R}$ is $\mathfrak{M}_L$-measurable if and only if it satisfies the condition

$$(1) \qquad \mu_L^*(I) = \mu_L^*(I \cap E) + \mu_L^*(I \cap E^c) \quad \text{for every } I \in \mathfrak{I}_o.$$

Consider an infinite open interval (a, ∞) in $\mathbb{R}$ where $a \in \mathbb{R}$. Then for an arbitrary $I \in \mathfrak{I}_o$ we have

$$I = I \cap \mathbb{R} = I \cap \{(a, \infty) \cup (a, \infty)^c\} = \{I \cap (a, \infty)\} \cup \{I \cap (a, \infty)^c\}.$$

Now $I \cap (a, \infty)$ is either an interval or $\emptyset$ and similarly $\{I \cap (a, \infty)^c\}$ is either an interval or $\emptyset$. Since $I \cap (a, \infty)$ and $\{I \cap (a, \infty)^c\}$ are disjoint and their union is equal to I, we have

$$(2) \qquad \ell(I) = \ell\big(I \cap (a, \infty)\big) + \ell\big(I \cap (a, \infty)^c\big).$$

According to Lemma 3.4, we have $\ell(I) = \mu_L(I)$, $\ell\big(I \cap (a, \infty)\big) = \mu_L\big(I \cap (a, \infty)\big)$, and $\ell\big(I \cap (a, \infty)^c\big) = \mu_L\big(I \cap (a, \infty)^c\big)$. Substituting these into (2), we have

$$(3) \qquad \mu_L(I) = \mu_L\big(I \cap (a, \infty)\big) + \mu_L\big(I \cap (a, \infty)^c\big) \quad \text{for every } I \in \mathfrak{I}_o.$$

This verifies condition (1) for (a, ∞) and shows that (a, ∞) is $\mathfrak{M}_L$-measurable.

By similar argument, we show that an infinite open interval $(-\infty, b)$, where $b \in \mathbb{R}$, is $\mathfrak{M}_L$-measurable. If $a, b \in \mathbb{R}$ and $a < b$, then $\mathbb{R} = (-\infty, b) \cup (a, \infty)$. Then since $(-\infty, b) \in \mathfrak{M}_L$, $(a, \infty) \in \mathfrak{M}_L$, and $\mathfrak{M}_L$ is a σ-algebra of subsets of $\mathbb{R}$, we have $\mathbb{R} \in \mathfrak{M}_L$. This shows that every infinite open interval in $\mathbb{R}$ is $\mathfrak{M}_L$-measurable. An arbitrary finite open interval is given by (a, b) where $a, b \in \mathbb{R}$ and $a < b$. Then $(a, b) = (-\infty, b) \cap (a, \infty)$. Since $(-\infty, b) \in \mathfrak{M}_L$, $(a, \infty) \in \mathfrak{M}_L$, and $\mathfrak{M}_L$ is a σ-algebra of subsets of $\mathbb{R}$, we have $(a, b) \in \mathfrak{M}_L$. Thus we have shown that every open interval in $\mathbb{R}$ is $\mathfrak{M}_L$-measurable.

2. Let J be an interval in $\mathbb{R}$ which is not an open interval. Then $J = \{a\} \cup I$, $J = I \cup \{b\}$, or $J = \{a\} \cup I \cup \{b\}$, where I is an open interval, $a, b \in \mathbb{R}$, $a < b$, and $\{I, \{a\}, \{b\}\}$ is a disjoint collection. Now $I \in \mathfrak{M}_L$ by our result above and $\{a\}, \{b\} \in \mathfrak{M}_L$ by Lemma 3.3. Then since $\mathfrak{M}_L$ is a σ-algebra of subsets of $\mathbb{R}$, we have $J \in \mathfrak{M}_L$. ∎

A set E in a topological space X is said to be dense in X if for every non-empty open set O in X we have $E \cap O \neq \emptyset$. In particular if E is a dense subset of $\mathbb{R}$ then every open interval (α, β) in $\mathbb{R}$ contains infinitely many points of E. Indeed the denseness of E in $\mathbb{R}$

implies that (α, β) contains a point $x_1 \in E$. Then the subinterval (α, x_1) of (α, β) contains a point $x_2 \in E$ and the subinterval (α, x_2) of (α, x_1) contains a point $x_3 \in E$ and so on. Thus (α, β) contains an infinite sequence $(x_n : n \in \mathbb{N})$ of distinct points in E.

Observation 3.7. If E is a null set in $(\mathbb{R}, \mathfrak{M}_L, \mu_L)$, then E^c is a dense subset of $\mathbb{R}$.

Proof. For every open interval I in $\mathbb{R}$, we have $\mu_L(I) > 0$. Thus if $E \in \mathfrak{M}_L$ and $\mu_L(E) = 0$, then by the monotonicity of μ_L on $\mathfrak{M}_L$, E cannot contain any open interval as a subset. This implies that $E^c \cap I \neq \emptyset$ for every open interval I and therefore E^c is dense in $\mathbb{R}$. ∎

Examples. (a) If Q is the set of all rational numbers and P is the set of all irrational numbers, then Q is a null set in $(\mathbb{R}, \mathfrak{M}_L, \mu_L)$ and $P \in \mathfrak{M}_L$ with $\mu_L(P) = \infty$.
(b) For every interval I in $\mathbb{R}$, we have $I \cap P \in \mathfrak{M}_L$ with $\mu_L(I \cap P) = \ell(I)$.

Proof. 1. The set Q of all rational numbers is a countable set and thus a null set in $(\mathbb{R}, \mathfrak{M}_L, \mu_L)$ by (b) of Lemma 3.3. The set P of all irrational numbers, being the complement of Q in $\mathbb{R}$, is a member of $\mathfrak{M}_L$. By the additivity of μ_L on $\mathfrak{M}_L$, we have $\mu_L(P) + \mu_L(Q) = \mu_L(\mathbb{R})$. Since $\mu_L(Q) = 0$ and $\mu_L(\mathbb{R}) = \ell(\mathbb{R}) = \infty$, we have $\mu_L(P) = \infty$.

2. Let I be an interval in $\mathbb{R}$. By Lemma 3.6, I is a member of $\mathfrak{M}_L$. This implies that $I \cap P$ and $I \cap Q$ are members of $\mathfrak{M}_L$. Since P and Q are disjoint and their union is $\mathbb{R}$, $I \cap P$ and $I \cap Q$ are disjoint and their union is I. By the additivity of μ_L on $\mathfrak{M}_L$, we have $\mu_L(I \cap P) + \mu_L(I \cap Q) = \mu_L(I) = \ell(I)$ by Lemma 3.4. Since $\mu_L(Q) = 0$, we have $\mu_L(I \cap Q) = 0$ by the monotonicity of μ_L on $\mathfrak{M}_L$. Thus $\mu_L(I \cap P) = \ell(I)$. ∎

[II] Some Properties of the Lebesgue Measure Space

We show that the Lebesgue measure $(\mathbb{R}, \mathfrak{M}_L, \mu_L)$ has the following properties:

(i) $(\mathbb{R}, \mathfrak{M}_L, \mu_L)$ is a complete measure space.
(ii) $(\mathbb{R}, \mathfrak{M}_L, \mu_L)$ is a σ-finite (but not finite) measure space.
(iii) $\mathfrak{B}_{\mathbb{R}} \subset \mathfrak{M}_L$.
(iv) $\mu_L(O) > 0$ for every nonempty open set O in $\mathbb{R}$.
(v) $(\mathbb{R}, \mathfrak{M}_L, \mu_L)$ has no atoms.
(vi) $(\mathbb{R}, \mathfrak{M}_L, \mu_L)$ is translation invariant.
(vii) $(\mathbb{R}, \mathfrak{M}_L, \mu_L)$ is positively homogeneous.
(viii) Non $\mathfrak{M}_L$-measurable sets exist.
(ix) The Lebesgue outer measure μ_L^* is a Borel regular outer measure.

Property (i) is immediate. Indeed since the measure space $(\mathbb{R}, \mathfrak{M}_L, \mu_L)$ is constructed by means of an outer measure μ_L^* it is a complete measure space according to Theorem 2.9. We prove the rest of the properties below.

Theorem 3.8. *The Lebesgue measure space $(\mathbb{R}, \mathfrak{M}_L, \mu_L)$ is a σ-finite, but not finite, measure space.*

Proof. Note that $\mu_L(\mathbb{R}) = \ell(\mathbb{R}) = \infty$ by Lemma 3.4. This shows that $(\mathbb{R}, \mathfrak{M}_L, \mu_L)$ is not a finite measure space. Let us show that $(\mathbb{R}, \mathfrak{M}_L, \mu_L)$ is a σ-finite measure space. Let $E_n = (-n, n)$ for $n \in \mathbb{N}$. Then $E_n \in \mathfrak{M}_L$ for $n \in \mathbb{N}$ and $\bigcup_{n\in\mathbb{N}} E_n = \mathbb{R}$ with $\mu_L(E_n) = \ell(E_n) = 2n < \infty$. Thus $(\mathbb{R}, \mathfrak{M}_L, \mu_L)$ is a σ-finite measure space. ∎

Theorem 3.9. *$\mathfrak{B}_{\mathbb{R}} \subset \mathfrak{M}_L$, that is, every Borel set in $\mathbb{R}$ is a Lebesgue measurable set.*

Proof. Every interval in $\mathbb{R}$ is a member of the σ-algebra $\mathfrak{M}_L$ by Lemma 3.6. Since every open set in $\mathbb{R}$ is a countable union of open intervals, it is a member of $\mathfrak{M}_L$. Thus if we let $\mathfrak{O}$ be the collection of all open sets in $\mathbb{R}$, then $\mathfrak{O} \subset \mathfrak{M}_L$. From this, it follows that $\sigma(\mathfrak{O}) \subset \sigma(\mathfrak{M}_L) = \mathfrak{M}_L$. Since $\mathfrak{B}_{\mathbb{R}} = \sigma(\mathfrak{O})$, we have $\mathfrak{B}_{\mathbb{R}} \subset \mathfrak{M}_L$. ∎

Observation 3.10. For every nonempty open set O in $\mathbb{R}$, we have $\mu_L(O) > 0$.

Proof. Every nonempty open set O in $\mathbb{R}$ contains an open interval I. Then we have $\mu_L(O) \geq \mu_L(I) = \ell(I) > 0$. ∎

Proposition 3.11. *Regarding atoms in $(\mathbb{R}, \mathfrak{M}_L)$ and atoms in $(\mathbb{R}, \mathfrak{M}_L, \mu_L)$, we have:*
(a) *In the Lebesgue measurable space $(\mathbb{R}, \mathfrak{M}_L)$, the only atoms are the singletons.*
(b) *The Lebesgue measure space $(\mathbb{R}, \mathfrak{M}_L, \mu_L)$ has no atoms. Indeed if $E \in \mathfrak{M}_L$ and $\mu_L(E) > 0$, then for every $\varepsilon > 0$ there exists $E_0 \subset E$, $E_0 \in \mathfrak{M}_L$, with $\mu_L(E_0) \in (0, \varepsilon]$.*

Proof. 1. For every $x \in \mathbb{R}$, we have $\{x\} \in \mathfrak{M}_L$ by (a) of Lemma 3.3. The only subsets of $\{x\}$ are $\emptyset$ and $\{x\}$. Thus $\{x\}$ is an atom.

Now let $E \in \mathfrak{M}_L$ and suppose E contains at least two points. Let $x \in E$. Then $\{x\} \in \mathfrak{M}_L$. But $\{x\} \neq \emptyset$ and $\{x\} \neq E$. Thus E is not an atom of the measurable space $(\mathbb{R}, \mathfrak{M}_L)$. This proves (a).

2. For $\varepsilon > 0$ arbitrarily given, let $I_n = \big((n-1)\varepsilon, n\varepsilon\big]$ for $n \in \mathbb{Z}$. Then $(I_n : n \in \mathbb{Z})$ is a disjoint sequence in $\mathfrak{I}_{oc}$ with $\bigcup_{n\in\mathbb{Z}} I_n = \mathbb{R}$. Let $E_n = E \cap I_n$ for $n \in \mathbb{Z}$. Then $(E_n : n \in \mathbb{Z})$ is a disjoint sequence in $\mathfrak{M}_L$ and $\bigcup_{n\in\mathbb{Z}} E_n = E$. By the countable additivity of μ_L on $\mathfrak{M}_L$ we have $\sum_{n\in\mathbb{Z}} \mu_L(E_n) = \mu_L\left(\bigcup_{n\in\mathbb{Z}} E_n\right) = \mu_L(E) > 0$. Thus there exists some $n_0 \in \mathbb{Z}$ such that $\mu_L(E_{n_0}) > 0$. Since $E_{n_0} = E \cap I_{n_0} \subset I_{n_0}$, we have $\mu_L(E_{n_0}) \leq \mu_L(I_{n_0}) = \varepsilon$ by the monotonicity of μ_L. Thus we have $\mu_L(E_{n_0}) \in (0, \varepsilon]$. This proves (b). ∎

Definition 3.12. *Let X be a linear space over the field of scalars $\mathbb{R}$.*
(a) *For $E \subset X$ and $x_0 \in X$, we write*

$$E + x_0 = \{x + x_0 : x \in E\} = \{y \in X : y = x + x_0 \text{ for some } x \in E\}$$

and call the set the x_0-translate of E.
(b) *For $\alpha \in \mathbb{R}$, we write*

$$\alpha E = \{\alpha x : x \in E\} = \{y \in X : y = \alpha x \text{ for some } x \in E\}$$

and call it the dilation of E by factor α.

(c) *For a collection $\mathfrak{E}$ of subsets of X, $x \in X$, and $\alpha \in \mathbb{R}$ we write $\mathfrak{E}+x = \{E+x : E \in \mathfrak{E}\}$ and $\alpha\mathfrak{E} = \{\alpha E : E \in \mathfrak{E}\}$.*

Observation 3.13. For translates of sets the following equalities hold:

$$(E + x_1) + x_2 = E + (x_1 + x_2),$$

$$(E + x)^c = E^c + x,$$

$$E_1 \subset E_2 \Rightarrow E_1 + x \subset E_2 + x.$$

For translates of collections of sets we have

$$\left(\bigcup_{\alpha \in A} E_\alpha\right) + x = \bigcup_{\alpha \in A}(E_\alpha + x),$$

$$\left(\bigcap_{\alpha \in A} E_\alpha\right) + x = \bigcap_{\alpha \in A}(E_\alpha + x).$$

For dilations of sets we have

$$\beta(\alpha E) = (\alpha\beta)E,$$

$$(\alpha E)^c = \alpha E^c.$$

Definition 3.14. *Given a measure space $(X, \mathfrak{A}, \mu)$ where X is a linear space.*
(a) *We say that σ-algebra $\mathfrak{A}$ is translation invariant if for every $E \in \mathfrak{A}$ and $x \in X$ we have $E + x \in \mathfrak{A}$.*
(b) *We say that the measure μ is translation invariant if $\mathfrak{A}$ is translation invariant and if for every $E \in \mathfrak{A}$ and $x \in X$ we have $\mu(E + x) = \mu(E)$.*
(c) *We call $(X, \mathfrak{A}, \mu)$ a translation invariant measure space if both $\mathfrak{A}$ and μ are translation invariant.*

Example. As an example of a σ-algebra of subsets of $\mathbb{R}$ that is not translation invariant, let $I_n = (n-1, n]$ for $n \in \mathbb{Z}$ and let $\mathfrak{A}$ be the collection of $\emptyset$ and all countable unions of members of the collection $\{I_n : n \in \mathbb{Z}\}$. It is easily verified that $\mathfrak{A}$ is a σ-algebra of subsets of $\mathbb{R}$. $\mathfrak{A}$ is not translation invariant since $I_n \in \mathfrak{A}$ but $I_n + \frac{1}{2} = \left(n - \frac{1}{2}, n + \frac{1}{2}\right]$ is not a member of $\mathfrak{A}$.

Lemma 3.15. (Translation Invariance of the Lebesgue Outer Measure) *For every $E \in \mathfrak{P}(\mathbb{R})$ and $x \in \mathbb{R}$, we have $\mu_L^*(E + x) = \mu_L^*(E)$.*

Proof. Note that for every $I \in \mathfrak{I}_o$ and $x \in \mathbb{R}$, we have $I + x \in \mathfrak{I}_o$ and $\ell(I + x) = \ell(I)$. Let $E \in \mathfrak{P}(\mathbb{R})$. Take an arbitrary sequence $(I_n : n \in \mathbb{N})$ in $\mathfrak{I}_o$ such that $\bigcup_{n\in\mathbb{N}} I_n \supset E$. For an arbitrary $x \in \mathbb{R}$, $(I_n + x : n \in \mathbb{N})$ is a sequence in $\mathfrak{I}_o$ with $\ell(I_n + x) = \ell(I_n)$ and $\bigcup_{n\in\mathbb{N}}(I_n+x) = \left(\bigcup_{n\in\mathbb{N}} I_n\right)+x \supset E+x$. Thus $\sum_{n\in\mathbb{N}} \ell(I_n) = \sum_{n\in\mathbb{N}} \ell(I_n+x) \geq \mu_L^*(E+x)$ by the fact that $(I_n + x : n \in \mathbb{N})$ is a covering sequence for $E + x$ and by the definition of $\mu_L^*(E+x)$ as an infimum. Since this holds for an arbitrary covering sequence $(I_n : n \in \mathbb{N})$

for E and since $\mu_L^*(E)$ is the infimum of $\sum_{n\in\mathbb{N}}\ell(I_n)$, we have $\mu_L^*(E) \geq \mu_L^*(E+x)$ for any $E \in \mathfrak{P}(\mathbb{R})$ and $x \in \mathbb{R}$.

Applying this result to $E + x$ and its translate $(E + x) - x = E$, we have the reverse inequality $\mu_L^*(E+x) \geq \mu_L^*(E.)$. Therefore $\mu_L^*(E+x) = \mu_L^*(E)$. ∎

Theorem 3.16. (Translation Invariance of the Lebesgue Measure Space) *The Lebesgue measure space* $(\mathbb{R}, \mathfrak{M}_L, \mu_L)$ *is translation invariant, that is, for every* $E \in \mathfrak{M}_L$ *and* $x \in \mathbb{R}$ *we have* $E + x \in \mathfrak{M}_L$ *and* $\mu_L(E+x) = \mu_L(E)$. *Let* $\mathfrak{M}_L + x := \{E + x : E \in \mathfrak{M}_L\}$. *Then* $\mathfrak{M}_L + x = \mathfrak{M}_L$ *for every* $x \in \mathbb{R}$.

Proof. Let $E \in \mathfrak{M}_L$ and $x \in \mathbb{R}$. To show that $E + x \in \mathfrak{M}_L$, we show that for every $A \in \mathfrak{P}(\mathbb{R})$, we have $\mu_L^*\big(A \cap (E+x)\big) + \mu_L^*\big(A \cap (E+x)^c\big) = \mu_L^*(A)$. By Lemma 3.15

$$\mu_L^*\big(A \cap (E+x)\big) + \mu_L^*\big(A \cap (E+x)^c\big)$$
$$= \mu_L^*\big(\{A \cap (E+x)\} - x\big) + \mu_L^*\big(\{A \cap (E+x)^c\} - x\big)$$
$$= \mu_L^*\big((A-x) \cap E\big) + \mu_L^*\big((A-x) + E^c\big)$$
$$= \mu_L^*(A-x) = \mu_L^*(A),$$

where the second equality is by Observation 3.13, the third equality is by the Carathéodory condition satisfied by our $E \in \mathfrak{M}_L$ with $A - x$ as a testing set, and the last equality is by Lemma 3.15. Thus $E + x \in \mathfrak{M}_L$. Then $\mu_L(E+x) = \mu_L^*(E+x) = \mu_L^*(E) = \mu_L(E)$, where the second equality is by Lemma 3.15.

For every $E \in \mathfrak{M}_L$ and $x \in \mathbb{R}$ we have $E + x \in \mathfrak{M}_L$ as we showed above. Thus $\mathfrak{M}_L + x \subset \mathfrak{M}_L$. Now $\mathfrak{M}_L = \mathfrak{M}_L + 0 = \mathfrak{M}_L + (-x) + x$ for any $x \in \mathbb{R}$. Since $-x \in \mathbb{R}$ we have $\mathfrak{M}_L + (-x) \subset \mathfrak{M}_L$. Then $\mathfrak{M}_L + (-x) + x \subset \mathfrak{M}_L + x$, that is, $\mathfrak{M}_L \subset \mathfrak{M}_L + x$. Therefore we have $\mathfrak{M}_L + x = \mathfrak{M}_L$. ∎

Lemma 3.17. *For* $E \in \mathfrak{P}(\mathbb{R})$ *and* $\alpha \in \mathbb{R}$, *let* $\alpha E = \{y \in \mathbb{R} : y = \alpha x \text{ for some } x \in E\}$. *Then* $\mu_L^*(\alpha E) = |\alpha|\mu_L^*(E)$.

Proof. Consider the collection $\mathfrak{I}_o$ of $\emptyset$ and all open intervals in $\mathbb{R}$. Let $S = (I_n : n \in \mathbb{N})$ be a sequence in $\mathfrak{I}_o$ and let $\mathcal{S}$ be the collection of all sequences in $\mathfrak{I}_o$. For $\alpha \in \mathbb{R}$, $\alpha \neq 0$, let us define a mapping M_α of $\mathcal{S}$ into $\mathcal{S}$ by setting $M_\alpha(S) = (\alpha I_n : n \in \mathbb{N})$ for $S = (I_n : n \in \mathbb{N})$. Clearly M_α is a one-to-one mapping of $\mathcal{S}$ onto $\mathcal{S}$ with inverse mapping given by $M_{1/\alpha}$. For an arbitrary $E \in \mathfrak{P}(\mathbb{R})$, let $\mathcal{S}_E$ be the subcollection of $\mathcal{S}$ consisting of sequences $S = (I_n : n \in \mathbb{N})$ such that $\bigcup_{n\in\mathbb{N}} I_n \supset E$. Note that $\bigcup_{n\in\mathbb{N}} I_n \supset E$ if and only if $\bigcup_{n\in\mathbb{N}} \alpha I_n \supset \alpha E$. Thus $M_\alpha(\mathcal{S}_E) = \mathcal{S}_{\alpha E}$ and in fact M_α is a one-to-one mapping of $\mathcal{S}_E$ onto $\mathcal{S}_{\alpha E}$. Let λ be a nonnegative extended real-valued set function on $\mathcal{S}$ defined by $\lambda(S) = \sum_{n\in\mathbb{N}}\ell(I_n)$ for $S = (I_n : n \in \mathbb{N})$. Then

$$(1) \qquad \lambda(M_\alpha(S)) = \sum_{n\in\mathbb{N}} \ell(\alpha I_n) = |\alpha| \sum_{n\in\mathbb{N}} \ell(I_n) = |\alpha|\lambda(S).$$

We have $\mu_L^*(E) = \inf_{S\in\mathcal{S}_E} \lambda(S)$ and $\mu_L^*(\alpha E) = \inf_{T\in\mathcal{S}_{\alpha E}} \lambda(T)$. By the one-to-one correspondence between $\mathcal{S}_E$ and $\mathcal{S}_{\alpha E}$ by M_α and by (1), we have

$$\inf_{T\in\mathcal{S}_{\alpha E}} \lambda(T) = \inf_{S\in\mathcal{S}_E} \lambda(M_\alpha(S)) = \inf_{S\in\mathcal{S}_E} |\alpha|\lambda(S) = |\alpha|\mu_L^*(E).$$

This shows that $\mu_L^*(\alpha E) = |\alpha|\mu_L^*(E)$ when $\alpha \neq 0$. When $\alpha = 0$, we have $\alpha E = \{0\}$ with $\mu_L^*(\{0\}) = 0$ so that $\mu_L^*(\alpha E) = |\alpha|\mu_L^*(E)$ holds trivially. ∎

Theorem 3.18. (Positive Homogeneity of the Lebesgue Measure Space) *For every set $E \in \mathfrak{M}_L$ and $\alpha \in \mathbb{R}$, we have $\alpha E \in \mathfrak{M}_L$ and $\mu_L(\alpha E) = |\alpha|\mu_L(E)$. For every $\alpha \in \mathbb{R}$, let $\alpha\mathfrak{M}_L := \{\alpha E : E \in \mathfrak{M}_L\}$. Then $\alpha\mathfrak{M}_L = \mathfrak{M}_L$ for every $\alpha \in \mathbb{R}$ such that $\alpha \neq 0$.*

Proof. The Theorem is trivially true when $\alpha = 0$. In what follows we assume $\alpha \neq 0$. Let $E \in \mathfrak{M}_L$. To show that $\alpha E \in \mathfrak{M}_L$, we show that $\mu_L^*(A) = \mu_L^*(A \cap \alpha E) + \mu_L^*\big(A \cap (\alpha E)^c\big)$ for every $A \in \mathfrak{P}(\mathbb{R})$. With $\frac{1}{\alpha}A$ as a testing set for $E \in \mathfrak{M}_L$, we have

(1) $$\mu_L^*\left(\tfrac{1}{\alpha}A\right) = \mu_L^*\left(\tfrac{1}{\alpha}A \cap E\right) + \mu_L^*\left(\tfrac{1}{\alpha}A \cap E^c\right).$$

By Lemma 3.17, we have

$$\mu_L^*\left(\tfrac{1}{\alpha}A\right) = \tfrac{1}{|\alpha|}\mu_L^*(A),$$
$$\mu_L^*\left(\tfrac{1}{\alpha}A \cap E\right) = \mu_L^*\left(\tfrac{1}{\alpha}A \cap \tfrac{1}{\alpha}\alpha E\right) = \tfrac{1}{|\alpha|}\mu_L^*(A \cap \alpha E),$$
$$\mu_L^*\left(\tfrac{1}{\alpha}A \cap E^c\right) = \mu_L^*\left(\tfrac{1}{\alpha}A \cap \tfrac{1}{\alpha}\alpha E^c\right) = \tfrac{1}{|\alpha|}\mu_L^*\big(A \cap (\alpha E)^c\big).$$

Substituting these equalities in (1) and multiplying the resulting equation by $|\alpha|$, we have $\mu_L^*(A) = \mu_L^*(A \cap \alpha E) + \mu_L^*(A \cap (\alpha E)^c)$. This shows that $\alpha E \in \mathfrak{M}_L$. Then we have $\mu_L(\alpha E) = \mu_L^*(\alpha E) = |\alpha|\mu_L^*(E) = |\alpha|\mu_L(E)$ by Lemma 3.17.

For every $E \in \mathfrak{M}_L$ and $\alpha \in \mathbb{R}$ we have $\alpha E \in \mathfrak{M}_L$ as we showed above. Thus $\alpha\mathfrak{M}_L \subset \mathfrak{M}_L$. Let $\alpha \in \mathbb{R}$ and $\alpha \neq 0$. Then $\mathfrak{M}_L = 1\mathfrak{M}_L = \alpha\frac{1}{\alpha}\mathfrak{M}_L \subset \alpha\mathfrak{M}_L$. Thus we have $\alpha\mathfrak{M}_L = \mathfrak{M}_L$. ∎

[III] Existence of Non Lebesgue Measurable Sets

Consider the interval $[0, 1)$ in $\mathbb{R}$. Define addition modulo 1 of $x, y \in [0, 1)$ by

$$x \overset{\circ}{+} y = \begin{cases} x + y & \text{if } x + y < 1, \\ x + y - 1 & \text{if } x + y \geq 1. \end{cases}$$

The operation $\overset{\circ}{+}$ takes a pair of numbers in $[0, 1)$ to a number in $[0, 1)$. It is commutative, that is, $x \overset{\circ}{+} y = y \overset{\circ}{+} x$, and associative, that is, $(x \overset{\circ}{+} y) \overset{\circ}{+} z = x \overset{\circ}{+} (y \overset{\circ}{+} z)$. For $E \subset [0, 1)$ and $y \in [0, 1)$, let $E \overset{\circ}{+} y = \big\{z \in [0, 1) : z = x \overset{\circ}{+} y \text{ for some } x \in E\big\}$, and call this set the y-translate of E modulo 1.

Lemma 3.19. *Let $E \subset [0, 1)$. If $E \in \mathfrak{M}_L$, then for every $y \in [0, 1)$, we have $E \overset{\circ}{+} y \in \mathfrak{M}_L$ and $\mu_L(E \overset{\circ}{+} y) = \mu_L(E)$.*

Proof. Let $E \in \mathfrak{M}_L$ and $y \in [0, 1)$. Then $[0, 1-y)$ and $[1-y, 1)$ are two disjoint subsets of $[0, 1)$ whose union is $[0, 1)$. Define two disjoint subsets of E by $E_1 = E \cap [0, 1-y)$ and $E_2 = E \cap [1-y, 1)$. Then $E_1 \cup E_2 = E$. Also $E_1, E_2 \in \mathfrak{M}_L$. Thus

(1) $$\mu_L(E) = \mu_L(E_1) + \mu_L(E_2).$$

Now $E_1 \overset{\circ}{+} y = E_1 + y \in \mathfrak{M}_L$ by the translation invariance of $(\mathbb{R}, \mathfrak{M}_L, \mu_L)$ and we have

(2) $$\mu_L(E_1 \overset{\circ}{+} y) = \mu_L(E_1 + y) = \mu_L(E_1).$$

On the other hand, $E_2 \overset{\circ}{+} y = E_2 + y - 1 \in \mathfrak{M}_L$ by the translation invariance of $(\mathbb{R}, \mathfrak{M}_L, \mu_L)$ and we have

(3) $$\mu_L(E_2 \overset{\circ}{+} y) = \mu_L(E_2 + y - 1) = \mu_L(E_2).$$

Now

(4) $$E \overset{\circ}{+} y = (E_1 \cup E_2) \overset{\circ}{+} y = (E_1 \overset{\circ}{+} y) \cup (E_2 \overset{\circ}{+} y).$$

The disjointness of E_1 and E_2 implies that of $E_1 \overset{\circ}{+} y$ and $E_2 \overset{\circ}{+} y$. Since $E_1 \overset{\circ}{+} y$ and $E_2 \overset{\circ}{+} y$ are members of $\mathfrak{M}_L$ as we showed above, we have $E \overset{\circ}{+} y \in \mathfrak{M}_L$. Then by (4), (2), (3), and (1), we have $\mu_L(E \overset{\circ}{+} y) = \mu_L(E_1 \overset{\circ}{+} y) + \mu_L(E_2 \overset{\circ}{+} y) = \mu_L(E_1) + \mu_L(E_2) = \mu_L(E)$. ∎

Theorem 3.20. *The interval* $[0, 1)$ *in* $\mathbb{R}$ *contains a non Lebesgue measurable set.*

Proof. For $x, y \in [0, 1)$, let us say that x and y are equivalent and write $x \sim y$ if $x - y$ is a rational number. Then

$$x \sim x$$
$$x \sim y \Rightarrow y \sim x,$$
$$x \sim y, y \sim z \Rightarrow x \sim z.$$

Thus $\sim$ is an equivalence relation among elements of $[0, 1)$. Let $\{E_\alpha\}$ be the collection of the equivalence classes with respect to this equivalence relation. $\{E_\alpha\}$ is a disjoint collection and $\bigcup_\alpha E_\alpha = [0, 1)$. Any two elements of $[0, 1)$ which are in one equivalence class differ by a rational number and any two elements of $[0, 1)$ which are not in the same equivalence class differ by an irrational number.

Let P be a subset of $[0, 1)$ constructed by picking an element from each E_α. (According to the Axiom of Choice, given a nonempty collection of nonempty sets it is possible to select an element from each set.) Let $\{r_n : n \in \mathbb{Z}_+\}$ be an enumeration of the rational numbers in $[0, 1)$ with $r_0 = 0$. Let

$$P_n = P \overset{\circ}{+} r_n \quad \text{for } n \in \mathbb{Z}_+.$$

Note that $P_0 = P \overset{\circ}{+} r_0 = P \overset{\circ}{+} 0 = P$. Let us show first that $\{P_n : n \in \mathbb{Z}_+\}$ is a disjoint collection, that is,

(1) $$m, n \in \mathbb{Z}_+, m \neq n \Rightarrow P_m \cap P_n = \emptyset.$$

Assume the contrary. Then $P_m \cap P_n \neq \emptyset$ and there exists $x \in P_m \cap P_n$. Since $x \in P_m$ we have $x = p_m \overset{\circ}{+} r_m$ for some $p_m \in P$. Similarly since $x \in P_n$ we have $x = p_n \overset{\circ}{+} r_n$ for some $p_n \in P$. Then $p_m \overset{\circ}{+} r_m = p_n \overset{\circ}{+} r_n$. Since $p_m \overset{\circ}{+} r_m$ is either $p_m + r_m$ or $p_m + r_m - 1$ and $p_n \overset{\circ}{+} r_n$ is either $p_n + r_n$ or $p_n + r_n - 1$, the difference $p_m - p_n$ is a rational number. Then $p_m, p_n \in E_\alpha$ for some $\alpha \in A$. But P contains only one element of E_α. Thus $p_m = p_n$. Then the equality $p_m \overset{\circ}{+} r_m = p_n \overset{\circ}{+} r_n$, which we established above, is reduced to $p_m \overset{\circ}{+} r_m = p_m \overset{\circ}{+} r_n$. This implies $r_m = r_n$ and then $m = n$, a contradiction. This proves (1).

Let us show next that

$$\text{(2)} \qquad \bigcup_{n\in\mathbb{Z}_+} P_n = [0, 1).$$

Since $P_n \subset [0, 1)$ for $n \in \mathbb{Z}_+$ we have $\bigcup_{n\in\mathbb{Z}_+} P_n \subset [0, 1)$. To prove the reverse inclusion we show that if $x \in [0, 1)$ then $x \in P_n$ for some $n \in \mathbb{Z}_+$. Now if $x \in [0, 1)$, then $x \in E_\alpha$ for some α. Since P contains one element from each equivalent class, there exists $p \in P$ such that $p \in E_\alpha$. Thus x and p differ by a rational number. Since both x and p are in $[0, 1)$, their difference is a rational number in $[0, 1)$, that is, their difference is r_n for some $n \in \mathbb{Z}_+$. Now if $x \geq p$, then $x = p + r_n$ so that $x \in P_n$. On the other hand, if $x < p$, then $p = x + r_n$ so that $x = p - r_n$. Let $r_m = 1 - r_n$, a rational number in $[0, 1)$. Then $x = p + r_m - 1 = p \overset{\circ}{+} r_m$ so that $x \in P_m$. Thus in any case x is in P_n for some $n \in \mathbb{Z}_+$. This shows that $[0, 1) \subset \bigcup_{n\in\mathbb{Z}_+} P_n$ and proves (2).

Finally to show that $P \notin \mathfrak{M}_L$, assume the contrary, that is, $P \in \mathfrak{M}_L$. Then by Lemma 3.19, $P_n = P \overset{\circ}{+} r_n \in \mathfrak{M}_L$ and $\mu_L(P_n) = \mu_L(P)$. By (2) and (1) we have

$$\text{(3)} \qquad 1 = \mu_L([0, 1)) = \mu_L\Big(\bigcup_{n\in\mathbb{Z}_+} P_n\Big) = \sum_{n\in\mathbb{Z}_+} \mu_L(P_n) = \sum_{n\in\mathbb{Z}_+} \mu_L(P).$$

Since $P \in \mathfrak{M}_L$, $\mu_L(P)$ is defined and $\mu_L(P) \geq 0$. If $\mu_L(P) = 0$, then $\sum_{n\in\mathbb{Z}_+} \mu_L(P) = 0$ so that (3) reduces to $1 = 0$, a contradiction. If $\mu_L(P) > 0$, then $\sum_{n\in\mathbb{Z}_+} \mu_L(P) = \infty$ so that (3) reduces to $1 = \infty$, a contradiction. This shows that P is not in $\mathfrak{M}_L$. ∎

[IV] Regularity of Lebesgue Outer Measure

Lemma 3.21. (Borel Regularity of the Lebesgue Outer Measure) *The Lebesgue outer measure μ_L^* on $\mathbb{R}$ has the following properties:*
(a) *For every $E \in \mathfrak{P}(\mathbb{R})$ and $\varepsilon > 0$, there exists an open set O in $\mathbb{R}$ such that $O \supset E$ and*

$$\mu_L^*(E) \leq \mu_L^*(O) \leq \mu_L^*(E) + \varepsilon.$$

(Note that the strict inequalities $\mu_L^*(E) < \mu_L^*(O)$ and $\mu_L^*(O) < \mu_L^*(E) + \varepsilon$ may not hold.)
(b) *For every $E \in \mathfrak{P}(\mathbb{R})$, there exists a G_δ-set G in $\mathbb{R}$ such that $G \supset E$ and*

$$\mu_L^*(G) = \mu_L^*(E).$$

(c) *The Lebesgue outer measure μ_L^* is a Borel regular outer measure.*

Proof. 1. Let $E \in \mathfrak{P}(\mathbb{R})$ and let $\varepsilon > 0$. If $\mu_L^*(E) = \infty$ then with the open set $\mathbb{R} \supset E$ we have $\mu_L^*(E) \le \mu_L^*(\mathbb{R}) = \infty = \infty + \varepsilon = \mu_L^*(E) + \varepsilon$. Thus (a) is trivially true in this case. Now consider the case $\mu_L^*(E) < \infty$. By the definition of $\mu_L^*(E)$ as an infimum there exists a sequence $(I_n : n \in \mathbb{N})$ in $\mathfrak{I}_o$ such that $\bigcup_{n\in\mathbb{N}} I_n \supset E$ and $\mu_L^*(E) \le \sum_{n\in\mathbb{N}} \ell(I_n) < \mu_L^*(E)+\varepsilon$. If we let $O = \bigcup_{n\in\mathbb{N}} I_n$, then O is an open set, $O \supset E$, and by the monotonicity and the countable subadditivity of μ_L^*, we have

$$\mu_L^*(E) \le \mu_L^*(O) = \mu_L^*\Big(\bigcup_{n\in\mathbb{N}} I_n\Big) \le \sum_{n\in\mathbb{N}} \mu_L^*(I_n) = \sum_{n\in\mathbb{N}} \ell(I_n) < \mu_L^*(E) + \varepsilon.$$

(Note that for $E \in \mathfrak{P}(\mathbb{R})$ such that $\mu_L^*(E) = \infty$, it is impossible to have an open set O in $\mathbb{R}$ such that $O \supset E$ and the strict inequality $\mu_L^*(E) \le \mu_L^*(O) < \mu_L^*(E) + \varepsilon$, that is, $\infty < \infty$.)

2. Let $E \in \mathfrak{P}(\mathbb{R})$. For $\varepsilon = \frac{1}{n}$ for $n \in \mathbb{N}$, there exists an open set $O_n \supset E$ such that $\mu_L^*(E) \le \mu_L^*(O_n) \le \mu_L^*(E) + \frac{1}{n}$ by (a) . Let $G = \bigcap_{n\in\mathbb{N}} O_n$. Then G is a G_δ- set and $G \supset E$. Since $G \subset O_n$ for every $n \in \mathbb{N}$, we have $\mu_L^*(E) \le \mu_L^*(G) \le \mu_L^*(O_n) \le \mu_L^*(E) + \frac{1}{n}$, by the monotonicity of μ_L^*. Since this holds for every $n \in \mathbb{N}$, we have $\mu_L^*(E) \le \mu_L^*(G) \le \mu_L^*(E)$ and therefore $\mu_L^*(E) = \mu_L^*(G)$.

3. Since $\mathfrak{B}_{\mathbb{R}} \subset \mathfrak{M}_L = \mathfrak{M}(\mu_L^*)$ by Theorem 3.9, μ_L^* is a Borel outer measure on $\mathbb{R}$. A G_δ-set is a member of $\mathfrak{B}_{\mathbb{R}}$. Thus μ_L^* is a Borel regular outer measure by (b). ∎

Theorem 3.22. *For $E \in \mathfrak{P}(\mathbb{R})$, the following conditions are all equivalent :*
(i) $E \in \mathfrak{M}_L$.
(ii) *For every $\varepsilon > 0$, there exists an open set $O \supset E$ with $\mu_L^*(O \setminus E) \le \varepsilon$.*
(iii) *There exists a G_δ-set $G \supset E$ with $\mu_L^*(G \setminus E) = 0$.*
(iv) *For every $\varepsilon > 0$, there exists a closed set $C \subset E$ with $\mu_L^*(E \setminus C) \le \varepsilon$.*
(v) *There exists an F_σ-set $F \subset E$ with $\mu_L^*(E \setminus F) = 0$.*

Proof. We show that (i) ⇒ (ii), (ii) ⇒ (iii), and (iii) ⇒ (i). This establishes (i)⇔(ii)⇔(iii). We then show that (i) ⇒ (iv) and (iv) ⇒ (i). Finally we show that (iv)⇒ (v) and (v) ⇒ (i). (Note that proving (iv) ⇒ (i) is actually superfluous.)

1. (i) ⇒ (ii). Assume that E satisfies (i). According to Lemma 3.21, for every $\varepsilon > 0$ there exists an open set $O \supset E$ with $\mu_L^*(E) \le \mu_L^*(O) \le \mu_L^*(E) + \varepsilon$. Since $E \in \mathfrak{M}_L$, with O as a testing set in the Carathéodory condition satisfied by E, we have $\mu_L^*(E) + \varepsilon \ge \mu_L^*(O) = \mu_L^*(O \cap E) + \mu_L^*(O \cap E^c) = \mu_L^*(E) + \mu_L^*(O \setminus E)$. If $\mu_L^*(E) < \infty$, then subtracting $\mu_L^*(E)$ from the last inequality we have $\mu_L^*(O \setminus E) \le \varepsilon$. If $\mu_L^*(E) = \infty$, let $E_n = E \cap (n-1, n]$ for $n \in \mathbb{Z}$. Then $(E_n : n \in \mathbb{Z})$ is a disjoint sequence in $\mathfrak{M}_L$ with $\bigcup_{n\in\mathbb{Z}} E_n = E$ and $\mu_L(E_n) \le \mu_L\big((n-1, n]\big) = 1$. Applying the result above to E_n for each $n \in \mathbb{Z}$, for every $\varepsilon > 0$ we have an open set $O_n \supset E_n$ such that $\mu_L(O_n \setminus E_n) \le 3^{-1}2^{-|n|}\varepsilon$.

If we let $O = \bigcup_{n\in\mathbb{Z}} O_n$, then O is an open set, $O \supset E$, and

$$\begin{aligned} O \setminus E &= \Big(\bigcup_{n\in\mathbb{Z}} O_n\Big) \setminus \Big(\bigcup_{n\in\mathbb{Z}} E_n\Big) = \Big(\bigcup_{n\in\mathbb{Z}} O_n\Big) \cap \Big(\bigcup_{n\in\mathbb{Z}} E_n\Big)^c \\ &= \bigcup_{n\in\mathbb{Z}} \Big\{O_n \cap \Big(\bigcup_{n\in\mathbb{Z}} E_n\Big)^c\Big\} = \bigcup_{n\in\mathbb{Z}} \Big\{O_n \setminus \Big(\bigcup_{n\in\mathbb{Z}} E_n\Big)\Big\} \\ &\subset \bigcup_{n\in\mathbb{Z}} (O_n \setminus E_n). \end{aligned}$$

Then we have

$$\begin{aligned} \mu_L^*(O \setminus E) &\le \mu_L^*\Big(\bigcup_{n\in\mathbb{Z}} (O_n \setminus E_n)\Big) \le \sum_{n\in\mathbb{Z}} \mu_L^*(O_n \setminus E) \\ &\le \sum_{n\in\mathbb{Z}} \frac{1}{3}\frac{\varepsilon}{2^{|n|}} = \frac{1}{3}\varepsilon + 2\sum_{n\in\mathbb{N}} \frac{1}{3}\frac{\varepsilon}{2^n} = \frac{1}{3}\varepsilon + \frac{2}{3}\varepsilon \\ &= \varepsilon. \end{aligned}$$

This shows that E satisfies (ii).

2. (ii) $\Rightarrow$ (iii). Assume that E satisfies (ii). Then for $\varepsilon = \frac{1}{n}$, $n \in \mathbb{N}$, there exists an open set $O_n \supset E$ with $\mu_L^*(O_n \setminus E) \le \frac{1}{n}$. Let $G = \bigcap_{n\in\mathbb{N}} O_n$, a G_δ-set containing E. Now $G \subset O_n$ implies $\mu_L^*(G \setminus E) \le \mu_L^*(O_n \setminus E) \le \frac{1}{n}$ for every $n \in \mathbb{N}$. Thus $\mu_L^*(G \setminus E) = 0$. This shows that E satisfies (iii).

3. (iii) $\Rightarrow$ (i). Assume that E satisfies (iii). Then there exists a G_δ-set $G \supset E$ with $\mu_L^*(G \setminus E) = 0$. Now $\mu_L^*(G \setminus E) = 0$ implies that $G \setminus E \in \mathfrak{M}(\mu_L^*) = \mathfrak{M}_L$ according to Lemma 2.6. Since $E \subset G$, we have $E = G \setminus (G \setminus E)$. Then the fact that G and $G \setminus E$ are in $\mathfrak{M}_L$ implies that E is in $\mathfrak{M}_L$.

4. (i) $\Rightarrow$ (iv). Assume that E satisfies (i). Then $E \in \mathfrak{M}_L$ and hence $E^c \in \mathfrak{M}_L$. As we showed in **1**, **2**, and **3**, (i), (ii), and (iii) are all equivalent. Thus applying (ii) to our $E^c \in \mathfrak{M}_L$, for every $\varepsilon > 0$ we have an open set $O \supset E^c$ with $\mu_L^*(O \setminus E^c) \le \varepsilon$. Now for any two sets A and B, we have $A \setminus B = A \cap B^c = B^c \cap (A^c)^c = B^c \setminus A^c$. Thus $O \setminus E^c = E \setminus O^c$ and then $\mu_L^*(E \setminus O^c) \le \varepsilon$. Now O^c is a closed set and $O^c \subset E$ since $O \supset E^c$. If we let $C = O^c$, then we have (iv) for E.

5. (iv) $\Rightarrow$ (i). Assume that E satisfies (iv). Then for every $\varepsilon > 0$, there exists a closed set $C \subset E$ with $\mu_L^*(E \setminus C) \le \varepsilon$. Now $E \setminus C = C^c \setminus E^c$ so that $\mu_L^*(C^c \setminus E^c) \le \varepsilon$. Also $C \subset E$ implies $C^c \supset E^c$. Let $O = C^c$. Then O is an open set, $O \supset E^c$ and $\mu_L^*(O \setminus E^c) \le \varepsilon$. This shows that E^c satisfies (ii). Since (ii) is equivalent to (i), we have $E^c \in \mathfrak{M}_L$. Then $E \in \mathfrak{M}_L$. Thus E satisfies (i).

6. (iv) $\Rightarrow$ (v). Assume that E satisfies (iv). Then for every $n \in \mathbb{N}$, there exists a closed set $C_n \subset E$ with $\mu_L^*(E \setminus C_n) \le \frac{1}{n}$. Let $C = \bigcup_{n\in\mathbb{N}} C_n$, an F_σ-set contained in E. Now $E \supset C \supset C_n$ so that $E \setminus C \subset E \setminus C_n$ and $\mu_L^*(E \setminus C) \le \mu_L^*(E \setminus C_n) \le \frac{1}{n}$ for every $n \in \mathbb{N}$. Thus $\mu_L^*(E \setminus C) = 0$. This shows that E satisfies (v).

7. (v) $\Rightarrow$ (i). Assume that E satisfies (v). Then there exists an F_σ-set $F \subset E$ with $\mu_L^*(E \setminus F) = 0$. Now $\mu_L^*(E \setminus F) = 0$ implies that $E \setminus F \in \mathfrak{M}(\mu_L^*) = \mathfrak{M}_L$ by Lemma 2.6.

Since $F \subset E$, we have $E = F \cup (E \setminus F)$. Since F is an F_σ-set, $F \in \mathfrak{B}_{\mathbb{R}} \subset \mathfrak{M}_L$. Thus $E \in \mathfrak{M}_L$. This shows that E satisfies (i). ∎

Remark 3.23. According to (b) of Lemma 3.21, for every $E \in \mathfrak{P}(\mathbb{R})$ there exists a G_δ-set $G \supset E$ with $\mu_L^*(G) = \mu_L^*(E)$. This does not imply $\mu_L^*(G \setminus E) = 0$. Since $G = E \cup (G \setminus E)$, we have $\mu_L^*(G) \le \mu_L^*(E) + \mu_L^*(G \setminus E)$ by the subadditivity of the outer measure μ_L^*. If $\mu_L^*(G) = \mu_L^*(E) < \infty$, then subtracting this real number from the last inequality we have $0 \le \mu_L^*(G \setminus E)$. On the other hand, as we showed in **3** in the Proof of Theorem 3.22, the equality $\mu_L^*(G \setminus E) = 0$ implies the $\mathfrak{M}_L$-measurability of E. Thus for any non $\mathfrak{M}_L$-measurable set E, we have $\mu_L^*(G \setminus E) > 0$. Theorem 2.13 is relevant here.

Next we derive from Theorem 3.22 some approximation theorems for Lebesgue measurable sets by open sets in terms of symmetric differences. The symmetric difference of two sets A and B is defined by $A \triangle B = (A \setminus B) \cup (B \setminus A)$.

Theorem 3.24. *Let $E \in \mathfrak{P}(\mathbb{R})$. Then $E \in \mathfrak{M}_L$ if and only if for every $\varepsilon > 0$ there exists an open set V with $\mu_L^*(E \triangle V) < \varepsilon$.*

Proof. 1. If $E \in \mathfrak{M}_L$, then by (ii) of Theorem 3.22 for every $\varepsilon > 0$ there exists an open set $V \supset E$ with $\mu_L^*(V \setminus E) < \varepsilon$. Since $V \supset E$, we have $V \setminus E = E \triangle V$. Thus $\mu_L^*(E \triangle V) < \varepsilon$.

2. Conversely suppose for every $\varepsilon > 0$, there exists an open set V with $\mu_L^*(E \triangle V) < \varepsilon$. To show that $E \in \mathfrak{M}_L$, we show that E satisfies (ii) of Theorem 3.22. Let $\varepsilon > 0$ be arbitrarily given. By our assumption there exists an open set V with $\mu_L^*(E \triangle V) < \frac{\varepsilon}{3}$. Consider the set $E \setminus V \subset E \triangle V$. By Lemma 3.21, there exists an open set $W \supset E \setminus V$ with $\mu_L^*(W) \le \mu_L^*(E \setminus V) + \frac{\varepsilon}{3} \le \mu_L^*(E \triangle V) + \frac{\varepsilon}{3} < \frac{2}{3}\varepsilon$. If we let $O = W \cup V$, then O is an open set containing E and

$$O \setminus E = (W \cup V) \setminus E \subset W \cup (V \setminus E) \subset W \cup (E \triangle V)$$

so that

$$\mu_L^*(O \setminus E) \le \mu_L^*(W) + \mu_L^*(E \triangle V) < \frac{2}{3}\varepsilon + \frac{1}{3}\varepsilon = \varepsilon.$$

Thus E satisfies (ii) of Theorem 3.22 and therefore $E \in \mathfrak{M}_L$. ∎

Theorem 3.25. *Let $E \in \mathfrak{M}_L$. If $\mu_L(E) < \infty$, then for every $\varepsilon > 0$, there exist finitely many disjoint finite open intervals $J_1, \ldots, J_m$ such that $\mu_L\left(E \triangle \bigcup_{n=1}^m J_n\right) < \varepsilon$.*

Proof. If $E \in \mathfrak{M}_L$, then by (ii) of Theorem 3.22 for an arbitrary $\varepsilon > 0$ there exists an open set $O \supset E$ such that

$$(1) \qquad \mu_L(O \setminus E) < \frac{\varepsilon}{3}.$$

If $\mu_L(E) < \infty$, then by the additivity of μ_L on $\mathfrak{M}_L$, we have

$$(2) \qquad \mu_L(O) = \mu_L(E) + \mu_L(O \setminus E) < \infty.$$

Now $\mu_L^*(O) = \mu_L(O) < \infty$. Thus by the definition of $\mu_L^*(O)$ as an infimum, there exists a sequence $(I_n : n \in \mathbb{N})$ in $\mathfrak{I}_o$ such that $\bigcup_{n\in\mathbb{N}} I_n \supset O$ and

$$(3) \qquad \sum_{n\in\mathbb{N}} \ell(I_n) < \mu_L^*(O) + \frac{\varepsilon}{3} < \infty.$$

The convergence of the series $\sum_{n\in\mathbb{N}} \ell(I_n)$ implies that there exists $N \in \mathbb{N}$ such that

$$(4) \qquad \sum_{n>N} \ell(I_n) < \frac{\varepsilon}{3}.$$

By the triangle inequality for symmetric differences, that is, $A\triangle B \subset (A\triangle C) \cup (C\triangle B)$ for any three sets A, B, and C, we have

$$E\triangle \bigcup_{n=1}^{N} I_n \subset (E\triangle O) \cup \Big(O\triangle \Big\{\bigcup_{n\in\mathbb{N}} I_n\Big\}\Big) \cup \Big(\Big\{\bigcup_{n\in\mathbb{N}} I_n\Big\}\triangle\Big\{\bigcup_{n=1}^{N} I_n\Big\}\Big).$$

By the monotonicity and finite subadditivity of μ_L we have

$$(5) \qquad \mu_L\Big(E\triangle \bigcup_{n=1}^{N} I_n\Big) \le \mu_L(E\triangle O) + \mu_L\Big(O\triangle \bigcup_{n\in\mathbb{N}} I_n\Big) + \mu_L\Big(\Big\{\bigcup_{n\in\mathbb{N}} I_n\Big\}\triangle\Big\{\bigcup_{n=1}^{N} I_n\Big\}\Big).$$

Since $E \subset O$, we have $E\triangle O = O \setminus E$ and then by (1) we have

$$(6) \qquad \mu_L(E\triangle O) = \mu_L(O\setminus E) < \frac{\varepsilon}{3}.$$

Since $O \subset \bigcup_{n\in\mathbb{N}} I_n$, we have $O\triangle \bigcup_{n\in\mathbb{N}} I_n = \big(\bigcup_{n\in\mathbb{N}} I_n\big) \setminus O$. Then we have

$$(7) \qquad \mu_L\Big(O\triangle \bigcup_{n\in\mathbb{N}} I_n\Big) = \mu_L\Big(\Big\{\bigcup_{n\in\mathbb{N}} I_n\Big\}\setminus O\Big) = \mu_L\Big(\bigcup_{n\in\mathbb{N}} I_n\Big) - \mu_L(O)$$
$$\le \sum_{n\in\mathbb{N}} \mu_L(I_n) - \mu_L(O) + \sum_{n\in\mathbb{N}} \ell(I_n) - \mu_L(O) < \frac{\varepsilon}{3}$$

where the second equality is by (3) of Lemma 1.25 which is applicable since $\mu_L(O) < \infty$ by (2) and the last equality is by (3). Finally since $\bigcup_{n=1}^{N} I_n \subset \bigcup_{n\in\mathbb{N}} I_n$ we have

$$\Big\{\bigcup_{n\in\mathbb{N}} I_n\Big\}\triangle\Big\{\bigcup_{n=1}^{N} I_n\Big\} = \Big\{\bigcup_{n\in\mathbb{N}} I_n\Big\}\setminus\Big\{\bigcup_{n=1}^{N} I_n\Big\} \subset \bigcup_{n>N} I_n,$$

and then

$$(8) \quad \mu_L\Big(\Big\{\bigcup_{n\in\mathbb{N}} I_n\Big\}\triangle\Big\{\bigcup_{n=1}^{N} I_n\Big\}\Big) \le \mu_L\Big(\bigcup_{n>N} I_n\Big) \le \sum_{n>N} \mu_L(I_n) = \sum_{n>N} \ell(I_n) < \frac{\varepsilon}{3}.$$

Using (6) - (8) in (5), we have $\mu_L\big(E\triangle\bigcup_{n=1}^N I_n\big) < \varepsilon$. Note that $\sum_{n\in\mathbb{N}} \ell(I_n) < \infty$ according to (3) and this implies that I_n is a finite interval for every $n \in \mathbb{N}$. Let us note that since $\{I_n : n = 1, \dots, N\}$ is a finite collection of finite open intervals, $\bigcup_{n=1}^N I_n$ is the union of finitely many disjoint finite open intervals $J_1, \dots, J_m$. ∎

Remark 3.26. Theorem 3.25 is not valid without the assumption that $\mu_L(E) < \infty$. For example, let $E = \bigcup_{n\in\mathbb{N}} E_n$ where $E_n = \big(n - \frac{1}{2}, n\big)$ for $n \in \mathbb{N}$. Then $E \in \mathfrak{M}_L$ and $\mu_L(E) = \infty$. Let $\{I_1, \dots, I_N\}$ be an arbitrary finite disjoint collection of finite open intervals in $\mathbb{R}$. Then there exists $N_0 \in \mathbb{N}$ such that $\bigcup_{n=1}^N I_n \subset (-N_0, N_0)$. This implies

$$E\triangle\bigcup_{n=1}^N I_n \supset E\setminus\bigcup_{n=1}^N I_n \supset E\setminus(-N_0, N_0) = \bigcup_{n\geq N_0+1} E_n.$$

Thus we have

$$\mu_L\Big(E\triangle\bigcup_{n=1}^N I_n\Big) \geq \mu_L\Big(\bigcup_{n\geq N_0+1} E_n\Big) = \sum_{n\geq N_0+1} \mu_L(E_n) = \sum_{n\geq N_0+1} \frac{1}{2} = \infty. \ \blacksquare$$

Since every open interval in $\mathbb{R}$ has a positive Lebesgue measure, every Lebesgue measurable set containing an open interval has a positive Lebesgue measure. The converse is false, that is, a Lebesgue measurable set with a positive Lebesgue measure need not contain an open interval. For instance the set of all irrational numbers contained in $(0, 1)$ has Lebesgue measure equal to 1 but this set contains no open interval. The next theorem shows that if $E \in \mathfrak{M}_L$ and $\mu_L(E) > 0$, then there exists a finite open interval I such that $\mu_L(E \cap I)$ is as close to $\mu_L(I)$ as we wish.

Theorem 3.27. *Let $E \in \mathfrak{M}_L$ with $\mu_L(E) > 0$. Then for every $\alpha \in (0, 1)$ there exists a finite open interval I such that $\alpha\,\mu_L(I) \leq \mu_L(E \cap I) \leq \mu_L(I)$.*

Proof. Consider first the case where $\mu_L(E) \in (0, \infty)$. Let $\alpha \in (0, 1)$. Then $\frac{1}{\alpha} = 1 + \eta$ with $\eta > 0$. Since $\mu_L(E) \in (0, \infty)$, we have $\eta\,\mu_L(E) \in (0, \infty)$. By Lemma 3.21, there exists an open set $O \supset E$ such that

$$\text{(1)} \qquad \mu_L(O) \leq \mu_L(E) + \eta\,\mu_L(E) = (1 + \eta)\mu_L(E) = \frac{1}{\alpha}\mu_L(E) < \infty.$$

Now since O is an open set in $\mathbb{R}$, it is the union of a disjoint sequence $(I_n : n \in \mathbb{N})$ of open intervals in $\mathbb{R}$. Then $\mu_L(O) = \sum_{n\in\mathbb{N}} \mu_L(I_n)$. Since $E \subset O$, we have

$$\mu_L(E) = \mu_L(E \cap O) = \mu_L\Big(E \cap \bigcup_{n\in\mathbb{N}} I_n\Big) = \sum_{n\in\mathbb{N}} \mu_L(E \cap I_n).$$

Using these expressions for $\mu_L(O)$ and $\mu_L(E)$ in (1), we have

$$\text{(2)} \qquad \sum_{n\in\mathbb{N}} \mu_L(I_n) \leq \frac{1}{\alpha}\sum_{n\in\mathbb{N}} \mu_L(E \cap I_n).$$

Therefore there exists at least one $n_0 \in \mathbb{N}$ such that $\mu_L(I_{n_0}) \leq \frac{1}{\alpha}\mu_L(E \cap I_{n_0})$. Since $\mu_L(O) < \infty$, I_n is a finite open interval for every $n \in \mathbb{N}$. Let $I = I_{n_0}$. Then I is a finite open interval with $\alpha\mu_L(I) \leq \mu_L(E \cap I) \leq \mu_L(I)$.

Now consider the case $\mu_L(E) = \infty$. By the σ-finiteness of the Lebesgue measure space $(\mathbb{R}, \mathfrak{M}_L, \mu_L)$, there exists a $\mathfrak{M}_L$-measurable subset E_0 of E with $\mu_L(E_0) \in (0, \infty)$. Then by our result above for $E \in \mathfrak{M}_L$ with $\mu_L(E) \in (0, \infty)$, there exists a finite open interval I such that $\alpha\,\mu_L(I) \leq \mu_L(E_0 \cap I) \leq \mu_L(E \cap I) \leq \mu_L(I)$. ∎

Definition 3.28. *For every $E \in \mathfrak{P}(\mathbb{R})$, let the difference set of E, $\Delta(E)$, be defined by*

$$\Delta(E) = \{z \in \mathbb{R} : z = x - y \text{ for some } x, y \in E\}.$$

Note that for every $E \in \mathfrak{P}(\mathbb{R})$, we have $0 \in \Delta(E)$.

Theorem 3.29. *If $E \in \mathfrak{M}_L$ and $\mu_L(E) > 0$, then $\Delta(E)$ contains an open interval.*

Proof. If $\mu_L(E) > 0$, then by Theorem 3.27 for every $\alpha \in (0, 1)$ there exists a finite open interval I_α such that

$$(1) \qquad \alpha\mu_L(I_\alpha) \leq \mu_L(E \cap I_\alpha).$$

Let $J_\alpha = \left(-\frac{\alpha}{2}\mu_L(I_\alpha), \frac{\alpha}{2}\mu_L(I_\alpha)\right)$. Let us show that for sufficiently large $\alpha \in (0, 1)$ we have $J_\alpha \subset \Delta(E)$. Let $z \in J_\alpha$ be arbitrarily chosen. We show that $z \in \Delta(E)$ by showing that $z = x - y$ for some $x, y \in E$. Now $(E \cap I_\alpha) \cup [(E \cap I_\alpha) + z] \subset I_\alpha \cup (I_\alpha + z)$. Since $z \in J_\alpha$, the set $I_\alpha \cup (I_\alpha + z)$ is an open interval with $\mu_L(I_\alpha \cup (I_\alpha + z)) \leq \{1 + \frac{\alpha}{2}\}\mu_L(I_\alpha)$. Thus we have

$$(2) \qquad \mu_L\left((E \cap I_\alpha) \cup [(E \cap I_\alpha) + z]\right) \leq \mu_L\left(I_\alpha \cup (I_\alpha + z)\right) \leq \left\{1 + \frac{\alpha}{2}\right\}\mu_L(I_\alpha).$$

By the translation invariance of $(\mathbb{R}, \mathfrak{M}_L, \mu_L)$, we have $\mu_L((E \cap I_\alpha) + z) = \mu_L(E \cap I_\alpha)$. Now if we assume that $E \cap I_\alpha$ and $(E \cap I_\alpha) + z$ are disjoint, then

$$(3) \qquad \begin{aligned} \mu_L\left((E \cap I_\alpha) \cup [(E \cap I_\alpha) + z]\right) &= \mu_L(E \cap I_\alpha) + \mu_L((E \cap I_\alpha) + z) \\ &= 2\,\mu_L(E \cap I_\alpha) \geq 2\alpha\,\mu_L(I_\alpha), \end{aligned}$$

by (1). Thus, if we assume that $E \cap I_\alpha$ and $(E \cap I_\alpha) + z$ are disjoint then by (2) and (3)

$$(4) \qquad 2\alpha \leq 1 + \frac{\alpha}{2}.$$

Now as $\alpha \uparrow 1$, we have $2\alpha \uparrow 2$ and $1 + \frac{\alpha}{2} \uparrow 1\frac{1}{2}$. Thus for sufficiently large $\alpha \in (0, 1)$, (4) is impossible. Indeed (4) is impossible for $\alpha = \frac{3}{4}$. Therefore with the choice $\alpha = \frac{3}{4}$, $E \cap I_\alpha$ and $(E \cap I_\alpha) + z$ cannot be disjoint. Let $x \in (E \cap I_\alpha) \cap [(E \cap I_\alpha) + z]$. Then we have $x \in E \cap I_\alpha$ as well as $x \in (E \cap I_\alpha) + z$. Thus $x = y + z$ for some $y \in E \cap I_\alpha$. Then $z = x - y$ where $x, y \in E$ so that $z \in \Delta(E)$. This shows that with $\alpha = \frac{3}{4}$, we have $J_\alpha \subset \Delta(E)$. ∎

[V] Lebesgue Inner Measure on $\mathbb{R}$

As we show next, the regularity of the Lebesgue outer measure μ_L^* on $\mathbb{R}$ implies that for every $E \in \mathfrak{P}(\mathbb{R})$, $\mu_L^*(E)$ is the infimum of the Lebesgue measures of all the open sets that contain E. We then define the Lebesgue inner measure $\mu_{*,L}(E)$ of E as the supremum of the Lebesgue measures of all the closed sets that are contained in E. We show that if $E \in \mathfrak{P}(\mathbb{R})$ and $\mu_L^*(E) < \infty$, then $E \in \mathfrak{M}_L$ if and only if $\mu_{*,L}(E) = \mu_L^*(E)$.

Notations. We write $\mathfrak{O}_{\mathbb{R}}$, $\mathfrak{C}_{\mathbb{R}}$, and $\mathfrak{K}_{\mathbb{R}}$ for the collections of the open sets, the closed sets, and the compact sets in $\mathbb{R}$ respectively.

Proposition 3.30. *For $E \in \mathfrak{P}(\mathbb{R})$, we have* $\mu_L^*(E) = \inf\{\mu_L(O) : O \supset E, O \in \mathfrak{O}_{\mathbb{R}}\}$.

Proof. Let $E \in \mathfrak{P}(\mathbb{R})$. If $O \in \mathfrak{O}_{\mathbb{R}}$ and $O \supset E$, then by the monotonicity of μ_L^* and by the fact that $O \in \mathfrak{M}_L$ we have $\mu_L^*(E) \le \mu_L^*(O) = \mu_L(O)$. Thus we have $\mu_L^*(E) \le \inf\{\mu_L(O) : O \supset E, O \in \mathfrak{O}_{\mathbb{R}}\}$. To prove the reverse inequality, let $\varepsilon > 0$ be arbitrarily given. According to Lemma 3.21, there exists $O \in \mathfrak{O}_{\mathbb{R}}$ such that $O \supset E$ and $\mu_L(O) \le \mu_L^*(E) + \varepsilon$. Thus $\inf\{\mu_L(O) : O \supset E, O \in \mathfrak{O}_{\mathbb{R}}\} \le \mu_L^*(E) + \varepsilon$. From the arbitrariness of $\varepsilon > 0$, we have $\inf\{\mu_L(O) : O \supset E, O \in \mathfrak{O}_{\mathbb{R}}\} \le \mu_L^*(E)$. ∎

Definition 3.31. *The Lebesgue inner measure of $E \in \mathfrak{P}(\mathbb{R})$ is defined by*

$$\mu_{*,L}(E) = \sup\{\mu_L(C) : C \subset E, C \in \mathfrak{C}_{\mathbb{R}}\}.$$

(Note that $\emptyset \subset E$ and $\emptyset \in \mathfrak{C}_{\mathbb{R}}$ so that the collection of all closed sets contained in E is nonempty.)

Observation 3.32. For every $C \in \mathfrak{C}_{\mathbb{R}}$, we have $\mu_{*,L}(C) = \mu_L(C)$.

Proof. This is an immediate consequence of Definition 3.31. ∎

We say that a topological space X is a σ-compact space if there exists a sequence $(K_n : n \in \mathbb{N})$ of compact sets in X such that $\bigcup_{n\in\mathbb{N}} K_n = X$. Since a finite union of compact sets is a compact set, if we let $K_n' = \bigcup_{i=1}^n K_i$ for $n \in \mathbb{N}$ then $(K_n' : n \in \mathbb{N})$ is an increasing sequence of compact sets in X with $\lim_{n\to\infty} K_n' = \bigcup_{n\in\mathbb{N}} K_n' = \bigcup_{n\in\mathbb{N}} K_n = X$. The intersection of a closed set and a compact set in a Hausdorff space is a compact set. Thus if X is a σ-compact Hausdorff space, then for every closed set C in X there exists an increasing sequence $(C_n : n \in \mathbb{N})$ of compact sets in X such that $\lim_{n\to\infty} C_n = C$. Indeed if we let $C_n = C \cap K_n'$, then the sequence $(C_n : n \in \mathbb{N})$ is such a sequence.

Consider the Hausdorff space $\mathbb{R}$. If we let $I_n = [-n, n]$ then $(I_n : n \in \mathbb{N})$ is an increasing sequence of compact sets in $\mathbb{R}$ and $\bigcup_{n\in\mathbb{N}} I_n = \mathbb{R}$. Thus $\mathbb{R}$ is a σ-compact Hausdorff space. Then for every closed set $C \in \mathbb{R}$ there exists an increasing sequence $(C_n : n \in \mathbb{N})$ of compact sets in $\mathbb{R}$ such that $C = \lim_{n\to\infty} C_n$. From this fact we derive the following expression for the Lebesgue inner measure on $\mathbb{R}$.

Proposition 3.33. *For every* $E \in \mathfrak{P}(\mathbb{R})$, $\mu_{*,L}(E) = \sup\{\mu_L(K) : K \subset E, K \in \mathfrak{K}_{\mathbb{R}}\}$.

Proof. Let $\alpha = \sup\{\mu_L(C) : C \subset E, C \in \mathfrak{C}_{\mathbb{R}}\}$ and $\beta = \sup\{\mu_L(K) : K \subset E, K \in \mathfrak{K}_{\mathbb{R}}\}$ for brevity. Since $\mathfrak{C}_{\mathbb{R}} \supset \mathfrak{K}_{\mathbb{R}}$ we have $\alpha \geq \beta$. It remains to prove $\alpha \leq \beta$.

Consider first the case $\alpha < \infty$. In this case, for every $\varepsilon > 0$ there exists $C_0 \in \mathfrak{C}_{\mathbb{R}}$ such that $C_0 \subset E$ and

$$\mu_L(C_0) > \alpha - \varepsilon.$$

Let $(K_n : n \in \mathbb{N})$ be an increasing sequence in $\mathfrak{K}_{\mathbb{R}}$ such that $\lim_{n\to\infty} K_n = \bigcup_{n\in\mathbb{N}} K_n = C_0$. We have $\lim_{n\to\infty} \mu_L(K_n) = \mu_L(C_0)$. Then

$$\beta \geq \sup_{n\in\mathbb{N}} \mu_L(K_n) = \lim_{n\to\infty} \mu_L(K_n) = \mu_L(C_0) > \alpha - \varepsilon.$$

Then by the arbitrariness of $\varepsilon > 0$, we have $\beta \geq \alpha$.

Next consider the case $\alpha = \infty$. In this case, for every $M > 0$ there exists $C_M \in \mathfrak{C}_{\mathbb{R}}$ such that $C_M \subset E$ and $\mu_L(C_M) > M$. Let $(K_n : n \in \mathbb{N})$ be an increasing sequence in $\mathfrak{K}_{\mathbb{R}}$ such that $\lim_{n\to\infty} K_n = \bigcup_{n\in\mathbb{N}} K_n = C_M$. Then $\lim_{n\to\infty} \mu_L(K_n) = \mu_L(C_M)$ so that

$$M \leq \mu_L(C_M) = \lim_{n\to\infty} \mu_L(K_n) = \sup_{n\in\mathbb{N}} \mu_L(K_n) \leq \beta.$$

By the arbitrariness of $M > 0$, we have $\beta = \infty = \alpha$. ∎

Theorem 3.34. *The Lebesgue inner measure* $\mu_{*,L}$ *on* $\mathbb{R}$ *has the following properties:*

1° *nonnegative extended real-valued :* $\mu_{*,L}(E) \in [0, \infty]$ *for every* $E \in \mathfrak{P}(\mathbb{R})$,

2° $\mu_{*,L}(\emptyset) = 0$,

3° *monotonicity :* $E_1, E_2 \in \mathfrak{P}(\mathbb{R})$, $E_1 \subset E_2 \Rightarrow \mu_{*,L}(E_1) \leq \mu_{*,L}(E_2)$,

4° *countable superadditivity :*

$$(E_n : n \in \mathbb{N}) \subset \mathfrak{P}(\mathbb{R}), \text{ disjoint } \Rightarrow \mu_{*,L}\big(\bigcup_{n\in\mathbb{N}} E_n\big) \geq \sum_{n\in\mathbb{N}} \mu_{*,L}(E_n),$$

5° *translation invariance:* $E \in \mathfrak{P}(\mathbb{R})$ *and* $x \in \mathbb{R} \Rightarrow \mu_{*,L}(E + x) = \mu_{*,L}(E)$,

6° *positive homogeneity:* $E \in \mathfrak{P}(\mathbb{R})$ *and* $\alpha \in \mathbb{R} \Rightarrow \mu_{*,L}(\alpha E) = |\alpha|\mu_{*,L}(E)$.

Proof. Properties 1° - 3° are immediate from Definition 3.31. Property 5° follows from the translation invariance of $(\mathbb{R}, \mathfrak{M}_L, \mu_L)$ applied to the closed sets C in Definition 3.31. Similarly 6° follows from the positive homogeneity of $(\mathbb{R}, \mathfrak{M}_L, \mu_L)$.

To prove 4°, let $(E_n : n \in \mathbb{N})$ be a disjoint sequence in $\mathfrak{P}(\mathbb{R})$. If $\mu_{*,L}(E_{n_0}) = \infty$ for some $n_0 \in \mathbb{N}$, then we have $\mu_{*,L}\big(\bigcup_{n\in\mathbb{N}} E_n\big) \geq \mu_{*,L}(E_{n_0}) = \infty$ by 3° so that $\mu_{*,L}\big(\bigcup_{n\in\mathbb{N}} E_n\big) = \infty = \sum_{n\in\mathbb{N}} \mu_{*,L}(E_n)$. Thus consider the case where $\mu_{*,L}(E_n) < \infty$ for every $n \in \mathbb{N}$. Let $N \in \mathbb{N}$ be arbitrarily fixed. Let $\varepsilon > 0$. Now for every $n \in \mathbb{N}$, since $\mu_{*,L}(E_n) < \infty$ there exists $C_n \in \mathfrak{C}_{\mathbb{R}}$ such that $C_n \subset E_n$ and $\mu_L(C_n) > \mu_{*,L}(E_n) - \frac{\varepsilon}{2^n}$ by Definition 3.31. Since $(E_n : n \in \mathbb{N})$ is a disjoint sequence, $(C_n : n \in \mathbb{N})$ is a disjoint sequence. For the closed set $\bigcup_{n=1}^N C_n$ contained in $\bigcup_{n=1}^N E_n$, we have by 3° , Observation

3.32, and the finite additivity of μ_L

$$\mu_{*,L}\Big(\bigcup_{n=1}^{N} E_n\Big) \geq \mu_{*,L}\Big(\bigcup_{n=1}^{N} C_n\Big) = \mu_L\Big(\bigcup_{n=1}^{N} C_n\Big)$$
$$= \sum_{n=1}^{N} \mu_L(C_n) > \sum_{n=1}^{N} \mu_{*,L}(E_n) - \varepsilon.$$

By the arbitrariness of $\varepsilon > 0$, we have $\mu_{*,L}\big(\bigcup_{n=1}^{N} E_n\big) \geq \sum_{n=1}^{N} \mu_{*,L}(E_n)$. Then by 3°, we have $\mu_{*,L}\big(\bigcup_{n\in\mathbb{N}} E_n\big) \geq \sum_{n=1}^{N} \mu_{*,L}(E_n)$. Since this holds for every $N \in \mathbb{N}$, we have $\mu_{*,L}\big(\bigcup_{n\in\mathbb{N}} E_n\big) \geq \sum_{n\in\mathbb{N}} \mu_{*,L}(E_n)$. This proves 4°. ∎

Remark 3.35. (a) Needless to say condition 4° in Theorem 3.34 does not hold without the assumption of the disjointness of $(E_n : n \in \mathbb{N})$. For instance if we let $E_1 = E_2 = [0, 1]$ and $E_3 = \emptyset$ for $n \geq 3$, then $\mu_{*,L}\big(\bigcup_{n\in\mathbb{N}} E_n\big) = \mu_{*,L}(E_1) = 1$, but $\sum_{n\in\mathbb{N}} \mu_{*,L}(E_n) = 2$.
(b) Let us observe also that the superadditivity of $\mu_{*,L}$ on $\mathfrak{P}(\mathbb{R})$ implies its finite superadditivity on $\mathfrak{P}(\mathbb{R})$, that is, if $(E_n : n = 1, \ldots, N)$ is disjoint finite sequence in $\mathfrak{P}(\mathbb{R})$ then $\mu_{*,L}\big(\bigcup_{n=1}^{N} E_n\big) \geq \sum_{n=1}^{N} \mu_{*,L}(E_n)$.

Observation 3.36. For every $E \in \mathfrak{P}(\mathbb{R})$, we have $\mu_{*,L}(E) \leq \mu_L^*(E)$.

Proof. Let $E \in \mathfrak{P}(\mathbb{R})$. Let C_0 be an arbitrary closed set in $\mathbb{R}$ such that $C_0 \subset E$ and let O_0 be an arbitrary open set in $\mathbb{R}$ such that $O_0 \supset E$. Then $C_0 \subset O_0$ so that $\mu_L(C_0) \leq \mu_L(O_0)$. Since this holds for an arbitrary $O_0 \in \mathfrak{O}_{\mathbb{R}}$ such that $O_0 \supset E$, we have

$$\mu_L(C_0) \leq \inf\big\{\mu_L(O) : O \supset E,\, O \in \mathfrak{O}_{\mathbb{R}}\big\} = \mu_L^*(E)$$

by Proposition 3.30. Since this holds for an arbitrary $C_0 \in \mathfrak{C}_{\mathbb{R}}$ such that $C_0 \subset E$, we have $\sup\{\mu_L(C); C \subset E, C \in \mathfrak{C}_{\mathbb{R}}\} \leq \mu_L^*(E)$, that is, $\mu_{*,L}(E) \leq \mu_L^*(E)$. ∎

The next theorem is a criterion for the Lebesgue measurability of a subset of $\mathbb{R}$ in terms of the Lebesgue inner measure.

Theorem 3.37. *Let $E \in \mathfrak{P}(\mathbb{R})$.*
(a) *If $E \in \mathfrak{M}_L$, then $\mu_{*,L}(E) = \mu_L^*(E)$.*
(b) *If $\mu_{*,L}(E) = \mu_L^*(E)$, then $E \in \mathfrak{M}_L$ provided that $\mu_L^*(E) < \infty$.*

Proof. 1. Suppose $E \in \mathfrak{M}_L$. Let $I_n = (n-1, n]$ and $E_n = E \cap I_n$ for $n \in \mathbb{Z}$. Consider the disjoint sequence $(E_n : n \in \mathbb{Z})$ in $\mathfrak{M}_L$ with $\mu_L(E_n) < \infty$ for every $n \in \mathbb{Z}$ and $\bigcup_{n\in\mathbb{Z}} E_n = E$. By (iv) of Theorem 3.22, for an arbitrary $\varepsilon > 0$ there exists $C_n \in \mathfrak{C}_{\mathbb{R}}$ such that $C_n \subset E_n$ and $\mu_L(E_n \setminus C_n) < 3^{-1}2^{-|n|}\varepsilon$ for every $n \in \mathbb{Z}$. Since $\mu_L(C_n) \leq \mu_L(E_n) < \infty$, we have

$\mu_L(E_n) - \mu_L(C_n) = \mu_L(E_n \setminus C_n)$ so that

$$\begin{aligned}\mu_L(E_n) &< \mu_L(C_n) + \frac{1}{3}\frac{\varepsilon}{2^{|n|}} \\ &\le \sup\{\mu_L(C) : C \subset E_n, C \in \mathfrak{C}_{\mathbb{R}}\} + \frac{1}{3}\frac{\varepsilon}{2^{|n|}} \\ &= \mu_{*,L}(E_n) + \frac{1}{3}\frac{\varepsilon}{2^{|n|}}.\end{aligned}$$

Then we have

$$\begin{aligned}\mu_L^*(E) &= \mu_L(E) = \mu_L\Big(\bigcup_{n\in\mathbb{Z}} E_n\Big) = \sum_{n\in\mathbb{Z}} \mu_L(E_n) \\ &< \sum_{n\in\mathbb{Z}} \Big\{\mu_{*,L}(E_n) + \frac{1}{3}\frac{\varepsilon}{2^{|n|}}\Big\} \le \mu_{*,L}\Big(\bigcup_{n\in\mathbb{Z}} E_n\Big) + \varepsilon \\ &= \mu_{*,L}(E) + \varepsilon\end{aligned}$$

where the second inequality is by 4° of Theorem 3.34 (countable superadditivity of μ_L^*). Thus by the arbitrariness of $\varepsilon > 0$, we have $\mu_L^*(E) \le \mu_{*,L}(E)$. On the other hand, $\mu_L^*(E) \ge \mu_{*,L}(E)$ by Observation 3.36. Therefore $\mu_{*,L}(E) = \mu_L^*(E)$. This proves (a).

2. Conversely suppose $\mu_{*,L}(E) = \mu_L^*(E)$ and $\mu_L^*(E) < \infty$. To show that $E \in \mathfrak{M}_L$, it suffices to show that for every $\varepsilon > 0$ there exists $O \in \mathfrak{O}_{\mathbb{R}}$ such that $O \supset E$ and $\mu_L^*(O \setminus E) < \varepsilon$ according to (ii) of Theorem 3.22. Now according to Proposition 3.30 we have $\mu_L^*(E) = \inf\{\mu_L(O) : O \supset E, O \in \mathfrak{O}_{\mathbb{R}}\}$. Thus there exists $O \in \mathfrak{O}_{\mathbb{R}}$ such that $O \supset E$ and $\mu_L(O) < \mu_L^*(E) + \frac{\varepsilon}{2}$. Since $\mu_{*,L}(E) = \sup\{\mu_L(C) : C \subset E, C \in \mathfrak{C}_{\mathbb{R}}\}$ and since $\mu_{*,L}(E) \le \mu_L^*(E) < \infty$ there exists $C \in \mathfrak{C}_{\mathbb{R}}$ such that $C \subset E$ and $\mu_{*,L}(E) - \frac{\varepsilon}{2} < \mu_L(C)$. Since $C \subset E$ and thus $O \setminus E \subset O \setminus C$, we have

$$\begin{aligned}\mu_L^*(O \setminus E) &\le \mu_L^*(O \setminus C) \\ &= \mu_L(O \setminus C) = \mu_L(O) - \mu_L(C) \\ &< \{\mu_L^*(E) + \tfrac{\varepsilon}{2}\} - \{\mu_{*,L}(E) - \tfrac{\varepsilon}{2}\} = \varepsilon,\end{aligned}$$

where the second equality is valid by the fact that $\mu_L(C) = \mu_L^*(C) \le \mu_L^*(E) < \infty$. ∎

Remark 3.38. Without the assumption $\mu_L^*(E) < \infty$, (b) of Theorem 3.37 is false, that is, if $E \in \mathfrak{P}(\mathbb{R})$ and if $\mu_L^*(E) = \infty$, then $\mu_{*,L}(E) = \mu_L^*(E)$ does not imply $E \in \mathfrak{M}_L$. This is shown by the existence of a set $E \in \mathfrak{P}(\mathbb{R})$, $E \notin \mathfrak{M}_L$, with $\mu_{*,L}(E) = \mu_L^*(E) = \infty$.

Example. By Theorem 3.20, there exists $E_0 \subset [0, 1)$ such that $E_0 \notin \mathfrak{M}_L$. Consider the set $E = E_0 \cup [1, \infty)$. Since $\mu_{*,L}\big([1, \infty)\big) = \infty$, we have $\mu_{*,L}(E) = \infty$ by the monotonicity of $\mu_{*,L}$. Thus by Observation 3.36, we have $\mu_{*,L}(E) = \mu_L^*(E) = \infty$. To show that $E \notin \mathfrak{M}_L$, note first that since E_0 and $[1, \infty)$ are disjoint we have $E_0 = E \setminus [1, \infty)$. Since $[1, \infty) \in \mathfrak{M}_L$, if $E \in \mathfrak{M}_L$, then we would have $E_0 \in \mathfrak{M}_L$. This is a contradiction. This shows that $E \notin \mathfrak{M}_L$.

The Lebesgue inner measure on $\mathbb{R}$ has the following regularity properties. Compare these properties with the regularity properties of the Lebesgue outer measure on $\mathbb{R}$ in Lemma 3.21.

Lemma 3.39. *Let $E \in \mathfrak{P}(\mathbb{R})$ with $\mu_L^*(E) < \infty$.*
(a) *For every $\varepsilon > 0$ there exists $C \in \mathfrak{C}_{\mathbb{R}}$ such that $C \subset E$ and*

$$\mu_{*,L}(E) - \varepsilon < \mu_L(C) \leq \mu_{*,L}(E).$$

(b) *There exists an F_σ-set F in $\mathbb{R}$ such that $F \subset E$ and $\mu_L(F) = \mu_{*,L}(E)$.*

Proof. 1. We have $\mu_{*,L}(E) \leq \mu_L^*(E) < \infty$ by Observation 3.36. It follows then from the definition of $\mu_{*,L}(E)$ that for every $\varepsilon > 0$ there exists $C \in \mathfrak{C}_{\mathbb{R}}$ such that $C \subset E$ and $\mu_{*,L}(E) - \varepsilon < \mu_L(C)$. By Observation 3.32 and by the monotonicity of $\mu_{*,L}$, we have $\mu_L(C) = \mu_{*,L}(C) \leq \mu_{*,L}(E)$. This proves (a).

2. For every $n \in \mathbb{N}$ there exists $C_n \in \mathfrak{C}_{\mathbb{R}}$ such that $C_n \subset E$ and $\mu_{*,L}(E) - \frac{1}{n} < \mu_L(C_n) \leq \mu_{*,L}(E)$ by (a). If we let $F = \bigcup_{n \in \mathbb{N}} C_n$, then F is an F_σ-set and $C_n \subset F \subset E$ for every $n \in \mathbb{N}$. Since $F \in \mathfrak{M}_L$ we have $\mu_{*,L}(F) = \mu_L(F)$ by (a) of Theorem 3.37. Thus $\mu_{*,L}(E) - \frac{1}{n} < \mu_L(C_n) \leq \mu_L(F) = \mu_{*,L}(F) \leq \mu_{*,L}(E)$. Since this holds for every $n \in \mathbb{N}$, we have $\mu_L(F) = \mu_{*,L}(E)$. ■

As an outer measure the Lebesgue outer measure μ_L^* is subadditive on $\mathfrak{P}(\mathbb{R})$. As we showed in Theorem 3.34, the Lebesgue inner measure $\mu_{*,L}$ is superadditive on $\mathfrak{P}(\mathbb{R})$. We show next some mixed inequalities.

Proposition 3.40. *Let $E_1, E_2 \in \mathfrak{P}(\mathbb{R})$ and $\mu_L^*(E_1 \cup E_2) < \infty$.*

(a) *We have $\mu_{*,L}(E_1 \cup E_2) \leq \mu_{*,L}(E_1) + \mu_L^*(E_2)$.*

(b) *If $E_1 \cap E_2 = \emptyset$, then $\mu_{*,L}(E_1) + \mu_L^*(E_2) \leq \mu_L^*(E_1 \cup E_2)$.*

Proof. 1. Since $\mu_L^*(E_1 \cup E_2) < \infty$, there exists an F_σ-set E such that $F \subset E_1 \cup E_2$ and $\mu_L(F) = \mu_{*,L}(E_1 \cup E_2)$ by Lemma 3.39. By Lemma 3.21, there exists a G_δ-set $G \supset E_2$ such that $\mu_L(G) = \mu_L^*(E_2)$. Now $F \subset (F \setminus G) \cup G$ and $F \setminus G \subset (E_1 \cup E_2) \setminus G \subset E_1$. Thus we have

$$\begin{aligned}\mu_{*,L}(E_1 \cup E_2) &= \mu_L(F) \leq \mu_L\big((F \setminus G) \cup G\big) \\ &\leq \mu_L(F \setminus G) + \mu_L(G) = \mu_L(F \setminus G) + \mu_L^*(E_2) \\ &= \mu_{*,L}(F \setminus G) + \mu_L^*(E_2) \leq \mu_{*,L}(E_1) + \mu_L^*(E_2),\end{aligned}$$

where the third equality is by (a) of Theorem 3.37 and the last inequality is by the monotonicity of $\mu_{*,L}$. This proves (a).

2. Assume that $E_1 \cap E_2 = \emptyset$. Since $\mu_L^*(E_1) \leq \mu_L^*(E_1 \cup E_2) < \infty$, there exists an F_σ-set $F \subset E_1$ such that $\mu_L(F) = \mu_{*,L}(E_1)$ by Lemma 3.39. By Lemma 3.21, there exists a G_δ-set $G \supset E_1 \cup E_2$ such that $\mu_L(G) = \mu_L^*(E_1 \cup E_2)$. Since $G \supset (E_1 \cup E_2) \supset F$, we have $G = F \cup (G \setminus F)$. Since $E_1 \cap E_2 = \emptyset$ and $F \subset E_1$, we have $G \setminus F \supset E_2$. Thus

$$\mu_L^*(E_1 \cup E_2) = \mu_L(G) = \mu_L(F) + \mu_L(G \setminus F)$$
$$= \mu_{*,L}(E_1) + \mu_L^*(G \setminus F) \geq \mu_{*,L}(E_1) + \mu_L^*(E_2). \blacksquare$$

The next theorem says that if E is a Lebesgue measurable set in $\mathbb{R}$ with finite Lebesgue measure and if we decompose the set into two parts in an arbitrary way, then the sum of the Lebesgue inner measure of one part and the Lebesgue outer measure of the other part is always equal to the Lebesgue measure of the set.

Theorem 3.41. *Let $E \in \mathfrak{P}(\mathbb{R})$. If $E \in \mathfrak{M}_L$ and $\mu_L(E) < \infty$, then for every $A \in \mathfrak{P}(\mathbb{R})$, we have $\mu_{*,L}(A \cap E) + \mu_L^*(A^c \cap E) = \mu_L(E)$.*

Proof. For an arbitrary $A \in \mathfrak{P}(\mathbb{R})$, $A \cap E$ and $A^c \cap E$ are two disjoint sets whose union is E so that $\mu_L^*\big((A \cap E) \cup (A^c \cap E)\big) = \mu_L^*(E) < \infty$. Thus by (a) and (b) of Proposition 3.40, we have $\mu_{*,L}(E) \leq \mu_{*,L}(A \cap E) + \mu_L^*(A^c \cap E) \leq \mu_L^*(E)$. But $E \in \mathfrak{M}_L$ and this implies $\mu_{*,L}(E) = \mu_L^*(E) = \mu_L(E)$ according to (a) of Theorem 3.37. Thus $\mu_{*,L}(A \cap E) + \mu_L^*(A^c \cap E) = \mu_L(E)$. $\blacksquare$

Lemma 3.42. *Let $\xi \in (0, 1)$ be an irrational number and let*

$$A = \{n + m\xi : m, n \in \mathbb{Z}\},$$
$$B = \{n + m\xi : m, n \in \mathbb{Z},\ n \text{ even}\},$$
$$C = \{n + m\xi : m, n \in \mathbb{Z},\ n \text{ odd}\}.$$

Then $A = B \cup C$, $B \cap C = \emptyset$, $C = B + 1$ (that is, the translate of B by $1 \in \mathbb{R}$), $0 \in B$, and A, B, and C are all countable dense subsets of $\mathbb{R}$.

Proof. Clearly A, B, and C are all countable subsets of $\mathbb{R}$. It remains to show that they are dense in $\mathbb{R}$. Let us observe that if $m', m'', n', n'' \in \mathbb{Z}$ and $(n', m') \neq (n'', m'')$ then

(1) $$n' + m'\xi \neq n'' + m''\xi.$$

To prove this, suppose $(n', m') \neq (n'', m'')$ but $n' + m'\xi = n'' + m''\xi$, that is,

(2) $$n' - n'' = (m'' - m')\xi.$$

Now since $(n', m') \neq (n'', m'')$ we have either $m' \neq m''$ or $n' \neq n''$ or both. If $m' \neq m''$ then by (2) we have $\xi = (n' - n'')/(m'' - m')$ which is a rational number. This contradicts the fact that ξ is an irrational number. On the other hand if $n' \neq n''$ then $n' - n'' \neq 0$ and then by (2) we have $m'' - m' \neq 0$ so that $\xi = (n' - n'')/(m'' - m')$, contradicting the fact that ξ is an irrational number. This proves (1).

To show that A is dense in $\mathbb{R}$, we show that every open interval I in $\mathbb{R}$ contains at least one point of A. Now for every $i \in \mathbb{N}$, there exists a unique $n_i \in \mathbb{Z}$ such that $n_i + i\xi \in [0, 1)$. Let $x_i = n_i + i\xi \in A$ for $i \in \mathbb{N}$. Let $k \in \mathbb{N}$ be so large that $\frac{1}{k} < \mu_L(I)$. Consider the collection $\{x_i : i = 1, \ldots, k+1\} \subset A$. Note that $x_1, \ldots, x_{k+1}$ are all distinct by (1). Let $x_i, i = 1, \ldots, k+1$, be enumerated in increasing order and let the result be labeled as $0 \le y_1 < \cdots < y_{k+1} < 1$. Now if $y_{j+1} - y_j \ge \frac{1}{k}$ for all $j = 1, \ldots, k$, then $y_{k+1} - y_1 \ge k \cdot \frac{1}{k} = 1$ which is impossible since both y_1 and y_{k+1} are in $[0, 1)$. Thus $y_{j+1} - y_j < \frac{1}{k}$ for at least one $j = 1, \ldots, k$. In other words, in the collection $\{x_i : i = 1, \ldots, k+1\}$, there exists at least one pair x_p and x_q such that $0 < x_p - x_q < \frac{1}{k}$. Consider all the integral multiples of $x_p - x_q$, that is, the collection $\{n(x_p - x_q) : n \in \mathbb{Z}\}$. This is a subcollection of A since

$$n(x_p - x_q) = n\{(n_p + p\xi) - (n_q + q\xi)\} = n(n_p - n_q) + n(p - q)\xi \in A.$$

The distance between two adjacent points in the collection $\{n(x_p - x_q) : n \in \mathbb{Z}\}$ is less than $\frac{1}{k}$. Since the length of the open interval I is greater than $\frac{1}{k}$, I must contain at least one point in the collection. Thus I contains at least one point of A. This proves the denseness of A in $\mathbb{R}$.

The proof for the denseness of B is similar. We need only to replace $[0, 1)$ in the argument above for the denseness of A by $[0, 2)$. The proof for the denseness of C follows from the fact that $C = B + 1$, that is, C is a translate of B by 1. ∎

Proposition 3.43. *Let $\xi \in (0, 1)$ be an irrational number and let $A = \{n + m\xi : m, n \in \mathbb{Z}\}$. For $x, y \in \mathbb{R}$, let us write $x \sim y$ if $x - y \in A$. Then $\sim$ is an equivalence relation, and if we let E_0 be a set consisting of an element from each of the equivalence classes with respect to the equivalence relation $\sim$, then $E_0 \notin \mathfrak{M}_L$.*

Proof. 1. For every $x \in \mathbb{R}$, we have $x \sim x$ since $x - x = 0 \in A$. Secondly for any $x, y \in \mathbb{R}$, $x \sim y$ implies $y \sim x$ since $a \in A$ implies $-a \in A$. Finally for any $x, y, z \in \mathbb{R}$, if $x \sim y$ and $y \sim z$ then $x \sim z$ since $a_1, a_2 \in A$ implies $a_1 + a_2 \in A$. Therefore $\sim$ is an equivalence relation. Let $\mathbb{R}$ be decomposed into the equivalence classes with respect to $\sim$. By the Axiom of Choice we can select one element from each of the equivalence classes. Let E_0 be the set of points thus selected. Let the elements of the countable set A be enumerated as $\{a_n : n \in \mathbb{Z}_+\}$ with $a_0 = 0$. Consider the collection of sets $\{E_0 + a_n : n \in \mathbb{Z}_+\}$.

To show that $E_0 + a_m$ and $E_0 + a_n$ are disjoint when $m \ne n$, suppose $(E_0 + a_m) \cap (E_0 + a_n) \ne \emptyset$. This implies that there exist $x_1, x_2 \in E_0$ such that $x_1 + a_m = x_2 + a_n$. Then $x_1 - x_2 = a_n - a_m \in A$. But $a_m \ne a_n$ so that $x_1 \ne x_2$. Thus we have two distinct elements x_1 and x_2 from E_0 such that $x_1 - x_2 \in A$, that is, $x_1 \sim x_2$, a contradiction to the definition of E_0.

Next we show that $\bigcup_{n \in \mathbb{Z}_+}(E_0 + a_n) = \mathbb{R}$. Every $y \in \mathbb{R}$ is in one of the equivalence classes with respect to $\sim$. Let x be the element in that equivalence class that has been selected to be an element of E_0. Then $y - x \in A$, say $y - x = a_n \in A$ for some $n \in \mathbb{Z}_+$. Thus $y = x + a_n \in E_0 + a_n$. Therefore $\bigcup_{n \in \mathbb{Z}_+}(E_0 + a_n) = \mathbb{R}$, that is, $E_0 + A = \mathbb{R}$.

2. To show that $E_0 \notin \mathfrak{M}_L$, let us show first that the assumption $E_0 \in \mathfrak{M}_L$ leads to $\mu_L(E_0) = 0$. Consider the difference set $\Delta(E_0)$ of E_0 as in Definition 3.28. Let $z \in \Delta(E_0)$. Then $z = x - y$ where $x, y \in E_0$. Now if $x = y$, then $z = 0 \in A$. If on the other hand $x \ne y$, then since $x, y \in E_0$, x and y do not belong to the same equivalence class with respect to the equivalence relation $\sim$ and thus $z = x - y \notin A$. Therefore we have

$\Delta(E_0) \cap A = \{0\}$, or $\emptyset$. Now if $E_0 \in \mathfrak{M}_L$ and $\mu_L(E_0) > 0$ then $\Delta(E_0)$ contains an open interval I by Theorem 3.29. Then since A is a dense subset of $\mathbb{R}$ the open interval I contains infinitely many points of A so that $\Delta(E_0) \cap A \supset I \cap A$ contains infinitely many points of A. This contradicts $\Delta(E_0) \cap A = \{0\}$, or $\emptyset$. Thus if $E_0 \in \mathfrak{M}_L$ then $\mu_L(E_0) = 0$.

3. Suppose $E_0 \in \mathfrak{M}_L$. Then $\mu_L(E_0) = 0$ as we showed in **2**. By the translation invariance of $(\mathbb{R}, \mathfrak{M}_L, \mu_L)$, we have $E_0 + a_n \in \mathfrak{M}_L$ and $\mu_L(E_0 + a_n) = \mu_L(E_0) = 0$ for $n \in \mathbb{Z}_+$. Since $\bigcup_{n \in \mathbb{Z}_+}(E_0 + a_n) = \mathbb{R}$ and $\{E_0 + a_n : n \in \mathbb{Z}_+\}$ is a disjoint collection, we have $\mu_L(\mathbb{R}) = \mu_L\big(\bigcup_{n \in \mathbb{Z}_+}(E_0 + a_n)\big) = \sum_{n \in \mathbb{Z}_+} \mu_L(E_0 + a_n) = 0$, a contradiction. Therefore $E_0 \notin \mathfrak{M}_L$. ■

By means of the non $\mathfrak{M}_L$-measurable set E_0 in Proposition 3.43 above, we construct next a set $M \in \mathfrak{P}(\mathbb{R})$ such that both $\mu_{*,L}(M) = 0$ and $\mu_{*,L}(M^c) = 0$. Then we have $\mu_{*,L}(M) + \mu_{*,L}(M^c) = 0 \neq \mu_{*,L}(\mathbb{R})$. This shows in particular that the Lebesgue inner measure $\mu_{*,L}$, though superadditive, is not an additive set function on $\mathfrak{P}(\mathbb{R})$.

Theorem 3.44. *There exists a set $M \in \mathfrak{P}(\mathbb{R})$ with the following properties:*

(1) $\mu_{*,L}(M) = \mu_{*,L}(M^c) = 0$,

(2) $\mu_{*,L}(M \cap E) = \mu_{*,L}(M^c \cap E) = 0$ *for every $E \in \mathfrak{P}(\mathbb{R})$,*

(3) $\mu_L^*(M \cap E) = \mu_L^*(M^c \cap E) = \mu_L(E)$ *for every $E \in \mathfrak{M}_L$ with $\mu_L(E) < \infty$.*

Proof. 1. Let A, B, and C be the three subsets of $\mathbb{R}$ defined in Lemma 3.42 and let E_0 be the non $\mathfrak{M}_L$-measurable subset of $\mathbb{R}$ defined in Proposition 3.43. Let $M = E_0 + B$. By the definition of $\mu_{*,L}$, to show that $\mu_{*,L}(M) = 0$, it suffices to show that for every closed set $F \subset M$ we have $\mu_L(F) = 0$. We show that actually for every $F \in \mathfrak{M}_L$ such that $F \subset M$, we have $\mu_L(F) = 0$. Let $\Delta(F)$ be the difference set of F as in Definition 3.28. Since $F \in \mathfrak{M}_L$, to show that $\mu_L(F) = 0$ it suffices to show that $\Delta(F)$ contains no open intervals in $\mathbb{R}$ according to Theorem 3.29. Since C is a dense subset of $\mathbb{R}$ by Lemma 3.42, every open interval in $\mathbb{R}$ contains some points of C. Thus to show that $\Delta(F)$ contains no open intervals, it suffices to show that $\Delta(F) \cap C = \emptyset$. To show $\Delta(F) \cap C = \emptyset$, let $z \in \Delta(F)$. Then $z = z_1 - z_2$ where $z_1, z_2 \in F \subset M = E_0 + B$ so that $z = (x_1 + b_1) - (x_2 + b_2) = (x_1 - x_2) + (b_1 - b_2)$ with $x_1, x_2 \in E_0$ and $b_1, b_2 \in B$. Now if $x_1 = x_2$, then $z = b_1 - b_2 \in B$ and thus $z \notin C$. On the other hand if $x_1 \neq x_2$, then by the definition of E_0 in Proposition 3.43, we have $x_1 - x_2 \notin A$. Now $x_1 - x_2 = z + (b_2 - b_1)$. Since $b_1, b_2 \in B$, we have $b_2 - b_1 \in B \subset A$. If $z \in A$, then since A is closed under addition, we have $x_1 - x_2 \in A$, a contradiction. Thus $z \notin A$. Then since $C \subset A$, we have $z \notin C$. Thus in any case, if $z \in \Delta(F)$ then $z \notin C$. This shows that $\Delta(F) \cap C = \emptyset$ and proves $\mu_{*,L}(M) = 0$.

To show that $\mu_{*,L}(M^c) = 0$, recall that we showed in **1** of the Proof of Proposition 3.43 that $E_0 + A = \mathbb{R}$. Thus we have

$$M^c = \mathbb{R} \setminus M = (E_0 + A) \setminus (E_0 + B)$$

$$= E_0 + C = (E_0 + B + 1) = M + 1.$$

By the translation invariance of $\mu_{*,L}$ on $\mathfrak{P}(\mathbb{R})$ by 5° of Theorem 3.34, we have the equalities $\mu_{*,L}(M^c) = \mu_{*,L}(M+1) = \mu_{*,L}(M) = 0$. This completes the proof of (1).

2. (2) follows from (1) by the monotonicity of $\mu_{*,L}$ on $\mathfrak{P}(\mathbb{R})$.

3. Let $E \in \mathfrak{M}_L$ and $\mu_L(E) < \infty$. According to Theorem 3.41, for every $A \in \mathfrak{P}(\mathbb{R})$, we have $\mu_{*,L}(A \cap E) + \mu_L^*(A^c \cap E) = \mu_L(E)$. With the choice $A = M^c$ we have $0 + \mu_L^*(M \cap E) = \mu_L(E)$ by (2) and similarly with the choice $A = M$ we have $0 + \mu_L^*(M^c \cap E) = \mu_L(E)$ by (b). This proves (3). ∎

Theorem 3.45. *Every $\mathfrak{M}_L$-measurable set E with $\mu_L(E) > 0$ contains a non $\mathfrak{M}_L$-measurable set.*

Proof. Let M be as in Theorem 3.44. Then by (2) and (3) of Theorem 3.44, we have $\mu_{*,L}(M \cap E) = 0$ and $\mu_L^*(M \cap E) = \mu_L(E) > 0$ and therefore $\mu_{*,L}(M \cap E) \neq \mu_L^*(M \cap E)$. This shows that $M \cap E \notin \mathfrak{M}_L$ by (a) of Theorem 3.37. ∎

Problems

Prob. 3.1. Let a decreasing sequence $(E_n : n \in \mathbb{N}) \subset \mathfrak{M}_L$ in the Lebesgue measure space $(\mathbb{R}, \mathfrak{M}_L, \mu_L)$ be given by $E_n = [n, \infty)$ for $n \in \mathbb{N}$.
(a) Find $\lim_{n\to\infty} E_n$ and $\mu_L\big(\lim_{n\to\infty} E_n\big)$.
(b) Find $\lim_{n\to\infty} \mu_L(E_n)$.

Prob. 3.2. Consider a sequence $(E_n : n \in \mathbb{N}) \subset \mathfrak{M}_L$ defined by $E_n = [0,1) \cup [n, n+1)$ when n is odd and $E_n = [0,1) \cup [n, n+2)$ when n is even.
(a) Show that $\lim_{n\to\infty} E_n$ exists and find $\lim_{n\to\infty} E_n$.
(b) Show that $\lim_{n\to\infty} \mu_L(E_n)$ does not exist.

Prob. 3.3. For each of the following sequences $(E_n : n \in \mathbb{N}) \subset \mathfrak{M}_L$,
(a) show that $\lim_{n\to\infty} E_n$ exists and find $\lim_{n\to\infty} E_n$,
(b) show that $\lim_{n\to\infty} \mu_L(E_n)$ exists and $\lim_{n\to\infty} \mu_L(E_n) \neq \mu_L\big(\lim_{n\to\infty} E_n\big)$.

(1) $E_n = [0,1) \cup [n, n+1)$ for $n \in \mathbb{N}$,

(2) $E_n = [0,1) \cup [n, 2n)$ for $n \in \mathbb{N}$,

(3) $E_n = [n, n+1)$ for $n \in \mathbb{N}$.

Prob. 3.4. Let $\mathfrak{J}$ be the collection of $\emptyset$ and all finite open intervals in $\mathbb{R}$. Define an outer measure μ^* on $\mathbb{R}$ by setting for each $E \in \mathfrak{P}(\mathbb{R})$

$$\mu^*(E) = \inf\big\{\textstyle\sum_{n\in\mathbb{N}} \ell(I_n) : (I_n : n \in \mathbb{N}) \subset \mathfrak{J}, \bigcup_{n\in\mathbb{N}} I_n \supset E\big\}.$$

Show that $\mu^* = \mu_o^*$ defined in Observation 3.2.

Prob. 3.5. Let $\mathfrak{E}$ be a disjoint collection of members of $\mathfrak{M}_L$ in the Lebesgue measure space $(\mathbb{R}, \mathfrak{M}_L, \mu_L)$. Show that if $\mu_L(E) > 0$ for every $E \in \mathfrak{E}$, then the collection $\mathfrak{E}$ is at most countable.

Prob. 3.6. For $E \in \mathfrak{M}_L$ with $\mu_L(E) < \infty$, define a real-valued function φ_E on $\mathbb{R}$ by setting

$$\varphi_E(x) = \mu_L\big(E \cap (-\infty, x]\big) \quad \text{for } x \in \mathbb{R}.$$

(a) Show that φ_E is an increasing function on $\mathbb{R}$.
(b) Show that $\lim_{x\to-\infty} \varphi_E(x) = 0$ and $\lim_{x\to\infty} \varphi_E(x) = \mu_L(E)$.
(c) Show that φ_E satisfies the Lipschitz condition on $\mathbb{R}$, that is,

$$|\varphi_E(x') - \varphi_E(x'')| \le |x' - x''| \quad \text{for } x', x'' \in \mathbb{R}.$$

(d) Show that φ_E is uniformly continuous on $\mathbb{R}$.

Prob. 3.7. Let $E \in \mathfrak{M}_L$ with $\mu_L(E) < \infty$. Show that for every $\alpha \in (0, 1)$, there exists a subset E_α of E such that $E_\alpha \in \mathfrak{M}_L$ and $\mu_L(E_\alpha) = \alpha\mu_L(E)$.
(Hint: Use the continuity of the function φ_E in Prob. 3.6.)
(Thus the Lebesgue measure space $(\mathbb{R}, \mathfrak{M}_L, \mu_L)$ not only does not have any atoms but in fact every Lebesgue-measurable set with finite measure has a measurable subset with an arbitrarily designated fraction of the measure. See Prob. 3.8 for $E \in \mathfrak{M}_L$ with $\mu_L(E) = \infty$.)

Prob. 3.8. Let $E \in \mathfrak{M}_L$ with $\mu_L(E) = \infty$. Show that for every $\lambda \in [0, \infty)$, there exists a subset E_λ of E such that $E_\lambda \in \mathfrak{M}_L$ and $\mu_L(E_\lambda) = \lambda$.

Prob. 3.9. Let $E \subset \mathbb{R}$ and $\mu_L^*(E) = 0$. Show that E^c is a dense subset of $\mathbb{R}$, that is, every open interval in $\mathbb{R}$ contains some points of E^c, in other words no open interval can be disjoint from E^c.

Prob. 3.10. For $E \subset \mathbb{R}$, define $-E$, called the reflexion of E with respect to the origin, by setting $-E = \{y \in \mathbb{R} : y = -x \text{ for some } x \in E\}$.
(a) Show $\mu_L^*(-E) = \mu_L^*(E)$.
(b) Show that if $E \in \mathfrak{M}_L$, then $-E \in \mathfrak{M}_L$ and $\mu_L(-E) = \mu_L(E)$.

Prob. 3.11. For $E \subset \mathbb{R}$ and $\alpha \in \mathbb{R}$, define αE, called the dilation of E by the factor α, by setting $\alpha E = \{y \in \mathbb{R} : y = \alpha x \text{ for some } x \in E\}$.
(a) Show $\mu_L^*(\alpha E) = |\alpha|\mu_L^*(E)$.
(b) Show that if $E \in \mathfrak{M}_L$, then $\alpha E \in \mathfrak{M}_L$ and $\mu_L(\alpha E) = |\alpha|\mu_L(E)$.

Prob. 3.12. Consider the measure space $(\mathbb{R}, \mathfrak{B}_{\mathbb{R}}, \mu_L)$.
(a) Show that if $E \in \mathfrak{B}_{\mathbb{R}}$ and $t \in \mathbb{R}$, then $E + t \in \mathfrak{B}_{\mathbb{R}}$ and $\mu_L(E + t) = \mu_L(E)$.
(a) Show that if $E \in \mathfrak{B}_{\mathbb{R}}$ and $\alpha \in \mathbb{R}$, then $\alpha E \in \mathfrak{B}_{\mathbb{R}}$ and $\mu_L(\alpha E) = |\alpha|\mu_L(E)$.

Prob. 3.13. Let f be a real-valued function on (a, b) such that f' exists and satisfies $|f'(x)| \le M$ for $x \in (a, b)$ for some $M \ge 0$. Show that for every subset E of (a, b) we have $\mu_L^*\big(f(E)\big) \le M\mu_L^*(E)$.

Prob. 3.14. From the interval $[0, 1]$ remove the open middle third $\left(\frac{1}{3}, \frac{2}{3}\right)$. Remove the open middle third from each of the two remaining intervals $\left[0, \frac{1}{3}\right]$ and $\left[\frac{2}{3}, 1\right]$. This leaves us four closed intervals. Remove the open middle third from each of the four. Continue this process of removal indefinitely. The resulting set T is called the Cantor ternary set.
(a) Show that T is a Borel set.
(b) Show that T is a compact set. (A set in $\mathbb{R}$ is compact if and only if it is bounded, that is, contained in a finite interval, and closed.)

(c) Show that T is nowhere dense. (A set A in a topological space $(X, \mathfrak{O})$ is called nowhere dense, or non-dense, if the interior of its closure is empty, that is, $\left(\overline{A}\right)^{\circ} = \emptyset$.)
(d) Show that T is a perfect set. (A set A in a topological space $(X, \mathfrak{O})$ is called perfect if $A = A'$ where A' is the derived set of A, that is, the set consisting of all the limit points of the set A. A point in a set A is called an isolated point of A if it is contained in an open set which contains no other point of A. Thus a set A is perfect if and only if it is closed and has no isolated point in it.)
(e) Show that T is an uncountable set and in fact a one-to-one correspondence between T and $[0, 1]$ can be established.
(Hint: Express every number in $[0, 1]$ as a ternary number, that is, $0.a_1a_2a_3\cdots$ where $a_i = 0, 1,$ *or* 2 for each i. Then T consists of those ternary numbers in which the numeral 1 never appears as a_i for any i. To establish a one-to-one correspondence between T and $[0, 1]$, express every number in $[0, 1]$ as a binary number.)
(f) Show that $\mu_L(T) = 0$.
(Thus the Cantor ternary set is an example of a null set in $(\mathbb{R}, \mathfrak{M}_L, \mu_L)$ which is an uncountable set.)

Prob. 3.15. Let Q be the set of all rational numbers in $\mathbb{R}$. For an arbitrary $\varepsilon > 0$, construct an open set O in $\mathbb{R}$ such that $O \supset Q$ and $\mu_L^*(O) \le \varepsilon$.

Prob. 3.16. Let Q be the set of all rational numbers in $\mathbb{R}$.
(a) Show that Q is a null set in $(\mathbb{R}, \mathfrak{B}_{\mathbb{R}}, \mu_L)$.
(b) Show that Q is a F_σ-set.
(c) Show that there exists a G_δ-set G such that $G \supset Q$ and $\mu_L(G) = 0$.
(d) Show that the set of all irrational numbers in $\mathbb{R}$ is a G_δ-set.

Prob. 3.17. Consider the measure space $\left(\mathbb{R}, \mathfrak{M}_L, \mu_L\right)$.
(a) Show that every null set in $\left(\mathbb{R}, \mathfrak{M}_L, \mu_L\right)$ is a subset of a G_δ set G which is itself a null set in $\left(\mathbb{R}, \mathfrak{M}_L, \mu_L\right)$.
(b) Show that every $E \in \mathfrak{M}_L$ can be written as $E = G \setminus N$ where G is a G_δ set and N is a null set in $\left(\mathbb{R}, \mathfrak{M}_L, \mu_L\right)$ contained in G.
(c) Show that every $E \in \mathfrak{M}_L$ can be written as $E = F \cup N$ where F is an F_σ set and N is a null set in $\left(\mathbb{R}, \mathfrak{M}_L, \mu_L\right)$ disjoint from F.

Prob. 3.18. Show that for every increasing sequence $(E_n : n \in \mathbb{N})$ of subsets of $\mathbb{R}$, we have $\lim_{n\to\infty} \mu_L^*(E_n) = \mu_L^*\left(\lim_{n\to\infty} E_n\right)$.

Prob. 3.19. (a) Show that the μ_L^*-measurability condition on $E \in \mathfrak{P}(\mathbb{R})$:

(1) $$\mu_L^*(A) = \mu_L^*(A \cap E) + \mu_L^*(A \cap E^c) \text{ for every } A \in \mathfrak{P}(\mathbb{R}),$$

is equivalent to the following condition:

(2) $$\mu_L^*(I) = \mu_L^*(I \cap E) + \mu_L^*(I \cap E^c) \text{ for every } I \in \mathfrak{I}_o.$$

(b) Let $\mathfrak{J}$ be the collection of all open intervals in $\mathbb{R}$ with rational endpoints. This is a countable subcollection of $\mathfrak{I}_o$. Show that condition (2) is equivalent to the following apparently, and only apparently, weaker condition :

(3) $$\mu_L^*(J) = \mu_L^*(J \cap E) + \mu_L^*(J \cap E^c) \text{ for every } J \in \mathfrak{J}.$$

Prob. 3.20. Let E be the set of all irrational numbers in the interval $(0, 1)$. Show that for every $\varepsilon \in (0, 1)$ there exists a closed set C in $\mathbb{R}$ such that $C \subset E$ and $\mu_L(C) > 1 - \varepsilon$.

Prob. 3.21. (a) Let $A \subset \mathbb{R}$. If there exists $\varepsilon \in (0, 1)$ such that $\mu_L^*(A \cap I) \leq \varepsilon \ell(I)$ for every open interval I then $\mu_L^*(A) = 0$.
(b) Let $E \in \mathfrak{M}_L$ with $\mu_L(E) > 0$. Then there exist $\eta \in (0, 1)$ and an open interval I such that $(E \cap I + h) \cap (E \cap I) \neq \emptyset$ for every $h \in \mathbb{R}$ with $|h| < \eta \ell(I)$.
(c) For $A \in \mathbb{R}$ let $\Delta A = \{z \in \mathbb{R} : z = x - y \text{ for some } x, y \in A\}$. If $E \in \mathfrak{M}_L$ and $\mu_L(E) > 0$ then ΔE contains an open interval.

Prob. 3.22. Consider the measure space $(\mathbb{R}, \mathfrak{M}_L, \mu_L)$. Show that if $E \in \mathfrak{M}_L$ and $\mu_L(E) > 0$ then the set $E + E$ contains an open interval.

Prob. 3.23. Let Q be the set of all rational numbers in $\mathbb{R}$. Show that for any $x \in \mathbb{R}$ we have $\lim_{t \to 0} \mathbf{1}_{Q+t}(x) \neq \mathbf{1}_Q(x)$ and indeed $\lim_{t \to 0} \mathbf{1}_{Q+t}(x)$ does not exist.

Prob. 3.24. Consider a measure space $(\mathbb{R}, \mathfrak{B}_{\mathbb{R}}, \mu)$. Suppose for every $E \in \mathfrak{B}_{\mathbb{R}}$ we have $\lim_{t \to 0} \mathbf{1}_{E+t}(x) = \mathbf{1}_E(x)$ for μ-a.e. $x \in \mathbb{R}$. Show that $\mu(\mathbb{R}) = 0$.

§4 Measurable Functions

[I] Measurability of Functions

Notations. For a set D we write $\{x \in D : \cdots\}$ for the subset of D consisting of those points x in D satisfying one or more conditions represented by $\cdots$. Note that $\{x \in D : \cdots\} = \emptyset$ when the defining conditions $\cdots$ are not satisfied by any $x \in D$. We also abbreviate $\{x \in D : \cdots\}$ as $\{D : \cdots\}$ when there is no ambiguity. Thus for example, if f is an extended real-valued function defined on a subset D of a set X and $\alpha \in \mathbb{R}$, then $\{x \in D : f(x) \le \alpha\}$ is the subset of D consisting of all $x \in D$ at which $f(x) \le \alpha$ holds. We also use the abbreviation $\{D : f \le \alpha\}$.

Definition 4.1. *Let $(X, \mathfrak{A})$ be an arbitrary measurable space and let $D \in \mathfrak{A}$. An extended real-valued function f defined on D is said to be $\mathfrak{A}$-measurable on D if it satisfies the condition that $\{x \in D : f(x) \le \alpha\} \in \mathfrak{A}$, that is, $f^{-1}([-\infty, \alpha]) \in \mathfrak{A}$, for every $\alpha \in \mathbb{R}$.*

In particular, when $(X, \mathfrak{A}) = (\mathbb{R}, \mathfrak{M}_L)$, $D \in \mathfrak{M}_L$, and f is $\mathfrak{M}_L$-measurable on D, we say that f is Lebesgue measurable. Similarly when $(X, \mathfrak{A}) = (\mathbb{R}, \mathfrak{B}_{\mathbb{R}})$, $D \in \mathfrak{B}_{\mathbb{R}}$, and f is $\mathfrak{B}_{\mathbb{R}}$-measurable on D, we say that f is Borel measurable.

Observation 4.2. **(a)** If $\mathfrak{A}_1$ and $\mathfrak{A}_2$ are two σ-algebras of subsets of a set X such that $\mathfrak{A}_1 \subset \mathfrak{A}_2$, then every $\mathfrak{A}_1$-measurable function is also a $\mathfrak{A}_2$-measurable function.
(b) Let $\mathfrak{A}_0$ be the smallest σ-algebra of subsets of a set X, that is, $\mathfrak{A}_0 = \{\emptyset, X\}$. Then an extended real-valued function f defined on X is $\mathfrak{A}_0$-measurable on X if and only if f is a constant function on X.
(c) With $\mathfrak{P}(X)$, the greatest σ-algebra of subsets of a set X consisting of all subsets of X, every extended real-valued function on an arbitrary $E \in \mathfrak{P}(X)$ is $\mathfrak{P}(X)$-measurable on E.

As an intermediate case to the two extreme cases (b) and (c) above, we have

Example. Let $\mathfrak{E} = \{[n, n+1) : n \in \mathbb{Z}\}$. The σ-algebra generated by $\mathfrak{E}$, $\sigma(\mathfrak{E})$, is the collection of all countable unions of members of $\mathfrak{E}$. An extended real-valued function defined on $\mathbb{R}$ is $\sigma(\mathfrak{E})$-measurable if and only if it is a right-continuous step function with jump discontinuity occurring at integers in $\mathbb{R}$ only.

For an arbitrary subset E of a set X, the identity function (or the characteristic function) of E is a function defined on X by

$$\mathbf{1}_E(x) = \begin{cases} 1 & \text{if } x \in E, \\ 0 & \text{if } x \in E^c. \end{cases}$$

Observation 4.3. Let $(X, \mathfrak{A})$ be a measurable space and let $E \in \mathfrak{P}(X)$. Then $\mathbf{1}_E$ is a $\mathfrak{A}$-measurable function on X if and only if $E \in \mathfrak{A}$.

Proof. Let $E \in \mathfrak{P}(X)$. Note that the domain of definition of $\mathbf{1}_E$ is $X \in \mathfrak{A}$. If $E \in \mathfrak{A}$ then

$$\{x \in X : \mathbf{1}_E(x) \le \alpha\} = \begin{cases} \emptyset \in \mathfrak{A} & \text{for } \alpha < 0, \\ E^c \in \mathfrak{A} & \text{for } 0 \le \alpha < 1, \\ X \in \mathfrak{A} & \text{for } \alpha \ge 1. \end{cases}$$

Thus $\{x \in X : \mathbf{1}_E(x) \le \alpha\} \in \mathfrak{A}$ for every $\alpha \in \mathbb{R}$ and $\mathbf{1}_E$ is $\mathfrak{A}$-measurable on X. Conversely if $\mathbf{1}_E$ is $\mathfrak{A}$-measurable on X, then we have $\{x \in X : \mathbf{1}_E(x) \le \alpha\} \in \mathfrak{A}$ for every $\alpha \in \mathbb{R}$ so that in particular $E^c = \{x \in X : \mathbf{1}_E(x) \le \frac{1}{2}\} \in \mathfrak{A}$ and then $E \in \mathfrak{A}$. ∎

Lemma 4.4. *Let $(X, \mathfrak{A})$ be a measurable space and f be an extended real-valued function defined on $D \in \mathfrak{A}$. Then the following conditions are all equivalent:*
(i) $\{x \in D : f(x) \le \alpha\} \in \mathfrak{A}$, *that is,* $f^{-1}([-\infty, \alpha]) \in \mathfrak{A}$, *for every* $\alpha \in \mathbb{R}$,
(ii) $\{x \in D : f(x) > \alpha\} \in \mathfrak{A}$, *that is,* $f^{-1}((\alpha, \infty]) \in \mathfrak{A}$, *for every* $\alpha \in \mathbb{R}$,
(iii) $\{x \in D : f(x) \ge \alpha\} \in \mathfrak{A}$, *that is,* $f^{-1}([\alpha, \infty]) \in \mathfrak{A}$, *for every* $\alpha \in \mathbb{R}$,
(iv) $\{x \in D : f(x) < \alpha\} \in \mathfrak{A}$, *that is,* $f^{-1}([-\infty, \alpha)) \in \mathfrak{A}$, *for every* $\alpha \in \mathbb{R}$.

Proof. 1. (i) $\Leftrightarrow$ (ii). Let $\alpha \in \mathbb{R}$ be fixed. Let D_1 and D_2 be the two sets in (i) and (ii) respectively. Then $D_1 \cap D_2 = \emptyset$ and $D_1 \cup D_2 = D \in \mathfrak{A}$ so that $D_1 = D \setminus D_2$ and $D_2 = D \setminus D_1$. Thus $D_1 \in \mathfrak{A}$ if and only if $D_2 \in \mathfrak{A}$.

2. (iii) $\Leftrightarrow$ (iv) by the same argument as in **1**.

3. (iv) $\Rightarrow$ (i). Note that for every $x \in D$ and $\alpha \in \mathbb{R}$, we have $f(x) \le \alpha$ if and only if $f(x) < \alpha + \frac{1}{n}$ for every $n \in \mathbb{N}$. Thus we have $\{D : f \le \alpha\} = \bigcap_{n \in \mathbb{N}} \{D : f < \alpha + \frac{1}{n}\}$. If f satisfies (iv), then every set in the intersection is in $\mathfrak{A}$ and then so is the countable intersection. This shows that f satisfies (i).

4. (ii) $\Rightarrow$ (iii). Note that we have $f(x) \ge \alpha$ if and only if $f(x) > \alpha - \frac{1}{n}$ for every $n \in \mathbb{N}$. Then by the same argument as in **3**, (ii) implies (iii). ∎

Corollary 4.5. *Let $(X, \mathfrak{A})$ be a measurable space and let f be an extended real-valued $\mathfrak{A}$-measurable function defined on $D \in \mathfrak{A}$. Then*
(a) $\{x \in D : f(x) = \alpha\} \in \mathfrak{A}$ *for every* $\alpha \in \overline{\mathbb{R}}$.
(b) $\{x \in D : f(x) \in \mathbb{R}\} \in \mathfrak{A}$.

Proof. For $\alpha \in \mathbb{R}$, we have

$$\{D : f = \alpha\} = \{D : f \le \alpha\} \setminus \{D : f < \alpha\} \in \mathfrak{A}$$

by (i) and (iv) of Lemma 4.4. We have also

$$\{D : f = \infty\} = D \setminus \{D : f < \infty\} = D \setminus \bigcup_{n \in \mathbb{N}} \{D : f \le n\} \in \mathfrak{A}$$

and

$$\{D : f = -\infty\} = D \setminus \{D : f > -\infty\} = D \setminus \bigcup_{n \in \mathbb{N}} \{D : f \ge -n\} \in \mathfrak{A}.$$

This proves (a). For (b), note that

$$\{D : f \in \mathbb{R}\} = D \setminus (\{D : f = \infty\} \cup \{D : f = -\infty\}) \in \mathfrak{A}$$

by (a). ■

Let us note that condition (a) in Corollary 4.5 is a necessary but not sufficient condition for the $\mathfrak{A}$-measurability of f on D, that is, an extended real-valued f on $D \in \mathfrak{A}$ satisfying condition (a) need not be $\mathfrak{A}$-measurable on D. Below is an example.

Example. Consider the measurable space $(\mathbb{R}, \mathfrak{M}_L)$. By Theorem 3.20 there exists a non $\mathfrak{M}_L$-measurable set $E \subset (0,1)$. Let f be a function defined on $(0,1) \in \mathfrak{M}_L$ by

$$f(x) = \begin{cases} x & \text{for } x \in E, \\ -x & \text{for } x \in (0,1) \setminus E. \end{cases}$$

For every $\alpha \in \overline{\mathbb{R}}$, the set $\{x \in (0,1) : f(x) = \alpha\}$ is either $\emptyset$ or a singleton. In any case it is a $\mathfrak{M}_L$-measurable set. Thus f satisfies condition (a) of Corollary 4.5. But f is a non $\mathfrak{M}_L$-measurable function since $\{x \in (0,1) : f(x) > 0\} = E \notin \mathfrak{M}_L$.

Let $(X, \mathfrak{A})$ and $(Y, \mathfrak{B})$ be measurable spaces. By Definition 1.38, a mapping f of a set $D \subset X$ into Y is $\mathfrak{A}/\mathfrak{B}$-measurable if $f^{-1}(\mathfrak{B}) \subset \mathfrak{A}$. This implies that $D = f^{-1}(Y) \in \mathfrak{A}$. Let us relate this definition of measurable mapping to $\mathfrak{A}$-measurability of a real-valued function f on a set $D \in \mathfrak{A}$ in a measurable space $(X, \mathfrak{A})$ as in Definition 4.1.

Theorem 4.6. *Let $(X, \mathfrak{A})$ be a measure space and let f be a real-valued function on a set $D \in \mathfrak{A}$. Let $\mathfrak{O}_{\mathbb{R}}$ and $\mathfrak{C}_{\mathbb{R}}$ be the collections of the open sets and the closed sets in $\mathbb{R}$ respectively.*
(a) *f is $\mathfrak{A}$-measurable on D if and only if f is a $\mathfrak{A}/\mathfrak{B}_{\mathbb{R}}$-measurable mapping of D into $\mathbb{R}$, that is, $f^{-1}(\mathfrak{B}_{\mathbb{R}}) \subset \mathfrak{A}$.*
(b) *f is $\mathfrak{A}$-measurable on D if and only if $f^{-1}(\mathfrak{O}_{\mathbb{R}}) \subset \mathfrak{A}$.*
(c) *f is $\mathfrak{A}$-measurable on D if and only if $f^{-1}(\mathfrak{C}_{\mathbb{R}}) \subset \mathfrak{A}$.*

Proof. 1. Let $\mathfrak{J}$ be the collection of all open intervals in $\mathbb{R}$. Since every open set O in $\mathbb{R}$ is a countable union of open intervals, we have $O \in \sigma(\mathfrak{J})$ and thus $\mathfrak{O}_{\mathbb{R}} \subset \sigma(\mathfrak{J})$ and then $\sigma(\mathfrak{O}_{\mathbb{R}}) \subset \sigma(\mathfrak{J})$. On the other hand, since $\mathfrak{J} \subset \mathfrak{O}_{\mathbb{R}}$, we have $\sigma(\mathfrak{J}) \subset \sigma(\mathfrak{O}_{\mathbb{R}})$. Therefore $\sigma(\mathfrak{J}) = \sigma(\mathfrak{O}_{\mathbb{R}}) = \mathfrak{B}_{\mathbb{R}}$.

Now suppose f is a real-valued $\mathfrak{A}$-measurable function defined on $D \in \mathfrak{A}$. For an arbitrary finite open interval (α, β) in $\mathbb{R}$, we have

$$f^{-1}\big((\alpha,\beta)\big) = f^{-1}\big((-\infty,\beta) \setminus (-\infty,\alpha]\big) = f^{-1}\big((-\infty,\beta)\big) \setminus f^{-1}\big((-\infty,\alpha]\big) \in \mathfrak{A}$$

by (iv) and (i) of Lemma 4.4. Similarly for an infinite open interval I in $\mathbb{R}$, we have $f^{-1}(I) \in \mathfrak{A}$ by (ii) and (iv) of Lemma 4.4 and Corollary 4.5. This shows that $f^{-1}(\mathfrak{J}) \subset \mathfrak{A}$. Then since $\sigma(\mathfrak{J}) = \mathfrak{B}_{\mathbb{R}}$, we have

$$f^{-1}(\mathfrak{B}_{\mathbb{R}}) = f^{-1}\big(\sigma(\mathfrak{J})\big) = \sigma\big(f^{-1}(\mathfrak{J})\big) \subset \sigma(\mathfrak{A}) = \mathfrak{A}$$

by Theorem 1.14. This shows that f is a $\mathfrak{A}/\mathfrak{B}_{\mathbb{R}}$-measurable mapping of D into $\mathbb{R}$.

Conversely if f is a $\mathfrak{A}/\mathfrak{B}_{\mathbb{R}}$-measurable mapping of D into $\mathbb{R}$, then we have $f^{-1}(B) \in \mathfrak{A}$ for every $B \in \mathfrak{B}_{\mathbb{R}}$ and in particular for every $\alpha \in \mathbb{R}$ we have $f^{-1}\big((-\infty, \alpha]\big) \in \mathfrak{A}$. This shows that f is a $\mathfrak{A}$-measurable function on D. This proves (a).

2. To prove (b), note that $f^{-1}(\mathfrak{O}_{\mathbb{R}}) \subset \mathfrak{A}$ if and only if $f^{-1}(\mathfrak{B}_{\mathbb{R}}) \subset \mathfrak{A}$. (Indeed since we have $\mathfrak{O}_{\mathbb{R}} \subset \mathfrak{B}_{\mathbb{R}}$, $f^{-1}(\mathfrak{B}_{\mathbb{R}}) \subset \mathfrak{A}$ implies $f^{-1}(\mathfrak{O}_{\mathbb{R}}) \subset \mathfrak{A}$. Conversely if $f^{-1}(\mathfrak{O}_{\mathbb{R}}) \subset \mathfrak{A}$, then we have $f^{-1}(\mathfrak{B}_{\mathbb{R}}) = f^{-1}\big(\sigma(\mathfrak{O}_{\mathbb{R}})\big) = \sigma\big(f^{-1}(\mathfrak{O}_{\mathbb{R}})\big) \subset \sigma(\mathfrak{A}) = \mathfrak{A}$ by Theorem 1.14.) Then (b) follows from (a).

3. (c) is proved in the same way as (b) using the fact that $\sigma(\mathfrak{C}_{\mathbb{R}}) = \mathfrak{B}_{\mathbb{R}}$. ∎

According to Theorem 4.6, among all the σ-algebras of subsets of $\mathbb{R}$, the Borel σ-algebra $\mathfrak{B}_{\mathbb{R}}$ has the special property that every real-valued function f on a set $D \in \mathfrak{A}$ in an arbitrary measurable space $(X, \mathfrak{A})$ is $\mathfrak{A}$-measurable on D if and only if f is a $\mathfrak{A}/\mathfrak{B}_{\mathbb{R}}$-measurable mapping of D into $\mathbb{R}$. This is a consequence of Definition 4.1 for $\mathfrak{A}$-measurability of a real-valued function in terms of the Borel sets $(-\infty, \alpha]$ for $\alpha \in \mathbb{R}$. In particular, a real-valued function f on a set $D \in \mathfrak{M}_L$ in the measurable space $(\mathbb{R}, \mathfrak{M}_L)$ is $\mathfrak{M}_L$-measurable on D if and only if f is a $\mathfrak{M}_L/\mathfrak{B}_{\mathbb{R}}$-measurable mapping of D into $\mathbb{R}$, and similarly a real-valued function f on a set $D \in \mathfrak{B}_{\mathbb{R}}$ in the measurable space $(\mathbb{R}, \mathfrak{B}_{\mathbb{R}})$ is $\mathfrak{B}_{\mathbb{R}}$-measurable on D if and only if f is a $\mathfrak{B}_{\mathbb{R}}/\mathfrak{B}_{\mathbb{R}}$-measurable mapping of D into $\mathbb{R}$.

Let us note that Theorem 4.6 is applicable to extended real-valued functions. Let f be an extended real-valued $\mathfrak{A}$-measurable function on a set $D \in \mathfrak{A}$ in a measurable space $(X, \mathfrak{A})$. By (b) of Corollary 4.5, $D_0 = \{D : f \in \mathbb{R}\}$ is a $\mathfrak{A}$-measurable set. Then by Theorem 4.6, the restriction f_0 of f to D_0 is a $\mathfrak{A}/\mathfrak{B}_{\mathbb{R}}$-measurable mapping of D_0 into $\mathbb{R}$.

Lemma 4.7. *Let $(X, \mathfrak{A})$ be a measurable space.*

(a) *If f is an extended real-valued $\mathfrak{A}$-measurable function on a set $D \in \mathfrak{A}$, then for every $D_0 \subset D$ such that $D_0 \in \mathfrak{A}$, the restriction of f to D_0 is a $\mathfrak{A}$-measurable function on D_0.*

(b) *Let $(D_n : n \in \mathbb{N})$ be a sequence in $\mathfrak{A}$ and let $D = \bigcup_{n\in\mathbb{N}} D_n$. Let f be an extended real-valued function on D. If the restriction of f to D_n is $\mathfrak{A}$-measurable on D_n for every $n \in \mathbb{N}$, then f is $\mathfrak{A}$-measurable on D.*

Proof. 1. To prove (a), note that $\{D_0 : f \le \alpha\} = \{D : f \le \alpha\} \cap D_0$ for every $\alpha \in \mathbb{R}$. If f is $\mathfrak{A}$-measurable on D, then $\{D : f \le \alpha\} \in \mathfrak{A}$ so that $\{D_0 : f \le \alpha\} \in \mathfrak{A}$.

2. To prove (b), note that $\{D : f \le \alpha\} = \{\bigcup_{n\in\mathbb{N}} D_n : f \le \alpha\} = \bigcup_{n\in\mathbb{N}} \{D_n : f \le \alpha\}$ for every $\alpha \in \mathbb{R}$. If f is $\mathfrak{A}$-measurable on D_n for every $n \in \mathbb{N}$, then $\{D_n : f \le \alpha\} \in \mathfrak{A}$ for every $n \in \mathbb{N}$ so that $\{D : f \le \alpha\} \in \mathfrak{A}$. ∎

Observation 4.8. Let $(X, \mathfrak{A})$ be a measurable space and let $D \in \mathfrak{A}$. Then every constant function on D is $\mathfrak{A}$-measurable on D, that is, if $f(x) = \gamma$ for every $x \in D$ for some $\gamma \in \overline{\mathbb{R}}$, then f is $\mathfrak{A}$-measurable on D.

Proof. If $f(x) = \gamma$ for every $x \in D$, then for every $\alpha \in \mathbb{R}$, we have

$$\{D : f \le \alpha\} = \{D : \gamma \le \alpha\} = \begin{cases} D \in \mathfrak{A} & \text{if } \alpha \ge \gamma, \\ \emptyset \in \mathfrak{A} & \text{if } \alpha < \gamma. \end{cases}$$

Thus f is $\mathfrak{A}$-measurable on D. ∎

Observation 4.9. Let $(X, \mathfrak{A})$ be a measurable space and let f be an extended real-valued function on a set $D \in \mathfrak{A}$. Let us decompose D into three subsets $D_0 = \{D : f \in \mathbb{R}\}$, $D_1 = \{D : f = -\infty\}$, and $D_2 = \{D : f = \infty\}$. Then we have

$$f \text{ is } \mathfrak{A}\text{-measurable on } D \Leftrightarrow \begin{cases} D_0, D_1, D_2 \in \mathfrak{A}, \\ f \text{ is } \mathfrak{A}\text{-measurable on each of} D_0, D_1, D_2. \end{cases}$$

Proof. If f is $\mathfrak{A}$-measurable on D then $D_0, D_1, D_2 \in \mathfrak{A}$ by (a) and (b) of Corollary 4.5 and f is $\mathfrak{A}$-measurable on each of D_0,D_1, and D_2 by (a) of Lemma 4.7. Conversely if $D_0, D_1, D_2 \in \mathfrak{A}$ and f is $\mathfrak{A}$-measurable on each of D_0,D_1, and D_2 then f is $\mathfrak{A}$-measurable on D by (b) of Lemma 4.7. ∎

Observation 4.10. Let $(X, \mathfrak{A})$ be a measurable space and let f be an extended real-valued function on a set $D \in \mathfrak{A}$ assuming at most countably many values $\gamma_n \in \overline{\mathbb{R}}$ for $n \in \mathbb{N}$. Let $D_n = \{D : f = \gamma_n\}$ for $n \in \mathbb{N}$. Then f is $\mathfrak{A}$-measurable on D if and only if $D_n \in \mathfrak{A}$ for every $n \in \mathbb{N}$.

Proof. If f is $\mathfrak{A}$-measurable on D, then $D_n \in \mathfrak{A}$ for every $n \in \mathbb{N}$ by (a) of Corollary 4.5. Conversely suppose $D_n \in \mathfrak{A}$ for every $n \in \mathbb{N}$. Then since f is constant on $D_n \in \mathfrak{A}$, f is $\mathfrak{A}$-measurable on D_n by Observation 4.8 for every $n \in \mathbb{N}$. Since $D = \bigcup_{n \in \mathbb{N}} D_n$, f is $\mathfrak{A}$-measurable on D by (b) of Lemma 4.7. ∎

[II] Operations with Measurable Functions

Addition, multiplication and product of extended real-valued functions entails addition and multiplication of extended real numbers.

Convention. For addition and multiplication in the extended real number system $\overline{\mathbb{R}}$ let us adopt the convention that $\frac{c}{\infty} = \frac{c}{-\infty} = 0$ for any $c \in \mathbb{R}$. Such expressions as $\infty - \infty$, $-\infty + \infty$, $(\pm\infty) \cdot 0$, $0 \cdot (\pm\infty)$, and $\frac{\pm\infty}{\pm\infty}$ remain undefined.

Theorem 4.11. *Let $(X, \mathfrak{A})$ be a measurable space and let f be an extended real-valued $\mathfrak{A}$-measurable function on a set $D \in \mathfrak{A}$. Then for every $c \in \mathbb{R}$, we have the domain of definition of the function cf, $\mathfrak{D}(cf) \in \mathfrak{A}$ and cf is a $\mathfrak{A}$-measurable function on $\mathfrak{D}(cf)$.*

Proof. Clearly $\mathfrak{D}(cf) \subset \mathfrak{D}(f) = D$. Let us find $\mathfrak{D}(cf)$ and show that it is a $\mathfrak{A}$-measurable set. If $c \in \mathbb{R}$ and $c \neq 0$, then $\mathfrak{D}(cf) = D \in \mathfrak{A}$. If $c = 0$, then $(cf)(x) = cf(x)$

is defined only if $f(x) \in \mathbb{R}$. so that $\mathfrak{D}(cf) = \{D : f \in \mathbb{R}\} \in \mathfrak{A}$ by (b) of Corollary 4.5. This shows that $\mathfrak{D}(cf) \in \mathfrak{A}$ for every $c \in \mathbb{R}$.

Let us show that cf is $\mathfrak{A}$-measurable on $\mathfrak{D}(cf) \in \mathfrak{A}$. If $c = 0$, then $(cf)(x) = 0$ for every $x \in \mathfrak{D}(cf)$ and thus cf is $\mathfrak{A}$-measurable on $\mathfrak{D}(cf)$ by Observation 4.8. If $c > 0$, then for every $\alpha \in \mathbb{R}$, we have $\{\mathfrak{D}(cf) : cf < \alpha\} = \{D : f < \frac{\alpha}{c}\} \in \mathfrak{A}$ by Lemma 4.4 so that cf is $\mathfrak{A}$-measurable on $\mathfrak{D}(cf)$. Similarly if $c < 0$, then we have $\{\mathfrak{D}(cf) : cf < \alpha\} = \{D : f > \frac{\alpha}{c}\} \in \mathfrak{A}$ by Lemma 4.4 so that cf is $\mathfrak{A}$-measurable on $\mathfrak{D}(cf)$. ∎

Theorem 4.12. *Let $(X, \mathfrak{A})$ be a measurable space and let f and g be two extended real-valued $\mathfrak{A}$-measurable function on a set $D \in \mathfrak{A}$. Then the domain of definition of the function $f + g$, $\mathfrak{D}(f + g) \in \mathfrak{A}$ and $f + g$ is a $\mathfrak{A}$-measurable function on $\mathfrak{D}(f + g)$.*

Proof. To find $\mathfrak{D}(f + g)$ which is a subset of D let us recall that $\infty - \infty$ and $-\infty + \infty$ are undefined. Thus let

$$
\begin{aligned}
D_0 &= \{D : f \in \mathbb{R}\} \cap \{D : g \in \mathbb{R}\} \in \mathfrak{A},\\
D_1 &= \{D : f = -\infty\} \cap \{D : g < \infty\} \in \mathfrak{A},\\
D_2 &= \{D : f = \infty\} \cap \{D : g > -\infty\} \in \mathfrak{A},\\
D_3 &= \{D : f < \infty\} \cap \{D : g = -\infty\} \in \mathfrak{A},\\
D_4 &= \{D : f > -\infty\} \cap \{D : g = \infty\} \in \mathfrak{A},\\
D_5 &= \{D : f = \infty\} \cap \{D : g = -\infty\} \in \mathfrak{A},\\
D_6 &= \{D : f = -\infty\} \cap \{D : g = \infty\} \in \mathfrak{A}.
\end{aligned}
$$

Then $\{D_k : k = 0, \ldots, 6\}$ is a disjoint collection of subsets of D and $\bigcup_{k=0}^{6} D_k = D$. Since $f(x) + g(x)$ is defined on each of $D_0, \ldots, D_4$ and undefined on D_5 and D_6, we have $\mathfrak{D}(f + g) = \bigcup_{k=0}^{4} D_k \in \mathfrak{A}$. To show the $\mathfrak{A}$-measurability of $f + g$ on $\mathfrak{D}(f + g)$, it suffices to show that $f + g$ is $\mathfrak{A}$-measurable on D_k for $k = 0, \ldots, 4$ according to (b) of Lemma 4.7. Since $f + g$ has the constant value $-\infty, \infty, -\infty$, and ∞ on D_1, D_2, D_3, and D_4 respectively, $f + g$ is $\mathfrak{A}$-measurable on each of these four sets by Observation 4.8.

Regarding D_0, we have $\{D_0 : f + g < \alpha\} = \{D_0 : f < \alpha - g\}$ for every $\alpha \in \mathbb{R}$. Let $\{r_n : n \in \mathbb{N}\}$ be the collection of all rational numbers in $\mathbb{R}$. Now if $f(x) < \alpha - g(x)$ then there exists a rational number r_n such that $f(x) < r_n < \alpha - g(x)$. Conversely if $f(x) < r_n < \alpha - g(x)$ for some rational number r_n, then $f(x) < \alpha - g(x)$. Thus

$$
\begin{aligned}
\{D_0 : f < \alpha - g\} &= \bigcup_{n \in \mathbb{N}} \{D_0 : f < r_n < \alpha - g\}\\
&= \bigcup_{n \in \mathbb{N}} [\{D_0 : f < r_n\} \cap \{D_0 : r_n < \alpha - g\}]\\
&= \bigcup_{n \in \mathbb{N}} [\{D_0 : f < r_n\} \cap \{D_0 : g < \alpha - r_n\}]\\
&\in \mathfrak{A}.
\end{aligned}
$$

(Regarding the first of the equalities above, note that for a rational number r_n such that $f(x) < r_n < \alpha - g(x)$ does not hold for any $x \in D_0$ we have $\{D_0 : f < r_n < \alpha - g\} = \emptyset$.) This shows that $f + g$ is $\mathfrak{A}$-measurable on D_0. ∎

Theorem 4.13. *Let $(X, \mathfrak{A})$ be a measurable space and let f and g be two extended real-valued $\mathfrak{A}$-measurable functions on a set $D \in \mathfrak{A}$. Then the domain of definition of the function fg, $\mathfrak{D}(fg) \in \mathfrak{A}$ and fg is a $\mathfrak{A}$-measurable function on $\mathfrak{D}(fg)$.*

Proof. Let us determine $\mathfrak{D}(fg) \subset D$. Recall that $\pm\infty \cdot 0$ and $0 \cdot \pm\infty$ are undefined. Let

$$D_1 = \big[\{D : f = \infty\} \cup \{D : f = -\infty\}\big] \cap \{D : g = 0\} \in \mathfrak{A},$$
$$D_2 = \big[\{D : g = \infty\} \cup \{D : g = -\infty\}\big] \cap \{D : f = 0\} \in \mathfrak{A}.$$

Then $\mathfrak{D}(fg) = D \setminus (D_1 \cup D_2) \in \mathfrak{A}$.

To show the $\mathfrak{A}$-measurability of fg on $\mathfrak{D}(fg)$, note that $\mathfrak{D}(fg)$ is the union of the following three sets:

$$E_0 = \{\mathfrak{D}(fg) : g = 0\} \in \mathfrak{A},$$
$$E_1 = \{\mathfrak{D}(fg) : g > 0\} \in \mathfrak{A},$$
$$E_2 = \{\mathfrak{D}(fg) : g < 0\} \in \mathfrak{A}.$$

Now $fg = 0$ on E_0 so that fg is $\mathfrak{A}$-measurable on E_0 by Observation 4.8. Let $\{r_n : n \in \mathbb{N}\}$ be the collection of all rational numbers. Let $\alpha \in \mathbb{R}$ be fixed. Then for $x \in E_1$ we have $f(x)g(x) < \alpha$, that is, $f(x) < \frac{\alpha}{g(x)}$, if and only if there exists r_n such that $f(x) < r_n < \frac{\alpha}{g(x)}$. Thus we have

$$\begin{aligned}\{E_1 : fg < \alpha\} &= \{E_1 : f < \tfrac{\alpha}{g}\} = \textstyle\bigcup_{n\in\mathbb{N}} \{E_1 : f < r_n < \tfrac{\alpha}{g}\} \\ &= \bigcup_{n\in\mathbb{N}} \big[\{E_1 : f < r_n\} \cap \{E_1 : r_n g < \alpha\}\big] \in \mathfrak{A},\end{aligned}$$

where $\{E_1 : r_n g < \alpha\} \in \mathfrak{A}$ since $r_n g$ is $\mathfrak{A}$-measurable on E_1 by Theorem 4.11. This proves the $\mathfrak{A}$-measurability of fg on E_1. The $\mathfrak{A}$-measurability of fg on E_2 is shown likewise. Then fg is $\mathfrak{A}$-measurable on $\mathfrak{D}(fg)$ by (b) of Lemma 4.7. ∎

Lemma 4.14. *Let $(X, \mathfrak{A})$ be a measurable space and let f be an extended real-valued $\mathfrak{A}$-measurable function on a set $D \in \mathfrak{A}$. Then the domain of definition of the function $\frac{1}{f}$, $\mathfrak{D}\big(\frac{1}{f}\big) \in \mathfrak{A}$ and $\frac{1}{f}$ is a $\mathfrak{A}$-measurable function on $\mathfrak{D}\big(\frac{1}{f}\big) \in \mathfrak{A}$.*

Proof. Let us note that $\mathfrak{D}\big(\frac{1}{f}\big) = D \setminus \{D : f = 0\} \in \mathfrak{A}$. To show the $\mathfrak{A}$-measurability of $\frac{1}{f}$ on $\mathfrak{D}\big(\frac{1}{f}\big)$, we show that $\big\{\mathfrak{D}\big(\frac{1}{f}\big) : \frac{1}{f} > \alpha\big\} \in \mathfrak{A}$ for every $\alpha \in \mathbb{R}$. Let us observe that

(1) $$0 < a < b \Rightarrow \tfrac{1}{a} > \tfrac{1}{b},$$

(2) $$a < b < 0 \Rightarrow \tfrac{1}{a} > \tfrac{1}{b},$$

(3) $$a < 0 < b \Rightarrow \tfrac{1}{a} < \tfrac{1}{b}.$$

1. If $\alpha = 0$, then $\{\mathfrak{D}(\frac{1}{f}) : \frac{1}{f} > 0\} = \{\mathfrak{D}(\frac{1}{f}) : f > 0\} \setminus \{\mathfrak{D}(\frac{1}{f}) : f = \infty\} \in \mathfrak{A}$.
2. If $\alpha > 0$, then $\{\mathfrak{D}(\frac{1}{f}) : \frac{1}{f} > \alpha\} = \{\mathfrak{D}(\frac{1}{f}) : 0 < f < \frac{1}{\alpha}\} \in \mathfrak{A}$.
3. If $\alpha < 0$, then

$$\{\mathfrak{D}(\tfrac{1}{f}) : \tfrac{1}{f} > \alpha\} = \{\mathfrak{D}(\tfrac{1}{f}) : \tfrac{1}{f} > \alpha, f > 0\} \cup \{\mathfrak{D}(\tfrac{1}{f}) : \tfrac{1}{f} > \alpha, f < 0\}.$$

Since $\alpha < 0$, we have

$$\{\mathfrak{D}(\tfrac{1}{f}) : \tfrac{1}{f} > \alpha, f > 0\} = \{\mathfrak{D}(\tfrac{1}{f}) : f > 0\} \in \mathfrak{A}.$$

On the other hand $\alpha < 0$ implies

$$\{\mathfrak{D}(\tfrac{1}{f}) : \tfrac{1}{f} > \alpha, f < 0\} = \{\mathfrak{D}(\tfrac{1}{f}) : f < \tfrac{1}{\alpha}, f < 0\} = \{\mathfrak{D}(\tfrac{1}{f}) : f < \tfrac{1}{\alpha}\} \in \mathfrak{A}.$$

This shows that $\{\mathfrak{D}(\frac{1}{f}) : \frac{1}{f} > \alpha\} \in \mathfrak{A}$ for the case $\alpha < 0$ and completes the proof of the $\mathfrak{A}$-measurability of $\frac{1}{f}$ on $\mathfrak{D}(\frac{1}{f})$. ∎

Theorem 4.15. *Let $(X, \mathfrak{A})$ be a measurable space and let f and g be two extended real-valued $\mathfrak{A}$-measurable function on a set $D \in \mathfrak{A}$. Then the domain of definition of the function $\frac{g}{f}$, $\mathfrak{D}(\frac{g}{f}) \in \mathfrak{A}$ and $\frac{g}{f}$ is a $\mathfrak{A}$-measurable function on $\mathfrak{D}(\frac{g}{f})$.*

Proof. Let

$$\begin{aligned} D_0 &= \{D : f = 0\}, \\ D_1 &= \{D : f = \infty\} \cap \{D : g = \infty\}, \\ D_2 &= \{D : f = \infty\} \cap \{D : g = -\infty\}, \\ D_3 &= \{D : f = -\infty\} \cap \{D : g = \infty\}, \\ D_4 &= \{D : f = -\infty\} \cap \{D : g = -\infty\}. \end{aligned}$$

By (a) of Corollary 4.5, $D_i \in \mathfrak{A}$ for $i = 0, \ldots, 4$. Then $\mathfrak{D}(\frac{g}{f}) = D \setminus (\bigcup_{i=0}^{4} D_i) \in \mathfrak{A}$. Since $\frac{g}{f} = (\frac{1}{f}) \cdot g$, the $\mathfrak{A}$-measurability of $\frac{g}{f}$ on $\mathfrak{D}(\frac{g}{f})$ follows from Lemma 4.14 and Theorem 4.13. ∎

Theorem 4.16. *Let $(X, \mathfrak{A})$ be a measurable space and let f and g be two extended real-valued $\mathfrak{A}$-measurable functions on a set $D \in \mathfrak{A}$. Then*

(1) $$\{D : f = g\} \in \mathfrak{A},$$

(2) $$\{D : f < g\} \in \mathfrak{A},$$

(3) $$\{D : f \le g\} \in \mathfrak{A},$$

(4) $$\{D : f \ne g\} \in \mathfrak{A}.$$

Proof. To prove (1), note that the set $\{D : f = g\}$ is the union of the three disjoint sets $\{D : f = g = -\infty\}$, $\{D : f = g = \infty\}$, and $\{D : f = g \in \mathbb{R}\}$. Now by (a) of

Corollary 4.5, $\{D : f = g = -\infty\} = \{D : f = -\infty\} \cap \{D : g = -\infty\} \in \mathfrak{A}$. Similarly $\{D : f = g = \infty\} \in \mathfrak{A}$. Finally, since $f(x) = g(x) \in \mathbb{R}$ if and only if $f(x) - g(x) = 0$, $\{D : f = g \in \mathbb{R}\} = \{D : f - g = 0\} = \{\mathfrak{D}(f - g) : f - g = 0\} \in \mathfrak{A}$ by Theorem 4.11, Theorem 4.12 and (a) of Corollary 4.5. Thus $\{D : f = g\}$ is the union of three $\mathfrak{A}$-measurable sets and is therefore $\mathfrak{A}$-measurable.

To prove (2), let $\{r_n : n \in \mathbb{N}\}$ be the collection of all rational numbers in $\mathbb{R}$. Then we have $\{D : f < g\} = \bigcup_{n\in\mathbb{N}} \{D : f < r_n < g\} = \bigcup_{n\in\mathbb{N}} [\{D : f < r_n\} \cap \{D : r_n < g\}] \in \mathfrak{A}$.

Next by (1) and (2), we have $\{D : f \le g\} = \{D : f = g\} \cup \{D : f < g\} \in \mathfrak{A}$. This proves (3).

Finally we have $\{D : f \ne g\} = D \setminus \{D : f = g\} \in \mathfrak{A}$ by (1). This proves (4). ∎

[III] Equality Almost Everywhere

Definition 4.17. *Given a measure space $(X, \mathfrak{A}, \mu)$. We say that two extended real-valued $\mathfrak{A}$-measurable functions f and g defined on a set $D \in \mathfrak{A}$ are equal almost everywhere and we write $f = g$ a.e. on D if there exists a null set N in $(X, \mathfrak{A}, \mu)$ such that $N \subset D$ and $f(x) = g(x)$ for $x \in D \setminus N$.*

Remark 4.18. In the definition above, $f = g$ a.e. on D if $f = g$ outside of a null set N of the measure space contained in D. This does not exclude the possibility that $f(x) = g(x)$ for some, and indeed for every, $x \in N$.

If f and g are $\mathfrak{A}$-measurable on D, then the set $D_0 = \{D : f \ne g\}$ is $\mathfrak{A}$-measurable by Theorem 4.16. If $f = g$ a.e. on D, that is, $f = g$ outside of a null set N, then $D_0 \subset N$ and thus by the monotonicity of μ we have $\mu(D_0) = 0$.

Remark 4.19. Almost everywhere equality of two functions are defined with respect to a given measure μ on a given σ-algebra of subsets of a set X. If more than one measures on a given σ-algebra $\mathfrak{A}$ are under consideration, or more than one σ-algebras of subsets of X are under consideration, we say that $f = g$, μ-a.e. on D, or even $f = g$, $(\mathfrak{A}, \mu)$-a.e. on D to indicate the measure μ and the σ-algebra $\mathfrak{A}$ to which the null set belongs.

Observation 4.20. Let $(X, \mathfrak{A}, \mu)$ be a complete measure space.

(a) Every extended real-valued function f defined on a null set N in $(X, \mathfrak{A}, \mu)$ is $\mathfrak{A}$-measurable on N.

(b) Let f and g be two extended real-valued functions defined on a set $D \in \mathfrak{A}$ such that $f = g$ a.e. on D. If f is $\mathfrak{A}$-measurable on D then so is g.

Proof. 1. Let f be an arbitrary extended real-valued function defined on a null set N. Now $N \in \mathfrak{A}$. To show that f is $\mathfrak{A}$-measurable on N, we show that for every $\alpha \in \mathbb{R}$, the set $\{N : f \le \alpha\}$ is a $\mathfrak{A}$-measurable set. But $(X, \mathfrak{A}, \mu)$ is a complete measure space and this implies that every subset of the null set N is a $\mathfrak{A}$-measurable set. Thus our set $\{N : f \le \alpha\}$ is a $\mathfrak{A}$-measurable set.

2. Suppose $f = g$ a.e. on D. Then there exists a null set N in the measure space

$(X, \mathfrak{A}, \mu)$ such that $N \subset D$ and $f = g$ on $D \setminus N$. If f is $\mathfrak{A}$-measurable on D, then f is $\mathfrak{A}$-measurable on the $\mathfrak{A}$-measurable subset $D \setminus N$ of D by (a) of Lemma 4.7. Since $f = g$ on $D \setminus N$, g is $\mathfrak{A}$-measurable on $D \setminus N$. On the other hand since N is a null set in a complete measure space, g is $\mathfrak{A}$-measurable on N by (a). Thus g is $\mathfrak{A}$-measurable on $D \setminus N$ and on N and therefore $\mathfrak{A}$-measurable on $(D \setminus N) \cup N = D$ by (b) of Lemma 4.7. ∎

[IV] Sequence of Measurable Functions

Let $(f_n : n \in \mathbb{N})$ be a sequence of extended real-valued functions on a subset D of a set X. For each $x \in D$, $\big(f_n(x) : n \in \mathbb{N}\big)$ is a sequence of extended real numbers. Thus $\liminf_{n\to\infty} f_n(x) = \lim_{n\to\infty} \big\{\inf_{k\geq n} f_k(x)\big\}$ and $\limsup_{n\to\infty} f_n(x) = \lim_{n\to\infty} \big\{\sup_{k\geq n} f_k(x)\big\}$ always exist in $\overline{\mathbb{R}}$. $\lim_{n\to\infty} f_n(x)$ exists in $\overline{\mathbb{R}}$ if and only if $\liminf_{n\to\infty} f_n(x) = \limsup_{n\to\infty} f_n(x)$.

We say that $\big(f_n(x) : n \in \mathbb{N}\big)$ converges if $\lim_{n\to\infty} f_n(x)$ exists and $\lim_{n\to\infty} f_n(x) \in \mathbb{R}$.

We say that $(f_n : n \in \mathbb{N})$ is a monotone sequence if $\big(f_n(x) : n \in \mathbb{N}\big)$ is a monotone sequence of extended real numbers for every $x \in D$.

The functions $\min_{n=1,\dots,N} f_n$, $\max_{n=1,\dots,N} f_n$, $\inf_{n\in\mathbb{N}} f_n$, $\sup_{n\in\mathbb{N}} f_n$, $\liminf_{n\in\mathbb{N}} f_n$, $\limsup_{n\in\mathbb{N}} f_n$, and $\lim_{n\in\mathbb{N}} f_n$ are defined pointwise. Thus

$$\big(\min_{n=1,\dots,N} f_n\big)(x) = \min_{n=1,\dots,N} f_n(x),$$

$$\big(\max_{n=1,\dots,N} f_n\big)(x) = \max_{n=1,\dots,N} f_n(x),$$

$$\big(\inf_{n\in\mathbb{N}} f_n\big)(x) = \inf_{n\in\mathbb{N}} f_n(x),$$

$$\big(\sup_{n\in\mathbb{N}} f_n\big)(x) = \sup_{n\in\mathbb{N}} f_n(x),$$

$$\big(\liminf_{n\in\mathbb{N}} f_n\big)(x) = \liminf_{n\in\mathbb{N}} f_n(x),$$

$$\big(\limsup_{n\in\mathbb{N}} f_n\big)(x) = \limsup_{n\in\mathbb{N}} f_n(x),$$

$$\big(\lim_{n\to\infty} f_n\big)(x) = \lim_{n\to\infty} f_n(x).$$

Theorem 4.21. *Let $(X, \mathfrak{A})$ be a measurable space and let $\big(f_n : n \in \mathbb{N}\big)$ be a monotone sequence of extended real-valued $\mathfrak{A}$-measurable functions on a set $D \in \mathfrak{A}$. Then $\lim_{n\to\infty} f_n$ exists on D and is $\mathfrak{A}$-measurable on D.*

Proof. If $(f_n : n \in \mathbb{N})$ is a monotone sequence on D, then $\big(f_n(x) : n \in \mathbb{N}\big)$ is a monotone sequence of extended real numbers so that $\lim_{n\to\infty} f_n(x)$ exists in $\overline{\mathbb{R}}$ for every $x \in D$. Thus $\lim_{n\to\infty} f_n$ exists on D.

Let us show the $\mathfrak{A}$-measurability of $\lim_{n\to\infty} f_n$ on D. If $(f_n : n \in \mathbb{N})$ is an increasing sequence, then for any $\alpha \in \mathbb{R}$ we have $\lim_{n\to\infty} f_n(x) > \alpha$ if and only if $f_n(x) > \alpha$ for some $n \in \mathbb{N}$. Thus $\{D : \lim_{n\to\infty} f_n > \alpha\} = \bigcup_{n\in\mathbb{N}} \{D : f_n > \alpha\} \in \mathfrak{A}$. This shows that $\lim_{n\to\infty} f_n$ is $\mathfrak{A}$-measurable on D by (ii) of Lemma 4.4. If $(f_n : n \in \mathbb{N})$ is a decreasing sequence, then $(-f_n : n \in \mathbb{N})$ is an increasing sequence and thus $\lim_{n\to\infty}(-f_n)$ is $\mathfrak{A}$-measurable on D by our result above. But $\lim_{n\to\infty}(-f_n) = -\lim_{n\to\infty} f_n$. Thus $-\lim_{n\to\infty} f_n$ is $\mathfrak{A}$-measurable on D. Then $\lim_{n\to\infty} f_n$ is $\mathfrak{A}$-measurable on D by Theorem 4.11. ∎

In the proof above of the $\mathfrak{A}$-measurability of the limit function $\lim_{n\to\infty} f_n$ of an increasing sequence $(f_n : n \in \mathbb{N})$, we applied criterion (ii) of Lemma 4.4 for $\mathfrak{A}$-measurability. Let us comment that criterion (iii) of Lemma 4.4 for $\mathfrak{A}$-measurability is not applicable here. For an increasing sequence $(f_n : n \in \mathbb{N})$ and an arbitrary $\alpha \in \mathbb{R}$, we have $\lim_{n\to\infty} f_n(x) > \alpha$ if and only if $f_n(x) > \alpha$ for some $n \in \mathbb{N}$. However

$$\lim_{n\to\infty} f_n(x) \geq \alpha \not\Leftrightarrow f_n(x) \geq \alpha \text{ for some } n \in \mathbb{N}.$$

Consider for instance a strictly increasing sequence $\big(f_n(x) : n \in \mathbb{N}\big)$ such that $f_n(x) < \alpha$ for every $n \in \mathbb{N}$ and $\lim_{n\to\infty} f_n(x) = \alpha$. We have $\lim_{n\to\infty} f_n(x) \geq \alpha$ but $f_n(x) < \alpha$ for every $n \in \mathbb{N}$.

Theorem 4.22. *Let $(X, \mathfrak{A})$ be a measurable space and let $(f_n : n \in \mathbb{N})$ be a sequence of extended real-valued $\mathfrak{A}$-measurable functions on a set $D \in \mathfrak{A}$.*
(a) *The functions $\min_{n=1,\dots,N} f_n$, $\max_{n=1,\dots,N} f_n$, $\inf_{n\in\mathbb{N}} f_n$, $\sup_{n\in\mathbb{N}} f_n$, $\liminf_{n\in\mathbb{N}} f_n$, and $\limsup_{n\in\mathbb{N}} f_n$ are $\mathfrak{A}$-measurable on D.*
(b) *Let $D_e = \{D : \lim_{n\to\infty} f_n \in \overline{\mathbb{R}}\}$. Then $D_e \in \mathfrak{A}$ and $\lim_{n\to\infty} f_n$ is $\mathfrak{A}$-measurable on D_e.*

Proof. 1. Let us show the $\mathfrak{A}$-measurability of $\min_{n=1,\dots,N} f_n$ on D. Let $\alpha \in \mathbb{R}$ and $x \in D$. Then $\min\{f_1(x), \dots, f_N(x)\} < \alpha$ if and only if $f_n(x) < \alpha$ for some $n = 1, \dots, N$. Thus

$$\{D : \min_{n=1,\dots,N} f_n < \alpha\} = \bigcup_{n=1}^{N} \{D : f_n < \alpha\} \in \mathfrak{A}$$

by (iv) of Lemma 4.4. This proves the $\mathfrak{A}$-measurability of $\min_{n=1,\dots,N} f_n$ on D. Similarly $\max\{f_1(x), \dots, f_N(x)\} > \alpha$ if and only if $f_n(x) > \alpha$ for some $n = 1, \dots, N$ so that

$$\{D : \max_{n=1,\dots,N} f_n > \alpha\} = \bigcup_{n=1}^{N} \{D : f_n > \alpha\} \in \mathfrak{A}$$

by (ii) of Lemma 4.4. This proves the $\mathfrak{A}$-measurability of $\max_{n=1,\dots,N} f_n$ on D.

2. Let us show that $\inf_{n\in\mathbb{N}} f_n$ is $\mathfrak{A}$-measurable on D. Let $\alpha \in \mathbb{R}$ and $x \in D$. Now $\inf_{n\in\mathbb{N}} f_n(x)$ is the greatest lower bound of $\{f_n(x) : n \in \mathbb{N}\}$. If the greatest lower bound of

$\{f_n(x) : n \in \mathbb{N}\}$ is less than α then α is not a lower bound of $\{f_n(x) : n \in \mathbb{N}\}$ so that $f_n(x) < \alpha$ for some $n \in \mathbb{N}$. Conversely if $f_n(x) < \alpha$ for some $n \in \mathbb{N}$ then since the greatest lower bound of $\{f_n(x) : n \in \mathbb{N}\}$ does not exceed $f_n(x)$, it is less than α. Therefore we have $\inf_{n\in\mathbb{N}} f_n(x) < \alpha$ if and only if $f_n(x) < \alpha$ for some $n \in \mathbb{N}$. Thus we have

$$\{D : \inf_{n\in\mathbb{N}} f_n < \alpha\} = \bigcup_{n\in\mathbb{N}} \{D : f_n < \alpha\} \in \mathfrak{A}$$

by (iv) of Lemma 4.4. This proves the $\mathfrak{A}$-measurability of $\inf_{n\in\mathbb{N}} f_n$ on D.

3. To prove the $\mathfrak{A}$-measurability of $\sup_{n\in\mathbb{N}} f_n$ on D, let us note that for sequence of extended real numbers $\{f_n(x) : n \in \mathbb{N}\}$ for $x \in D$ we have $\sup_{n\in\mathbb{N}} f_n(x) = -\inf_{n\in\mathbb{N}}\big(-f_n(x)\big)$. Now the $\mathfrak{A}$-measurability of f_n on D implies that of $-f_n$ by Theorem 4.11. Then $\inf_{n\in\mathbb{N}}\big(-f_n\big)$ is $\mathfrak{A}$-measurable on D by our result above. Then $-\inf_{n\in\mathbb{N}}\big(-f_n(x)\big)$ is $\mathfrak{A}$-measurable on D by Theorem 4.11. This shows that $\sup_{n\in\mathbb{N}} f_n$ is $\mathfrak{A}$-measurable on D.

4. Recall that $\liminf_{n\to\infty} f_n = \lim_{n\to\infty}\{\inf_{k\geq n} f_k\}$ where $(\inf_{k\geq n} f_k : n \in \mathbb{N})$ is an increasing sequence. By our result above, $\inf_{k\geq n} f_k$ is $\mathfrak{A}$-measurable on D for every $n \in \mathbb{N}$. Then by Theorem 4.21, $\lim_{n\to\infty}\{\inf_{k\geq n} f_k\}$ is $\mathfrak{A}$-measurable on D. This proves the $\mathfrak{A}$-measurability of $\liminf_{n\to\infty} f_n$ on D. The $\mathfrak{A}$-measurability of $\limsup_{n\to\infty} f_n$ on D is proved likewise.

5. To prove (b), note that since $D_e = \{D : \liminf_{n\to\infty} f_n = \limsup_{n\to\infty} f_n\}$ and since $\liminf_{n\to\infty} f_n$ and $\limsup_{n\to\infty} f_n$ are two $\mathfrak{A}$-measurable functions on D, we have $D_e \in \mathfrak{A}$ by (1) of Theorem 4.16. The $\mathfrak{A}$-measurability of $\liminf_{n\to\infty} f_n$ on D implies its $\mathfrak{A}$-measurability on D_e by (a) of Lemma 4.7. Then since $\lim_{n\to\infty} f_n = \liminf_{n\to\infty} f_n$ on D_e, $\lim_{n\to\infty} f_n$ is $\mathfrak{A}$-measurable on D_e. ∎

Theorem 4.23. *Let $(X, \mathfrak{A})$ be a measurable space and let $(f_n : n \in \mathbb{N})$ be a sequence of extended real-valued $\mathfrak{A}$-measurable functions on a set $D \in \mathfrak{A}$. Let*

(1) $$D_e = \{D : \lim_{n\to\infty} f_n \in \overline{\mathbb{R}}\},$$

(2) $$D_c = \{D : \lim_{n\to\infty} f_n \in \mathbb{R}\},$$

(3) $$D_\infty = \{D : \lim_{n\to\infty} f_n = \infty\},$$

(4) $$D_{-\infty} = \{D : \lim_{n\to\infty} f_n = -\infty\},$$

(5) $$D_{ne} = \{D : \lim_{n\to\infty} f_n \text{ does not exist}\},$$

so that D_e and D_{ne} are disjoint and $D_e \cup D_{ne} = D$; D_c, D_∞, and $D_{-\infty}$ are disjoint and $D_c \cup D_\infty \cup D_{-\infty} = D_e$. Then $D_e, D_c, D_\infty, D_{-\infty}, D_{ne} \in \mathfrak{A}$ and $\lim_{n\to\infty} f_n$ is $\mathfrak{A}$-measurable on each of D_e, D_c, D_∞, and $D_{-\infty}$.

Proof. By (b) of Theorem 4.22, $D_e \in \mathfrak{A}$ and $\lim_{n\to\infty} f_n$ is $\mathfrak{A}$-measurable on D_e. Then $D_c \in \mathfrak{A}$ and $D_\infty, D_{-\infty} \in \mathfrak{A}$ by (b) and (a) of Corollary 4.5 respectively. By (a) of Lemma 4.7, $\lim_{n\to\infty} f_n$ is $\mathfrak{A}$-measurable on D_c, D_∞, and $D_{-\infty}$. Finally $D_{ne} = D \setminus D_e \in \mathfrak{A}$. ∎

Proposition 4.24. *Let $(X, \mathfrak{A})$ be a measurable space and let $(f_n : n \in \mathbb{N})$ be a sequence of extended real-valued $\mathfrak{A}$-measurable functions on a set $D \in \mathfrak{A}$. Let D_c be the subset of D on which the sequence converges and let D_∞ and $D_{-\infty}$ be the subsets of D on which the limit of the sequence is equal to ∞ and $-\infty$ respectively. Then*

$$(1) \qquad D_c = \bigcap_{m\in\mathbb{N}} \bigcup_{N\in\mathbb{N}} \bigcap_{p\in\mathbb{N}} \left\{D : |f_{N+p} - f_N| < \tfrac{1}{m}\right\},$$

$$(2) \qquad D_\infty = \bigcap_{m\in\mathbb{N}} \bigcup_{N\in\mathbb{N}} \bigcap_{p\in\mathbb{N}} \{D : f_{N+p} \geq m\},$$

$$(3) \qquad D_{-\infty} = \bigcap_{m\in\mathbb{N}} \bigcup_{N\in\mathbb{N}} \bigcap_{p\in\mathbb{N}} \{D : f_{N+p} \leq -m\}.$$

Proof. 1. A sequence of real numbers $(a_n : n \in \mathbb{N})$ converges if and only if it is a Cauchy sequence, that is, it satisfies the condition that for every $\varepsilon > 0$ there exists $N \in \mathbb{N}$ such that $|a_n - a_n'| < \varepsilon$ for $n, n' \geq N$. If a sequence of extended real numbers $(\alpha_n : n \in \mathbb{N})$ converges then $\alpha_n \in \mathbb{R}$ for all but finitely many $n \in \mathbb{N}$ so that there exists $N \in \mathbb{N}$ such that $\alpha_n - \alpha_n'$ is defined for $n, n' \geq N$. Thus a sequence of extended real numbers converges if and only if it is a Cauchy sequence, that is, for every $\varepsilon > 0$ there exists $N \in \mathbb{N}$ such that for $n, n' \geq N$ the difference $\alpha_n - \alpha_n'$ is defined and $|\alpha_n - \alpha_n'| < \varepsilon$.

Now for each $x \in D$ we have a sequence of extended real numbers $(f_n(x) : n \in \mathbb{N})$ and $(f_n(x) : n \in \mathbb{N})$ converges if and only if it is a Cauchy sequence. Thus we have

$$\begin{aligned}
&(f_n(x) : n \in \mathbb{N}) \text{ converges} \\
\Leftrightarrow &\forall \varepsilon > 0, \exists N \in \mathbb{N} \text{ such that } |f_n(x) - f_n'(x)| < \varepsilon \text{ for } n, n' \geq N \\
\Leftrightarrow &\forall \varepsilon > 0, \exists N \in \mathbb{N} \text{ such that } |f_{N+p}(x) - f_N(x)| < \varepsilon \text{ for all } p \in \mathbb{N} \\
\Leftrightarrow &\forall m \in \mathbb{N}, \exists N \in \mathbb{N} \text{ such that } |f_{N+p}(x) - f_N(x)| < \tfrac{1}{m} \text{ for all } p \in \mathbb{N}
\end{aligned}$$

The expression (1) for D_c follows from this.

(Let us note that the set $\left\{x \in D : |f_{N+p}(x) - f_N(x)| < \frac{1}{m}\right\}$ in (1) is the subset of D consisting of all points $x \in D$ at which $|f_{N+p}(x) - f_N(x)|$ exists and is less than $\frac{1}{m}$. It may be an empty set. This is the case when $f_{N+p}(x) = f_N(x) = \infty$ for every $x \in D$ so that $|f_{N+p}(x) - f_N(x)|$ does not exist for any $x \in D$. However if $(f_n(x) : n \in \mathbb{N})$ converges at some $x \in D$, then there exists $N \in \mathbb{N}$ such that $f_n(x) \in \mathbb{R}$ for all $n \geq N$ and $\left\{x \in D : |f_{N+p}(x) - f_N(x)| < \frac{1}{m}\right\} \neq \emptyset$ for all $p \in \mathbb{N}$.)

2. Now $\lim_{n\to\infty} f_n(x) = \infty$ if and only if for every $m \in \mathbb{N}$ there exists $N \in \mathbb{N}$ such that $f_n(x) \geq m$ for every $n \geq N$, that is, $f_{N+p}(x) \geq m$ for every $p \in \mathbb{N}$. From this we have (2). Similarly for (3). ∎

Notations. For two extended real-valued functions f and g defined on a set D, we write $f \wedge g = \min\{f, g\}$ and $f \vee g = \max\{f, g\}$, that is, $(f \wedge g)(x) = \min\{f(x), g(x)\}$ and $(f \vee g)(x) = \max\{f(x), g(x)\}$ for $x \in D$.

Definition 4.25. *Let f be an extended real-valued function on a set D. The positive part f^+, the negative part f^-, and the absolute $|f|$ of f are nonnegative extended real-valued functions on D defined for $x \in D$ by setting*

$$\begin{aligned} f^+(x) &= (f \vee 0)(x) = \max\{f(x), 0\}, \\ f^-(x) &= -(f \wedge 0)(x) = -\min\{f(x), 0\}, \\ |f|(x) &= |f(x)|. \end{aligned}$$

Note that at any $x \in D$, at least one of $f^+(x)$ and $f^-(x)$ is equal to 0 so that the difference $f^+(x) - f^-(x)$ is always defined and moreover

$$f^+(x) - f^-(x) = \max\{f(x), 0\} + \min\{f(x), 0\} = f(x).$$

We have also

$$f^+(x) + f^-(x) = \max\{f(x), 0\} - \min\{f(x), 0\} = |f(x)| = |f|(x).$$

Proposition 4.26. *Let f be an extended real-valued $\mathfrak{A}$-measurable function on a set $D \in \mathfrak{A}$ where $\mathfrak{A}$ is a σ-algebra of subsets of set X. Then f^+, f^-, and $|f|$ are $\mathfrak{A}$-measurable functions on D.*

Proof. The $\mathfrak{A}$-measurability of f^+ and f^- is from Theorem 4.22. The $\mathfrak{A}$-measurability of $|f|$ is from Theorem 4.12. ∎

[V] Continuity and Borel and Lebesgue Measurability of Functions on $\mathbb{R}$

An extended real-valued function f on a subset D of $\mathbb{R}$ is said to be continuous at $x_0 \in D$ if $f(x_0) \in \mathbb{R}$ and if for every $\varepsilon > 0$ there exists $\delta > 0$ such that $|f(x) - f(x_0)| < \varepsilon$ for every $x \in (x_0 - \delta, x_0 + \delta) \cap D$. We say that f is continuous on D if f is continuous at every $x \in D$. It follows then that if f is continuous on D, then its restriction to a subset D_0 of D is continuous on D_0.

Let D_1 and D_2 be two disjoint subsets of $\mathbb{R}$. Let f_1 and f_2 be continuous on D_1 and D_2 respectively. If we define a function f on $D_1 \cup D_2$ by setting

$$f(x) = \begin{cases} f_1(x) & \text{for } x \in D_1, \\ f_2(x) & \text{for } x \in D_2, \end{cases}$$

then f may not be continuous on $D_1 \cup D_2$ and in fact it may be discontinuous at every $x \in D_1 \cup D_2$. (Thus if we merge two continuous functions, the result may not be a continuous function. Compare this with (b) of Lemma 4.7 according to which the result of merging countably many measurable functions is a measurable function.)

Example. Let Q and P be respectively the set of all rational numbers and the set of all irrational numbers in $\mathbb{R}$. We have $Q \cap P = \emptyset$ and $Q \cup P = \mathbb{R}$. Let f_1 be defined on Q by setting $f_1(x) = 1$ for every $x \in Q$, and let f_2 be defined on P by setting $f_2(x) = 0$ for every $x \in P$. Since f_1 is constant on Q, it is continuous on Q. Similarly f_2 is continuous on P. If we define a function f on $Q \cup P = \mathbb{R}$ by setting

$$f(x) = \begin{cases} f_1(x) = 1 & \text{for } x \in Q, \\ f_2(x) = 0 & \text{for } x \in P, \end{cases}$$

then f is discontinuous at every $x \in \mathbb{R}$.

Since f_1 is constant on $Q \in \mathfrak{B}_{\mathbb{R}}$, f_1 is $\mathfrak{B}_{\mathbb{R}}$-measurable on Q by Observation 4.8 and similarly f_2 is $\mathfrak{B}_{\mathbb{R}}$-measurable on P. Then f is $\mathfrak{B}_{\mathbb{R}}$-measurable on $Q \cup P = \mathbb{R}$ by (b) of Lemma 4.7. Actually the $\mathfrak{B}_{\mathbb{R}}$-measurability of f on $\mathbb{R}$ can be shown directly by observing that for every $\alpha \in \mathbb{R}$ we have

$$\{x \in \mathbb{R} : f(x) \le \alpha\} = \begin{cases} \emptyset \in \mathfrak{B}_{\mathbb{R}} & \text{for } \alpha \in (-\infty, 0), \\ P \in \mathfrak{B}_{\mathbb{R}} & \text{for } \alpha \in [0, 1), \\ \mathbb{R} \in \mathfrak{B}_{\mathbb{R}} & \text{for } \alpha \in [1, \infty). \end{cases} \blacksquare$$

Recall that a real-valued function f on a subset D of $\mathbb{R}$ is continuous on D if and only if for every open set O in $\mathbb{R}$ we have $f^{-1}(O) = D \cap V$ where V is an open set in $\mathbb{R}$.

Theorem 4.27. *Let f be a real-valued continuous function on a subset D of $\mathbb{R}$.*
(a) *If $D \in \mathfrak{B}_{\mathbb{R}}$, then f is $\mathfrak{B}_{\mathbb{R}}$-measurable, and hence $\mathfrak{M}_L$-measurable also, on D.*
(b) *If $D \in \mathfrak{M}_L$, then f is $\mathfrak{M}_L$-measurable on D.*

Proof. 1. Suppose $D \in \mathfrak{B}_{\mathbb{R}}$. To show that f is $\mathfrak{B}_{\mathbb{R}}$-measurable on D, it suffices to show that $f^{-1}(O) \in \mathfrak{B}_{\mathbb{R}}$ for every open set O in $\mathbb{R}$ by (b) of Theorem 4.6. Now the continuity of f on D implies that $f^{-1}(O) = D \cap V$ where V is an open set in $\mathbb{R}$ so that $f^{-1}(O) \in \mathfrak{B}_{\mathbb{R}}$. Thus f is $\mathfrak{B}_{\mathbb{R}}$-measurable on D. Since $\mathfrak{B}_{\mathbb{R}} \subset \mathfrak{M}_L$, f is also $\mathfrak{M}_L$-measurable on D by (a) of Observation 4.2.

2. If $D \in \mathfrak{M}_L$, then $f^{-1}(O) = D \cap V \in \mathfrak{M}_L$ so that f is $\mathfrak{M}_L$-measurable on D by (b) of Theorem 4.6. ∎

Theorem 4.28. *Let f be a real-valued $\mathfrak{M}_L$-measurable function on a set $D \in \mathfrak{M}_L$ and let g be a real-valued $\mathfrak{B}_{\mathbb{R}}$-measurable function on a set $E \in \mathfrak{B}_{\mathbb{R}}$ such that $E \supset f(D)$. Then the composite function $g \circ f$ is a real-valued $\mathfrak{M}_L$-measurable function on D.*

Proof. If f is a real-valued $\mathfrak{M}_L$-measurable function on $D \in \mathfrak{M}_L$ then f is a $\mathfrak{M}_L/\mathfrak{B}_{\mathbb{R}}$-measurable mapping of D into $\mathbb{R}$ by Theorem 4.6. Similarly if g is a real-valued $\mathfrak{B}_{\mathbb{R}}$-measurable function on $E \in \mathfrak{B}_{\mathbb{R}}$ then g is a $\mathfrak{B}_{\mathbb{R}}/\mathfrak{B}_{\mathbb{R}}$-measurable mapping of E into $\mathbb{R}$ by Theorem 4.6. Since $f(D) \subset E$ the composite function $g \circ f$ is defined on D and then by Theorem 1.40 (Chain Rule of Measurable Mappings) the mapping $g \circ f$ is a $\mathfrak{M}_L/\mathfrak{B}_{\mathbb{R}}$-measurable mapping of D into $\mathbb{R}$, that is, the function $g \circ f$ is a real-valued $\mathfrak{M}_L$-measurable function on D. ∎

Corollary 4.29. *Let f be a real-valued $\mathfrak{M}_L$-measurable function on a set $D \in \mathfrak{M}_L$ and let g be a real-valued continuous function on a set $E \in \mathfrak{B}_{\mathbb{R}}$ such that $E \supset f(D)$. Then the composite function $g \circ f$ is a real-valued $\mathfrak{M}_L$-measurable function on D.*

Proof. A real-valued continuous function g on a set $E \in \mathfrak{B}_{\mathbb{R}}$ is a $\mathfrak{B}_{\mathbb{R}}$-measurable function on E by (a) of Theorem 4.27. Thus the Corollary is a particular case of Theorem 4.28. ∎

Definition 4.30. *Given a measure space $(\mathbb{R}, \mathfrak{A}, \mu)$. Let f be an extended real-valued function on a set $D \in \mathfrak{A}$. We say that f is continuous a.e. on D (or to be more precise, $(\mathfrak{A}, \mu)$-a.e. on D), if there exists a null set N in $(\mathbb{R}, \mathfrak{A}, \mu)$ such that $N \subset D$ and f is continuous at every $x \in D \setminus N$.*

Theorem 4.31. *Let f be an extended real-valued function on a set $D \in \mathfrak{M}_L$. If f is continuous $(\mathfrak{M}_L, \mu_L)$-a.e. on D, then f is $\mathfrak{M}_L$-measurable on D.*

Proof. Since f is continuous a.e. on D, there exists a null set N in $(\mathbb{R}, \mathfrak{M}_L, \mu_L)$ such that $N \subset D$ and f is continuous at every $x \in D \setminus N$. The restriction f_1 of f to $D \setminus N$ is continuous at every $x \in D \setminus N$ and thus $\mathfrak{M}_L$-measurable on $D \setminus N$ by (b) of Theorem 4.27. On the other hand the restriction f_2 of f to N is $\mathfrak{M}_L$-measurable on N since every function on a null set of a complete measure space is measurable on the null set by (a) of Observation 4.20. The $\mathfrak{M}_L$-measurability of f_1 on $D \setminus N$ and the $\mathfrak{M}_L$-measurability of f_2 on N imply the $\mathfrak{M}_L$-measurability of f on $(D \setminus N) \cup N = D$ by (b) of Lemma 4.7. ∎

In general if $D \in \mathfrak{B}_{\mathbb{R}}$ or $D \in \mathfrak{M}_L$ and f is a continuous function on D, the image of D by f, $f(D)$, may not be in $\mathfrak{B}_{\mathbb{R}}$ or $\mathfrak{M}_L$. However we have the following special case.

Proposition 4.32. *Let f be a real-valued function with both $\mathfrak{D}(f)$ and $\mathfrak{R}(f)$ in $\mathfrak{B}_{\mathbb{R}}$. Suppose f is a homeomorphism (that is, f is continuous and one-to-one and its inverse function is also continuous). Then for every $\mathfrak{B}_{\mathbb{R}}$-measurable subset B of $\mathfrak{D}(f)$, we have $f(B) \in \mathfrak{B}_{\mathbb{R}}$.*

Proof. Let g be the inverse function of f. Then we have $\mathfrak{D}(g) = \mathfrak{R}(f) \in \mathfrak{B}_{\mathbb{R}}$ and $\mathfrak{R}(g) = \mathfrak{D}(f) \in \mathfrak{B}_{\mathbb{R}}$. Since g is continuous on $\mathfrak{D}(g) \in \mathfrak{B}_{\mathbb{R}}$, g is a $\mathfrak{B}_{\mathbb{R}}$-measurable function on $\mathfrak{D}(g)$ by (a) of Theorem 4.27. Then g is a $\mathfrak{B}_{\mathbb{R}}/\mathfrak{B}_{\mathbb{R}}$-measurable mapping from $\mathbb{R}$ to $\mathbb{R}$. Let $B \subset \mathfrak{D}(f)$ and $B \in \mathfrak{B}_{\mathbb{R}}$. Since f and g are one-to-one, we have $f(B) = g^{-1}(B)$. Since g is a $\mathfrak{B}_{\mathbb{R}}/\mathfrak{B}_{\mathbb{R}}$-measurable mapping, we have $g^{-1}(B) \in \mathfrak{B}_{\mathbb{R}}$. This shows that $f(B) \in \mathfrak{B}_{\mathbb{R}}$. ∎

[VI] Cantor Ternary Set and Cantor-Lebesgue Function

[VI.1] Construction of Cantor Ternary Set

Let $T_0 = [0, 1]$.

Let $I_{1,1} = \left(\frac{1}{3}, \frac{2}{3}\right)$, the open middle third of T_0. Let $G_1 = I_{1,1}$ and let $T_1 = T_0 \setminus G_1$.

Let $I_{2,1}$ and $I_{2,2}$ be the open middle thirds of the two closed intervals constituting T_1. Let G_2 be the union of $I_{1,1}$; $I_{2,1}, I_{2,2}$ and let $T_2 = T_0 \setminus G_2$.

Let $I_{3,1}, I_{3,2}, I_{3,3}$, and $I_{3,4}$ be the open middle thirds of the four closed intervals constituting T_2. Let G_3 be the union of $I_{1,1}$; $I_{2,1}, I_{2,2}$; $I_{3,1}, I_{3,2}, I_{3,3}, I_{3,4}$ and $T_3 = T_0 \setminus G_3$, and so on. In general let $I_{k,1}, \dots, I_{k,2^{k-1}}$ be the open middle thirds of the 2^{k-1} closed intervals constituting T_{k-1}. Let G_k be the union of $I_{1,1}; \dots; I_{k,1}, \dots, I_{k,2^{k-1}}$ and $T_k = T_0 \setminus G_k$. G_k is the union of $2^0 + 2^1 + 2^2 + \cdots + 2^{k-1} = 2^k - 1$ disjoint open intervals contained in T_0 and $(G_k : k \in \mathbb{N})$ is an increasing sequence of open sets contained in T_0. T_k is the union of 2^k disjoint closed intervals contained in T_0 and $(T_k : k \in \mathbb{N})$ is a decreasing sequence of closed sets contained in T_0. Let

$$G = \lim_{k\to\infty} G_k = \bigcup_{k\in\mathbb{N}} G_k \quad \text{and} \quad T = T_0 \setminus G.$$

We call T the Cantor ternary set. Note that $T \cap G = \emptyset$, $T \cup G = [0, 1]$ and

$$T = T_0 \setminus G = T_0 \cap G^c = T_0 \cap \Big(\bigcup_{k\in\mathbb{N}} G_k\Big)^c = T_0 \cap \Big(\bigcap_{k\in\mathbb{N}} G_k^c\Big)$$

$$= \bigcap_{k\in\mathbb{N}} (T_0 \cap G_k^c) = \bigcap_{k\in\mathbb{N}} (T_0 \setminus G_k) = \bigcap_{k\in\mathbb{N}} T_k = \lim_{k\to\infty} T_k.$$

Theorem 4.33. *The Cantor ternary set T has the following properties:*

(a) *T is a null set in the Borel measure space $(\mathbb{R}, \mathfrak{B}_{\mathbb{R}}, \mu_L)$.*

(b) *$G = [0, 1] \setminus T$ is a union of countably many disjoint open intervals in $\mathbb{R}$; G is dense in $[0, 1]$, and $\mu_L(G) = 1$.*

(c) *T is an uncountable set. Indeed the cardinality of T is equal to $\mathfrak{c}$, the continuum.*

(d) *T is a compact set in $\mathbb{R}$.*

(e) *T is a perfect set in $\mathbb{R}$, that is, T is identical with the set of all its limit points.*

(f) *T is nowhere dense in $\mathbb{R}$, that is, the interior of its closure, $(\overline{T})^\circ$, is an empty set.*

Proof. 1. Since T is a closed set, $T \in \mathfrak{B}_{\mathbb{R}}$. Since $(T_k : n \in \mathbb{N})$ is a decreasing sequence contained in $[0, 1]$ and $\lim_{k\to\infty} T_k = T$, we have $\mu_L(T) = \mu_L\big(\lim_{k\to\infty} T_k\big) = \lim_{k\to\infty} \mu_L(T_k)$. Now $\mu_L(T_0) = 1$, $\mu_L(T_1) = \frac{2}{3}$, $\mu_L(T_2) = \left(\frac{2}{3}\right)^2$, and so on and in general $\mu_L(T_k) = \left(\frac{2}{3}\right)^k$ for $k \in \mathbb{N}$. Thus $\mu_L(T) = \lim_{k\to\infty} \left(\frac{2}{3}\right)^k = 0$. This shows that T is a null set in $(\mathbb{R}, \mathfrak{B}_{\mathbb{R}}, \mu_L)$.

2. $G = [0, 1] \setminus T$ is a union of countably many disjoint open intervals in $\mathbb{R}$. Being a null set in $(\mathbb{R}, \mathfrak{B}_{\mathbb{R}}, \mu_L)$ by (a), T is a null set in $(\mathbb{R}, \mathfrak{M}_L, \mu_L)$. Then by Observation 3.6, T^c is dense in $\mathbb{R}$. Since $G = [0, 1] \cap T^c$, the denseness of T^c in $\mathbb{R}$ implies the denseness of G in $[0, 1]$. Since $G \cup T = [0, 1]$ and $\mu_L(T) = 0$, the additivity of μ_L on $\mathfrak{B}_{\mathbb{R}}$ implies that $\mu_L(G) = \mu_L([0, 1]) - \mu_L(T) = 1$.

3. T is the result of indefinitely iterated process of deleting open intervals from $[0, 1]$. In the first step an open interval is deleted from $[0, 1]$, leaving 2 disjoint closed intervals. In the second step an open interval is deleted from each of the 2 disjoint closed intervals, leaving 2^2 disjoint closed intervals. In the third step an open interval is deleted from each of the 2^2 disjoint closed intervals, leaving 2^3 disjoint closed intervals and so on indefinitely. The two endpoints of each of the 2^k disjoint closed intervals in the k-th step of deletion are never deleted in subsequent steps of deletion and are thus elements of T. Thus the cardinality of T is at least equal to $2^{\aleph_0} = \mathfrak{c}$, the continuum. Since $T \subset [0, 1]$ and the cardinality of $[0, 1]$ is equal to $\mathfrak{c}$, the cardinality of T is equal to $\mathfrak{c}$.

4. T is a bounded closed set and is thus a compact set in $\mathbb{R}$.

5. Let T' be the set of all limit points of T. Since T is a closed set, we have $T \supset T'$. It remains to show that $T \subset T'$. To show that every $x_0 \in T$ is a limit point of T, we show that $(x_0 - \delta, x_0 + \delta)$ contains at least one point of T other than x_0 itself for every $\delta > 0$. Now since $x_0 \in T$ and $T = \bigcap_{k\in\mathbb{N}} T_k$, we have $x_0 \in T_k$ for every $k \in \mathbb{N}$. Given $\delta > 0$, let $k \in \mathbb{N}$ be so large that $\frac{1}{3^k} < \delta$. Since $x_0 \in T_k$ and since T_k is the disjoint union of 2^k closed intervals each with length $\frac{1}{3^k}$, the one closed interval among the 2^k constituting T that contains x_0 is contained in $(x_0 - \delta, x_0 + \delta)$. Thus $(x_0 - \delta, x_0 + \delta)$ contains points of T other than x_0. Thus $T \subset T'$. This shows that $T = T'$.

6. Since T is a closed set, we have $\overline{T} = T$ and then $(\overline{T})^\circ = T^\circ$. To show that $T^\circ = \emptyset$, assume the contrary. Then T° is a non-empty open set in $\mathbb{R}$ so that $\mu_L(T^\circ) > 0$ by Observation 3.10. This contradicts the fact that $\mu_L(T) = 0$. Therefore $T^\circ = \emptyset$ and then $(\overline{T})^\circ = \emptyset$. ∎

If I is an open interval in $\mathbb{R}$ then $\mu_L(I) > 0$. A set $E \in \mathfrak{M}_L$ with $\mu_L(E) > 0$ need not contain an open interval. The set of all irrational numbers in an interval is such a set. Below we present a closed set F in $\mathbb{R}$ with $\mu_L(F) > 0$ and containing no open interval. Its construction resembles that of the Cantor ternary set.

Example. Let $\alpha \in (0, 1)$. We have $\sum_{n\in\mathbb{N}} \frac{1}{2^n}\alpha = \alpha$. From the closed interval $[0, 1]$ delete an open interval in the center with length $\frac{1}{2}\alpha$. From each of the two resulting closed intervals delete an open interval in the center with length $\frac{1}{2}\frac{1}{2^2}\alpha$. From each of the resulting closed intervals delete an open interval in the center with length $\frac{1}{4}\frac{1}{2^3}\alpha$ and so on. If we let G be the union of all the deleted open intervals, then G is an open set contained in $[0, 1]$ and $\mu_L(G) = \frac{1}{2}\alpha + \frac{1}{2^2}\alpha + \frac{1}{2^3}\alpha + \cdots = \alpha$. The set $F = [0, 1] \setminus G$ is a closed set and $\mu_L(F) = 1 - \alpha > 0$. The fact that F contains no open intervals can be shown as follows. After n-th step in the process of deleting open intervals, we have deleted an open set G_n with $\mu_L(G_n) = \{\frac{1}{2} + \frac{1}{2^2} + \cdots + \frac{1}{2^n}\}\alpha = \{1 - \frac{1}{2^n}\}\alpha$. We are left with a closed set F_n consisting of 2^n disjoint closed intervals each with length $\frac{1}{2^n}[1 - \{1 - \frac{1}{2^n}\}\alpha] \le \frac{1}{2^n}$. Note that $F = [0, 1] \setminus G = [0, 1] \cap \left(\bigcup_{n\in\mathbb{N}} G_n\right)^c = [0, 1] \cap \left(\bigcap_{n\in\mathbb{N}} F_n\right) = \bigcap_{n\in\mathbb{N}} F_n$. Now let I be an open interval. Then $\beta := \mu_L(I) > 0$. Let $n \in \mathbb{N}$ be so large that $\frac{1}{2^n} < \beta$. Since F_n is the union of disjoint closed intervals of length less than $\frac{1}{2^n}$, I cannot be contained in F_n. Then I cannot be contained in $F = \bigcap_{n\in\mathbb{N}} F_n$.

[VI.2] Cantor-Lebesgue Function

The open set G, defined in the construction of the Cantor set, is the union of the disjoint open intervals: $I_{1,1}$; $I_{2,1}, I_{2,2}$; $I_{3,1}, I_{3,2}, I_{3,3}, I_{3,4}$; ...; $I_{k,1}, \dots, I_{k,2^{k-1}}$; ..., with length $\ell(I_{k,j}) = 1/3^k$ for $j = 1, \dots, 2^{k-1}$ and $k \in \mathbb{N}$. Let us define a real-valued function τ_0 on G by setting

$$\tau_0(x) = \begin{cases} \frac{1}{2} & \text{for } x \text{ in } I_{1,1}, \\ \frac{1}{2^2}, \frac{3}{2^2} & \text{for } x \text{ in } I_{2,1},\ I_{2,2} \text{ respectively,} \\ \frac{1}{2^3}, \frac{3}{2^3}, \frac{5}{2^3}, \frac{7}{2^3} & \text{for } x \text{ in } I_{3,1},\ I_{3,2},\ I_{3,3},\ I_{3,4} \text{ respectively,} \\ \vdots & \\ \frac{1}{2^k}, \dots, \frac{2^k-1}{2^k} & \text{for } x \text{ in } I_{k,1}, \dots, I_{k,2^{k-1}} \text{ respectively,} \\ \vdots & \end{cases}$$

Thus defined, τ_0 is an increasing function on G. If x' and x'' are two points in G and if the distance between the two is less than $\frac{1}{3^k}$, then the difference between $\tau(x')$ and $\tau(x'')$ does not exceed $\frac{1}{2^k}$. Thus for every $\varepsilon > 0$, if $k \in \mathbb{N}$ is so large that $\frac{1}{2^k} < \varepsilon$, then

$$x', x'' \in G,\ |x' - x''| < \tfrac{1}{3^k} \Rightarrow |\tau_0(x') - \tau_0(x'')| \le \tfrac{1}{2^k} < \varepsilon.$$

Thus τ_0 is uniformly continuous on G. This implies that τ_0 has a unique continuous extension to $\overline{G}$. Since G is dense in $[0, 1]$ we have $\overline{G} = [0, 1]$. Let τ be the continuous extension of τ_0 to $[0, 1]$. We call this function the Cantor-Lebesgue function on $[0, 1]$.

Theorem 4.34. *The Cantor-Lebesgue function* τ *on* $[0, 1]$ *has the following properties:*
(a) τ *is continuous on* $[0, 1]$.
(b) τ *is increasing on* $[0, 1]$.
(c) $\tau(0) = 0$ *and* $\tau(1) = 1$.

Proof. By its definition, τ is continuous on $[0, 1]$. To show that τ is increasing on $[0, 1]$, let $x', x'' \in [0, 1]$ and $x' < x''$. Now since G is dense in $[0, 1]$, we can select two sequences $(a_n : n \in \mathbb{N})$ and $(b_n : n \in \mathbb{N})$ in G such that $a_n < b_n$ for every $n \in \mathbb{N}$ and $a_n \downarrow x'$ and $b_n \uparrow x''$. By the continuity of τ on $[0, 1]$, we have $\lim_{n\to\infty} \tau(a_n) = \tau(x')$ and $\lim_{n\to\infty} \tau(b_n) = \tau(x'')$. Since τ is increasing on G and since $a_n, b_n \in G$ and $a_n < b_n$, we have $\tau(a_n) \le \tau(b_n)$ for $n \in \mathbb{N}$. Thus $\lim_{n\to\infty} \tau(a_n) \le \lim_{n\to\infty} \tau(b_n)$, that is, $\tau(x') \le \tau(x'')$. This shows that τ is increasing on $[0, 1]$.

To show that $\tau(0) = 0$, consider the following sequence of intervals in the construction of T: $I_{1,1} = \left(\frac{1}{3}, \frac{2}{3}\right)$, $I_{2,1} = \left(\frac{1}{3^2}, \frac{2}{3^2}\right)$, $I_{3,1} = \left(\frac{1}{3^3}, \frac{2}{3^3}\right), \dots, I_{k,1} = \left(\frac{1}{3^k}, \frac{2}{3^k}\right), \dots$ Let x_k be the midpoint of $I_{k,1}$, that is, $x_k = \frac{3}{2}\frac{1}{3^k}$, for $k \in \mathbb{N}$. Then $\lim_{n\to\infty} x_k = 0$. From our definition of τ on G, we have $\tau(x_k) = \frac{1}{2^k}$. Then by the continuity of τ at $x = 0$, we have $\tau(0) = \tau\left(\lim_{k\to\infty} x_k\right) = \lim_{k\to\infty} \tau(x_k) = \lim_{k\to\infty} \frac{1}{2^k} = 0$. We show similarly that $\tau(1) = 1$. ∎

We use the Cantor ternary set and the Cantor-Lebesgue function to construct :
(a) a strictly increasing continuous function transforming a null set with Lebesgue

measure 0 into a set with positive Lebesgue measure, (Proposition 4.37),

(b) a continuous function transforming a Lebesgue measurable set into a non Lebesgue measurable set, (Proposition 4.38),

(c) a $\mathfrak{M}_L$-measurable function f and a $\mathfrak{M}_L$-measurable set E such that $f^{-1}(E)$ is not $\mathfrak{M}_L$-measurable, (Proposition 4.39),

(d) a Lebesgue measurable set which is not Borel measurable, (Proposition 4.40).

Proposition 4.35. *Let X be an arbitrary topological space and Y be a Hausdorff space. Let f be a continuous mapping of a compact set K in X into Y. If f is a one-to-one mapping, then f is a homeomorphism of K to $f(K)$.*

Proof. The compactness of K and the continuity of f imply that $f(K)$ is a compact set in Y. Consider the relative topology on K derived from the topology of X and the relative topology on $f(K)$ derived from the topology of Y. If C is a closed set in K then C is a compact set in K so that $f(C)$ is a compact set in $f(K)$. Since Y is a Hausdorff space, $f(K)$ is a Hausdorff space and this implies that the compact set $f(C)$ in $f(K)$ is a closed set in $f(K)$. Thus $f(C)$ is a closed set in $f(K)$ for every closed set C in K. This implies that $f(V)$ is an open set in $f(K)$ for every open set V in K. Now since f is a one-to-one mapping, an inverse mapping g exists. For every open set V in K, $g^{-1}(V) = f(V)$ which is an open set in $f(K)$. This shows that g is a continuous mapping of $f(X)$ to K. Therefore f is a homeomorphism of K to $f(K)$. ∎

Lemma 4.36. *Let $\varphi = \tau + \iota$ where τ is the Cantor-Lebesgue function on $[0, 1]$ and ι is the identity mapping of $[0, 1]$, that is, $\iota(x) = x$ for $x \in [0, 1]$. Then φ is a homeomorphism of $[0, 1]$ onto $[0, 2]$. Furthermore both φ and its inverse function ψ are strictly increasing functions on their respective domains of definition.*

Proof. Since both τ and ι are real-valued, continuous, and increasing on $[0, 1]$, so is $\varphi = \tau + \iota$. Since $\tau(0) = 0$, $\tau(1) = 1$, $\iota(0) = 0$, and $\iota(1) = 1$, we have $\varphi(0) = 0$ and $\varphi(1) = 2$. By the Intermediate Value Theorem for continuous functions, for every $\alpha \in (0, 2)$ there exists $x \in (0, 1)$ such that $\varphi(x) = \alpha$. Thus $\varphi\big([0, 1]\big) = [0, 2]$.

Since τ is increasing and ι is strictly increasing on $[0, 1]$, $\varphi = \tau + \iota$ is strictly increasing on $[0, 1]$. Thus φ maps $[0, 1]$ one-to-one onto $[0, 2]$. Then by Proposition 4.35, φ is a homeomorphism of $[0, 1]$ to $[0, 2]$. The inverse function ψ of φ is defined on $[0, 2]$ and maps $[0, 2]$ one-to-one onto $[0, 1]$. Since φ is increasing on $[0, 1]$, ψ is increasing on $[0, 2]$. Since ψ is one-to-one, it is strictly increasing. ∎

Proposition 4.37. *There exists a strictly increasing continuous function f on $[0, 1]$ and a compact set $K \subset [0, 1]$ such that K is a null set in $(\mathbb{R}, \mathfrak{B}_{\mathbb{R}}, \mu_L)$ and $\mu_L\big(f(K)\big) = 1$.*

Proof. We show that the function $\varphi = \tau + \iota$ in Lemma 4.36 and the Cantor ternary set T are such function and set. Let G be the open set defined in the construction of T. G is the union of countably many disjoint open intervals. Let $(J_n : n \in \mathbb{N})$ be an enumeration of the open intervals constituting G. Let c_n be the value of the Cantor-Lebesgue function on J_n for $n \in \mathbb{N}$. Then $\varphi(J_n) = (\iota + \tau)(J_n) = J_n + c_n$, the c_n-translate of J_n. Thus $\varphi(G) = \varphi\left(\bigcup_{n \in \mathbb{N}} J_n\right) = \bigcup_{n \in \mathbb{N}}(J_n + c_n) \in \mathfrak{B}_{\mathbb{R}}$. For $m \neq n$, if the open interval J_n is to the left of the open interval J_m, then $c_n < c_m$ since τ is an increasing function. Thus $J_n + c_n$ and $J_m + c_m$ remain disjoint. Then by the countable additivity of μ_L and by the translation invariance of $(\mathbb{R}, \mathfrak{M}_L, \mu_L)$, we have

$$\mu_L\big(\varphi(G)\big) = \sum_{n\in\mathbb{N}} \mu_L(J_n + c_n) = \sum_{n\in\mathbb{N}} \mu_L(J_n) = \mu_L(G) = 1.$$

Now

$$\varphi(T) = \varphi\big([0,1] \setminus G\big) = \varphi\big([0,1]\big) \setminus \varphi(G) = [0,2] \setminus \varphi(G) \in \mathfrak{B}_{\mathbb{R}}$$

and hence $\mu_L\big(\varphi(T)\big) = \mu_L\big([0,2]\big) - \mu_L\big(\varphi(G)\big) = 2 - 1 = 1$. Thus for the Cantor ternary set T, which is a compact set in $\mathbb{R}$ and a null set in $(\mathbb{R}, \mathfrak{B}_{\mathbb{R}}, \mu_L)$ and for the strictly increasing continuous function φ on $[0,1]$, we have $\mu_L\big(\varphi(T)\big) = 1$. ∎

Proposition 4.38. *There exist a set $E \in \mathfrak{M}_L$ and a continuous function f on an interval containing E such that $f(E) \notin \mathfrak{M}_L$.*

Proof. Let T be the Cantor ternary set and let the function φ and ψ be as in Lemma 4.36. We showed in the Proof of Proposition 4.37 that $\varphi(T) \in \mathfrak{B}_{\mathbb{R}}$ and $\mu_L\big(\varphi(T)\big) > 0$. Then by Theorem 3.45, there exists a set $A \subset \varphi(T)$ such that $A \notin \mathfrak{M}_L$. Now $\psi(A) \subset \psi\big(\varphi(T)\big) = T$. Since T is a null set in the complete measure space $\big(\mathbb{R}, \mathfrak{M}_L, \mu_L\big)$, its subset $\psi(A)$ is a null set in $\big(\mathbb{R}, \mathfrak{M}_L, \mu_L\big)$. Let $E = \psi(A)$ and $f = \varphi$. Then $E \subset [0,1]$, $E \in \mathfrak{M}_L$, f is a continuous function on $[0,1]$ and $f(E) = (\varphi \circ \psi)(A) = A \notin \mathfrak{M}_L$. ∎

If f is a real-valued $\mathfrak{M}_L$-measurable function on a set $D \in \mathfrak{M}_L$, then f is a $\mathfrak{M}_L/\mathfrak{B}_{\mathbb{R}}$-measurable mapping by (a) of Theorem 4.6 so that $f^{-1}(B) \in \mathfrak{M}_L$ for every $B \in \mathfrak{B}_{\mathbb{R}}$. However for $E \in \mathfrak{M}_L$, we may not have $f^{-1}(E) \in \mathfrak{M}_L$. An example is given in the next Proposition.

Proposition 4.39. *There exists a real-valued $\mathfrak{M}_L$-measurable function f for which we have $f^{-1}(E) \notin \mathfrak{M}_L$ for some $E \in \mathfrak{M}_L$.*

Proof. Let T, A, φ and ψ be as in Proposition 4.38. We have $A \subset \varphi(T)$, $A \notin \mathfrak{M}_L$ and $\psi(A) \in \mathfrak{M}_L$. Now ψ is a real-valued continuous function on $[0,2]$ and hence a $\mathfrak{M}_L$-measurable function on $[0,2]$. Since ψ is a one-to-one mapping, we have $\psi^{-1}\big(\psi(A)\big) = A \notin \mathfrak{M}_L$. Let $f = \psi$ and $E = \psi(A)$. Then f is a real-valued $\mathfrak{M}_L$-measurable function on $[0,2]$, $E \in \mathfrak{M}_L$ and $f^{-1}(E) = \psi^{-1}\big(\psi(A)\big) = A \notin \mathfrak{M}_L$. ∎

Proposition 4.40. *There exists $E \in \mathfrak{M}_L$ such that $E \notin \mathfrak{B}_{\mathbb{R}}$.*

Proof. Let T, A, φ and ψ be as in Proposition 4.38. We have $A \subset \varphi(T)$, $A \notin \mathfrak{M}_L$ and $\psi(A) \in \mathfrak{M}_L$. Let us show that $\psi(A) \notin \mathfrak{B}_{\mathbb{R}}$. Now since φ is a homeomorphism of $[0,1]$ onto $[0,2]$, for every $\mathfrak{B}_{\mathbb{R}}$-measurable subset B of $[0,1]$ we have $\varphi(B) \in \mathfrak{B}_{\mathbb{R}}$ by Proposition 4.32. Thus if $\psi(A) \in \mathfrak{B}_{\mathbb{R}}$, then $\varphi\big(\psi(A)\big) \in \mathfrak{B}_{\mathbb{R}}$. But $\varphi\big(\psi(A)\big) = A \notin \mathfrak{M}_L$ and consequently $\varphi\big(\psi(A)\big) \notin \mathfrak{B}_{\mathbb{R}}$. This shows that $\psi(A) \notin \mathfrak{B}_{\mathbb{R}}$. ∎

Problems

Prob. 4.1. Given a measurable space $(X, \mathfrak{A})$. Let f be an extended real-valued function on a set $D \in \mathfrak{A}$. Let Q be the collection of all rational numbers.
(a) Show that if $\{x \in D : f(x) < r\} \in \mathfrak{A}$ for every $r \in Q$ then f is $\mathfrak{A}$-measurable on D.
(b) What subsets of $\mathbb{R}$ other than Q have this property?
(c) Show that if f is $\mathfrak{A}$-measurable on D, then there exists a countable subcollection $\mathfrak{C}$ of $\mathfrak{A}$, depending on f, such that f is $\sigma(\mathfrak{C})$-measurable on D.

Prob. 4.2. Let $\mathfrak{E}$ be a collection of subsets of a set X. Consider $\sigma(\mathfrak{E})$, the smallest σ-algebra of subsets of X containing $\mathfrak{E}$. Let f be an extended real-valued $\sigma(\mathfrak{E})$-measurable function on X. Show that there exists a countable subcollection $\mathfrak{C}$ of $\mathfrak{E}$ such that f is $\sigma(\mathfrak{C})$-measurable on X.

Prob. 4.3. Show that the following functions defined on $\mathbb{R}$ are all Borel measurable, and hence Lebesgue measurable also, on $\mathbb{R}$:
(a) $f(x) = 0$ if x is rational and $f(x) = 1$ if x is irrational.
(b) $f(x) = x$ if x is rational and $f(x) = -x$ if x is irrational.
(c) $f(x) = \sin x$ if x is rational and $f(x) = \cos x$ if x is irrational.

Prob. 4.4. Let $f(x)$ be a real-valued increasing function on $\mathbb{R}$. Show that f is Borel measurable, and hence Lebesgue measurable also, on its domain of definition.

Prob. 4.5. Let f be an extended real-valued Borel measurable function on a set $D \in \mathfrak{B}_{\mathbb{R}}$. Show that if we redefine f arbitrarily on a countable subset D_0 of D the resulting function is still Borel measurable on D.

Prob. 4.6. Let D be a countable subset of $\mathbb{R}$. Show that every extended real-valued function f on D is Borel measurable on D.

Prob. 4.7. Let f be an extended real-valued function defined on an interval I in $\mathbb{R}$. Let Q be an arbitrary countable subset of $\mathbb{R}$. Show that if f is continuous at every $x \in I \setminus Q$, then f is Borel measurable on I.

Note that the countable set Q may be dense in I. For instance Q may be the set of all rational numbers in I. Below are some examples of functions of the type being considered.

(a) Let $(c_k : k \in \mathbb{N})$ be an arbitrary sequence of real numbers. Let f be a real-valued function on $\mathbb{R}$ defined by

$$f(x) = \begin{cases} \tan x & \text{for } x \neq (2k+1)\frac{\pi}{2} \text{ for } k \in \mathbb{Z}, \\ c_k & \text{for } x = (2k+1)\frac{\pi}{2} \text{ for } k \in \mathbb{Z}. \end{cases}$$

(b) f is a right-continuous step function on $\mathbb{R}$ defined as follows. Let $(x_k : k \in \mathbb{Z})$ be a sequence in $\mathbb{R}$ such that $x_k < x_{k+1}$ for $k \in \mathbb{Z}$ with $\lim_{k \to -\infty} x_k = -\infty$ and $\lim_{k \to \infty} x_k = \infty$. Let $(c_k : k \in \mathbb{Z})$ be a sequence of real numbers. Define f by $f(x) = c_k$ for $x \in [x_k, x_{k+1})$ for $k \in \mathbb{Z}$.
(c) Let Q be the collection of all rational numbers in $(0, \infty)$. Let each $x \in Q$ be expressed as $\frac{q}{p}$ where p and q are positive integers without common factors. Define a function f on $(0, \infty)$ by setting

$$f(x) = \begin{cases} \frac{1}{p} & \text{if } x \in Q \text{ and } x = \frac{q}{p}, \\ 0 & \text{if } x \in (0, \infty) \setminus Q. \end{cases}$$

Show that f is discontinuous at every rational $x \in (0, \infty)$ and continuous at every irrational $x \in (0, \infty)$.

Prob. 4.8. Let f be an extended real-valued $\mathfrak{M}_L$-measurable function on a set $D \in \mathfrak{M}_L$. Show that $\left(\sqrt{|f|}\right)$ is a $\mathfrak{M}_L$-measurable function on D.

Prob. 4.9. Consider the Lebesgue measure space $(\mathbb{R}, \mathfrak{M}_L, \mu_L)$. Let f be an extended real-valued function defined on $\mathbb{R}$. Consider the following two concepts:
(i) f is continuous a.e. on $\mathbb{R}$, that is, there exists a null set N in $(\mathbb{R}, \mathfrak{M}_L, \mu_L)$ such that f is continuous at every $x \in \mathbb{R} \setminus N$,
(ii) f is equal to a continuous function a.e. on $\mathbb{R}$, that is, there exist a continuous function g on $\mathbb{R}$ and a null set N in $(\mathbb{R}, \mathfrak{M}_L, \mu_L)$ such that $f(x) = g(x)$ for every $x \in \mathbb{R} \setminus N$.
To show that these are two different concepts:
(a) Construct a function which satisfies condition (ii) but not (i),
(b) Construct a function which satisfies condition (i) but not (ii).

Prob. 4.10. Let $(f_n : n \in \mathbb{N})$ and f be extended real-valued functions on a set D. Show that if $\lim_{n\to\infty} f_n(x) = f(x)$ at some $x \in D$, then we have both $\lim_{n\to\infty} f_n^+(x) = f^+(x)$ and $\lim_{n\to\infty} f_n^-(x) = f^-(x)$.

Prob. 4.11. Let f and g be two real-valued functions on $\mathbb{R}$.
(a) Show that if f and g are $\mathfrak{B}_{\mathbb{R}}$-measurable on $\mathbb{R}$, then so is $g \circ f$.
(b) Show that if f is $\mathfrak{M}_L$-measurable on $\mathbb{R}$ and g is $\mathfrak{B}_{\mathbb{R}}$-measurable on $\mathbb{R}$, then $g \circ f$ is $\mathfrak{M}_L$-measurable on $\mathbb{R}$.

Prob. 4.12. Let X be a set and let $(Y, \mathfrak{G})$ be a measurable space. Let T be a mapping from X to Y. Show that $T^{-1}(\mathfrak{G})$ is a σ-algebra of subsets of the set $\mathfrak{D}(T)$, and in particular, if $\mathfrak{D}(T) = X$ then $T^{-1}(\mathfrak{G})$ is a σ-algebra of subsets of the set X.

Prob. 4.13. Given two topological spaces $(X, \mathfrak{O}_X)$ and $(Y, \mathfrak{O}_Y)$. Let $\mathfrak{B}_X = \sigma(\mathfrak{O}_X)$, that is, the Borel σ-algebra in X, and similarly let $\mathfrak{B}_Y = \sigma(\mathfrak{O}_Y)$. Show that if T is a homeomorphism of $(X, \mathfrak{O}_X)$ onto $(Y, \mathfrak{O}_Y)$, then $T^{-1}(\mathfrak{B}_Y) = \mathfrak{B}_X$ and $T(\mathfrak{B}_X) = \mathfrak{B}_Y$.
(Hint: Show first that the fact that $\mathfrak{B}_Y$ is a σ-algebra of subsets of Y implies that $T^{-1}(\mathfrak{B}_Y)$ is a σ-algebra of subsets of X.)

Prologue to Prob. 4.14 to 4.19. Note that the infimum and the supremum of uncountably many $\mathfrak{M}_L$-measurable functions may not be $\mathfrak{M}_L$-measurable. To construct an example, let E be a non $\mathfrak{M}_L$-measurable set contained in $(0, 1)$ and define a real-valued function F on $(0, 1) \times (0, 1)$ by

$$F(x, y) = \begin{cases} 1 & \text{for } (x, y) \in (0, 1) \times (0, 1) \text{ with } x = y \in E, \\ 0 & \text{otherwise on } (0, 1) \times (0, 1). \end{cases}$$

For each $y \in (0, 1)$, define a function f_y on $(0, 1)$ by setting $f_y(x) = F(x, y)$ for $x \in (0, 1)$. Consider the uncountable collection of functions $\{f_y : y \in (0, 1)\}$. For each $y \in E$,

$$f_y(x) = \begin{cases} 1 & \text{when } x = y, \\ 0 & \text{when } x \in (0,1) \setminus \{y\}. \end{cases}$$

and for each $y \in (0,1) \setminus E$, we have

$$f_y(x) = 0 \text{ when } x \in (0,1).$$

Thus f_y is a $\mathfrak{M}_L$-measurable function on $(0,1)$ for each $y \in (0,1)$. But

$$\sup_{y\in(0,1)} f_y(x) = \begin{cases} 1 & \text{when } x \in E, \\ 0 & \text{when } x \in (0,1) \setminus E. \end{cases}$$

Since $E \notin \mathfrak{M}_L$, the function $\sup_{y\in(0,1)} f_y$ is not $\mathfrak{M}_L$-measurable on $(0,1)$.

Prob. 4.14. Let g be a continuous function on an interval I in $\mathbb{R}$. Let $(r_n : n \in \mathbb{N})$ be the collection of the rational numbers in I. Show that

$$\inf\{g(y) : y \in I\} = \inf\{g(r_n) : n \in \mathbb{N}\},$$

$$\sup\{g(y) : y \in I\} = \sup\{g(r_n) : n \in \mathbb{N}\}.$$

Prob. 4.15. Let $D \in \mathfrak{M}_L$ and let F be a real-valued function on $D \times (c,d)$ satisfying the conditions:
1°. For each $y \in (c,d)$, $F(\cdot, y)$ is a $\mathfrak{M}_L$-measurable function on D.
2°. For each $x \in D$, $F(x, \cdot)$ is a continuous function on (c,d).
Show that φ and ψ defined by $\varphi(x) = \inf_{y\in(c,d)} F(x,y)$ and $\psi(x) = \sup_{y\in(c,d)} F(x,y)$ for $x \in D$ are $\mathfrak{M}_L$-measurable on D.

Prob. 4.16. Let F be as in Prob. 4.15. Show that for if $y_0 = c$ or d, then $\liminf_{y\to y_0} F(\cdot, y)$ and $\limsup_{y\to y_0} F(\cdot, y)$ are $\mathfrak{M}_L$-measurable functions on D.

Prob. 4.17. Let F be as in Prob. 4.15. Show that if $y_0 = c$ or d, the subset of D on which the function $\lim_{y\to y_0} F(\cdot, y)$ exists is a $\mathfrak{M}_L$-measurable set and the function $\lim_{y\to y_0} F(\cdot, y)$ is $\mathfrak{M}_L$-measurable on the subset.

Prob. 4.18. Let f be a real-valued continuous function on $\mathbb{R}$. For $x \in \mathbb{R}$, let

$$\left(\tfrac{df}{dx}\right)(x) = \lim_{h\to 0} \tfrac{f(x+h)-f(x)}{h}$$

if the limit exists in $\overline{\mathbb{R}}$. (We say the f is differentiable at $x \in \mathbb{R}$ only if $\left(\frac{df}{dx}\right)(x) \in \mathbb{R}$.) Show that $\mathfrak{D}\left(\frac{df}{dx}\right) \in \mathfrak{B}_{\mathbb{R}}$ and $\frac{df}{dx}$ is $\mathfrak{B}_{\mathbb{R}}$-measurable on $\mathfrak{D}\left(\frac{df}{dx}\right)$.

Prob. 4.19. Let F be a real-valued continuous function on $[a,b] \times [c,d]$. Show that the function f defined on $[a,b]$ by $f(x) = \int_c^d F(x,y)\,dy$ for $x \in [a,b]$ is a $\mathfrak{M}_L$-measurable function on $[a,b]$.

Prob. 4.20. Let $(X, \mathfrak{A}, \mu)$ be a measure space and let f be an extended real-valued $\mathfrak{A}$-measurable function on X. Let $E_\lambda = \{X : |f| > \lambda\}$ and $\varphi(\lambda) = \mu(E_\lambda)$ for $\lambda \in [0,\infty)$.
(a) Show that φ is a nonnegative extended real-valued decreasing function on $[0,\infty)$.
(b) Show that if $\varphi(\lambda_0) < \infty$ for some $\lambda_0 \in [0,\infty)$, then φ is right-continuous on $[\lambda_0, \infty)$.

Prob. 4.21. Let $(X, \mathfrak{A}, \mu)$ be a measure space. Let $(f_n : n \in \mathbb{N})$ be an increasing sequence of nonnegative extended real-valued $\mathfrak{A}$-measurable functions on X and let $f = \lim_{n\to\infty} f_n$

on X. For $\lambda \in [0, \infty)$ let $E_n = \{X : f_n > \lambda\}$ for $n \in \mathbb{N}$ and let $E = \{X : f > \lambda\}$. Show that $\lim_{n\to\infty} E_n = E$ and $\lim_{n\to\infty} \mu(E_n) = \mu(E)$.

Prob. 4.22. Let us call a sequence $(a_n : n \in \mathbb{N})$ of extended real numbers eventually monotone if there exists $n_0 \in \mathbb{N}$ such that $(a_n : n > n_0)$ is monotone, that is, either increasing or decreasing.

Let $(X, \mathfrak{A}, \mu)$ be a measure space. Let $(f_n : n \in \mathbb{N})$ be a sequence of extended real-valued $\mathfrak{A}$-measurable functions on X such that $(f_n(x) : n \in \mathbb{N})$ is eventually monotone for every $x \in X$. Let $f = \lim_{n\to\infty} f_n$ on X.

For $\lambda \in [0, \infty)$ let $E_n = \{X : |f_n| > \lambda\}$ for $n \in \mathbb{N}$ and let $E = \{X : |f| > \lambda\}$.

(a) Show that $\lim_{n\to\infty} E_n = E$.

(b) Show that if $(X, \mathfrak{A}, \mu)$ is a finite measure space then $\lim_{n\to\infty} \mu(E_n) = \mu(E)$.

Prob. 4.23. Let $(X, \mathfrak{A}, \mu)$ be a measure space. Let $(f_n : n \in \mathbb{N})$ and f be extended real-valued $\mathfrak{A}$-measurable functions on a set $D \in \mathfrak{A}$ such that $\lim_{n\to\infty} f_n = f$ on D. Then for every $\alpha \in \mathbb{R}$ we have

(1) $$\mu\{D : f > \alpha\} \le \liminf_{n\to\infty} \mu\{D : f_n \ge \alpha\}$$

(2) $$\mu\{D : f < \alpha\} \le \liminf_{n\to\infty} \mu\{D : f_n \le \alpha\}.$$

§5 Completion of Measure Space

[I] Complete Extension and Completion of a Measure Space

By Definition 1.34 a measure space $(X, \mathfrak{A}, \mu)$ is called a complete measure space if every subset of a null set in $(X, \mathfrak{A}, \mu)$ is a member of $\mathfrak{A}$. According to Theorem 2.9, a measure space constructed by means of an outer measure is always a complete measure space. In particular the Lebesgue measure space $(\mathbb{R}, \mathfrak{M}_L, \mu_L)$ on $\mathbb{R}$ is a complete measure space.

Definition 5.1. *Given a measure space $(X, \mathfrak{A}, \mu)$. If there exists a complete measure space $(X, \mathfrak{F}, \nu)$ such that $\mathfrak{F} \supset \mathfrak{A}$ and $\nu = \mu$ on $\mathfrak{A}$, then we say that $(X, \mathfrak{F}, \nu)$ is a complete extension of $(X, \mathfrak{A}, \mu)$. If there exists a complete extension $(X, \mathfrak{F}_0, \nu_0)$ of $(X, \mathfrak{A}, \mu)$ such that for any complete extension $(X, \mathfrak{F}, \nu)$ of $(X, \mathfrak{A}, \mu)$ we have $\mathfrak{F} \supset \mathfrak{F}_0$ and $\nu = \nu_0$ on $\mathfrak{F}_0$, then we call $(X, \mathfrak{F}_0, \nu_0)$ a completion of $(X, \mathfrak{A}, \mu)$. A completion is thus a smallest complete extension.*

For an arbitrary measure space $(X, \mathfrak{A}, \mu)$, if a completion exists then it is unique. Indeed if $(X, \mathfrak{F}_1, \nu_1)$ and $(X, \mathfrak{F}_2, \nu_2)$ are two completions of $(X, \mathfrak{A}, \mu)$, then $\mathfrak{F}_1 \subset \mathfrak{F}_2$ and $\mathfrak{F}_2 \subset \mathfrak{F}_1$ so that $\mathfrak{F}_1 = \mathfrak{F}_2$ and since $\nu_2 = \nu_1$ on $\mathfrak{F}_1$ and $\nu_1 = \nu_2$ on $\mathfrak{F}_2$ we have $\nu_1 = \nu_2$ on $\mathfrak{F}_1 = \mathfrak{F}_2$. We show below that every measure space has a completion by constructing it.

Definition 5.2. *Given a measure space $(X, \mathfrak{A}, \mu)$. Let $\mathfrak{N}$ be the collection of all null sets in $(X, \mathfrak{A}, \mu)$ and $\overline{\mathfrak{N}}$ be the collection of all subsets of the members of $\mathfrak{N}$. Let $\overline{\mathfrak{A}}$ be the collection of all subsets of X of the type $E = A \cup C$ where $A \in \mathfrak{A}$ and $C \in \overline{\mathfrak{N}}$. We call $\overline{\mathfrak{A}}$ the completion of the σ-algebra $\mathfrak{A}$ with respect to the measure μ.*

Lemma 5.3. *Let $(X, \mathfrak{A}, \mu)$ be an arbitrary measure space. The completion $\overline{\mathfrak{A}}$ of the σ-algebra $\mathfrak{A}$ with respect to the measure μ is a σ-algebra of subsets of X. Indeed we have $\overline{\mathfrak{A}} = \sigma(\mathfrak{A} \cup \overline{\mathfrak{N}})$.*

Proof. Let us show that $\overline{\mathfrak{A}}$ is a σ-algebra of subsets of X. First of all, since $\emptyset \in \mathfrak{N} \subset \overline{\mathfrak{N}}$, we have $X = X \cup \emptyset \in \overline{\mathfrak{A}}$. To show that $\overline{\mathfrak{A}}$ is closed under complementations, let $E \in \overline{\mathfrak{A}}$. Then $E = A \cup C$ where $A \in \mathfrak{A}$ and $C \subset B \in \mathfrak{N}$. Now $E^c = (A \cup C)^c = A^c \cap C^c$. Since $C = B \setminus (B \setminus C) = B \cap (B \setminus C)^c$, we have $C^c = B^c \cup (B \setminus C)$. Then $E^c = A^c \cap [B^c \cup (B \setminus C)] = (A^c \cap B^c) \cup [A^c \cap (B \setminus C)]$. Since $A, B \in \mathfrak{A}$, we have $A^c \cap B^c \in \mathfrak{A}$. On the other hand, $A^c \cap (B \setminus C) \subset B \setminus C \subset B$ so that $A^c \cap (B \setminus C) \in \overline{\mathfrak{N}}$. Thus $E^c \in \overline{\mathfrak{A}}$. To show that $\overline{\mathfrak{A}}$ is closed under countable unions, let $E_n \in \overline{\mathfrak{A}}$ be given by $E_n = A_n \cup C_n$ where $A_n \in \mathfrak{A}$ and $C_n \subset B_n \in \mathfrak{N}$ for $n \in \mathbb{N}$. Then $\bigcup_{n\in\mathbb{N}} E_n = \bigcup_{n\in\mathbb{N}}(A_n \cup C_n) = \left(\bigcup_{n\in\mathbb{N}} A_n\right) \cup \left(\bigcup_{n\in\mathbb{N}} C_n\right)$. Now $\bigcup_{n\in\mathbb{N}} A_n \in \mathfrak{A}$. Also $\bigcup_{n\in\mathbb{N}} C_n \subset \bigcup_{n\in\mathbb{N}} B_n$. Since a countable union of null sets is a null set in any measure space, $\bigcup_{n\in\mathbb{N}} B_n \in \mathfrak{N}$ and thus $\bigcup_{n\in\mathbb{N}} C_n \in \overline{\mathfrak{N}}$. Thus $\bigcup_{n\in\mathbb{N}} E_n \in \overline{\mathfrak{A}}$. This completes the proof that $\overline{\mathfrak{A}}$ is a σ-algebra of subsets of X.

If $A \in \mathfrak{A}$, then $A = A \cup \emptyset$. Since $\emptyset \in \overline{\mathfrak{N}}$, we have $A \in \overline{\mathfrak{A}}$. Thus $\mathfrak{A} \subset \overline{\mathfrak{A}}$. If $C \in \overline{\mathfrak{N}}$, then since $\emptyset \in \mathfrak{A}$, we have $C = \emptyset \cup C \in \overline{\mathfrak{A}}$. Thus $\overline{\mathfrak{N}} \subset \overline{\mathfrak{A}}$. Therefore $\mathfrak{A} \cup \overline{\mathfrak{N}} \subset \overline{\mathfrak{A}}$ and then $\sigma(\mathfrak{A} \cup \overline{\mathfrak{N}}) \subset \sigma(\overline{\mathfrak{A}}) = \overline{\mathfrak{A}}$.

To show that $\overline{\mathfrak{A}}$ is the smallest σ-algebra of subsets of X containing $\mathfrak{A} \cup \overline{\mathfrak{N}}$, note that if $\mathfrak{F}$ is a σ-algebra of subsets of X containing $\mathfrak{A} \cup \overline{\mathfrak{N}}$, then it contains any set of the type $A \cup C$ where $A \in \mathfrak{A}$ and $C \in \overline{\mathfrak{N}}$. Thus $\mathfrak{F}$ contains $\overline{\mathfrak{A}}$. Therefore $\overline{\mathfrak{A}}$ is the smallest σ-algebra of subsets of X containing $\mathfrak{A}$ and $\overline{\mathfrak{N}}$. ∎

Theorem 5.4. (Existence of the Completion of a Measure Space) *Let $(X, \mathfrak{A}, \mu)$ be an arbitrary measure space. Let $\mathfrak{N}$ be the collection of all null sets in $(X, \mathfrak{A}, \mu)$ and $\overline{\mathfrak{N}}$ be the collection of all subsets of the members of $\mathfrak{N}$. Let $\overline{\mathfrak{A}}$ be the completion of the σ-algebra $\mathfrak{A}$ with respect to the measure μ, that is, $\overline{\mathfrak{A}} = \sigma\left(\mathfrak{A} \cup \overline{\mathfrak{N}}\right)$. Let us extend the domain of definition of the set function μ from $\mathfrak{A}$ to $\overline{\mathfrak{A}}$ by setting*

$$(1) \qquad \mu(E) = \mu(A),$$

for $E \in \overline{\mathfrak{A}}$ given as $E = A \cup C$ with $A \in \mathfrak{A}$ and $C \in \overline{\mathfrak{N}}$. Then μ is a measure on $\overline{\mathfrak{A}}$ and the measure space $\left(X, \overline{\mathfrak{A}}, \mu\right)$ is the completion of $(X, \mathfrak{A}, \mu)$.

Proof. 1. Let us show that the set function μ on the σ-algebra $\overline{\mathfrak{A}}$ is well-defined, that is, $\mu(E)$ defined by (1) does not depend on the representation of E as the union of a set in $\mathfrak{A}$ and a set in $\overline{\mathfrak{N}}$. Thus let $E = A_1 \cup C_1$ and $E = A_2 \cup C_2$ where $A_1, A_2 \in \mathfrak{A}$ and $C_1, C_2 \in \overline{\mathfrak{N}}$. Since $C_1, C_2 \in \overline{\mathfrak{N}}$, we have $C_1 \subset B_1$ and $C_2 \subset B_2$ where $B_1, B_2 \in \mathfrak{N}$. Then by $A_1 \cup C_1 = A_2 \cup C_2$, we have $A_1 \cup C_1 \cup B_1 \cup B_2 = A_2 \cup C_2 \cup B_1 \cup B_2$. Since $C_1 \subset B_1$ and $C_2 \subset B_2$, the last equality reduces to $A_1 \cup B_1 \cup B_2 = A_2 \cup B_1 \cup B_2$ and thus $\mu(A_1 \cup B_1 \cup B_2) = \mu(A_2 \cup B_1 \cup B_2)$. Now $\mu(A_1) \le \mu(A_1 \cup B_1 \cup B_2) \le \mu(A_1) + \mu(B_1) + \mu(B_2) = \mu(A_1)$ and therefore $\mu(A_1) = \mu(A_1 \cup B_1 \cup B_2)$. Similarly we have $\mu(A_2) = \mu(A_2 \cup B_1 \cup B_2)$. This shows that $\mu(A_1) = \mu(A_2)$, proving the independence of $\mu(E)$ from the representation $E = A \cup C$.

2. Let us show that μ is a measure on the σ-algebra $\overline{\mathfrak{A}}$. First of all, we have $\mu(E) \in [0, \infty]$ for any $E \in \overline{\mathfrak{A}}$ by (1). Secondly, since $\emptyset = \emptyset \cup \emptyset$ where $\emptyset \in \mathfrak{A}$ and $\emptyset \in \mathfrak{N} \subset \overline{\mathfrak{N}}$, we have $\mu(\emptyset) = 0$ by (1). To show the countable additivity of μ on $\overline{\mathfrak{A}}$, let $(E_n : n \in \mathbb{N})$ be a disjoint sequence in $\overline{\mathfrak{A}}$. Let $E_n = A_n \cup C_n$ where $A_n \in \mathfrak{A}$ and $C_n \subset B_n \in \mathfrak{N}$ for $n \in \mathbb{N}$. The disjointness of $(E_n : n \in \mathbb{N})$ implies that of $(A_n : n \in \mathbb{N})$ and that of $(C_n : n \in \mathbb{N})$. Let $A = \bigcup_{n \in \mathbb{N}} A_n \in \mathfrak{A}$, $B = \bigcup_{n \in \mathbb{N}} B_n \in \mathfrak{N}$ and $C = \bigcup_{n \in \mathbb{N}} C_n$. Then $C \subset B$ so that $C \in \overline{\mathfrak{N}}$. Now $\bigcup_{n \in \mathbb{N}} E_n = \bigcup_{n \in \mathbb{N}} (A_n \cup C_n) = \left(\bigcup_{n \in \mathbb{N}} A_n\right) \cup \left(\bigcup_{n \in \mathbb{N}} C_n\right) = A \cup C$. Thus by (1), we have $\mu\left(\bigcup_{n \in \mathbb{N}} E_n\right) = \mu(A \cup C) = \mu(A) = \mu\left(\bigcup_{n \in \mathbb{N}} A_n\right) = \sum_{n \in \mathbb{N}} \mu(A_n) = \sum_{n \in \mathbb{N}} \mu(E_n)$. This proves the countable additivity of μ on $\overline{\mathfrak{A}}$ and completes the proof that μ is a measure on $\overline{\mathfrak{A}}$.

3. To show that the measure space $\left(X, \overline{\mathfrak{A}}, \mu\right)$ is a complete measure space, we show that if E is a null set in $\left(X, \overline{\mathfrak{A}}, \mu\right)$, that is, $E \in \overline{\mathfrak{A}}$ and $\mu(E) = 0$, then every subset E_0 of E is a member of $\overline{\mathfrak{A}}$. Now $E = A \cup C$ where $A \in \mathfrak{A}$ and $C \subset B \in \mathfrak{N}$. By (1), we have $\mu(A) = \mu(E) = 0$. Since $B \in \mathfrak{N}$, we have $\mu(B) = 0$. Thus $\mu(A \cup B) \le \mu(A) + \mu(B) = 0$ so that $\mu(A \cup B) = 0$. Then since $A \cup B \in \mathfrak{A}$ and $\mu(A \cup B) = 0$, we have $A \cup B \in \mathfrak{N}$. Now $E_0 \subset E = A \cup C \subset A \cup B \in \mathfrak{N}$. Thus $E_0 \in \overline{\mathfrak{N}} \subset \overline{\mathfrak{A}}$.

4. Since $\mathfrak{A} \subset \overline{\mathfrak{A}}$, $\left(X, \overline{\mathfrak{A}}, \mu\right)$ is a complete extension of $(X, \mathfrak{A}, \mu)$.

5. To show that $\left(X, \overline{\mathfrak{A}}, \mu\right)$ is the completion of $(X, \mathfrak{A}, \mu)$, let $(X, \mathfrak{F}, \nu)$ be an arbitrary complete extension of $(X, \mathfrak{A}, \mu)$. Then the σ-algebra $\mathfrak{F}$ contains $\mathfrak{A}$ and $\overline{\mathfrak{N}}$. This implies that $\mathfrak{F} \supset \sigma(\mathfrak{A} \cup \overline{\mathfrak{N}}) = \overline{\mathfrak{A}}$ by Lemma 5.3. Next let us show that $\nu = \mu$ on $\overline{\mathfrak{A}}$. For any $E \in \overline{\mathfrak{A}}$ given by $E = A \cup C$ where $A \in \mathfrak{A}$ and $C \subset B \in \mathfrak{N}$, let us use the representation $E = A \cup (C \setminus A)$. Note that $C \setminus A \subset C \subset B \in \mathfrak{N}$ so that $C \setminus A \in \overline{\mathfrak{N}}$ and $A \cap (C \setminus A) = \emptyset$. Then we have $\nu(E) = \nu(A \cup (C \setminus A)) = \nu(A) + \nu(C \setminus A)$. Since $A \in \mathfrak{A}$ we have $\nu(A) = \mu(A) = \mu(E)$. On the other hand, since $C \setminus A \subset B \in \mathfrak{N}$ and $\nu(C \setminus A) \leq \nu(B) = \mu(B) = 0$, we have $\nu(C \setminus A) = 0$. Thus $\nu(E) = \mu(E)$. This shows that $\nu = \mu$ on $\overline{\mathfrak{A}}$. Therefore $(X, \overline{\mathfrak{A}}, \mu)$ is the smallest complete extension of $(X, \mathfrak{A}, \mu)$. ∎

For an arbitrary measure space $(X, \mathfrak{A}, \mu)$, a completion exists by Theorem 5.4. With $\overline{\mathfrak{A}}$, the completion of $\mathfrak{A}$ with respect to μ as in Definition 5.2 and with μ extended to $\overline{\mathfrak{A}}$ by (1) of Theorem 5.4, $(X, \overline{\mathfrak{A}}, \mu)$ is a completion of $(X, \mathfrak{A}, \mu)$. The completion of a measure space is unique as we noted above. This justifies representation of the completion of $(X, \mathfrak{A}, \mu)$ by our construction $(X, \overline{\mathfrak{A}}, \mu)$.

Observation 5.5. Consider the completion $(X, \overline{\mathfrak{A}}, \mu)$ of a measure space $(X, \mathfrak{A}, \mu)$.

(a) Every null set in $(X, \overline{\mathfrak{A}}, \mu)$ is a subset of a null set in $(X, \mathfrak{A}, \mu)$.

(b) Let $\overline{\overline{\mathfrak{A}}}$ be the completion of the σ-algebra $\overline{\mathfrak{A}}$ with respect to the measure μ on $\overline{\mathfrak{A}}$. Then we have $\overline{\overline{\mathfrak{A}}} = \overline{\mathfrak{A}}$.

Proof. 1. Let E be a null set in $(X, \overline{\mathfrak{A}}, \mu)$. Then $E \in \overline{\mathfrak{A}}$ so that $E = A \cup C$ with $A \in \mathfrak{A}$ and $C \subset B$ where B is a null set in $(X, \mathfrak{A}, \mu)$ and $\mu(E) = \mu(A)$. Since $\mu(E) = 0$ we have $\mu(A) = 0$ and then $A \cup B$ is a null set in $(X, \mathfrak{A}, \mu)$. Thus $E = A \cup C \subset A \cup B$ which is a null set in $(X, \mathfrak{A}, \mu)$. This proves (a).

2. By Definition 5.2, $\overline{\overline{\mathfrak{A}}}$ is the collection of all subsets E of X of the type $E = A \cup C$ where $A \in \overline{\mathfrak{A}}$ and C is a subset of a null set in $(X, \overline{\mathfrak{A}}, \mu)$. Since $(X, \overline{\mathfrak{A}}, \mu)$ is a complete measure space, $C \in \overline{\mathfrak{A}}$. Thus $E \in \overline{\mathfrak{A}}$. This shows that $\overline{\overline{\mathfrak{A}}} \subset \overline{\mathfrak{A}}$. On the other hand since $\overline{\overline{\mathfrak{A}}}$ is the completion of $\overline{\mathfrak{A}}$ with respect to μ, we have $\overline{\overline{\mathfrak{A}}} \supset \overline{\mathfrak{A}}$. Therefore $\overline{\overline{\mathfrak{A}}} = \overline{\mathfrak{A}}$. ∎

Let $(X, \overline{\mathfrak{A}}, \mu)$ be the completion of a measure space $(X, \mathfrak{A}, \mu)$. Since $\mathfrak{A} \subset \overline{\mathfrak{A}}$, a $\overline{\mathfrak{A}}$-measurable function f on a set $D \in \mathfrak{A} \subset \overline{\mathfrak{A}}$ may not be $\mathfrak{A}$-measurable on D. The following theorem shows that there exists a null set N in $(X, \mathfrak{A}, \mu)$ such that f is $\mathfrak{A}$-measurable on $D \setminus N$.

Theorem 5.6. *Let $(X, \overline{\mathfrak{A}}, \mu)$ be the completion of a measure space $(X, \mathfrak{A}, \mu)$. Let f be an extended real-valued $\overline{\mathfrak{A}}$-measurable function on a set $D \in \mathfrak{A} \subset \overline{\mathfrak{A}}$. Then there exist a null set N in $(X, \mathfrak{A}, \mu)$ and an extended real-valued $\mathfrak{A}$-measurable function g on D such that $f = g$ on $D \setminus N$. (In other words, if f is an extended real-valued $\overline{\mathfrak{A}}$-measurable function on a set $D \in \mathfrak{A}$, then there exists a null set N in $(X, \mathfrak{A}, \mu)$ such that f is $\mathfrak{A}$-measurable on $D \setminus N$.)*

Proof. Let $\{r_n : n \in \mathbb{N}\}$ be the collection of all rational numbers. For every $n \in \mathbb{N}$, let $E_n = \{D : f < r_n\}$. By the $\overline{\mathfrak{A}}$-measurability of f on D, we have $E_n \in \overline{\mathfrak{A}}$. Then $E_n = A_n \cup C_n$ where $A_n \in \mathfrak{A}$ and C_n is a subset of a null set B_n in $(X, \mathfrak{A}, \mu)$. If we let

$N = \bigcup_{n \in \mathbb{N}} B_n$, then N is a null set in $(X, \mathfrak{A}, \mu)$. Define an extended real-valued function g on D by setting

$$g(x) = \begin{cases} f(x) & \text{for } x \in D \setminus N, \\ 0 & \text{for } x \in N. \end{cases}$$

It remains to show that g is $\mathfrak{A}$-measurable on D. Now for every $n \in \mathbb{N}$, we have

$$(1) \qquad \{D : g < r_n\} = \{D \setminus N : g < r_n\} \cup \{N : g < r_n\}.$$

Note that $\{N : g < r_n\} = \emptyset$ or N according as $r_n \leq 0$ or $r_n > 0$. In any case we have

$$(2) \qquad \{N : g < r_n\} \in \mathfrak{A}.$$

On the other hand we have

$$(3) \qquad \begin{aligned} \{D \setminus N : g < r_n\} &= \{D \setminus N : f < r_n\} = (D \setminus N) \cap \{D : f < r_n\} \\ &= (D \setminus N) \cap (A_n \cup C_n) = (D \setminus N) \cap A_n \in \mathfrak{A}, \end{aligned}$$

where the last equality is by the fact that $C_n \subset B_n \subset N$. By (1), (2), and (3) we have

$$(4) \qquad \{D : g < r_n\} \in \mathfrak{A} \quad \text{for every } n \in \mathbb{N}.$$

Now let $\alpha \in \mathbb{R}$. There exists an increasing sequence of rational numbers $(r_k : k \in \mathbb{N})$ such that $r_k \uparrow \alpha$. Then $\{D : g < \alpha\} = \bigcup_{k \in \mathbb{N}} \{D : g < r_k\} \in \mathfrak{A}$ by (4). This proves the $\mathfrak{A}$-measurability of g on D. ∎

[II] Completion of the Borel Measure Space to the Lebesgue Measure Space

The Borel measure space $(\mathbb{R}, \mathfrak{B}_{\mathbb{R}}, \mu_L)$ on $\mathbb{R}$ is obtained by restricting the Lebesgue measure μ_L on the Lebesgue σ-algebra $\mathfrak{M}_L$ to the Borel σ-algebra $\mathfrak{B}_{\mathbb{R}} \subset \mathfrak{M}_L$. Thus $(\mathbb{R}, \mathfrak{M}_L, \mu_L)$ is a complete extension of $(\mathbb{R}, \mathfrak{B}_{\mathbb{R}}, \mu_L)$. We show next that $(\mathbb{R}, \mathfrak{M}_L, \mu_L)$ is actually the completion of $(\mathbb{R}, \mathfrak{B}_{\mathbb{R}}, \mu_L)$.

Theorem 5.7. *Consider the Lebesgue σ-algebra $\mathfrak{M}_L$ and the Borel σ-algebra $\mathfrak{B}_{\mathbb{R}}$ on $\mathbb{R}$.*
(a) *Let $E \in \mathfrak{P}(\mathbb{R})$. Then $E \in \mathfrak{M}_L$ if and only if $E = A \cup C$ where $A \in \mathfrak{B}_{\mathbb{R}}$ and C is a subset of a set $B \in \mathfrak{B}_{\mathbb{R}}$ with $\mu_L(B) = 0$.*
(b) *The Lebesgue measure space $(\mathbb{R}, \mathfrak{M}_L, \mu_L)$ is the completion of the Borel measure space*
$(\mathbb{R}, \mathfrak{B}_{\mathbb{R}}, \mu_L)$.

Proof. 1. If $E \in \mathfrak{M}_L$ then by Theorem 3.22, there exists a G_δ-set $G \supset E$ with $\mu_L(G \setminus E) = 0$ and there exists an F_σ-set $F \subset E$ with $\mu_L(E \setminus F) = 0$. Then $F \subset E \subset G$ and $\mu_L(G \setminus F) = \mu_L(G \setminus E) + \mu_L(E \setminus F) = 0$. Now $E = F \cup (E \setminus F)$ where $F \in \mathfrak{B}_{\mathbb{R}}$, $E \setminus F \subset G \setminus F$, and $G \setminus F \in \mathfrak{B}_{\mathbb{R}}$ with $\mu_L(G \setminus F) = 0$.

Conversely suppose $E = A \cup C$ where $A \in \mathfrak{B}_{\mathbb{R}}$ and C is a subset of a set $B \in \mathfrak{B}_{\mathbb{R}}$ with $\mu_L(B) = 0$. Now $A, B \in \mathfrak{B}_{\mathbb{R}} \subset \mathfrak{M}_L$. Since C is a subset of $B \in \mathfrak{M}_L$ with $\mu_L(B) = 0$, we have $C \in \mathfrak{M}_L$ by the completeness of the measure space $(\mathbb{R}, \mathfrak{M}_L, \mu_L)$. Thus $E \in \mathfrak{M}_L$. This completes the proof of (a).

2. To prove (b) let $(\mathbb{R}, \overline{\mathfrak{B}_{\mathbb{R}}}, \mu_L)$ be the completion of $(\mathbb{R}, \mathfrak{B}_{\mathbb{R}}, \mu_L)$. By Definition 5.2, $\overline{\mathfrak{B}_{\mathbb{R}}}$ consists of subsets of $\mathbb{R}$ of the type $E = A \cup C$ where $A \in \mathfrak{B}_{\mathbb{R}}$ and C is a subset of a set $B \in \mathfrak{B}_{\mathbb{R}}$ with $\mu_L(B) = 0$. Thens by (a) we have $\overline{\mathfrak{B}_{\mathbb{R}}} = \mathfrak{M}_L$. ∎

Corollary 5.8. *The Borel measure space $(\mathbb{R}, \mathfrak{B}_{\mathbb{R}}, \mu_L)$ is not a complete measure space.*

Proof. If $(\mathbb{R}, \mathfrak{B}_{\mathbb{R}}, \mu_L)$ were a complete measure space, then it would be equal to its completion which is the Lebesgue measure space $(\mathbb{R}, \mathfrak{M}_L, \mu_L)$ by Theorem 5.7. But according to Proposition 4.40 we have $\mathfrak{M}_L \neq \mathfrak{B}_{\mathbb{R}}$. ∎

Proposition 5.9. *Let f be an extended real-valued $\mathfrak{M}_L$-measurable function on a set $D \in \mathfrak{B}_{\mathbb{R}}$. Then there exist a null set N in $(\mathbb{R}, \mathfrak{B}_{\mathbb{R}}, \mu_L)$ and an extended real-valued $\mathfrak{B}_{\mathbb{R}}$-measurable function g on D such that $g = f$ on $D \setminus N$ so that f is $\mathfrak{B}_{\mathbb{R}}$-measurable on $D \setminus N$.*

Proof. According to Theorem 5.7, $(\mathbb{R}, \mathfrak{M}_L, \mu_L)$ is the completion of $(\mathbb{R}, \mathfrak{B}_{\mathbb{R}}, \mu_L)$. Then the Proposition is a particular case of Theorem 5.6. ∎

§6 Convergence a.e. and Convergence in Measure

[I] Convergence a.e.

Definition 6.1. *Given a measure space $(X, \mathfrak{A}, \mu)$. Let $(f_n : n \in \mathbb{N})$ be a sequence of extended real-valued $\mathfrak{A}$-measurable functions on a set $D \in \mathfrak{A}$. We say that $\lim_{n\to\infty} f_n$ exists a.e. on D if there exists a null set N in $(X, \mathfrak{A}, \mu)$ such that $N \subset D$ and $\lim_{n\to\infty} f_n(x)$ exists for every $x \in D \setminus N$. We say that $(f_n : n \in \mathbb{N})$ converges a.e. on D if $\lim_{n\to\infty} f_n(x)$ exists and $\lim_{n\to\infty} f_n(x) \in \mathbb{R}$ for every $x \in D \setminus N$.*

Note that in the definition above for the existence of $\lim_{n\to\infty} f_n$ a.e. on D, it is possible that $\lim_{n\to\infty} f_n(x)$ exists at some, and indeed at every, $x \in N$. If we let D_e be the subset of D consisting of only those points $x \in D$ for which $\lim_{n\to\infty} f_n(x)$ exists, then $D_e \in \mathfrak{A}$ by Theorem 4.22 and we have $D \setminus N \subset D_e$ and $D \setminus D_e \subset N$. The extended real-valued function $\lim_{n\to\infty} f_n$ is $\mathfrak{A}$-measurable on D_e by Theorem 4.22.

Lemma 6.2. *Given a measure space $(X, \mathfrak{A}, \mu)$. Let $(f_n : n \in \mathbb{N})$ be a sequence of extended real-valued $\mathfrak{A}$-measurable functions on a set $D \in \mathfrak{A}$. If for every $\eta > 0$ there exists a $\mathfrak{A}$-measurable subset E of D with $\mu(E) < \eta$ such that $\lim_{n\to\infty} f_n(x)$ exists (resp. $f_n(x)$ converges) for every $x \in D \setminus E$, then $\lim_{n\to\infty} f_n(x)$ exists (resp. $f_n(x)$ converges) a.e. on D.*

Proof. Under the assumption of the condition in the Lemma, for every $k \in \mathbb{N}$ there exists a $\mathfrak{A}$-measurable subset E_k of D with $\mu(E_k) < \frac{1}{k}$ such that $\lim_{n\to\infty} f_n(x)$ exists (resp. $f_n(x)$ converges) for every $x \in D \setminus E_k$. Let $E = \bigcap_{n\in\mathbb{N}} E_k$. Then $E \subset D$, $E \in \mathfrak{A}$, and $\mu(E) = \mu\left(\bigcap_{k\in\mathbb{N}} E_k\right) \le \mu(E_k) < \frac{1}{k}$ for every $k \in \mathbb{N}$. This implies that $\mu(E) = 0$ so that E is a null set in $(X, \mathfrak{A}, \mu)$.

Let us show that $\lim_{n\to\infty} f_n(x)$ exists (resp. $f_n(x)$ converges) for every $x \in D \setminus E$. Now $D \setminus E = D \cap E^c = D \cap \left(\bigcap_{k\in\mathbb{N}} E_k\right)^c = D \cap \left(\bigcup_{k\in\mathbb{N}} E_k^c\right) = \bigcup_{k\in\mathbb{N}} (D \cap E_k^c) = \bigcup_{k\in\mathbb{N}} (D \setminus E_k)$. Thus if $x \in D \setminus E$ then $x \in D \setminus E_k$ for some $k \in \mathbb{N}$. Then $\lim_{n\to\infty} f_n(x)$ exists (resp. $f_n(x)$ converges). ∎

Proposition 6.3. (Uniqueness of Limit a.e.) *Given a measure space $(X, \mathfrak{A}, \mu)$. Let $(f_n : n \in \mathbb{N})$ be a sequence of extended real-valued $\mathfrak{A}$-measurable functions on a set $D \in \mathfrak{A}$. Let g_1 and g_2 be two extended real-valued $\mathfrak{A}$-measurable functions on D. If $\lim_{n\to\infty} f_n = g_1$ a.e. on D and $\lim_{n\to\infty} f_n = g_2$ a.e. on D also, then $g_1 = g_2$ a.e. on D.*

Proof. Under the assumption of the Proposition, there exist two null sets N_1 and N_2 of $(X, \mathfrak{A}, \mu)$ such that $N_1, N_2 \subset D$, $\lim_{n\to\infty} f_n = g_1$ on $D \setminus N_1$, and $\lim_{n\to\infty} f_n = g_2$ on $D \setminus N_2$. Let $N = N_1 \cup N_2$. Then N is a subset of D and a null set of $(X, \mathfrak{A}, \mu)$ and $g_1 = \lim_{n\to\infty} f_n = g_2$ on $D \setminus N$. ∎

Proposition 6.4. *Given a measure space $(X, \mathfrak{A}, \mu)$. Let $(f_n : n \in \mathbb{N})$ be a sequence of extended real-valued $\mathfrak{A}$-measurable functions on a set $D \in \mathfrak{A}$ and let f be a real-valued $\mathfrak{A}$-measurable function on D. Then*

(1)
$$\left\{D : \lim_{n\to\infty} f_n = f\right\} = \bigcap_{m\in\mathbb{N}} \bigcup_{N\in\mathbb{N}} \bigcap_{p\in\mathbb{N}} \left\{D : |f_{N+p} - f| < \tfrac{1}{m}\right\},$$

and

(2)
$$\{D : f_n \not\to f\} = \bigcup_{m\in\mathbb{N}} \bigcap_{N\in\mathbb{N}} \bigcup_{p\in\mathbb{N}} \left\{D : |f_{N+p} - f| \geq \tfrac{1}{m}\right\}.$$

Proof. The set $\left\{D : \lim_{n\to\infty} f_n = f\right\}$ consists of every $x \in D$ for which the sequence of extended real numbers $\left(f_n(x) : n \in \mathbb{N}\right)$ converges to $f(x)$. Now for each $x \in D$, $\left(f_n(x) : n \in \mathbb{N}\right)$ converges to $f(x)$ if and only if for every $m \in \mathbb{N}$ there exists $N \in \mathbb{N}$ such that $|f_{N+p}(x) - f(x)| < \frac{1}{m}$ for every $p \in \mathbb{N}$. Equation (1) is a set theoretic statement of this fact. To prove (2), note that the set $\{x \in D : f_n(x) \not\to f(x)\}$ consists of every $x \in D$ for which either $\lim_{n\to\infty} f_n(x)$ does not exist or $\lim_{n\to\infty} f_n(x)$ exists but is not equal to $f(x)$. Now we have

$$\begin{aligned}
\{D : f_n \not\to f\} &= D \setminus \left\{D : \lim_{n\to\infty} f_n = f\right\} \\
&= D \setminus \bigcap_{m\in\mathbb{N}} \bigcup_{N\in\mathbb{N}} \bigcap_{p\in\mathbb{N}} \left\{D : |f_{N+p} - f| < \tfrac{1}{m}\right\} \\
&= D \cap \left[\bigcap_{m\in\mathbb{N}} \bigcup_{N\in\mathbb{N}} \bigcap_{p\in\mathbb{N}} \left\{D : |f_{N+p} - f| < \tfrac{1}{m}\right\}\right]^c \\
&= D \cap \left[\bigcup_{m\in\mathbb{N}} \bigcap_{N\in\mathbb{N}} \bigcup_{p\in\mathbb{N}} \left\{D : |f_{N+p} - f| < \tfrac{1}{m}\right\}^c\right] \\
&= \bigcup_{m\in\mathbb{N}} \bigcap_{N\in\mathbb{N}} \bigcup_{p\in\mathbb{N}} \left[D \cap \left\{D : |f_{N+p} - f| < \tfrac{1}{m}\right\}^c\right] \\
&= \bigcup_{m\in\mathbb{N}} \bigcap_{N\in\mathbb{N}} \bigcup_{p\in\mathbb{N}} \left\{D : |f_{N+p} - f| \geq \tfrac{1}{m}\right\}. \ \blacksquare
\end{aligned}$$

The following theorem is a criterion for convergence a.e. of a sequence of measurable functions. It is fundamental in that several convergence theorems will be derived from it.

Theorem 6.5. *Given a measure space $(X, \mathfrak{A}, \mu)$. Let $(f_n : n \in \mathbb{N})$ be a sequence of extended real-valued $\mathfrak{A}$-measurable functions on a set $D \in \mathfrak{A}$ and let f be a real-valued $\mathfrak{A}$-measurable function on D. Then*

$$\begin{aligned}
&(f_n : n \in \mathbb{N}) \text{ converges to } f \text{ a.e. on } D \\
&\Leftrightarrow \mu\left(\limsup_{n\to\infty} \left\{D : |f_n - f| \geq \tfrac{1}{m}\right\}\right) = 0 \text{ for every } m \in \mathbb{N}.
\end{aligned}$$

Proof. $(f_n : n \in \mathbb{N})$ converges to f a.e. on D if and only if $\mu\{D : f_n \not\to f\} = 0$. By (2) of Proposition 6.4, this last condition is equivalent to

$$(1) \qquad \mu\Big(\bigcup_{m\in\mathbb{N}}\bigcap_{N\in\mathbb{N}}\bigcup_{p\in\mathbb{N}}\{D : |f_{N+p} - f| \geq \tfrac{1}{m}\}\Big) = 0.$$

By the monotonicity of μ and by the fact that a countable union of null sets is always a null set, (1) is equivalent to the condition that

$$(2) \qquad \mu\Big(\bigcap_{N\in\mathbb{N}}\bigcup_{p\in\mathbb{N}}\{D : |f_{N+p} - f| \geq \tfrac{1}{m}\}\Big) = 0 \quad \text{for every } m \in \mathbb{N}.$$

Condition (2) is equivalent to the condition that

$$(3) \qquad \mu\Big(\bigcap_{N\in\mathbb{N}}\bigcup_{n>N}\{D : |f_n - f| \geq \tfrac{1}{m}\}\Big) = 0 \quad \text{for every } m \in \mathbb{N}$$

that is, $\mu\big(\limsup_{n\to\infty}\{D : |f_n - f| \geq \frac{1}{m}\}\big) = 0$ for every $m \in \mathbb{N}$. ∎

The necessary and sufficient condition in Theorem 6.5 for convergence almost everywhere of a sequence of measurable functions leads to the question as to when the equality $\mu\big(\limsup_{n\to\infty} A_n\big) = 0$ holds for a sequence $(A_n : n \in \mathbb{N})$ of $\mathfrak{A}$-measurable sets in a measure space $(X, \mathfrak{A}, \mu)$. Recall that according to Lemma 1.7, $\limsup_{n\to\infty} A_n$ consists of every x in X at which infinitely many A_n intersect. If we make each A_n smaller, the set of infinite intersections becomes smaller. How small must the sets A_n be to ensure $\mu\big(\limsup_{n\to\infty} A_n\big) = 0$? The Borel-Cantelli Lemma gives a sufficient condition for this to take place.

Theorem 6.6. (Borel-Cantelli Lemma) *Given a measure space $(X, \mathfrak{A}, \mu)$. For any sequence $(A_n : n \in \mathbb{N})$ in $\mathfrak{A}$ with $\sum_{n\in\mathbb{N}} \mu(A_n) < \infty$, we have $\mu\big(\limsup_{n\to\infty} A_n\big) = 0$.*

Proof. By definition, $\limsup_{n\to\infty} A_n = \bigcap_{n\in\mathbb{N}}\bigcup_{k\geq n} A_k$. If we let $E_n = \bigcup_{k\geq n} A_k$ for $n \in \mathbb{N}$, then $(E_n : n \in \mathbb{N})$ is a decreasing sequence in $\mathfrak{A}$ and $\limsup_{n\to\infty} A_n = \lim_{n\to\infty} E_n$. Now for $E_1 = \bigcup_{k\geq 1} A_k$, we have $\mu(E_1) = \mu\big(\bigcup_{n\in\mathbb{N}} A_n\big) \leq \sum_{n\in\mathbb{N}} \mu(A_n) < \infty$. Thus $(E_n : n \in \mathbb{N})$ is a decreasing sequence in $\mathfrak{A}$ and $\mu(E_1) < \infty$. Then $\mu\big(\lim_{n\to\infty} E_n\big) = \lim_{n\to\infty} \mu(E_n)$ by Theorem 1.26. Thus we have

$$\mu\Big(\limsup_{n\to\infty} A_n\Big) = \mu\big(\lim_{n\to\infty} E_n\big) = \lim_{n\to\infty} \mu(E_n) = \lim_{n\to\infty} \mu\Big(\bigcup_{k\geq n} A_k\Big) \leq \lim_{n\to\infty} \sum_{k\geq n} \mu(A_k).$$

Since $\sum_{n\in\mathbb{N}} \mu(A_n) < \infty$ implies $\lim_{n\to\infty} \sum_{k\geq n} \mu(A_k) = 0$, we have $\mu\big(\limsup_{n\to\infty} A_n\big) = 0$.
∎

Combining Theorem 6.5 with the Borel-Cantelli Lemma, we obtain the following sufficient conditions for a.e. convergence of a sequence:

Theorem 6.7. *Given a measure space $(X, \mathfrak{A}, \mu)$. Let $(f_n : n \in \mathbb{N})$ be a sequence of extended real-valued $\mathfrak{A}$-measurable functions on a set $D \in \mathfrak{A}$ and let f be a real-valued $\mathfrak{A}$-measurable function on D. Suppose there exists a sequence $(\varepsilon_n : n \in \mathbb{N})$ of positive numbers such that*

1° $\lim_{n\to\infty} \varepsilon_n = 0$,

2° $\sum_{n\in\mathbb{N}} \mu\{D : |f_n - f| \geq \varepsilon_n\} < \infty$.

Then $(f_n : n \in \mathbb{N})$ converges to f a.e. on D.

Proof. If we let $A_n = \{D : |f_n - f| \geq \varepsilon_n\}$ for $n \in \mathbb{N}$, then $\sum_{n\in\mathbb{N}} \mu(A_n) < \infty$ by 2°. Thus by the Borel-Cantelli Lemma (Theorem 6.6), we have $\mu\big(\limsup_{n\to\infty} A_n\big) = 0$, that is,

$$(1) \qquad \mu\Big(\limsup_{n\to\infty}\{D : |f_n - f| \geq \varepsilon_n\}\Big) = 0.$$

Let $m \in \mathbb{N}$ be arbitrarily given. By 1°, there exists $N \in \mathbb{N}$ such that $\varepsilon_n < \frac{1}{m}$ for $n \geq N$. This implies that $\{D : |f_n - f| \geq \varepsilon_n\} \supset \big\{D : |f_n - f| \geq \frac{1}{m}\big\}$ for $n \geq N$. Then we have

$$\limsup_{n\to\infty}\{D : |f_n - f| \geq \varepsilon_n\} \supset \limsup_{n\to\infty}\big\{D : |f_n - f| \geq \tfrac{1}{m}\big\},$$

by applying Lemma 1.7 for the limit superior of a sequence of sets. Then we have

$$(2) \qquad \mu\Big(\limsup_{n\to\infty}\{D : |f_n - f| \geq \varepsilon_n\}\Big) \geq \mu\Big(\limsup_{n\to\infty}\big\{D : |f_n - f| \geq \tfrac{1}{m}\big\}\Big).$$

By (1) and (2) we have

$$\mu\Big(\limsup_{n\to\infty}\big\{D : |f_n - f| \geq \tfrac{1}{m}\big\}\Big) = 0 \quad \text{for every } m \in \mathbb{N}.$$

By Theorem 6.5 this shows that $(f_n : n \in \mathbb{N})$ converges to f a.e. on D. ∎

Remark 6.8. Conditions 1° and 2° are sufficient conditions for $(f_n : n \in \mathbb{N})$ to converge to f a.e. on D but they are not necessary. Below is such an example.

Example. In $(\mathbb{R}, \mathfrak{M}_L, \mu_L)$, let $D = (0, 1)$ and let $f_n(x) = 1$ for $x \in \big(0, \frac{1}{n}\big)$ and $f_n(x) = 0$ for $x \in \big[\frac{1}{n}, 1\big)$ for $n \in \mathbb{N}$ and let $f(x) = 0$ for $x \in (0, 1)$. Then $(f_n : n \in \mathbb{N})$ converges to f everywhere on D. Let $(\varepsilon_n : n \in \mathbb{N})$ be an arbitrary sequence of positive numbers such that $\lim_{n\to\infty} \varepsilon_n = 0$. Let us show that $\sum_{n\in\mathbb{N}} \mu_L\{(0, 1) : |f_n - f| \geq \varepsilon_n\} = \infty$. Let $N \in \mathbb{N}$ be so large that $\varepsilon_n < 1$ for $n \geq N$. Then

$$\sum_{n\in\mathbb{N}} \mu_L\{(0, 1) : |f_n - f| \geq \varepsilon_n\} \geq \sum_{n\geq N} \mu_L\{(0, 1) : |f_n - f| \geq \varepsilon_n\}$$
$$= \sum_{n\geq N} \mu_L\big(0, \tfrac{1}{n}\big) = \sum_{n\geq N} \frac{1}{n} = \infty.$$

Theorem 6.9. *Given a measure space $(X, \mathfrak{A}, \mu)$. Let $(g_n : n \in \mathbb{N})$ be a sequence of real-valued $\mathfrak{A}$-measurable functions on a set $D \in \mathfrak{A}$. Suppose there exists a sequence $(\eta_n : n \in \mathbb{N})$ of positive numbers such that*

1° $\displaystyle\sum_{n\in\mathbb{N}} \eta_n < \infty,$

2° $\displaystyle\sum_{n\in\mathbb{N}} \mu\{x \in D : |g_n(x)| \geq \eta_n\} < \infty.$

Then the series $\sum_{n\in\mathbb{N}} g_n$ converges absolutely a.e. on D and hence converges a.e. on D.

Proof. Let $A_n = \{x \in D : |g_n(x)| \geq \eta_n\}$ for $n \in \mathbb{N}$. By 2° we have $\sum_{n\in\mathbb{N}} \mu(A_n) < \infty$ and thus by Theorem 6.6 we have $\mu\left(\limsup_{n\to\infty} A_n\right) = 0$. Let $E = \limsup_{n\to\infty} A_n$. Then E is a subset of D and is also a null set in $(X, \mathfrak{A}, \mu)$. Also, for every $x \in D$, we have $x \in E$ if and only if $x \in A_n$ for infinitely many $n \in \mathbb{N}$. Then

$$\begin{aligned} x \in D \setminus E &\Leftrightarrow x \in A_n \text{ for at most finitely many } n \in \mathbb{N} \\ &\Leftrightarrow \text{there exists } N(x) \in \mathbb{N} \text{ such that } x \notin A_n \text{ for } n > N(x) \\ &\Leftrightarrow \text{there exists } N(x) \in \mathbb{N} \text{ such that } |g_n(x)| < \eta_n \text{ for } n > N(x). \end{aligned}$$

Thus for $x \in D \setminus E$, we have

$$\sum_{n\in\mathbb{N}} |g_n(x)| = \sum_{n=1}^{N(x)} |g_n(x)| + \sum_{n\geq N(x)+1} |g_n(x)| \geq \sum_{n=1}^{N(x)} |g_n(x)| + \sum_{n\geq N(x)+1} \eta_n.$$

Since g_n is real-valued for every $n \in \mathbb{N}$, we have $\sum_{n=1}^{N(x)} |g_n(x)| < \infty$. By 1° we have $\sum_{n\geq N(x)+1} \eta_n \leq \sum_{n\in\mathbb{N}} \eta_n < \infty$ also. Thus for $x \in D \setminus E$, we have $\sum_{n\in\mathbb{N}} |g_n(x)| < \infty$, that is, the series $\sum_{n\in\mathbb{N}} g_n(x)$ converges absolutely and hence converges for every $x \in D \setminus E$, that is, a.e. on D. ∎

[II] Almost Uniform Convergence

We say that a sequence of extended real-valued functions $(f_n : n \in \mathbb{N})$ converges uniformly on a set D to a real-valued function f if for every $\varepsilon > 0$ there exists $N \in \mathbb{N}$, depending on ε but not on $x \in D$, such that $|f_n(x) - f(x)| < \varepsilon$ for all $x \in D$ when $n \geq N$, or equivalently, for every $m \in \mathbb{N}$ there exists $N \in \mathbb{N}$ such that $|f_n(x) - f(x)| < \frac{1}{m}$ for all $x \in D$ when $n \geq N$.

Definition 6.10. *Given a measure space $(X, \mathfrak{A}, \mu)$. Let $(f_n : n \in \mathbb{N})$ be a sequence of extended real-valued $\mathfrak{A}$-measurable functions on a set $D \in \mathfrak{A}$ and let f be a real-valued $\mathfrak{A}$-measurable function on D. We say that $(f_n : n \in \mathbb{N})$ converges almost uniformly on D to f if for every $\eta > 0$ there exists a $\mathfrak{A}$-measurable subset E of D such that $\mu(E) < \eta$ and $(f_n : n \in \mathbb{N})$ converges uniformly on $D \setminus E$ to f.*

If $(f_n : n \in \mathbb{N})$ converges almost uniformly on D to f, then $(f_n : n \in \mathbb{N})$ satisfies the condition in Lemma 6.2 and thus $(f_n : n \in \mathbb{N})$ converges to f a.e. on D. Egoroff's

Theorem proves that the converse holds provided that $\mu(D) < \infty$, that is, convergence a.e. on D implies almost uniform convergence on D provided that $\mu(D) < \infty$. We specialize the necessary and sufficient condition for convergence a.e. in Theorem 6.5 to the case $\mu(D) < \infty$ in the next Proposition and then derive Egoroff's Theorem from it.

Proposition 6.11. *Given a measure space $(X, \mathfrak{A}, \mu)$. Let $(f_n : n \in \mathbb{N})$ be a sequence of extended real-valued $\mathfrak{A}$-measurable functions on a set $D \in \mathfrak{A}$ and let f be a real-valued $\mathfrak{A}$-measurable function on D. Suppose*

1° *$(f_n : n \in \mathbb{N})$ converges to f a.e. on D,*
2° *$\mu(D) < \infty$.*

For $m \in \mathbb{N}$ and $n \in \mathbb{N}$, let

$$D_n(m) = \bigcup_{k \geq n} \left\{x \in D : |f_k(x) - f(x)| \geq \tfrac{1}{m}\right\}. \tag{1}$$

Then we have

$$\lim_{n \to \infty} \mu\big(D_n(m)\big) = 0 \quad \textit{for every } m \in \mathbb{N}, \tag{2}$$

and consequently

$$\lim_{n \to \infty} \mu\left\{D : |f_n - f| \geq \tfrac{1}{m}\right\} = 0 \quad \textit{for every } m \in \mathbb{N}. \tag{3}$$

Proof. Condition 1° is equivalent to $\mu\big(\limsup_{n \to \infty} \{D : |f_n - f| \geq \frac{1}{m}\}\big) = 0$ for every $m \in \mathbb{N}$ by Theorem 6.5, in other words, $\mu\big(\bigcap_{n \in \mathbb{N}} \bigcup_{k \geq n} \{D : |f_k - f| \geq \frac{1}{m}\}\big) = 0$ for every $m \in \mathbb{N}$. Thus we have

$$\mu\Big(\bigcap_{n \in \mathbb{N}} D_n(m)\Big) = 0 \quad \text{for every } m \in \mathbb{N}. \tag{4}$$

Now $D_n(m) \subset D$, $D_n(m) \in \mathfrak{A}$, and $\big(D_n(m) : n \in \mathbb{N}\big)$ is a decreasing sequence for every $m \in \mathbb{N}$. Thus $\bigcap_{n \in \mathbb{N}} D_n(m) = \lim_{n \to \infty} D_n(m)$ for every $m \in \mathbb{N}$. Since $D_n(m) \subset D$ and $\mu(D) < \infty$, we have $\mu\big(\bigcap_{n \in \mathbb{N}} D_n(m)\big) = \mu\big(\lim_{n \to \infty} D_n(m)\big) = \lim_{n \to \infty} \mu\big(D_n(m)\big)$ by Theorem 1.26 for every $m \in \mathbb{N}$. But according to (4), $\mu\big(\bigcap_{n \in \mathbb{N}} D_n(m)\big) = 0$ for every $m \in \mathbb{N}$. Therefore $\lim_{n \to \infty} \mu\big(D_n(m)\big) = 0$ for every $m \in \mathbb{N}$. This proves (2). Then since $\{D : |f_n - f| \geq \frac{1}{m}\} \subset D_n(m)$, (3) follows from (2) by the monotonicity of μ. ∎

Theorem 6.12. **(D. E. Egoroff)** *Given a measure space $(X, \mathfrak{A}, \mu)$. Let $D \in \mathfrak{A}$ and $\mu(D) < \infty$. Let $(f_n : n \in \mathbb{N})$ be a sequence of extended real-valued $\mathfrak{A}$-measurable functions on D and let f be a real-valued $\mathfrak{A}$-measurable function on D. If $(f_n : n \in \mathbb{N})$ converges to f a.e. on D, then $(f_n : n \in \mathbb{N})$ converges to f almost uniformly on D.*

Proof. For $m \in \mathbb{N}$ and $n \in \mathbb{N}$, let

(1)
$$D_n(m) = \bigcup_{k \geq n} \left\{x \in D : |f_k(x) - f(x)| \geq \tfrac{1}{m}\right\}.$$

Then by (2) of Proposition 6.11, we have

(2)
$$\lim_{n\to\infty} \mu\big(D_n(m)\big) = 0 \quad \text{for every } m \in \mathbb{N}.$$

Let $\eta > 0$ be arbitrarily given. By (2), for every $m \in \mathbb{N}$ there exists $N(m) \in \mathbb{N}$ such that

$$\mu\big(D_{N(m)}(m)\big) < \frac{\eta}{2^m}.$$

Consider the sequence $\big(D_{N(m)}(m) : m \in \mathbb{N}\big)$ of $\mathfrak{A}$-measurable subsets of D. Let

$$E = \bigcup_{m\in\mathbb{N}} D_{N(m)}(m).$$

Then $E \subset D$, $E \in \mathfrak{A}$, and

$$\mu(E) = \mu\Big(\bigcup_{m\in\mathbb{N}} D_{N(m)}(m)\Big) \leq \sum_{m\in\mathbb{N}} \mu\big(D_{N(m)}(m)\big) < \sum_{m\in\mathbb{N}} \frac{\eta}{2^m} = \eta.$$

Consider $D \setminus E$. If $x \in D \setminus E$, then $x \notin E$ and thus $x \notin D_{N(m)}(m)$ for every $m \in \mathbb{N}$. Then by (1), we have $|f_k(x) - f(x)| < \frac{1}{m}$ for $k \geq N(m)$ for every $m \in \mathbb{N}$. This shows that $(f_n : n \in \mathbb{N})$ converges to f uniformly on $D \setminus E$. ∎

Remark 6.13. Theorem 6.12 does not hold without the assumption $\mu(D) < \infty$.

Example. Consider $(\mathbb{R}, \mathfrak{M}_L, \mu_L)$. Let $D = [0, \infty)$ with $\mu_L(D) = \infty$. Let $f_n(x) = \frac{1}{n}x$ for $x \in D$ for $n \in \mathbb{N}$ and let $f(x) = 0$ for $x \in D$. Then the sequence $(f_n : n \in \mathbb{N})$ converges to f everywhere on D. To show that $(f_n : n \in \mathbb{N})$ does not converge to f almost uniformly on D, we show that for some $\eta > 0$ there does not exist a $\mathfrak{M}_L$-measurable subset E of D with $\mu_L(E) < \eta$ such that $(f_n : n \in \mathbb{N})$ converges to f uniformly on $D \setminus E$. For our example, we show that actually for any $\mathfrak{M}_L$-measurable subset E of D with $\mu_L(E) < \infty$, $(f_n : n \in \mathbb{N})$ does not converge to f uniformly on $D \setminus E$. Let E be a $\mathfrak{M}_L$-measurable subset of D with $\mu_L(E) < \infty$. Let $\mu_L(E) < M$ where $M \in (0, \infty)$. Let $D_k = \big[(k-1)M, kM\big)$ for $k \in \mathbb{N}$. Then $(D_k : k \in \mathbb{N})$ is a disjoint sequence and $\bigcup_{k\in\mathbb{N}} D_k = D$. Since $\mu_L(E) < M$ and $\mu_L(D_k) = M$, there exists $x_k \in D_k$ such that $x_k \notin E$ for each $k \in \mathbb{N}$. We have $\lim_{n\to\infty} x_k = \infty$. To show that $(f_n : n \in \mathbb{N})$ does not converge to f uniformly on the subset $\{x_k : k \in \mathbb{N}\}$ of $D \setminus E$, note that given $\varepsilon > 0$ for any $N \in \mathbb{N}$, however large, and $n \geq N$ there exists $k \in \mathbb{N}$ such that $\frac{1}{n}x_k > \varepsilon$ so that $|f_n(x_k) - f(x_k)| = f_n(x_k) = \frac{1}{n}x_k > \varepsilon$. This shows that $(f_n : n \in \mathbb{N})$ does not converge to f uniformly on set $\{x_k : k \in \mathbb{N}\}$. Thus $(f_n : n \in \mathbb{N})$ does not converge to f on the set $D \setminus E$ containing the set $\{x_k : k \in \mathbb{N}\}$.

[III] Convergence in Measure

Definition 6.14. *Given a measure space $(X, \mathfrak{A}, \mu)$. Let $(f_n : n \in \mathbb{N})$ be a sequence of extended real-valued $\mathfrak{A}$-measurable functions on a set $D \in \mathfrak{A}$. We say that $(f_n : n \in \mathbb{N})$ converges in measure μ on D if there exists a real-valued $\mathfrak{A}$-measurable function f on D such that for every $\varepsilon > 0$ we have*

$$(1) \qquad \lim_{n\to\infty} \mu\{D : |f_n - f| \geq \varepsilon\} = 0,$$

that is, for every $\varepsilon > 0$ and $\eta > 0$ there exists $N_{\varepsilon,\eta} \in \mathbb{N}$ such that

$$(2) \qquad \mu\{D : |f_n - f| \geq \varepsilon\} < \eta \quad \text{for } n \geq N_{\varepsilon,\eta}.$$

We write $f_n \xrightarrow{\mu} f$ on D for this convergence.

Remark 6.15. (a) If (1) of Definition 6.14 holds for every $\varepsilon \in (0, \varepsilon_0]$ for some $\varepsilon_0 > 0$, then (1) holds for every $\varepsilon > 0$. This follows from the fact that for $\varepsilon > \varepsilon_0$, we have $\{D : |f_n - f| \geq \varepsilon\} \subset \{D : |f_n - f| \geq \varepsilon_0\}$ so that $\mu\{D : |f_n - f| \geq \varepsilon\} \leq \mu\{D : |f_n - f| \geq \varepsilon_0\}$ and then $\lim_{n\to\infty} \mu\{D : |f_n - f| \geq \varepsilon\} \leq \lim_{n\to\infty} \mu\{D : |f_n - f| \geq \varepsilon_0\} = 0$.

(b) It is clear from (2) of Definition 6.14 that a sequence obtained by dropping finitely many terms from $(f_n : n \in \mathbb{N})$ or by adding finitely many terms to $(f_n : n \in \mathbb{N})$ converges in measure if and only if $(f_n : n \in \mathbb{N})$ converges in measure.

Observation 6.16. Condition (1) in Definition 6.14 is equivalent to the following condition

$$(3) \qquad \lim_{n\to\infty} \mu\{D : |f_n - f| \geq \tfrac{1}{m}\} = 0 \quad \text{for every } m \in \mathbb{N}.$$

Proof. Clearly if (1) holds then (3) holds. Conversely assume that (3) holds. Given $\varepsilon > 0$, let $m \in \mathbb{N}$ be such that $\frac{1}{m} < \varepsilon$. Then $\{D : |f_n - f| \geq \varepsilon\} \subset \{D : |f_n - f| \geq \frac{1}{m}\}$ and $\mu\{D : |f_n - f| \geq \varepsilon\} \leq \mu\{D : |f_n - f| \geq \frac{1}{m}\}$ by the monotonicity of μ. Thus (3) implies (1). Therefore (1) and (3) are equivalent. ∎

Example 1. In $(\mathbb{R}, \mathfrak{M}_L, \mu_L)$, let $(f_n : n \in \mathbb{N})$ be a sequence of functions on $D = [0, \infty)$ defined by

$$f_n(x) = \begin{cases} n & \text{for } x \in \left(n - \frac{1}{n}, n\right], \\ 0 & \text{for } x \in D \setminus \left(n - \frac{1}{n}, n\right], \end{cases}$$

for $n \in \mathbb{N}$. Let $f = 0$ on D. For $\varepsilon \in (0, 1]$, we have

$$\lim_{n\to\infty} \mu_L\{D : |f_n - f| \geq \varepsilon\} = \lim_{n\to\infty} \mu_L\left(\left(n - \tfrac{1}{n}, n\right]\right) = \lim_{n\to\infty} \frac{1}{n} = 0.$$

This shows that $(f_n : n \in \mathbb{N})$ converges to f on D in measure μ_L by (a) of Remark 6.15.

Example 2. In $(\mathbb{R}, \mathfrak{M}_L, \mu_L)$, let $D = [0, \infty)$ and let

$$f_n(x) = \begin{cases} 1 & \text{for } x \in (n-1, n], \\ 0 & \text{for } x \in D \setminus (n-1, n] \end{cases}$$

for $n \in \mathbb{N}$ and let $f = 0$ on D. For $\varepsilon \in (0, 1]$, we have

$$\lim_{n\to\infty} \mu_L\{D : |f_n - f| \geq \varepsilon\} = \mu_L\big((n-1, n]\big) = \lim_{n\to\infty} 1 = 1.$$

Thus $(f_n : n \in \mathbb{N})$ does not converge to f on D in measure μ_L.

Proposition 6.17. *Given a measure space $(X, \mathfrak{A}, \mu)$. Let $(f_n : n \in \mathbb{N})$ be a sequence of extended real-valued $\mathfrak{A}$-measurable functions on a set $D \in \mathfrak{A}$ and let f be a real-valued $\mathfrak{A}$-measurable function on D. Then $f_n \xrightarrow{\mu} f$ on D if and only if for every $\delta > 0$, there exists $N_\delta \in \mathbb{N}$ such that*

$$(4) \qquad \mu\{D : |f_n - f| \geq \delta\} < \delta \quad \textit{for } n \geq N_\delta.$$

Proof. Let us show that (4) is equivalent to (2) of Definition 1.14. Clearly (2) implies (4). To show that (4) implies (2) let $\varepsilon > 0$ and $\eta > 0$ be arbitrarily given and let $\delta = \varepsilon \wedge \eta$. By (4) there exists $N_\delta \in \mathbb{N}$ such that $\mu\{D : |f_n - f| \geq \delta\} < \delta$ for $n \geq N_\delta$. Now $\varepsilon \geq \delta$ implies $\{D : |f_n - f| \geq \varepsilon\} \subset \{D : |f_n - f| \geq \delta\}$ and then we have

$$\mu\{D : |f_n - f| \geq \varepsilon\} \leq \mu\{D : |f_n - f| \geq \delta\} < \delta \leq \eta \quad \text{for } n \geq N_\delta.$$

This shows that (4) implies (2). ∎

Corollary 6.18. $f_n \xrightarrow{\mu} f$ *on D if and only if for every $m \in \mathbb{N}$, there exists $N_m \in \mathbb{N}$ such that*

$$(5) \qquad \mu\{D : |f_n - f| \geq \tfrac{1}{m}\} < \tfrac{1}{m} \quad \textit{for } n \geq N_m.$$

Proof. Let us show that (5) is equivalent to (4) of Proposition 6.17. Clearly (4) implies (5). To show that (5) implies (4), let $\delta > 0$ be arbitrarily given. Let $m(\delta) \in \mathbb{N}$ be such that $\frac{1}{m(\delta)} < \delta$. Then $\{D : |f_n - f| \geq \delta\} \subset \left\{D : |f_n - f| \geq \frac{1}{m(\delta)}\right\}$. By (5) there exists $N_{m(\delta)} \in \mathbb{N}$ such that $\mu\left\{D : |f_n - f| \geq \frac{1}{m(\delta)}\right\} < \frac{1}{m(\delta)}$ for $n \geq N_{m(\delta)}$. Then we have

$$\mu\{D : |f_n - f| \geq \delta\} \leq \mu\left\{D : |f_n - f| \geq \frac{1}{m(\delta)}\right\} < \frac{1}{m(\delta)} < \delta \quad \text{for } n \geq N_{m(\delta)}.$$

This shows that (5) implies (4). ∎

Corollary 6.19. *Given a measure space $(X, \mathfrak{A}, \mu)$. Let $(f_n : n \in \mathbb{N})$ be a sequence of extended real-valued $\mathfrak{A}$-measurable functions on a set $D \in \mathfrak{A}$ and let f be a real-valued*

$\mathfrak{A}$-measurable function on D. Suppose there exist two sequences of positive numbers $(\alpha_n : n \in \mathbb{N})$ and $(\beta_n : n \in \mathbb{N})$ such that

1° $\lim_{n\to\infty} \alpha_n = 0$ *and* $\lim_{n\to\infty} \beta_n = 0$,

2° $\mu\{D : |f_n - f| \geq \alpha_n\} < \beta_n$ *for all every* $n \in \mathbb{N}$.

Then $(f_n : n \in \mathbb{N})$ converges to f on D in measure μ.

Proof. Assume 1° and 2°. Let $\varepsilon > 0$ be arbitrarily given. Let $N \in \mathbb{N}$ be so large that for $n \geq N$ we have $\alpha_n < \varepsilon$ and $\beta_n < \varepsilon$. Then $\{D : |f_n - f| \geq \varepsilon\} \subset \{D : |f_n - f| \geq \alpha_n\}$ so that $\mu\{D : |f_n - f| \geq \varepsilon\} \leq \mu\{D : |f_n - f| \geq \alpha_n\} < \beta_n < \varepsilon$ for $n \geq N$. This shows that $(f_n : n \in \mathbb{N})$ converges to f on D in measure μ according to Proposition 6.17. ∎

Lemma 6.20. *Given a measure space $(X, \mathfrak{A}, \mu)$. Let f and g be two real-valued $\mathfrak{A}$-measurable functions on a set $D \in \mathfrak{A}$. Let ε, ε_1, and ε_2 be positive numbers such that $\varepsilon = \varepsilon_1 + \varepsilon_2$. Then for an arbitrary extended real-valued $\mathfrak{A}$-measurable function h on D, we have*

$$(1) \qquad \{D : |f - g| \geq \varepsilon\} \subset \{D : |f - h| \geq \varepsilon_1\} \cup \{D : |h - g| \geq \varepsilon_2\},$$

and

$$(2) \qquad \mu\{D : |f - g| \geq \varepsilon\} \leq \mu\{D : |f - h| \geq \varepsilon_1\} + \mu\{D : |h - g| \geq \varepsilon_2\}.$$

Proof. Let the three sets in (1) be denoted by A, A_1, and A_2. If $x \in D$ and $x \notin A_1 \cup A_2$, then $x \notin A_1$ and $x \notin A_2$ so that $|f(x) - h(x)| < \varepsilon_1$ and $|h(x) - g(x)| < \varepsilon_2$ and then

$$|f(x) - g(x)| \leq |f(x) - h(x)| + |h(x) - g(x)| < \varepsilon_1 + \varepsilon_2 = \varepsilon$$

so that $x \notin A$. Thus if $x \in D \setminus (A_1 \cup A_2)$ then $x \in D \setminus A$. This shows that $D \setminus (A_1 \cup A_2) \subset D \setminus A$, that is, $A_1 \cup A_2 \supset A$. This proves (1). Then (2) follows from (1) by the monotonicity and subadditivity of μ. ∎

Theorem 6.21. (Uniqueness of Limit of Convergence in Measure) *Given a measure space $(X, \mathfrak{A}, \mu)$. Let $(f_n : n \in \mathbb{N})$ be a sequence of extended real-valued $\mathfrak{A}$-measurable functions on a set $D \in \mathfrak{A}$ and let f and g be two real-valued $\mathfrak{A}$-measurable functions on D. If $f_n \xrightarrow{\mu} f$ on D and $f_n \xrightarrow{\mu} g$ on D also, then $f = g$ a.e. on D.*

Proof. Suppose $f_n \xrightarrow{\mu} f$ on D and $f_n \xrightarrow{\mu} g$ on D also. Then for every $\varepsilon > 0$, we have

$$(1) \qquad \begin{cases} \lim_{n\to\infty} \mu\{D : |f_n - f| \geq \varepsilon\} = 0, \\ \lim_{n\to\infty} \mu\{D : |f_n - g| \geq \varepsilon\} = 0. \end{cases}$$

To show that $f = g$ a.e. on D, let us assume the contrary, that is, $\mu\{D : f \neq g\} > 0$. Then since $f(x) \neq g(x)$ if and only if $|f(x) - g(x)| > 0$, we have

$$(2) \qquad \mu\{D : |f - g| > 0\} > 0.$$

Now since $\{D : |f - g| > 0\} = \bigcup_{k \in \mathbb{N}} \{D : |f - g| \geq \frac{1}{k}\}$, we have

(3) $$\mu\{D : |f - g| > 0\} \leq \sum_{k \in \mathbb{N}} \mu\left\{D : |f - g| \geq \tfrac{1}{k}\right\}.$$

By (2), the left side of (3) is positive. Then not all of the summands on the right side are equal to 0. Thus there exists some $k_0 \in \mathbb{N}$ such that

(4) $$\mu\left\{D : |f - g| \geq \tfrac{1}{k_0}\right\} > 0.$$

By Lemma 6.20, for every $n \in \mathbb{N}$ we have

$$\mu\left\{D : |f - g| \geq \tfrac{1}{k_0}\right\} \leq \mu\left\{D : |f - f_n| \geq \tfrac{1}{2k_0}\right\} + \mu\left\{D : |f_n - g| \geq \tfrac{1}{2k_0}\right\}.$$

Letting $n \to \infty$ on the right side of the last inequality, we have $\mu\left\{D : |f - g| \geq \frac{1}{k_0}\right\} = 0$ by (1). This contradicts (4). ∎

Theorem 6.22. (H. Lebesgue) *Given a measure space $(X, \mathfrak{A}, \mu)$. Let $(f_n : n \in \mathbb{N})$ be a sequence of extended real-valued $\mathfrak{A}$-measurable functions on a set $D \in \mathfrak{A}$ and let f be a real-valued $\mathfrak{A}$-measurable function on D. Suppose*

1° *$(f_n : n \in \mathbb{N})$ converges to f a.e. on D,*

2° *$\mu(D) < \infty$.*

Then $(f_n : n \in \mathbb{N})$ converges to f in measure μ on D.

Proof. According to Proposition 6.11, conditions 1° and 2° imply that for every $m \in \mathbb{N}$ we have $\lim_{n\to\infty} \mu\{D : |f_n - f| \geq \frac{1}{m}\} = 0$. But this is convergence of $(f_n : n \in \mathbb{N})$ to f on D in measure by Observation 6.16. ∎

Remark 6.23. Theorem 6.22 does not hold without the condition $\mu(D) < \infty$.

Example. In $(\mathbb{R}, \mathfrak{M}_L, \mu_L)$, let $D = [0, \infty)$ with $\mu_L(D) = \infty$. Let $f_n(x) = \frac{1}{n}x$ for $x \in D$ and $n \in \mathbb{N}$ and let $f = 0$ on D. Then $(f_n : n \in \mathbb{N})$ converges to f everywhere on D but $(f_n : n \in \mathbb{N})$ does not converge to f on D in measure μ_L since for every $\varepsilon > 0$ we have $\mu_L\{D : |f_n - f| \geq \varepsilon\} = \mu_L\{D : f_n \geq \varepsilon\} = \mu_L([n\varepsilon, \infty)) = \infty$ for every $n \in \mathbb{N}$.

Theorem 6.24. (F. Riesz) *Given a measure space $(X, \mathfrak{A}, \mu)$. Let $(f_n : n \in \mathbb{N})$ be a sequence of extended real-valued $\mathfrak{A}$-measurable functions on a set $D \in \mathfrak{A}$ and let f be a real-valued $\mathfrak{A}$-measurable function on D. If $(f_n : n \in \mathbb{N})$ converges to f in measure μ on D, then there exists a subsequence $(f_{n_k} : k \in \mathbb{N})$ which converges to f, μ-a.e. on D.*

Proof. If $(f_n : n \in \mathbb{N})$ converges to f on D in measure, then by Proposition 6.17, for every $\varepsilon > 0$ there exists $N_\varepsilon \in \mathbb{N}$ such that $\mu\{D : |f_n - f| \geq \varepsilon\} < \varepsilon$ for $n \geq N_\varepsilon$. Then for $\varepsilon = \frac{1}{2}$, there exists $n_1 \in \mathbb{N}$ such that

$$\mu\{D : |f_{n_1} - f| \geq \tfrac{1}{2}\} < \tfrac{1}{2},$$

and for $\varepsilon = \frac{1}{2^2}$, there exists $n_2 \in \mathbb{N}$, $n_2 > n_1$, such that

$$\mu\{D : |f_{n_2} - f| \geq \tfrac{1}{2^2}\} < \tfrac{1}{2^2},$$

and so on. In general for every $k \in \mathbb{N}$, there exists $n_k \in \mathbb{N}$, $n_k > n_{k-1}$, such that

$$\mu\{D : |f_{n_k} - f| \geq \tfrac{1}{2^k}\} < \tfrac{1}{2^k}.$$

Then since $\lim_{k\to\infty} \frac{1}{2^k} = 0$ and since $\sum_{k\in\mathbb{N}} \mu\{D : |f_{n_k} - f| \geq \frac{1}{2^k}\} < \sum_{k\in\mathbb{N}} \frac{1}{2^k} = 1 < \infty$, the sequence $(f_{n_k} : k \in \mathbb{N})$ converges to f a.e. on D by Theorem 6.7. ∎

Example. (A Sequence Converging in Measure but Diverging Everywhere)
1. Consider the Lebesgue measure space $(\mathbb{R}, \mathfrak{M}_L, \mu_L)$. For $\alpha \in \mathbb{R}$, let $[\alpha]$ be the greatest integer not exceeding α and let $(\alpha)_{mod\,1} = \alpha - [\alpha]$. Thus $(\alpha)_{mod\,1} \in [0, 1)$ for every $\alpha \in \mathbb{R}$ and $(\alpha)_{mod\,1} = 0$ if and only if α is an integer. Let us define a sequence $(\alpha_n : n \in \mathbb{N})$ in $D = [0, 1)$ by setting for every $n \in \mathbb{N}$

$$\alpha_n = \left(1 + \frac{1}{2} + \frac{1}{3} + \cdots + \frac{1}{n}\right)_{mod\,1}.$$

For instance, we have

$$\alpha_1 = (1)_{mod\,1} = 0,$$

$$\alpha_2 = \left(1 + \frac{1}{2}\right)_{mod\,1} = \left(\frac{3}{2}\right)_{mod\,1} = \frac{1}{2},$$

$$\alpha_3 = \left(1 + \frac{1}{2} + \frac{1}{3}\right)_{mod\,1} = \left(\frac{11}{6}\right)_{mod\,1} = \frac{5}{6},$$

$$\alpha_4 = \left(1 + \frac{1}{2} + \frac{1}{3} + \frac{1}{4}\right)_{mod\,1} = \left(\frac{25}{12}\right)_{mod\,1} = \frac{1}{12},$$

and so on. Define a sequence of subsets of D, $(D_n : n \in \mathbb{N})$, by setting

$$D_n = \begin{cases} [\alpha_n, \alpha_{n+1}) & \text{if } \alpha_n < \alpha_{n+1}, \\ [\alpha_n, 1) \cup [0, \alpha_{n+1}) & \text{if } \alpha_n > \alpha_{n+1}. \end{cases}$$

Note that $\mu_L(D_n) = \frac{1}{n+1}$ for $n \in \mathbb{N}$. Since $\sum_{n\in\mathbb{N}} \frac{1}{n} = \infty$, the endpoint α_{n+1} of D_n traverses $D = [0, 1)$ infinitely many times as $n \to \infty$. Define a sequence of functions $(f_n : n \in \mathbb{N})$ on $D = [0, 1)$ by setting $f_n(x) = 1$ for $x \in D_n$ and $f_n(x) = 0$ for $x \in D \setminus D_n$, and let $f(x) = 0$ for $x \in D$.

2. For $\varepsilon \in (0, 1]$, we have

$$\lim_{n\to\infty} \mu_L\{D : |f_n - f| \geq \varepsilon\} = \lim_{n\to\infty} \mu_L(D_n) = \lim_{n\to\infty} \frac{1}{n+1} = 0.$$

This shows that $(f_n : n \in \mathbb{N})$ converges to f on D in measure μ_L.

3. Let us show that $(f_n(x) : n \in \mathbb{N})$ diverges for every $x \in D$. For each $x \in D$, we have $x \in D_n$ for infinitely many $n \in \mathbb{N}$ and $x \notin D_n$ for infinitely many $n \in \mathbb{N}$ also. Thus $f_n(x) = 1$ for infinitely many $n \in \mathbb{N}$ and $f_n(x) = 0$ for infinitely many $n \in \mathbb{N}$ also. This shows $(f_n(x) : n \in \mathbb{N})$ diverges.

4. To select a subsequence $(f_{n_k} : n \in \mathbb{N})$ of $(f_n : n \in \mathbb{N})$ which converges to f a.e. on D, let us decompose D into a disjoint sequence of subintervals $(A_m : m \in \mathbb{N})$ of the type $[a, b)$ with $\mu_L(A_m) = \frac{1}{2^m}$. Thus

$$A_1 = \left[0, \tfrac{1}{2}\right), A_2 = \left[\tfrac{1}{2}, \tfrac{1}{2} + \tfrac{1}{2^2}\right), A_3 = \left[\tfrac{1}{2} + \tfrac{1}{2^2}, \tfrac{1}{2} + \tfrac{1}{2^2} + \tfrac{1}{2^3}\right), \ldots .$$

Since $\lim_{n\to\infty} \mu_L(D_n) = \lim_{n\to\infty} \frac{1}{n+1} = 0$, we can select $n_1 \in \mathbb{N}$ such that $D_{n_1} \subset A_1$. Then select $n_2 \in \mathbb{N}$, $n_2 > n_1$, such that $D_{n_2} \subset A_2$ and so on. Consider the subsequence $(f_{n_k} : k \in \mathbb{N})$ of $(f_n : n \in \mathbb{N})$. For each $x \subset D$, we have $x \in A_m$ for some $m \in \mathbb{N}$. Then for $k > m$ we have $f_{n_k}(x) = 0$ and thus $\lim_{x\to\infty} f_{n_k}(x) = 0 = f(x)$. Thus our subsequence $(f_{n_k} : k \in \mathbb{N})$ converges to f everywhere on D.

5. Note also that the sequence $(f_n : n \in \mathbb{N})$ not only converges to f on D in measure but it satisfies the stronger condition $\lim_{n\to\infty} \mu_L\{D : |f_n - f| > 0\} = 0$. Indeed we have $\{D : |f_n - f| > 0\} = D_n$ and $\mu_L(D_n) = \frac{1}{n+1}$.

[IV] Cauchy Sequences in Convergence in Measure

Definition 6.25. *Given a measure space $(X, \mathfrak{A}, \mu)$. Let $(f_n : n \in \mathbb{N})$ be a sequence of real-valued $\mathfrak{A}$-measurable functions on a set $D \in \mathfrak{A}$. We say that the sequence is a Cauchy sequence with respect to convergence in measure μ on D if for every $\varepsilon > 0$ and $\eta > 0$ there exists $N_{\varepsilon,\eta} \in \mathbb{N}$ such that*

$$(1) \qquad \mu\{x \in D : |f_m(x) - f_n(x)| \geq \varepsilon\} < \eta \quad \text{for } m, n \geq N_{\varepsilon,\eta},$$

or equivalently, for every $\delta > 0$ there exists $N_\delta \in \mathbb{N}$ such that

$$(2) \qquad \mu\{x \in D : |f_m(x) - f_n(x)| \geq \delta\} < \delta \quad \text{for } m, n \geq N_\delta.$$

(The equivalence of (1) and (2) can be proved by the same argument as in Proposition 6.17.)

Observation 6.26. If $(f_n : n \in \mathbb{N})$ is a sequence of real-valued $\mathfrak{A}$-measurable functions and if $f_n \xrightarrow{\mu} f$ on D, then $(f_n : n \in \mathbb{N})$ is a Cauchy sequence with respect to convergence in measure μ on D.

Proof. If $f_n \xrightarrow{\mu} f$ on D, then by Proposition 6.17 for every $\varepsilon > 0$ there exists $N_\varepsilon > 0$ such that $\mu\left\{D : |f_n - f| \geq \frac{\varepsilon}{2}\right\} < \frac{\varepsilon}{2}$ for $n \geq N_\varepsilon$. Then by Lemma 6.20 we have

$$\begin{aligned}\mu\{D : |f_m - f_n| \geq \varepsilon\} &\leq \mu\left\{D : |f_m - f| \geq \tfrac{\varepsilon}{2}\right\} + \mu\left\{D : |f_n - f| \geq \tfrac{\varepsilon}{2}\right\} \\ &< \frac{\varepsilon}{2} + \frac{\varepsilon}{2} = \varepsilon \quad \text{for } m, n \geq N_\varepsilon.\end{aligned}$$

This shows that $(f_n : n \in \mathbb{N})$ is a Cauchy sequence with respect to convergence in measure μ on D. ∎

Theorem 6.27. *Given a measure space $(X, \mathfrak{A}, \mu)$. Let $(f_n : n \in \mathbb{N})$ be a sequence of real-valued $\mathfrak{A}$-measurable functions on a set $D \in \mathfrak{A}$. If $(f_n : n \in \mathbb{N})$ is a Cauchy sequence with respect to convergence in measure μ on D, then there exists a real-valued $\mathfrak{A}$-measurable function on D such that $f_n \overset{\mu}{\to} f$ on D.*

Proof. 1. If $(f_n : n \in \mathbb{N})$ is a Cauchy sequence with respect to convergence in measure on D, then by (2) of Definition 6.25 for every $\delta > 0$ there exists $N_\delta \in \mathbb{N}$ such that

$$\mu\{D : |f_m - f_n| \ge \delta\} < \delta \quad \text{for } m, n \ge N_\delta.$$

Then for $\delta = \frac{1}{2}$, there exists $n_1 \in \mathbb{N}$ such that we have $\mu\{D : |f_{n_1+p} - f_{n_1}| \ge \frac{1}{2}\} < \frac{1}{2}$ for all $p \in \mathbb{N}$. Next for $\delta = \frac{1}{2^2}$, there exists $n_2 \in \mathbb{N}$, $n_2 > n_1$, such that we have $\mu\{D : |f_{n_2+p} - f_{n_2}| \ge \frac{1}{2^2}\} < \frac{1}{2^2}$ for all $p \in \mathbb{N}$, and so on. In general for $\delta = \frac{1}{2^k}$ there exists $n_k \in \mathbb{N}$, $n_k > n_{k-1}$, such that

$$(1) \qquad \mu\{D : |f_{n_k+p} - f_{n_k}| \ge \tfrac{1}{2^k}\} < \tfrac{1}{2^k} \quad \text{for all } p \in \mathbb{N}.$$

Thus for the subsequence $(f_{n_k} : n \in \mathbb{N})$ of $(f_n : n \in \mathbb{N})$, since $n_{k+1} = n_k + p$ for some $p \in \mathbb{N}$, we have

$$(2) \qquad \mu\{D : |f_{n_{k+1}} - f_{n_k}| \ge \tfrac{1}{2^k}\} < \tfrac{1}{2^k} \quad \text{for } k \in \mathbb{N}.$$

For brevity, let us write $g_k = f_{n_k}$ for $k \in \mathbb{N}$. Let us show that the sequence $(g_k : k \in \mathbb{N})$ converges a.e. on D. According to Lemma 6.2, it suffices to show that for every $\eta > 0$ there exists a $\mathfrak{A}$-measurable subset E of D with $\mu(E) < \eta$ such that $(g_k(x) : k \in \mathbb{N})$ converges for every $x \in D \setminus E$. Let $D_j = \{D : |g_{j+1} - g_j| \ge \frac{1}{2^j}\}$ for $j \in \mathbb{N}$ and let $E_j = \bigcup_{\ell \ge j} D_\ell$ for $j \in \mathbb{N}$. By (2), we have $\mu(D_j) < \frac{1}{2^j}$ and then

$$\mu(E_j) \le \sum_{\ell \ge j} \mu(D_\ell) < \sum_{\ell \ge j} \frac{1}{2^\ell} < \frac{1}{2^{j-1}} \sum_{\ell \in \mathbb{N}} \frac{1}{2^\ell} = \frac{1}{2^{j-1}}.$$

For $\eta > 0$ arbitrarily given, let $j \in \mathbb{N}$ be so large that $\frac{1}{2^{j-1}} < \eta$ so that $\mu(E_j) < \eta$. Let us show that $(g_k(x) : k \in \mathbb{N})$ converges for every $x \in D \setminus E_j$. Let $\varepsilon > 0$ be arbitrarily given. Let $N \in \mathbb{N}$ be so large that $\frac{1}{2^{N-1}} < \varepsilon$ and $N \ge j$ also. If $x \in D \setminus E_j$, then $x \notin D_\ell$ for $\ell \ge j$ so that by (2) we have

$$(3) \qquad |g_{N+p}(x) - g_N(x)| \le \sum_{\ell=N}^{N+p-1} |g_{\ell+1}(x) - g_\ell(x)|$$

$$< \sum_{\ell=N}^{N+p-1} \frac{1}{2^\ell} < \frac{1}{2^{N-1}} \sum_{\ell \in \mathbb{N}} \frac{1}{2^\ell} = \frac{1}{2^{N-1}} < \varepsilon \quad \text{for } p \in \mathbb{N}.$$

This shows that $(g_k(x) : k \in \mathbb{N})$ is a Cauchy sequence of real numbers and hence it is a convergent sequence. Therefore we have shown that for the arbitrarily given $\eta > 0$, we have a $\mathfrak{A}$-measurable subset E_j of D such that $\mu(E_j) < \eta$ and $(g_k(x) : k \in \mathbb{N})$ converges for $x \in D \setminus E_j$. Thus by Lemma 6.2, $(g_k : k \in \mathbb{N})$ converges a.e. on D. Let $D_c = \{x \in D : \lim_{k\to\infty} g_k(x) \in \mathbb{R}\}$. Then $D_c \in \mathfrak{A}$ by Theorem 4.23 and $\mu(D \setminus D_c) = 0$ since $(g_k : k \in \mathbb{N})$ converges a.e. on D. Let us define a real-valued $\mathfrak{A}$-measurable function f on D by setting $f(x) = \lim_{k\to\infty} g_k(x)$ for $x \in D_c$ and $f(x) = 0$ for $x \in D \setminus D_c$.

2. Let us show next that $g_k \xrightarrow{\mu} f$ on D. (Note that Theorem 6.22 is not applicable here since we do not assume $\mu(D) < \infty$.) In (3) we showed that for every $j \in \mathbb{N}$ we have

$$|g_{N+p}(x) - g_N(x)| < \tfrac{1}{2^{N-1}} \quad \text{for } N \geq j,\ p \in \mathbb{N} \text{ and } x \in D \setminus E_j.$$

Letting $p \to \infty$ and recalling $\lim_{n\to\infty} g_k(x) = f(x)$ for x in $D_c \supset D \setminus E_j$, we have

$$|f(x) - g_N(x)| \leq \tfrac{1}{2^{N-1}} \quad \text{for } N \geq j \text{ and } x \in D \setminus E_j.$$

This implies that $\{D : |g_N - f| > \frac{1}{2^{N-1}}\} \subset E_j$ for $N \geq j$ and then we have

$$\mu\{D : |g_N - f| > \tfrac{1}{2^{N-1}}\} \leq \mu(E_j) < \tfrac{1}{2^{j-1}} \quad \text{for } N \geq j.$$

For an arbitrary $\varepsilon > 0$, let $j \in \mathbb{N}$ be so large that $\frac{1}{2^{j-2}} < \varepsilon$. Then for $N \geq j$, we have

$$\begin{aligned}\mu\{D : |g_N - f| \geq \varepsilon\} &\leq \mu\{D : |g_N - f| \geq \tfrac{1}{2^{j-2}}\} \leq \mu\{D : |g_N - f| > \tfrac{1}{2^{j-1}}\} \\ &\leq \mu\{D : |g_N - f| > \tfrac{1}{2^{N-1}}\} < \frac{1}{2^{j-1}} < \frac{1}{2^{j-2}} < \varepsilon.\end{aligned}$$

This shows that $g_k \xrightarrow{\mu} f$ on D by Proposition 6.17.

3. Finally to show that $f_n \xrightarrow{\mu} f$ on D, note that for every $\varepsilon > 0$ and every f_{n_k} we have

$$\mu\{D : |f_n - f| \geq \varepsilon\} \leq \mu\left\{D : |f_n - f_{n_k}| \geq \tfrac{\varepsilon}{2}\right\} + \mu\left\{D : |f_{n_k} - f| \geq \tfrac{\varepsilon}{2}\right\},$$

by Lemma 6.20. Now since $(f_n : n \in \mathbb{N})$ is a Cauchy sequence with respect to convergence in measure, there exists $N'_\varepsilon \in \mathbb{N}$ such that $\mu\left\{D : |f_n - f_{n_k}| \geq \frac{\varepsilon}{2}\right\} < \frac{\varepsilon}{2}$ for $n, n_k \geq N'_\varepsilon$. On the other hand since $f_{n_k} \xrightarrow{\mu} f$ on D, there exists $N''_\varepsilon \in \mathbb{N}$ such that $\mu\left\{D : |f_{n_k} - f| \geq \frac{\varepsilon}{2}\right\} < \frac{\varepsilon}{2}$ for $n_k \geq N''_\varepsilon$. If we let $N_\varepsilon = \max\{N'_\varepsilon, N''_\varepsilon\}$, then $\mu\{D : |f_n - f| \geq \varepsilon\} < \frac{\varepsilon}{2} + \frac{\varepsilon}{2} = \varepsilon$ for $n \geq N_\varepsilon$. This shows that $f_n \xrightarrow{\mu} f$ on D by Proposition 6.17. ∎

Remark 6.28. **(a)** Let $(f_n : n \in \mathbb{N})$ be a sequence of real-valued $\mathfrak{A}$-measurable functions on a set $D \in \mathfrak{A}$ in a measure space $(X, \mathfrak{A}, \mu)$. By Observation 6.26 and Theorem 6.27, the sequence converges to a real-valued $\mathfrak{A}$-measurable function on D in measure μ if and only if it is a Cauchy sequence with respect to convergence in measure μ.

(b) Referring to Definition 6.25, the definition of Cauchy sequence for convergence in measure is restricted to sequences of real-valued (but not extended real-valued) measurable

functions. The reason for this restriction is as follows. For given $\varepsilon > 0$ and $\eta > 0$, consider the set $D_{m,n} = \{D : |f_m - f_n| \geq \varepsilon\}$. If $f_n(x) = \infty$ for all $x \in D$ and $n \in \mathbb{N}$, then the difference $f_m(x) - f_n(x)$ is undefined for any $m, n \in \mathbb{N}$ and $x \in D$. Then $D_{m,n} = \emptyset$ since by definition $D_{m,n}$ is the set of all points in D at which $|f_m(x) - f_n(x)|$ is defined and is not less than ε. Thus we have $\mu(D_{m,n}) = 0 < \eta$ for all $m, n \in \mathbb{N}$. Yet the sequence $(f_n : n \in \mathbb{N})$ does not converge in measure on D to a real-valued measurable function.
(c) If f_n is real-valued a.e. on D for every $n \in \mathbb{N}$, that is, there exists a subset E_n of D which is a null set in $(X, \mathfrak{A}, \mu)$ such that f_n is real-valued on $D \setminus E_n$, then $E = \bigcup_{n\in\mathbb{N}} E_n$ is a null set contained in D and by restricting f_n to $D \setminus E$ for every $n \in \mathbb{N}$, we have a sequence of real-valued measurable functions on $D \setminus E$ to which we can apply Theorem 6.27.

[V] Approximation by Step Functions and Continuous Functions

[V.1] Approximation of Lebesgue Measurable Functions by Step Functions

Definition 6.29. *By a right-continuous step function on a finite interval $[a, b) \in \mathfrak{I}_{co}$, we mean a function f on $[a, b)$ of the form $f = \sum_{k=1}^{n} c_k \cdot \mathbf{1}_{I_k}$ where $(I_k : k = 1, \ldots, n)$ is a finite disjoint sequence in $\mathfrak{I}_{co}$ with $\bigcup_{k=1}^{n} I_k = [a, b)$. By a right-continuous step function on $\mathbb{R}$ or on an infinite interval $[a, \infty) \in \mathfrak{I}_{co}$, we mean a function f on $\mathbb{R}$ or $[a, \infty)$ such that its restriction to any finite subinterval of the class $\mathfrak{I}_{co}$ is a right-continuous step function on the finite subinterval.*

Let us note that the set of the endpoints of the intervals in the domain of definition of a right-continuous step function on $\mathbb{R}$ or on $[a, \infty)$ has no limit points in $\mathbb{R}$. In particular the set of points of discontinuity of a right-continuous step function is a countable set having no limit points in $\mathbb{R}$. Let us note also that a right-continuous step function on an infinite interval may be an unbounded function and its range is a countable subset of $\mathbb{R}$.

Observation 6.30. If $f_1, \ldots, f_n$ are right-continuous step functions on a common domain of definition, then $\alpha_1 f_1 + \cdots + \alpha_n f_n$ with $\alpha_1, \ldots, \alpha_n \in \mathbb{R}$ is a right-continuous step function. If f_1 and f_2 are right-continuous step functions on the same domain of definition, then $f_1 \vee f_2$ and $f_1 \wedge f_2$ are right-continuous step functions.

Lemma 6.31. *Given a measure space $(X, \mathfrak{A}, \mu)$. Let f be an extended real-valued $\mathfrak{A}$-measurable function on a set $D \in \mathfrak{A}$ with $\mu(D) < \infty$. If f is real-valued a.e. on D, then for every $\varepsilon > 0$ there exist a $\mathfrak{A}$-measurable subset E of D with $\mu(E) < \varepsilon$ and a constant $B > 0$ such that $|f| \leq B$ on $D \setminus E$.*

Proof. Let $D_n = \{D : |f| \leq n\} = f^{-1}([-n, n])$ for $n \in \mathbb{N}$. Then $(D_n : n \in \mathbb{N})$ is an increasing sequence in $\mathfrak{A}$ with $\lim_{n\to\infty} D_n = \bigcup_{n\in\mathbb{N}} D_n = \{D : |f| < \infty\}$ so that $\lim_{n\to\infty} \mu(D_n) = \mu\{D : |f| < \infty\} = \mu(D) < \infty$ since $\mu\{D : |f| = \infty\} = 0$. Thus for an arbitrarily given $\varepsilon > 0$, there exists $N \in \mathbb{N}$ such that $\mu(D \setminus D_N) = \mu(D) - \mu(D_N) < \varepsilon$. Let $E = D \setminus D_N$. Then $\mu(E) < \varepsilon$ and for $x \in D \setminus E = D_N$ we have $|f(x)| \leq N$. Let $B = N$. ∎

Remark 6.32. Lemma 6.31 does not hold without the condition $\mu(D) < \infty$.

Example 1. In $(\mathbb{R}, \mathfrak{M}_L, \mu_L)$, let $D = [0, \infty)$ and let $f(x) = x$ for $x \in D$. We have $\mu_L(D) = \infty$. Let us show that not only for some $\varepsilon > 0$ but in fact for every $\varepsilon > 0$ it is impossible to have a $\mathfrak{M}_L$-measurable subset E of D with $\mu_L(E) < \varepsilon$ such that f is bounded on $D \setminus E$. Let E be a $\mathfrak{M}_L$-measurable subset of D with $\mu_L(E) = c > 0$. Let $I_n = \big[2(n-1)c, 2nc\big)$ for $n \in \mathbb{N}$. Then $(I_n : n \in \mathbb{N})$ is a disjoint sequence of intervals and $\bigcup_{n \in \mathbb{N}} I_n = D$. Now since $\mu_L(I_n) = 2c$ and $\mu_L(E) = c$, I_n is not a subset of E and therefore there exists $x_n \in I_n \setminus E$ for every $n \in \mathbb{N}$. Consider the subset $\{x_n : n \in \mathbb{N}\}$ of $D \setminus E$. Since $x_n \in I_n$, we have $f(x_n) \geq 2(n-1)c$ for $n \in \mathbb{N}$ and thus for any constant $B > 0$, we have $f(x_n) > B$ for sufficiently large $n \in \mathbb{N}$. Thus f is not bounded on $D \setminus E$.

Example 2. In $(\mathbb{R}, \mathfrak{M}_L, \mu_L)$, let $D = \mathbb{R} \setminus \{k\pi : k \in \mathbb{Z}\}$ and let $f(x) = |\csc x|$ for $x \in D$. We have $\mu_L(D) = \infty$. Let us show that for any constant $B > 0$ however large, it is impossible to have a $\mathfrak{M}_L$-measurable subset E of D with finite measure such that $|f| \leq B$ on $D \setminus E$. Let $c \in \big(0, \frac{\pi}{2}\big)$ be such that $\csc c = B$. Then $\{D : |f| > B\} = \bigcup_{k \in \mathbb{Z}} (k\pi - c, k\pi + c)$ so that $\mu_L\{D : |f| > B\} = \infty$. Thus any $\mathfrak{M}_L$-measurable subset E of D with $\mu_L(E) < \infty$ cannot have the set $\{D : |f| > B\}$ as a subset and f cannot satisfy the condition $|f| \leq B$ on $D \setminus E$.

Theorem 6.33. *Let f be an extended real-valued $\mathfrak{M}_L$-measurable function on $D = [a, b)$. Suppose f is real-valued a.e. on D. Then for every $\varepsilon > 0$ there exist a $\mathfrak{M}_L$-measurable subset E of D with $\mu_L(E) < \varepsilon$ and a right-continuous step function s on D such that $|f - s| < \varepsilon$ on $D \setminus E$. Moreover if $M_1 \leq f \leq M_2$ on D for two constants $M_1, M_2 \in \mathbb{R}$ such that $M_1 < M_2$, then s can be so chosen that $M_1 \leq s \leq M_2$ on D.*

Proof. 1. Consider first the case where f is nonnegative extended real-valued on D. Since $\mu_L(D) < \infty$, according to Lemma 6.31 for an arbitrary $\varepsilon > 0$ there exists a $\mathfrak{M}_L$-measurable subset D_{ub} of D with $\mu_L(D_{ub}) < \frac{\varepsilon}{2}$ and there exists $B > 0$ such that $f(x) \in [0, B)$ for $x \in D \setminus D_{ub}$. Let $D_b = D \setminus D_{ub}$. Then

(1) $$\mu_L(D_b) = \mu_L(D) - \mu_L(D_{ub}) > \mu_L(D) - \tfrac{\varepsilon}{2}.$$

Let $N \in \mathbb{N}$ be so large that $\frac{B}{N} < \varepsilon$. Let

(2) $$D_n = \big\{D_b : f \in \big[n\tfrac{B}{N}, (n+1)\tfrac{B}{N}\big)\big\} \quad \text{for } n = 0, \dots, N-1.$$

Note that $D_n = f^{-1}\big(\big[n\frac{B}{N}, (n+1)\frac{B}{N}\big)\big) \cap D_b \in \mathfrak{M}_L$ for $n = 0, \dots, N-1$. Let $D_N = D_{ub}$. Then $\{D_0, \dots, D_N\}$ is a disjoint collection and

(3) $$\bigcup_{n=0}^{N-1} D_n = D_b \quad \text{and} \quad \bigcup_{n=0}^{N} D_n = D.$$

For each $n = 0, \dots, N$, we have $\mu_L(D_n) < \infty$ so that by Theorem 3.25 there exists a union V_n of finitely many finite open intervals such that

(4) $$\mu_L(D_n \triangle V_n) < \frac{\varepsilon}{2(N+1)^2}.$$

To each open interval in the union V_n adjoin its left endpoint so that V_n is a union of intervals of the class $\mathfrak{I}_{co}$. This does not affect the inequality (4). For $n = 0, \ldots, N$, let us define a function s_n on D by

(5)
$$s_n(x) = \begin{cases} n\frac{B}{N} & \text{for } x \in V_n \cap D, \\ 0 & \text{for } x \in D \setminus V_n. \end{cases}$$

Thus defined, s_n is a right-continuous step function on D. Let

(6)
$$s(x) = \sum_{n=0}^{N} s_n(x) \quad \text{for } x \in D.$$

Then s is a right-continuous step function on D by Observation 6.30. By (6), we have

(7)
$$s(x) = s_n(x) \quad \text{if } x \in V_n \cap D \text{ and } x \notin V_k \text{ for } k \neq n.$$

For each $n = 0, \ldots, N$, let

(8)
$$\begin{aligned} D_n^* &= (V_n \cap D_n) \setminus \bigcup_{k \neq n} V_k \\ &= (V_n \cap D_n) \setminus \bigcup_{k \neq n} \left[(V_k \setminus D_k) \cup (V_k \cap D_k)\right] \\ &= (V_n \cap D_n) \setminus \left[\bigcup_{k \neq n} (V_k \setminus D_k) \cup \bigcup_{k \neq n} (V_k \cap D_k)\right] \\ &= (V_n \cap D_n) \setminus \bigcup_{k \neq n} (V_k \setminus D_k), \end{aligned}$$

where the last equality is from the fact that if $k \neq n$, then D_n and D_k are disjoint so that $V_n \cap D_n$ and $V_k \cap D_k$ are disjoint. Now we have

(9)
$$\begin{aligned} \mu_L(D_n^*) &\geq \mu_L(V_n \cap D_n) - \mu_L\Big(\bigcup_{k \neq n} (V_k \setminus D_k)\Big) \\ &\geq \mu_L(V_n \cap D_n) - \sum_{k \neq n} \mu_L(V_k \setminus D_k) \\ &\geq \mu_L(D_n) - \mu_L(V_n \triangle D_n) - \sum_{k \neq n} \mu_L(V_k \triangle D_k) \\ &= \mu_L(D_n) - \sum_{k=0}^{N} \mu_L(V_k \triangle D_k) \\ &> \mu_L(D_n) - \frac{\varepsilon}{2(N+1)}, \end{aligned}$$

where the last inequality is by (4). Let

$$D^* = \bigcup_{n=0}^{N-1} D_n^* \quad \text{and} \quad E = D \setminus D^*. \tag{10}$$

Since $\{D_n : n = 0, \ldots, N\}$ is a disjoint collection and $D_n^* \subset D_n$ for $n = 0, \ldots, N$, $\{D_n^* : n = 0, \ldots, N\}$ is a disjoint collection. Thus we have

$$\begin{aligned} \mu_L(D^*) &= \sum_{n=0}^{N-1} \mu_L(D_n^*) > \sum_{n=0}^{N-1} \mu_L(D_n) - \frac{\varepsilon}{2}\frac{N}{N+1} \\ &> \mu_L(D_b) - \frac{\varepsilon}{2} > \mu_L(D) - \varepsilon, \end{aligned}$$

where the first inequality is by (9), the second inequality is by (3) and the last inequality is by (1). Thus we have $\mu_L(E) = \mu_L(D) - \mu_L(D^*) < \varepsilon$.

Let us show that $|f - s| < \varepsilon$ on $D \setminus E$. If $x \in D \setminus E = D^*$, then $x \in D_n^*$ for some $n = 1, \ldots, N-1$. Then since $x \in D_n^* \subset D_n$ by (8) and $n \leq N-1$, we have $f(x) \in \left[n\frac{B}{N}, (n+1)\frac{B}{N}\right)$ by (2). On the other hand, since $x \in D_n^*$, we have $x \in V_n \cap D_n$ and $x \notin V_k$ for $k \neq n$ according to (8). Thus $s(x) = s_n(x) = n\frac{B}{N}$ by (7) and (5). Therefore $|f(x) - s(x)| < \frac{B}{N} < \varepsilon$. This proves that $|f - s| < \varepsilon$ on $D \setminus E$.

2. If $0 \leq f \leq M$ on D for some constant $M \geq 0$, let $s' = s \wedge M$. Then s' is a right-continuous step function on D such that $0 \leq s' \leq M$ and $|f - s'| < \varepsilon$ on $D \setminus E$.

3. Let f be extended real-valued on D and let $f = f^+ - f^-$. By our result in **1**, for $\varepsilon > 0$ there exists a $\mathfrak{M}_L$-measurable subset E_1 of D with $\mu_L(E_1) < \frac{\varepsilon}{2}$ and there exists a right-continuous step function s_1 on D such that $|f^+ - s_1| < \frac{\varepsilon}{2}$ on $D \setminus E_1$ and similarly there exists a $\mathfrak{M}_L$-measurable subset E_2 of D with $\mu_L(E_2) < \frac{\varepsilon}{2}$ and there exists a right-continuous step function s_2 on D such that $|f^- - s_2| < \frac{\varepsilon}{2}$ on $D \setminus E_2$. If we let $E = E_1 \cup E_2$ and $s = s_1 - s_2$, then E is a $\mathfrak{M}_L$-measurable subset of D with $\mu_L(E) \leq \mu_L(E_1) + \mu_L(E_2) < \varepsilon$ and s is a right-continuous step function on D. For $x \in D \setminus E$ we have

$$\begin{aligned} |f(x) - s(x)| &= \left|\{f^+(x) - f^-(x)\} - \{s_1(x) - s_2(x)\}\right| \\ &\leq |f^+(x) - s_1(x)| + |f^-(x) - s_2(x)| < \frac{\varepsilon}{2} + \frac{\varepsilon}{2} = \varepsilon. \end{aligned}$$

4. Suppose $M_1 \leq f \leq M_2$ on D for two constants $M_1, M_2 \in \mathbb{R}$ such that $M_1 < M_2$. Consider the function $g = f - M_1$ on D. Now $0 \leq g \leq M_2 - M_1$ on D. For every $\varepsilon > 0$, by our result in **2** there exists a $\mathfrak{M}_L$-measurable subset E of D with $\mu_L(E) < \varepsilon$ and there exists a right-continuous step function t on D such that $0 \leq t \leq M_2 - M_1$ on D and $|g - t| < \varepsilon$ on $D \setminus E$. Let $s = t + M_1$ on D. Then s is a right-continuous step function on D such that $M_1 \leq s \leq M_2$ on D, and $|f - s| = |\{g + M_1\} - \{t + M_1\}| = |g - t| < \varepsilon$ on $D \setminus E$. ∎

In Theorem 6.33 the domain of definition of the $\mathfrak{M}_L$-measurable function f is assumed to be a finite interval so that it has a finite Lebesgue measure. With this assumption, Lemma

6.31 and Theorem 3.25 are applicable in constructing a step function approximating f. Once the existence of such a step function is proved for a $\mathfrak{M}_L$-measurable function defined on a finite interval, we can use this result to show that a $\mathfrak{M}_L$-measurable function defined on an infinite interval can be approximated by a step function. This is done in the next Corollary.

Corollary 6.34. *Let f be an extended real-valued $\mathfrak{M}_L$-measurable function on $D = [a, \infty)$ or $\mathbb{R}$. Suppose f is real-valued a.e. on D. Then for every $\varepsilon > 0$ there exist a $\mathfrak{M}_L$-measurable subset E of D with $\mu_L(E) < \varepsilon$ and a right-continuous step function s on D such that $|f - s| < \varepsilon$ on $D \setminus E$. Moreover if $M_1 \le f \le M_2$ on D for two constants $M_1, M_2 \in \mathbb{R}$ such that $M_1 < M_2$, then s can be so chosen that $M_1 \le s \le M_2$ on D.*

Proof. The Corollary can be proved in the same way for the two cases $D = [a, \infty)$ and $D = \mathbb{R}$. Let us consider the case $D = \mathbb{R}$. Let $\varepsilon > 0$ be arbitrarily given. Let $I_n = [n-1, n)$ for $n \in \mathbb{Z}$. Then $\{I_n : n \in \mathbb{Z}\}$ is a disjoint collection of finite intervals of the class $\mathfrak{I}_{co}$ with $\bigcup_{n\in\mathbb{Z}} I_n = \mathbb{R}$. By Theorem 6.33 there exist a $\mathfrak{M}_L$-measurable subset E_n of I_n with $\mu_L(E_n) < 2^{-|n|}\frac{\varepsilon}{3}$ and a right-continuous step function s_n on I_n such that $|f - s_n| < 2^{-|n|}\frac{\varepsilon}{3}$ on $I_n \setminus E_n$ for every $n \in \mathbb{Z}$. If $M_1 \le f \le M_2$ on D, then s_n can be so chosen that $M_1 \le s_n \le M_2$ on I_n according to Theorem 6.33. If we let $E = \bigcup_{n\in\mathbb{Z}} E_n$ then E is a $\mathfrak{M}_L$-measurable subset of D with

$$\mu_L(E) = \mu_L\Big(\bigcup_{n\in\mathbb{Z}} E_n\Big) = \sum_{n\in\mathbb{Z}} \mu_L(E_n) < \sum_{n\in\mathbb{Z}} \frac{1}{2^{|n|}}\frac{\varepsilon}{3} = \varepsilon.$$

Let s be the right-continuous step function on $\mathbb{R}$ defined by setting $s = s_n$ on I_n for $n \in \mathbb{Z}$. Then $|f - s| = |f - s_n| < \varepsilon$ on $I_n \setminus E_n$ for $n \in \mathbb{Z}$ so that $|f - s| < \varepsilon$ on $\mathbb{R} \setminus E$. If $M_1 \le f \le M_2$ on $\mathbb{R}$, then as noted above s_n can be so chosen that $M_1 \le s_n \le M_2$ on I_n for $n \in \mathbb{Z}$ and consequently $M_1 \le s \le M_2$ on $\mathbb{R}$. ■

Theorem 6.35. *Let f be an extended real-valued $\mathfrak{M}_L$-measurable function on $D = [a, b)$, $[a, \infty)$, or $\mathbb{R}$. Suppose f is real-valued a.e. on D. Then there exists a sequence $(s_n : n \in \mathbb{N})$ of right-continuous step functions on D which converges to f on D in measure μ_L and also converges to f a.e. on D. Moreover if $M_1 \le f \le M_2$ on D for two constants $M_1, M_2 \in \mathbb{R}$ such that $M_1 < M_2$, then the sequence $(s_n : n \in \mathbb{N})$ can be so chosen that $M_1 \le s_n \le M_2$ on D for every $n \in \mathbb{N}$.*

Proof. By Theorem 6.33 and Corollary 6.34, for every $k \in \mathbb{N}$ there exists a $\mathfrak{M}_L$-measurable subset E_k of D with $\mu_L(E_k) < \frac{1}{k}$ and there exists a right-continuous step function s_k on D such that $|s_k - f| < \frac{1}{k}$ on $D \setminus E_k$. This implies $\big\{D : |s_k - f| \ge \frac{1}{k}\big\} \subset E_k$ so that we have $\mu_L\big\{D : |s_k - f| \ge \frac{1}{k}\big\} < \frac{1}{k}$ for every $k \in \mathbb{N}$. According to Corollary 6.18, this shows that the sequence $(s_k : k \in \mathbb{N})$ converges to f on D in measure. Then by Theorem 6.24 (Riesz), there exists a subsequence $(s_{k_\ell} : \ell \in \mathbb{N})$ which converges to f a.e. on D. As a subsequence of $(s_k : k \in \mathbb{N})$, the sequence $(s_{k_\ell} : \ell \in \mathbb{N})$ also converges to f on D in measure. If $M_1 \le f \le M_2$ on D, then s_k can be so chosen that $M_1 \le s_k \le M_2$ on D for every $k \in \mathbb{N}$ according to Theorem 6.33 and Corollary 6.34 and thus $M_1 \le s_{k_\ell} \le M_2$ on D. ■

[V.2] Approximation of Lebesgue Measurable Functions by Continuous Functions

Theorem 6.36. (E. Borel) *Let f be an extended real-valued $\mathfrak{M}_L$-measurable function on $\mathbb{R}$ which is real-valued a.e. on $\mathbb{R}$. Then for every $\varepsilon > 0$ there exist a $\mathfrak{M}_L$-measurable subset E of $\mathbb{R}$ with $\mu_L(E) < \varepsilon$ and a continuous real-valued function g on $\mathbb{R}$ such that $|f - g| < \varepsilon$ on $\mathbb{R} \setminus E$. Moreover if $M_1 \le f \le M_2$ on $\mathbb{R}$ for two constants $M_1, M_2 \in \mathbb{R}$ such that $M_1 < M_2$, then g can be so chosen that $M_1 \le g \le M_2$ on $\mathbb{R}$.*

Proof. Let $\varepsilon > 0$ be arbitrarily given. According to Corollary 6.34, there exist a $\mathfrak{M}_L$-measurable subset E_0 of $\mathbb{R}$ with $\mu_L(E_0) < \frac{\varepsilon}{2}$ and a right-continuous step function s on $\mathbb{R}$ such that $|f - s| < \frac{\varepsilon}{2}$ on $\mathbb{R} \setminus E_0$. Let s be given by $s = \sum_{n \in \mathbb{Z}} c_n \cdot \mathbf{1}_{I_n}$ where $c_n \in \mathbb{R}$ for $n \in \mathbb{Z}$ and $\{I_n : n \in \mathbb{Z}\}$ is a disjoint collection of finite intervals of the class $\mathfrak{I}_{co}$ with $\bigcup_{n \in \mathbb{Z}} I_n = \mathbb{R}$. We number the intervals I_n in such a way that I_n is to the left of I_{n+1} on the real line $\mathbb{R}$ for every $n \in \mathbb{Z}$. This is possible since the set of the endpoints of the intervals in the domain of definition of a right-continuous step functions on $\mathbb{R}$ does not have any limit points in $\mathbb{R}$ as we noted following Definition 6.29. Let $I_n = [a_n, a_{n+1})$ where $a_n < a_{n+1}$ for $n \in \mathbb{Z}$. The set $\{a_n : n \in \mathbb{Z}\}$ contains all the points of discontinuity of the step function s in $\mathbb{R}$. Now for every $n \in \mathbb{Z}$, a_n is the right endpoint of the interval $I_{n-1} = [a_{n-1}, a_n)$ and the left endpoint of the interval $I_n = [a_n, a_{n+1})$. Let us define

$$\delta_n = \frac{1}{4} \min\left\{\frac{1}{2^{|n|}} \frac{\varepsilon}{3}, \ell(I_{n-1}), \ell(I_n)\right\},$$

$$I_n^* = [a_n + \delta_n, a_{n+1} - \delta_{n+1}] \subset I_n,$$

$$J_n = [a_n - \delta_n, a_n + \delta_n].$$

Then $\mathbb{R} = \cdots \cup J_{n-1} \cup I_{n-1}^* \cup J_n \cup I_n^* \cup J_{n+1} \cup I_{n+1}^* \cup \cdots$, where J_{n-1} is to the left of I_{n-1}^*, I_{n-1}^* is to the left of J_n, J_n is to the left of I_n^* and so on. We define a continuous real-valued function g on $\mathbb{R}$ by setting g be equal to s on I_n^* and letting g to be linear on J_n assuming the value of s at the two endpoints of J_n for every $n \in \mathbb{Z}$. Thus for $x \in I_n^*$ we define $g(x) = c_n$ and for $x \in J_n$ we define $g(x)$ by

$$\frac{g(x) - c_{n-1}}{x - (a_n - \delta_n)} = \frac{c_n - c_{n-1}}{(a_n + \delta_n) - (a_n - \delta_n)},$$

that is,

$$g(x) = \frac{c_n - c_{n-1}}{2\delta_n}\{x - (a_n - \delta_n)\} + c_{n-1}.$$

Let $E = E_0 \cup \left(\bigcup_{n \in \mathbb{Z}} J_n\right)$. Then E is a $\mathfrak{M}_L$-measurable subset of $\mathbb{R}$ with

$$\mu_L(E) \le \mu_L(E_0) + \sum_{n \in \mathbb{Z}} \mu_L(J_n) < \frac{\varepsilon}{2} + \sum_{n \in \mathbb{Z}} 2\delta_n < \frac{\varepsilon}{2} + \frac{1}{2} \sum_{n \in \mathbb{Z}} \frac{1}{2^{|n|}} \frac{\varepsilon}{3} = \varepsilon,$$

and $g = s$ on $\mathbb{R} \setminus E$. Then $|f - g| \le |f - s| + |s - g| = |f - s| < \varepsilon$ on $\mathbb{R} \setminus E$.

If $M_1 \le f \le M_2$ on $\mathbb{R}$, then s can be so chosen that $M_1 \le s \le M_2$ on $\mathbb{R}$ according to Corollary 6.34. Then for the continuous real-valued function g defined above, we have $M_1 \le g \le M_2$ on $\mathbb{R}$. ∎

Theorem 6.37. *Let f be an extended real-valued $\mathfrak{M}_L$-measurable function on $\mathbb{R}$. Suppose f is real-valued a.e. on $\mathbb{R}$. Then there exists a sequence $(g_n : n \in \mathbb{N})$ of continuous real-valued functions on $\mathbb{R}$ which converges to f on $\mathbb{R}$ in measure μ_L and also converges to f a.e. on $\mathbb{R}$. Moreover if $M_1 \le f \le M_2$ on $\mathbb{R}$ for two constants $M_1, M_2 \in \mathbb{R}$ such that $M_1 < M_2$, then the sequence $(g_n : n \in \mathbb{N})$ can be so chosen that $M_1 \le g_n \le M_2$ on $\mathbb{R}$ for every $n \in \mathbb{N}$.*

Proof. This Theorem can be proved in the same way as Theorem 6.35. ∎

(Let us remark that if f is a $\mathfrak{M}_L$-measurable but not $\mathfrak{B}_{\mathbb{R}}$-measurable real-valued function on $\mathbb{R}$ then there does not exist any sequence $(g_n : n \in \mathbb{N})$ of continuous functions on $\mathbb{R}$ which converges to f everywhere on $\mathbb{R}$ since the existence of such a sequence implies that f is $\mathfrak{B}_{\mathbb{R}}$-measurable on $\mathbb{R}$ by (b) of Theorem 4.22.)

A continuous real-valued function f defined on an open set G in a topological space X may not have a continuous extension to X, that is, there may not be any continuous real-valued function g on X such that $g = f$ on G. For example, if $X = \mathbb{R}$ and $G = (-\infty, 0) \cup (0, \infty)$ and f is defined by $f(x) = x^{-1}$ for $x \in G$, then f is continuous on G but has no continuous extension to X. On the other hand, if f is a continuous real-valued function defined on a closed set F in a metric space X, or more generally in a normal topological space, then f has a continuous extension to X. In the next Proposition, we prove this for the particular case $X = \mathbb{R}$ by an elementary construction.

Proposition 6.38. *Let F be a closed set in $\mathbb{R}$. If f is a continuous real-valued function defined on F, then f has a continuous extension to $\mathbb{R}$, that is, there exists a continuous real-valued function g on $\mathbb{R}$ such that $g = f$ on F.*

Proof. 1. Let $G = F^c$. Then G is an open set in $\mathbb{R}$ so that $G = \bigcup_{k\in\mathbb{N}} I_k$ where $(I_k : k \in \mathbb{N})$ is a countable collection of disjoint open intervals in $\mathbb{R}$. Let $I_k = (a_k, b_k)$ for $k \in \mathbb{N}$. Let us define a real-valued function g on $\mathbb{R}$ by setting $g = f$ on F and setting

$$g(x) = \frac{f(b_k) - f(a_k)}{b_k - a_k}(x - a_k) + f(a_k) \quad \text{for } x \in [a_k, b_k] \text{ for } k \in \mathbb{N},$$

that is, g is linear on the closed interval $[a_k, b_k] \supset I_k$. Note that a_k and b_k are in F so that $f(a_k)$ and $f(b_k)$ are defined and $g(a_k) = f(a_k)$ and $g(b_k) = f(b_k)$. Let us show that g is continuous at every $x_0 \in \mathbb{R}$. Now $G \cap F = \emptyset$ and $G \cup F = \mathbb{R}$ and every $x_0 \in \mathbb{R}$ is either in G or in F.

2. If $x_0 \in G$, then $x_0 \in (a_k, b_k)$ for some $k \in \mathbb{N}$. Since g is linear on (a_k, b_k), g is continuous at x_0. Suppose $x_0 \in F$. To show that g is continuous at x_0, we show that g is both left-continuous at x_0 and right-continuous at x_0. Let us prove the right-continuity of g at x_0. There are three cases to be considered regarding the position of x_0 in F:
(a) There exists $\eta > 0$ such that $[x_0, x_0 + \eta) \subset F$.
(b) There exists $\eta > 0$ such that $(x_0, x_0 + \eta) \subset G$.
(c) For every $\eta > 0$, $(x_0, x_0 + \eta)$ contains points from both F and G.

Case (a). Since f is continuous on F, it is continuous on $[x_0, x_0 + \eta)$ and in particular right-continuous at x_0. Then since $g = f$ on $[x_0, x_0 + \eta)$, g is right-continuous at x_0.

Case (b). The fact that $x_0 \in F$ and $(x_0, x_0 + \eta) \subset G$ implies that there exists $k \in \mathbb{N}$ such that $(x_0, x_0 + \eta) \subset (a_k, b_k)$ and $x_0 = a_k$. Then since g is linear on $[a_k, b_k]$, g is right-continuous at $x_0 = a_k$.

Case (c). By the continuity of f at x_0, for every $\varepsilon > 0$ there exists $\delta > 0$ such that

$$(1) \qquad |f(x) - f(x_0)| < \frac{\varepsilon}{3} \quad \text{for } x \in [x_0, x_0 + \delta) \cap F.$$

Now since $(x_0, x_0 + \eta) \cap F \neq \emptyset$ for every $\eta > 0$, there exists $x_1 \in (x_0, x_0 + \delta) \cap F$. Let us write $x_1 = x_0 + \delta_0$ with $\delta_0 \in (0, \delta)$. By (1) and by the fact that $g = f$ on F, we have

$$(2) \qquad |g(x) - g(x_0)| < \frac{\varepsilon}{3} \quad \text{for } x \in [x_0, x_0 + \delta_0] \cap F.$$

Consider $[x_0, x_0 + \delta_0] \cap G$. If $x \in [x_0, x_0 + \delta_0] \cap G$, then since $x_0, x_0 + \delta_0 \notin G$ and $x \in G$, there exists $k \in \mathbb{N}$ such that $x \in (a_k, b_k) \subset [x_0, x_0 + \delta_0] \subset [x_0, x_0 + \delta)$. Since $a_k, b_k \in [x_0, x_0 + \delta) \cap F$, (1) implies that $|f(a_k) - f(x_0)| < \frac{\varepsilon}{3}$ and $|f(b_k) - f(x_0)| < \frac{\varepsilon}{3}$. Since g is linear on $[a_k, b_k]$, for our $x \in (a_k, b_k)$ we have either $f(a_k) \le g(x) \le f(b_k)$ or $f(a_k) \ge g(x) \ge f(b_k)$ and in any case we have

$$|g(x) - f(a_k)| \le |f(a_k) - f(b_k)| \le |f(a_k) - f(x_0)| + |f(x_0) - f(b_k)| < \frac{2\varepsilon}{3}.$$

Then

$$|g(x) - g(x_0)| = |g(x) - f(x_0)| \le |g(x) - f(a_k)| + |f(a_k) - f(x_0)| < \frac{2\varepsilon}{3} + \frac{\varepsilon}{3}\varepsilon = \varepsilon.$$

Thus we have shown

$$(3) \qquad |g(x) - g(x_0)| < \varepsilon \quad \text{for } x \in [x_0, x_0 + \delta_0] \cap G.$$

Since $F \cup G = \mathbb{R}$, (2) and (3) imply that $|g(x) - g(x_0)| < \varepsilon$ for $x \in [x_0, x_0 + \delta_0]$. This proves the right-continuity of g at x_0. The left-continuity of g at x_0 is proved likewise. ∎

Theorem 6.39. (N. N. Lusin) *Let f be an extended real-valued $\mathfrak{M}_L$-measurable function on $\mathbb{R}$ which is real-valued a.e. on $\mathbb{R}$. Then for every $\varepsilon > 0$ there exist an open set G in $\mathbb{R}$ with $\mu_L(G) < \varepsilon$ and a continuous real-valued function g on $\mathbb{R}$ such that $g = f$ on $\mathbb{R} \setminus G$.*

Proof. Let $\varepsilon > 0$. According to Theorem 6.36 (Borel), for every $n \in \mathbb{N}$ there exist a $\mathfrak{M}_L$-measurable subset E_n of $\mathbb{R}$ with $\mu_L(E_n) < \frac{\varepsilon}{2}\frac{1}{2^n}$ and a continuous real-valued function f_n on $\mathbb{R}$ such that $|f_n - f| < \frac{\varepsilon}{2}\frac{1}{2^n}$ on $\mathbb{R} \setminus E_n$. Let $E = \bigcup_{n \in \mathbb{N}} E_n$. Then we have

$$|f_n - f| < \frac{\varepsilon}{2}\frac{1}{2^n} \quad \text{on } \mathbb{R} \setminus E \text{ for every } n \in \mathbb{N}.$$

This shows that the sequence $(f_n : n \in \mathbb{N})$ of continuous functions converges to f uniformly on $\mathbb{R} \setminus E$. The continuity of f_n and the uniform convergence of f_n to f on $\mathbb{R} \setminus E$ implies that f is continuous on $\mathbb{R} \setminus E$. Now

$$\mu_L(E) = \mu_L\Big(\bigcup_{n \in \mathbb{N}} E_n\Big) \le \sum_{n \in \mathbb{N}} \mu_L(E_n) < \sum_{n \in \mathbb{N}} \frac{\varepsilon}{2}\frac{1}{2^n} = \frac{\varepsilon}{2}.$$

According to (a) of Lemma 3.21, there exists an open set G in $\mathbb{R}$ such that $G \supset E$ and $\mu_L(G) \leq \mu_L(E) + \frac{\varepsilon}{2} < \varepsilon$. Since f is continuous on $\mathbb{R} \setminus E$, it is continuous on the closed set $\mathbb{R} \setminus G$. Then by Proposition 6.38 there exists a continuous real-valued function g on $\mathbb{R}$ such that $g = f$ on $\mathbb{R} \setminus G$. ∎

Problems

Assumptions. For problems in this section, unless otherwise stated, it is assumed that $(f_n : n \in \mathbb{N})$ is a sequence of extended real-valued $\mathfrak{A}$-measurable functions on a set $D \in \mathfrak{A}$ in a measure space $(X, \mathfrak{A}, \mu)$ and f is a real-valued $\mathfrak{A}$-measurable function on D. Similarly for $(g_n : n \in \mathbb{N})$ and g.

Prob. 6.1. Let $(X, \mathfrak{A}, \mu)$ be a measure space and let $D \in \mathfrak{A}$. Let $(f_n : n \in \mathbb{N})$ and f be extended real-valued $\mathfrak{A}$-measurable functions on D and let f be a real-valued on D. For an arbitrary $\varepsilon > 0$ let $D_{\varepsilon,n} = \{D : |f_n - f| \geq \varepsilon\}$ for $n \in \mathbb{N}$. Then $\lim_{n\to\infty} f_n = f$ a.e. on D if and only if $\mu\left(\limsup_{n\to\infty} D_{\varepsilon,n}\right) = 0$ for every $\varepsilon > 0$.

Prob. 6.2. Let $(X, \mathfrak{A}, \mu)$ be a measure space. Let $(f_n : n \in \mathbb{N})$ be a sequence of extended real-valued $\mathfrak{A}$-measurable functions on X and let f be an extended real-valued $\mathfrak{A}$-measurable function which is finite a.e. on X. Suppose $\lim_{n\to\infty} f_n = f$ a.e. on X. Let $\alpha \in [0, \mu(X))$ be arbitrarily chosen. Show that for every $\varepsilon > 0$ there exists $N \in \mathbb{N}$ such that $\mu\{X : |f_n - f| < \varepsilon\} \geq \alpha$ for $n \geq N$.

Prob. 6.3. Let $(f_n : n \in \mathbb{N})$ be a sequence of real-valued continuous functions on $\mathbb{R}$. Let E be the set of all $x \in \mathbb{R}$ such that the sequence $(f_n(x) : n \in \mathbb{N})$ of real numbers converges. Show that E is an $F_{\sigma\delta}$-set, that is, the intersection of countably many F_σ-sets.
(Hint: Use the expression $E = \bigcap_{m\in\mathbb{N}} \bigcup_{N\in\mathbb{N}} \bigcap_{p\in\mathbb{N}} \{x \in \mathbb{R} : |f_{N+p}(x) - f_N(x)| \leq \frac{1}{m}\}$.)

Prob. 6.4. Conditions 1° and 2° in Theorem 6.7 are sufficient for $(f_n : n \in \mathbb{N})$ to converge to f a.e. on D, but they are not necessary. To show this, consider the following example: In $(\mathbb{R}, \mathfrak{M}_L, \mu_L)$, let $D = (0, 1)$ and let $f_n(x) = 1$ for $x \in \left(0, \frac{1}{n}\right)$ and $f(x) = 0$ for $x \in \left[\frac{1}{n}, 1\right)$ for $n \in \mathbb{N}$ and let $f(x) = 0$ for $x \in (0, 1)$. Then $(f_n : n \in \mathbb{N})$ converges to f everywhere on D. Show that for any sequence of positive numbers $(\varepsilon_n : n \in \mathbb{N})$ such that $\lim_{n\to\infty} \varepsilon_n = 0$, we have $\sum_{n\in\mathbb{N}} \mu_L\{x \in (0, 1) : |f_n(x) - f(x)| \geq \varepsilon_n\} = \infty$.

Prob. 6.5. To show that Egoroff's Theorem holds only on sets with finite measures, we have the following example:
Consider $(\mathbb{R}, \mathfrak{M}_L, \mu_L)$. Let $D = [0, \infty)$ with $\mu_L(D) = \infty$. Let $f_n(x) = \frac{x}{n}$ for $x \in D$ for $n \in \mathbb{N}$ and let $f(x) = 0$ for $x \in D$. Then the sequence $(f_n : n \in \mathbb{N})$ converges to f everywhere on D. To show that $(f_n : n \in \mathbb{N})$ does not converge to f almost uniformly on D, we show that for some $\eta > 0$ there does not exist a $\mathfrak{M}_L$-measurable subset E of D with $\mu_L(E) < \eta$ such that $(f_n : n \in \mathbb{N})$ converges to f uniformly on $D \setminus E$. Show that actually for any $\mathfrak{M}_L$-measurable subset E of D with $\mu_L(E) < \infty$, $(f_n : n \in \mathbb{N})$ does not converge to f uniformly on $D \setminus E$.

Prob. 6.6. By definition, $f_n \xrightarrow{\mu} f$ on D if for every $\varepsilon > 0$

$$\lim_{n\to\infty} \mu\{x \in D : |f_n(x) - f(x)| \geq \varepsilon\} = 0.$$

Show that this condition is equivalent to the condition that for every $\varepsilon > 0$

$$\lim_{n\to\infty} \mu\{x \in D : |f_n(x) - f(x)| > \varepsilon\} = 0.$$

Prob. 6.7. (a) Show that the condition

$$\lim_{n\to\infty} \mu\{x \in D : |f_n(x) - f(x)| > 0\} = 0$$

implies that $f_n \xrightarrow{\mu} f$ on D.
(b) Show by example that the converse of (a) is false.
(c) Show that the condition in (a) implies that for a.e. $x \in D$ we have $f_n(x) = f(x)$ for infinitely many $n \in \mathbb{N}$.

Prob. 6.8. Let $(a_n : n \in \mathbb{N})$ be a sequence of extended real numbers. Show that the sequence converges to a real number a if and only if every subsequence $(a_{n_k} : k \in \mathbb{N})$ of the sequence $(a_n : n \in \mathbb{N})$ has a subsequence $(a_{n_{k_\ell}} : \ell \in \mathbb{N})$ converging to a.

Prob. 6.9. (a) Show that if every subsequence $(f_{n_k} : k \in \mathbb{N})$ of a sequence $(f_n : n \in \mathbb{N})$ has a subsequence $(f_{n_{k_\ell}} : \ell \in \mathbb{N})$ which converges to f on D in measure, then the sequence $(f_n : n \in \mathbb{N})$ itself converges to f on D in measure.
(b) Show that if $\mu(D) < \infty$ and if every subsequence $(f_{n_k} : k \in \mathbb{N})$ of the sequence $(f_n : n \in \mathbb{N})$ has a subsequence $(f_{n_{k_\ell}} : \ell \in \mathbb{N})$ which converges to f a.e. on D, then the sequence $(f_n : n \in \mathbb{N})$ itself converges in measure to f on D.

Prob. 6.10. Show that if $f_n \xrightarrow{\mu} f$ on D, then $cf_n \xrightarrow{\mu} cf$ on D for every $c \in \mathbb{R}$. (Assume that f_n is real-valued for every $n \in \mathbb{N}$ to make multiplication $cf_n(x)$ of $f_n(x)$ by $c \in \mathbb{R}$ including $c = 0$ possible at every $x \in D$. Alternately, consider only $c \neq 0$ and permit f_n to be extended real-valued.)

Prob. 6.11. Show that if $f_n \xrightarrow{\mu} f$ on D and $g_n \xrightarrow{\mu} g$ on D, then $f_n + g_n \xrightarrow{\mu} f + g$ on D. (Assume that f_n and g_n are real-valued for every $n \in \mathbb{N}$ to make the addition $f_n(x)+g_n(x)$ possible at every $x \in D$.)

Prob. 6.12. Show that if $f_n \xrightarrow{\mu} f$ on D and $g_n \xrightarrow{\mu} g$ on D and if $\mu(D) < \infty$, then $f_n g_n \xrightarrow{\mu} fg$ on D. (Assume that both f_n and g_n are real-valued for every $n \in \mathbb{N}$ so that the multiplication $f_n(x)g_n(x)$ is possible at every $x \in D$.)

Prob. 6.13. Show that if $f_n \xrightarrow{\mu} f$ on D and if $\mu(D) < \infty$, then for every real-valued continuous function F on $\mathbb{R}$ we have $F \circ f_n \xrightarrow{\mu} F \circ f$ on D. (Assume that f_n is real-valued for every $n \in \mathbb{N}$ to make the composition $(F \circ f_n)(x) = F(f_n(x))$ possible at every $x \in D$.)

Prob. 6.14. The following are two modifications of Prob. 6.13:
(a) Show that if we remove the condition $\mu(D) < \infty$, we still have $F \circ f_n \xrightarrow{\mu} F \circ f$ on D if we assume that F is uniformly continuous on $\mathbb{R}$.
(b) Show that if we remove the condition $\mu(D) < \infty$, we still have $F \circ f_n \xrightarrow{\mu} F \circ f$ on D

if we assume that the sequence $(f_n : n \in \mathbb{N})$ is uniformly bounded.

Prob. 6.15. Referring to Prob. 1.32, show that $([\mathfrak{A}], \rho^*)$ is a complete metric space, that is, if a sequence $\{[A_n] : n \in \mathbb{N}\}$ in $[\mathfrak{A}]$ is a Cauchy sequence with respect to the metric ρ^* then there exists $[A] \in [\mathfrak{A}]$ to which the sequence converges with respect the metric ρ^*.

Prob. 6.16. An extended real valued function f on $\mathbb{R}$ is said to be left-continuous at $x_0 \in \mathbb{R}$ if $f(x_0) \in \mathbb{R}$ and $\lim_{x \uparrow x_0} f(x) = f(x_0)$. Similarly f is said to be right-continuous at $x_0 \in \mathbb{R}$ if $f(x_0) \in \mathbb{R}$ and $\lim_{x \downarrow x_0} f(x) = f(x_0)$.
(a) Show that if f is left-continuous everywhere on $\mathbb{R}$, then f is $\mathfrak{B}_{\mathbb{R}}$-measurable on $\mathbb{R}$.
(b) Prove a parallel statement for a right-continuous function.
(Hint: For (a), construct a sequence of right-continuous step functions which converges to f everywhere on $\mathbb{R}$.)

Prob. 6.17. Let f be an extended real valued function on $\mathbb{R}$.
(a) Show that if $\lim_{x \uparrow x_0} f(x)$ exists in $\overline{\mathbb{R}}$ for every $x_0 \in \mathbb{R}$, then the collection of points $x_0 \in \mathbb{R}$ at which $\lim_{x \uparrow x_0} f(x) \neq f(x_0)$ is a countable collection.
(b) Prove a parallel statement for $\lim_{x \downarrow x_0} f(x)$.

Prob. 6.18. Let f be an extended real valued function on $\mathbb{R}$.
(a) Show that if $\lim_{x \uparrow x_0} f(x)$ exists in $\overline{\mathbb{R}}$ for every $x_0 \in \mathbb{R}$, then f is $\mathfrak{B}_{\mathbb{R}}$- measurable on $\mathbb{R}$.
(b) Prove a parallel statement for $\lim_{x \downarrow x_0} f(x)$.

Prob. 6.19. Let $(X, \mathfrak{A}, \mu)$ be a finite measure space. Let $(f_n : n \in \mathbb{N})$ be an arbitrary sequence of real-valued $\mathfrak{A}$-measurable functions on X. Show that for every $\varepsilon > 0$ there exist $E \in \mathfrak{A}$ with $\mu(E) < \varepsilon$ and a sequence of positive real numbers $(a_n : n \in \mathbb{N})$ such that $\lim_{n \to \infty} a_n f_n = 0$ on $X \setminus E$.

Prob. 6.20. Let $(X, \mathfrak{A}, \mu)$ be a finite measure space. Let $(f_n : n \in \mathbb{N})$ be an arbitrary sequence of real-valued $\mathfrak{A}$-measurable functions on X. Show there exists a sequence of positive real numbers $(a_n : n \in \mathbb{N})$ such that $\lim_{n \to \infty} a_n f_n = 0$, μ-a.e. on X.

Chapter 2

The Lebesgue Integral

§7 Integration of Bounded Functions on Sets of Finite Measure

[I] Integration of Simple Functions

Definition 7.1. *Given a measure space $(X, \mathfrak{A}, \mu)$. A function φ is called a simple function if it satisfies the following conditions:*
1° $\mathfrak{D}(\varphi) \in \mathfrak{A}$,
2° *φ is $\mathfrak{A}$-measurable on $\mathfrak{D}(\varphi)$,*
3° *φ assumes only finitely many real values, that is, $\mathfrak{R}(\varphi)$ is a finite subset of $\mathbb{R}$.*
(Note that $\mu(\mathfrak{D}(\varphi)) = \infty$ is permissible but ∞ and $-\infty$ are not permissible values for a simple function φ.)

Lemma 7.2. *If φ_1 and φ_2 are simple functions on a set $D \in \mathfrak{A}$ and if $c_1, c_2 \in \mathbb{R}$, then $c_1\varphi_1 + c_2\varphi_2$ is a simple function on D.*

Proof. The $\mathfrak{A}$-measurability of $c_1\varphi_1 + c_2\varphi_2$ on D follows from that of φ_1 and φ_2 by Theorem 4.12. Since φ_1 and φ_2 each assume only finitely many real values, $c_1\varphi_1 + c_2\varphi_2$ assumes only finitely many real values. In fact if $\{a_i : i = 1, \dots, m\}$ and $\{b_j : j = 1, \dots, n\}$ are the values of φ_1 and φ_2 respectively, then the values of $c_1\varphi_1 + c_2\varphi_2$ are given by $\{c_1 a_i + c_2 b_j : i = 1, \dots, m;\ j = 1, \dots, n\}$ which is a finite set in $\mathbb{R}$. Thus $c_1\varphi_1 + c_2\varphi_2$ is a simple function on D. ∎

Definition 7.3. *Let φ be a simple function on a set $D \in \mathfrak{A}$ in a measure space $(X, \mathfrak{A}, \mu)$. Let $\{a_i : i = 1, \dots, n\}$ be the set of the distinct values assumed by φ on D and let $D_i = \{x \in D : \varphi(x) = a_i\}$ for $i = 1, \dots, n$. $\{D_i : i = 1, \dots, n\}$ is a disjoint collection in $\mathfrak{A}$ and $\bigcup_{i=1}^n D_i = D$. The expression $\varphi(x) = \sum_{i=1}^n a_i \mathbf{1}_{D_i}(x)$ for $x \in D$, or briefly, $\varphi = \sum_{i=1}^n a_i \mathbf{1}_{D_i}$, is called the canonical representation of φ.*

Note that if $D_i \in \mathfrak{A}$ and $a_i \in \mathbb{R}$ for $i = 1, \dots, n$ and if we let $D = \bigcup_{i=1}^n D_i$ and define $\varphi = \sum_{i=1}^n a_i \mathbf{1}_{D_i}$ on D then φ is a simple function on D. However $\sum_{i=1}^n a_i \mathbf{1}_{D_i}$ may not be the canonical representation of φ since $\{D_i : i = 1, \ldots n\}$ may not be a disjoint collection and since $\{a_i : i = 1, \ldots, n\}$ may not be a collection of distinct real numbers.

Convention. We adopt the convention that $0 \cdot \pm\infty = \pm\infty \cdot 0 = 0$ if one the two factors is the constant value of a measurable function on a measurable set and the other is the measure of the set.

Let $\varphi = \sum_{i=1}^{n} a_i \mathbf{1}_{D_i}$ be the canonical representation of a simple function on a set $D \in \mathfrak{A}$ of a measure space $(X, \mathfrak{A}, \mu)$. By our convention above, $a_i \mu(D_i)$ is defined and exists in the extended real number system $\overline{\mathbb{R}}$ for $i = 1, \ldots, n$ but $\sum_{i=1}^{n} a_i \mu(D_i)$ may not exist since $\infty - \infty$ is undefined.

Definition 7.4. *Let $\varphi = \sum_{i=1}^{n} a_i \mathbf{1}_{D_i}$ be the canonical representation of a simple function on a set $D \in \mathfrak{A}$ in a measure space $(X, \mathfrak{A}, \mu)$. The Lebesgue integral of φ on D with respect to μ is defined by*

$$\int_D \varphi(x)\, \mu(dx) = \sum_{i=1}^{n} a_i \mu(D_i),$$

provided that the sum exists in $\overline{\mathbb{R}}$. In this case we say that φ is Lebesgue semi-integrable on D with respect to μ, or simply, μ semi-integrable on D. We say that φ is Lebesgue integrable on D with respect to μ, or simply μ-integrable on D, only when $\int_D \varphi(x)\, \mu(dx)$ is a real number. The notation $\int_D \varphi\, d\mu$ is an abbreviation for $\int_D \varphi(x)\, \mu(dx)$.

Let $D \in \mathfrak{A}$ and let $\{D_i : i = 1, \ldots, n\}$ be a disjoint collection in $\mathfrak{A}$ such that $\bigcup_{i=1}^{n} D_i = D$. Let $a_1, \ldots, a_n \in \mathbb{R}$. If we define a function φ on D by setting $\varphi = \sum_{i=1}^{n} a_i \mathbf{1}_{D_i}$ on D then φ is a simple function on D. However $\sum_{i=1}^{n} a_i \mathbf{1}_{D_i}$ may not be the canonical representation of φ since $a_1, \ldots, a_n$ are not assumed to be distinct. Nevertheless Definition 7.4 implies that if $\int_D \varphi\, d\mu$ exists then it is given by

$$\int_D \varphi\, d\mu = \sum_{i=1}^{n} a_i \mu(D_i).$$

To show this, consider for instance the case that $a_1, \ldots, a_{n_1}$ are distinct and $a_{n-1} = a_n$. In this case we let $E = D_{n-1} \cup D_n$. Then $\sum_{i=1}^{n-2} a_i \mathbf{1}_{D_i} + a_{n-1} \mathbf{1}_E$ is the canonical representation of φ so that by Definition 7.4 we have

$$\begin{aligned}\int_D \varphi\, d\mu &= \sum_{i=1}^{n-2} a_i \mu(D_i) + a_{n-1}\mu(E) = \sum_{i=1}^{n-2} a_i \mu(D_i) + a_{n-1}\mu(D_{n-1}) + a_{n-1}\mu(D_n) \\ &= \sum_{i=1}^{n} a_i \mu(D_i).\end{aligned}$$

Examples. Consider $(\mathbb{R}, \mathfrak{B}_{\mathbb{R}}, \mu_L)$. Let Q and P be respectively the sets of all rational numbers and irrational numbers in $\mathbb{R}$. We have $P \cap Q = \emptyset$, $P \cup Q = \mathbb{R}$ and $P,\ Q \in \mathfrak{B}_{\mathbb{R}}$.

Let us define a simple function φ_1 on $[0,1] \in \mathfrak{B}_{\mathbb{R}}$ by setting $\varphi_1(x) = 0$ for $x \in [0,1] \cap Q$ and $\varphi_1(x) = 1$ for $x \in [0,1] \cap P$. With $\mu_L([0,1] \cap Q) = 0$ and $\mu_L([0,1] \cap P) = 1$, we have $\int_{[0,1]} \varphi_1 \, d\mu_L = 0 \cdot 0 + 1 \cdot 1 = 1$ and thus φ_1 is μ-integrable on $[0,1]$. Next let us define a simple function φ_2 on $\mathbb{R}$ by setting $\varphi_2(x) = 0$ for $x \in Q$ and $\varphi_2(x) = 1$ for $x \in P$. In this case we have $\mu_L(Q) = 0$ and $\mu_L(P) = \infty$ so that $\int_{\mathbb{R}} \varphi_2 \, d\mu_L = 0 \cdot 0 + 1 \cdot \infty = \infty$ and thus φ_2 is μ semi-integrable on $\mathbb{R}$. As another example, if we define a simple function φ_3 on $[0,\infty) \in \mathfrak{B}_{\mathbb{R}}$ by setting $\varphi_3(x) = -1$ for $x \in \bigcup_{k \in \mathbb{Z}_+} [2k+1, 2k+2)$ and $\varphi_3(x) = 1$ for $x \in \bigcup_{k \in \mathbb{Z}_+} [2k, 2k+1)$, then since $\mu_L\big(\bigcup_{k \in \mathbb{Z}_+} [2k+1, 2k+2)\big) = \infty$ and $\mu_L\big(\bigcup_{k \in \mathbb{Z}_+} [2k, 2k+1)\big) = \infty$ also, and since $-1 \cdot \infty + 1 \cdot \infty$ is undefined, $\int_{[0,\infty)} \varphi_3 \, d\mu_L$ does not exist.

Observation 7.5. Let φ be a simple function on a set $D \in \mathfrak{A}$ in a measure space $(X, \mathfrak{A}, \mu)$.
1° If $\mu(D) = 0$, then $\int_D \varphi \, d\mu = 0$.
2° If $\varphi \equiv 0$ on D, then $\int_D \varphi \, d\mu = 0$.
3° If $\varphi \geq 0$ on D, then $\int_D \varphi \, d\mu \in [0, \infty]$.
4° If $\varphi \leq 0$ on D, then $\int_D \varphi \, d\mu \in [-\infty, 0]$.
5° If $\mu(D) < \infty$, then $\int_D \varphi \, d\mu \in \mathbb{R}$.
6° φ is μ-integrable on D if and only if $\mu\{D : \varphi \neq 0\} < \infty$.
7° If $\varphi \geq 0$ on D, then $\int_D \varphi \, d\mu = 0$ if and only if $\mu\{x \in D : \varphi(x) \neq 0\} = 0$.
8° If $\varphi \geq 0$ on D, $E \subset D$, and $E \in \mathfrak{A}$, then $\int_E \varphi \, d\mu \leq \int_D \varphi \, d\mu$.

Observation 7.6. Let φ be a simple function on a set $D \in \mathfrak{A}$ in a measure space $(X, \mathfrak{A}, \mu)$. Let $\{E_1, \ldots, E_k\}$ be a disjoint collection in $\mathfrak{A}$ such that $\bigcup_{j=1}^k E_j = D$. Then φ is a simple function on E_j for $j = 1, \ldots, k$ and if φ is μ semi-integrable on D then so is φ on E_j for $j = 1, \ldots, k$ and furthermore $\int_D \varphi \, d\mu = \sum_{j=1}^k \int_{E_j} \varphi \, d\mu$.

Proof. Since $E_j \in \mathfrak{A}$ and $E_j \subset D$, the $\mathfrak{A}$-measurability of φ on D implies the $\mathfrak{A}$-measurability of the restriction of φ to E_j by Lemma 4.7. Since φ assumes only finitely many real values on D, its restriction to E_j assumes only finitely many real values on E_j. Thus the restriction of φ to E_j is a simple function on E_j for every $j = 1, \ldots, k$. Now let $\varphi = \sum_{i=1}^n a_i \mathbf{1}_{D_i}$ be the canonical representation of φ. Consider the disjoint collection $\{D_i \cap E_j : i = 1, \ldots, n;\ j = 1, \ldots, k\}$ in $\mathfrak{A}$ with $\bigcup_{j=1}^k (D_i \cap E_j) = D_i$ for $i = 1, \ldots, n$ and $\bigcup_{i=1}^n \bigcup_{j=1}^k (D_i \cap E_j) = D$. The canonical representation of φ on E_j is given by $\varphi = \sum_{i=1}^n a_i \mathbf{1}_{D_i \cap E_j}$. Now

$$\int_D \varphi \, d\mu = \sum_{i=1}^n a_i \, \mu(D_i) = \sum_{i=1}^n a_i \Big[\sum_{j=1}^k \mu(D_i \cap E_j) \Big]$$
$$= \sum_{j=1}^k \Big[\sum_{i=1}^n a_i \, \mu(D_i \cap E_j) \Big] = \sum_{j=1}^k \int_{E_j} \varphi \, d\mu. \ \blacksquare$$

Lemma 7.7. *Let φ_1, φ_2, and φ be simple functions on a set $D \in \mathfrak{A}$ in a measure space $(X, \mathfrak{A}, \mu)$. Assume that φ_1, φ_2, and φ are μ semi-integrable on D.*

(a) *If $\varphi_1 = \varphi_2$ a.e. on D, then $\int_D \varphi_1\,d\mu = \int_D \varphi_2\,d\mu$.*
(b) *If $\varphi_1 \le \varphi_2$ on D, then $\int_D \varphi_1\,d\mu \le \int_D \varphi_2\,d\mu$.*
(c) *If $\varphi(x) \in [M_1, M_2]$ for $x \in D$, then $\int_D \varphi\,d\mu \in [M_1\mu(D), M_2\mu(D)]$.*
(d) *$c\varphi$ is μ semi-integrable on D for every $c \in \mathbb{R}$.*
If $c \ne 0$ then we have $\int_D c\varphi\,d\mu = c\int_D \varphi\,d\mu$.
If $c = 0$ then $\int_D c\varphi\,d\mu = 0$ but $c\int_D \varphi\,d\mu$ is undefined in case $\int_D \varphi\,d\mu = \pm\infty$.
(e) *If φ is μ-integrable on D, then so is $c\varphi$ for every $c \in \mathbb{R}$ and moreover we have $\int_D c\varphi\,d\mu = c\int_D \varphi\,d\mu$.*
(f) *If $\varphi_1 \ge 0$ and $\varphi_2 \ge 0$ on D, then $\varphi_1 + \varphi_2$ is μ semi-integrable on D and moreover $\int_D \{\varphi_1 + \varphi_2\}\,d\mu = \int_D \varphi_1\,d\mu + \int_D \varphi_2\,d\mu$.*
(g) *If at least one of φ_1 and φ_2 is μ-integrable on D, then $\varphi_1 + \varphi_2$ is μ semi-integrable on D and moreover we have $\int_D \{\varphi_1 + \varphi_2\}\,d\mu = \int_D \varphi_1\,d\mu + \int_D \varphi_2\,d\mu$.*
In particular, if both φ_1 and φ_2 are μ-integrable on D, then so is $\varphi_1 + \varphi_2$.
(h) *If $D_1, D_2 \in \mathfrak{A}$, $D_1 \cap D_2 = \emptyset$ and $D_1 \cup D_2 = D$, then $\int_D \varphi\,d\mu = \int_{D_1} \varphi\,d\mu + \int_{D_2} \varphi\,d\mu$.*

Proof. 1. To prove (a), suppose $\varphi_1 = \varphi_2$ on $D \setminus E$ where E is a subset of D which is a null set in $(X, \mathfrak{A}, \mu)$. By Observation 7.6, we have $\int_D \varphi_1\,d\mu = \int_{D\setminus E} \varphi_1\,d\mu + \int_E \varphi_1\,d\mu$ and $\int_D \varphi_2\,d\mu = \int_{D\setminus E} \varphi_2\,d\mu + \int_E \varphi_2\,d\mu$. On $D \setminus E$ we have $\varphi_1 = \varphi_2$ so that we have $\int_{D\setminus E} \varphi_1\,d\mu = \int_{D\setminus E} \varphi_2\,d\mu$. Since $\mu(E) = 0$, we have $\int_E \varphi_1\,d\mu = 0$ and $\int_E \varphi_2\,d\mu = 0$ by 1° of Observation 7.5. Thus we have $\int_D \varphi_1\,d\mu = \int_D \varphi_2\,d\mu$.

2. (b), (c), (d), and (e) are immediate from Definition 7.4.

3. To prove (f), let $\varphi_1 = \sum_{i=1}^m a_i \mathbf{1}_{E_i}$ and $\varphi_2 = \sum_{j=1}^n b_j \mathbf{1}_{F_j}$ be the canonical representations of φ_1 and φ_2 respectively. If we let $G_{i,j} = E_i \cap F_j$, then we have a disjoint collection $\{G_{i,j} : i = 1, \ldots, m;\ j = 1, \ldots, n\}$ in $\mathfrak{A}$ with $\bigcup_{i=1}^m \bigcup_{j=1}^n G_{i,j} = D$. We have

$$\int_D \varphi_1\,d\mu = \sum_{i=1}^m a_i \mu(E_i) = \sum_{i=1}^m a_i \Big[\sum_{j=1}^n \mu(G_{i,j})\Big] = \sum_{i=1}^m \sum_{j=1}^n a_i \mu(G_{i,j})$$

and similarly

$$\int_D \varphi_2\,d\mu = \sum_{j=1}^n b_j \mu(F_j) = \sum_{j=1}^n b_j \Big[\sum_{i=1}^m \mu(G_{i,j})\Big] = \sum_{i=1}^m \sum_{j=1}^n b_j \mu(G_{i,j}).$$

On the other hand since $\varphi_1 + \varphi_2$ assumes the value $a_i + b_j$ on the set $G_{i,j}$ for $i = 1, \ldots, m$ and $j = 1, \ldots, n$, we have $\int_D \{\varphi_1 + \varphi_2\}\,d\mu = \sum_{i=1}^m \sum_{j=1}^n (a_i + b_j)\mu(G_{i,j})$. Thus we have $\int_D \{\varphi_1 + \varphi_2\}\,d\mu = \int_D \varphi_1\,d\mu + \int_D \varphi_2\,d\mu$.

4. To prove (g), let $\varphi_1 = \sum_{i=1}^m a_i \mathbf{1}_{E_i}$ and $\varphi_2 = \sum_{j=1}^n b_j \mathbf{1}_{F_j}$ be the canonical representations of φ_1 and φ_2 respectively. Let $G_{i,j} = E_i \cap F_j$ as in **3.** We have the equalities $\int_D \varphi_1\,d\mu = \sum_{i=1}^m \sum_{j=1}^n a_i \mu(G_{i,j})$ and $\int_D \varphi_2\,d\mu = \sum_{i=1}^m \sum_{j=1}^n b_j \mu(G_{i,j})$. Let us show that if at least one of φ_1 and φ_2 is μ-integrable on D, then the simple function $\varphi_1 + \varphi_2$ is μ semi-integrable on D. Let us assume that φ_2 is μ-integrable on D. Since $\varphi_1 + \varphi_2$ assumes the value $a_i + b_j$ on $G_{i,j}$, to show that $\varphi_1 + \varphi_2$ is μ semi-integrable on D we show that $\sum_{i=1}^m \sum_{j=1}^n (a_i + b_j)\mu(G_{i,j})$ exists in $\overline{\mathbb{R}}$. Suppose among the $m \times n$ summands in $\sum_{i=1}^m \sum_{j=1}^n (a_i + b_j)\mu(G_{i,j})$, we have $(a_i + b_j)\mu(G_{i,j}) = \infty$ for some (i, j)

and $(a_k + b_\ell)\mu(G_{k,\ell}) = -\infty$ for some (k, ℓ). Now since φ_2 is μ-integrable on D and thus $b_j\mu(G_{i,j})$ and $b_\ell\mu(G_{k,\ell})$ are finite, we have $a_i\mu(G_{i,j}) = \infty$ and $a_k\mu(G_{k,\ell}) = -\infty$ and then $a_i\mu(E_i) = \infty$ and $a_k\mu(E_k) = -\infty$. This is impossible since φ_1 is μ semi-integrable on D and $\sum_{i=1}^m a_i\mu(E_i)$ exists in $\overline{\mathbb{R}}$. This shows that it is impossible that one of the $m \times n$ summands assumes the value ∞ and another the value $-\infty$. Therefore $\sum_{i=1}^m \sum_{j=1}^n (a_i + b_j)\mu(G_{i,j})$ exists in $\overline{\mathbb{R}}$.

5. (h) is a particular case of Observation 7.6. ∎

[II] Integration of Bounded Functions on Sets of Finite Measure

Given a measure space $(X, \mathfrak{A}, \mu)$. Let D be a $\mathfrak{A}$-measurable set with $\mu(D) < \infty$. Then every simple function on D is μ-integrable on D with respect to μ. Let f be a bounded real-valued function on D, say $|f(x)| \le M$ for $x \in D$ for some constant $M \ge 0$. Let φ and ψ be simple functions on D such that $\varphi(x) \le f(x)$ and $f(x) \le \psi(x)$ for $x \in D$. (Such simple functions always exist. For instance $\varphi(x) = -M$ and $\psi(x) = M$ for $x \in D$ will do.) Now since $\varphi \le \psi$ on D, we have $\int_D \varphi\, d\mu \le \int_D \psi\, d\mu$.

Observation 7.8. Let f be a bounded real-valued function on a set $D \in \mathfrak{A}$ with $\mu(D) < \infty$ in a measure space $(X, \mathfrak{A}, \mu)$. Suppose $f(x) \in [M_1, M_2]$ for some constants $M_1, M_2 \in \mathbb{R}$ such that $M_1 < M_2$. Then we have

$$M_1\mu(D) \le \sup_{\varphi \le f} \int_D \varphi\, d\mu \le \inf_{f \le \psi} \int_D \psi\, d\mu \le M_2\mu(D),$$

where $\sup_{\varphi \le f}$ is the supremum on the collection of all simple functions φ on D such that $\varphi \le f$ on D and $\inf_{f \le \psi}$ is the infimum on the collection of all simple functions ψ on D such that $f \le \psi$ on D.

Proof. For every φ and ψ, we have $\varphi \le f \le \psi$ and thus $\int_D \varphi\, d\mu \le \int_D \psi\, d\mu$ by (b) of Lemma 7.7. Then with an arbitrarily fixed φ, we have $\int_D \varphi\, d\mu \le \inf_{f \le \psi} \int_D \psi\, d\mu$. Since this holds for every φ, we have $\sup_{\varphi \le f} \int_D \varphi\, d\mu \le \inf_{f \le \psi} \int_D \psi\, d\mu$. ∎

Theorem 7.9. *Let f be a bounded real-valued function on a set $D \in \mathfrak{A}$ with $\mu(D) < \infty$ in a measure space $(X, \mathfrak{A}, \mu)$.*

(a) *If f is $\mathfrak{A}$-measurable on D, then*

$$\sup_{\varphi \le f} \int_D \varphi\, d\mu = \inf_{f \le \psi} \int_D \psi\, d\mu. \tag{1}$$

(b) *Conversely if (1) holds, then f is $\mathfrak{A}$-measurable on D provided that $(X, \mathfrak{A}, \mu)$ is a complete measure space.*

Proof. 1. Let us assume that f is $\mathfrak{A}$-measurable on D. Since f is bounded on D, there exists $M > 0$ such that $f(x) \in (-M, M]$ for $x \in D$. For each $n \in \mathbb{N}$, let us divide the interval $(-M, M]$ into $2n$ equal subintervals of length $\frac{M}{n}$ given by $I_{n,k} = \left((k-1)\frac{M}{n}, k\frac{M}{n}\right]$ for $k = -n+1, \ldots, n$. Let $D_{n,k} = \{D : f \in I_{n,k}\} = f^{-1}(I_{n,k})$. Since f is a real-valued $\mathfrak{A}$-measurable function, it is a $\mathfrak{A}/\mathfrak{B}_{\mathbb{R}}$-measurable mapping by Theorem 4.6. Then since $I_{n,k} \in \mathfrak{B}_{\mathbb{R}}$, we have $f^{-1}(I_{n,k}) \in \mathfrak{A}$. Thus $\{D_{n,k} : k = -n+1, \ldots, n\}$ is a disjoint collection in $\mathfrak{A}$ and $\bigcup_{k=-n+1}^{n} D_{n,k} = D$. For each $n \in \mathbb{N}$, let us define two simple functions φ_n and ψ_n on D by setting

$$\begin{cases} \varphi_n(x) = (k-1)\dfrac{M}{n} & \text{for } x \in D_{n,k} \text{ for } k = -n+1, \ldots, n, \\ \psi_n(x) = k\dfrac{M}{n} & \text{for } x \in D_{n,k} \text{ for } k = -n+1, \ldots, n. \end{cases}$$

We have $\varphi_n(x) \le f(x) \le \psi_n(x)$ for $x \in D$. Now by Observation 7.8, we have

$$\begin{aligned} 0 \le \inf_{f \le \psi} \int_D \psi \, d\mu - \sup_{\varphi \le f} \int_D \varphi \, d\mu &\le \int_D \psi_n \, d\mu - \int_D \varphi_n \, d\mu \\ &= \int_D \{\psi_n - \varphi_n\} d\mu = \sum_{k=-n+1}^{n} \int_{D_{n,k}} \{\psi_n - \varphi_n\} d\mu \\ &= \sum_{k=-n+1}^{n} \left\{k\frac{M}{n} - (k-1)\frac{M}{n}\right\} \mu(D_{n,k}) \\ &= \sum_{k=-n+1}^{n} \frac{M}{n} \mu(D_{n,k}) = \frac{M}{n} \mu(D) \end{aligned}$$

for an arbitrary $n \in \mathbb{N}$. Letting $n \to \infty$, we have $\inf_{f \le \psi} \int_D \psi \, d\mu - \sup_{\varphi \le f} \int_D \varphi \, d\mu = 0$. This proves (1).

2. Conversely suppose (1) holds. Assume also that $(X, \mathfrak{A}, \mu)$ is a complete measure space. Let us show that f is $\mathfrak{A}$-measurable. Now for every $n \in \mathbb{N}$, there exist a simple function $\varphi_n \le f$ and a simple function $\psi_n \ge f$ such that

$$\sup_{\varphi \le f} \int_D \varphi \, d\mu - \frac{1}{n} < \int_D \varphi_n \, d\mu$$

and

$$\inf_{f \le \psi} \int_D \psi \, d\mu + \frac{1}{n} > \int_D \psi_n \, d\mu.$$

Subtracting the first inequality from the second and applying (1) we have

$$(2) \qquad \int_D \psi_n \, d\mu - \int_D \varphi_n \, d\mu \le \frac{2}{n}.$$

If we define two functions g and h on D by setting $g = \sup_{n \in \mathbb{N}} \varphi_n$ and $h = \inf_{n \in \mathbb{N}} \psi_n$, then g and h are $\mathfrak{A}$-measurable on D by Theorem 4.22. Also from $\varphi_n \le f \le \psi_n$ on D for every $n \in \mathbb{N}$, we have $g \le f \le h$ on D. If we show that $g = h$ a.e. on D, then we have $f = g$ a.e. on D. Then the fact that g is $\mathfrak{A}$-measurable on D and $(X, \mathfrak{A}, \mu)$ is a complete measure space implies that f is $\mathfrak{A}$-measurable on D according to (b) of Observation 4.20. Thus it remains to show that $g = h$ a.e. on D. Now

$$\{D : g \neq h\} = \{D : h - g > 0\} = \bigcup_{k \in \mathbb{N}} \{D : h - g \geq \tfrac{1}{k}\}$$

so that

$$\mu\{D : g \neq h\} \leq \sum_{k \in \mathbb{N}} \mu\{D : h - g \geq \tfrac{1}{k}\}.$$

If we show that

$$\text{(3)} \qquad \mu\{D : h - g \geq \tfrac{1}{k}\} = 0 \quad \text{for every } k \in \mathbb{N},$$

then $\mu\{D : g \neq h\} = 0$ and $g = h$ a.e. on D. To prove (3), let $k \in \mathbb{N}$ be fixed. Since $\varphi_n \leq g \leq f \leq h \leq \psi_n$ for $n \in \mathbb{N}$, we have $\{D : h - g \geq \frac{1}{k}\} \subset \{D : \psi_n - \varphi_n \geq \frac{1}{k}\}$ and then $\mu\{D : h - g \geq \frac{1}{k}\} \leq \mu\{D : \psi_n - \varphi_n \geq \frac{1}{k}\}$. Thus we have

$$\begin{aligned}
&\frac{1}{k}\mu\left\{D : h - g \geq \frac{1}{k}\right\} \leq \frac{1}{k}\mu\left\{D : \psi_n - \varphi_n \geq \frac{1}{k}\right\} \\
&= \int_{\{D : \psi_n - \varphi_n \geq \frac{1}{k}\}} \frac{1}{k}\, d\mu \leq \int_{\{D : \psi_n - \varphi_n \geq \frac{1}{k}\}} \{\psi_n(x) - \varphi_n(x)\}\, d\mu \\
&\leq \int_D \{\psi_n - \varphi_n\}\, d\mu \leq \frac{2}{n},
\end{aligned}$$

where the second inequality is from the fact that $\frac{1}{k} \leq \psi_n - \varphi_n$ on $\{D : \psi_n - \varphi_n \geq \frac{1}{k}\}$ so that (b) of Lemma 7.7 is applicable, the third inequality is by 8° of Observation 7.5, and the last inequality is by (2). Since this holds for every $n \in \mathbb{N}$, we have $\frac{1}{k}\mu\{D : h - g \geq \frac{1}{k}\} = 0$ and then multiplying by k we have $\mu\{D : h - g \geq \frac{1}{k}\} = 0$. This proves (3). ∎

If f is a bounded real-valued $\mathfrak{A}$-measurable function on a set $D \in \mathfrak{A}$ with $\mu(D) < \infty$ in a measure space $(X, \mathfrak{A}, \mu)$, then $\sup_{\varphi \leq f} \int_D \varphi\, d\mu = \inf_{f \leq \psi} \int_D \psi\, d\mu \in \mathbb{R}$ by Theorem 7.9 and Observation 7.8. This is the basis of the following definition.

Definition 7.10. *Given a measure space $(X, \mathfrak{A}, \mu)$. Let f be a bounded real-valued $\mathfrak{A}$-measurable function on a set $D \in \mathfrak{A}$ with $\mu(D) < \infty$. We define the Lebesgue integral of f on D with respect to μ by*

$$\int_D f(x)\mu(dx) = \sup_{\varphi \leq f} \int_D \varphi\, d\mu = \inf_{f \leq \psi} \int_D \psi\, d\mu \in \mathbb{R}.$$

In other words, with the collection Φ of all simple functions φ on D such that $\varphi \leq f$ we let

$$\int_D f(x)\mu(dx) = \sup\left\{\int_D \varphi\, d\mu : \varphi \in \Phi\right\}.$$

We say that f is Lebesgue integrable on D with respect to μ or simply μ-integrable on D.

If $D \in \mathfrak{A}$ with $\mu(D) < \infty$ then a simple function f defined on D is also a bounded real-valued $\mathfrak{A}$-measurable function on a set with finite measure so that $\int_D f\,d\mu$ is defined by both Definitions 7.4 and 7.10. Thus we need to verify that the two definitions of $\int_D f\,d\mu$ are consistent for such a function. Let us write $I(f)$ for the integral of f on D in Definition 7.10, reserving the notation $\int_D f\,d\mu$ for the integral of a simple function f as in Definition 7.4. Thus $I(f) = \sup\left\{\int_D \varphi\,d\mu : \varphi \in \Phi\right\}$. Now if f is itself a simple function on D, then $f \in \Phi$ and moreover for every $\varphi \in \Phi$ we have $\varphi \le f$ so that $\int_D \varphi\,d\mu \le \int_D f\,d\mu$ by (b) of Lemma 7.7. Thus $I(f) = \sup\left\{\int_D \varphi\,d\mu : \varphi \in \Phi\right\} = \int_D f\,d\mu$. This shows the consistency of Definitions 7.10 and 7.4.

Lemma 7.11. *Given a measure space $(X, \mathfrak{A}, \mu)$. Let f_1 and f_2 be bounded real-valued $\mathfrak{A}$-measurable functions on a set $D \in \mathfrak{A}$ with $\mu(D) < \infty$. If $f_1 = f_2$ a.e. on D, then $\int_D f_1\,d\mu = \int_D f_2\,d\mu$.*

Proof. For $i = 1$ and 2, let Φ_i be the collection of all simple functions φ_i on D such that $\varphi_i \le f_i$. By Definition 7.10, we have

$$(1) \qquad \int_D f_i\,d\mu = \sup\left\{\int_D \varphi_i\,d\mu : \varphi_i \in \Phi_i\right\}.$$

Let us show that corresponding to every $\varphi_1 \in \Phi_1$ there exists $\varphi_2 \in \Phi_2$ such that $\int_D \varphi_1\,d\mu = \int_D \varphi_2\,d\mu$. Now since $f_1 = f_2$ a.e. on D, there exists a subset D_0 of D which is a null set in $(X, \mathfrak{A}, \mu)$ such that $f_1 = f_2$ on $D \setminus D_0$. Since f_1 and f_2 are bounded on D, there exists $M > 0$ such that $f_1(x), f_2(x) \in [-M, M]$ for $x \in D$. Let us define a simple function φ_2 on D by setting $\varphi_2(x) = \varphi_1(x)$ for $x \in D \setminus D_0$ and $\varphi_2(x) = -M$ for $x \in D_0$. Then $\varphi_2 \le f_2$ on D so that $\varphi_2 \in \Phi_2$. By (h) of Lemma 7.7 and 1° of Observation 7.5, we have

$$\begin{aligned}\int_D \varphi_2\,d\mu &= \int_{D\setminus D_0} \varphi_2\,d\mu + \int_{D_0} \varphi_2\,d\mu = \int_{D\setminus D_0} \varphi_1\,d\mu + 0\\ &= \int_{D\setminus D_0} \varphi_1\,d\mu + \int_{D_0} \varphi_1\,d\mu = \int_D \varphi_1\,d\mu.\end{aligned}$$

Thus corresponding to every $\varphi_1 \in \Phi_1$ there exists $\varphi_2 \in \Phi_2$ such that $\int_D \varphi_1\,d\mu = \int_D \varphi_2\,d\mu$. Then the collection of real numbers $\left\{\int_D \varphi_1\,d\mu : \varphi_1 \in \Phi_1\right\}$ is a subcollection of the collection $\left\{\int_D \varphi_2\,d\mu : \varphi_2 \in \Phi_2\right\}$ and therefore we have

$$(2) \qquad \sup\left\{\int_D \varphi_1\,d\mu : \varphi_1 \in \Phi_1\right\} \le \sup\left\{\int_D \varphi_2\,d\mu : \varphi_2 \in \Phi_2\right\}.$$

By (1) and (2), we have $\int_D f_1\,d\mu \le \int_D f_2\,d\mu$. Interchanging the roles of f_1 and f_2 in the argument above we have $\int_D f_2\,d\mu \le \int_D f_1\,d\mu$. Therefore $\int_D f_1\,d\mu = \int_D f_2\,d\mu$. ■

Lemma 7.12. *Given a measure space $(X, \mathfrak{A}, \mu)$. Let f, f_1, and f_2 be bounded real-valued $\mathfrak{A}$-measurable functions on a set $D \in \mathfrak{A}$ with $\mu(D) < \infty$. Then*

$$(1) \qquad \int_D cf\,d\mu = c\int_D f\,d\mu \quad \text{for } c \in \mathbb{R},$$

and

$$(2) \qquad \int_D \{f_1 + f_2\}\,d\mu = \int_D f_1\,d\mu + \int_D f_2\,d\mu.$$

Proof. 1. To prove (1), note that if $c = 0$, then $cf = 0$ on D and thus $\int_D cf\,d\mu = 0$. On the other hand since $\int_D f\,d\mu \in \mathbb{R}$, if $c = 0$ then $c\int_D f\,d\mu = 0$. This proves (1) for the case $c = 0$. Next consider the case $c > 0$. Now

$$(3) \qquad \begin{aligned} \int_D cf\,d\mu &= \sup_{\varphi \le cf} \int_D \varphi\,d\mu = c \sup_{\varphi \le cf} \left\{ \frac{1}{c} \int_D \varphi\,d\mu \right\} \\ &= c \sup_{\varphi \le cf} \int_D \frac{1}{c}\varphi\,d\mu = c \sup_{\varphi/c \le f} \int_D \frac{1}{c}\varphi\,d\mu \\ &= c \sup_{\varphi \le f} \int_D \varphi\,d\mu = c \int_D f\,d\mu. \end{aligned}$$

This proves (1) for the case $c > 0$. When $c = -1$, we have

$$(4) \qquad \begin{aligned} \int_D (-f)\,d\mu &= \sup_{\varphi \le -f} \int_D \varphi\,d\mu = -\inf_{\varphi \le -f} \left\{ -\int_D \varphi\,d\mu \right\} \\ &= -\inf_{f \le -\varphi} \int_D (-\varphi)\,d\mu = -\inf_{f \le \varphi} \int_D \varphi\,d\mu \\ &= -\int_D f\,d\mu, \end{aligned}$$

where the last equality is by Definition 7.10. This proves (1) for the case $c = -1$. When $c < 0$, recalling the results above for the cases $c = -1$ and $c > 0$, we have

$$(5) \qquad \int_D cf\,d\mu = \int_D (-|c|)f\,d\mu = -\int_D |c|f\,d\mu = -|c|\int_D f\,d\mu = c\int_D f\,d\mu.$$

This proves (1) for the case $c < 0$. With (3), (4), and (5), we have (1).

2. To prove (2), let φ_1 and φ_2 be simple functions on D such that $\varphi_1 \le f_1$ and $\varphi_2 \le f_2$. By (g) of Lemma 7.7, we have

$$\int_D \varphi_1\,d\mu + \int_D \varphi_2\,d\mu = \int_D \{\varphi_1 + \varphi_2\}\,d\mu \le \sup_{\varphi \le f_1 + f_2} \int_D \varphi\,d\mu,$$

since $\varphi_1 + \varphi_2$ is one simple function on D such that $\varphi_1 + \varphi_2 \le f_1 + f_2$. Then we have

$$\int_D \varphi_1\,d\mu + \sup_{\varphi_2 \le f_2} \int_D \varphi_2\,d\mu \le \sup_{\varphi \le f_1 + f_2} \int_D \varphi\,d\mu,$$

and

$$\sup_{\varphi_1 \le f_1} \int_D \varphi_1\,d\mu + \sup_{\varphi_2 \le f_2} \int_D \varphi_2\,d\mu \le \sup_{\varphi \le f_1 + f_2} \int_D \varphi\,d\mu,$$

that is,

$$(6)\qquad \int_D f_1\,d\mu + \int_D f_2\,d\mu \le \int_D \{f_1 + f_2\}\,d\mu.$$

Similarly for simple functions ψ_1 and ψ_2 on D such that $f_1 \le \psi_1$ and $f_2 \le \psi_2$ we have $f_1 + f_2 \le \psi_1 + \psi_2$ so that

$$\int_D \psi_1\,d\mu + \int_D \psi_2\,d\mu = \int_D \{\psi_1 + \psi_2\}\,d\mu \ge \inf_{f_1+f_2\le\psi} \int_D \psi\,d\mu$$

and then

$$\inf_{f_1\le\psi_1} \int_D \psi_1\,d\mu + \inf_{f_2\le\psi_2} \int_D \psi_2\,d\mu \ge \inf_{f_1+f_2\le\psi} \int_D \psi\,d\mu,$$

that is,

$$(7)\qquad \int_D f_1\,d\mu + \int_D f_2\,d\mu \ge \int_D \{f_1 + f_2\}\,d\mu.$$

By (6) and (7), we have (2). ∎

Lemma 7.13. *Given a measure space $(X, \mathfrak{A}, \mu)$. Let f and g be bounded real-valued $\mathfrak{A}$-measurable functions on a set $D \in \mathfrak{A}$ with $\mu(D) < \infty$. Then*

$$(1)\qquad f \le g \text{ on } D \Rightarrow \int_D f\,d\mu \le \int_D g\,d\mu.$$

If $|f| \le M$ on D where $M \ge 0$, then

$$(2)\qquad \left|\int_D f\,d\mu\right| \le \int_D |f|\,d\mu \le M\,\mu(D).$$

Proof. 1. Suppose $f \le g$ on D. If φ is a simple function on D such that $\varphi \le f$ then $\varphi \le g$ also. Thus we have $\sup_{\varphi\le f} \int_D \varphi\,d\mu \le \sup_{\varphi\le g} \int_D \varphi\,d\mu$, that is, $\int_D f\,d\mu \le \int_D g\,d\mu$. This proves (1).

2. Since $f \le |f|$ and $-f \le |f|$ also, we have $\int_D f\,d\mu \le \int_D |f|\,d\mu$ by (1) and similarly $\int_D (-f)\,d\mu \le \int_D |f|\,d\mu$, that is, $-\int_D f\,d\mu \le \int_D |f|\,d\mu$ by Lemma 7.12. Therefore we have $\left|\int_D f\,d\mu\right| \le \int_D |f|\,d\mu$. Also if $|f| \le M$, then $\int_D |f|\,d\mu \le \int_D M\,d\mu = M\mu(D)$ by (1). ∎

Lemma 7.14. *Given a measure space $(X, \mathfrak{A}, \mu)$. Let f and g be bounded real-valued $\mathfrak{A}$-measurable functions on a set $D \in \mathfrak{A}$ with $\mu(D) < \infty$.*
(a) *If $f \ge 0$ a.e. on D and $\int_D f\,d\mu = 0$, then $f = 0$ a.e. on D.*
(b) *If $f \le g$ a.e. on D and $\int_D f\,d\mu = \int_D g\,d\mu$, then $f = g$ a.e. on D.*

Proof. 1. Consider first the special case that $f \ge 0$ on D. Let $D_0 = \{x \in D : f(x) = 0\}$ and $D_1 = \{x \in D : f(x) > 0\}$. Then we have $D_0, D_1 \in \mathfrak{A}$, $D_0 \cap D_1 = \emptyset$ and $D_0 \cup D_1 = D$. Let us show that

$$(1)\qquad f = 0 \text{ a.e. on } D \Leftrightarrow \mu(D_1) = 0.$$

Suppose $f = 0$ a.e. on D. Then by Definition 4.17 there exists a null set E in $(X, \mathfrak{A}, \mu)$ such that $E \subset D$ and $f = 0$ on $D \setminus E$. Then by our definition of D_0 we have $D \setminus E \subset D_0 = D \setminus D_1$. Thus $D_1 \subset E$ and consequently $\mu(D_1) \leq \mu(E) = 0$ so that $\mu(D_1) = 0$. Conversely suppose $\mu(D_1) = 0$. Then D_1 is a null set in $(X, \mathfrak{A}, \mu)$ and $D_1 \subset D$. Now $f = 0$ on $D_0 = D \setminus D_1$. Thus $f = 0$ a.e. on D. This completes the proof of (1).

Let us note that if $\mu(D) = 0$, that is, D is a null set in $(X, \mathfrak{A}, \mu)$, then $\mu(D_1) \leq \mu(D) = 0$ so that $f = 0$ a.e. on D by (1).

Now let us show that if $f \geq 0$ a.e. on D and $\int_D f \, d\mu = 0$, then $f = 0$ a.e. on D. By our remark above we need only to consider the case $\mu(D) \in (0, \infty)$. To show that $f = 0$ a.e. on D, assume the contrary, that is, the statement "$f = 0$ a.e. on D" is false. By (1) this implies that $\mu(D_1) > 0$. Now

$$D_1 = \{D : f > 0\} = \bigcup_{k \in \mathbb{N}} \{D : f \geq \tfrac{1}{k}\}$$

and then

$$0 < \mu(D_1) \leq \sum_{k \in \mathbb{N}} \mu\{D : f \geq \tfrac{1}{k}\}.$$

Thus there exist $k_0 \in \mathbb{N}$ such that $\mu\{D : f \geq \frac{1}{k_0}\} > 0$. Let us define a simple function φ on D by setting

$$\varphi(x) = \begin{cases} \dfrac{1}{k_0} & \text{for } x \in \{D : f \geq \frac{1}{k_0}\}, \\ 0 & \text{for } x \in D \setminus \{D : f \geq \frac{1}{k_0}\}. \end{cases}$$

Then $\varphi \leq f$ on D. Thus we have

$$\int_D f \, d\mu \geq \int_D \varphi \, d\mu = \frac{1}{k_0} \mu\left\{D : f \geq \frac{1}{k_0}\right\} > 0.$$

This contradicts the assumption that $\int_D f \, d\mu = 0$. Therefore we have $f = 0$ a.e. on D.

Finally consider the general case that $f \geq 0$ a.e. on D and $\int_D f \, d\mu = 0$. Then there exists a null set E in $(X, \mathfrak{A}, \mu)$ such that $E \subset D$ and $f \geq 0$ on $D \setminus E$. By Lemma 7.15

$$0 = \int_D f \, d\mu = \int_E f \, d\mu + \int_{D \setminus E} f \, d\mu = \int_{D \setminus E} f \, d\mu.$$

Thus $f \geq 0$ on $D \setminus E$ and $\int_{D \setminus E} f \, d\mu = 0$. By our result above this implies that $f = 0$ a.e. on $D \setminus E$. Thus there exists a null set F in $(X, \mathfrak{A}, \mu)$ such that $F \subset D \setminus E$ and $f = 0$ on $(D \setminus E) \setminus F = D \setminus (E \cup F)$. As the union of two null sets, $E \cup F$ is a null set in $(X, \mathfrak{A}, \mu)$. Thus $f = 0$ a.e. on D.

2. To prove (b), note that if $f \leq g$ a.e. on D then $g - f \geq 0$ a.e. on D. If in addition we have $\int_D f \, d\mu = \int_D g \, d\mu$, then $\int_D \{g - f\} \, d\mu = 0$ by Lemma 7.12. Then by (a), we have $g - f = 0$ a.e. on D, that is, $f = g$ a.e. on D. ∎

Lemma 7.15. *Given a measure space $(X, \mathfrak{A}, \mu)$. Let f be a bounded real-valued $\mathfrak{A}$-measurable functions on a set $D \in \mathfrak{A}$ with $\mu(D) < \infty$. Let $(D_n : n \in \mathbb{N})$ be a disjoint sequence in $\mathfrak{A}$ such that $\bigcup_{n \in \mathbb{N}} D_n = D$. Then $\int_D f \, d\mu = \sum_{n \in \mathbb{N}} \int_{D_n} f \, d\mu$.*

Proof. Let φ be an arbitrary simple function on D such that $\varphi \leq f$. Let $\varphi = \sum_{i=1}^{p} a_i \mathbf{1}_{E_i}$ be its canonical representation. Let φ_n be the simple function on D_n which is the restriction of φ to D_n. Its canonical representation is given by $\varphi_n = \sum_{i=1}^{p} a_i \mathbf{1}_{E_i \cap D_n}$. Note that for each $i = 1, \ldots, p$, $\{E_i \cap D_n : n \in \mathbb{N}\}$ is a disjoint collection in $\mathfrak{A}$ with $\bigcup_{n \in \mathbb{N}} (E_i \cap D_n) = E_i$ so that $\mu(E_i) = \sum_{n \in \mathbb{N}} \mu(E_i \cap D_n)$. Now

$$\begin{aligned}
\int_D \varphi \, d\mu &= \sum_{i=1}^{p} a_i \, \mu(E_i) = \sum_{i=1}^{p} a_i \left[\sum_{n \in \mathbb{N}} \mu(E_i \cap D_n) \right] \\
&= \sum_{n \in \mathbb{N}} \left[\sum_{i=1}^{p} a_i \, \mu(E_i \cap D_n) \right] = \sum_{n \in \mathbb{N}} \int_{D_n} \varphi_n \, d\mu \\
&\leq \sum_{n \in \mathbb{N}} \int_{D_n} f \, d\mu,
\end{aligned}$$

where the third equality is by the fact that if $\sum_{n \in \mathbb{N}} \alpha_{1,n}, \ldots, \sum_{n \in \mathbb{N}} \alpha_{p,n}$ are convergent series of real numbers then $\sum_{n \in \mathbb{N}} \{\alpha_{1,n} + \cdots + \alpha_{p,n}\} = \sum_{n \in \mathbb{N}} \alpha_{1,n} + \cdots + \sum_{n \in \mathbb{N}} \alpha_{p,n}$, and the last inequality is from the fact that φ_n is just one simple function on D_n such that $\varphi_n \leq f$ so that $\int_{D_n} \varphi_n \, d\mu \leq \sup_{\varphi \leq f} \int_{D_n} \varphi \, d\mu = \int_{D_n} f \, d\mu$ for every $n \in \mathbb{N}$. Thus we have

$$\int_D f \, d\mu = \sup_{\varphi \leq f} \int_D \varphi \, d\mu \leq \sum_{n \in \mathbb{N}} \int_{D_n} f \, d\mu.$$

Similarly by starting with a simple function ψ on D such that $f \leq \psi$ we obtain

$$\int_D f \, d\mu = \inf_{f \leq \psi} \int_D \psi \, d\mu \geq \sum_{n \in \mathbb{N}} \int_{D_n} f \, d\mu.$$

Therefore we have $\int_D f \, d\mu = \sum_{n \in \mathbb{N}} \int_{D_n} f \, d\mu$. ∎

Theorem 7.16. (Bounded Convergence Theorem) *Given a measure space $(X, \mathfrak{A}, \mu)$. Let $(f_n : n \in \mathbb{N})$ be a uniformly bounded sequence of real-valued $\mathfrak{A}$-measurable functions on a set $D \in \mathfrak{A}$ with $\mu(D) < \infty$. Let f be a bounded real-valued $\mathfrak{A}$-measurable function on D. If $(f_n : n \in \mathbb{N})$ converges to f a.e. on D, then*

(1) $$\lim_{n \to \infty} \int_D |f_n - f| \, d\mu = 0,$$

and in particular

(2) $$\lim_{n \to \infty} \int_D f_n \, d\mu = \int_D f \, d\mu.$$

Proof. By the uniform boundedness of $(f_n : n \in \mathbb{N})$ on D, there exists $M > 0$ such that $|f_n(x)| \leq M$ for all $x \in D$ and $n \in \mathbb{N}$. Since f is also bounded on D, we assume $M > 0$ has been so chosen that $|f(x)| \leq M$ for all $x \in D$. Now since $(f_n : n \in \mathbb{N})$ converges to f a.e. on D and since $\mu(D) < \infty$, according to Theorem 6.12 (Egoroff), for every $\eta > 0$ there exists a $\mathfrak{A}$-measurable subset E of D with $\mu(E) < \eta$ such that $(f_n : n \in \mathbb{N})$ converges to f uniformly on $D \setminus E$, that is, for every $\varepsilon > 0$ there exists $N \in \mathbb{N}$ such that $|f_n(x) - f(x)| < \varepsilon$ for all $x \in D \setminus E$ when $n \geq N$. Then for $n \geq N$, we have by Lemma 7.15

$$\begin{aligned}\int_D |f_n - f| \, d\mu &= \int_{D\setminus E} |f_n - f| \, d\mu + \int_E |f_n - f| \, d\mu \\ &\leq \int_{D\setminus E} \varepsilon \, d\mu + \int_E 2M \, d\mu \\ &\leq \varepsilon \, \mu(D \setminus E) + 2M \, \mu(E) \\ &< \varepsilon \mu(D) + 2M\eta.\end{aligned}$$

Since this holds for all $n \geq \mathbb{N}$, we have $\limsup_{n\to\infty} \int_D |f_n - f| \, d\mu \leq \varepsilon\mu(D) + 2M\eta$. Then since this holds for every $\eta > 0$ and $\varepsilon > 0$, we have $\limsup_{n\to\infty} \int_D |f_n - f| \, d\mu = 0$. Now $\int_D |f_n - f| \, d\mu \geq 0$ for every $n \in \mathbb{N}$ so that $\liminf_{n\to\infty} \int_D |f_n - f| \, d\mu \geq 0$. Then we have

$$0 \leq \liminf_{n\to\infty} \int_D |f_n - f| \, d\mu \leq \limsup_{n\to\infty} \int_D |f_n - f| \, d\mu = 0$$

so that $\liminf_{n\to\infty} \int_D |f_n - f| \, d\mu = \limsup_{n\to\infty} \int_D |f_n - f| \, d\mu = 0$. This equality implies that we have $\lim_{n\to\infty} \int_D |f_n - f| \, d\mu = 0$. This proves (1).

To derive (2) from (1), note that by Lemmas 7.12 and 7.13, we have

$$\text{(3)} \qquad \left| \int_D f_n \, d\mu - \int_D f \, d\mu \right| = \left| \int_D \{f_n - f\} \, d\mu \right| \leq \int_D |f_n - f| \, d\mu.$$

Applying (1) to the last member of (3), we have $\lim_{n\to\infty} \left| \int_D f_n \, d\mu - \int_D f \, d\mu \right| = 0$. Now for a sequence $(a_n : n \in \mathbb{N})$ of real numbers $\lim_{n\to\infty} |a_n| = 0$ implies $\lim_{n\to\infty} a_n = 0$. Thus we have $\lim_{n\to\infty} \left\{ \int_D f_n \, d\mu - \int_D f \, d\mu \right\} = 0$. This implies (2). ∎

Given a sequence of extended real numbers $(a_n : n \in \mathbb{N})$. If there exists a real number a such that every subsequence $(a_{n_k} : k \in \mathbb{N})$ has a subsequence $\left(a_{n_{k_\ell}} : \ell \in \mathbb{N}\right)$ converging to a, then the sequence $(a_n : n \in \mathbb{N})$ converges to a. Using this fact we can replace the convergence a.e. condition on the sequence of functions in Theorem 7.16 by the weaker condition of convergence in measure. Recall that since $\mu(D) < \infty$, convergence a.e. on D implies convergence on D in measure according to Theorem 6.22 (Lebesgue).

Corollary 7.17. (Bounded Convergence Theorem under Convergence in Measure) *If the condition that $(f_n : n \in \mathbb{N})$ converges to f a.e. on D in Theorem 7.16 is replaced by the weaker condition that $(f_n : n \in \mathbb{N})$ converges to f on D in measure, the conclusions (1) and (2) are still valid.*

Proof. Assume that $(f_n : n \in \mathbb{N})$ converges to f on D in measure. We are to show that $\lim_{n\to\infty} \int_D |f_n - f| \, d\mu = 0$. Consider the sequence of real numbers $(a_n : n \in \mathbb{N})$ defined by $a_n = \int_D |f_n - f| \, d\mu$ for $n \in \mathbb{N}$. Take an arbitrary subsequence $(a_{n_k} : k \in \mathbb{N})$. Consider the sequence of functions $(f_{n_k} : k \in \mathbb{N})$. Since $(f_n : n \in \mathbb{N})$ converges to f on D in measure, the subsequence $(f_{n_k} : k \in \mathbb{N})$ converges to f on D in measure too. Then by Theorem 6.23 (Riesz), there exists a subsequence $\left(f_{n_{k_\ell}} : \ell \in \mathbb{N}\right)$ which converges to f a.e. on D. Thus by Theorem 7.16, we have $\lim_{n\to\infty} \int_D |f_{n_{k_\ell}} - f| \, d\mu = 0$, that is, the subsequence $\left(a_{n_{k_\ell}} : \ell \in \mathbb{N}\right)$ of the arbitrary subsequence $(a_{n_k} : k \in \mathbb{N})$ of $(a_n : n \in \mathbb{N})$ converges to 0. Therefore the sequence $(a_n : n \in \mathbb{N})$ converges to 0, that is, $\lim_{n\to\infty} \int_D |f_n - f| \, d\mu = 0$. ∎

[III] Riemann Integrability

Let $I = [a, b] \subset \mathbb{R}$. Let $\mathcal{P} = \{x_0, \dots, x_n\}$ where $a = x_0 < \cdots < x_n = b$ and consider a partition of I into subintervals $I_k = [x_{k-1}, x_k]$ for $k = 1, \dots, n$. We call $x_0, \dots, x_n$ the partition points in $\mathcal{P}$. The mesh of $\mathcal{P}$ is defined by $|\mathcal{P}| = \max\{\ell(I_k) : k = 1, \dots, n\}$. Let $\mathfrak{P}$ be the collection of all partitions of I. If $\mathcal{P}_1, \mathcal{P}_2 \in \mathfrak{P}$ and $\mathcal{P}_1 \subset \mathcal{P}_2$, then we say that $\mathcal{P}_2$ is a refinement of $\mathcal{P}_1$.

Definition 7.18. *Let f be a bounded real-valued function on I. For $\mathcal{P} \in \mathfrak{P}$ given by $\mathcal{P} = \{x_0, \dots, x_n\}$, let $I_k = [x_{k-1}, x_k]$ and $\xi_k \in I_k$ be arbitrarily chosen for $k = 1, \dots, n$. The Riemann sum of f corresponding to $\mathcal{P}$ and the selection $\xi = (\xi_k : k = 1, \dots, n)$ is defined by*

$$S(f, \mathcal{P}, \xi) = \sum_{k=1}^{n} f(\xi_k)\ell(I_k).$$

We say that f is Riemann integrable on I if there exists $J \in \mathbb{R}$ such that for every $\varepsilon > 0$ there exists $\delta > 0$ such that

$$\left|S(f, \mathcal{P}, \xi) - J\right| < \varepsilon \quad \text{for every } \mathcal{P} \in \mathfrak{P} \text{ with } |\mathcal{P}| < \delta.$$

In this case, J is called the Riemann integral of f on I and we write $\int_a^b f(x)\,dx$ for J.

Definition 7.19. *Let f be a bounded real-valued function on $I = [a, b]$ and let $|f(x)| \le M$ for $x \in I$ for some $M \ge 0$. With $\mathcal{P} \in \mathfrak{P}$ given by $\mathcal{P} = \{x_0, \dots, x_n\}$, let $I_k = [x_{k-1}, x_k]$, $m_k = \inf_{x\in I_k} f(x)$ and $M_k = \sup_{x\in I_k} f(x)$ for $k = 1, \dots, n$. The lower and upper Darboux sums of f corresponding to a partition $\mathcal{P}$ are defined by*

$$(1) \qquad \underline{S}(f, \mathcal{P}) = \sum_{k=1}^{n} m_k \ell(I_k) \quad \text{and} \quad \overline{S}(f, \mathcal{P}) = \sum_{k=1}^{n} M_k \ell(I_k).$$

Note that

$$(2) \qquad -M\ell(I) \le \underline{S}(f, \mathcal{P}) \le \overline{S}(f, \mathcal{P}) \le M\ell(I),$$

and

(3) $$\mathcal{P}_1 \subset \mathcal{P}_2 \Rightarrow \underline{S}(f, \mathcal{P}_1) \le \underline{S}(f, \mathcal{P}_2) \text{ and } \overline{S}(f, \mathcal{P}_1) \ge \overline{S}(f, \mathcal{P}_2).$$

Let

(4) $$\underline{S}(f) = \sup_{\mathcal{P} \in \mathfrak{P}} \underline{S}(f, \mathcal{P}) \quad \text{and} \quad \overline{S}(f) = \inf_{\mathcal{P} \in \mathfrak{P}} \overline{S}(f, \mathcal{P}).$$

We call $\underline{S}(f)$ *and* $\overline{S}(f)$ *the lower and upper Darboux integrals of* f *on* I. *Clearly*

(5) $$-M\ell(I) \le \underline{S}(f) \le \overline{S}(f) \le M\ell(I).$$

Lemma 7.20. *Let* f *be a bounded real-valued function on* $I = [a, b]$. *Then* $\underline{S}(f) = \overline{S}(f)$ *if and only if for every* $\varepsilon > 0$ *there exists* $\mathcal{P}_0 \in \mathfrak{P}$ *such that*

(1) $$\overline{S}(f, \mathcal{P}_0) - \underline{S}(f, \mathcal{P}_0) < \varepsilon.$$

Proof. 1. Suppose for every $\varepsilon > 0$ there exists $\mathcal{P}_0 \in \mathfrak{P}$ such that (1) holds. Then

$$\begin{aligned} \overline{S}(f) &= \inf_{\mathcal{P} \in \mathfrak{P}} \overline{S}(f, \mathcal{P}) \le \overline{S}(f, \mathcal{P}_0) = \overline{S}(f, \mathcal{P}_0) - \underline{S}(f, \mathcal{P}_0) + \underline{S}(f, \mathcal{P}_0) \\ &< \varepsilon + \underline{S}(f, \mathcal{P}_0) \le \varepsilon + \sup_{\mathcal{P} \in \mathfrak{P}} \underline{S}(f, \mathcal{P}) = \varepsilon + \underline{S}(f). \end{aligned}$$

By the arbitrariness of $\varepsilon > 0$, we have $\overline{S}(f) \le \underline{S}(f)$. On the other hand, $\overline{S}(f) \ge \underline{S}(f)$ for any bounded real-valued function f on I. Therefore $\underline{S}(f) = \overline{S}(f)$.

2. Conversely suppose $\underline{S}(f) = \overline{S}(f)$. Since $\underline{S}(f) = \sup_{\mathcal{P} \in \mathfrak{P}} \underline{S}(f, \mathcal{P})$, for every $\varepsilon > 0$ there exists $\mathcal{P}_1 \in \mathfrak{P}$ such that

$$\underline{S}(f, \mathcal{P}_1) > \underline{S}(f) - \tfrac{\varepsilon}{2},$$

and similarly there exists $\mathcal{P}_2 \in \mathfrak{P}$ such that

$$\overline{S}(f, \mathcal{P}_2) < \overline{S}(f) + \tfrac{\varepsilon}{2}.$$

If we let $\mathcal{P}_0 = \mathcal{P}_1 \cup \mathcal{P}_2 \in \mathfrak{P}$, then $\underline{S}(f, \mathcal{P}_1) \le \underline{S}(f, \mathcal{P}_0)$ and $\overline{S}(f, \mathcal{P}_2) \ge \overline{S}(f, \mathcal{P}_0)$ by (3) in Definition 7.19 so that

$$\overline{S}(f, \mathcal{P}_0) - \underline{S}(f, \mathcal{P}_0) \le \overline{S}(f, \mathcal{P}_2) - \underline{S}(f, \mathcal{P}_1) < \{\overline{S}(f) + \tfrac{\varepsilon}{2}\} - \{\underline{S}(f) - \tfrac{\varepsilon}{2}\} = \varepsilon. \blacksquare$$

Proposition 7.21. *Let* f *be a bounded real-valued function on* $I = [a, b]$. *Then we have* $\underline{S}(f) = \overline{S}(f)$ *if and only if for every* $\varepsilon > 0$ *there exists* $\delta > 0$ *such that*

(1) $$\overline{S}(f, \mathcal{P}) - \underline{S}(f, \mathcal{P}) < \varepsilon \quad \text{for every } \mathcal{P} \in \mathfrak{P} \text{ with } |\mathcal{P}| < \delta.$$

Proof. 1. Suppose for every $\varepsilon > 0$ there exists $\delta > 0$ such that (1) holds. Take any $\mathcal{P}_0 \in \mathfrak{P}$ such that $|\mathcal{P}_0| < \delta$. Then $\overline{S}(f, \mathcal{P}_0) - \underline{S}(f, \mathcal{P}_0) < \varepsilon$ so that $\underline{S}(f) = \overline{S}(f)$ by Lemma 7.20.

2. Conversely suppose $\underline{S}(f) = \overline{S}(f)$. Let $\varepsilon > 0$ be arbitrarily given. Then by Lemma 7.20, there exists $\mathcal{P}^* \in \mathfrak{P}$ such that

$$\overline{S}(f, \mathcal{P}^*) - \underline{S}(f, \mathcal{P}^*) < \tfrac{\varepsilon}{2}. \tag{2}$$

Let $\mathcal{P}^* = \{a_0, \dots, a_N\}$ where $a = a_0 < \cdots < a_N = b$. Since f is bounded on I, there exists $M > 0$ such that $|f(x)| \le M$ for $x \in I$. Let $\delta = \frac{\varepsilon}{8NM}$. To verify that our δ satisfies (1), let $\mathcal{P} \in \mathfrak{P}$ be such that $|\mathcal{P}| < \delta$. Let $\mathcal{P} = \{x_0, \dots, x_n\}$ where $a = x_0 < \cdots < x_n = b$. Consider $\mathcal{P}_0, \mathcal{P}_1, \dots, \mathcal{P}_N \in \mathfrak{P}$ defined by

$$\mathcal{P}_0 = \mathcal{P},\ \mathcal{P}_1 = \mathcal{P} \cup \{a_1\},\ \mathcal{P}_2 = \mathcal{P} \cup \{a_1, a_2\}, \dots, \mathcal{P}_N = \mathcal{P} \cup \{a_1, \dots, a_N\} = \mathcal{P} \cup \mathcal{P}^*.$$

Comparing $\mathcal{P}_0$ and $\mathcal{P}_1$, we have $a_1 \in [x_{k-1}, x_k]$ for some $k = 1, \dots, n$. Thus we have

$$\begin{aligned}
&\underline{S}(f, \mathcal{P}_1) - \underline{S}(f, \mathcal{P}_0)\\
=&\Big\{\inf_{[x_{k-1}, a_1]} f\Big\}\ell([x_{k-1}, a_1]) + \Big\{\inf_{[a_1, x_k]} f\Big\}\ell([a_1, x_k]) - \Big\{\inf_{[x_{k-1}, x_k]} f\Big\}\ell([x_{k-1}, x_k])\\
\le& M\big\{(a_1 - x_{k-1}) + (x_k - a_1)\big\} + M(x_k - x_{k-1})\\
=&2M(x_k - x_{k-1}) \le 2M|\mathcal{P}|.
\end{aligned}$$

By the same argument as above, we obtain $\underline{S}(f, \mathcal{P}_k) - \underline{S}(f, \mathcal{P}_{k-1}) \le 2M|\mathcal{P}|$ for each $k = 1, \dots, n$. Therefore we have

$$\begin{aligned}
\underline{S}(f, \mathcal{P}_N) - \underline{S}(f, \mathcal{P}_0) &= \sum_{k=1}^{N} \{\underline{S}(f, \mathcal{P}_k) - \underline{S}(f, \mathcal{P}_{k-1})\}\\
&\le 2NM|\mathcal{P}| < 2NM\frac{\varepsilon}{8NM} = \frac{\varepsilon}{4}.
\end{aligned}$$

Since $\mathcal{P}_N = \mathcal{P} \cup \mathcal{P}^* \supset \mathcal{P}^*$, we have $\underline{S}(f, \mathcal{P}_N) \ge \underline{S}(f, \mathcal{P}^*)$ by (3) in Definition 7.19. Thus

$$\underline{S}(f, \mathcal{P}^*) - \underline{S}(f, \mathcal{P}) \le \underline{S}(f, \mathcal{P}_N) - \underline{S}(f, \mathcal{P}_0) < \tfrac{\varepsilon}{4}.$$

By similar argument we have $\overline{S}(f, \mathcal{P}) - \overline{S}(f, \mathcal{P}^*) < \frac{\varepsilon}{4}$. By these two inequalities, we have

$$\begin{aligned}
\overline{S}(f, \mathcal{P}) - \underline{S}(f, \mathcal{P}) &< \{\overline{S}(f, \mathcal{P}^*) + \tfrac{\varepsilon}{4}\} - \{\underline{S}(f, \mathcal{P}^*) - \tfrac{\varepsilon}{4}\}\\
&= \overline{S}(f, \mathcal{P}^*) - \underline{S}(f, \mathcal{P}^*) + \tfrac{\varepsilon}{2}\\
&< \varepsilon \quad \text{by (2)}.
\end{aligned}$$

This verifies (1) for our δ. ∎

Theorem 7.22. *A bounded real-valued function f on $I = [a, b] \subset \mathbb{R}$ is Riemann integrable on I if and only if $\underline{S}(f) = \overline{S}(f)$. In this case we have $\int_a^b f(x)\,dx = \underline{S}(f) = \overline{S}(f)$.*

Proof. 1. Suppose $\underline{S}(f) = \overline{S}(f)$. To show that f is Riemann integrable on I and moreover $\int_a^b f(x)\,dx = \underline{S}(f)$, we show that for every $\varepsilon > 0$ there exists $\delta > 0$ such that

$$(1) \qquad \left| S(f, \mathcal{P}, \xi) - \underline{S}(f) \right| < \varepsilon \quad \text{for every } \mathcal{P} \in \mathfrak{P} \text{ with } |\mathcal{P}| < \delta.$$

Now since $\underline{S}(f) = \overline{S}(f)$, by Proposition 7.21 for an arbitrary $\varepsilon > 0$ there exists $\delta > 0$ such that

$$(2) \qquad \left| \overline{S}(f, \mathcal{P}) - \underline{S}(f, \mathcal{P}) \right| < \varepsilon \quad \text{for every } \mathcal{P} \in \mathfrak{P} \text{ with } |\mathcal{P}| < \delta.$$

Let $\mathcal{P} \in \mathfrak{P}$ with $|\mathcal{P}| < \delta$ be given by $\mathcal{P} = \{x_0, \ldots, x_n\}$ where $a = x_0 < \cdots < x_n = b$. Let $\xi_k \in [x_{k-1}, x_k]$ for $k = 1, \ldots, n$. Since $\inf_{[x_{k-1}, x_k]} f \leq f(\xi_k) \leq \sup_{[x_{k-1}, x_k]} f$, we have

$$(3) \qquad \underline{S}(f, \mathcal{P}) \leq S(f, \mathcal{P}, \xi) \leq \overline{S}(f, \mathcal{P}).$$

By (2) we have

$$(4) \qquad \overline{S}(f, \mathcal{P}) < \underline{S}(f, \mathcal{P}) + \varepsilon \leq \underline{S}(f) + \varepsilon,$$

and similarly by (2) we have

$$(5) \qquad \underline{S}(f, \mathcal{P}) > \overline{S}(f, \mathcal{P}) - \varepsilon \geq \overline{S}(f) - \varepsilon.$$

By (3), (4), and (5) and by the fact that $\underline{S}(f) = \overline{S}(f)$, we have

$$\underline{S}(f) - \varepsilon < S(f, \mathcal{P}, \xi) < \underline{S}(f) + \varepsilon.$$

This proves (1).

2. Conversely suppose f is Riemann integrable on I. Then there exists $J \in \mathbb{R}$ such that for every $\varepsilon > 0$ there exists $\delta > 0$ such that

$$(6) \qquad \left| S(f, \mathcal{P}, \xi) - J \right| < \varepsilon \quad \text{for every } \mathcal{P} \in \mathfrak{P} \text{ with } |\mathcal{P}| < \delta$$

and in this case $\int_a^b f(x)\,dx = J$. Let $\varepsilon > 0$ be arbitrarily given. Let $\mathcal{P} \in \mathfrak{P}$ with $|\mathcal{P}| < \delta$ be given by $\mathcal{P} = \{x_0, \ldots, x_n\}$ where $a = x_0 < \cdots < x_n = b$. Now there exists $\xi_k \in [x_{k-1}, x_k]$ such that $f(\xi_k) < \inf_{[x_{k-1}, x_k]} f + \varepsilon$. With this choice of ξ_k for $k = 1, \ldots, n$, we have

$$(7) \qquad S(f, \mathcal{P}, \xi) \leq \underline{S}(f, \mathcal{P}) + \varepsilon(b - a).$$

Then by (7) and (6), we have

$$\underline{S}(f) \geq \underline{S}(f, \mathcal{P}) \geq S(f, \mathcal{P}, \xi) - \varepsilon(b - a) > J - \varepsilon - \varepsilon(b - a).$$

By the arbitrariness of $\varepsilon > 0$, we have $\underline{S}(f) \geq J$. By similar argument we have $\overline{S}(f) \leq J$ also. Thus $\overline{S}(f) \leq J \leq \underline{S}(f)$. Since $\underline{S}(f) \leq \overline{S}(f)$, we have $\underline{S}(f) = \overline{S}(f) = J = \int_a^b f(x)\,dx$. ■

Theorem 7.23. *If f is a continuous real-valued function on $I = [a,b]$, then $\underline{\mathcal{S}}(f) = \overline{\mathcal{S}}(f)$ and consequently f is Riemann integrable on I.*

Proof. If f is continuous on the compact set I, then it is uniformly continuous on I. Thus for every $\varepsilon > 0$ there exists $\delta > 0$ such that

$$|f(x') - f(x'')| < \frac{\varepsilon}{b-a} \quad \text{for any } x', x'' \in [a,b] \text{ such that } |x' - x''| < \delta.$$

Let $\mathcal{P}_0 \in \mathfrak{P}$ be such that $|\mathcal{P}_0| < \delta$. Let $\mathcal{P}_0 = \{x_0, \dots, x_n\}$ where $a = x_0 < \cdots < x_n = b$. Then

$$\begin{aligned}\overline{\mathcal{S}}(f, \mathcal{P}_0) - \underline{\mathcal{S}}(f, \mathcal{P}_0) &= \sum_{k=1}^{n} \Big\{ \sup_{[x_{k-1},x_k]} f - \inf_{[x_{k-1},x_k]} f \Big\} (x_k - x_{k-1}) \\ &\le \frac{\varepsilon}{b-a}(b-a) = \varepsilon.\end{aligned}$$

From this it follows that

$$\overline{\mathcal{S}}(f) - \underline{\mathcal{S}}(f) = \inf_{\mathcal{P} \in \mathfrak{P}} \overline{\mathcal{S}}(f, \mathcal{P}) - \sup_{\mathcal{P} \in \mathfrak{P}} \underline{\mathcal{S}}(f, \mathcal{P}) \le \varepsilon.$$

Since this holds for every $\varepsilon > 0$, we have $\underline{\mathcal{S}}(f) = \overline{\mathcal{S}}(f)$. Then f is Riemann integrable on I by Theorem 7.22. ∎

To relate the Riemann integral to the Lebesgue integral, we show in Theorem 7.27 below that if a bounded real-valued function f on $I = [a,b]$ is Riemann integrable on I, then it is Lebesgue integrable on I with respect to μ_L in the sense of Definition 7.10 and moreover we have the equality $\int_a^b f(x)\,dx = \int_{[a,b]} f \, d\mu_L$. Regarding the Riemann integrability of a bounded real-valued function f on $I = [a,b]$, we show in Theorem 7.28 that f is Riemann integrable if and only if $\mu_L^*(E) = 0$ where E is the set of all points of discontinuity of f in I. In preparation for these results, let us review the limit inferior and the limit superior of a function and then introduce the lower envelope and the upper envelope of a function.

For $x_0 \in \mathbb{R}$ and $\delta > 0$, let $U(x_0, \delta) = (x_0 - \delta, x_0 + \delta)$ and $U_0(x_0, \delta) = U(x_0, \delta) \setminus \{x_0\}$, that is, $U(x_0, \delta) = \{x \in \mathbb{R} : |x - x_0| < \delta\}$ and $U_0(x_0, \delta) = \{x \in \mathbb{R} : 0 < |x - x_0| < \delta\}$.

Let f be an extended real-valued function on a set $D \in \mathbb{R}$ and let $x_0 \in \overline{D}$. The limit inferior and limit superior of f as x approaches x_0 are defined by

$$\liminf_{x \to x_0} f(x) = \lim_{\delta \to 0} \inf_{U_0(x_0,\delta) \cap D} f \quad \text{and} \quad \limsup_{x \to x_0} f(x) = \lim_{\delta \to 0} \sup_{U_0(x_0,\delta) \cap D} f.$$

Note that $\inf_{U_0(x_0,\delta)\cap D} f \uparrow$ and $\sup_{U_0(x_0,\delta)\cap D} f \downarrow$ as $\delta \downarrow 0$ so that $\liminf_{x\to x_0} f(x)$ and $\limsup_{x \to x_0} f(x)$ always exist in $\overline{\mathbb{R}}$ and $\liminf_{x\to x_0} f(x) \le \limsup_{x\to x_0} f(x)$. We have

$$\lim_{x \to x_0} f(x) \text{ exists in } \overline{\mathbb{R}} \Leftrightarrow \liminf_{x\to x_0} f(x) = \limsup_{x \to x_0} f(x)$$

and

$$f \text{ is continuous at } x_0 \Leftrightarrow \begin{cases} f(x_0) \in \mathbb{R}, \\ \liminf_{x\to x_0} f(x) = \limsup_{x\to x_0} f(x) = f(x_0). \end{cases}$$

Definition 7.24. *Let f be an extended real-valued function on a set $D \subset \mathbb{R}$. Let $x_0 \in \overline{D}$. We define the lower envelope and upper envelope of f at x_0 by*

$$f_*(x_0) = \lim_{\delta\to 0} \inf_{U(x_0,\delta)\cap D} f \quad \textit{and} \quad f^*(x_0) = \lim_{\delta\to 0} \sup_{U(x_0,\delta)\cap D} f.$$

(Note that $U_0(x_0, \delta)$ in the definition of $\liminf_{x\to x_0} f(x)$ and $\limsup_{x\to x_0} f(x)$ is now replaced by $U(x_0, \delta)$. Note also that since $\inf_{U(x_0,\delta)\cap D} f \uparrow$ and $\sup_{U(x_0,\delta)\cap D} f \downarrow$ as $\delta \downarrow 0$, $f_*(x_0)$ and $f^*(x_0)$ always exist in $\overline{\mathbb{R}}$.)

Observation 7.25. For an extended real-valued function f on a set $D \subset \mathbb{R}$ and for $x_0 \in \overline{D}$, we have

(1) $$f_*(x_0) \le \liminf_{x\to x_0} f(x) \le \limsup_{x\to x_0} f(x) \le f^*(x_0)$$

(2) $$f_*(x_0) \le f(x_0) \le f^*(x_0)$$

(3) $$f \text{ is continuous at } x_0 \in D \Leftrightarrow \begin{cases} f(x_0) \in \mathbb{R}, \\ f_*(x_0) = f^*(x_0). \end{cases}$$

Proof. (1) and (2) are immediate from the definitions. Let us prove (3). Suppose f is continuous at $x_0 \in D$. Then for every $\varepsilon > 0$ there exists $\eta > 0$ such that $f(x_0) - \varepsilon < f(x) < f(x_0) + \varepsilon$ for $x \in U(x_0, \eta) \cap D$. Since $\inf_{U(x_0,\delta)\cap D} f \uparrow$ as $\delta \downarrow 0$, we have

$$f(x_0) - \varepsilon \le \inf_{U(x_0,\eta)\cap D} f \le \lim_{\delta\to 0} \inf_{U(x_0,\delta)\cap D} f = f_*(x_0).$$

Similarly we have $f^*(x_0) \le f(x_0) + \varepsilon$. Thus we have $f^*(x_0) - f_*(x_0) \le 2\varepsilon$. By the arbitrariness of $\varepsilon > 0$, we have $f_*(x_0) = f^*(x_0)$.

Conversely if $f(x_0) \in \mathbb{R}$ and $f_*(x_0) = f^*(x_0)$, then we have $f_*(x_0) = f^*(x_0) = f(x_0)$ by (2). Then by (1) we have $\liminf_{x\to x_0} f(x) = \limsup_{x\to x_0} f(x) = f(x_0)$ and this proves the continuity of f at x_0. ∎

We show next that whether f is $\mathfrak{M}_L$-measurable or not, its lower envelope f_* and upper envelope f^* are always $\mathfrak{B}_{\mathbb{R}}$-measurable. Moreover the Lebesgue integrals of f_* and f^* on I with respect to μ_L are equal to the lower and upper Darboux sums $\underline{S}(f)$ and $\overline{S}(f)$ respectively.

Lemma 7.26. *Let f be a bounded real-valued function on $I = [a, b]$.*
(a) *The lower envelope f_* and the upper envelope f^* of f are bounded real-valued*

$\mathfrak{B}_{\mathbb{R}}$*-measurable functions on* I.
(b) $\int_I f_* \, d\mu_L = \underline{S}(f)$ *and* $\int_I f^* \, d\mu_L = \overline{S}(f)$.

Proof. **1.** Since f is a bounded real-valued function on I, f_* and f^* are bounded real-valued functions on I. Let us prove the $\mathfrak{B}_{\mathbb{R}}$-measurability of f^* on I. Consider the upper Darboux integral $\overline{S}(f) = \inf_{\mathfrak{P}} \overline{S}(f, \mathcal{P})$ of f on I. For every $m \in \mathbb{N}$, there exists $\mathcal{P}_m \in \mathfrak{P}$ such that

$$(1) \qquad \overline{S}(f) \le \overline{S}(f, \mathcal{P}_m) < \overline{S}(f) + \tfrac{1}{m}.$$

Then for the sequence of partitions $(\mathcal{P}_m : m \in \mathbb{N})$, we have

$$(2) \qquad \lim_{m\to\infty} \overline{S}(f, \mathcal{P}_m) = \overline{S}(f).$$

Let us add partition points to $\mathcal{P}_m$ if necessary so that $|\mathcal{P}_m| \le \frac{1}{m}$. Then $\lim_{m\to\infty} |\mathcal{P}_m| = 0$. Since adding partition points to $\mathcal{P}_m$ does not increase $\overline{S}(f, \mathcal{P}_m)$, (1) remains valid and so does (2). Let $\mathcal{P}_m$ be given by $\mathcal{P}_m = \{x_{m,0}, x_{m,1}, \dots, x_{m,n(m)}\}$ and let $I_{m,k} = [x_{m,k-1}, x_{m,k}]$ for $k = 1, \dots, n(m)$. For each $m \in \mathbb{N}$, let us define a function ψ_m on I by

$$(3) \qquad \psi_m = \sum_{k=1}^{n(m)} \sup_{I_{m,k}} f \cdot \mathbf{1}_{I_{m,k}}.$$

Since intervals are $\mathfrak{B}_{\mathbb{R}}$-measurable sets, ψ_m is a simple function on $(\mathbb{R}, \mathfrak{B}_{\mathbb{R}})$. However the expression (3) is not its canonical representation since the intervals $I_{m,k}$ are not disjoint. Rather, (3) is the sum of $m(n)$ simple functions each in its canonical representation $\sup_{I_{m,k}} f \cdot \mathbf{1}_{I_{m,k}}$. Let E be the countable set consisting of all the partition points in $(\mathcal{P}_m : m \in \mathbb{N})$. Let us show

$$(4) \qquad \lim_{m\to\infty} \psi_m(x) = f^*(x) \quad \text{for } x \in I \setminus E.$$

Let $x_0 \in I \setminus E$. Then for each $m \in \mathbb{N}$, there exists a subinterval in the partition $\mathcal{P}_m$ which contains x_0 in its interior, that is, there exists $k(m) \in \mathbb{N}$, $1 \le k(m) \le n(m)$, such that $x_0 \in I^{\circ}_{m,k(m)}$. Let $\delta_m > 0$ be so small that $U(x_0, \delta_m) \subset I^{\circ}_{m,k(m)}$. Then

$$(5) \qquad f^*(x_0) = \lim_{\delta\to 0} \sup_{U(x_0,\delta)} f \le \sup_{U(x_0,\delta_m)} f \le \sup_{I^{\circ}_{m,k(m)}} f \le \sup_{I_{m,k(m)}} f = \psi_m(x_0),$$

where the last inequality is from the definition of ψ_m by (3) and the fact that x_0 is an interior point in a subinterval in the partition and is not a point common to two adjacent subintervals. Since $\sup_{U(x_0,\delta)\cap I} f \downarrow f^*(x_0)$ as $\delta \downarrow 0$, for an arbitrary $\varepsilon > 0$ there exists $\delta > 0$ such that

$$(6) \qquad f^*(x_0) \le \sup_{U(x_0,\delta)\cap I} f < f^*(x_0) + \varepsilon.$$

Since $\lim_{m\to\infty} |\mathcal{P}_m| = 0$ and since $x_0 \in I^{\circ}_{m,k(m)}$ for some $k(m) \in \mathbb{N}$ for every $m \in \mathbb{N}$, there exists $M > 0$ such that for $m \ge M$ we have

$$(7) \qquad I_{m,k(m)} \subset U(x_0, \delta) \cap I.$$

By (5), (7), and (6), we have for $m \geq M$

$$f^*(x_0) \leq \psi_m(x_0) = \sup_{I_{m,k(m)}} f \leq \sup_{U(x_0,\delta)\cap I} f < f^*(x_0) + \varepsilon.$$

This proves (4).

Being a countable set, $E \in \mathfrak{B}_{\mathbb{R}}$ and then $I \setminus E \in \mathfrak{B}_{\mathbb{R}}$. Since ψ_m is a $\mathfrak{B}_{\mathbb{R}}$-measurable function on I, it is $\mathfrak{B}_{\mathbb{R}}$-measurable on $I \setminus E$ by (a) of Lemma 4.7. Then (4) implies that f^* is a $\mathfrak{B}_{\mathbb{R}}$-measurable function on $I \setminus E$ by Theorem 4.22. Since E is a countable set, any extended real-valued function on E is $\mathfrak{B}_{\mathbb{R}}$-measurable on E. (If g is an extended real-valued function on a countable set E in $\mathbb{R}$, then for any $\alpha \in \mathbb{R}$ the set $\{x \in E : g(x) \leq \alpha\}$, being a subset of E, is a countable set and is therefore a $\mathfrak{B}_{\mathbb{R}}$-measurable set. This shows that g is a $\mathfrak{B}_{\mathbb{R}}$-measurable function on E.) In particular f^* is $\mathfrak{B}_{\mathbb{R}}$-measurable on E. The $\mathfrak{B}_{\mathbb{R}}$-measurability of f^* on $I \setminus E$ and on E implies the $\mathfrak{B}_{\mathbb{R}}$-measurability of f^* on I by (b) of Lemma 4.7. The $\mathfrak{B}_{\mathbb{R}}$-measurability of f_* on I is proved likewise.

2. Let us prove $\int_I f^* \, d\mu_L = \overline{S}(f)$. As we showed above, f^* is a bounded real-valued $\mathfrak{B}_{\mathbb{R}}$-measurable function on $I \in \mathfrak{B}_{\mathbb{R}}$. Thus $\int_I f^* \, d\mu_L$ exists by Definition 7.10. Consider the sequence of simple functions $(\psi_m : m \in \mathbb{N})$ on $(\mathbb{R}, \mathfrak{B}_{\mathbb{R}})$ defined by (3). Then

$$\int_I \psi_m \, d\mu_L = \sum_{k=1}^{n(m)} \int_I \sup_{I_{m,k}} f \cdot \mathbf{1}_{I_{m,k}} \, d\mu_L = \sum_{k=1}^{n(m)} \sup_{I_{m,k}} f \cdot \ell(I_{m,k}) = \overline{S}(f, \mathcal{P}_m),$$

which is the upper Darboux sum of f corresponding to the partition $\mathcal{P}_m$. Then by (2)

$$\lim_{m\to\infty} \int_I \psi_m \, d\mu_L = \lim_{m\to\infty} \overline{S}(f, \mathcal{P}_m) = \overline{S}(f). \tag{8}$$

On the other hand, since f is bounded on I we have $|f(x)| \leq M$ for $x \in I$ for some $M \geq 0$. Then by (3), we have $|\psi_m(x)| \leq 2M$ for $x \in I$ and $m \in \mathbb{N}$. According to (4), we have $\lim_{m\to\infty} \psi_m = f^*$ on $I \setminus E$, that is, a.e. on I. Thus by Theorem 7.16 (Bounded Convergence Theorem), we have

$$\lim_{m\to\infty} \int_I \psi_m \, d\mu_L = \int_I f^* \, d\mu_L. \tag{9}$$

With (8) and (9) we have $\int_I f^* \, d\mu_L = \overline{S}(f)$. The equality $\int_I f_* \, d\mu_L = \underline{S}(f)$ is proved similarly. ■

Theorem 7.27. *Let f be a bounded real-valued function on $I = [a, b]$. If f is Riemann integrable on I, then f is $\mathfrak{M}_L$-measurable and Lebesgue integrable on I with respect to μ_L and moreover $\int_a^b f(x)\,dx = \int_I f \, d\mu_L$.*

Proof. If f is Riemann integrable on I, then the lower and upper Darboux integrals of f on I are equal, that is, $\underline{S}(f) = \overline{S}(f)$, by Theorem 7.22. Then by Lemma 7.26, we have $\int_I f_* \, d\mu_L = \int_I f^* \, d\mu_L$. Since f_* and f^* are $\mathfrak{B}_{\mathbb{R}}$-measurable on I, they are $\mathfrak{M}_L$-measurable on I. Since $f_* \leq f^*$ on I, the equality of the Lebesgue integrals of f_* and f^*

implies that $f_* = f^*$, $(\mathfrak{M}_L, \mu_L)$-a.e. on I by (b) of Lemma 7.14. Then since $f_* \le f \le f^*$ on I by Observation 7.25, we have $f = f_* = f^*$, $(\mathfrak{M}_L, \mu_L)$-a.e. on I. The completeness of the measure space $(\mathbb{R}, \mathfrak{M}_L, \mu_L)$ then implies that f is $\mathfrak{M}_L$-measurable on I according to (b) of Observation 4.20. Now that f is $\mathfrak{M}_L$-measurable on I and $f = f^*$, $(\mathfrak{M}_L, \mu_L)$-a.e. on I, we have $\int_I f \, d\mu_L = \int_I f^* \, d\mu_L$ by Lemma 7.11. But $\int_I f^* \, d\mu_L = \overline{S}(f) = \int_a^b f(x) \, dx$ by Lemma 7.26 and Theorem 7.22. Thus $\int_a^b f(x) \, dx = \int_I f \, d\mu_L$. ∎

Theorem 7.28. *Let f be a bounded real-valued function on $I = [a, b]$ and let E be the set of all points of discontinuity of f in I. Then*

$$\begin{aligned} & f \text{ is Riemann integrable on } I \\ & \Leftrightarrow f_* = f^*, (\mathfrak{M}_L, \mu_L)\text{-a.e. on } I \\ & \Leftrightarrow \mu_L(E) = 0. \end{aligned}$$

Proof. **1.** If f is Riemann integrable on I, then as we saw in the Proof of Theorem 7.27, $f_* = f^*$, $(\mathfrak{M}_L, \mu_L)$-a.e. on I. Conversely if $f_* = f^*$, $(\mathfrak{M}_L, \mu_L)$-a.e. on I, then $\int_I f_* \, d\mu_L = \int_I f^* \, d\mu_L$ by Lemma 7.11. Then we have $\underline{S}(f) = \overline{S}(f)$ by Lemma 7.26. But this implies the Riemann integrability of f on I by Theorem 7.22.

2. By Observation 7.25, f is continuous at $x_0 \in I$ if and only if $f_*(x_0) = f^*(x_0)$. Thus $f_* = f^*$, $(\mathfrak{M}_L, \mu_L)$-a.e. on I if and only if $\mu_L(E) = 0$. ∎

Example. Let Q be the set of all rational numbers and P be the set of all irrational numbers in $\mathbb{R}$. Let f be a bounded real-valued function on $[0, 1]$ defined by $f(x) = 0$ if $x \in [0, 1] \cap Q$ and $f(x) = 1$ if $x \in [0, 1] \cap P$. The set E of all points of discontinuity of f in $[0, 1]$. Then $E = [0, 1]$ and $\mu_L(E) = 1$. Thus by Theorem 7.28, f is not Riemann integrable on $[0, 1]$. On the other hand since $[0, 1] \cap Q, [0, 1] \cap P \in \mathfrak{B}_{\mathbb{R}}$, f is a simple function on $(\mathbb{R}, \mathfrak{B}_{\mathbb{R}})$ and its Lebesgue integral with respect to μ_L is given by $\int_{[0,1]} f \, d\mu_L = 0 \cdot \mu_L([0, 1] \cap Q) + 1 \cdot \mu_L([0, 1] \cap P) = 1$.

Problems

Prob. 7.1. Let f be an extended real-valued $\mathfrak{A}$-measurable function on a set $D \in \mathfrak{A}$ in a measure space $(X, \mathfrak{A}, \mu)$. For $M_1, M_2 \in \mathbb{R}$, $M_1 < M_2$, let the truncation of f at M_1 and M_2 be defined by

$$g(x) = \begin{cases} M_1 & \text{if } f(x) < M_1, \\ f(x) & \text{if } f(x) \in [M_1, M_2], \\ M_2 & \text{if } f(x) > M_2. \end{cases}$$

Show that g is $\mathfrak{A}$-measurable on D.

Prob. 7.2. Let f be a bounded real-valued $\mathfrak{A}$-measurable function on a set $D \in \mathfrak{A}$ in a measure space $(X, \mathfrak{A}, \mu)$.

(a) Show that there exists an increasing sequence of simple functions on D, $(\varphi_n : n \in \mathbb{N})$, such that $\varphi_n \uparrow f$ uniformly on D.

(b) Show that there exists a decreasing sequence of simple functions on D, $(\psi_n : n \in \mathbb{N})$, such that $\psi_n \downarrow f$ uniformly on D.

Prob. 7.3. Let f be a real-valued $\mathfrak{A}$-measurable function on a set $D \in \mathfrak{A}$ in a measure space $(X, \mathfrak{A}, \mu)$. Suppose f is not bounded on D.
(a) Show that it is not possible that for an arbitrary $\varepsilon > 0$ there exists a simple function φ on D such that $|\varphi(x) - f(x)| < \varepsilon$ for all $x \in D$.
(b) Show that there may not be any sequence of simple function which converges to f uniformly on D.

Prob. 7.4. Given a measure space $(X, \mathfrak{A}, \mu)$. Let us call a function φ an elementary function if it satisfies the following conditions:
1° $\mathfrak{D}(\varphi) \in \mathfrak{A}$,
2° φ is $\mathfrak{A}$-measurable on $\mathfrak{D}(\varphi)$,
3° φ assumes only countably many real values, that is, $\mathfrak{R}(\varphi)$ is a countable subset of $\mathbb{R}$.
Let f be a real-valued $\mathfrak{A}$-measurable function on a set $D \in \mathfrak{A}$.
(a) Show that there exists an increasing sequence of elementary functions on D, $(\varphi_n : n \in \mathbb{N})$, such that $\varphi_n \uparrow f$ uniformly on D.
(b) Show that there exists a decreasing sequence of elementary functions on D, $(\psi_n : n \in \mathbb{N})$, such that $\psi_n \downarrow f$ uniformly on D.

Prob. 7.5. Given a measure space $(X, \mathfrak{A}, \mu)$. Let f be a nonnegative real-valued function on a set $D \in \mathfrak{A}$. Let Φ be the collection of all nonnegative elementary functions φ such that $0 \le \varphi \le f$ on D and let Ψ be the collection of all nonnegative elementary functions φ such that $0 \le f \le \psi$ on D.
(a) Show that $\sup_{\varphi\in\Phi} \int_D \varphi \, d\mu \le \inf_{\psi\in\Phi} \int_D \psi \, d\mu$.
(b) Show that in f is a measurable on D then $\int_D f \, d\mu = \sup_{\varphi\in\Phi} \int_D \varphi \, d\mu = \inf_{\psi\in\Phi} \int_D \psi \, d\mu$.
(c) Show that if $\sup_{\varphi\in\Phi} \int_D \varphi \, d\mu = \inf_{\psi\in\Phi} \int_D \psi \, d\mu$ then f is $\mathfrak{A}$-measurable on D, provided that $(X, \mathfrak{A}, \mu)$ is a complete measure space.

Prob. 7.6. Given a measure space $(X, \mathfrak{A}, \mu)$. Let f be a bounded real-valued $\mathfrak{A}$-measurable function on a set $D \in \mathfrak{A}$ with $\mu(D) < \infty$. Suppose $|f(x)| \le M$ for $x \in D$ for some constant $M > 0$.
(a) Show that if $\int_D f \, d\mu = M\mu(D)$, then $f = M$ a.e. on D.
(b) Show that if $f < M$ a.e. on D and if $\mu(D) > 0$, then $\int_D f \, d\mu < M\mu(D)$.

Prologue to Prob. 7.7 and Prob. 7.8. Recall that for a bounded real-valued $\mathfrak{A}$-measurable function on a set $D \in \mathfrak{A}$ with $\mu(D) < \infty$ in a measure space $(X, \mathfrak{A}, \mu)$, the Lebesgue integral of f on D is defined by $\int_D f \, d\mu = \sup_{\varphi\le f} \int_D \varphi \, d\mu = \inf_{\psi\ge f} \int_D \psi \, d\mu$, where φ and ψ are simple functions on D. The integral of f thus defined can be computed as the limit of a sequence of integrals of simple functions.

Prob. 7.7. Let f be a bounded real-valued $\mathfrak{A}$-measurable function on a set $D \in \mathfrak{A}$ with $\mu(D) < \infty$ satisfying $f(x) \in (-M, M]$ for $x \in D$ for some $M > 0$.
For each $n \in \mathbb{N}$, let

$$D_{n,k} = \{x \in D : f(x) \in ((k-1)\tfrac{M}{n}, k\tfrac{M}{n}]\} \text{ for } k = -n+1, \cdots, n.$$

Define simple functions φ_n and ψ_n on D by setting

$$\varphi_n = \sum_{k=-n+1}^{n} (k-1)\tfrac{M}{n} \cdot \mathbf{1}_{D_{n,k}}$$

and

$$\psi_n = \sum_{k=-n+1}^{n} k\tfrac{M}{n} \cdot \mathbf{1}_{D_{n,k}}.$$

Show that $\lim_{n\to\infty} \int_D \varphi_n \, d\mu = \int_D f \, d\mu$ and $\lim_{n\to\infty} \int_D \psi_n \, d\mu = \int_D f \, d\mu$.

Prob. 7.8. Let f be a bounded real-valued $\mathfrak{A}$-measurable function on a set $D \in \mathfrak{A}$ with $\mu(D) < \infty$ satisfying $f(x) \in (-M, M]$ for $x \in D$ for some $M > 0$.
Let $\mathcal{P} = (y_0, \ldots, y_p)$ be a partition of $[-M, M]$, that is, $-M = y_0 < \cdots < y_p = M$.
Let $I_k = [y_{k-1}, y_k]$ for $k = 1, \ldots, p$ and let $|\mathcal{P}| = \max_{k=1,\ldots,p} \ell(I_k)$.
Let $E_k = \{x \in D : f(x) \in (y_{k-1}, y_k]\}$ for $k = 1, \ldots, p$.
Let $\varphi(\mathcal{P})$ and $\psi(\mathcal{P})$ be simple functions on D defined by $\varphi(\mathcal{P}) = \sum_{k=1}^{p} y_{k-1} \cdot \mathbf{1}_{E_k}$ and $\psi(\mathcal{P}) = \sum_{k=1}^{p} y_k \cdot \mathbf{1}_{E_k}$.
Let $(\mathcal{P}_n : n \in \mathbb{N})$ be a sequence of partitions such that $\lim_{n\to\infty} |\mathcal{P}_n| = 0$.
Show that $\lim_{n\to\infty} \int_D \varphi(\mathcal{P}_n) \, d\mu = \int_D f \, d\mu$ and $\lim_{n\to\infty} \int_D \psi(\mathcal{P}_n) \, d\mu = \int_D f \, d\mu$.

Prob. 7.9. Consider a sequence of functions $(f_n : n \in \mathbb{N})$ defined on $[0, 1]$ by setting

$$f_n(x) = \frac{nx}{1+n^2x^2} \text{ for } x \in [0, 1].$$

(a) Show that $(f_n : n \in \mathbb{N})$ is a uniformly bounded sequence on $[0, 1]$ and evaluate

$$\lim_{n\to\infty} \int_{[0,1]} \frac{nx}{1+n^2x^2} \, \mu_L(dx).$$

(b) Show that $(f_n : n \in \mathbb{N})$ does not converge uniformly on $[0, 1]$. (This makes justification of $\lim_{n\to\infty} \int_{[0,1]} f_n(x) \, dx = \int_{[0,1]} \lim_{n\to\infty} f_n(x) \, dx$ by uniform convergence inapplicable.)

Prob. 7.10. Given a measure space $(X, \mathfrak{A}, \mu)$. Let f_n, $n \in \mathbb{N}$, and f be extended real-valued $\mathfrak{A}$-measurable functions on a set $D \in \mathfrak{A}$ with $\mu(D) < \infty$ and assume that f is real-valued a.e. on D. Show that $f_n \overset{\mu}{\to} f$ on D if and only if

$$\lim_{n\to\infty} \int_D \frac{|f_n - f|}{1 + |f_n - f|} \, \mu(dx) = 0.$$

Prob. 7.11. Let $(X, \mathfrak{A}, \mu)$ be a finite measure space. Let Φ be the set of all extended real-valued $\mathfrak{A}$-measurable functions on X where we identify functions that are equal μ-a.e. on X. Let

$$\rho(f, g) = \int_X \frac{|f-g|}{1+|f-g|} \, d\mu \quad \text{for } f, g \in \Phi.$$

(a) Show that ρ is a metric on Φ.
(b) Show that Φ is complete with respect to the metric ρ.

Prob. 7.12. Given a measure space $(X, \mathfrak{A}, \mu)$. Let $(f_n : n \in \mathbb{N})$ be a uniformly bounded sequence of real-valued $\mathfrak{A}$-measurable functions and let f be a bounded real-valued $\mathfrak{A}$-measurable function on $D \in \mathfrak{A}$ with $\mu(D) < \infty$.
Show that if $f_n \overset{\mu}{\to} f$ on D then $\lim_{n\to\infty} \int_D |f_n - f| \, d\mu = 0$ and $\lim_{n\to\infty} \int_D f_n \, d\mu = \int_D f \, d\mu$.
This is the Bounded Convergence Theorem under Convergence in Measure. Construct a

direct proof without using the subsequence argument as in Corollary 7.17.

Prob. 7.13. Let Q be the collection of all rational numbers in $(0, \infty)$. Let each $x \in Q$ be expressed as $\frac{q}{p}$ where p and q are positive integers without common factors. Define a function f on $[0, 1]$ by setting

$$f(x) = \begin{cases} \frac{1}{p} & \text{if } x \in [0,1] \cap Q \text{ and } x = \frac{q}{p}, \\ 0 & \text{if } x \in [0,1] \setminus Q. \end{cases}$$

Show that f is Riemann integrable on $[0, 1]$ and evaluate $\int_0^1 f(x)\,dx$.

Prob. 7.14. Let f be an extended real-valued function on $\mathbb{R}$. Define two extended real-valued functions φ and ψ on $\mathbb{R}$ by letting $\varphi(x) = \liminf_{y \to x} f(y)$ and $\psi(x) = \limsup_{y \to x} f(y)$ for $x \in \mathbb{R}$. Show that φ and ψ are $\mathfrak{B}_{\mathbb{R}}$-measurable on $\mathbb{R}$. (Note that f itself is not assumed to be $\mathfrak{B}_{\mathbb{R}}$-measurable on $\mathbb{R}$.)

Prob. 7.15. Let f be a real-valued function on $\mathbb{R}$. Assume that f is locally bounded on $\mathbb{R}$ in the sense that for every $x \in \mathbb{R}$ there exists $\eta > 0$ such that f is bounded on $U(x, \eta) = (x - \eta, x + \eta)$. Then at every $x \in \mathbb{R}$, $\sup_{U(x,\delta)} f - \inf_{U(x,\delta)} f$ is a nonnegative real number for sufficiently small $\delta > 0$ and decreases as $\delta \downarrow 0$. The nonnegative real-valued function $\omega(x) = \lim_{\delta \to 0} \left\{ \sup_{U(x,\delta)} f - \inf_{U(x,\delta)} f \right\}$ for $x \in \mathbb{R}$ is call the oscillation of f at x.

(a) Show that f is continuous at $x \in \mathbb{R}$ if and only if $\omega(x) = 0$.

(b) Show that for every $\varepsilon > 0$, $\omega^{-1}([\varepsilon, \infty)) = \{x \in \mathbb{R} : \omega(x) \geq \varepsilon\}$ is a closed set in $\mathbb{R}$.

(c) Show that ω need not be continuous on $\mathbb{R}$. (Consider the function f defined by $f(x) = 0$ for $x \in (-\infty, 0)$ and $f(x) = 1$ for $x \in [0, \infty)$.)

(d) Show that the set of all points of continuity of f is a G_δ-set and the set of all points of discontinuity of f is an F_σ-set.

§8 Integration of Nonnegative Functions

[I] Lebesgue Integral of Nonnegative Functions

Definition 8.1. *Given a measure space $(X, \mathfrak{A}, \mu)$. Let f be a nonnegative extended real-valued $\mathfrak{A}$-measurable function on a set $D \in \mathfrak{A}$. We define the Lebesgue integral of f on D with respect to μ by*

$$\int_D f \, d\mu = \sup_{0 \le \varphi \le f} \int_D \varphi \, d\mu \in [0, \infty],$$

where the supremum is on the collection of all nonnegative simple functions φ on D such that $0 \le \varphi \le f$. Thus $\int_D f \, d\mu$ exists for every nonnegative extended real-valued measurable function and we say that every nonnegative extended real-valued measurable function is Lebesgue semi-integrable on D with respect to μ, or simply μ semi-integrable on D. We say that the function is Lebesgue integrable with respect to μ, or simply μ-integrable, only when $\int_D f \, d\mu \in \mathbb{R}$, that is, $\int_D f \, d\mu < \infty$.

The definition given above for a nonnegative extended real-valued $\mathfrak{A}$-measurable function f is consistent with the definition $\int_D f \, d\mu = \sup_{\varphi \le f} \int_D \varphi \, d\mu$ in Definition 7.10 for a bounded real-valued $\mathfrak{A}$-measurable function f defined on $D \in \mathfrak{A}$ with $\mu(D) < \infty$. Since a simple function assumes only finitely many real values, a simple function is necessarily a bounded function on its domain of definition. A nonnegative extended real-valued function need not be bounded and therefore there may not be any simple function ψ such that $\psi \ge f$. Thus the equality $\int_D f \, d\mu = \sup_{f \le \psi} \int_D \varphi \, d\mu$ for a bounded real-valued $\mathfrak{A}$-measurable function f does not exist for a nonnegative extended real-valued $\mathfrak{A}$-measurable function f. This fact has the consequence that while the integral of a nonnegative extended real-valued function can be approximated by integrals of simple functions from below (see Lemma 8.6), it cannot be so approximated from above.

Lemma 8.2. *Given a measure space $(X, \mathfrak{A}, \mu)$. Let f, f_1, and f_2 be nonnegative extended real-valued $\mathfrak{A}$-measurable functions on a set $D \in \mathfrak{A}$.*

(a) *If $\int_D f \, d\mu < \infty$, then $f < \infty$ a.e. on D.*
(b) *If $\int_D f \, d\mu = 0$, then $f = 0$ a.e. on D.*
(c) *If D_0 is a $\mathfrak{A}$-measurable subset of D, then $\int_{D_0} f \, d\mu \le \int_D f \, d\mu$.*
(d) *If $f > 0$ a.e. on D and $\int_D f \, d\mu = 0$, then $\mu(D) = 0$.*
(e) *If $f_1 \le f_2$ on D, then $\int_D f_1 \, d\mu \le \int_D f_2 \, d\mu$.*
(f) *If $f_1 = f_2$ a.e. on D, then $\int_D f_1 \, d\mu = \int_D f_2 \, d\mu$.*

Proof. 1. To prove (a), suppose $\int_D f \, d\mu < \infty$ but the statement that $f < \infty$ for a.e. on D is false. Then we have $\mu(E) > 0$ where $E = \{D : f = \infty\} \in \mathfrak{A}$. For each $n \in \mathbb{N}$, define a nonnegative simple function φ_n on D by $\varphi_n = n \cdot \mathbf{1}_E + 0 \cdot \mathbf{1}_{D \setminus E}$. Then $\varphi_n \le f$ on E and $\varphi_n \le f$ on $D \setminus E$ also so that $\varphi_n \le f$ on D. Since $\{\varphi_n : n \in \mathbb{N}\}$ is a subcollection of the collection of all nonnegative simple functions φ on D such that $0 \le \varphi \le f$, we have

$$\int_D f \, d\mu = \sup_{0 \le \varphi \le f} \int_D \varphi \, d\mu \ge \sup_{n \in \mathbb{N}} \int_D \varphi_n \, d\mu = \sup_{n \in \mathbb{N}} n\mu(E) = \infty$$

since $\mu(E) > 0$. This contradicts the assumption $\int_D f \, d\mu < \infty$.

2. We say that $f = 0$ a.e. on D if there exists a subset E of D such that E is a null set in $(X, \mathfrak{A}, \mu)$ and $f = 0$ on $D \setminus E$. Consider the case $\mu(D) = 0$. Let f be an arbitrary nonnegative extended real-valued $\mathfrak{A}$-measurable function on D and let $E = \{D : f \neq 0\}$. Then $E \in \mathfrak{A}$ and $\mu(E) = 0$ since $E \subset D$. Now $f = 0$ on $D \setminus E$ so that $f = 0$ a.e. on D. Thus to prove (b), we need only to consider the case $\mu(D) > 0$. Suppose $\int_D f \, d\mu = 0$ but the statement that $f = 0$ a.e. on D is false. Then for the set $E = \{D : f > 0\}$, we have $\mu(E) > 0$. Now since $f > 0$ on E, we have $E = \bigcup_{n \in \mathbb{N}} E_n$ where $E_n = \left\{D : f \geq \frac{1}{n}\right\}$. The sequence $(E_n : n \in \mathbb{N})$ is an increasing sequence in $\mathfrak{A}$ so that $\lim_{n \to \infty} E_n = \bigcup_{n \in \mathbb{N}} E_n = E$. By Theorem 1.26, we have $\lim_{n \to \infty} \mu(E_n) = \mu(E) > 0$. This implies that there exists $n_0 \in \mathbb{N}$ such that $\mu(E_{n_0}) > 0$. Let us define a nonnegative simple function φ_0 on D by $\varphi_0 = \frac{1}{n_0} \mathbf{1}_{E_{n_0}} + 0 \cdot \mathbf{1}_{D \setminus E_{n_0}}$. Then $0 \leq \varphi_0 \leq f$ on D and

$$\int_D f \, d\mu = \sup_{0 \leq \varphi \leq f} \int_D \varphi \, d\mu \geq \int_D \varphi_0 \, d\mu = \frac{1}{n_0} \mu(E_{n_0}) > 0,$$

which contradicts the assumption $\int_D f \, d\mu = 0$.

3. To prove (c), let D_0 be a $\mathfrak{A}$-measurable subset of D. Corresponding to each nonnegative simple function φ_0 on D_0 such that $0 \leq \varphi_0 \leq f$ on D_0, let us define a nonnegative simple function φ_1 on D by

$$\varphi_1(x) = \begin{cases} \varphi_0(x) & \text{for } x \in D_0, \\ 0 & \text{for } x \in D \setminus D_0. \end{cases}$$

Then $0 \leq \varphi_1 \leq f$ on D and $\int_{D_0} \varphi_0 \, d\mu = \int_D \varphi_1 \, d\mu$. Thus we have

$$\sup_{0 \leq \varphi_0 \leq f} \int_{D_0} \varphi_0 \, d\mu = \sup_{0 \leq \varphi_1 \leq f} \int_D \varphi_1 \, d\mu \leq \sup_{0 \leq \varphi \leq f} \int_D \varphi \, d\mu,$$

that is, $\int_{D_0} f \, d\mu \leq \int_D f \, d\mu$.

4. To prove (d), let $D_0 = \{x \in D : f(x) = 0\}$ and $D_k = \left\{x \in D : f(x) \geq \frac{1}{k}\right\}$ for $k \in \mathbb{N}$. Now $\mu(D_0) > 0$ would contradict our assumption that $f > 0$ a.e. on D. Thus $\mu(D_0) = 0$. Now $D_0 \cup D_k \uparrow D$ so that $\lim_{k \to \infty} \mu(D_0 \cup D_k) = \mu(D)$. If $\mu(D) > 0$, then there exists $k_0 \in \mathbb{N}$ such that $\mu(D_0 \cup D_{k_0}) > 0$. Since $\mu(D_0) = 0$, we have $\mu(D_{k_0}) > 0$. Then by (c) we have $\int_D f \, d\mu \geq \int_{D_{k_0}} f \, d\mu \geq \frac{1}{k_0} \mu(D_{k_0}) > 0$. This is a contradiction. Thus $\mu(D) = 0$.

5. Let us prove (e). For $i = 1$ and 2, let Φ_i be the collection of all nonnegative simple functions φ on D such that $0 \leq \varphi \leq f_i$. Since $f_1 \leq f_2$ on D, if a simple function φ is such that $0 \leq \varphi \leq f_1$ on D, then $0 \leq \varphi \leq f_2$ on D also. Thus $\Phi_1 \subset \Phi_2$. Then

$$\left\{\int_D \varphi \, d\mu : \varphi \in \Phi_1\right\} \subset \left\{\int_D \varphi \, d\mu : \varphi \in \Phi_2\right\}$$

so that

$$\int_D f_1 \, d\mu = \sup\left\{\int_D \varphi \, d\mu : \varphi \in \Phi_1\right\} \leq \sup\left\{\int_D \varphi \, d\mu : \varphi \in \Phi_2\right\} = \int_D f_2 \, d\mu.$$

6. If $f_1 = f_2$ a.e. on D, then by the same argument as in the Proof of Lemma 7.11 for bounded real-valued measurable functions on a set with finite measure we have $\int_D f_1\,d\mu = \int_D f_2\,d\mu$. ■

Lemma 8.3. *Given a measure space $(X, \mathfrak{A}, \mu)$. Let f be a nonnegative extended real-valued $\mathfrak{A}$-measurable function on a set $D \in \mathfrak{A}$. Then for every $c > 0$, we have*

$$\int_D cf\,d\mu = c\int_D f\,d\mu.$$

If f is nonnegative real-valued, then for $c = 0$ we have $cf = 0$ on D and $\int_D cf\,d\mu = 0$ but $c\int_D f\,d\mu$ does not exist when $\int_D f\,d\mu = \infty$.

Proof. For $c > 0$, the equality is proved in the same way as in Lemma 7.12 for bounded real-valued measurable functions on a set of finite measure. ■

In Lemma 7.12, we proved the equality $\int_D\{f_1 + f_2\}\,d\mu = \int_D f_1\,d\mu + \int_D f_2\,d\mu$ for two bounded real-valued measurable functions f_1 and f_2 on a set D with $\mu(D) < \infty$. By using the fact that $\int_D f\,d\mu = \sup_{\varphi\le f}\int_D \varphi\,d\mu = \inf_{f\le\psi}\int_D \psi\,d\mu$ for a bounded real-valued measurable function f. The argument used there is no longer available when f_1 and f_2 are nonnegative extended real-valued measurable functions since for an unbounded nonnegative function f no simple function ψ such that $f \le \psi$ exists. Instead, we derive the equality $\int_D\{f_1 + f_2\}\,d\mu = \int_D f_1\,d\mu + \int_D f_2\,d\mu$ in Proposition 8.7 below by applying the Monotone Convergence Theorem.

[II] Monotone Convergence Theorem

Proposition 8.4. *Given a measure space $(X, \mathfrak{A}, \mu)$. Let φ be a nonnegative simple function on X. If we define a set function ν on $\mathfrak{A}$ by setting $\nu(A) = \int_A \varphi\,d\mu$ for $A \in \mathfrak{A}$, then ν is a measure on $\mathfrak{A}$.*

Proof. Let $\varphi = \sum_{i=1}^m a_i \cdot \mathbf{1}_{D_i}$ be the canonical representation of φ. Since φ is defined on X, $\{D_i : i = 1, \dots, m\}$ is a disjoint collection in $\mathfrak{A}$ with $\bigcup_{i=1}^m D_i = X$. The restriction of φ to A is given by $\varphi = \sum_{i=1}^m a_i \cdot \mathbf{1}_{D_i\cap A}$. Thus $\nu(A) = \int_A \varphi\,d\mu = \sum_{i=1}^m a_i\,\mu(D_i \cap A) \in [0, \infty]$. Since $\mu(\emptyset) = 0$, we have $\nu(\emptyset) = \int_\emptyset \varphi\,d\mu = 0$ by 1° of Observation 7.5. Thus it remains to show that ν is countably additive on $\mathfrak{A}$. Let $(A_n : n \in \mathbb{N})$ be a disjoint sequence in $\mathfrak{A}$. Then

$$\begin{aligned}\nu\Big(\bigcup_{n\in\mathbb{N}} A_n\Big) &= \int_{\bigcup_{n\in\mathbb{N}} A_n} \varphi\,d\mu = \sum_{i=1}^m a_i\,\mu\Big(D_i \cap \bigcup_{n\in\mathbb{N}} A_n\Big) = \sum_{i=1}^m a_i\,\mu\Big(\bigcup_{n\in\mathbb{N}} (D_i \cap A_n)\Big)\\ &= \sum_{i=1}^m a_i\Big[\sum_{n\in\mathbb{N}} \mu(D_i \cap A_n)\Big] = \sum_{n\in\mathbb{N}}\Big[\sum_{i=1}^m a_i\,\mu(D_i \cap A_n)\Big] = \sum_{n\in\mathbb{N}}\int_{A_n} \varphi\,d\mu = \sum_{n\in\mathbb{N}} \nu(A_n),\end{aligned}$$

where the fifth equality is by the fact that for series of nonnegative extended real numbers $\sum_{n\in\mathbb{N}} \alpha_{1,n}, \dots, \sum_{n\in\mathbb{N}} \alpha_{p,n}$ we have $\sum_{n\in\mathbb{N}}\{\alpha_{1,n} + \cdots + \alpha_{p,n}\} = \sum_{n\in\mathbb{N}} \alpha_{1,n} + \cdots +$

$\sum_{n\in\mathbb{N}} \alpha_{p,n}$. This proves the countable additivity of ν on $\mathfrak{A}$ and completes the proof that ν is a measure on $\mathfrak{A}$. ∎

Theorem 8.5. (Monotone Convergence Theorem) *Given a measure space $(X, \mathfrak{A}, \mu)$. Let $(f_n : n \in \mathbb{N})$ be an increasing sequence of nonnegative extended real-valued $\mathfrak{A}$-measurable functions on a set $D \in \mathfrak{A}$ and let $f = \lim_{n\to\infty} f_n$ on D. Then $\lim_{n\to\infty} \int_D f_n \, d\mu = \int_D f \, d\mu$.*

Proof. 1. Since $(f_n : n \in \mathbb{N})$ is an increasing sequence of nonnegative functions on D, $\lim_{n\to\infty} f_n(x)$ exists in $[0, \infty]$ for every $x \in D$ and thus $f = \lim_{n\to\infty} f_n$ is a nonnegative extended real-valued function on D. f is $\mathfrak{A}$-measurable on D by Theorem 4.22. Since $f_n \leq f$ on D, we have $\int_D f_n \, d\mu \leq \int_D f \, d\mu$ for every $n \in \mathbb{N}$ by (e) of Lemma 8.2. Also $f_n \leq f_{n+1}$ implies $\int_D f_n \, d\mu \leq \int_D f_{n+1} \, d\mu$ and thus $\left(\int_D f_n \, d\mu : n \in \mathbb{N}\right)$ is an increasing sequence of nonnegative extended real numbers bounded above by $\int_D f \, d\mu$ and consequently $\lim_{n\to\infty} \int_D f_n \, d\mu \leq \int_D f \, d\mu$.

2. Let us prove the reverse inequality $\lim_{n\to\infty} \int_D f_n \, d\mu \geq \int_D f \, d\mu$. Let φ be an arbitrary nonnegative simple function on D such that $0 \leq \varphi \leq f$. With $\alpha \in (0, 1)$ arbitrarily fixed, we have $0 \leq \alpha\varphi \leq \varphi \leq f$ on D. Let us define a sequence $(E_n : n \in \mathbb{N})$ of subsets of D by

$$(1) \qquad E_n = \{x \in D : f_n(x) \geq \alpha\varphi(x)\} \quad \text{for } n \in \mathbb{N}.$$

Since f_n and $\alpha\varphi$ are $\mathfrak{A}$-measurable on D, we have $E_n \in \mathfrak{A}$ by Theorem 4.16. Now $f_n \leq f_{n+1}$ implies $E_n \subset E_{n+1}$ for every $n \in \mathbb{N}$ so that $(E_n : n \in \mathbb{N})$ is an increasing sequence in $\mathfrak{A}$. Since $E_n \subset D$ for every $n \in \mathbb{N}$, we have $\bigcup_{n\in\mathbb{N}} E_n \subset D$. Let us show that actually $\bigcup_{n\in\mathbb{N}} E_n = D$. To show this, we show that if $x \in D$ then $x \in E_n$ for some $n \in \mathbb{N}$. Let $x \in D$. If $f(x) = 0$, then since $0 \leq \varphi \leq f$ we have $\varphi(x) = 0$ also. Since $0 \leq f_n \leq f$, we have $f_n(x) = 0$ for every $n \in \mathbb{N}$. Thus $f_n(x) = 0 = \alpha\varphi(x)$ for every $n \in \mathbb{N}$ so that $x \in E_n$ for every $n \in \mathbb{N}$. On the other hand, if $f(x) > 0$ then since $0 \leq \varphi \leq f$ and $\alpha \in (0, 1)$ we have $f(x) > \alpha\varphi(x)$. Since $f_n(x) \uparrow f(x)$, there exists $N(x) \in \mathbb{N}$ such that $f_{N(x)}(x) \geq \alpha\varphi(x)$. Thus $x \in E_{N(x)}$. This completes the proof that if $x \in D$ then $x \in E_n$ for some $n \in \mathbb{N}$. Therefore $D \subset \bigcup_{n\in\mathbb{N}} E_n$ so that $D = \bigcup_{n\in\mathbb{N}} E_n$.

Let us define a set function ν on $\mathfrak{A}$ by setting $\nu(A) = \int_A \varphi \, d\mu$ for $A \in \mathfrak{A}$. Then by Proposition 8.4, ν is a measure on $(X, \mathfrak{A})$. Now

$$(2) \qquad \int_D f_n \, d\mu \geq \int_{E_n} f_n \, d\mu \geq \int_{E_n} \alpha\varphi \, d\mu = \alpha \int_{E_n} \varphi \, d\mu = \alpha\nu(E_n),$$

where the second inequality is from (1). Since $(E_n : n \in \mathbb{N})$ is an increasing sequence, we have $\lim_{n\to\infty} E_n = \bigcup_{n\in\mathbb{N}} E_n = D$. Then by Theorem 1.26 (Monotone Convergence), we have $\lim_{n\to\infty} \nu(E_n) = \nu\left(\lim_{n\to\infty} E_n\right) = \nu(D)$. Letting $n \to \infty$ in (2), we have

$$\lim_{n\to\infty} \int_D f_n \, d\mu \geq \alpha \lim_{n\to\infty} \nu(E_n) = \alpha\nu(D).$$

Since this holds for every $\alpha \in (0, 1)$, letting $\alpha \uparrow 1$ we have

$$\lim_{n\to\infty} \int_D f_n \, d\mu \geq \nu(D) = \int_D \varphi \, d\mu.$$

Since this holds for an arbitrary nonnegative simple function φ on D such that $0 \leq \varphi \leq f$, we have

$$\lim_{n\to\infty} \int_D f_n \, d\mu \geq \sup_{0\leq\varphi\leq f} \int_D \varphi \, d\mu = \int_D f \, d\mu.$$

This proves the required reverse inequality. ∎

Theorem 8.5 is not valid for a decreasing sequence. As a counter example in the measure space $(\mathbb{R}, \mathfrak{M}_L, \mu_L)$, let $(f_n : n \in \mathbb{N})$ be a decreasing sequence of nonnegative real-valued $\mathfrak{M}_L$-measurable functions on $\mathbb{R}$ defined by $f_n = \mathbf{1}_{[n,\infty)}$ for $n \in \mathbb{N}$. Then for every $n \in \mathbb{N}$ we have $\int_{\mathbb{R}} f_n \, d\mu_L = \infty$ so that $\lim_{n\to\infty} \int_{\mathbb{R}} f_n \, d\mu_L = \infty$ but $\lim_{n\to\infty} f_n = 0$ on $\mathbb{R}$ and thus $\int_{\mathbb{R}} \lim_{n\to\infty} f_n \, d\mu_L = 0$.

Lemma 8.6. *Given a measure space $(X, \mathfrak{A}, \mu)$. Let f be a nonnegative extended real-valued $\mathfrak{A}$-measurable function on X. Then there exists an increasing sequence of nonnegative simple functions $(\varphi_n : n \in \mathbb{N})$ on X such that*

1° $\varphi_n \uparrow f$ *on* X,

2° $\varphi_n \uparrow f$ *uniformly on an arbitrary subset E of X on which f is bounded,*

3° $\lim_{n\to\infty} \int_D \varphi_n \, d\mu = \int_D f \, d\mu$ *for every* $D \in \mathfrak{A}$.

Moreover if $(X, \mathfrak{A}, \mu)$ is a σ-finite measure space then the nonnegative simple function φ_n can be chosen to be μ-integrable for every $n \in \mathbb{N}$.

Proof. 1. The range of f is contained in $[0, \infty]$. For every $n \in \mathbb{N}$, decompose the interval $[0, n)$ into $n2^n$ subintervals of equal length $\frac{1}{2^n}$ given by $I_{n,k} = \left[\frac{k-1}{2^n}, \frac{k}{2^n}\right)$ for $k = 1, \ldots, n2^n$, and let $J_n = [n, \infty]$. Thus we have a decomposition of $[0, \infty]$ into a class of sets

$$\text{(1)} \qquad \mathfrak{I}_n = \left\{I_{n,k}, k = 1, \ldots, n2^n, J_n\right\}.$$

Let $E_{n,k} = f^{-1}(I_{n,k})$ for $k = 1, \ldots, n2^n$, and let $F_n = f^{-1}(J_n)$. Consider the collection of subsets of X:

$$\text{(2)} \qquad \mathfrak{E}_n = \left\{E_{n,k}, k = 1, \ldots, n2^n, F_n\right\}.$$

By the $\mathfrak{A}$-measurability of f on X, $\mathfrak{E}_n$ is a collection in $\mathfrak{A}$. The disjointness of the collection $\mathfrak{I}_n$ implies the disjointness of the collection $\mathfrak{E}_n$. Since the union of the members of $\mathfrak{I}_n$ is equal to $[0, \infty]$, the union of members of $\mathfrak{E}_n$ is equal to X. Thus $\mathfrak{E}_n$ is a decomposition of X into disjoint $\mathfrak{A}$-measurable subsets of X. Let us define a nonnegative simple function φ_n on X by setting

$$\text{(3)} \qquad \varphi_n = \sum_{k=1}^{n2^n} \frac{k-1}{2^n} \cdot \mathbf{1}_{E_{n,k}} + n \cdot \mathbf{1}_{F_n}.$$

Clearly we have $0 \leq \varphi_n \leq f$ on X for every $n \in \mathbb{N}$.

2. Let us show that $(\varphi_n : n \in \mathbb{N})$ is an increasing sequence of functions on X, that is, for every $x \in X$, we have $\varphi_n(x) \leq \varphi_{n+1}(x)$ for every $n \in \mathbb{N}$. Consider the transition from the decomposition $\mathfrak{I}_n$ to the decomposition $\mathfrak{I}_{n+1}$ of $[0, \infty]$ and the resulting transition from the decomposition $\mathfrak{E}_n$ to the decomposition $\mathfrak{E}_{n+1}$ of X.

The interval $I_{n,k} = \left[\frac{k-1}{2^n}, \frac{k}{2^n}\right)$ in $\mathfrak{I}_n$ is decomposed into two intervals $I_{n+1,2k-1} = \left[\frac{2k-2}{2^{n+1}}, \frac{2k-1}{2^{n+1}}\right)$ and $I_{n+1,2k} = \left[\frac{2k-1}{2^{n+1}}, \frac{2k}{2^{n+1}}\right)$ in $\mathfrak{I}_{n+1}$. Thus $E_{n,k} = f^{-1}(I_{n,k})$ in $\mathfrak{E}_n$ is decomposed into two sets $E_{n+1,2k-1} = f^{-1}(I_{n+1,2k-1})$ and $E_{n+1,2k} = f^{-1}(I_{n+1,2k})$ in $\mathfrak{E}_{n+1}$.

On the other hand, the set $J_n = [n, \infty]$ in $\mathfrak{I}_n$ is decomposed into $I_{n+1,n2^{n+1}+1}, \dots, I_{n+1,(n+1)2^{n+1}}$, and $J_{n+1} = [n+1, \infty]$, a total of $2^{n+1}+1$ sets in $\mathfrak{I}_{n+1}$. Thus $F_n = f^{-1}(J_n)$ in $\mathfrak{E}_n$ is decomposed into $E_{n+1,n2^{n+1}+1}, \dots, E_{n+1,(n+1)2^{n+1}}$, and F_{n+1} in $\mathfrak{E}_{n+1}$.

Let $x \in X$. Since the union of the members of the disjoint collection $\mathfrak{E}_n$ is equal to X, either $x \in E_{n,k}$ for some $k = 1, \dots, n2^n$ or $x \in F_n$. If $x \in E_{n,k}$, then $\varphi_n(x) = \frac{k-1}{2^n}$. Also, if $x \in E_{n,k}$, then either $x \in E_{n+1,2k-1}$ so that $\varphi_{n+1}(x) = \frac{2k-2}{2^{n+1}} = \frac{k-1}{2^n} = \varphi_n(x)$, or $x \in E_{n+1,2k}$ so that $\varphi_{n+1}(x) = \frac{2k-1}{2^{n+1}} > \frac{2k-2}{2^{n+1}} = \frac{k-1}{2^n} = \varphi_n(x)$. On the other hand, if $x \in F_n$, then $\varphi_n(x) = n$. Now if $x \in F_n$, then either $x \in E_{n+1,k}$ for some $k = n2^{n+1}+1, \dots, (n+1)2^{n+1}$ or $x \in F_{n+1}$. If $x \in E_{n+1,k}$, then $\varphi_{n+1}(x) = \frac{k-1}{2^{n+1}} \geq \frac{n2^{n+1}}{2^{n+1}} = n = \varphi_n(x)$. If $x \in F_{n+1}$, then $\varphi_{n+1}(x) = n+1 > n = \varphi_n(x)$.

Thus we have verified that for every $x \in X$, we have $\varphi_n(x) \leq \varphi_{n+1}(x)$ for every $n \in \mathbb{N}$.

3. Let us show that $\lim_{n\to\infty} \varphi_n(x) = f(x)$ for every $x \in X$. Let $x \in X$. Consider first the case $f(x) = \infty$. Then for every $n \in \mathbb{N}$, we have $f(x) \in [n, \infty] = J_n$ so that $x \in f^{-1}(J_n) = F_n$. Then by (3), we have $\varphi_n(x) = n$ so that $\lim_{n\to\infty} \varphi_n(x) = \infty = f(x)$. Next consider the case $f(x) \in [0, \infty)$. In this case there exists $N \in \mathbb{N}$ such that $f(x) \in [0, N)$. Then for any $n \geq N$, we have $f(x) \in [0, n)$. Since the intervals $I_{n,k}$ for $k = 1, \dots, n2^n$ are disjoint and their union is equal to $[0, n)$, we have $f(x) \in I_{n,k} = \left[\frac{k-1}{2^n}, \frac{k}{2^n}\right)$ for some $k = 1, \dots, n2^n$. Then $x \in f^{-1}(I_{n,k}) = E_{n,k}$ so that $\varphi_n(x) = \frac{k-1}{2^n}$ by (3). Thus we have $0 \leq f(x) - \varphi_n(x) \leq \frac{1}{2^n}$. Since this holds for every $n \geq N$, we have $\lim_{n\to\infty} \varphi_n(x) = f(x)$.

4. Suppose f is bounded on a subset E of X. Then there exists $N \in \mathbb{N}$ such that $f(x) \in [0, N)$ for every $x \in E$. Then for every $n \geq N$, we have $f(x) \notin [n, \infty] = J_n$ so that $x \notin f^{-1}(J_n) = F_n$. Since $\mathfrak{E}_n$ is a disjoint collection of sets whose union is equal to X, x is contained in $E_{n,k} = f^{-1}(I_{n,k})$ for some $k = 1, \dots, n2^n$. Then $f(x) \in I_{n,k} = \left[\frac{k-1}{2^n}, \frac{k}{2^n}\right)$. Now $x \in E_{n,k}$ implies $\varphi_n(x) = \frac{k-1}{2^n}$ by (3). Thus we have $0 \leq f(x) - \varphi_n(x) \leq \frac{1}{2^n}$ for every $x \in E$ when $n \geq N$. This shows that $\lim_{n\to\infty} \varphi_n(x) = f(x)$ uniformly on E.

5. 3° follows from 1° and Theorem 8.5 (Monotone Convergence).

6. Suppose $(X, \mathfrak{A}, \mu)$ is a σ-finite measure space. Then there exists an increasing sequence $(A_n : n \in \mathbb{N})$ in $\mathfrak{A}$ such that $\lim_{n\to\infty} A_n = \bigcup_{n\in\mathbb{N}} A_n = X$ and $\mu(A_n) < \infty$ for $n \in \mathbb{N}$. If we let $\psi_n = \varphi_n \mathbf{1}_{A_n}$ for $n \in \mathbb{N}$ then $(\psi_n : n \in \mathbb{N})$ is an increasing sequence of μ-integrable nonnegative simple functions on X. Let us show that $\psi_n \uparrow f$ on X. Let $x \in X$. Since $(A_n : n \in \mathbb{N})$ is an increasing sequence in $\mathfrak{A}$ with $\lim_{n\to\infty} A_n = X$, there exists $N \in \mathbb{N}$ such that $x \in A_n$ for $n \geq N$. Then $\psi_n(x) = \varphi_n(x)\mathbf{1}_{A_n}(x) = \varphi_n(x)$ for $n \geq N$. Since $\varphi_n(x) \uparrow f(x)$ we have $\psi_n(x) \uparrow f(x)$. This shows that $\psi_n \uparrow f$ on X. ∎

Proposition 8.7. *Given a measure space $(X, \mathfrak{A}, \mu)$. Let $D \in \mathfrak{A}$.*
(a) *If $f_1, \dots, f_N$ are nonnegative extended real-valued $\mathfrak{A}$-measurable functions defined on*

D, then we have

$$\int_D \Big\{\sum_{n=1}^N f_n\Big\} d\mu = \sum_{n=1}^N \Big\{\int_D f_n \, d\mu\Big\}.$$

(b) *If* $(f_n : n \in \mathbb{N})$ *is a sequence of nonnegative extended real-valued* $\mathfrak{A}$*-measurable functions defined on D, then we have*

$$\int_D \Big\{\sum_{n\in\mathbb{N}} f_n\Big\} d\mu = \sum_{n\in\mathbb{N}} \Big\{\int_D f_n \, d\mu\Big\}.$$

Proof. Let f_1 and f_2 be two nonnegative extended real-valued $\mathfrak{A}$-measurable functions defined on D. By Lemma 8.6, there exist two increasing sequences of nonnegative simple functions $(\varphi_{1,n} : n \in \mathbb{N})$ and $(\varphi_{2,n} : n \in \mathbb{N})$ on X such that $\varphi_{1,n} \uparrow f_1$ and $\varphi_{2,n} \uparrow f_2$. Then for the increasing sequence of nonnegative simple functions $(\varphi_{1,n} + \varphi_{2,n} : n \in \mathbb{N})$, we have $\varphi_{1,n} + \varphi_{2,n} \uparrow f_1 + f_2$. Thus by Theorem 8.5 (Monotone Convergence), we have

$$(1) \qquad \lim_{n\to\infty} \int_D \{\varphi_{1.n} + \varphi_{2,n}\} d\mu = \int_D \{f_1 + f_2\} d\mu.$$

Now

$$(2) \qquad \lim_{n\to\infty} \int_D \{\varphi_{1.n} + \varphi_{2,n}\} d\mu = \lim_{n\to\infty} \Big\{\int_D \varphi_{1.n} \, d\mu + \int_D \varphi_{2.n} \, d\mu\Big\}$$

$$= \lim_{n\to\infty} \int_D \varphi_{1.n} \, d\mu + \lim_{n\to\infty} \int_D \varphi_{2.n} \, d\mu = \int_D f_1 \, d\mu + \int_D f_2 \, d\mu$$

where the first equality is by (f) of Lemma 7.7 and the last equality is by Theorem 8.5. By (1) and (2) we have

$$(3) \qquad \int_D \{f_1 + f_2\} d\mu = \int_D f_1 \, d\mu + \int_D f_2 \, d\mu.$$

If $f_1, \dots, f_N$ are nonnegative extended real-valued $\mathfrak{A}$-measurable functions defined on D, then by iterated application of (3), we have

$$(4) \qquad \int_D \Big\{\sum_{n=1}^N f_n\Big\} d\mu = \sum_{n=1}^N \Big\{\int_D f_n d\mu\Big\}.$$

If $(f_n : n \in \mathbb{N})$ is a sequence of nonnegative extended real-valued $\mathfrak{A}$-measurable functions defined on D, then for every $N \in \mathbb{N}$, (4) holds. Now the sum of the series $\sum_{n\in\mathbb{N}} f_n$ is the limit of the sequence of partial sums $\big(\sum_{n=1}^N f_n : N \in \mathbb{N}\big)$. Since f_n is nonnegative for every $n \in \mathbb{N}$, the sequence of partial sums is an increasing sequence. Thus by Theorem 8.5 and (4), we have

$$\int_D \Big\{\sum_{n\in\mathbb{N}} f_n\Big\} d\mu = \int_D \Big\{\lim_{N\to\infty} \sum_{n=1}^N f_n\Big\} d\mu = \lim_{N\to\infty} \int_D \Big\{\sum_{n=1}^N f_n\Big\} d\mu$$

$$= \lim_{N\to\infty} \Big\{\sum_{n=1}^N \int_D f_n \, d\mu\Big\} = \sum_{n\in\mathbb{N}} \Big\{\int_D f_n \, d\mu\Big\}. \ \blacksquare$$

Definition 8.8. *Given a measure space $(X, \mathfrak{A}, \mu)$. Let f be a nonnegative extended real-valued $\mathfrak{A}$-measurable function defined only on $D \setminus N$ where $D, N \in \mathfrak{A}$, $N \subset D$ and N is a null set in $(X, \mathfrak{A}, \mu)$. We write $\int_D f \, d\mu$ for $\int_D \tilde{f} \, d\mu$ where $\tilde{f}$ is a nonnegative extended real-valued $\mathfrak{A}$-measurable function defined on D by setting*

$$\tilde{f}(x) = \begin{cases} f(x) & \text{for } x \in D \setminus N, \\ 0 & \text{for } x \in N. \end{cases}$$

(The fact that $\int_D \tilde{f} \, d\mu = \int_{D \setminus N} \tilde{f} \, d\mu + \int_N \tilde{f} \, d\mu = \int_{D \setminus N} f \, d\mu$ will follow from Proposition 8.11 below. The definition of $\int_D f \, d\mu$ for a function f which is defined only a.e. on D is for the convenience of not having to write $\int_{D \setminus N} f \, d\mu$.)

Corollary 8.9. *Given a measure space $(X, \mathfrak{A}, \mu)$. Let f_1 and f_2 be nonnegative extended real-valued $\mathfrak{A}$-measurable functions on a set $D \in \mathfrak{A}$. If $f_1 \le f_2$ on D and if we assume that f_1 is μ-integrable on D, then $f_2 - f_1$ is defined a.e. on D and moreover we have $\int_D \{f_2 - f_1\} \, d\mu = \int_D f_2 \, d\mu - \int_D f_1 \, d\mu$.*

Proof. If f_1 is μ-integrable on D, then f_1 is real-valued a.e. on D by (a) of Lemma 8.2. Thus there exists a null set N in $(X, \mathfrak{A}, \mu)$ such that $N \subset D$ and $f(x) \in [0, \infty)$ for $x \in D \setminus N$. Then $f_2 - f_1$ is defined on $D \setminus N$ and $f_1 + \{f_2 - f_1\}$ is defined on $D \setminus N$ and $f_1 + \{f_2 - f_1\} = f_2$. For the integrals $\int_D \{f_2 - f_1\} \, d\mu$ and $\int_D \{f_1 + \{f_2 - f_1\}\} \, d\mu$ in the sense of Definition 8.8 we have

$$\int_D f_2 \, d\mu = \int_D \{f_1 + \{f_2 - f_1\}\} \, d\mu = \int_D f_1 \, d\mu + \int_D \{f_2 - f_1\} \, d\mu,$$

by (f) of Lemma 8.2 and by (a) of Proposition 8.7. Subtracting the real number $\int_D f_1 \, d\mu$ from both sides, we have $\int_D f_2 \, d\mu - \int_D f_1 \, d\mu = \int_D \{f_2 - f_1\} \, d\mu$. ∎

Remark 8.10. Corollary 8.9 may not hold if f_1 is not μ-integrable. In this case we have $\int_D f_1 \, d\mu = \infty$ and consequently $\int_D f_2 \, d\mu = \infty$ also so that $\int_D f_2 \, d\mu - \int_D f_1 \, d\mu = \infty - \infty$ which is undefined.

Proposition 8.11. *Given a measure space $(X, \mathfrak{A}, \mu)$. Let f be a nonnegative extended real-valued $\mathfrak{A}$-measurable function on a set $D \in \mathfrak{A}$.*
(a) *If $\{D_1, \ldots, D_N\}$ is a disjoint collection in $\mathfrak{A}$ such that $\bigcup_{n=1}^N D_n = D$, then we have*

$$\int_D f \, d\mu = \sum_{n=1}^{N} \int_{D_n} f \, d\mu.$$

(b) *If $(D_n : n \in \mathbb{N})$ is a disjoint sequence in $\mathfrak{A}$ such that $\bigcup_{n \in \mathbb{N}} D_n = D$, then we have*

$$\int_D f \, d\mu = \sum_{n \in \mathbb{N}} \int_{D_n} f \, d\mu.$$

(c) *If* $(E_n : n \in \mathbb{N})$ *is an increasing sequence in* $\mathfrak{A}$ *such that* $\lim_{n\to\infty} E_n = D$, *then we have*

$$\int_D f\,d\mu = \lim_{n\to\infty} \int_{E_n} f\,d\mu.$$

Proof. 1. Let g be a nonnegative extended real-valued $\mathfrak{A}$-measurable function on a set $D \in \mathfrak{A}$. Suppose $A, B \in \mathfrak{A}$, $A \cap B = \emptyset$, $A \cup B = D$, and $g = 0$ on B. Then we have

$$(1)\qquad \int_D g\,d\mu = \int_A g\,d\mu.$$

This can be proved as follows. According to Lemma 8.6, there exists an increasing sequence $(\varphi_n : n \in \mathbb{N})$ of nonnegative simple functions such that $\varphi_n \uparrow g$ on D. Since $0 \le \varphi_n \le g$ on D and since $g = 0$ on B, we have $\varphi_n = 0$ on B so that $\int_B \varphi_n\,d\mu = 0$ for every $n \in \mathbb{N}$. Then by Observation 7.6, we have

$$\int_D \varphi_n\,d\mu = \int_A \varphi_n\,d\mu + \int_B \varphi_n\,d\mu = \int_A \varphi_n\,d\mu.$$

Now $\varphi_n \uparrow g$ on D and in particular $\varphi_n \uparrow g$ on A. Thus by Theorem 8.5, we have

$$\int_D g\,d\mu = \lim_{n\to\infty} \int_D \varphi_n\,d\mu = \lim_{n\to\infty} \int_A \varphi_n\,d\mu = \int_A g\,d\mu.$$

This proves (1).

2. To prove (a), let f be a nonnegative extended real-valued $\mathfrak{A}$-measurable function on $D \in \mathfrak{A}$ and let $\{D_1, \dots, D_N\}$ be a disjoint collection in $\mathfrak{A}$ with $\bigcup_{n=1}^N D_n = D$. For $n = 1, \dots, N$, let us definition a function on D by setting

$$(2)\qquad f_{D_n}(x) = \begin{cases} f(x) & \text{for } x \in D_n, \\ 0 & \text{for } x \in D \setminus D_n. \end{cases}$$

Then $f_{D_1}, \dots, f_{D_N}$ are nonnegative extended real-valued $\mathfrak{A}$-measurable functions on D and $\sum_{n=1}^N f_{D_n} = f$ on D. Thus by Proposition 8.7 and by (1), we have

$$\int_D f\,d\mu = \sum_{n=1}^N \int_D f_{D_n}\,d\mu = \sum_{n=1}^N \int_{D_n} f\,d\mu.$$

3. Let us prove (c) next. Thus let $(E_n : n \in \mathbb{N})$ be an increasing sequence in $\mathfrak{A}$ with $\lim_{n\to\infty} E_n = D$. For each $n \in \mathbb{N}$, let us define a nonnegative extended real-valued $\mathfrak{A}$-measurable function f_{E_n} by (2). Then $(f_{E_n} : n \in \mathbb{N})$ is an increasing sequence and $f_{E_n} \uparrow f$ on D. Thus by Theorem 8.5 and by (1), we have

$$\int_D f\,d\mu = \lim_{n\to\infty} \int_D f_{E_n}\,d\mu = \lim_{n\to\infty} \int_{E_n} f\,d\mu.$$

4. Finally to prove (b), let $(D_n : n \in \mathbb{N})$ be a sequence of disjoint members of $\mathfrak{A}$ such that $\bigcup_{n\in\mathbb{N}} D_n = D$. Let $E_n = \bigcup_{k=1}^n D_k$ for $n \in \mathbb{N}$. Then $(E_n : n \in \mathbb{N})$ is an increasing sequence in $\mathfrak{A}$ and $\lim_{n\to\infty} E_n = \bigcup_{n\in\mathbb{N}} E_n = D$. Then by (c) and (a), we have

$$\begin{aligned}\int_D f\,d\mu &= \lim_{n\to\infty}\int_{E_n} f\,d\mu = \lim_{n\to\infty}\int_{\bigcup_{k=1}^n D_k} f\,d\mu \\ &= \lim_{n\to\infty}\sum_{k=1}^n \int_{D_k} f\,d\mu = \sum_{n\in\mathbb{N}}\int_{D_n} f\,d\mu. \quad\blacksquare\end{aligned}$$

Proposition 8.12. *Given a measure space $(X, \mathfrak{A}, \mu)$. Let f be a nonnegative extended real-valued $\mathfrak{A}$-measurable function on X. If we define a set function ν on $\mathfrak{A}$ by setting $\nu(E) = \int_E f\,d\mu$ for $E \in \mathfrak{A}$, then ν is a measure on $(X, \mathfrak{A})$.*

Proof. Clearly $\nu(E) \in [0, \infty]$ for every $E \in \mathfrak{A}$ and $\nu(\emptyset) = 0$. The countable additivity of ν on $\mathfrak{A}$ follows from (b) of Proposition 8.11. ∎

Theorem 8.13. (Fatou's Lemma) *Given a measure space $(X, \mathfrak{A}, \mu)$. Then for every sequence $(f_n : n \in \mathbb{N})$ of nonnegative extended real-valued $\mathfrak{A}$-measurable functions on a set $D \in \mathfrak{A}$, we have*

(1) $$\int_D \liminf_{n\to\infty} f_n\,d\mu \le \liminf_{n\to\infty}\int_D f_n\,d\mu.$$

In particular if $f = \lim_{n\to\infty} f_n$ exists a.e. on D, then

(2) $$\int_D f\,d\mu \le \liminf_{n\to\infty}\int_D f_n\,d\mu.$$

Proof. By definition, $\liminf_{n\to\infty} f_n = \lim_{n\to\infty}\inf_{k\ge n} f_k$. Since $\left(\inf_{k\ge n} f_k : n \in \mathbb{N}\right)$ is an increasing sequence of nonnegative extended real-valued $\mathfrak{A}$-measurable functions on D, Theorem 8.5 implies that

$$\begin{aligned}\int_D \liminf_{n\to\infty} f_n\,d\mu &= \int_D \lim_{n\to\infty}\inf_{k\ge n} f_k\,d\mu = \lim_{n\to\infty}\int_D \inf_{k\ge n} f_k\,d\mu \\ &= \liminf_{n\to\infty}\int_D \inf_{k\ge n} f_k\,d\mu \le \liminf_{n\to\infty}\int_D f_n\,d\mu,\end{aligned}$$

where the third equality is from the fact that $\left(\int_D \inf_{k\ge n} f_k\,d\mu : n \in \mathbb{N}\right)$ is an increasing sequence in $\overline{\mathbb{R}}$ so that its limit exists and the limit is equal to the limit inferior and the last inequality is from the fact that $\inf_{k\ge n} f_k \le f_n$. This proves (1). If $f = \lim_{n\to\infty} f_n$ exists a.e. on D, then $f = \liminf_{n\to\infty} f_n$ a.e. on D and thus (2) follows from (1) by Definition 8.8. ∎

Theorem 8.14. *Let $(X, \mathfrak{A}, \mu)$ be a measure space and let $(f_n : n \in \mathbb{N})$ be a sequence of nonnegative extended real-valued $\mathfrak{A}$-measurable functions on a set $D \in \mathfrak{A}$. Suppose*

$\lim_{n\to\infty} f_n = f$ *exists* μ*-a.e. on* D *and moreover* $f_n \le f$, μ*-a.e. on* D *for all* $n \in \mathbb{N}$. *Then*

$$\int_D f\,d\mu = \lim_{n\to\infty} \int_D f_n\,d\mu.$$

Proof. By the hypothesis of the theorem there exists a null set N in the measure space $(X, \mathfrak{A}, \mu)$ such that $N \subset D$, $\lim_{n\to\infty} f_n = f$, and $f_n \le f$ for all $n \in \mathbb{N}$ on $D \setminus N$. Then f is a nonnegative extended real-valued $\mathfrak{A}$-measurable function on $D \setminus N$ by Theorem 4.22 and $\int_D f\,d\mu$ exists in $[0, \infty]$ by Definition 8.8. Now $f = \lim_{n\to\infty} f_n = \liminf_{n\to\infty} f_n$ on $D \setminus N$ so that by Theorem 8.13 (Fatou's Lemma) we have

$$\begin{aligned}\int_D f\,d\mu &= \int_D \liminf_{n\to\infty} f_n\,d\mu \le \liminf_{n\to\infty} \int_D f_n\,d\mu \\ &\le \limsup_{n\to\infty} \int_D f_n\,d\mu \le \int_D f\,d\mu\end{aligned}$$

where the last inequality is from the fact that $\int_D f_n\,d\mu \le \int_D f\,d\mu$ for all $n \in \mathbb{N}$. Thus we have

$$\liminf_{n\to\infty} \int_D f_n\,d\mu = \limsup_{n\to\infty} \int_D f_n\,d\mu = \int_D f\,d\mu$$

and therefore $\lim_{n\to\infty} \int_D f_n\,d\mu = \int_D f\,d\mu$. ∎

[III] Approximation of the Integral by Truncation

Let us consider first a truncation of the range of an extended real-valued function.

Definition 8.15. *Given a measure space* $(X, \mathfrak{A}, \mu)$. *Let* f *be an extended real-valued* $\mathfrak{A}$*-measurable on a set* $D \in \mathfrak{A}$. *For a constant* $M > 0$, *define a bounded function* $f^{[M]}$ *on* D *by setting*

$$f^{[M]}(x) = \begin{cases} f(x) & \text{if } f(x) \in [-M, M], \\ M & \text{if } f(x) > M, \\ -M & \text{if } f(x) < -M. \end{cases}$$

Let us call $f^{[M]}$ *the truncation of (the range of)* f *at level* M.

Observe that $f^{[M]} = \max\big\{-M, \min\{M, f\}\big\}$ and thus $f^{[M]}$ is $\mathfrak{A}$-measurable on D by Theorem 4.22. Note also that for the sequence $\big(f^{[n]} : n \in \mathbb{N}\big)$, we have $\lim_{n\to\infty} f^{[n]} = f$ on D. Indeed if $f(x) \in \mathbb{R}$ then there exist $N \in \mathbb{N}$ such that $|f(x)| \le N$ and then $f^{[n]}(x) = f(x)$ for every $n \ge N$ so that $\lim_{n\to\infty} f^{[n]}(x) = f(x)$; if $f(x) = \infty$, then $f^{[n]}(x) = n$ for every $n \in \mathbb{N}$ so that $\lim_{n\to\infty} f^{[n]}(x) = \lim_{n\to\infty} n = \infty = f(x)$ and similarly if $f(x) = -\infty$, then $f^{[n]}(x) = -n$ for every $n \in \mathbb{N}$ so that $\lim_{n\to\infty} f^{[n]}(x) = \lim_{n\to\infty} -n = -\infty = f(x)$. In particular if f is nonnegative then $f^{[n]} \uparrow f$ on D.

Lemma 8.16. *Given a measure space $(X, \mathfrak{A}, \mu)$. Let f be a μ-integrable nonnegative extended real-valued $\mathfrak{A}$-measurable on a set $D \in \mathfrak{A}$. Then for every $\varepsilon > 0$, there exists a constant $M > 0$ such that*

$$\int_D \{f - f^{[M]}\}\, d\mu = \int_D f\, d\mu - \int_D f^{[M]}\, d\mu < \varepsilon.$$

Proof. For any $M > 0$, we have $0 \leq f^{[M]} \leq f$ on D so that the μ-integrability of f on D implies that of $f^{[M]}$ by (e) of Lemma 8.2. Then $\int_D \{f - f^{[M]}\}\, d\mu = \int_D f\, d\mu - \int_D f^{[M]}\, d\mu$ by Corollary 8.9.

For $n \in \mathbb{N}$, consider $f^{[n]}$, the truncation of f at level n. Then $(f^{[n]} : n \in \mathbb{N})$ is an increasing sequence of nonnegative real-valued $\mathfrak{A}$-measurable functions such that $f^{[n]} \uparrow f$ on D. Thus by Theorem 8.5, we have $\lim_{n\to\infty} \int_D f^{[n]}\, d\mu = \int_D f\, d\mu$. Since $\int_D f\, d\mu < \infty$, for every $\varepsilon > 0$ there exists $N \in \mathbb{N}$ such that $\int_D f\, d\mu - \varepsilon < \int_D f^{[n]}\, d\mu$ for $n \geq N$. Thus $\int_D f\, d\mu - \int_D f^{[N]}\, d\mu < \varepsilon$. ■

Next let us consider a truncation of the domain of definition of a function.

Definition 8.17. *Let $(X, \mathfrak{A}, \mu)$ be a measure space and let f be an extended real-valued $\mathfrak{A}$-measurable function on a set $D \in \mathfrak{A}$. Let $E \in \mathfrak{A}$ and $E \subset D$. Let an extended real-valued $\mathfrak{A}$-measurable function on D be defined by setting*

$$f_E(x) = \begin{cases} f(x) & \text{for } x \in E, \\ 0 & \text{for } x \in D \setminus E. \end{cases}$$

We call f_E the truncation of f at the set E.

Note that if $(E_n : n \in \mathbb{N})$ is an increasing sequence of $\mathfrak{A}$-measurable subsets of D such that $\lim_{n\to\infty} E_n = D$, then $\lim_{n\to\infty} f_{E_n} = f$ on D. This can be shown as follows. If $x \in D$, then since $D = \bigcup_{n\in\mathbb{N}} E_n$ there exists $N \in \mathbb{N}$ such that $x \in E_N$. Then $x \in E_n$ for every $n \geq N$ so that $f_{E_n}(x) = f(x)$ for every $n \geq N$. Hence $\lim_{n\to\infty} f_{E_n}(x) = f(x)$ for every $x \in D$. In particular if f is nonnegative then $f_{E_n} \uparrow f$ on D.

Lemma 8.18. *Given a measure space $(X, \mathfrak{A}, \mu)$. Let f be a μ-integrable nonnegative extended real-valued $\mathfrak{A}$-measurable function on a σ-finite set $D \in \mathfrak{A}$. Then for every $\varepsilon > 0$, there exists a $\mathfrak{A}$-measurable subset E of D with $\mu(E) < \infty$ such that*

(1) $$\int_D \{f - f_E\}\, d\mu = \int_D f\, d\mu - \int_D f_E\, d\mu < \varepsilon.$$

In particular, when $(X, \mathfrak{A}, \mu) = (\mathbb{R}, \mathfrak{M}_L, \mu_L)$, there exists a constant $M > 0$ such that

(2) $$\int_D \{f - f_{D\cap[-M,M]}\}\, d\mu_L = \int_D f\, d\mu_L - \int_D f_{D\cap[-M,M]}\, d\mu_L < \varepsilon.$$

Proof. 1. For any $E \subset D$ such that $E \in \mathfrak{A}$, we have $0 \leq f_E \leq f$ on D so that the μ-integrability of f on D implies that of f_E. Then $\int_D \{f - f_E\}\,d\mu = \int_D f\,d\mu - \int_D f_E\,d\mu$ by Corollary 8.9.

Since D is a σ-finite set, according to Lemma 1.31, there exists an increasing sequence $(E_n : n \in \mathbb{N})$ in $\mathfrak{A}$ such that $\lim_{n\to\infty} E_n = D$ and $\mu(E_n) < \infty$ for every $n \in \mathbb{N}$. Then $(f_{E_n} : n \in \mathbb{N})$ is an increasing sequence of nonnegative extended real-valued $\mathfrak{A}$-measurable functions on D such that $f_{E_n} \uparrow f$ on D, we have $\lim_{n\to\infty} \int_D f_{E_n}\,d\mu = \int_D f\,d\mu$ by Theorem 8.5. Since $\int_D f\,d\mu < \infty$, for an arbitrary $\varepsilon > 0$ there exists $N \in \mathbb{N}$ such that we have $\int_D f\,d\mu - \int_D f_{E_n}\,d\mu < \varepsilon$ for $n \geq N$. Let $E = E_N$ and we have the inequality in (1).

2. When $(X, \mathfrak{A}, \mu) = (\mathbb{R}, \mathfrak{M}_L, \mu_L)$ and $D \in \mathfrak{M}_L$, we let $E_n = D \cap [-n, n]$ for $n \in \mathbb{N}$. Then $(E_n : n \in \mathbb{N})$ is an increasing sequence in $\mathfrak{M}_L$ such that $\lim_{n\to\infty} E_n = D$ and $\mu_L(E_n) < \infty$ for every $n \in \mathbb{N}$. Then (2) follows by the same argument as in **1.** ■

The next Theorem shows that if f is a nonnegative extended real-valued $\mathfrak{A}$-measurable function on a σ-finite set $D \in \mathfrak{A}$ and if $\int_D f\,d\mu < \infty$, then $\int_D f\,d\mu$ can be approximated by the integral of a bounded nonnegative real-valued $\mathfrak{A}$-measurable function which vanishes outside of a set with finite measure.

Theorem 8.19. *Given a measure space $(X, \mathfrak{A}, \mu)$. Let f be a μ-integrable nonnegative extended real-valued $\mathfrak{A}$-measurable function on a σ-finite set $D \in \mathfrak{A}$. Then for every $\varepsilon > 0$, there exist a constant $M > 0$ and a $\mathfrak{A}$-measurable subset E of D with $\mu(E) < \infty$ such that*

(1) $$\int_D \{f - (f^{[M]})_E\}\,d\mu = \int_D f\,d\mu - \int_D (f^{[M]})_E\,d\mu < \varepsilon.$$

In particular, when $(X, \mathfrak{A}, \mu) = (\mathbb{R}, \mathfrak{M}_L, \mu_L)$, there exists a constant $M > 0$ such that

(2) $$\int_D \left\{f - (f^{[M]})_{D\cap[-M,M]}\right\} d\mu_L = \int_D f\,d\mu_L - \int_D (f^{[M]})_{D\cap[-M,M]}\,d\mu_L < \varepsilon.$$

Proof. 1. For every $M > 0$ and $E \in \mathfrak{A}$ such that $E \subset D$, we have $0 \leq (f^{[M]})_E \leq f^{[M]} \leq f$ on D. Thus the μ-integrability of f on D implies that of f_{E_n} and $(f^{[M]})_E$. Then by Corollary 8.9, we have $\int_D \{f - (f^{[M]})_E\}\,d\mu = \int_D f\,d\mu - \int_D (f^{[M]})_E\,d\mu$.

Let $\varepsilon > 0$. The μ-integrability of f on D implies according to Lemma 8.16 that there exists $M > 0$ such that

(3) $$\int_D f\,d\mu - \int_D f^{[M]}\,d\mu < \frac{\varepsilon}{2}.$$

The μ-integrability of $f^{[M]}$ on D implies according to Lemma 8.18 the existence of a $\mathfrak{A}$-measurable subset E of D with $\mu(E) < \infty$ such that

(4) $$\int_D f^{[M]}\,d\mu - \int_D (f^{[M]})_E\,d\mu < \frac{\varepsilon}{2}.$$

Adding (3) and (4), we have the inequality in (1).

2. When $(X, \mathfrak{A}, \mu) = (\mathbb{R}, \mathfrak{M}_L, \mu_L)$, by Lemma 8.16 there exists $M_1 > 0$ such that

$$(5) \qquad \int_D f \, d\mu_L - \int_D f^{[M_1]} \, d\mu_L < \frac{\varepsilon}{2},$$

and by Lemma 8.17, there exists $M_2 > 0$ such that

$$(6) \qquad \int_D f^{[M_1]} \, d\mu_L - \int_D (f^{[M_1]})_{D \cap [-M_2, M_2]} \, d\mu_L < \frac{\varepsilon}{2}.$$

Adding (5) and (6), we have

$$(7) \qquad \int_D f \, d\mu_L - \int_D (f^{[M_1]})_{D \cap [-M_2, M_2]} \, d\mu_L < \varepsilon.$$

Now $\int_D (f^{[M_1]})_{D \cap [-M_2, M_2]} \, d\mu_L$ increases if either one of M_1 and M_2 increases. Thus replacing M_1 and M_2 in (7) with $M = \max\{M_1, M_2\}$, we have

$$\int_D (f^{[M_1]})_{D \cap [-M_2, M_2]} \, d\mu_L \leq \int_D (f^{[M]})_{D \cap [-M, M]} \, d\mu_L.$$

Therefore we have

$$\int_D f \, d\mu_L - \int_D (f^{[M]})_{D \cap [-M, M]} \, d\mu_L \leq \int_D f \, d\mu_L - \int_D (f^{[M_1]})_{D \cap [-M_2, M_2]} \, d\mu_L < \varepsilon.$$

This proves the inequality in (2). ∎

Theorem 8.20. (Uniform Absolute Continuity of the Integral with Respect to the Measure) *Given a measure space $(X, \mathfrak{A}, \mu)$. Let f be a μ-integrable nonnegative extended real-valued $\mathfrak{A}$-measurable on a set $D \in \mathfrak{A}$. Then for every $\varepsilon > 0$, there exists $\delta > 0$ such that*

$$\int_E f \, d\mu < \varepsilon$$

for every $\mathfrak{A}$-measurable subset E of D with $\mu(E) < \delta$.

Proof. By Lemma 8.16, for every $\varepsilon > 0$ there exists $M > 0$ such that $\int_D \{f - f^{[M]}\} \, d\mu < \frac{\varepsilon}{2}$. Let $\delta < \frac{\varepsilon}{2M}$. Then for every $\mathfrak{A}$-measurable subset E of D with $\mu(E) < \delta$, we have $\int_E f^{[M]} \, d\mu \leq M\mu(E) < \frac{\varepsilon}{2}$. Thus

$$\int_E f \, d\mu - \int_E f^{[M]} \, d\mu = \int_E \{f - f^{[M]}\} \, d\mu \leq \int_D \{f - f^{[M]}\} \, d\mu < \frac{\varepsilon}{2},$$

and then

$$\int_E f \, d\mu \leq \int_E f^{[M]} \, d\mu + \frac{\varepsilon}{2} < \frac{\varepsilon}{2} + \frac{\varepsilon}{2} = \varepsilon. \ ∎$$

Remark 8.21. Theorem 8.20 is not valid if the function f is only μ semi-integrable, but not μ-integrable, on D.

Example. In the measure space $(\mathbb{R}, \mathfrak{M}_L, \mu_L)$, let $D = [0, \infty)$ and let $f(x) = x$ for $x \in D$. Then $\int_D f \, d\mu_L = \infty$ so that f is not μ_L-integrable on D. Now let $E_n = \left[n, n + \frac{1}{n}\right)$ for $n \in \mathbb{N}$. We have $\int_{E_n} f \, d\mu_L = \int_n^{n+\frac{1}{n}} x \, dx = \frac{1}{2}\left[x^2\right]_n^{n+\frac{1}{n}} = 1 + \frac{1}{2n^2}$. Thus if $\varepsilon = 1$ for instance, then for any $\delta > 0$ however small, if we take $n \in \mathbb{N}$ so large that $\frac{1}{n} < \delta$ then $\mu_L(E_n) = \frac{1}{n} < \delta$ but $\int_{E_n} f \, d\mu_L = 1 + \frac{1}{2n^2} > \varepsilon$.

Problems

Prob. 8.1. Given a measure space $(X, \mathfrak{A}, \mu)$. Let f be a nonnegative extended real-valued $\mathfrak{A}$-measurable function on a set $D \in \mathfrak{A}$. Let Φ be the collection of all nonnegative simple functions φ on D such that $0 \le \varphi \le f$ on D and let Θ be the collection of all Lebesgue integrable nonnegative simple functions ϑ on D such that $0 \le \vartheta \le f$ on D. Recall that by our definition of the Lebesgue integral of f, we have $\int_D f \, d\mu = \sup_{\varphi \in \Phi} \int_D \varphi \, d\mu$. Note that since $\Theta \subset \Phi$, we have $\sup_{\vartheta \in \Theta} \int_D \vartheta \, d\mu \le \sup_{\varphi \in \Phi} \int_D \varphi \, d\mu$. Show that if $(X, \mathfrak{A}, \mu)$ is a σ-finite measure space, then $\sup_{\vartheta \in \Theta} \int_D \vartheta \, d\mu = \sup_{\varphi \in \Phi} \int_D \varphi \, d\mu$.

Prob. 8.2. Consider the Lebesgue measure space $(\mathbb{R}, \mathfrak{M}_L, \mu_L)$.
(a) Construct nonnegative real-valued $\mathfrak{M}_L$-measurable functions $(f_n : n \in \mathbb{N})$ and f on $\mathbb{R}$ such that $(f_n : n \in \mathbb{N})$ converges to f uniformly on $\mathbb{R}$ but $\lim_{n \to \infty} \int_{\mathbb{R}} f_n \, d\mu_L \neq \int_{\mathbb{R}} f \, d\mu_L$.
(b) Construct nonnegative real-valued $\mathfrak{M}_L$-measurable functions $(f_n : n \in \mathbb{N})$ and f on $\mathbb{R}$ such that $f_n \downarrow f$ as $n \to \infty$ on $\mathbb{R}$ but $\lim_{n \to \infty} \int_{\mathbb{R}} f_n \, d\mu_L \neq \int_{\mathbb{R}} f \, d\mu_L$.

Prob. 8.3. Given a measure space $(X, \mathfrak{A}, \mu)$. Let $(f_n : n \in \mathbb{N})$ and f be extended real-valued $\mathfrak{A}$-measurable functions on a set $D \in \mathfrak{A}$ and assume that f is real-valued a.e. on D. Suppose there exists a sequence of positive numbers $(\varepsilon_n : n \in \mathbb{N})$ such that
1° $\sum_{n \in \mathbb{N}} \varepsilon_n < \infty$,
2° $\int_D |f_n - f|^p \, d\mu < \varepsilon_n$ for every $n \in \mathbb{N}$ for some fixed $p \in (0, \infty)$.
Show that the sequence $(f_n : n \in \mathbb{N})$ converges to f a.e. on D. (Note that no μ-integrability of f_n, f, $|f_n|^p$, and $|f|^p$ on D is assumed.)

Prob. 8.4. Given a measure space $(X, \mathfrak{A}, \mu)$. Let $(f_n : n \in \mathbb{N})$ and f be extended real-valued $\mathfrak{A}$-measurable functions on a set $D \in \mathfrak{A}$ and assume that f is real-valued a.e. on D. Suppose there exists a sequence of positive numbers $(\varepsilon_n : n \in \mathbb{N})$ such that
1° $\lim_{n \to \infty} \varepsilon_n = 0$,
2° $\int_D |f_n - f|^p \, d\mu < \varepsilon_n$ for every $n \in \mathbb{N}$ for some fixed $p \in (0, \infty)$.
Show that the sequence $(f_n : n \in \mathbb{N})$ has a subsequence $(f_{n_k} : k \in \mathbb{N})$ which converges to f a.e. on D.

Prob. 8.5. Given a measure space $(X, \mathfrak{A}, \mu)$. Let $(f_n : n \in \mathbb{N})$ and f be extended real-valued $\mathfrak{A}$-measurable functions on a set $D \in \mathfrak{A}$ and assume that f is real-valued a.e. on D. Suppose $\lim_{n \to \infty} \int_D |f_n - f|^p \, d\mu = 0$ for some fixed $p \in (0, \infty)$. Show that the sequence $(f_n : n \in \mathbb{N})$ converges to f on D in measure.

Prob. 8.6. Let $(X, \mathfrak{A}, \mu)$ be a measure space and let f be an extended real-valued $\mathfrak{A}$-

measurable function on X such that $\int_X |f|^p \, d\mu < \infty$ for some $p \in (0, \infty)$. Show that $\lim_{\lambda \to \infty} \lambda^p \mu\{X : |f| \geq \lambda\} = 0$.

Prob. 8.7. Let $(X, \mathfrak{A}, \mu)$ be a σ-finite measure space. Let f be an extended real-valued $\mathfrak{A}$-measurable function on X. Show that for every $p \in (0, \infty)$ we have

$$\int_X |f|^p \, d\mu = \int_{[0,\infty)} p\lambda^{p-1} \mu\{X : |f| > \lambda\} \, \mu_L(d\lambda).$$

Prob. 8.8. Given a measure space $(X, \mathfrak{A}, \mu)$. Let f be a nonnegative extended real-valued $\mathfrak{A}$-measurable function on a set $D \in \mathfrak{A}$ with $\mu(D) < \infty$. Suppose $f > 0$ a.e. on D.
(a) Show that for every $\delta > 0$ there exists $\eta > 0$ such that for every $\mathfrak{A}$-measurable subset E of D with $\mu(E) \geq \delta$, we have $\int_E f \, d\mu \geq \eta$.
(b) Show that (a) does not hold without the assumption $\mu(D) < \infty$.

Prob. 8.9. Given a measure space $(X, \mathfrak{A}, \mu)$ with $\mu(X) < \infty$. Let f be a nonnegative extended real-valued $\mathfrak{A}$-measurable function on X such that $f > 0$, μ-a.e. on X. Let $(E_n : n \in \mathbb{N})$ be a sequence in $\mathfrak{A}$ such that $\lim_{n \to \infty} \int_{E_n} f \, d\mu = 0$. Prove the following:
(a) $\lim_{n \to \infty} \mu(E_n) = 0$.
(b) $\lim_{n \to \infty} \mu(E_n) = 0$ does not hold without the assumption that $f > 0$, μ-a.e. on X.
(c) $\lim_{n \to \infty} \mu(E_n) = 0$ does not hold without the assumption that $\mu(X) < \infty$.

Prob. 8.10. Given a measure space $(X, \mathfrak{A}, \mu)$. Let f be a nonnegative extended real-valued $\mathfrak{A}$-measurable function on a set $D \in \mathfrak{A}$ with $\mu(D) < \infty$. Let $D_n = \{x \in D : f(x) \geq n\}$ for $n \in \mathbb{N}$. Show that $\int_D f \, d\mu < \infty$ if and only if $\sum_{n \in \mathbb{N}} \mu(D_n) < \infty$.

Prob. 8.11. Given a measure space $(X, \mathfrak{A}, \mu)$ with $\mu(X) < \infty$. Let f be a nonnegative real-valued $\mathfrak{A}$-measurable function on X. Show that f in μ-integrable on X if and only if

$$\sum_{n \in \mathbb{Z}_+} 2^n \mu\{x \in X : f(x) > 2^n\} < \infty.$$

Prob. 8.12. Given a measure space $(X, \mathfrak{A}, \mu)$. Let f and g be two nonnegative extended real-valued $\mathfrak{A}$-measurable function on a set $D \in \mathfrak{A}$ such that $f \leq g$ a.e. on D.
(a) Show that if $\int_D f \, d\mu = \int_D g \, d\mu < \infty$, then $f = g$ a.e. on D.
(b) Show by constructing a counter example that if the condition $\int_D g \, d\mu < \infty$ in (a) is removed then the conclusion is not valid.

Prob. 8.13. Given a measure space $(X, \mathfrak{A}, \mu)$. Let f be a real-valued $\mathfrak{A}$-measurable function on a set $D \in \mathfrak{A}$ with $\mu(D) \in (0, \infty)$ such that $f(x) \in [0, 1)$ for every $x \in D$.
(a) Show that $\int_D f \, d\mu < \mu(D)$.
(b) Show by counter example that without the condition $\mu(D) < \infty$, the conclusion in (a) is not valid.

Prob. 8.14. **(a)** Let $\{c_{n,i} : n \in \mathbb{N}, i \in \mathbb{N}\}$ be an array of nonnegative extended real numbers, that is, $c_{n,i} \in [0, \infty]$ for $n \in \mathbb{N}$ and $i \in \mathbb{N}$. Show that

$$\liminf_{n\to\infty} \sum_{i\in\mathbb{N}} c_{n,i} \geq \sum_{i\in\mathbb{N}} \liminf_{n\to\infty} c_{n,i}.$$

(b) Show that if $(c_{n,i} : n \in \mathbb{N})$ is an increasing sequence for each $i \in \mathbb{N}$ then

$$\lim_{n\to\infty} \sum_{i\in\mathbb{N}} c_{n,i} = \sum_{i\in\mathbb{N}} \lim_{n\to\infty} c_{n,i}.$$

§9 Integration of Measurable Functions

[I] Lebesgue Integral of Measurable Functions

If f is an extended real-valued $\mathfrak{A}$-measurable function on a set $D \in \mathfrak{A}$ in a measure space $(X, \mathfrak{A}, \mu)$, then $f = f^+ - f^-$ where f^+ and f^- are the positive and negative parts of f, that is, $f^+ = \max\{f, 0\}$ and $f^- = -\min\{f, 0\}$. Since f^+ and f^- are nonnegative extended real-valued $\mathfrak{A}$-measurable functions on D, $\int_D f^+ \, d\mu$ and $\int_D f^- \, d\mu$ exist in $[0, \infty]$ by Definition 8.1, but $\int_D f^+ \, d\mu - \int_D f^- \, d\mu$ may not exist in $\overline{\mathbb{R}}$.

Definition 9.1. *Given a measure space $(X, \mathfrak{A}, \mu)$. Let f be an extended real-valued $\mathfrak{A}$-measurable function on a set $D \in \mathfrak{A}$. If $\int_D f^+ \, d\mu - \int_D f^- \, d\mu$ exists in $\overline{\mathbb{R}}$, then we say that f is Lebesgue semi-integrable on D with respect to μ, or simply μ semi-integrable on D, and define $\int_D f \, d\mu = \int_D f^+ \, d\mu - \int_D f^- \, d\mu$. We say that f is Lebesgue integrable on D with respect to μ, or simply μ-integrable on D, only when $\int_D f \, d\mu \in \mathbb{R}$.*

For the class of bounded measurable functions on a set with finite measure and for the class of nonnegative extended real-valued measurable functions the Lebesgue integral has been defined by Definition 7.10 and by Definition 8.1 respectively. Let us reconcile these definitions with Definition 9.1 for the class of extended real-valued measurable functions which includes the last two classes of functions as particular cases. To do this, let $I_0(f)$, $I(f)$, and $J(f)$ be the integrals of f in the sense of Definition 7.10, Definition 8.1, and Definition 9.1 respectively.

If f is a nonnegative extended real-valued $\mathfrak{A}$-measurable function on a set $D \in \mathfrak{A}$, then $f = f^+ - f^-$ where $f^+ = f$ and $f^- = 0$. Then $J(f) = I(f^+) - I(f^-) = I(f) - I(0) = I(f)$. Thus Definition 8.1 is consistent with Definition 9.1. If f is a bounded real-valued $\mathfrak{A}$-measurable function on a set $D \in \mathfrak{A}$ with $\mu(D) < \infty$, then

$$\begin{aligned} I_0(f) &= I_0(f^+ - f^-) = I_0(f^+) - I_0(f^-) \\ &= \sup_{\varphi \le f^+} \int_D \varphi \, d\mu - \sup_{\varphi \le f^-} \int_D \varphi \, d\mu \\ &= \sup_{0 \le \varphi \le f^+} \int_D \varphi \, d\mu - \sup_{0 \le \varphi \le f^-} \int_D \varphi \, d\mu \\ &= I(f^+) - I(f^-) = J(f), \end{aligned}$$

where the second equality is by Lemma 7.12. This shows that Definition 7.10 is consistent with Definition 9.1.

Observation 9.2. (a) f is μ semi-integrable on D if and only if at least one of $\int_D f^+ \, d\mu$ and $\int_D f^- \, d\mu$ is finite.
(b) f is μ-integrable on D if and only if both $\int_D f^+ \, d\mu$ and $\int_D f^- \, d\mu$ are finite.
(c) If f is μ semi-integrable on D, then $|\int_D f \, d\mu| \le \int_D |f| \, d\mu$.

(d) f is μ-integrable on D if and only if $|\int_D f\,d\mu| < \infty$.
(e) f is μ-integrable on D if and only if $|f|$ is.
(f) If f is μ-integrable on D, then $|f| < \infty$ a.e. on D, that is, f is real-valued a.e. on D.
(g) If f is μ-integrable on D, then the set $\{x \in D : f(x) \neq 0\}$ is a σ-finite set.

Proof. 1. (a) and (b) are immediate from Definition 9.1.

2. To prove (c), note that if $\int_D f\,d\mu$ exists then

$$\begin{aligned}\int_D f\,d\mu &= \int_D f^+\,d\mu - \int_D f^-\,d\mu \le \int_D f^+\,d\mu + \int_D f^-\,d\mu \\ &= \int_D \{f^+ + f^-\}\,d\mu = \int_D |f|\,d\mu,\end{aligned}$$

by Proposition 8.7. Similarly

$$-\int_D f\,d\mu = -\int_D f^+\,d\mu + \int_D f^-\,d\mu \le \int_D f^+\,d\mu + \int_D f^-\,d\mu = \int_D |f|\,d\mu.$$

Thus $|\int_D f\,d\mu| \le \int_D |f|\,d\mu$.

3. To prove (d), suppose f is μ-integrable on D. Then by (b) we have

$$\left|\int_D f\,d\mu\right| = \left|\int_D f^+\,d\mu - \int_D f^-\,d\mu\right| \le \left|\int_D f^+\,d\mu\right| + \left|\int_D f^-\,d\mu\right| < \infty.$$

Conversely if $|\int_D f\,d\mu| < \infty$, that is, $|\int_D f^+\,d\mu - \int_D f^-\,d\mu| < \infty$, then we have $\int_D f^+\,d\mu - \int_D f^-\,d\mu \in \mathbb{R}$, that is, f is μ-integrable on D.

4. To prove (e), suppose f is μ-integrable on D. Then we have $\int_D f^+\,d\mu < \infty$ and $\int_D f^-\,d\mu < \infty$ by (b). This then implies

$$\int_D |f|\,d\mu = \int_D \{f^+ + f^-\}\,d\mu = \int_D f^+\,d\mu + \int_D f^-\,d\mu < \infty$$

so that $|f|$ is μ-integrable on D. Conversely suppose $|f|$ is μ-integrable on D, that is, $\int_D |f|\,d\mu < \infty$. Now since $0 \le f^+ \le |f|$ and $0 \le f^- \le |f|$ on D, we have both $\int_D f^+\,d\mu \le \int_D |f|\,d\mu < \infty$ and $\int_D f^-\,d\mu \le \int_D |f|\,d\mu < \infty$ by (e) of Lemma 8.2. This shows that f is μ-integrable on D by (b).

5. To prove (f), note that if f is μ-integrable on D then so is $|f|$ by (e). Then $|f| < \infty$ a.e. on D by (a) of Lemma 8.2.

6. To prove (g), note that if we define $D_k = \left\{x \in D : |f(x)| \ge \frac{1}{k}\right\}$ for $k \in \mathbb{N}$ then we have $\left\{x \in D : f(x) \neq 0\right\} = \left\{x \in X : |f(x)| > 0\right\} = \bigcup_{k\in\mathbb{N}} D_k$. Now for each $k \in \mathbb{N}$, we have the estimate $\frac{1}{k}\mu(D_k) \le \int_{D_k} |f|\,d\mu \le \int_D |f|\,d\mu < \infty$ so that $\mu(D_k) < \infty$. This shows that the set $\left\{x \in D : f(x) \neq 0\right\}$ is a σ-finite set. ∎

Lemma 9.3. *Given a measure space $(X, \mathfrak{A}, \mu)$. Let f be an extended real-valued $\mathfrak{A}$-measurable function on a set $D \in \mathfrak{A}$. If $\int_E f\,d\mu$ exists and $\int_E f\,d\mu \ge 0$ for every $\mathfrak{A}$-measurable subset E of D, then $f \ge 0$ a.e. on D.*

Proof. Suppose the statement that $f \geq 0$ a.e. on D is false. Then for the set $E = \{x \in D : f(x) < 0\}$, we have $\mu(E) > 0$. Now $E = \bigcup_{k \in \mathbb{N}} E_k$ where $E_k = \left\{x \in D : f(x) \leq -\frac{1}{k}\right\}$ for $k \in \mathbb{N}$. Thus $\sum_{k \in \mathbb{N}} \mu(E_k) \geq \mu\left(\bigcup_{k \in \mathbb{N}} E_k\right) = \mu(E) > 0$ so that there exists $k_0 \in \mathbb{N}$ such that $\mu(E_{k_0}) > 0$. Then $\int_{E_{k_0}} f \, d\mu \leq -\frac{1}{k_0}\mu(E_{k_0}) < 0$, contradicting the assumption that $\int_E f \, d\mu \geq 0$ for every $\mathfrak{A}$-measurable subset E of D. ■

Observation 9.4. Let f and g be two extended real-valued functions on a set D. If $f \leq g$ on D, then $f^+ \leq g^+$ and $f^- \geq g^-$ on D.

Proof. If $f \leq g$, then $f^+ = \max\{f, 0\} \leq \max\{g, 0\} = g^+$. Also $f \leq g$ implies that $\min\{f, 0\} \leq \min\{g, 0\}$ and then $f^- = -\min\{f, 0\} \geq -\min\{g, 0\} = g^-$. ■

Lemma 9.5. *Given a measure space $(X, \mathfrak{A}, \mu)$. Let f and g be two extended real-valued $\mathfrak{A}$-measurable functions on a set $D \in \mathfrak{A}$. Suppose $f \leq g$ on D. If one of f and g is μ-integrable on D, then the other is at least μ semi-integrable on D.*

Proof. If f is μ-integrable on D, then we have $\int_D f^+ \, d\mu < \infty$ and $\int_D f^- \, d\mu < \infty$. Since $f \leq g$ on D we have $f^- \geq g^-$ on D by Observation 9.4. This implies that we have $\int_D g^- \, d\mu \leq \int_D f^- \, d\mu < \infty$. This proves the μ semi-integrability of g on D by (a) of Observation 9.2.

On the other hand, if we suppose that g is μ-integrable on D, then $\int_D g^+ \, d\mu < \infty$ and $\int_D g^- \, d\mu < \infty$. Since $f \leq g$ on D, we have $f^+ \leq g^+$ on D by Observation 9.4. This implies $\int_D f^+ \, d\mu \leq \int_D g^+ \, d\mu < \infty$. Then $\int_D f^+ \, d\mu < \infty$ implies the μ semi-integrability of f on D. ■

Lemma 9.6. *Given a measure space $(X, \mathfrak{A}, \mu)$. Let f and g be two extended real-valued $\mathfrak{A}$-measurable functions on a set $D \in \mathfrak{A}$. Suppose $f \leq g$ on D.*
(a) *If f is μ semi-integrable on D and $\int_D f \, d\mu \neq -\infty$, then g is μ semi-integrable on D.*
(b) *If g is μ semi-integrable on D and $\int_D g \, d\mu \neq \infty$, then f is μ semi-integrable on D.*

Proof. 1. Let us prove (a). Now if f is μ semi-integrable on D and $\int_D f \, d\mu \neq -\infty$, then since $\int_D f \, d\mu = \int_D f^+ \, d\mu - \int_D f^- \, d\mu$ we have $\int_D f^- \, d\mu < \infty$. Since $f \leq g$ on D implies $f^- \geq g^-$ on D by Observation 9.4, we have $\int_D g^- \, d\mu \leq \int_D f^- \, d\mu < \infty$. Thus g is μ semi-integrable on D by (a) of Observation 9.2.

2. To prove (b), suppose g is μ semi-integrable on D and $\int_D g \, d\mu \neq \infty$. Then since $\int_D g \, d\mu = \int_D g^+ \, d\mu - \int_D g^- \, d\mu$, we have $\int_D g^+ \, d\mu < \infty$. Since $f \leq g$ on D implies $f^+ \leq g^+$ on D by Observation 9.4, we have $\int_D f^+ \, d\mu \leq \int_D g^+ \, d\mu < \infty$. This shows that f is μ semi-integrable on D by (a) of Observation 9.2. ■

Lemma 9.7. *Given a measure space $(X, \mathfrak{A}, \mu)$. Let f and g be two extended real-valued $\mathfrak{A}$-measurable functions on a set $D \in \mathfrak{A}$. If f and g are both μ semi-integrable on D and if $f \leq g$ on D, then $\int_D f \, d\mu \leq \int_D g \, d\mu$.*

Proof. $f \leq g$ on D implies $f^+ \leq g^+$ and $f^- \geq g^-$ on D by Observation 9.4. Thus we have $\int_D f^+ \, d\mu \leq \int_D g^+ \, d\mu$ and $\int_D f^- \, d\mu \geq \int_D g^- \, d\mu$. Subtracting the second inequality from the first side by side we have $\int_D f^+ \, d\mu - \int_D f^- \, d\mu \le \int_D g^+ \, d\mu - \int_D g^- \, d\mu$, where the two differences exist since f and g are μ semi-integrable on D. Thus we have $\int_D f \, d\mu \le \int_D g \, d\mu$. ∎

Lemma 9.8. *Given a measure space $(X, \mathfrak{A}, \mu)$. Let f and g be two extended real-valued $\mathfrak{A}$-measurable functions on a set $D \in \mathfrak{A}$. If $f = g$ a.e. on D and if f is μ semi-integrable on D, then so is g and $\int_D f \, d\mu = \int_D g \, d\mu$.*

Proof. If $f = g$ a.e. on D then $f^+ = g^+$ and $f^- = g^-$ a.e. on D and therefore we have $\int_D f^+ \, d\mu = \int_D g^+ \, d\mu$ and $\int_D f^- \, d\mu = \int_D g^- \, d\mu$ by (f) of Lemma 8.2. Thus if $\int_D f^+ \, d\mu - \int_D f^- \, d\mu$ exists then so does $\int_D g^+ \, d\mu - \int_D g^- \, d\mu$ and the two differences are equal. ∎

Lemma 9.9. *Given a measure space $(X, \mathfrak{A}, \mu)$. Let f be an extended real-valued $\mathfrak{A}$-measurable function on a set $D \in \mathfrak{A}$. Let D_0 be an arbitrary $\mathfrak{A}$-measurable subset of D.*

(a) *If f is μ semi-integrable on D, then f is μ semi-integrable on D_0. In other words, if f is not μ semi-integrable on D_0, then f is not μ semi-integrable on D.*

(b) *If f is μ-integrable on D, then f is μ-integrable on D_0. In other words, if f is not μ-integrable on D_0, then f is not μ-integrable on D.*

Proof. 1. We have $\int_D f^+ \, d\mu \ge \int_{D_0} f^+ \, d\mu \ge 0$ and $\int_D f^- \, d\mu \ge \int_{D_0} f^- \, d\mu \ge 0$ by (c) of Lemma 8.2. If f is μ semi-integrable on D, then at least one of $\int_D f^+ \, d\mu$ and $\int_D f^- \, d\mu$ is finite so that at least one of $\int_{D_0} f^+ \, d\mu$ and $\int_{D_0} f^- \, d\mu$ is finite. Thus f is μ semi-integrable on D_0.

2. If f is μ-integrable on D, then both $\int_D f^+ \, d\mu$ and $\int_D f^- \, d\mu$ are finite so that both $\int_{D_0} f^+ \, d\mu$ and $\int_{D_0} f^- \, d\mu$ are finite and then f is μ-integrable on D_0. ∎

Lemma 9.10. *Given a measure space $(X, \mathfrak{A}, \mu)$. Let f be an extended real-valued $\mathfrak{A}$-measurable function on a set $D \in \mathfrak{A}$. Suppose f is μ semi-integrable on D.*

(a) *If $\{D_1, \dots, D_N\}$ is a disjoint collection in $\mathfrak{A}$ such that $\bigcup_{n=1}^N D_n = D$, then we have*

$$\int_D f \, d\mu = \sum_{n=1}^N \int_{D_n} f \, d\mu.$$

(b) *If $\{D_n : n \in \mathbb{N}\}$ is a disjoint collection in $\mathfrak{A}$ such that $\bigcup_{n \in \mathbb{N}} D_n = D$, then we have*

$$\int_D f \, d\mu = \sum_{n \in \mathbb{N}} \int_{D_n} f \, d\mu.$$

(c) *If $(E_n : n \in \mathbb{N})$ is an increasing sequence in $\mathfrak{A}$ such that $\lim_{n\to\infty} E_n = D$, then we have*

$$\int_D f \, d\mu = \lim_{n\to\infty} \int_{E_n} f \, d\mu.$$

Proof. 1. The μ semi-integrability of f on D implies its μ semi-integrability on D_n by Lemma 9.9. Then (a) and (b) follow from (a) and (b) of Proposition 8.11 respectively.

2. To prove (b) for instance, note that $\int_D f\,d\mu = \int_D f^+\,d\mu - \int_D f^-\,d\mu$ and then we have $\int_D f^+\,d\mu = \sum_{n\in\mathbb{N}} \int_{D_n} f^+\,d\mu$ and $\int_D f^-\,d\mu = \sum_{n\in\mathbb{N}} \int_{D_n} f^-\,d\mu$ according to (b) of Proposition 8.11. Thus

$$\begin{aligned}\int_D f\,d\mu &= \sum_{n\in\mathbb{N}} \int_{D_n} f^+\,d\mu - \sum_{n\in\mathbb{N}} \int_{D_n} f^-\,d\mu \\ &= \sum_{n\in\mathbb{N}} \left\{ \int_{D_n} f^+\,d\mu - \int_{D_n} f^-\,d\mu \right\},\end{aligned}$$

where the second equality is from the fact that if $(a_n : n \in \mathbb{N})$ and $(b_n : n \in \mathbb{N})$ are two sequences of nonnegative extended real numbers such that at least one of $\sum_{n\in\mathbb{N}} a_n$ and $\sum_{n\in\mathbb{N}} b_n$ is finite then $\sum_{n\in\mathbb{N}}(a_n - b_n) = \sum_{n\in\mathbb{N}} a_n - \sum_{n\in\mathbb{N}} b_n$. Thus

$$\begin{aligned}\int_D f\,d\mu &= \lim_{n\to\infty} \sum_{k=1}^{n} \left\{ \int_{D_k} f^+\,d\mu - \int_{D_k} f^-\,d\mu \right\} \\ &= \lim_{n\to\infty} \sum_{k=1}^{n} \int_{D_k} f\,d\mu = \sum_{n\in\mathbb{N}} \int_{D_n} f\,d\mu.\end{aligned}$$

3. To prove (c), Let $D_1 = E_1$ and $D_n = E_n \setminus E_{n-1}$ for $n \geq 2$. Then $\{D_n : n \in \mathbb{N}\}$ is a disjoint collection in $\mathfrak{A}$ with $\bigcup_{k=1}^{n} D_k = E_n$ for $n \in \mathbb{N}$ and $\bigcup_{n\in\mathbb{N}} D_n = \bigcup_{n\in\mathbb{N}} E_n = D$. Thus by (b) and (a) we have

$$\begin{aligned}\int_D f\,d\mu &= \sum_{n\in\mathbb{N}} \int_{D_n} f\,d\mu = \lim_{n\to\infty} \sum_{k=1}^{n} \int_{D_k} f\,d\mu \\ &= \lim_{n\to\infty} \int_{\bigcup_{k=1}^{n} D_k} f\,d\mu = \lim_{n\to\infty} \int_{E_n} f\,d\mu.\end{aligned}$$

This proves (c). ∎

For the integral of a measurable function defined a.e. on a measurable set, the following definition is convenient.

Definition 9.11. *Given a measure space $(X, \mathfrak{A}, \mu)$. Let f be an extended real-valued $\mathfrak{A}$-measurable function on $D \setminus N$ where $D, N \in \mathfrak{A}$, $N \subset D$, and N is a null set in $(X, \mathfrak{A}, \mu)$. Suppose f is μ semi-integrable on $D \setminus N$. We write $\int_D f\,d\mu$ for $\int_D \tilde{f}\,d\mu$ where $\tilde{f}$ is an extended real-valued $\mathfrak{A}$-measurable function on D defined by*

$$\tilde{f} = \begin{cases} f & \text{on } D \setminus N, \\ 0 & \text{on } N. \end{cases}$$

(Note that $\int_D \tilde{f}\,d\mu = \int_{D\setminus N} \tilde{f}\,d\mu + \int_N \tilde{f}\,d\mu = \int_{D\setminus N} f\,d\mu + 0$ by (a) of Lemma 9.10.)

Lemma 9.12. *Given a measure space $(X, \mathfrak{A}, \mu)$. Let f be an extended real-valued $\mathfrak{A}$-measurable function on a set $D \in \mathfrak{A}$.*
(a) *If f is μ semi-integrable on D, then so is $-f$ and $\int_D (-f)\,d\mu = -\int_D f\,d\mu$.*
(b) *If f is μ semi-integrable on D, then so is cf for every $c \neq 0$ and $\int_D cf\,d\mu = c\int_D f\,d\mu$.*
(c) *If f is μ-integrable on D, then $0f$ is defined a.e. on D and $\int_D 0\cdot f\,d\mu = 0\cdot\int_D f\,d\mu$.*

Proof. (a) follows from the fact that $(-f)^+ = f^-$ and $(-f)^- = f^+$. (b) follows from (a) and Lemma 8.3. (c) follows from the fact that if f is μ-integrable on D, then f is real-valued a.e. on D. ∎

It is usually shown that if f and g are both μ-integrable, then so is $f+g$ and moreover $\int_D\{f+g\}\,d\mu = \int_D f\,d\mu + \int_D g\,d\mu$. Here we show that the equality holds even when f and g are only μ semi-integrable as long as the sum $\int_D f\,d\mu + \int_D g\,d\mu$ exists in $\overline{\mathbb{R}}$.

Observation 9.13. Let $a, b \in \overline{\mathbb{R}}$ be such that $a+b$ exists in $\overline{\mathbb{R}}$. Then $\max\{a+b, 0\} \leq \max\{a, 0\} + \max\{b, 0\}$.

Proof. If $a < 0$ and $b < 0$ also, then $\max\{a+b, 0\} = 0 = \max\{a, 0\} + \max\{b, 0\}$. If only one of a and b is negative, say $a < 0$ and $b \geq 0$, then $\max\{a+b, 0\} \leq \max\{b, 0\} = \max\{a, 0\} + \max\{b, 0\}$. If $a \geq 0$ and $b \geq 0$ also, then $\max\{a+b, 0\} = a+b = \max\{a, 0\} + \max\{b, 0\}$. ∎

Theorem 9.14. *Given a measure space $(X, \mathfrak{A}, \mu)$. Let f and g be two extended real-valued $\mathfrak{A}$-measurable functions on a set $D \in \mathfrak{A}$. Suppose f and g are both μ semi-integrable on D. If $\int_D f\,d\mu + \int_D g\,d\mu$ exists in $\overline{\mathbb{R}}$, then*
(a) *$f+g$ is defined a.e. on D,*
(b) *$f+g$ is μ semi-integrable on D,*
(c) *$\int_D\{f+g\}\,d\mu = \int_D f\,d\mu + \int_D g\,d\mu$.*
In particular if both f and g are μ-integrable on D then $f+g$ is μ-integrable on D and $\int_D\{f+g\}\,d\mu = \int_D f\,d\mu + \int_D g\,d\mu$.

Proof. 1. Suppose $\int_D f\,d\mu + \int_D g\,d\mu$ exists in $\overline{\mathbb{R}}$. Then either

(1) $$\int_D f\,d\mu, \int_D g\,d\mu \in (-\infty, \infty],$$

or

(2) $$\int_D f\,d\mu, \int_D g\,d\mu \in [-\infty, \infty).$$

Let us consider case (1). Case (2) can be treated likewise. Now if we assume (1), then we have

(3) $$\int_D f^+\,d\mu, \int_D g^+\,d\mu \in [0, \infty],$$

and

$$\int_D f^- \, d\mu, \int_D g^- \, d\mu \in [0, \infty). \tag{4}$$

By (a) of Lemma 8.2, (4) implies that $f^-, g^- \in [0, \infty)$ a.e. on D. Then $f, g \in (-\infty, \infty]$ a.e. on D so that $f + g$ is defined and $f + g \in (-\infty, \infty]$ a.e. on D.

2. Let us show that $f + g$ is μ semi-integrable on D. By Observation 9.13, we have

$$\begin{cases} (f+g)^+ \leq f^+ + g^+ & \text{a.e. on } D, \\ (f+g)^- = \left(-\{f+g\}\right)^+ \leq (-f)^+ + (-g)^+ = f^- + g^- & \text{a.e. on } D. \end{cases}$$

Then recalling Definition 9.11, by (e) of Lemma 8.2, (a) of Proposition 8.7, (3) and (4), we have

$$\int_D (f+g)^+ \, d\mu \leq \int_D f^+ \, d\mu + \int_D g^+ \, d\mu \in [0, \infty], \tag{5}$$

and similarly

$$\int_D (f+g)^- \, d\mu \leq \int_D f^- \, d\mu + \int_D g^- \, d\mu \in [0, \infty). \tag{6}$$

By (5) and (6), we have

$$\int_D \{f+g\} \, d\mu = \int_D (f+g)^+ \, d\mu - \int_D (f+g)^- \, d\mu \in (-\infty, \infty]. \tag{7}$$

This proves the μ semi-integrability of $f + g$ on D.

3. To prove that the integral of the sum is equal to the sum of the integrals, note that

$$(f+g)^+ - (f+g)^- = f + g = \{f^+ - f^-\} + \{g^+ - g^-\} \quad \text{a.e. on } D.$$

Since (4) implies that f^- and g^- are finite a.e. on D and (6) implies that $(f+g)^-$ is finite a.e. on D, transposition of f^-, g^-, and $(f+g)^-$ in the last equality is possible a.e. on D and thus we have

$$(f+g)^+ + f^- + g^- = f^+ + g^+ + (f+g)^- \quad \text{a.e. on } D.$$

Then recalling Definition 9.11, we have by (a) of Proposition 8.7

$$\begin{aligned} &\int_D (f+g)^+ \, d\mu + \int_D f^- \, d\mu + \int_D g^- \, d\mu \\ &= \int_D f^+ \, d\mu + \int_D g^+ \, d\mu + \int_D (f+g)^- \, d\mu. \end{aligned}$$

Since $\int_D f^- \, d\mu$, $\int_D g^- \, d\mu$, and $\int_D (f+g)^- \, d\mu$ are finite, we may transpose them in the last equality. Then we have

$$\begin{aligned} &\int_D (f+g)^+ \, d\mu - \int_D (f+g)^- \, d\mu \\ &= \int_D f^+ \, d\mu - \int_D f^- \, d\mu + \int_D g^+ \, d\mu - \int_D g^- \, d\mu, \end{aligned}$$

that is, $\int_D \{f + g\} \, d\mu = \int_D f \, d\mu + \int_D g \, d\mu$.

4. If both f and g are μ-integrable on D then $\int_D f \, d\mu + \int_D g \, d\mu$ exists in $\mathbb{R}$ so that we have $\int_D \{f + g\} \, d\mu = \int_D f \, d\mu + \int_D g \, d\mu \in \mathbb{R}$ by (c). ∎

In Theorem 9.14, the existence of $\int_D f \, d\mu + \int_D g \, d\mu$ in $\overline{\mathbb{R}}$ is a sufficient condition for the existence of $\int_D \{f + g\} \, d\mu$ in $\overline{\mathbb{R}}$. This condition however is not a necessary condition. For instance, if f is a nonnegative real-valued $\mathfrak{A}$-measurable function on D with $\int_D f \, d\mu = \infty$ and $g = -f$, then $f + g = 0$ on D and thus $\int_D \{f + g\} \, d\mu = 0$ but $\int_D f \, d\mu + \int_D g \, d\mu = \infty - \infty$ which does not exist.

Corollary 9.15. *Given a measure space $(X, \mathfrak{A}, \mu)$. Let f and g be two extended real-valued $\mathfrak{A}$-measurable functions on a set $D \in \mathfrak{A}$. Suppose f and g are both μ semi-integrable on D. If $\int_D f \, d\mu - \int_D g \, d\mu$ exists in $\overline{\mathbb{R}}$, then*
(a) *$f - g$ is defined a.e. on D,*
(b) *$f - g$ is μ semi-integrable on D,*
(c) *$\int_D \{f - g\} \, d\mu = \int_D f \, d\mu - \int_D g \, d\mu$.*
In particular if both f and g are μ-integrable on D then $f - g$ is μ-integrable on D and $\int_D \{f - g\} \, d\mu = \int_D f \, d\mu - \int_D g \, d\mu$.

Proof. Since $\int_D g \, d\mu$ exists, $\int_D (-g) \, d\mu$ exists and $\int_D (-g) \, d\mu = -\int_D g \, d\mu$. Then since $\int_D f \, d\mu - \int_D g \, d\mu$ exists, $\int_D f \, d\mu + \int_D (-g) \, d\mu$ exists. Thus by Theorem 9.14, $f + (-g)$ is defined a.e. on D, $\int_D \{f + (-g)\} \, d\mu$ exists and $\int_D \{f + (-g)\} \, d\mu = \int_D f \, d\mu + \int_D (-g) \, d\mu$. This proves the Corollary. ∎

Proposition 9.16. *Given a measure space $(X, \mathfrak{A}, \mu)$. Let f and g be extended real-valued $\mathfrak{A}$-measurable functions on a set $D \in \mathfrak{A}$.*
(a) *If f and g are μ-integrable on D and $\int_E f \, d\mu = \int_E g \, d\mu$ for every $E \in \mathfrak{A}$ such that $E \subset D$, then $f = g$ a.e. on D.*
(b) *If f and g are μ semi-integrable on D and $\int_E f \, d\mu = \int_E g \, d\mu$ for every $E \in \mathfrak{A}$ such that $E \subset D$, then $f = g$ a.e. on D provided that D is a σ-finite set.*

Proof. 1. Consider first the case where f and g are two μ-integrable real-valued $\mathfrak{A}$-measurable functions on D. If the statement that $f = g$ a.e. on D is false, then at least one of the two sets $E = \{x \in D : f(x) < g(x)\}$ and $F = \{x \in D : f(x) > g(x)\}$ has a positive measure. Consider the case $\mu(E) > 0$. Now since both f and g are real-valued on D, we have $E = \bigcup_{k \in \mathbb{N}} E_k$ where $E_k = \left\{x \in D : g(x) - f(x) \geq \frac{1}{k}\right\}$. Then $0 < \mu(E) \leq \sum_{k \in \mathbb{N}} \mu(E_k)$. Thus there exists $k_0 \in \mathbb{N}$ such that $\mu(E_{k_0}) > 0$ so that $\int_{E_{k_0}} \{g - f\} \, d\mu \geq \frac{1}{k_0} \mu(E_{k_0}) > 0$. Now $\int_{E_{k_0}} \{g - f\} \, d\mu = \int_{E_{k_0}} g \, d\mu - \int_{E_{k_0}} f \, d\mu$ by Corollary 9.15. Thus we have $\int_{E_{k_0}} g \, d\mu \geq \int_{E_{k_0}} f \, d\mu + \frac{1}{k_0} \mu(E_{k_0}) > \int_{E_{k_0}} f \, d\mu$. This is a contradiction. Thus $\mu(E) = 0$. Similarly $\mu(F) = 0$. This shows that $f = g$ a.e. on D.

Now consider the general case that f and g are two μ-integrable extended real-valued $\mathfrak{A}$-measurable functions on D. The μ-integrability of f and g implies that f and g are real-valued a.e. on D by (f) of Observation 9.2. Thus there exists a null set D_0 contained in D such that f and g are real-valued on $D \setminus D_0$. Let two functions $\tilde{f}$ and $\tilde{g}$ be defined by setting $\widetilde{f} = f$ on $D \setminus D_0$ and $\widetilde{f} = 0$ on D_0, and similarly $\widetilde{g} = g$ on $D \setminus D_0$ and $\widetilde{g} = 0$ on D_0. Then $\widetilde{f}$ and $\widetilde{g}$ are real-valued on D and $\int_E \widetilde{f}\,d\mu = \int_E f\,d\mu = \int_E g\,d\mu = \int_E \widetilde{g}\,d\mu$ for every $E \in \mathfrak{A}$ such that $E \subset D$ so that by our result above we have $\widetilde{f} = \widetilde{g}$ a.e. on D. Then since $f = \widetilde{f}$ a.e. on D and $g = \widetilde{g}$ a.e. on D, we have $f = g$ a.e. on D.

2. Suppose $D \in \mathfrak{A}$ is a σ-finite set and f and g are μ semi-integrable on D. Suppose $\int_E f\,d\mu = \int_E g\,d\mu$ for every $E \in \mathfrak{A}$ such that $E \subset D$. The σ-finiteness of D implies according to Lemma 1.31 that there exists a disjoint sequence $(A_n : n \in \mathbb{N})$ in $\mathfrak{A}$ such that $\bigcup_{n\in\mathbb{N}} A_n = D$ and $\mu(A_n) < \infty$ for every $n \in \mathbb{N}$. Since a countable union of null sets is a null set, to show that $f = g$ a.e. on D, it suffices to show that $f = g$ a.e. on A_n for every $n \in \mathbb{N}$. Now suppose the statement that $f = g$ a.e. on A_n is false for some $n \in \mathbb{N}$. Then at least one of the two sets $E = \{A_n : f < g\}$ and $F = \{A_n : f > g\}$ has a positive measure. Suppose $\mu(E) > 0$. Now E is the union of three mutually disjoint $\mathfrak{A}$-measurable sets given by

$$E^{(1)} = \{A_n : -\infty < f < g < \infty\},$$

$$E^{(2)} = \{A_n : -\infty < f < g = \infty\},$$

$$E^{(3)} = \{A_n : -\infty = f < g\}.$$

If $\mu(E) > 0$ then at least one of the three sets $E^{(1)}$, $E^{(2)}$, and $E^{(3)}$ has a positive measure.

Consider first the case $\mu(E^{(1)}) > 0$. Now $E^{(1)} = \bigcup_{m\in\mathbb{N}} \bigcup_{k\in\mathbb{N}} \bigcup_{\ell\in\mathbb{N}} E^{(1)}_{m,k,\ell}$ where

$$E^{(1)}_{m,k,\ell} = \{A_n : -m \le f, f + \tfrac{1}{k} \le g, g \le \ell\}.$$

Since $0 < \mu(E^{(1)}) \le \sum_{m\in\mathbb{N}} \sum_{k\in\mathbb{N}} \sum_{\ell\in\mathbb{N}} \mu(E^{(1)}_{m,k,\ell})$, there exist $m_0, k_0, \ell_0 \in \mathbb{N}$ such that $\mu(E^{(1)}_{m_0,k_0,\ell_0}) > 0$. For brevity, let $E^* = E^{(1)}_{m_0,k_0,\ell_0}$. Then $\int_{E^*}\{g - f\}\,d\mu \ge \frac{1}{k_0}\mu(E^*) > 0$. Since $-m_0 \le f \le g \le \ell_0$ on E^* and $\mu(E^*) \le \mu(A_n) < \infty$, f and g are μ-integrable on E^* and then $\int_{E^*}\{g - f\}\,d\mu = \int_{E^*} g\,d\mu - \int_{E^*} f\,d\mu$ by Corollary 9.15. This then implies $\int_{E^*} g\,d\mu - \int_{E^*} f\,d\mu \ge \frac{1}{k_0}\mu(E^*) > 0$ so that $\int_{E^*} g\,d\mu > \int_{E^*} f\,d\mu$ which contradicts the assumption that $\int_E f\,d\mu = \int_E g\,d\mu$ for every $E \in \mathfrak{A}$ such that $E \subset D$.

Consider next the case $\mu(E^{(2)}) > 0$. Now $E^{(2)} = \bigcup_{\ell\in\mathbb{N}} E^{(2)}_\ell$ where

$$E^{(2)}_\ell = \{A_n : -\infty < f \le \ell, g = \infty\}.$$

Since $0 < \mu(E^{(2)}) \le \sum_{\ell\in\mathbb{N}} \mu(E^{(2)}_\ell)$, there exists $\ell_0 \in \mathbb{N}$ such that $\mu(E^{(2)}_{\ell_0}) > 0$. For brevity let us write $E^* = E^{(2)}_{\ell_0}$. Then since $g = \infty$ on E^* and $\mu(E^*) > 0$ we have $\int_{E^*} g\,d\mu = \infty$. On the other hand, $\int_{E^*} f\,d\mu \le \ell_0\mu(E^*) \le \ell_0\mu(A_n) < \infty$. Thus $\int_{E^*} f\,d\mu \ne \int_{E^*} g\,d\mu$, contradicting the assumption that $\int_E f\,d\mu = \int_E g\,d\mu$ for every $E \in \mathfrak{A}$ such that $E \subset D$.

Finally consider next the case $\mu(E^{(3)}) > 0$. Now $E^{(3)} = \bigcup_{\ell\in\mathbb{N}} E^{(3)}_m$ where

$$E^{(3)}_m = \{A_n : f = -\infty, g \ge -m\}.$$

Then $0 < \mu(E^{(3)}) \le \sum_{m\in\mathbb{N}} \mu(E^{(3)}_m)$ so that there exists $m_0 \in \mathbb{N}$ such that $\mu(E^{(3)}_{m_0}) > 0$. For brevity let us write $E^* = E^{(3)}_{m_0}$. Then $\int_{E^*} f\,d\mu = -\infty$ and $\int_{E^*} g\,d\mu \ge -m_0\mu(E^*) > -\infty$ since $\mu(E^*) \le \mu(A_n) < \infty$. Thus $\int_{E^*} f\,d\mu \ne \int_{E^*} g\,d\mu$. This is a contradiction.

Thus we have shown that the assumption $\mu(E) > 0$ leads to a contradiction. Similarly the assumption $\mu(F) > 0$ leads to a contradiction. Therefore $f = g$ a.e. on A_n for every $n \in \mathbb{N}$ and hence $f = g$ a.e. on D. ∎

Let us observe that (b) of Proposition 9.16 is not valid without the σ-finiteness of the set $D \in \mathfrak{A}$. As a counter example, consider a measure space $(X, \mathfrak{A}, \mu)$ where X is an arbitrary nonempty set, $\mathfrak{A} = \{\emptyset, X\}$, and $\mu(\emptyset) = 0$ and $\mu(X) = \infty$. Let $f = 1$ on X and $g = 2$ on X. Then $\int_\emptyset f\,d\mu = 0 = \int_\emptyset g\,d\mu$ and $\int_X f\,d\mu = \infty = \int_X g\,d\mu$ so that f and g are μ semi-integrable on X. We have $\int_E f\,d\mu = \int_E g\,d\mu$ for every $E \in \mathfrak{A}$. But $f(x) \neq g(x)$ for any $x \in X$.

[II] Convergence Theorems

The Monotone Convergence Theorem (Theorem 8.5) for an increasing sequence of nonnegative extended real-valued functions is extended next for a monotone sequence of extended real-valued functions.

Theorem 9.17. (Generalized Monotone Convergence Theorem) *Let $(X, \mathfrak{A}, \mu)$ be an arbitrary measure space. Let $(f_n : n \in \mathbb{N})$ be a monotone sequence of extended real-valued $\mathfrak{A}$-measurable functions on a set $D \in \mathfrak{A}$ and let $f = \lim_{n\to\infty} f_n$.*

(a) *If $(f_n : n \in \mathbb{N})$ is an increasing sequence and there exists a μ-integrable extended real-valued $\mathfrak{A}$-measurable function g such that $f_n \geq g$ on D for every $n \in \mathbb{N}$, then f_n, $n \in \mathbb{N}$, and f are μ semi-integrable on D and $\lim_{n\to\infty} \int_D f_n\,d\mu = \int_D f\,d\mu$.*

(b) *If $(f_n : n \in \mathbb{N})$ is a decreasing sequence and there exists a μ-integrable extended real-valued $\mathfrak{A}$-measurable function g such that $f_n \leq g$ on D for every $n \in \mathbb{N}$, then f_n, $n \in \mathbb{N}$, and f are μ semi-integrable on D and $\lim_{n\to\infty} \int_D f_n\,d\mu = \int_D f\,d\mu$.*

Proof. 1. Assume the hypothesis in (a). Since $g \leq f_n$ and g is μ-integrable on D, f_n is μ-semi-integrable on D by Lemma 9.5. Similarly for f. By the μ-integrability of g on D, there exists a subset E of D which is a null set in $(X, \mathfrak{A}, \mu)$ such that g is real-valued on $D \setminus E$ by (f) of Observation 9.2. Then $f_n - g$ is defined on $D \setminus E$. Now $(f_n - g : n \in \mathbb{N})$ is an increasing sequence of nonnegative extended real-valued $\mathfrak{A}$-measurable functions on $D \setminus E$ and thus by Theorem 8.5 (Monotone Convergence Theorem), we have

$$\lim_{n\to\infty} \int_{D\setminus E} \{f_n - g\}\,d\mu = \int_{D\setminus E} \{f - g\}\,d\mu.$$

By the μ-integrability of g, we have $\int_{D\setminus E} g\,d\mu \in \mathbb{R}$. Thus $\int_{D\setminus E} f_n\,d\mu - \int_{D\setminus E} g\,d\mu$ exists in $\overline{\mathbb{R}}$ and this implies that $\int_{D\setminus E}\{f_n - g\}\,d\mu = \int_{D\setminus E} f_n\,d\mu - \int_{D\setminus E} g\,d\mu$ by Corollary 9.15. Similarly we have $\int_{D\setminus E}\{f - g\}\,d\mu = \int_{D\setminus E} f\,d\mu - \int_{D\setminus E} g\,d\mu$. Therefore we have

$$\lim_{n\to\infty} \int_{D\setminus E} f_n\,d\mu - \int_{D\setminus E} g\,d\mu = \int_{D\setminus E} f\,d\mu - \int_{D\setminus E} g\,d\mu$$

and thus $\lim_{n\to\infty} \int_{D\setminus E} f_n\,d\mu = \int_{D\setminus E} f\,d\mu$. Then we have $\lim_{n\to\infty} \int_D f_n\,d\mu = \int_D f\,d\mu$ by Definition 9.11.

2. To prove (b), assume the hypothesis in (b). By Lemma 9.5, f_n, $n \in \mathbb{N}$, and f are μ semi-integrable on D. Now $(-f_n : n \in \mathbb{N})$ is an increasing sequence and $-f_n \geq -g$ on D for every $n \in \mathbb{N}$. Since $-g$ is μ-integrable, (a) implies $\lim_{n\to\infty} \int_D (-f_n)\, d\mu = \int_D (-f)\, d\mu$ and thus $\lim_{n\to\infty} \int_D f_n\, d\mu = \int_D f\, d\mu$. ■

Remark 9.18. (a) In Theorem 9.17, if some entry in the monotone sequence $(f_n : n \in \mathbb{N})$, say f_{n_0} for some $n_0 \in \mathbb{N}$, is μ-integrable, then f_{n_0} can serve as the bounding μ-integrable function g for the monotone sequence $(f_n : n \geq n_0)$. Thus $\lim_{n\to\infty} \int_D f_n\, d\mu = \int_D f\, d\mu$.

(b) Theorem 9.17 does not hold without the existence of the μ-integrable lower bounding function g for an increasing sequence $(f_n : n \in \mathbb{N})$ or the μ-integrable upper bounding function g for a decreasing sequence $(f_n : n \in \mathbb{N})$. For example, in $(\mathbb{R}, \mathfrak{M}_L, \mu_L)$, let $f_n = -\frac{1}{n}$ on $\mathbb{R}$ for $n \in \mathbb{N}$ and let $f = 0$ on $\mathbb{R}$. Then $(f_n : n \in \mathbb{N})$ is an increasing sequence and $\lim_{n\to\infty} f_n = f$ on $\mathbb{R}$. We have $\int_{\mathbb{R}} f_n\, d\mu_L = -\infty$ for every $n \in \mathbb{N}$ while $\int_{\mathbb{R}} f\, d\mu_L = 0$.

Fatou's Lemma (Theorem 8.13) for a sequence of nonnegative extended real-valued measurable functions is extended to a sequence of extended real-valued measurable functions as follows.

Theorem 9.19. (Generalized Fatou's Lemma) *Given a measure space $(X, \mathfrak{A}, \mu)$. Let $(f_n : n \in \mathbb{N})$ be a sequence of extended real-valued $\mathfrak{A}$-measurable functions on a set $D \in \mathfrak{A}$.*

(a) *If there exists a μ-integrable extended real-valued $\mathfrak{A}$-measurable function g such that $f_n \geq g$ on D for every $n \in \mathbb{N}$, then f_n, $n \in \mathbb{N}$, and $\liminf_{n\to\infty} f_n$ are μ semi-integrable on D and*

(1) $$\int_D \liminf_{n\to\infty} f_n\, d\mu \leq \liminf_{n\to\infty} \int_D f_n\, d\mu.$$

In particular if $f = \lim_{n\to\infty} f_n$ exists a.e. on D, then f is μ semi-integrable on D and

(2) $$\int_D f\, d\mu \leq \liminf_{n\to\infty} \int_D f_n\, d\mu.$$

(b) *If there exists a μ-integrable extended real-valued $\mathfrak{A}$-measurable function g such that $f_n \leq g$ on D for every $n \in \mathbb{N}$, then f_n, $n \in \mathbb{N}$, and $\limsup_{n\to\infty} f_n$ are μ semi-integrable on D and*

(3) $$\int_D \limsup_{n\to\infty} f_n\, d\mu \geq \limsup_{n\to\infty} \int_D f_n\, d\mu.$$

In particular if $f = \lim_{n\to\infty} f_n$ exists a.e. on D, then f is μ semi-integrable on D and

(4) $$\int_D f\, d\mu \geq \limsup_{n\to\infty} \int_D f_n\, d\mu.$$

Proof. 1. To prove (1), assume the hypothesis in (a). Since $f_n \geq g$ on D for every $n \in \mathbb{N}$, we have $\liminf_{n\to\infty} f_n \geq g$ on D. Then by Lemma 9.5, f_n, $n \in \mathbb{N}$, and $\liminf_{n\to\infty} f_n$ are μ semi-integrable on D. Now $\liminf_{n\to\infty} f_n = \lim_{n\to\infty} \inf_{k\geq n} f_k$ and $\big(\inf_{k\geq n} f_k : n \in \mathbb{N}\big)$ is an increasing sequence of extended real-valued $\mathfrak{A}$-measurable functions on D satisfying $\inf_{k\geq n} f_k \geq g$ on D for every $n \in \mathbb{N}$. Thus by (a) of Theorem 9.17 (Generalized Monotone Convergence Theorem), we have

$$\begin{aligned}\int_D \liminf_{n\to\infty} f_n \, d\mu &= \int_D \lim_{n\to\infty} \inf_{k\geq n} f_k \, d\mu = \lim_{n\to\infty} \int_D \inf_{k\geq n} f_k \, d\mu \\ &= \liminf_{n\to\infty} \int_D \inf_{k\geq n} f_k \, d\mu \leq \liminf_{n\to\infty} \int_D f_n \, d\mu.\end{aligned}$$

This proves (1). Now if $f = \lim_{n\to\infty} f_n$ exists a.e. on D, then $f = \liminf_{n\to\infty} f_n$ a.e. on D. Then $\int_D f \, d\mu = \int_D \liminf_{n\to\infty} f_n \, d\mu$ by Definition 9.11. Substituting this equality in (1), we have (2).

2. To prove (3), assume the hypothesis of (b). Since $f_n \leq g$ on D for every $n \in \mathbb{N}$, we have $\limsup_{n\to\infty} f_n \leq g$ on D. Then f_n, $n \in \mathbb{N}$, and $\limsup_{n\to\infty} f_n$ are μ semi-integrable on D by Lemma 9.5. For the sequence $(-f_n : n \in \mathbb{N})$, we have $-f_n \geq -g$ on D for every $n \in \mathbb{N}$ and $-g$ is μ-integrable on D. Applying (1) to the sequence $(-f_n : n \in \mathbb{N})$, we have

(5) $$\int_D \liminf_{n\to\infty}(-f_n) \, d\mu \leq \liminf_{n\to\infty} \int_D (-f_n) \, d\mu = \liminf_{n\to\infty} \Big(- \int_D f_n \, d\mu \Big).$$

Now $\liminf_{n\to\infty}(-a_n) = -\limsup_{n\to\infty} a_n$ for an arbitrary sequence $(a_n : n \in \mathbb{N})$ in $\overline{\mathbb{R}}$. Thus

$$\int_D \liminf_{n\to\infty}(-f_n) \, d\mu = -\int_D \limsup_{n\to\infty} f_n \, d\mu$$

and

$$\liminf_{n\to\infty} \Big(- \int_D f_n \, d\mu \Big) = -\limsup_{n\to\infty} \int_D f_n \, d\mu.$$

Using these two equalities in (5), we obtain (3). If $f = \lim_{n\to\infty} f_n$ exists a.e. on D, then we have $f = \limsup_{n\to\infty} f_n$ a.e. on D. Then $\int_D f \, d\mu = \int_D \limsup_{n\to\infty} f_n \, d\mu$ by Definition 9.11. Substituting this equality in (3), we have (4). ∎

The Dominated Convergence Theorem which we prove next contains the Bounded Convergence Theorem (Theorem 7.16) as a particular case.

Theorem 9.20. (Lebesgue's Dominated Convergence Theorem) *Given a measure space $(X, \mathfrak{A}, \mu)$. Let $(f_n : n \in \mathbb{N})$ be a sequence of extended real-valued $\mathfrak{A}$-measurable functions on a set $D \in \mathfrak{A}$ such that $|f_n| \leq g$ on D for every $n \in \mathbb{N}$ for some μ-integrable nonnegative* *extended real-valued $\mathfrak{A}$-measurable function g on D. If $f = \lim_{n\to\infty} f_n$ exists a.e. on D, then f is μ-integrable on D and furthermore*

$$\lim_{n\to\infty}\int_D f_n\,d\mu = \int_D f\,d\mu, \tag{1}$$

and

$$\lim_{n\to\infty}\int_D |f_n - f|\,d\mu = 0. \tag{2}$$

Proof. 1. Since $|f_n| \le g$ on D, we have $-g \le f_n \le g$ on D for every $n \in \mathbb{N}$. Since g is μ-integrable on D, $-g$ is also μ-integrable on D. Thus by Theorem 9.19 (Generalized Fatou's Lemma), f_n, $n \in \mathbb{N}$, and f are μ semi-integrable on D and

$$\int_D f\,d\mu \le \liminf_{n\to\infty}\int_D f_n\,d\mu, \tag{3}$$

as well as

$$\int_D f\,d\mu \ge \limsup_{n\to\infty}\int_D f_n\,d\mu. \tag{4}$$

Since $\liminf_{n\to\infty} \int_D f_n\,d\mu \le \limsup_{n\to\infty} \int_D f_n\,d\mu$, (3) and (4) imply

$$\liminf_{n\to\infty}\int_D f_n\,d\mu = \limsup_{n\to\infty}\int_D f_n\,d\mu = \int_D f_n\,d\mu.$$

Thus $\lim_{n\to\infty} \int_D f_n\,d\mu = \int_D f\,d\mu \in \overline{\mathbb{R}}$. This proves (1).

2. Let us show that f is not only μ semi-integrable but in fact μ-integrable on D, that is, $\int_D f\,d\mu \in \mathbb{R}$. Since $-g \le f_n \le g$ on D, we have $-\int_D g\,d\mu \le \int_D f_n\,d\mu \le \int_D g\,d\mu$ for every $n \in \mathbb{N}$. Thus $\int_D f\,d\mu = \lim_{n\to\infty} \int_D f_n\,d\mu \in \left[-\int_D g\,d\mu, \int_D g\,d\mu\right]$ and therefore we have $\int_D f\,d\mu \in \mathbb{R}$.

3. To prove (2), we apply (1) to the sequence $(|f_n - f| : n \in \mathbb{N})$ as follows. Since $f = \lim_{n\to\infty} f_n$ a.e. on D, there exists a null set E_1 in $(X, \mathfrak{A}, \mu)$ contained in D such that $f = \lim_{n\to\infty} f_n$ on $D \setminus E_1$. Since f is μ-integrable on D, f is real-valued a.e. on D by (f) of Observation 9.2 so that there exists a null set E_2 in $(X, \mathfrak{A}, \mu)$ contained in D such that f is real-valued on $D \setminus E_2$. Let $E = E_1 \cup E_2$, a null set in $(X, \mathfrak{A}, \mu)$ contained in D. Now f is real-valued on $D \setminus E$ so that $f_n - f$ is defined on $D \setminus E$ for $n \in \mathbb{N}$. Since $-g \le f_n \le g$ on D for $n \in \mathbb{N}$ and since $\lim_{n\to\infty} f_n = f$ on $D \setminus E$, we have $-g \le f \le g$ on $D \setminus E$. Then $|f_n - f| \le |f_n| + |f| \le 2g$ on $D \setminus E$. Since $f = \lim_{n\to\infty} f_n$ on $D \setminus E$, we have $\lim_{n\to\infty} f_n - f = 0$ on $D \setminus E$ and consequently $\lim_{n\to\infty} |f_n - f| = 0$ on $D \setminus E$. Thus applying (1) to the sequence $(|f_n - f| : n \in \mathbb{N})$ which is dominated by the μ-integrable function $2g$ and converges to 0 on $D \setminus E$, we have $\lim_{n\to\infty} \int_{D\setminus E} |f_n - f|\,d\mu = \int_{D\setminus E} 0\,d\mu = 0$. Then by Definition 9.11, we have $\lim_{n\to\infty} \int_D |f_n - f|\,d\mu = 0$. ∎

Theorem 9.20 still holds if the condition that $|f_n| \le g$ on D for every $n \in \mathbb{N}$ is replaced by the condition that $|f_n| \le g$ a.e. on D for each $n \in \mathbb{N}$. In this case we redefine f_n to be equal to 0 on the null set on which $|f_n| \le g$ does not hold. The redefined functions $\widetilde{f}_n$ now satisfy $|\widetilde{f}_n| \le g$ on D and we still have $\lim_{n\to\infty} \widetilde{f}_n = f$ a.e. on D. Then by Theorem 9.20, we have $\lim_{n\to\infty} \int_D \widetilde{f}_n \, d\mu = \int_D f \, d\mu$. But $\int_D \widetilde{f}_n \, d\mu = \int_D f_n \, d\mu$. Thus we have $\lim_{n\to\infty} \int_D f_n \, d\mu = \int_D f \, d\mu$. Similarly $\lim_{n\to\infty} \int_D |f_n - f| \, d\mu = 0$.

[III] Convergence Theorems under Convergence in Measure

Let $(a_n : n \in \mathbb{N})$ be a sequence in $\overline{\mathbb{R}}$. By definition, $\liminf_{n\to\infty} a_n$ is the limit of the increasing sequence $\left(\inf_{k\ge n} a_k : n \in \mathbb{N}\right)$ in $\overline{\mathbb{R}}$, and $\limsup_{n\to\infty} a_n$ is the limit of the decreasing sequence $\left(\sup_{k\ge n} a_k : n \in \mathbb{N}\right)$ in $\overline{\mathbb{R}}$. It follows then that if $\liminf_{n\to\infty} a_n < c$ for some $c \in \mathbb{R}$, then $a_n < c$ for infinitely many $n \in \mathbb{N}$, and if $\limsup_{n\to\infty} a_n > c$, then $a_n > c$ for infinitely many $n \in \mathbb{N}$.

Theorem 9.21. (Fatou's Lemma under Convergence in Measure) *Given a measure space $(X, \mathfrak{A}, \mu)$. Let $(f_n : n \in \mathbb{N})$ be a sequence of extended real-valued $\mathfrak{A}$-measurable functions on a set $D \in \mathfrak{A}$ converging in measure to an extended real-valued $\mathfrak{A}$-measurable function f which is real-valued a.e. on D.*

(a) *If $f_n \ge g$ on D for every $n \in \mathbb{N}$ for some μ-integrable extended real-valued $\mathfrak{A}$-measurable function g on D, then f_n, $n \in \mathbb{N}$, and f are all μ semi-integrable on D and*

$$(1) \qquad \int_D f \, d\mu \le \liminf_{n\to\infty} \int_D f_n \, d\mu.$$

(b) *If $f_n \le g$ on D for every $n \in \mathbb{N}$ for some μ-integrable extended real-valued $\mathfrak{A}$-measurable function g on D, then f_n, $n \in \mathbb{N}$, and f are all μ semi-integrable on D and*

$$(2) \qquad \int_D f \, d\mu \ge \limsup_{n\to\infty} \int_D f_n \, d\mu.$$

Proof. Assume the hypothesis in (a). Since $f_n \ge g$ on D, the μ-integrability of g on D implies the μ semi-integrability of f_n on D, that is, $\int_D f_n \, d\mu$ exists in $\overline{\mathbb{R}}$, for every $n \in \mathbb{N}$ by Lemma 9.5. The convergence in measure of $(f_n : n \in \mathbb{N})$ to f on D implies the existence of a subsequence $(f_{n_k} : k \in \mathbb{N})$ which converges to f a.e. on D by Theorem 6.24. Then $f \ge g$ a.e. on D and this implies the μ semi-integrability of f on D by Lemma 9.5. Thus we have proved the existence of the integrals $\int_D f_n \, d\mu$ for $n \in \mathbb{N}$ and $\int_D f \, d\mu$.

To prove the inequality (1), let $a_n = \int_D f_n \, d\mu$ for $n \in \mathbb{N}$ and $a = \int_D f \, d\mu$. To show $a \le \liminf_{n\to\infty} a_n$, let us assume the contrary. Then there exists $c \in \mathbb{R}$ such that $a > c > \liminf_{n\to\infty} a_n$. This implies that $a_n < c$ for infinitely many $n \in \mathbb{N}$ and thus there exists a subsequence $(a_{n_k} : k \in \mathbb{N})$ such that $a_{n_k} < c$ for every $k \in \mathbb{N}$. The convergence in measure of $(f_n : n \in \mathbb{N})$ to f on D implies the convergence in measure of the subsequence

$(f_{n_k} : k \in \mathbb{N})$ to f on D. This then implies the existence of a subsequence $(f_{n_{k_\ell}} : \ell \in \mathbb{N})$ which converges to f a.e. on D by Theorem 6.24. Then by (2) of Theorem 9.19, we have $\int_D f\,d\mu \leq \liminf_{\ell\to\infty} \int_D f_{n_{k_\ell}}\,d\mu$, that is, $a \leq \liminf_{\ell\to\infty} a_{n_{k_\ell}}$. But $a_{n_k} < c < a$ for every $k \in \mathbb{N}$ and thus $\liminf_{\ell\to\infty} a_{n_{k_\ell}} \leq c < a$. This is a contradiction. This completes the proof of (a). (b) is proved by applying (a) to the sequence $(-f_n : n \in \mathbb{N})$. ∎

Theorem 9.22. (Dominated Convergence Theorem under Convergence in Measure) *Given a measure space $(X, \mathfrak{A}, \mu)$. Let $(f_n : n \in \mathbb{N})$ be a sequence of extended real-valued $\mathfrak{A}$-measurable functions on a set $D \in \mathfrak{A}$ such that $|f_n| \leq g$ on D for every $n \in \mathbb{N}$ for some μ-integrable extended real-valued $\mathfrak{A}$-measurable function g on D. If $(f_n : n \in \mathbb{N})$ converges in measure to an extended real-valued $\mathfrak{A}$-measurable function f which is real-valued a.e. on D, then f is μ-integrable on D and furthermore $\lim_{n\to\infty} \int_D f_n\,d\mu = \int_D f\,d\mu$ and $\lim_{n\to\infty} \int_D |f_n - f|\,d\mu = 0$.*

Proof. The converges of the sequence $(f_n : n \in \mathbb{N})$ to f in measure on D implies the existence of a subsequence $(f_{n_k} : n \in \mathbb{N})$ which converges to f a.e. on D. This implies the μ-integrability of f on D by Theorem 9.20. To show $\lim_{n\to\infty} \displaystyle\int_D f_n\,d\mu = \int_D f\,d\mu$, we use the fact that for an arbitrary sequence $(a_n : n \in \mathbb{N})$ in $\overline{\mathbb{R}}$ if there exists $a \in \overline{\mathbb{R}}$ such that every subsequence $(a_{n_k} : k \in \mathbb{N})$ has a subsequence $(a_{n_{k_\ell}} : \ell \in \mathbb{N})$ such that $\lim_{\ell\to\infty} a_{n_{k_\ell}} = a$ then $\lim_{n\to\infty} a_n = a$. Thus let $a_n = \int_D f_n\,d\mu$ for $n \in \mathbb{N}$ and $a = \int_D f\,d\mu$. Let $(a_{n_k} : n \in \mathbb{N})$ be an arbitrary subsequence of $(a_n : n \in \mathbb{N})$. The convergence in measure of $(f_n : n \in \mathbb{N})$ to f on D implies the convergence in measure of $(f_{n_k} : k \in \mathbb{N})$ to f on D and thus there exists a subsequence $(f_{n_{k_\ell}} : \ell \in \mathbb{N})$ which converges to f a.e. on D by Theorem 6.24. Then by Theorem 9.20, we have $\lim_{\ell\to\infty} \int_D f_{n_{k_\ell}}\,d\mu = \int_D f\,d\mu$, that is, $\lim_{\ell\to\infty} a_{n_{k_\ell}} = a$. Thus $\lim_{n\to\infty} a_n = a$, that is, $\lim_{n\to\infty} \int_D f_n\,d\mu = \int_D f\,d\mu$. By Theorem 9.20, we have also $\lim_{\ell\to\infty} \int_D |f_{n_{k_\ell}} - f|\,d\mu = 0$. Then $\lim_{n\to\infty} \int_D |f_n - f|\,d\mu = 0$ follows by the same argument as above. ∎

[IV] Approximation of the Integral by Truncation

For an extended real-valued function f and $M > 0$, recall the truncation $f^{[M]}$ of f at M defined by Definition 8.15. Let us consider an approximation of f by $f^{[M]}$ in terms of the integral.

Lemma 9.23. *Given a measure space $(X, \mathfrak{A}, \mu)$. Let f be a μ-integrable extended real-valued $\mathfrak{A}$-measurable function on a set $D \in \mathfrak{A}$. Then for every $\varepsilon > 0$, there exists a constant $M > 0$ such that*

$$\left|\int_D f\,d\mu - \int_D f^{[M]}\,d\mu\right| \leq \int_D |f - f^{[M]}|\,d\mu < \varepsilon.$$

Proof. For every $M > 0$, we have $|f^{[M]}| \le |f|$ on D. Then the μ-integrability of f on D implies that of $f^{[M]}$. Thus by (c) of Corollary 9.15 and (c) of Observation 9.2, we have

$$\left| \int_D f \, d\mu - \int_D f^{[M]} \, d\mu \right| = \left| \int_D \{f - f^{[M]}\} \, d\mu \right| \le \int_D |f - f^{[M]}| \, d\mu.$$

For every $n \in \mathbb{N}$, we have $|f - f^{[n]}| \le |f| + |f^{[n]}| \le 2|f|$. Also $\lim_{n\to\infty} f^{[n]} = f$ on D implies $\lim_{n\to\infty} |f - f^{[n]}| = 0$ on D. Applying Theorem 9.20 (Lebesgue's Dominated Convergence Theorem) to the sequence $\left(|f - f^{[n]}| : n \in \mathbb{N}\right)$ of functions dominated by the μ-integrable function $2|f|$, we have

$$\lim_{n\to\infty} \int_D |f - f^{[n]}| \, d\mu = \int_D \lim_{n\to\infty} |f - f^{[n]}| \, d\mu = 0.$$

Thus for every $\varepsilon > 0$, there exists $M \in \mathbb{N}$ such that $\int_D |f - f^{[M]}| \, d\mu < \varepsilon$. ■

Recall Definition 8.17 for f_E where E is a subset of the domain of definition D of the function f.

Lemma 9.24. *Given a measure space $(X, \mathfrak{A}, \mu)$. Let $D \in \mathfrak{A}$ be a σ-finite set. Let f be a μ-integrable extended real-valued $\mathfrak{A}$-measurable function on D. Then for every $\varepsilon > 0$, there exist a $\mathfrak{A}$-measurable subset E of D with $\mu(E) < \infty$ such that*

(1) $$\left| \int_D f \, d\mu - \int_D f_E \, d\mu \right| \le \int_D |f - f_E| \, d\mu < \varepsilon.$$

In particular, when $(X, \mathfrak{A}, \mu) = \left(\mathbb{R}, \mathfrak{M}_L, \mu_L\right)$, there exists a constant $M > 0$ such that

(2) $$\left| \int_D f \, d\mu_L - \int_D f_{D\cap[-M,M]} \, d\mu_L \right| \le \int_D |f - f_{D\cap[-M,M]}| \, d\mu_L < \varepsilon.$$

Proof. For every $E \subset D$ such that $E \in \mathfrak{A}$, we have $|f_E| \le |f|$ on D. Then the μ-integrability of f on D implies that of f_E. Thus by (c) of Corollary 9.15 and (c) of Observation 9.2, we have

$$\left| \int_D f \, d\mu - \int_D f_E \, d\mu \right| = \left| \int_D \{f - f_E\} \, d\mu \right| \le \int_D |f - f_E| \, d\mu.$$

Since D is a σ-finite set there exists an increasing sequence $(E_n : n \in \mathbb{N})$ in $\mathfrak{A}$ such that $\lim_{n\to\infty} E_n = D$ and $\mu(E_n) < \infty$ for every $n \in \mathbb{N}$ by Lemma 1.31. For the sequence $(f_{E_n} : n \in \mathbb{N})$, we have $\lim_{n\to\infty} f_{E_n} = f$ on D. Since f is μ-integrable on D, f is real-valued a.e. on D. Thus $f - f_{E_n}$ is defined a.e. on D. Furthermore $\lim_{n\to\infty} |f - f_{E_n}| = 0$ and $|f - f_{E_n}| \le |f| + |f_{E_n}| \le 2|f|$ a.e. on D for every $n \in \mathbb{N}$. Thus by Theorem 9.20, we have $\lim_{n\to\infty} \int_D |f - f_{E_n}| \, d\mu = 0$. Then for every $\varepsilon > 0$ there exists $M \in \mathbb{N}$ such that $\lim_{n\to\infty} \int_D |f - f_{E_n}| \, d\mu < \varepsilon$ for $n \ge M$. With $E = E_M$, we have the inequality in (1). When $(X, \mathfrak{A}, \mu) = \left(\mathbb{R}, \mathfrak{M}_L, \mu_L\right)$, we let $E_n = D \cap [-n, n]$. Then (2) follows from (1). ■

Theorem 9.25. *Given a measure space $(X, \mathfrak{A}, \mu)$. Let $D \in \mathfrak{A}$ be a σ-finite set. Let f be a μ-integrable extended real-valued $\mathfrak{A}$-measurable on D. Then for every $\varepsilon > 0$ there exist a constant $M > 0$ and a $\mathfrak{A}$-measurable subset E of D with $\mu(E) < \infty$ such that*

$$(1) \qquad \left| \int_D f \, d\mu - \int_D (f^{[M]})_E \, d\mu \right| \leq \int_D |f - (f^{[M]})_E| \, d\mu < \varepsilon.$$

In particular, when $(X, \mathfrak{A}, \mu) = (\mathbb{R}, \mathfrak{M}_L, \mu_L)$, there exists a constant $M > 0$ such that

$$(2) \quad \left| \int_D f \, d\mu_L - \int_D (f^{[M]})_{D\cap[-M,M]} \, d\mu_L \right| \leq \int_D |f - (f^{[M]})_{D\cap[-M,M]}| \, d\mu_L < \varepsilon.$$

Proof. The Theorem follows from Lemma 9.23 and Lemma 9.24 by the same argument as in the Proof of Theorem 8.19 for nonnegative functions. ∎

Theorem 9.26. (Uniform Absolute Continuity of the Integral with Respect to the Measure) *Given a measure space $(X, \mathfrak{A}, \mu)$. Let f be a μ-integrable extended real-valued $\mathfrak{A}$-measurable function on a set $D \in \mathfrak{A}$. Then for every $\varepsilon > 0$, there exists $\delta > 0$ such that*

$$\left| \int_E f \, d\mu \right| \leq \int_E |f| \, d\mu < \varepsilon,$$

for every $\mathfrak{A}$-measurable subset E of D with $\mu(E) < \delta$.

Proof. The first inequality is by (c) of Observation 9.2. If f is μ-integrable on D then $|f|$ is μ-integrable on D by (e) of Observation 9.2. Then by applying Theorem 8.20 to the μ-integrable nonnegative extended real-valued $\mathfrak{A}$-measurable function $|f|$ on D we complete the proof. ∎

Observation 9.27. Let $(X, \mathfrak{A}, \mu)$ be a measure space and let f be an extended real-valued $\mathfrak{A}$-measurable function on a set $D \in \mathfrak{A}$. Let $(E_n : n \in \mathbb{N})$ be an increasing sequence in $\mathfrak{A}$ such that $\lim_{n\to\infty} E_n = \bigcup_{n\in\mathbb{N}} E_n = D$.

(a) Suppose f is μ semi-integrable on D, that is, $\int_D f \, d\mu$ exists in $\overline{\mathbb{R}}$. Then we have $\int_D f \, d\mu = \lim_{n\to\infty} \int_{E_n} f \, d\mu$ by (c) of Lemma 9.10. Thus if $\int_D f \, d\mu$ exists, then it can be computed as the limit of a sequence of extended real numbers $\left(\int_{E_n} f \, d\mu : n \in \mathbb{N}\right)$.

(b) Conversely suppose f is such that $\int_{E_n} f \, d\mu$ exists in $\overline{\mathbb{R}}$ for every $n \in \mathbb{N}$ and the sequence of extended real numbers $\left(\int_{E_n} f \, d\mu : n \in \mathbb{N}\right)$ has $\lim_{n\to\infty} \int_{E_n} f \, d\mu \in \overline{\mathbb{R}}$. This does not imply the existence of $\int_D f \, d\mu$ in $\overline{\mathbb{R}}$, not even if $\lim_{n\to\infty} \int_{E_n} f \, d\mu \in \mathbb{R}$. See Example below.

(c) Now $\left(\int_{E_n} |f| \, d\mu \ : \ n \in \mathbb{N}\right)$ is an increasing sequence of nonnegative extended real numbers so that $\lim_{n\to\infty} \int_{E_n} |f| \, d\mu$ exists in $[0, \infty]$. In Theorem 9.28 below, we show that if $\lim_{n\to\infty} \int_{E_n} |f| \, d\mu < \infty$, then f is μ-integrable on D.

Example. In $(\mathbb{R}, \mathfrak{M}_L, \mu_L)$, let $D = [0, \infty)$ and $E_n = \left[0, \sum_{k=1}^n \frac{1}{k}\right)$ for $n \in \mathbb{N}$. Then $(E_n : n \in \mathbb{N})$ is an increasing sequence in $\mathfrak{M}_L$ with $\lim_{n\to\infty} E_n = \bigcup_{n\in\mathbb{N}} E_n = D$. With $E_0 := \emptyset$, let $D_n = E_n \setminus E_{n-1}$ for $n \in \mathbb{N}$. Then $(D_n : n \in \mathbb{N})$ is a sequence of disjoint intervals with $\bigcup_{n\in\mathbb{N}} D_n = \bigcup_{n\in\mathbb{N}} E_n = D$ and $\mu_L(D_n) = \frac{1}{n}$ for $n \in \mathbb{N}$. Let f be a real-valued function on D defined by setting $f(x) = (-1)^n$ for $x \in D_n$ for $n \in \mathbb{N}$. Then

$$\int_{E_n} f \, d\mu_L = \int_{\bigcup_{k=1}^n D_k} f \, d\mu_L = \sum_{k=1}^n \int_{D_k} f \, d\mu_L = \sum_{k=1}^n (-1)^k \frac{1}{k},$$

and thus

$$\lim_{n\to\infty} \int_{E_n} f \, d\mu_L = \lim_{n\to\infty} \sum_{k=1}^n (-1)^k \frac{1}{k} = \sum_{n\in\mathbb{N}} (-1)^n \frac{1}{n} \in \mathbb{R}.$$

On the other hand we have

$$\int_D f^+ \, d\mu_L = \sum_{n\in\mathbb{N}, n \text{ even}} \mu_L(D_n) = \frac{1}{2} + \frac{1}{4} + \frac{1}{6} + \cdots = \infty,$$

and

$$\int_D f^- \, d\mu_L = \sum_{n\in\mathbb{N}, n \text{ odd}} \mu_L(D_n) = 1 + \frac{1}{3} + \frac{1}{5} + \cdots = \infty.$$

Therefore $\int_D f \, d\mu = \int_D f^+ \, d\mu - \int_D f^- \, d\mu$ does not exist.

Theorem 9.28. *Given a measure space $(X, \mathfrak{A}, \mu)$. Let f be an extended real-valued $\mathfrak{A}$-measurable function on a set $D \in \mathfrak{A}$. Suppose there exists an increasing sequence $(E_n : n \in \mathbb{N})$ in $\mathfrak{A}$ with $\lim_{n\to\infty} E_n = \bigcup_{n\in\mathbb{N}} E_n = D$ such that $\lim_{n\to\infty} \int_{E_n} |f| \, d\mu < \infty$. Then f is μ-integrable on D and moreover $\int_D f \, d\mu = \lim_{n\to\infty} \int_{E_n} f \, d\mu$.*

Proof. Let $(E_n : n \in \mathbb{N})$ be an increasing sequence in $\mathfrak{A}$ such that $\lim_{n\to\infty} E_n = \bigcup_{n\in\mathbb{N}} E_n = D$. Let us define a nonnegative extended real-valued $\mathfrak{A}$-measurable function $|f|_{E_n}$ on D by setting

$$|f|_{E_n}(x) = \begin{cases} |f|(x) & \text{for } x \in E_n, \\ 0 & \text{for } x \in D \setminus E_n. \end{cases}$$

Then $(|f|_{E_n} : n \in \mathbb{N})$ is an increasing sequence and $\lim_{n\to\infty} |f|_{E_n} = |f|$ on D. Thus by Theorem 8.5 (Monotone Convergence Theorem), we have

$$(1) \qquad \int_D |f| \, d\mu = \lim_{n\to\infty} \int_D |f|_{E_n} \, d\mu.$$

Now we have

$$(2) \qquad \int_D |f|_{E_n} \, d\mu = \int_{E_n} |f|_{E_n} \, d\mu + \int_{D\setminus E_n} |f|_{E_n} \, d\mu = \int_{E_n} |f| \, d\mu + \int_{D\setminus E_n} 0 \, d\mu.$$

Thus by (1) and (2), we have $\int_D |f|\, d\mu = \lim_{n\to\infty} \int_{E_n} |f|\, d\mu < \infty$. This proves that $|f|$ is μ-integrable on D. Then f is μ-integrable on D by (e) of Observation 9.2. Then we have $\int_D f\, d\mu = \lim_{n\to\infty} \int_{E_n} f\, d\mu$ by (c) of Lemma 9.10. ∎

Remark 9.29. Theorem 9.28 offers the possibility of calculating the Lebesgue integral of a real-valued function on a finite or infinite interval in $\mathbb{R}$ as an improper Riemann integral. Let f be a bounded real-valued function defined on a finite closed interval $[a, b] \subset \mathbb{R}$. If f is Riemann integrable on $[a, b]$ then f is μ_L-integrable on $[a, b]$ and moreover we have the equality $\int_a^b f(x)\, dx = \int_{[a,b]} f\, d\mu_L$. (See Theorem 7.27.) Regarding the Riemann integrability of a function, it is known that a bounded real-valued function f on $[a, b]$ is Riemann integrable on $[a, b]$ if and only if the subset E of $[a, b]$ consisting of all points of discontinuity of f has $\mu_L(E) = 0$. (See Theorem 7.28.)

If a real-valued function f defined on $[a, \beta)$ where $-\infty < a < \beta \leq \infty$ is such that f is Riemann integrable on $[a, c]$ for every $c \in [a, \beta)$ and if $\lim_{c\to\beta} \int_a^c f(x)\, dx$ exists, then we call this limit the improper Riemann integral of f on $[a, \beta)$ and write $\int_a^\beta f(x)\, dx$ for this limit, that is, we define $\int_a^\beta f(x)\, dx = \lim_{c\to\beta} \int_a^c f(x)\, dx$.

Similarly for f defined on $(\alpha, b]$ where $-\infty \leq \alpha < b < \infty$, $c \in (\alpha, b]$ we define $\int_\alpha^b f(x)\, dx = \lim_{c\to\alpha} \int_c^b f(x)\, dx$.

Let f be a real-valued function on an interval $[a, \beta)$ where $-\infty < a < \beta \leq \infty$ and suppose f is continuous a.e. on $[a, \beta)$ and bounded on $[a, c]$ for every $c \in [a, \beta)$. Then f is Riemann integrable on $[a, c]$ and $\int_a^c f(x)\, dx = \int_{[a,c]} f\, d\mu_L$ for every $c \in [a, \beta)$.

Suppose the improper Riemann integral $\int_a^\beta f(x)\, dx = \lim_{c\to\beta} \int_a^c f(x)\, dx$ exists. Then for every increasing sequence $(c_n : n \in \mathbb{N})$ in $[a, \beta)$ such that $\lim_{n\to\infty} c_n = \beta$, we have the equality $\int_a^\beta f(x)\, dx = \lim_{n\to\infty} \int_a^{c_n} f(x)\, dx$. If $\lim_{n\to\infty} \int_a^{c_n} |f|(x)\, dx < \infty$, that is, $\lim_{n\to\infty} \int_{[a,c_n]} |f|\, d\mu_L < \infty$, then according to Theorem 9.28, f is μ_L-integrable on $[a, \beta)$ and

$$\int_{[a,\beta)} f\, d\mu_L = \lim_{n\to\infty} \int_{[a,c_n]} f\, d\mu_L = \lim_{n\to\infty} \int_a^{c_n} f(x)\, dx = \int_a^\beta f(x)\, dx.$$

Theorem 9.30. *If f is a μ_L-integrable extended real-valued $\mathfrak{M}_L$-measurable function on $\mathbb{R}$, then for every $\varepsilon > 0$ there exists a continuous real-valued function g on $\mathbb{R}$ such that $\int_{\mathbb{R}} |f - g|\, d\mu_L < \varepsilon$. Moreover g can be so chosen that it vanishes outside of a finite closed interval.*

Proof. Let $\varepsilon > 0$ be arbitrarily given. By (2) of Theorem 9.25, there exists $M > 0$ such that

(1) $$\int_{\mathbb{R}} \left|f - (f^{[M]})_{[-M,M]}\right| d\mu_L < \frac{\varepsilon}{4}.$$

Now $(f^{[M]})_{[-M,M]}$ is a real-valued $\mathfrak{M}_L$-measurable function with $|(f^{[M]})_{[-M,M]}| \le M$ on $\mathbb{R}$. Thus by Theorem 6.36, there exist a $\mathfrak{M}_L$-measurable subset E of $\mathbb{R}$ with $\mu_L(E) < \frac{\varepsilon}{16M}$ and a continuous real-valued function g_0 on $\mathbb{R}$ such that

$$(2)\qquad \begin{cases} \left|(f^{[M]})_{[-M,M]} - g_0\right| < \dfrac{\varepsilon}{16M} & \text{on } \mathbb{R} \setminus E, \\ |g_0| \le M \quad \text{on } \mathbb{R}. \end{cases}$$

Then

$$(3)\qquad \int_{[-M,M]} \left|(f^{[M]})_{[-M,M]} - g_0\right| d\mu_L \le \int_{[-M,M]\setminus E} \left|(f^{[M]})_{[-M,M]} - g_0\right| d\mu_L$$
$$+ \int_{[-M,M]\cap E} \left|(f^{[M]})_{[-M,M]}\right| d\mu_L + \int_{[-M,M]\cap E} |g_0|\, d\mu_L$$
$$\le \frac{\varepsilon}{16M} 2M + M\frac{\varepsilon}{16M} + M\frac{\varepsilon}{16M} = \frac{\varepsilon}{4}.$$

Let us define a continuous real-valued function g on $\mathbb{R}$ which is equal to g_0 on $[-M, M]$ and vanishes outside of a finite closed interval by setting

$$(4)\qquad g(x) = \begin{cases} g_0(x) & \text{for } x \in [-M, M], \\ 0 & \text{for } x \in \left[-M - \dfrac{\varepsilon}{4M}, M + \dfrac{\varepsilon}{4M}\right]^c, \\ \text{linear} & \text{on } \left[-M - \dfrac{\varepsilon}{4M}, -M\right] \text{ and on } \left[M, M + \dfrac{\varepsilon}{4M}\right]. \end{cases}$$

Then

$$(5)\qquad \int_{[-M,M]^c} |g|\, d\mu_L = \int_{[-M-\frac{\varepsilon}{4M}, -M]} |g|\, d\mu_L + \int_{[M, M+\frac{\varepsilon}{4M}]} |g|\, d\mu_L \le 2 \cdot \frac{M}{2}\frac{\varepsilon}{4M} = \frac{\varepsilon}{4}.$$

Therefore we have

$$\int_{\mathbb{R}} |f - g|\, d\mu_L \le \int_{\mathbb{R}} \left|f - (f^{[M]})_{[-M,M]}\right| d\mu_L + \int_{\mathbb{R}} \left|(f^{[M]})_{[-M,M]} - g\right| d\mu_L$$
$$\le \frac{\varepsilon}{4} + \int_{[-M,M]} \left|(f^{[M]})_{[-M,M]} - g_0\right| d\mu_L + \int_{[-M,M]^c} \left|(f^{[M]})_{[-M,M]} - g\right| d\mu_L$$
$$\le \frac{\varepsilon}{4} + \frac{\varepsilon}{4} + \int_{[-M,M]^c} \left|(f^{[M]})_{[-M,M]}\right| d\mu_L + \int_{[-M,M]^c} |g|\, d\mu_L$$
$$< \frac{\varepsilon}{2} + 0 + \frac{\varepsilon}{4} < \varepsilon,$$

by (1), (3), (4), and (5). ∎

[V] Translation and Linear Transformation of the Lebesgue Integral on $\mathbb{R}$

Consider a measure space $(X, \mathfrak{A}, \mu)$. If the set X is endowed with algebraic structure then it is pertinent to consider the effect of an algebraic operation in X on an integral on X. In particular consider the Lebesgue measure space $(\mathbb{R}, \mathfrak{M}_L, \mu_L)$ on $\mathbb{R}$ which is a linear space over the scalar field of the real number system by addition and multiplication in the real number system. For $E \subset \mathbb{R}$ and $h \in \mathbb{R}$ the translate of E by h is defined by

$$E + h = \{x + h : x \in E\} = \{y \in \mathbb{R} : y = x + h \text{ for some } x \in E\}.$$

According to Theorem 3.16 (Translation Invariance of $(\mathbb{R}, \mathfrak{M}_L, \mu_L)$), the measure space $(\mathbb{R}, \mathfrak{M}_L, \mu_L)$ is translation invariant, that is, for every $E \in \mathfrak{M}_L$ and $h \in \mathbb{R}$ we have $E + h \in \mathfrak{M}_L$ and $\mu_L(E + h) = \mu_L(E)$. Let us consider the effect of a translation on an integral on $(\mathbb{R}, \mathfrak{M}_L, \mu_L)$.

Theorem 9.31. (Translation Invariance of the Lebesgue Integral on $\mathbb{R}$) *Let f be an extended real-valued $\mathfrak{M}_L$-measurable function on a set $D \in \mathfrak{M}_L$. Let $h \in \mathbb{R}$ and let g be a function on $D - h$ defined by $g(x) = f(x + h)$ for $x \in D - h$. Then g is $\mathfrak{M}_L$-measurable on $D - h$ and moreover*

$$(1) \qquad \int_D f(x)\,\mu_L(dx) = \int_{D-h} f(x + h)\,\mu_L(dx),$$

in the sense that if one of the two integrals exists then so does the other and the two are equal. If f is defined and $\mathfrak{M}_L$-measurable on $\mathbb{R}$, then

$$(2) \qquad \int_{\mathbb{R}} f(x)\,\mu_L(dx) = \int_{\mathbb{R}} f(x + h)\,\mu_L(dx),$$

in the same sense as above. In particular if f is nonnegative extended real-valued $\mathfrak{M}_L$-measurable, then (1) and (2) always hold.

Proof. Let T be a translation on $\mathbb{R}$ by an element $h \in \mathbb{R}$, that is, T is a mapping of $\mathbb{R}$ into $\mathbb{R}$ defined by $T(x) = x + h$ for $x \in \mathbb{R}$. T maps $\mathbb{R}$ one-to-one onto $\mathbb{R}$ and its inverse mapping is given by $T^{-1}(x) = x - h$ for $x \in \mathbb{R}$, a translation by $-h \in \mathbb{R}$. Thus for every $E \in \mathfrak{M}_L$, we have $T^{-1}(E) = E - h$. Now $E - h \in \mathfrak{M}_L$ by Theorem 3.16. Thus T is a $\mathfrak{M}_L/\mathfrak{M}_L$-measurable mapping of $\mathbb{R}$ into $\mathbb{R}$.

Let f be an extended real-valued $\mathfrak{M}_L$-measurable function on a set $D \in \mathfrak{M}_L$. To show that $g = f \circ T$ is a $\mathfrak{M}_L$-measurable function on $D - h \in \mathfrak{M}_L$, it suffices according to Lemma 4.4 to show that $g^{-1}([-\infty, c]) \in \mathfrak{M}_L$ for every $c \in \mathbb{R}$. Now

$$g^{-1}([-\infty, c]) = (f \circ T)^{-1}([-\infty, c]) = T^{-1}(f^{-1}([-\infty, c])).$$

Since f is $\mathfrak{M}_L$-measurable, $f^{-1}([-\infty, c]) \in \mathfrak{M}_L$. Then since T is $\mathfrak{M}_L/\mathfrak{M}_L$-measurable, $T^{-1}(f^{-1}([-\infty, c])) \in \mathfrak{M}_L$. This shows the $\mathfrak{M}_L$-measurability of $f \circ T$.

To prove (1), recall that for every $E \in \mathfrak{M}_L$ and $h \in \mathbb{R}$, we have $\mu_L(E - h) = \mu_L(E)$

by Theorem 3.16. Consider first the case $f = \mathbf{1}_E$ where $E \in \mathfrak{M}_L$ and $E \subset D$. Then

$$\int_D f(x)\,\mu_L(dx) = \int_D \mathbf{1}_E(x)\,\mu_L(dx)$$
$$= \mu_L(E \cap D) = \mu_L\big((E \cap D) - h\big) = \mu_L\big((E-h) \cap (D-h)\big)$$
$$= \int_{D-h} \mathbf{1}_{E-h}(x)\,\mu_L(dx) = \int_{D-h} \mathbf{1}_E(x+h)\,\mu_L(dx)$$
$$= \int_{D-h} f(x+h)\,\mu_L(dx),$$

where the third equality is by Theorem 3.16 (Translation Invariance of $(\mathbb{R}, \mathfrak{M}_L, \mu_L)$) and the fifth equality is from the fact that $\mathbf{1}_{E-h}(\cdot) = \mathbf{1}_E(\cdot + h)$. This shows that (1) holds for the case $f = \mathbf{1}_E$. Then by the linearity of the integrals with respect to their integrands, (1) holds when f is a nonnegative simple function on D. Indeed if f is a nonnegative simple function on D, say $f = \sum_{i=1}^k c_i \mathbf{1}_{E_i}$ where $E_i \in \mathfrak{M}_L$, $E_i \subset D$, and $c_i \geq 0$ for $i = 1, \dots, k$, then by Theorem 9.14 and Lemma 9.12 we have

$$\int_D f(x)\,\mu_L(dx) = \int_D \sum_{i=1}^k c_i \mathbf{1}_{E_i}(x)\,\mu_L(dx)$$
$$= \sum_{i=1}^k c_i \int_D \mathbf{1}_{E_i}(x)\,\mu_L(dx) = \sum_{i=1}^k c_i \int_{D-h} \mathbf{1}_{E_i}(x+h)\,\mu_L(dx)$$
$$= \int_{D-h} \sum_{i=1}^k c_i \mathbf{1}_{E_i}(x+h)\,\mu_L(dx) = \int_{D-h} f(x+h)\,\mu_L(dx).$$

If f is a nonnegative extended real-valued $\mathfrak{M}_L$-measurable function on D, then there exists an increasing sequence $(\varphi_n : n \in \mathbb{N})$ of nonnegative simple functions on D such that $\varphi_n \uparrow f$ on D as $n \to \infty$ by Lemma 8.6. Applying our result above to the nonnegative simple function φ_n we have $\int_D \varphi_n(x)\,\mu_L(dx) = \int_{D-h} \varphi_n(x+h)\,\mu_L(dx)$ for every $n \in \mathbb{N}$. Letting $n \to \infty$ and applying the Monotone Convergence Theorem (Theorem 8.5), we have (1) for this case. If f is an extended real-valued $\mathfrak{M}_L$-measurable function on D, then by decomposing $f = f^+ - f^-$ and by applying the result above for nonnegative functions to f^+ and f^- we have

$$\begin{cases} \int_D f^+(x)\,\mu_L(dx) = \int_{D-h} f^+(x+h)\,\mu_L(dx), \\ \int_D f^-(x)\,\mu_L(dx) = \int_{D-h} f^-(x+h)\,\mu_L(dx). \end{cases}$$

If not both $\int_D f^+(x)\,\mu_L(dx)$ and $\int_D f^-(x)\,\mu_L(dx)$ are equal to ∞, then subtracting the second equality from the first equality we have (1). Since $\mathbb{R} - h = \mathbb{R}$, (2) is a particular case of (1). ∎

For $E \subset \mathbb{R}$ and $\alpha \in \mathbb{R}$ the dilation of E by the factor α is defined by setting

$$\alpha E = \{\alpha x : x \in E\} = \{y \in \mathbb{R} : y = \alpha x \text{ for some } x \in E\}.$$

According to Theorem 3.18, if $E \in \mathfrak{M}_L$ then $\alpha E \in \mathfrak{M}_L$ and $\mu_L(\alpha E) = |\alpha|\mu_L(E)$. Let us consider the effect of a linear mapping of $\mathbb{R}$ into $\mathbb{R}$ on an integral on $(\mathbb{R}, \mathfrak{M}_L, \mu_L)$ when the linear mapping is defined by $T(x) = \alpha x$ for $x \in \mathbb{R}$ with a fixed real number α.

Theorem 9.32. (Linear Transformation of the Lebesgue Integral in $\mathbb{R}$) *Let f be an extended real-valued $\mathfrak{M}_L$-measurable function on a set $D \in \mathfrak{M}_L$. For $\alpha \in \mathbb{R}$, $\alpha \neq 0$, let g be a function on $\frac{1}{\alpha}D$ defined by $g(x) = f(\alpha x)$ for $x \in \frac{1}{\alpha}D$. Then g is $\mathfrak{M}_L$-measurable on $\frac{1}{\alpha}D$ and moreover*

(1) $$\int_D f(x)\,\mu_L(dx) = |\alpha| \int_{\frac{1}{\alpha}D} f(\alpha x)\,\mu_L(dx),$$

in the sense that the existence of one of the two integrals implies that of the other and the equality of the two. If f is defined and $\mathfrak{M}_L$-measurable on $\mathbb{R}$, then

(2) $$\int_{\mathbb{R}} f(x)\,\mu_L(dx) = |\alpha| \int_{\mathbb{R}} f(\alpha x)\,\mu_L(dx),$$

and in particular

(3) $$\int_{\mathbb{R}} f(x)\,\mu_L(dx) = \int_{\mathbb{R}} f(-x)\,\mu_L(dx),$$

in the same sense as above. In particular if f is nonnegative extended real-valued and $\mathfrak{M}_L$-measurable, then (1), (2) and (3) always hold.

Proof. With $\alpha \in \mathbb{R}$, $\alpha \neq 0$, a linear transformation T of $\mathbb{R}$ into $\mathbb{R}$ defined by $T(x) = \alpha x$ for $x \in \mathbb{R}$ maps $\mathbb{R}$ one-to-one onto $\mathbb{R}$ with inverse mapping given by $T^{-1}(x) = \frac{1}{\alpha}x$ for $x \in \mathbb{R}$. Then for every $E \in \mathfrak{M}_L$, we have $T^{-1}(E) = \frac{1}{\alpha}E \in \mathfrak{M}_L$ by Theorem 3.18. This shows that T is a $\mathfrak{M}_L/\mathfrak{M}_L$-measurable mapping of $\mathbb{R}$ into $\mathbb{R}$.

The fact that if f is an extended real-valued $\mathfrak{M}_L$-measurable function on a set $D \in \mathfrak{M}_L$ then $g = f \circ T$ is a $\mathfrak{M}_L$-measurable function on the set $\frac{1}{\alpha}D \in \mathfrak{M}_L$ follows from the $\mathfrak{M}_L$-measurability of f and the $\mathfrak{M}_L/\mathfrak{M}_L$-measurability of T by the same argument as in the Proof of Theorem 9.31.

To prove (1), let us recall that if $E \in \mathfrak{M}_L$ and $\alpha \in \mathbb{R}$, $\alpha \neq 0$, then $\mu_L(\frac{1}{\alpha}E) = |\alpha|\mu_L(E)$ by Theorem 3.18. Now if $f = \mathbf{1}_E$ where $E \in \mathfrak{M}_L$ and $E \subset D$ then

$$\begin{aligned}\int_D f(x)\,\mu_L(dx) &= \int_D \mathbf{1}_E(x)\,\mu_L(dx) = \mu_L(E \cap D) = |\alpha|\mu_L\big(\tfrac{1}{\alpha}[E \cap D]\big)\\ &= |\alpha| \int_{\frac{1}{\alpha}D} \mathbf{1}_{\frac{1}{\alpha}E}(x)\,\mu_L(dx) = |\alpha| \int_{\frac{1}{\alpha}D} \mathbf{1}_E(\alpha x)\,\mu_L(dx)\\ &= |\alpha| \int_{\frac{1}{\alpha}D} f(\alpha x)\,\mu_L(dx).\end{aligned}$$

Starting with this particular case for f we reach the general case by the same argument as in the Proof of Theorem 9.31. ∎

Theorem 9.33. (Affine Transformation of the Lebesgue Integral in $\mathbb{R}$) *Let D be a $\mathfrak{M}_L$-measurable subset of $\mathbb{R}$. Then for $\alpha \in \mathbb{R}$, $\alpha \neq 0$, and $\beta \in \mathbb{R}$, the set $\frac{1}{\alpha}D - \frac{\beta}{\alpha}$ is a $\mathfrak{M}_L$-measurable subset of $\mathbb{R}$. Let f be an extended real-valued $\mathfrak{M}_L$-measurable function defined on D. Let g be a function on $\frac{1}{\alpha}D - \frac{\beta}{\alpha}$ defined by $g(x) = f(\alpha x + \beta)$ for $x \in \frac{1}{\alpha}D - \frac{\beta}{\alpha}$. Then g is $\mathfrak{M}_L$-measurable on $\frac{1}{\alpha}D - \frac{\beta}{\alpha}$ and moreover*

$$(1) \qquad \int_D f(x)\,\mu_L(dx) = |\alpha| \int_{\frac{1}{\alpha}D - \frac{\beta}{\alpha}} f(\alpha x + \beta)\,\mu_L(dx),$$

in the sense that the existence of one of the two integrals implies that of the other and the equality of the two. If f is defined and $\mathfrak{M}_L$-measurable on $\mathbb{R}$, then

$$(2) \qquad \int_{\mathbb{R}} f(x)\,\mu_L(dx) = |\alpha| \int_{\mathbb{R}} f(\alpha x + \beta)\,\mu_L(dx),$$

in the same sense as above. In particular if f is nonnegative extended real-valued and $\mathfrak{M}_L$-measurable, then (1) and (2) always hold.

Proof. With $\alpha \in \mathbb{R}$ such that $\alpha \neq 0$, a mapping $T_1(x) = \alpha x$ for $x \in \mathbb{R}$ is a one-to-one mapping of $\mathbb{R}$ onto $\mathbb{R}$ and T_1 is $\mathfrak{M}_L/\mathfrak{M}_L$-measurable by Theorem 3.18. With $\beta \in \mathbb{R}$, a mapping $T_2(x) = x + \beta$ for $x \in \mathbb{R}$ is a one-to-one mapping of $\mathbb{R}$ onto $\mathbb{R}$ and T_2 is $\mathfrak{M}_L/\mathfrak{M}_L$-measurable by Theorem 3.16. Let $T = T_2 \circ T_1$, that is, $T(x) = (T_2 \circ T_1)(x) = \alpha x + \beta$ for $x \in \mathbb{R}$, is a one-to-one mapping of $\mathbb{R}$ onto $\mathbb{R}$ and T is $\mathfrak{M}_L/\mathfrak{M}_L$-measurable by Theorem 1.40 (Chain Rule).

Let $D \in \mathfrak{M}_L$. Since T is a $\mathfrak{M}_L/\mathfrak{M}_L$-measurable mapping, we have $T^{-1}(D) \in \mathfrak{M}_L$. But $T^{-1}(D) = (T_2 \circ T_1)^{-1}(D) = T_1^{-1}\big(T_2^{-1}(D)\big) = T_1^{-1}(D - \beta) = \frac{1}{\alpha}D - \frac{\beta}{\alpha}$. This shows that $\frac{1}{\alpha}D - \frac{\beta}{\alpha}$ is a $\mathfrak{M}_L$-measurable subset of $\mathbb{R}$.

Let f be an extended real-valued $\mathfrak{M}_L$-measurable function on D. Let g be an extended real-valued function on $\frac{1}{\alpha}D - \frac{\beta}{\alpha}$ defined by setting $g = f \circ T$ on $\frac{1}{\alpha}D - \frac{\beta}{\alpha}$. We have $g(x) = (f \circ T)(x) = f(\alpha x + \beta)$ for $x \in \frac{1}{\alpha}D - \frac{\beta}{\alpha}$. Let us show the $\mathfrak{M}_L$-measurability of g on $\frac{1}{\alpha}D - \frac{\beta}{\alpha}$. For every $c \in \mathbb{R}$, we have

$$g^{-1}\big([-\infty, c]\big) = (f \circ T)^{-1}\big([-\infty, c]\big) = T^{-1}\big(f^{-1}\big([-\infty, c]\big)\big).$$

Now since f is $\mathfrak{M}_L$-measurable, we have $f^{-1}\big([-\infty, c]\big) \in \mathfrak{M}_L$. Then since T is $\mathfrak{M}_L/\mathfrak{M}_L$-measurable, we have $T^{-1}\big(f^{-1}\big([-\infty, c]\big)\big) \in \mathfrak{M}_L$. Thus g is a $\mathfrak{M}_L$-measurable function.

To prove (1), note that

$$\begin{aligned}
\int_D f(x)\,\mu_L(dx) &= |\alpha| \int_{\frac{1}{\alpha}D} f(\alpha x)\,\mu_L(dx) \\
&= |\alpha| \int_{\frac{1}{\alpha}D - \frac{\beta}{\alpha}} f\big(\alpha\big(x + \tfrac{\beta}{\alpha}\big)\big)\,\mu_L(dx) \\
&= |\alpha| \int_{\frac{1}{\alpha}D - \frac{\beta}{\alpha}} f(\alpha x + \beta)\,\mu_L(dx),
\end{aligned}$$

where the first equality is by (1) of Theorem 9.32 and the second equality is by (1) of Theorem 9.31. This proves (1). Then since $\frac{1}{\alpha}\mathbb{R} - \frac{\beta}{\alpha} = \mathbb{R}$, we have (2). ∎

[VI] Integration by Image Measure

Theorem 9.34. *Given a measure space $(X, \mathfrak{A}, \mu)$ and a measurable space $(Y, \mathfrak{B})$. Let Φ be a $\mathfrak{A}/\mathfrak{B}$-measurable mapping of X into Y and let ν be the image measure of μ on $\mathfrak{B}$ by the mapping Φ, that is, $\nu = \mu \circ \Phi^{-1}$. If f is an extended real-valued $\mathfrak{B}$-measurable function on a set $D \in \mathfrak{B}$, then $f \circ \Phi$ is $\mathfrak{A}$-measurable on the set $\Phi^{-1}(D) \in \mathfrak{A}$ and*

$$(1) \qquad \int_{\Phi^{-1}(D)} (f \circ \Phi)(x)\, \mu(dx) = \int_D f(y)\, \nu(dy),$$

in the sense that the existence of one of the two integrals implies that of the other and the equality of the two. In particular, if f is nonnegative then (1) holds.

Proof. Let f be an extended real-valued $\mathfrak{B}$-measurable function on a set $D \in \mathfrak{B}$. The $\mathfrak{A}/\mathfrak{B}$-measurability of Φ implies that $\Phi^{-1}(D) \in \mathfrak{A}$. To show the $\mathfrak{A}$-measurability of $f \circ \Phi$ on $\Phi^{-1}(D)$, it suffices according to Lemma 4.4 to show that for every $\alpha \in \mathbb{R}$, $(f \circ \Phi)^{-1}([-\infty, \alpha]) \in \mathfrak{A}$. Now $(f \circ \Phi)^{-1}([-\infty, \alpha]) = \Phi^{-1}(f^{-1}([-\infty, \alpha]))$. The $\mathfrak{B}$-measurability of f implies that $f^{-1}([-\infty, \alpha]) \in \mathfrak{B}$ and then the $\mathfrak{A}/\mathfrak{B}$-measurability of Φ implies that $\Phi^{-1}(f^{-1}([-\infty, \alpha])) \in \mathfrak{A}$.

To prove (1), consider first the case $D = Y$ so that $\Phi^{-1}(D) = X$. Thus we are to prove

$$(2) \qquad \int_X (f \circ \Phi)(x)\, \mu(dx) = \int_Y f(y)\, \nu(dy).$$

To prove (2), let us start with the case $f = \mathbf{1}_E$ with $E \in \mathfrak{B}$. In this case we have on the one hand $\int_X (f \circ \Phi)(x)\, \mu(dx) = \int_X \mathbf{1}_E(\Phi(x))\, \mu(dx) = \int_X \mathbf{1}_{\Phi^{-1}(E)}(x)\, \mu(dx) = \mu(\Phi^{-1}(E)) = \nu(E)$ and on the other hand $\int_Y f(y)\, \nu(dy) = \int_Y \mathbf{1}_E(y)\, \nu(dy) = \nu(E)$ so that (2) holds. If f is a nonnegative simple function on Y, then (2) holds by our result above and by the linearity of the integrals with respect to their integrands. If f is a nonnegative extended real-valued $\mathfrak{B}$-measurable function on Y, then by Lemma 8.6 there exists an increasing sequence of nonnegative simple functions $(\varphi_n : n \in \mathbb{N})$ such that $\varphi_n \uparrow f$ on Y as $n \to \infty$. By our result above, we have $\int_X (\varphi_n \circ \Phi)(x)\, \mu(dx) = \int_Y \varphi_n(y)\, \nu(dy)$ for every $n \in \mathbb{N}$. Letting $n \to \infty$, we have (2) holding for our f by Theorem 8.5 (Monotone Convergence Theorem). Finally let f be an arbitrary extended real-valued $\mathfrak{B}$-measurable function on Y. Decomposing $f = f^+ - f^-$ and applying our result above to each of the two nonnegative extended real-valued $\mathfrak{B}$-measurable functions f^+ and f^-, we have

$$(3) \qquad \int_X (f^+ \circ \Phi)(x)\, \mu(dx) = \int_Y f^+(y)\, \nu(dy),$$

$$(4) \qquad \int_X (f^- \circ \Phi)(x)\, \mu(dx) = \int_Y f^-(y)\, \nu(dy).$$

Now suppose $\int_Y f\, d\nu$ exists, that is, $\int_Y f^+\, d\nu - \int_Y f^-\, d\nu$ exists in $\overline{\mathbb{R}}$. Then by (3) and (4), $\int_X f^+ \circ \Phi\, d\mu - \int_X f^- \circ \Phi\, d\mu$ exists in $\overline{\mathbb{R}}$ so that $\int_X f \circ \Phi\, d\mu$ exists and moreover we have $\int_X (f \circ \Phi)(x)\, \mu(dx) = \int_Y f(y)\, \nu(dy)$. Conversely if $\int_X (f \circ \Phi)(x)\, \mu(dx)$ exists then by (3) and (4), $\int_Y f(y)\, \nu(dy)$ exists and $\int_X (f \circ \Phi)(x)\, \mu(dx) = \int_Y f(y)\, \nu(dy)$. This proves (2).

Now let f be an extended real-valued $\mathfrak{B}$-measurable function on a set $D \in \mathfrak{B}$. Let the domain of definition of f be extended to Y by setting $f = 0$ on D^c. Then f is an extended real-valued $\mathfrak{B}$-measurable function on Y and so is the product $\mathbf{1}_D f$. Thus by (2), we have

$$(5) \qquad \int_X (\mathbf{1}_D f \circ \Phi)(x)\, \mu(dx) = \int_Y (\mathbf{1}_D f)(y)\, \nu(dy).$$

On the right side of (5) we have $\int_Y \mathbf{1}_D f\, d\nu = \int_D f\, d\nu$ and on the left side of (5) we have

$$\int_X (\mathbf{1}_D f \circ \Phi)\, d\mu = \int_X (\mathbf{1}_D \circ \Phi)(f \circ \Phi)\, d\mu$$
$$= \int_X \mathbf{1}_{\Phi^{-1}(D)}(f \circ \Phi)\, d\mu = \int_{\Phi^{-1}(D)} (f \circ \Phi)\, d\mu.$$

This proves (1). ∎

Problems

Prob. 9.1. Let $(X, \mathfrak{A}, \mu)$ be a measure space and let f and g be extended real-valued $\mathfrak{A}$-measurable and μ-integrable functions on X. Show that if $\int_E f\, d\mu = \int_E g\, d\mu$ for every $E \in \mathfrak{A}$ then $f = g$ a.e. on X.

Prob. 9.2. Let $(X, \mathfrak{A}, \mu)$ be a σ-finite measure space and let f and g be extended real-valued $\mathfrak{A}$-measurable functions on X. Show that if $\int_E f\, d\mu = \int_E g\, d\mu$ for every $E \in \mathfrak{A}$ then $f = g$ a.e. on X.
(Note that with the σ-finiteness of $(X, \mathfrak{A}, \mu)$, the μ-integrability of f and g is no longer assumed.)

Prob. 9.3. In Prob. 9.2, instead of the σ-finiteness of the measure space $(X, \mathfrak{A}, \mu)$ let us assume the weaker condition that for every $E \in \mathfrak{A}$ with $\mu(E) > 0$ there exists $E_0 \in \mathfrak{A}$ such that $E_0 \subset E$ and $\mu(E_0) \in (0, \infty)$.
(a) Show that if $\int_E f\, d\mu = \int_E g\, d\mu$ for every $E \in \mathfrak{A}$ then $f = g$ a.e. on X.
(b) Find a measure space $(X, \mathfrak{A}, \mu)$ for which the conclusion of (a) is not valid.

Prob. 9.4. Let $(X, \mathfrak{A}, \mu)$ be a measure space. Show that if there exists an extended real-valued valued $\mathfrak{A}$-measurable function f on X such that $f > 0$ μ-a.e. on X and f is μ-integrable on X, then $(X, \mathfrak{A}, \mu)$ is a σ-finite measure space. (In other words, if $(X, \mathfrak{A}, \mu)$ is not σ-finite, then there does not exist an extended real-valued $\mathfrak{A}$-measurable function f on X such that $f > 0$ μ-a.e. on X and f is μ-integrable on X.)

Prob. 9.5. Given a measure space $(X, \mathfrak{A}, \mu)$. Let f be an extended real-valued $\mathfrak{A}$-measurable and μ-integrable function on X. Let $(E_n : n \in \mathbb{N})$ be a sequence in $\mathfrak{A}$ such that $\lim_{n\to\infty} \mu(E_n) = 0$. Show that $\lim_{n\to\infty} \int_{E_n} f\, d\mu = 0$.

Prob. 9.6. Given a measure space $(X, \mathfrak{A}, \mu)$. Let f be an extended real-valued $\mathfrak{A}$-measurable and μ-integrable function on X. Let $E_n = \{x \in X : |f(x)| \geq n\}$ for $n \in \mathbb{N}$. Show that $\lim_{n\to\infty} \mu(E_n) = 0$.

Prob. 9.7. Given a measure space $(X, \mathfrak{A}, \mu)$ and an extended real-valued $\mathfrak{A}$-measurable function f on X. Suppose that for some increasing sequence $(A_n : n \in \mathbb{N})$ in $\mathfrak{A}$ with $\bigcup_{n \in \mathbb{N}} A_n = X$, we have $\lim_{n\to\infty} \int_{A_n} |f| \, d\mu < \infty$.
(a) Show that f is μ-integrable on X.
(b) Show that for every increasing sequence $(B_n : n \in \mathbb{N})$ in $\mathfrak{A}$ with $\bigcup_{n\in\mathbb{N}} B_n = X$, we have $\lim_{n\to\infty} \int_{B_n} f \, d\mu = \int_X f \, d\mu$.

Prob. 9.8. Given a measure space $(X, \mathfrak{A}, \mu)$ and an extended real-valued $\mathfrak{A}$-measurable function f on X. The existence of an increasing sequence $(A_n : n \in \mathbb{N})$ in $\mathfrak{A}$ with $\bigcup_{n\in\mathbb{N}} A_n = X$ such that $\lim_{n\to\infty} \int_{A_n} f \, d\mu$ exists does not imply the existence of $\int_X f \, d\mu$ even when the limit is finite. Show this by constructing an example.

Prob. 9.9. Construct a real-valued continuous and μ_L-integrable function f on $[0, \infty)$ for which $\lim_{x\to\infty} f(x)$ does not exist.

Prob. 9.10. Let f be an extended real-valued $\mathfrak{M}_L$-measurable and μ_L integrable function on the interval $[0, \infty)$. Show that if f is uniformly continuous on $[a, \infty)$ for some $a \geq 0$ then $\lim_{x\to\infty} f(x) = 0$.

Prob. 9.11. Construct extended real-valued $\mathfrak{M}_L$-measurable and μ_L-integrable functions $(f_n : n \in \mathbb{N})$ and f on a set $D \in \mathfrak{M}_L$ such that

$$\lim_{n\to\infty} f_n = f \quad \text{a.e. on } D,$$

$$\lim_{n\to\infty} \int_D f_n \, d\mu_L = \int_D f \, d\mu_L,$$

but for some $E \subset D$, $E \in \mu_L$, we have

$$\lim_{n\to\infty} \int_E f_n \, d\mu_L \neq \int_E f \, d\mu_L.$$

Prob. 9.12. Let $(X, \mathfrak{A}, \mu)$ be a measure space and let $(f_n : n \in \mathbb{N})$, f and g be extended real-valued $\mathfrak{A}$-measurable and μ-integrable functions on a set $D \in \mathfrak{A}$. Suppose that
1° $\lim_{n\to\infty} f_n = f$ a.e. on D,
2° $\lim_{n\to\infty} \int_D f_n \, d\mu = \int_D \, d\mu$,
3° either $f_n \geq g$ on D for all $n \in \mathbb{N}$ or $f_n \leq g$ on D for all $n \in \mathbb{N}$.
Show that for every $E \subset D$ such that $E \in \mathfrak{A}$ we have $\lim_{n\to\infty} \int_E f_n \, d\mu = \int_E \, d\mu$.
(Note that as a particular case of condition 3°, we have the condition that either $f_n \geq 0$ on D for all $n \in \mathbb{N}$ or $f_n \leq 0$ on D for all $n \in \mathbb{N}$.)

Prob. 9.13. Given a measure space $(X, \mathfrak{A}, \mu)$. Let $(f_n : n \in \mathbb{N})$ be an increasing sequence of extended real-valued $\mathfrak{A}$-measurable functions on a set $D \in \mathfrak{A}$ and let $f = \lim_{n\to\infty} f_n$. Suppose
1° $\int_D |f_1| \, d\mu < \infty$,
2° $\sup_{n\in\mathbb{N}} \int_D f_n \, d\mu < \infty$.
Show that $\lim_{n\to\infty} \int_D f_n \, d\mu = \int_D f \, d\mu$ and $\int_D |f| \, d\mu < \infty$.

Prob. 9.14. Given a measure space $(X, \mathfrak{A}, \mu)$. Show that for an arbitrary extended real-valued $\mathfrak{A}$-measurable function on f on a set $D \in \mathfrak{A}$, there exists a sequence of simple functions $(\varphi_n : n \in \mathbb{N})$ on D such that

1° $|\varphi_n| \le |f|$ on D for every $n \in \mathbb{N}$ and $\lim_{n\to\infty} \varphi_n = f$ on D,

2° $\lim_{n\to\infty} \int_D \varphi_n \, d\mu = \int_D f \, d\mu$, provided that f is μ semi-integrable on D.

Prob. 9.15. Given a measure space $(X, \mathfrak{A}, \mu)$. Let f_n, $n \in \mathbb{N}$, and f be extended real-valued $\mathfrak{A}$-measurable functions on a set $D \in \mathfrak{A}$ such that $\lim_{n\to\infty} f_n = f$ a.e. on D. Suppose there exists a nonnegative extended real-valued $\mathfrak{A}$-measurable function g on D such that

1° $|f_n| \le g$ on D for every $n \in \mathbb{N}$,

2° $\int_D g^p \, d\mu < \infty$ for fixed $p \in [1, \infty)$.

Show that $\lim_{n\to\infty} \int_D |f_n - f|^p \, d\mu = 0$.

Prob. 9.16. Assume the hypothesis in Theorem 9.20 (Lebesgue Dominated Convergence Theorem). Then $g + f_n \ge 0$ and $g - f_n \ge 0$ on D for every $n \in \mathbb{N}$ so that $(g + f_n : n \in \mathbb{N})$ and $(g - f_n : n \in \mathbb{N})$ are two sequences of nonnegative extended real-valued $\mathfrak{A}$-measurable functions on D and $\lim_{n\to\infty} \{g + f_n\} = g + f$ and $\lim_{n\to\infty} \{g - f_n\} = g - f$ a.e. on D. Apply Fatou's Lemma for nonnegative functions (Theorem 8.13) to construct an alternate proof to Theorem 9.20.

Prologue to Prob. 9.17. Prob. 9.17 is a very useful extension of Theorem 9.20 (Lebesgue Dominated Convergence Theorem). It can be proved by the same kind of argument as in Prob. 9.16. For examples of application of this extension we have Prob. 9.18, Prob. 9.19, Prob. 9.20, and Prob. 9.21 below.

Prob. 9.17. (An Extension of the Dominated Convergence Theorem) Prove the following:
Given a measure space $(X, \mathfrak{A}, \mu)$. Let $(f_n : n \in \mathbb{N})$, $(g_n : n \in \mathbb{N})$, f, and g be extended real-valued $\mathfrak{A}$-measurable functions on a set $D \in \mathfrak{A}$. Suppose

1° $\lim_{n\to\infty} f_n = f$ and $\lim_{n\to\infty} g_n = g$ a.e. on D,

2° $(g_n : n \in \mathbb{N})$ and g are all μ-integrable on D and $\lim_{n\to\infty} \int_D g_n \, d\mu = \int_D g \, d\mu$,

3° $|f_n| \le g_n$ on D for every $n \in \mathbb{N}$.

Then f is μ-integrable on D and $\lim_{n\to\infty} \int_D f_n \, d\mu = \int_D f \, d\mu$.
(Hint: Apply Fatou's Lemma (Theorem 8.13) to the two sequences of nonnegative functions $(g_n + f_n : n \in \mathbb{N})$ and $(g_n - f_n : n \in \mathbb{N})$.)

Prob. 9.18. Given a measure space $(X, \mathfrak{A}, \mu)$. Let $(f_n : n \in \mathbb{N})$ and f be extended real-valued $\mathfrak{A}$-measurable functions on a set $D \in \mathfrak{A}$. Suppose

1° $\lim_{n\to\infty} f_n = f$ a.e. on D,

2° $f_n, n \in \mathbb{N}$, and f are all μ-integrable on D.

(a) Show that if $\lim_{n\to\infty} \int_D |f_n| \, d\mu = \int_D |f| \, d\mu$, then $\lim_{n\to\infty} \int_D f_n \, d\mu = \int_D f \, d\mu$.

(b) Show that the converse of (a) is false by constructing a counter example.

Prob. 9.19. Given a measure space $(X, \mathfrak{A}, \mu)$. Let $(f_n : n \in \mathbb{N})$ and f be extended real-valued $\mathfrak{A}$-measurable functions on a set $D \in \mathfrak{A}$. Suppose

1° $\lim_{n\to\infty} f_n = f$ a.e. on D,

2° $(f_n : n \in \mathbb{N})$ and f are all μ-integrable on D.

3° $\lim_{n\to\infty} \int_D |f_n|\,d\mu = \int_D |f|\,d\mu$.

(a) Show that for every $\mathfrak{A}$-measurable subset E of D, we have $\lim_{n\to\infty} \int_E |f_n|\,d\mu = \int_E |f|\,d\mu$ and $\lim_{n\to\infty} \int_E f_n\,d\mu = \int_E f\,d\mu$.

(b) Show by constructing a counter example that if condition 3° is replaced by the condition that $\lim_{n\to\infty} \int_D f_n\,d\mu = \int_D f\,d\mu$, then (a) does not hold.

Prob. 9.20. Let us replace condition 3° in Prob. 9.19 with

4° $\lim_{n\to\infty} \int_D f_n\,d\mu = \int_D f\,d\mu$.

5° there exists a μ-integrable extended real-valued $\mathfrak{A}$-measurable function g on D such that $f_n \geq g$ (or $f_n \leq g$) on D for $n \in \mathbb{N}$.

Show that for every $\mathfrak{A}$-measurable subset E of D, we have $\lim_{n\to\infty} \int_E f_n\,d\mu = \int_E f\,d\mu$.

Prob. 9.21. Given a measure space $(X, \mathfrak{A}, \mu)$. Let $(f_n : n \in \mathbb{N})$ and f be extended real-valued $\mathfrak{A}$-measurable functions on a set $D \in \mathfrak{A}$. Assume that $(f_n : n \in \mathbb{N})$ and f are all μ-integrable on D.

(a) Show that if $\lim_{n\to\infty} \int_D |f_n - f|\,d\mu = 0$, then

1° $(f_n : n \in \mathbb{N})$ converges to f on D in measure,

2° $\lim_{n\to\infty} \int_D |f_n|\,d\mu = \int_D |f|\,d\mu$.

(b) Show that if

3° $\lim_{n\to\infty} f_n = f$ a.e. on D,

4° $\lim_{n\to\infty} \int_D |f_n|\,d\mu = \int_D |f|\,d\mu$,

then $\lim_{n\to\infty} \int_D |f_n - f|\,d\mu = 0$.

(c) Show by constructing a counter example that if f is only μ semi-integrable on D, then (b) does not hold.

Prob. 9.22. (Dominated Convergence and Convergence in Measure) Given a measure space $(X, \mathfrak{A}, \mu)$. Let$(f_n : n \in \mathbb{N})$ and f be extended real-valued $\mathfrak{A}$-measurable functions on a set $D \in \mathfrak{A}$ and assume that f is real-valued a.e. on D. According to Theorem 6.22 (Lebesgue), if $(f_n : n \in \mathbb{N})$ converges to f a.e. on D and if $\mu(D) < \infty$, then $(f_n : n \in \mathbb{N})$ converges to f on D in measure.

Let us remove the condition $\mu(D) < \infty$ and instead assume that $|f_n| \leq g$ on D for every $n \in \mathbb{N}$ with a μ-integrable extended real-valued $\mathfrak{A}$-measurable function g on D. Show that $(f_n : n \in \mathbb{N})$ converges to f on D in measure.

Prob. 9.23. Given a measure space $(X, \mathfrak{A}, \mu)$.

(a) Show that there exists a μ-integrable extended real-valued $\mathfrak{A}$-measurable function f on X such that $\mu\{x \in X : f(x) = 0\} = 0$ if and only if $(X, \mathfrak{A}, \mu)$ is a σ-finite measure space.

(b) If f is a μ-integrable extended real-valued $\mathfrak{A}$-measurable function on X, then the set $\{x \in X : f(x) \neq 0\}$ is a σ-finite set, that is, the union of countably many members of $\mathfrak{A}$, each with a finite measure.

Prob. 9.24. Given a measure space $(X, \mathfrak{A}, \mu)$. Let $(f_n : n \in \mathbb{N})$ be a sequence of extended real-valued $\mathfrak{A}$-measurable functions on a set $D \in \mathfrak{A}$. Show that if
1° $\sum_{n\in\mathbb{N}} \int_D |f_n| \, d\mu < \infty$,
then
2° the series $\sum_{n\in\mathbb{N}} f_n$ converges absolutely a.e. on D,
3° $\int_D \{\sum_{n\in\mathbb{N}} f_n\} \, d\mu = \sum_{n\in\mathbb{N}} \int_D f_n \, d\mu$.

Prob. 9.25. With $0 < a < b$, let $(f_n : n \in \mathbb{N})$ be a sequence of functions defined by setting

$$f_n(x) = ae^{-nax} - be^{-nbx} \text{ for } x \in [0, \infty).$$

(a) Show that $\int_{[0,\infty)} f_n \, d\mu_L = 0$ for every $n \in \mathbb{N}$.
(b) Compute $\int_{[0,\infty)} |f_n| \, d\mu_L$ for every $n \in \mathbb{N}$.
(c) Show that $\sum_{n\in\mathbb{N}} \int_{[0,\infty)} |f_n| \, d\mu_L = \infty$.
(d) Compute $\sum_{n\in\mathbb{N}} f_n$.
(e) Show that $\int_{[0,\infty)} \{\sum_{n\in\mathbb{N}} f_n\} \, d\mu_L$ does not exist.

Prob. 9.26. If f and g are two μ-integrable extended real-valued $\mathfrak{A}$-measurable functions on set $D \in \mathfrak{A}$ in a measure space $(X, \mathfrak{A}, \mu)$, then the product fg may not be μ-integrable, and in fact it may not even be μ semi-integrable, on D. To show this, construct the following examples.
(a) Construct a μ-integrable function f such that f^2 is μ semi-integrable but not μ-integrable.
(b) Construct two μ-integrable functions f and g such that fg is not μ semi-integrable.

Prob. 9.27. Let f and g be two μ-integrable extended real-valued $\mathfrak{A}$-measurable functions on set $D \in \mathfrak{A}$ in a measure space $(X, \mathfrak{A}, \mu)$. Show that if f is bounded on D, then fg is μ-integrable on D.

Prob. 9.28. (H. Lebesgue) Given a measure space $(X, \mathfrak{A}, \mu)$. Suppose f is an extended real-valued $\mathfrak{A}$-measurable function on a set $D \in \mathfrak{A}$ such that $\int_D fg \, d\mu$ exists in $\overline{\mathbb{R}}$ for every μ-integrable extended real-valued $\mathfrak{A}$-measurable function g on D. Show that f is bounded a.e. on D, that is, there exists a constant $M > 0$ such that $\mu\{x \in D : |f(x)| \geq M\} = 0$.

Prob. 9.29. (First Mean Value Theorem of the Lebesgue Integral) Given a measure space $(X, \mathfrak{A}, \mu)$. Let f and g be two extended real-valued $\mathfrak{A}$-measurable functions on a set $D \in \mathfrak{A}$. Suppose f is bounded on D and g is μ-integrable on D.
(a) Show that fg and $f|g|$ are μ-integrable on D.
(b) Show if $m = \inf_{x\in D} f(x)$ and $M = \sup_{x\in D} f(x)$ and $\int_D |g| \, d\mu > 0$, then there exists $c \in [m, M]$ such that $\int_D f|g| \, d\mu = c \int_D |g| \, d\mu$.

Prob. 9.30. Let f be an extended real-valued $\mathfrak{M}_L$-measurable function on $[0, \infty)$ such that
1° f is μ_L-integrable on every finite subinterval of $[0, \infty)$,
2° $\lim_{x\to\infty} f(x) = c \in \mathbb{R}$.
Show that $\lim_{a\to\infty} \frac{1}{a} \int_{[0,a]} f \, d\mu_L = c$.

Prob. 9.31. Let f be a μ_L-integrable nonnegative extended real-valued $\mathfrak{M}_L$-measurable function on $\mathbb{R}$. Show that if the function $g(x) = \sum_{n\in\mathbb{N}} f(x+n)$ for $x \in \mathbb{R}$ is μ_L-integrable on $\mathbb{R}$ then $f = 0$, μ_L-a.e. on $\mathbb{R}$.

Prob. 9.32. Let f be an extended real-valued $\mathfrak{M}_L$-measurable function on $\mathbb{R}$. For $x \in \mathbb{R}$ and $r > 0$ let $B_r(x) = \{y \in \mathbb{R} : |y - x| < r\}$.
With $r > 0$ fixed, define a function g on $\mathbb{R}$ by setting

$$g(x) = \int_{B_r(x)} f(y)\, \mu_L(dy) \text{ for } x \in \mathbb{R}.$$

(a) Suppose f is locally μ_L-integrable on $\mathbb{R}$, that is, f is μ_L-integrable on every bounded $\mathfrak{M}_L$-measurable set in $\mathbb{R}$. Show that g is a real-valued continuous function on $\mathbb{R}$.
(b) Show that if f is μ_L-integrable on $\mathbb{R}$ then g is uniformly continuous on $\mathbb{R}$.

Prob. 9.33. Let f be a μ_L-integrable extended real-valued $\mathfrak{M}_L$-measurable function on $\mathbb{R}$ in the measure space $(\mathbb{R}, \mathfrak{M}_L, \mu_L)$. Show that

$$\lim_{h\to 0} \int_{\mathbb{R}} |f(x+h) - f(x)| \mu_L(dx) = 0.$$

(Note that we may not have $\lim_{h\to 0} f(x+h) = f(x)$ as f may not be continuous at any $x \in \mathbb{R}$.)

Prob. 9.34. Let f be a μ_L-integrable extended real-valued $\mathfrak{M}_L$-measurable function on $\mathbb{R}$ in the measure space $(\mathbb{R}, \mathfrak{M}_L, \mu_L)$. Show that

$$\lim_{h\to\infty} \int_{\mathbb{R}} |f(x+h) - f(x)| \mu_L(dx) = 2 \int_{\mathbb{R}} |f(x)| \mu_L(dx)$$

and similarly for $h \to -\infty$.

Prologue to Prob. 9.35. Given a measure space $(X, \mathfrak{A}, \mu)$. Let f be a complex valued function on a set $D \in \mathfrak{A}$. Let $\Re f$ and $\Im f$ be the real and imaginary parts of f respectively. We say that f is $\mathfrak{A}$-measurable, μ semi-integrable, or μ-integrable on D if both $\Re f$ and $\Im f$ are $\mathfrak{A}$-measurable, μ semi-integrable, or μ-integrable on D. If f is μ semi-integrable on D, that is, both $\Re f$ and $\Im f$ are μ semi-integrable on D, then we define $\int_D f\, d\mu = \int_D \Re f\, d\mu + i \int_D \Im f\, d\mu$.

Prob. 9.35. Let f be a μ_L-integrable extended real-valued $\mathfrak{M}_L$-measurable function on $\mathbb{R}$.
(a) Show that for every $y \in \mathbb{R}$, the complex valued function $e^{ixy} f(x)$ for $x \in \mathbb{R}$ is $\mathfrak{M}_L$-measurable and μ_L-integrable on $\mathbb{R}$.
(b) The Fourier transform of f is a complex valued function $\hat{f}$ on $\mathbb{R}$ defined by setting

$$\hat{f}(y) = \int_{\mathbb{R}} e^{ixy} f(x) \mu_L(dx) \quad \text{for } y \in \mathbb{R}.$$

Show that $\hat{f}$ is bounded on $\mathbb{R}$ and in fact $\sup_{y\in\mathbb{R}} |\hat{f}(y)| \le \int_{\mathbb{R}} |f(x)| \mu_L(dx)$.
(c) Show that $\hat{f}$ defined in (b) is continuous on $\mathbb{R}$.

Prob. 9.36. (Riemann-Lebesgue Theorem) Let f and $\hat{f}$ be as in Prob. 9.35. Show that $\lim_{y\to\infty} \hat{f}(y) = 0$ and $\lim_{y\to-\infty} \hat{f}(y) = 0$.
(Hint: Show first that by the translation invariance of the Lebesgue integral in $(\mathbb{R}, \mathfrak{M}_L, \mu_L)$,

$$\hat{f}(y) = -\int_{\mathbb{R}} e^{i(x+\pi/y)y} f(x) \mu_L(dx) = -\int_{\mathbb{R}} e^{ixy} f\left(x - \tfrac{\pi}{y}\right) \mu_L(dx).$$

Then use this expression in $2\hat{f}$.)

Prob. 9.37. Let f be an extended real-valued function on $\mathbb{R}$. Let $\alpha \in \mathbb{R}$, $\alpha \neq 0$ and let g be an extended real-valued function on $\mathbb{R}$ defined by $g(x) = f(\alpha x)$ for $x \in \mathbb{R}$.
(a) Show that g is $\mathfrak{M}_L$-measurable on $\mathbb{R}$ if and only if f is.
(b) Assume that f is $\mathfrak{M}_L$-measurable on $\mathbb{R}$. Show that

$$\int_{\mathbb{R}} f(\alpha x)\mu_L(dx) = \frac{1}{|\alpha|}\int_{\mathbb{R}} f(x)\mu_L(dx)$$

in the sense that the existence of one side implies that of the other and the equality of the two.

Prob. 9.38. (a) Given a measure space $(X, \mathfrak{A}, \mu)$. Let f be a μ-integrable extended real-valued $\mathfrak{A}$-measurable function on a set $D \in \mathfrak{A}$. Let $(D_n : n \in \mathbb{N})$ be a sequence of $\mathfrak{A}$-measurable subsets of D such that $\lim_{n\to\infty} \mu(D_n) = 0$. (Note that $(D_n : n \in \mathbb{N})$ is not assumed to be a decreasing sequence.) Show that $\lim_{n\to\infty} \int_{D_n} f\,d\mu = 0$.
(b) Show by constructing a counter example that the conclusion in (a) does not hold if f is only μ semi-integrable on D.

Prob. 9.39. Find a real-valued continuous function f on $\mathbb{R}$ such that $\lim_{a\to\infty} \int_{-a}^{a} f(x)\,dx$ converges but $\int_{\mathbb{R}} f\,d\mu_L$ does not exist.

Prob. 9.40. Evaluate $\int_{[0,\infty)} xe^{-x^2}\,\mu_L(dx)$.

Prob. 9.41. Let $f(x) = e^{-x^2}$ for $x \in [0, \infty)$.
(a) Show that f is μ-integrable on $[0, \infty)$.
(b) Evaluate $\int_{[0,\infty)} e^{-nx^2} \sin nx\,\mu_L(dx)$.

Prob. 9.42. From the improper Riemann integral $\frac{1}{\sqrt{2\pi t}}\int_{-\infty}^{\infty} \exp\{-\frac{x^2}{2t}\}dx = 1$ for $t > 0$, we have $\frac{1}{\sqrt{2\pi t}}\int_{\mathbb{R}} \exp\{-\frac{x^2}{2t}\}\,\mu_L(dx) = 1$ for $t > 0$.
(a) Show that for $t > 0$ and $y \in \mathbb{R}$ we have

$$\frac{1}{\sqrt{2\pi t}}\int_{\mathbb{R}} \exp\Big\{-\frac{(x-y)^2}{2t}\Big\}\,\mu_L(dx) = 1.$$

(b) Let f be a bounded real valued $\mathfrak{M}_L$-measurable function on $\mathbb{R}$. Show that if f is continuous at some $y \in \mathbb{R}$, then

$$\lim_{t\downarrow 0} \frac{1}{\sqrt{2\pi t}}\int_{\mathbb{R}} \exp\Big\{-\frac{(x-y)^2}{2t}\Big\} f(x)\,\mu_L(dx) = f(y).$$

Prob. 9.43. Show that the function f for $x \in (0, \infty)$ defined by

$$f(x) = \frac{1}{1+x^2}\log(1-e^{-x})$$

is Lebesgue integrable on $(0, \infty)$.

Prob. 9.44. Let f be a real valued function on $\mathbb{R}$ defined by

$$f(x) = \begin{cases} \frac{1}{\sqrt{x}} & \text{for } x \in (0, 1) \\ 0 & \text{for } x \in \mathbb{R} - (0, 1). \end{cases}$$

Let $(r_n : n \in \mathbb{N})$ be an arbitrary enumeration of the rational numbers in $\mathbb{R}$ and set for $x \in \mathbb{R}$

$$g(x) = \sum_{n\in\mathbb{N}} 2^{-n} f(x - r_n)$$

(a) Show that g is Lebesgue integrable on $\mathbb{R}$ and thus $g < \infty$ a.e. on $\mathbb{R}$.
(b) Show that g is discontinuous at every $x \in \mathbb{R}$ and unbounded on every interval and that it remains so after any redefinition on a null set.
(c) Show that $g^2 < \infty$ a.e. on $\mathbb{R}$ but g^2 is not Lebesgue integrable on any interval.

Prob. 9.45. Let $D \in \mathfrak{M}_L$ and f be an extended real valued function on $D \times [a, b]$ such that $f(\cdot, t)$ is a $\mathfrak{M}_L$-measurable function on D for every $t \in [a, b]$. Suppose
1° $\lim_{t \to t_0} f(x, t) = f(x, t_0)$ for every $x \in D$ for some $t_0 \in [a, b]$.
2° there exists an extended positive valued Lebesgue integrable function g on D such that $|f(x, t)| \leq g(x)$ for $(x, t) \in D \times [a, b]$.
Show that $f(\cdot, t)$ is Lebesgue integrable on D for every $t \in [a, b]$ and moreover we have $\lim_{t \to t_0} \int_D f(x, t)\, \mu_L(dx) = \int_D f(x, t_0)\, \mu_L(dx)$.
(This is the Dominated Convergence Theorem for the continuous approach $t \to t_0$.)

Prob. 9.46. Let $D \in \mathfrak{M}_L$ and f be a real valued function on $D \times [a, b]$ such that $f(\cdot, t)$ is a $\mathfrak{M}_L$-measurable function on D for every $t \in [a, b]$ and $f(x, \cdot)$ is a continuous on $[a, b]$ for every $x \in D$. Suppose that there exists an extended positive valued Lebesgue integrable function g on D such that $|f(x, t)| \leq g(x)$ for $(x, t) \in D \times [a, b]$. Show that the function F on $[a, b]$ defined by $F(t) = \int_D f(x, t)\, \mu_L(dx)$ for $t \in [a, b]$ is real valued and continuous on $[a, b]$.

Prob. 9.47. Let $a_{n,k} \geq 0$ for $n, k \in \mathbb{N}$. For $\alpha, \beta \in \mathbb{R}$, write $\alpha \wedge \beta$ for $\min\{\alpha, \beta\}$.
(a) Show that

$$\lim_{n \to \infty} \sum_{k \in \mathbb{N}} \frac{1}{2^k} (a_{n,k} \wedge 1) = 0$$

if and only if $\lim_{n \to \infty} a_{n,k} = 0$ for every $k \in \mathbb{N}$.
(b) Show that

$$\lim_{n \to \infty} \sum_{k \in \mathbb{N}} \frac{1}{2^k} \frac{a_{n,k}}{1 + a_{n,k}} = 0$$

if and only if $\lim_{n \to \infty} a_{n,k} = 0$ for every $k \in \mathbb{N}$.

§10 Signed Measures

[I] Signed Measure Spaces

Definition 10.1. *Given a measurable space $(X, \mathfrak{A})$. A set function λ on $\mathfrak{A}$ is called a signed measure on $\mathfrak{A}$ if it satisfies the following conditions:*
1° $\lambda(E) \in (-\infty, \infty]$ *for every* $E \in \mathfrak{A}$ *or* $\lambda(E) \in [-\infty, \infty)$ *for every* $E \in \mathfrak{A}$,
2° $\lambda(\emptyset) = 0$,
3° *countable additivity: for every disjoint sequence* $(E_n : n \in \mathbb{N})$ *in* $\mathfrak{A}$, $\sum_{n\in\mathbb{N}} \lambda(E_n)$ *exists in* $\overline{\mathbb{R}}$ *and* $\sum_{n\in\mathbb{N}} \lambda(E_n) = \lambda\left(\bigcup_{n\in\mathbb{N}} E_n\right)$.
If λ is a signed measure on $\mathfrak{A}$, the triple $(X, \mathfrak{A}, \lambda)$ is called a signed measure space.

Thus a measure μ on a measurable space $(X, \mathfrak{A})$ is a signed measure satisfying the condition that $\mu(E) \in [0, \infty]$ for every $E \in \mathfrak{A}$. For emphasis, a measure and a measure space will often be called a positive measure and a positive measure space respectively.

Remark 10.2. If λ is a set function defined on $\mathfrak{A}$ satisfying conditions 1° and 2° of Definition 10.1 and if $(E_n : n \in \mathbb{N})$ is a disjoint sequence in $\mathfrak{A}$, the sum $\sum_{n\in\mathbb{N}} \lambda(E_n)$ may not exist in $\overline{\mathbb{R}}$. Condition 3° requires that the sum $\sum_{n\in\mathbb{N}} \lambda(E_n)$ exists in $\overline{\mathbb{R}}$ and is equal to $\lambda\left(\bigcup_{n\in\mathbb{N}} E_n\right)$. Now if $(E_1, \dots, E_N)$ is a finite sequence in $\mathfrak{A}$, then since at most one of the two infinite values ∞ and $-\infty$ is assumed by λ, the sum $\sum_{k=1}^{N} \lambda(E_k)$ always exist in $\overline{\mathbb{R}}$. If $(E_1, \dots, E_N)$ is a disjoint finite sequence, then $(E_1, \dots, E_N, \emptyset, \emptyset, \dots)$ is a disjoint infinite sequence and since $\lambda(\emptyset) = 0$, condition 3° in Definition 10.1 implies:
4° finite additivity: for every disjoint finite sequence $(E_1, \dots, E_N)$ in $\mathfrak{A}$, $\sum_{k=1}^{N} \lambda(E_k)$ exists in $\overline{\mathbb{R}}$ and $\sum_{k=1}^{N} \lambda(E_k) = \lambda\left(\bigcup_{k=1}^{N} E_k\right)$.

Let $(X, \mathfrak{A}, \mu)$ be a measure space and let f be a nonnegative extended real-valued $\mathfrak{A}$-measurable function on X. If we define a set function ν on $\mathfrak{A}$ by setting $\nu(E) = \int_E f\, d\mu$ for $E \in \mathfrak{A}$ then ν is a measure on $(X, \mathfrak{A})$ according to Proposition 8.12. We show next that if we remove the nonnegativity condition on f and instead assume that f is μ semi-integrable on X then ν as defined above is a signed measure on $(X, \mathfrak{A})$.

Proposition 10.3. *Let $(X, \mathfrak{A}, \mu)$ be a measure space. Let f be a μ semi-integrable extended real-valued $\mathfrak{A}$-measurable function on X. Let us define a set function λ on $\mathfrak{A}$ by setting $\lambda(E) = \int_E f\, d\mu$ for $E \in \mathfrak{A}$. Then λ is a signed measure on $\mathfrak{A}$.*

Proof. Note that the μ semi-integrability of f on X implies the μ semi-integrability of f on every $E \in \mathfrak{A}$ by Lemma 9.9 so that $\lambda(E) = \int_E f\, d\mu$ exists for every $E \in \mathfrak{A}$. Now since f is μ semi-integrable on X, at least one of $\int_X f^+\, d\mu$ and $\int_X f^-\, d\mu$ is finite. Suppose $\int_X f^-\, d\mu < \infty$. Then for every $E \in \mathfrak{A}$ we have $0 \le \int_E f^-\, d\mu \le \int_X f^-\, d\mu < \infty$. Thus $\lambda(E) = \int_E f^+\, d\mu - \int_E f^-\, d\mu \in (-\infty, \infty]$. Similarly we have $\lambda(E) \in [-\infty, \infty)$ for every $E \in \mathfrak{A}$ when $\int_X f^+\, d\mu < \infty$. Thus condition 1° of Definition 10.1 is satisfied. Since $\lambda(\emptyset) = \int_\emptyset f\, d\mu = 0$, condition 2° is satisfied. Let $(E_n : n \in \mathbb{N})$ be a disjoint sequence in $\mathfrak{A}$ and let $E = \bigcup_{n\in\mathbb{N}} E_n$. Then by (b) of Lemma 9.10, we have

$$\lambda(E) = \int_E f\,d\mu = \int_{\bigcup_{n\in\mathbb{N}} E_n} f\,d\mu = \sum_{n\in\mathbb{N}} \int_{E_n} f\,d\mu = \sum_{n\in\mathbb{N}} \lambda(E_n).$$

This shows that condition 3° is satisfied. Therefore λ is a signed measure on $\mathfrak{A}$. ∎

The converse of Proposition 10.3, that is, the possibility of representing a signed measure λ on a measurable space $(X, \mathfrak{A})$ as integral of an extended real-valued $\mathfrak{A}$-measurable function f on X with respect to a measure μ on $(X, \mathfrak{A})$ will be proved in Proposition 10.24.

Comparison of two measures μ and ν on a measurable space $(X, \mathfrak{A})$ leads to consideration of a set function $\mu - \nu$ on $\mathfrak{A}$. We show next that $\mu - \nu$ is a signed measure on $(X, \mathfrak{A})$ provided that at least one of the two measures μ and ν is a finite measure so that $\mu - \nu$ can be defined.

Proposition 10.4. *Let $(X, \mathfrak{A})$ be a measurable space and let μ and ν be two measures on $\mathfrak{A}$ at least one of which is finite. Then the set function λ on $\mathfrak{A}$ defined by $\lambda = \mu - \nu$ is a signed measure on $\mathfrak{A}$.*

Proof. Let us show that λ satisfies conditions 1°, 2°, and 3° of Definition 10.1. Suppose ν is a finite measure on $\mathfrak{A}$. Then for every $E \in \mathfrak{A}$ we have $\lambda(E) = \mu(E) - \nu(E) \in (-\infty, \infty]$. Secondly we have $\lambda(\emptyset) = \mu(\emptyset) - \nu(\emptyset) = 0$. Finally if $(E_n : n \in \mathbb{N})$ is a disjoint sequence in $\mathfrak{A}$ and if we let $E = \bigcup_{n\in\mathbb{N}} E_n$, then by the countable additivity of μ and ν we have $\lambda(E) = \mu(E) - \nu(E) = \sum_{n\in\mathbb{N}} \mu(E_n) - \sum_{n\in\mathbb{N}} \nu(E_n)$. Now $\sum_{n\in\mathbb{N}} \nu(E_n) = \nu(E) < \infty$ implies that

$$(1) \qquad \sum_{n\in\mathbb{N}} \mu(E_n) - \sum_{n\in\mathbb{N}} \nu(E_n) = \sum_{n\in\mathbb{N}} \{\mu(E_n) - \nu(E_n)\} = \sum_{n\in\mathbb{N}} \lambda(E_n).$$

Indeed if we have $\sum_{n\in\mathbb{N}} \mu(E_n) < \infty$ also then we have two convergent series of real numbers $\sum_{n\in\mathbb{N}} \mu(E_n)$ and $\sum_{n\in\mathbb{N}} \nu(E_n)$ and this fact implies (1). If on the other hand we have $\sum_{n\in\mathbb{N}} \mu(E_n) = \infty$, then

$$\begin{aligned}\sum_{n\in\mathbb{N}} \{\mu(E_n) - \nu(E_n)\} &= \lim_{n\to\infty} \sum_{k=1}^{n} \{\mu(E_n) - \nu(E_n)\} = \lim_{n\to\infty} \Big\{\sum_{k=1}^{n} \mu(E_n) - \sum_{k=1}^{n} \nu(E_n)\Big\} \\ &= \lim_{n\to\infty} \sum_{k=1}^{n} \mu(E_n) - \lim_{n\to\infty} \sum_{k=1}^{n} \nu(E_n) = \sum_{n\in\mathbb{N}} \mu(E_n) - \sum_{n\in\mathbb{N}} \nu(E_n)\end{aligned}$$

so that (1) holds. Then by (1), we have

$$\sum_{n\in\mathbb{N}} \lambda(E_n) = \sum_{n\in\mathbb{N}} \mu(E_n) - \sum_{n\in\mathbb{N}} \nu(E_n) = \mu(E) - \nu(E) = \lambda(E).$$

This proves the countable additivity of λ on $\mathfrak{A}$ and completes the proof that λ is a signed measure on $(X, \mathfrak{A})$ for the case that ν is a finite measure on $(X, \mathfrak{A})$. Similarly for the case that μ is a finite measure on $(X, \mathfrak{A})$. ∎

The converse of Proposition 10.4, that is, the possibility of representing a signed measure λ on a measurable space $(A, \mathfrak{A})$ as the difference of two measures μ and ν on $(A, \mathfrak{A})$ will be proved by the Jordan decomposition theorem (Theorem 10.21) below.

Unlike a positive measure space $(X, \mathfrak{A}, \mu)$, a signed measure space $(X, \mathfrak{A}, \lambda)$ does not have the monotonicity property, that is, $E, F \in \mathfrak{A}$ and $E \subset F$ do not imply $\lambda(E) \leq \lambda(F)$. For instance, if $E_1, E_2 \in \mathfrak{A}$, $E_1 \cap E_2 = \emptyset$, $\lambda(E_1) > 0$, and $\lambda(E_2) < 0$, and if we let $E = E_1 \cup E_2$, then $E_1 \subset E$ but $\lambda(E) = \lambda(E_1) + \lambda(E_2) < \lambda(E_1)$.

Example. Let us consider the measure space $\big([0, 2\pi], \mathfrak{M}_L \cap [0, 2\pi], \mu_L\big)$ where we define $\mathfrak{M}_L \cap [0, 2\pi] := \{A \cap [0, 2\pi] : A \in \mathfrak{M}_L\}$, a σ-algebra of subsets of $[0, 2\pi]$. If we define a set function λ on $\mathfrak{M}_L \cap [0, 2\pi]$ by setting $\lambda(E) = \int_E \sin x \, \mu_L(dx)$ for $E \in \mathfrak{M}_L \cap [0, 2\pi]$, then λ is a signed measure on $\big([0, 2\pi], \mathfrak{M}_L \cap [0, 2\pi]\big)$ by Proposition 10.3. Now

$$\lambda\big([0, \pi]\big) = \int_{[0,\pi]} \sin x \, \mu_L(dx) = \int_0^{\pi} \sin x \, dx = 2,$$

$$\lambda\big(\big[0, \tfrac{3}{2}\pi\big]\big) = \int_{\left[0,\frac{3}{2}\pi\right]} \sin x \, \mu_L(dx) = \int_0^{\frac{3}{2}\pi} \sin x \, dx = 1,$$

$$\lambda\big([0, 2\pi]\big) = \int_{[0,2\pi]} \sin x \, \mu_L(dx) = \int_0^{2\pi} \sin x \, dx = 0.$$

Thus we have $\lambda\big([0, \pi]\big) > \lambda\big(\big[0, \tfrac{3}{2}\pi\big]\big) > \lambda\big([0, 2\pi]\big)$, while $[0, \pi] \subset \big[0, \tfrac{3}{2}\pi\big] \subset [0, 2\pi]$.

In the absence of the monotonicity property in a signed measure space $(X, \mathfrak{A}, \lambda)$, we have the following partial monotonicity property.

Lemma 10.5. *Given a signed measure space* $(X, \mathfrak{A}, \lambda)$. *Let* $E_1, E_2 \in \mathfrak{A}$ *and* $E_1 \subset E_2$.
(a) *If* $\lambda(E_2) \in \mathbb{R}$, *then* $\lambda(E_1) \in \mathbb{R}$.
(b) *If* $\lambda(E_1) = \infty$, *then* $\lambda(E_2) = \infty$.
(c) *If* $\lambda(E_1) = -\infty$, *then* $\lambda(E_2) = -\infty$.

Proof. 1. Let $E_0 = E_2 \setminus E_1$. By the finite additivity of λ, we have $\lambda(E_0) + \lambda(E_1) = \lambda(E_2)$. If $\lambda(E_2) \in \mathbb{R}$, then $\lambda(E_0), \lambda(E_1) \in \mathbb{R}$.

2. If $\lambda(E_1) = \infty$, then $\lambda(E_2) \notin \mathbb{R}$ for otherwise we would have $\lambda(E_1) \in \mathbb{R}$ by (a). Thus $\lambda(E_2) \in \{-\infty, \infty\}$. Since $\lambda(E_1) = \infty$ and since λ cannot assume both the infinite values ∞ and $-\infty$, we have $\lambda(E_2) = \infty$. This proves (b). Similarly for (c). ■

Consider a sequence of objects $(A_n : n \in \mathbb{N})$. Let φ be a one-to-one mapping of $\mathbb{N}$ onto $\mathbb{N}$. Let us call the sequence $(B_n : n \in \mathbb{N})$ where $B_n = A_{\varphi^{-1}(n)}$ for $n \in \mathbb{N}$ a renumbering of the sequence $(A_n : n \in \mathbb{N})$ corresponding to φ.

Let $(X, \mathfrak{A}, \lambda)$ be a signed measure space. Let $(E_n : n \in \mathbb{N})$ be a disjoint sequence in $\mathfrak{A}$ and let $(F_n : n \in \mathbb{N})$ be an arbitrary renumbering of the sequence $(E_n : n \in \mathbb{N})$. We have

$\bigcup_{n\in\mathbb{N}} E_n = \bigcup_{n\in\mathbb{N}} F_n$. By the countable additivity of λ on $\mathfrak{A}$, we have

$$(1) \qquad \sum_{n\in\mathbb{N}} \lambda(E_n) = \lambda\Big(\bigcup_{n\in\mathbb{N}} E_n\Big) = \lambda\Big(\bigcup_{n\in\mathbb{N}} F_n\Big) = \sum_{n\in\mathbb{N}} \lambda(F_n).$$

Suppose $\lambda\big(\bigcup_{n\in\mathbb{N}} E_n\big) \in \mathbb{R}$. Then by Lemma 10.5, we have $\lambda(E_n) \in \mathbb{R}$ and $\lambda(F_n) \in \mathbb{R}$ for every $n \in \mathbb{N}$. Let $(u_n : n \in \mathbb{N}) = (\lambda(E_n) : n \in \mathbb{N})$ and $(v_n : n \in \mathbb{N}) = (\lambda(F_n) : n \in \mathbb{N})$. Then (1) tells us that the sequence of real numbers $(u_n : n \in \mathbb{N})$ is such that for an arbitrary renumbering $(v_n : n \in \mathbb{N})$ of $(u_n : n \in \mathbb{N})$ we have $\sum_{n\in\mathbb{N}} v_n = \sum_{n\in\mathbb{N}} u_n \in \mathbb{R}$. Not every sequence of real numbers has this property.

If $(a_n : n \in \mathbb{N})$ is a sequence of nonnegative real numbers and if $(b_n : n \in \mathbb{N})$ is an arbitrary renumbering of $(a_n : n \in \mathbb{N})$, then for an arbitrary $n_1 \in \mathbb{N}$ there exists $n_2 \in \mathbb{N}$ such that $\sum_{k=1}^{n_1} a_k \le \sum_{k=1}^{n_2} b_k$ and similarly for an arbitrary $n_1 \in \mathbb{N}$ there exists $n_2 \in \mathbb{N}$ such that $\sum_{k=1}^{n_1} b_k \le \sum_{k=1}^{n_2} a_k$. From this it follows that $\sum_{n\in\mathbb{N}} a_n = \sum_{n\in\mathbb{N}} b_n$. For an arbitrary sequence $(u_n : n \in \mathbb{N})$ of real numbers, we have the following :

Lemma 10.6. *Let $(u_n : n \in \mathbb{N})$ be a sequence of real numbers.*
(a) *For every renumbering $(v_n : n \in \mathbb{N})$ of $(u_n : n \in \mathbb{N})$ the series $\sum_{n\in\mathbb{N}} v_n$ converges if and only if the series $\sum_{n\in\mathbb{N}} u_n$ converges absolutely, that is, $\sum_{n\in\mathbb{N}} |u_n| < \infty$.*
(b) *If $\sum_{n\in\mathbb{N}} |u_n| < \infty$, then $\sum_{n\in\mathbb{N}} u_n = \sum_{n\in\mathbb{N}} v_n$ for every renumbering $(v_n : n \in \mathbb{N})$ of $(u_n : n \in \mathbb{N})$.*

Proof. 1. Suppose $\sum_{n\in\mathbb{N}} |u_n| < \infty$. Let $(v_n : n \in \mathbb{N})$ be an arbitrary renumbering of $(u_n : n \in \mathbb{N})$. Then $(|v_n| : n \in \mathbb{N})$ is a renumbering of the sequence of nonnegative real numbers $(|u_n| : n \in \mathbb{N})$ so that $\sum_{n\in\mathbb{N}} |u_n| < \infty$ implies $\sum_{n\in\mathbb{N}} |v_n| < \infty$. Since absolute convergence of a series implies convergence, the series $\sum_{n\in\mathbb{N}} v_n$ converges.

2. Suppose for every renumbering $(v_n : n \in \mathbb{N})$ of $(u_n : n \in \mathbb{N})$, the series $\sum_{n\in\mathbb{N}} v_n$ converges. To show $\sum_{n\in\mathbb{N}} |u_n| < \infty$, let us assume the contrary, that is, $\sum_{n\in\mathbb{N}} |u_n| = \infty$. Let $(a_i : i \in \mathbb{N})$ and $(b_j : j \in \mathbb{N})$ be respectively the subsequence consisting of all the nonnegative entries and the subsequence consisting of all the negative entries in $(u_n : n \in \mathbb{N})$ and let $A = \sum_{i\in\mathbb{N}} a_i$ and $B = \sum_{j\in\mathbb{N}} b_j$. Then $A + |B| = \sum_{n\in\mathbb{N}} |u_n| = \infty$ and thus at least one of A and $|B|$ is equal to ∞. Suppose $A = \infty$. In this case let us consider a renumbering $(v_n : n \in \mathbb{N})$ of $(u_n : n \in \mathbb{N})$ given by

$$a_1, \dots, a_{k_1}, b_1, a_{k_1+1}, \dots, a_{k_2}, b_2, a_{k_2+1}, \dots, a_{k_3}, b_3, \dots,$$

where $k_1, k_2, k_3, \dots$ are so chosen that the partial sum of $\sum_{n\in\mathbb{N}} v_n$ from a_1 through a_{k_n} is greater than n for every $n \in \mathbb{N}$. This is possible since $A = \infty$ and b_j is finite for every $j \in \mathbb{N}$. Thus the sequence of the partial sums of $\sum_{n\in\mathbb{N}} v_n$ is unbounded and hence $\sum_{n\in\mathbb{N}} v_n$ does not converge. This is a contradiction. Similarly if $|B| = \infty$, that is, $B = -\infty$, we consider a renumbering $(v_n : n \in \mathbb{N})$ of $(u_n : n \in \mathbb{N})$ given by

$$b_1, \dots, b_{k_1}, a_1, b_{k_1+1}, \dots, b_{k_2}, a_2, b_{k_2+1}, \dots, b_{k_3}, a_3, \dots,$$

where $k_1, k_2, k_3, \dots$ are so chosen that the partial sum of $\sum_{n\in\mathbb{N}} v_n$ from b_1 through b_{k_n} is less than $-n$ for every $n \in \mathbb{N}$. This is possible since $B = -\infty$ and a_i is finite for every $i \in \mathbb{N}$. Thus the sequence of the partial sums of $\sum_{n\in\mathbb{N}} v_n$ is unbounded and hence $\sum_{n\in\mathbb{N}} v_n$ does not converge and we have a contradiction. This shows that $\sum_{n\in\mathbb{N}} |u_n| < \infty$ and completes the proof of (a).

3. To prove (b), suppose $\sum_{n\in\mathbb{N}} |u_n| < \infty$. In this case $A + |B| = \sum_{n\in\mathbb{N}} |u_n| < \infty$ so that $A < \infty$ and $|B| < \infty$. Let us show that $\sum_{n\in\mathbb{N}} u_n = A + B$. For $n \in \mathbb{N}$, let $S_n = \sum_{k=1}^n u_k$, $A_n = \sum_{i=1}^n a_i$, and $B_n = \sum_{j=1}^n b_j$. We have $\lim_{n\to\infty} A_n = A \in [0, \infty)$ and $\lim_{n\to\infty} B_n = B \in (-\infty, 0]$. Thus for an arbitrary $\varepsilon > 0$, there exists $M \in \mathbb{N}$ such that $|A_n - A| < \frac{\varepsilon}{2}$ and $|B_n - B| < \frac{\varepsilon}{2}$ for $n \geq M$. Let $N \in \mathbb{N}$ be so large that $\{a_1, \ldots, a_M\} \cup \{b_1, \ldots, b_M\} \subset \{u_1, \ldots, u_N\}$. Then for $n \geq N$, we have $\{u_1, \ldots, u_n\} = \{a_1, \ldots, a_p\} \cup \{b_1, \ldots, b_q\}$ with some $p, q \geq M$. Thus for $n \geq N$ we have

$$\begin{aligned}|S_n - (A + B)| &= |(A_p + B_q) - (A + B)| \\ &\leq |A_p - A| + |B_q - B| \\ &< \frac{\varepsilon}{2} + \frac{\varepsilon}{2} = \varepsilon.\end{aligned}$$

This shows that $\lim_{n\to\infty} S_n = A + B$ and hence $\sum_{n\in\mathbb{N}} u_n = A + B$.

Let $(v_n : n \in \mathbb{N})$ be an arbitrary renumbering of $(u_n : n \in \mathbb{N})$. Let A' be the sum of the nonnegative entries and B' be the sum of the negative entries in $(v_n : n \in \mathbb{N})$. Now $\sum_{n\in\mathbb{N}} |u_n| < \infty$ implies that $\sum_{n\in\mathbb{N}} |v_n| < \infty$. Thus by our result above, we have $\sum_{n\in\mathbb{N}} v_n = A' + B'$. But $A' = A$ and $B' = B$. Thus $\sum_{n\in\mathbb{N}} v_n = A + B = \sum_{n\in\mathbb{N}} u_n$. This proves (b). ∎

Proposition 10.7. *Let $(X, \mathfrak{A}, \lambda)$ be a signed measure space. Let $(E_n : n \in \mathbb{N})$ be a disjoint sequence in $\mathfrak{A}$. Then*

$$\lambda\Big(\bigcup_{n\in\mathbb{N}} E_n\Big) \in \mathbb{R} \Leftrightarrow \sum_{n\in\mathbb{N}} |\lambda(E_n)| < \infty.$$

In other words, the series $\sum_{n\in\mathbb{N}} \lambda(E_n)$ converges if and only if it converges absolutely.

Proof. 1. By the countable additivity of λ, we have $\sum_{n\in\mathbb{N}} \lambda(E_n) = \lambda\big(\bigcup_{n\in\mathbb{N}} E_n\big) \in \overline{\mathbb{R}}$. Thus we have

$$\lambda\Big(\bigcup_{n\in\mathbb{N}} E_n\Big) \in \mathbb{R} \Leftrightarrow \sum_{n\in\mathbb{N}} \lambda(E_n) \in \mathbb{R}.$$

2. If $\sum_{n\in\mathbb{N}} |\lambda(E_n)| < \infty$, that is, the series $\sum_{n\in\mathbb{N}} \lambda(E_n)$ converges absolutely, then it converges so that $\sum_{n\in\mathbb{N}} \lambda(E_n) \in \mathbb{R}$. Thus we have $\lambda\left(\bigcup_{n\in\mathbb{N}} E_n\right) \in \mathbb{R}$.

3. Conversely suppose $\lambda\left(\bigcup_{n\in\mathbb{N}} E_n\right) \in \mathbb{R}$. Let $(F_n : n \in \mathbb{N})$ be an arbitrary renumbering of the sequence $(E_n : n \in \mathbb{N})$. Then $(F_n : n \in \mathbb{N})$ is a disjoint sequence in $\mathfrak{A}$ with $\bigcup_{n\in\mathbb{N}} F_n = \bigcup_{n\in\mathbb{N}} E_n$ so that $\sum_{n\in\mathbb{N}} \lambda(F_n) = \lambda\left(\bigcup_{n\in\mathbb{N}} F_n\right) = \lambda\left(\bigcup_{n\in\mathbb{N}} E_n\right) \in \mathbb{R}$ by the countable additivity of λ. This shows that for an arbitrary renumbering $(\lambda(F_n) : n \in \mathbb{N})$ of the sequence $(\lambda(E_n) : n \in \mathbb{N})$ the series $\sum_{n\in\mathbb{N}} \lambda(F_n)$ converges. Then by (a) of Lemma 10.6, we have $\sum_{n\in\mathbb{N}} |\lambda(E_n)| < \infty$. ∎

Example. Let $(X, \mathfrak{A}, \lambda)$ be a signed measure space and let $(E_n : n \in \mathbb{N})$ be a disjoint sequence in $\mathfrak{A}$. According to Proposition 10.7, the series $\sum_{n\in\mathbb{N}} \lambda(E_n)$ converges if and only if it converges absolutely. For a sequence of real numbers $(u_n : n \in \mathbb{N})$ the absolute convergence of the series $\sum_{n\in\mathbb{N}} u_n$ implies the convergence of the series but the converse does not hold. For instance the series $\sum_{n\in\mathbb{N}}(-1)^n\frac{1}{n}$ converges but it does not converge absolutely. This implies that there does not exist a signed measure space $(X, \mathfrak{A}, \lambda)$ such that for a disjoint sequence $(E_n : n \in \mathbb{N})$ in $\mathfrak{A}$ we have $\lambda(E_n) = (-1)^n\frac{1}{n}$ for $n \in \mathbb{N}$.

Let $(X, \mathfrak{A}, \mu)$ be a measure space and let $(E_n : n \in \mathbb{N})$ be a monotone sequence in $\mathfrak{A}$. According to Theorem 1.26, if $(E_n : n \in \mathbb{N})$ is an increasing sequence then we have $\lim_{n\to\infty} \mu(E_n) = \mu\big(\lim_{n\to\infty} E_n\big)$ and if $(E_n : n \in \mathbb{N})$ is a decreasing sequence and $\mu(E_1) < \infty$ then we have $\lim_{n\to\infty} \mu(E_n) = \mu\big(\lim_{n\to\infty} E_n\big)$. We show next that similar convergence theorem holds in a signed measure space $(X, \mathfrak{A}, \lambda)$.

Theorem 10.8. *Let $(X, \mathfrak{A}, \lambda)$ be a signed measure space.*
(a) *If $(E_n : n \in \mathbb{N})$ is an increasing sequence in $\mathfrak{A}$, then $\lim_{n\to\infty} \lambda(E_n) = \lambda\big(\lim_{n\to\infty} E_n\big)$.*
(b) *If $(E_n : n \in \mathbb{N})$ is a decreasing sequence in $\mathfrak{A}$, and if $\lambda(E_1) \in \mathbb{R}$, then we have $\lim_{n\to\infty} \lambda(E_n) = \lambda\big(\lim_{n\to\infty} E_n\big)$.*

Proof. 1. If $(E_n : n \in \mathbb{N})$ is an increasing sequence in $\mathfrak{A}$, then $\lim_{n\to\infty} E_n = \bigcup_{n\in\mathbb{N}} E_n$. Now if $\lambda(E_{n_0}) = \infty$ for some $n_0 \in \mathbb{N}$, then since $E_{n_0} \subset E_m \subset \bigcup_{n\in\mathbb{N}} E_n$ for every $m \geq n_0$, we have $\lambda(E_m) = \infty$ for every $m \geq n_0$ as well as $\lambda\left(\bigcup_{n\in\mathbb{N}} E_n\right) = \infty$ by (b) of Lemma 10.5. Thus $\lim_{n\to\infty} \lambda(E_n) = \infty = \lambda\big(\lim_{n\to\infty} E_n\big)$. Similarly if $\lambda(E_{n_0}) = -\infty$ for some $n_0 \in \mathbb{N}$, we have by means of (c) of Lemma 10.5 the equality $\lim_{n\to\infty} \lambda(E_n) = -\infty = \lambda\big(\lim_{n\to\infty} E_n\big)$.

Now consider the case $\lambda(E_n) \in \mathbb{R}$ for every $n \in \mathbb{N}$. Let $F_1 = E_1$ and $F_n = E_n \setminus E_{n-1}$ for $n \geq 2$. Then $(F_n : n \in \mathbb{N})$ is a disjoint sequence in $\mathfrak{A}$ with $\bigcup_{n\in\mathbb{N}} F_n = \bigcup_{n\in\mathbb{N}} E_n$. By the countable additivity of λ, we have $\sum_{n\in\mathbb{N}} \lambda(F_n) = \lambda\left(\bigcup_{n\in\mathbb{N}} F_n\right) = \lambda\left(\bigcup_{n\in\mathbb{N}} E_n\right)$. Since $E_{n-1} \cup F_n = E_n$ and $E_{n-1} \cap F_n = \emptyset$ for $n \geq 2$, we have by the finite additivity of λ, $\lambda(E_{n-1}) + \lambda(F_n) = \lambda(E_n)$ and thus $\lambda(F_n) = \lambda(E_n) - \lambda(E_{n-1})$. Then we have

$$
\begin{aligned}
\lambda\Big(\bigcup_{n\in\mathbb{N}} E_n\Big) &= \lambda\Big(\bigcup_{n\in\mathbb{N}} F_n\Big) = \lambda(F_1) + \sum_{n\geq 2} \lambda(F_n) = \lambda(E_1) + \sum_{n\geq 2} \{\lambda(E_n) - \lambda(E_{n-1})\} \\
&= \lambda(E_1) + \lim_{n\to\infty} \sum_{k\geq 2}^{n} \{\lambda(E_k) - \lambda(E_{k-1})\} = \lambda(E_1) + \lim_{n\to\infty} \{\lambda(E_n) - \lambda(E_1)\} \\
&= \lim_{n\to\infty} \lambda(E_n).
\end{aligned}
$$

2. Let $(E_n : n \in \mathbb{N})$ be a decreasing sequence in $\mathfrak{A}$ with $\lambda(E_1) \in \mathbb{R}$. In this case, we have $\lambda(E_n) \in \mathbb{R}$ for every $n \in \mathbb{R}$ by (a) of Lemma 10.5. Now $\lim_{n\to\infty} E_n = \bigcap_{n\in\mathbb{N}} E_n$ and

$$E_1 \setminus \bigcap_{n\in\mathbb{N}} E_n = E_1 \cap \Big(\bigcap_{n\in\mathbb{N}} E_n\Big)^c = E_1 \cap \Big(\bigcup_{n\in\mathbb{N}} E_n^c\Big)$$
$$= \bigcup_{n\in\mathbb{N}} (E_1 \cap E_n^c) = \bigcup_{n\in\mathbb{N}} (E_1 \setminus E_n) = \lim_{n\to\infty} (E_1 \setminus E_n),$$

since $(E_1 \setminus E_n : n \in \mathbb{N})$ is an increasing sequence. Thus we have

$$\lambda\Big(E_1 \setminus \bigcap_{n\in\mathbb{N}} E_n\Big) = \lambda\Big(\lim_{n\to\infty} (E_1 \setminus E_n)\Big) = \lim_{n\to\infty} \lambda(E_1 \setminus E_n)$$
$$= \lim_{n\to\infty} \{\lambda(E_1) - \lambda(E_n)\} = \lambda(E_1) - \lim_{n\to\infty} \lambda(E_n),$$

where the second equality is by (a) and the third equality is by the finite additivity of λ and the fact that $\lambda(E_n) \in \mathbb{R}$. On the other hand, by the finite additivity of λ we have $\lambda\left(E_1 \setminus \bigcap_{n\in\mathbb{N}} E_n\right) = \lambda(E_1) - \lambda\left(\bigcap_{n\in\mathbb{N}} E_n\right)$. Therefore we have

$$\lambda(E_1) - \lim_{n\to\infty} \lambda(E_n) = \lambda(E_1) - \lambda\Big(\bigcap_{n\in\mathbb{N}} E_n\Big).$$

Subtracting $\lambda(E_1) \in \mathbb{R}$, we have

$$\lim_{n\to\infty} \lambda(E_n) = \lambda\Big(\bigcap_{n\in\mathbb{N}} E_n\Big) = \lambda\Big(\lim_{n\to\infty} E_n\Big). \ \blacksquare$$

[II] Decomposition of Signed Measures

Definition 10.9. *Given a signed measure space $(X, \mathfrak{A}, \lambda)$.*
(a) *$E \in \mathfrak{A}$ is called a positive set for λ if $\lambda(E_0) \geq 0$ for every $E_0 \in \mathfrak{A}$ such that $E_0 \subset E$.*
(b) *$E \in \mathfrak{A}$ is called a negative set for λ if $\lambda(E_0) \leq 0$ for every $E_0 \in \mathfrak{A}$ such that $E_0 \subset E$.*
(c) *$E \in \mathfrak{A}$ is called a null set for λ if $\lambda(E_0) = 0$ for every $E_0 \in \mathfrak{A}$ such that $E_0 \subset E$.*
(Note that the definition of a null set in a signed measure space $(X, \mathfrak{A}, \lambda)$ is consistent with Definition 1.32 for a null set in a positive measure space $(X, \mathfrak{A}, \mu)$.)

Observation 10.10. As immediate consequences of Definition 10.9, we have the following:
(a) Every $\mathfrak{A}$-measurable subset of a positive set for λ is itself a positive set for λ. Similarly every $\mathfrak{A}$-measurable subset of a negative set for λ is itself a negative set for λ, and every $\mathfrak{A}$-measurable subset of a null set for λ is itself a null set for λ.
(b) A set is a null set for λ if and only if it is both a positive set and a negative set for λ.
(c) If E_1 is a positive set for λ and E_2 is a negative set for λ, then $E_1 \cap E_2$ is a null set for λ. (This is a consequence of (a) and (b).)

Example. Consider the example of a signed measure space $(X, \mathfrak{A}, \lambda)$ in Proposition 10.3 defined by $\lambda(E) = \int_E f\, d\mu$ for $E \in \mathfrak{A}$ where f is a μ semi-integrable extended real-valued $\mathfrak{A}$-measurable function f on X and μ is a positive measure on $(X, \mathfrak{A})$. If $E \in \mathfrak{A}$

and $f \geq 0$ μ-a.e. on E, then E is a positive set for λ, and if $f \leq 0$ μ-a.e. on E, then E is a negative set for λ. If $f = 0$ μ-a.e. on E, then E is a null set for λ.
Consider for instance the measure space $([0, 2\pi], \mathfrak{M}_L \cap [0, 2\pi], \mu_L)$ and a signed measure λ on $([0, 2\pi], \mathfrak{M}_L \cap [0, 2\pi])$ defined by $\lambda(E) = \int_E \sin x \, \mu_L(dx)$ for $E \in \mathfrak{M}_L \cap [0, 2\pi]$. The set $[0, \pi]$ is a positive set for λ and the set $[\pi, 2\pi]$ is a negative set for λ.

Lemma 10.11. *Given a signed measure space $(X, \mathfrak{A}, \lambda)$.*
(a) *If $(A_n : n \in \mathbb{N})$ is a sequence of positive sets for λ, then $\bigcup_{n\in\mathbb{N}} A_n$ is a positive set for λ.*
(b) *If $(B_n : n \in \mathbb{N})$ is a sequence of negative sets for λ, then $\bigcup_{n\in\mathbb{N}} B_n$ is a negative set for λ.*
(c) *If $(E_n : n \in \mathbb{N})$ is a sequence of null sets for λ, then $\bigcup_{n\in\mathbb{N}} E_n$ is a null set for λ.*

Proof. Let $(A_n : n \in \mathbb{N})$ be a sequence of positive sets for λ. To show that $\bigcup_{n\in\mathbb{N}} A_n$ is a positive set for λ, we show that for every $E \in \mathfrak{A}$ such that $E \subset \bigcup_{n\in\mathbb{N}} A_n$ we have $\lambda(E) \geq 0$. Let $A'_1 = A_1$ and $A'_n = A_n \setminus \bigcup_{k=1}^{n-1} A_k$ for $n \geq 2$. Then $(A'_n : n \in \mathbb{N})$ is a disjoint sequence in $\mathfrak{A}$ and $\bigcup_{n\in\mathbb{N}} A'_n = \bigcup_{n\in\mathbb{N}} A_n$. Then $E = E \cap \bigcup_{n\in\mathbb{N}} A_n = E \cap \bigcup_{n\in\mathbb{N}} A'_n = \bigcup_{n\in\mathbb{N}} (E \cap A'_n)$. By the countable additivity of λ, we have $\lambda(E) = \sum_{n\in\mathbb{N}} \lambda(E \cap A'_n)$. Since A'_n is a $\mathfrak{A}$-measurable subset of a positive set A_n, it is a positive set. Then its $\mathfrak{A}$-measurable subset $E \cap A'_n$ is a positive set and in particular $\lambda(E \cap A'_n) \geq 0$ for every $n \in \mathbb{N}$. Thus $\lambda(E) \geq 0$. This shows that $\bigcup_{n\in\mathbb{N}} A_n$ is a positive set. Similarly for a sequence of negative sets and a sequence of null sets. ∎

Let $(X, \mathfrak{A}, \lambda)$ be an arbitrary signed measure space. There exists at least one positive set for λ. Indeed the empty set $\emptyset$ is a null set for λ and hence both a positive set and a negative set for λ. If $(A_n : n \in \mathbb{N})$ is a sequence of positive sets for λ, then $\bigcup_{n\in\mathbb{N}} A_n$ is a positive set for λ by Lemma 10.11. Does a greatest positive set for λ exist? The collection $\{A_\gamma : \gamma \in \Gamma\}$ of all positive sets for λ may be an uncountable collection and $\bigcup_{\gamma\in\Gamma} A_\gamma$ may not be in $\mathfrak{A}$ with $\lambda\big(\bigcup_{\gamma\in\Gamma} A_\gamma\big)$ undefined. So, rather than pursuing the idea of a greatest positive set let us define a maximal positive set.

Definition 10.12. *Let $(X, \mathfrak{A}, \lambda)$ be a signed measure space.*
(a) *Let us call a positive set A for λ a maximal positive set if for every positive set A' for λ the set $A' \setminus A$ is a null set for λ.*
(b) *Let us call a negative set B for λ a maximal negative set if for every negative set B' for λ the set $B' \setminus B$ is a null set for λ.*
(We show in Corollary 10.15 below that for an arbitrary signed measure space a maximal positive set and a maximal negative set always exist.)

Definition 10.13. *Given a signed measure space $(X, \mathfrak{A}, \lambda)$. A pair $\{A, B\}$ of $\mathfrak{A}$-measurable sets such that $A \cap B = \emptyset$, $A \cup B = X$, A is a positive set and B is a negative set for λ is called a Hahn decomposition of $(X, \mathfrak{A}, \lambda)$.*

Theorem 10.14. (Hahn Decomposition of Signed Measure Spaces) *For an arbitrary signed measure space $(X, \mathfrak{A}, \lambda)$, a Hahn decomposition exists and is unique up to null sets of λ, that is, there exist a positive set A and a negative set B for λ such that $A \cap B = \emptyset$ and*

$A \cup B = X$, and moreover if A' and B' are another such pair, then $A \triangle A'$ and $B \triangle B'$ are null sets for λ.

Proof. 1. Since λ is a signed measure on $(X, \mathfrak{A})$, we have either $\lambda(E) \in [-\infty, \infty)$ for every $E \in \mathfrak{A}$, or $\lambda(E) \in (-\infty, \infty]$ for every $E \in \mathfrak{A}$. Let us consider the former case. In this case, if $E \in \mathfrak{A}$ and $\lambda(E) \neq -\infty$ then $\lambda(E) \in \mathbb{R}$.

For every $C \in \mathfrak{A}$ there exists $C' \in \mathfrak{A}$ such that $C' \subset C$ and $\lambda(C') \geq \lambda(C)$. (The choice $C' := C$ will do.) Let us show that if $C \in \mathfrak{A}$ and $\lambda(C) \in \mathbb{R}$ then for every $\varepsilon > 0$ there exists $C' \in \mathfrak{A}$ such that $C' \subset C$ with $\lambda(C') \geq \lambda(C)$ and moreover $\lambda(E) > -\varepsilon$ for every $E \in \mathfrak{A}$ such that $E \subset C'$. Suppose for some $\varepsilon > 0$ no such subset C' of C exists. Then there exists $E_1 \in \mathfrak{A}$ such that $E_1 \subset C$ and $\lambda(E_1) \leq -\varepsilon$. By the finite additivity of λ we have $\lambda(E_1) + \lambda(C \setminus E_1) = \lambda(C)$. Since $\lambda(C) \in \mathbb{R}$ and since E_1 and $C \setminus E_1$ are $\mathfrak{A}$-measurable subsets of C, we have $\lambda(E_1), \lambda(C \setminus E_1) \in \mathbb{R}$ by Lemma 10.5. Thus $\lambda(C \setminus E_1) = \lambda(C) - \lambda(E_1) > \lambda(C)$ since $-\lambda(E_1) \geq \varepsilon > 0$. Now since $C \setminus E_1 \in \mathfrak{A}$, $C \setminus E_1 \subset C$ and $\lambda(C \setminus E_1) \geq \lambda(C)$, by the assumed non existence of the set C' there exists $E_2 \in \mathfrak{A}$ such that $E_2 \subset C \setminus E_1$ and $\lambda(E_2) \leq -\varepsilon$. Thus continuing, we have a disjoint sequence $(E_n : n \in \mathbb{N})$ of $\mathfrak{A}$-measurable subsets of C with $\lambda(E_n) \leq -\varepsilon$ for every $n \in \mathbb{N}$. Let $E = \bigcup_{n \in \mathbb{N}} E_n$. By the countable additivity of λ, we have $\lambda(E) = \sum_{n \in \mathbb{N}} \lambda(E_n) = -\infty$. But the fact that $E \subset C$ and $\lambda(C) \in \mathbb{R}$ implies that $\lambda(E) \in \mathbb{R}$ by Lemma 10.5. This is a contradiction. This proves the existence of a $\mathfrak{A}$-measurable subset C' of C such that $\lambda(C') \geq \lambda(C)$ and $\lambda(E) > -\varepsilon$ for every $\mathfrak{A}$-measurable subset E of C'.

2. Let us show that if $C \in \mathfrak{A}$ and $\lambda(C) \in \mathbb{R}$, then there exists a positive set $A \subset C$ with $\lambda(A) \geq \lambda(C)$. Let us define inductively a decreasing sequence $(C_n : n \in \mathbb{N})$ in $\mathfrak{A}$ with $\lambda(C_n) \in \mathbb{R}$ for $n \in \mathbb{N}$ as follows. Let $C_1 = C$. Suppose we have selected $C_1, \ldots, C_{n-1}$. By our result in **1** there exists $C_n \subset C_{n-1}$ such that $C_n \in \mathfrak{A}$ with $\lambda(C_n) \geq \lambda(C_{n-1})$ and $\lambda(E) \geq -\frac{1}{n}$ for every $\mathfrak{A}$-measurable subset E of C_n. With the decreasing sequence of sets $(C_n : n \in \mathbb{N})$ in $\mathfrak{A}$ thus selected, let $A = \bigcap_{n \in \mathbb{N}} C_n$. Since $(C_n : n \in \mathbb{N})$ is a decreasing sequence and since $\lambda(C_1) = \lambda(C) \in \mathbb{R}$, we have $\lambda(A) = \lim_{n \to \infty} \lambda(C_n)$ by (b) of Theorem 10.8. Since $(\lambda(C_n) : n \in \mathbb{N})$ is an increasing sequence of real numbers and since $\lambda(C_1) = \lambda(C)$, we have $\lim_{n \to \infty} \lambda(C_n) \geq \lambda(C)$. Therefore $\lambda(A) \geq \lambda(C)$. To show that A is a positive set, let E be an arbitrary $\mathfrak{A}$-measurable subset of A. Then $E \subset A = \bigcap_{n \in \mathbb{N}} C_n$ so that $E \subset C_n$ for every $n \in \mathbb{N}$. This implies that $\lambda(E) \geq -\frac{1}{n}$ for every $n \in \mathbb{N}$ and thus $\lambda(E) \geq 0$. This shows that A is a positive set.

3. Let $\alpha = \sup\{\lambda(E) : E \in \mathfrak{A}\}$. Since $\lambda(\emptyset) = 0$, we have $\alpha \in [0, \infty]$. Then we can select $(C_n : n \in \mathbb{N})$ in $\mathfrak{A}$ such that $\lambda(C_n) \in \mathbb{R}$ for every $n \in \mathbb{N}$ and $\lambda(C_n) \uparrow \alpha$. Then by our result in **2**, for every $n \in \mathbb{N}$ there exists a positive set A_n for λ such that $A_n \subset C_n$ and $\lambda(A_n) \geq \lambda(C_n)$. Let $A = \bigcup_{n \in \mathbb{N}} A_n \in \mathfrak{A}$. By the definition of α we have $\lambda(A) \leq \alpha$. Let us show that actually $\lambda(A) = \alpha$. Let $F_n = \bigcup_{k=1}^{n} A_k$ for $n \in \mathbb{N}$. Then F_n is a positive set for λ by (a) of Lemma 10.11. Now $(F_n : n \in \mathbb{N})$ is an increasing sequence in $\mathfrak{A}$ with $\bigcup_{n \in \mathbb{N}} F_n = \bigcup_{n \in \mathbb{N}} A_n = A$ so that $\lambda(A) = \lambda\big(\lim_{n \to \infty} F_n\big) = \lim_{n \to \infty} \lambda(F_n)$ by (a) of Theorem 10.8. On the other hand $F_n = (F_n \setminus A_n) \cup A_n$ and the finite additivity of λ on $\mathfrak{A}$ implies $\lambda(F_n) = \lambda(F_n \setminus A_n) + \lambda(A_n)$. Since $F_n \setminus A_n$ is a $\mathfrak{A}$-measurable subset of the positive set F_n for λ, $F_n \setminus A_n$ is a positive set for λ by (a) of Observation 10.10. Thus $\lambda(F_n \setminus A_n) \geq 0$.

This implies $\lambda(F_n) \geq \lambda(A_n)$. Thus we have

$$\lambda(A) = \lim_{n\to\infty} \lambda(F_n) \geq \limsup_{n\to\infty} \lambda(A_n) \geq \limsup_{n\to\infty} \lambda(C_n) = \lim_{n\to\infty} \lambda(C_n) = \alpha.$$

Therefore we have $\lambda(A) = \alpha$. (Since $\lambda(E) \in [-\infty, \infty)$ for every $E \in \mathfrak{A}$, we have $\alpha = \lambda(A) < \infty$.)

Now let $B = A^c$. We have $A \cap B = \emptyset$ and $A \cup B = X$. To show that $\{A, B\}$ is a Hahn decomposition for $(X, \mathfrak{A}, \lambda)$, it remains to show that B is a negative set for λ. Let $E \in \mathfrak{A}$ and $E \subset B$. If $\lambda(E) > 0$, then by the finite additivity of λ on $\mathfrak{A}$ we have $\lambda(A \cup E) = \lambda(A) + \lambda(E) > \lambda(A) = \alpha$, contradicting the definition of α. Thus $\lambda(E) \leq 0$. This shows that B is a negative set for λ. Thus we have proved the existence of a Hahn decomposition for a signed measure space $(X, \mathfrak{A}, \lambda)$ in which $\lambda(E) \in [-\infty, \infty)$ for every $E \in \mathfrak{A}$.

4. Consider a signed measure space $(X, \mathfrak{A}, \lambda)$ in which $\lambda(E) \in (-\infty, \infty]$ for every $E \in \mathfrak{A}$. Then $-\lambda$ is a signed measure on $(X, \mathfrak{A})$ such that $-\lambda(E) \in [-\infty, \infty)$ for every $E \in \mathfrak{A}$. Thus by our result above there exists a Hahn decomposition $\{A, B\}$ for the signed measure space $(X, \mathfrak{A}, -\lambda)$. Then $\{B, A\}$ is a Hahn decomposition for $(X, \mathfrak{A}, \lambda)$.

5. Let us prove the uniqueness of a Hahn decomposition. Suppose $\{A, B\}$ and $\{A', B'\}$ are two Hahn decompositions of a signed measure space $(X, \mathfrak{A}, \lambda)$. Since A is a positive set for λ, its $\mathfrak{A}$-measurable subset $A \setminus A'$ is a positive set for λ by (a) of Observation 10.10. On the other hand $A \setminus A' = A \cap (A')^c = A \cap B'$ which is a $\mathfrak{A}$-measurable subset of a negative set B' for λ and hence a negative set for λ by (a) of Observation 10.10. Thus $A \setminus A'$ is both a positive set and a negative set for λ and hence a null set for λ by (b) of Observation 10.10. Similarly $A' \setminus A$ is a null set for λ. Then $A \triangle A' = (A \setminus A') \cup (A' \setminus A)$ is a null set for λ by Lemma 10.11. Similarly $B \triangle B$ is a null set for λ. ∎

Corollary 10.15. *In an arbitrary signed measure space $(X, \mathfrak{A}, \lambda)$, a maximal positive set and a maximal negative set for λ always exist. Indeed if $\{A, B\}$ is a Hahn decomposition for $(X, \mathfrak{A}, \lambda)$, then A is a maximal positive set and B is a maximal negative set for λ.*

Proof. Let $\{A, B\}$ be a Hahn decomposition for $(X, \mathfrak{A}, \lambda)$. Let A' be an arbitrary positive set for λ. Then $A' \setminus A$, being a $\mathfrak{A}$-measurable subset of a positive set A', is a positive set for λ by (a) of Observation 10.10. On the other hand, $A' \setminus A = A' \cap A^c = A' \cap B$ being a $\mathfrak{A}$-measurable subset of a negative set B is a negative set for λ by (a) of Observation 10.10. Thus $A' \setminus A$ is both a positive set and a negative set for λ and is therefore a null set for λ by (b) of Observation 10.10. This shows that A is a maximal positive set for λ. Similarly B is a maximal negative set for λ. ∎

Definition 10.16. *Two signed measures λ_1 and λ_2 on a measurable space $(X, \mathfrak{A})$ are said to be mutually singular and we write $\lambda_1 \perp \lambda_2$ if there exist two $\mathfrak{A}$-measurable sets C_1 and C_2 such that $C_1 \cap C_2 = \emptyset$, $C_1 \cup C_2 = X$, C_1 is a null set for λ_2 and C_2 is a null set for λ_1.*

Definition 10.17. *Given a signed measure space $(X, \mathfrak{A}, \lambda)$. If there exist two positive measures μ and ν, at least one of which is finite, on the measurable space $(X, \mathfrak{A})$ such that $\mu \perp \nu$ and $\lambda = \mu - \nu$, then $\{\mu, \nu\}$ is called a Jordan decomposition of λ.*

(Note that the requirement that at least one of the two positive measures μ and ν is finite is to ensure that $\mu - \nu$ is defined.)

A Hahn decomposition and a Jordan decomposition of a signed measure space are related. The existence of one implies that of the other as the next proposition shows.

Proposition 10.18. *Let $(X, \mathfrak{A}, \lambda)$ be a signed measure space.*

(a) *If $\{A, B\}$ is a Hahn decomposition of $(X, \mathfrak{A}, \lambda)$ and if we define two set functions μ and ν on $\mathfrak{A}$ by $\mu(E) = \lambda(E \cap A)$ and $\nu(E) = -\lambda(E \cap B)$ for $E \in \mathfrak{A}$, then $\{\mu, \nu\}$ is a Jordan decomposition for λ.*

(b) *If $\{\mu, \nu\}$ is a Jordan decomposition of λ and C_1 and C_2 are two $\mathfrak{A}$-measurable subsets of X such that $C_1 \cap C_2 = \emptyset$, $C_1 \cup C_2 = X$, C_1 is a null set for ν and C_2 is a null set for μ, then $\{C_1, C_2\}$ is a Hahn decomposition of $(X, \mathfrak{A}, \lambda)$.*

Proof. 1. Let us prove (a). Every $\mathfrak{A}$-measurable subset of the positive set A for λ is a positive set for λ and every $\mathfrak{A}$-measurable subset of the negative set B for λ is a negative set for λ by Observation 10.10. From this it follows that μ and ν are positive measures on $(X, \mathfrak{A})$. Since λ cannot assume both ∞ and $-\infty$, we have either $\lambda(A) \in [0, \infty)$ or $\lambda(B) \in (-\infty, 0]$. Thus at least one of μ and ν is a finite positive measure. Also by the finite additivity of λ, we have $\lambda(E) = \lambda(E \cap A) + \lambda(E \cap B) = \mu(E) - \nu(E)$ for every $E \in \mathfrak{A}$ so that $\lambda = \mu - \nu$. To show that $\mu \perp \nu$, we show that A is a null set for ν and B is a null set for μ. Now $\nu(A) = -\lambda(A \cap B) = -\lambda(\emptyset) = 0$ so that A is a null set for the positive measure ν. Similarly $\mu(B) = \lambda(B \cap A) = \lambda(\emptyset) = 0$ so that B is a null set for the positive measure μ. This shows that $\mu \perp \nu$. Therefore $\{\mu, \nu\}$ is a Jordan decomposition for λ.

2. Let us prove (b). We show that C_1 is a positive set for λ and C_2 is a negative set for λ. Now since $\{\mu, \nu\}$ is a Jordan decomposition for $(X, \mathfrak{A}, \lambda)$, we have $\lambda(E) = \mu(E) - \nu(E)$ for every $E \in \mathfrak{A}$. To show that C_1 is a positive set for λ, let $E \in \mathfrak{A}$ and $E \subset C_1$. Since C_1 is a null set for the positive measure ν, its subset E is a null set for ν. Thus $\nu(E) = 0$ and consequently $\lambda(E) = \mu(E) \geq 0$. This shows that C_1 is a positive set for λ. To show that C_2 is a negative set for λ, let $E \in \mathfrak{A}$ and $E \subset C_2$. Since C_2 is a null set for the positive measure μ, its subset E is a null set for μ. Thus $\mu(E) = 0$ and consequently $\lambda(E) = -\nu(E) \leq 0$. This shows that C_2 is a negative set for λ. ∎

Observation 10.19. Given a signed measure space $(X, \mathfrak{A}, \lambda)$. Let $C, C' \in \mathfrak{A}$ be such that $C \Delta C'$ is a null set for λ. Then for every $E \in \mathfrak{A}$, we have $\lambda(E \cap C) = \lambda(E \cap C')$.

Proof. Since $C = (C \cap C') \cup (C \setminus C')$, we have $E \cap C = \{E \cap (C \cap C')\} \cup \{E \cap (C \setminus C')\}$ and thus $\lambda(E \cap C) = \lambda\big(E \cap (C \cap C')\big) + \lambda\big(E \cap (C \setminus C')\big)$. Since $C \Delta C'$ is a null set for λ, its subset $C \setminus C'$ is a null set for λ and then so is its subset $E \cap (C \setminus C')$. Thus $\lambda\big(E \cap (C \setminus C')\big) = 0$ and hence $\lambda(E \cap C) = \lambda\big(E \cap (C \cap C')\big)$. Interchanging the roles of C and C', we have $\lambda(E \cap C') = \lambda\big(E \cap (C' \cap C)\big)$. Thus we have $\lambda(E \cap C) = \lambda(E \cap C')$. ∎

Lemma 10.20. *Let $(X, \mathfrak{A}, \lambda)$ be a signed measure space. Let $\{A, B\}$ and $\{A', B'\}$ be two*

Hahn decompositions of $(X, \mathfrak{A}, \lambda)$. *Then* $\lambda(E \cap A) = \lambda(E \cap A')$ *and* $\lambda(E \cap B) = \lambda(E \cap B')$ *for every* $E \in \mathfrak{A}$.

Proof. By the uniqueness of the Hahn decomposition (Theorem 10.14), $A \triangle A'$ and $B \triangle B'$ are null sets for λ. Then the Lemma follows from Observation 10.19. ∎

Theorem 10.21. (Jordan Decomposition of Signed Measures) *Given a signed measure space* $(X, \mathfrak{A}, \lambda)$. *A Jordan decomposition for* $(X, \mathfrak{A}, \lambda)$ *exists and is unique, that is, there exists a unique pair* $\{\mu, \nu\}$ *of positive measures on* $(X, \mathfrak{A})$, *at least one of which is finite, such that* $\mu \perp \nu$ *and* $\lambda = \mu - \nu$. *Moreover with an arbitrary Hahn decomposition* $\{A, B\}$ *of* $(X, \mathfrak{A}, \lambda)$, *if we define two set functions* μ *and* ν *on* $\mathfrak{A}$ *by setting*

$$\mu(E) = \lambda(E \cap A) \quad \text{and} \quad \nu(E) = -\lambda(E \cap B) \quad \text{for } E \in \mathfrak{A},$$

then $\{\mu, \nu\}$ *is a Jordan decomposition for* $(X, \mathfrak{A}, \lambda)$.

Proof. 1. By Theorem 10.14, a Hahn decomposition $\{A, B\}$ of $(X, \mathfrak{A}, \lambda)$ exists. According to (a) of Proposition 10.18, if we define two set functions μ and ν on $\mathfrak{A}$ by $\mu(E) = \lambda(E \cap A)$ and $\nu(E) = -\lambda(E \cap B)$ for $E \in \mathfrak{A}$, then $\{\mu, \nu\}$ is a Jordan decomposition for λ. Therefore a Jordan decomposition for $(X, \mathfrak{A}, \lambda)$ exists.

2. To prove the uniqueness of the Jordan decomposition, let $\{\mu, \nu\}$ be the Jordan decomposition for $(X, \mathfrak{A}, \lambda)$ defined in **1** and let $\{\mu', \nu'\}$ be a Jordan decomposition for $(X, \mathfrak{A}, \lambda)$. Then there exist $A', B' \in \mathfrak{A}$ such that $A' \cap B' = \emptyset$, $A' \cup B' = X$, A' is a null set for ν' and B' is a null set for μ'. By (b) of Proposition 10.18, $\{A', B'\}$ is a Hahn decomposition for $(X, \mathfrak{A}, \lambda)$. Then by Lemma 10.20, we have $\lambda(E \cap A) = \lambda(E \cap A')$ and $\lambda(E \cap B) = \lambda(E \cap B')$ for every $E \in \mathfrak{A}$. Then for an arbitrary $E \in \mathfrak{A}$, we have

$$\begin{aligned}
\mu(E) &= \lambda(E \cap A) = \lambda(E \cap A') \\
&= \mu'(E \cap A') - \nu'(E \cap A') \quad \text{since } \lambda = \mu' - \nu' \\
&= \mu'(E \cap A') \quad \text{since } A' \text{ and } E \cap A' \text{ are null sets for } \nu' \\
&= \mu'(E \cap A') + \mu'(E \cap B') \quad \text{since } B' \text{ and } E \cap B' \text{ are null sets for } \mu' \\
&= \mu'(E) \quad \text{by the finite additivity of } \mu'.
\end{aligned}$$

Thus $\mu = \mu'$ on $\mathfrak{A}$. Similarly $\nu = \nu'$ on $\mathfrak{A}$. This proves the uniqueness of the Jordan decomposition as well as the expressions (1) for μ and ν. ∎

Definition 10.22. *Given a signed measure space* $(X, \mathfrak{A}, \lambda)$. *Let* μ *and* ν *be the unique positive measures on* $\mathfrak{A}$, *at least one of which is finite, such that* $\mu \perp \nu$ *and* $\lambda = \mu - \nu$. *Let us call* μ *and* ν *the positive and negative parts of* λ *and write* λ^+ *for* μ *and* λ^- *for* ν. *The total variation of* λ *is a positive measure* $\|\lambda\|$ *on* $\mathfrak{A}$ *defined by*

(1) $$\|\lambda\|(E) = \lambda^+(E) + \lambda^-(E) \quad \text{for } E \in \mathfrak{A}.$$

The positive and negative parts, λ^+ *and* λ^-, *are also called the positive and negative variations of* λ. *For every* $E \in \mathfrak{A}$, *we have*

(2) $$\|\lambda\|(E) = \lambda^+(E) + \lambda^-(E) \geq |\lambda^+(E) - \lambda^-(E)| = |\lambda(E)|.$$

The next proposition explains the terminology 'total variation' for $\|\lambda\|$.

Proposition 10.23. *Given a signed measure space* $(X, \mathfrak{A}, \lambda)$. *For* $E \in \mathfrak{A}$, *let*

$$\alpha_E = \sup\left\{\sum_{k=1}^n |\lambda(E_k)| : \{E_1, \dots, E_n\} \subset \mathfrak{A}, \textit{ disjoint}, \bigcup_{k=1}^n E_k = E, n \in \mathbb{N}\right\}.$$

Then $\|\lambda\|(E) = \alpha_E$.

Proof. Let $E \in \mathfrak{A}$. Then for any finite disjoint collection $\{E_1, \dots, E_n\}$ in $\mathfrak{A}$ such that $\bigcup_{k=1}^n E_k = E$, we have

$$\begin{aligned}\sum_{k=1}^n |\lambda(E_k)| &= \sum_{k=1}^n |\lambda^+(E_k) - \lambda^-(E_k)| \le \sum_{k=1}^n \{|\lambda^+(E_k)| + |\lambda^-(E_k)|\} \\ &= \sum_{k=1}^n \{\lambda^+(E_k) + \lambda^-(E_k)\} = \sum_{k=1}^n \|\lambda\|(E_k) = \|\lambda\|(E).\end{aligned}$$

Thus we have $\alpha_E \le \|\lambda\|(E)$.

To prove the reverse inequality, let $\{A, B\}$ be an arbitrary Hahn decomposition of $(X, \mathfrak{A}, \lambda)$. Let $E_1 = E \cap A$ and $E_2 = E \cap B$. Then $\{E_1, E_2\}$ is a disjoint collection in $\mathfrak{A}$ such that $E_1 \cup E_2 = E$ so that by the definition of α_E we have

$$\alpha_E \ge |\lambda(E_1)| + |\lambda(E_2)| = |\lambda(E \cap A)| + |\lambda(E \cap B)|.$$

If we define $\mu(E) = \lambda(E \cap A)$ and $\nu(E) = -\lambda(E \cap B)$ for $E \in \mathfrak{A}$, then $\{\mu, \nu\}$ is a Jordan decomposition for $(X, \mathfrak{A}, \lambda)$ by (a) of Proposition 10.18. By the uniqueness of the Jordan decomposition (Theorem 10.21), we have $\lambda^+(E) = \lambda(E \cap A)$ and $\lambda^-(E) = -\lambda(E \cap B)$. Now since A is a positive set for λ, its subset $E \cap A$ is a positive set for λ by (a) of Observation 10.10 so that $|\lambda(E \cap A)| = \lambda(E \cap A) = \lambda^+(E)$. Similarly since B is a negative set for λ so is its subset $E \cap B$ so that $\lambda(E \cap B) \le 0$. Then $|\lambda(E \cap B)| = -\lambda(E \cap B) = \lambda^-(E)$. Thus we have $\alpha_E \ge \lambda^+(E) + \lambda^-(E) = \|\lambda\|(E)$. Therefore we have $\|\lambda\|(E) = \alpha_E$. ∎

The next proposition is the converse of Proposition 10.3. It shows that an arbitrary signed measure can be represented as an integral.

Proposition 10.24. *Let* $(X, \mathfrak{A}, \lambda)$ *be a signed measure space. Let* $\{A, B\}$ *be a Hahn decomposition of* $(X, \mathfrak{A}, \lambda)$ *and let* $\|\lambda\|$ *be the total variation of* λ. *Then we have*

$$\lambda(E) = \int_E \{\mathbf{1}_A - \mathbf{1}_B\}\, d\|\lambda\| \quad \textit{for } E \in \mathfrak{A}.$$

Proof. Let $\{A, B\}$ be a Hahn decomposition of $(X, \mathfrak{A}, \lambda)$. By Theorem 10.21, the positive variation λ^+ and the negative variation λ^- of the signed measure λ are given by

$$(1) \qquad \lambda^+(F) = \lambda(A \cap F) \quad \text{and} \quad \lambda^-(F) = -\lambda(B \cap F) \quad \text{for } F \in \mathfrak{A}.$$

Then for every $E \in \mathfrak{A}$, we have

$$\begin{aligned}\int_E \{\mathbf{1}_A - \mathbf{1}_B\}\, d|\lambda| &= \int_X \mathbf{1}_{A\cap E}\, d|\lambda| - \int_X \mathbf{1}_{B\cap E}\, d|\lambda| = |\lambda|(A\cap E) - |\lambda|(B\cap E)\\ &= \lambda^+(A\cap E) + \lambda^-(A\cap E) - \lambda^+(B\cap E) - \lambda^-(B\cap E)\\ &= \lambda(A\cap E) - \lambda(B\cap A\cap E) - \lambda(A\cap B\cap E) + \lambda(B\cap E)\\ &= \lambda(A\cap E) + \lambda(B\cap E) = \lambda(E),\end{aligned}$$

where the third equality is by $|\lambda| = \lambda^+ + \lambda^-$, the fourth equality is by (1), the fifth equality is by $A \cap B = \emptyset$, and the sixth equality is by $A \cup B = X$. ∎

The next proposition relates the positive variation, the negative variation, and the total variation of a signed measure obtained by integrating a function to the positive part, the negative part, and the absolute value of the function.

Proposition 10.25. *Let f be a μ semi-integrable extended real-valued $\mathfrak{A}$-measurable function on a measure space $(X, \mathfrak{A}, \mu)$. Let us define a signed measure λ and three positive measures μ_1, μ_2, and μ_3 on $\mathfrak{A}$ by setting*

$$\begin{aligned}\lambda(E) &= \int_E f\, d\mu,\\ \mu_1(E) &= \int_E f^+\, d\mu,\\ \mu_2(E) &= \int_E f^-\, d\mu,\\ \mu_3(E) &= \int_E |f|\, d\mu,\end{aligned}$$

for $E \in \mathfrak{A}$. Then μ_1, μ_2, and μ_3 are respectively the positive variation, the negative variation, and the total variation of λ.

Proof. Note that $\lambda(E) = \int_E f\, d\mu = \int_E f^+\, d\mu - \int_E f^-\, d\mu = \mu_1(E) - \mu_2(E)$ for every $E \in \mathfrak{A}$ so that $\lambda = \mu_1 - \mu_2$. If we let $A = \{X : f \geq 0\}$ and $B = \{X : f < 0\}$, then $A, B \in \mathfrak{A}$, $A \cap B = \emptyset$, $A \cup B = X$, $\mu_1(B) = 0$, and $\mu_2(A) = 0$ so that $\mu_1 \perp \mu_2$. This shows that $\{\mu_1, \mu_2\}$ is a Jordan decomposition of λ and μ_1 and μ_2 are the positive variation and the negative variation of λ. Then for every $E \in \mathfrak{A}$, we have

$$\mu_3(E) = \int_E |f|\, d\mu = \int_E f^+\, d\mu + \int_E f^-\, d\mu = \mu_1(E) + \mu_2(E) = |\lambda|(E).$$

This shows that μ_3 is the total variation of λ. ∎

Observation 10.26. Given a signed measure space $(X, \mathfrak{A}, \lambda)$.
(a) A set $E \in \mathfrak{A}$ is a null set for λ if and only if E is a null set for $|\lambda|$.
(b) If a set $E \in \mathfrak{A}$ is a null set for λ, then E is a null set for both λ^+ and λ^-.

Proof. 1. Let us prove (a). According to Theorem 10.21, with an arbitrary Hahn decomposition $\{A, B\}$ of $(X, \mathfrak{A}, \lambda)$, the positive and negative parts of λ, λ^+ and λ^-, are given by

$$(1) \qquad \lambda^+(E) = \lambda(E \cap A) \quad \text{and} \quad \lambda^-(E) = -\lambda(E \cap B) \quad \text{for } E \in \mathfrak{A}.$$

Then we have

$$(2) \qquad \|\lambda\|(E) = \lambda^+(E) + \lambda^-(E) = \lambda(E \cap A) - \lambda(E \cap B) \quad \text{for } E \in \mathfrak{A}.$$

If E is a null set for λ, then so are its subsets $E \cap A$ and $E \cap B$ by (a) of Observation 10.10 so that $\|\lambda\|(E) = 0$ by (2) and this shows that E is a null set for the positive measure $\|\lambda\|$. Conversely if E is a null set for $\|\lambda\|$, then for every $\mathfrak{A}$-measurable subset E_0 of E we have $|\lambda(E_0)| \leq \|\lambda\|(E_0) = 0$ so that $\lambda(E_0) = 0$. This shows that E is a null set for λ.

2. Suppose a set $E \in \mathfrak{A}$ is a null set for λ. Then E is a null set for $\|\lambda\|$ by (a) so that $\|\lambda\|(E) = 0$. Now since λ^+, λ^-, and $\|\lambda\|$ are all positive measures and $\|\lambda\| = \lambda^+ + \lambda^-$, $\|\lambda\|(E) = 0$ implies both $\lambda^+(E) = 0$ and $\lambda^-(E) = 0$. This shows that E is a null set for both λ^+ and λ^-. ■

Proposition 10.27. Let λ_1 and λ_2 be two signed measures on a measurable space $(X, \mathfrak{A})$. Then we have
(a) $\lambda_1 \perp \lambda_2 \Leftrightarrow \|\lambda_1\| \perp \lambda_2 \Leftrightarrow \lambda_1 \perp \|\lambda_2\| \Leftrightarrow \|\lambda_1\| \perp \lambda_2 \Leftrightarrow \|\lambda_1\| \perp \|\lambda_2\|$,
(b) $\|\lambda_1\| \perp \lambda_2 \Leftrightarrow \lambda_1^+ \perp \lambda_2$ and $\lambda_1^- \perp \lambda_2$.

Proof. 1. The equivalence $\lambda_1 \perp \lambda_2 \Leftrightarrow \|\lambda_1\| \perp \lambda_2$ is an immediate consequence of the fact that a $\mathfrak{A}$-measurable set E is a null set of λ_1 if and only if E is a null set for $\|\lambda_1\|$. Similarly for $\lambda_1 \perp \lambda_2 \Leftrightarrow \lambda_1 \perp \|\lambda_2\|$. Applying this to the pair $\|\lambda_1\|$ and λ_2, we have $\|\lambda_1\| \perp \lambda_2 \Leftrightarrow \|\lambda_1\| \perp \|\lambda_2\|$.

2. Suppose $\|\lambda_1\| \perp \lambda_2$. Then there exist $C_1, C_2 \in \mathfrak{A}$ such that $C_1 \cap C_2 = \emptyset$, $C_1 \cup C_2 = X$, C_1 is a null set for λ_2, and C_2 is a null set for $\|\lambda_1\|$. This implies that C_2 is a null set for both λ_1^+ and λ_1^-. This shows that $\lambda_1^+ \perp \lambda_2$ and $\lambda_1^- \perp \lambda_2$.

Conversely suppose $\lambda_1^+ \perp \lambda_2$ and $\lambda_1^- \perp \lambda_2$. Then there exist $C_1^+, C_2^+ \in \mathfrak{A}$ such that $C_1^+ \cap C_2^+ = \emptyset$, $C_1^+ \cup C_2^+ = X$, C_1^+ is a null set for λ_2 and C_2^+ is a null set for λ_1^+, and there exist $C_1^-, C_2^- \in \mathfrak{A}$ such that $C_1^- \cap C_2^- = \emptyset$, $C_1^- \cup C_2^- = X$, C_1^- is a null set for λ_2 and C_2^- is a null set for λ_1^-. Consider $C_2^+ \cap C_2^-$. Since $\lambda_1^+(C_2^+) = 0$ and λ_1^+ is a positive measure, we have $\lambda_1^+(C_2^+ \cap C_2^-) = 0$. Similarly we have $\lambda_1^-(C_2^+ \cap C_2^-) = 0$. Therefore we have $\|\lambda_1\|(C_2^+ \cap C_2^-) = \lambda_1^+(C_2^+ \cap C_2^-) + \lambda_1^-(C_2^+ \cap C_2^-) = 0$. This shows that $C_2^+ \cap C_2^-$ is a null set for the positive measure $\|\lambda_1\|$. Consider the set

$$X \setminus (C_2^+ \cap C_2^-) = (C_2^+ \cap C_2^-)^c = (C_2^+)^c \cup (C_2^-)^c = C_1^+ \cup C_1^-.$$

Since C_1^+ and C_1^- are null sets for λ_2, so is $C_1^+ \cup C_1^-$. Therefore we have shown that $C_2^+ \cap C_2^-$ is a null set for the positive measure $\|\lambda_1\|$ and its complement $X \setminus (C_2^+ \cap C_2^-)$ is a null set for λ_2. Thus $\|\lambda_1\| \perp \lambda_2$. ■

Theorem 10.28. (a) *Let λ be a signed measure on a measurable space $(X, \mathfrak{A})$. Then for every $c \in \mathbb{R}$, $c\lambda$ is a signed measure on $(X, \mathfrak{A})$ and moreover we have $\|c\lambda\| = |c|\,\|\lambda\|$.*

(b) *Let λ_1 and λ_2 be signed measures on a measurable space $(X, \mathfrak{A})$. Suppose $\lambda_1 + \lambda_2$ is defined on $(X, \mathfrak{A})$. (This is the case if $\lambda_1(E), \lambda_2(E) \in (-\infty, \infty]$ for every $E \in \mathfrak{A}$, or $\lambda_1(E), \lambda_2(E) \in [-\infty, \infty)$ for every $E \in \mathfrak{A}$.) Then $\lambda_1 + \lambda_2$ is a signed measure on $(X, \mathfrak{A})$ and moreover we have $\|\lambda_1 + \lambda_2\| \le \|\lambda_1\| + \|\lambda_2\|$.*

Proof. 1. It is easily verified that $c\lambda$ and $\lambda_1 + \lambda_2$ satisfy conditions 1°, 2°, and 3° in Definition 10.1. Thus $c\lambda$ and $\lambda_1 + \lambda_2$ are signed measures on $(X, \mathfrak{A})$.

2. For every signed measure λ on $(X, \mathfrak{A})$, we have by Proposition 10.23

(1) $$\|\lambda\|(E) = \sup\Big\{\sum_{k=1}^{n} |\lambda(E_k)| : \{E_1, \dots, E_n\} \subset \mathfrak{A},\ \text{disjoint}, \bigcup_{k=1}^{n} E_k = E, n \in \mathbb{N}\Big\}.$$

It follows from (1) that $\|c\lambda\| = |c|\,\|\lambda\|$ for every $c \in \mathbb{R}$.

3. Let λ_1 and λ_2 be signed measures on $(X, \mathfrak{A})$ such that $\lambda_1 + \lambda_2$ is defined on $(X, \mathfrak{A})$. Then by (1) we have

(2)
$$\begin{aligned}
&\|\lambda_1 + \lambda_2\|(E) \\
&= \sup\Big\{\sum_{k=1}^{n} |\lambda_1(E_k) + \lambda_2(E_k)| : \{E_1, \dots, E_n\} \subset \mathfrak{A},\ \text{disjoint}, \bigcup_{k=1}^{n} E_k = E, n \in \mathbb{N}\Big\} \\
&\le \sup\Big\{\sum_{k=1}^{n} |\lambda_1(E_k)| + \sum_{k=1}^{n} |\lambda_2(E_k)| : \{E_1, \dots, E_n\} \subset \mathfrak{A},\ \text{disjoint}, \bigcup_{k=1}^{n} E_k = E, n \in \mathbb{N}\Big\}.
\end{aligned}$$

Since $\sum_{k=1}^{n} |\lambda_1(E_k)| \le \|\lambda_1\|$ and $\sum_{k=1}^{n} |\lambda_2(E_k)| \le \|\lambda_2\|$, we have

(3) $$\|\lambda_1 + \lambda_2\| \le \|\lambda_1\| + \|\lambda_2\|. \blacksquare$$

[III] Integration on a Signed Measure Space

Definition 10.29. *Let $(X, \mathfrak{A}, \lambda)$ be a signed measure space.*

(a) *We say that $(X, \mathfrak{A}, \lambda)$ is a complete signed measure space if every subset of a null set for λ is $\mathfrak{A}$-measurable.*

(b) *We say that $(X, \mathfrak{A}, \lambda)$ is a finite signed measure space if $\lambda(X) \in \mathbb{R}$.*

(c) *We say that $(X, \mathfrak{A}, \lambda)$ is a σ-finite signed measure space if there exists a sequence $(E_n : n \in \mathbb{N})$ in $\mathfrak{A}$ such that $\bigcup_{n\in\mathbb{N}} E_n = X$ and $\lambda(E_n) \in \mathbb{R}$ for every $n \in \mathbb{N}$.*

Regarding (b), note that if $\lambda(X) \in \mathbb{R}$, then $\lambda(E) \in \mathbb{R}$ for every $E \in \mathfrak{A}$ by Lemma 10.5. Thus λ is a finite signed measure if and only if $\lambda(E) \in \mathbb{R}$ for every $E \in \mathfrak{A}$, or equivalently, if and only if $|\lambda(E)| < \infty$ for every $E \in \mathfrak{A}$.

Regarding (c), note that if we let $F_1 = E_1$ and $F_n = E_n \setminus \bigcup_{k=1}^{n-1} E_k$ for $n \ge 2$, then $(F_n : n \in \mathbb{N})$ is a disjoint sequence in $\mathfrak{A}$, $\bigcup_{n\in\mathbb{N}} F_n = \bigcup_{n\in\mathbb{N}} E_n = X$, and since F_n is a subset of E_n we have $\lambda(F_n) \in \mathbb{R}$ by Lemma 10.5 for every $n \in \mathbb{N}$.

If we let $G_n = \bigcup_{k=1}^{n} F_k$ for $n \in \mathbb{N}$, then $(G_n : n \in \mathbb{N})$ is an increasing sequence in $\mathfrak{A}$ with $\bigcup_{n\in\mathbb{N}} G_n = \bigcup_{n\in\mathbb{N}} F_n = X$ and $\lambda(G_n) = \sum_{k=1}^{n} \lambda(F_k) \in \mathbb{R}$ for every $n \in \mathbb{N}$.

Observation 10.30. Given a signed measure space $(X, \mathfrak{A}, \lambda)$.
(a) $(X, \mathfrak{A}, \lambda)$ is a complete signed measure space if and only if $(X, \mathfrak{A}, |\lambda|)$ is complete.
(b) $(X, \mathfrak{A}, \lambda)$ is a finite signed measure space if and only if $(X, \mathfrak{A}, |\lambda|)$ is finite.
(c) $(X, \mathfrak{A}, \lambda)$ is a σ-finite signed measure space if and only if $(X, \mathfrak{A}, |\lambda|)$ is σ-finite.
(d) If μ and ν are two σ-finite measures on $(X, \mathfrak{A})$, one of which is finite, then the signed measure $\lambda = \mu - \nu$ is σ-finite.

Proof. (a), (b), and (c) are immediate consequences of Definition 10.29. Let us prove (d). Now if μ and ν are two σ-finite measures on $(X, \mathfrak{A})$, then there exist two increasing sequences $(E_n : n \in \mathbb{N})$ and $(F_n : n \in \mathbb{N})$ in $\mathfrak{A}$ such that $\lim_{n\to\infty} E_n = \bigcup_{n\in\mathbb{N}} E_n = X$, $\mu(E_n) < \infty$ for every $n \in \mathbb{N}$, $\lim_{n\to\infty} F_n = \bigcup_{n\in\mathbb{N}} F_n = X$, and $\nu(F_n) < \infty$ for every $n \in \mathbb{N}$. If we let $G_n = E_n \cap F_n$ for $n \in \mathbb{N}$, then $(G_n : n \in \mathbb{N})$ is an increasing sequence in $\mathfrak{A}$. Let us show that $\lim_{n\to\infty} G_n = X$. Let $x \in X$. Then $x \in E_{n_1}$ for some $n_1 \in \mathbb{N}$ and $x \in F_{n_2}$ for some $n_2 \in \mathbb{N}$. Let $n_0 = \max\{n_1, n_2\}$. Then $x \in E_{n_0}$ and $x \in F_{n_0}$ so that $x \in G_{n_0}$. This shows that every $x \in X$ is in G_n for some $n \in \mathbb{N}$. Thus $X \subset \bigcup_{n\in\mathbb{N}} G_n$ and therefore $\lim_{n\to\infty} G_n = \bigcup_{n\in\mathbb{N}} G_n = X$. Now $\mu(G_n) \le \mu(E_n) < \infty$ and $\nu(G_n) \le \nu(F_n) < \infty$. Then $\lambda(G_n) = \mu(G_n) - \nu(G_n) \in \mathbb{R}$ for every $n \in \mathbb{N}$. This shows that λ is a σ-finite signed measure on $(X, \mathfrak{A})$. ∎

If μ and ν are two positive measures on a measurable space $(X, \mathfrak{A})$, then the set function $\mu + \nu$ on $\mathfrak{A}$ is a positive measure on $\mathfrak{A}$. Let f be an extended real-valued $\mathfrak{A}$-measurable function on X. If all the three integrals $\int_X f\,d\mu$, $\int_X f\,d\nu$, and $\int_X f\,d(\mu+\nu)$ exist and the sum $\int_X f\,d\mu + \int_X f\,d\nu$ also exist, does the equality $\int_X f\,d\mu + \int_X f\,d\nu = \int_X f\,d(\mu+\nu)$? Of particular interest is a signed measure space $(X, \mathfrak{A}, \lambda)$ for which we have $|\lambda| = \lambda^+ + \lambda^-$. Does the equality $\int_X f\,d\lambda^+ + \int_X f\,d\lambda^- = \int_X f\,d|\lambda|$ hold?

Proposition 10.31. *Let μ, ν, and $\mu + \nu$ be three positive measures on a measurable space $(X, \mathfrak{A})$. Let f be extended real-valued $\mathfrak{A}$-measurable function on X. Then we have*

$$\int_X f\,d\mu + \int_X f\,d\nu = \int_X f\,d(\mu+\nu),$$

in the sense that the existence of one side implies that of the other and the equality of the two.

Proof. It is easily verified that the equality holds if $f = \mathbf{1}_E$ where $E \in \mathfrak{A}$. By the linearity of the integrals, it follows that the equality holds when f is a nonnegative simple function. If f is a nonnegative extended real-valued $\mathfrak{A}$-measurable function on X, then there exists an increasing sequence $(\varphi_n : n \in \mathbb{N})$ of nonnegative simple function such that $\varphi_n \uparrow f$ on X according to Lemma 8.6. Applying the Monotone Convergence Theorem (Theorem 8.5), we have the equality for our f. If f is an extended real-valued function, then $f = f^+ - f^-$ and we apply the result above for nonnegative functions to f^+ and f^-. ∎

Definition 10.32. *Given a signed measure space $(X, \mathfrak{A}, \lambda)$. Let λ^+ and λ^- be the positive and negative parts of λ. Let f be an extended real-valued $\mathfrak{A}$-measurable function on X.*

We define

$$\int_X f\,d\lambda = \int_X f\,d\lambda^+ - \int_X f\,d\lambda^-,$$

if the two integrals $\int_X f\,d\lambda^+$ and $\int_X f\,d\lambda^-$ and their difference $\int_X f\,d\lambda^+ - \int_X f\,d\lambda^-$ all exists in $\overline{\mathbb{R}}$. If $\int_X f\,d\lambda$ exists in $\overline{\mathbb{R}}$, we say that f is λ semi-integrable on X. If $\int_X f\,d\lambda$ exists in $\mathbb{R}$, we say that f is λ-integrable on X.

Proposition 10.33. *Given a signed measure space $(X, \mathfrak{A}, \lambda)$. Let f be an extended real-valued $\mathfrak{A}$-measurable function on X. If f is λ semi-integrable on X, then we have*

$$\left|\int_X f\,d\lambda\right| \le \int_X |f|\,d|\lambda|.$$

Proof. By Definition 10.32 we have

$$\begin{aligned}\left|\int_X f\,d\lambda\right| &= \left|\int_X f\,d\lambda^+ - \int_X f\,d\lambda^-\right| \le \left|\int_X f\,d\lambda^+\right| + \left|\int_X f\,d\lambda^-\right| \\ &\le \int_X |f|\,d\lambda^+ + \int_X |f|\,d\lambda^- = \int_X |f|\,d|\lambda|\end{aligned}$$

where the last equality is by Proposition 10.31 and the fact that $\lambda^+ + \lambda^- = |\lambda|$. ∎

Proposition 10.34. *Let μ and ν be two positive measures, at least one of which is finite, on a measurable space $(X, \mathfrak{A})$. Let λ be a signed measure on $(X, \mathfrak{A})$ defined by setting $\lambda = \mu - \nu$. Let f be an extended real-valued $\mathfrak{A}$-measurable function on X. Then we have*

$$\int_X f\,d\lambda = \int_X f\,d\mu - \int_X f\,d\nu,$$

in the sense that the existence of one side implies that of the other and the equality of the two.

Proof. Recalling the definition of $\int_X f\,d\lambda$ by Definition 10.32, we are to show

$$(1) \qquad \int_X f\,d\lambda^+ - \int_X f\,d\lambda^- = \int_X f\,d\mu - \int_X f\,d\nu,$$

in the sense that the existence of one side implies that of the other and the equality of the two. Now since $\lambda^+ - \lambda^- = \lambda = \mu - \nu$, we have

$$(2) \qquad \lambda^+(E) - \lambda^-(E) = \lambda(E) = \mu(E) - \nu(E) \quad \text{for every } E \in \mathfrak{A}.$$

Let $\varphi = \sum_{i=1}^k a_i \mathbf{1}_{E_i}$ be a simple function on $(X, \mathfrak{A})$ in its canonical expression with $a_i \in \mathbb{R}$

and $E_i \in \mathfrak{A}$ for $i = 1, \ldots, k$. Then we have

$$(3) \qquad \int_X \varphi\, d\lambda^+ - \int_X \varphi\, d\lambda^- = \sum_{i=1}^k a_i \lambda^+(E_i) - \sum_{i=1}^k a_i \lambda^-(E_i)$$

$$= \sum_{i=1}^k a_i \{\lambda^+(E_i) - \lambda^-(E_i)\} = \sum_{i=1}^k a_i \{\mu(E_i) - \nu(E_i)\}$$

$$= \sum_{i=1}^k a_i \mu(E_i) - \sum_{i=1}^k a_i \nu(E_i) = \int_X \varphi\, d\mu - \int_X \varphi\, d\nu.$$

Let f be a nonnegative extended real-valued $\mathfrak{A}$-measurable function on X. According to Lemma 8.6, there exists an increasing sequence of simple functions on $(X, \mathfrak{A})$, $(\varphi_n : n \in \mathbb{N})$, such that $\varphi_n \uparrow f$ on X. Applying (3) to φ_n, we have

$$(4) \qquad \int_X \varphi_n\, d\lambda^+ - \int_X \varphi_n\, d\lambda^- = \int_X \varphi_n\, d\mu - \int_X \varphi_n\, d\nu.$$

Letting $n \to \infty$ in (4) and applying the Monotone Convergence Theorem (Theorem 8.5), we have (1) for a nonnegative extended real-valued $\mathfrak{A}$-measurable function f on X in the sense that the existence of one side implies that of the other and the equality of the two.

If f is an extended real-valued $\mathfrak{A}$-measurable function on X, let $f = f^+ - f^-$. Applying the result above to the two nonnegative extended real-valued $\mathfrak{A}$-measurable functions f^+ and f^-, we have

$$(5) \qquad \int_X f^+\, d\lambda^+ - \int_X f^+\, d\lambda^- = \int_X f^+\, d\mu - \int_X f^+\, d\nu,$$

$$(6) \qquad \int_X f^-\, d\lambda^+ - \int_X f^-\, d\lambda^- = \int_X f^-\, d\mu - \int_X f^-\, d\nu,$$

in the sense that the existence of one side implies that of the other and the equality of the two. Subtracting (6) from (5) and assuming we do not encounter $\infty - \infty$ or $-\infty + \infty$, we have (1). ∎

Theorem 10.35. *Let λ be a signed measure on a measurable space $(X, \mathfrak{A})$. Let $c \in \mathbb{R}$ and consider the signed measure $c\lambda$ on $(X, \mathfrak{A})$. Let f be an extended real-valued $\mathfrak{A}$-measurable function on X. Then we have*

$$\int_X f\, d(c\lambda) = c \int_X f\, d\lambda,$$

in the sense that the existence of one side implies that of the other and the equality of the two.

Proof. Let us write $(c\lambda)^+$ and $(c\lambda)^-$ for the positive and negative parts of the signed measure $c\lambda$ on $(X, \mathfrak{A})$. By Definition 10.32, we have

$$(1) \qquad \int_X f\, d(c\lambda) = \int_X f\, d(c\lambda)^+ - \int_X f\, d(c\lambda)^-.$$

We have also

$$(2) \qquad c\lambda = c\{\lambda^+ - \lambda^-\} = c\lambda^+ - c\lambda^-.$$

1. Consider the case $c \geq 0$. Now $c \geq 0$ implies that $c\lambda^+$ and $c\lambda^-$ are measures on $(X, \mathfrak{A})$. Since λ^+ and λ^- are the positive and negative parts of the signed measure λ, we have $\lambda^+ \perp \lambda^-$ on $(X, \mathfrak{A})$ and this implies that $c\lambda^+ \perp c\lambda^-$ on $(X, \mathfrak{A})$. Then by the uniqueness of Jordan decomposition of a signed measure (Theorem 10.21), the equality (2) implies that $(c\lambda)^+ = c\lambda^+$ and $(c\lambda)^- = c\lambda^-$. Substituting these in (1), we have

$$(3) \qquad \begin{aligned} \int_X f\, d(c\lambda) &= \int_X f\, d(c\lambda^+) - \int_X f\, d(c\lambda^-) \\ &= c\Big\{\int_X f\, d\lambda^+ - \int_X f\, d\lambda^-\Big\} \\ &= c\int_X f\, d\lambda. \end{aligned}$$

This proves the theorem for the case $c \geq 0$.

2. Consider the case $c = -1$. Now $-\lambda = -\{\lambda^+ - \lambda^-\} = \lambda^- - \lambda^+$. Since λ^- and λ^+ are measures on $(X, \mathfrak{A})$ and $\lambda^- \perp \lambda^+$ on $(X, \mathfrak{A})$, uniqueness of Jordan decomposition implies that $(-\lambda)^+ = \lambda^-$ and $(-\lambda)^- = \lambda^+$. Substituting these in (1) for $c = -1$, we have

$$(4) \qquad \int_X f\, d(-\lambda) = \int_X f\, d\lambda^- - \int_X f\, d\lambda^+ = -\Big\{\int_X f\, d\lambda^+ - \int_X f\, d\lambda^-\Big\} = -\int_X f\, d\lambda.$$

This proves the theorem for the case $c = -1$.

3. Consider the case $c < 0$. Now if $c < 0$, then $c = -|c|$. Thus we have

$$(5) \qquad \begin{aligned} \int_X f\, d(c\lambda) &= \int_X f\, d\big(-|c|\lambda\big) = -\int_X f\, d\big(|c|\lambda\big) \\ &= -|c|\int_X f\, d\lambda = c\int_X f\, d\lambda, \end{aligned}$$

where the second equality is by (4) and the third equality is by (3). This proves the theorem for the case $c < 0$. ∎

Theorem 10.36. *Let λ_1 and λ_2 be two signed measures, at least one of which is finite, on a measurable space $(X, \mathfrak{A})$. Let $\lambda = \lambda_1 + \lambda_2$. Let f be an extended real-valued $\mathfrak{A}$-measurable function on X. Then we have*

$$\int_X f\, d\lambda = \int_X f\, d\lambda_1 + \int_X f\, d\lambda_2,$$

in the sense that the existence of one side implies that of the other and the equality of the two.

Proof. Let $\lambda_1 = \lambda_1^+ - \lambda_1^-$ and $\lambda_2 = \lambda_2^+ - \lambda_2^-$ be the Jordan decompositions of λ_1 and λ_2. We have then

$$(1) \qquad \lambda = \lambda_1 + \lambda_2 = \{\lambda_1^+ + \lambda_2^+\} - \{\lambda_1^- + \lambda_2^-\}.$$

Recall that for every signed measure, either its positive part or its negative part is finite and for a finite signed measure both the positive part and the negative part are finite. Then since at least one of λ_1 and λ_2 is a finite signed measure, at least three of the four positive measures λ_1^+, λ_1^-, λ_2^+, and λ_2^- are finite. Thus the expression (1) is defined. Now

$$(2) \qquad \begin{aligned} \int_X f\,d\lambda &= \int_X f\,d\{\lambda_1^+ + \lambda_2^+\} - \int_X f\,d\{\lambda_1^- + \lambda_2^-\} \\ &= \int_X f\,d\lambda_1^+ + \int_X f\,d\lambda_2^+ - \int_X f\,d\lambda_1^- - \int_X f\,d\lambda_2^- \\ &= \int_X f\,d\lambda_1^+ - \int_X f\,d\lambda_1^- + \int_X f\,d\lambda_2^+ - \int_X f\,d\lambda_2^- \\ &= \int_X f\,d\lambda_1 + \int_X f\,d\lambda_2, \end{aligned}$$

where the first equality is by (1) and Proposition 10.34, the second equality is by Proposition 10.31, and the last equality is by Definition 10.32. ∎

Problems

Prob. 10.1. Let $(X, \mathfrak{A}, \mu)$ be a measure space and let f be an extended real-valued $\mathfrak{A}$-measurable function on X. Suppose $\int_X f\,d\mu$ exists.
(a) Show that $\int_E f\,d\mu$ exists for every $E \in \mathfrak{A}$.
(b) Show that

$$\begin{array}{lll} \int_X f\,d\mu \in \mathbb{R} & \Rightarrow \int_E f\,d\mu \in \mathbb{R} & \text{for every } E \in \mathfrak{A}, \\ \int_X f\,d\mu = \infty & \Rightarrow \int_E f\,d\mu \neq -\infty & \text{for every } E \in \mathfrak{A}, \\ \int_X f\,d\mu = -\infty & \Rightarrow \int_E f\,d\mu \neq \infty & \text{for every } E \in \mathfrak{A}. \end{array}$$

Prob. 10.2. Let $(X, \mathfrak{A}, \lambda)$ be a signed measure space. Let f be an extended real-valued $\mathfrak{A}$-measurable function on a set $A \in \mathfrak{A}$. Let $A_0 \in \mathfrak{A}$ and $A_0 \subset A$.
(a) Show that if $\int_A f\,d\lambda$ exists in $\overline{\mathbb{R}}$, then so does $\int_{A_0} f\,d\lambda$.
(b) Show that if $\int_A f\,d\lambda$ exists in $\mathbb{R}$, then so does $\int_{A_0} f\,d\lambda$.

Prob. 10.3. Given a measure space $(X, \mathfrak{A}, \mu)$. Let f be a μ semi-integrable extended real-valued $\mathfrak{A}$-measurable function on X. Define a signed measure λ on $(X, \mathfrak{A})$ by setting $\lambda(E) = \int_E f\,d\mu$ for $E \in \mathfrak{A}$.
(a) Find a Hahn decomposition of $(X, \mathfrak{A}, \lambda)$ in terms of f.
(b) Find the positive and negative variations λ^+ and λ^- of λ in terms of f.
(c) Find the total variation $\|\lambda\|$ of λ in terms of f.

Prob. 10.4. Consider the measure space $(X, \mathfrak{A}, \mu) = \big([0, 2\pi], \mathfrak{M}_L \cap [0, 2\pi], \mu_L\big)$. Define a signed measure λ on $([0, 2\pi], \mathfrak{M}_L \cap [0, 2\pi])$ by setting $\lambda(E) = \int_E \sin x\, \mu_L(dx)$ for $E \in \mathfrak{M}_L \cap [0, 2\pi]$. Let $C = \left[\frac{4}{3}\pi, \frac{5}{3}\pi\right]$. Let $\varepsilon > 0$ be arbitrarily given. Find a $\mathfrak{M}_L \cap [0, 2\pi]$-measurable subset C' of C such that $\lambda(C') \geq \lambda(C)$ and $\lambda(E) > -\varepsilon$ for every $\mathfrak{M}_L \cap [0, 2\pi]$-measurable subset E of C'.

Prob. 10.5. Given a signed measure space $(X, \mathfrak{A}, \lambda)$.
(a) Show that if $E \in \mathfrak{A}$ and $\lambda(E) > 0$, then there exists a subset E_0 of E which is a positive

set for λ with $\lambda(E_0) \geq \lambda(E)$.
(a) Show that if $E \in \mathfrak{A}$ and $\lambda(E) < 0$, then there exists a subset E_0 of E which is a negative set for λ with $\lambda(E_0) \leq \lambda(E)$.

Prob. 10.6. Let $(X, \mathfrak{A}, \lambda)$ be a signed measure space. Prove the following statements:
(a) If A is a maximal positive set for λ, then A^c is a maximal negative set for λ.
(b) If B is a maximal negative set for λ, then B^c is a maximal positive set for λ.
(c) A is a maximal positive set for λ if and only if A^c is a maximal negative set for λ.
(d) B is a maximal negative set for λ if and only if B^c is a maximal positive set for λ.

Prob. 10.7. Let μ and ν be two positive measures on a measurable space $(X, \mathfrak{A})$. Suppose $\mu \perp \nu$. Show that if ρ is a positive measure on $(X, \mathfrak{A})$ and if ρ is nontrivial in the sense that $\rho(X) > 0$, then $(\mu + \rho) \perp (\nu + \rho)$ does not hold.

Prob. 10.8. Let $(X, \mathfrak{A}, \lambda)$ be a signed measure space. Show that for the positive and negative variations λ^+ and λ^- of λ, we have

$$\lambda^+(E) = \sup\{\lambda(E_0) : E_0 \subset E, E_0 \in \mathfrak{A}\},$$
$$\lambda^-(E) = -\inf\{\lambda(E_0) : E_0 \subset E, E_0 \in \mathfrak{A}\}.$$

Prob. 10.9. Given a signed measure space $(X, \mathfrak{A}, \lambda)$. For $E \in \mathfrak{A}$, let

$$\alpha_E = \sup\left\{\sum_{k=1}^n |\lambda(E_k)| : \{E_1, \ldots, E_n\} \subset \mathfrak{A}, \text{ disjoint}, \bigcup_{k=1}^n E_k = E, n \in \mathbb{N}\right\},$$

and

$$\beta_E = \sup\left\{\sum_{n\in\mathbb{N}} |\lambda(E_n)| : \{E_n : n \in \mathbb{N}\} \subset \mathfrak{A}, \text{ disjoint}, \bigcup_{n\in\mathbb{N}} E_n = E\right\}.$$

Show that $\alpha_E = \beta_E$.

Prob. 10.10. Let μ and ν be two positive measures on a measurable space $(X, \mathfrak{A})$. Suppose for every $\varepsilon > 0$, there exists $E \in \mathfrak{A}$ such that $\mu(E) < \varepsilon$ and $\nu(E^c) < \varepsilon$. Show that $\mu \perp \nu$.

§11 Absolute Continuity of a Measure

[I] The Radon-Nikodym Derivative

Let $(X, \mathfrak{A}, \mu)$ be a measure space. Let f be a μ semi-integrable extended real-valued $\mathfrak{A}$-measurable function on X. The set function λ on $\mathfrak{A}$ defined by $\lambda(E) = \int_E f \, d\mu$ for $E \in \mathfrak{A}$ is a signed measure on $(X, \mathfrak{A})$ as we showed in Proposition 10.3. In particular if f is nonnegative then λ is a positive measure on $(X, \mathfrak{A})$ by Proposition 8.12. Conversely if λ is a signed measure on $(X, \mathfrak{A})$, does there exist an extended real-valued $\mathfrak{A}$-measurable function f on X such that $\lambda(E) = \int_E f \, d\mu$ for every $E \in \mathfrak{A}$? If such a function exists, then $\lambda(E) = 0$ for every $E \in \mathfrak{A}$ with $\mu(E) = 0$. In the Radon-Nikodym Theorem we show that when μ and λ are σ-finite then this necessary condition is also a sufficient condition for the existence of such a function f.

Definition 11.1. (Radon-Nikodym Derivative) *Let μ be a positive measure and λ be a signed measure on a measurable space $(X, \mathfrak{A})$. If there exists an extended real-valued $\mathfrak{A}$-measurable function f on X such that $\lambda(E) = \int_E f \, d\mu$ for every $E \in \mathfrak{A}$, then f is called a Radon-Nikodym derivative of λ with respect to μ and we write $\frac{d\lambda}{d\mu}$ for it.*

Let us note that if a Radon-Nikodym derivative f of λ with respect to μ exists, then $\int_X f \, d\mu = \lambda(X) \in \overline{\mathbb{R}}$ and thus f is necessarily μ semi-integrable but it need not be μ-integrable on X.

Proposition 11.2. (a) *Let f be a Radon-Nikodym derivative of a signed measure λ with respect to a positive measure μ on a measurable space $(X, \mathfrak{A})$. If g is an extended real-valued $\mathfrak{A}$-measurable function on X such that $f = g$, μ-a.e. on X, then g too is a Radon-Nikodym derivative of λ with respect to μ.*

(b) (Uniqueness of the Radon-Nikodym Derivative) *Let μ be a σ-finite positive measure and λ be a signed measure on a measurable space $(X, \mathfrak{A})$. If two extended real-valued $\mathfrak{A}$-measurable functions f and g are Radon-Nikodym derivatives of λ with respect to μ, then $f = g$, μ-a.e. on X.*

Proof. 1. To prove (a), note that if g is an extended real-valued $\mathfrak{A}$-measurable function on X such that $f = g$, μ-a.e. on X then for every $E \in \mathfrak{A}$ we have $\int_E g \, d\mu = \int_E f \, d\mu = \lambda(E)$ by Lemma 9.8. This shows that g is a Radon-Nikodym derivative of λ with respect to μ.

2. Let us prove (b). Now if two extended real-valued $\mathfrak{A}$-measurable functions f and g are Radon-Nikodym derivatives of λ with respect to μ, then f and g are μ semi-integrable on X and $\int_E f \, d\mu = \lambda(E) = \int_E g \, d\mu$ for every $E \in \mathfrak{A}$. Since μ is σ-finite, this implies that $f = g$, μ-a.e. on X by Proposition 9.16. ∎

Let μ be a positive measure and λ be a signed measure on a measurable space $(X, \mathfrak{A})$. Consider the set Φ of all extended real-valued $\mathfrak{A}$-measurable functions on X. The relation that $f = g$, μ-a.e. on X for $f, g \in \Phi$ is an equivalence relation. Consider the collection of all equivalence classes with respect to this equivalence relation. If $f \in \Phi$ is a Radon-Nikodym derivative of λ with respect to μ, then every member in the equivalence class to which f belongs is a Radon-Nikodym derivative of λ with respect to μ as we showed in (a) of Proposition 11.2. This equivalence class is also called a Radon-Nikodym derivative of λ with respect to μ. In this terminology each member of the equivalence class is called a version, or a representative, of the Radon-Nikodym derivative. When we speak of a Radon-Nikodym derivative, whether a function or an equivalence class of functions is meant should be clear from the context.

Remark 11.3. The uniqueness of the Radon-Nikodym derivative of λ with respect to μ in (b) of Proposition 11.2 does not hold without the σ-finiteness of μ.

Example. For an arbitrary nonempty set X, let $\mathfrak{A} = \{\emptyset, X\}$. Let μ be a positive measure on $(X, \mathfrak{A})$ defined by setting $\mu(\emptyset) = 0$ and $\mu(X) = \infty$. The measure μ is not σ-finite. Now $\lambda(\emptyset) = 0 = \int_\emptyset 1\,d\mu$ and $\lambda(X) = \mu(X) = \int_X 1\,d\mu$. This shows that the function $f_1 = 1$ on X is a Radon-Nikodym derivative of λ with respect to μ. Next let $c \in \mathbb{R}$, $c > 0$, and let $f_c = c$ on X. Then we have $\lambda(\emptyset) = 0 = \int_\emptyset c\,d\mu$ and $\lambda(X) = \mu(X) = \infty = \int_X c\,d\mu$. This shows that the function $f_c = c$ on X is a Radon-Nikodym derivative of λ with respect to μ. Consider the two Radon-Nikodym derivatives f_1 and f_c of λ with respect to μ. If $c \neq 1$ then $f_1 \neq f_c$ on X and in particular we do not have $f_1 = f_c$, μ-a.e. on X.

[II] Absolute Continuity of a Signed Measure Relative to a Positive Measure

Definition 11.4. (Absolute Continuity) *Let μ be a positive measure and λ be a signed measure on a measurable space $(X, \mathfrak{A})$. We say that λ is absolutely continuous with respect to μ and write $\lambda \ll \mu$ if for every $E \in \mathfrak{A}$ with $\mu(E) = 0$ we have $\lambda(E) = 0$, or equivalently, every null set for the positive measure μ is a null set for the signed measure λ. (Note that since a positive measure has the monotonicity property, $\mu(E) = 0$ for some $E \in \mathfrak{A}$ implies $\mu(E_0) = 0$ for every $E_0 \in \mathfrak{A}$ such that $E_0 \subset E$.)*

Definition 11.5. (Lebesgue Decomposition) *Let μ be a positive measure and λ be a signed measure on a measurable space $(X, \mathfrak{A})$. If there exist two signed measures λ_a and λ_s on $(X, \mathfrak{A})$ such that $\lambda_a \ll \mu$, $\lambda_s \perp \mu$ and $\lambda = \lambda_a + \lambda_s$, then we call $\{\lambda_a, \lambda_s\}$ a Lebesgue decomposition of λ with respect to μ. We call λ_a and λ_s the absolutely continuous part and the singular part of λ with respect to μ.*

Remark 11.6. If $\{\lambda_a, \lambda_s\}$ is a Lebesgue decomposition of λ with respect to μ, then we have $\lambda_a \perp \lambda_s$ also. Thus a Lebesgue decomposition of λ is a pair of signed measures $\{\lambda_a, \lambda_s\}$ such that $\lambda = \lambda_a + \lambda_s$, $\lambda_a \perp \lambda_s$, $\lambda_a \ll \mu$, and $\lambda_s \perp \mu$.

Proof. Since $\lambda_s \perp \mu$, there exist $C_1, C_2 \in \mathfrak{A}$ such that $C_1 \cap C_2 = \emptyset$, $C_1 \cup C_2 = X$, C_1 is a null set for μ, and C_2 is a null set for λ_s. Since $\lambda_a \ll \mu$, every null set for μ is a null set for λ_a and in particular C_1 is a null set for λ_a. Therefore we have $\lambda_a \perp \lambda_s$. ∎

Observation 11.7. Let μ and ν be two positive measures on a measurable space $(X, \mathfrak{A})$. In general there does not exist a positive number $c > 0$ such that $c\mu \le \nu$ on $\mathfrak{A}$, not even when μ is a finite positive measure.

Example 1. Let A and B be two non-empty subset of a set X such that $A \cap B = \emptyset$ and $A \cup B = X$. Then $\mathfrak{A} = \{\emptyset, A, B, X\}$ is a σ-algebra of subsets of X. Let us define two set functions μ and ν on $\mathfrak{A}$ by setting

$$\mu(\emptyset) = 0, \quad \mu(A) = 1, \quad \mu(B) = 0, \quad \mu(x) = 1;$$
$$\nu(\emptyset) = 0, \quad \nu(A) = 0, \quad \nu(B) = 1, \quad \mu(x) = 1.$$

Then μ and ν are two finite positive measures on $(X, \mathfrak{A})$ and moreover $\mu \perp \nu$ on $(X, \mathfrak{A})$. Now no matter how small $c > 0$ may be, we have $c\mu(A) = c > 0 = \nu(A)$. Thus there does not exist $c > 0$ such that $c\mu \le \nu$ on $\mathfrak{A}$. Similarly there does not exist $c > 0$ such that $c\nu \le \mu$ on $\mathfrak{A}$.

Example 2. Consider the measure space $(X, \mathfrak{A}, \mu) = \big([0,1], \mathfrak{M}_L \cap [0,1], \mu_L\big)$. Let f be a function on $[0, 1]$ defined by

$$f(x) = \begin{cases} \left|x \sin \frac{1}{x}\right| & \text{for } x \in (0, 1], \\ 0 & \text{for } x = 0. \end{cases}$$

Let us define a set function ν on $\mathfrak{A}$ by setting $\nu(E) = \int_E f \, d\mu_L$ for $E \in \mathfrak{E}$. Then ν is a positive measure on $(X, \mathfrak{A})$ by Proposition 8.12. Now since $\lim_{x \to 0} f(x) = 0$, for every $c > 0$ there exists $a \in (0, 1]$ such that $f(x) < \frac{c}{2}$ for $x \in [0, a]$. Then we have

$$\nu([0,a]) = \int_{[0,a]} f \, d\mu_L \le \frac{c}{2}\mu_L([0,a]) < c\mu_L([0,a]).$$

This shows that there does not exist $c > 0$ such that $c\mu_L \le \nu$ on $\mathfrak{A}$.

For a measurable space $(X, \mathfrak{A})$ and for a set $E \in \mathfrak{A}$, we write $\mathfrak{A} \cap E = \{A \cap E : A \in \mathfrak{A}\}$, that is, the collection of all $\mathfrak{A}$-measurable subsets of E.

Lemma 11.8. *Let μ be a finite positive measure and ν be a positive measure on a measurable space $(X, \mathfrak{A})$. Suppose μ and ν are not mutually singular. Then there exists $c > 0$ and $E \in \mathfrak{A}$ with $\mu(E) > 0$ such that $c\mu \le \nu$ on $\mathfrak{A} \cap E$, (that is, $c\mu(E_0) \le \nu(E_0)$ for every $E_0 \in \mathfrak{A}$ such that $E_0 \subset E$, in other words, E is a positive set for the signed measure $\nu - c\mu$).*

Proof. For every $n \in \mathbb{N}$, $\frac{1}{n}\mu$ is a finite positive measure on $(X, \mathfrak{A})$ and thus $\nu - \frac{1}{n}\mu$ is a signed measure on $(X, \mathfrak{A})$ by Proposition 10.4. Let $\{A_n, B_n\}$ be a Hahn decomposition for the signed measure space $\big(X, \mathfrak{A}, \nu - \frac{1}{n}\mu\big)$ for each $n \in \mathbb{N}$. Let $A = \bigcup_{n \in \mathbb{N}} A_n$ and $B = \bigcap_{n \in \mathbb{N}} B_n$. Note that $A^c = \big(\bigcup_{n \in \mathbb{N}} A_n\big)^c = \bigcap_{n \in \mathbb{N}} B_n = B$. Thus $A \cap B = \emptyset$ and $A \cup B = X$. Since $B \subset B_n$ for every $n \in \mathbb{N}$ and since B_n is a negative set for $\nu - \frac{1}{n}\mu$, B is a negative set for $\nu-\frac{1}{n}\mu$ for every $n \in \mathbb{N}$. Then $(\nu-\frac{1}{n}\mu)(B) \leq 0$, that is, $0 \leq \nu(B) \leq \frac{1}{n}\mu(B)$ for every $n \in \mathbb{N}$. Thus $\nu(B) = 0$. Now if $\mu(A) = 0$, then $\mu \perp \nu$. On the other hand, if $\mu(A) > 0$, then since $\mu(A) \leq \sum_{n\in\mathbb{N}} \mu(A_n)$ there exists $n_0 \in \mathbb{N}$ such that $\mu(A_{n_0}) > 0$. Now $\{A_{n_0}, B_{n_0}\}$ being a Hahn decomposition of $(X, \mathfrak{A}, \nu - \frac{1}{n_0}\mu)$, the set A_{n_0} is a positive set for $\nu - \frac{1}{n_0}\mu$. Let us select $E = A_{n_0}$ and $c = \frac{1}{n_0}$. Then $\mu(E) > 0$, $c > 0$ and $E = A_{n_0}$ is a positive set for the signed measure $\nu - c\mu = \nu - \frac{1}{n_0}\mu$. ∎

Let us note that if λ_1 and λ_2 are two signed measures on a measurable space $(X, \mathfrak{A})$ and if $\lambda_1(E) + \lambda_2(E)$ exists in $\overline{\mathbb{R}}$ for every $E \in \mathfrak{A}$, then $\lambda_1 + \lambda_2$ is defined and is a signed measure on $(X, \mathfrak{A})$. This is the case when one of λ_1 and λ_2 does not assume the value ∞ or when one of λ_1 and λ_2 does not assume the value $-\infty$.

Lemma 11.9. (a) *Let λ_1 and λ_2 be signed measures and μ be a positive measure on a measurable space $(X, \mathfrak{A})$. If $\lambda_1 \ll \mu$, $\lambda_2 \ll \mu$, and $\lambda_1+\lambda_2$ is defined, then $\{\lambda_1+\lambda_2\} \ll \mu$.*
(b) *Let λ_1, λ_2, and λ be signed measures on a measurable space $(X, \mathfrak{A})$. If $\lambda_1 \perp \lambda$, $\lambda_2 \perp \lambda$, and $\lambda_1 + \lambda_2$ is defined, then $\{\lambda_1 + \lambda_2\} \perp \lambda$.*

Proof. (a) is immediate. Let us prove (b). If $\lambda_1 \perp \lambda$, then there exist $A_1, B_1 \in \mathfrak{A}$ such that $A_1 \cap B_1 = \emptyset$, $A_1 \cup B_1 = X$, A_1 is a null set for λ and B_1 is a null set for λ_1. Similarly if $\lambda_2 \perp \lambda$, then there exist $A_2, B_2 \in \mathfrak{A}$ such that $A_2 \cap B_2 = \emptyset$, $A_2 \cup B_2 = X$, A_2 is a null set for λ and B_2 is a null set for λ_2. Consider $A_1 \cup A_2$ and $(A_1 \cup A_2)^c = A_1^c \cap A_2^c = B_1 \cap B_2$. Since A_1 and A_2 are null sets for λ, $A_1 \cup A_2$ is a null set for λ by Lemma 10.11. Since B_1 is a null set for λ_1, so is its $\mathfrak{A}$-measurable subset $B_1 \cap B_2$. Similarly $B_1 \cap B_2$ is a null set for λ_2. Let E be an arbitrary $\mathfrak{A}$-measurable subset of $B_1 \cap B_2$. Then $\lambda_1(E) = 0$ and $\lambda_2(E) = 0$ so that $\{\lambda_1 + \lambda_2\}(E) = \lambda_1(E) + \lambda_2(E) = 0$. Thus $B_1 \cap B_2$ is a null set for $\lambda_1 + \lambda_2$. Therefore $A_1 \cup A_2$ is a null set for λ and its complement $B_1 \cap B_2$ is a null set for $\lambda_1 + \lambda_2$. This shows that $\{\lambda_1 + \lambda_2\} \perp \lambda$. ∎

Theorem 11.10. (Existence of Lebesgue Decomposition)(Lebesgue-Radon-Nikodym) *Let μ be a σ-finite positive measure and λ be a σ-finite signed measure on a measurable space $(X, \mathfrak{A})$. Then there exist two signed measures λ_a and λ_s on $(X, \mathfrak{A})$ such that $\lambda_a \ll \mu$, $\lambda_s \perp \mu$, $\lambda = \lambda_a + \lambda_s$ and λ_a is defined by $\lambda_a(E) = \int_E f\,d\mu$ for $E \in \mathfrak{A}$ where f is an extended real-valued $\mathfrak{A}$-measurable function on X.*

Proof. 1. Consider the case that both μ and λ are finite positive measures. Let Φ be a collection of nonnegative extended real-valued $\mathfrak{A}$-measurable functions φ on X defined by

$$\Phi = \Big\{\varphi : \varphi \geq 0 \text{ on } X \text{ and } \int_E \varphi\,d\mu \leq \lambda(E) \text{ for every } E \in \mathfrak{A}\Big\}.$$

Since the identically vanishing function on X is such a function, $\Phi \neq \emptyset$. Let us observe that if $\varphi_1, \varphi_2 \in \Phi$, then $\max\{\varphi_1, \varphi_2\} \in \Phi$. Indeed if we let $A = \{x \in X : \varphi_1(x) > \varphi_2(x)\}$, then for every $E \in \mathfrak{A}$ we have

$$\begin{aligned}\int_E \max\{\varphi_1, \varphi_2\}\, d\mu &= \int_{E\cap A} \max\{\varphi_1, \varphi_2\}\, d\mu + \int_{E\setminus A} \max\{\varphi_1, \varphi_2\}\, d\mu \\ &= \int_{E\cap A} \varphi_1\, d\mu + \int_{E\setminus A} \varphi_2\, d\mu \\ &\leq \lambda(E\cap A) + \lambda(E\setminus A) = \lambda(E).\end{aligned}$$

This shows that $\max\{\varphi_1, \varphi_2\} \in \Phi$. It follows by induction that for every finite collection $\{\varphi_1, \ldots, \varphi_n\} \subset \Phi$, we have $\max\{\varphi_1, \ldots, \varphi_n\} \in \Phi$. Let

$$\alpha = \sup_{\varphi\in\Phi} \int_X \varphi\, d\mu.$$

Let us show that there exists $f \in \Phi$ such that

$$\int_X f\, d\mu = \alpha.$$

Note that since $\int_X \varphi\, d\mu \leq \lambda(X)$ for every $\varphi \in \Phi$, we have $0 \leq \alpha \leq \lambda(X) < \infty$. By the definition of α there exists a sequence $(\varphi_n : n \in \mathbb{N})$ in Φ such that $\lim_{n\to\infty} \int_X \varphi_n\, d\mu = \alpha$. Let $f = \sup_{n\in\mathbb{N}} \varphi_n$. To show $f \in \Phi$, let $\psi_n = \max\{\varphi_1, \ldots, \varphi_n\}$ for $n \in \mathbb{N}$. Then $(\psi_n : n \in \mathbb{N})$ is an increasing sequence in Φ. Now since $\psi_n \geq \varphi_n$ for every $n \in \mathbb{N}$, we have $\sup_{n\in\mathbb{N}} \psi_n \geq \sup_{n\in\mathbb{N}} \varphi_n = f$. On the other hand, $\psi_n = \max\{\varphi_1, \ldots, \varphi_n\} \leq \sup_{n\in\mathbb{N}} \varphi_n = f$ for every $n \in \mathbb{N}$ so that $\sup_{n\in\mathbb{N}} \psi_n \leq f$. Thus $f = \sup_{n\in\mathbb{N}} \psi_n$. Then since $(\psi_n : n \in \mathbb{N})$ is an increasing sequence, we have $\psi_n \uparrow f$ on X. Then by the Monotone Convergence Theorem (Theorem 8.5) and by the fact that $\int_E \psi_n\, d\mu \leq \lambda(E)$ for every $E \in \mathfrak{A}$, we have

$$(1) \qquad \int_E f\, d\mu = \lim_{n\to\infty} \int_E \psi_n\, d\mu \leq \lambda(E).$$

This shows that $f \in \Phi$. To show that $\int_X f\, d\mu = \alpha$, note that from the fact that $\psi_n \geq \varphi_n$ on X, we have $\int_X f\, d\mu = \lim_{n\to\infty} \int_X \psi_n\, d\mu \geq \lim_{n\to\infty} \int_X \varphi_n\, d\mu = \alpha$. On the other hand, since $f \in \Phi$ we have $\int_X f\, d\mu \leq \alpha$. Thus $\int_X f\, d\mu = \alpha$.

With our function f define a set function λ_a on $\mathfrak{A}$ by

$$(2) \qquad \lambda_a(E) = \int_E f\, d\mu \quad \text{for } E \in \mathfrak{A}.$$

Since $f \geq 0$ on X, λ_a is a positive measure on $\mathfrak{A}$. Since $\lambda_a(X) = \int_X f\, d\mu = \alpha < \infty$, λ_a is a finite measure. For every $E \in \mathfrak{A}$ with $\mu(E) = 0$, we have $\int_E f\, d\mu = 0$ and thus $\lambda_a \ll \mu$. By (1) and (2), $\lambda_a(E) \leq \lambda(E)$ for every $E \in \mathfrak{A}$ and thus $\lambda \geq \lambda_a$ on $\mathfrak{A}$. Since both λ and λ_a are finite measures, if we let $\lambda_s = \lambda - \lambda_a$ then λ_s is defined and is a finite positive measure on $\mathfrak{A}$. Let us show that $\lambda_s \perp \mu$. Suppose not. Then by Lemma 11.8, there exist a set $E_0 \in \mathfrak{A}$ with $\mu(E_0) > 0$ and a constant $c > 0$ such that $\lambda_s \geq c\mu$ on $\mathfrak{A} \cap E_0$.

Consider the nonnegative extended real-valued $\mathfrak{A}$-measurable function $f + c\mathbf{1}_{E_0}$ on X. Let us show that this function is a member of Φ. Now for every $E \in \mathfrak{A}$, we have

$$\int_E \{f + c\mathbf{1}_{E_0}\}\,d\mu = \int_E f\,d\mu + c\mu(E \cap E_0) \le \lambda_a(E) + \lambda_s(E \cap E_0)$$
$$\le \lambda_a(E) + \lambda_s(E) = \lambda(E).$$

This shows that $f + c\mathbf{1}_{E_0} \in \Phi$. Now $\int_X \{f + c\mathbf{1}_{E_0}\}\,d\mu = \alpha + c\mu(E_0) > \alpha$ since $\mu(E_0) > 0$. This contradicts the definition of α. Thus we have $\lambda_s \perp \mu$.

2. Next consider the case that both μ and λ are σ-finite positive measures on $(X, \mathfrak{A})$. Now the σ-finiteness of μ implies that there exists a disjoint sequence $(F_m : m \in \mathbb{N})$ in $\mathfrak{A}$ such that $\bigcup_{m\in\mathbb{N}} F_m = X$ with $\mu(F_m) < \infty$ for every $m \in \mathbb{N}$ and similarly the σ-finiteness of λ implies that there exists a disjoint sequence $(G_n : n \in \mathbb{N})$ in $\mathfrak{A}$ such that $\bigcup_{n\in\mathbb{N}} G_n = X$ with $\mu(G_n) < \infty$ for every $n \in \mathbb{N}$. Then $\{F_m \cap G_n : m \in \mathbb{N}, n \in \mathbb{N}\}$ is a disjoint collection in $\mathfrak{A}$, $\bigcup_{m\in\mathbb{N}} \bigcup_{n\in\mathbb{N}} F_m \cap G_n = X$ with $\mu(F_m \cap G_n) < \infty$ and $\lambda(F_m \cap G_n) < \infty$ for every $m \in \mathbb{N}$ and $n \in \mathbb{N}$. Let $(X_n : n \in \mathbb{N})$ be a renumbering of the countable collection $\{F_m \cap G_n : m \in \mathbb{N}, n \in \mathbb{N}\}$. Then $(X_n : n \in \mathbb{N})$ is a disjoint sequence in $\mathfrak{A}$, $\bigcup_{n\in\mathbb{N}} X_n = X$ with $\mu(X_n) < \infty$ and $\lambda(X_n) < \infty$ for every $n \in \mathbb{N}$.

Let μ_n and λ_n be the restrictions of μ and λ to the measurable space $(X_n, \mathfrak{A} \cap X_n)$ for $n \in \mathbb{N}$. Then μ_n and λ_n are finite positive measures on the measurable space $(X_n, \mathfrak{A} \cap X_n)$. Thus by our result in **1** we have

$$(3) \qquad \lambda_n = \lambda_{n,a} + \lambda_{n,s};\ \lambda_{n,a} \ll \mu_n;\ \lambda_{n,s} \perp \mu_n \quad \text{on } (X_n, \mathfrak{A} \cap X_n),$$

$$(4) \qquad \lambda_{n,a}(E) \int_E f_n\,d\mu_n \quad \text{for } E \in \mathfrak{A} \cap X_n$$

where f_n is a nonnegative extended real-valued $\mathfrak{A} \cap X_n$-measurable function on X_n.

Let us define two set functions λ_a and λ_s on $\mathfrak{A}$ by setting

$$(5) \qquad \lambda_a(E) = \sum_{n\in\mathbb{N}} \lambda_{n,a}(E \cap X_n) \quad \text{for } E \in \mathfrak{A},$$

$$(6) \qquad \lambda_s(E) = \sum_{n\in\mathbb{N}} \lambda_{n,s}(E \cap X_n) \quad \text{for } E \in \mathfrak{A}.$$

It follows immediately that λ_a and λ_s are two positive measures on $(X, \mathfrak{A})$. To show the countable additivity of λ_a on $\mathfrak{A}$ for instance, let $(E_k : k \in \mathbb{N})$ be a disjoint sequence in $\mathfrak{A}$. Then we have

$$\lambda_a\Big(\bigcup_{k\in\mathbb{N}} E_k\Big) = \sum_{n\in\mathbb{N}} \lambda_{n,a}\Big(\bigcup_{k\in\mathbb{N}} E_k \cap X_n\Big) = \sum_{n\in\mathbb{N}} \sum_{k\in\mathbb{N}} \lambda_{n,a}(E_k \cap X_n)$$
$$= \sum_{k\in\mathbb{N}} \sum_{n\in\mathbb{N}} \lambda_{n,a}(E_k \cap X_n) = \sum_{k\in\mathbb{N}} \lambda_a(E_k).$$

It remains to show that for the two positive measures λ_a and λ_s on $(X, \mathfrak{A})$ defined respectively by (5) and (6) we have

$$(7) \qquad \lambda = \lambda_a + \lambda_s;\ \lambda_a \ll \mu;\ \lambda_s \perp \mu \quad \text{on } (X, \mathfrak{A}).$$

Now for every $E \in \mathfrak{A}$ we have

$$\begin{aligned}\lambda(E) &= \sum_{n\in\mathbb{N}} \lambda(E\cap X_n) = \sum_{n\in\mathbb{N}} \lambda_n(E\cap X_n)\\ &= \sum_{n\in\mathbb{N}} \{\lambda_{n,a}(E\cap X_n) + \lambda_{n,s}(E\cap X_n)\}\\ &= \sum_{n\in\mathbb{N}} \lambda_{n,a}(E\cap X_n) + \sum_{n\in\mathbb{N}} \lambda_{n,s}(E\cap X_n)\\ &= \lambda_a(E) + \lambda_s(E)\end{aligned}$$

where the third equality is by (3) and the last equality is by (5) and (6). This shows that we have $\lambda = \lambda_a + \lambda_s$ on $(X, \mathfrak{A})$.

Let us show that $\lambda_a \ll \mu$ on $(X, \mathfrak{A})$. Let E be a null set in $(X, \mathfrak{A}, \mu)$. Then $\mu(E) = 0$ and this implies $\mu_n(E \cap X_n) = \mu(E \cap X_n) = 0$ for every $n \in \mathbb{N}$. Since $\lambda_{n,a} \ll \mu_n$ on $(X_n, X_n \cap \mathfrak{A})$ by (3), we have $\lambda_{n,a}(E \cap X_n) = 0$ for every $n \in \mathbb{N}$. Then by (5) we have $\lambda_a(E) = \sum_{n\in\mathbb{N}} \lambda_{n,a}(E \cap X_n) = 0$. Thus E is a null set in $(X, \mathfrak{A}, \lambda_a)$. This shows that $\lambda_a \ll \mu$ on $(X, \mathfrak{A})$.

Let us show that $\lambda_s \perp \mu$ on $(X, \mathfrak{A})$. By (3) we have $\lambda_{n,s} \perp \mu_n$ on $(X_n, \mathfrak{A} \cap X_n)$ for every $n \in \mathbb{N}$. Thus there exist $A_n, B_n \in \mathfrak{A} \cap X_n$ such that $A_n \cap B_n = \emptyset$, $A_n \cup B_n = X_n$, B_n is a null set for $\lambda_{n,s}$, and A_n is a null set for μ_n. Let $A = \bigcup_{n\in\mathbb{N}} A_n$ and $B = \bigcup_{n\in\mathbb{N}} B_n$. Then $A, B \in \mathfrak{A}$. Since $(X_n : n \in \mathbb{N})$ is a disjoint sequence and A_n and B_n are disjoint subsets of X_n for every $n \in \mathbb{N}$, $\{A_n, B_n : n \in \mathbb{N}\}$ is a disjoint collection. Then we have

$$\begin{aligned}A\cap B &= \Big(\bigcup_{n\in\mathbb{N}} A_n\Big) \cap \Big(\bigcup_{n\in\mathbb{N}} B_n\Big) = \bigcup_{n\in\mathbb{N}} \Big(A_n \cap \Big(\bigcup_{n\in\mathbb{N}} B_n\Big)\Big) = \bigcup_{n\in\mathbb{N}} \emptyset = \emptyset,\\ A\cup B &= \Big(\bigcup_{n\in\mathbb{N}} A_n\Big) \cup \Big(\bigcup_{n\in\mathbb{N}} B_n\Big) = \bigcup_{n\in\mathbb{N}} (A_n \cup B_n) = \bigcup_{n\in\mathbb{N}} X_n = X.\end{aligned}$$

Now $\mu(A) = \sum_{n\in\mathbb{N}} \mu(A \cap X_n) = \sum_{n\in\mathbb{N}} \mu_n(A_n) = 0$ since A_n is a null set for μ_n. Also by (6) we have $\lambda_s(B) = \sum_{n\in\mathbb{N}} \lambda_{n,s}(B \cap X_n) = \sum_{n\in\mathbb{N}} \lambda_{n,s}(B_n) = 0$ since B_n is a null set for $\lambda_{n,s}$. This shows that $\lambda_s \perp \mu$ on $(X, \mathfrak{A})$. This completes the proof of (7).

Let us show that for the positive measure λ_a on $(X, \mathfrak{A})$ defined by (5) there exists a nonnegative extended real-valued $\mathfrak{A}$-measurable function f on X such that

$$\lambda_a(E) = \int_E f\,d\mu \quad \text{for } E \in \mathfrak{A}. \tag{8}$$

With the nonnegative extended real-valued $\mathfrak{A} \cap X_n$-measurable function f_n on X_n defined in (4) for $n \in \mathbb{N}$, let us define a function f on $X = \bigcup_{n\in\mathbb{N}} X_n$ by setting

$$f(x) = f_n(x) \quad \text{for } x \in X_n.$$

Then f is a nonnegative extended real-valued $\mathfrak{A}$-measurable function on X. Moreover for every $E \in \mathfrak{A}$ we have

$$\begin{aligned}\int_E f\,d\mu &= \int_{E\cap\bigcup_{n\in\mathbb{N}} X_n} f\,d\mu = \sum_{n\in\mathbb{N}}\int_{E\cap X_n} f\,d\mu \\ &= \sum_{n\in\mathbb{N}}\int_{E\cap X_n} f_n\,d\mu_n = \sum_{n\in\mathbb{N}}\lambda_{n,a}(E) \\ &= \lambda_a(E)\end{aligned}$$

where the fourth equality is by (4) and the last equality is by (5). This proves (8).

3. Finally consider the case that μ is a σ-finite positive measure and λ is a σ-finite signed measure. Let $\lambda = \lambda^+ - \lambda^-$ be the Jordan decomposition of λ. Here λ^+ and λ^- are two positive measures at least one of which is a finite measure. The other is at least σ-finite since the σ-finiteness of λ implies that of $|\lambda|$ according to Observation 10.30 and this implies the σ-finiteness of both λ^+ and λ^-.

Let us consider the case that λ^+ is finite and λ^- is σ-finite. By (7) we have

(9) $$\lambda^+ = \lambda_a^+ + \lambda_s^+;\ \lambda_a^+ \ll \mu;\ \lambda_s^+ \perp \mu$$

(10) $$\lambda^- = \lambda_a^- + \lambda_s^-;\ \lambda_a^- \ll \mu;\ \lambda_s^- \perp \mu$$

and by (8) we have

(11) $$\lambda_a^+(E) = \int_E f_1\,d\mu \quad \text{for } E \in \mathfrak{A}$$

(12) $$\lambda_a^-(E) = \int_E f_2\,d\mu \quad \text{for } E \in \mathfrak{A}$$

where f_1 and f_2 are nonnegative extended real-valued $\mathfrak{A}$-measurable functions on X. Now by (9) and (10) we have

$$\lambda = \lambda^+ - \lambda^- = \{\lambda_a^+ + \lambda_s^+\} - \{\lambda_a^- + \lambda_s^-\} = \{\lambda_a^+ - \lambda_a^-\} + \{\lambda_s^+ - \lambda_s^-\}.$$

Note that finiteness of λ^+ implies that of λ_a^+ and λ_s^+ so that the differences $\lambda_a^+ - \lambda_a^-$ and $\lambda_s^+ - \lambda_s^-$ are defined. Let us define

(13) $$\lambda_a = \lambda_a^+ - \lambda_a^-;\ \lambda_s = \lambda_s^+ - \lambda_s^-.$$

Then $\lambda = \lambda_a + \lambda_s$. Also $\lambda_a^+ \ll \mu$ and $\lambda_a^- \ll \mu$ imply $\lambda_a = \lambda_a^+ - \lambda_a^- \ll \mu$ by (a) of Lemma 11.9 and $\lambda_s^+ \perp \mu$ and $\lambda_s^- \perp \mu$ imply $\lambda_s = \lambda_s^+ - \lambda_s^- \perp \mu$ by (b) of Lemma 11.9.

Let $f = f_1 - f_2$. Now since λ^+ is a finite signed measure and $\lambda^+ = \lambda_a^+ + \lambda_s^+$, λ_a^+ is a finite signed measure. Then by (11) we have $\int_X f_1\,d\mu = \lambda_a^+(X) \in \mathbb{R}$ so that f_1 is real-valued μ-a.e. on X. Thus $f = f_1 - f_2$ is defined μ-a.e. on X and is an extended real-valued $\mathfrak{A}$-measurable function. Then for every $E \in \mathfrak{A}$ we have

$$\begin{aligned}\int_E f\,d\mu &= \int_E \{f_1 - f_2\}\,d\mu = \int_E f_1\,d\mu - \int_E f_2\,d\mu \\ &= \lambda_a^+(E) - \lambda_a^-(E) = \lambda_a(E)\end{aligned}$$

where the third equality is by (11) and (12) and the last equality is by (13). This completes the proof that $\lambda = \lambda_a + \lambda_s$, $\lambda_a \ll \mu$, $\lambda_s \perp \mu$ and $\lambda_a(E) = \int_E f\,d\mu$ for $E \in \mathfrak{A}$ where f is an extended real-valued $\mathfrak{A}$-measurable function on X for the case that λ^+ is a finite and λ^- is a σ-finite positive measure. The case that λ^+ is a σ-finite and λ^- is a finite positive measure is treated similarly. ∎

Observation 11.11. Let μ be a positive measure and λ be a signed measure on a measurable space $(X, \mathfrak{A})$.

(a) $\lambda \ll \mu \Leftrightarrow \lambda^+ \ll \mu$ and $\lambda^- \ll \mu$.

(b) $\|\lambda\| \ll \mu \Leftrightarrow \lambda^+ \ll \mu$ and $\lambda^- \ll \mu$.

(c) $\|\lambda\| \ll \mu \Leftrightarrow \lambda \ll \mu$.

Proof. 1. Suppose $\lambda \ll \mu$. Let us show that $\lambda^+ \ll \mu$ and $\lambda^- \ll \mu$. Let $\{A, B\}$ be an arbitrary Hahn decomposition of $(X, \mathfrak{A}, \lambda)$. By Theorem 10.21, $\lambda^+(E) = \lambda(E \cap A)$ and $\lambda^-(E) = -\lambda(E \cap B)$ for every $E \in \mathfrak{A}$. If $E \in \mathfrak{A}$ and $\mu(E) = 0$, then by the monotonicity of the positive measure μ, we have $\mu(E \cap A) = 0$ and $\mu(E \cap B) = 0$. Since $\lambda \ll \mu$, this implies that $\lambda(E \cap A) = 0$ and $\lambda(E \cap B) = 0$. Then $\lambda^+(E) = 0$ and $\lambda^-(E) = 0$. This shows that $\lambda^+ \ll \mu$ and $\lambda^- \ll \mu$.

Conversely suppose $\lambda^+ \ll \mu$ and $\lambda^- \ll \mu$. Then for every $E \in \mathfrak{A}$ with $\mu(E) = 0$, we have $\lambda^+(E) = 0$ and $\lambda^-(E) = 0$ and hence $|\lambda(E)| \leq \|\lambda\|(E) = \lambda^+(E) + \lambda^-(E) = 0$, implying $\lambda(E) = 0$. This shows that $\lambda \ll \mu$.

2. (b) is immediate from the fact that $\|\lambda\| = \lambda^+ + \lambda^-$ and the fact that λ^+, λ^-, and $\|\lambda\|$ are all positive measures. (c) is a combination of (a) and (b). ∎

Observation 11.12. Let μ be a positive measure and λ be a signed measure on a measurable space $(X, \mathfrak{A})$. If $\lambda \perp \mu$ and $\lambda \ll \mu$, then $\lambda = 0$.

Proof. Since $\lambda \perp \mu$ if and only if $\|\lambda\| \perp \mu$ by Proposition 10.27 and since $\lambda \ll \mu$ if and only if $\|\lambda\| \ll \mu$ by Observation 11.11, it suffices to show that if $\|\lambda\| \perp \mu$ and $\|\lambda\| \ll \mu$ then $\lambda = 0$.

Now since $\|\lambda\| \perp \mu$, there exist $C_1, C_2 \in X$ such that $C_1 \cap C_2 = \emptyset$, $C_1 \cup C_2 = X$, $\mu(C_1) = 0$, and $\|\lambda\|(C_2) = 0$. Then $\|\lambda\| \ll \mu$ implies that $\|\lambda\|(C_1) = 0$. Thus we have $\|\lambda\|(X) = \|\lambda\|(C_1) + \|\lambda\|(C_2) = 0$. Then for every $E \in \mathfrak{A}$, we have $\|\lambda\|(E) = 0$ by the monotonicity of the positive measure $\|\lambda\|$. Since $|\lambda(E)| \leq \|\lambda\|(E)$ for every $E \in \mathfrak{A}$, the fact that $\|\lambda\|(E) = 0$ for every $E \in \mathfrak{A}$ implies that $\lambda(E) = 0$ for every $E \in \mathfrak{A}$, that is, $\lambda = 0$. ∎

Theorem 11.13. (Uniqueness of the Lebesgue Decomposition) *Let μ be a positive measure and λ be a σ-finite signed measure on a measurable space $(X, \mathfrak{A})$. If a Lebesgue decomposition of λ with respect to μ exists, then it is unique.*

Proof. Suppose a Lebesgue decomposition of λ with respect to μ exists, that is, there exist two signed measures λ_a and λ_s on $\mathfrak{A}$ such that $\lambda_a \ll \mu$, $\lambda_s \perp \mu$ and $\lambda = \lambda_a + \lambda_s$. Let λ'_a and λ'_s be another such pair of signed measures on $\mathfrak{A}$. Let us show that $\lambda_a = \lambda'_a$ and $\lambda_s = \lambda'_s$.

1. Consider first the case that λ is a finite signed measure. In this case λ_a, λ_s, λ'_a, and λ'_s are all finite signed measures on $\mathfrak{A}$ so that $\lambda_a - \lambda'_a$ and $\lambda_s - \lambda'_s$ are defined. From $\lambda_a + \lambda_s = \lambda = \lambda'_a + \lambda'_s$, we have $\lambda_a - \lambda'_a = \lambda'_s - \lambda_s$. Since $\lambda_a \ll \mu$ and $\lambda'_a \ll \mu$ we have $\lambda_a - \lambda'_a \ll \mu$ by (a) of Lemma 11.9. Since $\lambda_s \perp \mu$ and $\lambda'_s \perp \mu$ we have $\lambda'_s - \lambda_s \perp \mu$ by (b) of Lemma 11.9. If we let $\nu = \lambda_a - \lambda'_a = \lambda'_s - \lambda_s$, then $\nu \ll \mu$ and $\nu \perp \mu$ so that $\nu = 0$ by Observation 11.12. Thus $\lambda_a = \lambda'_a$ and $\lambda_s = \lambda'_s$.

2. Consider the case that λ is a σ-finite signed measure. In this case there exists a disjoint sequence $(E_n : n \in \mathbb{N})$ in $\mathfrak{A}$ such that $\bigcup_{n\in\mathbb{N}} E_n = X$ and $\lambda(E_n) \in \mathbb{R}$ for every $n \in \mathbb{R}$. For every $n \in \mathbb{N}$, consider the finite signed measure space $(E_n, \mathfrak{A} \cap E_n, \lambda_n)$ where λ_n is the restriction of λ to $\mathfrak{A} \cap E_n$. Let $\lambda_{n,a}$, $\lambda_{n,s}$, $\lambda'_{n,a}$, and $\lambda'_{n,s}$ be the restrictions of λ_a, λ_s, λ'_a, and λ'_s to $\mathfrak{A} \cap E_n$, and let μ_n be the restriction of μ to $\mathfrak{A} \cap E_n$. From $\lambda_a \ll \mu$ and $\lambda_s \perp \mu$, it follows immediately that $\lambda_{n,a} \ll \mu_n$ and $\lambda_{n,s} \perp \mu_n$ for every $n \in \mathbb{N}$. Similarly from $\lambda'_a \ll \mu$ and $\lambda'_s \perp \mu$, we have $\lambda'_{n,a} \ll \mu_n$ and $\lambda'_{n,s} \perp \mu_n$ for every $n \in \mathbb{N}$. Thus by **1** we have $\lambda_{n,a} = \lambda'_{n,a}$ and $\lambda_{n,s} = \lambda'_{n,s}$ for every $n \in \mathbb{N}$. Then $\lambda_a = \lambda'_a$ and $\lambda_s = \lambda'_s$. ∎

Theorem 11.14. (Radon-Nikodym Theorem) *Let μ be a σ-finite positive measure and λ be a σ-finite signed measure on a measurable space $(X, \mathfrak{A})$. If $\lambda \ll \mu$, then the Radon-Nikodym derivative of λ with respect to μ exists, that is, there exists an extended real-valued $\mathfrak{A}$-measurable function f on X such that $\lambda(E) = \int_E f\,d\mu$ for every $E \in \mathfrak{A}$.*

Proof. According to Theorem 11.10, there exist two signed measures λ_a and λ_s on $(X, \mathfrak{A})$ such that $\lambda = \lambda_a + \lambda_s$, $\lambda_a \ll \mu$, $\lambda_s \perp \mu$ and moreover $\lambda_a(E) = \int_E f\,d\mu$ for $E \in \mathfrak{A}$ where f is an extended real-valued $\mathfrak{A}$-measurable function on X. Now suppose $\lambda \ll \mu$. If we write $\lambda = \lambda + 0$, then since $\lambda \ll \mu$ and $0 \perp \mu$, $\{\lambda, 0\}$ is a Lebesgue decomposition of λ with respect to μ. Then the uniqueness of the Lebesgue decomposition by Theorem 11.13 implies that we have $\lambda = \lambda_a$. Therefore $\lambda(E) = \lambda_a(E) = \int_E f\,d\mu$ for every $E \in \mathfrak{A}$. ∎

We showed above that if μ is a σ-finite positive measure, λ is a σ-finite signed measure and $\lambda \ll \mu$ on a measurable space $(X, \mathfrak{A})$, then the Radon-Nikodym derivative $\frac{d\lambda}{d\mu}$ exists. We show next that $\frac{d\lambda}{d\mu}$ exists without the assumption that λ is σ-finite. For this we require a decomposition of the measurable space $(X, \mathfrak{A})$ as follows.

Lemma 11.15. *Let μ and ν be two positive measures on a measurable space $(X, \mathfrak{A})$. If $\mu(X) < \infty$, $\nu(X) = \infty$, and $\nu \ll \mu$, then there exists a disjoint sequence $(X_n : n \in \mathbb{Z}_+)$ in $\mathfrak{A}$ with $\bigcup_{n\in\mathbb{Z}_+} X_n = X$ such that*
1° $\nu(X_n) < \infty$ *for* $n \in \mathbb{N}$,
2° *for every $A \in \mathfrak{A} \cap X_0$, we have either $\mu(A) = \nu(A) = 0$ or $\mu(A) > 0$ and $\nu(A) = \infty$.*

Proof. 1. Let

(1) $$\mathfrak{E} = \{E \in \mathfrak{A} : \nu(E) < \infty\},$$

(2) $$\alpha = \sup_{E\in\mathfrak{E}} \mu(E) \le \mu(X) < \infty.$$

Then there exists a sequence $(E_n : n \in \mathbb{N})$ in $\mathfrak{E}$ such that $\lim_{n\to\infty} \mu(E_n) = \alpha$. Since

$\mathfrak{E}$ is closed under finite unions, we can select an increasing sequence $(E_n : n \in \mathbb{N})$ in $\mathfrak{E}$ such that $\lim_{n\to\infty} \mu(E_n) = \alpha$. Let $G = \lim_{n\to\infty} E_n = \bigcup_{n\in\mathbb{N}} E_n \in \mathfrak{A}$. Then we have $\mu(G) = \lim_{n\to\infty} \mu(E_n) = \alpha$. With the increasing sequence $(E_n : n \in \mathbb{N}) \subset \mathfrak{E} \subset \mathfrak{A}$, let $X_1 = E_1$ and $X_n = E_n \setminus (E_1 \cup \cdots \cup E_{n-1})$ for $n \geq 2$. Then $(X_n : n \in \mathbb{N})$ is a disjoint sequence in $\mathfrak{E} \subset \mathfrak{A}$ with $\bigcup_{n\in\mathbb{N}} X_n = \bigcup_{n\in\mathbb{N}} E_n = G$ and $\nu(X_n) \leq \nu(E_n) < \infty$ for $n \in \mathbb{N}$.

2. Let

$$X_0 = G^c = \Big(\bigcup_{n\in\mathbb{N}} E_n\Big)^c = \Big(\bigcup_{n\in\mathbb{N}} X_n\Big)^c.$$

Let us show that X_0 satisfies condition 2°. Let $A \in \mathfrak{A} \cap X_0$. Now either $\nu(A) = \infty$ or $\nu(A) < \infty$. If $\nu(A) = \infty$, then since $\nu \ll \mu$ we cannot have $\mu(A) = 0$ so that $\mu(A) > 0$. Suppose $\nu(A) < \infty$. Then $A \in \mathfrak{E}$ by (1). Since $E_n \in \mathfrak{E}$ and $\mathfrak{E}$ is closed under finite unions, we have $E_n \cup A \in \mathfrak{E}$ and then $\mu(E_n \cup A) \leq \alpha$ by (2) for every $n \in \mathbb{N}$. Now since $G = \lim_{n\to\infty} E_n$ for the increasing sequence $(E_n : n \in \mathbb{N})$, we have $G \cup A = \lim_{n\to\infty} (E_n \cup A)$. This implies then that

$$\mu(G \cup A) = \lim_{n\to\infty} \mu(E_n \cup A) \leq \alpha. \tag{3}$$

Since $A \subset X_0 = G^c$, we have $A \cap G = \emptyset$. Thus we have

$$\mu(G \cup A) = \mu(G) + \mu(A) = \alpha + \mu(A). \tag{4}$$

By (3) and (4), we have $\alpha + \mu(A) \leq \alpha$ and hence $\mu(A) = 0$. Then since $\nu \ll \mu$, we have $\nu(A) = 0$. This shows that if $\nu(A) < \infty$ then $\mu(A) = \nu(A) = 0$. ∎

Theorem 11.16. (Radon-Nikodym Theorem) *Let μ be a σ-finite positive measure and λ be a signed measure on a measurable space $(X, \mathfrak{A})$. If $\lambda \ll \mu$, then the Radon-Nikodym derivative of λ with respect to μ, $\frac{d\lambda}{d\mu}$, exists.*

Proof. Let μ be a σ-finite positive measure and λ be a signed measure on a measurable space $(X, \mathfrak{A})$ such that $\lambda \ll \mu$. If λ is a finite signed measure then $\frac{d\lambda}{d\mu}$ exists by Theorem 11.14. Thus it remains to prove the existence of $\frac{d\lambda}{d\mu}$ when λ is an infinite signed measure. Below we prove the existence of $\frac{d\lambda}{d\mu}$ first for the case that μ is a finite positive measure and λ is an infinite signed measure and then for the case that μ is a σ-finite positive measure and λ is an infinite signed measure.

1. Consider the case that μ is a finite positive measure and λ is an infinite signed measure on $(X, \mathfrak{A})$. Since λ is an infinite signed measure on $(X, \mathfrak{A})$, either there exists $E \in \mathfrak{A}$ such that $\lambda(E) = \infty$ or there exists $E \in \mathfrak{A}$ such that $\lambda(E) = -\infty$. In the former case we have $\lambda(X) = \infty$ and in the latter case we have $\lambda(X) = -\infty$ by Lemma 10.5. Let us start with the case $\lambda(X) = \infty$. By Theorem 10.21 (Jordan decomposition), there exist two positive measures ν and ν' on $(X, \mathfrak{A})$, at least one of which is finite, such that $\lambda = \nu - \nu'$. Since $\lambda(X) = \infty$, ν' is a finite positive measure and $\nu(X) = \infty$. Now $\lambda \ll \mu$ implies $\nu \ll \mu$ and $\nu' \ll \mu$ by (a) of Observation 11.11. Applying Lemma 11.15 to the finite positive measure μ and the infinite positive measure ν, we have a decomposition of X into a disjoint sequence $(X_n : n \in \mathbb{Z}_+)$ in $\mathfrak{A}$ with $\bigcup_{n\in\mathbb{Z}_+} X_n = X$ such that

1° $\nu(X_n) < \infty$ for $n \in \mathbb{N}$,
2° for every $A \in \mathfrak{A} \cap X_0$, we have either $\mu(A) = \nu(A) = 0$ or $\mu(A) > 0$ and $\nu(A) = \infty$.
For each $n \in \mathbb{N}$, on the measurable space $(X_n, \mathfrak{A} \cap X_n)$ we have $\mu(X_n) < \infty$, $\nu(X_n) < \infty$ and $\nu \ll \mu$. Then by Theorem 11.14, the Radon-Nikodym derivative of ν with respect to μ exists, that is, there exists a nonnegative extended real-valued $\mathfrak{A} \cap X_n$-measurable function f_n on X_n such that

$$\nu(E) = \int_E f_n \, d\mu \quad \text{for } E \in \mathfrak{A} \cap X_n.$$

Let $f_0 = \infty$ on X_0. Then by 2° we have

$$\nu(E) = \int_E f_0 \, d\mu \quad \text{for } E \in \mathfrak{A} \cap X_0.$$

Let us define a nonnegative extended real-valued $\mathfrak{A}$-measurable function g on X by setting

$$g(x) = f_n(x) \quad \text{for } x \in X_n, n \in \mathbb{Z}_+.$$

Then for any $E \in \mathfrak{A}$ we have

$$\begin{aligned} \nu(E) &= \sum_{n \in \mathbb{Z}_+} \nu(E \cap X_n) = \sum_{n \in \mathbb{Z}_+} \int_{E \cap X_n} f_n \, d\mu \\ &= \int_{\bigcup_{n \in \mathbb{Z}_+}(E \cap X_n)} g \, d\mu = \int_E g \, d\mu. \end{aligned}$$

This shows that g is the Radon-Nikodym derivative of ν with respect to μ.

Since ν' is a finite measure and $\nu' \ll \mu$, the Radon-Nikodym derivative of ν' with respect to μ exists by Theorem 11.14, that is, there exists a nonnegative extended real-valued $\mathfrak{A}$-measurable function h on X such that

$$\nu'(E) = \int_E h \, d\mu \quad \text{for } E \in \mathfrak{A}.$$

Note that since $\nu'(X) < \infty$, h is finite μ-a.e. on X. Then $g - h$ is defined μ-a.e. on X. Now for every $E \in \mathfrak{A}$, we have

$$\lambda(E) = \nu(E) - \nu'(E) = \int_E g \, d\mu - \int_E h \, d\mu = \int_E \{g - h\} \, d\mu.$$

Thus $g - h$ is the Radon-Nikodym derivative of λ with respect to μ. This proves the existence of $\frac{d\lambda}{d\mu}$ for the case that μ is a finite positive measure and λ is an infinite signed measure with $\lambda(X) = \infty$.

If λ is an infinite signed measure with $\lambda(X) = -\infty$, then $-\lambda$ is an infinite signed measure with $(-\lambda)(X) = \infty$. Thus by our result above, the Radon-Nikodym derivative of $-\lambda$ with respect to μ exists, that is, there exists an extended real-valued $\mathfrak{A}$-measurable function g on X such that $(-\lambda)(E) = \int_E g \, d\mu$ for $E \in \mathfrak{A}$. Then for every $E \in \mathfrak{A}$ we have $\int_E -g \, d\mu = -\int_E g \, d\mu = -(-\lambda)(E) = \lambda(E)$. Thus $-g$ is the Radon-Nikodym derivative of λ with respect to μ.

2. Now let μ be a σ-finite positive measure and λ be an infinite signed measure on $(X, \mathfrak{A})$. Now σ-finiteness of μ implies that there exists a disjoint sequence $(Y_n : n \in \mathbb{N})$ in $\mathfrak{A}$ such that $\bigcup_{n\in\mathbb{N}} Y_n = X$ and $\mu(Y_n) < \infty$ for $n \in \mathbb{N}$. For each $n \in \mathbb{N}$, μ is a finite positive measure and λ is a signed measure on the measurable space $(Y_n, \mathfrak{A} \cap Y_n)$ and $\lambda \ll \mu$. If λ is a finite signed measure on $(Y_n, \mathfrak{A} \cap Y_n)$ then the Radon-Nikodym derivative of λ with respect to μ exists by Theorem 11.14 and if λ is an infinite signed measure on $(Y_n, \mathfrak{A} \cap Y_n)$ then the Radon-Nikodym derivative of λ with respect to μ exists as we showed in **1** above. Denote the Radon-Nikodym derivative of λ with respect to μ on $(Y_n, \mathfrak{A} \cap Y_n)$ by f_n. Let us define an extended real-valued $\mathfrak{A}$-measurable function f on X by setting

$$f(x) = f_n(x) \quad \text{for } x \in Y_n \text{ for } n \in \mathbb{N}.$$

Then for any $E \in \mathfrak{A}$ we have

$$\begin{aligned}\lambda(E) &= \sum_{n\in\mathbb{N}} \lambda(E \cap Y_n) = \sum_{n\in\mathbb{N}} \int_{E\cap Y_n} f_n \, d\mu \\ &= \int_{\bigcup_{n\in\mathbb{N}}(E\cap Y_n)} f \, d\mu = \int_E f \, d\mu.\end{aligned}$$

This shows that f is the Radon-Nikodym derivative of λ with respect to μ. ∎

Remark 11.17. Let μ be a positive measure and λ be a signed measure on a measurable space $(X, \mathfrak{A})$ such that $\lambda \ll \mu$. Without the assumption that μ is σ-finite the Radon-Nikodym derivative $\frac{d\lambda}{d\mu}$ may not exist. See the following example.

Example. Consider a measurable space $(X, \mathfrak{A})$ where $\mathfrak{A} = \{\emptyset, X\}$. The only extended real-valued $\mathfrak{A}$-measurable functions on X are the constant functions on X, that is, $f(x) = \gamma$ for $x \in X$ where $\gamma \in \overline{\mathbb{R}}$. Let μ and λ be two measures on $(X, \mathfrak{A})$ defined by setting: $\mu(\emptyset) = 0$, $\mu(X) = \infty$, $\lambda(\emptyset) = 0$, and $\lambda(X) = 1$. The measure μ is not σ-finite. The only null set in $(X, \mathfrak{A}, \mu)$ is $\emptyset$ and we have $\lambda(\emptyset) = 0$. Thus $\lambda \ll \mu$. Now if f is an extended real-valued $\mathfrak{A}$-measurable function on X then $f = \gamma$ for some $\gamma \in \overline{\mathbb{R}}$. Then $\int_X f \, d\mu = \int_X \gamma \, d\mu = \gamma \mu(X) = \gamma \cdot \infty \neq 1 = \lambda(X)$. Thus f is not a Radon-Nikodym derivative of λ with respect to μ. This shows that no extended real-valued $\mathfrak{A}$-measurable function on X can be a Radon-Nikodym derivative of λ with respect to μ and therefore $\frac{d\lambda}{d\mu}$ does not exist.

[III] Properties of the Radon-Nikodym Derivative

Let μ be a positive measure and λ be a signed measure on a measurable space $(X, \mathfrak{A})$. According to Definition 11.1, a Radon-Nikodym derivative of λ with respect to μ, if it exists, is an extended real-valued $\mathfrak{A}$-measurable function $\frac{d\lambda}{d\mu}$ on X such that $\lambda(E) = \int_E \frac{d\lambda}{d\mu} \, d\mu$ for every $E \in \mathfrak{A}$. Let us list first the more immediate properties of a Radon-Nikodym deivative.

Observation 11.18. Let μ be a positive measure and λ be a signed measure on $(X, \mathfrak{A})$.
(a) If $\lambda \ll \mu$ and if μ is σ-finite, then $\frac{d\lambda}{d\mu}$ exists.

(b) If $\frac{d\lambda}{d\mu}$ exists and if μ is σ-finite, then $\frac{d\lambda}{d\mu}$ is unique up to null sets of μ.
(c) If $\frac{d\lambda}{d\mu}$ exists, then $\lambda \ll \mu$.

Proof. (a) is by Theorem 11.16. (b) is by Proposition 11.2. Let us prove (c). Now if $\frac{d\lambda}{d\mu}$ exists, then we have $\lambda(E) = \int_E \frac{d\lambda}{d\mu}\, d\mu$ for every $E \in \mathfrak{A}$. Then for every $E \in \mathfrak{A}$ such that $\mu(E) = 0$ we have $\lambda(E) = 0$. This shows that $\lambda \ll \mu$. ∎

Observation 11.19. Let μ be a positive measure and λ be a signed measure on a measurable space $(X, \mathfrak{A})$. Suppose the Radon-Nikodym derivative $\frac{d\lambda}{d\mu}$ of λ with respect to μ exists.
(a) $\frac{d\lambda}{d\mu}$ is μ semi-integrable on X.
(b) If λ is a positive measure, then $\frac{d\lambda}{d\mu} \geq 0$ μ-a.e. on X.
(c) If λ is a finite signed measure, then $\frac{d\lambda}{d\mu}$ is μ-integrable on X.
(d) If λ is a σ-finite signed measure, then $\frac{d\lambda}{d\mu}$ is real-valued μ-a.e. on X.
(e) If λ is a σ-finite signed measure, then $\frac{d\lambda}{d\mu}$ has a version which is real-valued on X.

Proof. 1. By Definition 11.1 we have $\int_X \frac{d\lambda}{d\mu}\, d\mu = \lambda(X) \in \overline{\mathbb{R}}$. Thus $\frac{d\lambda}{d\mu}$ is μ semi-integrable on X.

2. Suppose λ is a positive measure. Then we have $\int_E \frac{d\lambda}{d\mu}\, d\mu = \lambda(E) \geq 0$ for every $E \in \mathfrak{A}$. Thus by Lemma 9.3, we have $\frac{d\lambda}{d\mu} \geq 0$, μ-a.e. on X.

3. Suppose λ is a finite signed measure. Then $\int_X \frac{d\lambda}{d\mu}\, d\mu = \lambda(X) \in \mathbb{R}$. This shows that $\frac{d\lambda}{d\mu}$ is μ-integrable on X.

4. Suppose λ is a σ-finite signed measure. This implies that there exists a disjoint sequence $(E_n : n \in \mathbb{N})$ in $\mathfrak{A}$ such that $\bigcup_{n\in\mathbb{N}} E_n = X$ and $\lambda(E_n) \in \mathbb{R}$ for $n \in \mathbb{N}$. Then $\int_{E_n} \frac{d\lambda}{d\mu}\, d\mu = \lambda(E_n) \in \mathbb{R}$, that is, $\frac{d\lambda}{d\mu}$ is μ-integrable on E_n. This implies that $\frac{d\lambda}{d\mu}$ is real-valued μ-a.e. on E_n by (f) of Observation 9.2. Since this holds for every $n \in \mathbb{N}$ and since a countable union of null sets is a null set, $\frac{d\lambda}{d\mu}$ is real-valued μ-a.e. on X. ∎

5. If λ is a σ-finite measure, then $\frac{d\lambda}{d\mu}$ is real-valued μ-a.e. on X by (d). Let E_0 be the null set in $(X, \mathfrak{A}, \mu)$ on which $\frac{d\lambda}{d\mu}$ assumes the values ∞ and $-\infty$. Let us define a function on X by setting

$$g(x) = \begin{cases} \frac{d\lambda}{d\mu}(x) & \text{for } x \in X \setminus E_0, \\ 0 & \text{for } x \in E_0. \end{cases}$$

Then g is a real-valued $\mathfrak{A}$-measurable function on X and $\frac{d\lambda}{d\mu} = g$, μ-a.e. on X so that g is a Radon-Nikodym derivative of λ with respect to μ by (a) of Proposition 11.2. ∎

Thus in the theorems to follow we assume the σ-finiteness of μ to ensure the uniqueness of $\frac{d\lambda}{d\mu}$ up to null sets in $(X, \mathfrak{A}, \mu)$ and we assume the σ-finiteness of λ to ensure that $\frac{d\lambda}{d\mu}$ is real-valued μ-a.e. on X and has a real-valued version on X.

Lemma 11.20. *Let μ be a σ-finite positive measure and λ be a σ-finite signed measure on a measurable space $(X, \mathfrak{A})$ and let $\lambda = \lambda^+ - \lambda^-$ be the Jordan decomposition of λ. Suppose the Radon-Nikodym derivative $\frac{d\lambda}{d\mu}$ of λ with respect to μ exists.*

(a) *The Radon-Nikodym derivatives of λ^+ and λ^- with respect to μ, $\frac{d\lambda^+}{d\mu}$ and $\frac{d\lambda^-}{d\mu}$ exist. Furthermore, with an arbitrary Hahn decomposition $\{A, B\}$ of $(X, \mathfrak{A}, \lambda)$, we have*

$$(1)\qquad \begin{cases} \frac{d\lambda^+}{d\mu} = \mathbf{1}_A \cdot \frac{d\lambda}{d\mu} & \mu\text{-a.e. on } X, \\ \frac{d\lambda^-}{d\mu} = -\mathbf{1}_B \cdot \frac{d\lambda}{d\mu} & \mu\text{-a.e. on } X. \end{cases}$$

(b) *For the positive and negative parts $\left[\frac{d\lambda}{d\mu}\right]^+$ and $\left[\frac{d\lambda}{d\mu}\right]^-$ of the function $\frac{d\lambda}{d\mu}$, we have*

$$(2)\qquad \begin{cases} \frac{d\lambda^+}{d\mu} = \left[\frac{d\lambda}{d\mu}\right]^+ & \mu\text{-a.e. on } X, \\ \frac{d\lambda^-}{d\mu} = \left[\frac{d\lambda}{d\mu}\right]^- & \mu\text{-a.e. on } X. \end{cases}$$

Proof. 1. With an arbitrary Hahn decomposition $\{A, B\}$ of $(X, \mathfrak{A}, \lambda)$, the positive part λ^+ and the negative part λ^- of λ are given, according to Theorem 10.21, by

$$\lambda^+(E) = \lambda(E \cap A) \quad \text{and} \quad \lambda^-(E) = -\lambda(E \cap B) \quad \text{for every } E \in \mathfrak{A}.$$

Now if $\frac{d\lambda}{d\mu}$ exists, then since λ is a σ-finite signed measure, $\frac{d\lambda}{d\mu}$ is real-valued μ-a.e. on X by (d) of Observation 11.19. Thus $\mathbf{1}_A \cdot \frac{d\lambda}{d\mu}$ and $\mathbf{1}_B \cdot \frac{d\lambda}{d\mu}$ are defined μ-a.e. on X. Therefore for every $E \in \mathfrak{A}$, we have

$$\lambda^+(E) = \lambda(E \cap A) = \int_{E\cap A} \frac{d\lambda}{d\mu}\, d\mu = \int_E \mathbf{1}_A \cdot \frac{d\lambda}{d\mu}\, d\mu$$

and similarly

$$\lambda^-(E) = -\lambda(E \cap B) = -\int_{E\cap B} \frac{d\lambda}{d\mu}\, d\mu = -\int_E \mathbf{1}_B \cdot \frac{d\lambda}{d\mu}\, d\mu.$$

This shows that $\mathbf{1}_A \cdot \frac{d\lambda}{d\mu}$ and $-\mathbf{1}_B \cdot \frac{d\lambda}{d\mu}$ are Radon-Nikodym derivatives of λ^+ and λ^- with respect to μ. Since μ is a σ-finite positive measure, $\frac{d\lambda}{d\mu}$ is unique up to null sets in $(X, \mathfrak{A}, \mu)$ by (b) of Proposition 11.2. Thus (1) holds.

2. Let $A' = \left\{x \in X : \left(\frac{d\lambda}{d\mu}\right)(x) \geq 0\right\}$ and $B' = \left\{x \in X : \left(\frac{d\lambda}{d\mu}\right)(x) < 0\right\}$ so that

$$(3)\qquad \begin{cases} \left[\frac{d\lambda}{d\mu}\right]^+ = \mathbf{1}_{A'} \cdot \frac{d\lambda}{d\mu} & \mu\text{-a.e. on } X, \\ \left[\frac{d\lambda}{d\mu}\right]^- = -\mathbf{1}_{B'} \cdot \frac{d\lambda}{d\mu} & \mu\text{-a.e. on } X. \end{cases}$$

Now $A', B' \in \mathfrak{A}$, $A' \cap B' = \emptyset$, and $A' \cup B' = X$. Let us show that $\{A', B'\}$ is a Hahn decomposition of $(X, \mathfrak{A}, \lambda)$. For this, we show that A' is a positive set and B' is a negative set for λ. Now for every $\mathfrak{A}$-measurable subset E of A', we have $\lambda(E) = \int_E \frac{d\lambda}{d\mu}\, d\mu \geq 0$ since $\frac{d\lambda}{d\mu} \geq 0$ on $E \subset A'$. Thus A' is a positive set for λ. Likewise B' is a negative set for λ. Thus $\{A', B'\}$ is a Hahn decomposition of $(X, \mathfrak{A}, \lambda)$. Then by (a) and (3), we have

$$\begin{cases} \frac{d\lambda^+}{d\mu} = \mathbf{1}_{A'} \cdot \frac{d\lambda}{d\mu} = \left[\frac{d\lambda}{d\mu}\right]^+ & \mu\text{-a.e. on } X, \\ \frac{d\lambda^-}{d\mu} = -\mathbf{1}_{B'} \cdot \frac{d\lambda}{d\mu} = \left[\frac{d\lambda}{d\mu}\right]^- & \mu\text{-a.e. on } X. \end{cases}$$

This completes the proof. ∎

Theorem 11.21. *Let μ be a σ-finite positive measure and λ be a σ-finite signed measure on a measurable space $(X, \mathfrak{A})$. Suppose the Radon-Nikodym derivative $\frac{d\lambda}{d\mu}$ of λ with respect to μ exists. Then for every extended real-valued $\mathfrak{A}$-measurable function f on X, we have*

$$\int_X f\,d\lambda = \int_X f\cdot\frac{d\lambda}{d\mu}\,d\mu,$$

in the sense that the existence of one side implies that of the other and the equality.

Proof. 1. Consider first the case that λ is a σ-finite positive measure. Let $f = \mathbf{1}_E$ where $E \in \mathfrak{A}$. Since λ is σ-finite, $\frac{d\lambda}{d\mu}$ is real-valued μ-a.e. on X by (d) of Observation 11.19. Thus $\mathbf{1}_E \cdot \frac{d\lambda}{d\mu}$ is defined μ-a.e. on X. Then we have

$$\begin{aligned}\int_X f\,d\lambda &= \int_X \mathbf{1}_E\,d\lambda = \lambda(E) = \int_E \frac{d\lambda}{d\mu}\,d\mu\\ &= \int_X \mathbf{1}_E\cdot\frac{d\lambda}{d\mu}\,d\mu = \int_X f\cdot\frac{d\lambda}{d\mu}\,d\mu.\end{aligned}$$

If f is a nonnegative simple function, then by the result above and by the linearity of the integral, we have $\int_X f\,d\lambda = \int_X f\cdot\frac{d\lambda}{d\mu}\,d\mu$. If f is a nonnegative extended real-valued $\mathfrak{A}$-measurable function on X, then according to Lemma 8.6, there exists an increasing sequence of nonnegative simple functions $(\varphi_n : n \in \mathbb{N})$ such that $\varphi_n \uparrow f$ on X. By our result above, we have for every $n \in \mathbb{N}$

$$(1) \qquad \int_X \varphi_n\,d\lambda = \int_X \varphi_n\cdot\frac{d\lambda}{d\mu}\,d\mu.$$

Now since λ is a positive measure, we have $\frac{d\lambda}{d\mu} \geq 0$, μ-a.e. on X by (b) of Observation 11.19. Then $\varphi_n \uparrow f$ on X implies that $\varphi_n\cdot\frac{d\lambda}{d\mu} \uparrow f\cdot\frac{d\lambda}{d\mu}$, μ-a.e. on X. Thus by the Monotone Convergence Theorem (Theorem 8.5), we have

$$(2) \qquad \lim_{n\to\infty}\int_X \varphi_n\cdot\frac{d\lambda}{d\mu}\,d\mu = \int_X f\cdot\frac{d\lambda}{d\mu}\,d\mu.$$

Also by the Monotone Convergence Theorem, we have

$$(3) \qquad \lim_{n\to\infty}\int_X \varphi_n\,d\lambda = \int_X f\,d\lambda.$$

Letting $n \to \infty$ in (1), we have $\int_X f\,d\lambda = \int_X f\cdot\frac{d\lambda}{d\mu}\,d\mu$ by (2) and (3).

If f is an extended real-valued $\mathfrak{A}$-measurable function on X, let us write $f = f^+ - f^-$. We have $\int_X f^+\,d\lambda = \int_X f^+\cdot\frac{d\lambda}{d\mu}\,d\mu$ and $\int_X f^-\,d\lambda = \int_X f^-\cdot\frac{d\lambda}{d\mu}\,d\mu$ by our result above for nonnegative functions. Thus if one of the two differences $\int_X f^+\,d\lambda - \int_X f^-\,d\lambda$ and $\int_X f^+\cdot\frac{d\lambda}{d\mu}\,d\mu - \int_X f^-\cdot\frac{d\lambda}{d\mu}\,d\mu$ exists in $\overline{\mathbb{R}}$, then so does the other and the two are equal, that is, $\int_X f\,d\lambda = \int_X f\cdot\frac{d\lambda}{d\mu}\,d\mu$.

2. Consider the case that λ is a σ-finite signed measure. Let $\lambda = \lambda^+ - \lambda^-$ be the Jordan decomposition of λ. Then λ^+ and λ^- are σ-finite positive measures. The existence of $\frac{d\lambda}{d\mu}$ implies that $\left[\frac{d\lambda}{d\mu}\right]^+$ and $\left[\frac{d\lambda}{d\mu}\right]^-$ are the Radon-Nikodym derivatives of λ^+ and λ^- with respect to μ according to Lemma 11.20. Since λ^+ and λ^- are positive measures, by our result in **1**, we have for every extended real-valued $\mathfrak{A}$-measurable function f on X

$$\int_X f\,d\lambda^+ = \int_X f \cdot \frac{d\lambda^+}{d\mu}\,d\mu = \int_X f \cdot \left[\frac{d\lambda}{d\mu}\right]^+ d\mu,$$

in the sense that the existence of one side implies that of the other and the equality of the two, and similarly

$$\int_X f\,d\lambda^- = \int_X f \cdot \frac{d\lambda^-}{d\mu}\,d\mu = \int_X f \cdot \left[\frac{d\lambda}{d\mu}\right]^- d\mu.$$

If one of the differences $\int_X f\,d\lambda^+ - \int_X f\,d\lambda^-$ and $\int_X f \cdot \left[\frac{d\lambda}{d\mu}\right]^+ d\mu - \int_X f \cdot \left[\frac{d\lambda}{d\mu}\right]^- d\mu$ exists in $\overline{\mathbb{R}}$, then so does the other and moreover the two differences are equal, that is, $\int_X f\,d\lambda = \int_X f \cdot \frac{d\lambda}{d\mu}\,d\mu$. ∎

Theorem 11.22. (a) *Let μ be a σ-finite positive measure and λ be a signed measure on a measurable space $(X, \mathfrak{A})$. If $\frac{d\lambda}{d\mu}$ exists, then for every $c \in \mathbb{R}$, $c \neq 0$, $\frac{d(c\lambda)}{d\mu}$ exists and furthermore*

(1) $$\frac{d(c\lambda)}{d\mu} = c \cdot \frac{d\lambda}{d\mu} \quad \mu\text{-a.e. on } X.$$

(b) *Let μ be a σ-finite positive measure and let λ_1 and λ_2 be two σ-finite signed measures on a measurable space $(X, \mathfrak{A})$ such that $\lambda_1 + \lambda_2$ is defined. If $\frac{d\lambda_1}{d\mu}$ and $\frac{d\lambda_2}{d\mu}$ exist, then so does $\frac{d(\lambda_1+\lambda_2)}{d\mu}$ and furthermore*

(2) $$\frac{d(\lambda_1 + \lambda_2)}{d\mu} = \frac{d\lambda_1}{d\mu} + \frac{d\lambda_2}{d\mu} \quad \mu\text{-a.e. on } X.$$

Proof. 1. To prove (a), note that if $\frac{d\lambda}{d\mu}$ exists then for $c \in \mathbb{R}$, $c \neq 0$, and $E \in \mathfrak{A}$ we have

$$\int_E c\frac{d\lambda}{d\mu}\,d\mu = c\int_E \frac{d\lambda}{d\mu}\,d\mu = c\lambda(E) = (c\lambda)(E).$$

Thus $c \cdot \frac{d\lambda}{d\mu}$ is a Radon-Nikodym derivative of $c\lambda$ with respect to μ. The σ-finiteness of μ implies the uniqueness of a Radon-Nikodym derivative with respect to μ up to null sets in $(X, \mathfrak{A}, \mu)$ by (b) of Proposition 11.2. Thus $\frac{d(c\lambda)}{d\mu} = c \cdot \frac{d\lambda}{d\mu}$, μ-a.e. on X.

2. To prove (b), note first that if $\frac{d\lambda_1}{d\mu}$ and $\frac{d\lambda_2}{d\mu}$ exist then the σ-finiteness of λ_1 and λ_2 implies that $\frac{d\lambda_1}{d\mu}$ and $\frac{d\lambda_2}{d\mu}$ are real-valued μ-a.e. on X according to (d) of Observation 11.19 and thus $\frac{d\lambda_1}{d\mu} + \frac{d\lambda_2}{d\mu}$ is defined μ-a.e. on X. Now for every $E \in \mathfrak{A}$, we have

$$\begin{aligned}\int_E \left\{\frac{d\lambda_1}{d\mu} + \frac{d\lambda_2}{d\mu}\right\} d\mu &= \int_E \frac{d\lambda_1}{d\mu}\,d\mu + \int_E \frac{d\lambda_2}{d\mu}\,d\mu = \lambda_1(E) + \lambda_2(E) \\ &= (\lambda_1 + \lambda_2)(E).\end{aligned}$$

Thus $\frac{d\lambda_1}{d\mu} + \frac{d\lambda_2}{d\mu}$ is a Radon-Nikodym derivative of $\lambda_1 + \lambda_2$ with respect to μ. The σ-finiteness of μ implies the uniqueness of a Radon-Nikodym derivative with respect to μ up to null sets in $(X, \mathfrak{A}, \mu)$ by (b) of Proposition 11.2. Thus $\frac{d(\lambda_1+\lambda_2)}{d\mu} = \frac{d\lambda_1}{d\mu} + \frac{d\lambda_2}{d\mu}$, μ-a.e. on X. ∎

Theorem 11.23. (Chain Rule for Radon-Nikodym Derivatives) *Let μ_1 and μ_2 be two σ-finite positive measures and λ be a σ-finite signed measure on a measurable space $(X, \mathfrak{A})$. If $\frac{d\mu_2}{d\mu_1}$ and $\frac{d\lambda}{d\mu_2}$ exist, then $\frac{d\lambda}{d\mu_1}$ exists and*

$$\frac{d\lambda}{d\mu_1} = \frac{d\lambda}{d\mu_2} \cdot \frac{d\mu_2}{d\mu_1} \quad \mu\text{-a.e. on } X.$$

Proof. Since μ_2 is σ-finite, $\frac{d\mu_2}{d\mu_1}$ has a real-valued version on X by (d) of Proposition 11.19. Similarly since λ is σ-finite, $\frac{d\lambda}{d\mu_2}$ has a real-valued version on X. Thus the product $\frac{d\lambda}{d\mu_2} \cdot \frac{d\mu_2}{d\mu_1}$ is defined on X. Let $E \in \mathfrak{A}$. Applying Theorem 11.21 to the real-valued $\mathfrak{A}$-measurable function $\mathbf{1}_E \cdot \frac{d\lambda}{d\mu_2}$, we have

$$\begin{aligned}\int_E \frac{d\lambda}{d\mu_2} \cdot \frac{d\mu_2}{d\mu_1} d\mu_1 &= \int_X \mathbf{1}_E \cdot \frac{d\lambda}{d\mu_2} \cdot \frac{d\mu_2}{d\mu_1} d\mu_1 = \int_X \mathbf{1}_E \cdot \frac{d\lambda}{d\mu_2} d\mu_2 \\ &= \int_E \frac{d\lambda}{d\mu_2} d\mu_2 = \lambda(E).\end{aligned}$$

This shows that $\frac{d\lambda}{d\mu_2} \cdot \frac{d\mu_2}{d\mu_1}$ is a Radon-Nikodym derivative of λ with respect to μ_1. The σ-finiteness of μ_1 implies the uniqueness of $\frac{d\lambda}{d\mu_1}$ by (b) of Proposition 11.2. Thus we have $\frac{d\lambda}{d\mu_1} = \frac{d\lambda}{d\mu_2} \cdot \frac{d\mu_2}{d\mu_1}$, μ_1-a.e. on X. ∎

Definition 11.24. *Let μ_1 and μ_2 be two positive measures on a measurable space $(X, \mathfrak{A})$. We say that μ_1 and μ_2 are equivalent and write $\mu_1 \sim \mu_2$ if $\mu_1 \ll \mu_2$ and $\mu_2 \ll \mu_1$.*

If $\mu_1 \sim \mu_2$, then a set $E \in \mathfrak{A}$ is a null set for μ_1 if and only if it is a null set for μ_2. Thus 'μ_1-a.e. on X' and 'μ_2-a.e. on X' are equivalent, that is, a statement is valid μ_1-a.e. on X if and only if it is valid μ_2-a.e. on X.

Observation 11.25. If μ_1 and μ_2 are two σ-finite positive measures on a measurable space $(X, \mathfrak{A})$ and if $\mu_1 \sim \mu_2$, then $\frac{d\mu_1}{d\mu_2}$ and $\frac{d\mu_2}{d\mu_1}$ exist by Theorem 11.13. Moreover $\frac{d\mu_1}{d\mu_2}$ is unique up to null sets in $(X, \mathfrak{A}, \mu_2)$ and $\frac{d\mu_2}{d\mu_1}$ is unique up to null sets in $(X, \mathfrak{A}, \mu_1)$ by (b) of Proposition 11.2. Furthermore $\frac{d\mu_1}{d\mu_2}$ is real-valued μ_2-a.e. on X and $\frac{d\mu_2}{d\mu_1}$ is real-valued μ_1-a.e. on X by (d) of Observation 11.19 so that $\frac{d\mu_1}{d\mu_2} \cdot \frac{d\mu_2}{d\mu_1}$ is defined μ_2-a.e. on X and μ_1-a.e. on X. We show next that $\frac{d\mu_1}{d\mu_2} \cdot \frac{d\mu_2}{d\mu_1} = 1$, μ_2-a.e. on X and μ_1-a.e. on X.

Corollary 11.26. *Let μ_1 and μ_2 be two σ-finite positive measures on a measurable space $(X, \mathfrak{A})$. If $\frac{d\mu_2}{d\mu_1}$ and $\frac{d\mu_1}{d\mu_2}$ both exist, then $\mu_1 \sim \mu_2$ and $\frac{d\mu_2}{d\mu_1} \cdot \frac{d\mu_1}{d\mu_2} = 1$, μ_1-a.e. on X and μ_2-a.e. on X.*

Proof. The existence of $\frac{d\mu_2}{d\mu_1}$ implies that $\mu_2 \ll \mu_1$ and the existence of $\frac{d\mu_1}{d\mu_2}$ implies that $\mu_1 \ll \mu_2$. Thus $\mu_1 \sim \mu_2$. In particular a statement is valid μ_1-a.e. on X if and only if it is valid μ_2-a.e. on X. Apply Theorem 11.23 with $\lambda := \mu_1$. Then the existence of $\frac{d\mu_2}{d\mu_1}$ and $\frac{d\mu_1}{d\mu_2}$ implies $\frac{d\mu_1}{d\mu_1} = \frac{d\mu_1}{d\mu_2} \cdot \frac{d\mu_2}{d\mu_1}$, μ_1-a.e. on X. But $\frac{d\mu_1}{d\mu_1} = 1$, μ_1-a.e. on X. Thus $\frac{d\mu_1}{d\mu_2} \cdot \frac{d\mu_2}{d\mu_1} = 1$, μ_1-a.e. on X. ∎

According to the Jordan decomposition theorem (Theorem 10.21), every signed measure λ on a measurable space $(X, \mathfrak{A})$ can be decomposed as $\lambda = \lambda^+ - \lambda^-$ where λ^+ and λ^- are two mutually singular positive measures on $(X, \mathfrak{A})$. The total variation measure of λ is defined by $|\lambda| = \lambda^+ + \lambda^-$. Now let μ_1 and μ_2 be two σ-finite positive measures on $(X, \mathfrak{A})$ at least one of which is finite. Then $\nu := \mu_1 - \mu_2$ is a σ-finite signed measure on $(X, \mathfrak{A})$. For the total variation measure $|\nu|$ of ν we have the following estimate.

Theorem 11.27. *Let μ_1 and μ_2 be two σ-finite measures on a measurable space $(X, \mathfrak{A})$ one of which is finite. Let $\nu = \mu_1 - \mu_2$. Then for the total variation $|\nu|$ of the σ-finite signed measure ν on $(X, \mathfrak{A})$, we have $|\nu| \le \mu_1 + \mu_2$, that is, $|\nu|(E) \le \mu_1(E) + \mu_2(E)$ for every $E \in \mathfrak{A}$.*

Proof. Since at least one of μ_1 and μ_2 is a finite measure on $(X, \mathfrak{A})$, $\nu = \mu_1 - \mu_2$ is defined and is a signed measure on $(X, \mathfrak{A})$ by Proposition 10.4. The σ-finiteness of ν follows from that of μ_1 and μ_2 by Observation 10.30. Let $\lambda = \mu_1 + \mu_2$, a σ-finite measure on $(X, \mathfrak{A})$. Then for every $E \in \mathfrak{A}$ such that $\lambda(E) = 0$, we have $\mu_1(E) = 0$ and $\mu_2(E) = 0$ and then $\nu(E) = 0$ also. Thus $\mu_1 \ll \lambda$, $\mu_2 \ll \lambda$ and $\nu \ll \lambda$ on $(X, \mathfrak{A})$. By Theorem 11.14, the Radon-Nikodym derivatives $\frac{d\mu_1}{d\lambda}$, $\frac{d\mu_2}{d\lambda}$ and $\frac{d\nu}{d\lambda}$ exist. By (d) of Observation 11.19, $\frac{d\mu_1}{d\lambda}$, $\frac{d\mu_2}{d\lambda}$, and $\frac{d\nu}{d\lambda}$ are all real-valued λ-a.e. on X. Now for every $E \in \mathfrak{A}$, we have

$$\nu(E) = \mu_1(E) - \mu_2(E) = \int_E \frac{d\mu_1}{d\lambda}\, d\lambda - \int_E \frac{d\mu_2}{d\lambda}\, d\lambda = \int_E \left\{\frac{d\mu_1}{d\lambda} - \frac{d\mu_2}{d\lambda}\right\} d\lambda.$$

This shows that $\left\{\frac{d\mu_1}{d\lambda} - \frac{d\mu_2}{d\lambda}\right\}$ is a Radon-Nikodym derivative of ν with respect to λ. Thus by the uniqueness of Radon-Nikodym derivative ((b) of Proposition 11.2), we have

(1) $$\frac{d\nu}{d\lambda} = \frac{d\mu_1}{d\lambda} - \frac{d\mu_2}{d\lambda} \quad \lambda\text{-a.e. on } (X, \mathfrak{A}).$$

Let $A = \left\{x \in X : \frac{d\nu}{d\lambda}(x) \ge 0\right\}$ and $B = \left\{x \in X : \frac{d\nu}{d\lambda}(x) < 0\right\}$. Then $A, B \in \mathfrak{A}$, $A \cap B = \emptyset$ and $A \cup B = X$. Let us define two measures ν_1 and ν_2 on $(X, \mathfrak{A})$ by setting for every $E \in \mathfrak{A}$

(2) $$\nu_1(E) = \int_E \mathbf{1}_A \cdot \frac{d\nu}{d\lambda}\, d\lambda$$

and

(3) $$\nu_2(E) = \int_E \mathbf{1}_B \cdot \left\{-\frac{d\nu}{d\lambda}\right\} d\lambda.$$

Let us show that the pair $\{\nu_1, \nu_2\}$ is a Jordan decomposition of the signed measure ν. Now for every $E \in \mathfrak{A}$, we have

$$\nu_1(E) - \nu_2(E) = \int_E \mathbf{1}_A \cdot \frac{d\nu}{d\lambda}\, d\lambda + \int_E \mathbf{1}_B \cdot \frac{d\nu}{d\lambda}\, d\lambda$$
$$= \nu(E \cap A) + \nu(E \cap B) = \nu(E).$$

This shows that $\nu = \nu_1 - \nu_2$. Let us show $\nu_1 \perp \nu_2$ on $(X, \mathfrak{A})$. Now

$$\nu_1(B) = \int_B \mathbf{1}_A \cdot \frac{d\nu}{d\lambda}\, d\lambda = \int_{A \cap B} \frac{d\nu}{d\lambda}\, d\lambda = \int_{\emptyset} \frac{d\nu}{d\lambda}\, d\lambda = 0$$

so that B is a null set for the measure ν_1. Similarly A is a null set for the measure ν_2. Thus $\nu_1 \perp \nu_2$ by Definition 10.16. This proves that $\{\nu_1, \nu_2\}$ is a Jordan decomposition of ν.

The total variation $\|\nu\|$ of ν is given by $\|\nu\| = \nu_1 + \nu_2$. Thus for every $E \in \mathfrak{A}$, we have $\|\nu\|(E) = \nu_1(E) + \nu_2(E)$. Recalling (2) and (3), we have

$$\text{(4)} \qquad \|\nu\|(E) = \int_E \mathbf{1}_A \cdot \frac{d\nu}{d\lambda}\, d\lambda + \int_E \mathbf{1}_B \cdot \left\{-\frac{d\nu}{d\lambda}\right\} d\lambda = \int_E \left|\frac{d\nu}{d\lambda}\right| d\lambda.$$

By (1) we have $\left|\frac{d\nu}{d\lambda}\right| = \left|\frac{d\mu_1}{d\lambda} - \frac{d\mu_2}{d\lambda}\right| \le \frac{d\mu_1}{d\lambda} + \frac{d\mu_2}{d\lambda}$ by the nonnegativity of $\frac{d\mu_1}{d\lambda}$ and $\frac{d\mu_2}{d\lambda}$. Substituting this in (4), we have

$$\|\nu\|(E) \le \int_E \left\{\frac{d\mu_1}{d\lambda} + \frac{d\mu_2}{d\lambda}\right\} d\lambda = \mu_1(E) + \mu_2(E).$$

This completes the proof. ∎

Problems

Prob. 11.1. Consider the Lebesgue measure space $(\mathbb{R}, \mathfrak{M}_L, \mu_L)$. Let ν be the counting measure on $\mathfrak{M}_L$, that is, ν is defined by setting $\nu(E)$ to be equal to the number of elements in $E \in \mathfrak{M}_L$ if E is a finite set and $\nu(E) = \infty$ if E is an infinite set.
(a) Show that $\mu_L \ll \nu$ but $\frac{d\mu_L}{d\nu}$ does not exist.
(b) Show that ν does not have a Lebesgue decomposition with respect to μ_L.

Prob. 11.2. Let μ and ν be two positive measures on a measurable space $(X, \mathfrak{A})$.
(a) Show that if for every $\varepsilon > 0$ there exists $\delta > 0$ such that $\nu(E) < \varepsilon$ for every $E \in \mathfrak{A}$ with $\mu(E) < \delta$, then $\nu \ll \mu$.
(b) Show that if ν is a finite positive measure, then the converse of (a) holds.
(c) Show that (a) is still valid if ν is replaced by a signed measure λ, and that (b) is still valid if ν is replaced by a finite signed measure λ.

Prob. 11.3. Let μ and ν be two positive measures on a measurable space $(X, \mathfrak{A})$. Suppose $d\nu/d\mu$ exists so that $\nu \ll \mu$.
(a) Show that if $\frac{d\nu}{d\mu} > 0$, μ-a.e. on X, then $\mu \ll \nu$ and thus $\mu \sim \nu$.

(b) Show that if $\frac{d\nu}{d\mu} > 0$, μ-a.e. on X and if μ and ν are σ-finite, then $\frac{d\mu}{d\nu}$ exists and $\frac{d\mu}{d\nu} = \left(\frac{d\nu}{d\mu}\right)^{-1}$ μ- and ν-a.e. on X.

Prob. 11.4. Let $(X, \mathfrak{A}, \mu)$ be a measure space. Assume that there exists a $(0, \infty)$-valued $\mathfrak{A}$-measurable function f on X satisfying the condition that $\mu\{x \in X : f(x) \leq n\} < \infty$ for every $n \in \mathbb{N}$.
(a) Show that the existence of such a function f implies that μ is a σ-finite measure.
(b) Define a positive measure ν on $\mathfrak{A}$ by setting $\nu(E) = \int_E f \, d\mu$ for $E \in \mathfrak{A}$. Show that ν is a σ-finite measure.
(c) Show that $\frac{d\mu}{d\nu}$ exists and $\frac{d\mu}{d\nu} = 1/f$, μ- and ν-a.e. on X.

Prob. 11.5. Let $(X, \mathfrak{A}, \lambda)$ be a measure space. Let f and g be extended real-valued $\mathfrak{A}$-measurable functions on X such that $\int_X f \, d\lambda$ and $\int_X g \, d\lambda$ exist. Define two signed measures μ and ν on $(X, \mathfrak{A})$ by $\mu(E) = \int_E f \, d\lambda$ and $\nu(E) = \int_E g \, d\lambda$ for $E \in \mathfrak{A}$ so that $\frac{d\mu}{d\lambda} = f$ and $\frac{d\nu}{d\lambda} = g$, λ-a.e. on X and $\mu \ll \lambda$ and $\nu \ll \lambda$.
Let $\mathfrak{N}_\mu$ and $\mathfrak{N}_\nu$ be the collections of the null sets with respect to μ and ν respectively.
Let $N_f = \{x \in X : f(x) = 0\}$ and $N_g = \{x \in X : g(x) = 0\}$.
(a) Find a necessary and sufficient condition on f for μ to be σ-finite.
(b) Find a necessary and sufficient condition for $E \in \mathfrak{A}$ to be in $\mathfrak{N}_\mu$ in terms of N_f.
(c) Find a necessary and sufficient condition for $\nu \ll \mu$ in terms of N_f and N_g.
(d) Let λ and μ be σ-finite measures and ν be a σ-finite signed measure. Assume that $\nu \ll \mu$. Find $\frac{d\nu}{d\mu}$ in terms of f and g.
(e) Find a necessary and sufficient condition for $\mu \sim \nu$ in terms of N_f and N_g.
(f) Find a necessary and sufficient condition for $\mu \perp \nu$ in terms of N_f and N_g.
(g) Assume that μ and ν are σ-finite signed measures so that the Lebesgue decomposition $\nu = \nu_a + \nu_s$ exists where $\nu_a \ll \mu$ and $\nu_s \perp \mu$. Find ν_a and ν_s.

Prob. 11.6. Given a measurable space $(X, \mathfrak{A})$. For $A \in \mathfrak{A}$, let us write $\mathfrak{A} \cap A$ for the σ-algebra of subsets of A consisting of subsets of A of the type $E \cap A$ where $E \in \mathfrak{A}$.

Let μ and ν be σ-finite positive measures on $(X, \mathfrak{A})$. Show that there exist $A, B \in \mathfrak{A}$ such that $A \cap B = \emptyset$, $A \cup B = X$, $\mu \sim \nu$ on $(A, \mathfrak{A} \cap A)$ and $\mu \perp \nu$ on $(B, \mathfrak{A} \cap B)$. (Hint: Note that if we define $\lambda = \mu + \nu$, then $\mu \ll \lambda$ and $\nu \ll \lambda$ also.)

Prob. 11.7. (A variant of the Lebesgue Decomposition Theorem) Let μ and ν be σ-finite positive measures on a measurable space $(X, \mathfrak{A})$. Show that there exists a nonnegative extended real-valued $\mathfrak{A}$-measurable function φ on X and a set $A_0 \in \mathfrak{A}$ with $\mu(A_0) = 0$ such that $\nu(E) = \int_E \varphi \, d\mu + \nu(E \cap A_0)$ for every $E \in \mathfrak{A}$.

Chapter 3

Differentiation and Integration

§12 Monotone Functions and Functions of Bounded Variation

[I] The Derivative

To define the right-derivative and the left-derivative of an extended real-valued function defined on an interval in $\mathbb{R}$ let us review the definition of the right limit and the left limit of a function.

Let f be an extended real-valued function on (a, b).

1. If there exists $\alpha \in \mathbb{R}$ such that for every $\varepsilon > 0$ there exists $\delta > 0$ such that $|f(x)-\alpha| < \varepsilon$ for $x \in (a, a+\delta)$ then we say that α is the right limit of f as $x \downarrow a$, that is, as x tends to a from the right, and write $\lim_{x \downarrow a} f(x) = \alpha$.

If for every $M > 0$ there exists $\delta > 0$ such that $f(x) \geq M$ for $x \in (a, a+\delta)$ then we say that ∞ is the right hand limit of f as $x \downarrow a$ and write $\lim_{x \downarrow a} f(x) = \infty$.

If for every $M > 0$ there exists $\delta > 0$ such that $f(x) \leq -M$ for $x \in (a, a+\delta)$ then we say that $-\infty$ is the right hand limit of f as $x \downarrow a$ and write $\lim_{x \downarrow a} f(x) = -\infty$.

2. If $\lim_{x \downarrow a} f(x)$ exists in $\overline{\mathbb{R}}$, then for every sequence $(x_n : n \in \mathbb{N})$ in (a, b) such that $\lim_{n \to \infty} x_n = a$ we have $\lim_{n \to \infty} f(x_n) = \lim_{x \downarrow a} f(x)$.

Conversely if for every sequence $(x_n : n \in \mathbb{N})$ in (a, b) such that $\lim_{n \to \infty} x_n = a$ the limit $\lim_{n \to \infty} f(x_n)$ exists in $\overline{\mathbb{R}}$ then the limit is the same for all such sequences $(x_n : n \in \mathbb{N})$ in (a, b) and moreover $\lim_{x \downarrow a} f(x)$ exists and $\lim_{x \downarrow a} f(x) = \lim_{n \to \infty} f(x_n)$.

3. The left limit of f as $x \uparrow b$, that is, as x tends to b from the left, is defined likewise. Thus for instance $\lim_{x \uparrow b} f(x) = \beta \in \mathbb{R}$ if for every $\varepsilon > 0$ there exists $\delta > 0$ such that $|f(x) - \beta| < \varepsilon$ for $x \in (b - \delta, b)$.

4. Let f be an extended real-valued function on $[a, b]$. Let $x \in [a, b)$ and $f(x) \in \mathbb{R}$. We define the right-derivative of f at x as follows. Let $\delta > 0$ be so small that $x + \delta < b$. Define a function g on $(0, \delta)$ by setting

$$g(h) = \frac{f(x+h) - f(x)}{h} \quad \text{for } h \in (0, \delta).$$

The right-derivative of f at x is defined by

$$(D_r f)(x) = \lim_{h \downarrow 0} g(h),$$

provided the limit exists in $\overline{\mathbb{R}}$.

Definition 12.1. *Let f be an extended real-valued function on $[a, b]$.*
(a) *If $x \in [a, b)$, $f(x) \in \mathbb{R}$, and*

$$(D_r f)(x) := \lim_{h \downarrow 0} \frac{f(x+h) - f(x)}{h}$$

exists in $\overline{\mathbb{R}}$, then we call $(D_r f)(x)$ the right-derivative of f at x and say that f is right semi-differentiable at x. We say that f is right-differentiable at x only if $(D_r f)(x) \in \mathbb{R}$.
(b) *If $x \in (a, b]$, $f(x) \in \mathbb{R}$, and*

$$(D_\ell f)(x) := \lim_{h \uparrow 0} \frac{f(x+h) - f(x)}{h}$$

exists in $\overline{\mathbb{R}}$, then we call $(D_\ell f)(x)$ the left-derivative of f at x and say that f is left semi-differentiable at x. We say that f is left-differentiable at x only if $(D_\ell f)(x) \in \mathbb{R}$.
(c) *If $x \in (a, b)$, $f(x) \in \mathbb{R}$, and both $(D_\ell f)(x)$ and $(D_r f)(x)$ exist in $\overline{\mathbb{R}}$ and are equal, we let $(Df)(x)$, or as an alternate notation $f'(x)$, be the common value. Thus*

$$f'(x) = (Df)(x) = \lim_{h \to 0} \frac{f(x+h) - f(x)}{h}.$$

We call $f'(x)$ the derivative of f at x and say that f is semi-differentiable at x. We say that f is differentiable at x only when $f'(x) \in \mathbb{R}$.
(d) *Let us agree to say that f' exists on $[a, b]$ if $D_r f$ exists on $[a, b)$, $D_\ell f$ exists on $(a, b]$, and $D_\ell f = D_r f$ on (a, b).*

If a real-valued function f on (a, b) is differentiable at some $x \in (a, b)$, then f is continuous at x. The converse is false. In fact there exists a continuous function on $\mathbb{R}$ which is not differentiable at any $x \in \mathbb{R}$. To show such an example, let us prove a necessary condition for differentiability.

Proposition 12.2. *Let f be a real-valued function on (a, b). Suppose f is differentiable at some $x \in (a, b)$. Then for any two sequences $(a_n : n \in \mathbb{N})$ and $(b_n : n \in \mathbb{N})$ such that $a_n \uparrow x$ and $b_n \downarrow x$ as $n \to \infty$, we have*

$$\lim_{n \to \infty} \frac{f(b_n) - f(a_n)}{b_n - a_n} = f'(x).$$

Proof. For $n \in \mathbb{N}$, let $\lambda_n = (b_n - x)/(b_n - a_n)$. Then we have $\lambda_n \in [0, 1]$ and $1 - \lambda_n = (x - a_n)/(b_n - a_n)$. Now

$$f'(x) = \lambda_n f'(x) + (1 - \lambda_n) f'(x) \tag{1}$$

and

$$\begin{aligned} \frac{f(b_n) - f(a_n)}{b_n - a_n} &= \frac{f(b_n) - f(x)}{b_n - a_n} + \frac{f(x) - f(a_n)}{b_n - a_n} \\ &= \lambda_n \frac{f(b_n) - f(x)}{b_n - x} + (1 - \lambda_n) \frac{f(x) - f(a_n)}{x - a_n}. \end{aligned} \tag{2}$$

Subtracting (1) from (2), we have

$$\frac{f(b_n) - f(a_n)}{b_n - a_n} - f'(x) = \lambda_n A_n + (1 - \lambda_n) B_n,$$

where

$$A_n = \frac{f(b_n) - f(x)}{b_n - x} - f'(x) \quad \text{and} \quad B_n = \frac{f(x) - f(a_n)}{x - a_n} - f'(x).$$

By the differentiability of f at x, we have $\lim_{n\to\infty} A_n = 0$ and $\lim_{n\to\infty} B_n = 0$. Then since $(\lambda_n : n \in \mathbb{N})$ and $(1 - \lambda_n : n \in \mathbb{N})$ are bounded sequences of real numbers, we have $\lim_{n\to\infty} \lambda_n A_n = 0$ and $\lim_{n\to\infty} (1 - \lambda_n) B_n = 0$. Therefore we have

$$\lim_{n\to\infty} \left\{ \frac{f(b_n) - f(a_n)}{b_n - a_n} - f'(x) \right\} = 0.$$

This completes the proof. ∎

Example. (A Continuous Nowhere Differentiable Function on $\mathbb{R}$) Let φ be a real-valued function on $\mathbb{R}$ defined by setting $\varphi(x) = |x|$ for $x \in [-1, 1]$ and then setting $\varphi(x + 2) = \varphi(x)$ for $x \in \mathbb{R}$. Define a sequence $(\varphi_n : n \in \mathbb{N})$ of functions on $\mathbb{R}$ by setting $\varphi_n(x) = \varphi(4^n x)$ for $x \in \mathbb{R}$ and $n \in \mathbb{N}$. Let

$$f(x) = \sum_{n\in\mathbb{Z}_+} \left(\tfrac{3}{4}\right)^n \varphi_n(x) \quad \text{for } x \in \mathbb{R}.$$

Then f is continuous at every $x \in \mathbb{R}$ and not differentiable at any $x \in \mathbb{R}$.

Proof. The function φ is continuous and periodic with period 2 and $|\varphi| \leq 1$ on $\mathbb{R}$. The series of functions in the definition of f converges uniformly on $\mathbb{R}$ by the Weierstrass Test in Calculus and thus f is a continuous function on $\mathbb{R}$. Let us show that f is not differentiable at an arbitrary $x \in \mathbb{R}$. For a fixed $m \in \mathbb{Z}_+$, there exists $k \in \mathbb{Z}$ such that

$$k \leq 4^m x \leq k + 1.$$

Let

$$a_m = 4^{-m}k \quad \text{and} \quad b_m = 4^{-m}(k+1).$$

Consider $4^n a_m$ and $4^n b_m$ for $n \in \mathbb{Z}_+$. For $n > m$, we have $4^n b_m - 4^n a_m = 4^n 4^{-m}$ which is an even integer so that $|\varphi(4^n b_m) - \varphi(4^n a_m)| = 0$ by the fact that φ is periodic with period 2. Next if $n = m$, then $4^n a_m$ and $4^n b_m$ are integers and $4^n b_m - 4^n a_m = 1$ so that $|\varphi(4^n b_m) - \varphi(4^n a_m)| = 1$. Finally if $n < m$, then $4^n a_m$ and $4^n b_m$ are not integers and furthermore no integers lie between them so that by the definition of φ and by the fact that $4^n b_m - 4^n a_m = 4^{n-m}$, we have $|\varphi(4^n b_m) - \varphi(4^n a_m)| = 4^{n-m}$. Summarizing these results, we have

$$|\varphi_n(b_m) - \varphi_n(a_m)| = |\varphi(4^n b_m) - \varphi(4^n a_m)| = \begin{cases} 0 & \text{for } n > m, \\ 4^{n-m} & \text{for } n \le m. \end{cases}$$

Thus we have

$$\left|\frac{f(b_m) - f(a_m)}{b_m - a_m}\right| = \left|\sum_{n \in \mathbb{Z}_+} \left(\frac{3}{4}\right)^n \frac{\varphi_n(b_m) - \varphi_n(a_m)}{b_m - a_m}\right| = \left|\sum_{n=0}^{m} \left(\frac{3}{4}\right)^n \frac{\varphi_n(b_m) - \varphi_n(a_m)}{b_m - a_m}\right|.$$

Applying the inequality $\big||\alpha| - |\beta|\big| \le |\alpha - \beta|$ for $\alpha, \beta \in \mathbb{R}$, we have

$$\begin{aligned}
\left|\frac{f(b_m) - f(a_m)}{b_m - a_m}\right| &\ge \left(\frac{3}{4}\right)^m \left|\frac{\varphi_m(b_m) - \varphi_m(a_m)}{b_m - a_m}\right| - \sum_{n=0}^{m-1} \left(\frac{3}{4}\right)^n \left|\frac{\varphi_n(b_m) - \varphi_n(a_m)}{b_m - a_m}\right| \\
&= \left(\frac{3}{4}\right)^m \frac{1}{4^{-m}} - \sum_{n=0}^{m-1} \left(\frac{3}{4}\right)^n \frac{1}{4^{-m}} 4^{n-m} = 3^m - \sum_{n=0}^{m-1} 3^n \\
&= 3^m - \frac{3^m - 1}{2} > \frac{3^m}{2}.
\end{aligned}$$

Thus we have $\lim_{m \to \infty} |f(b_m) - f(a_m)|/|b_m - a_m| = \infty$. Since $a_m \uparrow x$ and $b_m \downarrow x$ as $m \to \infty$, this implies that f is not differentiable at x by Proposition 12.2. ∎

If f is a differentiable function on $[a, b]$, then the real-valued function f' need not be continuous on $[a, b]$. (For example, the function $f(x) = x^2 \sin 1/x$ for $x \in [-1, 1] \setminus \{0\}$ and $f(0) = 0$ is differentiable on $[-1, 1]$ but f' is discontinuous at $x = 0$.) Nevertheless f', like a continuous function, has the intermediate value property.

Theorem 12.3. (Intermediate Value Theorem for Derivatives) *Let f be a real-valued function on $[a, b]$. If f is differentiable on $[a, b]$ and if $f'(x_1) \ne f'(x_2)$ for some $x_1, x_2 \in [a, b]$, $x_1 < x_2$, then for every $c \in \mathbb{R}$ between $f'(x_1)$ and $f'(x_2)$ there exists $x_0 \in (x_1, x_2)$ such that $f'(x_0) = c$.*

Proof. Let us assume $f'(x_1) < f'(x_2)$. Let $g(x) = f(x) - cx$ for $x \in [a, b]$. Then g is a real-valued continuous function on $[a, b]$ and in particular on $[x_1, x_2]$ so that it assumes a minimum on $[x_1, x_2]$ at some point in $[x_1, x_2]$, that is, there exists $x_0 \in [x_1, x_2]$ such that

$g(x_0) \leq g(x)$ for $x \in [x_1, x_2]$. Let us show that actually $x_0 \in (x_1, x_2)$. Now $g'(x) = f'(x) - c$ so that $g'(x_1) < 0 < g'(x_2)$. Since

$$g'(x_1) = (D_r g)(x_1) = \lim_{h \downarrow 0} \frac{g(x_1 + h) - g(x_1)}{h} < 0,$$

there exists $\delta > 0$ such that $g(x_1 + h) - g(x_1) < 0$ for $h \in (0, \delta)$. Thus there exists $\xi \in (x_1, x_2)$ such that $g(\xi) - g(x_1) < 0$, that is, $g(\xi) < g(x_1)$. Thus $x_0 \neq x_1$. Similarly

$$g'(x_2) = (D_\ell g)(x_2) = \lim_{h \uparrow 0} \frac{g(x_2 + h) - g(x_2)}{h} = \lim_{h \downarrow 0} \frac{g(x_2) - g(x_2 - h)}{h} > 0$$

so that there exists $\delta > 0$ such that $g(x_2) - g(x_2 - h) > 0$ for $h \in (0, \delta)$. Then there exists $\xi \in (x_1, x_2)$ such that $g(x_2) - g(\xi) > 0$, that is, $g(x_2) > g(\xi)$. This shows that $x_0 \neq x_2$. Therefore $x_0 \in (x_1, x_2)$. Then since g is differentiable at x_0 and since g assumes a minimum at x_0, we have $g'(x_0) = 0$. Then $f'(x_0) = c$. ∎

Example. If we define a real-valued function f on $[-1, 1]$ by setting $f(x) = 0$ for $x \in [-1, 0)$ and $f(x) = 1$ for $x \in [0, 1]$, then f does not have the intermediate value property so that by Theorem 12.3, f is not the derivative of any function on $[-1, 1]$.

Regarding the measurability of the derivatives of a measurable function, we have the following

Proposition 12.4. *Let f be a real-valued $\mathfrak{M}_L$-measurable function on $[a, b]$. Let E be a $\mathfrak{M}_L$-measurable subset of $[a, b]$. If $D_r f$ exists on E, then $D_r f$ is a $\mathfrak{M}_L$-measurable function on E. The same holds for $D_\ell f$ and f'.*

Proof. Let us extend the definition of f to $\mathbb{R}$ by setting $f = 0$ on $\mathbb{R} \setminus [a, b]$. Let us define a sequence of real-valued function on $\mathbb{R}$ by setting $g_n(x) = n\left\{f\left(x + \frac{1}{n}\right) - f(x)\right\}$ for $x \in \mathbb{R}$. The $\mathfrak{M}_L$-measurability of f on $\mathbb{R}$ implies that of g_n by Theorem 9.31 (Translation Invariance). Let $D_e = \left\{\mathbb{R} : \lim_{n \to \infty} g_n \text{ exists in } \overline{\mathbb{R}}\right\}$. By Theorem 4.23, $D_e \in \mathfrak{M}_L$ and $\lim_{n \to \infty} g_n$ is $\mathfrak{M}_L$-measurable on D_e. Now if $(D_r f)(x)$ exists for some $x \in [a, b)$, then

$$(D_r f)(x) = \lim_{h \downarrow 0} \frac{f(x + h) - f(x)}{h} = \lim_{n \to \infty} \frac{f\left(x + \frac{1}{n}\right) - f(x)}{\frac{1}{n}} = \lim_{n \to \infty} g_n(x).$$

Thus if $D_r f$ exists on a $\mathfrak{M}_L$-measurable subset E of $[a, b)$, then $E \subset D_e$ and moreover $D_r f = \lim_{n \to \infty} g_n$ on E so that $D_r f$ is $\mathfrak{M}_L$-measurable on E. ∎

We shall show in §13 that if f is an absolutely continuous function on $[a, b]$ then f' exists a.e. on $[a, b]$ and $\int_{[a,b]} f' \, d\mu_L = f(b) - f(a)$. We shall show also that if f is a real-valued function on $[a, b]$ and f' exists and is finite everywhere on $[a, b]$ and is μ_L-integrable on $[a, b]$, then $\int_{[a,b]} f' \, d\mu_L = f(b) - f(a)$.

To define the Dini derivates let us review the definition of the right limit inferior, the right limit superior, the left limit inferior, and the left limit superior of an extended real-valued function defined on an interval in $\mathbb{R}$. Let f be an extended real-valued function on (a, b).

1. The right limit inferior and the right limit superior of f as $x \downarrow a$ are defined by

$$\liminf_{x \downarrow a} f(x) = \lim_{\delta \downarrow 0} \inf_{(a,a+\delta)} f(x)$$

and

$$\limsup_{x \downarrow a} f(x) = \lim_{\delta \downarrow 0} \sup_{(a,a+\delta)} f(x).$$

Let us note that $\inf_{(a,a+\delta)} f(x)$ increases as $\delta \downarrow 0$ so that $\lim_{\delta \downarrow 0} \inf_{(a,a+\delta)} f(x)$ exists in $\overline{\mathbb{R}}$. Similarly $\sup_{(a,a+\delta)} f(x)$ decreases as $\delta \downarrow 0$ so that $\lim_{\delta \downarrow 0} \sup_{(a,a+\delta)} f(x)$ exists in $\overline{\mathbb{R}}$. Thus $\liminf_{x \downarrow a} f(x)$ and $\liminf_{x \downarrow a} f(x)$ always exist. Moreover $\lim_{x \downarrow a} f(x)$ exists if and only if $\liminf_{x \downarrow a} f(x) = \liminf_{x \downarrow a} f(x)$ and when this is the case we have $\lim_{x \downarrow a} f(x) = \liminf_{x \downarrow a} f(x) = \liminf_{x \downarrow a} f(x)$.

2. There exists a decreasing sequence $(x_n : n \in \mathbb{N})$ in (a, b) such that $x_n \downarrow a$ and $\lim_{n \to \infty} f(x_n) = \liminf_{x \downarrow a} f(x)$. Similarly there exists a decreasing sequence $(x_n : n \in \mathbb{N})$ in (a, b) such that $x_n \downarrow a$ and $\lim_{n \to \infty} f(x_n) = \limsup_{x \downarrow a} f(x)$.

(Let us prove the first of the two statements for instance. Let $\gamma = \liminf_{x \downarrow a} f(x) \in \overline{\mathbb{R}}$, that is, $\gamma = \lim_{\delta \downarrow 0} \inf_{(a,a+\delta)} f(x)$. Since $\inf_{(a,a+\delta)} f(x)$ increases as $\delta \downarrow 0$, for an arbitrary sequence $(\delta_k : k \in \mathbb{N})$ of positive numbers such that $\delta_k \downarrow 0$ we have

$$\lim_{k \to \infty} \inf_{(a,a+\delta_k)} f(x) = \lim_{\delta \downarrow 0} \inf_{(a,a+\delta)} f(x) = \gamma.$$

Select $x_k \in (a, a + \delta_k)$ such that $\inf_{(a,a+\delta_k)} f(x) \le f(x_k) \le \inf_{(a,a+\delta_k)} f(x) + \frac{1}{k}$. Then since $\lim_{k \to \infty} \inf_{(a,a+\delta_k)} f(x) = \gamma$ and $\lim_{k \to \infty} \inf_{(a,a+\delta_k)} f(x) + \frac{1}{k} = \gamma$ also, we have $\lim_{k \to \infty} f(x_k) = \gamma$. Now $x_k \in (a, a + \delta_k)$ and $\delta_k \downarrow 0$ so that $x_k > a$ and $\lim_{k \to \infty} x_k = a$. Then we can select a subsequence $(x_{k_n} : n \in \mathbb{N})$ of $(x_k : k \in \mathbb{N})$ such that $x_{k_n} \downarrow a$. Let $(x_n : n \in \mathbb{N})$ be the sequence $(x_{k_n} : n \in \mathbb{N})$. Then $(x_n : n \in \mathbb{N})$ is a decreasing sequence in (a, b) such that $x_n \downarrow a$ and $\lim_{n \to \infty} f(x_n) = \lim_{n \to \infty} f(x_{k_n}) = \lim_{k \to \infty} f(x_k) = \gamma$.)

3. The left limit inferior and the left limit superior of f as $x \uparrow b$ are defined by

$$\liminf_{x \uparrow b} f(x) = \lim_{\delta \downarrow 0} \inf_{(b-\delta,b)} f(x)$$

and

$$\limsup_{x \uparrow b} f(x) = \lim_{\delta \downarrow 0} \sup_{(b-\delta,b)} f(x).$$

4. Let f be an extended real-valued function on $[a, b]$. Let $x \in [a, b)$ and $f(x) \in \mathbb{R}$. We define the lower-right Dini derivate of f at x as follows. Let $\delta > 0$ be so small that $x + \delta < b$. Define a function g on $(0, \delta)$ by setting

$$g(h) = \frac{f(x+h) - f(x)}{h} \quad \text{for } h \in (0, \delta).$$

The lower-right Dini derivate of f at x is defined by $(D_+f)(x) = \liminf_{h \downarrow 0} g(h)$.

Definition 12.5. *Let f be an extended real-valued function on $[a, b]$.*
(a) *If $x \in [a, b)$ and $f(x) \in \mathbb{R}$, we define*

$$(D_+f)(x) = \liminf_{h \downarrow 0} \frac{f(x+h) - f(x)}{h} \in \overline{\mathbb{R}},$$

$$(D^+f)(x) = \limsup_{h \downarrow 0} \frac{f(x+h) - f(x)}{h} \in \overline{\mathbb{R}}.$$

(b) *If $x \in (a, b]$ and $f(x) \in \mathbb{R}$, we define*

$$(D_-f)(x) = \liminf_{h \uparrow 0} \frac{f(x+h) - f(x)}{h} \in \overline{\mathbb{R}},$$

$$(D^-f)(x) = \limsup_{h \uparrow 0} \frac{f(x+h) - f(x)}{h} \in \overline{\mathbb{R}}.$$

We call $(D_+f)(x)$, $(D^+f)(x)$, $(D_-f)(x)$, and $(D^-f)(x)$ respectively the lower-right, upper-right, lower-left, and upper-left Dini derivates of f at x.

Note that the four Dini derivates always exist in $\overline{\mathbb{R}}$ and $(D_-f)(x) \le (D^-f)(x)$ and $(D_+f)(x) \le (D^+f)(x)$.
f is right semi-differentiable at x if and only if $(D_+f)(x) = (D^+f)(x) \in \overline{\mathbb{R}}$.
f is right-differentiable at x if and only if $(D_+f)(x) = (D^+f)(x) \in \mathbb{R}$.
f is left semi-differentiable at x if and only if $(D_-f)(x) = (D^-f)(x) \in \overline{\mathbb{R}}$.
f is left-differentiable at x if and only if $(D_-f)(x) = (D^-f)(x) \in \mathbb{R}$.
f is semi-differentiable at x if and only if the four Dini derivates of f at x are all equal.
f is differentiable at x if and only if the four Dini derivates of f at x are all finite and equal.

[II] Differentiability of Monotone Functions

A real-valued function f on an interval I in $\mathbb{R}$ is called an increasing (or non-decreasing) function if $f(x_1) \le f(x_2)$ for every pair $x_1, x_2 \in I$ such that $x_1 < x_2$, and a decreasing (or non-increasing) function if $f(x_1) \ge f(x_2)$. If f is a decreasing function, then $-f$ is an increasing function and thus we need not treat decreasing functions separately. A function is called a monotone function if it is either an increasing function or a decreasing function.

Let f be a real-valued increasing function on an open interval I in $\mathbb{R}$. Let a be an arbitrary point in I. Since $f(x) \uparrow$ as $x \uparrow a$ and since $f(x) \le f(a)$ for $x \in I$ such that

$x < a$, $f(a-) := \lim_{x \uparrow a} f(x)$ exists in $\mathbb{R}$. Similarly $f(a+) := \lim_{x \downarrow a} f(x)$ exists in $\mathbb{R}$. Also $f(a) \in [f(a-), f(a+)]$. Thus a real-valued increasing function f on open interval I is continuous at $a \in I$ if and only if $f(a-) = f(a+)$ and discontinuous at a if and only if $f(a-) < f(a+)$. If f is discontinuous at a, we call $\eta(a) := f(a+) - f(a-) > 0$, the jump of f at a.

Theorem 12.6. *Let f be a real-valued increasing function on an open interval I in $\mathbb{R}$. Then f has at most countably many points of discontinuity in I.*

Proof. Let a_1 and a_2 be two points of discontinuity of f in I and assume $a_1 < a_2$. Consider the two open intervals $J(a_1) = \big(f(a_1-), f(a_1+)\big)$ and $J(a_2) = \big(f(a_2-), f(a_2+)\big)$. To show that $J(a_1)$ and $J(a_2)$ are disjoint, let us take an arbitrary point $x_0 \in (a_1, a_2)$. From the fact that f is increasing on I, we have $f(a_1+) \le f(x_0) \le f(a_2-)$. Thus the two open intervals $J(a_1)$ and $J(a_2)$ are disjoint.

Let A be the set of all points of discontinuity of f in I and consider the collection of the open intervals $\{J(a) : a \in A\}$ where $J(a) = \big(f(a-), f(a+)\big)$. From $J(a)$ select arbitrarily a rational number r_a. Since $\{J(a) : a \in A\}$ is a collection of pairwise disjoint sets, $\{r_a : a \in A\}$ is a collection of distinct rational numbers. Since there are only countably many rational numbers, A is at most a countable set. ∎

If f is a real-valued increasing function on an open interval I in $\mathbb{R}$, then f is continuous except possibly at countably many points in I. Since continuity of f at $x \in I$ implies $f(x) = f(x+)$, we have $f(x) = f(x+)$ except at countably many $x \in I$. Thus if we define a real-valued function g on I by setting $g(x) = f(x+)$ for $x \in I$, then $g = f$ except at countably many points in I. Since f is an increasing function, we have $f(x_1+) \le f(x_2+)$ for any $x_1, x_2 \in I$ such that $x_1 < x_2$. Thus g is an increasing function on I. Furthermore g is right-continuous on I. (This can be shown as follows. Let $a \in I$ be arbitrarily chosen. By the definition of $f(a+)$, for every $\varepsilon > 0$ there exists $\delta > 0$ such that $f(x) - f(a+) < \varepsilon$ for $x \in (a, a+\delta)$. Then $f(x+) - f(a+) \le \varepsilon$, that is, $g(x) - g(a) \le \varepsilon$, for $x \in (a, a+\delta)$. This proves the right-continuity of g at a.) Our function g is thus a real-valued right-continuous increasing function on I obtained by redefining f at countably many points at most. Let us call g the right-continuous modification of f.

According to Theorem 12.6, if f is a real-valued increasing function on an open interval I in $\mathbb{R}$, then the set of the points of discontinuity of f is a countable subset of I. This countable subset of I may have limit points in I. In fact it may even be a dense subset of I. We construct some examples below.

Example 1. Let $E = \{\xi_n : n \in \mathbb{N}\}$ be a countable dense subset of $\mathbb{R}$ (for instance the set of all rational numbers) and let $(\alpha_n : n \in \mathbb{N})$ be a sequence of positive numbers such that $\alpha := \sum_{n \in \mathbb{N}} \alpha_n < \infty$. Let us define a function f on $\mathbb{R}$ by setting

$$f(x) = \sum_{\{n \in \mathbb{N} : \xi_n \le x\}} \alpha_n \quad \text{for } x \in \mathbb{R}.$$

Then the function f has the following properties:

1° f is a real-valued right-continuous increasing function on $\mathbb{R}$.

2° E is the set of points of discontinuity of f and the jump of f at ξ_n is equal to α_n for every $n \in \mathbb{N}$.

3° $0 \le f \le \alpha$ on $\mathbb{R}$.

Proof. It is immediate from the definition that f is a nonnegative real valued increasing function on $\mathbb{R}$ which is bounded above by α. Let us show that f is right-continuous at every $x_0 \in \mathbb{R}$ by showing that for every $\varepsilon > 0$ there exists $\delta > 0$ such that $f(x) - f(x_0) < \varepsilon$ for $x \in (x_0, x_0 + \delta)$. Let $\varepsilon > 0$ be arbitrarily given. Now since $\sum_{n \in \mathbb{N}} \alpha_n < \infty$, there exists $N \in \mathbb{N}$ such that $\sum_{n > N} \alpha_n < \varepsilon$. Let $\delta > 0$ be so small that the interval $(x_0, x_0 + \delta)$ does not contain any point in the finite collection $\{\xi_1, \ldots, \xi_N\}$. Then $f(x) - f(x_0) \le \sum_{n > N} \alpha_n < \varepsilon$ for every $x \in (x_0, x_0 + \delta)$. This proves the right-continuity of f at x_0.

Let us show that for every $n_0 \in \mathbb{N}$, f is discontinuous at ξ_{n_0} with jump at ξ_{n_0} given by $f(\xi_{n_0}) - f(\xi_{n_0}-) = \alpha_{n_0}$. Let $\mathbb{N}(\xi_{n_0}) = \{n \in \mathbb{N} : \xi_n < \xi_{n_0}\}$ and let $\beta = \sum_{n \in \mathbb{N}(\xi_{n_0})} \alpha_n$. Then by the definition of f, we have

$$f(\xi_{n_0}) = \sum_{\{n \in \mathbb{N} : \xi_n \le \xi_{n_0}\}} \alpha_n = \sum_{n \in \mathbb{N}(\xi_{n_0})} \alpha_n + \alpha_{n_0} = \beta + \alpha_{n_0}. \tag{1}$$

For every $x < \xi_{n_0}$ we have $f(x) \le \sum_{n \in \mathbb{N}(\xi_{n_0})} \alpha_n = \beta$ and thus

$$f(\xi_{n_0}-) = \lim_{x \uparrow \xi_{n_0}} f(x) \le \beta. \tag{2}$$

For an arbitrary $\varepsilon > 0$, since $\sum_{n \in \mathbb{N}(\xi_{n_0})} \alpha_n = \beta < \infty$, there exists a finite subcollection $\mathbb{N}_0$ of $\mathbb{N}(\xi_{n_0})$ such that $\sum_{n \in \mathbb{N}_0} \alpha_n > \beta - \varepsilon$. Since $\mathbb{N}_0$ is a finite collection, there exists $x < \xi_{n_0}$ such that $x > \xi_n$ for $n \in \mathbb{N}_0$. Then $f(x) \ge \sum_{n \in \mathbb{N}_0} \alpha_n > \beta - \varepsilon$. This shows that $f(\xi_{n_0}-) = \lim_{x \uparrow \xi_{n_0}} f(x) \ge \beta - \varepsilon$. Then by the arbitrariness of $\varepsilon > 0$ we have $f(\xi_{n_0}-) \ge \beta$. Therefore $f(\xi_{n_0}-) = \beta$ by (2). Substituting this in (1), we have $f(\xi_{n_0}) - f(\xi_{n_0}-) = \alpha_{n_0}$.

To show the continuity of f at every $x_0 \in \mathbb{R} \setminus E$, let $\mathbb{N}(x_0) = \{n \in \mathbb{N} : \xi_n < x_0\}$ and $\beta = \sum_{n \in \mathbb{N}(x_0)} \alpha_n$. Then $f(x_0) = \sum_{n \in \mathbb{N}(x_0)} \alpha_n = \beta$. By the same argument as above, we have $f(x_0-) = \beta$ and thus $f(x_0) = f(x_0-)$, proving the continuity of f at x_0. ∎

Example 2. In Example 1, let us assume that $E \cap \mathbb{Z} = \emptyset$. For $x \in \mathbb{R}$, let $[x]$ be the greatest integer not exceeding x. Let f be the function in Example 1 and let g be defined by

$$g(x) = f(x) + [x] \quad \text{for } x \in \mathbb{R}.$$

Then the function g has the following properties:

1° g is a real-valued right-continuous increasing function on $\mathbb{R}$.

2° $E \cup \mathbb{Z}$ is the set of the points of discontinuity of g. The jump of g at ξ_n is equal to α_n for every $n \in \mathbb{N}$ and the jump of g at an integer is equal to 1.

3° $\lim_{x \to -\infty} g(x) = -\infty$ and $\lim_{x \to \infty} g(x) = \infty$.

Example 3. Consider a finite open interval (a, b) in $\mathbb{R}$. The real-valued function φ defined on (a, b) by $\varphi(x) = \frac{\pi}{b-a}(x - a) - \frac{\pi}{2}$ for $x \in (a, b)$ is a homeomorphism of (a, b) onto

$\left(-\frac{\pi}{2}, \frac{\pi}{2}\right)$. The function $\psi(y) = \tan y$ for $y \in \left(-\frac{\pi}{2}, \frac{\pi}{2}\right)$ is a homeomorphism of $\left(-\frac{\pi}{2}, \frac{\pi}{2}\right)$ onto $\mathbb{R}$ and thus the function $\vartheta = \psi \circ \varphi$ is a homeomorphism of (a, b) onto $\mathbb{R}$. With the function g in Example 2, let $h = g \circ \vartheta$. Then h is a real-valued right-continuous increasing function on (a, b) with $\lim_{x \downarrow a} h(x) = -\infty$ and $\lim_{x \uparrow b} h(x) = \infty$, and the set of points of discontinuity is a dense subset of (a, b).

We show next that a real-valued increasing function f is differentiable a.e. on $[a, b]$. For this we require the Vitali Covering Theorem.

Definition 12.7. *Let E be a subset of $\mathbb{R}$. A collection $\mathfrak{V}$ of finite closed intervals is called a Vitali cover of E if for every $x \in E$ and $\varepsilon > 0$ there exists $I \in \mathfrak{V}$ such that $x \in I$ and $\ell(I) < \varepsilon$. (Singletons are not admissible as finite closed intervals.)*

The collection of all finite closed intervals in $\mathbb{R}$ is a Vitali cover for any subset of $\mathbb{R}$. For an arbitrary subset E of $\mathbb{R}$, the collection $\left\{\left[x - \frac{1}{n}, x + \frac{1}{n}\right] : n \in \mathbb{N}, x \in E\right\}$ is an example of Vitali cover of E. So is the collection $\left\{\left[x, x + \frac{1}{n}\right] : n \in \mathbb{N}, x \in E\right\}$.

Observation 12.8. Let $\mathfrak{V}$ be a Vitali cover of a subset E of $\mathbb{R}$. For an arbitrary open set O such that $O \supset E$, let $\mathfrak{V}_O$ be the subcollection of $\mathfrak{V}$ consisting of all members of $\mathfrak{V}$ which are contained in O. Then $\mathfrak{V}_O$ is a Vitali cover of E.

Proof. To show that $\mathfrak{V}_O$ is a Vitali cover of E, we show that for every $x \in E$ and $\varepsilon > 0$ there exists $I \in \mathfrak{V}_O$ such that $x \in I$ and $\ell(I) < \varepsilon$. Now if $x \in E$, then since $E \subset O$ and O is an open set there exists $\delta > 0$ such that $(x - \delta, x + \delta) \subset O$. Since $\mathfrak{V}$ is a Vitali cover of E, there exists $I \in \mathfrak{V}$ such that $x \in I$ and $\ell(I) < \min\left\{\frac{1}{2}\delta, \varepsilon\right\}$. Then $I \subset (x - \delta, x + \delta) \subset O$ so that $I \in \mathfrak{V}_O$. Also $\ell(I) < \varepsilon$. This shows that $\mathfrak{V}_O$ is a Vitali cover of E. ∎

Theorem 12.9. (Vitali Covering Theorem) *Let $\mathfrak{V}$ be a Vitali cover of an arbitrary subset E of $\mathbb{R}$. Then there exists a countable disjoint subcollection $\{I_n : n \in \mathbb{N}\}$ of $\mathfrak{V}$ such that*

(1)
$$\mu_L^*\left(E \setminus \bigcup_{n \in \mathbb{N}} I_n\right) = 0.$$

If $\mu_L^*(E) < \infty$, then the disjoint collection $\{I_n : n \in \mathbb{N}\}$ can be so chosen that

(2)
$$\sum_{n \in \mathbb{N}} \ell(I_n) < \infty,$$

and for every $\varepsilon > 0$ there exists $N \in \mathbb{N}$ such that

(3)
$$\mu_L^*\left(E \setminus \bigcup_{n=1}^{N} I_n\right) < \varepsilon.$$

Proof. **1.** Let us assume first that $\mu_L^*(E) < \infty$. By Lemma 3.21, there exists an open set $O \supset E$ such that $\mu_L^*(E) \le \mu_L(O) \le \mu_L^*(E) + 1 < \infty$. Let $\mathfrak{V}_O$ be the subcollection of $\mathfrak{V}$ consisting of those members of $\mathfrak{V}$ which are contained in O. By Observation 12.8, $\mathfrak{V}_O$ is a Vitali cover of E. Let $I_1 \in \mathfrak{V}_O$ be arbitrarily chosen. If $\mu_L^*(E \setminus I_1) = 0$, then the collection $\{I_1\}$ satisfies conditions (1), (2), and (3). If $\mu_L^*(E \setminus I_1) > 0$, then we select a disjoint subcollection $\{I_n : n \in \mathbb{N}\}$ of $\mathfrak{V}_O$ inductively as follows. Suppose for some $n \in \mathbb{N}$ we have selected a disjoint subcollection $\{I_1, \dots, I_n\}$ of $\mathfrak{V}_O$. If $\mu_L^*\big(E \setminus \bigcup_{k=1}^n I_k\big) = 0$, then the collection $\{I_1, \dots, I_n\}$ satisfies conditions (1), (2), and (3). If $\mu_L^*\big(E \setminus \bigcup_{k=1}^n I_k\big) > 0$, then we select $I_{n+1} \in \mathfrak{V}_O$ as follows.

For brevity let

(4) $$F_n = \bigcup_{k=1}^n I_k \quad \text{and} \quad O_n = O \setminus F_n.$$

Then F_n is a closed set in $\mathbb{R}$ and $O_n = O \cap F_n^c$ is an open set in $\mathbb{R}$ and moreover we have

(5) $$F_n \cap O_n = \emptyset \quad \text{and} \quad F_n \cup O_n = O.$$

Let

(6) $$\mathfrak{V}_{O_n} = \{I \in \mathfrak{V}_O : I \subset O_n\}.$$

Since $\mathfrak{V}_O$ is a Vitali cover of E, it is certainly a Vitali cover of the subset $E \setminus F_n$ of E. Since $E \setminus F_n \subset O \setminus F_n = O_n$, the collection $\mathfrak{V}_{O_n}$ defined by (6) is a Vitali cover of $E \setminus F_n$ by Lemma 12.8. In particular, $\mathfrak{V}_{O_n} \neq \emptyset$. Let

(7) $$d_n = \sup\{\ell(I) : I \in \mathfrak{V}_{O_n}\}.$$

Since $\mathfrak{V}_{O_n} \neq \emptyset$ and $\ell(I) > 0$ for every $I \in \mathfrak{V}_{O_n}$, we have $d_n > 0$. On the other hand, for every $I \in \mathfrak{V}_{O_n}$ we have $I \subset O_n \subset O$ so that $\ell(I) \le \mu_L(O) < \infty$ and then $d_n \le \mu_L(O) < \infty$. Let us select $I_{n+1} \in \mathfrak{V}_{O_n}$ such that

(8) $$\ell(I_{n+1}) > \tfrac{1}{2} d_n.$$

Since $I_{n+1} \subset O_n$, we have $I_{n+1} \cap F_n = \emptyset$. Thus $\{I_1, \dots, I_n, I_{n+1}\}$ is a disjoint collection in $\mathfrak{V}_O$. If $\mu_L^*\big(\bigcup_{k=1}^{n+1} I_k\big) = 0$, then conditions (1), (2), and (3) are satisfied by our collection $\{I_1, \dots, I_n, I_{n+1}\}$ and we are done. If $\mu_L^*\big(\bigcup_{k=1}^{n+1} I_k\big) > 0$, then we repeat the selection process above to select $I_{n+2} \in \mathfrak{V}_{O_{n+1}}$. If this selection process does not terminate in finitely many steps then we have a disjoint collection $\{I_n : n \in \mathbb{N}\}$ in $\mathfrak{V}_O$ satisfying condition (8). By the disjointness of the collection $\{I_n : n \in \mathbb{N}\}$ we have

$$\sum_{n \in \mathbb{N}} \ell(I_n) = \sum_{n \in \mathbb{N}} \mu_L(I_n) = \mu_L\Big(\bigcup_{n \in \mathbb{N}} I_n\Big) \le \mu_L(O) < \infty$$

so that condition (2) is satisfied. Let us show that the collection $\{I_n : n \in \mathbb{N}\}$ satisfies (1).

Let $F = \bigcup_{n \in \mathbb{N}} I_n$. If $\mu_L^*(E \setminus F) = 0$, then we have (1). Suppose $\mu_L^*(E \setminus F) > 0$. For each $n \in \mathbb{N}$, let J_n be a closed interval with the same midpoint as I_n but with $\ell(J_n) =$

$5\ell(I_n)$. Then $\sum_{n\in\mathbb{N}} \ell(J_n) = 5\sum_{n\in\mathbb{N}} \ell(I_n) < \infty$ and this implies $\lim_{p\to\infty} \sum_{n\geq p} \ell(J_n) = 0$. Thus

$$\lim_{p\to\infty} \mu_L\Big(\bigcup_{n\geq p} J_n\Big) \leq \lim_{p\to\infty} \sum_{n\geq p} \ell(J_n) = 0. \tag{9}$$

If we show that

$$E \setminus F \subset \bigcup_{n\geq p} J_n \quad \text{for every } p \in \mathbb{N}, \tag{10}$$

then $\mu_L^*(E \setminus F) \leq \mu_L\big(\bigcup_{n\geq p} J_n\big)$ for every $p \in \mathbb{N}$ and then by (9) we have

$$\mu_L^*(E \setminus F) \leq \lim_{p\to\infty} \mu_L\Big(\bigcup_{n\geq p} J_n\Big) = 0$$

and condition (1) is satisfied. Thus to prove (1) it suffices to prove (10).

Let $p \in \mathbb{N}$ be arbitrarily fixed. Now since $\mu_L^*(E \setminus F) > 0$, we have $E \setminus F \neq \emptyset$. Let $x \in E \setminus F$ be arbitrarily chosen. We show that there exists $I^* \in \mathfrak{V}_O$ such that

$$x \in I^* \subset J_q \quad \text{for some } q > p. \tag{11}$$

Then we have $x \in \bigcup_{n\geq p} J_n$ so that $E \setminus F \subset \bigcup_{n\geq p} J_n$, proving (10). Let us prove the existence of such $I^* \in \mathfrak{V}_O$. Now since $x \in E\setminus F$, we have $x \notin F_n$ so that $x \in O\setminus F_n = O_n$ for every $n \in \mathbb{N}$. Then $x \in \bigcap_{n\in\mathbb{N}} O_n$ and in particular $x \in O_p$. Now since $\mathfrak{V}_O$ is a Vitali cover of E, it is a Vitali cover of the subset $\{x\}$ of E. Now since O_p is an open set and $\{x\} \subset O_p$, $\mathfrak{V}_{O_p} = \{I \in \mathfrak{V}_O : I \subset O_p\}$ is a Vitali cover of $\{x\}$ by Observation 12.8. Thus there exists $I^* \in \mathfrak{V}_{O_p}$ such that $x \in I^* \subset O_p$.

It remains to show that $I^* \subset J_q$ for some $q > p$. For this let us note first that since $x \in O_n = O\setminus F_n$ and $F_n \supset I_n$ we have $x \notin I_n$ for every $n \in \mathbb{N}$. Then since $x \in I^*$, we have $I^* \neq I_n$ for every $n \in \mathbb{N}$. Let us show that there exists $N \in \mathbb{N}$ such that $I^* \cap F_N \neq \emptyset$. Now by (2) we have $\sum_{n\in\mathbb{N}} \ell(I_n) < \infty$ so that $\lim_{n\to\infty} \ell(I_n) = 0$. This implies that $\lim_{n\to\infty} d_n = 0$ by (8) so that for sufficiently large $N \in \mathbb{N}$ we have $d_N < \ell(I^*)$. Then by (7) we have $I^* \notin \mathfrak{V}_{O_N}$ and hence $I^* \not\subset O_N$ by (6) so that $I^* \cap F_N \neq \emptyset$.

Let $q = \min\{n \in \mathbb{N} : I^* \cap F_n \neq \emptyset\}$. Now since $I^* \subset O_p$ we have $I^* \cap F_p = \emptyset$. Since $(F_n : n \in \mathbb{N})$ is an increasing sequence, this implies $q > p$. By the definition of q, we have $I^* \cap F_q \neq \emptyset$ but $I^* \cap F_{q-1} = \emptyset$. This implies $I^* \cap I_q \neq \emptyset$ and $I^* \subset O_{q-1}$. The last inclusion and (6), (7) and (8) imply $\ell(I^*) \leq d_{q-1} < 2\ell(I_q)$. Now since $I^* \cap I_q \neq \emptyset$ and since the closed interval J_q has the same midpoint as I_q and $\ell(J_q) = 5\ell(I_q)$, we have $I^* \subset J_q$. Then since $q > p$ we have $I^* \subset \bigcup_{n\geq p} J_n$. Since $x \in I^*$, we have $x \in \bigcup_{n\geq p} J_n$. By the arbitrariness of $x \in E \setminus F$, we have $E \setminus F \subset \bigcup_{n\geq p} J_n$. This proves (10) and completes the proof that our disjoint collection $\{I_n : n \in \mathbb{N}\}$ satisfies condition (1).

Let us prove (3). Since $\sum_{n\in\mathbb{N}} \ell(I_n) < \infty$, for every $\varepsilon > 0$ there exists $N \in \mathbb{N}$ such that $\sum_{n\geq N+1} \ell(I_n) < \varepsilon$. Now we have $F_N = F \setminus \bigcup_{n\geq N+1} I_n = F \cap \big(\bigcup_{n\geq N+1} I_n\big)^c$ so that $F_N^c = F^c \cup \big(\bigcup_{n\geq N+1} I_n\big)$. Thus we have

$$E \cap F_N^c = (E \cap F^c) \cup \Big(E \cap \bigcup_{n\geq N+1} I_n\Big) \subset (E \setminus F) \cup \Big(\bigcup_{n\geq N+1} I_n\Big).$$

Since $\mu_L^*(E \setminus F) = 0$ by (1), we have

$$\mu_L^*(E \cap F_N^c) \le \mu_L^*(E \setminus F) + \mu_L^*\Big(\bigcup_{n \ge N+1} I_n\Big) \le \sum_{n \ge N+1} \ell(I_n) < \varepsilon.$$

This proves (3) for our disjoint collection $\{I_n : n \in \mathbb{N}\}$ in $\mathfrak{V}_O$.

2. Let E be an arbitrary subset of $\mathbb{R}$. Let $E_n = E \cap (n, n+1)$ for $n \in \mathbb{Z}$, the collection of all integers. Since $\mathfrak{V}$ is a Vitali cover of E, $\mathfrak{V}$ is a Vitali cover of E_n for every $n \in \mathbb{Z}$. Let $\mathfrak{V}_{(n,n+1)}$ be the subcollection of $\mathfrak{V}$ consisting of those members of $\mathfrak{V}$ which are contained in the open interval $(n, n+1)$. According to Observation 12.8, $\mathfrak{V}_{(n,n+1)}$ is a Vitali cover of E_n. Since $\mu_L^*(E_n) < \infty$, by our result in **1** there exists a countable disjoint subcollection $\{I_{n,k} : k \in \mathbb{N}\}$ of $\mathfrak{V}_{(n,n+1)}$ such that $\mu_L^*\big(E_n \setminus \bigcup_{k \in \mathbb{N}} I_{n,k}\big) = 0$. Now $\{I_{n,k} : k \in \mathbb{N}, n \in \mathbb{Z}\}$ is a countable disjoint subcollection of $\mathfrak{V}$ such that

$$\bigcup_{n \in \mathbb{Z}} E_n \setminus \bigcup_{n \in \mathbb{Z}} \bigcup_{k \in \mathbb{N}} I_{n,k} = \bigcup_{n \in \mathbb{Z}} \Big(E_n \setminus \bigcup_{n \in \mathbb{Z}} \bigcup_{k \in \mathbb{N}} I_{n,k}\Big) \subset \bigcup_{n \in \mathbb{Z}} \Big(E_n \setminus \bigcup_{k \in \mathbb{N}} I_{n,k}\Big),$$

and

$$\mu_L^*\Big(\bigcup_{n \in \mathbb{Z}} E_n \setminus \bigcup_{n \in \mathbb{Z}} \bigcup_{k \in \mathbb{N}} I_{n,k}\Big) \le \sum_{n \in \mathbb{Z}} \mu_L^*\Big(E_n \setminus \bigcup_{k \in \mathbb{N}} I_{n,k}\Big) = 0.$$

Now $E \subset \big\{\bigcup_{n \in \mathbb{Z}} E_n\big\} \cup \mathbb{Z}$ so that

$$E \setminus \bigcup_{n \in \mathbb{N}} \bigcup_{k \in \mathbb{N}} I_{n,k} \subset \Big\{\bigcup_{n \in \mathbb{N}} E_n \setminus \bigcup_{n \in \mathbb{N}} \bigcup_{k \in \mathbb{N}} I_{n,k}\Big\} \cup \mathbb{Z}.$$

Then we have

$$\mu_L^*\Big(E \setminus \bigcup_{n \in \mathbb{N}} \bigcup_{k \in \mathbb{N}} I_{n,k}\Big) \le \mu_L^*\Big(\bigcup_{n \in \mathbb{N}} E_n \setminus \bigcup_{n \in \mathbb{N}} \bigcup_{k \in \mathbb{N}} I_{n,k}\Big) + \mu_L^*(\mathbb{Z}) = 0.$$

This proves (1). ∎

Theorem 12.10. (H. Lebesgue) *Let f be a real-valued increasing function on $[a, b]$. Then the derivative f' exists and is nonnegative on $(a, b) \setminus E$ where E is a null set in $(\mathbb{R}, \mathfrak{M}_L, \mu_L)$ contained in (a, b). Furthermore f' is $\mathfrak{M}_L$-measurable and μ_L-integrable on $(a, b) \setminus E$ with $\int_{[a,b]} f' \, d\mu_L \le f(b) - f(a)$. In particular f' is real-valued, that is f is differentiable, $(\mathfrak{M}_L, \mu_L)$-a.e. on $[a, b]$.*

Proof. 1. Since f is an increasing function, if $f'(x)$ exists at some $x \in (a, b)$ then $f'(x) \ge 0$. To show that f' exists a.e. on (a, b), we show that the four Dini derivates are equal a.e. on (a, b). Since $D_- f \le D^- f$ and $D_+ f \le D^+ f$, it suffices to show that

$$D^- f \le D_+ f \le D^+ f \le D_- f \quad \text{a.e. on } (a, b).$$

Since the second inequality holds on (a, b), it remains to show

(1) $$D^- f \le D_+ f \quad \text{a.e. on } (a, b)$$

and

(2) $$D^+ f \le D_- f \quad \text{a.e. on } (a,b).$$

Since each of the two inequalities asserts that the limit superior from one direction is dominated by the limit inferior from the other direction, they can be proved exactly in the same way. Thus let us prove (2) for instance. To prove (2) we let

$$E = \{x \in (a,b) : (D^+ f)(x) > (D_- f)(x)\}$$

and show that $\mu_L^*(E) = 0$. Now since f is an increasing function the four Dini derivates are all nonnegative. Let Q be the collection of all rational numbers. For $u, v \in Q$ such that $0 < u < v$ let

(3) $$E_{u,v} = \{x \in (a,b) : (D_- f)(x) < u < v < (D^+ f)(x)\}.$$

Then $E = \bigcup_{u,v \in Q, 0<u<v} E_{u,v}$. By the countable subadditivity of μ_L^*, we have

$$\mu_L^*(E) \le \sum_{u,v \in Q, 0<u<v} \mu_L^*(E_{u,v}).$$

Thus to show that $\mu_L^*(E) = 0$, it suffice to show that $\mu_L^*(E_{u,v}) = 0$ for every pair $u, v \in Q$ such that $0 < u < v$. Let $\gamma = \mu_L^*(E_{u,v})$. Since $E_{u,v} \subset (a,b)$, we have $\gamma \in [0, \infty)$. We outline the two steps in showing $\gamma = 0$.

Step 1. Using the fact that $(D_- f)(x) < u$ for every $x \in E_{u,v}$, we show that for an arbitrary $\varepsilon > 0$ the set $E_{u,v}$ can be approximated by a finite collection of disjoint bounded and closed intervals $\{I_1, \dots, I_p\}$ such that $\sum_{m=1}^p \ell(I_m) < u(\gamma + \varepsilon)$ by means of the Vitali Covering Theorem.

Step 2. By the fact that $(D^+ f)(x) > v$ for every $x \in E_{u,v}$, we show that some subset of $E_{u,v}$ can be approximated by a finite collection of disjoint bounded and closed intervals $\{J_1, \dots, J_q\}$ each of which is contained in one in the collection $\{I_1, \dots, I_p\}$ such that $v(\gamma - 2\varepsilon) < \sum_{n=1}^q \ell(J_n)$ by means of the Vitali Covering Theorem. Then we have $v(\gamma - 2\varepsilon) < \sum_{n=1}^q \ell(J_n) \le \sum_{m=1}^p \ell(I_m) < u(\gamma + \varepsilon)$. By the arbitrariness of $\varepsilon > 0$, we have $v\gamma \le u\gamma$. This implies $\gamma = 0$ since $u < v$.

Since $\mu_L^*(E_{u,v}) \le \mu_L((a,b)) < \infty$, by Lemma 3.21 for every $\varepsilon > 0$ there exists an open set $O \supset E$ such that

(4) $$\mu_L(O) < \mu_L^*(E_{u,v}) + \varepsilon = \gamma + \varepsilon.$$

For each $x \in E_{u,v}$, we have $(D_- f)(x) < u$. Now

$$(D_- f)(x) = \liminf_{h \uparrow 0} \frac{f(x+h) - f(x)}{h} = \liminf_{-k \uparrow 0} \frac{f(x-k) - f(x)}{-k}$$
$$= \liminf_{k \downarrow 0} \frac{f(x) - f(x-k)}{k} = \liminf_{h \downarrow 0} \frac{f(x) - f(x-h)}{h}.$$

Recall that for an extended real-valued function φ on an open interval containing 0, we have

$$\liminf_{h \downarrow 0} \varphi(h) = \lim_{\delta \downarrow 0} \inf_{h \in (0,\delta)} \varphi(h).$$

Now $\inf_{h \in (0,\delta)} \varphi(h) \uparrow$ as $\delta \downarrow 0$. This implies that there exists a sequence of positive numbers $(h_m : m \in \mathbb{N})$ such that

$$h_m \downarrow 0 \text{ and } \varphi(h_m) \uparrow \liminf_{h \downarrow 0} \varphi(h).$$

Since $(D_- f)(x) = \liminf_{h \downarrow 0} \{f(x) - f(x-h)\}/h$, there exists a sequence of positive numbers $(h_{x,m} : m \in \mathbb{N})$ such that

$$(5) \qquad \begin{cases} h_{x,m} \downarrow 0, [x - h_{x,m}, x] \subset O & \text{for } m \in \mathbb{N}, \\ \{f(x) - f(x - h_{x,m})\}/h_{x,m} \uparrow (D_- f)(x) < u \end{cases}$$

so that

$$(6) \qquad f(x) - f(x - h_{x,m}) < u h_{x,m} \quad \text{for } m \in \mathbb{N}.$$

Now $\mathfrak{V} = \{[x - h_{x,m}, x] : m \in \mathbb{N}, x \in E_{u,v}\}$ is a Vitali cover of $E_{u,v}$. Thus by Theorem 12.9 (Vitali Covering Theorem), there exists a finite disjoint collection of members of $\mathfrak{V}$ which we denote by $I_1 = [x_1 - h_1, x_1], \dots, I_p = [x_p - h_p, x_p]$ where $x_1, \dots, x_p \in E_{u,v}$, $x_1 < \cdots < x_p$ and $h_1, \dots, h_p > 0$ such that

$$(7) \qquad \mu_L^*\Big(E_{u,v} \setminus \bigcup_{m=1}^{p} I_m\Big) < \varepsilon.$$

Let I_m° be the interior of I_m for $m = 1, \dots, p$ and let

$$(8) \qquad E_{u,v}^* = E_{u,v} \cap \bigcup_{m=1}^{p} I_m^\circ.$$

For an extended real-valued function ψ on an open interval containing 0, we have

$$\limsup_{k \downarrow 0} \psi(k) = \lim_{\delta \downarrow 0} \sup_{k \in (0,\delta)} \psi(k).$$

Since $\sup_{k \in (0,\delta)} \psi(k) \downarrow$ as $\delta \downarrow 0$ there exists a sequence of positive numbers $(k_n : n \in \mathbb{N})$ such that

$$k_n \downarrow 0 \text{ and } \psi(k_n) \downarrow \limsup_{k \downarrow 0} \psi(k).$$

Now for every $y \in E_{u,v}^* \subset E_{u,v}$, we have $(D^+ f)(y) > v$ by the definition of $E_{u,v}$ by (3). Since $(D^+ f)(y) = \limsup_{h \downarrow 0} \{f(y+h) - f(y)\}/h$, there exists a sequence of positive numbers $(k_{y,n} : n \in \mathbb{N})$ such that

$$(9) \qquad \begin{cases} k_{y,n} \downarrow 0, [y, y + k_{y,n}] \subset I_m^\circ & \text{for } n \in \mathbb{N} \text{ for some } m = 1, \dots, p, \\ \{f(y + k_{y,n}) - f(y)\}/k_{y,n} \downarrow (D^+ f)(y) > v \end{cases}$$

so that

$$(10) \qquad f(y + k_{y,n}) - f(y) > v k_{y,n} \quad \text{for } n \in \mathbb{N}.$$

Then the collection $\mathfrak{W} = \{[y, y+k_{y,n}] : n \in \mathbb{N}, y \in E^*_{u,v}\}$ is a Vitali cover of $E^*_{u,v}$ so that by Theorem 12.9, there exists a finite disjoint collection of members of $\mathfrak{W}$ which we denote by $J_1 = [y_1, y_1+k_1], \ldots, J_q = [y_q, y_q+k_q]$ where $y_1, \ldots, y_q \in E^*_{u,v}$, $y_1 < \cdots < y_q$ and $k_1, \ldots, k_q > 0$ such that

$$\mu_L^*\Big(E^*_{u,v} \setminus \bigcup_{n=1}^{q} J_n\Big) < \varepsilon. \tag{11}$$

For our $I_m = [x_m - h_m, x_m]$ and $J_n = [y_n, y_n + k_n]$, let $\Delta f(I_m) = f(x_m) - f(x_m - h_m)$ and $\Delta f(J_n) = f(y_n + k_n) - f(y_n)$. Since $\{J_1, \ldots, J_q\}$ is a disjoint collection and each J_n is contained in some I_m°, and since f is an increasing function, we have

$$\sum_{n=1}^{q} \Delta f(J_n) \le \sum_{m=1}^{p} \Delta f(I_m). \tag{12}$$

From (10) we have $\Delta f(J_n) > v\ell(J_n)$ for $n = 1, \ldots, q$ and similarly from (6) we have $\Delta f(I_m) < u\ell(I_m)$ for $m = 1, \ldots, p$. Thus by (12) and (4), we have

$$\begin{aligned} v\sum_{n=1}^{q} \ell(J_n) &\le \sum_{n=1}^{q} \Delta f(J_n) \le \sum_{m=1}^{p} \Delta f(I_m) < u\sum_{m=1}^{p} \ell(I_m) \\ &= u\,\mu_L\Big(\bigcup_{m=1}^{p} I_m\Big) \le u\,\mu_L(O) < u(\gamma + \varepsilon). \end{aligned} \tag{13}$$

Now if we let N be the set of the endpoints of $I_1, \ldots, I_p$, then

$$\begin{aligned} E_{u,v} &= \Big(E_{u,v} \setminus \bigcup_{m=1}^{p} I_m\Big) \cup \Big(E_{u,v} \cap \bigcup_{m=1}^{p} I_m\Big) \\ &\subset \Big(E_{u,v} \setminus \bigcup_{m=1}^{p} I_m\Big) \cup \Big(E_{u,v} \cap \bigcup_{m=1}^{p} I_m^\circ\Big) \cup N \\ &= \Big(E_{u,v} \setminus \bigcup_{m=1}^{p} I_m\Big) \cup E^*_{u,v} \cup N \\ &\subset \Big(E_{u,v} \setminus \bigcup_{m=1}^{p} I_m\Big) \cup \Big(E^*_{u,v} \setminus \bigcup_{n=1}^{q} J_n\Big) \cup \Big(\bigcup_{n=1}^{q} J_n\Big) \cup N, \end{aligned}$$

where the second equality is by (8). Thus by the subadditivity of μ_L^* and by (7) and (11),

$$\gamma = \mu_L^*(E_{u,v}) < \varepsilon + \varepsilon + \sum_{n=1}^{q} \ell(J_n) + 0 = 2\varepsilon + \sum_{n=1}^{q} \ell(J_n).$$

Thus $\gamma - 2\varepsilon < \sum_{n=1}^{q} \ell(J_n)$, and then by (13) we have

$$v(\gamma - 2\varepsilon) < v\sum_{n=1}^{q} \ell(J_n) < u(\gamma + \varepsilon).$$

By the arbitrariness of $\varepsilon > 0$, we have $v\gamma \leq u\gamma$. Since $v > u$, we have $\gamma = 0$.

2. Thus we have shown that there exists a null set E in $(\mathbb{R}, \mathfrak{M}_L, \mu_L)$ contained in (a, b) such that f' exists on $(a, b) \setminus E$. Since $(a, b) \setminus E \in \mathfrak{M}_L$, f' is $\mathfrak{M}_L$-measurable on $(a, b) \setminus E$ by Proposition 12.4.

Let us extend the domain of definition of f from $[a, b]$ to $\mathbb{R}$ by setting $f(x) = f(a)$ for $x \in (-\infty, a)$ and $f(x) = f(b)$ for $x \in (b, \infty)$. Then f is a real-valued increasing function on $\mathbb{R}$ so that it is $\mathfrak{B}_{\mathbb{R}}$-measurable and hence $\mathfrak{M}_L$-measurable on $\mathbb{R}$. If we define a sequence $(g_n : n \in \mathbb{N})$ of nonnegative real-valued functions on $\mathbb{R}$ by

$$g_n(x) = \frac{f\left(x + \frac{1}{n}\right) - f(x)}{\frac{1}{n}} \quad \text{for } x \in \mathbb{R},$$

then g_n is $\mathfrak{M}_L$-measurable on $\mathbb{R}$ by Theorem 9.31 (Translation Invariance). At any point $x \in [a, b]$ at which $f'(x)$ exists, we have

$$f'(x) = \lim_{h \to 0} \frac{f(x+h) - f(x)}{h} = \lim_{n \to \infty} \frac{f\left(x + \frac{1}{n}\right) - f(x)}{\frac{1}{n}} = \lim_{n \to \infty} g_n(x).$$

Thus by Fatou's Lemma (Theorem 8.13), we have

$$(14) \qquad \int_{(a,b)\setminus E} f' \, d\mu_L \leq \liminf_{n \to \infty} \int_{(a,b)\setminus E} g_n \, d\mu_L.$$

Now since $\{a\} \cup E \cup \{b\}$ is a null set in $(\mathbb{R}, \mathfrak{M}_L, \mu_L)$, we have by Definition 9.11

$$\begin{aligned}
\int_{(a,b)\setminus E} g_n \, d\mu_L &= \int_{[a,b]} g_n \, d\mu_L = \int_{[a,b]} \frac{f\left(x + \frac{1}{n}\right) - f(x)}{\frac{1}{n}} \, \mu_L(dx) \\
&= n\left\{ \int_{\left[a+\frac{1}{n}, b+\frac{1}{n}\right]} f(x) \, \mu_L(dx) - \int_{[a,b]} f(x) \, \mu_L(dx) \right\} \\
&= n\left\{ \int_{\left(b, b+\frac{1}{n}\right]} f(x) \, \mu_L(dx) - \int_{\left[a, a+\frac{1}{n}\right)} f(x) \, \mu_L(dx) \right\} \\
&\leq n\left\{ f(b)\frac{1}{n} - f(a)\frac{1}{n} \right\} = f(b) - f(a),
\end{aligned}$$

where the third equality is by Theorem 9.31, the fourth equality is by (a) of Lemma 9.10, and the inequality is by the fact that f is bounded above by $f(b)$ and bounded below by $f(a)$ on $\mathbb{R}$. Using this estimate in (14), we have $\int_{(a,b)\setminus E} f' \, d\mu_L \leq f(b) - f(a)$. Then by Definition 9.11 we have $\int_{[a,b]} f' \, d\mu_L = \int_{(a,b)\setminus E} f' \, d\mu_L \leq f(b) - f(a) < \infty$. Finally by (f) of Observation 9.2, f' is real-valued, that is, f is differentiable, $(\mathfrak{M}_L, \mu_L)$-a.e. on $[a, b]$. ■

Remark 12.11. The equality $\int_{[a,b]} f' \, d\mu_L = f(b) - f(a)$ does not hold in general.
(a) For example the Cantor-Lebesgue function τ is a real-valued continuous increasing function on $[0, 1]$ with $f(0) = 0$ and $f(1) = 1$; τ is constant on each of a collection of

disjoint open intervals whose union G has $\mu_L(G) = 1$. Thus the derivative τ' exists and is equal to 0 on G. Then $\int_{[0,1]} \tau' \, d\mu_L = 0 \neq \tau(1) - \tau(0)$.
(b) For a simpler example, let f be a real-valued increasing function on $[0, 1]$ defined by setting $f = 0$ on $\left[0, \frac{1}{2}\right)$ and $f = 1$ on $\left[\frac{1}{2}, 1\right]$. The derivative f' exists and is equal to 0 on $(0, 1)$ except at $x = \frac{1}{2}$. Thus $\int_{[0,1]} f' \, d\mu_L = 0 \neq f(1) - f(0)$.
(c) In §13 we show that whether f is an increasing function or not, if f' exists and is finite everywhere on $[a, b]$ and if f' is μ_L-integrable on $[a, b]$ then the equality $\int_{[a,b]} f' \, d\mu_L = f(b) - f(a)$ holds. We show also that if f is absolutely continuous on $[a, b]$ then the equality holds.

[III] Functions of Bounded Variation

Definition 12.12. *Let $[a, b] \subset \mathbb{R}$ with $a \leq b$. A partition of $[a, b]$ is a finite ordered set $\mathcal{P} = \{a = x_0 \leq \cdots \leq x_n = b\}$. Let $\mathfrak{P}_{a,b}$ be the collection of all partitions of $[a, b]$. For a real-valued function f on $[a, b]$, we define the variation of f on $[a, b]$ corresponding to a partition $\mathcal{P} = \{a = x_0 \leq \cdots \leq x_n = b\}$ by*

$$V_a^b(f, \mathcal{P}) = \sum_{k=1}^{n} |f(x_k) - f(x_{k-1})| \in [0, \infty).$$

We define the total variation of f on $[a, b]$ by

$$V_a^b(f) = \sup_{\mathcal{P} \in \mathfrak{P}_{a,b}} V_a^b(f, \mathcal{P}) \in [0, \infty].$$

We say that f is a function of bounded variation on $[a, b]$, or simply a BV function on $[a, b]$, if $V_a^b(f) < \infty$. We write $BV([a, b])$ for the collection of all BV functions on $[a, b]$.

If $\mathcal{P}, \mathcal{P}' \in \mathfrak{P}_{a,b}$ and $\mathcal{P} \subset \mathcal{P}'$ then we say that $\mathcal{P}'$ is a refinement of $\mathcal{P}$. If $\mathcal{P}, \mathcal{P}' \in \mathfrak{P}_{a,b}$, then $\mathcal{P} \cup \mathcal{P}' \in \mathfrak{P}_{a,b}$ and moreover $\mathcal{P} \cup \mathcal{P}'$ is a refinement of both $\mathcal{P}$ and $\mathcal{P}'$.

Observation 12.13. (a) If $\mathcal{P}, \mathcal{P}' \in \mathfrak{P}_{a,b}$ and if $\mathcal{P} \subset \mathcal{P}'$, then $V_a^b(f, \mathcal{P}) \leq V_a^b(f, \mathcal{P}')$.
(b) If $f \in BV([a, b])$, then f is a bounded function on $[a, b]$ and in fact for every $x \in [a, b]$
we have $f(x) \in \left[f(a) - V_a^b(f), f(a) + V_a^b(f)\right]$.
(c) If $f \in BV([a, b])$, then $-f \in BV([a, b])$ and $V_a^b(-f) = V_a^b(f)$.
(d) If f is a real-valued monotone function on $[a, b]$, then $f \in BV([a, b])$ and in this case we have $V_a^b(f) = |f(b) - f(a)|$.

Proof. 1. Let $\mathcal{P}, \mathcal{P}' \in \mathfrak{P}_{a,b}$ and $\mathcal{P} \subset \mathcal{P}'$. Let n be the number of partition points in $\mathcal{P}$ and let m be that of $\mathcal{P}'$. We have $n \leq m$. If $n = m$ then $\mathcal{P} = \mathcal{P}'$ and $V_a^b(f, \mathcal{P}) = V_a^b(\mathcal{P}')$. Consider the case $n < m$. Then $m = n + p$ where $p \in \mathbb{N}$. Let $\xi_1, \ldots \xi_p$ be the p partition points in $\mathcal{P}'$ that are not in $\mathcal{P}$. Let $\mathcal{P}_0 = \mathcal{P}$. Let $\mathcal{P}_1$ be the partition of $[a, b]$ obtained by adding ξ_1 to $\mathcal{P}_0$. Let $\mathcal{P}_2$ be the partition of $[a, b]$ obtained by adding ξ_2 to $\mathcal{P}_1$ and so on so that $\mathcal{P}_p = \mathcal{P}'$ is the partition of $[a, b]$ obtained by adding ξ_p to $\mathcal{P}_{p-1}$. Thus we have a chain of partitions of $[a,b]$, $\mathcal{P} = \mathcal{P}_0 \subset \mathcal{P}_1 \subset \cdots \subset \mathcal{P}_{p-1} \subset \mathcal{P}_p = \mathcal{P}'$, each having one more partition point than its predecessor. Let $\mathcal{P} = \mathcal{P}_0 = \{a = x_0 \le x_1 \le x_2 \le \cdots \le x_n = b\}$. Then $\xi_1 \in (x_{k_0-1}, x_{k_0})$ for some $k_0 \le n$. Now we have $V_a^b(f, \mathcal{P}_0) = \sum_{k=1}^n |f(x_k) - f(x_{k-1})|$ and $V_a^b(f, \mathcal{P}_1)$ is obtained by replacing the summand $|f(x_{k_0}) - f(x_{k_0-1})|$ in $V_a^b(f, \mathcal{P}_0)$ with $|f(\xi_1) - f(x_{k_0-1})| + |f(x_{k_0}) - f(\xi_1)|$ which is greater than or equal to $|f(x_{k_0}) - f(x_{k_0-1})|$ by the triangle inequality in $\mathbb{R}$. Thus we have $V_a^b(f, \mathcal{P}_0) \le V_a^b(f, \mathcal{P}_1)$. By the same argument we have $V_a^b(f, \mathcal{P}_1) \le V_a^b(f, \mathcal{P}_2)$, $V_a^b(f, \mathcal{P}_2) \le V_a^b(f, \mathcal{P}_3), \ldots, V_a^b(f, \mathcal{P}_{p-1}) \le V_a^b(f, \mathcal{P}_p)$. Thus we have $V_a^b(f, \mathcal{P}) \le V_a^b(f, \mathcal{P}')$.

2. To prove (b), suppose $f \in BV([a,b])$. For an arbitrary $x \in [a,b]$, consider the partition $\mathcal{P} = \{a \le x \le b\}$ of $[a,b]$. Then we have

$$|f(x) - f(a)| \le |f(x) - f(a)| + |f(b) - f(x)| = V_a^b(f, \mathcal{P}) \le V_a^b(f) < \infty.$$

From this we have $f(a) - V_a^b(f) \le f(x) \le f(a) + V_a^b(f)$ for every $x \in [a,b]$.

3. To prove (c), suppose $f \in BV([a,b])$ and let $\mathcal{P} = \{a = x_0 \le \cdots \le x_n = b\}$ be an arbitrary partition of $[a,b]$. Then

$$\begin{aligned} V_a^b(-f, \mathcal{P}) &= \sum_{k=1}^n \big|(-f)(x_k) - (-f)(x_{k-1})\big| \\ &= \sum_{k=1}^n |f(x_k) - f(x_{k-1})| = V_a^b(f, \mathcal{P}), \end{aligned}$$

and then

$$V_a^b(-f) = \sup_{\mathcal{P} \in \mathfrak{P}_{a,b}} V_a^b(-f, \mathcal{P}) = \sup_{\mathcal{P} \in \mathfrak{P}_{a,b}} V_a^b(f, \mathcal{P}) = V_a^b(f).$$

4. To prove (d), let f be a real-valued monotone function on $[a,b]$. Then for an arbitrary partition $\mathcal{P} = \{a = x_0 \le \cdots \le x_n = b\}$ of $[a,b]$, the differences $f(x_k) - f(x_{k-1})$ for $k = 1, \ldots, n$ are all nonnegative if f is an increasing function and all nonpositive if f is a decreasing function. Therefore we have

$$V_a^b(f, \mathcal{P}) = \sum_{k=1}^n |f(x_k) - f(x_{k-1})| = \left| \sum_{k=1}^n \{f(x_k) - f(x_{k-1})\} \right| = |f(b) - f(a)|,$$

and then

$$V_a^b(f) = \sup_{\mathcal{P} \in \mathfrak{P}_{a,b}} V_a^b(f, \mathcal{P}) = |f(b) - f(a)|. \ \blacksquare$$

Lemma 12.14. *Let $f_1, f_2 \in BV([a,b])$ and $c_1, c_2 \in \mathbb{R}$. Then $c_1 f_1 + c_2 f_2 \in BV([a,b])$ and $V_a^b(c_1 f_1 + c_2 f_2) \le |c_1| V_a^b(f_1) + |c_2| V_a^b(f_2)$.*

Proof. For an arbitrary partition $\mathcal{P} = \{a = x_0 \le \cdots \le x_n = b\}$ of $[a, b]$, we have

$$V_a^b(c_1 f_1 + c_2 f_2, \mathcal{P}) = \sum_{k=1}^{n} \left|(c_1 f_1 + c_2 f_2)(x_k) - (c_1 f_1 + c_2 f_2)(x_{k-1})\right|$$

$$\le |c_1| \sum_{k=1}^{n} |f_1(x_k) - f_1(x_{k-1})| + |c_2| \sum_{k=1}^{n} |f_2(x_k) - f_2(x_{k-1})|$$

$$= |c_1| V_a^b(f_1, \mathcal{P}) + |c_2| V_a^b(f_2, \mathcal{P}) \le |c_1| V_a^b(f_1) + |c_2| V_a^b(f_2).$$

Thus we have

$$V_a^b(c_1 f_1 + c_2 f_2) = \sup_{\mathcal{P} \in \mathfrak{P}_{a,b}} V_a^b(c_1 f_1 + c_2 f_2, \mathcal{P}) \le |c_1| V_a^b(f_1) + |c_2| V_a^b(f_2). \ \blacksquare$$

Lemma 12.15. (a) *If $f \in BV([a, b])$, then for every closed subinterval $[a_0, b_0]$ of $[a, b]$, we have $f \in BV([a_0, b_0])$ and $V_{a_0}^{b_0}(f) \le V_a^b(f)$.*

(b) *Let $c \in (a, b)$. If $f \in BV([a, c])$ and $f \in BV([c, b])$, then $f \in BV([a, b])$ and moreover $V_a^b(f) = V_a^c(f) + V_c^b(f)$.*

Proof. 1. To prove (a), let $\mathcal{P} = \{a_0 = x_0 \le \cdots \le x_n = b_0\}$ be an arbitrary partition of $[a_0, b_0]$. Consider the partition $\mathcal{P}' = \{a \le a_0 = x_0 \le \cdots \le x_n = b_0 \le b\}$ of $[a, b]$. Then

$$V_{a_0}^{b_0}(f, \mathcal{P}) = \sum_{k=1}^{n} |f(x_k) - f(x_{k-1})|$$

$$\le |f(x_0) - f(a)| + \sum_{k=1}^{n} |f(x_k) - f(x_{k-1})| + |f(b) - f(x_n)|$$

$$= V_a^b(f, \mathcal{P}') \le V_a^b(f) < \infty.$$

Thus we have $V_{a_0}^{b_0}(f) = \sup_{\mathcal{P} \in \mathfrak{P}_{a_0,b_0}} V_{a_0}^{b_0}(f, \mathcal{P}) \le V_a^b(f) < \infty$ so that $f \in BV([a_0, b_0])$.

2. To prove (b), let $c \in (a, b)$ and suppose $f \in BV([a, c])$ and $f \in BV([c, b])$. Let $\mathcal{P}$ be an arbitrary partition of $[a, b]$ and let $\mathcal{P}'$ be the refinement of $\mathcal{P}$ by adding c as a partition point to $\mathcal{P}$. By (a) of Observation 12.13, we have

(1) $$V_a^b(f, \mathcal{P}) \le V_a^b(f, \mathcal{P}').$$

Let $\mathcal{P}_1$ and $\mathcal{P}_2$ be the restrictions of $\mathcal{P}'$ to $[a, c]$ and $[c, b]$ respectively. Then we have

(2) $$V_a^b(f, \mathcal{P}') = V_a^c(f, \mathcal{P}_1) + V_c^b(f, \mathcal{P}_2) \le V_a^c(f) + V_c^b(f).$$

By (1) and (2), we have

(3) $$V_a^b(f) = \sup_{\mathcal{P} \in \mathfrak{P}_{a,b}} V_a^b(f, \mathcal{P}) \le V_a^c(f) + V_c^b(f) < \infty.$$

This shows that $f \in BV([a,b])$.

To show $V_a^b(f) = V_a^c(f) + V_c^b(f)$, it remains to show that $V_a^b(f) \geq V_a^c(f) + V_c^b(f)$. Let $\mathcal{P}_1$ and $\mathcal{P}_2$ be arbitrary partitions of $[a,c]$ and $[c,b]$ respectively. Then $\mathcal{P}_1 \cup \mathcal{P}_2$ is a partition of $[a,b]$ and $V_a^b(f, \mathcal{P}_1 \cup \mathcal{P}_2) = V_a^c(f, \mathcal{P}_1) + V_c^b(f, \mathcal{P}_2)$. Thus we have

$$V_a^b(f) = \sup_{\mathcal{P} \in \mathfrak{P}_{a,b}} V_a^b(f, \mathcal{P}) \geq V_a^b(f, \mathcal{P}_1 \cup \mathcal{P}_2) = V_a^c(f, \mathcal{P}_1) + V_c^b(f, \mathcal{P}_2).$$

Since this holds for arbitrary partitions $\mathcal{P}_1$ and $\mathcal{P}_2$ of $[a,c]$ and $[c,b]$ respectively, we have

$$(4) \qquad V_a^b(f) \geq V_a^c(f) + V_c^b(f).$$

This completes the proof. ∎

Let $f \in BV([a,b])$. Then for any $x \in [a,b]$ we have $f \in BV([a,x])$ by (a) of Lemma 12.15 so that $V_a^x(f) < \infty$. Thus $V_a^x(f)$ is a nonnegative real-valued function of $x \in [a,b]$.

Definition 12.16. *Let $f \in BV([a,b])$. The total variation function v_f of f on $[a,b]$ is defined by $v_f(x) = V_a^x(f)$ for $x \in [a,b]$.*

Observation 12.17. Let $f \in BV([a,b])$. The total variation function v_f of f is a nonnegative real-valued increasing function on $[a,b]$ with $v_f(a) = 0$. Indeed for $a \leq x' < x'' \leq b$, we have

$$(1) \qquad v_f(x'') - v_f(x') = V_{x'}^{x''}(f) \geq 0.$$

Proof. If $f \in BV([a,b])$, then for any $x \in [a,b]$ we have $f \in BV([a,x])$ with $V_a^x(f) \leq V_a^b(f)$ by (a) of Lemma 12.15. Thus v_f is a nonnegative real-valued function on $[a,b]$. To show that v_f is an increasing function on $[a,b]$, let $a \leq x' < x'' \leq b$. By (b) of Lemma 12.15 we have $V_a^{x'}(f) + V_{x'}^{x''}(f) = V_a^{x''}(f)$, that is, $v_f(x') + V_{x'}^{x''}(f) = v_f(x'')$ so that $v_f(x'') - v_f(x') = V_{x'}^{x''}(f) \geq 0$. ∎

Theorem 12.18. (Jordan Decomposition of Functions of Bounded Variation) *Let f be a real-valued function on $[a,b]$. Then $f \in BV([a,b])$ if and only if there exist two real-valued increasing functions g_1 and g_2 on $[a,b]$ such that $f = g_1 - g_2$ on $[a,b]$.*

Proof. 1. Suppose there exist two real-valued increasing functions g_1 and g_2 on $[a,b]$ such that $f = g_1 - g_2$ on $[a,b]$. Then $g_1, g_2 \in BV([a,b])$ by (d) of Observation 12.13 and thus $f \in BV([a,b])$ by Lemma 12.14.

2. Conversely suppose $f \in BV([a,b])$. Consider the total variation function v_f of f on $[a,b]$. Let $g_1 = v_f$ and $g_2 = v_f - f$. Then $f = g_1 - g_2$. Now g_1 is a real-valued increasing function on $[a,b]$ by Observation 12.17. It remains to show that g_2 is a real-valued increasing function on $[a,b]$. Thus let $a \leq x' < x'' \leq b$ and let $\mathcal{P} = \{x', x''\}$, a partition of $[x', x'']$. Then we have

$$g_2(x'') - g_2(x') = \{v_f(x'') - v_f(x')\} - \{f(x'') - f(x')\}$$
$$\geq V_{x'}^{x''}(f) - |f(x'') - f(x')| = V_{x'}^{x''}(f) - V_{x'}^{x''}(f, \mathcal{P}) \geq 0$$

by (1) of Observation 12.17. This shows that g_2 is an increasing function on $[a, b]$. ∎

Definition 12.19. *Let $f \in BV([a, b])$. The expression $f = g_1 - g_2$ where g_1 and g_2 are two real-valued increasing functions on $[a, b]$ is called a Jordan decomposition of f.*

Remark 12.20. A Jordan decomposition of $f \in BV([a, b])$ is not unique. In fact if $f = g_1 - g_2$ is a Jordan decomposition, then with an arbitrary real-valued increasing function h on $[a, b]$, $f = \{g_1+h\}-\{g_2+h\}$ is another Jordan decomposition. In the Proof of Theorem 12.18, we showed that $v_f - f$ is an increasing function on $[a, b]$. Actually $v_f + f$ is also an increasing function on $[a, b]$. This can be shown as follows. Note that $f \in BV([a, b])$ implies that $-f \in BV([a, b])$. This implies that $v_{-f} - (-f)$ is an increasing function. But $v_{-f} = v_f$ by (c) of Observation 12.13. Thus $v_f + f$ is an increasing function on $[a, b]$. Then we have yet another Jordan decomposition $f = \frac{1}{2}\{v_f + f\} - \frac{1}{2}\{v_f - f\}$.

Theorem 12.21. *Let $f \in BV([a, b])$. Then we have:*
(a) *f is $\mathfrak{B}_{\mathbb{R}}$-measurable on $[a, b]$.*
(b) *f' exists $(\mathfrak{M}_L, \mu_L)$-a.e. on $[a, b]$ and is $\mathfrak{M}_L$-measurable and μ_L-integrable on $[a, b]$.*

Proof. By Theorem 12.18, $f = g_1 - g_2$ where g_1 and g_2 are real-valued and increasing on $[a, b]$. Then g_1 and g_2 are $\mathfrak{B}_{\mathbb{R}}$-measurable on $[a, b]$ and so is f. By Theorem 12.10 (Lebesgue), g_1' and g_2' exist $(\mathfrak{M}_L, \mu_L)$-a.e. on $[a, b]$ and are $\mathfrak{M}_L$-measurable and μ_L-integrable on $[a, b]$. Then so does $f' = g_1' - g_2'$. ∎

Theorem 12.22. *Let $f \in BV([a, b])$ where $a < b$. Then the total variation function v_f of f is continuous at $x_0 \in [a, b]$ if and only if f is continuous at x_0. (Continuity at a is understood to be right-continuity at a and continuity at b is understood to be left-continuity at b.)*

Proof. 1. Suppose f is continuous at $x_0 \in [a, b]$. Let us show that if $x_0 < b$, then v_f is right-continuous at x_0. Let $\varepsilon > 0$ be arbitrarily given. Then there exists a partition $\mathcal{P}$ of $[x_0, b]$ such that

(1) $$V_{x_0}^b(f, \mathcal{P}) > V_{x_0}^b(f) - \varepsilon.$$

Let $\mathcal{P} = \{x_0 < x_1 \leq x_2 \leq \cdots \leq x_n = b\}$. Since f is continuous at x_0, there exists $\delta > 0$ such that $|f(\xi) - f(x_0)| < \varepsilon$ for $\xi \in [x_0, x_0 + \delta)$. Let us take δ so small that $x_0 + \delta < x_1$. With an arbitrarily chosen $\xi \in (x_0, x_0 + \delta)$, let $\mathcal{P}'$ be the refinement of $\mathcal{P}$ by adding ξ as a partition point for $[x_0, b]$, that is, $\mathcal{P}' = \{x_0 < \xi < x_1 \leq x_2 \leq \cdots \leq x_n = b\}$. Then by (a) of Observation 12.13 we have

(2) $$V_{x_0}^b(f, \mathcal{P}') \geq V_{x_0}^b(f, \mathcal{P}).$$

We have also

$$(3)\qquad V_{x_0}^b(f,\mathcal{P}') = |f(\xi)-f(x_0)| + |f(x_1)-f(\xi)| + \sum_{k=2}^{n}|f(x_k)-f(x_{k-1})|$$
$$< \varepsilon + |f(x_1)-f(\xi)| + \sum_{k=2}^{n}|f(x_k)-f(x_{k-1})| \le \varepsilon + V_{\xi}^b(f).$$

By (1), (2), and (3), we have

$$(4)\qquad V_{x_0}^b(f) < V_{x_0}^b(f,\mathcal{P}) + \varepsilon \le V_{x_0}^b(f,\mathcal{P}') + \varepsilon \le V_{\xi}^b(f) + 2\varepsilon.$$

By Definition 12.16, (b) of Lemma 12.15, and (4), we have

$$v_f(\xi) - v_f(x_0) = V_a^{\xi}(f) - V_a^{x_0}(f) = V_{x_0}^{\xi}(f)$$
$$= V_{x_0}^b(f) - V_{\xi}^b(f) < 2\varepsilon \quad \text{for } \xi \in (x_0, x_0+\delta).$$

This proves that if f is continuous at x_0 and $x_0 < b$, then v_f is right-continuous at x_0. We show similarly that if f is continuous at x_0 and $x_0 > a$, then v_f is left-continuous at x_0. Thus if f is continuous at $x_0 \in [a,b]$, then v_f is continuous at x_0.

2. Conversely suppose v_f is continuous at $x_0 \in [a,b]$. Let us show that if $x_0 < b$, then f is right-continuous at x_0. Let $\varepsilon > 0$ be arbitrarily given. The continuity of v_f at x_0 implies that there exists $\delta > 0$ such that $|v_f(x) - v_f(x_0)| < \varepsilon$ for $x \in [x_0, x_0+\delta)$. Now for any $x \in (x_0, x_0+\delta)$, with the partition $\mathcal{P} = \{x_0 < x\}$ of $[x_0, x]$ we have

$$|f(x) - f(x_0)| = V_{x_0}^x(f,\mathcal{P}) \le V_{x_0}^x(f) = v_f(x) - v_f(x_0) < \varepsilon.$$

This shows that f is right-continuous at x_0. We show similarly that if v_f is continuous at $x_0 \in [a,b]$ and if $x_0 > a$ then f is left-continuous at x_0. Thus if v_f is continuous at $x_0 \in [a,b]$, then f is continuous at x_0. ∎

Corollary 12.23. *Let $f \in BV([a,b])$. If f is continuous on $[a,b]$, then in a Jordan decomposition $f = g_1 - g_2$ of f, the real-valued increasing functions g_1 and g_2 can be chosen to be continuous functions on $[a,b]$.*

Proof. Consider the Jordan decomposition $f = g_1 - g_2$ of f where $g_1 = v_f$ and $g_2 = v_f - f$ as in the Proof of Theorem 12.18. If f is continuous on $[a,b]$, then v_f is continuous on $[a,b]$ by Theorem 12.22. Then $g_1 = v_f$ and $g_2 = v_f - f$ are continuous functions on $[a,b]$. ∎

Problems

Prob. 12.1. Let $f, g \in BV([a,b])$.
(a) Show that $cf \in BV([a,b])$ and $V_a^b(cf) = |c|V_a^b(f)$ for every $c \in \mathbb{R}$.
(b) Show that $f+g \in BV([a,b])$ and $V_a^b(f+g) \le V_a^b(f) + V_a^b(g)$.
(b) Construct an example for which $V_a^b(f+g) \ne V_a^b(f) + V_a^b(g)$.

Prob. 12.2. Let $f \in BV([a,b])$. Show that if $f \geq c$ on $[a,b]$ for some constant $c > 0$, then $1/f \in BV([a,b])$.

Prob. 12.3. Let $f, g \in BV([a,b])$. Show that $fg \in BV([a,b])$ and

$$V_a^b(fg) \leq \sup_{[a,b]} |f| \cdot V_a^b(g) + \sup_{[a,b]} |g| \cdot V_a^b(f).$$

Prob. 12.4. Let f be a real-valued function on $[a,b]$ satisfying a Lipschitz condition, that is, there exists a constant $M > 0$ such that $|f(x') - f(x'')| \leq M|x' - x''|$ for every $x', x'' \in [a,b]$. Show that $f \in BV([a,b])$ and $V_a^b(f) \leq M(b-a)$.

Prob. 12.5. Let f be a real-valued function on $[a,b]$. Suppose f is continuous on $[a,b]$ and is differentiable on (a,b) with $|f'| \leq M$ for some constant $M > 0$. Show that $f \in BV([a,b])$ and $V_a^b(f) \leq M(b-a)$.

Prob. 12.6. According to Prob. 12.5, continuity of f on $[a,b]$ and existence and boundedness of the derivative f' on (a,b) imply that f is a BV function. These conditions are sufficient but not necessary for f to be a BV function. To show this find a real-valued function f on $[a,b]$ satisfying the following conditions:
1° f is continuous on $[a,b]$.
2° f' exists on (a,b).
3° f' is not bounded on (a,b).
4° $f \in BV([a,b])$.

Prob. 12.7. Let f be a polygonal function on $[0,1]$ defined as follows. Let $a_n = \frac{1}{n}$ for $n \in \mathbb{N}$ and let $b_n = \frac{1}{2}(a_n + a_{n+1})$ for $n \in \mathbb{N}$. Let f be defined by setting

$$\begin{aligned} &f(0) = 0, \\ &f(a_n) = 1 \quad \text{for } n \in \mathbb{N}, \\ &f(b_n) = 0 \quad \text{for } n \in \mathbb{N}, \end{aligned}$$

f is linear on $[a_{n+1}, b_n]$ and on $[b_n, a_n]$ for $n \in \mathbb{N}$.
Show that $f \notin BV([0,1])$. Show also that f is continuous at every $x \in (0,1]$ and is discontinuous at $x = 0$.

Prob. 12.8. Let f be a real-valued function on $\left[0, \frac{2}{\pi}\right]$ defined by

$$f(x) = \begin{cases} \sin \frac{1}{x} & \text{for } x \in \left(0, \frac{2}{\pi}\right], \\ 0 & \text{for } x = 0. \end{cases}$$

Show that $f \notin BV\left(\left[0, \frac{2}{\pi}\right]\right)$.

Prob. 12.9. Let $f(x) = \sin x$ for $x \in [0, 2\pi]$.
(a) Find the total variation function v_f of f on $[0, 2\pi]$.
(b) Find a Jordan decomposition of f.

Prob. 12.10. Let f be a real-valued function on $[a,b]$ such that $f \in BV([a,x])$ for every $x \in [a,b)$ and consider the total variation function $v_f(x) = V_a^x(f)$ for $x \in [a,b)$.
(a) Show that $f \in BV([a,b])$ if and only if v_f is bounded on $[a,b)$, or equivalently,

$$\begin{aligned} &\lim_{x \to b} v_f(x) < \infty \quad \Rightarrow \quad f \in BV([a,b]) \\ &\lim_{x \to b} v_f(x) = \infty \quad \Rightarrow \quad f \notin BV([a,b]). \end{aligned}$$

(b) Show that if $\lim_{x \to b} v_f(x) < \infty$ and if f is continuous at b, then $\lim_{x \to b} v_f(x) = V_a^b(f)$.
(c) Show by constructing a counter example that the equality in (b) does not hold without the continuity of f at b.

Prob. 12.11. For each $p \in (0, \infty)$, let f_p be a real-valued continuous function on $\left[0, \frac{2}{\pi}\right]$ defined by

$$f(x) = \begin{cases} x^p \sin \frac{1}{x} & \text{for } x \in \left(0, \frac{2}{\pi}\right], \\ 0 & \text{for } x = 0. \end{cases}$$

(a) Show that $f \in BV\left(\left[0, \frac{2}{\pi}\right]\right)$ if $p \in (1, \infty)$.
(b) Show that $f \notin BV\left(\left[0, \frac{2}{\pi}\right]\right)$ if $p \in (0, 1]$.

Prob. 12.12. Let $f \in BV([a, b])$ and $g \in BV([c, d])$. Assume that $f([a, b]) \subset [c, d]$ so that $g \circ f$ is defined on $[a, b]$. Is $g \circ f$ a BV function on $[a, b]$?

Prob. 12.13. Let f be a real-valued continuous and BV function on $[0, 1]$. Show that

$$\lim_{n \to \infty} \sum_{i=1}^{n} \left| f\left(\tfrac{i}{n}\right) - f\left(\tfrac{i-1}{n}\right) \right|^2 = 0.$$

Prob. 12.14. Let $(f_i : i \in \mathbb{N})$ and f be real-valued functions on an interval $[a, b]$ such that $\lim_{i \to \infty} f_i(x) = f(x)$ for $x \in [a, b]$. Show that $V_a^b(f) \le \liminf_{i \to \infty} V_a^b(f_i)$.

Prob. 12.15. Let $(f_i : i \in \mathbb{N})$ and f be real-valued functions on an interval $[a, b]$ such that $\lim_{i \to \infty} f_i(x) = f(x)$ for $x \in [a, b]$. Show that if $f_i \in BV([a, b])$ for every $i \in \mathbb{N}$ then $f \in BV([a, b])$.

Prob. 12.16. Let f be real-valued and differentiable everywhere on (a, b). Show that the derivative f' is a Borel measurable function on (a, b).

§13 Absolutely Continuous Functions

[I] Absolute Continuity

Definition 13.1. *We say that a collection of closed intervals in* $\mathbb{R}$ *is non-overlapping if their interiors are disjoint. A real-valued function* f *on a finite closed interval* $[a, b]$ *is said to be absolutely continuous on* $[a, b]$ *if for every* $\varepsilon > 0$ *there exists* $\delta > 0$ *such that for any finite collection of non-overlapping closed intervals* $\{[a_k, b_k] : k = 1, \dots, n\}$ *contained in* $[a, b]$ *with* $\sum_{k=1}^{n}(b_k - a_k) < \delta$, *we have* $\sum_{k=1}^{n} |f(b_k) - f(a_k)| < \varepsilon$.

Note that unlike continuity, absolute continuity is not a pointwise property of a function; rather it is a property of a function on a finite closed interval.

Theorem 13.2. *Let* f *be an absolutely continuous real-valued function on* $[a, b]$. *Then*
(a) f *is uniformly continuous on* $[a, b]$.
(b) f *is a function of bounded variation on* $[a, b]$.

Proof. 1. To show the uniform continuity of f on $[a, b]$, let $\varepsilon > 0$ be arbitrarily given and let $\delta > 0$ be as in Definition 13.1. Thus if $x', x'' \in [a, b]$ and $|x' - x''| < \delta$, then $|f(x') - f(x'')| < \varepsilon$. This proves the uniform continuity of f on $[a, b]$.

2. To show that f is a BV function on $[a, b]$, let $\varepsilon > 0$ be arbitrarily given and let $\delta > 0$ be as in Definition 13.1. Let $N \in \mathbb{N}$ be so large that $\frac{1}{N}(b - a) < \delta$ and let $\mathcal{P}_0$ be the partition of $[a, b]$ into N closed subintervals $I_1, \dots, I_N$ with equal length $\frac{1}{N}(b - a)$.

Let $\mathcal{P}$ be an arbitrary partition of $[a, b]$. Consider the partition $\mathcal{P} \cup \mathcal{P}_0$ of $[a, b]$. It partitions each I_k into finitely many closed subintervals $I_{k,1}, \dots, I_{k,n(k)}$ with $n(k) \in \mathbb{N}$ for $k = 1, \dots, N$. Let $I_{k,j} = [a_{k,j}, b_{k,j}]$ for $j = 1, \dots, n(k)$ and $k = 1, \dots, N$. Then

$$V_a^b(f, \mathcal{P}) \le V_a^b(f, \mathcal{P} \cup \mathcal{P}_0) = \sum_{k=1}^{N} \sum_{j=1}^{n(k)} \left| f(b_{k,j}) - f(a_{k,j}) \right|.$$

Now since $\sum_{j=1}^{n(k)}(b_{k,j} - a_{k,j}) = \frac{1}{N}(b - a) < \delta$, we have $\sum_{j=1}^{n(k)} \left| f(b_{k,j}) - f(a_{k,j}) \right| < \varepsilon$. Thus $V_a^b(f, \mathcal{P}) \le N\varepsilon$ and then $V_a^b(f) = \sup_{\mathcal{P} \in \mathfrak{P}_{a,b}} V_a^b(f, \mathcal{P}) \le N\varepsilon < \infty$. This shows that f is a BV function on $[a, b]$. ∎

The converse of Theorem 13.2 is false, that is, a continuous function of bounded variation on $[a, b]$ may not be absolutely continuous on $[a, b]$. We shall show that the Cantor-Lebesgue function is such a function.

Corollary 13.3. *If* f *is an absolutely continuous real-valued function on* $[a, b]$, *then* f *is* $\mathfrak{B}_{\mathbb{R}}$*-measurable on* $[a, b]$*; the derivative* f' *exists* $(\mathfrak{M}_L, \mu_L)$*-a.e. on* $[a, b]$ *and is* $\mathfrak{M}_L$*-measurable and* μ_L*-integrable on* $[a, b]$.

Proof. An absolutely continuous real-valued function on $[a, b]$ is a function of bounded variation on $[a, b]$ by Theorem 13.2. Thus the Corollary follows from Theorem 12.21. ∎

We show in Theorem 13.17 that actually $\int_{[a,b]} f' \, d\mu_L = f(b) - f(a)$ for an absolutely continuous function f on $[a, b]$.

The derivative of a function can be used to estimate the range of the function in terms of its domain of definition. As a simple example, let f be a real-valued continuous function on a finite closed interval $I = [a, b]$ with derivative f' such that $|f'| \leq M$ on (a, b) for some constant $M > 0$. Since I is a compact set in $\mathbb{R}$ and f is continuous, $f(I)$ is a compact set in $\mathbb{R}$ and thus $f(I) \in \mathfrak{B}_{\mathbb{R}}$. Since f is bounded on I, $\inf_I f$ and $\sup_I f$ are finite. By the continuity of f on I, there exist $x_1, x_2 \in I$ such that $f(x_1) = \inf_I f$ and $f(x_2) = \sup_I f$. By the Mean Value Theorem, we have $f(x_2) - f(x_1) \leq M|x_1 - x_2| \leq M\mu_L(I)$. Now $f(I) \subset [f(x_1), f(x_2)]$. Thus $\mu_L\big(f(I)\big) \leq f(x_2) - f(x_1) \leq M\mu_L(I)$, which estimates the Lebesgue measure of the set $f(I)$ in terms of the Lebesgue measure of I. The next Theorem gives such an estimate for $f(E)$ where f is a continuous function on $[a, b]$ and E is an arbitrary $\mathfrak{M}_L$-measurable subset of $[a, b]$ on which f' exists and is bounded.

Lemma 13.4. *Let f be a real-valued function defined on a finite closed interval $I = [a, b]$. Let E be an arbitrary subset of I such that $f'(x)$ exists and $|f'(x)| < M$ for some constant $M > 0$ for every $x \in E$. Then $\mu_L^*\big(f(E)\big) \leq M\mu_L^*(E)$.*

Proof. Let $x \in E$. Then $f'(x)$ exists and $|f'(x)| < M$, that is,

$$|f'(x)| = \left| \lim_{y \to x} \frac{f(y) - f(x)}{y - x} \right| < M.$$

Thus for every $x \in E$ there exists $\delta > 0$ such that

$$|f(y) - f(x)| < M|y - x| \quad \text{for } y \in (x - \delta, x + \delta) \cap [a, b].$$

For every $n \in \mathbb{N}$, let

(1) $$E_n = \left\{x \in E : |f(y) - f(x)| < M|y - x| \text{ for } y \in \left(x - \tfrac{1}{n}, x - \tfrac{1}{n}\right)\right\}.$$

Then every $x \in E$ is contained in E_n with sufficiently large $n \in \mathbb{N}$ so that $E \subset \bigcup_{n\in\mathbb{N}} E_n$. On the other hand $E_n \subset E$ for every $n \in \mathbb{N}$ and $(E_n : n \in \mathbb{N})$ is an increasing sequence of subsets of E. Thus we have $E = \bigcup_{n\in\mathbb{N}} E_n = \lim_{n\to\infty} E_n$. By Lemma 3.21 and Remark 2.16 the Lebesgue outer measure μ_L^* on $\mathbb{R}$ is a regular outer measure. Then by Theorem 2.11, $E_n \uparrow E$ implies

(2) $$\mu_L^*(E) = \lim_{n\to\infty} \mu_L^*(E_n).$$

We have also $f(E) = f\big(\bigcup_{n\in\mathbb{N}} E_n\big) = \bigcup_{n\in\mathbb{N}} f(E_n)$. Since $(E_n : n \in \mathbb{N})$ is an increasing sequence, $\big(f(E_n) : n \in \mathbb{N}\big)$ is an increasing sequence. Thus $\bigcup_{n\in\mathbb{N}} f(E_n) = \lim_{n\to\infty} f(E_n)$. Therefore $f(E) = \lim_{n\to\infty} f(E_n)$. Then $f(E_n) \uparrow f(E)$ implies by Theorem 2.11 that

(3) $$\mu_L^*\big(f(E)\big) = \lim_{n\to\infty} \mu_L^*\big(f(E_n)\big).$$

For each $n \in \mathbb{N}$, let us estimate $\mu_L^*\big(f(E_n)\big)$ by $\mu_L^*(E_n)$. Let $\varepsilon > 0$ be arbitrarily given and let $(I_{n,k} : k \in \mathbb{N})$ be a sequence of open intervals such that

(4) $$\ell(I_{n,k}) < 1/n \quad \text{for } k \in \mathbb{N},$$

(5) $$E_n \subset \bigcup_{k\in\mathbb{N}} I_{n,k},$$

(6) $$\sum_{k\in\mathbb{N}} \ell(I_{n,k}) \le \mu_L^*(E_n) + \varepsilon.$$

Then we have

(7) $$E_n = E_n \cap \Big(\bigcup_{k\in\mathbb{N}} I_{n,k}\Big) = \bigcup_{k\in\mathbb{N}} (E_n \cap I_{n,k})$$

and

(8) $$f(E_n) = f\Big(\bigcup_{k\in\mathbb{N}} (E_n \cap I_{n,k})\Big) = \bigcup_{k\in\mathbb{N}} f(E_n \cap I_{n,k}).$$

If $x_1, x_2 \in E_n \cap I_{n,k}$ then $|f(x_1) - f(x_2)| < M|x_1 - x_2| \le M\ell(I_{n,k})$ by (4) and (1). Thus $f(E_n \cap I_{n,k})$ is contained in an interval of length equal to $M\ell(I_{n,k})$ and hence

(9) $$\mu_L^*\big(f(E_n \cap I_{n,k})\big) \le M\ell(I_{n,k}).$$

By (8), (9), and (6) we have

(10) $$\begin{aligned}\mu_L^*\big(f(E_n)\big) &= \mu_L^*\Big(\bigcup_{k\in\mathbb{N}} f(E_n \cap I_{n,k})\Big) \le \sum_{k\in\mathbb{N}} \mu_L^*\big(f(E_n \cap I_{n,k})\big) \\ &\le \sum_{k\in\mathbb{N}} M\ell(I_{n,k}) < M\{\mu_L^*(E_n) + \varepsilon\}.\end{aligned}$$

Then by (3), (10), and (2) we have

$$\mu_L^*\big(f(E)\big) = \lim_{n\to\infty} \mu_L^*\big(f(E_n)\big) < M\{\lim_{n\to\infty} \mu_L^*(E_n) + \varepsilon\} = M\{\mu_L^*(E) + \varepsilon\}.$$

Since this holds for an arbitrary $\varepsilon > 0$, we have $\mu_L^*\big(f(E)\big) \le M\mu_L^*(E)$. ∎

Theorem 13.5. *Let f be a real-valued function on $[a, b]$. Let E be a $\mathfrak{M}_L$-measurable subset of $[a, b]$. Suppose the derivative f' of f exists at every $x \in E$. Then we have $\mu_L^*\big(f(E)\big) \le \int_E |f'| \, d\mu_L$.*

Proof. The existence of f' on $E \in \mathfrak{M}_L$ implies that f' is $\mathfrak{M}_L$-measurable on E by Proposition 12.4. Consider first the case that f' is bounded on E. Let $M \in \mathbb{N}$ be such that $|f'| < M$ on E. For every $n \in \mathbb{N}$, let

$$E_{n,k} = \big\{x \in E : |f'(x)| \in \big[\tfrac{k-1}{2^n}, \tfrac{k}{2^n}\big)\big\} \quad \text{for } k = 1, \ldots, 2^n M.$$

Then $\{E_{n,k} : k \in \mathbb{N}\}$ is a disjoint collection in $\mathfrak{M}_L$ by the $\mathfrak{M}_L$-measurability of f' on E and $\bigcup_{k=1}^{2^n M} E_{n,k} = E$. Now $f(E) = f\left(\bigcup_{k=1}^{2^n M} E_{n,k}\right) = \bigcup_{k=1}^{2^n M} f(E_{n,k})$ so that we have $\mu_L^*\big(f(E)\big) \leq \sum_{k=1}^{2^n M} \mu_L^*\big(f(E_{n,k})\big)$. Now $|f'| < \frac{k}{2^n}$ on $E_{n,k}$ implies that we have $\mu_L^*\big(f(E_{n,k})\big) \leq \frac{k}{2^n}\mu_L(E_{n,k})$ by Lemma 13.4. Thus we have

$$\mu_L^*\big(f(E)\big) \leq \sum_{k=1}^{2^n M} \frac{k}{2^n}\mu_L(E_{n,k}) = \sum_{k=1}^{2^n M} \frac{k-1}{2^n}\mu_L(E_{n,k}) + \sum_{k=1}^{2^n M} \frac{1}{2^n}\mu_L(E_{n,k}).$$

Since $|f'| \geq \frac{k-1}{2^n}$ on $E_{n,k}$, we have $\int_{E_{n,k}} |f'|\,d\mu_L \geq \frac{k-1}{2^n}\mu_L(E_{n,k})$. Then we have

$$\mu_L^*\big(f(E)\big) \leq \sum_{k=1}^{2^n M} \int_{E_{n,k}} |f'|\,d\mu_L + \sum_{k=1}^{2^n M} \frac{1}{2^n}\mu_L(E_{n,k}) = \int_E |f'|\,d\mu_L + \frac{1}{2^n}\mu_L(E).$$

Since this holds for every $n \in \mathbb{N}$, we have $\mu_L^*(f(E)) \leq \int_E |f'|\,d\mu_L$. This proves the Theorem for the particular case that f' is bounded on E.

For the general case, let $E_n = \{x \in E : |f'(x)| \in [n-1, n)\}$ for $n \in \mathbb{N}$. Then $\{E_n : n \in \mathbb{N}\}$ is a disjoint collection in $\mathfrak{M}_L$ and $E = \bigcup_{n\in\mathbb{N}} E_n$. By our result above, we have $\mu_L^*\big(f(E_n)\big) \leq \int_{E_n} |f'|\,d\mu_L$ for every $n \in \mathbb{N}$. Now $f(E) = f\big(\bigcup_{n\in\mathbb{N}} E_n\big) = \bigcup_{n\in\mathbb{N}} f(E_n)$. Then by the countable subadditivity of μ_L^* we have

$$\mu_L^*\big(f(E)\big) \leq \sum_{n\in\mathbb{N}} \mu_L^*\big(f(E_n)\big) \leq \sum_{n\in\mathbb{N}} \int_{E_n} |f'|\,d\mu_L = \int_E |f'|\,d\mu_L.$$

This completes the proof. ∎

[II] Banach-Zarecki Criterion for Absolute Continuity

Definition 13.6. *Let f be a real-valued function on $[a, b]$. We say that f satisfies Lusin's Condition (N) on $[a, b]$ if for every subset E of $[a, b]$ which is a null set in $(\mathbb{R}, \mathfrak{M}_L, \mu_L)$, the set $f(E)$ is also a null set in $(\mathbb{R}, \mathfrak{M}_L, \mu_L)$.*

Lemma 13.7. *If a real-valued function f is absolutely continuous on $[a, b]$, then for every $\varepsilon > 0$ there exists $\delta > 0$ such that for every countable collection $\big\{[a_n, b_n] : n \in \mathbb{N}\big\}$ of non-overlapping closed intervals contained in $[a, b]$ such that $\sum_{n\in\mathbb{N}}(b_n - a_n) < \delta$, we have $\sum_{n\in\mathbb{N}} |f(b_n) - f(a_n)| < \varepsilon$ and $\sum_{n\in\mathbb{N}} \big\{\sup_{[a_k,b_k]} f - \inf_{[a_k,b_k]} f\big\} < \varepsilon$.*

Proof. Let $\varepsilon > 0$. By the absolute continuity of f on $[a, b]$, there exists $\delta > 0$ such that for any finite collection of non-overlapping closed intervals $\big\{[a_k, b_k] : k = 1, \ldots, n\big\}$ contained in $[a, b]$ with $\sum_{k=1}^n (b_k - a_k) < \delta$, we have $\sum_{k=1}^n |f(b_k) - f(a_k)| < \frac{\varepsilon}{2}$. Take an arbitrary countable collection $\big\{[a_n, b_n] : n \in \mathbb{N}\big\}$ of non-overlapping closed intervals contained in $[a, b]$ such that $\sum_{n\in\mathbb{N}}(b_n - a_n) < \delta$. Then for every $N \in \mathbb{N}$, the collection $\{[a_k, b_k] : k = 1, \ldots, N\}$ is a finite collection of non-overlapping closed intervals contained in $[a, b]$ with $\sum_{k=1}^N (b_k - a_k) < \delta$ so that $\sum_{k=1}^N |f(b_k) - f(a_k)| < \frac{\varepsilon}{2}$. Since this holds for every $N \in \mathbb{N}$, we have $\sum_{n\in\mathbb{N}} |f(b_n) - f(a_n)| = \lim_{N\to\infty} \sum_{k=1}^N |f(b_k) - f(a_k)| \leq \frac{\varepsilon}{2} < \varepsilon$.

If f is absolutely continuous on $[a, b]$, then f is continuous on $[a, b]$, that is, f is continuous at every $x \in [a, b]$. (Actually the absolute continuity of f on $[a, b]$ implies its uniform continuity on $[a, b]$ as we show in Theorem 13.2.) Now the continuity of f on $[a_n, b_n]$ implies that there exist $\alpha_n, \beta_n \in [a_n, b_n]$ such that $f(\alpha_n) = \inf_{[a_n, b_n]} f$ and $f(\beta_n) = \sup_{[a_n, b_n]} f$. Since $[\alpha_n \wedge \beta_n, \alpha_n \vee \beta_n] \subset [a_n, b_n]$, we have a countable collection of non-overlapping closed intervals $\{[\alpha_n \wedge \beta_n, \alpha_n \vee \beta_n] : n \in \mathbb{N}\}$ contained in $[a, b]$ with $\sum_{n\in\mathbb{N}}(\alpha_n \vee \beta_n - \alpha_n \wedge \beta_n) \leq \sum_{n\in\mathbb{N}}(b_n - a_n) < \delta$. Then our result above implies that $\sum_{n\in\mathbb{N}} |f(\alpha_n \vee \beta_n) - f(\alpha_n \wedge \beta_n)| < \varepsilon$, that is, $\sum_{n\in\mathbb{N}} \{\sup_{[a_k, b_k]} f - \inf_{[a_k, b_k]} f\} < \varepsilon$. ■

Theorem 13.8. (Banach-Zarecki) *Let f be a real-valued function on $[a, b]$. Then f is absolutely continuous on $[a, b]$ if and only if it satisfies the following three conditions:*
1° *f is continuous on $[a, b]$.*
2° *f is of bounded variation on $[a, b]$.*
3° *f satisfies Condition (N) on $[a, b]$.*

Proof. 1. Suppose f is absolutely continuous on $[a, b]$. Then by Theorem 13.2, f is continuous and of bounded variation on $[a, b]$. It remains to show that f satisfies Condition (N) on $[a, b]$. Since $\mu_L^*(\{f(a), f(b)\}) = 0$, it suffices to show that for every subset E of (a, b) with $\mu_L^*(E) = 0$, we have $\mu_L^*(f(E)) = 0$.

Now let $E \subset (a, b)$ with $\mu_L^*(E) = 0$. Let $\varepsilon > 0$ be arbitrarily given. According to Lemma 13.7, the absolute continuity of f on $[a, b]$ implies that there exists $\delta > 0$ such that for every countable collection $\{I_n : n \in \mathbb{N}\}$ of non-overlapping closed intervals contained in $[a, b]$ with $\sum_{n\in\mathbb{N}} \ell(I_n) < \delta$ we have $\sum_{n\in\mathbb{N}} \{\sup_{I_n} f - \inf_{I_n} f\} < \varepsilon$. Now by Lemma 3.21, there exists an open set O containing E and contained in (a, b) such that $\mu_L(O) < \mu_L^*(E) + \delta = \delta$. Let $O = \bigcup_{n\in\mathbb{N}}(a_n, b_n)$ where $\{(a_n, b_n) : n \in \mathbb{N}\}$ is a disjoint collection of open intervals. If we let $I_n = [a_n, b_n]$, then $\{I_n : n \in \mathbb{N}\}$ is a countable collection of non-overlapping closed intervals contained in $[a, b]$ such that $\sum_{n\in\mathbb{N}} \ell(I_n) = \sum_{n\in\mathbb{N}}(b_n - a_n) = \mu_L(O) < \delta$. Then $\sum_{n\in\mathbb{N}} \{\sup_{[a_n, b_n]} f - \inf_{[a_n, b_n]} f\} < \varepsilon$. Since $E \subset O = \bigcup_{n\in\mathbb{N}}(a_n, b_n)$, we have $f(E) \subset \bigcup_{n\in\mathbb{N}} f((a_n, b_n))$. This implies that $\mu_L^*(f(E)) \leq \sum_{n\in\mathbb{N}} \mu_L^*(f((a_n, b_n)))$. But $f((a_n, b_n)) \subset [\inf_{[a_n, b_n]} f, \sup_{[a_n, b_n]} f]$ so that $\mu_L^*(f((a_n, b_n))) \leq \sup_{[a_n, b_n]} f - \inf_{[a_n, b_n]} f$ for every $n \in \mathbb{N}$. Thus we have $\mu_L^*(f(E)) \leq \sum_{n\in\mathbb{N}} \{\sup_{[a_n, b_n]} f - \inf_{[a_n, b_n]} f\} < \varepsilon$. By the arbitrariness of $\varepsilon > 0$, we have $\mu_L^*(f(E)) = 0$. This shows that f satisfies Condition (N) on $[a, b]$.

2. Conversely suppose f satisfies conditions 1°, 2°, and 3°. Let us show that f is absolutely continuous on $[a, b]$. Now since f is a function of bounded variation on $[a, b]$, its derivative f' exists a.e. on $[a, b]$, is $\mathfrak{M}_L$-measurable, and μ_L-integrable on $[a, b]$ by Theorem 12.21. By Theorem 9.26 (Absolute Uniform Continuity of the Integral), the μ_L-integrability of f' implies that for an arbitrary $\varepsilon > 0$ there exists $\delta > 0$ such that

(1) $$\int_F |f'| \, d\mu_L < \varepsilon \quad \text{for every } F \subset [a, b], F \in \mathfrak{M}_L, \text{ with } \mu_L(F) < \delta.$$

Let $I_k = [a_k, b_k]$, where $k = 1, \ldots, n$, be non-overlapping closed intervals contained in $[a, b]$ with $\sum_{k=1}^n (b_k - a_k) < \delta$. Since f' exists a.e. on $[a, b]$, the subset of $[a, b]$ on which f' exists is a $\mathfrak{M}_L$-measurable set. If we let $D_k = \{x \in I_k : f'(x) \text{ exists}\}$ then $D_k \in \mathfrak{M}_L$

and $E_k := I_k \setminus D_k \in \mathfrak{M}_L$ for $k = 1, \ldots, n$. Since f' exists a.e. on $[a, b]$, E_k is a null set in $(\mathbb{R}, \mathfrak{M}_L, \mu_L)$ and thus by Condition (N) we have $\mu_L^*(f(E_k)) = 0$ for $k = 1, \ldots, n$. Since f is a continuous function, the Intermediate Value Theorem implies that $f(I_k)$ is an interval. Since I_k is a compact set in $\mathbb{R}$, the continuity of f implies that $f(I_k)$ is a compact set in $\mathbb{R}$. Thus $f(I_k)$ is a finite closed interval. Therefore $f(I_k) = [\min_{I_k} f, \max_{I_k} f]$. Then

$$(2) \qquad |f(b_k) - f(a_k)| \le \max_{I_k} f - \min_{I_k} f = \mu_L(f(I_k)).$$

Since $I_k = D_k \cup E_k$, we have $f(I_k) = f(D_k) \cup f(E_k)$. Then by Theorem 13.5, we have

$$(3) \qquad \mu_L(f(I_k)) \le \mu_L^*(f(D_k)) + \mu_L^*(f(E_k)) \le \int_{D_k} |f'| \, d\mu_L.$$

Now $D_k \subset I_k$ for $k = 1, \ldots, n$ and $\{I_k : k = 1, \ldots, n\}$ is a collection of non-overlapping finite closed intervals. This implies that the inter section of any two members of the collection $\{D_k : k = 1, \ldots, n\}$ is either $\emptyset$ or a singleton. Then by (2) and (3) we have

$$(4) \qquad \sum_{k=1}^{n} |f(b_k) - f(a_k)| \le \sum_{k=1}^{n} \int_{D_k} |f'| \, d\mu_L = \int_{\bigcup_{k=1}^{n} D_k} |f'| \, d\mu_L.$$

Now

$$(5) \qquad \mu_L\Big(\bigcup_{k=1}^{n} D_k\Big) = \sum_{k=1}^{n} \mu_L(D_k) = \sum_{k=1}^{n} \mu_L(I_k) = \sum_{k=1}^{n} (b_k - a_k) < \delta.$$

Then applying (1) to the last integral in (4), we have $\sum_{k=1}^{n} |f(b_k) - f(a_k)| < \varepsilon$. This proves the absolute continuity of f on $[a, b]$. ∎

Example. The Cantor-Lebesgue function τ is continuous and increasing on $[0, 1]$ and is thus a continuous BV function on $[0, 1]$. However it is not absolutely continuous on $[0, 1]$ since it does not satisfy Condition (N) on $[a, b]$. Recall that τ maps $[0, 1]$ onto $[0, 1]$ and it maps the open set G contained in $[0, 1]$ with $\mu_L(G) = 1$ in the construction of the Cantor ternary set T in [VI] of §4 onto a countable subset $F = \{\frac{1}{2}; \frac{1}{2^2}, \frac{3}{2^2}; \frac{1}{2^3}, \frac{3}{2^3}, \frac{5}{2^3}, \frac{7}{2^3}; \ldots\}$ of $[0, 1]$. Thus for the Cantor ternary set $T = [0, 1] \setminus G$, which is a null set in $(\mathbb{R}, \mathfrak{M}_L, \mu_L)$, we have $\tau(T) = [0, 1] \setminus F$ which is not a null set in $(\mathbb{R}, \mathfrak{M}_L, \mu_L)$.

Let E be a $\mathfrak{M}_L$-measurable subset of $[a, b]$. For a real-valued continuous function f on $[a, b]$, the set $f(E)$ need not be $\mathfrak{M}_L$-measurable. (See Proposition 4.38.) We show next that if f satisfies Condition (N) in addition, then $f(E)$ is $\mathfrak{M}_L$-measurable.

Proposition 13.9. *Let f be a real-valued continuous function on $[a, b]$. If f satisfies Condition (N) on $[a, b]$, then for every $E \in \mathfrak{M}_L$, $E \subset [a, b]$, we have $f(E) \in \mathfrak{M}_L$. In particular if f is a real-valued absolutely continuous function on $[a, b]$, then for every $E \in \mathfrak{M}_L$, $E \subset [a, b]$, we have $f(E) \in \mathfrak{M}_L$.*

Proof. Let $E \subset [a,b]$ and $E \in \mathfrak{M}_L$. By Theorem 3.22, there exists an F_σ-set $F \subset E$ such that $\mu_L(E \setminus F) = 0$. Since F is an F_σ-set contained in $[a,b]$, we have $F = \bigcup_{n\in\mathbb{N}} C_n$ where $(C_n : n \in \mathbb{N})$ is a sequence of closed sets contained in $[a,b]$. Thus C_n is a compact set and the continuity of f implies that $f(C_n)$ is a compact set so that $f(C_n) \in \mathfrak{M}_L$ for $n \in \mathbb{N}$. Now $E = F \cup (E \setminus F) = \left(\bigcup_{n\in\mathbb{N}} C_n\right) \cup (E \setminus F)$ so that $f(E) = \left(\bigcup_{n\in\mathbb{N}} f(C_n)\right) \cup f(E \setminus F)$. Since $\mu_L(E \setminus F) = 0$, Condition (N) on f implies $\mu_L^*\big(f(E \setminus F)\big) = 0$. Since $(\mathbb{R}, \mathfrak{M}_L, \mu_L)$ is a complete measure space, this implies that $f(E \setminus F) \in \mathfrak{M}_L$. Thus $f(E)$ is a countable union of members of $\mathfrak{M}_L$ and is therefore a member of $\mathfrak{M}_L$. ∎

[III] Singular Functions

Definition 13.10. *A real-valued function f on $[a,b]$ is said to be singular on $[a,b]$ if its derivative f' exists and $f' = 0$, $(\mathfrak{M}_L, \mu_L)$-a.e. on $[a,b]$.*

Example. The Cantor-Lebesgue function τ on $[0,1]$ is constant on each of the disjoint countable collection of open intervals which constitute an open set $G \subset [0,1]$ which has $\mu_L(G) = 1$. Thus the derivative τ' exists and $\tau' = 0$ on G. This shows that τ is a singular function on $[0,1]$.

Proposition 13.11. *If f is a real-valued function which is both absolutely continuous and singular on $[a,b]$, then f is constant on $[a,b]$.*

Proof. 1. Let us show that if f is both absolutely continuous and singular on $[a,b]$, then $\mu_L^*\big(f([a,b])\big) = 0$.

The singularity of f on $[a,b]$ implies that there exists a null set E in $(\mathbb{R}, \mathfrak{M}_L, \mu_L)$ contained in $[a,b]$ such that f' exists and $f' = 0$ on $D = [a,b] \setminus E$. Then $[a,b] = D \cup E$ and $f\big([a,b]\big) = f(D) \cup f(E)$ so that $\mu_L^*\big(f([a,b])\big) \le \mu_L^*\big(f(D)\big) + \mu_L^*\big(f(E)\big)$. Since $D \in \mathfrak{M}_L$ and $f' = 0$ on D, we have $\mu_L^*\big(f(D)\big) \le \int_D |f'|\, d\mu_L = 0$ by Theorem 13.5 so that $\mu_L^*\big(f(D)\big) = 0$. On the other hand, since f is absolutely continuous on $[a,b]$, it satisfies Condition (N) on $[a,b]$ by Theorem 13.8. Then $\mu_L(E) = 0$ implies $\mu_L^*\big(f(E)\big) = 0$. Therefore $\mu_L^*\big(f([a,b])\big) = 0$.

2. Now since f is continuous on $[a,b]$, if it were not constant on $[a,b]$ then $f\big([a,b]\big)$ would contain an open interval by the Intermediate Value Theorem for continuous functions, and then we would have $\mu_L^*\big(f([a,b])\big) > 0$, a contradiction. ∎

[IV] Indefinite Integrals

Definition 13.12. *Let f be a μ_L-integrable extended real-valued $\mathfrak{M}_L$-measurable function on $[a,b]$. By an indefinite integral of f on $[a,b]$ we mean a real-valued function F on $[a,b]$ defined by $F(x) = \int_{[a,x]} f\, d\mu_L + c$ for $x \in [a,b]$ where c is an arbitrarily selected real number.*

For an arbitrary indefinite integral F of f as defined above, $F(a) = \int_{\{a\}} f\, d\mu_L + c = c$. Also if F_1 and F_2 are two arbitrary indefinite integrals of f, then $F_1 - F_2$ is a constant function on $[a, b]$.

Lemma 13.13. *Let f be a μ_L-integrable extended real-valued $\mathfrak{M}_L$-measurable function on $[a, b]$. If some, and hence every, indefinite integral of f is a constant function on $[a, b]$, then $f = 0$ a.e. on $[a, b]$.*

Proof. 1. Let us show first that if an indefinite integral of f is constant on $[a, b]$, then $\int_O f \, d\mu_L = 0$ for every open set $O \subset (a, b)$ and then $\int_C f \, d\mu_L = 0$ for every closed set $C \subset (a, b)$. Let F be an indefinite integral of f on $[a, b]$. Then $F(x) = \int_{[a,x]} f \, d\mu_L + c$ for $x \in [a, b]$ where $c \in \mathbb{R}$. If F is a constant function on $[a, b]$, then for any $a \leq x' < x'' \leq b$, we have

$$(1) \qquad \int_{(x',x'')} f \, d\mu_L = \int_{(x',x'']} f \, d\mu_L = \int_{[a,x'']} f \, d\mu_L - \int_{[a,x']} f \, d\mu_L$$
$$= \{F(x'') - c\} - \{F(x') - c\} = 0.$$

Let O be an open set in $\mathbb{R}$ contained in (a, b). Then O is the union of countably many disjoint open intervals (x_n', x_n''), $n \in \mathbb{N}$, contained in (a, b). By (b) of Lemma 9.10 and (1),

$$(2) \qquad \int_O f \, d\mu_L = \int_{\bigcup_{n\in\mathbb{N}}(x_n', x_n'')} f \, d\mu_L = \sum_{n\in\mathbb{N}} \int_{(x_n', x_n'')} f \, d\mu_L = 0.$$

Let C be a closed set in $\mathbb{R}$ contained in (a, b). Then $O := (a, b) \setminus C = (a, b) \cap C^c$ is an open set in $\mathbb{R}$ contained in (a, b) so that by (2) we have

$$(3) \qquad \int_C f \, d\mu_L = \int_C f \, d\mu_L + \int_O f \, d\mu_L = \int_{(a,b)} f \, d\mu_L = 0.$$

2. To show that $f = 0$ a.e. on (a, b), let us note that $\{(a, b) : f \neq 0\}$ is the union of two disjoint $\mathfrak{M}_L$-measurable sets $\{(a, b) : f > 0\}$ and $\{(a, b) : f < 0\}$. This implies then

$$\mu_L\{(a, b) : f \neq 0\} = \mu_L\{(a, b) : f > 0\} + \mu_L\{(a, b) : f < 0\}.$$

Suppose the statement $f = 0$ a.e. on (a, b) is false. Then $\mu_L\{(a, b) : f \neq 0\} > 0$ and thus at least one of $\mu_L\{(a, b) : f > 0\}$ and $\mu_L\{(a, b) : f < 0\}$ is positive. Consider the case $\mu_L\{(a, b) : f > 0\} > 0$, that is, $\gamma := \mu_L(E) > 0$ where $E = \{(a, b) : f > 0\}$. Let $E_k = \{(a, b) : f \geq \frac{1}{k}\}$ for $k \in \mathbb{N}$. Then $(E_k : k \in \mathbb{N})$ is an increasing sequence in $\mathfrak{M}_L$ with $\lim_{k\to\infty} E_k = \bigcup_{k\in\mathbb{N}} E_k = E$ and thus $\mu_L(E_k) \uparrow \mu_L(E)$. Then there exists $k_0 \in \mathbb{N}$ such that $\mu_L(E_{k_0}) \geq \frac{1}{2}\gamma$. Since $E_{k_0} \in \mathfrak{M}_L$, by Theorem 3.22 there exists a closed set C in $\mathbb{R}$ such that $C \subset E_{k_0}$ and $\mu_L(E_{k_0} \setminus C) \leq \frac{1}{4}\gamma$. Then we have

$$\mu_L(C) = \mu_L(E_{k_0}) - \mu_L(E_{k_0} \setminus C) > \frac{1}{2}\gamma - \frac{1}{4}\gamma = \frac{1}{4}\gamma > 0.$$

Since $C \subset E_{k_0}$, we have $f \geq \frac{1}{k_0}$ on C. Then $\int_C f \, d\mu_L \geq \frac{1}{k_0}\frac{\gamma}{4} > 0$, contradicting (3). Similarly we get a contradiction if we assume $\mu_L\{(a, b) : f < 0\} > 0$. This shows that $f = 0$ a.e. on (a, b). Hence $f = 0$ a.e. on $[a, b]$. ∎

Lemma 13.14. *Let f be a μ_L-integrable extended real-valued $\mathfrak{M}_L$-measurable function on $[a, b]$ and let F be an indefinite integral of f on $[a, b]$. If f is continuous at $x_0 \in [a, b]$, then F is differentiable at x_0 and $F'(x_0) = f(x_0)$.*

Proof. By definition $F(x) = \int_{[a,x]} f \, d\mu_L + c$ for $x \in [a, b]$ where $c \in \mathbb{R}$. Suppose f is continuous at $x_0 \in [a, b]$. With $h > 0$, we have

$$\begin{aligned}\frac{1}{h}\{F(x_0+h) - F(x_0)\} - f(x_0) &= \frac{1}{h}\Big\{\int_{[a,x_0+h]} f \, d\mu_L - \int_{[a,x_0]} f \, d\mu_L\Big\} - f(x_0) \\ &= \frac{1}{h}\int_{(x_0,x_0+h]} f \, d\mu_L - \frac{1}{h}\int_{(x_0,x_0+h]} f(x_0) \, d\mu_L \\ &= \frac{1}{h}\int_{(x_0,x_0+h]} \{f - f(x_0)\} \, d\mu_L,\end{aligned}$$

and therefore

$$(1) \qquad \Big|\frac{1}{h}\{F(x_0+h) - F(x_0)\} - f(x_0)\Big| \leq \frac{1}{h}\int_{(x_0,x_0+h]} |f - f(x_0)| \, d\mu_L.$$

Now the continuity of f at x_0 implies that for every $\varepsilon > 0$ there exists $\delta > 0$ such that $|f(x) - f(x_0)| < \varepsilon$ for $x \in (x_0 - \delta, x_0 + \delta) \cap [a, b]$. Let $h > 0$ be so small that $h < \delta$. Then $\frac{1}{h}\int_{(x_0,x_0+h]} |f - f(x_0)| \, d\mu_L \leq \frac{1}{h}\varepsilon h = \varepsilon$. Substituting this in (1) and letting $h \downarrow 0$ we have

$$\limsup_{h\downarrow 0} \big|\tfrac{1}{h}\{F(x_0+h) - F(x_0)\} - f(x_0)\big| \leq \varepsilon.$$

By the arbitrariness of $\varepsilon > 0$, we have

$$\limsup_{h\downarrow 0} \big|\tfrac{1}{h}\{F(x_0+h) - F(x_0)\} - f(x_0)\big| = 0.$$

This implies that the limit inferior is also equal to 0 and then

$$\lim_{h\downarrow 0} \big|\tfrac{1}{h}\{F(x_0+h) - F(x_0)\} - f(x_0)\big| = 0,$$

and thus

$$\lim_{h\downarrow 0} \big\{\tfrac{1}{h}\{F(x_0+h) - F(x_0)\} - f(x_0)\big\} = 0,$$

that is, the right-derivative of f at x_0 exists and $(D_r F)(x_0) = f(x_0) \in \mathbb{R}$. Similarly the left-derivative of f at x_0 exists and $(D_l F)(x_0) = f(x_0) \in \mathbb{R}$. This shows that F is differentiable at x_0 and $F'(x_0) = f(x_0)$. ∎

Theorem 13.15. (Lebesgue Differentiation Theorem for the Indefinite Integral) *Let F be an indefinite integral of a μ_L-integrable extended real-valued $\mathfrak{M}_L$-measurable function f on $[a, b]$. Then F is absolutely continuous on $[a, b]$ and the derivative F' exists a.e. on $[a, b]$ and is $\mathfrak{M}_L$-measurable and μ_L-integrable on $[a, b]$. Furthermore $F' = f$ a.e. on $[a, b]$.*

Proof. Since the difference between any two indefinite integrals of f is a constant function, it suffices to prove the Theorem for the particular indefinite integral F_0 of f defined by

(1) $$F_0(x) = \int_{[a,x]} f \, d\mu_L \quad \text{for } x \in [a,b]$$

with the arbitrary constant c in Definition 13.12 being equal to 0. We have $F_0(a) = 0$.

1. To prove the absolute continuity of F_0 on $[a,b]$, let $\varepsilon > 0$ be arbitrarily given. By Theorem 9.26, there exists $\delta > 0$ such that $\int_E |f| \, d\mu_L < \varepsilon$ whenever $E \subset [a,b]$, $E \in \mathfrak{M}_L$, and $\mu_L(E) < \delta$. Let $\{[a_k, b_k] : k = 1, \ldots, n\}$ be an arbitrary finite collection of non-overlapping closed intervals contained in $[a,b]$ with $\sum_{k=1}^n (b_k - a_k) < \delta$. Then

$$\sum_{k=1}^n |F_0(b_k) - F_0(a_k)| = \sum_{k=1}^n \left| \int_{[a,b_k]} f \, d\mu_L - \int_{[a,a_k]} f \, d\mu_L \right|$$
$$\leq \sum_{k=1}^n \int_{(a_k,b_k]} |f| \, d\mu_L = \int_{\bigcup_{k=1}^n (a_k,b_k]} |f| \, d\mu_L < \varepsilon,$$

since $\mu_L\left(\bigcup_{k=1}^n (a_k, b_k]\right) = \sum_{k=1}^n (b_k - a_k) < \delta$. This proves the absolute continuity of F_0 on $[a,b]$. Then by Corollary 13.3, F_0' exists a.e. on $[a,b]$ and is $\mathfrak{M}_L$-measurable and μ_L-integrable on $[a,b]$.

2. Let us prove $F_0' = f$ a.e. on $[a,b]$. Now since F_0' and f are μ_L-integrable on $[a,b]$, so is $F_0' - f$. If we show

(2) $$\int_{[a,x]} \{F_0' - f\} \, d\mu_L = 0 \quad \text{for } x \in [a,b],$$

then every indefinite integral of $F_0' - f$ is a constant function on $[a,b]$, and this implies according to Lemma 13.13 that $F_0' - f = 0$ a.e. on $[a,b]$, that is, $F_0' = f$ a.e. on $[a,b]$ and we are done. Thus it suffices to prove (2). Note that (2) is equivalent to the equality $\int_{[a,x]} F_0' \, d\mu_L = \int_{[a,x]} f \, d\mu_L$ for $x \in [a,b]$, that is,

(3) $$\int_{[a,x]} F_0' \, d\mu_L = F_0(x) \quad \text{for } x \in [a,b].$$

Thus it remains to prove (3).

3. Let us consider first the case that f is a bounded real-valued $\mathfrak{M}_L$-measurable function on $[a,b]$, say $|f| \leq M$ on $[a,b]$ for some $M > 0$. Let us extended the domain of definition of f from $[a,b]$ to $[a, b+1]$ by setting $f(x) = 0$ for $x \in (b, b+1]$. Then the domain of definition of F_0 is extended from $[a,b]$ to $[a, b+1]$ and $F_0(x) = F_0(b)$ for $x \in (b, b+1]$. Then F_0 is continuous and hence $\mathfrak{M}_L$-measurable on $[a, b+1]$. Thus for every $n \in \mathbb{N}$, $F_0\left(x + \frac{1}{n}\right)$ is a $\mathfrak{M}_L$-measurable function of $x \in [a,b]$ according to Theorem 9.31 (Translation Invariance). Let us define a sequence $(G_n : n \in \mathbb{N})$ of real-valued $\mathfrak{M}_L$-measurable functions on $[a,b]$ by setting

(4) $$G_n(x) = \frac{F_0\left(x + \frac{1}{n}\right) - F_0(x)}{\frac{1}{n}} \quad \text{for } x \in [a,b].$$

Since F_0' exists a.e. on $[a, b]$, we have $\lim_{n\to\infty} G_n = F_0'$ a.e. on $[a, b]$. Now

$$|G_n(x)| = n\big|F_0\big(x+\tfrac{1}{n}\big) - F_0(x)\big| = n\Big|\int_{[a,x+\frac{1}{n}]} f\,d\mu_L - \int_{[a,x]} f\,d\mu_L\Big|$$
$$\leq n\int_{(x,x+\frac{1}{n}]} |f|\,d\mu_L \leq nM\frac{1}{n} = M \quad \text{for } x \in [a, b] \text{ and } n \in \mathbb{N}.$$

Thus $(G_n : n \in \mathbb{N})$ is a bounded sequence of real-valued $\mathfrak{M}_L$-measurable functions on $[a, b]$. Then by the Bounded Convergence Theorem (Theorem 7.16), we have

$$(5) \quad \int_{[a,x]} F_0'\,d\mu_L = \lim_{n\to\infty}\int_{[a,x]} G_n\,d\mu_L = \lim_{n\to\infty} n\int_{[a,x]} \big\{F_0\big(t+\tfrac{1}{n}\big) - F_0(t)\big\}\mu_L(dt)$$
$$= \lim_{n\to\infty} n\Big\{\int_{[a+\frac{1}{n},x+\frac{1}{n}]} F_0(t)\mu_L(dt) - \int_{[a,x]} F_0(t)\mu_L(dt)\Big\}$$
$$= \lim_{n\to\infty} n\Big\{\int_{(x,x+\frac{1}{n}]} F_0(t)\mu_L(dt) - \int_{[a,a+\frac{1}{n})} F_0(t)\mu_L(dt)\Big\},$$

where the third equality is by Theorem 9.31. Now F_0 is a real-valued continuous function on $[a, b+1]$ and is thus μ_L-integrable on $[a, b+1]$. Then according to Lemma 13.14, its indefinite integral defined by $\Phi_0(x) = \int_{[a,x]} F_0\,d\mu_L$ for $x \in [a, b+1]$ is differentiable with $\Phi_0'(x) = F_0(x)$ for $x \in [a, b+1]$. This implies

$$\lim_{n\to\infty} n\big\{\Phi_0\big(x+\tfrac{1}{n}\big) - \Phi_0(x)\big\} = F_0(x) \quad \text{for } x \in [a, b].$$

But from the definition of Φ_0 we have

$$\Phi_0\big(x+\tfrac{1}{n}\big) - \Phi_0(x) = \int_{[a,x+\frac{1}{n}]} F_0\,d\mu_L - \int_{[a,x]} F_0\,d\mu_L = \int_{(x,x+\frac{1}{n}]} F_0\,d\mu_L.$$

Thus we have

$$\lim_{n\to\infty} n\int_{(x,x+\frac{1}{n}]} F_0\,d\mu_L = F_0(x) \quad \text{for } x \in [a, b].$$

Applying this to the two integrals in the last member of (5), we have

$$\int_{[a,x]} F_0'\,d\mu_L = F_0(x) - F_0(a) = F_0(x) \quad \text{for } x \in [a, b].$$

This proves (3) for the case that f is a bounded real-valued $\mathfrak{M}_L$-measurable function on $[a, b]$.

4. Next consider the case that f is a μ_L-integrable nonnegative extended real-valued $\mathfrak{M}_L$-measurable function on $[a, b]$. In this case, the indefinite integral F_0 of f is a real-valued increasing function on $[a, b]$ so that by Theorem 12.10 (Lebesgue), the derivative F_0' exists a.e. on $[a, b]$ and is $\mathfrak{M}_L$-measurable and μ_L-integrable on $[a, b]$, and furthermore

$$(6) \quad \int_{[a,x]} F_0'\,d\mu_L \leq F_0(x) - F_0(a) = \int_{[a,x]} f\,d\mu_L \quad \text{for } x \in [a, b].$$

If we show that $F_0' \geq f$ a.e. on $[a,b]$, then we have $\int_{[a,x]} F_0' \, d\mu_L \geq \int_{[a,x]} f \, d\mu_L$ for $x \in [a,b]$ and this and (6) imply $\int_{[a,x]} F_0' \, d\mu_L = \int_{[a,x]} f \, d\mu_L = F_0(x)$ for $x \in [a,b]$, which is (3).

To show that $F_0' \geq f$ a.e. on $[a,b]$, consider the truncation of f at level n, that is, $f^{[n]}(x) = \min\{f(x), n\}$ for $x \in [a,b]$ and $n \in \mathbb{N}$. Consider the indefinite integral F_n of $f^{[n]}$ on $[a,b]$ defined by $F_n(x) = \int_{[a,x]} f^{[n]} \, d\mu_L$ for $x \in [a,b]$. Since $f^{[n]}$ is a bounded real-valued $\mathfrak{M}_L$-measurable function on $[a,b]$, we have $F_n' = f^{[n]}$ a.e. on $[a,b]$ by our result in **3**. Consider the indefinite integral G_n of the nonnegative extended real-valued $\mathfrak{M}_L$-measurable function $f - f^{[n]}$ on $[a,b]$ defined by $G_n(x) = \int_{[a,x]} \{f - f^{[n]}\} \, d\mu_L$ for $x \in [a,b]$. The nonnegativity of $f - f^{[n]}$ implies that G_n is a real-valued increasing function on $[a,b]$ and thus by Theorem 12.10 (Lebesgue) the derivative G_n' exists and $G_n' \geq 0$ a.e. on $[a,b]$. Now for $x \in [a,b]$ we have

$$F_0(x) = \int_{[a,x]} f \, d\mu_L = \int_{[a,x]} \{f - f^{[n]}\} \, d\mu_L + \int_{[a,x]} f^{[n]} \, d\mu_L = G_n(x) + F_n(x)$$

and therefore

$$F_0'(x) = G_n'(x) + F_n'(x) = G_n'(x) + f^{[n]}(x) \geq f^{[n]}(x) \quad \text{for a.e. } x \in [a,b].$$

Then since $f^{[n]} \uparrow f$ on $[a,b]$, we have $F_0' \geq f$ a.e. on $[a,b]$.

5. Finally let f be an arbitrary μ_L-integrable extended real-valued $\mathfrak{M}_L$-measurable function on $[a,b]$. We let $f = f^+ - f^-$ and apply the result in **4** to f^+ and f^-. Then F_0' exists a.e. on $[a,b]$ and is $\mathfrak{M}_L$-measurable and μ_L-integrable on $[a,b]$ and $F_0' = f$ a.e. on $[a,b]$. ∎

Lemma 13.16. *Let f be an absolutely continuous real-valued function on $[a,b]$. If $f' = 0$ a.e. on $[a,b]$ then f is constant on $[a,b]$.*

Proof. To show that f is constant on $[a,b]$ we show that for every $c \in (a,b]$ we have $f(c) = f(a)$. Now since $f' = 0$ a.e. on $[a,b]$, we have $f' = 0$ a.e. on (a,c). Thus there exists a null set F in $(\mathbb{R}, \mathfrak{M}_L, \mu_L)$ contained in (a,c) such that $f' = 0$ on $E = (a,c) \setminus F$. We have $E \in \mathfrak{M}_L$ with $\mu_L(E) = \mu_L\big((a,c)\big) = c - a < \infty$.

Let $\varepsilon > 0$ and $\eta > 0$ be arbitrarily given. Let $\delta > 0$ be a positive number corresponding to our $\varepsilon > 0$ in Definition 13.1 for the absolute continuity of f on $[a,c]$. Now at every $x \in E$ we have $f'(x) = 0$ so that there exists $h_x > 0$ such that

$$(1) \qquad |f(x+h) - f(x)| < \eta h \quad \text{for every } h \in (0, h_x).$$

Then the collection of closed intervals $\mathfrak{V} = \big\{[x, x+h] : h \in (0, h_x), x \in E\big\}$ is a Vitali cover of E. Since $\mu_L(E) < \infty$, by the Vitali Covering Theorem (Theorem 12.5) there exists a finite disjoint subcollection $\{I_k : k = 1, \ldots, N\}$ of $\mathfrak{V}$ such that

$$(2) \qquad \mu_L\Big(E \setminus \bigcup_{k=1}^{N} I_k\Big) < \delta.$$

Let $I_k = [x_k, y_k]$ with $x_k \in E \subset (a, c)$ for $k = 1, \dots, N$ and let $\{I_k : k = 1, \dots, N\}$ be so numbered that

$$(3) \qquad y_0 = a < x_1 < y_1 < x_2 < y_2 < \cdots < x_N < y_N < c = x_{N+1}.$$

Since $E \subset (a, c)$ and $\mu_L(E) = \mu_L\big((a, c)\big)$, (2) implies

$$(4) \qquad \mu_L\Big((a, c) \setminus \bigcup_{k=1}^{N} I_k\Big) < \delta.$$

Now

$$(a, c) \setminus \bigcup_{k=1}^{N} I_k = (y_0, x_1) \cup (y_1, x_2) \cup (y_2, x_3) \cup \cdots \cup (y_N, x_{N+1}).$$

Thus by (4) we have $\sum_{k=0}^{N}(x_{k+1} - y_k) < \delta$. Then by the absolute continuity of f on $[a, c]$ we have

$$(5) \qquad \sum_{k=0}^{N} \big|f(x_{k+1}) - f(y_k)\big| < \varepsilon.$$

On the other hand, (1) and the disjointness of the collection $\{I_k : k = 1, \dots, N\}$ imply

$$(6) \qquad \sum_{k=1}^{N} \big|f(y_k) - f(x_k)\big| < \eta \sum_{k=1}^{N} (y_k - x_k) \le \eta(c - a).$$

Therefore by (5) and (6) we have

$$\begin{aligned} |f(c) - f(a)| &= \Big|\sum_{k=0}^{N} \{f(x_{k+1}) - f(y_k)\} + \sum_{k=1}^{N} \{f(y_k) - f(x_k)\}\Big| \\ &\le \sum_{k=0}^{N} \big|\{f(x_{k+1}) - f(y_k)\}\big| + \sum_{k=1}^{N} \big|\{f(y_k) - f(x_k)\}\big| \\ &< \varepsilon + \eta(c - a). \end{aligned}$$

Since this holds for every $\varepsilon > 0$ and $\eta > 0$, we have $|f(c) - f(a)| = 0$, that is, we have $f(c) = f(a)$. ∎

Theorem 13.17. *Let f be an absolutely continuous real-valued function on $[a, b]$. Then*

$$\int_{[a,x]} f' \, d\mu_L = f(x) - f(a) \quad \textit{for } x \in [a, b].$$

Thus an absolutely continuous function is an indefinite integral of its derivative.

Proof. f' exists a.e. on $[a, b]$ and is $\mathfrak{M}_L$-measurable and μ_L-integrable on $[a, b]$ according to Corollary 13.3. Consider the indefinite integral of f' defined by $\varphi(x) = \int_{[a,x]} f' \, d\mu_L$

for $x \in [a, b]$. By Theorem 13.15, φ is absolute continuous on $[a, b]$ and the derivative φ' exists and $\varphi' = f'$ a.e. on $[a, b]$. Since φ and f are absolute continuous on $[a, b]$, $\varphi - f$ is absolute continuous on $[a, b]$. Moreover $(\varphi - f)' = \varphi' - f' = 0$ a.e. on $[a, b]$. Thus by Lemma 13.16, $\varphi - f$ is constant on $[a, b]$. Therefore there exists $c \in \mathbb{R}$ such that $\varphi(x) - f(x) = c$ for every $x \in [a, b]$. Then $\int_{[a,x]} f' \, d\mu_L - f(x) = c$ for every $x \in [a, b]$. With $x = a$, we have $c = \int_{\{a\}} f' \, d\mu_L - f(a) = -f(a)$. Thus $\int_{[a,x]} f' \, d\mu_L = f(x) - f(a)$ for every $x \in [a, b]$. ∎

As a characterization of an absolutely continuous function we have the following:

Theorem 13.18. *A real-valued function f on $[a, b]$ is absolutely continuous on $[a, b]$ if and only if it satisfies the following conditions:*
1° *f' exists a.e. on $[a, b]$.*
2° *f' is $\mathfrak{M}_L$-measurable and μ_L-integrable on $[a, b]$.*
3° *$\int_{[a,x]} f' \, d\mu_L = f(x) - f(a)$ for every $x \in [a, b]$.*

Proof. If f is absolutely continuous on $[a, b]$, then 1° and 2° hold by Corollary 13.3 and 3° holds by Theorem 13.17.

Conversely if 1°, 2°, and 3° hold, then f is an indefinite integral of a μ_L-integrable extended real-valued $\mathfrak{M}_L$-measurable function on $[a, b]$, namely f', and is thus absolutely continuous on $[a, b]$ by Theorem 13.15. ∎

The following theorem is contained in Theorem 13.15 and Theorem 13.17. We state it here because of its simplicity.

Theorem 13.19. *A real-valued function f on $[a, b]$ is absolute continuous on $[a, b]$ if and only if it is an indefinite integral of a μ_L-integrable extended real-valued $\mathfrak{M}_L$-measurable function on $[a, b]$.*

Proof. If f is absolutely continuous on $[a, b]$ then f is an indefinite integral of its derivative by Theorem 13.17. Conversely if f is an indefinite integral of a μ_L-integrable $\mathfrak{M}_L$-measurable extended real-valued function on $[a, b]$ then f is absolutely continuous on $[a, b]$ by Theorem 13.15. ∎

Lebesgue's decomposition of a real-valued increasing function f on $[a, b]$ as a sum of an absolutely continuous function and a singular function can be derived from the existence and the integrability of f' (Theorem 12.10) and Theorem 13.15 as follows.

Theorem 13.20. (Lebesgue Decomposition of Increasing Functions) (a) *For every real-valued increasing function f on $[a, b]$, there exist an absolutely continuous increasing function g and a singular increasing function h on $[a, b]$ such that $f = g + h$ on $[a, b]$. Moreover the decomposition is unique up to constants.*
(b) *If f is a function of bounded variation on $[a, b]$, then there exist an absolutely continuous function g and a singular function of bounded variation h on $[a, b]$ such that $f = g + h$ on $[a, b]$. The decomposition is unique up to constants.*

Proof. 1. If f is a real-valued increasing function on $[a, b]$, then by Theorem 12.10 the derivative f' exists and is nonnegative a.e. on $[a, b]$ and furthermore f' is $\mathfrak{M}_L$-measurable and μ_L-integrable on $[a, b]$. Then an indefinite integral of the μ_L-integrable nonnegative function f' defined by $g(x) = \int_{[a,b]} f' \, d\mu_L$ for $x \in [a, b]$ is a real-valued increasing function on $[a, b]$. By Theorem 13.15, g is absolute continuous on $[a, b]$ and $g' = f'$ a.e. on $[a, b]$. If we define a real-valued function h on $[a, b]$ by setting $h = f - g$, then $h' = f' - g' = 0$ a.e. on $[a, b]$ so that h is a singular function on $[a, b]$. Let us show that h is an increasing function on $[a, b]$. Let $x_1, x_2 \in [a, b]$ and $x_1 < x_2$. By Theorem 12.10, we have $\int_{[x_1,x_2]} f' \, d\mu_L \leq f(x_2) - f(x_1)$. Now $f' = g' + h' = g'$ a.e. on $[a, b]$. Thus $\int_{[x_1,x_2]} f' \, d\mu_L = \int_{[x_1,x_2]} g' \, d\mu_L = g(x_2) - g(x_1)$ by Theorem 13.17 since g is absolutely continuous on $[a, b]$. Therefore we have $g(x_2) - g(x_1) \leq f(x_2) - f(x_1)$, that is, $(f - g)(x_1) \leq (f - g)(x_2)$, in other words, $h(x_1) \leq h(x_2)$. This shows that h is an increasing function on $[a, b]$. Thus $f = g + h$ is a decomposition we seek.

To prove the uniqueness of the decomposition up to constants, suppose that we have $f = g_1 + h_1 = g_2 + h_2$ where g_1 and g_2 are absolutely continuous increasing functions and h_1 and h_2 are singular increasing functions on $[a, b]$. Then $f_0 := g_1 - g_2 = h_2 - h_1$. Now $g_1 - g_2$ is an absolutely continuous function and $h_2 - h_1$ is a singular function on $[a, b]$. Thus f_0 being both absolutely continuous and singular is constant on $[a, b]$ by Proposition 13.11. Therefore $g_1 - g_2$ and $h_2 - h_1$ are constant on $[a, b]$.

2. If f is a function of bounded variation on $[a, b]$, then $f = f_1 - f_2$ where f_1 and f_2 are real-valued increasing functions on $[a, b]$ by Theorem 12.18. Then (b) follows by applying (a) to f_1 and f_2. ∎

If f is an extended real-valued $\mathfrak{M}_L$-measurable and μ_L-integrable function on $[a, b]$ then its indefinite integral F on $[a, b]$ is an absolutely continuous function on $[a, b]$ and in particular it is a function of bounded variation on $[a, b]$. We show next that the total variation of F on $[a, b]$ is equal to the integral of $|f|$ on $[a, b]$.

Theorem 13.21. *Let f be an extended real-valued $\mathfrak{M}_L$-measurable and μ_L-integrable function on $[a, b]$ and let F be an indefinite integral of f on $[a, b]$ defined by*

$$F(x) = \int_{[a,x]} f(t) \, \mu_L(dt) + c \quad \text{for } x \in [a, b]$$

where $c \in \mathbb{R}$. Then $V_a^b(F)$, the total variation of F on $[a, b]$, is given by

$$V_a^b(F) = \int_{[a,b]} |f(t)| \, \mu_L(dt).$$

Proof. Let $\mathfrak{P}_{a,b}$ be the collection of all partitions of the interval $[a, b]$. Take an arbitrary partition $\mathcal{P} = \{a = x_0 \leq \cdots \leq x_n = b\}$. Then for the variation of F on $[a, b]$

corresponding to $\mathcal{P}$ we have

$$V_a^b(F,\mathcal{P}) = \sum_{k=1}^n |F(x_k) - F(x_{k-1})| = \sum_{k=1}^n \int_{[x_{k-1},x_k]} f\,d\mu_L$$
$$\leq \sum_{k=1}^n \int_{[x_{k-1},x_k]} |f|\,d\mu_L = \int_{[a,b]} |f|\,d\mu_L.$$

Thus we have

$$V_a^b(F) = \sup_{\mathcal{P}\in\mathfrak{P}_{a,b}} V_a^b(F,\mathcal{P}) \leq \int_{[a,b]} |f|\,d\mu_L.$$

It remains to prove the reverse inequality.

Let $E_1 = \{t \in (a,b) : f(t) \geq 0\}$ and $E_2 = \{t \in (a,b) : f(t) < 0\}$. Then we have $E_1, E_2 \in \mathfrak{M}_L$, $E_1 \cap E_2 = \emptyset$, $E_1 \cup E_2 = (a,b)$, and

(1) $$\int_{[a,b]} |f|\,d\mu_L = \int_{E_1} f\,d\mu_L - \int_{E_2} f\,d\mu_L.$$

Since f is μ_L-integrable on (a,b), by Theorem 9.26 (Uniform Absolute Continuity of Integral with Respect to Measure) for every $\varepsilon > 0$ there exists $\delta > 0$ such that for every $E \subset (a,b)$, $E \in \mathfrak{M}_L$ with $\mu_L(E) \leq \delta$ we have

(2) $$\left|\int_E f\,d\mu_L\right| \leq \int_E |f|\,d\mu_L < \varepsilon.$$

Since $E_i \in \mathfrak{M}_L$, according to Theorem 3.22 there exists a closed set C_i such that $C_i \subset E_i$ and $\mu_L(E_i \setminus C_i) \leq \delta$ for $i = 1, 2$. Thus by (2) we have

(3) $$\begin{cases} \int_{E_1} f\,d\mu_L = \int_{C_1} f\,d\mu_L + \int_{E_1\setminus C_1} f\,d\mu_L < \int_{C_1} f\,d\mu_L + \varepsilon, \\ \int_{E_2} f\,d\mu_L = \int_{C_2} f\,d\mu_L + \int_{E_2\setminus C_2} f\,d\mu_L > \int_{C_2} f\,d\mu_L - \varepsilon. \end{cases}$$

By (1) and (3) we have

(4) $$\int_{[a,b]} |f|\,d\mu_L < \int_{C_1} f\,d\mu_L - \int_{C_2} f\,d\mu_L + 2\varepsilon.$$

Since $C_i \subset E_i$ and $E_1 \cap E_2 = \emptyset$ we have $C_1 \cap C_2 = \emptyset$. Then there exists an open set O_i such that $O_i \supset E_i$ and $O_1 \cap O_2 = \emptyset$. We may assume without loss of generality that $O_i \subset (a,b)$. (If $O_i \not\subset (a,b)$, take $O_i \cap (a,b)$.) Note that while $O_i \subset (a,b)$ we may not have $O_i \subset E_i$. Since $C_i \in \mathfrak{M}_L$, by Theorem 3.22 there exists an open set G_i such that $G_i \supset C_i$ and $\mu_L(G_i \setminus C_i) \leq \delta$ for $i = 1, 2$. Let $V_i = O_i \cap G_i$ for $i = 1, 2$. Then V_i is an open set and $V_i \subset O_i \subset (a,b)$. Since $O_1 \cap O_2 = \emptyset$, we have $V_1 \cap V_2 = \emptyset$. Since $C_i \subset O_i$

and $C_i \subset G_i$ also, we have $C_i \subset V_i$. We have also $\mu_L(V_i \setminus C_i) \le \mu_L(G_i \setminus C_i) \le \delta$. Thus by (2) we have

$$(5)\qquad \begin{cases} \displaystyle\int_{C_1} f\,d\mu_L = \int_{V_1} f\,d\mu_L + \int_{V_1\setminus C_1} f\,d\mu_L < \int_{V_1} f\,d\mu_L + \varepsilon, \\ \displaystyle\int_{C_2} f\,d\mu_L = \int_{V_2} f\,d\mu_L + \int_{V_2\setminus C_2} f\,d\mu_L > \int_{V_2} f\,d\mu_L - \varepsilon. \end{cases}$$

By (4) and (5) we have

$$(6)\qquad \int_{[a,b]} |f|\,d\mu_L < \int_{V_1} f\,d\mu_L - \int_{V_2} f\,d\mu_L + 4\varepsilon.$$

Since V_1 is an open set in $\mathbb{R}$, we have $V_1 = \bigcup_{k\in\mathbb{N}}(a_k, b_k)$ where $\big((a_k, b_k) : k \in \mathbb{N}\big)$ is a disjoint sequence of open intervals in $\mathbb{R}$. Thus we have $\mu_L(V_1) = \sum_{k\in\mathbb{N}} \mu_L\big((a_k, b_k)\big)$. Then since $\mu_L(V_1) < \infty$ there exists $m \in \mathbb{N}$ such that

$$\mu_L\Big(V_1 \setminus \bigcup_{k=1}^{m}(a_k, b_k)\Big) = \mu_L(V_1) - \sum_{k=1}^{m} \mu_L\big((a_k, b_k)\big) < \delta.$$

Then by (2) we have

$$(7)\qquad \begin{aligned} \int_{V_1} f\,d\mu_L &= \int_{\bigcup_{k=1}^{m}(a_k,b_k)} f\,d\mu_L + \int_{V_1\setminus\bigcup_{k=1}^{m}(a_k,b_k)} f\,d\mu_L \\ &< \sum_{k=1}^{m} \int_{(a_k,b_k)} f\,d\mu_L + \varepsilon \\ &= \sum_{k=1}^{m} \{F(b_k) - F(a_k)\} + \varepsilon. \end{aligned}$$

Similarly since V_2 is an open set in $\mathbb{R}$, we have $V_2 = \bigcup_{\ell\in\mathbb{N}}(c_\ell, d_\ell)$ where $\big((c_\ell, d_\ell) : \ell \in \mathbb{N}\big)$ is a disjoint sequence of open intervals in $\mathbb{R}$. Thus there exists $n \in \mathbb{N}$ such that

$$\mu_L\Big(V_2 \setminus \bigcup_{\ell=1}^{n}(c_\ell, d_\ell)\Big) = \mu_L(V_2) - \sum_{\ell=1}^{n} \mu_L\big((c_\ell, d_\ell)\big) < \delta.$$

Then by (2) we have

$$(8)\qquad \begin{aligned} \int_{V_2} f\,d\mu_L &= \int_{\bigcup_{\ell=1}^{n}(c_\ell,d_\ell)} f\,d\mu_L + \int_{V_2\setminus\bigcup_{\ell=1}^{n}(c_\ell,d_\ell)} f\,d\mu_L \\ &> \sum_{\ell=1}^{n} \int_{(c_\ell,d_\ell)} f\,d\mu_L - \varepsilon \\ &= \sum_{\ell=1}^{n} \{F(b_\ell) - F(a_\ell)\} - \varepsilon. \end{aligned}$$

Substituting (7) and (8) in (6) we have

$$\int_{[a,b]} |f|\,d\mu_L < \sum_{k=1}^{m}\{F(b_k)-F(a_k)\} - \sum_{\ell=1}^{n}\{F(b_\ell)-F(a_\ell)\} + 6\varepsilon$$
$$\leq \sum_{k=1}^{m}\big|F(b_k)-F(a_k)\big| + \sum_{\ell=1}^{n}\big|F(b_\ell)-F(a_\ell)\big| + 6\varepsilon.$$

Since $\big((a_k,b_k) : k\in\mathbb{N}\big)$ is a disjoint sequence with $\bigcup_{k\in\mathbb{N}}(a_k,b_k) = V_1$, $\bigcup_{\ell\in\mathbb{N}}(c_\ell,d_\ell)$ is a disjoint sequence with $\bigcup_{\ell\in\mathbb{N}}(c_\ell,d_\ell) = V_2$ and since $V_1\cap V_2=\emptyset$, we have $(a_k,b_k)\cap(c_\ell,d_\ell)=\emptyset$ for any $k,\ell\in\mathbb{N}$. Let $\mathcal{P}_0$ be the partition of $[a,b]$ with $a_k, b_k : k=1,\dots,m$ and $c_\ell, d_\ell : \ell=1,\dots,n$ as the partition points. Then

$$V_a^b(F,\mathcal{P}_0) \geq \sum_{k=1}^{m}\big|F(b_k)-F(a_k)\big| + \sum_{\ell=1}^{n}\big|F(b_\ell)-F(a_\ell)\big|.$$

Thus

$$\int_{[a,b]} |f|\,d\mu_L < V_a^b(F,\mathcal{P}_0) + 6\varepsilon \leq V_a^b(F) + 6\varepsilon.$$

Since this holds for an arbitrary $\varepsilon>0$, we have $\int_{[a,b]}|f|\,d\mu_L \leq V_a^b(F)$. This completes the proof that $V_a^b(F) = \int_{[a,b]}|f|\,d\mu_L$. ■

Theorem 13.22. *Let f be a real-valued absolutely continuous function on $[a,b]$. Then*

$$V_a^b(f) = \int_{[a,b]} |f'(t)|\,\mu_L(dt).$$

Proof. If f is a real-valued absolutely continuous function on $[a,b]$, then by Corollary 13.3 the derivative f' exists $(\mathfrak{M}_L,\mu_L)$-a.e. on $[a,b]$ and is $\mathfrak{M}_L$-measurable and μ_L-integrable on $[a,b]$ and moreover by Theorem 13.17 we have

$$f(x) = \int_{[a,x]} f'(t)\,\mu_L(dt) + f(a) \quad \text{for } x\in[a,b].$$

Then by Theorem 13.21, we have $V_a^b(f) = \int_{[a,b]}|f'(t)|\,\mu_L(dt)$. ■

[V] Calculation of the Lebesgue Integral by Means of the Derivative

[V.1] The Fundamental Theorem of Calculus

Let us consider the problem of recapturing a function by integrating its derivative. According to the Fundamental Theorem of Calculus, if a real-valued function f is differentiable everywhere on $[a,b]$ and its derivative f' is Riemann integrable on $[a,b]$, then $\int_a^b f'(x)\,dx = f(b)-f(a)$. According to Theorem 7.28, f' is Riemann integrable on $[a,b]$ if and only if f' is bounded and continuous a.e. on $[a,b]$. In this way, the Fundamental Theorem of Calculus relies on the continuity of f'. In contrast to this, if f is an absolutely continuous function on $[a, b]$, then according to Theorem 13.17, f' exists a.e. and is μ_L-integrable on $[a, b]$ with $\int_{[a,b]} f'\, d\mu_L = f(b) - f(a)$. In the next two theorems we show that if, instead of assuming the absolutely continuity of f, we assume that f is differentiable everywhere on $[a, b]$ and f' is bounded on $[a, b]$ (as in Theorem 13.23 below), or if f is continuous on $[a, b]$, f' exists a.e. on $[a, b]$, f' is μ_L-integrable on $[a, b]$, and the upper-right Dini derivate $D^+ f > -\infty$ on $[a, b)$ (as in Theorem 13.26 below), then $\int_{[a,b]} f'\, d\mu_L = f(b) - f(a)$.

Theorem 13.23. **(a)** *If a real-valued function f is differentiable everywhere on $[a, b]$ and f' is a bounded function on $[a, b]$, then f is absolutely continuous on $[a, b]$ and thus $\int_{[a,b]} f'\, d\mu_L = f(b) - f(a)$.*
(b) *If a real-valued function f on $[a, b]$ satisfies the Lipschitz condition on $[a, b]$, that is, there exists a constant $M > 0$ such that $|f(x') - f(x'')| \le M|x' - x''|$ for every $x', x'' \in [a, b]$, then f is absolutely continuous on $[a, b]$ so that $\int_{[a,b]} f'\, d\mu_L = f(b) - f(a)$.*

Proof. 1. The existence of f' on $[a, b] \in \mathfrak{M}_L$ implies that f' is $\mathfrak{M}_L$-measurable on $[a, b]$ by Proposition 12.4. Suppose $|f'| \le M$ on $[a, b]$ for some constant $M > 0$. Then by the Mean Value Theorem we have $|f(x'') - f(x')| = |f'(\xi)(x'' - x')| \le M(x'' - x')$ where $\xi \in (x', x'')$ for any $x', x'' \in [a, b]$ such that $x' < x''$. Given $\varepsilon > 0$, let $\delta \in \left(0, \frac{\varepsilon}{M}\right)$. Then for any finite collection of non-overlapping closed intervals $\{[a_k, b_k] : k = 1, \dots, n\}$ contained in $[a, b]$ with $\sum_{k=1}^n (b_k - a_k) < \delta$, we have $\sum_{k=1}^n |f(b_k) - f(a_k)| \le M \sum_{k=1}^n (b_k - a_k) \le M\delta < \varepsilon$. This shows that f is absolutely continuous on $[a, b]$. Then $\int_{[a,b]} f'\, d\mu_L = f(b) - f(a)$ by Theorem 13.17.

2. Suppose a real-valued function f satisfies the Lipschitz condition with coefficient $M > 0$ on $[a, b]$. For an arbitrary $\varepsilon > 0$, let $\delta \in \left(0, \frac{\varepsilon}{M}\right)$. Then for any finite collection $\{[a_k, b_k] : k = 1, \dots, n\}$ of non-overlapping closed intervals contained in $[a, b]$ such that $\sum_{k=1}^n (b_k - a_k) < \delta$, we have $\sum_{k=1}^n |f(b_k) - f(a_k)| \le M \sum_{k=1}^n (b_k - a_k) \le M\delta < \varepsilon$. This shows that f is absolutely continuous on $[a, b]$. Then by Theorem 13.17, we have $\int_{[a,b]} f'\, d\mu_L = f(b) - f(a)$. ∎

Regarding the next lemma, let us note that while a real-valued increasing function on $[a, b]$ has at most countably many points of discontinuity, at a point of continuity of f we may have $f' = \infty$. For example, if $f(x) = \sqrt{x}$ for $x \in [0, 1]$ and $f(x) = -\sqrt{|x|}$ for $x \in [-1, 0)$, then f is continuous at $x = 0$ and $f'(0) = \infty$.

Lemma 13.24. *Let $E \subset [a, b]$ be a null set in $(\mathbb{R}, \mathfrak{M}_L, \mu_L)$. Then for every $\varepsilon > 0$, there exists an absolutely continuous increasing function f on $[a, b]$ such that $f(b) - f(a) \le \varepsilon$ and $f' = \infty$ on E.*

Proof. Given $\varepsilon > 0$. Let $(\varepsilon_n : n \in \mathbb{N})$ be a decreasing sequence of positive numbers such that $\sum_{n \in \mathbb{N}} \varepsilon_n = \varepsilon$. Since $\mu_L(E) = 0$, for every $n \in \mathbb{N}$ there exists an open set $O_n \supset E$ such that $\mu_L(O_n) < \varepsilon_n$ by Lemma 3.21. Let $V_n = \bigcap_{k=1}^n O_k$ for $n \in \mathbb{N}$. Then $(V_n : n \in \mathbb{N})$ is a decreasing sequence of open sets with $\mu_L(V_n) < \varepsilon_n$ for $n \in \mathbb{N}$ and $E \subset \bigcap_{n \in \mathbb{N}} V_n$. Let $V = \bigcap_{n \in \mathbb{N}} V_n$. Then $\mu_L(V) \le \mu_L(V_n) \le \varepsilon_n$ for every $n \in \mathbb{N}$ so that $\mu_L(V) = 0$.

Let $g_n = \sum_{k=1}^n \mathbf{1}_{V_k}$ for $n \in \mathbb{N}$. Then $(g_n : n \in \mathbb{N})$ is an increasing sequence of nonnegative real-valued $\mathfrak{M}_L$-measurable and μ_L-integrable functions on $\mathbb{R}$. Let $g = \lim_{n\to\infty} g_n$. Then g is a nonnegative extended real-valued $\mathfrak{M}_L$-measurable function on $\mathbb{R}$. The μ_L-integrability of g on $\mathbb{R}$ follows from Theorem 8.5 (Monotone Convergence Theorem). Indeed we have

$$(1) \qquad \int_{\mathbb{R}} g \, d\mu_L = \lim_{n\to\infty} \int_{\mathbb{R}} g_n \, d\mu_L = \lim_{n\to\infty} \int_{\mathbb{R}} \sum_{k=1}^n \mathbf{1}_{V_k} \, d\mu_L$$
$$= \lim_{n\to\infty} \sum_{k=1}^n \mu_L(V_k) \le \lim_{n\to\infty} \sum_{k=1}^n \varepsilon_k = \varepsilon.$$

Consider the indefinite integrals of the μ_L-integrable functions g_n, $n \in \mathbb{N}$, and g defined by

$$\begin{cases} f_n(x) = \int_{[a,x]} g_n \, d\mu_L & \text{for } x \in [a,b] \text{ for } n \in \mathbb{N}, \\ f(x) = \int_{[a,x]} g \, d\mu_L & \text{for } x \in [a,b]. \end{cases}$$

Then f_n, $n \in \mathbb{N}$, and f are all real-valued increasing functions on $[a,b]$ and furthermore absolutely continuous with $f_n' = g_n$, $n \in \mathbb{N}$, and $f' = g$ a.e. on $[a,b]$ by Theorem 13.15. In particular by Theorem 13.17 and by (1) we have

$$f(b) - f(a) = \int_{[a,b]} f' \, d\mu_L = \int_{[a,b]} g \, d\mu_L \le \int_{\mathbb{R}} g \, d\mu_L \le \varepsilon.$$

It remains to show that $f' = \infty$ on E. Since $E \subset V \cap [a,b]$, let us show that $f' = \infty$ on $V \cap [a,b]$. If we show that for every $x \in V \cap [a,b]$, the right-lower Dini derivate $(D_+ f)(x) = \infty$ and the left-lower Dini derivate $(D_- f)(x) = \infty$, then the right-derivative $(D_r f)(x) = \infty$ and the left-derivative $(D_\ell f)(x) = \infty$ and thus $f'(x) = \infty$.

Now $(D_+ f)(x) = \liminf_{h\downarrow 0} h^{-1}\{f(x+h) - f(x)\}$. Let $x \in V \cap [a,b]$. For $h > 0$ and an arbitrary $n \in \mathbb{N}$,

$$\frac{f(x+h) - f(x)}{h} = \frac{1}{h} \int_{[x,x+h]} g \, d\mu_L \ge \frac{1}{h} \int_{[x,x+h]} g_n \, d\mu_L = \frac{f_n(x+h) - f_n(x)}{h}.$$

Since the function $\mathbf{1}_{V_k}$ is continuous at $x \in V_k$ and since $V_n \subset V_k$ for $k = 1, \ldots, n$, the function $g_n = \sum_{k=1}^n \mathbf{1}_{V_k}$ is continuous at $x \in V_n$. Thus by Lemma 13.14, we have $f_n'(x) = g_n(x) = n$ for $x \in V_n \cap [a,b]$. This implies that there exists $\delta > 0$ such that $h^{-1}\{f_n(x+h) - f_n(x)\} \ge n - 1$ for $h \in (0, \delta)$ for $x \in V_n \cap [a,b]$. Thus $(D_+ f)(x) \ge n - 1$ for $x \in V_n \cap [a,b]$ and in particular for $x \in V \cap [a,b]$. Since this holds for an arbitrary $n \in \mathbb{N}$, we have $(D_+ f)(x) = \infty$ for $x \in V \cap [a,b]$.

The fact that $(D_- f)(x) = \infty$ for $x \in V \cap [a,b]$ can be shown by the same argument observing that

$$(D_- f)(x) = \liminf_{h\uparrow 0} \frac{f(x+h) - f(x)}{h} = \liminf_{h\downarrow 0} \frac{f(x-h) - f(x)}{-h}$$

and with $h > 0$ we have

$$\frac{f(x-h)-f(x)}{-h} = \frac{f(x)-f(x-h)}{h}$$
$$= \frac{1}{h}\int_{[x-h,x]} g\,d\mu_L \geq \frac{1}{h}\int_{[x-h,x]} g_n\,d\mu_L$$
$$= \frac{f_n(x)-f_n(x-h)}{h}.$$

The rest of the argument is the same as above. ∎

If f is a real-valued continuous function on $[a, b]$ and if f' exists and is nonnegative on (a, b) then f is an increasing function on $[a, b]$. This is an immediate consequence of the Mean Value Theorem. Now if f is a real-valued function on $[a, b]$ then its four Dini derivates always exist whether f' exists or not. We show next that if the upper-right Dini derivate D^+f is nonnegative on $[a, b)$ then f is an increasing function on $[a, b]$.

Lemma 13.25. *Let f be a real-valued continuous function on $[a, b]$.*
(a) *If the upper-right Dini derivate $D^+f \geq 0$ everywhere on $[a, b)$, then f is an increasing function on $[a, b]$.*
(b) *If $D^+f \geq 0$ a.e. on $[a, b)$ and $D^+f > -\infty$ everywhere on $[a, b)$, then f is an increasing function on $[a, b]$.*

Proof. 1. To prove (a), let us assume that $D^+f \geq 0$ everywhere on $[a, b)$. To show that f is an increasing function on $[a, b]$, let $\alpha, \beta \in [a, b]$ and $\alpha < \beta$. To show that $f(\alpha) \leq f(\beta)$, assume the contrary. Then $f(\beta) - f(\alpha) < 0$ so that there exists $\varepsilon > 0$ such that

(1) $$f(\beta) - f(\alpha) < -\varepsilon(\beta - \alpha).$$

Let g be a real-valued continuous function on $[\alpha, \beta]$ defined by

(2) $$g(x) = f(x) - f(\alpha) + \varepsilon(x - \alpha) \quad \text{for } x \in [\alpha, \beta].$$

By (2) and (1), we have $g(\beta) < 0$. Now $(x-\alpha)^{-1}g(x) = (x-\alpha)^{-1}\{f(x) - f(\alpha)\} + \varepsilon$ for $x \in (\alpha, \beta]$ so that $\limsup_{x\downarrow\alpha}(x-\alpha)^{-1}g(x) = (D^+f)(\alpha) + \varepsilon \geq \varepsilon$. This implies that there exists $x_0 \in (\alpha, \beta]$ such that $g(x_0) \geq \frac{1}{2}\varepsilon(x_0 - \alpha) > 0$. Then by the Intermediate Value Theorem applied to the continuous function g, there exists $x_1 \in (x_0, \beta)$ such that $g(x_1) = 0$. Let $\xi = \max\{x \in (x_0, \beta) : g(x) = 0\}$. By the continuity of g, we have $g(\xi) = 0$ and $g(x) < 0$ for $x \in (\xi, \beta]$. This implies that $(D^+g)(\xi) \leq 0$. But (2) implies that $(D^+g)(\xi) = (D^+f)(\xi) + \varepsilon$. Thus $(D^+f)(\xi) \leq -\varepsilon < 0$. This contradicts the assumption that $D^+f \geq 0$ on $[a, b]$. Therefore $f(\alpha) \leq f(\beta)$ and f is an increasing function on $[a, b]$.

2. To prove (b), assume that $D^+f \geq 0$ a.e. on $[a, b)$ and $D^+f > -\infty$ everywhere on $[a, b)$. Let $\alpha, \beta \in [a, b]$, $\alpha < \beta$, and $E = \{[\alpha, \beta] : D^+f < 0\}$. Then by our assumption, $\mu_L(E) = 0$ and $D^+f \geq 0$ on $[\alpha, \beta) \setminus E$. Let $\varepsilon > 0$. By Lemma 13.24, there exists an absolutely continuous increasing function g on $[\alpha, \beta]$ such that $g(\beta) - g(\alpha) \leq \varepsilon$ and

$g' = \infty$ on E. Let $h = f + g$ on $[\alpha, \beta]$. Then h is a real-valued continuous function on $[\alpha, \beta]$. To apply (a) to h on $[\alpha, \beta]$, let us show that $D^+ h \geq 0$ everywhere on $[\alpha, \beta)$. Now for two arbitrary real-valued functions φ_1 and φ_2 on a set D, we have $\sup_D\{\varphi_1 + \varphi_2\} \geq \sup_D \varphi_1 + \inf_D \varphi_2$. Thus we have

$$(3) \qquad (D^+h)(x) = \limsup_{t \downarrow 0} \frac{h(x+t) - h(x)}{t}$$
$$= \lim_{\delta \downarrow 0} \sup_{0<t<\delta} \left\{ \frac{f(x+t) - f(x)}{t} + \frac{g(x+t) - g(x)}{t} \right\}$$
$$\geq (D^+ f)(x) + (D_+ g)(x).$$

Since $D^+ f > -\infty$ everywhere on $[a, b)$ and since $g' = \infty$ and hence $D_+ g = \infty$ on E, (3) implies that $D^+ h = \infty$ on E. On the other hand since g is an increasing function on $[\alpha, \beta]$, we have $t^{-1}\{g(x+t) - g(x)\} \geq 0$ for $t > 0$ so that by the second equality in (3) we have

$$(D^+h)(x) \geq \lim_{\delta \downarrow 0} \sup_{0<t<\delta} \frac{h(x+t) - h(x)}{t} = (D^+ f)(x) \geq 0$$

for $x \in [\alpha, \beta) \setminus E$. Therefore $D^+ h \geq 0$ everywhere on $[\alpha, \beta)$. Thus by (a), h is an increasing function on $[\alpha, \beta]$. Then

$$0 \leq h(\beta) - h(\alpha) = \{f(\beta) - f(\alpha)\} + \{g(\beta) - g(\alpha)\} \leq f(\beta) - f(\alpha) + \varepsilon.$$

Thus $f(\beta) - f(\alpha) \geq -\varepsilon$. By the arbitrariness of $\varepsilon > 0$, we have $f(\beta) \geq f(\alpha)$. Since this holds for every $\alpha, \beta \in [a, b]$ such that $\alpha < \beta$, f is an increasing function on $[a, b]$. ∎

Theorem 13.26. *Let f be a real-valued continuous function on $[a, b]$ such that f' exists a.e. on $[a, b]$ and f' is μ_L-integrable on $[a, b]$. If the upper-right Dini derivate $D^+ f > -\infty$ on $[a, b)$, then we have $\int_{[a,x]} f' \, d\mu_L = f(x) - f(a)$ for every $x \in [a, b]$. In particular, f is absolutely continuous on $[a, b]$.*

Proof. The μ_L-integrability of f' on $[a, b]$ implies that f' is real-valued a.e. on $[a, b]$. Then there exists a null set E in $(\mathbb{R}, \mathfrak{M}_L, \mu_L)$ contained in $[a, b]$ such that f' is a real-valued $\mathfrak{M}_L$-measurable function on $[a, b] \setminus E$ by Proposition 12.4. Let us write $f' = 0$ on E. (This is just a matter of convenience in notation. Then f' need not be the derivative of f at $x \in E$.) Let us define two sequences of real-valued $\mathfrak{M}_L$-measurable functions $(g_n : n \in \mathbb{N})$ and $(h_n : n \in \mathbb{N})$ on $[a, b]$ by

$$(1) \qquad \begin{cases} g_n(x) = \min\{f'(x), n\} & \text{for } x \in [a, b], \\ h_n(x) = \max\{f'(x), -n\} & \text{for } x \in [a, b]. \end{cases}$$

Then we have

$$(2) \qquad \begin{cases} g_n \leq f', \quad |g_n| \leq |f'|, \quad \lim_{n \to \infty} g_n = f' \quad \text{on } [a, b], \\ h_n \geq f', \quad |h_n| \leq |f'|, \quad \lim_{n \to \infty} h_n = f' \quad \text{on } [a, b]. \end{cases}$$

The μ_L-integrability of f' on $[a, b]$ implies the μ_L-integrability of g_n and h_n on $[a, b]$. Now f' is μ_L-integrable on $[a, b]$ and hence μ-integrable on $[a, x]$ for every $x \in [a, b]$. Then by Theorem 9.20 (Dominated Convergence Theorem) we have

(3)
$$\begin{cases} \lim_{n\to\infty} \int_{[a,x]} g_n \, d\mu_L = \int_{[a,x]} f' \, d\mu_L & \text{for } x \in [a, b], \\ \lim_{n\to\infty} \int_{[a,x]} h_n \, d\mu_L = \int_{[a,x]} f' \, d\mu_L & \text{for } x \in [a, b]. \end{cases}$$

Let G_n and H_n be indefinite integrals of g_n and h_n respectively defined by

(4)
$$\begin{cases} G_n(x) = \int_{[a,x]} g_n \, d\mu_L & \text{for } x \in [a, b], \\ H_n(x) = \int_{[a,x]} h_n \, d\mu_L & \text{for } x \in [a, b]. \end{cases}$$

Let us show that for every $n \in \mathbb{N}$, the two real-valued continuous functions $f - G_n$ and $H_n - f$ on $[a, b]$ are increasing functions by showing that they satisfy the hypothesis in (b) of Lemma 13.25, that is, $D^+(f - G_n) \geq 0$ a.e. on $[a, b)$ and $D^+(f - G_n) > -\infty$ everywhere on $[a, b)$, and similarly for $D^+(H_n - f)$. Now for any two real-valued functions φ_1 and φ_2 on a set D, we have $\sup_D\{\varphi_1 - \varphi_2\} \geq \sup_D \varphi_1 - \sup_D \varphi_2$. Thus for every $x \in [a, b)$ we have

(5)
$$D^+(f - G_n)(x) \geq (D^+ f)(x) - (D^+ G_n)(x).$$

By (4) and (1) we have

(6)
$$\begin{aligned} (D^+ G_n)(x) &= \limsup_{t\downarrow 0} \frac{G_n(x+t) - G_n(x)}{t} \\ &= \limsup_{t\downarrow 0} \frac{1}{t} \int_{[x,x+t]} g_n \, d\mu_L \\ &\leq \frac{1}{t} nt = n. \end{aligned}$$

By (5) and (6) and by the fact that $D^+ f > -\infty$ on $[a, b)$, we have $D^+(f - G_n)(x) > -\infty$ for every $x \in [a, b)$. By Theorem 13.15, G_n' exists and $G_n' = g_n$ a.e. on $[a, b]$ and hence $D^+ G_n = g_n$ a.e. on $[a, b)$. Thus by (5) and (1), we have $D^+(f - G_n) \geq f' - g_n \geq 0$ on $[a, b) \setminus E$. This verifies that $f - G_n$ satisfies the hypothesis of (b) of Lemma 13.25 and therefore $f - G_n$ is an increasing function on $[a, b]$. Then for every $x \in [a, b]$, we have $(f - G_n)(x) \geq (f - G_n)(a)$, that is, $f(x) - f(a) \geq \int_{[a,x]} g_n \, d\mu_L$. Letting $n \to \infty$ and recalling (3), we have

(7)
$$f(x) - f(a) \geq \int_{[a,x]} f' \, d\mu_L \quad \text{for } x \in [a, b].$$

Similarly for every $x \in [a, b)$ we have

(8)
$$D^+(H_n - f)(x) \geq (D^+ H_n)(x) - (D^+ f)(x)$$

and by (1) we have

$$(9) \qquad (D^+H_n)(x) = \limsup_{t \downarrow 0} \frac{H_n(x+t) - H_n(x)}{t}$$
$$= \limsup_{t \downarrow 0} \frac{1}{t} \int_{[x,x+t]} h_n \, d\mu_L$$
$$\geq \frac{1}{t}(-n)t = -n.$$

From (8) and (9) and the fact that $D^+f > -\infty$ on $[a, b)$, we have $D^+(H_n - f)(x) > -\infty$ for all $x \in [a, b)$ and $D^+(H_n - f) \geq h_n - f' \geq 0$ on $[a, b) \setminus E$ and thus by (b) of Lemma 13.25, $H_n - f$ is an increasing function on $[a, b]$. From this we derive by the same argument as above that

$$(10) \qquad f(x) - f(a) \leq \int_{[a,x]} f' \, d\mu_L \quad \text{for } x \in [a, b].$$

With (7) and (10), we have $f(x) - f(a) = \int_{[a,x]} f' \, d\mu_L$ for $x \in [a, b]$. Now that f is an indefinite integral of the μ_L-integrable function f', f is absolutely continuous on $[a, b]$ by Theorem 13.15. ∎

Corollary 13.27. *Let f be a real-valued continuous and increasing function on $[a, b]$. If $D^+f > -\infty$ on $[a, b)$, then f is absolutely continuous on $[a, b]$.*

Proof. If f is a real-valued increasing function on $[a, b]$, then by Theorem 12.10, f' exists a.e. on $[a, b]$ and $0 \leq \int_{[a,b]} f' \, d\mu_L \leq f(b) - f(a) < \infty$ so that f' is μ_L-integrable on $[a, b]$. If we assume that $D^+f > -\infty$ on $[a, b)$, then the conditions in Theorem 13.26 are all satisfied and thus f is absolutely continuous on $[a, b]$. ∎

Let g be a real-valued continuous function on $[a, b]$. If f is also a real-valued continuous function on $[a, b]$ and if $f = g$ a.e. on $[a, b]$, then $f = g$ on $[a, b]$. (Indeed the set $E = \{[a, b] : f \neq g\}$ is a null set in $(\mathbb{R}, \mathfrak{M}_L, \mu_L)$ and thus E^c is a dense subset of $\mathbb{R}$ by Observation 3.6. Then for every $x \in [a, b]$, there exists a sequence $(x_n : n \in \mathbb{N})$ in $[a, b] \cap E^c$ such that $\lim_{n\to\infty} x_n = x$. By the continuity of f and g, we have $\lim_{n\to\infty} f(x_n) = f(x)$ and $\lim_{n\to\infty} g(x_n) = g(x)$. Since $x_n \in [a, b] \cap E^c$, we have $f(x_n) = g(x_n)$ for every $n \in \mathbb{N}$. Then $f(x) = g(x)$.)

Let g be a real-valued continuous function on $[a, b]$. Let h be a real-valued function on $[a, b]$ such that $h = g$ a.e. on $[a, b]$. We show in the next Corollary that if h is the derivative of some differentiable continuous function on $[a, b]$, then $h = g$ on $[a, b]$.

Corollary 13.28. *Let h and g be two real-valued function h on $[a, b]$. Suppose*
1° *$h = g$ a.e. on $[a, b]$,*
2° *$h = f'$ for some differentiable continuous function f on $[a, b]$,*
3° *g is continuous on $[a, b]$.*
Then $h = g$ on $[a, b]$.

Proof. Condition 3° implies that g is μ_L-integrable on $[a, b]$. Then by 1°, h is μ_L-integrable on $[a, b]$. By 2°, the continuous function f is differentiable on $[a, b]$ and its derivative h is μ_L-integrable on $[a, b]$. Since $f' = h$ we have $D^+ f = f' = h > -\infty$ on $[a, b)$. Thus f satisfies all the conditions in Theorem 13.26 so that by Theorem 13.26 and 1° we have

$$f(x) - f(a) = \int_{[a,x]} h \, d\mu_L = \int_{[a,x]} g \, d\mu_L \quad \text{for every } x \in [a, b].$$

This shows that f is an indefinite integral of g. Then 3° implies that $f' = g$ on $[a, b]$ by Lemma 13.14. Therefore $h = g$ on $[a, b]$ by 2° . ∎

Example. Let h be a real-valued function on $[-1, 1]$ defined by $h(x) = 1$ for $x = 0$ and $h(x) = 0$ for $x \in [-1, 1] \setminus \{0\}$, and let $g = 0$ on $[-1, 1]$. We have $h = g$ a.e. on $[-1, 1]$. If h were the derivative of some real-valued continuous function on $[-1, 1]$, then by Corollary 13.28, h would be equal to g on $[-1, 1]$ contradicting the fact that $h(0) \neq g(0)$. Therefore h is not the derivative of any function on $[-1, 1]$.

[V.2] Integration by Parts

If two real-valued functions f and g are differentiable on $[a, b]$ and the derivatives f' and g' are Riemann integrable on $[a, b]$, then fg' and $f'g$ are Riemann integrable on $[a, b]$ and $\int_a^b f(x)g'(x)\,dx + \int_a^b f'(x)g(x)\,dx = f(x)g(x) - f(a)g(a)$. This is the Integration by Parts Formula for the Riemann integral. A corresponding formula for the Lebesgue integral may be stated as follows.

Theorem 13.29. *Let f and g be two real-valued absolutely continuous functions on $[a, b]$. Then fg' and $f'g$ are μ_L-integrable on $[a, b]$ and for every $x \in [a, b]$ we have*

$$\int_{[a,x]} fg' \, d\mu_L + \int_{[a,x]} f'g \, d\mu_L = f(x)g(x) - f(a)g(a).$$

Proof. The absolute continuity of f and g implies that of fg. Thus by Theorem 13.17, the derivatives f', g' and $(fg)'$ exist a.e. on $[a, b]$ and are $\mathfrak{M}_L$-measurable and μ_L-integrable on $[a, b]$, and

$$(1) \qquad \int_{[a,x]} (fg)' \, d\mu_L = f(x)g(x) - f(a)g(a)$$

for $x \in [a, b]$. Now $(fg)' = fg' + f'g$ wherever both f' and g' exist, that is, a.e. on $[a, b]$. Then for every $x \in [a, b]$ we have

$$(2) \qquad \int_{[a,x]} (fg)' \, d\mu_L = \int_{[a,x]} \{fg' + f'g\} \, d\mu_L.$$

Since g' is μ_L-integrable on $[a, b]$ and since f is absolutely continuous on $[a, b]$ and is hence bounded on $[a, b]$, fg' is μ_L-integrable on $[a, b]$. Similarly for $f'g$. Thus for every

$x \in [a, b]$ we have by Proposition 9.14

(3) $$\int_{[a,x]} \{fg' + f'g\}\, d\mu_L = \int_{[a,x]} fg'\, d\mu_L + \int_{[a,x]} f'g\, d\mu_L.$$

With (1), (2), and (3), we are done. ∎

Corollary 13.30. *Let f and g be two real-valued absolutely continuous functions on $[a, b]$ and let F and G be arbitrary indefinite integrals of f and g on $[a, b]$ respectively. Then Fg and fG are μ_L-integrable on $[a, b]$ and for every $x \in [a, b]$ we have*

$$\int_{[a,x]} Fg\, d\mu_L + \int_{[a,x]} fG\, d\mu_L = F(x)G(x) - F(a)G(a).$$

Proof. If f and g are absolutely continuous functions on $[a, b]$, then f and g are μ_L-integrable on $[a, b]$. If F and G are arbitrary indefinite integrals of f and g respectively on $[a, b]$, then F and G are absolutely continuous on $[a, b]$ and furthermore $F' = f$ and $G' = g$ a.e. on $[a, b]$ by Theorem 13.15. Applying Theorem 13.29 to the absolutely continuous functions F and G, we have the Corollary. ∎

[V.3] Change of Variables

A standard Change of Variable Theorem for the Riemann integral states that if f is a real-valued continuous function on $[\alpha, \beta]$ and if g is a real-valued increasing function with continuous derivative g' on $[a, b]$ and $g(a) = \alpha$ and $g(b) = \beta$, then we have the equality $\int_\alpha^\beta f(y)\, dy = \int_a^b (f \circ g)(x) g'(x)\, dx$. Note that the continuity of g' on $[a, b]$ implies that g is absolutely continuous on $[a, b]$. In Theorem 13.32 below, we prove a corresponding Change of Variable Theorem for the Lebesgue integral which does not require the continuity of g'.

Lemma 13.31. *Let g be an absolutely continuous increasing function on $[a, b]$ which is not constant on $[a, b]$ and let $g([a, b]) = [\alpha, \beta]$. Let $D = \{[a, b] : g' > 0\}$.*

(a) *If $G \subset [\alpha, \beta]$ is a G_δ-set, then $\mu_L(G) = \int_{g^{-1}(G)} g'\, d\mu_L$.*

(b) *If $E \subset [\alpha, \beta]$, $E \in \mathfrak{M}_L$, $\mu_L(E) = 0$, then $g^{-1}(E) \cap D \in \mathfrak{M}_L$, $\mu_L\big(g^{-1}(E) \cap D\big) = 0$.*

(c) *If $E \subset [\alpha, \beta]$ and $E \in \mathfrak{M}_L$, then $g^{-1}(E) \cap D \in \mathfrak{M}_L$, $\mathbf{1}_{g^{-1}(E)} g'$ is $\mathfrak{M}_L$-measurable on $[a, b]$ and $\mu_L(E) = \int_{[a,b]} \mathbf{1}_{g^{-1}(E)} g'\, d\mu_L$.*

(Let us note that since $[a, b]$ is a compact set and g is a continuous mapping, $g([a, b])$ is a compact set. The continuity of g also implies that $g([a, b])$ is an interval by the Intermediate Value Theorem. Thus $g([a, b])$ is a finite closed interval which we denote by $[\alpha, \beta]$. Since g is an increasing function, we have $g(a) = \alpha$ and $g(b) = \beta$. The fact that g is an increasing function implies also that the derivative g' exists and $g' \geq 0$ a.e. on $[a, b]$, g' is $\mathfrak{M}_L$-measurable and μ_L-integrable on $[a, b]$ by Theorem 12.10. The absolute continuity of g implies that $\int_{[a,b]} g'\, d\mu_L = g(b) - g(a) = \beta - \alpha$ by Theorem 13.17. The

$\mathfrak{M}_L$-measurability of g implies that $g^{-1}(E) \in \mathfrak{M}_L$ for $E \in \mathfrak{B}_{\mathbb{R}}$, but for $E \in \mathfrak{M}_L$ we may not have $g^{-1}(E) \in \mathfrak{M}_L$ and $\mathbf{1}_{g^{-1}(E)}$ may not be a $\mathfrak{M}_L$-measurable function. Part (c) of the Lemma asserts that $\mathbf{1}_{g^{-1}(E)} g'$ is a $\mathfrak{M}_L$-measurable function.)

Proof. **1.** To prove (a), as a particular case of a G_δ-set consider first an open interval contained in $[\alpha, \beta]$, say $(\alpha_0, \beta_0) \subset [\alpha, \beta]$. Since g is a continuous mapping, $g^{-1}\big((\alpha_0, \beta_0)\big)$ is an open set contained in $[a, b]$. The monotonicity of g implies that the open set $g^{-1}\big((\alpha_0, \beta_0)\big)$ is an open interval, rather than a union of two or more disjoint open intervals. The continuity of g implies that $g^{-1}\big(\{\alpha_0\}\big)$ is a closed set contained in $[a, b]$ and similarly $g^{-1}\big(\{\beta_0\}\big)$ is a closed set contained in $[a, b]$. Let $a_0 = \max g^{-1}\big(\{\alpha_0\}\big)$ and $b_0 = \min g^{-1}\big(\{\beta_0\}\big)$. By the continuity of g we have $g(a_0) = \alpha_0$ and $g(b_0) = \beta_0$ and thus $g^{-1}\big((\alpha_0, \beta_0)\big) = (a_0, b_0)$. Now

$$
(1) \qquad \mu_L\big((\alpha_0, \beta_0)\big) = \beta_0 - \alpha_0 = g(b_0) - g(a_0) = \int_{[a_0, b_0]} g' \, d\mu_L
$$
$$
= \int_{(a_0, b_0)} g' \, d\mu_L = \int_{g^{-1}((\alpha_0, \beta_0))} g' \, d\mu_L
$$

where the third equality is by Theorem 13.17.

Next, let O be an open set contained in $[\alpha, \beta]$. Then $O = \bigcup_{n \in \mathbb{N}} (\alpha_n, \beta_n)$ where $\big((\alpha_n, \beta_n) : n \in \mathbb{N}\big)$ is a disjoint sequence of open intervals contained in $[\alpha, \beta]$. Then

$$
(2) \quad \mu_L(O) = \sum_{n \in \mathbb{N}} \mu_L\big((\alpha_n, \beta_n)\big) = \sum_{n \in \mathbb{N}} \int_{g^{-1}((\alpha_n, \beta_n))} g' \, d\mu_L = \int_{\bigcup_{n \in \mathbb{N}} g^{-1}((\alpha_n, \beta_n))} g' \, d\mu_L
$$
$$
= \int_{g^{-1}(\bigcup_{n \in \mathbb{N}} (\alpha_n, \beta_n))} g' \, d\mu_L = \int_{g^{-1}(O)} g' \, d\mu_L
$$

where the second equality is by (1) and the third equality is by the disjointness of the sequence $\big(g^{-1}\big((\alpha_n, \beta_n)\big) : n \in \mathbb{N}\big)$ implied by the disjointness of $\big((\alpha_n, \beta_n) : n \in \mathbb{N}\big)$.

If G is a G_δ-set, then $G = \bigcap_{n \in \mathbb{N}} O_n$ where $(O_n : n \in \mathbb{N})$ is a sequence of open sets. If $G \subset [\alpha, \beta]$, then O_n may not be contained in $[\alpha, \beta]$ but we can assume $O_n \subset (\alpha - 1, \beta + 1)$ by taking the intersection of O_n with $(\alpha - 1, \beta + 1)$ if necessary. To be able to apply (2) to the open sets $O_n \subset (\alpha - 1, \beta + 1)$, we extend g to an absolutely continuous increasing function g_1 on $[a - 1, b + 1]$ with $g_1\big([a - 1, b + 1]\big) = [\alpha - 1, \beta + 1]$. Let g_1 be defined by setting $g_1 = g$ on $[a, b]$, $g_1(x) = x + \alpha - a$ for $x \in [a - 1, a]$, and $g_1(x) = x + \beta - b$ for $x \in [b, b + 1]$. The function g_1 thus defined is absolutely continuous and increasing on $[a-1, b+1]$ with $g_1(a-1) = \alpha - 1$, $g_1(b+1) = \beta + 1$, and $g_1\big([a-1, b+1]\big) = [\alpha - 1, \beta + 1]$. Now let $V_n = \bigcap_{k=1}^n O_k$ for $n \in \mathbb{N}$. Then $(V_n : n \in \mathbb{N})$ is a decreasing sequence of open sets contained in $(\alpha - 1, \beta + 1)$ with $\bigcap_{n \in \mathbb{N}} V_n = G$. Since $\mu_L\big((\alpha - 1, \beta + 1)\big) < \infty$, we have $\mu_L(G) = \lim_{n \to \infty} \mu_L(V_n)$. Applying (2) to our absolutely continuous increasing function g_1 on $[a - 1, b + 1]$, we have $\mu_L(V_n) = \int_{g_1^{-1}(V_n)} g_1' \, d\mu_L$ for $n \in \mathbb{N}$. Now since $(V_n : n \in \mathbb{N})$ is a decreasing sequence, so is $\big(g_1^{-1}(V_n) : n \in \mathbb{N}\big)$. Thus we have

$$
\lim_{n \to \infty} g_1^{-1}(V_n) = \bigcap_{n \in \mathbb{N}} g_1^{-1}(V_n) = g_1^{-1}\Big(\bigcap_{n \in \mathbb{N}} V_n\Big) = g_1^{-1}(G).
$$

This implies $\lim_{n\to\infty} \mathbf{1}_{g_1^{-1}(V_n)}(x) = \mathbf{1}_{g_1^{-1}(G)}(x)$ for $x \in [a-1, b+1]$. Now since we have $0 \le \mathbf{1}_{g_1^{-1}(V_n)} g_1' \le g_1'$ and g_1' is μ_L-integrable on $[a-1, b+1]$, we can apply the Dominated Convergence Theorem (Theorem 9.20) to have

$$
\begin{aligned}
(3) \qquad \mu_L(G) &= \lim_{n\to\infty} \mu_L(V_n) = \lim_{n\to\infty} \int_{g_1^{-1}(V_n)} g_1' \, d\mu_L \\
&= \lim_{n\to\infty} \int_{[a-1,b+1]} \mathbf{1}_{g_1^{-1}(V_n)} g_1' \, d\mu_L = \int_{[a-1,b+1]} \mathbf{1}_{g_1^{-1}(G)} g_1' \, d\mu_L \\
&= \int_{g_1^{-1}(G)} g_1' \, d\mu_L = \int_{g^{-1}(G)} g' \, d\mu_L .
\end{aligned}
$$

This proves (a).

2. Let $E \subset [\alpha, \beta]$, $E \in \mathfrak{M}_L$, and $\mu_L(E) = 0$. By Theorem 3.22, there exists a G_δ-set $G \supset E$ with $\mu_L(G \setminus E) = 0$. Intersecting with $(\alpha - 1, \beta + 1)$ if necessary, we may take $G \subset (\alpha - 1, \beta + 1)$. Now $\mu_L(G \setminus E) = 0$ implies $\mu_L(G) = \mu_L(E) + \mu_L(G \setminus E) = 0$. Then by (a), we have $\int_{g_1^{-1}(G)} g_1' \, d\mu_L = \mu_L(G) = 0$. Let $D_1 = \{[a-1, b+1] : g_1' > 0\}$. Since $g_1' \ge 0$, we have $\int_{g_1^{-1}(G)\cap D_1} g_1' \, d\mu_L \le \int_{g_1^{-1}(G)} g_1' \, d\mu_L$ so that $\int_{g_1^{-1}(G)\cap D_1} g_1' \, d\mu_L = 0$. Since $g_1' > 0$ on $g_1^{-1}(G) \cap D_1$, this implies $\mu_L\big(g_1^{-1}(G) \cap D_1\big) = 0$. Now $E \subset G$ implies $g^{-1}(E) \cap D \subset g_1^{-1}(G) \cap D_1$. Therefore $g^{-1}(E) \cap D$ is a subset of a null set $g_1^{-1}(G) \cap D_1$ in the complete measure space $(\mathbb{R}, \mathfrak{M}_L, \mu_L)$. Then $g^{-1}(E) \cap D \in \mathfrak{M}_L$ with $\mu_L(g^{-1}(E) \cap D) = 0$. This proves (b).

3. Let $E \subset [\alpha, \beta]$ and $E \in \mathfrak{M}_L$. As in **2** above, there exists a G_δ-set G such that $E \subset G \subset (\alpha - 1, \beta + 1)$ with $\mu_L(G \setminus E) = 0$. Then

$$
(4) \qquad \mu_L(E) = \mu_L(G) = \int_{g_1^{-1}(G)} g_1' \, d\mu_L = \int_{g_1^{-1}(G)\cap D_1} g_1' \, d\mu_L
$$

where the second equality is from (3) and the last equality follows from the definition of the set D_1. Since $G \setminus E \subset [\alpha - 1, \beta + 1]$, $G \setminus E \in \mathfrak{M}_L$, and $\mu_L(G \setminus E) = 0$, applying (b) to the absolutely continuous increasing function g_1 on $[a-1, b+1]$ and the null set $G \setminus E$, we have $g_1^{-1}(G \setminus E) \cap D_1 \in \mathfrak{M}_L$ and $\mu_L\big(g_1^{-1}(G \setminus E) \cap D_1\big) = 0$. Since $E = G \setminus (G \setminus E)$, we have $g_1^{-1}(E) = g_1^{-1}(G) \setminus g_1^{-1}(G \setminus E)$ and thus

$$
(5) \qquad g_1^{-1}(E) \cap D_1 = \big[g_1^{-1}(G) \cap D_1\big] \setminus \big[g_1^{-1}(G \setminus E) \cap D_1\big].
$$

Since $G \in \mathfrak{B}_{\mathbb{R}}$, we have $g_1^{-1}(G) \in \mathfrak{M}_L$ by the $\mathfrak{M}_L$-measurability of g_1. Thus by (5), we have $g_1^{-1}(E) \cap D_1 \in \mathfrak{M}_L$. Since $\mu_L\big(g_1^{-1}(G \setminus E) \cap D_1\big) = 0$, we have

$$
\mu_L\big(g_1^{-1}(E) \cap D_1\big) = \mu_L\big(g_1^{-1}(G) \cap D_1\big).
$$

Since $g_1^{-1}(E) \cap D_1 \subset g_1^{-1}(G) \cap D_1$ and $g_1' \ge 0$, the equality of the measures of the two sets implies $\int_{g_1^{-1}(E)\cap D_1} g_1' \, d\mu_L = \int_{g_1^{-1}(G)\cap D_1} g_1' \, d\mu_L$. Using this in (4), we have

$$
\begin{aligned}
\mu_L(E) &= \int_{g_1^{-1}(E)\cap D_1} g_1' \, d\mu_L = \int_{g^{-1}(E)\cap D} g' \, d\mu_L \\
&= \int_{[a,b]} \mathbf{1}_{g^{-1}(E)} \mathbf{1}_D g' \, d\mu_L = \int_{[a,b]} \mathbf{1}_{g^{-1}(E)} g' \, d\mu_L . \quad \blacksquare
\end{aligned}
$$

Theorem 13.32. *Let f be an extended real-valued $\mathfrak{M}_L$-measurable function on $[\alpha, \beta]$. If g is an absolutely continuous increasing function on $[a, b]$ with $g(a) = \alpha$ and $g(b) = \beta$, then*

$$(1) \qquad \int_{[\alpha,\beta]} f(y)\,\mu_L(dy) = \int_{[a,b]} (f \circ g)(x) g'(x)\,\mu_L(dx)$$

in the sense that the existence of one of the two integrals implies that of the other and the equality of the two.

Proof. Consider the case $f = \mathbf{1}_E$ where E is a $\mathfrak{M}_L$-measurable subset of $[\alpha, \beta]$. Then by (c) of Lemma 13.31, we have

$$\begin{aligned}\int_{[\alpha,\beta]} f\,d\mu_L &= \int_{[\alpha,\beta]} \mathbf{1}_E\,d\mu_L = \mu_L(E) = \int_{[a,b]} \mathbf{1}_{g^{-1}(E)} g'\,d\mu_L \\ &= \int_{[a,b]} \mathbf{1}_E\big(g(x)\big) g'(x)\,\mu_L(dx) \\ &= \int_{[a,b]} (f \circ g)(x) g'(x)\,\mu_L(dx).\end{aligned}$$

This shows that (1) holds for this particular case. Then (1) holds when f is a nonnegative simple function on $[\alpha, \beta]$ by our result above and by the linearity of the integrals with respect to the integrands. If f is a nonnegative extended real-valued $\mathfrak{M}_L$-measurable function on $[\alpha, \beta]$, then by Lemma 8.6, there exists an increasing sequence of nonnegative simple functions $(\varphi_n : n \in \mathbb{N})$ such that $\varphi_n \uparrow f$ on $[\alpha, \beta]$. Since g is absolutely continuous and increasing on $[a, b]$, g' exists and $g' \geq 0$ a.e. on $[a, b]$. Define $g' = 0$ on the exceptional null set so that $g' \geq 0$ on $[a, b]$. Then we have $(\varphi_n \circ g) g' \uparrow (f \circ g) g'$ on $[\alpha, \beta]$. By Theorem 8.5 (Monotone Convergence Theorem), we have $\lim_{n\to\infty} \int_{[\alpha,\beta]} \varphi_n\,d\mu_L = \int_{[\alpha,\beta]} f\,d\mu_L$ and $\lim_{n\to\infty} \int_{[a,b]} (\varphi_n \circ g) g'\,d\mu_L = \int_{[a,b]} (f \circ g) g'\,d\mu_L$. Since φ_n is a nonnegative simple function, we have $\int_{[\alpha,\beta]} \varphi_n\,d\mu_L = \int_{[a,b]} (\varphi_n \circ g) g'\,d\mu_L$ for every $n \in \mathbb{N}$. Thus we have $\int_{[\alpha,\beta]} f\,d\mu_L = \int_{[a,b]} (f \circ g) g'\,d\mu_L$. This shows that (1) holds when f is a nonnegative extended real-valued $\mathfrak{M}_L$-measurable function on $[\alpha, \beta]$. When f is an extended real-valued $\mathfrak{M}_L$-measurable function on $[\alpha, \beta]$, we decompose f as $f = f^+ - f^-$ and apply the result above to f^+ and f^-. ∎

[VI] Length of Rectifiable Curves

Definition 13.33. *Let $f_1, \ldots, f_m$ be real-valued functions on $[a, b]$. Consider a mapping x of $[a, b]$ into $\mathbb{R}^m$ defined by $x(t) = \big(f_1(t), \ldots, f_m(t)\big) \in \mathbb{R}^m$ for $t \in [a, b]$. We call the subset $C = \{x(t) : t \in [a, b]\}$ of $\mathbb{R}^m$ a curve in $\mathbb{R}^m$ defined by the functions $f_1, \ldots, f_m$.*

Definition 13.34. *Let C be a curve in $\mathbb{R}^m$ defined by m real-valued functions $f_1, \ldots, f_m$ on $[a, b]$. Let $\mathcal{P} = \{a = t_0 \leq \cdots \leq t_n = b\}$ be a partition of $[a, b]$ and let $\mathfrak{P}_{a,b}$ be the collection of all partitions of $[a, b]$. Consider the points $x(t_0), \ldots, x(t_n) \in C$ and the line segments $\overline{x(t_0), x(t_1)}, \ldots, \overline{x(t_{n-1}), x(t_n)}$ in $\mathbb{R}^m$. Let*

$$(1) \qquad P(\mathcal{P}) = \overline{x(t_0), x(t_1)} \cup \cdots \cup \overline{x(t_{n-1}), x(t_n)}.$$

We call $P(\mathcal{P})$ the polygonal line inscribed to the curve C by the partition $\mathcal{P}$ of $[a, b]$. The length of the polygonal line $P(\mathcal{P})$, written $\ell(P(\mathcal{P}))$, is given by

$$(2) \qquad \ell(P(\mathcal{P})) = \sum_{k=1}^{n} \ell\big(\overline{x(t_{k-1}), x(t_k)}\big) = \sum_{k=1}^{n} \sqrt{\sum_{i=1}^{m} |f_i(t_k) - f_i(t_{k-1})|^2}.$$

We define the length of the curve C by setting

$$(3) \qquad \ell(C) = \sup_{\mathcal{P} \in \mathfrak{P}_{a,b}} \ell(P(\mathcal{P})).$$

We say that the curve C is rectifiable if $\ell(C) < \infty$.

Observation 13.35. For $a_1, \ldots, a_m \in \mathbb{R}$, we have

$$\max_{i=1,\ldots,m} |a_i| \le \sqrt{\sum_{i=1}^{m} |a_i|^2} \le \sum_{i=1}^{m} |a_i|.$$

Proof. For each $i = 1, \ldots, m$, we have $|a_i|^2 \le \sum_{j=1}^{m} |a_j|^2$. Thus $|a_i| \le \sqrt{\sum_{j=1}^{m} |a_j|^2}$ for $i = 1, \ldots, m$ so that $\max_{i=1,\ldots,m} |a_i| \le \sqrt{\sum_{i=1}^{m} |a_i|^2}$. On the other hand we have $\sum_{i=1}^{m} |a_i|^2 \le \sum_{i=1}^{m} \sum_{j=1}^{m} |a_i||a_j| = \big\{\sum_{i=1}^{m} |a_i|\big\}^2$ so that $\sqrt{\sum_{i=1}^{m} |a_i|^2} \le \sum_{i=1}^{m} |a_i|$. ∎

Lemma 13.36. *Let C be a curve in $\mathbb{R}^m$ defined by m real-valued functions $f_1, \ldots, f_m$ on $[a, b]$. Then for the length $\ell(P(\mathcal{P}))$ of the polygonal line $P(\mathcal{P})$ inscribed to C by a partition $\mathcal{P}$ of $[a, b]$ we have*

$$\max_{i=1,\ldots,m} V_a^b(f_i, \mathcal{P}) \le \ell(P(\mathcal{P})) \le \sum_{i=1}^{m} V_a^b(f_i, \mathcal{P})$$

where $V_a^b(f_i, \mathcal{P})$ is the variation of the function f_i corresponding to the partition $\mathcal{P}$ of $[a, b]$.

Proof. Let a partition of $[a, b]$ be given by $\mathcal{P} = \{a = t_0 \le \cdots \le t_n = b\}$. Let

$$\Delta_k^i = f_i(t_k) - f(t_{k-1}) \quad \text{for } k = 1, \ldots, n; i = 1, \ldots, m.$$

Then by (2) of Definition 13.35, we have

$$(1) \qquad \ell(P(\mathcal{P})) = \sum_{k=1}^{n} \sqrt{\sum_{i=1}^{m} |\Delta_k^i|^2}.$$

By Observation 13.35 we have for each fixed $k = 1, \ldots, n$

$$\max_{i=1,\ldots,m} |\Delta_k^i| \le \sqrt{\sum_{i=1}^{m} |\Delta_k^i|^2} \le \sum_{i=1}^{m} |\Delta_k^i|$$

and then summing over $k = 1, \ldots, n$ and recalling (1) we have

$$(2) \qquad \sum_{k=1}^{n} \max_{i=1,\ldots,m} |\Delta_k^i| \leq \ell\big(P(\mathcal{P})\big) \leq \sum_{k=1}^{n} \Big\{ \sum_{i=1}^{m} |\Delta_k^i| \Big\}.$$

Now for each fixed $i = 1, \ldots, m$, we have

$$V_a^b\big(f_i, \mathcal{P}\big) = \sum_{k=1}^{n} |f_i(t_k) - f(t_{k-1})| = \sum_{k=1}^{n} |\Delta_k^i| \leq \sum_{k=1}^{n} \max_{i=1,\ldots,m} |\Delta_k^i| \leq \ell\big(P(\mathcal{P})\big)$$

so that we have

$$(3) \qquad \max_{i=1,\ldots,m} V_a^b\big(f_i, \mathcal{P}\big) \leq \ell\big(P(\mathcal{P})\big).$$

On the other hand, from (2) we have

$$(4) \qquad \ell\big(P(\mathcal{P})\big) \leq \sum_{k=1}^{n} \Big\{ \sum_{i=1}^{m} |\Delta_k^i| \Big\} = \sum_{i=1}^{m} \sum_{k=1}^{n} \{ |f_i(t_k) - f(t_{k-1})| \} = \sum_{i=1}^{m} V_a^b\big(f_i, \mathcal{P}\big).$$

With (3) and (4) the proof is complete. ∎

Theorem 13.37. *Let C be a curve in $\mathbb{R}^m$ defined by m real-valued functions $f_1, \ldots, f_m$ on $[a, b]$. Then C is rectifiable, that is $\ell(C) < \infty$, if and only if $f_1, \ldots, f_m \in BV\big([a, b]\big)$.*

Proof. By Lemma 13.36, for any partition $\mathcal{P}$ of $[a, b]$ we have

$$\max_{i=1,\ldots,m} V_a^b\big(f_i, \mathcal{P}\big) \leq \ell\big(P(\mathcal{P})\big) \leq \sum_{i=1}^{m} V_a^b\big(f_i, \mathcal{P}\big).$$

Then

$$(1) \qquad \sup_{\mathcal{P} \in \mathfrak{P}_{a,b}} \max_{i=1,\ldots,m} V_a^b\big(f_i, \mathcal{P}\big) \leq \sup_{\mathcal{P} \in \mathfrak{P}_{a,b}} \ell\big(P(\mathcal{P})\big) \leq \sup_{\mathcal{P} \in \mathfrak{P}_{a,b}} \sum_{i=1}^{m} V_a^b\big(f_i, \mathcal{P}\big).$$

Now $\sum_{i=1}^{m} V_a^b\big(f_i, \mathcal{P}\big) \leq \sum_{i=1}^{m} \sup_{\mathcal{P} \in \mathfrak{P}_{a,b}} V_a^b\big(f_i, \mathcal{P}\big)$ so that
(2)

$$\sup_{\mathcal{P} \in \mathfrak{P}_{a,b}} \sum_{i=1}^{m} V_a^b\big(f_i, \mathcal{P}\big) \leq \sup_{\mathcal{P} \in \mathfrak{P}_{a,b}} \sum_{i=1}^{m} \sup_{\mathcal{P} \in \mathfrak{P}_{a,b}} V_a^b\big(f_i, \mathcal{P}\big) = \sum_{i=1}^{m} \sup_{\mathcal{P} \in \mathfrak{P}_{a,b}} V_a^b\big(f_i, \mathcal{P}\big) = \sum_{i=1}^{m} V_a^b\big(f_i\big).$$

By (1) and (2) and recalling that $\ell(C) = \sup_{\mathcal{P} \in \mathfrak{P}_{a,b}} \ell\big(P(\mathcal{P})\big)$ by Definition 13.34, we have

$$(3) \qquad \ell(C) \leq \sum_{i=1}^{m} V_a^b\big(f_i\big).$$

Now if $f_1, \dots f_m \in BV([a,b])$, then $V_a^b(f_i) < \infty$ for $i = 1, \dots, m$ and $\ell(C) < \infty$ by (2). Conversely if $\ell(C) = \sup_{\mathcal{P}\in\mathfrak{P}_{a,b}} \ell(P(\mathcal{P})) < \infty$, then (1) implies

$$\sup_{\mathcal{P}\in\mathfrak{P}_{a,b}} \max_{i=1,\dots,m} V_a^b(f_i, \mathcal{P}) < \infty.$$

This implies that we have $\sup_{\mathcal{P}\in\mathfrak{P}_{a,b}} V_a^b(f_i, \mathcal{P}) < \infty$ for $i = 1, \dots, m$, that is, $V_a^b(f_i) < \infty$ for $i = 1, \dots, m$ so that $f_1, \dots, f_m \in BV([a,b])$. ∎

Let C be a curve in $\mathbb{R}^m$ defined by m real-valued functions $f_1, \dots, f_m$ on $[a,b]$. In Theorem 15.35, we showed that $\ell(C) < \infty$ if and only if $f_1, \dots, f_m \in BV([a,b])$. We show below that if $f_1, \dots, f_m$ are absolutely continuous on $[a,b]$ then we have

$$\ell(C) = \int_{[a,b]} \sqrt{\sum_{i=1}^m (f_i')^2}\, d\mu_L.$$

Proposition 13.38. *Let f be a real-valued function which is differentiable (that is, the derivative f' exists and is finite) $(\mathfrak{M}_L, \mu_L)$-a.e. on $[a,b]$. For $n \in \mathbb{N}$, let*

(1) $$t_{n,k} = a + k\frac{b-a}{2^n} \quad \text{for } k = 0, 1, \dots, 2^n$$

and let

(2) $$J_{n,k} = (t_{n,k-1}, t_{n,k}) \quad \text{for } k = 1, \dots, 2^n.$$

Define a sequence of functions $(g_n : n \in \mathbb{N})$ by setting

(3) $$g_n(t) = \begin{cases} \dfrac{f(t_{n,k}) - f(t_{n,k-1})}{t_{n,k} - t_{n,k-1}} & \text{for } t \in J_{n,k};\ k = 1, \dots, 2^n, \\ 0 & \text{for } t = t_{n,k};\ k = 1, \dots, 2^n. \end{cases}$$

Then

(4) $$\lim_{n\to\infty} g_n = f' \quad \text{a.e. on } [a,b].$$

Proof. According to Proposition 12.2, if f is a real-valued function on (a,b) and differentiable at some $t \in (a,b)$, then for any two sequences $(a_n : n \in \mathbb{N})$ and $(b_n : n \in \mathbb{N})$ in (a,b) such that $a_n \uparrow t$ and $b_n \downarrow t$ we have

(5) $$f'(t) = \lim_{n\to\infty} \frac{f(b_n) - f(a_n)}{b_n - a_n}.$$

Suppose f is differentiable a.e. on $[a,b]$. Then there exists a null set N in the measure space $(\mathbb{R}, \mathfrak{M}_L, \mu_L)$ such that $N \subset [a,b]$ and f is differentiable at every $t \in [a,b] \setminus N$. Let $Q_n = \{t_{n,k} : k = 0, 1, \dots, 2^n\}$ for $n \in \mathbb{N}$ and let $Q = \bigcup_{n\in\mathbb{N}} Q_n$. Then Q is a countable subset of $[a, b]$ and is therefore a null set in $(\mathbb{R}, \mathfrak{M}_L, \mu_L)$. Now $N \cup Q \subset [a, b]$ is a null set in $(\mathbb{R}, \mathfrak{M}_L, \mu_L)$. Consider $[a, b] \setminus \{N \cup Q\}$. To prove (4) we show that $\lim_{n\to\infty} g_n(t) = f'(t)$ for every $t \in [a, b] \setminus \{N \cup Q\}$.

Let $\mathfrak{J}_n = \{J_{n,k} : k = 1, \dots, 2^n\}$ for $n \in \mathbb{N}$. Then $\mathfrak{J}_n$ is a disjoint collection of open intervals for each $n \in \mathbb{N}$. Moreover if $n', n'' \in \mathbb{N}$ and $n' < n''$ and $J' \in \mathfrak{J}_{n'}$ and $J'' \in \mathfrak{J}_{n''}$, then either $J' \cap J'' = \emptyset$ or $J' \supset J''$. Now let $t \in [a, b] \setminus \{N \cup Q\}$. Then for every $n \in \mathbb{N}$ we have $t \in J_{n,k(n)}$ for some $k(n) = 1, \dots, 2^n$. For brevity let us write (a_n, b_n) for $J_{n,k(n)} = (t_{n,k(n)-1}, t_{n,k(n)})$. Now since $t \in (a_n, b_n)$ for every $n \in \mathbb{N}$, for $n', n'' \in \mathbb{N}$ such that $n' < n''$ we have $(a_{n'}, b_{n'}) \cap (a_{n''}, b_{n''}) \neq \emptyset$ and hence $(a_{n'}, b_{n'}) \supset (a_{n''}, b_{n''})$. Thus $((a_n, b_n) : n \in \mathbb{N})$ is a decreasing sequence of open intervals. Then since $b_n - a_n = \frac{b-a}{2^n}$ we have $\lim_{n\to\infty}(b_n - a_n) = 0$. Thus we have $\bigcap_{n\in\mathbb{N}}(a_n, b_n) = \{t\}$ and this implies that $a_n \uparrow t$ and $b_n \downarrow t$. Therefore (5) holds for our sequence $((a_n, b_n) : n \in \mathbb{N})$. But we have

$$(6) \qquad \frac{f(b_n) - f(a_n)}{b_n - a_n} = \frac{f(t_{n,k(n)}) - f(t_{n,k(n)-1})}{t_{n,k(n)} - t_{n,k(n)-1}} = g_n(t) \quad \text{for } t \in J_{n,k(n)}.$$

By (5) and (6), we have $f'(t) = \lim_{n\to\infty} g_n(t)$ for $t \in [a, b] \setminus \{N \cup Q\}$. ∎

Theorem 13.39. *Let C be a curve in $\mathbb{R}^m$ defined by m real-valued functions $f_1, \dots, f_m$ on $[a, b]$. If $f_1, \dots, f_m \in BV([a, b])$, then*

$$\ell(C) \geq \int_{[a,b]} \sqrt{\sum_{i=1}^{m} (f_i')^2}\, d\mu_L.$$

Proof. If $f_i \in BV([a, b])$ then by Theorem 12.21 the derivative f_i' exists $(\mathfrak{M}_L, \mu_L)$-a.e. on $[a, b]$ and f_i' is $\mathfrak{M}_L$-measurable and μ_L-integrable on $[a, b]$. The μ_L-integrability of f_i' implies that f_i' is finite, that is, f_i is differentiable, $(\mathfrak{M}_L, \mu_L)$-a.e. on $[a, b]$.

For $n \in \mathbb{N}$, let

$$(1) \qquad t_{n,k} = a + k\frac{b-a}{2^n} \quad \text{for } k = 0, 1, \dots, 2^n,$$

$$(2) \qquad J_{n,k} = (t_{n,k-1}, t_{n,k}) \quad \text{for } k = 1, \dots, 2^n,$$

$$(3) \qquad \mathcal{P}_n = \{a = t_{n,0} < \cdots < t_{n,2^n} = b\}.$$

For each $i = 1, \dots, m$, define a sequence of functions $(g_{i,n} : n \in \mathbb{N})$ by

$$(4) \qquad g_{i,n}(t) = \begin{cases} \dfrac{f_i(t_{n,k}) - f_i(t_{n,k-1})}{t_{n,k} - t_{n,k-1}} & \text{for } t \in J_{n,k};\ k = 1, \dots, 2^n, \\ 0 & \text{for } t = t_{n,k};\ k = 1, \dots, 2^n. \end{cases}$$

Then by Proposition 13.38, $\lim_{n\to\infty} g_{i,n} = f_i'$, $(\mathfrak{M}_L, \mu_L)$-a.e. on $[a, b]$. Thus we have

$$\lim_{n\to\infty} \sqrt{\sum_{i=1}^{m} (g_{i,n})^2} = \sqrt{\sum_{i=1}^{m} (f_i')^2} \quad (\mathfrak{M}_L, \mu_L)\text{-a.e. on } [a, b].$$

Then by Fatou's Lemma (Theorem 8.13) we have

$$\int_{[a,b]} \sqrt{\sum_{i=1}^{m} (f_i')^2}\, d\mu_L \le \liminf_{n\to\infty} \int_{[a,b]} \sqrt{\sum_{i=1}^{m} (g_{i,n})^2}\, d\mu_L \le \sup_{n\in\mathbb{N}} \int_{[a,b]} \sqrt{\sum_{i=1}^{m} (g_{i,n})^2}\, d\mu_L .$$

Now

$$\begin{aligned}\int_{[a,b]} \sqrt{\sum_{i=1}^{m} (g_{i,n})^2}\, d\mu_L &= \sum_{k=1}^{2^n} \int_{J_{n,k}} \sqrt{\sum_{i=1}^{m} (g_{i,n})^2}\, d\mu_L = \sum_{k=1}^{2^n} \sqrt{\sum_{i=1}^{m} (g_{i,n})^2}(t_{n,k} - t_{n,k_1}) \\ &= \sum_{k=1}^{2^n} \sqrt{\sum_{i=1}^{m} \big| f_i(t_{n,k}) - f_i(t_{n,k-1})\big|^2} = \ell\big(P(\mathcal{P}_n)\big)\end{aligned}$$

where the second equality is by the fact that the function $g_{i,n}$ is constant on the interval $J_{n,k}$ and the third equality is by (4). Thus we have

$$\int_{[a,b]} \sqrt{\sum_{i=1}^{m} (f_i')^2}\, d\mu_L \le \sup_{n\in\mathbb{N}} \ell(P(\mathcal{P}_n)) \le \sup_{\mathcal{P}\in\mathfrak{P}_{a,b}} \ell\big(P(\mathcal{P})\big) = \ell(C). \ \blacksquare$$

Lemma 13.40. *Let $a_1, \dots, a_m \in \mathbb{R}$. Then there exist $\alpha_1, \dots, \alpha_m \in \mathbb{R}$ with $\sum_{i=1}^m \alpha_i^2 = 1$ such that*

$$(1) \qquad \sqrt{\sum_{i=1}^{m} a_i^2} = \sum_{i=1}^{m} \alpha_i a_i .$$

Proof. If $a_1, \dots, a_m$ are all equal to 0, then with arbitrary $\alpha_1, \dots, \alpha_m \in \mathbb{R}$ the equality (1) holds trivially. Suppose not all $a_1, \dots, a_m$ are equal to 0. Then

$$S := \sqrt{\sum_{i=1}^{m} a_i^2} > 0.$$

Let $\alpha_i = \frac{a_i}{S}$ for $i = 1, \dots, m$. Then we have $\sum_{i=1}^m \alpha_i^2 = S^{-2} \sum_{i=1}^m a_i^2 = S^{-2} S^2 = 1$ and

$$\sqrt{\sum_{i=1}^{m} a_i^2} = S = \Big\{\sum_{i=1}^{m} \alpha_i^2\Big\} S = \sum_{i=1}^{m} \alpha_i (\alpha_i S) = \sum_{i=1}^{m} \alpha_i a_i . \ \blacksquare$$

Theorem 13.41. *Let C be a curve in $\mathbb{R}^m$ defined by m real-valued functions $f_1, \dots, f_m$ on $[a, b]$. If $f_1, \dots, f_m$ are absolutely continuous on $[a, b]$, then*

$$\ell(C) = \int_{[a,b]} \sqrt{\sum_{i=1}^{m} (f_i')^2}\, d\mu_L .$$

Proof. If $f_1, \ldots, f_m$ are absolutely continuous on $[a, b]$ then $f_1, \ldots, f_m \in BV([a, b])$ by Theorem 13.3. Then by Theorem 13.39 we have

$$(1) \qquad \ell(C) \geq \int_{[a,b]} \sqrt{\sum_{i=1}^{m} (f_i')^2}\, d\mu_L.$$

Thus it remains to show that

$$(2) \qquad \ell(C) \leq \int_{[a,b]} \sqrt{\sum_{i=1}^{m} (f_i')^2}\, d\mu_L.$$

By Definition 13.34 we have

$$(3) \qquad \ell(C) = \sup_{\mathcal{P} \in \mathfrak{P}_{a,b}} \ell\big(P(\mathcal{P})\big)$$

where $P(\mathcal{P})$ is a polygonal line inscribed to the curve C by a partition $\mathcal{P}$ of $[a, b]$. Let

$$\mathcal{P} = \{a = t_0 < \cdots < t_n = b\},$$

$$I_k = [t_{k-1}, t_k] \quad \text{for } k = 1, \ldots, n,$$

$$\Delta_{i,k} = f_i(t_k) - f_i(t_{k-1}) \quad \text{for } k = 1, \ldots, n;\ i = 1, \ldots, m.$$

Then

$$(4) \qquad \ell\big(P(\mathcal{P})\big) = \sum_{k=1}^{n} \sqrt{\sum_{i=1}^{m} |\Delta_{i,k}|^2} = \sum_{k=1}^{n} S_k$$

where

$$(5) \qquad S_k = \sqrt{\sum_{i=1}^{m} |\Delta_{i,k}|^2} \quad \text{for } k = 1, \ldots, n.$$

By Lemma 13.40 there exist $\alpha_1, \ldots, \alpha_m \in \mathbb{R}$ with $\sum_{i=1}^{m} \alpha_i^2 = 1$ such that

$$S_k = \sum_{i=1}^{m} \alpha_i \Delta_{i,k}.$$

Now since f_i is absolutely continuous on $[a, b]$ we have $\Delta_{i,k} = \int_{I_k} f_i'\, d\mu_L$ by Theorem 13.17. Thus we have

$$S_k = \sum_{i=1}^{m} \alpha_i \Big\{ \int_{I_k} f_i'\, d\mu_L \Big\} = \int_{I_k} \Big\{ \sum_{i=1}^{m} \alpha_i f_i' \Big\}\, d\mu_L.$$

Now for $a_1, \ldots, a_m, b_1, \ldots, b_m \in \mathbb{R}$ we have $|\sum_{i=1}^m a_i b_i| \le \sqrt{\sum_{i=1}^m a_i^2}\sqrt{\sum_{i=1}^m b_i^2}$. (For a proof of this inequality, see for instance Lemma 15.13.) Thus we have

$$\Big|\sum_{i=1}^m \alpha_i f_i'\Big| \le \sqrt{\sum_{i=1}^m \alpha_i^2}\sqrt{\sum_{i=1}^m (f_i')^2} = \sqrt{\sum_{i=1}^m (f_i')^2}.$$

Then

$$(6) \qquad S_k = \int_{I_k}\Big\{\sum_{i=1}^m \alpha_i f_i'\Big\}\,d\mu_L \le \int_{I_k}\Big|\sum_{i=1}^m \alpha_i f_i'\Big|\,d\mu_L \le \int_{I_k}\sqrt{\sum_{i=1}^m (f_i')^2}\,d\mu_L.$$

Substituting (6) in (4) we have

$$(7) \qquad \ell\big(P(\mathcal{P})\big) \le \sum_{k=1}^n \int_{I_k}\sqrt{\sum_{i=1}^m (f_i')^2}\,d\mu_L = \int_{[a,b]}\sqrt{\sum_{i=1}^m (f_i')^2}\,d\mu_L.$$

Since (7) holds for an arbitrary partition $\mathcal{P}$ of $[a, b]$, we have from (3)

$$\ell(C) = \sup_{\mathcal{P}\in\mathfrak{P}_{a,b}} \ell\big(P(\mathcal{P})\big) \le \int_{[a,b]}\sqrt{\sum_{i=1}^m (f_i')^2}\,d\mu_L.$$

This proves (2). ∎

Problems

Prob. 13.1. Let f and g be real-valued absolutely continuous functions on $[a, b]$. Show that the following functions are absolutely continuous on $[a, b]$:
(a) cf where $c \in \mathbb{R}$,
(b) $f + g$,
(c) fg,
(d) $1/f$ provided that $f(x) \ne 0$ for $x \in [a, b]$.

Prob. 13.2. Let f be a real-valued function on $[a, b]$. Let $c \in (a, b)$. Show that if f is absolutely continuous on $[a, c]$ and $[c, b]$ then f is absolutely continuous on $[a, b]$.

Prob. . Let f be a real-valued function on $[a, b]$ satisfying the Lipschitz condition on $[a, b]$, that is, there exists $K > 0$ such that for any $x', x'' \in [a, b]$ we have

$$|f(x') - f(x'')| \le K|x' - x''|.$$

Show that f is absolutely continuous on $[a, b]$.

Prob. 13.4. Show that if f is continuous on $[a, b]$ and f' exists on (a, b) and satisfies $|f'(x)| \le M$ for $x \in (a, b)$ with some $M > 0$, then f satisfies the Lipschitz condition and thus absolutely continuous on $[a, b]$.

Prob. 13.5. Under the same hypothesis on f as in Prob. 13.4, show that for every $E \subset [a,b]$ we have $\mu_L^*(f(E)) \le M\mu_L^*(E)$.

Prob. 13.6. Let f be an absolutely continuous function on $[a,b]$. Let $Z = \{x \in [a,b] : f'(x) = 0\}$. Show that $\mu_L(f(Z)) = 0$.

Prob. 13.7. Find a real-valued function f on $[a,b]$ such that f is continuous on $[a,b]$, absolutely continuous on $[a+\eta, b]$ for every $\eta \in (0, b-a)$, but not absolutely continuous on $[a,b]$.

Prob. 13.8. Let f be a real-valued function on $[a,b]$ such that f is absolutely continuous on $[a+\eta, b]$ for every $\eta \in (0, b-a)$. Show that if f is continuous and of bounded variation on $[a,b]$, then f is absolutely continuous on $[a,b]$.

Prob. 13.9. Let f be a real-valued continuous on $[0,1]$. Suppose $f \in BV\big([0,1]\big)$ and f' exists and $|f'(x)| < \frac{1}{x}$ for $x \in (0,1]$. Show that f is absolutely continuous on $[0,1]$.

Prob. 13.10. Find an absolutely continuous function on $[a,b]$ which does not satisfy the Lipschitz condition on $[a,b]$.

Prob. 13.11. For each $p \in (0,\infty)$, let f_p be a real-valued continuous function on $\left[0, \frac{2}{\pi}\right]$ defined by

$$f_p(x) = \begin{cases} x^p \sin\frac{1}{x} & \text{for } x \in \left(0, \frac{2}{\pi}\right], \\ 0 & \text{for } x = 0. \end{cases}$$

Show that f_p is absolutely continuous on $\left[0, \frac{2}{\pi}\right]$ if and only if $p \in (1,\infty)$.

Prob. 13.12. Let f be an absolutely continuous function on $[a,b]$ and g be an absolutely continuous function on $[c,d]$. Show that if f is monotone on $[a,b]$ and $f([a,b]) \subset [c,d]$, then $g \circ f$ is absolutely continuous on $[a,b]$ by verifying the definition of absolute continuity for $g \circ f$.

Prob. 13.13. Let f and g be two functions defined on $[0,1]$ by

$$f(x) = \sqrt{x} \quad \text{for } x \in [0,1],$$

and

$$g(x) = \begin{cases} x^2|\sin\frac{1}{x}| & \text{for } x \in (0,1], \\ 0 & \text{for } x = 0. \end{cases}$$

(a) Show that f and g are absolutely continuous on $[0,1]$.
(b) Show that $g \circ f$ is absolutely continuous on $[0,1]$.
(c) Show that $f \circ g$ is not absolutely continuous on $[0,1]$.
(Hint for (c)): Show that $f \circ g$ is not of bounded variation on $[0,1]$.)

Prob. 13.14. Let f be an absolutely continuous function on $[a,b]$ and g be an absolutely continuous function on $[c,d]$. Assume that $f([a,b]) \subset [c,d]$. Show that $g \circ f$ is absolutely continuous on $[a,b]$ if and only if $g \circ f$ is of bounded variation on $[a,b]$.

Prob. 13.15. For $p \in (1,\infty)$, let

$$f(x) = \begin{cases} x^p \sin \frac{1}{x} & \text{for } x \in \left(0, \frac{2}{\pi}\right], \\ 0 & \text{for } x = 0. \end{cases}$$

(a) Calculate f_p' on $\left[0, \frac{2}{\pi}\right]$.
(b) Show that f_p' is Lebesgue integrable on $\left[0, \frac{2}{\pi}\right]$.
(c) Show that when $p \in (2, \infty)$, f_p' is continuous on $\left[0, \frac{2}{\pi}\right]$.
(d) Show that when $p \in (1, 2]$, f_p' is not continuous on $\left[0, \frac{2}{\pi}\right]$. Discuss the nature of discontinuity.
(e) Show that when $p \in [2, \infty)$, f_p' is bounded on $\left[0, \frac{2}{\pi}\right]$.
(f) Show that when $p \in (1, 2)$, f_p' is not bounded on $\left[0, \frac{2}{\pi}\right]$.

Prob. 13.16. Let

$$f(x) = \begin{cases} x^2 \cos \frac{\pi}{x^2} & \text{for } x \in \left(0, \frac{2}{\pi}\right], \\ 0 & \text{for } x = 0. \end{cases}$$

(a) Calculate f' on $\left[0, \frac{2}{\pi}\right]$.
(b) Show that f' is not μ_L-integrable on $\left[0, \frac{2}{\pi}\right]$.
(Hint: With $\alpha_n = \left(2n + \frac{1}{2}\right)^{-1/2}$ and $\beta_n = (2n)^{-1/2}$, $n \in \mathbb{N}$, compute $\int_{[\alpha_n, \beta_n]} f' \, d\mu_L$.)

Prob. 13.17. Let $D \subset [a, b]$ and $D \in \mathfrak{M}_L$. Show that for every $\lambda \in [0, 1]$, there exists $c \in [a, b]$ such that $\mu_L\left(D \cap [a, c]\right) = \lambda \mu_L(D)$.
(Hint: Consider the Lebesgue integral of $\mathbf{1}_{D \cap [a,x]}$ for $x \in [a, b]$.)

Prob. 13.18. Let $D \subset \mathbb{R}$, $D \in \mathfrak{M}_L$, and $\mu_L(D) < \infty$. Show that for every $\lambda \in (0, 1)$, there exists $\xi \in \mathbb{R}$ such that $\mu_L\left(D \cap (-\infty, \xi]\right) = \lambda \mu_L(D)$.

Prob. 13.19. The average density of set $E \subset \mathbb{R}$, $E \in \mathfrak{M}_L$, over an interval $(x - h, x + h)$, where $x \in \mathbb{R}$ and $h > 0$, is defined by $(2h)^{-1} \mu_L\left(E \cap (x - h, x + h)\right)$ and the density of E at a point $x \in \mathbb{R}$ is defined by

$$p_E(x) = \lim_{h \downarrow 0} \frac{1}{2h} \mu_L\left(E \cap (x - h, x + h)\right).$$

Show that

$$p_E(x) = \begin{cases} 1 & \text{for a.e. } x \in E, \\ 0 & \text{for a.e. } x \in E^c. \end{cases}$$

§14 Convex Functions

[I] Continuity and Differentiability of a Convex Function

Geometrically a convex function is a real-valued function on an interval I in $\mathbb{R}$ such that for any two points $x_1 < x_2$ in I the chord joining the two points $(x_1, f(x_1))$ and $(x_2, f(x_2))$ on the graph of f is on or above the graph of f on the interval $[x_1, x_2]$. An analytic definition of a convex function is given as follows.

Definition 14.1. *A real-valued function f on an interval I in $\mathbb{R}$ is said to be convex on I if for any $x_1, x_2 \in I$ and $\lambda \in [0, 1]$, we have $f\big(\lambda x_1 + (1-\lambda)x_2\big) \le \lambda f(x_1) + (1-\lambda) f(x_2)$. Let us call this condition the convexity condition.*

The equivalence of this definition to the geometric definition will be proved below. We shall show that a convex function f on an open interval I is continuous on I, differentiable everywhere except at countably many points in I, the derivative is an increasing function on the subset of I on which it exists, and furthermore f is absolutely continuous on every finite closed interval contained in I.

Observation 14.2. With $x_1, x_2 \in \mathbb{R}$ such that $x_1 < x_2$, the real-valued function φ on $[0, 1]$ defined by $\varphi(\lambda) = \lambda x_1 + (1-\lambda)x_2$ for $\lambda \in [0, 1]$ is continuous and strictly decreasing on $[0, 1]$, mapping $[0, 1]$ one-to-one onto $[x_1, x_2]$ with $\varphi(0) = x_2$ and $\varphi(1) = x_1$. This is immediate from the fact that $\varphi(\lambda) = (x_1 - x_2)\lambda + x_2$ for $\lambda \in [0, 1]$.

Lemma 14.3. (a) *Let f be a real-valued function on an interval I in $\mathbb{R}$. For $x_1, x_2 \in I$ such that $x_1 < x_2$, and $\lambda \in [0, 1]$, let $\xi = \lambda x_1 + (1-\lambda)x_2$. Then the following conditions are all equivalent:*

(i) $$f(\xi) \le \lambda f(x_1) + (1-\lambda) f(x_2).$$

(ii) $$f(\xi) \le \frac{x_2 - \xi}{x_2 - x_1} f(x_1) + \frac{\xi - x_1}{x_2 - x_1} f(x_2).$$

(iii) $$f(\xi) \le \frac{f(x_2) - f(x_1)}{x_2 - x_1}(\xi - x_1) + f(x_1).$$

(iv) $$\frac{f(\xi) - f(x_1)}{\xi - x_1} \le \frac{f(x_2) - f(\xi)}{x_2 - \xi} \quad \text{for } \xi \ne x_1, x_2.$$

(b) *A real-valued function f on an interval I in $\mathbb{R}$ is a convex function if and only if for any $x_1, x_2 \in I$ such that $x_1 < x_2$ the chord joining the two points $(x_1, f(x_1))$ and $(x_2, f(x_2))$ on the graph of f is on or above the graph of f on the interval $[x_1, x_2]$.*

Proof. 1. To show the equivalence of (i), (ii), and (iii), we show that the right sides of (i), (ii), and (iii) are all equal. Note that by $\xi = \lambda x_1 + (1-\lambda)x_2 = (x_1 - x_2)\lambda + x_2$, we have

$\lambda = \{x_2 - \xi\}/\{x_2 - x_1\}$ and $1 - \lambda = \{\xi - x_1\}/\{x_2 - x_1\}$. Then

$$\lambda f(x_1) + (1-\lambda) f(x_2) = \frac{x_2 - \xi}{x_2 - x_1} f(x_1) + \frac{\xi - x_1}{x_2 - x_1} f(x_2).$$

This shows that the right side of (i) is equal to that of (ii).

Next, writing $x_2 - \xi = (x_2 - x_1) + (x_1 - \xi)$, we have by simple computation

$$\frac{x_2 - \xi}{x_2 - x_1} f(x_1) + \frac{\xi - x_1}{x_2 - x_1} f(x_2) = \frac{f(x_2) - f(x_1)}{x_2 - x_1}(\xi - x_1) + f(x_1).$$

This shows that the right side of (ii) is equal to that of (iii).

To show the equivalence of (ii) and (iv), we multiply both sides of (ii) by $x_2 - x_1$ to obtain $(x_2 - x_1) f(\xi) \leq (x_2 - \xi) f(x_1) + (\xi - x_1) f(x_2)$. Adding $-(\xi - x_1) f(\xi)$, we have $(x_2 - x_1) f(\xi) - (\xi - x_1) f(\xi) \leq (x_2 - \xi) f(x_1) + (\xi - x_1)\{f(x_2) - f(\xi)\}$ so that we have $(x_2 - \xi)\{f(\xi) - f(x_1)\} \leq (\xi - x_1)\{f(x_2) - f(\xi)\}$. For $\xi \neq x_1, x_2$, by dividing by the factors $x_2 - \xi$ and $\xi - x_1$ we obtain (iv). Conversely starting with (iv) and retracing the calculation above, we have (ii) for $\xi \neq x_1, x_2$. But (ii) holds trivially for $\xi = x_1$ and for $\xi = x_2$. This proves the equivalence of (ii) and (iv).

2. Let f be a real-valued function on an interval I in $\mathbb{R}$. Let $x_1, x_2 \in I$ and $x_1 < x_2$. The equation of the chord through the two points $(x_1, f(x_1))$ and $(x_2, f(x_2))$ is given by

$$g(\xi) = \frac{f(x_2) - f(x_1)}{x_2 - x_1}(\xi - x_1) + f(x_1) \quad \text{for } \xi \in [x_1, x_2].$$

According to (a), the convexity condition (i) is equivalent to condition (iii). Therefore f is convex on I if and only if $f(\xi) \leq g(\xi)$. ∎

The next Proposition gives some fundamental inequalities for convex functions. Continuity and differentiability of a convex function are consequences of these inequalities.

Proposition 14.4. *Let f be a convex function on an interval I in $\mathbb{R}$. Then for any $u, v, w \in I$ such that $u < v < w$, we have*

$$\frac{f(v) - f(u)}{v - u} \leq \frac{f(w) - f(u)}{w - u} \leq \frac{f(w) - f(v)}{w - v}.$$

Proof. Since $v \in (u, w)$, there exists $\lambda \in (0, 1)$ such that $v = \lambda u + (1-\lambda) w$. Solving this equation for λ, we have $\lambda = \{w - v\}/\{w - u\}$ and then $1 - \lambda = \{v - u\}/\{w - u\}$. By the convexity condition satisfied by f, we have

$$f(v) \leq \lambda f(u) + (1-\lambda) f(w) = \frac{w - v}{w - u} f(u) + \frac{v - u}{w - u} f(w)$$

and then

(1) $$(w - u) f(v) \leq (w - v) f(u) + (v - u) f(w).$$

To derive the first inequality of the Proposition, let us write (1) as

$$(w-u)f(v) \le \{(w-u)-(v-u)\}f(u)+(v-u)f(w).$$

Then we have

$$(w-u)\{f(v)-f(u)\} \le (v-u)\{f(w)-f(u)\}.$$

Cross-multiplying, we have the first inequality of the Proposition.

To derive the second inequality of the Proposition, rewrite (1) as

$$(w-u)f(v) \le (w-v)f(u)+\{(w-u)-(w-v)\}f(w)$$

and then

$$(w-u)\{f(v)-f(w)\} \le (w-v)\{f(u)-f(w)\}.$$

Cross-multiplying and then multiplying by -1, we have the second inequality of the Proposition. ■

Theorem 14.5. *Let f be a convex function on an open interval I in $\mathbb{R}$. Then we have:*
(a) *f is both left- and right-differentiable at every $x_0 \in I$, that is,*

(1) $$(D_\ell f)(x_0) = \lim_{x\uparrow x_0} \frac{f(x)-f(x_0)}{x-x_0} \in \mathbb{R},$$

(2) $$(D_r f)(x_0) = \lim_{x\downarrow x_0} \frac{f(x)-f(x_0)}{x-x_0} \in \mathbb{R}.$$

Furthermore for any $x_1, x_2 \in I$ such that $x_1 < x_2$, we have

(3) $$(D_r f)(x_1) \le \frac{f(x_2)-f(x_1)}{x_2-x_1} \le (D_\ell f)(x_2).$$

(b) *f is continuous on I.*
(c) *$D_\ell f \le D_r f$ on I.*
(d) *$D_\ell f$ and $D_r f$ are real-valued increasing functions on I.*
(e) *f is differentiable, that is, the derivative f' exists and is finite, everywhere except at countably many points in I. If f' exists at $x_1, x_2 \in I$, $x_1 < x_2$, then $f'(x_1) \le f'(x_2)$.*

Proof. 1. To prove the right-differentiability of f at any $x_0 \in I$, let $x_1, x_2 \in I$ be such that $x_0 < x_1 < x_2$. Identifying x_0, x_1, x_2 respectively with u, v, w in the first inequality in Proposition A.4, we have $\{f(x_1)-f(x_0)\}/\{x_1-x_0\} \le \{f(x_2)-f(x_0)\}/\{x_2-x_0\}$. This shows that $\{f(x)-f(x_0)\}/\{x-x_0\} \downarrow$ as $x \downarrow x_0$ so that the right-derivative of f at x_0, that is, $(D_r f)(x_0) = \lim_{x\downarrow x_0}\{f(x)-f(x_0)\}/\{x-x_0\}$ exists in $\overline{\mathbb{R}}$ and moreover we have $(D_r f)(x_0) \le \{f(x)-f(x_0)\}/\{x-x_0\}$ for any $x > x_0$. This proves the first inequality in (3). Let $u, x \in I$ be such that $u < x_0 < x$. Identifying u, x_0, x respectively with u, v, w in Proposition A.4, we have $\{f(x_0)-f(u)\}/\{x_0-u\} \le \{f(x)-f(x_0)\}/\{x-x_0\}$. Thus for $x > x_0$, $\{f(x)-f(x_0)\}/\{x-x_0\}$ is bounded below by the constant $\{f(x_0)-f(u)\}/\{x_0-u\}$. Therefore we have $(D_r f)(x_0) = \lim_{x\downarrow x_0}\{f(x)-f(x_0)\}/\{x-x_0\} \in \mathbb{R}$ and f is right-differentiable at x_0.

Similarly, to prove the left-differentiability of f at x_0, let $x_1, x_2 \in I$ be such that $x_2 < x_1 < x_0$. Identifying x_2, x_1, x_0 respectively with u, v, w in the second inequality in Proposition 14.4, we have $\{f(x_0) - f(x_2)\}/\{x_0 - x_2\} \le \{f(x_0) - f(x_1)\}/\{x_0 - x_1\}$. Thus $\{f(x_0) - f(x)\}/\{x_0 - x\} \uparrow$ as $x \uparrow x_0$. Therefore the left-derivative $(D_\ell f)(x_0) = \lim_{x \uparrow x_0} \{f(x) - f(x_0)\}/\{x - x_0\}$ exists in $\overline{\mathbb{R}}$ and $(D_\ell f)(x_0) \ge \{f(x_0) - f(x)\}/\{x_0 - x\}$ for any $x < x_0$, proving the second inequality in (3). Next let $x, w \in I$ be such that $x < x_0 < w$. Identifying x, x_0, w respectively with u, v, w in Proposition 14.4, we have $\{f(x_0) - f(x)\}/\{x_0 - x\} \le \{f(w) - f(x_0)\}/\{w - x_0\}$. Thus for $x < x_0$, the quotient $\{f(x_0) - f(x)\}/\{x_0 - x\}$ is bounded above by the constant $\{f(w) - f(x_0)\}/\{w - x_0\}$ and therefore $(D_\ell f)(x_0) = \lim_{x \uparrow x_0} \{f(x) - f(x_0)\}/\{x - x_0\} \in \mathbb{R}$. This proves the left-differentiability of f at x_0.

2. To prove (b), let $x_0 \in I$. By (1), f is left-differentiable at x_0 so that f is left-continuous at x_0. By (2) f is right-continuous at x_0. Therefore f is continuous at x_0.

3. To prove (c), let $x_0 \in I$. Let $x', x'' \in I$ be such that $x' < x_0 < x''$. Identifying x', x_0, x'' respectively with u, v, w in Proposition 14.4, we have $\{f(x_0) - f(x')\}/\{x_0 - x'\} \le \{f(x'') - f(x_0)\}/\{x'' - x_0\}$. Thus we have

$$(D_\ell f)(x_0) = \lim_{x' \uparrow x_0} \frac{f(x_0) - f(x')}{x_0 - x'} \le \frac{f(x'') - f(x_0)}{x'' - x_0}$$

and then

$$(D_\ell f)(x_0) \le \lim_{x'' \downarrow x_0} \frac{f(x'') - f(x_0)}{x'' - x_0} = (D_r f)(x_0).$$

This proves (c).

4. Let us prove (d). To show that $D_r f$ is an increasing function on I, let $x_1, x_2 \in I$ be such that $x_1 < x_2$. Let $x_1 < x' < \alpha < \beta < x_2 < x''$ be points in I. By Proposition 14.4, we have

$$\frac{f(x') - f(x_1)}{x' - x_1} \le \frac{f(\alpha) - f(x')}{\alpha - x'} \le \frac{f(\beta) - f(\alpha)}{\beta - \alpha} \le \frac{f(x_2) - f(\beta)}{x_2 - \beta} \le \frac{f(x'') - f(x_2)}{x'' - x_2}.$$

Letting $x' \downarrow x_1$ in the first quotient and letting $x'' \downarrow x_2$ in the last quotient above, we have

$$(D_r f)(x_1) \le \frac{f(\beta) - f(\alpha)}{\beta - \alpha} \le (D_r f)(x_2).$$

This shows that $D_r f$ is an increasing function on I. Similarly, to show that $D_\ell f$ is an increasing function on I we let $x' < x_1 < \alpha < \beta < x'' < x_2$ be points in I. Applying Proposition 14.4 as above, we obtain in this case

$$\frac{f(x_1) - f(x')}{x_1 - x'} \le \frac{f(\beta) - f(\alpha)}{\beta - \alpha} \le \frac{f(x_2) - f(x'')}{x_2 - x''}.$$

Letting $x' \uparrow x_1$ and $x'' \uparrow x_2$, we have

$$(D_\ell f)(x_1) \le \frac{f(\beta) - f(\alpha)}{\beta - \alpha} \le (D_\ell f)(x_2).$$

Thus $D_\ell f$ is an increasing function on I.

5. To prove (e), note that since $D_\ell f$ is a real-valued increasing function on I by (d), there exists a countable subset E_ℓ of I such that $D_\ell f$ is continuous at every $x \in I \setminus E_\ell$. Similarly there exists a countable subset E_r of I such that $D_r f$ is continuous at every $x \in I \setminus E_r$. Let $E = E_\ell \cup E_r$. Then E is a countable subset of I and both $D_\ell f$ and $D_r f$ are continuous at every $x \in I \setminus E$. Let $x_0 \in I \setminus E$. Then for $x \in I$ such that $x > x_0$, we have $(D_\ell f)(x_0) \leq (D_r f)(x_0) \leq (D_\ell f)(x)$ by (c) and by (3). Since $D_\ell f$ is an increasing function, $\lim_{x \downarrow x_0} (D_\ell f)(x)$ exists and we have $(D_r f)(x_0) \leq \lim_{x \downarrow x_0} (D_\ell f)(x)$. Now the continuity of $D_\ell f$ at x_0 implies that $\lim_{x \downarrow x_0} (D_\ell f)(x) = (D_\ell f)(x_0)$. Therefore we have $(D_\ell f)(x_0) \leq (D_r f)(x_0) \leq (D_\ell f)(x_0)$ and thus $(D_\ell f)(x_0) = (D_r f)(x_0)$ This shows that f is differentiable at x_0. Thus f is differentiable except at countably many points in I.

Finally if f' exists at $x_1, x_2 \in I$ and $x_1 < x_2$, then $f'(x_1) = (D_\ell f)(x_1) \leq (D_\ell f)(x_2) = f'(x_2)$ by (c). ∎

Remark 14.6. In Theorem 14.5, the domain of definition of the convex function is an open interval in $\mathbb{R}$. If f is a convex function defined on a finite interval which includes its endpoints, say $[a, b]$, then the right-derivative $(D_r f)(a)$ and the left-derivative $(D_\ell f)(b)$ exist but may not be finite and f may not be right-continuous at a and left-continuous at b.

Proof. Applying the first inequality in Proposition 14.4 as in **1** of the Proof of Theorem 14.5, we show that for $x > a$ we have $\{f(x) - f(a)\}/\{x - a\} \downarrow$ as $x \downarrow a$ so that $(D_r f)(a) = \lim_{x \downarrow a} \{f(x) - f(a)\}/\{x - a\}$ exists in $\overline{\mathbb{R}}$. However since there does not exist any point $u \in [a, b]$ such that $u < a$, it is not possible to use Proposition 14.4 to show that the quotient $\{f(x) - f(a)\}/\{x - a\}$ is bounded below for $x > a$ as it was done in the Proof of Theorem 14.5. Thus it may occur that $(D_r f)(a) = -\infty$ so that f is not right-differentiable at a and f may not be right-continuous at a. Similarly $(D_\ell f)(b)$ exists in $\overline{\mathbb{R}}$ and we may have $(D_\ell f)(b) = \infty$ so that f is not left-differentiable at b and f may not be left-continuous at b. ∎

Example 1. Let $f(x) = -\sqrt{1 - x^2}$ for $x \in [-1, 1]$. Then f is convex on $[-1, 1]$ by (b) of Lemma 14.3; f is right-continuous at $x = -1$ but $(D_r f)(-1) = -\infty$; f is left-continuous at $x = 1$ but $(D_\ell f)(1) = \infty$.

Example 2. Let f be defined on $[-1, 1]$ by setting $f(x) = 0$ for $x \in (-1, 1)$ and $f(-1) = f(1) = 1$. Then f is convex on $[-1, 1]$ by (b) of Lemma 14.3; f is not right-continuous at $x = -1$ and $(D_r f)(-1) = -\infty$; f is not left-continuous at $x = 1$ and $(D_\ell f)(1) = \infty$.

Remark 14.7. If f is a convex function on an interval I in $\mathbb{R}$, then f is a convex function on every subinterval of the interval. On the other hand it may not be possible to extend f as a convex function beyond I, that is, there may not exist a convex function g on an interval J such that $J \supset I$, $J \neq I$, and $g = f$ on I. Here are some examples.

(a) Consider the convex function f on $[-1, 1]$ in Example 1 above. Suppose there exists a convex function g on $(a, b) \supset [-1, 1]$ such that $g = f$ on $[-1, 1]$. By (a) of Theorem 14.5, $D_r g$ is a real-valued function on (a, b). But the equality $g = f$ on $[-1, 1]$ implies that $D_r g(-1) = D_r f(-1) = -\infty$, a contradiction. This shows that f cannot be extended as a convex function beyond $x = -1$. Similarly f cannot be extended as a convex function beyond $x = 1$.

(b) Next, consider the convex function f on $[-1, 1]$ in Example 2 above. Suppose there exists a convex function g on $(a, b) \supset [-1, 1]$ such that $g = f$ on $[-1, 1]$. By (b) of Theorem 14.5, g is continuous on (a, b) and in particular g is right-continuous at $x = -1$. Since $g = f$ on $[-1, 1]$, f is right-continuous at $x = -1$. This is a contradiction. Thus f cannot be extended as a convex function beyond $x = -1$. Similarly f cannot be extended as a convex function beyond $x = 1$.

(c) For an example of a convex function on an open interval which cannot be extended, consider $f(x) = \sec x$ for $x \in \left(-\frac{\pi}{2}, \frac{\pi}{2}\right)$. By (b) of Lemma 14.3, f is a convex function on $\left(-\frac{\pi}{2}, \frac{\pi}{2}\right)$. Let us show that f cannot be extended as a convex function beyond $\left(-\frac{\pi}{2}, \frac{\pi}{2}\right)$. Suppose there exists a convex function g on $\left[-\frac{\pi}{2}, \frac{\pi}{2}\right]$ such that $g = f$ on $\left(-\frac{\pi}{2}, \frac{\pi}{2}\right)$. Since g is a convex function, g is real-valued and thus in particular $g\left(-\frac{\pi}{2}\right) \in \mathbb{R}$. Consider the chord joining the two points $\left(-\frac{\pi}{2}, g\left(-\frac{\pi}{2}\right)\right)$ and $(0, 1)$ on the graph of g. This chord is not on or above the graph of g on the interval $\left[-\frac{\pi}{2}, 0\right]$ since $\lim_{x \downarrow -\pi/2} g(x) = \lim_{x \downarrow -\pi/2} f(x) = \infty$. Thus g is not a convex function on $\left[-\frac{\pi}{2}, \frac{\pi}{2}\right]$ by (b) of Lemma 14.3.

Theorem 14.8. *Let f be a convex function on an open interval I in $\mathbb{R}$. Then we have:*
(a) *For every finite closed interval $[\alpha, \beta] \subset I$, if we let $M = \max\left\{|(D_r f)(\alpha)|, |(D_\ell f)(\beta)|\right\}$ then f satisfies a Lipschitz condition with coefficient M on $[\alpha, \beta]$, that is,*

$$(1) \qquad |f(x') - f(x'')| \le M|x' - x''| \quad \text{for every } x', x'' \in [\alpha, \beta].$$

(b) *For every $x_0 \in I$ and every real number $m \in \left[(D_\ell f)(x_0), (D_r f)(x_0)\right]$, we have*

$$(2) \qquad f(x) \ge m(x - x_0) + f(x_0) \quad \text{for } x \in I.$$

(Thus a line passing through a point $(x_0, f(x_0))$ on the graph of f with a slope m which is an intermediate value of the left- and right-derivative of f at x_0 is on or below the graph of f. In particular, if f is differentiable at a point $x_0 \in I$, then the tangent line to the graph of f at $(x_0, f(x_0))$ is on or below the graph of f.)
(c) *There exists a countable collection $\{g_n : n \in \mathbb{N}\}$ of affine functions, $g_n(x) = \alpha_n x + \beta_n$ for $x \in I$ with $\alpha_n, \beta_n \in \mathbb{R}$ for $n \in \mathbb{N}$, such that*

$$(3) \qquad f(x) = \sup_{n \in \mathbb{N}} g_n(x) \quad \text{for } x \in I.$$

Proof. 1. Let $[\alpha, \beta] \subset I$ and let $\alpha \le x_1 < x_2 \le \beta$. By (3) of Theorem 14.5 and by the fact that $D_r f$ and $D_\ell f$ are increasing functions on I according to (d) of Theorem 14.5, we have

$$(D_r f)(\alpha) \le (D_r f)(x_1) \le \frac{f(x_2) - f(x_1)}{x_2 - x_1} \le (D_\ell f)(x_2) \le (D_\ell f)(\beta).$$

If we let $M = \max\left\{|(D_r f)(\alpha)|, |(D_\ell f)(\beta)|\right\}$, then $-M \le (D_r f)(\alpha)$ and $(D_\ell f)(\beta) \le M$ so that $-M \le \{f(x_2) - f(x_1)\}/\{x_2 - x_1\} \le M$ and then $|f(x_2) - f(x_1)| \le M|x_2 - x_1|$. This proves (a).

2. Let $x_0 \in I$ and let $m \in \big[(D_\ell f)(x_0), (D_r f)(x_0)\big]$. Let $x \in I$. If $x < x_0$, then by (3) of Theorem 14.5, we have $\{f(x) - f(x_0)\}/\{x - x_0\} \leq (D_\ell f)(x_0) \leq m$. Then since $x - x_0 < 0$, we have $f(x) - f(x_0) \geq m(x - x_0)$. On the other hand, if $x > x_0$, then by (3) of Theorem 14.5, we have $\{f(x) - f(x_0)\}/\{x - x_0\} \geq (D_r f)(x_0) \geq m$ so that $f(x) - f(x_0) \geq m(x - x_0)$. Thus we have shown that for $x \in I$ such that $x \neq x_0$, we have $f(x) - f(x_0) \geq m(x - x_0)$. This inequality holds trivially for $x = x_0$. Thus we have $f(x) \geq m(x - x_0) + f(x_0)$ for $x \in I$.

3. To prove (c), consider the collection of affine functions $\{g_\xi : \xi \in I\}$ defined by

(4) $$g_\xi(x) = (D_r f)(\xi)(x - \xi) + f(\xi) \quad \text{for } x \in I.$$

By (b), we have $f(x) \geq g_\xi(x)$ for all $x \in I$ for each $\xi \in I$ so that $f(x) \geq \sup_{\xi \in I} g_\xi(x)$ for $x \in I$. At each $x \in I$, if we choose $\xi = x$, then $g_\xi(x) = f(x)$. Thus we have

(5) $$f(x) = \sup_{\xi \in I} g_\xi(x) \quad \text{for } x \in I.$$

Let $(r_n : n \in \mathbb{N})$ be an arbitrary enumeration of the rational numbers in I. By (4), we have

(6) $$g_{r_n}(r_n) = f(r_n).$$

By (5) and by the fact that $\{g_{r_n} : n \in \mathbb{N}\}$ is a subcollection of $\{g_\xi : \xi \in I\}$, we have

(7) $$f(x) \geq \sup_{n \in \mathbb{N}} g_{r_n}(x) \quad \text{for } x \in I.$$

We claim that

(8) $$f(x) \leq \sup_{n \in \mathbb{N}} g_{r_n}(x) \quad \text{for } x \in I.$$

To prove (8), let $x_0 \in I$ be arbitrarily fixed. Let $\delta > 0$ be such that $[x_0 - \delta, x_0 + \delta] \subset I$. Since $D_r f$ is a real-valued increasing function on I by (d) of Theorem 14.5, if we let $m_1 = (D_r f)(x_0 - \delta)$ and $m_2 = (D_r f)(x_0 + \delta)$, then we have

(9) $$(D_r f)(\xi) \in [m_1, m_2] \quad \text{for } \xi \in [x_0 - \delta, x_0 + \delta].$$

Let $(r_{n_k} : k \in \mathbb{N})$ be a subsequence of $(r_n : n \in \mathbb{N})$ contained in $[x_0 - \delta, x_0 + \delta]$ such that $\lim_{k\to\infty} r_{n_k} = x_0$. Since f is continuous on I by (b) of Theorem 14.5, we have $\lim_{k\to\infty} f(r_{n_k}) = f(x_0)$. By (6), $g_{r_{n_k}}(r_{n_k}) = f(r_{n_k})$. Thus we have $\lim_{k\to\infty} g_{r_{n_k}}(r_{n_k}) = f(x_0)$. By (4), $g_{r_{n_k}}(r_{n_k}) - g_{r_{n_k}}(x_0) = (D_r f)(r_{n_k})(r_{n_k} - x_0)$. Since $r_{n_k} \in [x_0 - \delta, x_0 + \delta]$, we have $(D_r f)(r_{n_k}) \in [m_1, m_2]$ by (9). Thus if we let $M = \max\big\{|m_1|, |m_2|\big\}$, then we have $|g_{r_{n_k}}(r_{n_k}) - g_{r_{n_k}}(x_0)| \leq M|r_{n_k} - x_0|$ and then $g_{r_{n_k}}(x_0) \geq g_{r_{n_k}}(r_{n_k}) - M|r_{n_k} - x_0|$. Then since $\lim_{k\to\infty} r_{n_k} = x_0$ and $\lim_{k\to\infty} g_{r_{n_k}}(r_{n_k}) = f(x_0)$, we have $\limsup_{k\to\infty} g_{r_{n_k}}(x_0) \geq f(x_0)$. Thus

$$\sup_{n \in \mathbb{N}} g_{r_n}(x_0) \geq \sup_{k \in \mathbb{N}} g_{r_{n_k}}(x_0) \geq \limsup_{k\to\infty} g_{r_{n_k}}(x_0) \geq f(x_0).$$

By the arbitrariness of $x_0 \in I$, we have (8). Then with (7) and (8), we have (3). ∎

Lemma 14.9. (Chain Rule for Dini Derivates) *Let f be a strictly increasing convex function on an open interval I in $\mathbb{R}$ and let g be a real-valued function on the open interval $J = f(I)$. Then*

$$(1)\qquad D^+(g \circ f)(x) = (D^+g)(f(x))(D^+f)(x) \quad \text{for } x \in I.$$

The same equality holds for $D_+(g \circ f)$, $D^-(g \circ f)$, and $D_-(g \circ f)$.

Proof. Note that the convexity of f implies its continuity on I by Theorem 14.5. Thus f is a strictly increasing continuous function on I and this implies that $J = f(I)$ is an open interval in $\mathbb{R}$. Let $x_0 \in I$ be arbitrarily fixed. Let us define a function h on J by setting

$$(2)\qquad h(y) = \begin{cases} \dfrac{g(y) - g(f(x_0))}{y - f(x_0)} & \text{for } y \in J \setminus \{f(x_0)\}, \\ c & \text{for } y = f(x_0) \end{cases}$$

where c is an arbitrary real number. From (2), we have

$$g(y) - g(f(x_0)) = h(y)\{y - f(x_0)\} \quad \text{for } y \in J.$$

Note in particular that the equality holds at $y = f(x_0) \in J$ since both sides are equal to 0. Now since $f(I) = J$, the last equality implies

$$g\big(f(x)\big) - g\big(f(x_0)\big) = h\big(f(x)\big)\{f(x) - f(x_0)\} \quad \text{for } x \in I$$

so that for $x \in I$, $x \neq x_0$, we have

$$(3)\qquad \frac{(g \circ f)(x) - (g \circ f)(x_0)}{x - x_0} = h\big(f(x)\big)\frac{f(x) - f(x_0)}{x - x_0}.$$

Since f is a strictly increasing continuous function and $f(x) \downarrow f(x_0)$ as $x \downarrow x_0$, we have by (2)

$$(4)\qquad \limsup_{x \downarrow x_0} h(f(x)) = \limsup_{x \downarrow x_0} \frac{g\big(f(x)\big) - g\big(f(x_0)\big)}{f(x) - f(x_0)} = (D^+g)\big(f(x_0)\big).$$

The convexity of f on I implies that f is right-differentiable on I by Theorem 14.5. Then

$$(5)\qquad \lim_{x \downarrow x_0} \frac{f(x) - f(x_0)}{x - x_0} = (D_r f)(x_0) = (D^+f)(x_0).$$

Applying (4) and (5) to (3), we have

$$\begin{aligned} D^+(g \circ f)(x_0) &= \limsup_{x \downarrow x_0} \frac{(g \circ f)(x) - (g \circ f)(x_0)}{x - x_0} \\ &= \limsup_{x \downarrow x_0} \left[h\big(f(x)\big)\frac{f(x) - f(x_0)}{x - x_0}\right] \\ &= (D^+g)\big(f(x_0)\big)(D^+f)(x_0). \end{aligned}$$

This proves (1).

By considering $\liminf_{x \downarrow x_0} h(f(x)$ and by the equality $(D_r f)(x_0) = (D_+ f)(x_0)$, we have the equality (1) for $D_+(g \circ f)$. Then by considering $\limsup_{x \uparrow x_0} h(f(x)$ and $\liminf_{x \uparrow x_0} h(f(x)$, we obtain the equality (1) for $D^-(g \circ f)$ and $D_-(g \circ f)$. ∎

Theorem 14.10. *Let f be a real-valued continuous function on an open interval I in $\mathbb{R}$.*

(a) *If f is a convex function on I, then all the four Dini derivates $D^+ f$, $D_+ f$, $D^- f$, and $D_- f$ are real-valued increasing functions on I.*

(b) *Conversely if one of the four Dini derivates of f is a real-valued increasing function on I, then f is a convex function on I.*

Proof. 1. If f is a convex function on I, then the right-derivative $D_r f$ and the left-derivative D_ℓ exist, are real-valued and increasing on I by Theorem 14.5. The existence of $D_r f$ and D_ℓ implies $D_r f = D^+ f = D_+ f$ and $D_\ell = D^- f = D_- f$. Therefore $D^+ f$, $D_+ f$, $D^- f$, and $D_- f$ are real-valued increasing functions on I.

2. Conversely suppose g is a real-valued continuous function on an open interval I such that one of the Dini derivates, say $D^+ g$, is a real-valued increasing function on I. To show that g is a convex function on I, we show that for any $x_1, x_2 \in I$ such that $x_1 > x_2$ we have

(1) $$g(\lambda x_1 + (1-\lambda)x_2) \le \lambda g(x_1) + (1-\lambda) g(x_2) \quad \text{for every } \lambda \in [0,1].$$

Consider a function f on $[0, 1]$ defined by

(2) $$f(\lambda) = \lambda x_1 + (1-\lambda)x_2 \quad \text{for } \lambda \in [0,1].$$

Since $x_1 > x_2$, f is a strictly increasing function on $[0, 1]$. Since the graph of f is a line segment, f is a convex function on $[0, 1]$ by (b) of Lemma 14.3. Let us define a function φ on $[0, 1]$ by setting

(3) $$\varphi(\lambda) = (g \circ f)(\lambda) - \lambda g(x_1) - (1-\lambda) g(x_2) \quad \text{for } \lambda \in [0,1].$$

Then (1) is equivalent to

(4) $$\varphi(\lambda) \le 0 \quad \text{for every } \lambda \in [0,1].$$

Now φ is a real-valued continuous function on $[0, 1]$ with $\varphi(0) = \varphi(1) = 0$. From (3) and by Lemma 14.9, we have

$$\begin{aligned}(D^+\varphi)(\lambda) &= D^+(g \circ f)(\lambda) - g(x_1) + g(x_2) \\ &= (D^+ g)\big(f(\lambda)\big)(D^+ f)(\lambda) - g(x_1) + g(x_2) \\ &= \{x_1 - x_2\}(D^+ g)\big(f(\lambda)\big) - g(x_1) + g(x_2).\end{aligned}$$

Since $D^+ g$ and f are real-valued increasing functions, $(D^+ g)\big(f(\cdot)\big)$ is a real-valued increasing function on $[0, 1]$. Therefore $D^+\varphi$ is a real-valued increasing function on $[0, 1]$.

The real-valued continuous function φ has a maximum point λ_1 and a minimum point λ_2 in $[0, 1]$. If φ vanishes identically on $[0, 1]$ then (3) holds and we are done. If on the other hand φ is not identically vanishing on $[0, 1]$ then $\lambda_1 \neq \lambda_2$. Then since $\varphi(0) = \varphi(1)$, at least one of λ_1 and λ_2 is in $(0, 1)$. Consider the case $\lambda_1 \in (0, 1)$. Now since $\lambda_1 \in (0, 1)$ is a maximum point for φ, we have $(D^+\varphi)(\lambda_1) \leq 0$. Then since $D^+\varphi$ is an increasing function on $[0, 1]$, we have $D^+\varphi \leq 0$ on $[0, \lambda_1]$. This implies that φ is decreasing on $[0, \lambda_1]$ so that $\varphi(\lambda_1) \leq \varphi(0) = 0$. Then $\varphi(\lambda) \leq \varphi(\lambda_1) \leq 0$ for every $\lambda \in [0, 1]$ so that (4) holds. The case $\lambda_2 \in (0, 1)$ is treated similarly. ∎

[II] Monotonicity and Absolute Continuity of a Convex Function

Theorem 14.11. *Let f be a convex function on an open interval I in $\mathbb{R}$. Then either f is monotone on I or else there exists $x_0 \in I$ such that f is decreasing on $I \cap (-\infty, x_0]$ and increasing on $I \cap [x_0, \infty)$.*

Proof. Suppose f in not monotone on I. Then there exist $x_1, x_2, x_3 \in I$, $x_1 < x_2 < x_3$, such that $f(x_1) > f(x_2)$ and $f(x_2) < f(x_3)$. (Note that it is impossible to have $f(x_1) < f(x_2)$ and $f(x_2) > f(x_3)$ since this would contradict the fact that the chord joining the two points $(x_1, f(x_1))$ and $(x_3, f(x_3))$ on the graph of f is on or above the graph of f on the interval $[x_1, x_3]$.)

Now since f is continuous on I by (b) of Theorem 14.5 and in particular continuous on the compact set $[x_1, x_3]$, there exists a minimum point of f on $[x_1, x_3]$, that is, there exists $x_0 \in [x_1, x_3]$ such that $f(x) \geq f(x_0)$ for every $x \in [x_1, x_3]$. Note that since $f(x_1) > f(x_2)$ and $f(x_2) < f(x_3)$, we have $x_0 \neq x_1$ and $x_0 \neq x_3$ so that $x_0 \in (x_1, x_3)$. Let us show that x_0 is actually a minimum point of f on the entire interval I. Let $x \in I$ and $x < x_1$. Then $x < x_1 < x_0$ and therefore by the second inequality in Proposition 14.4, we have $\{f(x_0) - f(x)\}/\{x_0 - x\} \leq \{f(x_0) - f(x_1)\}/\{x_0 - x_1\}$. Since $f(x_0) - f(x_1) < 0$, the inequality above implies that $f(x_0) - f(x) < 0$. Thus $f(x) > f(x_0)$ for every $x \in I$ such that $x < x_1$. Similarly for $x \in I$ such that $x > x_3$, we have $x_0 < x_3 < x$ so that by the first inequality in Proposition 14.4, we have $\{f(x_3) - f(x_0)\}/\{x_3 - x_0\} \leq \{f(x) - f(x_0)\}/\{x - x_0\}$. Since $f(x_3) - f(x_0) > 0$, we have $f(x) - f(x_0) > 0$. Thus $f(x) > f(x_0)$ for every $x \in I$ such that $x > x_3$. This shows that x_0 is a minimum point of f on I.

To show that f is decreasing on $I \cap (-\infty, x_0]$, let $x', x'' \in I \cap (-\infty, x_0]$ such that $x' < x'' < x_0$. Then we have $\{f(x'') - f(x')\}/\{x'' - x'\} \leq \{f(x_0) - f(x')\}/\{x_0 - x'\}$ by the first inequality in Proposition 14.4. Since $f(x_0) - f(x') \leq 0$, we have $f(x'') - f(x') \leq 0$, that is, $f(x') \geq f(x'')$. This shows that f is decreasing on $I \cap (-\infty, x_0]$. Similarly to show that f is increasing on $I \cap [x_0, \infty)$, let $x', x'' \in I \cap (-\infty, x_0]$ be such that $x' < x'' < x_0$. By the second inequality in Proposition 14.4, we have $\{f(x'') - f(x_0)\}/\{x'' - x_0\} \leq \{f(x'') - f(x')\}/\{x'' - x'\}$. Since $f(x'') - f(x_0) \geq 0$, we have $f(x'') - f(x') \geq 0$, that is, $f(x') \leq f(x'')$. This shows that f is increasing on $I \cap [x_0, \infty)$. ∎

Proposition 14.12. *If f is a convex function on an open interval I, then f is absolutely continuous on every finite closed interval $[\alpha, \beta]$ contained in I.*

Proof. If f is convex on an open interval I, then f satisfies a Lipschitz condition on every finite closed interval $[\alpha, \beta]$ contained in I by (a) of Theorem 14.8. Then by (b) of Theorem 13.19, f is absolutely continuous on $[\alpha, \beta]$. ∎

Theorem 14.13. *Let f be a convex function on $[a, b]$. If f is right-continuous at a and left-continuous at b, then f is absolutely continuous on $[a, b]$.*

Proof. If f is convex on $[a, b]$, then it is convex on the open subinterval (a, b) so that f is continuous on (a, b) by (b) of Theorem 14.5. Then by the right-continuity of f at a and the left-continuity of f at b, f is continuous on $[a, b]$. To show that f is absolutely continuous on $[a, b]$, we show that for every $\varepsilon > 0$ there exists $\delta > 0$ such that for any finite collection of non overlapping closed intervals $[\alpha_j, \beta_j]$, $j = 1, \dots, n$, contained in $[a, b]$ with $\sum_{j=1}^n (\beta_j - \alpha_j) < \delta$, we have $\sum_{j=1}^n |f(\beta_j) - f(\alpha_j)| < \varepsilon$.

Let $\varepsilon > 0$ be arbitrarily given. By the continuity of f on $[a, b]$, there exists $\eta > 0$ such that

$$2\eta < b - a,$$
$$|f(x) - f(a)| < \tfrac{\varepsilon}{3} \quad \text{for } x \in [a, a + \eta],$$
$$|f(x) - f(b)| < \tfrac{\varepsilon}{3} \quad \text{for } x \in [b - \eta, b].$$

According to Theorem 14.11, the convexity of f on (a, b) implies that either f is monotone on (a, b) or else there exists $x_0 \in (a, b)$ such that f is increasing on $(a, x_0]$ and decreasing on $[x_0, b)$. Let us take $\eta > 0$ so small that f is monotone on each of the two intervals $(a, a + \eta]$ and $[b - \eta, b)$. Then by the continuity of f on $[a, b]$, f is monotone on each of the two intervals $[a, a + \eta]$ and $[b - \eta, b]$. Now by (a) of Theorem 14.8, there exists $M > 0$ such that

$$|f(x') - f(x'')| \le M|x' - x''| \quad \text{for } x', x'' \in [a + \eta, b - \eta].$$

Let $\delta < \frac{\varepsilon}{3M}$. Let $\{[\alpha_j, \beta_j] : j = 1, \cdots, n\}$ be an arbitrary finite collection of non overlapping closed intervals contained in $[a, b]$ with $\sum_{j=1}^n (\beta_j - \alpha_j) < \frac{\delta}{2}$. By subdividing one or two members in this collection or by adding one or two more non overlapping closed intervals of sufficiently small lengths to this collection, we have a finite collection of non overlapping closed intervals $\{[a_k, b_k] : k = 1, \cdots, m\}$ contained in $[a, b]$ such that $\sum_{k=1}^m (b_k - a_k) < \delta$ and $a + \eta$ is the right endpoint of a member and $b - \eta$ is the left endpoint of another member of the collection. Let us assume that the members in this collection are so numbered that $a_1 < a_2 < \cdots < a_m$. Then we have $b_p = a + \eta$ and $a_q = b - \eta$ for some $p, q \in \mathbb{N}$ such that $1 \le p < q \le m$. Clearly $\sum_{j=1}^n |f(\beta_j) - f(\alpha_j)| \le \sum_{k=1}^m |f(b_k) - f(a_k)|$. Let us estimate the last sum. By the fact that f is monotone on $[a, a + \eta]$ and $b_p = a + \eta$, we have

$$\text{(1)} \qquad \sum_{k=1}^{p} |f(b_k) - f(a_k)| = \left| \sum_{k=1}^{p} \{f(b_k) - f(a_k)\} \right| \le |f(b_p) - f(a)| < \frac{\varepsilon}{3}$$

and similarly by the fact that f is monotone on $[b - \eta, b]$ and $a_q = b - \eta$, we have

$$\text{(2)} \qquad \sum_{k=q}^{m} |f(b_k) - f(a_k)| = \left| \sum_{k=q}^{m} \{f(b_k) - f(a_k)\} \right| \le |f(b) - f(a_q)| < \frac{\varepsilon}{3}.$$

We have also

(3) $$\sum_{k=p+1}^{q-1} |f(b_k) - f(a_k)| \le M \sum_{k=p+1}^{q-1} (b_k - a_k) \le M\delta < \frac{\varepsilon}{3}.$$

Then by (1), (2), and (3), we have

$$\sum_{j=1}^{n} |f(\beta_j) - f(\alpha_j)| \le \sum_{k=1}^{m} |f(b_k) - f(a_k)| < \frac{\varepsilon}{3} + \frac{\varepsilon}{3} + \frac{\varepsilon}{3} = \varepsilon.$$

This proves the absolute continuity of f on $[a, b]$. ∎

Theorem 14.14. *Let f be a real-valued function on an open interval I in $\mathbb{R}$. Suppose*
1° *f is absolutely continuous on every finite closed interval contained in I.*
2° *f' is an increasing function on the subset of I on which it exists.*
Then f is a convex function on I.

Proof Since I is an open interval, there exists an increasing sequence $(J_n : n \in \mathbb{N})$ of finite closed intervals contained in I such that $\bigcup_{n\in\mathbb{N}} J_n = I$. Now condition 1° implies that f' exists a.e. on J_n for each $n \in \mathbb{N}$. Hence f' exists a.e. on I. Let E be the subset of I on which f' exists. By 2°, if $x_1, x_2 \in E$ and $x_1 < x_2$, then $f'(x_1) \le f'(x_2)$. By Lemma 14.3, to show that f is a convex function on i, it suffices to show that for any $x_1, \xi, x_2 \in I$ such that $x_1 < \xi < x_2$, we have

(1) $$\frac{f(\xi) - f(x_1)}{\xi - x_1} \le \frac{f(x_2) - f(\xi)}{x_2 - \xi}.$$

Now the absolute continuity of f on each of the two finite closed intervals $[x_1, \xi]$ and $[\xi, x_2]$ contained in I implies by Theorem 13.16 that $f(\xi) - f(x_1) = \int_{[x_1,\xi]} f'\, d\mu_L$ and $f(x_2) - f(\xi) = \int_{[\xi,x_2]} f'\, d\mu_L$. Let us define

$$\alpha_1 = \inf_{[x_1,\xi]\cap E} f', \quad \beta_1 = \sup_{[x_1,\xi]\cap E} f', \quad \alpha_2 = \inf_{[\xi,x_2]\cap E} f', \quad \beta_2 = \sup_{[\xi,x_2]\cap E} f'.$$

Then by the fact that f' is an increasing function on E and $x_1 < \xi < x_2$, we have

(2) $$\alpha_1 \le \beta_1 \le \alpha_2 \le \beta_2.$$

Now $\alpha_1(\xi - x_1) \le \int_{[x_1,\xi]} f'\, d\mu_L \le \beta_1(\xi - x_1)$ and $\alpha_2(x_2 - \xi) \le \int_{[\xi,x_2]} f'\, d\mu_L \le \beta_2(x_2 - \xi)$ by the definition of α_1, β_1, α_2, and β_2 so that

$$\alpha_1 \le \frac{\int_{[x_1,\xi]} f'\, d\mu_L}{\xi - x_1} \le \beta_1 \quad \text{and} \quad \alpha_2 \le \frac{\int_{[x_1,\xi]} f'\, d\mu_L}{x_2 - \xi} \le \beta_2,$$

that is,

$$\alpha_1 \le \frac{f(\xi) - f(x_1)}{\xi - x_1} \le \beta_1 \quad \text{and} \quad \alpha_2 \le \frac{f(x_2) - f(\xi)}{x_2 - \xi} \le \beta_2.$$

Then by (2), we have (1). ∎

[III] Jensen's Inequality

Lemma 14.15. *Let f be a convex function on an open interval I in $\mathbb{R}$. Then for any $x_1, \ldots, x_n \in I$ and $\lambda_1, \ldots, \lambda_n \in [0,1]$ such that $\lambda_1 + \cdots + \lambda_n = 1$, we have*

$$(1) \qquad f(\lambda_1 x_1 + \cdots + \lambda_n x_n) \le \lambda_1 f(x_1) + \cdots + \lambda_n f(x_n).$$

Proof. Let the n points $x_1, \ldots, x_n$ in I be so numbered that $x_1 \le \cdots \le x_n$. Let $\lambda_1, \ldots, \lambda_n \in [0,1]$ be such that $\lambda_1 + \cdots + \lambda_n = 1$. If we let $\xi = \lambda_1 x_1 + \cdots + \lambda_n x_n$, then $\xi \in [x_1, x_n] \subset I$. Take an arbitrary real number $m \in \big[(D_\ell f)(\xi), (D_r f)(\xi)\big]$. By (2) of Theorem 14.8, we have $f(x) \ge m(x - \xi) + f(\xi)$ for $x \in I$. Then

$$\begin{aligned}\sum_{k=1}^{n} \lambda_k f(x_k) &\ge \sum_{k=1}^{n} \lambda_k \{m(x_k - \xi) + f(\xi)\} \\ &= m \sum_{k=1}^{n} \lambda_k (x_k - \xi) + f(\xi) \\ &= f(\lambda_1 x_1 + \cdots + \lambda_n x_n). \ \blacksquare\end{aligned}$$

Let us observe that for $n = 2$, the inequality (1) in Lemma 14.15 is the convexity condition and in fact (1) can be derived directly from the convexity condition by induction on n and furthermore it is not necessary to assume that the interval I is an open one in this derivation. The proof given above is of interest in that the proof for the Jensen's Inequality below resembles this proof.

Lemma 14.15 has the following equivalent statement: Let f be a convex function on an open interval I in $\mathbb{R}$. Then for any $x_1, \ldots, x_n \in I$ and $t_1, \ldots, t_n \in [0, \infty)$ such that $t_1 + \cdots + t_n > 0$, we have

$$f\left(\frac{t_1 x_1 + \cdots + t_n x_n}{t_1 + \cdots + t_n}\right) \le \frac{t_1 f(x_1) + \cdots + t_n f(x_n)}{t_1 + \cdots + t_n}.$$

This form resembles more closely Jensen's Inequality.

Theorem 14.16. (Jensen's Inequality) *Let $(X, \mathfrak{A}, \mu)$ be a measure space. Let g be a real-valued $\mathfrak{A}$-measurable and μ-integrable function on a set $D \in \mathfrak{A}$ with $\mu(D) \in (0, \infty)$. If f is a convex function on an open interval I in $\mathbb{R}$ and if $g(D) \subset I$, then*

$$(1) \qquad f\left(\frac{1}{\mu(D)} \int_D g \, d\mu\right) \le \frac{1}{\mu(D)} \int_D f \circ g \, d\mu.$$

Proof. Being a real-valued $\mathfrak{A}$-measurable function on $D \in \mathfrak{A}$, g is a $\mathfrak{A}/\mathfrak{B}_{\mathbb{R}}$-measurable mapping from $(X, \mathfrak{A})$ to $(\mathbb{R}, \mathfrak{B}_{\mathbb{R}})$ by Theorem 4.6. The convexity of f on the open interval I implies that f is a real-valued continuous function on I by (b) of Theorem 14.5 and then

f is a $\mathfrak{B}_{\mathbb{R}}/\mathfrak{B}_{\mathbb{R}}$-measurable mapping from $(\mathbb{R}, \mathfrak{B}_{\mathbb{R}})$ to $(\mathbb{R}, \mathfrak{B}_{\mathbb{R}})$ by Theorem 4.27 and Theorem 4.6 and therefore $f \circ g$ is a $\mathfrak{A}/\mathfrak{B}_{\mathbb{R}}$-measurable mapping from $(X, \mathfrak{A})$ to $(\mathbb{R}, \mathfrak{B}_{\mathbb{R}})$ by Theorem 1.40, that is, $f \circ g$ is a real-valued $\mathfrak{A}$-measurable function on D.

Let the open interval I be given by $I = (a, b)$ where $-\infty \le a < b \le \infty$. We have $\int_D g\,d\mu \in \mathbb{R}$ by the μ-integrability of g on D. The assumption $g(D) \subset (a, b)$ implies $a < g < b$ on D and hence $\int_D g\,d\mu \in (a\mu(D), b\mu(D))$. Let

$$\text{(2)} \qquad \xi = \frac{1}{\mu(D)} \int_D g\,d\mu \in (a, b).$$

Then the convexity of f on (a, b) implies that if we take an arbitrary real number m in the range $[(D_\ell f)(\xi), (D_r f)(\xi)]$ then by (b) of Theorem 14.8 we have $f(x) \ge m(x-\xi)+f(\xi)$ for every $x \in (a, b)$. Since $g(y) \in (a, b)$ for every $y \in D$, we have

$$f(g(y)) \ge m\{g(y) - \xi\} + f(\xi) \quad \text{for } y \in D.$$

Integrating with respect to μ and recalling the definition of ξ by (2), we have

$$\int_D f \circ g\,d\mu \ge m\left\{\int_D g\,d\mu - \xi\mu(D)\right\} + f(\xi)\mu(D) = f(\xi)\mu(D).$$

Dividing both sides by $\mu(D)$, we have the inequality (1). ∎

For $r \in (1, \infty)$, the function $f(x) = x^r$ is a convex function for $x \in [0, \infty)$ which can be extended to be a convex function on $\mathbb{R}$ by setting $f(x) = 0$ for $x \in (-\infty, 0)$. Applying Theorem 14.16 to this convex function, we have the following Proposition.

Proposition 14.17. *Let $(X, \mathfrak{A}, \mu)$ be a measure space. Let g be an extended real-valued $\mathfrak{A}$-measurable function on a set $D \in \mathfrak{A}$ with $\mu(D) \in (0, \infty)$. Then for any $0 < p < q < \infty$, we have*

$$\text{(1)} \qquad \left[\frac{1}{\mu(D)} \int_D |g|^p\,d\mu\right]^{1/p} \le \left[\frac{1}{\mu(D)} \int_D |g|^q\,d\mu\right]^{1/q}.$$

Proof. If $0 < p < q < \infty$, then $q/p \in (1, \infty)$. Define a real-valued function f on $\mathbb{R}$ by

$$f(x) = \begin{cases} x^{q/p} & \text{for } x \in [0, \infty), \\ 0 & \text{for } x \in (-\infty, 0). \end{cases}$$

Then f is differentiable everywhere on $\mathbb{R}$ and its derivative is given by

$$f'(x) = \begin{cases} \frac{q}{p} x^{(q/p)-1} & \text{for } x \in [0, \infty), \\ 0 & \text{for } x \in (-\infty, 0). \end{cases}$$

Thus f' is a real-valued increasing function on $\mathbb{R}$ and this implies according to (b) of Theorem 14.10 that f is a convex function on $\mathbb{R}$.

For every $n \in \mathbb{N}$, let $g_n = |g| \wedge n$ on D. Then $|g_n|^p \leq n^p$ on D. Since $\mu(D) < \infty$, $|g_n|^p$ is μ-integrable on D. Also $f \circ |g_n|^p = |g_n|^q$ on D. Thus by Theorem 14.16, we have

$$\left[\frac{1}{\mu(D)} \int_D |g_n|^p \, d\mu\right]^{q/p} \leq \frac{1}{\mu(D)} \int_D |g_n|^q \, d\mu$$

and then

$$\text{(2)} \qquad \left[\frac{1}{\mu(D)} \int_D |g_n|^p \, d\mu\right]^{1/p} \leq \left[\frac{1}{\mu(D)} \int_D |g_n|^q \, d\mu\right]^{1/q}.$$

Since $|g_n| \uparrow |g|$ on D as $n \to \infty$, we have $|g_n|^p \uparrow |g|^p$ and $|g_n|^q \uparrow |f|^q$. By Theorem 8.5 (Monotone Convergence Theorem), we have $\lim_{n\to\infty} \int_D |g_n|^p \, d\mu = \int_D |g|^p \, d\mu$. Applying these to (2), we have (1). ■

Chapter 4

The Classical Banach Spaces

§15 Normed Linear Spaces

[I] Banach Spaces

We consider linear spaces over the field $\mathbb{K}$ where $\mathbb{K}$ is either the field of real numbers $\mathbb{R}$ or the field of complex numbers $\mathbb{C}$.

Definition 15.1. *A linear space over the field $\mathbb{K} = \mathbb{R}$ or $\mathbb{C}$ is a set V with a mapping of $V \times V$ into V: $(u, v) \mapsto u + v$, called addition and a mapping of $\mathbb{K} \times V$ into V: $(\alpha, v) \mapsto \alpha v$, called scalar multiplication, which satisfy the following conditions:*
1° $u + v = v + u$ *for* $u, v \in V$,
2° $u + (v + w) = (u + v) + w$ *for* $u, v, w \in V$,
3° *there exists* $0 \in V$ *such that* $u + 0 = u$ *for every* $u \in V$,
4° *for each* $v \in V$ *there exists some* $-v \in V$ *such that* $v + (-v) = 0$,
5° $(\alpha\beta)v = \alpha(\beta v)$ *for* $\alpha, \beta \in \mathbb{K}$ *and* $v \in V$,
6° $\alpha(u + v) = \alpha u + \alpha v$ *for* $\alpha \in \mathbb{K}$ *and* $u, v \in V$,
7° $(\alpha + \beta)v = \alpha v + \beta v$ *for* $\alpha, \beta \in \mathbb{K}$ *and* $v \in V$,
8° $1 \cdot v = v$ *for every* $v \in V$,
9° $(-1)v = -v$ *for every* $v \in V$.
Note that $0 \cdot v = 0$ *for every* $v \in V$ *follows from* 7°, 8°, 9°, *and* 4°.
A linear space over the field $\mathbb{R}$ is called a real linear space.
A linear space over the field $\mathbb{C}$ is called a complex linear space.

Definition 15.2. *Let V be a linear space over the field $\mathbb{K}$. If a subset M of V satisfies the following conditions:*
1° $u + v \in M$ *for* $u, v \in M$,
2° $\alpha v \in M$ *for* $v \in M$ *and* $\alpha \in \mathbb{K}$,
then we call M a linear subspace of V.
(Note that by satisfying conditions 1° and 2°, M is itself a linear space over the field $\mathbb{K}$.)

Remark 15.3. Every complex linear space is also a real linear space. Indeed since $\mathbb{R} \subset \mathbb{C}$

if V satisfies the defining conditions for a complex linear space it certainly satisfies the defining conditions for a real linear space; in other words if we ignore the possibility of multiplying elements of V by complex numbers that are not real numbers and restrict to multiplication by real numbers only then V is a real linear space.

Remark 15.4. The set $\mathbb{K}$ is a linear space over the field $\mathbb{K}$. Thus $\mathbb{R}$ is a real linear space and $\mathbb{C}$ is a complex linear space.

Definition 15.5. *Let V be a linear space over the field of scalars $\mathbb{K}$. A real-valued function $\|\cdot\|$ on V is called a norm on V if it satisfies the following conditions:*
1° *non-negativity: $\|v\| \in [0, \infty)$ for $v \in V$.*
2° *$\|v\| = 0$ if and only if $v = 0 \in V$.*
3° *positive homogeneity: $\|\alpha v\| = |\alpha|\|v\|$ for $v \in V$ and $\alpha \in \mathbb{K}$.*
4° *triangle inequality: $\|u + v\| \leq \|u\| + \|v\|$ for $u, v \in V$.*
A linear space V with a norm $\|\cdot\|$ defined on it is called a normed linear space and we write $(V, \|\cdot\|)$ for it.

Observation 15.6. Let $\|\cdot\|$ be a norm on a linear space V. Then $\big|\|u\| - \|v\|\big| \leq \|u - v\|$ for $u, v \in V$.

Proof. By the triangle inequality of the norm, we have $\|u\| = \|u-v+v\| \leq \|u-v\|+\|v\|$ and thus $\|u\| - \|v\| \leq \|u - v\|$. Interchanging the roles of u and v, we have $\|v\| - \|u\| \leq \|v - u\| = \|u - v\|$ by the positive homogeneity of the norm. With $\|u\| - \|v\| \leq \|u - v\|$ and $-\{\|u\| - \|v\|\} \leq \|u - v\|$, we have $\big|\|u\| - \|v\|\big| \leq \|u - v\|$. ∎

Lemma 15.7. *Let $(V, \|\cdot\|)$ be a normed linear space. If we define a function ρ on $V \times V$ by $\rho(u, v) = \|u - v\|$ for $u, v \in V$, then ρ is a translation invariant metric on V, that is,*
1° *non-negativity: $\rho(u, v) \in [0, \infty)$ for $u, v \in V$.*
2° *$\rho(u, v) = 0$ if and only if $u = v$.*
3° *symmetry: $\rho(u, v) = \rho(v, u)$ for $u, v \in V$.*
4° *triangle inequality: $\rho(u, v) \leq \rho(u, w) + \rho(w, v)$ for $u, v, w \in V$.*
5° *translation invariance: $\rho(u, v) = \rho(u + w, v + w)$ for $u, v, w \in V$.*

Proof.

$$\begin{aligned}
&\rho(u, g) = \|u - v\| \in [0, \infty),\\
&\rho(u, v) = 0 \Leftrightarrow \|u - v\| = 0 \Leftrightarrow u - v = 0 \Leftrightarrow u = v,\\
&\rho(u, v) = \|u - v\| = \|v - u\| = \rho(v, u),\\
&\rho(u, v) = \|u - v\| = \|u - w + w - v\| \leq \|u - w\| + \|w - v\| = \rho(u, w) + \rho(w, v),\\
&\rho(u + w, v + w) = \|(u + w) - (v + w)\| = \|u - v\| = \rho(u, v). \ \blacksquare
\end{aligned}$$

Thus a norm $\|\cdot\|$ on a linear space V introduces a metric $\rho(u, v) = \|u - v\|$ and a metric topology to V. Unless otherwise specified, we understand the topology on V to be the metric topology by the metric derived from the norm $\|\cdot\|$.

Sometimes it is convenient to use a letter such as ν rather than the symbol $\|\cdot\|$ for a norm on a linear space V. Then we write $\nu(v)$ for $v \in V$ and $\nu(u-v)$ for the metric derived from the norm ν.

Definition 15.8. **(a)** *Let $(V, \|\cdot\|)$ be a normed linear space. Let $(v_n : n \in \mathbb{N})$ be a sequence in V and let $v \in V$. We say that the sequence converges in the norm to v if the sequence converges to v in the metric topology on V derived from the norm $\|\cdot\|$, that is, $\lim_{n\to\infty} \|v_n - v\| = 0$. We say that a sequence in V converges in the norm if there exists an element in V to which the sequence converges in the norm.*

We write $\lim_{n\to\infty} v_n = v$ for $\lim_{n\to\infty} \|v_n - v\| = 0$.

(b) *A sequence $(v_n : n \in \mathbb{N})$ in a normed linear space $(V, \|\cdot\|)$ is called a Cauchy sequence with respect to the norm $\|\cdot\|$ if it is a Cauchy sequence with respect to the metric derived from the norm, that is, if for every $\varepsilon > 0$ there exists $N \in \mathbb{N}$ such that $\|v_m - v_n\| < \varepsilon$ for $m, n \geq N$.*

(c) *We say that a normed linear space $(V, \|\cdot\|)$ is complete with respect to the norm $\|\cdot\|$ if for every Cauchy sequence $(v_n : n \in \mathbb{N})$ with respect to the norm there exists $v \in V$ such that $\lim_{n\to\infty} \|v_n - v\| = 0$.*

(d) *We call a normed linear space $(V, \|\cdot\|)$ a Banach space if $(V, \|\cdot\|)$ is complete with respect to the norm $\|\cdot\|$. We call a Banach space $(V, \|\cdot\|)$ a real Banach space or a complex Banach space according as the field of scalars of the normed linear space $(V, \|\cdot\|)$ is $\mathbb{R}$ or $\mathbb{C}$.*

Triangle inequality of a norm implies the following:

Proposition 15.9. *Let $(V, \|\cdot\|)$ be a normed linear space. Let $(v_n : n \in \mathbb{N}) \subset V$ and $v \in V$.*

(a) (Uniqueness of Limit) *If $\lim_{n\to\infty} v_n = v$ and $\lim_{n\to\infty} v_n = w \in V$ also, then $v = w$.*

(b) *If $(v_n : n \in \mathbb{N})$ converges in the norm then $(v_n : n \in \mathbb{N})$ is a Cauchy sequence with respect to the norm.*

(c) *If $\lim_{n\to\infty} v_n = v$ then $\lim_{n\to\infty} \|v_n\| = \|v\|$, that is, $\lim_{n\to\infty} \|v_n\| = \|\lim_{n\to\infty} v_n\|$.*

Proof. 1. Suppose $\lim_{n\to\infty} v_n = v$ and $\lim_{n\to\infty} v_n = w$ also. By the triangle inequality of the norm we have $\|v - w\| \leq \|v - v_n\| + \|w - v_n\|$ for every $n \in \mathbb{N}$. Letting $n \to \infty$ on the right side of the last inequality we have $\|v - w\| \leq 0$ and thus $\|v - w\| = 0$ and $v = w$.

2. Suppose $\lim_{n\to\infty} v_n = v$, that is, $\lim_{n\to\infty} \|v_n - v\| = 0$. Now by the triangle inequality we have $\|v_m - v_n\| \leq \|v_m - v\| + \|v_n - v\|$ for $m, n \in \mathbb{N}$. Since $\lim_{m\to\infty} \|v_m - v\| = 0$ and $\lim_{n\to\infty} \|v_n - v\|$, for every $\varepsilon > 0$ there exists $N \in \mathbb{N}$ such that $\|v_m - v\| < \frac{\varepsilon}{2}$ for $m \geq N$ and $\|v_n - v\| < \frac{\varepsilon}{2}$ for $n \geq N$. Then for $m, n \geq N$ we have $\|v_m - v_n\| \leq \|v_m - v\| + \|v_n - v\| < \varepsilon$ for $m, n \in \mathbb{N}$. This shows that $(v_n : n \in \mathbb{N})$ is a Cauchy sequence with respect to the norm.

3. Suppose $\lim_{n\to\infty} v_n = v$, that is, $\lim_{n\to\infty} \|v_n - v\| = 0$. By Observation 15.6, we have $\big|\|v_n\| - \|v\|\big| \leq \|v_n - v\|$ and then $\lim_{n\to\infty} \big|\|v_n\| - \|v\|\big| \leq \lim_{n\to\infty} \|v_n - v\| = 0$. Thus we have $\lim_{n\to\infty} \big|\|v_n\| - \|v\|\big| = 0$ and this implies $\lim_{n\to\infty} \|v_n\| = \|v\|$. ∎

Proposition 15.10. *Let $(V, \|\cdot\|)$ be a Banach space. If M is a linear subspace of V and M is a closed set in V then $(M, \|\cdot\|)$ is a Banach space.*

Proof. If M is a linear subspace of V then $(M, \|\cdot\|)$ is a normed linear space. Suppose M is also a closed set in V. To show that $(M, \|\cdot\|)$ is a Banach space we show that M is complete with respect to the norm $\|\cdot\|$. Thus let $(v_n : n \in \mathbb{N})$ be a Cauchy sequence in M with respect to $\|\cdot\|$. Then $(v_n : n \in \mathbb{N})$ is a Cauchy sequence in V with respect to $\|\cdot\|$. Since V is complete with respect to $\|\cdot\|$ there exists $v \in V$ such that $\lim_{n\to\infty} \|v_n - v\| = 0$. Then since $(v_n : n \in \mathbb{N}) \subset M$ and since M is a closed set in V, the convergence $\lim_{n\to\infty} \|v_n - v\| = 0$ implies that $v \in M$. Thus every Cauchy sequence in M with respect to $\|\cdot\|$ converges to an element in M, that is, M is complete with respect to $\|\cdot\|$. This shows that $(M, \|\cdot\|)$ is a Banach space. ∎

Definition 15.11. *Let ν_1 and ν_2 be two norms on a linear space V. We say that ν_1 and ν_2 are equivalent if there exist $\alpha, \beta > 0$ such that $\nu_1(v) \le \alpha\nu_2(v)$ and $\nu_2(v) \le \beta\nu_1(v)$ for $v \in V$.*

Proposition 15.12. *Let ν_1 and ν_2 be two equivalent norms on a linear space V.*

(a) *If a sequence $(v_n : n \in \mathbb{N})$ in V is a Cauchy sequence with respect to one of ν_1 and ν_2, then it is a Cauchy sequence with respect to the other.*

(b) *If a sequence $(v_n : n \in \mathbb{N})$ in V converges in the norm ν_1 to some $v \in V$, then $(v_n : n \in \mathbb{N})$ converges in the norm ν_2 to $v \in V$ and vice versa.*

(c) *If V is a Banach space with respect to one of ν_1 and ν_2, then V is a Banach space with respect to the other.*

Proof. The proofs of these statements are immediate from the fact that for the two metrics $\rho_1(u, v) = \nu_1(u - v)$ and $\rho_2(u, v) = \nu_2(u - v)$, we have $\rho_1(u, v) \le \alpha\rho_2(u, v)$ and $\rho_2(u, v) \le \beta\rho_1(u, v)$. ∎

[II] Banach Spaces on $\mathbb{R}^k$

$\mathbb{R}^k$ consisting of $x = (x_1, \dots, x_k)$ where $x_1, \dots, x_k \in \mathbb{R}$ is a linear space over the field of scalars $\mathbb{R}$ if addition and scalar multiplication are defined by $x + y = (x_1 + y_1, \dots, x_k + y_k)$ and $\alpha x = (\alpha x_1, \dots, \alpha x_k)$ for $x, y \in \mathbb{R}^k$ and $\alpha \in \mathbb{R}$. Let us prove algebraic forms of Hölder and Minkowski Inequalities with exponent 2.

Lemma 15.13. *For $a_i, b_i \in \mathbb{R}$ for $i = 1, \dots, k$, we have*

$$(1) \qquad \sum_{i=1}^{k} |a_i b_i| \le \left[\sum_{i=1}^{k} a_i^2\right]^{1/2} \left[\sum_{i=1}^{k} b_i^2\right]^{1/2},$$

and

$$(2) \qquad \left[\sum_{i=1}^{k} (a_i + b_i)^2\right]^{1/2} \le \left[\sum_{i=1}^{k} a_i^2\right]^{1/2} + \left[\sum_{i=1}^{k} b_i^2\right]^{1/2}.$$

Proof. To prove (1), let us prove first

$$\left[\sum_{i=1}^{k} a_i b_i\right]^2 \le \left[\sum_{i=1}^{k} a_i^2\right]\left[\sum_{i=1}^{k} b_i^2\right]. \tag{3}$$

Now

$$\left[\sum_{i=1}^{k} a_i b_i\right]^2 = \sum_{i=1}^{k}\sum_{j=1}^{k} a_i b_i a_j b_j = \sum_{i=1}^{k} a_i^2 b_i^2 + 2 \sum_{1\le i<j\le k} a_i b_i a_j b_j,$$

and

$$\left[\sum_{i=1}^{k} a_i^2\right]\left[\sum_{i=1}^{k} b_i^2\right] = \sum_{i=1}^{k}\sum_{j=1}^{k} a_i^2 b_j^2 = \sum_{i=1}^{k} a_i^2 b_i^2 + \sum_{1\le i<j\le k} a_i^2 b_j^2 + \sum_{1\le i<j\le k} a_j^2 b_i^2,$$

and then

$$\begin{aligned}\left[\sum_{i=1}^{k} a_i^2\right]\left[\sum_{i=1}^{k} b_i^2\right] - \left[\sum_{i=1}^{k} a_i b_i\right]^2 &= \sum_{1\le i<j\le k} \{a_i^2 b_j^2 + a_j^2 b_i^2 - 2a_i b_i a_j b_j\} \\ &= \sum_{1\le i<j\le k} \{a_i b_j - a_j b_i\}^2 \ge 0.\end{aligned}$$

This proves (3). Since (3) holds for any $a_i, b_i \in \mathbb{R}$, $i = 1, \dots, k$, it holds for $|a_i|, |b_i| \in \mathbb{R}$, $i = 1, \dots, k$. Thus we have $\left[\sum_{i=1}^{k} |a_i||b_i|\right]^2 \le \left[\sum_{i=1}^{k} |a_i|^2\right]\left[\sum_{i=1}^{k} |b_i|^2\right]$. Taking the square roots, we have (1).

To prove (2), note that

$$\begin{aligned}\sum_{i=1}^{k}\{a_i + b_i\}^2 &= \sum_{i=1}^{k} a_i^2 + \sum_{i=1}^{k} b_i^2 + 2\sum_{i=1}^{k} a_i b_i \\ &\le \sum_{i=1}^{k} a_i^2 + \sum_{i=1}^{k} b_i^2 + 2\left[\sum_{i=1}^{k} a_i^2\right]^{1/2}\left[\sum_{i=1}^{k} b_i^2\right]^{1/2} \quad \text{by (1)} \\ &= \left\{\left[\sum_{i=1}^{k} a_i^2\right]^{1/2} + \left[\sum_{i=1}^{k} b_i^2\right]^{1/2}\right\}^2.\end{aligned}$$

Taking the square roots, we have (2). ∎

Theorem 15.14. *On $\mathbb{R}^k$, let us define functions $\|\cdot\|_1$, $\|\cdot\|_2$, and $\|\cdot\|_u$ by*

$$\|x\|_1 = \textstyle\sum_{i=1}^{k} |x_i| \qquad \text{for } x \in \mathbb{R}^k,$$

$$\|x\|_2 = \left\{\textstyle\sum_{i=1}^{k} |x_i|^2\right\}^{1/2} \qquad \text{for } x \in \mathbb{R}^k,$$

$$\|x\|_u = \max_{i=1,\dots,k} |x_i| \qquad \text{for } x \in \mathbb{R}^k.$$

(a) *$\|\cdot\|_1$, $\|\cdot\|_2$, and $\|\cdot\|_u$ are norms on $\mathbb{R}^k$.*
(b) *The norms $\|\cdot\|_1$, $\|\cdot\|_2$, and $\|\cdot\|_u$ satisfy the following inequalities*

(1) $$\|x\|_1 \geq \|x\|_2 \geq \|x\|_u \quad \textit{for } x \in \mathbb{R}^k$$

and

(2) $$\frac{\|x\|_1}{k} \leq \frac{\|x\|_2}{k^{1/2}} \leq \|x\|_u \quad \textit{for } x \in \mathbb{R}^k.$$

Thus any two of the three norms is a pair of equivalent norms.
(c) *$\mathbb{R}^k$ is a Banach space with respect to each one of $\|\cdot\|_1$, $\|\cdot\|_2$, and $\|\cdot\|_u$.*

Proof. 1. It follows immediately from its definition that each one of $\|\cdot\|_1$, $\|\cdot\|_2$, and $\|\cdot\|_u$ satisfies conditions 1°, 2°, and 3° of Definition 15.1. Regarding condition 4°, triangle inequality, note that for $x, y \in \mathbb{R}^k$ given by $x = (x_1, \dots, x_k)$ and $y = (y_1, \dots, y_k)$, we have

$$\begin{aligned} \|x+y\|_1 &= \sum_{i=1}^k |x_i + y_i| \leq \sum_{i=1}^k \{|x_i| + |y_i|\} \\ &= \sum_{i=1}^k |x_i| + \sum_{i=1}^k |y_i| = \|x\|_1 + \|y\|_1, \end{aligned}$$

and

$$\begin{aligned} \|x+y\|_u &= \max_{i=1,\dots,k} |x_i + y_i| \leq \max_{i=1,\dots,k} \{|x_i| + |y_i|\} \\ &\leq \max_{i=1,\dots,k} |x_i| + \max_{i=1,\dots,k} |y_i| = \|x\|_u + \|y\|_u. \end{aligned}$$

This shows that $\|\cdot\|_1$ and $\|\cdot\|_u$ satisfy the triangle inequality. The fact that $\|\cdot\|_2$ satisfies the triangle inequality follows from (2) of Lemma 15.13 .

2. To prove (1), note that $\|x\|_1^2 = \left[\sum_{i=1}^k |x_i|\right]^2 \geq \sum_{i=1}^k |x_i|^2 = \|x\|_2^2$ and thus we have $\|x\|_1 \geq \|x\|_2$. Also $\|x\|_2^2 = \sum_{i=1}^k |x_i|^2 \geq \max_{i=1,\dots,k} |x_i|^2 = \left[\max_{i=1,\dots,k} |x_i|\right]^2$ so that $\|x\|_2 \geq \max_{i=1,\dots,k} |x_i| = \|x\|_u$. This proves (1).

To prove (2), let $x = (x_1, \dots, x_k) \in \mathbb{R}^k$. Applying (1) of Lemma 15.13 with $a_i = x_i$ and $b_i = 1$ for $i = 1, \dots, k$, we have $\sum_{i=1}^k |x_i| \leq \left[\sum_{i=1}^k |x_i|^2\right]^{1/2} k^{1/2}$. Dividing by k, we have

$$\frac{1}{k}\sum_{i=1}^k |x_i| \leq \frac{1}{k^{1/2}}\left[\sum_{i=1}^k |x_i|^2\right]^{1/2} \leq \frac{1}{k^{1/2}}\left[k \max_{i=1,\dots,k} |x_i|^2\right]^{1/2} = \max_{i=1,\dots,k} |x_i|.$$

This proves (2).

3. Since the three norms $\|\cdot\|_1$, $\|\cdot\|_2$, and $\|\cdot\|_u$ are equivalent, if we show that $\mathbb{R}^k$ is a Banach space with respect to one of the three, then $\mathbb{R}^k$ is a Banach space with respect to each of the other two according to Lemma 15.12. Let us show that $\mathbb{R}^k$ is a Banach space with respect to $\|\cdot\|_1$. Let $(x^{(n)} : n \in \mathbb{N})$ be a Cauchy sequence in $\mathbb{R}^k$ with respect to the norm $\|\cdot\|_1$. Then for every $\varepsilon > 0$, there exists $N \in \mathbb{N}$ such that $\|x^{(m)} - x^{(n)}\|_1 < \varepsilon$ for $m, n \geq N$. Then for each $i = 1, \ldots, k$, we have

$$\left|x_i^{(m)} - x_i^{(n)}\right| \leq \sum_{j=1}^{k} \left|x_j^{(m)} - x_j^{(n)}\right| = \left\|x^{(m)} - x^{(n)}\right\|_1 < \varepsilon \quad \text{for } m, n \geq N.$$

This shows that $(x_i^{(n)} : n \in \mathbb{N})$ is a Cauchy sequence in $\mathbb{R}$. Thus there exists $x_i \in \mathbb{R}$ such that $\lim_{n\to\infty} \left|x_i^{(n)} - x_i\right| = 0$. Let $x = (x_1, \ldots, x_k) \in \mathbb{R}^k$. Then

$$\lim_{n\to\infty} \left\|x^{(n)} - x\right\|_1 = \lim_{n\to\infty} \sum_{i=1}^{k} \left|x_i^{(n)} - x_i\right| = \sum_{i=1}^{k} \lim_{n\to\infty} \left|x_i^{(n)} - x_i\right| = 0.$$

This shows that $\mathbb{R}^k$ is complete with respect to the norm $\|\cdot\|_1$. Therefore $\mathbb{R}^k$ is a Banach space with respect to the norm $\|\cdot\|_1$. ■

[III] The Space of Continuous Functions $C([a, b])$

Let $C([a, b])$ be the collection of all real-valued continuous functions defined on $[a, b]$. If we define addition and scalar multiplication in $C([a, b])$ by pointwise addition and scalar multiplication on $[a, b]$, that is, $(f+g)(t) = f(t)+g(t)$ for $t \in [a, b]$ for $f, g \in C([a, b])$, and $(\alpha f)(t) = \alpha f(t)$ for $t \in [a, b]$ for $f \in C([a, b])$ and $\alpha \in \mathbb{R}$, then $C([a, b])$ is a linear space over the field of scalars $\mathbb{R}$ in which the origin is the identically vanishing function on $[a, b]$.

Theorem 15.15. *On $C([a, b])$, define $\|\cdot\|_1$, $\|\cdot\|_2$, and $\|\cdot\|_u$ by*

$$\begin{aligned} \|f\|_1 &= \textstyle\int_{[a,b]} |f| \, d\mu_L, \\ \|f\|_2 &= \left[\textstyle\int_{[a,b]} |f|^2 \, d\mu_L \right]^{1/2}, \\ \|f\|_u &= \max_{t\in[a,b]} |f(t)|. \end{aligned}$$

(a) *$\|\cdot\|_1$, $\|\cdot\|_2$, and $\|\cdot\|_u$ are norms on $C([a, b])$, called L^1 norm, L^2 norm, and uniform norm respectively.*

(b) *The norms $\|\cdot\|_1$, $\|\cdot\|_2$, and $\|\cdot\|_u$ satisfy the inequality that for $f \in C([a, b])$*

$$\frac{\|f\|_1}{b-a} \leq \frac{\|f\|_2}{(b-a)^{1/2}} \leq \|f\|_u.$$

(c) *$C([a, b])$ is a Banach space with respect to the norm $\|\cdot\|_u$.*

Proof. 1. It follows immediately from the definition that $\|\cdot\|_1$, $\|\cdot\|_2$, and $\|\cdot\|_u$ satisfy conditions 1°, 2°, and 3° in Definition 15.5. Let us verify for instance that if $\|f\|_1 = 0$ for some $f \in C([a, b])$ then $f = 0 \in C([a, b])$. Now if $\|f\|_1 = 0$, that is, $\int_{[a,b]} |f| \, d\mu_L = 0$, then $f = 0$ a.e. on $[a, b]$. Thus there exists a null set E in the measure space $(\mathbb{R}, \mathfrak{M}_L, \mu_L)$ such that $E \subset [a, b]$ and $f = 0$ on $[a, b] \cap E^c$. Now since every open interval in $\mathbb{R}$ has positive Lebesgue measure, the null set E cannot contain any open interval. Thus every open interval in $\mathbb{R}$ contains some point in E^c. This shows that E^c is a dense set in $\mathbb{R}$. Then for every $x \in [a, b]$ there exists a sequence $(x_n : n \in \mathbb{N})$ in $[a, b] \cap E^c$ such that $\lim_{n\to\infty} x_n = x$. The continuity of f on $[a, b]$ and the fact that $f = 0$ on $[a, b] \cap E^c$ imply that $f(x) = 0$. This shows that $f = 0$ on $[a, b]$. Thus $f = 0 \in C([a, b])$.

Condition 4° in Definition 15.5, the triangle inequality, is verified for $\|\cdot\|_u$ and $\|\cdot\|_1$ as follows:

$$\begin{aligned}\|f+g\|_u &= \max_{t\in[a,b]} |f(t)+g(t)| \le \max_{t\in[a,b]} \{|f(t)|+|g(t)|\} \\ &\le \max_{t\in[a,b]} |f(t)| + \max_{t\in[a,b]} |g(t)| = \|f\|_u + \|g\|_u,\end{aligned}$$

and

$$\begin{aligned}\|f+g\|_1 &= \int_{[a,b]} |f+g|\, d\mu_L \le \int_{[a,b]} \{|f|+|g|\}\, d\mu_L \\ &= \int_{[a,b]} |f|\, d\mu_L + \int_{[a,b]} |g|\, d\mu_L = \|f\|_1 + \|g\|_1.\end{aligned}$$

2. It remains to verify the triangle inequality for $\|\cdot\|_2$. Let us note that for every $h \in C([a, b])$, we have $\int_{[a,b]} h\, d\mu_L = \int_a^b h(t)\, dt$. For each $k \in \mathbb{N}$, let us partition $[a, b]$ into k equal subintervals with partition points $a = t_{k,0} < \cdots < t_{k,k} = b$, let $\ell_k = (b-a)/k$, and write $h_{k,i} = h(t_{k,i})$ for $i = 1, \dots, k$. Then

$$(1) \qquad \int_a^b h(t)\, dt = \lim_{k\to\infty} \sum_{i=1}^k h_{k,i}\ell_k.$$

For $f, g \in C([a, b])$, let $f_{k,i} = f(t_{k,i})$ and $g_{k,i} = g(t_{k,i})$ for $i = 1, \dots, k$ and $k \in \mathbb{N}$. By (2) of Lemma 15.13, we have

$$(2) \qquad \Big[\sum_{i=1}^k |f_{k,i} + g_{k,i}|^2 \ell_k\Big]^{1/2} \le \Big[\sum_{i=1}^k |f_{k,i}|^2 \ell_k\Big]^{1/2} + \Big[\sum_{i=1}^k |g_{k,i}|^2 \ell_k\Big]^{1/2}.$$

Letting $k \to \infty$ in (2) and applying (1) to our $|f+g|^2, |f|^2, |g|^2 \in C([a, b])$, we have

$$\Big[\int_a^b |f(t)+g(t)|^2\, dt\Big]^{1/2} \le \Big[\int_a^b |f(t)|^2\, dt\Big]^{1/2} + \Big[\int_a^b |g(t)|^2\, dt\Big]^{1/2}.$$

This shows that $\|f+g\|_2 \le \|f\|_2 + \|g\|_2$.

3. With $a_i = f_{k,i}\ell_k^{1/2}$ and $b_i = \ell_k^{1/2}$ for $i = 1, \dots, k$ in (1) of Lemma 15.13, we have

$$(3) \qquad \sum_{i=1}^k |f_{k,i}|\ell_k = \sum_{i=1}^k |f_{k,i}|\ell_k^{1/2}\ell_k^{1/2} \le \Big[\sum_{i=1}^k |f_{k,i}|^2 \ell_k\Big]^{1/2} (b-a)^{1/2}.$$

By (1) we have

$$\lim_{k\to\infty}\sum_{i=1}^{k}|f_{k,i}|\ell_k=\int_a^b|f(x)|\,dx=\int_{[a,b]}|f|\,d\mu_L$$

and similarly we have

$$\lim_{k\to\infty}\sum_{i=1}^{k}|f_{k,i}|^2\ell_k=\int_a^b|f(x)|^2\,dx=\int_{[a,b]}|f|^2\,d\mu_L.$$

Letting $k\to\infty$ in (3), we have

$$\int_{[a,b]}|f|\,d\mu_L\le(b-a)^{1/2}\left[\int_{[a,b]}|f|^2\,d\mu_L\right]^{1/2},$$

that is, $\|f\|_1(b-a)^{-1}\le\|f\|_2(b-a)^{-1/2}$. From $\int_{[a,b]}|f|^2\,d\mu_L\le\|f\|_u^2(b-a)$, we have $\|f\|_2(b-a)^{-1/2}\le\|f\|_u$. This proves (b).

4. Let us show that $C([a,b])$ is a Banach space with respect to the norm $\|\cdot\|_u$. Let $(f_n : n\in\mathbb{N})$ be a Cauchy sequence in $C([a,b])$ with respect to the norm $\|\cdot\|_u$. Then for every $\varepsilon>0$, there exists $N\in\mathbb{N}$ such that $\|f_m-f_n\|_u<\varepsilon$ for $m,n\ge N$ and therefore

$$(4)\qquad |f_m(t)-f_n(t)|\le\|f_m-f_n\|_u<\varepsilon\quad\text{for all } t\in[a,b].$$

This shows that for every $t\in[a,b]$, $(f_n(t) : n\in\mathbb{N})$ is a Cauchy sequence of real numbers and therefore converges to a real number $f(t)$. By (4) we have $\left|\lim_{m\to\infty}f_m(t)-f_n(t)\right|\le\varepsilon$ for all $n\ge N$, that is,

$$(5)\qquad |f(t)-f_n(t)|\le\varepsilon\quad\text{for all } t\in[a,b].$$

Thus the sequence $(f_n : n\in\mathbb{N})$ converges uniformly to f on $[a,b]$ so that f is continuous on $[a,b]$, that is, $f\in C([a,b])$. From (5), we have $\max_{t\in[a,b]}|f(t)-f_n(t)|\le\varepsilon$ for $n\ge N$. This shows that $\lim_{n\to\infty}\|f-f_n\|_u=0$. ■

[IV] A Criterion for Completeness of a Normed Linear Space

Observation 15.16. Given a metric space (S,ρ). If a Cauchy sequence $(v_n : n\in\mathbb{N})$ in S has a subsequence $(v_{n_k} : k\in\mathbb{N})$ which converges to an element $v\in S$, then the sequence $(v_n : n\in\mathbb{N})$ converges to v.

Proof. Let $\varepsilon>0$. Since $(v_n : n\in\mathbb{N})$ is a Cauchy sequence, there exists $N_1\in\mathbb{N}$ such that $\rho(v_n,v_m)<\frac{\varepsilon}{2}$ for $n,m\ge N_1$. Since $\lim_{k\to\infty}\rho(v_{n_k},v)=0$, there exists $N_2\in\mathbb{N}$ such that $\rho(v_{n_k},v)<\frac{\varepsilon}{2}$ if $k\ge N_2$. Let $N=\max\{N_1,N_2\}$. Then for $n\ge N$, with an arbitrary $k\ge N$, we have $\rho(v_n,v)\le\rho(v_n,v_{n_k})+\rho(v_{n_k},v)<\frac{\varepsilon}{2}+\frac{\varepsilon}{2}=\varepsilon$. This shows that $\lim_{n\to\infty}\rho(v_n,v)=0$. ■

Definition 15.17. *Given a sequence $(v_n : n \in \mathbb{N})$ in a normed linear space $(V, \|\cdot\|)$. Consider the sequence $(w_n : n \in \mathbb{N})$ in V defined by $w_n = \sum_{k=1}^n v_k$ for $n \in \mathbb{N}$. If the sequence $(w_n : n \in \mathbb{N})$ converges in the norm, that is, if there exists $w \in V$ such that $\lim_{n\to\infty} \|w_n - w\| = 0$, then we say that the series $\sum_{n\in\mathbb{N}} v_n$ converges in the norm to the sum w and write $\sum_{n\in\mathbb{N}} v_n = w$. When no such w exists in V, we say that the series $\sum_{n\in\mathbb{N}} v_n$ diverges. We call $(w_n : n \in \mathbb{N})$ the sequence of partial sums of the series $\sum_{n\in\mathbb{N}} v_n$.*

Theorem 15.18. *Given a normed linear space $(V, \|\cdot\|)$. The space V is complete with respect to the norm $\|\cdot\|$ if and only if for every sequence $(v_n : n \in \mathbb{N})$ in V such that $\sum_{n\in\mathbb{N}} \|v_n\| < \infty$, the series $\sum_{n\in\mathbb{N}} v_n$ converges in the norm.*

Proof. 1. Suppose V is complete with respect to the norm $\|\cdot\|$. Let $(v_n : n \in \mathbb{N})$ be a sequence in V with $\sum_{n\in\mathbb{N}} \|v_n\| < \infty$. Let us show that the series $\sum_{n\in\mathbb{N}} v_n$ converges, that is, for $w_n = \sum_{k=1}^n v_k$ for $n \in \mathbb{N}$ there exists $w \in V$ such that $\lim_{n\to\infty} \|w_n - w\| = 0$. Now since $\sum_{n\in\mathbb{N}} \|v_n\| < \infty$, for every $\varepsilon > 0$ there exists $N \in \mathbb{N}$ such that $\sum_{k\geq N} \|v_k\| < \varepsilon$. Then for $m > n \geq N$ we have

$$\|w_m - w_n\| = \Big\|\sum_{k=1}^m v_k - \sum_{k=1}^n v_k\Big\| = \Big\|\sum_{k=n+1}^m v_k\Big\|$$
$$\leq \sum_{k=n+1}^m \|v_k\| \leq \sum_{k\geq n+1} \|v_k\| \leq \sum_{k\geq N} \|v_k\| < \varepsilon.$$

This shows that $(w_n : n \in \mathbb{N})$ is a Cauchy sequence in V. The completeness of V with respect to the norm $\|\cdot\|$ implies then that there exists $w \in V$ such that $\lim_{n\to\infty} \|w_n - w\| = 0$.

2. Conversely suppose for every sequence $(v_n : n \in \mathbb{N})$ in V with $\sum_{n\in\mathbb{N}} \|v_n\| < \infty$, the series $\sum_{n\in\mathbb{N}} v_n$ converges in the norm. Let us show that V is complete with respect to the norm $\|\cdot\|$. Let $(v_n : n \in \mathbb{N})$ be a Cauchy sequence in V with respect to the norm. To show that the sequence converges in the norm, it suffices to show that there exists a subsequence which converges to some element in V in the norm according to Observation 15.16. Now since $(v_n : n \in \mathbb{N})$ is a Cauchy sequence, there exists $N_1 < N_2 < \cdots$ such that for every $k \in \mathbb{N}$ we have

$$(1) \qquad \|v_m - v_n\| < \frac{1}{2^k} \quad \text{for } m, n \geq N_k.$$

Let us show that the subsequence $(v_{N_k} : k \in \mathbb{N})$ of $(v_n ; n \in \mathbb{N})$ converges in the norm. Consider a sequence $(u_i : i \in \mathbb{N})$ in V defined by $u_1 = v_{N_1}$ and $u_i = v_{N_i} - v_{N_{i-1}}$ for $i \geq 2$. Then we have

$$(2) \qquad \sum_{i=1}^k u_i = v_{N_1} + \sum_{i=2}^k \{v_{N_i} - v_{N_{i-1}}\} = v_{N_k} \quad \text{for } k \in \mathbb{N},$$

By the definition of the sequence $(u_i : i \in \mathbb{N})$ and by (1) we have

$$\sum_{i\in\mathbb{N}} \|u_i\| = \|u_1\| + \sum_{i\geq 2} \|v_{N_i} - v_{N_{i-1}}\| \leq \|u_1\| + \sum_{i\geq 2} \frac{1}{2^{i-1}} = \|u_1\| + 1 < \infty.$$

Thus by our assumption the series $\sum_{i\in\mathbb{N}} u_i$ converges to some element $u \in V$ in the norm, that is, $\lim_{k\to\infty} \| \sum_{i=1}^{k} u_i - u\| = 0$. Substituting (2), we have $\lim_{k\to\infty} \|v_{N_k} - u\| = 0$. ∎

[V] Hilbert Spaces

Definition 15.19. *Given a linear space V over $\mathbb{C}$ (resp. $\mathbb{R}$). A complex-valued (resp. real-valued) function $\langle\cdot,\cdot\rangle$ on $V \times V$ is called an inner product on V if it satisfies the following conditions:*

1° $\langle v, w\rangle \in \mathbb{C}$ *(resp. $\mathbb{R}$) for $v, w \in V$,*
2° $\langle v, v\rangle \in [0,\infty)$ *for every $v \in V$,*
3° $\langle v, v\rangle = 0$ *if and only if $v = 0 \in V$,*
4° $\langle \alpha_1 v_1 + \alpha_2 v_2, w\rangle = \alpha_1 \langle v_1, w\rangle + \alpha_2 \langle v_2, w\rangle$ *for $v_1, v_2, w \in V$, $\alpha_1, \alpha_2 \in \mathbb{C}$ (resp. $\mathbb{R}$),*
5° $\langle v, w\rangle = \overline{\langle w, v\rangle}$ *for $v, w \in V$, where $\overline{z}$ is the complex conjugate of $z \in \mathbb{C}$.*

A linear space V on which an inner product $\langle\cdot,\cdot\rangle$ is defined is called an inner product space and we write $\big(V, \langle\cdot,\cdot\rangle\big)$ for it.

Observation 15.20. For an inner product $\langle\cdot,\cdot\rangle$ on a linear space V, we have

$$\langle v, \beta_1 w_1 + \beta_2 w_2\rangle = \overline{\beta_1}\,\langle v, w_1\rangle + \overline{\beta_2}\,\langle v, w_2\rangle\,,$$

for $v, w_1, w_2 \in V$ and $\beta_1, \beta_2 \in \mathbb{C}$. This is an immediate consequence of 4° and 5° in Definition 15.19.

Definition 15.21. *On an inner product space $\big(V, \langle\cdot,\cdot\rangle\big)$, we define a function $\|\cdot\|_h$ on V by setting $\|v\|_h = \{\langle v, v\rangle\}^{1/2}$ for $v \in V$.*

To show that $\|\cdot\|_h$ defined above is a norm and in particular satisfies the triangle inequality, we prove first the Schwarz's Inequality.

Theorem 15.22. (Schwarz's Inequality) *For the function $\|\cdot\|_h$ on an inner product space $\big(V, \langle\cdot,\cdot\rangle\big)$, we have $|\langle v, w\rangle| \le \|v\|_h\|w\|_h$ for $v, w \in V$. The equality holds if and only if v and w are linearly dependent.*

Proof. 1. Consider the case $\langle v, w\rangle = 0$. In this case, $|\langle v, w\rangle| \le \|v\|_h\|w\|_h$ holds trivially. If f and g are linearly dependent, say $v = \lambda w$ where $\lambda \in \mathbb{C}$ (resp. $\mathbb{R}$), then $\|\lambda w\|_h = |\lambda|\|w\|_h$ by Definition 15.21 and conditions 4° and 5° of Definition 15.19, and

$$|\langle v, w\rangle| = |\langle \lambda w, w\rangle| = |\lambda|\,\langle w, w\rangle = |\lambda|\|w\|_h^2 = \|\lambda w\|_h\|w\|_h = \|v\|_h\|w\|_h.$$

Conversely if $|\langle v, w\rangle| = \|v\|_h\|w\|_h$, then since $\langle v, w\rangle = 0$, at least one of $\|v\|_h$ and $\|w\|_h$ is equal to 0, that is, at least one of $\langle v, v\rangle$ and $\langle w, w\rangle$ is equal to 0. Thus according to condition 3° in Definition 15.19, at least one of v and w is equal to $0 \in V$ and therefore v and w are linearly dependent.

2. Assume that $\langle v, w\rangle \neq 0$. Let $\vartheta = \overline{\langle v, w\rangle} | \langle v, w\rangle |^{-1}$ so that $\vartheta \in \mathbb{C}$ with $|\vartheta| = 1$. Then for every $\lambda \in \mathbb{R}$, we have by 2°, 4°, and 5° of Definition 15.19

$$(1)\quad 0 \leq \langle \vartheta v + \lambda w, \vartheta v + \lambda w\rangle = \vartheta\overline{\vartheta} \langle v, v\rangle + \vartheta\lambda \langle v, w\rangle + \lambda\overline{\vartheta} \langle w, v\rangle + \lambda^2 \langle w, w\rangle$$
$$= |\vartheta|^2\|v\|_h^2 + 2\lambda\Re\{\vartheta \langle v, w\rangle\} + \lambda^2\|w\|_h^2 \leq \|v\|_h^2 + 2\lambda| \langle v, w\rangle | + \lambda^2\|w\|_h^2,$$

since $\vartheta \langle v, w\rangle = \overline{\langle v, w\rangle} \langle v, w\rangle | \langle v, w\rangle |^{-1} = | \langle v, w\rangle |$. Thus we have a polynomial of degree 2 in the real variable $\lambda$with real coefficients which does not assume negative values. Then its discriminant cannot be positive, that is,

$$(2)\qquad | \langle v, w\rangle |^2 - \|v\|_h^2\|w\|_h^2 \leq 0.$$

This inequality is equivalent to $| \langle v, w\rangle | \leq \|v\|_h\|w\|_h$. Furthermore the equality in (2) holds if and only if our second degree polynomial in λ has a real double root λ_0, that is,

$$(3)\qquad \lambda_0^2\|w\|_h^2 + 2\lambda_0| \langle v, w\rangle | + \|v\|_h^2 = 0,$$

which, in view of (1), is equivalent to $\langle \vartheta v + \lambda_0 w, \vartheta v + \lambda_0 w\rangle = 0$, that is, $\vartheta v + \lambda_0 w = 0$, in other words, v and w are linearly dependent. ∎

Corollary 15.23. *For the function* $\|\cdot\|_h$ *on an inner product space* $(V, \langle\cdot,\cdot\rangle)$, *we have* $\|v + w\|_h \leq \|v\|_h + \|w\|_h$ *for* $v, w \in V$.

Proof. If $v, w \in V$ then $\|v + w\|_h^2 = \langle v + w, v + w\rangle = \|v\|_h^2 + \langle v, w\rangle + \langle w, v\rangle + \|w\|_h^2$. But $\langle v, w\rangle + \langle w, v\rangle = 2\Re \langle v, w\rangle \leq 2| \langle v, w\rangle | \leq 2\|v\|_h^2\|w\|_h^2$ by Theorem 15.22. Therefore we have $\|v+w\|_h^2 \leq \|v\|_h^2+2\|v\|_h^2\|w\|_h^2+\|w\|_h^2 = \{\|v\|_h+\|w\|_h\}^2$, and then $\|v+w\|_h \leq \|v\|_h + \|w\|_h$. ∎

Theorem 15.24. *The function* $\|\cdot\|_h$ *on an inner product space* $(V, \langle\cdot,\cdot\rangle)$ *as defined by Definition 15.21 is a norm on the linear space* V.

Proof. By condition 2° of Definition 15.19, $\|v\|_h \in [0, \infty)$ for every $v \in V$ and by condition 3° of Definition 15.19, $\|v\|_h = 0$ if and only if $v = 0 \in V$. For $v \in V$ and $\alpha \in \mathbb{C}$, we have $\|\alpha v\|_h = [\langle \alpha v, \alpha v\rangle]^{1/2} = [\alpha\overline{\alpha} \langle v, v\rangle]^{1/2} = |\alpha|\|v\|_h$. The triangle inequality for $\|\cdot\|_h$ is given by Corollary 15.23. ∎

Definition 15.25. *The norm* $\|\cdot\|_h = \{\langle\cdot,\cdot\rangle\}^{1/2}$ *on an inner product space* $(V, \langle\cdot,\cdot\rangle)$ *is called a Hilbert norm. If* V *is complete with respect to the Hilbert norm, then* V *is called a Hilbert space.*

[VI] Bounded Linear Mappings of Normed Linear Spaces

Notations. For a linear space V over the field $\mathbb{K}$ where $\mathbb{K} = \mathbb{R}$ or $\mathbb{K} = \mathbb{C}$ we write $(V, \mathbb{K})$ to indicate that $\mathbb{K}$ is the field of scalars for the linear space. We write $(W, \mathbb{K}')$ for another linear space W whose field of scalars is $\mathbb{K}'$ where $\mathbb{K}' = \mathbb{R}$ or $\mathbb{K}' = \mathbb{C}$.

We write $(V, \mathbb{K}, \|\cdot\|)$ for a normed linear space when a norm $\|\cdot\|$ is defined on a linear space $(V, \mathbb{K})$.

Example. Consider a normed linear space $(V, \mathbb{K}, \|\cdot\|)$. Now $\mathbb{K}$ is a linear space over the field of scalars $\mathbb{K}$. For $\mathbb{K} = \mathbb{R}$ the absolute value $|\cdot|$ of a real number is a norm on $\mathbb{K}$. For $\mathbb{K} = \mathbb{C}$ the modulus $|\cdot|$ of a complex number is a norm on $\mathbb{K}$. Thus $(\mathbb{K}, \mathbb{K}, |\cdot|)$ is a particular case of a normed linear space over the field of scalars $\mathbb{K}$.

Definition 15.26. (a) *Let $(V, \mathbb{K})$ and $(W, \mathbb{K}')$ be two linear spaces. A mapping L of V into W is called a linear mapping if it satisfies the following conditions:*
1° *homogeneity: $L(\gamma v) = \gamma L(v)$ for $v \in V$ and $\gamma \in \mathbb{K}$,*
2° *additivity: $L(v_1 + v_2) = L(v_1) + L(v_2)$ for $v_1, v_2 \in V$.*
Let $\mathfrak{L}\{V, W\}$ be the collection of all linear mappings of the linear space $(V, \mathbb{K})$ into the linear space $(W, \mathbb{K}')$.
(b) *For the particular case that $(W, \mathbb{K}') = (\mathbb{K}', \mathbb{K}')$ a linear mapping L of the linear space $(V, \mathbb{K})$ into the linear space $(\mathbb{K}', \mathbb{K}')$ is called a $\mathbb{K}'$-valued linear functional on $(V, \mathbb{K})$. We write $\mathfrak{L}\{V, \mathbb{K}'\}$ for the collection of all $\mathbb{K}'$-valued linear functionals on $(V, \mathbb{K})$.*

Note that for $0 \in V$ we have $L(0) = L(0v) = 0L(v) = 0$ with an arbitrary $v \in V$. Note also that conditions 1° and 2° are equivalent to the single condition:

$$L(\gamma_1 v_1 + \gamma_2 v_2) = \gamma_1 L(v_1) + \gamma_2 L(v_2) \quad \text{for } v_1, v_2 \in V \text{ and } \gamma_1, \gamma_2 \in \mathbb{K}.$$

Remark 15.27. In the definition of a linear mapping of a linear space $(V, \mathbb{K})$ into a linear space $(W, \mathbb{K}')$ it is not required that $\mathbb{K} = \mathbb{K}'$. However a linear mapping of a linear space $(V, \mathbb{C})$ into a linear space $(W, \mathbb{R})$ may not exist since the homogeneity condition 1° of Definition 15.26

$$L(\gamma v) = \gamma L(v) \quad \text{for } v \in V \text{ and } \gamma \in \mathbb{C}$$

may be meaningless in this case. Indeed if L is a mapping of $(V, \mathbb{C})$ into $(W, \mathbb{R})$ then $L(\gamma v) \in W$ for $v \in V$ and $\gamma \in \mathbb{C}$ but $\gamma L(v)$ is undefined since the field of scalars for W is $\mathbb{R}$ and not $\mathbb{C}$. To ensure existence of linear mappings we assume that either $\mathbb{K} = \mathbb{K}'$ or $\mathbb{K} = \mathbb{R}$ and $\mathbb{K}' = \mathbb{C}$.

[VI.1] Continuous Linear Mappings

Let us consider continuity of a linear mapping L of a normed linear space into another.

Example. Consider the normed linear space $(\mathbb{R}^n, \mathbb{R}, \|\cdot\|_2)$ where $\|x\|_2 = \left\{\sum_{i=1}^n \xi_i^2\right\}^{1/2}$ for $x = (\xi_1, \dots, \xi_n) \in \mathbb{R}^n$ and the normed linear space $(\mathbb{R}, \mathbb{R}, |\cdot|)$. Let L be a linear mapping of $(\mathbb{R}^n, \mathbb{R}, \|\cdot\|_2)$ into $(\mathbb{R}, \mathbb{R}, |\cdot|)$. Then there exist $\lambda_1, \dots, \lambda_n \in \mathbb{R}$ such that

(1) $$L(x) = \lambda_1 \xi_1 + \cdots + \lambda_n \xi_n \quad \text{for } x = (\xi_1, \dots, \xi_n) \in \mathbb{R}.$$

Indeed if we let $e_1 = (1, 0, \dots, 0)$, $e_2 = (0, 1, 0, \dots, 0), \dots, e_n = (0, \dots, 0, 1)$, then $\{e_1, e_2, \dots, e_n\}$ is a basis for $\mathbb{R}$ and for every $x = (\xi_1, \dots, \xi_n) \in \mathbb{R}^n$ we have the expression

$x = \xi_1 e_1 + \cdots + \xi_n e_n$. Let $\lambda_1 = L(e_1), \ldots, \lambda_n = L(e_n)$. Then by the linearity of L we have

(2) $$L(x) = L(\xi_1 e_1 + \cdots + \xi_n e_n) = \xi_1 L(e_1) + \cdots + \xi_n L(e_n) = \lambda_1 \xi_1 + \cdots + \lambda_n \xi_n.$$

Let

(3) $$M = \left\{ \sum_{i=1}^{n} |\lambda_i|^2 \right\}^{1/2} \in [0, \infty).$$

Then for every $x = (\xi_1, \ldots, \xi_n) \in \mathbb{R}^n$, we have

$$|L(x)| = |\lambda_1 \xi_1 + \cdots + \lambda_n \xi_n| \leq \left\{ \sum_{i=1}^{n} |\lambda_i|^2 \right\}^{1/2} \left\{ \sum_{i=1}^{n} |\xi_i|^2 \right\}^{1/2}$$

so that

(4) $$|L(x)| \leq M \|x\|_2 \quad \text{for every } x \in \mathbb{R}^n.$$

Now (4) implies that L is uniformly continuous on $\mathbb{R}^n$. Indeed for an arbitrary $\varepsilon > 0$ let $\delta = \frac{\varepsilon}{M+1}$. Then for any $x', x'' \in \mathbb{R}^n$ such that $\|x' - x''\|_2 < \delta$, we have

$$|L(x') - L(x'')| = |L(x' - x'')| \leq M\|x' - x''\|_2 < M\delta = M\frac{\varepsilon}{M+1} \leq \varepsilon.$$

Example. A linear functional L on an arbitrary normed linear space $\big(X, \mathbb{K}, \|\cdot\|\big)$ may be discontinuous at every point in X. Let us consider fro instance the normed linear space $\big(C([a,b]), \mathbb{R}, \|\cdot\|_1\big)$ of real-valued continuous functions f on $[a,b]$ with the norm defined by $\|f\|_1 = \int_{[a,b]} |f| \, d\mu_L$ as in Theorem 15.15. If we take an arbitrary $c \in [a,b]$ and define a functional L on $C([a,b])$ by $L(f) = f(c)$ for $f \in C([a,b])$, then for every $\alpha, \beta \in \mathbb{R}$ and $f, g \in C([a,b])$ we have

$$L(\alpha f + \beta g) = (\alpha f + \beta g)(c) = \alpha f(c) + \beta g(c) = \alpha L(f) + \beta L(g).$$

Thus L is a linear functional on $C([a,b])$. However L is discontinuous at every f in $C([a,b])$. To show this, let $\varepsilon > 0$. For any $\delta > 0$, a real-valued continuous function g on $[a,b]$ such that $|f(c) - g(c)| > \varepsilon$ but $\|f - g\|_1 = \int_{[a,b]} |f - g| \, d\mu_L < \delta$ can be easily constructed.

Theorem 15.28. *Let* $\big(V, \mathbb{K}, \|\cdot\|\big)$ *and* $\big(W, \mathbb{K}', \|\cdot\|'\big)$ *be two normed linear spaces where either* $\mathbb{K} = \mathbb{K}'$ *or* $\mathbb{K} = \mathbb{R}$ *and* $\mathbb{K}' = \mathbb{C}$. *Let* L *be a linear mapping of* V *into* W. *If* L *is continuous at some* $v_0 \in V$ *then* L *is continuous at every* $v \in V$ *and in fact* L *is uniformly continuous on* V.

Proof. If L is continuous at $v_0 \in V$ then for every $\varepsilon > 0$ there exists $\delta > 0$ such that

(1) $$\|L(v) - L(v_0)\|' < \varepsilon \quad \text{for } f \in V \text{ with } \|v - v_0\| < \delta.$$

Let us show that L is uniformly continuous on V, that is, for every $\varepsilon > 0$ there exists $\delta > 0$ such that

$$(2) \qquad \|L(v') - L(v'')\|' < \varepsilon \quad \text{for } v', v'' \in V \text{ such that } \|v' - v''\| < \delta.$$

Let $\varepsilon > 0$ be arbitrarily given and let $\delta > 0$ be as in (1). Let $v', v'' \in V$ be such that $\|v' - v''\| < \delta$. Then by the linearity of L, we have

$$\big\|L(v') - L(v'')\big\|' = \big\|L(v' - (v'' - v_0) - v_0)\big\|' = \big\|L\big(v' - (v'' - v_0)\big) - L(v_0)\big\|' < \varepsilon,$$

by (1) since $\big\|\big(v' - (v'' - v_0)\big) - v_0\big\| = \|v' - v''\| < \delta$. This proves the uniform continuity of L on V. ∎

[VI.2] Bounded Linear Mappings

Definition 15.29. (a) *Let $\big(V, \mathbb{K}, \|\cdot\|\big)$ and $\big(W, \mathbb{K}', \|\cdot\|'\big)$ be two normed linear spaces where either $\mathbb{K} = \mathbb{K}'$ or $\mathbb{K} = \mathbb{R}$ and $\mathbb{K}' = \mathbb{C}$. A linear mapping L of V into W is called a bounded linear mapping if there exists a real number $M \geq 0$ such that*

$$\|L(v)\|' \leq M\|v\| \quad \textit{for all } v \in V.$$

Such constant $M \geq 0$, if it exists, is called a bound for the bounded linear mapping L. We write $\mathfrak{B}\{V, W\}$ for the collection of all bounded linear mappings of the normed linear space $\big(V, \mathbb{K}, \|\cdot\|\big)$ into the normed linear space $\big(W, \mathbb{K}', \|\cdot\|'\big)$. For $L \in \mathfrak{B}\{V, W\}$, we define

$$\|L\|_{V,W} = \inf\big\{M \geq 0 : \|L(v)\|' \leq M\|v\| \textit{ for all } v \in V\big\} \in [0, \infty).$$

(b) *For the particular case $\big(W, \mathbb{K}', \|\cdot\|'\big) = \big(\mathbb{K}', \mathbb{K}', |\cdot|\big)$ a linear mapping L of V into $\mathbb{K}'$ is called a $\mathbb{K}'$-valued bounded linear functional, abbreviated as b.l.f, if there exists $M \geq 0$ such that*

$$|L(v)| \leq M\|v\| \quad \textit{for all } v \in V.$$

We write $\mathfrak{B}\{V, \mathbb{K}'\}$ for the collection of all $\mathbb{K}'$-valued bounded linear functionals on the normed linear space $\big(V, \mathbb{K}, \|\cdot\|\big)$. For $L \in \mathfrak{B}\{V, \mathbb{K}'\}$ we define

$$\|L\|_{V,\mathbb{K}'} = \inf\big\{M \geq 0 : |L(v)| \leq M\|v\| \textit{ for all } v \in V\big\} \in [0, \infty).$$

We show next that the infimum of all bounds of a bounded linear mapping is itself a bound for the linear mapping.

Proposition 15.30. *Let $\big(V, \mathbb{K}, \|\cdot\|\big)$ and $\big(W, \mathbb{K}', \|\cdot\|'\big)$ be two normed linear spaces where either $\mathbb{K} = \mathbb{K}'$ or $\mathbb{K} = \mathbb{R}$ and $\mathbb{K}' = \mathbb{C}$. Let L be a bounded linear mapping of V into W. Then*

$$\|L(v)\|' \leq \|L\|_{V,W}\|v\| \quad \textit{for every } v \in V.$$

If L is not the identically vanishing, that is, if there exists $v \in V$ such that $L(v) \neq 0 \in W$, then $\|L\|_{V,W} > 0$.

Proof. 1. By the definition of $\|L\|_{V,W}$ as the infimum of all bounds for L, for every $k \in \mathbb{N}$ there exists a bound M_k for L such that $\|L\|_{V,W} \le M_k < \|L\|_{V,W} + \frac{1}{k}$. Since M_k is a bound for L, we have for every $v \in V$

$$\|L(v)\|' \le M_k\|v\| < \{\|L\|_{V,W} + \tfrac{1}{k}\}\|v\|.$$

Since this holds for every $k \in \mathbb{N}$, we have $\|L(v)\|' \le \|L\|_{V,W}\|v\|$ for every $v \in V$.

2. Suppose $L(v_0) \neq 0 \in W$ for some $v_0 \in V$. Then $0 < \|L(v_0)\|' \le \|L\|_{V,W}\|v_0\|$. Thus $\|L\|_{V,W} > 0$. ∎

[VI.3] Equivalence of Continuity and Boundedness of a Linear Mapping

Proposition 15.31. *Let $(V, \mathbb{K}, \|\cdot\|)$ and $(W, \mathbb{K}', \|\cdot\|')$ be two normed linear spaces where either $\mathbb{K} = \mathbb{K}'$ or $\mathbb{K} = \mathbb{R}$ and $\mathbb{K}' = \mathbb{C}$. Let L be a linear mapping of V into W. Then L is a bounded linear mapping if and only if L is continuous at $0 \in V$.*

Proof. 1. Suppose L is a bounded linear mapping. To show that L is continuous at $0 \in V$, we show that for every $\varepsilon > 0$ there exists $\delta > 0$ such that $\|L(v) - L(0)\|' < \varepsilon$ for every $v \in V$ such that $\|v - 0\| < \delta$, that is, $\|L(v)\|' < \varepsilon$ for every $v \in V$ such that $\|v\| < \delta$. Now according to Proposition 15.30 we have $\|L(v)\|' \le \|L\|_{V,W}\|v\|$ for every $v \in V$. If $\|L\|_{V,W} = 0$ then $\|L(v)\|' = 0$, that is, $L(v) = 0 \in W$ for every $v \in V$ so that L is a constant mapping and hence continuous on V. If $\|L\|_{V,W} > 0$ then for an arbitrarily given $\varepsilon > 0$ let $\delta = (\|L\|_{V,W})^{-1}\varepsilon$. Then for $v \in V$ such that $\|v\| < \delta$ we have

$$\|L(v)\|' \le \|L\|_{V,W}\|v\| < \|L\|_{V,W}\delta = \|L\|_{V,W}(\|L\|_{V,W})^{-1}\varepsilon = \varepsilon$$

so that L is continuous at $0 \in V$.

2. Conversely suppose L is continuous at $0 \in V$. Then for every $\varepsilon > 0$ there exists $\delta > 0$ such that $\|L(v)\|' < \varepsilon$ for every $v \in V$ with $\|v\| < \delta$.

Now let $v \in V$ and $v \neq 0 \in V$. For the element $\frac{\delta}{2\|v\|}v \in V$ we have $\left\|\frac{\delta}{2\|v\|}v\right\| = \frac{\delta}{2} < \delta$ so that $\left\|L\left(\frac{\delta}{2\|v\|}v\right)\right\|' < \varepsilon$, that is, $\frac{\delta}{2\|v\|}\|L(v)\|' < \varepsilon$ so that $\|L(v)\|' \le 2\frac{\varepsilon}{\delta}\|v\|$ for every $v \in V$, $v \neq 0 \in V$. On the other hand for $v = 0 \in V$, the last inequality holds trivially. Thus $\|L(v)\|' \le 2\frac{\varepsilon}{\delta}\|v\|$ for all $v \in V$. This shows that $2\frac{\varepsilon}{\delta}$ is a bound for the linear mapping L and then L is a bounded linear mapping. ∎

Theorem 15.32. *Let $(V, \mathbb{K}, \|\cdot\|)$ and $(W, \mathbb{K}', \|\cdot\|')$ be two normed linear spaces where either $\mathbb{K} = \mathbb{K}'$ or $\mathbb{K} = \mathbb{R}$ and $\mathbb{K}' = \mathbb{C}$. Let L be a linear mapping of V into W. Then the following three conditions are equivalent:*

1° *L is continuous at $0 \in V$.*

2° *L is uniformly continuous on V.*

3° *L is a bounded linear mapping.*

Proof. By Theorem 15.28, 1° and 2° are equivalent. By Proposition 15.31, 1° and 3° are equivalent. ∎

[VI.4] Infimum of the Bounds of a Linear Mapping

Definition 15.33. *Let (X, ρ) be a metric space. For $x_0 \in X$ and $r > 0$ we define*

$$B_r(x_0) = \{x \in X : \rho(x, x_0) < r\},$$
$$\overline{B}_r(x_0) = \{x \in X : \rho(x, x_0) \le r\},$$
$$S_r(x_0) = \{x \in X : \rho(x, x_0) = r\}.$$

The alternate notations $B(x_0, r)$, $\overline{B}(x_0, r)$, and $S(x_0, r)$ for these sets will also be used. In particular in a normed linear space $(V, \mathbb{K}, \|\cdot\|)$, for $v_0 \in V$ and $r > 0$ we have

$$B_r(v_0) = \{v \in V : \|v - v_0\| < r\},$$
$$\overline{B}_r(v_0) = \{v \in V : \|v - v_0\| \le r\},$$
$$S_r(v_0) = \{v \in V : \|v - v_0\| = r\}.$$

Observation 15.34. $B_r(x_0)$ is an open set. We call it an open ball with center at x_0 and radius r. It is easy to show that $\overline{B}_r(x_0)$ is the closure of $B_r(x_0)$. Thus $\overline{B}_r(x_0)$ is a closed set. We call it a closed ball with center at x_0 and radius r. Then $S_r(x_0) = \overline{B}_r(x_0) \setminus B_r(x_0)$ is a closed set. We call it a spherical hypersurface with center at x_0 and radius r. Also a non-empty open set in a metric space (X, ρ) is a union of open balls. (Indeed if O is a non-empty open set then for every $x \in X$ there exists $r_x > 0$ such that $B_{r_x}(x) \subset O$ so that $\bigcup_{x \in O} B_{r_x}(x) \subset O$. On the other hand for every $x \in O$ we have $x \in B_{r_x}(x)$ so that $O = \bigcup_{x \in O}\{x\} \subset \bigcup_{x \in O} B_{r_x}(x)$. Therefore we have $O = \bigcup_{x \in O} B_{r_x}(x)$.)

Lemma 15.35. *Let $(V, \mathbb{K}, \|\cdot\|)$ and $(W, \mathbb{K}', \|\cdot\|')$ be two normed linear spaces where either $\mathbb{K} = \mathbb{K}'$ or $\mathbb{K} = \mathbb{R}$ and $\mathbb{K}' = \mathbb{C}$. Let L be a linear mapping of V into W. Then*

$$\sup_{v \in S_1(0)} \|L(v)\|' = \sup_{v \in \overline{B}_1(0)} \|L(v)\|'.$$

Proof. Since $S_1(0) \subset \overline{B}_1(0)$ we have

$$\sup_{v \in S_1(0)} \|L(v)\|' \le \sup_{v \in \overline{B}_1(0)} \|L(v)\|'.$$

It remains to prove the reverse inequality. Let $v \in \overline{B}_1(0)$ and $v \neq 0 \in V$. If we let $u = \frac{1}{\|v\|} v$ then $\|u\| = 1$ so that $u \in S_1(0)$. Now

$$\|L(u)\|' = \left\|L\left(\tfrac{1}{\|v\|} v\right)\right\|' = \tfrac{1}{\|v\|}\|L(v)\|' \ge \|L(v)\|'$$

since $v \in \overline{B}_1(0)$ and $\|v\| \le 1$. This shows that for every $v \in \overline{B}_1(0)$ and $v \neq 0 \in V$ there exists $u \in S_1(0)$ such that $\|L(v)\|' \le \|L(u)\|'$. Also for $v = 0 \in V$ and for any $u \in S_1(0)$ we have $\|L(v)\|' = \|0\|' = 0$ so that the inequality $\|L(v)\|' \le \|L(u)\|'$ holds trivially. Thus we have

$$\sup_{v \in \overline{B}_1(0)} \|L(v)\|' \le \sup_{v \in S_1(0)} \|L(v)\|'.$$

This completes the proof. ■

Theorem 15.36. *Let* $(V, \mathbb{K}, \|\cdot\|)$ *and* $(W, \mathbb{K}', \|\cdot\|')$ *be two normed linear spaces where either* $\mathbb{K} = \mathbb{K}'$ *or* $\mathbb{K} = \mathbb{R}$ *and* $\mathbb{K}' = \mathbb{C}$. *Let* L *be a linear mapping of* V *into* W. *Then* L *is a bounded linear mapping if and only if*

$$\sup_{v \in S_1(0)} \|L(v)\|' < \infty. \tag{1}$$

Moreover if L *is a bounded linear mapping then*

$$\|L\|_{V,W} = \sup_{v \in S_1(0)} \|L(v)\|'. \tag{2}$$

Proof. 1. Suppose L is a bounded linear mapping. Then there exists $M \geq 0$ such that

$$\|L(v)\|' \leq M\|v\| \quad \text{for all } v \in V.$$

Then we have

$$\sup_{v \in S_1(0)} \|L(v)\|' \leq \sup_{v \in S_1(0)} M\|v\| = M < \infty.$$

Conversely suppose $\sup_{v \in S_1(0)} \|L(v)\|' < \infty$. Now for an arbitrary $v \in V$ such that $v \neq 0 \in V$, if we let $u = \frac{1}{\|v\|} v$ then $\|u\| = 1$ so that $u \in S_1(0)$. Then we have

$$\|L(v)\|' = \left\|L\left(\|v\| \tfrac{1}{\|v\|} v\right)\right\|' = \|v\| \, \|L(u)\|' \leq \|v\| \sup_{w \in S_1(0)} \|L(w)\|'.$$

For $v = 0 \in V$ the inequality $\|L(v)\|' \leq \|v\| \sup_{w \in S_1(0)} \|L(w)\|'$ holds trivially. Thus

$$\|L(v)\|' \leq \|v\| \sup_{w \in S_1(0)} \|L(w)\|' \quad \text{for all } v \in V. \tag{3}$$

This shows that the real number $\sup_{v \in S_1(0)} \|L(v)\|' \geq 0$ is a bound for the linear mapping L and L is a bounded linear mapping.

2. To prove (2) suppose L is a bounded linear mapping. By Proposition 15.30 we have $\|L(v)\|' \leq \|L\|_{V,W} \|v\|$ for all $v \in V$. Thus we have

$$\sup_{v \in S_1(0)} \|L(v)\|' \leq \|L\|_{V,W} \sup_{v \in S_1(0)} \|v\| = \|L\|_{V,W}.$$

Since L is a bounded linear mapping we have $\sup_{v \in S_1(0)} \|L(v)\|' < \infty$ as we showed in **1**. Then (3) shows that the real number $\sup_{v \in S_1(0)} \|L(v)\|'$ is a bound for the linear mapping L. Since $\|L\|_{V,W}$ is the infimum of all bounds of L we have $\|L\|_{V,W} \leq \sup_{v \in S_1(0)} \|L(v)\|'$. Thus $\|L\|_{V,W} = \sup_{v \in S_1(0)} \|L(v)\|'$. ■

[VI.5] Normed Linear Space of Bounded Linear Mappings

Theorem 15.37. *Given two normed linear spaces* $(V, \mathbb{K}, \|\cdot\|)$ *and* $(W, \mathbb{K}', \|\cdot\|')$ *where either* $\mathbb{K} = \mathbb{K}'$ *or* $\mathbb{K} = \mathbb{R}$ *and* $\mathbb{K}' = \mathbb{C}$. *Consider the collection* $\mathfrak{B}\{V, W\}$ *of all bounded* *linear mappings of V into W. Let us define addition and scalar multiplication in $\mathfrak{B}\{V, W\}$ by setting*

(1) $$(L_1 + L_2)(v) = L_1(v) + L_2(v) \quad \textit{for } v \in V \textit{ and } L_1, L_2 \in \mathfrak{B}\{V, W\}$$

and

(2) $$(\gamma L)(v) = \gamma L(v) \quad \textit{for } v \in V, L \in \mathfrak{B}\{V, W\} \textit{ and } \gamma \in \mathbb{K}'.$$

Then $\mathfrak{B}\{V, W\}$ is a linear space over the field of scalars $\mathbb{K}'$ and $\|\cdot\|_{V,W}$ is a norm on this linear space.

Proof. 1. Let $L_1, L_2 \in \mathfrak{B}\{V, W\}$. Let us show that $L_1 + L_2$ defined by (1) is a member of $\mathfrak{B}\{V, W\}$. Let $v_1, v_2 \in V$ and $\gamma_1, \gamma_2 \in \mathbb{K}$. Then

$$\begin{aligned}(L_1 + L_2)(\gamma_1 v_1 + \gamma_2 v_2) &= L_1(\gamma_1 v_1 + \gamma_2 v_2) + L_2(\gamma_1 v_1 + \gamma_2 v_2)\\ &= \{\gamma_1 L_1(v_1) + \gamma_2 L_1(v_2)\} + \{\gamma_1 L_2(v_1) + \gamma_2 L_2(v_2)\}\\ &= \gamma_1(L_1 + L_2)(v_1) + \gamma_2(L_1 + L_2)(v_2)\end{aligned}$$

where the first equality is by (1), the second equality is by the linearity of L_1 and L_2 and the third equality is by (1) again. This shows that $L_1 + L_2$ is a linear mapping of V into W. To show that $L_1 + L_2$ is a bounded linear mapping note that for every $v \in V$ we have

$$\begin{aligned}\|(L_1 + L_2)(v)\|' &= \|L_1(v) + L_2(v)\|' \le \|L_1(v)\|' + \|L_2(v)\|'\\ &\le \|L_1\|_{V,W}\|v\| + \|L_2\|_{V,W}\|v\| = \{\|L_1\|_{V,W} + \|L_2\|_{V,W}\}\|v\|.\end{aligned}$$

Thus $\|L_1\|_{V,W} + \|L_2\|_{V,W}$ is a bound for the linear mapping $L_1 + L_2$ and $L_1 + L_2$is a bounded linear mapping. This shows that $(L_1 + L_2) \in \mathfrak{B}\{V, W\}$.

Let $L \in \mathfrak{B}\{V, W\}$ and $\gamma \in \mathbb{K}'$. Let us show that γL defined by (2) is a member of $\mathfrak{B}\{V, W\}$. Let $v_1, v_2 \in V$ and $\gamma_1, \gamma_2 \in \mathbb{K}$. Then

$$\begin{aligned}(\gamma L)(\gamma_1 v_1 + \gamma_2 v_2) &= \gamma L(\gamma_1 v_1 + \gamma_2 v_2) = \gamma\{\gamma_1 L(v_1) + \gamma_2 L(v_2)\}\\ &= \gamma_1(\gamma L)(v_1) + \gamma_2(\gamma L)(v_2)\end{aligned}$$

where the first equality is by (2) and the second equality is by the linearity of L. This shows that γL is a linear mapping of V into W. To show that γL is a bounded linear mapping we note that for every $v \in V$ we have

$$\|(\gamma L)(v)\|' = \|(\gamma L(v)\|' = |\gamma|\,\|L(v)\|' \le |\gamma|\,\|L\|_{V,W}\|v\|$$

so that $|\gamma|\,\|L\|_{V,W}$ is a bound for γL and L is a bounded linear mapping. This shows that $\gamma L \in \mathfrak{B}\{V, W\}$.

With addition and scalar multiplication thus defined, the fact that $\mathfrak{B}\{V, W\}$ is a linear space over the field of scalars $\mathbb{K}'$ is easily established by verifying conditions 1° to 9° in Definition 15.1. Note that $0 \in \mathfrak{B}\{V, W\}$ is the identically vanishing linear mapping of V into W, that is, the mapping L such that $L(v) = 0 \in W$ for every $v \in V$.

2. Let us show that $\|L\|_{V,W}$ for $L \in \mathfrak{B}\{V, W\}$ is a norm on the linear space $\mathfrak{B}\{V, W\}$. Now by Definition 15.29 we have

$$\|L\|_{V,W} = \inf\left\{M \ge 0 : \|L(v)\|' \le M\|v\| \text{ for all } v \in V\right\} \in [0, \infty).$$

If $L = 0 \in \mathfrak{B}\{V, W\}$, the identically vanishing mapping of V into W, then $\|L\|_{V,W} = 0$. On the other hand by Proposition 15.30 if $L \neq 0 \in \mathfrak{B}\{V, W\}$ then $\|L\|_{V,W} > 0$. Thus $\|L\|_{V,W} = 0$ if and only if $L = 0 \in \mathfrak{B}\{V, W\}$.

For $L \in \mathfrak{B}\{V, W\}$ and $\gamma \in \mathbb{K}'$ we have by Theorem 15.36

$$\|\gamma L\|_{V,W} = \sup_{v \in S_1(0)} \left\|(\gamma L)(v)\right\|' = |\gamma| \sup_{v \in S_1(0)} \|L(v)\|' = |\gamma|\|L\|_{V,W}.$$

This proves the positive homogeneity of $\|\cdot\|_{V,W}$. To prove the triangle inequality for $\|\cdot\|_{V,W}$, let $L_1, L_2 \in \mathfrak{B}\{V, W\}$. Then by (2) of Theorem 15.36 we have

$$\begin{aligned}\|L_1 + L_2\|_{V,W} &= \sup_{v \in S_1(0)} \left\|(L_1 + L_2)(v)\right\|' \le \sup_{v \in S_1(0)} \left\{\|L_1(v)\|' + \|L_2(v)\|'\right\} \\ &\le \sup_{v \in S_1(0)} \left\{\|L_1\|_{V,W} + \|L_2\|_{V,W}\right\}\|v\| = \|L_1\|_{V,W} + \|L_2\|_{V,W},\end{aligned}$$

where the first inequality is by (1) and the triangle inequality of the norm $\|\cdot\|'$ and the second inequality is by Proposition 15.30. This proves the triangle inequality for $\|\cdot\|_{V,W}$ and completes the proof that $\|\cdot\|_{V,W}$ is a norm on the linear space $\mathfrak{B}\{V, W\}$. ∎

Theorem 15.38. *Let $(V, \mathbb{K}, \|\cdot\|)$ be a normed linear space and $(W, \mathbb{K}', \|\cdot\|')$ be a Banach space where either $\mathbb{K} = \mathbb{K}'$ or $\mathbb{K} = \mathbb{R}$ and $\mathbb{K}' = \mathbb{C}$. Then the linear space $\mathfrak{B}\{V, W\}$ over the field $\mathbb{K}'$ of all bounded linear mappings of V into W is a Banach space with respect to the norm $\|\cdot\|_{V,W}$.*

In particular the linear space $\mathfrak{B}\{V, \mathbb{K}'\}$ over the field $\mathbb{K}'$ of all $\mathbb{K}'$-valued bounded linear functionals on V is a Banach space with respect to the norm $\|\cdot\|_{V,\mathbb{K}'}$.

Proof. Let us show that $\mathfrak{B}\{V, W\}$ is complete with respect to the norm $\|\cdot\|_{V,W}$. Let $(L_n : n \in \mathbb{N})$ be a Cauchy sequence in $\mathfrak{B}\{V, W\}$ with respect to the norm $\|\cdot\|_{V,W}$, that is, for every $\varepsilon > 0$ there exists $N \in \mathbb{N}$ such that

$$(1) \qquad \|L_m - L_n\|_{V,W} < \varepsilon \quad \text{for } m, n \ge N.$$

Then for every $v \in V$ we have

$$(2) \qquad \begin{aligned}\|L_m(v) - L_n(v)\|' &= \|(L_m - L_n)(v)\|' \\ &\le \|L_m - L_n\|_{V,W}\|v\| < \varepsilon\|v\| \quad \text{for } m, n \ge N\end{aligned}$$

so that $(L_n(v) : n \in \mathbb{N})$ is a Cauchy sequence in W with respect to the norm $\|\cdot\|'$. Since W is complete with respect to the norm $\|\cdot\|'$ there exists an element in W, depending on v, to which the sequence $(L_n(v) : n \in \mathbb{N})$ converges. Let us denote this element in W as $L(v)$. Thus we define

$$(3) \qquad \lim_{n \to \infty} L_n(v) = L(v) \quad \text{for } v \in V.$$

This defines a mapping L of V into W.

Let us show that the mapping L of V into W thus defined is a linear mapping. Let $v_1, v_2 \in V$ and $\gamma_1, \gamma_2 \in \mathbb{K}$. Then

$$\begin{aligned} L(\gamma_1 v_1 + \gamma_2 v_2) &= \lim_{n\to\infty} L_n(\gamma_1 v_1 + \gamma_2 v_2) \\ &= \gamma_1 \lim_{n\to\infty} L_n(v_1) + \gamma_1 \lim_{n\to\infty} L_n(v_2) \\ &= \gamma_1 L(v_1) + \gamma_1 L(v_2). \end{aligned}$$

Let us show next that the linear mapping L is bounded. By the triangle inequality for the norm $\|\cdot\|_{V,W}$ we have the inequality $\big|\|L_m\|_{V,W} - \|L_n\|_{V,W}\big| \le \|L_m - L_n\|_{V,W}$ for every $m, n \in \mathbb{N}$. Then by (1) the sequence $(\|L_n\|_{V,W} : n \in \mathbb{N})$ is a Cauchy sequence of real numbers so that there exists $M \in \mathbb{R}$ such that $\lim_{n\to\infty} \|L_n\|_{V,W} = M$. Since $\|L_n\|_{V,W} \ge 0$ for every $n \in \mathbb{N}$ we have $M \ge 0$. By (3) for every $v \in V$ we have $\lim_{n\to\infty} \|L_n(v) - L(v)\|' = 0$. This implies $\|L(v)\|' = \lim_{n\to\infty} \|L_n(v)\|' \le \lim_{n\to\infty} \|L_n(v)\|_{V,W}\|v\| = M\|v\|$. This shows that M is a bound for the linear mapping L and thus L is a bounded linear mapping of V into W, that is, $L \in \mathfrak{B}\{V, W\}$.

Finally let us show that $\lim_{n\to\infty} \|L_n - L\|_{V,W} = 0$. For every $v \in V$ we have by (3)

$$\begin{aligned} (4) \qquad & \|L_n(v) - L(v)\|' = \big\|L_n(v) - \lim_{m\to\infty} L_m(v)\big\|' \\ &= \big\|\lim_{m\to\infty} \{L_n(v) - L_m(v)\}\big\|' = \lim_{m\to\infty} \big\|L_n(v) - L_m(v)\big\|' \\ &\le \limsup_{m\to\infty} \big\|L_n(v) - L_m(v)\big\|_{V,W} \|v\|, \end{aligned}$$

where the third equality is by applying (c) of Proposition 15.9 to the normed linear space $(W, \mathbb{K}', \|\cdot\|')$. Since $(L_n : n \in \mathbb{N})$ is a Cauchy sequence in $\mathfrak{B}\{V, W\}$, for every $\varepsilon > 0$ there exists $N \in \mathbb{N}$ such that $\|L_n(v) - L_m(v)\|_{V,W} < \varepsilon$ for $m, n \ge N$. Then for $n \ge N$ we have $\limsup_{m\to\infty} \|L_n(v) - L_m(v)\|_{V,W} \le \varepsilon$. Substituting this in (4), we have for $n \ge N$

$$\|L_n(v) - L(v)\|' \le \varepsilon\|v\|$$

and then

$$\|L_n - L\|_{V,W} = \sup_{v \in S_1(0)} \|L_n(v) - L(v)\|' \le \varepsilon.$$

This shows that $\lim_{n\to\infty} \|L_n - L\|_{V,W} = 0$.

In particular since $(W, \mathbb{K}', \|\cdot\|') := (\mathbb{K}', \mathbb{K}', |\cdot|)$ is a Banach space, the linear space $\mathfrak{B}\{V, \mathbb{K}'\}$ over the field $\mathbb{K}'$ of all $\mathbb{K}'$-valued bounded linear functionals on V is a Banach space by our result above. ∎

[VI.6] Dual of a Normed Linear Space

Let $(V, \mathbb{K}, \|\cdot\|)$ and $(W, \mathbb{K}', \|\cdot\|')$ be two normed linear spaces where either $\mathbb{K} = \mathbb{K}'$ or $\mathbb{K} = \mathbb{R}$ and $\mathbb{K}' = \mathbb{C}$. Consider the collection $\mathfrak{B}\{V, W\}$ of all bounded linear mappings of

V into W. In Definition 15.29, we defined a function $\|\cdot\|_{V,W}$ on the linear space $\mathfrak{B}\{V, W\}$ over the field of scalars $\mathbb{K}'$ by setting for every $L \in \mathfrak{B}\{V, W\}$

$$\|L\|_{V,W} = \inf\left\{M \geq 0 : \|L(v)\|' \leq M\|v\| \text{ for all } v \in V\right\}.$$

We showed in Theorem 15.37 that $\|\cdot\|_{V,W}$ is a norm on the linear space $\mathfrak{B}\{V, W\}$. We showed in Theorem 15.36 that if we let $S_1(0) = \left\{v \in V : \|v\| = 1\right\}$, then

$$\|L\|_{V,W} = \sup_{v \in S_1(0)} \|L(v)\|'.$$

We showed in Theorem 15.38 that if $\left(W, \mathbb{K}', \|\cdot\|'\right)$ is a Banach space, that is, W is complete with respect to the norm $\|\cdot\|'$, then $\mathfrak{B}\{V, W\}$ is a Banach space with respect to the norm $\|\cdot\|_{V,W}$.

Let us consider the particular case that $\left(W, \mathbb{K}', \|\cdot\|'\right) = \left(\mathbb{K}, \mathbb{K}, |\cdot|\right)$, a Banach space. Thus we consider the linear space $\mathfrak{B}\{V, \mathbb{K}\}$ over the field of scalars $\mathbb{K}$ of all $\mathbb{K}$-valued bounded linear functionals on V. Let us write V^* for this linear space $\mathfrak{B}\{V, \mathbb{K}\}$. If for every $f \in V^*$ we define

$$\|f\|_* = \inf\left\{M \geq 0 : |f(v)| \leq M\|v\| \text{ for all } v \in V\right\},$$

then $\|\cdot\|_*$ is a norm on V^* and V^* is a Banach space with respect to the norm $\|\cdot\|_*$. As in the general case, if we let $S_1(0) = \left\{v \in V : \|v\| = 1\right\}$, then

$$\|f\|_* = \sup_{v \in S_1(0)} |f(v)|.$$

Definition 15.39. *Let $\left(V, \mathbb{K}, \|\cdot\|\right)$ be a normed linear space where $\mathbb{K} = \mathbb{R}$ or $\mathbb{K} = \mathbb{C}$. Let V^* be the linear space over the field of scalars $\mathbb{K}$ of all $\mathbb{K}$-valued bounded linear functionals on V. Let $\|\cdot\|_*$ be a norm on V^* defined for every $f \in V^*$ by*

$$\|f\|_* = \inf\left\{M \geq 0 : |f(v)| \leq M\|v\| \textit{ for all } v \in V\right\}.$$

We call the Banach space $\left(V^*, \mathbb{K}, \|\cdot\|_*\right)$ the dual normed linear space of the normed linear $\left(V, \mathbb{K}, \|\cdot\|\right)$.

In a normed linear space $\left(V, \mathbb{K}, \|\cdot\|\right)$ and its dual space $\left(V^*, \mathbb{K}, \|\cdot\|_*\right)$ if we let $S_1(0) = \left\{v \in V : \|v\| = 1\right\}$, then as we showed above we have $\|f\|_* = \sup_{v \in S_1(0)} |f(v)|$ for every $f \in V^*$. We show in Proposition that if we let $S_1^*(0) = \left\{f \in V^* : \|f\|_* = 1\right\}$, then we have $\|v\| = \sup_{f \in S_1^*(0)} |f(v)|$ for every $v \in V$.

[VII] Baire Category Theorem

Definition 15.40. *Let (X, d) be a metric space. The diameter of a subset E of X is defined by* $\operatorname{diam}(E) = \sup_{x,y \in E} d(x, y)$.

Theorem 15.41. (Cantor Intersection Theorem) *Let (X, d) be a complete metric space and let $(F_n : n \in \mathbb{N})$ be a decreasing sequence of non-empty closed sets in X.*

If $\lim_{n\to\infty} \operatorname{diam}(F_n) = 0$, then there exist $x_0 \in X$ such that $\bigcap_{n\in\mathbb{N}} F_n = \{x_0\}$.

Proof. 1. Let us show that $\bigcap_{n\in\mathbb{N}} F_n$ contains at most one point. Suppose $a, b \in \bigcap_{n\in\mathbb{N}} F_n$ and $a \neq b$. Then $\delta := d(a, b) > 0$. Now $a, b \in \bigcap_{n\in\mathbb{N}} F_n$ implies that $a, b \in F_n$ for every $n \in \mathbb{N}$. Then $\operatorname{diam}(F_n) = \sup_{x,y\in F_n} d(x, y) \geq d(a, b) = \delta$ for every $n \in \mathbb{N}$. This contradicts the fact that $\lim_{n\to\infty} \operatorname{diam}(F_n) = 0$. Therefore $a = b$. This shows that $\bigcap_{n\in\mathbb{N}} F_n$ contains at most one point.

2. Let us show that $\bigcap_{n\in\mathbb{N}} F_n$ contains at least one point. Now $F_n \neq \emptyset$ for every $n \in \mathbb{N}$. Pick $x_n \in F_n$ arbitrarily for each $n \in \mathbb{N}$. To show that $(x_n : n \in \mathbb{N})$ is a Cauchy sequence in X with respect to the metric d, we show that for every $\varepsilon > 0$ there exists $N \in \mathbb{N}$ such that $d(x_m, x_n) < \varepsilon$ for $m, n \geq N$. Now since $\lim_{n\to\infty} \operatorname{diam}(F_n) = 0$, there exists $N \in \mathbb{N}$ such that $d(F_n) < \varepsilon$ for $n \geq N$. Let $m, n \geq N$. We may assume without loss of generality that $m \geq n \geq N$. Then $F_m \subset F_n$ so that $x_m, x_n \in F_n$. This implies that $d(x_m, x_n) \leq \sup_{x,y\in F_n} d(x, y) = \operatorname{diam}(F_n) < \varepsilon$. This shows that $(x_n : n \in \mathbb{N})$ is a Cauchy sequence in X with respect to d. Then by the completeness of the metric space (X, d), there exists $x_0 \in X$ such that $\lim_{n\to\infty} d(x_n, x_0) = 0$. It remains to show that $x_0 \in \bigcap_{n\in\mathbb{N}} F_n$. For every $k \in \mathbb{N}$, consider the sequence $(x_n : n \geq k)$ and its range $E_k = \{x_n : n \geq k\}$. Since $(F_n : n \in \mathbb{N})$ is a decreasing sequence and since $x_n \in F_n$ for every $n \in \mathbb{N}$, we have $E_k \subset F_k$. Now the convergence of $(x_n : n \geq k)$ to x_0 implies that $x_0 \in \overline{E}_k$. Then $x_0 \in \overline{E}_k \subset \overline{F}_k = F_k$ since F_k is a closed set. Thus $x_0 \in F_k$ for every $k \in \mathbb{N}$ and therefore $x_0 \in \bigcap_{n\in\mathbb{N}} F_n$. ∎

Definition 15.42. *Let (X, d) be a metric space and let E be a subset of X.*
(a) *We say that E is dense in X if $\overline{E} = X$.*
(b) *We say that E is nowhere dense in X, or non-dense in X, if $\left(\overline{E}\right)^\circ = \emptyset$.*

Remark 15.43. A set E in a metric space (X, d) cannot be both dense and non-dense in X. In fact, if E is dense in X then $\overline{E} = X$ so that $\left(\overline{E}\right)^\circ = X^\circ = X \neq \emptyset$ so that E is not non-dense in X.

Proposition 15.44. *Let (X, d) be a metric space and let E be a subset of X.*
(a) *E is dense in X if and only if for every non-empty open set G in X, we have $G \cap E \neq \emptyset$.*
(b) *E is dense in X if and only if for every open ball B in X, we have $B \cap E \neq \emptyset$.*

Proof. For an arbitrary set E in X let us write E' for the set of all limit points of E. Then $\overline{E} = E \cup E'$.

1. Let us prove (a). Suppose E is dense in X, that is, $\overline{E} = X$. Let G be a non-empty open set in X. Since $\overline{E} = X$, G contains a point $x \in \overline{E} = E \cup E'$. If $x \in E$, then we have $G \cap E \neq \emptyset$. If $x \in E'$, then the open set G containing x must contain at least one point of E other than x so that $G \cap E \neq \emptyset$.

Conversely suppose that for every non-empty open set G in X we have $G \cap E \neq \emptyset$. Let $x \in X$. If $x \in E$ then $x \in \overline{E}$. If $x \notin E$, take an arbitrary open set G containing x. Since

$G \cap E \neq \emptyset$, G contains a point of E distinct from x. Thus x is a limit point of E. This shows that if $x \notin E$ then $x \in E' \subset E \cup E' = \overline{E}$. Therefore every $x \in X$ is contained in $\overline{E}$. This shows that $\overline{E} = X$, that is, E is dense in X.

2. Let us prove (b). If E is dense in X, then by (a) for every non-empty open set G in X we have $G \cap E \neq \emptyset$ so that in particular for every open ball B in X we have $B \cap E \neq \emptyset$. Conversely suppose for every open ball B in X we have $B \cap E \neq \emptyset$. Let G be an arbitrary non-empty open set in X. Then G contains an open ball B. Since $B \cap E \neq \emptyset$ and since $G \supset B$, we have $G \cap E \neq \emptyset$. Then E is dense in X by (a). ∎

Proposition 15.45. *Let (X, d) be a metric space and let E be a subset of X.*
(a) *E is non-dense in X if and only if every non-empty open set G in X contains a non-empty*
open set G_0 such that $G_0 \cap E = \emptyset$.
(b) *E is non-dense in X if and only if every open ball B in X contains an open ball B_0 such that $B_0 \cap E = \emptyset$.*

Proof. 1. Let us prove (a). Suppose E is non-dense in X, that is, $\left(\overline{E}\right)^\circ = \emptyset$. Let G be an arbitrary non-empty open set in X. Then $G \not\subset \overline{E}$ for otherwise we have $\left(\overline{E}\right)^\circ \supset G \neq \emptyset$, a contradiction. Now $G \not\subset \overline{E}$ implies that $G \setminus \overline{E} \neq \emptyset$. Let $G_0 = G \setminus \overline{E} \neq \emptyset$. We have $G_0 \cap \overline{E} = \emptyset$ and thus $G_0 \cap E = \emptyset$. To show that G_0 is an open set, note that $G_0 = G \setminus \overline{E} = G \cap \left(\overline{E}\right)^c$. Since G and $\left(\overline{E}\right)^c$ are open sets, G_0 is an open set. Thus we have shown that every non-empty open set G in X contains a non-empty open set G_0 such that $G_0 \cap E = \emptyset$.

Conversely suppose every non-empty open set G in X contains a non-empty open set G_0 such that $G_0 \cap E = \emptyset$. To show that E is a non-dense set in X, assume the contrary, that is, $\left(\overline{E}\right)^\circ \neq \emptyset$. Then $G := \left(\overline{E}\right)^\circ$ is a non-empty open set. Let G_0 be an arbitrary non-empty open set contained in G. Then $G_0 \subset G = \left(\overline{E}\right)^\circ \subset \overline{E} = E \cup E'$. Thus an arbitrary point x in G_0 is either in E or in E' or in both. If $x \in E$, then $G_0 \cap E \neq \emptyset$. If on the other hand $x \in E'$, then G_0 is an open set containing a limit point x of E so that it must contain a point of E and then we have $G_0 \cap E \neq \emptyset$. Thus in any case for any non-empty open set G_0 contained in G we have $G_0 \cap E \neq \emptyset$, a contradiction.

2. (b) follows from (a) by the fact that an open ball is a non-empty open set and every non-empty open set contains an open ball.

If E is non-dense in X, then since every open ball B is a non-empty open set, B contains a non-empty open set G_0 such that $G_0 \cap E = \emptyset$ by (a). Since G_0 is a non-empty open set, it contains an open ball B_0. Then $G_0 \cap E = \emptyset$ implies $B_0 \cap E = \emptyset$.

Conversely suppose every open ball B in X contains an open ball B_0 such that $B_0 \cap E = \emptyset$. Let G be an arbitrary non-empty open set. Then G contains an open ball B and B contains an open ball B_0 such that $B_0 \cap E = \emptyset$. Thus every non-empty open set G contains a non-empty open set $G_0 := B_0$ such that $G_0 \cap E = \emptyset$. Then by (a), E is a non-dense set in X. ∎

Definition 15.46. *Let (X, d) be a metric space. A subset E of X is called a set of the first category if E is a countable union of non-dense sets in X. A subset E of X which is not a* *set of the first category is called a set of the second category.*

Theorem 15.47. (Baire Category Theorem) *Let (X, d) be a complete metric space. Then X is a set of the second category.*

Proof. To show that X is a set of the second category, we show that X is not a countable union of non-dense sets in X. We show this by showing that for any countable collection of non-dense sets in X, $\{E_n : n \in \mathbb{N}\}$, we have $\bigcup_{n\in\mathbb{N}} E_n \neq X$, that is, there exists $x \in X$ such that $x \notin E_n$ for every $n \in \mathbb{N}$.

Let us write $B_r(x) = \{y \in X : d(x, y) < r\}$ and $C_r(x) = \{y \in X : d(x, y) \leq r\}$ for an open ball and a closed ball in (X, d).

Let $B_{r_0}(x_0)$ be an arbitrary open ball in (X, d). Since E_1 is a non-dense set in X, the open ball $B_{r_0}(x_0)$ contains an open ball $B_{r_1}(x_1)$ such that $B_{r_1}(x_1) \cap E_1 = \emptyset$ by Proposition 15.45. We may choose $r_1 < 1$.

Consider $B_{r_1/2}(x_1) \subset C_{r_1/2}(x_1) \subset B_{r_1}(x_1)$. Since E_2 is a non-dense set in X, the open ball $B_{r_1/2}(x_1)$ contains an open ball $B_{r_2}(x_2)$ such that $B_{r_2}(x_2) \cap E_2 = \emptyset$ by Proposition 15.45. We may choose $r_2 < \frac{1}{2}$.

Consider $B_{r_2/2}(x_2) \subset C_{r_2/2}(x_2) \subset B_{r_2}(x_2)$. Since E_3 is a non-dense set in X, the open ball $B_{r_2/2}(x_2)$ contains an open ball $B_{r_3}(x_3)$ such that $B_{r_3}(x_3) \cap E_3 = \emptyset$ by Proposition 15.45. We may choose $r_3 < \frac{1}{3}$.

Thus continuing indefinitely, we have sequences $(B_{r_n}(x_n) : n \in \mathbb{N})$, $(C_{r_n/2}(x_n) : n \in \mathbb{N})$, and $(B_{r_n/2}(x_n) : n \in \mathbb{N})$, such that $B_{r_n}(x_n) \cap E_n = \emptyset$ and $r_n < \frac{1}{n}$ for $n \in \mathbb{N}$ and

$$\begin{aligned} &B_{r_1}(x_1) \supset C_{r_1/2}(x_1) \supset B_{r_1/2}(x_1) \\ &\supset B_{r_2}(x_2) \supset C_{r_2/2}(x_2) \supset B_{r_2/2}(x_2) \\ &\supset B_{r_3}(x_3) \supset C_{r_3/2}(x_3) \supset B_{r_3/2}(x_3) \\ &\vdots \end{aligned}$$

Now $C_{r_1/2}(x_1) \supset C_{r_2/2}(x_2) \supset C_{r_3/2}(x_3) \supset \cdots$ so that $(C_{r_n/2}(x_n) : n \in \mathbb{N})$ is a decreasing sequence of closed sets. Moreover $\lim_{n\to\infty} \operatorname{diam}(C_{r_n/2}(x_n)) = \lim_{n\to\infty} r_n \leq \lim_{n\to\infty} \frac{1}{n} = 0$. Since (X, d) is a complete metric space, this implies that there exists $x \in X$ such that $\bigcap_{n\in\mathbb{N}} C_{r_n/2}(x_n) = \{x\}$ by Theorem 15.41. Now $x \in C_{r_n/2}(x_n) \subset B_{r_n}(x_n)$ for every $n \in \mathbb{N}$. Then since $B_{r_n}(x_n) \cap E_n = \emptyset$ for every $n \in \mathbb{N}$, we have $x \notin E_n$ for every $n \in \mathbb{N}$. Therefore $x \notin \bigcup_{n\in\mathbb{N}} E_n$ and $\bigcup_{n\in\mathbb{N}} E_n \neq X$. ∎

[VIII] Uniform Boundedness Theorems

Theorem 15.48. *Let (X, d) be a complete metric space. Let $\mathfrak{F}$ be a collection of real-valued continuous functions on X with the property that for every $x \in X$ there exists $M_x \geq 0$ such that*

$$(1) \qquad |f(x)| \leq M_x \quad \textit{for all } f \in \mathfrak{F}.$$

Then there exist a non-empty open set O in X and a constant $M \geq 0$ such that

(2) $$|f(x)| \leq M \quad \textit{for all } x \in O \textit{ and all } f \in \mathfrak{F}.$$

Proof. Let us classify the points of X as follows. For each $k \in \mathbb{N}$ we let

(3) $$E_k = \{x \in X : |f(x)| \leq k \text{ for all } f \in \mathfrak{F}\},$$

and for each $k \in \mathbb{N}$ and $f \in \mathfrak{F}$ we let

(4) $$E_{k,f} = \{x \in X : |f(x)| \leq k\} = |f|^{-1}([0,k]).$$

Then we have

(5) $$E_k = \bigcap_{f \in \mathfrak{F}} E_{k,f}.$$

Let $x \in X$ be arbitrarily chosen. By (1) there exists $M_x \geq 0$ such that $|f(x)| \leq M_x$ for all $f \in \mathfrak{F}$. With $k_x \in \mathbb{N}$ such that $k_x \geq M_x$, we have $|f(x)| \leq k_x$ for all $f \in \mathfrak{F}$ so that $x \in E_{k_x}$. Thus every $x \in X$ is in E_k for some $k \in \mathbb{N}$ so that we have

(6) $$X = \bigcup_{k \in \mathbb{N}} E_k.$$

Let us show that E_k is a closed set in X for every $k \in \mathbb{N}$. Let $f \in \mathfrak{F}$. Since f is a real-valued continuous function on X, $|f|$ is a real-valued continuous function on X. Then $E_{k,f}$ as given by (4) is the preimage of the closed set $[0,k]$ in $\mathbb{R}$ under the continuous mapping $|f|$. Thus $E_{k,f}$ is a closed set in X. Then E_k as given by (5) is the intersection of a collection of closed sets in X and is therefore a closed set in X. Since X is a complete metric space, according to Theorem 15.47 (Baire Category Theorem) X is a set of the second category, that is, X is not a countable union of non-dense sets. Then (6) implies that there exist $k_0 \in \mathbb{N}$ such that E_{k_0} is not a non-dense set, that is, $\left(\overline{E}_{k_0}\right)^\circ \neq \emptyset$. Since E_{k_0} is a closed set we have $\overline{E}_{k_0} = E_{k_0}$. Thus we have $E_{k_0}^\circ \neq \emptyset$. Let $O = E_{k_0}^\circ$ and $M = k_0$. Then by (3) we have $|f(x)| \leq k_0 = M$ for all $x \in O$ and all $f \in \mathfrak{F}$. ∎

Theorem 15.49. (Uniform Boundedness Theorem) *Let $(V, \mathbb{K}, \|\cdot\|)$ be a Banach space and $(W, \mathbb{K}', \|\cdot\|')$ be a normed linear space where either $\mathbb{K} = \mathbb{K}'$ or $\mathbb{K} = \mathbb{R}$ and $\mathbb{K}' = \mathbb{C}$. Consider the collection $\mathfrak{B}\{V, W\}$ of all bounded linear mappings of V into W and let $\mathfrak{B}_0\{V, W\}$ be a subcollection of $\mathfrak{B}\{V, W\}$ such that for every $v \in V$ there exists $M_v \geq 0$ such that*

(1) $$\|L(v)\|' \leq M_v \quad \textit{for all } L \in \mathfrak{B}_0\{V, W\}.$$

Then there exists a constant $B \geq 0$ such that

(2) $$\|L\|_{V,W} \leq B \quad \textit{for all } L \in \mathfrak{B}_0\{V, W\}.$$

Proof. For each $L \in \mathfrak{B}_0\{V, W\}$ let us define a real-valued function f_L on V by setting

(3) $$f_L(v) = \|L(v)\|' \quad \text{for } v \in V.$$

Since L is a bounded linear mapping of V into W, L is a continuous mapping of V into W by Theorem 15.32. On the normed linear space $(W, \mathbb{K}', \|\cdot\|')$ the mapping $w \mapsto \|w\|'$ is a real-valued continuous function. (This is an immediate consequence of Observation 15.2.) Thus as a composition of two continuous mappings, L and $\|\cdot\|'$, f_L is a real-valued continuous function on V. Let $\mathfrak{F} = \{f_L : L \in \mathfrak{B}_0\{V, W\}\}$. This collection of nonnegative continuous functions on V has the property that for each $v \in V$ we have $f_L(v) = \|L(v)\|' \le M_v$ for some $M_v \ge 0$ by (1). Thus by Theorem 15.48 there exist a non-empty open set O in V and a constant $M \ge 0$ such that

$$f_L(v) \le M \quad \text{for all } v \in O \text{ and all } f_L \in \mathfrak{F},$$

that is, recalling (3),

(4) $$\|L(v)\|' \le M \quad \text{for all } v \in O \text{ and all } L \in \mathfrak{B}_0\{V, W\}.$$

Now since O is a non-empty open set in V there exist $v_0 \in O$ and $r_0 > 0$ such that

$$\overline{B}_{r_0}(v_0) = \{v \in V : \|v - v_0\| \le r_0\} \subset O.$$

Let $v \in V$ be such that $\|v\| = r_0$. Then we have $\|(v + v_0) - v_0\| = \|v\| = r_0$ so that $v + v_0 \in \overline{B}_{r_0}(v_0) \subset O$. Let $L \in \mathfrak{B}_0\{V, W\}$. Now by (4) and (1) we have

$$\|L(v)\|' = \|L(v + v_0) - L(v_0)\|' \le \|L(v + v_0)\|' + \|L(v_0)\|' \le M + M_{v_0}.$$

Thus for every $v \in V$ with $\|v\| = r_0$ we have

$$\left\| L\Big(\frac{1}{r_0} v\Big) \right\|' = \frac{1}{r_0}\|L(v)\|' \le \frac{M + M_{v_0}}{r_0}.$$

This implies that for every $u \in S_1(0)$ we have $\|L(u)\|' \le \frac{M+M_{v_0}}{r_0}$. Then

$$\|L\|_{V,W} = \sup_{u \in S_1(0)} \|L(u)\|' \le \frac{M + M_{v_0}}{r_0}.$$

Let $B = \frac{M+M_{v_0}}{r_0}$. Then we have $\|L\|_{V,W} \le B$ for all $L \in \mathfrak{B}_0\{V, W\}$. ∎

Corollary 15.50. *Let $(V, \mathbb{K}, \|\cdot\|)$ be a Banach space. Let $\mathfrak{B}_0\{V, W\}$ be a collection of bounded linear functionals on V with the property that for every $v \in V$ there exists a constant $M_v \ge 0$ such that $|L(v)| \le M_v$ for all $L \in \mathfrak{B}_0\{V, W\}$. Then there exists a constant $B \ge 0$ such that $\|L\|_* \le B$ for all $L \in \mathfrak{B}_0\{V, W\}$.*

Proof. The Corollary is a particular case of Theorem 15.49 in which we have $(W, \mathbb{K}', \|\cdot\|') = (\mathbb{K}', \mathbb{K}', |\cdot|)$. ∎

Theorem 15.51. (Banach-Steinhaus Theorem) *Let $(V, \mathbb{K}, \|\cdot\|)$ be a Banach space and $(W, \mathbb{K}', \|\cdot\|')$ be a normed linear spaces where either $\mathbb{K} = \mathbb{K}'$ or $\mathbb{K} = \mathbb{R}$ and $\mathbb{K}' = \mathbb{C}$. Let $(L_n : n \in \mathbb{N})$ be a sequence of bounded linear mappings of V into W such that the sequence $(L_n(v) : n \in \mathbb{N})$ converges in W for every $v \in V$. Then the mapping L of V into W defined by setting $L(v) := \lim_{n\to\infty} L_n(v)$ for $v \in V$ is a bounded linear mapping.*

Proof. The fact that L is a linear mapping follows immediately from the definition of L. Indeed for $v_1, v_2 \in V$ and $\gamma_1, \gamma_2 \in \mathbb{K}$ we have

$$\begin{aligned} L(\gamma_1 v_1 + \gamma_2 v_2) &= \lim_{n\to\infty} L_n(\gamma_1 v_1 + \gamma_2 v_2) \\ &= \gamma_1 \lim_{n\to\infty} L_n(v_1) + \gamma_2 \lim_{n\to\infty} L_n(v_2) \\ &= \gamma_1 L(v_1) + \gamma_2 L(v_2). \end{aligned}$$

Let us show that L is bounded. Since the sequence $(L_n(v) : n \in \mathbb{N})$ converges in W for every $v \in V$, we have $\sup_{n\in\mathbb{N}} \|L_n(v)\|' < \infty$. Thus by Theorem 15.49 there exists a constant $B \geq 0$ such that $\|L_n\|_{V,W} \leq B$ for all $n \in \mathbb{N}$. Then for every $v \in V$ we have

$$\|L(v)\|' = \lim_{n\to\infty} \|L_n(v)\|' \leq \limsup_{n\to\infty} \|L_n\|_{V,W} \|v\| \leq B\|v\|.$$

This shows that B is a bound for the linear mapping L and thus L is a bounded linear mapping. ∎

[IX] Open Mapping Theorem

Definition 15.52. *Let $(V, \mathbb{K})$ be a linear space over the field $\mathbb{K}$ where $\mathbb{K} = \mathbb{R}$ or $\mathbb{K} = \mathbb{C}$. For $A, B \subset V$ and $\alpha \in \mathbb{K}$, we define*

$$\begin{aligned} \alpha A &= \{\alpha u : u \in A\}. \\ A + B &= \{u + v : u \in A, v \in B\} \\ A - B &= \{u - v : u \in A, v \in B\}. \end{aligned}$$

Note that it follows from the definitions above that $A - B = A + (-B)$. Note also that for any non-empty subset A of V we have $0 \in A - A$.

Observation 15.53. Let $(V, \mathbb{K}, \|\cdot\|)$ be a normed linear space.
(a) For every $\lambda \in \mathbb{K}$, $\lambda \neq 0$, we have $\lambda B_r(0) = B_{|\lambda| r}(0)$.
(b) For every $v_0 \in V$ we have $v_0 + B_r(0) = B_r(v_0)$.
(c) For every $u_0, v_0 \in V$ we have $u_0 + B_r(v_0) = B_r(u_0 + v_0)$.
(d) For every $\lambda \in \mathbb{K}$, $\lambda \neq 0$, we have $\lambda B_r(v_0) = B_{|\lambda| r}(\lambda v_0)$.

Proof. 1. Let $\lambda \in \mathbb{K}$, $\lambda \neq 0$. We have $\lambda B_r(0) = \{\lambda u : u \in B_r(0)\}$. For $u \in B_r(0)$ we have $\|u\| < r$. Then $\|\lambda u\| = |\lambda| \, \|u\| < |\lambda| r$ so that $\lambda B_r(0) \subset B_{|\lambda| r}(0)$.

Conversely if $v \in B_{|\lambda|r}(0)$ then $\|v\| < |\lambda|r$ so that $\|\frac{1}{\lambda}v\| = \frac{1}{|\lambda|}\|v\| < \frac{1}{|\lambda|}|\lambda|r = r$, that is, $\frac{1}{\lambda}v \in B_r(0)$. Then $v = \lambda\frac{1}{\lambda}v \in \lambda B_r(0)$. Thus $B_{|\lambda|r}(0) \subset \lambda B_r(0)$. Therefore we have $\lambda B_r(0) = B_{|\lambda|r}(0)$.

2. If $v \in (v_0 + B_r(0))$ then $v = v_0 + u$ where $u \in B_r(0)$ with $\|u\| < r$. Then $v - v_0 = u$ so that $\|v - v_0\| = \|u\| < r$. Thus $v \in B_r(v_0)$. This shows that $(v_0 + B_r(0)) \subset B_r(v_0)$.

Conversely if $v \in B_r(v_0)$ then $\|v - v_0\| < r$ and $v - v_0 \in B_r(0)$. Now we have $v = v_0 + v - v_0 \in (v_0 + B_r(0))$. This shows that $B_r(v_0) \subset (v_0 + B_r(0))$. Therefore we have $v_0 + B_r(0) = B_r(v_0)$.

3. To prove (c) note that by (b) we have $B_r(v_0) = v_0 + B_r(0)$. Then

$$u_0 + B_r(v_0) = u_0 + \{v_0 + B_r(0)\} = \{u_0 + v_0\} + B_r(0) = B_r(u_0 + v_0)$$

where the last equality is by (b).

4. To prove (d) note that $\lambda B_r(v_0) = \lambda\{v_0 + B_r(0)\} = \lambda v_0 + \lambda B_r(0) = B_{|\lambda|r}(\lambda v_0)$ where the first equality is by (b) and the last equality is by (a) and (b). ∎

Observation 15.54. Let $(V, \mathbb{K}, \|\cdot\|)$ be a normed linear space.
(a) If A is an open set in V then for every $\alpha \in \mathbb{K}$, $\alpha \neq 0$, the set αA is an open set in V.
(b) If A is an arbitrary subset of V and B is a non-empty open set in V then $A + B$ and $A - B$ are open sets in V.

Proof. 1. Let us prove (a). Let A be an open set in V. If $A = \emptyset$ then for any $\alpha \in \mathbb{K}$ we have $\alpha A = \emptyset$. Suppose A is a non-empty open set. Then by Observation 15.34, A is a union of open balls and in fact $A = \bigcup_{v \in A} B_{r_v}(v)$ where $r_v > 0$. Then for any $\alpha \in \mathbb{K}$, $\alpha \neq 0$, we have

$$\alpha A = \alpha \bigcup_{v \in A} B_{r_v}(v) = \bigcup_{v \in A} \alpha B_{r_v}(v) = \bigcup_{v \in A} B_{|\alpha| r_v}(\alpha v)$$

by (d) of Observation 15.53. Then as a union of open balls, αA is an open set.

2. Let us prove (b). Let B be a non-empty open set in V. Then as we noted above, we can write $B = \bigcup_{v \in B} B_{r_v}(v)$ where $r_v > 0$. For any $u \in V$ we have

$$u + B = u + \bigcup_{v \in B} B_{r_v}(v) = \bigcup_{v \in B} \{u + B_{r_v}(v)\} = \bigcup_{v \in B} B_{r_v}(u + v)$$

by (c) of Observation 15.53. Then we have

$$A + B = \bigcup_{u \in A} \{u + B\} = \bigcup_{u \in A} \left\{u + \bigcup_{v \in B} B_{r_v}(v)\right\} = \bigcup_{u \in A} \bigcup_{v \in B} B_{r_v}(u + v)$$

by (c) of Observation 15.53. Then as a union of open balls, $A + B$ is an open set. Regarding $A - B$, note that $A - B = A + (-B)$. Now $-B$ is an open set by (a). Then by our result above, $A + (-B)$ is an open set. This shows that $A - B$ is an open set. ∎

Observation 15.55. Let $(V, \mathbb{K}, \|\cdot\|)$ be a normed linear space.
(a) For $A \subset V$ and $\alpha \in \mathbb{K}$ we have $\overline{\alpha A} = \alpha \overline{A}$.
(b) For $A, B \subset V$ we have $\overline{A + B} \supset \overline{A} + \overline{B}$ and $\overline{A - B} \supset \overline{A} - \overline{B}$.

Proof. 1. Let us show $\overline{\alpha A} = \alpha \overline{A}$. Let $v \in \overline{\alpha A}$. Then there exists a sequence $(v_n : n \in \mathbb{N})$ in αA such that $\lim_{n\to\infty} v_n = v$. Now since $v_n \in \alpha A$ there exists $u_n \in A$ such that $v_n = \alpha u_n$. Then $v = \lim_{n\to\infty} v_n = \lim_{n\to\infty} \alpha u_n = \alpha \lim_{n\to\infty} u_n$. Since $u_n \in A$ for every $n \in \mathbb{N}$, we have $\lim_{n\to\infty} u_n \in \overline{A}$. Thus $v \in \alpha \overline{A}$. This shows that $\overline{\alpha A} \subset \alpha \overline{A}$.

Conversely suppose $v \in \alpha \overline{A}$. Then $v = \alpha u$ for some $u \in \overline{A}$. Since $u \in \overline{A}$ there exists a sequence $(u_n : n \in \mathbb{N})$ in A such that $\lim_{n\to\infty} u_n = u$. Then $v = \alpha u = \alpha \lim_{n\to\infty} u_n = \lim_{n\to\infty} \alpha u_n$. Since $\alpha u_n \in \alpha A$ for every $n \in \mathbb{N}$ we have $v = \lim_{n\to\infty} \alpha u_n \in \overline{\alpha A}$. This shows that $\alpha \overline{A} \subset \overline{\alpha A}$. Therefore we have $\overline{\alpha A} = \alpha \overline{A}$.

2. Let us show that for every $A, B \subset V$, we have

(1) $$\overline{A} + \overline{B} \subset \overline{A + B}.$$

Let $w \in \overline{A} + \overline{B}$. Then $w = u + v$ where $u \in \overline{A}$ and $v \in \overline{B}$. Since $u \in \overline{A}$, there exists a sequence $(u_n : n \in \mathbb{N})$ in A such that $\lim_{n\to\infty} u_n = u$. Similarly there exists a sequence $(v_n : n \in \mathbb{N})$ in B such that $\lim_{n\to\infty} v_n = v$. Now $w = \lim_{n\to\infty} u_n + \lim_{n\to\infty} v_n = \lim_{n\to\infty} \{u_n + v_n\}$ Since $u_n + v_n \in A + B$ for every $n \in \mathbb{N}$ we have $w \in \overline{A + B}$. This shows that $\overline{A} + \overline{B} \subset \overline{A + B}$. Then we have

$$\overline{A} - \overline{B} = \overline{A} + (-\overline{B}) = \overline{A} + \overline{(-B)} \subset \overline{A + (-B)} = \overline{A - B}$$

where the second equality is by (a) and the set inclusion is by (1). ∎

Lemma 15.56. *Let $(V, \mathbb{K}, \|\cdot\|)$ and $(W, \mathbb{K}, \|\cdot\|')$ be two Banach spaces over the same filed of scalars $\mathbb{K}$. Let L be a bounded linear mapping of V into W.*

Let $A_a(0) = \{v \in V : \|v\| < a\}$, an open ball in V with center at $0 \in V$ and radius $a > 0$. If $\overline{L(A_a(0))}$ contains an open ball in W with center at $0 \in W$, then we have

$$\overline{L(A_a(0))} \subset L(A_{2a}(0)).$$

Proof. Suppose $\overline{L(A_a(0))}$ contains an open ball in W with center at $0 \in W$ given by $B_b(0) = \{w \in W : \|w\|' < b\}$ where $b > 0$. To show that $\overline{L(A_a(0))} \subset L(A_{2a}(0))$, we show that if $w \in \overline{L(A_a(0))}$ then $w \in L(A_{2a}(0))$, that is, if $w \in \overline{L(A_a(0))}$ then there exists $v \in A_{2a}(0)$ such that $L(v) = w$. Let $w \in \overline{L(A_a(0))}$ be arbitrarily selected. We construct $v \in V$ such that $v \in A_{2a}(0)$ and $L(v) = w$ as follows.

Now $w \in \overline{L(A_a(0))}$ implies that for our $b > 0$ we can select $v_1 \in A_a(0)$ such that

(1) $$\|w - L(v_1)\|' < \frac{b}{2}.$$

Then $\|2w - 2L(v_1)\|' < b$ so that $2w - 2L(v_1) \in B_b(0) \subset \overline{L(A_a(0))}$. Then we can select $v_2 \in A_a(0)$ such that $\|2w - 2L(v_1) - L(v_2)\|' < \frac{b}{2}$, that is,

(2) $$\left\|w - L(v_1) - \frac{1}{2}L(v_2)\right\|' < \frac{b}{2^2}.$$

Now $4w - 4L(v_1) - 2L(v_2) \in B_b(0) \subset \overline{L(A_a(0))}$. Then we can select $v_3 \in A_a(0)$ such that $\|4w - 4L(v_1) - 2L(v_2) - L(v_3)\|' < \frac{b}{2}$ so that

$$\text{(3)} \qquad \Big\|w - L(v_1) - \frac{1}{2}L(v_2) - \frac{1}{2^2}L(v_3)\Big\|' < \frac{b}{2^3}.$$

Thus proceeding we have a sequence $(v_n : n \in \mathbb{N})$ in $A_a(0)$ such that

$$\text{(4)} \qquad \Big\|w - L\Big(\sum_{i=1}^{n} \frac{1}{2^{i-1}} v_i\Big)\Big\|' = \Big\|w - \sum_{i=1}^{n} \frac{1}{2^{i-1}} L(v_i)\Big\|' < \frac{b}{2^n}$$

for every $n \in \mathbb{N}$. Then letting $n \to \infty$, we have

$$\text{(5)} \qquad \lim_{n\to\infty} \Big\|w - L\Big(\sum_{i=1}^{n} \frac{1}{2^{i-1}} v_i\Big)\Big\|' = 0.$$

Consider the sequence $\big(\sum_{i=1}^{n} \frac{1}{2^{i-1}} v_i : n \in \mathbb{N}\big) \subset V$. For $m > n$, we have

$$\Big\|\sum_{i=n}^{m} \frac{1}{2^{i-1}} v_i\Big\| \le \sum_{i=n}^{m} \frac{1}{2^{i-1}} \|v_i\| < a \sum_{i=n}^{m} \frac{1}{2^{i-1}}.$$

Since $\sum_{i=n}^{m} \frac{1}{2^{i-1}} = 2$, for $\varepsilon > 0$ there exists $N \in \mathbb{N}$ such that $\sum_{i \ge N} \frac{1}{2^{i-1}} < \frac{\varepsilon}{a}$. Thus we have $\big\|\sum_{i=n}^{m} \frac{1}{2^{i-1}} v_i\big\| < \varepsilon$ for $m > n \ge N$. This shows that $\big(\sum_{i=1}^{n} \frac{1}{2^{i-1}} v_i : n \in \mathbb{N}\big)$ is a Cauchy sequence in V. Then the completeness of the Banach space V implies that there exists $v \in V$ such that

$$v = \lim_{n\to\infty} \sum_{i=1}^{n} \frac{1}{2^{i-1}} v_i = \sum_{i \in \mathbb{N}} \frac{1}{2^{i-1}} v_i.$$

Let us show that $v \in A_{2a}(0)$. Now we have

$$\begin{aligned} \|v\| &= \Big\|\sum_{i\in\mathbb{N}} \frac{1}{2^{i-1}} v_i\Big\| = \Big\|\lim_{n\to\infty} \sum_{i=1}^{n} \frac{1}{2^{i-1}} v_i\Big\| \\ &= \lim_{n\to\infty} \Big\|\sum_{i=1}^{n} \frac{1}{2^{i-1}} v_i\Big\| \quad \text{by (c) of Proposition 15.9} \\ &\le \lim_{n\to\infty} \sum_{i=1}^{n} \frac{1}{2^{i-1}} \|v_i\| < \lim_{n\to\infty} \sum_{i=1}^{n} \frac{1}{2^{i-1}} a = 2a. \end{aligned}$$

This shows that $v \in A_{2a}(0)$.

It remains to show that $L(v) = w$. Now

$$L(v) = L\Big(\sum_{i\in\mathbb{N}} \frac{1}{2^{i-1}} v_i\Big) = L\Big(\lim_{n\to\infty} \sum_{i=1}^{n} \frac{1}{2^{i-1}} v_i\Big) = \lim_{n\to\infty} L\Big(\sum_{i=1}^{n} \frac{1}{2^{i-1}} v_i\Big)$$

by the continuity of L. Then

$$\|w - L(v)\|' = \Big\|w - \lim_{n\to\infty} L\Big(\sum_{i=1}^{n} \frac{1}{2^{i-1}} v_i\Big)\Big\|' = \Big\|\lim_{n\to\infty} \Big\{w - L\Big(\sum_{i=1}^{n} \frac{1}{2^{i-1}} v_i\Big)\Big\}\Big\|'$$
$$= \lim_{n\to\infty} \Big\|w - L\Big(\sum_{i=1}^{n} \frac{1}{2^{i-1}} v_i\Big)\Big\|' = 0$$

where the third equality is by (c) of Proposition 15.9 and the last equality is by (5). Thus we have $w = L(v)$. ∎

Theorem 15.57. (Open Mapping Theorem) *Let $(V, \mathbb{K}, \|\cdot\|)$ and $(W, \mathbb{K}, \|\cdot\|')$ be two Aanach spaces over the same field of scalars $\mathbb{K}$. Let L be a bounded linear mapping of V onto W. Then L is an open mapping, that is, for every open set G in V, $L(G)$ is an open set in W.*

Proof. Let us write $A_r(v_0)$ for an open ball in V with center at $v_0 \in V$ and radius $r > 0$ and $B_r(w_0)$ for an open ball in W with center at $w_0 \in W$ and radius $r > 0$, that is,

$$A_r(v_0) = \{v \in V : \|v - v_0\| < r\},$$
$$B_r(w_0) = \{w \in W : \|w - w_0\|' < r\}.$$

1. Let us show first that the image $L(A_1(0))$ of the open ball $A_1(0)$ in V contains an open ball in W with center at $0 \in W$, that is, there exists $a > 0$ such that

$$(1) \qquad B_a(0) \subset L(A_1(0)).$$

Now for every $v \in V$ we have $\|v\| < \infty$ so that $\|v\| < n$ for some $n \in \mathbb{N}$ and thus $v \in A_n(0)$. Therefore we have $V = \bigcup_{n\in\mathbb{N}} A_n(0)$. Now since L maps V onto W we have

$$W = L(V) = L\Big(\bigcup_{n\in\mathbb{N}} A_n(0)\Big) = \bigcup_{n\in\mathbb{N}} L(A_n(0)).$$

Since W is a complete metric space, according to Theorem 15.47 (Baire Category Theorem) at least one in the sequence $(L(A_n(0)) : n \in \mathbb{N})$ is not a non-dense set in W. Thus there exists $n_0 \in \mathbb{N}$ such that the interior of $\overline{L(A_{n_0}(0))}$ is non-empty. Then there exists a non-empty open set H in W such that

$$H \subset \overline{L(A_{n_0}(0))}.$$

Now if $u, v \in A_{n_0}(0)$ then $\|u - v\| \le \|u\| + \|v\| < 2n_0$ so that $u - v \in A_{2n_0}(0)$. Therefore $A_{2n_0}(0) \supset A_{n_0}(0) - A_{n_0}(0)$. Thus we have

$$L(A_{2n_0}(0)) \supset L(A_{n_0}(0) - A_{n_0}(0)) = L(A_{n_0}(0)) - L(A_{n_0}(0))$$

where the equality is by the linearity of L. Then

$$\overline{L(A_{2n_0}(0))} \supset \overline{L(A_{n_0}(0) - A_{n_0}(0))} \supset \overline{L(A_{n_0}(0))} - \overline{L(A_{n_0}(0))} \supset H - H$$

where the second set inclusion is by Observation 15.55. By Observation 15.54, $H - H$ is an open set in W. Since $H - H$ contains $0 \in W$ there exists $r > 0$ such that

$$B_r(0) \subset H - H \subset \overline{L(A_{2n_0}(0))}.$$

Then by Lemma 15.56 we have

$$B_r(0) \subset \overline{L(A_{2n_0}(0))} \subset L(A_{4n_0}(0)).$$

By Observation 15.54 we have

$$B_{r/4n_0}(0) = \frac{1}{4n_0} B_r(0) \subset \frac{1}{4n_0} L(A_{4n_0}(0)) = L\Big(\frac{1}{4n_0} A_{4n_0}(0)\Big) = L(A_1(0)).$$

If we let $a = \frac{r}{4n_0}$ then we have (1).

2. For the open set $\emptyset$ in V we have $L(\emptyset) = \emptyset$, an open set in W. Let G be a non-empty open set in V. To show that $L(G)$ is an open set in W we show that for every $w \in L(G)$ there exists $b > 0$ such that

$$(2) \qquad B_b(w) \subset L(G).$$

Let $w \in L(G)$. Then there exists $v \in G$ such that $w = L(v)$. Since $v \in G$ and G is an open set in V there exists $r > 0$ such that $A_r(v) \subset G$, that is, $v + A_r(0) \subset G$ by Observation 15.53. Then $L(G) \supset L(v + A_r(0)) = L(v) + L(A_r(0))$. By (1) there exists $a > 0$ such that $B_a(0) \subset L(A_1(0))$ and thus

$$B_{ar}(0) = r B_a(0) \subset r L(A_1(0)) = L(r A_1(0)) = L(A_r(0))$$

by Observation 15.53. Thus $L(G) \supset L(v) + B_{ar}(0) = B_{ar}(L(v))$. Then since $w = L(v)$ we have $B_{ar}(w) \subset L(G)$. If we let $b = ar$ then we have (2). ∎

Corollary 15.58. *Let $(V, \mathbb{K}, \|\cdot\|)$ and $(W, \mathbb{K}, \|\cdot\|')$ be two Banach spaces over the same field of scalars $\mathbb{K}$. Let L be a one-to-one continuous linear mapping of V onto W. Then the inverse mapping of L is a continuous mapping of W onto V.*

Proof. Let Λ be the inverse mapping of L. Now Λ maps W one-to-one onto V. Let us show that Λ is a linear mapping. Let $w_1, w_2 \in W$ and $\alpha_1, \alpha_2 \in \mathbb{K}$. Since L maps V one-to-one onto W there exist unique $v_1, v_2 \in V$ such that $L(v_1) = w_1$ and $L(v_2) = w_2$, that is, $v_1 = \Lambda(w_1)$ and $v_2 = \Lambda(w_2)$. Then

$$\begin{aligned}\Lambda(\alpha_1 w_1 + \alpha_2 w_2) &= \Lambda(\alpha_1 L(v_1) + \alpha_2 L(v_2)) = \Lambda(L(\alpha_1 v_1 + \alpha_2 v_2)) \\ &= \alpha_1 v_1 + \alpha_2 v_2 = \alpha_1 \Lambda(w_1) + \alpha_2 \Lambda(w_2).\end{aligned}$$

This shows that Λ is a linear mapping of W into V.

To show that Λ is a continuous mapping we show that for every open set G in V, $\Lambda^{-1}(G)$ is an open set in W. But $\Lambda^{-1}(G) = L(G)$ and according to Theorem 15.57, $L(G)$ is an open set in W. ∎

Definition 15.59. *Let X and Y be two sets and let T be a mapping of a subset D of X into Y. The subset of $X \times Y$ defined by $\{(x, T(x)) \in X \times Y : x \in D\}$ is called the graph of the mapping T.*

Proposition 15.60. *Given two normed linear spaces $(V, \mathbb{K}, \|\cdot\|)$ and $(W, \mathbb{K}, \|\cdot\|')$ over the same field of scalars $\mathbb{K}$.*
(a) *Let us define addition and scalar multiplication on the product space $V \times W$ by*

(1) $$(v_1, w_1) + (v_2, w_2) = (v_1 + v_2, w_1 + w_2) \quad \text{for } (v_1, w_1), (v_2, w_2) \in V \times W,$$

(2) $$\gamma\,(v, w) = (\gamma\, v, \gamma\, w) \quad \text{for } (v, w) \in V \times W \text{ and } \gamma \in \mathbb{K}.$$

Then $V \times W$ is a linear space over the field of scalars $\mathbb{K}$.
(b) *Let us define a function $\|\cdot\|''$ on $V \times W$ by setting*

(3) $$\|(v, w)\|'' = \|v\| + \|w\|' \quad \text{for } (v, w) \in V \times W.$$

Then $\|\cdot\|''$ is a norm on the linear space $V \times W$.
(c) *$V \times W$ is complete with respect to the norm $\|\cdot\|''$ if and only if V and W are complete with respect to the norms $\|\cdot\|$ and $\|\cdot\|'$ respectively.*
(d) *If $(V, \mathbb{K}, \|\cdot\|)$ and $(W, \mathbb{K}, \|\cdot\|')$ are Banach spaces then $(V \times W, \mathbb{K}, \|\cdot\|'')$ is a Banach space.*

Proof. 1. It is readily verified that, with addition and scalar multiplication defined by (1) and (2), $V \times W$ satisfies conditions 1° - 9° of Definition 15.1. Thus $V \times W$ is a linear space over the field $\mathbb{K}$.

2. Similarly it is easily verified that $\|\cdot\|''$ defined by (3) satisfies conditions 1° - 4° of Definition 15.1. Thus $\|\cdot\|''$ is a norm on the linear space $V \times W$.

3. Let us note that for $(v_1, w_1), (v_2, w_2) \in V \times W$ we have by (1) and (3)

(4) $$\|(v_1, w_1) - (v_2, w_2)\|'' = \|(v_1 - v_2, w_1 - w_2)\|'' = \|v_1 - v_2\| + \|w_1 - w_2\|'.$$

Thus we have

(5) $$\|v_1 - v_2\|, \|w_1 - w_2\|' \leq \|(v_1, w_1) - (v_2, w_2)\|'' = \|v_1 - v_2\| + \|w_1 - w_2\|'.$$

Now consider arbitrary sequences $(v_n : n \in \mathbb{N})$ in V, $(w_n : n \in \mathbb{N})$ in W, and $((v_n, w_n) : n \in \mathbb{N})$ in $V \times W$. It follows from (5) that $((v_n, w_n) : n \in \mathbb{N})$ is a Cauchy sequence with respect to the norm $\|\cdot\|''$ if and only if $(v_n : n \in \mathbb{N})$ is a Cauchy sequence with respect to the norm $\|\cdot\|$ and $(w_n : n \in \mathbb{N})$ is a Cauchy sequence with respect to the norm $\|\cdot\|'$.

Suppose V is complete with respect to $\|\cdot\|$ and W is complete with respect to $\|\cdot\|'$. To show that $V \times W$ is complete with respect to $\|\cdot\|''$, let $((v_n, w_n) : n \in \mathbb{N})$ be a Cauchy sequence in $V \times W$ with respect to $\|\cdot\|''$. Then as we noted above $(v_n : n \in \mathbb{N})$ is a Cauchy sequence in V with respect to $\|\cdot\|$ and $(w_n : n \in \mathbb{N})$ is a Cauchy sequence in W with respect to $\|\cdot\|'$. By the completeness of V with respect to $\|\cdot\|$ and the completeness of W with respect to $\|\cdot\|'$ there exist $v \in V$ and $w \in W$ such that $\lim_{n\to\infty} \|v_n - v\| = 0$ and $\lim_{n\to\infty} \|w_n - w\| = 0$. Then by (4) we have

$$\lim_{n\to\infty} \|(v_n, w_n) - (v, w)\|'' = \lim_{n\to\infty} \{\|v_n - v\| + \|w_n - w\|'\} = 0.$$

This shows that $V \times W$ is complete with respect to $\|\cdot\|''$.

Conversely suppose $V \times W$ is complete with respect to $\|\cdot\|''$. To show that V is complete with respect to $\|\cdot\|$ and W is complete with respect to $\|\cdot\|'$, let $(v_n : n \in \mathbb{N})$ be a Cauchy sequence in V with respect to $\|\cdot\|$ and $(w_n : n \in \mathbb{N})$ be a Cauchy sequence in W with respect to $\|\cdot\|'$. Then as we noted above $\big((v_n, w_n) : n \in \mathbb{N}\big)$ is a Cauchy sequence in $V \times W$ with respect to $\|\cdot\|''$. By the completeness of $V \times W$ with respect to $\|\cdot\|''$ there exists $(v, w) \in V \times W$ such that $\lim_{n\to\infty} \|(v_n, w_n) - (v, w)\|'' = 0$. Then by (5) we have

$$\lim_{n\to\infty} \|v_n - v\| \le \lim_{n\to\infty} \|(v_n, w_n) - (v, w)\|'' = 0.$$

This shows that V is complete with respect to $\|\cdot\|$. Similarly W is complete with respect to $\|\cdot\|'$.

4. If $(V, \mathbb{K}, \|\cdot\|)$ and $(W, \mathbb{K}, \|\cdot\|')$ are Banach spaces then V is complete with respect to the norm $\|\cdot\|$ and W is complete with respect to the norm $\|\cdot\|'$. Then by (c), $V \times W$ is complete with respect to the norm $\|\cdot\|''$, that is, $(V \times W, \mathbb{K}, \|\cdot\|'')$ is a Banach space. ∎

Theorem 15.61. (Closed Graph Theorem) *Let $(V, \mathbb{K}, \|\cdot\|)$ and $(W, \mathbb{K}, \|\cdot\|')$ be two Banach spaces over the same field of scalars $\mathbb{K}$. Let L be a linear mapping of V into W. If the graph of L is a closed set in the Banach space $(V \times W, \mathbb{K}, \|\cdot\|'')$ where $\|(v, w)\|'' = \|v\| + \|w\|'$ for $(v, w) \in V \times W$, then L is a continuous mapping of V into W.*

Proof. Let G be the graph of L, that is, $G = \big\{\big(v, L(v)\big) \in V \times W : v \in V\big\}$. Then G is a linear subspace of $V \times W$. Indeed if $\big(v_1, L(v_1)\big), \big(v_2, L(v_2)\big) \in G$ and $\gamma_1, \gamma_2 \in \mathbb{K}$ then by (1) and (2) of Proposition B.9 we have

$$\begin{aligned}\gamma_1\big(v_1, L(v_1)\big) + \gamma_2\big(v_2, L(v_2)\big) &= \big(\gamma_1 v_1 + \gamma_2 v_2, \gamma_1 L(v_1) + \gamma_2 L(v_2)\big)\\ &= \big(\gamma_1 v_1 + \gamma_2 v_2, L(\gamma_1 v_1 + \gamma_2 v_2)\big) \in G.\end{aligned}$$

If we assume that G is a closed set in the Banach space $(V \times W, \mathbb{K}, \|\cdot\|'')$ then $(G, \mathbb{K}, \|\cdot\|'')$ is a Banach space by Proposition 15.10.

Let us define two mappings P_1 and P_2 of $G \subset V \times W$ into V and W respectively by

$$\begin{aligned}P_1\big(v, L(v)\big) &= v \quad &&\text{for } \big(v, L(v)\big) \in G,\\ P_2\big(v, L(v)\big) &= L(v) \quad &&\text{for } \big(v, L(v)\big) \in G.\end{aligned}$$

Clearly P_1 and P_2 are linear mappings. Moreover we have

$$\big\|P_1\big(v, L(v)\big)\big\| = \|v\| \le \|v\| + \|L(v)\|' = \big\|\big(v, L(v)\big)\big\|'',$$

$$\big\|P_2\big(v, L(v)\big)\big\|' = \|L(v)\|' \le \|v\| + \|L(v)\|' = \big\|\big(v, L(v)\big)\big\|''.$$

This shows that the constant 1 is a bound for the linear mapping P_1 and for the linear mapping P_2. Thus P_1 and P_2 are bounded linear mappings. This implies that P_1 and P_2 are continuous linear mappings by Theorem 15.32.

Now P_1 is a one-to-one mapping of the Banach space G onto the Banach space V. Thus the continuity of L implies that its inverse mapping Q is a continuous mapping of V

onto G by Corollary 15.58. Now $L = P_2 \circ Q$. Then the continuity of the mapping Q of V onto G and the continuity of the mapping P_2 of G into W implies that L is a continuous mapping of V into W. ∎

[X] Hahn-Banach Extension Theorems

[X.1] Hahn-Banach Extension Theorem for Real Linear Spaces

Let P be a set. A partial ordering on P is a relation $\preceq$ between elements of P which satisfies the following conditions:
1° reflexivity: $x \preceq x$.
2° antisymmetry: $x \preceq y$ and $y \preceq x$ imply $x = y$.
3° transitivity: $x \preceq y$ and $y \preceq z$ imply $x \preceq z$.
$\preceq$ is called a linear, or total, ordering if it satisfies the additional condition :
4° trichotomy: $x, y \in P$ implies $x \preceq y$ or $y \preceq x$.
If for every $A \subset P$, $A \neq \emptyset$, there exists $a \in A$ such that $a \preceq x$ for every $x \in A$ then $\preceq$ is called a well ordering.
We call $(P, \preceq)$ a partially ordered set. If $x \preceq y$ and $x \neq y$ then we write $x \prec y$.
We write $x \succeq y$ for $y \preceq x$ and $x \succ y$ for $y \prec x$.

Let $(P, \preceq)$ be a partially ordered set. Let $A \subset P$.
If there exists $u \in P$ such that $x \preceq u$ for every $x \in A$ then we call u an upper bound for A.
If there exists $m \in P$ such that if $x \in P$ and $m \preceq x$ then $m = x$ and m is called a maximal element in P.
A lower bound for a subset A of P and a minimal element for P are defined likewise.
A subset C of P which is linearly ordered is called a chain in P.

Zorn's Lemma. *Every non-empty partially ordered set in which every chain has an upper bound has a maximal element.*

Zorn's Lemma is equivalent to the Axiom of Choice in set theory which states that for any non-empty collection of non-empty sets a set can be formed by selecting exactly one element from each set in the collection. For a proof of the equivalence we refer to [11].

Definition 15.62. *Let $(V, \mathbb{R})$ be a real linear space. A function p on V is called a sublinear functional on V if it satisfies the following conditions:*
1° *real-valued: $p(v) \in \mathbb{R}$ for every $v \in V$.*
2° *homogeneity: $p(\alpha v) = \alpha p(v)$ for $v \in V$ and $\alpha \geq 0$.*
3° *subadditivity: $p(u + v) \leq p(u) + p(v)$ for $u, v \in V$.*

Remark 15.63. (a) Conditions 1° to 4° in Definition 15.1 for a norm on a linear space imply conditions 1° - 3° in Definition 15.62. Thus a norm on a real linear space is a sublinear functional on the real linear space.

(b) If p is a sublinear functional on a real linear space $(V, \mathbb{R})$ then for $u, v \in V$ and $\alpha \in [0, 1]$

$$p\big(\alpha u + (1-\alpha)v\big) \leq \alpha p(u) + (1-\alpha)p(v).$$

Thus a sublinear functional is a convex function.

Theorem 15.64. (Hahn-Banach Theorem for Real Linear Spaces) *Let $(V, \mathbb{R})$ be a real linear space and let p be a sublinear functional on V. Let M be a linear subspace of V and let f be a real-valued linear functional on M such that $f(v) \leq p(v)$ for $v \in M$. Then there exists a real-valued linear functional g on V such that $g(v) = f(v)$ for $v \in M$ and $g(v) \leq p(v)$ for $v \in V$.*

Proof. 1. If $M = V$ there is nothing to prove. Thus we assume $M \neq V$. Let $v_0 \in V$ and $v_0 \notin M$. Let V_0 be the linear subspace of V spanned by M and v_0, that is,

(1) $$V_0 = \{u + \lambda v_0 : u \in M, \lambda \in \mathbb{R}\}.$$

Let us show that there exists a real-valued linear functional f_0 on V_0 such that $f_0(v) = f(v)$ for $v \in M$ and $f_0(v) \leq p(v)$ for $v \in V_0$.

With an arbitrary $c \in \mathbb{R}$ let us define a functional f_0 on V_0 by setting

(2) $$f_0(v) = f_0(u + \lambda v_0) = f(u) + \lambda c \quad \text{for } v = u + \lambda v_0 \in V_0.$$

Let us note that since $v_0 \notin M$ the representation of $v \in V_0$ as $v = u + \lambda c$ with $u \in M$ and $\lambda \in \mathbb{R}$ is unique and therefore the definition of f_0 by (2) is well defined.

Now f_0 is a real-valued functional on V_0. Let us show that f_0 is a linear functional on V_0. Let $v_1, v_2 \in V_0$ and $\alpha_1, \alpha_2 \in \mathbb{R}$. Now $v_1 = u_1 + \lambda_1 v_0$ and $v_2 = u_2 + \lambda_2 v_0$ for some $u_1, u_2 \in M$ and $\lambda_1, \lambda_2 \in \mathbb{R}$. Thus we have

$$\begin{aligned} f_0(\alpha_1 v_1 + \alpha_2 v_2) &= f_0\big((\alpha_1 u_1 + \alpha_2 u_2) + (\alpha_1\lambda_1 + \alpha_2\lambda_2)v_0\big) \\ &= f(\alpha_1 u_1 + \alpha_2 u_2) + (\alpha_1\lambda_1 + \alpha_2\lambda_2)c \\ &= \alpha_1\{f(u_1) + \lambda_1 c\} + \alpha_2\{f(u_2) + \lambda_2 c\} \\ &= \alpha_1 f_0(v_1) + \alpha_2 f_0(v_2). \end{aligned}$$

This shows that f_0 is a linear functional on V_0.

Clearly $f_0(v) = f(v)$ for $v \in M$. Let us show that the constant $c \in \mathbb{R}$ in the definition of f_0 by (2) can be so chosen that $f_0(v) \leq p(v)$ for $v \in V_0$, that is,

(3) $$f(u) + \lambda c \leq p(u + \lambda v_0) \quad \text{for all } u \in M \text{ and } \lambda \in \mathbb{R}.$$

Note that if $\lambda = 0$ then (3) reduces to $f(u) \leq p(u)$ for all $u \in M$ which is valid by our assumption on f. Thus dividing (3) by $\lambda \neq 0$ and applying the linearity of f on M and condition 2° in Definition 15.62 we obtain a set of conditions on c:

$$\begin{cases} f\big(\frac{u}{\lambda}\big) + c \leq p\big(\frac{u}{\lambda} + v_0\big) & \text{for all } u \in M \text{ and } \lambda > 0, \\ f\big(-\frac{u}{\lambda}\big) - c \leq p\big(-\frac{u}{\lambda} - v_0\big) & \text{for all } u \in M \text{ and } \lambda < 0, \end{cases}$$

which is equivalent to the condition (3) on c. This set of conditions on c is equivalent to the set of conditions:

$$\begin{cases} f(u) + c \le p(u + v_0) & \text{for all } u \in M, \\ f(u) - c \le p(u - v_0) & \text{for all } u \in M. \end{cases}$$

This set of conditions is further equivalent to the condition :

$$(4) \qquad f(u) - p(u - v_0) \le c \le -f(u) + p(u + v_0) \quad \text{for all } u \in M.$$

Let us show that $c \in \mathbb{R}$ satisfying condition (4) exists. Now for any $u', u'' \in M$ we have

$$\begin{aligned} f(u') + f(u'') &= f(u' + u'') \le p(u' + u'') \\ &= p(u' - v_0 + u'' + v_0) \\ &\le p(u' - v_0) + p(u'' + v_0). \end{aligned}$$

Thus we have $f(u') - p(u' - v_0) \le -f(u'') + p(u'' + v_0)$ for all $u', u'' \in M$ and therefore

$$\sup_{u \in M} \{f(u) - p(u - v_0)\} \le \inf_{u \in M} \{-f(u) + p(u + v_0)\}.$$

Thus if we select $c \in \big[\sup_{u \in M} \{f(u) - p(u - v_0)\}, \inf_{u \in M} \{-f(u) + p(u + v_0)\}\big]$, then c satisfies condition (4). With $c \in \mathbb{R}$ thus selected we have condition (3) holding for all $u \in M$ and $\lambda \in \mathbb{R}$, that is, $f_0(v) \le p(v)$ for $v \in V_0$.

2. Let V_α be a linear subspace of V containing M and let f_α be a real-valued linear functional on V_α such that $f_\alpha = f$ on M and $f_\alpha \le p$ on V_α. Let $\mathfrak{F} = \{(V_\alpha, f_\alpha) : \alpha \in A\}$ be the collection of all such pairs (V_α, f_α). Note that $\mathfrak{F} \neq \emptyset$ since with V_0 and f_0 constructed in **1** we have such a pair. Let us introduce a partial ordering in $\mathfrak{F}$ by defining $(V_{\alpha_1}, f_{\alpha_2}) \preceq (V_{\alpha_1}, f_{\alpha_2})$ if $(V_{\alpha_1}, f_{\alpha_1}), (V_{\alpha_2}, f_{\alpha_2}) \in \mathfrak{F}$ are such that $V_{\alpha_1} \subset V_{\alpha_2}$ and $f_{\alpha_1} = f_{\alpha_2}$ on V_{α_1}. It is easily verified that $\preceq$ satisfies the reflexivity condition, the symmetry condition, and the transitivity condition so that $\preceq$ is indeed a partial ordering on $\mathfrak{F}$.

Let $\mathfrak{C} \subset \mathfrak{F}$ be a chain, that is, if $(V_{\alpha_1}, f_{\alpha_1}), (V_{\alpha_2}, f_{\alpha_2}) \in \mathfrak{C}$ then we have either $(V_{\alpha_1}, f_{\alpha_1}) \preceq (V_{\alpha_2}, f_{\alpha_2})$ or $(V_{\alpha_2}, f_{\alpha_2}) \preceq (V_{\alpha_1}, f_{\alpha_1})$. Let us show that $\mathfrak{C}$ has an upper bound, that is, there exist $(V_\beta, f_\beta) \in \mathfrak{F}$ such that $(V_\alpha, f_\alpha) \preceq (V_\beta, f_\beta)$ for every $(V_\alpha, f_\alpha) \in \mathfrak{C}$. We construct V_β and f_β as follows. First let $V_\beta = \bigcup_{(V_\alpha, f_\alpha) \in \mathfrak{C}} V_\alpha$. Let us show that V_β is a linear subspace of V containing M. Let $v_1, v_2 \in V_\beta$ and $c_1, c_2 \in \mathbb{R}$. There exist $(V_{\alpha_1}, f_{\alpha_1}), (V_{\alpha_2}, f_{\alpha_2}) \in \mathfrak{C}$ such that $v_1 \in V_{\alpha_1}$ and $v_2 \in V_{\alpha_2}$. Since $\mathfrak{C}$ is a chain, either $(V_{\alpha_1}, f_{\alpha_1}) \preceq (V_{\alpha_2}, f_{\alpha_2})$ or $(V_{\alpha_2}, f_{\alpha_2}) \preceq (V_{\alpha_1}, f_{\alpha_1})$. Let $(V_{\alpha_3}, f_{\alpha_3})$ be the greater of the two. Then $v_1, v_2 \in V_{\alpha_3}$ so that $c_1 v_1 + c_2 v_2 \in V_{\alpha_3} \subset V_\beta$. This shows that V_β is a linear subspace of V. We have also $M \subset V_{\alpha_3} \subset V_\beta$.

Let us define a real-valued functional f_β on V_β as follows. Let $v \in V_\beta$. Then $v \in V_\alpha$ for some $(V_\alpha, f_\alpha) \in \mathfrak{C}$. We define $f_\beta(v) := f_\alpha(v)$. Let us show that this definition of f_β is consistent. Suppose $v \in V_{\alpha_1}$ and $v \in V_{\alpha_2}$ for some $(V_{\alpha_1}, f_{\alpha_1}), (V_{\alpha_2}, f_{\alpha_2}) \in \mathfrak{C}$. Since $\mathfrak{C}$ is a chain, either $(V_{\alpha_1}, f_{\alpha_1}) \preceq (V_{\alpha_2}, f_{\alpha_2})$ or $(V_{\alpha_2}, f_{\alpha_2}) \preceq (V_{\alpha_1}, f_{\alpha_1})$. We assume without loss of generality that $(V_{\alpha_1}, f_{\alpha_1}) \preceq (V_{\alpha_2}, f_{\alpha_2})$. Then $V_{\alpha_1} \subset V_{\alpha_2}$ and $f_{\alpha_1} = f_{\alpha_2}$ on V_{α_1} so that $f_\beta(v) := f_{\alpha_1}(v)$ and $f_\beta(v) := f_{\alpha_2}(v)$ are equal.

To show that $(V_\beta, f_\beta) \in \mathcal{F}$ we verify that f_β is a linear functional on V_β, $f_\beta = f$ on M, and $f_\beta \le p$ on V_β. To prove the linearity of f_β on V_β note that if $v_1, v_2 \in V_\beta$ then as we pointed out above there exists $(V_{\alpha_3}, f_{\alpha_3}) \in \mathcal{C}$ such that $v_1, v_2 \in V_{\alpha_3}$. Then for any $c_1, c_2 \in \mathbb{R}$ we have

$$\begin{aligned} f_\beta(c_1v_1 + c_2v_2) &= f_{\alpha_3}(c_1v_1 + c_2v_2) = c_1 f_{\alpha_3}(v_1) + c_2 f_{\alpha_3}(v_2) \\ &= c_1 f_\beta(v_1) + c_2 f_\beta(v_2). \end{aligned}$$

This shows that f_β is a linear functional on V_β. If $v \in M$ then for an arbitrary $(V_\alpha, f_\alpha) \in \mathcal{C}$ we have $v \in M \subset V_\alpha$ so that $f_\beta(v) = f_\alpha(v) = f(v)$. Thus $f_\beta = f$ on M. Let $v \in V_\beta$. Then there exists $(V_\alpha, f_\alpha) \in \mathcal{C}$ such that $v \in V_\alpha$. Then $f_\beta(v) = f_\alpha(v) \le p(v)$. This shows that $f_\beta \le p$ on V_β and completes the verification that $(V_\beta, f_\beta) \in \mathcal{F}$.

For any $(V_\alpha, f_\alpha) \in \mathcal{C}$ it is clear from our definition of V_β and f_β that $V_\alpha \subset V_\beta$ and $f_\beta = f_\alpha$ on V_α. Thus $(V_\alpha, f_\alpha) \preceq (V_\beta, f_\beta)$.

We have shown above that every chain in $\mathcal{F}$ has an upper bound. Thus by Zorn's Lemma there exists a maximal element in $\mathcal{F}$. Let (V_γ, f_γ) be a maximal element in $\mathcal{F}$. Let us show that $V_\gamma = V$. Suppose $V_\gamma \ne V$. Then there exists $v_0 \in V$ such that $v_0 \notin V_\gamma$. Let V_0 be the linear subspace of V spanned by V_γ and v_0. Then as we showed in **1** there exists a real-valued linear functional f_0 on V_0 such that $f_0 = f_\gamma$ on V_γ and $f_0 \le p$ on V_0. Since $M \subset V_\gamma \subset V_0$ and $f_\gamma = f$ on M we have $f_0 = f_\gamma = f$ on M. Thus $(V_0, f_0) \in \mathcal{F}$. Since $V_\gamma \subset V_0$ and $f_\gamma = f_0$ on V_γ we have $(V_\gamma, f_\gamma) \preceq (V_0, f_0)$. But $V_\gamma \ne V_0$ so that $(V_\gamma, f_\gamma) \ne (V_0, f_0)$. This contradicts the maximality of (V_γ, f_γ) in $\mathcal{F}$. Thus we have $V_\gamma = V$. Therefore f_γ is a real-valued linear functional on V such that $f_\gamma = f$ on M and $f_\gamma \le p$ on V. ∎

We apply the Hahn-Banach Theorem for real linear spaces to obtain an extension theorem for real-valued bounded linear functionals on a real normed linear space.

Theorem 15.65. *Let $(V, \mathbb{R}, \|\cdot\|)$ be a real normed linear space and let M be a linear subspace of V. Let f be a real-valued bounded linear functional on the real normed linear space $(M, \mathbb{R}, \|\cdot\|)$ and let $\|f\|_*$ be its norm. Then there exists a real-valued bounded linear functional g on V such that $g = f$ on M and $\|g\|_* = \|f\|_*$.*

Proof. Let us define a real-valued functional p on V by setting

$$(1) \qquad p(v) = \|f\|_*\|v\| \quad \text{for } v \in V.$$

Then for any $v \in V$ and $\alpha \ge 0$ we have

$$p(\alpha v) = \|f\|_*\|\alpha v\| = \|f\|_*\alpha\|v\| = \alpha p(v),$$

and for any $u, v \in V$ we have

$$p(u+v) = \|f\|_*\|u+v\| \le \|f\|_*\{\|u\| + \|v\|\} = p(u) + p(v).$$

Thus p satisfies conditions 1° - 3° of Definition 15.62 so that p is a sublinear functional on the real linear space V. We have also, by Proposition 15.30

$$f(v) \le |f(v)| \le \|f\|_*\|v\| = p(v) \quad \text{for } v \in M.$$

Thus by Theorem 15.64 (Hahn-Banach), there exists a real-valued linear functional g on V such that $g = f$ on M and $g \leq p$ on V. Now since $g(v) \leq p(v)$ for every $v \in V$ we have $g(v) = -g(-v) \geq -p(-v)$ so that

(2) $$-p(-v) \leq g(v) \leq p(v) \quad \text{for } v \in V.$$

Now (1) implies that $-p(-v) = -\|f\|_*\| - v\| = -\|f\|_*\|v\|$ for $v \in V$. Substituting this and (1) in (2), we have

$$-\|f\|_*\|v\| \leq g(v) \leq \|f\|_*\|v\| \quad \text{for } v \in V$$

and hence

(3) $$|g(v)| \leq \|f\|_*\|v\| \quad \text{for } v \in V.$$

This shows that g is a bounded linear functional on V and $\|g\|_* \leq \|f\|_*$. On the other hand by Theorem 15.36 we have

$$\begin{aligned} \|g\|_* &= \sup_{v \in V, \|v\|=1} |g(v)| \geq \sup_{v \in M, \|v\|=1} |g(v)| \\ &= \sup_{v \in M, \|v\|=1} |f(v)| = \|f\|_*. \end{aligned}$$

Therefore we have $\|g\|_* = \|f\|_*$. ■

[X.2] Hahn-Banach Extension Theorem for Complex Linear Spaces

Observation 15.66. Let $(V, \mathbb{C})$ be a complex linear space and let f be a complex-valued linear functional on V satisfying the homogeneity and additivity conditionss in Definition 15.26. Let $f = f_1 + if_2$ where f_1 and f_2 are the real and the imaginary part of f.
(a) f_1 and f_2 are related by

(1) $$\begin{cases} f_1(v) = f_2(iv) & \text{for } v \in V, \\ f_2(v) = -f_1(iv) & \text{for } v \in V. \end{cases}$$

(b) For $j = 1, 2$ the real-valued functional f_j on the complex linear space V satisfies the additivity condition but, unless f_j is identically vanishing on V, f_j does not satisfy the homogeneity condition and thus f_j is not a linear functional on V.
(c) As we noted in Remark 15.3, a complex linear space is also a real linear space. If we regard V as a real linear space then f_1 and f_2 are real-valued linear functionals on the real linear space V.

Proof. 1. To prove (1) note that for every $v \in V$ we have

(2) $$f(iv) = f_1(iv) + if_2(iv)$$

and by the homogeneity of f we have

(3) $$f(iv) = if(v) = i\{f_1(v) + if_2(v)\} = if_1(v) - f_2(v).$$

Equating the real part of (2) to that of (3) we have the second equation in (1), and equating the imaginary part of (2) to that of (3) we have the first equation in (1).

2. To show that f_1 and f_2 satisfy the additivity condition on V, let $u, v \in V$. Then

$$f(u+v) = f_1(u+v) + if_2(u+v). \tag{4}$$

On the other hand by the additivity condition on f we have

$$f(u+v) = f(u) + f(v) = \{f_1(u) + f_1(v)\} + i\{f_2(u) + f_2(v)\}. \tag{5}$$

Equating the real part of (3) to that of (4) and equating the imaginary part of (3) to that of (4) we have

$$f_1(u+v) = f_1(u) + f_1(v),$$
$$f_2(u+v) = f_2(u) + f_2(v).$$

This proves the additivity of f_1 and f_2.

3. Let $j = 1$ or 2. Suppose f_j is not identically vanishing on V. To show that f_j does not satisfy the homogeneity condition we show that there exist $v \in V$ and $\gamma \in \mathbb{C}$ such that

$$f_j(\gamma v) \neq \gamma f_j(v).$$

Now for any $v \in V$ and $\gamma \in \mathbb{C}$, $\gamma = \alpha + i\beta$ where $\alpha, \beta \in \mathbb{R}$, we have

$$f(\gamma v) = f_1(\gamma v) + if_2(\gamma v). \tag{6}$$

On the other hand by the homogeneity of f we have

$$\begin{aligned} f(\gamma v) &= \gamma f(v) = \{\alpha + i\beta\}\{f_1(v) + if_2(v)\} \\ &= \{\alpha f_1(v) - \beta f_2(v)\} + i\{\beta f_1(v) + \alpha f_2(v)\}. \end{aligned} \tag{7}$$

Equating the real part of (6) to that of (7) and equating the imaginary part of (6) to that of (7) we have

$$\begin{cases} f_1(\gamma v) = \alpha f_1(v) - \beta f_2(v) = \gamma f_1(v) - i\beta f_1(v) - \beta f_2(v), \\ f_2(\gamma v) = \beta f_1(v) + \alpha f_2(v) = \gamma f_2(v) - i\beta f_2(v) + \beta f_1(v). \end{cases} \tag{8}$$

If f_1 is not identically vanishing on V we can select $v \in V$ such that $f_1(v) \neq 0$. If we select $\beta \neq 0$ then $f_1(\gamma v) \neq \gamma f_1(v)$. If f_2 is not identically vanishing on V we can select $v \in V$ such that $f_2(v) \neq 0$. If we select $\beta \neq 0$ then $f_2(\gamma v) \neq \gamma f_2(v)$.

4. Let us regard V as a real linear space. Let us show that f_1 and f_2 are real-valued linear functionals on the real linear space V. We have verified the additivity of f_1 and f_2 in **2** above. To verify the homogeneity of f_1 and f_2 on the real linear space V, let $v \in V$ and $\alpha \in \mathbb{R}$. Then we have

$$f(\alpha v) = f_1(\alpha v) + if_2(\alpha v) \tag{9}$$

and by the homogeneity of f on the complex linear space V and by the fact that $\alpha \in \mathbb{R} \subset \mathbb{C}$ we have

$$f(\alpha v) = \alpha f(v) = \alpha f_1(v) + i\alpha f_2(v). \tag{10}$$

From (9) and (10) we have

$$f_1(\alpha v) = \alpha f_1(v),$$
$$f_2(\alpha v) = \alpha f_2(v).$$

This verifies the homogeneity of f_1 and f_2 on the real linear space V. Thus f_1 and f_2 are real-valued linear functionals on the real linear space V. ∎

Lemma 15.67. *Let $(V, \mathbb{C})$ be a complex linear space. Let f be a complex-valued functional on V satisfying the following conditions :*

$$f(u + v) = f(u) + f(v) \quad \text{for } u, v \in V \tag{1}$$

$$f(\alpha v) = \alpha f(v) \quad \text{for } v \in V, \alpha \in \mathbb{R} \tag{2}$$

$$f(iv) = if(v) \quad \text{for } v \in V. \tag{3}$$

Then f is a complex-valued linear functional on V.

Proof. By (1), f satisfies the additivity condition 2° of Definition 15.26. It remains to verify that f satisfies the homogeneity condition 1° of Definition 15.26, that is,

$$f(\gamma v) = \gamma f(v) \quad \text{for } v \in V, \gamma \in \mathbb{C}. \tag{4}$$

Let $v \in V$ and let $\gamma \in \mathbb{C}$ written as $\gamma = \alpha + i\beta$ where $\alpha, \beta \in \mathbb{R}$. Then by (1) we have

$$f(\gamma v) = f(\alpha v + i\beta v) = f(\alpha v) + f(i\beta v).$$

By (2) we have $f(\alpha v) = \alpha f(v)$. By (3) and (2) we have $f(i\beta v) = if(\beta v) = i\beta f(v)$. Thus we have $f(\gamma v) = \alpha f(v) = i\beta f(v) = \gamma f(v)$. This proves (4). ∎

Definition 15.68. *Let $(V, \mathbb{C})$ be a complex linear space. A functional p on V is called a sublinear functional on a V if it satisfies the following conditions:*
1° *real-valued: $p(v) \in \mathbb{R}$ for every $v \in V$.*
2° *positive homogeneity: $p(\gamma v) = |\gamma| p(v)$ for $v \in V$ and $\gamma \in \mathbb{C}$.*
3° *subadditivity: $p(u + v) \le p(u) + p(v)$ for $u, v \in V$.*

Remark 15.69. (a) A norm on a complex linear space $(V, \mathbb{C})$ satisfies conditions 1° - 3° of Definition 15.68 and is therefore a sublinear functional on V.
(b) As we noted in Remark 15.3 a complex linear space $(V, \mathbb{C})$ can be regarded as a real linear space. Since conditions 1° - 3° in Definition 15.68 imply conditions 1° - 3° in Definition 15.62 a sublinear functional p on a complex linear space V is a sublinear functional on the real linear space V.

The following extension of the Hahn-Banach Theorem for real linear spaces to complex linear spaces is due to H. F. Bohnenblust and A. Sobdczyk.

Theorem 15.70. (Hahn-Banach Theorem for Complex Linear Spaces) *Let $(V, \mathbb{C})$ be a complex linear space and let p be a sublinear functional on V. Let M be a linear subspace of V and let f be a complex-valued linear functional on M such that $|f| \leq p$ on M. Then there exists a complex-valued linear functional h on V such that $h = f$ on M and $|h| \leq p$ on V.*

Proof. Let us write $f = f_1 + if_2$ where f_1 and f_2 are the real and the imaginary parts of f. Let us regard the complex linear spaces V and M as real linear spaces. By (c) of Observation 15.66, f_1 and f_2 are real-valued linear functionals on the real linear space M. Now

$$|f_1(v)| \leq \left\{|f_1(v)|^2 + |f_2(v)|^2\right\}^{1/2} = |f(v)| \leq p(v) \quad \text{for } v \in M.$$

The sublinear functional p on the complex linear space V is also a sublinear functional on the real linear space V by (b) of Remark 15.69. Thus by Theorem 15.64 (Hahn-Banach) there exists a real-valued linear functional g on the real linear space V such that $g = f_1$ on M and $g \leq p$ on V. Thus we have

(1) $$g(u + v) = g(u) + g(v) \quad \text{for } u, v \in V,$$

(2) $$g(\alpha v) = \alpha g(v) \quad \text{for } v \in V, \alpha \in \mathbb{R},$$

(3) $$g(v) = f_1(v) \quad \text{for } v \in M,$$

(4) $$g(v) \leq p(v) \quad \text{for } v \in V.$$

Let us define a complex-valued functional h on the complex linear space V by setting

(5) $$h(v) = g(v) - ig(iv) \quad \text{for } v \in V.$$

Let us show that h is a complex-valued linear functional on the complex linear space V by verifying conditions (1) - (3) of Lemma 15.67. First of all, for $u, v \in V$ we have

$$\begin{aligned} h(u + v) &= g(u + v) - ig\big(i(u + v)\big) = g(u) + g(v) - i\{g(iu) + g(iv)\} \\ &= \{g(u) - ig(iu)\} + \{g(v) - ig(iv)\} = h(u) + h(v), \end{aligned}$$

where the second equality is by (1). This verifies condition (1) of Lemma 15.67. Secondly, for $v \in V$ and $\alpha \in \mathbb{R}$, we have

$$h(\alpha v) = g(\alpha v) - ig(i\alpha v) = \alpha g(v) - i\alpha g(iv) = \alpha h(v),$$

where the second equality is by (2). This verifies condition (2) of Lemma 15.67. Finally, for $v \in V$ we have

$$\begin{aligned} h(iv) &= g(iv) - ig(-v) = g(iv) + ig(v) \\ &= i\{-ig(iv) + g(v)\} = ih(v), \end{aligned}$$

where the second equality is by (2). This verifies condition (3) of Lemma 15.67. Thus by Lemma 15.67, h is a complex-valued linear functional on the complex linear space V.

To show that $h = f$ on M, note that for $v \in M$ we have

$$\begin{aligned} g(iv) + if_2(iv) &= f_1(iv) + if_2(iv) = f(iv) = if(v) \\ &= i\{f_1(v) + if_2(v)\} = ig(v) - f_2(v), \end{aligned}$$

where the first equality is by (3). Since g is real-valued the chain of equalities above implies that $g(iv) = -f_2(v)$. Thus we have

$$h(v) = g(v) - ig(iv) = f_1(v) + if_2(v) = f(v).$$

This shows that $h = f$ on M. It remains to show that $|h| \le p$ on V. For $v \in V$ let $\gamma = e^{i\vartheta}$ be such that $\gamma h(v) = |h(v)|$. Then $|h(v)| = \gamma h(v) = h(\gamma v) = g(\gamma v) - ig(i\gamma v)$. Since $|h(v)| \in \mathbb{R}$ and g is real-valued the last equality implies

$$|h(v)| = g(\gamma v) \le p(\gamma v) = |\gamma| p(v) = p(v),$$

by (4) and condition 2° of Definition 15.68. ∎

As in the real case, we apply the Hahn-Banach Theorem for complex linear spaces to obtain an extension theorem for complex-valued bounded linear functionals on a complex normed linear space.

Theorem 15.71. *Let $(V, \mathbb{C}, \|\cdot\|)$ be a complex normed linear space. Let M be a linear subspace of V and let f be a complex-valued bounded linear functional on the complex normed linear space $(M, \mathbb{C}, \|\cdot\|)$ with norm $\|f\|_*$. Then there exists a complex-valued bounded linear functional h on V such that $h = f$ on M and with norm $\|h\|_* = \|f\|_*$.*

Proof. Let us define a functional on V by setting

$$p(v) = \|f\|_* \|v\| \quad \text{for } v \in V.$$

Then for $v \in V$ and $\gamma \in \mathbb{C}$ we have

$$p(\gamma v) = \|f\|_* \|\gamma v\| = \|f\|_* |\gamma| \|v\| = |\gamma| p(v),$$

and for $u, v \in V$ we have

$$p(u+v) = \|f\|_* \|u+v\| \le \|f\|_* \{\|u\| + \|v\|\} = p(u) + p(v).$$

Thus p satisfies conditions 1° - 3° of Definition 15.68 and is therefore a sublinear functional on the complex linear space V. Now since f is a complex-valued bounded linear functional on the complex normed linear space $(M, \mathbb{C}, \|\cdot\|)$ we have

$$|f(v)| \le \|f\|_* \|v\| = p(v) \quad \text{for } v \in M.$$

Then by Theorem 15.70 there exists a complex-valued linear functional h on V such that $h = f$ on M and $|h| \le p$ on V. Thus we have $|h(v)| \le \|f\|_* \|v\|$ for $v \in V$. This shows that h is a bounded linear functional on V and $\|f\|_*$ is a bound for h. Thus $\|h\|_* \le \|f\|_*$. On the other hand we have

$$\|h\|_* = \sup_{v\in V, \|v\|=1} |h(v)| \ge \sup_{v\in M, \|v\|=1} |h(v)| = \sup_{v\in M, \|v\|=1} |f(v)| = \|f\|_*.$$

Thus we have $\|h\|_* = \|f\|_*$. ∎

Applying Theorem 15.65 and Theorem 15.71 we obtain the following extension theorem for bounded linear functionals on a normed linear space.

Theorem 15.72. *Let $(V, \mathbb{K}, \|\cdot\|)$ be a normed linear space over the field of scalars $\mathbb{K}$ where $\mathbb{K} = \mathbb{R}$ or $\mathbb{K} = \mathbb{C}$. Let M be a linear subspace of V, $M \neq V$, and let $v_0 \in V$ be such that the distance between v_0 and M is positive, that is, $d = \inf_{u\in M} \|u - v_0\| > 0$. Then there exists a $\mathbb{K}$-valued bounded linear functional f on V such that $f = 0$ on M, $f(v_0) = d$, and $\|f\|_* = 1$.*

Proof. Let V_0 be the linear subspace of V spanned by M and v_0, that is,

(1) $$V_0 = \{u + \gamma v_0 : u \in M, \gamma \in \mathbb{K}\}.$$

Let us define a $\mathbb{K}$-valued functional f_0 on V_0 by setting

(2) $$f_0(v) = f_0(u + \gamma v_0) = \gamma d \quad \text{for } u \in M \text{ and } \gamma \in \mathbb{K}.$$

Since $d = \inf_{u\in M} \|u - v_0\| > 0$, $v_0 \notin M$ so that the representation of $v \in V_0$ by $v = u + \gamma v_0$ with $u \in M$ and $\gamma \in \mathbb{K}$ is unique and therefore f_0 is well defined by (2).

It is easily verified that f_0 is a linear functional on V_0. Also $f = 0$ on M. Indeed for $v \in M \subset V_0$ we have $v = u + 0v_0$ with $u \in M$ so that by (2) we have

(3) $$f_0(v) = f_0(u + 0v_0) = 0 \cdot d = 0 \quad \text{for } v \in M.$$

Since $v_0 \in V_0$ and $v_0 = 0 + 1v_0$, we have by (2)

(4) $$f_0(v_0) = f_0(0 + 1v_0) = 1 \cdot d = d.$$

Let us show that f_0 is a bounded linear functional on V_0 with $\|f_0\|_* = 1$. According to Theorem 15.36 it suffices to show that if we let $S = \{v \in V_0 : \|v\| = 1\}$, then we have

(5) $$\sup_{v\in S} |f_0(v)| = 1.$$

Now

$$S = \left\{\frac{v}{\|v\|} : v \in V_0, v \neq 0\right\} = \left\{\frac{u + \gamma v_0}{\|u + \gamma v_0\|} : u \in M, \gamma \in \mathbb{K}, u + \gamma v_0 \neq 0\right\}.$$

Thus for $v \in S$ we have

(6) $$|f_0(v)| = \left|f_0\left(\frac{u + \gamma v_0}{\|u + \gamma v_0\|}\right)\right| = \frac{|f_0(u + \gamma v_0)|}{\|u + \gamma v_0\|} = \frac{|\gamma| d}{\|u + \gamma v_0\|}.$$

Since $v_0 \neq 0 \in V_0$, we have $\frac{v_0}{\|v_0\|} \in S$. By (4) we have

$$f_0\left(\frac{v_0}{\|v_0\|}\right) = \frac{f_0(v_0)}{\|v_0\|} = \frac{d}{\|v_0\|} > 0.$$

Thus we have

$$\text{(7)} \qquad \sup_{v \in S} |f_0(v)| = \sup_{v \in S, f_0(v) \neq 0} |f_0(v)|.$$

Now $f_0(v) \neq 0$ for $v \in S$ implies $\gamma \neq 0$ by (6). Then by (6) and (7) we have

$$\begin{aligned}\sup_{v \in S} |f_0(v)| &= \sup_{u \in M, \gamma \neq 0} \frac{|\gamma| d}{\|u + \gamma v_0\|} = \sup_{u \in M, \gamma \neq 0} \frac{d}{\|\frac{u}{\gamma} + v_0\|} \\ &= \sup_{w \in M} \frac{d}{\|v_0 - w\|} = \frac{d}{\inf_{w \in M} \|v_0 - w\|} = \frac{d}{d} \\ &= 1.\end{aligned}$$

Thus we have $\|f_0\|_* = 1$.

We have just shown that f_0 is a $\mathbb{K}$-valued bounded linear functional on V_0 such that $f_0 = 0$ on M, $f_0(v_0) = d$, and $\|f_0\|_* = 1$. Then by Theorem 15.65 if $\mathbb{K} = \mathbb{R}$ and by Theorem 15.71 if $\mathbb{K} = \mathbb{C}$, there exists a $\mathbb{K}$-valued bounded linear functional f on V such that $f = f_0$ on V_0 and $\|f\|_* = \|f_0\|_* = 1$. Since $M \subset V_0$ and $f_0 = 0$ on M by (3), we have $f = f_0 = 0$ on M. Since $v_0 \in V_0$, we have $f(v_0) = f_0(v_0) = d$ by (4). ∎

Corollary 15.73. *Let $(V, \mathbb{K}, \|\cdot\|)$ be a normed linear space over the field of scalars $\mathbb{K}$ where $\mathbb{K} = \mathbb{R}$ or $\mathbb{K} = \mathbb{C}$. If $v_0 \in V$ and $v_0 \neq 0 \in V$ then there exists a $\mathbb{K}$-valued bounded linear functional f on V such that $f(v_0) = \|v_0\|$ and $\|f\|_* = 1$.*

Proof. Consider the linear subspace M of V consisting of $0 \in V$ only, that is, $M = \{0\}$. Since $v_0 \neq 0$ we have $d = \inf_{u \in M} \|u - v_0\| = \|v_0\| > 0$. Thus by Corollary 15.72, there exists a $\mathbb{K}$-valued bounded linear functional f on V such that $f(v_0) = d = \|v_0\|$ and $\|f\|_* = 1$. ∎

[X.3] Bounded Linear Mappings of a Dual Space into a Dual Space

Proposition 15.74 . *Let $(V, \mathbb{K}, \|\cdot\|)$ be a normed linear space over the field of scalars $\mathbb{K}$ where $\mathbb{K} = \mathbb{R}$ or $\mathbb{K} = \mathbb{C}$. Let $(V^*, \mathbb{K}, \|\cdot\|_*)$ be the dual space of $(V, \mathbb{K}, \|\cdot\|)$ and let $S_1^*(0) = \{f \in V^* : \|f\|_* = 1\}$. Then for every $v \in V$ we have*

$$\text{(1)} \qquad \|v\| = \sup_{f \in S_1^*(0)} |f(v)|.$$

Consequently for each fixed $v \in V$ the mapping of V^* into $\mathbb{K}$ defined by $f \mapsto f(v)$ for $f \in V^*$ is a $\mathbb{K}$-valued bounded linear functional on V^* with norm equal to $\|v\|$.

Proof. For every $f \in S_1^*(0)$ we have $|f(v)| \le \|f\|_* \|v\| = \|v\|$ for all $v \in V$ so that

(2) $$\sup_{f \in S_1^*(0)} |f(v)| \le \|v\| \quad \text{for each } v \in V.$$

If $v = 0 \in V$, then $\|v\| = 0$ and $f(v) = 0$ for every $f \in V^*$ so that (1) holds trivially. If $v \in V$ and $v \ne 0 \in V$, then according to Corollary 15.73 there exists $f \in S_1^*(0)$ such that $f(v) = \|v\|$. This fact and (2) imply (1). ∎

Observation 15.75. Let A and B be two arbitrary indexing sets and let $c_{\alpha,\beta} \in \mathbb{R}$ for $\alpha \in A$ and $\beta \in B$. Then

(1) $$\sup_{\alpha \in A} \Big\{ \sup_{\beta \in B} c_{\alpha,\beta} \Big\} = \sup_{\alpha \in A, \beta \in B} c_{\alpha,\beta} = \sup_{\beta \in B} \Big\{ \sup_{\alpha \in A} c_{\alpha,\beta} \Big\}.$$

Proof. For each $\alpha \in A$ we have $c_{\alpha,\beta} \le \sup_{\beta \in B} c_{\alpha,\beta}$. Thus

(2) $$\sup_{\alpha \in A, \beta \in B} c_{\alpha,\beta} \le \sup_{\alpha \in A, \beta \in B} \Big\{ \sup_{\beta \in B} c_{\alpha,\beta} \Big\} = \sup_{\alpha \in A} \Big\{ \sup_{\beta \in B} c_{\alpha,\beta} \Big\}.$$

On the other hand $\sup_{\beta B} c_{\alpha,\beta} \le \sup_{\alpha \in A, \beta \in B} c_{\alpha,\beta}$ so that

(3) $$\sup_{\alpha \in A} \Big\{ \sup_{\beta B} c_{\alpha,\beta} \Big\} \le \sup_{\alpha \in A, \beta \in B} c_{\alpha,\beta}.$$

By (2) and (3) we have the first equality in (1). Interchanging the roles of A and B we have the second equality in (1) by the same argument. ∎

Theorem 15.76. *Let $(V, \mathbb{K}, \|\cdot\|)$ and $(W, \mathbb{K}, \|\cdot\|')$ be two normed linear spaces over the same field of scalars $\mathbb{K}$ where $\mathbb{K} = \mathbb{R}$ or $\mathbb{K} = \mathbb{C}$. Let $(V^*, \mathbb{K}, \|\cdot\|_*)$ and $(W^*, \mathbb{K}, \|\cdot\|'_*)$ be their respective dual spaces. Let $L \in \mathfrak{B}\{V, W\}$. Define a mapping L^* of W^* into V^* by*

(1) $$L^*(g) = g \circ L \quad \text{for } g \in W^*.$$

Then $L^* \in \mathfrak{B}\{W^*, V^*\}$, that is, L^* is a bounded linear mapping of W^* into V^*, and

(2) $$\|L^*\|_{W^*, V^*} = \|L\|_{V,W}.$$

Proof. Let us show first that L^* defined by (1) maps W^* into V^*, that is, $L^*(g) \in V^*$ for every $g \in W^*$. Now for $v \in V$ we have

$$L^*(g)(v) = (g \circ L)(v) = g\big(L(v)\big) \in \mathbb{K}.$$

This shows that $L^*(g)$ is a $\mathbb{K}$-valued function on V. The linearity of $L^*(g)$ on V follows from the linearity of g and L. We have moreover

$$|L^*(g)(v)| = \big|g\big(L(v)\big)\big| \le \|g\|'_* \|L(v)\|' \le \|g\|'_* \|L\|_{V,W} \|v\| \quad \text{for all } v \in V.$$

Thus the real number $\|g\|'_* \|L\|_{V,W} \geq 0$ is a bound for the $\mathbb{K}$-valued linear functional $L^*(g)$ on V so that $L^*(g)$ is a bounded linear functional on V, that is, $L^*(g) \in V^*$. This shows that L^* maps W^* into V^*.

To prove the linearity of L^* on W^*, let $g_1, g_2 \in W^*$ and $\gamma_1, \gamma_2 \in \mathbb{K}$. Then we have

$$\begin{aligned} L^*(\gamma_1 g_1 + \gamma_2 g_2) &= (\gamma_1 g_1 + \gamma_2 g_2) \circ L = \gamma_1 (g_1 \circ L) + \gamma_2 (g_2 \circ L) \\ &= \gamma_1 L^*(g_1) + \gamma_2 L^*(g_2). \end{aligned}$$

Let us show that L^* is a bounded linear mapping of W^* into V^*. According to Theorem 15.36, the linear mapping L^* is bounded if and only if

$$(3) \qquad \sup_{g \in W^*, \|g\|'_* = 1} \|L^*(g)\|_* < \infty$$

and moreover when (3) is valid then (2) also holds. Thus it remains to show that (3) holds. Now by Theorem 15.36, we have

$$(4) \qquad \|L\|_{V,W} = \sup_{v \in V, \|v\| = 1} \|L(v)\|'.$$

Applying Proposition 15.74 to the normed linear space $(W, \mathbb{K}, \|\cdot\|')$, we have

$$(5) \qquad \|L(v)\|' = \sup_{g \in W^*, \|g\|'_* = 1} \big|g\big(L(v)\big)\big|.$$

Substituting (5) in (4), we have

$$(6) \qquad \begin{aligned} \|L\|_{V,W} &= \sup_{v \in V, \|v\| = 1} \Big\{ \sup_{g \in W^*, \|g\|'_* = 1} \big|g\big(L(v)\big)\big| \Big\} \\ &= \sup_{g \in W^*, \|g\|'_* = 1} \Big\{ \sup_{v \in V, \|v\| = 1} \big|g\big(L(v)\big)\big| \Big\} \\ &= \sup_{g \in W^*, \|g\|'_* = 1} \Big\{ \sup_{v \in V, \|v\| = 1} \big|L^*(g)(v)\big| \Big\} \\ &= \sup_{g \in W^*, \|g\|'_* = 1} \|L^*(g)\|_*, \end{aligned}$$

where the second equality is by Observation 15.75. Since $\|L\|_{V,W} < \infty$, (6) shows that (3) holds. ∎

Definition 15.77. *Let $(V, \mathbb{K}, \|\cdot\|)$ and $(W, \mathbb{K}, \|\cdot\|')$ be two normed linear spaces over the same field of scalars $\mathbb{K}$ where $\mathbb{K} = \mathbb{R}$ or $\mathbb{K} = \mathbb{C}$. Let $(V^*, \mathbb{K}, \|\cdot\|_*)$ and $(W^*, \mathbb{K}, \|\cdot\|'_*)$ be their respective dual spaces. Let L be a bounded linear mapping of V into W. A bounded linear mapping L^* of W^* into V^* is called an adjoint of L if it satisfies the condition:*

$$L^*(g)(v) = (g \circ L)(v) \quad \textit{for all } v \in V \textit{ and } g \in W^*.$$

Remark 15.78. Such bounded linear mapping L^* exists by Theorem 15.76. Let us show that such bounded linear mapping L^* is unique. Suppose L_1^* and L_2^* are two such bounded linear mappings of W^* into V^*. Then we have

$$L_1^*(g)(v) = (g \circ L)(v) = L_2^*(g)(v) \quad \text{for all } v \in V \text{ and } g \in W^*.$$

Since $L_1^*(g), L_2^*(g) \in V^*$, the equality above for all $v \in V$ implies that $L_1^*(g) = L_2^*(g)$ as elements of V^*. Then since $L_1^*(g) = L_2^*(g)$ for all $g \in W^*$, L_1^* and L_2^* as mappings of W^* are equal.

[XI] Semicontinuous Functions

Definition 15.79. *Let f be an extended real-valued function defined on a subset D of a topological space X. Let $x_0 \in D$.*
(a) *We say that f is lower semicontinuous at x_0 if $f(x_0) \neq -\infty$ and for every $\gamma \in \mathbb{R}$ such that $\gamma < f(x_0)$ there exists an open set $U \ni x_0$ such that $\gamma < f(x)$ for all $x \in U \cap D$. We say that f is lower semicontinuous on D if f is lower semicontinuous at every $x_0 \in D$.*
(b) *We say that f is upper semicontinuous at x_0 if $f(x_0) \neq \infty$ and for every $\gamma \in \mathbb{R}$ such that $\gamma > f(x_0)$ there exists an open set $U \ni x_0$ such that $\gamma > f(x)$ for all $x \in U \cap D$. We say that f is upper semicontinuous on D if f is upper semicontinuous at every $x_0 \in D$.*

Remark 15.80. If f is a real-valued function defined on a subset D of a topological space X then f is lower semicontinuous at $x_0 \in D$ if and only if for every $\varepsilon > 0$ there exists an open set $U \ni x_0$ such that $f(x_0) - \varepsilon < f(x)$ for all $x \in U \cap D$ and f is upper semicontinuous at $x_0 \in D$ if and only if for every $\varepsilon > 0$ there exists an open set $U \ni x_0$ such that $f(x) < f(x_0) + \varepsilon$ for all $x \in U \cap D$.

Observation 15.81.
(a) f is continuous at $x_0 \in D$ if and only if f is both lower and upper semicontinuous at x_0.
(b) If f is lower semicontinuous at x_0, then cf is lower semicontinuous at x_0 if $c > 0$ and cf is upper semicontinuous at x_0 if $c < 0$.
(c) If f is upper semicontinuous at x_0, then cf is upper semicontinuous at x_0 if $c > 0$ and cf is lower semicontinuous at x_0 if $c < 0$.
(d) Let f be a $(-\infty, \infty]$-valued function on D. Then f is lower semicontinuous on D if and only if for every $\gamma \in \mathbb{R}$, $f^{-1}\big((\gamma, \infty]\big) = V \cap D$ where V is an open set in X.
(e) Let f be a $[-\infty, \infty)$-valued function on D. Then f is upper semicontinuous on D if and only if for every $\gamma \in \mathbb{R}$, $f^{-1}\big([-\infty, \gamma)\big) = V \cap D$ where V is an open set in X.
(f) In particular if O is an open set, then $\mathbf{1}_O$ is a lower semicontinuous function on X. If C is a closed set, then $\mathbf{1}_C$ is an upper semicontinuous function on X.

Proof. (a), (b), and (c) are immediate from Definition 15.79.

Let us prove (d). Let f be $(-\infty, \infty]$-valued function D. Suppose for every $\gamma \in \mathbb{R}$, $f^{-1}\big((\gamma, \infty]\big) = V \cap D$ where V is an open set in X. For an arbitrary $x_0 \in D$, let $\gamma \in \mathbb{R}$ and $\gamma < f(x_0)$. By our assumption, $f^{-1}\big((\gamma, \infty]\big) = V \cap D$ where V is an open set in X.

Then for any $x \in V \cap D$, we have $f(x) \in (\gamma, \infty]$, that is, $f(x) > \gamma$. This proves the lower semicontinuity of f at x_0. By the arbitrariness of $x_0 \in D$, f is lower semicontinuous on D.

Conversely suppose f is lower semicontinuous on D. Let $\gamma \in \mathbb{R}$ be arbitrarily given and let $E = f^{-1}\big((\gamma, \infty]\big) \subset D$. Let x_0 be an arbitrary point in E. Then $f(x_0) \in (\gamma, \infty]$ so that $f(x_0) > \gamma$. By the lower semicontinuity of f at $x_0 \in D$ there exists an open set $U_{x_0} \ni x_0$ such that $f(x) > \gamma$ for every $x \in U_{x_0} \cap D$. Then $f(U_{x_0} \cap D) \subset (\gamma, \infty]$ so that $U_{x_0} \cap D \subset E$. Thus $\bigcup_{x_0 \in E} \big(U_{x_0} \cap D\big) = E$. Let $V = \bigcup_{x_0 \in E} U_{x_0}$. Then V is an open set in X and $E = V \cap D$, that is, $f^{-1}\big((\gamma, \infty]\big) = V \cap D$.

(e) is proved in the same way as (d). (f) follows from (d) and (e). ∎

Proposition 15.82. *Let $(X, \mathfrak{B}_X)$ be a measurable space where X is a topological space and $\mathfrak{B}_X$ is the Borel σ-algebra of subsets of X. Let f be a real-valued lower (resp. upper) semicontinuous function on X. Then f is a $\mathfrak{B}_X/\mathfrak{B}_{\mathbb{R}}$-measurable mapping of X into $\mathbb{R}$.*

Proof. Let $\mathfrak{J}$ be the collection of all open intervals in $\mathbb{R}$ of the type (γ, ∞) where $\gamma \in \mathbb{R}$ and let $\mathfrak{O}$ be the collection of all open sets in $\mathbb{R}$. It is easily verified that $\sigma(\mathfrak{J}) = \sigma(\mathfrak{O}) = \mathfrak{B}_{\mathbb{R}}$. To show that a real-valued function f on X is a $\mathfrak{B}_X/\mathfrak{B}_{\mathbb{R}}$-measurable mapping of X into $\mathbb{R}$, it suffices to show that $f^{-1}(\mathfrak{J}) \subset \mathfrak{B}_X$ according to Theorem 1.41. Now if f is a real-valued lower semicontinuous function on X then $f^{-1}\big((\gamma, \infty)\big)$ is an open set in X by (d) of Observation 15.81 and thus a member of $\mathfrak{B}_X$. Thus for every $J \in \mathfrak{J}$ we have $f^{-1}(J) \in \mathfrak{B}_X$ and then $f^{-1}(\mathfrak{J}) \subset \mathfrak{B}_X$. This shows that a real-valued lower semicontinuous function f on X is a $\mathfrak{B}_X/\mathfrak{B}_{\mathbb{R}}$-measurable mapping of X into $\mathbb{R}$. The $\mathfrak{B}_X/\mathfrak{B}_{\mathbb{R}}$-measurability of a real-valued upper semicontinuous function f on X is proved similarly by applying (e) of Observation 15.81. ∎

Proposition 15.83. (Sequential Criterion for Semicontinuity on a Metric Space) *Let f be an extended real-valued function on a metric space (X, d).*

(a) *f is lower semicontinuous at $x_0 \in X$ if and only if for every sequence $(x_n : n \in \mathbb{N})$ in X such that $\lim_{n\to\infty} x_n = x_0$, we have $\liminf_{n\to\infty} f(x_n) \geq f(x_0)$.*

(b) *f is upper semicontinuous at $x_0 \in X$ if and only if for every sequence $(x_n : n \in \mathbb{N})$ in X such that $\lim_{n\to\infty} x_n = x_0$, we have $\limsup_{n\to\infty} f(x_n) \leq f(x_0)$.*

Proof. 1. Suppose f is lower semicontinuous at $x_0 \in X$. Let $(x_n : n \in \mathbb{N})$ be a sequence in X such that $\lim_{n\to\infty} x_n = x_0$. Since f is lower semicontinuous at x_0, for every $\gamma \in \mathbb{R}$ such that $\gamma < f(x_0)$ there exists an open set $U \ni x_0$ such that $f(x) > \gamma$ for all $x \in U$. Since $\lim_{n\to\infty} x_n = x_0$, there exists $N \in \mathbb{N}$ such that $x_n \in U$ for $n \geq N$ and thus $f(x_n) > \gamma$ for $n \geq N$. This implies $\liminf_{n\to\infty} f(x_n) \geq \gamma$. Since this holds for every $\gamma < f(x_0)$ we have $\liminf_{n\to\infty} f(x_n) \geq f(x_0)$.

2. Conversely suppose for every sequence $(x_n : n \in \mathbb{N})$ in X such that $\lim_{n\to\infty} x_n = x_0$ we have $\liminf_{n\to\infty} f(x_n) \geq f(x_0)$. To show that f is lower semicontinuous at x_0, assume the contrary. Then there exists $\gamma < f(x_0)$ such that every open set $U \ni x_0$ contains a point $x \in X$ such that $\gamma \geq f(x)$. Thus the open ball $B\big(x_0, \frac{1}{n}\big)$ contains a point $x_n \in X$

such that $\gamma \geq f(x_n)$ for $n \in \mathbb{N}$. The sequence $(x_n : n \in \mathbb{N})$ then has $\lim_{n\to\infty} x_n = x_0$ and $\liminf_{n\to\infty} f(x_n) \leq \gamma < f(x_0)$. This contradicts the assumption that for every sequence $(x_n : n \in \mathbb{N})$ in X such that $\lim_{n\to\infty} x_n = x_0$ we have $\liminf_{n\to\infty} f(x_n) \geq f(x_0)$. Therefore f is lower semicontinuous at x_0. This completes the proof of (a). (b) is proved similarly. ∎

Theorem 15.84. *Let $\{f_\alpha : \alpha \in A\}$ be an arbitrary collection of extended real-valued functions on a set D in a topological space X.*
(a) *If f_α is lower semicontinuous at $x_0 \in D$ for every $\alpha \in A$, then $f = \sup_{\alpha\in A} f_\alpha$ is lower semicontinuous at x_0.*
(b) *If f_α is upper semicontinuous at $x_0 \in D$ for every $\alpha \in A$, then $f = \inf_{\alpha\in A} f_\alpha$ is upper semicontinuous at x_0.*

Proof. Let us prove (a). Thus let $\gamma \in \mathbb{R}$ and $\gamma < f(x_0)$. Since $f(x_0) = \sup_{\alpha\in A} f_\alpha(x_0)$, there exists $\alpha_0 \in A$ such that $\gamma < f_{\alpha_0}(x_0)$. By the lower semicontinuity of f_{α_0} at x_0, there exists an open set $U \ni x_0$ such that $\gamma < f_{\alpha_0}(x)$ for all $x \in U \cap D$. Then we have $f(x) = \sup_{\alpha\in A} f_\alpha(x) \geq f_{\alpha_0}(x) > \gamma$ for all $x \in U \cap D$. This proves the lower semicontinuity of f at x_0. (b) is proved similarly. ∎

Definition 15.85. *A real-valued function p on a linear space $(V, \mathbb{K})$ where $\mathbb{K} = \mathbb{R}$ or $\mathbb{K} = \mathbb{C}$ is called a seminorm if it satisfies the following conditions :*
1° $p(v) \in [0, \infty)$ *for* $v \in V$,
2° $p(\gamma v) = |\gamma| p(v)$ *for* $v \in V$ *and* $\gamma \in \mathbb{K}$,
3° $p(u+v) \leq p(u) + p(v)$ *for* $u, v \in V$.

Every norm on a linear space $(V, \mathbb{K})$ satisfies conditions 1° - 3° in Definition 15.85 and is therefore a seminorm on $(V, \mathbb{K})$. Note that from 2° we have $p(0) = 0$. However $p(f) = 0$ does not imply $f = 0$. Thus a seminorm may not be a norm.

In $\mathbb{R}^k$, if we define $p(x) = |x_i|$ with some fixed $i = 1, \ldots, k$ for $x = (x_1, \ldots, x_k) \in \mathbb{R}^k$, then p is a seminorm on $\mathbb{R}^k$.

In $C([a, b])$, if we let $[c, d] \subset [a, b]$ and define $p(f) = \int_c^d |f(t)|\,dt$ for $f \in C([a, b])$, then p is a seminorm on $C([a, b])$.

For a bounded linear functional f on a normed linear space $(V, \mathbb{K}, \|\cdot\|)$ where $\mathbb{K} = \mathbb{R}$ or $\mathbb{K} = \mathbb{C}$, if we define $p(v) = |f(v)|$ for $v \in V$ then p is a seminorm on V.

Proposition 15.86. *Let $\{f_\alpha : \alpha \in A\}$ be an arbitrary subset of the dual space $(V^*, \mathbb{K}, \|\cdot\|_*)$ of a normed linear space $(V, \mathbb{K}, \|\cdot\|)$ where $\mathbb{K} = \mathbb{R}$ or $\mathbb{K} = \mathbb{C}$ such that*

(1) $$\sup_{\alpha\in A} |f_\alpha(v)| < \infty \quad \textit{for } v \in V.$$

Define a nonnegative real-valued function p on V by setting

(2) $$p(v) = \sup_{\alpha\in A} |f_\alpha(v)| \quad \textit{for } v \in V.$$

Then p is a lower semicontinuous seminorm on V.

Proof. By (1), we have $p(v) \in [0, \infty)$ for every $v \in V$. Also for $v \in V$ and $\gamma \in \mathbb{K}$, we have

$$p(\gamma v) = \sup_{\alpha \in A} |f_\alpha(\gamma v)| = |\gamma| \sup_{\alpha \in A} |f_\alpha(v)| = |\gamma| p(v),$$

and for $u, v \in V$ we have

$$p(u + v) = \sup_{\alpha \in A} |f_\alpha(u + v)| \leq \sup_{\alpha \in A} \{|f_\alpha(u)| + |f_\alpha(v)|\} \leq p(v) + p(v).$$

This shows that p is a seminorm on V. Finally the continuity of f_α on V implies that of $|f_\alpha|$ and the continuity of $|f_\alpha|$ on V implies its lower semicontinuity on V. Then by Theorem 15.84, p is lower semicontinuous on V. ∎

Lemma 15.87. *Let p be a seminorm on a normed linear space $(V, \mathbb{K}, \|\cdot\|)$ where $\mathbb{K} = \mathbb{R}$ or $\mathbb{K} = \mathbb{C}$. Then the following two conditions on p are equivalent:*
1° *p is bounded above on some open ball in V,*
2° *there exists a constant $M \geq 0$ such that $p(v) \leq M\|v\|$ for all $v \in V$.*

Proof. 1. Suppose there exist an open ball, say $B(v_0, r)$, and a constant $C \geq 0$ such that $p(v) \leq C$ for all $v \in B(v_0, r)$. Then for every $v \in B(0, r)$, we have $v + v_0 \in B(v_0, r)$ so that $p(v + v_0) \leq C$ and then $p(v) = p(v + v_0 - v_0) \leq p(v + v_0) + p(v_0) \leq 2C$.

Now let $v \in V$ and $v \neq 0 \in V$. Then $\left\|\{\frac{r}{2}\|v\|^{-1}\}v\right\| = \frac{r}{2}$ so that $\frac{r}{2}\|v\|^{-1}v \in B(0, r)$ and then we have $p\left(\frac{r}{2}\|v\|^{-1}v\right) \leq 2C$ and thus $p(v) \leq \frac{4C}{r}\|v\|$. When $v = 0 \in V$, this inequality is trivially satisfied. Thus with $M = \frac{4C}{r}$, we have $p(v) \leq M\|v\|$ for every $v \in V$.

2. Conversely suppose there exists a constant $M \geq 0$ such that $p(v) \leq M\|v\|$ for all $v \in V$. Take an arbitrary open ball, say $B(v_0, r)$. Then for every $v \in B(v_0, r)$, we have $p(v) \leq M\|v\| = M\|v - v_0 + v_0\| \leq M\{\|v - v_0\| + \|v_0\|\} < M\{r + \|v_0\|\}$. Thus p is bounded above on $B(v_0, r)$ by the constant $M\{r + \|v_0\|\}$. ∎

Proposition 15.88. *Let p be a seminorm on a Banach space $(V, \mathbb{K}, \|\cdot\|)$ where $\mathbb{K} = \mathbb{R}$ or $\mathbb{K} = \mathbb{C}$. Then the following two conditions on p are equivalent:*
1° *there exists a constant $M \geq 0$ such that $p(v) \leq M\|v\|$ for all $v \in V$,*
2° *p is lower semicontinuous on V.*

Proof. 1. Suppose there exists a constant $M \geq 0$ such that $p(v) \leq M\|v\|$ for all $v \in V$. To show that p is lower semicontinuous at an arbitrary $v_0 \in V$, let $\varepsilon > 0$ be arbitrarily given. Let $r > 0$ be such that $Mr < \varepsilon$. Then for every $v \in B(v_0, r)$, we have $p(v_0 - v) \leq M\|v_0 - v\| < Mr < \varepsilon$ and hence $p(v_0) \leq p(v_0 - v) + p(v) < \varepsilon + p(v)$, that is, $p(v_0) - \varepsilon < p(v)$ for $v \in B(v_0, r)$, proving the lower semicontinuity of p at v_0.

2. Conversely suppose p is lower semicontinuous. To prove the existence of a constant $M \geq 0$ such that $p(v) \leq M\|v\|$ for all $v \in V$, assume the contrary. Then by Lemma 15.87, p is not bounded above on any open ball. Let $\varepsilon > 0$ be arbitrarily fixed. Take an arbitrary open ball $B(0, r_0)$. Since p is not bounded above on $B(0, r_0)$, there exists

$v_1 \in B(0, r_0)$ such that $p(v_1) > 1 + \varepsilon$. By the lower semicontinuity of p at v_1, there exists an open ball $B(v_1, 2r_1)$ such that $p(v) > 1$ for all $v \in B(v_1, 2r_1)$ and thus $p(v) > 1$ for all $v \in \overline{B(v_1, r_1)}$. Let us take r_1 so small that $r_1 < \min\left\{\frac{1}{2}\{r_0 - \|v_1\|\}, 1\right\}$. Now since p is not bounded above on $B(v_1, r_1)$, there exists $v_2 \in B(v_1, r_1)$ such that $p(v_2) > 2 + \varepsilon$. Then since p is lower semicontinuous at v_2, there exists an open ball $B(v_2, 2r_2)$ such that $p(v) > 2$ for all $v \in B(v_2, 2r_2)$ and thus $p(v) > 2$ for all $v \in \overline{B(v_2, r_2)}$. Let us take r_2 so small that $r_2 < \min\left\{\frac{1}{2}\{r_1 - \|v_2\|\}, \frac{1}{2}\right\}$. Thus continuing we have a sequence of open balls $(B(v_n, r_n) : n \in \mathbb{N})$ such that $\overline{B(v_{n+1}, r_{n+1})} \subset B(v_n, r_n)$, $r_n \in \left(0, \frac{1}{n}\right)$, and $p(v) > n$ for $v \in \overline{B(v_n, r_n)}$ for $n \in \mathbb{N}$. Then $(v_n : n \in \mathbb{N})$ is a Cauchy sequence in V since for $m \geq n$ we have $v_m \in B(v_n, r_n)$ so that $\|v_m - v_n\| < r_n < \frac{1}{n}$. By the completeness of V there exists $v \in V$ such that $\lim_{n\to\infty} \|v_n - v\| = 0$. Let $N \in \mathbb{N}$. Then for $n \geq N$ we have $v_n \in B(v_N, r_N)$ so that $v = \lim_{n\to\infty} v_n \in \overline{B(v_N, r_N)}$. Thus $p(v) > N$. Since this holds for every $N \in \mathbb{N}$, we have $p(v) = \infty$, contradicting the fact that p is real-valued. ∎

Theorem 15.89. *Let $(f_n : n \in \mathbb{N})$ be a sequence in the dual space $(V^*, \mathbb{K}, \|\cdot\|_*)$ of a Banach space $(V, \mathbb{K}, \|\cdot\|)$ where $\mathbb{K} = \mathbb{R}$ or $\mathbb{K} = \mathbb{C}$. Suppose the sequence $(f_n(v) : n \in \mathbb{N})$ in $\mathbb{K}$ converges for every $v \in V$. Define a $\mathbb{K}$-valued function f on V by setting*

(1) $$f(v) = \lim_{n\to\infty} f_n(v) \quad \text{for } v \in V.$$

Then $f \in V^*$. Moreover $(\|f_n\|_* : n \in \mathbb{N})$ is a bounded sequence of nonnegative real numbers and

(2) $$\|f\|_* \leq \liminf_{n\to\infty} \|f_n\|_*.$$

Proof. For every $v \in V$, we have $f(v) \in \mathbb{K}$ by (1). The linearity of f follows immediately from the linearity of f_n for $n \in \mathbb{N}$. Now since $f_n \in V^*$, we have $|f_n(v)| \leq \|f_n\|_* \|v\|$ for every $v \in V$. Thus from (1) for every $v \in V$ we have

(3) $$|f(v)| = \lim_{n\to\infty} |f_n(v)| = \liminf_{n\to\infty} |f_n(v)| \leq \left(\liminf_{n\to\infty} \|f_n\|_*\right) \|v\|.$$

Now if we show that $(\|f_n\|_* : n \in \mathbb{N})$ is a bounded sequence of nonnegative real numbers, then $\liminf_{n\to\infty} \|f_n\|_*$ is a nonnegative real number so that it is a bound for the linear functional f by (3) and thus f is a bounded linear functional on V with $\|f\|_* \leq \liminf_{n\to\infty} \|f_n\|_*$.

To show that $(\|f_n\|_* : n \in \mathbb{N})$ is a bounded sequence, let us define a function p on V by

$$p(v) = \sup_{n\in\mathbb{N}} |f_n(v)| \quad \text{for } v \in V.$$

Since $(f_n(v) : n \in \mathbb{N})$ is a convergent sequence in $\mathbb{K}$, we have $\sup_{n\in\mathbb{N}} |f_n(v)| < \infty$ for every $v \in V$. This implies that p is a lower semicontinuous seminorm on V by Proposition 15.86. The lower semicontinuity of the seminorm p implies the existence of a constant $M \geq 0$ such that $p(v) \leq M\|v\|$ for every $v \in V$ according to Proposition 15.88. Then by the definition of p, we have $|f_n(v)| \leq p(v) \leq M\|v\|$ for every $v \in V$ so that M is a bound of the b.l.f. f_n and thus $\|f_n\|_* \leq M$. Since this holds for every $n \in \mathbb{N}$, the sequence $(\|f_n\|_* : n \in \mathbb{N})$ is bounded by M. ∎

§16 The L^p Spaces

[I] The $\mathcal{L}^p$ Spaces for $p \in (0, \infty)$

Let us define the extended complex number system $\overline{\mathbb{C}}$ as the set of all extended complex numbers $\zeta = \xi + i\eta$ where $\xi, \eta \in \overline{\mathbb{R}}$. We regard $\overline{\mathbb{R}}$ as a subset of $\overline{\mathbb{C}}$ identifying $\xi \in \overline{\mathbb{R}}$ with $\xi + i0 \in \overline{\mathbb{C}}$. If $\zeta_1, \zeta_2 \in \overline{\mathbb{C}}$, then $\zeta_1 + \zeta_2$, $\zeta_1\zeta_2$, and $0\zeta_1$ may be undefined unless $\zeta_1, \zeta_2 \in \mathbb{C}$.

By an extended complex-valued function we mean a function $f = f_1 + if_2$ where f_1 and f_2 are extended real-valued functions. We call f_1 and f_2 the real and imaginary parts of f and write $\Re f$ and $\Im f$ for them. Thus $f = \Re f + i\Im f$. We regard an extended real-valued function f as a particular case of extended complex-valued functions with $\Im f = 0$.

Definition 16.1. *Given a measure space $(X, \mathfrak{A}, \mu)$. Let f be an extended complex-valued function on a set $D \in \mathfrak{A}$.*

(a) *We say that f is $\mathfrak{A}$-measurable on D, μ semi-integrable on D, or μ-integrable on D according as both $\Re f$ and $\Im f$ are.*

(b) *When f is μ semi-integrable on D, that is, when both $\int_D \Re f\, d\mu$ and $\int_D \Im f\, d\mu$ exist in*

$\overline{\mathbb{R}}$, we let $\int_D f\, d\mu = \int_D \Re f\, d\mu + i \int_D \Im f\, d\mu \in \overline{\mathbb{C}}$.

(c) *When f in μ-integrable on D, that is, when both $\int_D \Re f\, d\mu$ and $\int_D \Im f\, d\mu$ exist in $\mathbb{R}$, we have $\int_D f\, d\mu \in \mathbb{C}$. In this case, since $\left|\int_D \Re f\, d\mu\right|, \left|\int_D \Im f\, d\mu\right| < \infty$, we have*

$$\left|\int_D f\, d\mu\right| = \left\{\left|\int_D \Re f\, d\mu\right|^2 + \left|\int_D \Im f\, d\mu\right|^2\right\}^{1/2} < \infty.$$

Observation 16.2. Let f be an extended complex-valued $\mathfrak{A}$-measurable function on a set $D \in \mathfrak{A}$ in a measure space $(X, \mathfrak{A}, \mu)$. If f is μ-integrable on D, then $f(x) \in \mathbb{C}$ for a.e. $x \in D$ and thus $|f| < \infty$ a.e. on D.

Proof. If f is μ-integrable on D, then by Definition 16.1, both $\Re f$ and $\Im f$ are μ-integrable on D. Then by Observation 9.2, $(\Re f)(x), (\Im f)(x) \in \mathbb{R}$ for a.e. $x \in D$ and thus $f(x) \in \mathbb{C}$ for a.e. $x \in D$. ■

Observation 16.3. Let f be an extended complex-valued $\mathfrak{A}$-measurable function on a set $D \in \mathfrak{A}$ in a measure space $(X, \mathfrak{A}, \mu)$. Then f is μ-integrable on D if and only if $|f|$ is μ-integrable on D.

Proof. By $f = \Re f + i\Im f$, we have $|f| = \left\{|\Re f|^2 + |\Im f|^2\right\}^{1/2}$. This implies that

$$(1) \qquad |\Re f|, |\Im f| \le |f| \le |\Re f| + |\Im f| \quad \text{on } D.$$

Integrating over D we have

$$(2) \qquad \int_D |\Re f|\, d\mu, \int_D |\Im f|\, d\mu \le \int_D |f|\, d\mu \le \int_D |\Re f|\, d\mu + \int_D |\Im f|\, d\mu.$$

Thus we have

$$\begin{aligned} & f \text{ is } \mu\text{-integrable on } D \\ \Leftrightarrow & \left|\int_D \Re f\,d\mu\right|, \left|\int_D \Im f\,d\mu\right| < \infty \quad \text{by Definition 16.1} \\ \Leftrightarrow & \int_D |\Re f|\,d\mu, \int_D |\Im f|\,d\mu < \infty \quad \text{by (e) of Observation 9.2} \\ \Leftrightarrow & \int_D |f|\,d\mu < \infty \quad \text{by (2).} \ \blacksquare \end{aligned}$$

Definition 16.4. *Let f be an extended complex-valued $\mathfrak{A}$-measurable function on a set $D \in \mathfrak{A}$ in a measure space $(X, \mathfrak{A}, \mu)$. Let $p \in (0, \infty)$. We say that f is p-th order μ-integrable on D if the nonnegative extended real-valued $\mathfrak{A}$-measurable function $|f|^p$ on D is μ-integrable on D, that is, $\int_D |f|^p\,d\mu < \infty$.*

Let us note that if f is p-th order μ-integrable on D for some $p \in (0, \infty)$, that is, $\int_D |f|^p\,d\mu < \infty$, then $|f|^p < \infty$ a.e. on D according to (f) of Observation 9.2 and thus $f(x) \in \mathbb{C}$ for a.e. $x \in D$.

Observation 16.5. Let f be an extended complex-valued $\mathfrak{A}$-measurable function on a set $D \in \mathfrak{A}$ in a measure space $(X, \mathfrak{A}, \mu)$. Let $p \in (0, \infty)$. Then f is p-th order μ-integrable on D if and only if both $\Re f$ and $\Im f$ are p-th order μ-integrable on D.

Proof. In the Proof of Observation 16.3 we showed

(1) $$|\Re f|, |\Im f| \le |f| \le |\Re f| + |\Im f| \quad \text{on } D.$$

Now for $\alpha, \beta \in [0, \infty]$, we have

(2) $$\{\alpha + \beta\}^p \le \left[2\max\{\alpha, \beta\}\right]^p \le 2^p\left\{\alpha^p + \beta^p\right\}.$$

By (1) and (2), we have

$$|\Re f|^p, |\Im f|^p \le |f|^p \le 2^p\left\{|\Re f|^p + |\Im f|^p\right\}.$$

Thus we have

$$\int_D |f|^p\,d\mu < \infty \Leftrightarrow \int_D |\Re f|^p\,d\mu < \infty \text{ and } \int_D |\Im f|^p\,d\mu < \infty. \ \blacksquare$$

Let f and g be two extended complex-valued $\mathfrak{A}$-measurable functions on a set $D \in \mathfrak{A}$ in a measure space $(X, \mathfrak{A}, \mu)$. Then the pointwise addition $f + g$, multiplication fg, and scalar multiplication $0f$ may be undefined on a $\mathfrak{A}$-measurable subset E of D with $\mu(E) > 0$. If we assume that $|f| < \infty$ and $|g| < \infty$ a.e. on D, that is, $f(x) \in \mathbb{C}$ and $g(x) \in \mathbb{C}$ for a.e. $x \in D$, then $f + g$, fg, and $0f$ are defined a.e. on D. The condition that $|f| < \infty$ and $|g| < \infty$ a.e. on D is satisfied if for instance f and g are p-th order μ-integrable on D for some $p \in (0, \infty)$ as we noted above.

Definition 16.6. *Let f and g be two extended complex-valued $\mathfrak{A}$-measurable functions on a set $D \in \mathfrak{A}$ in a measure space $(X, \mathfrak{A}, \mu)$ such that $|f| < \infty$ and $|g| < \infty$ a.e. on D. On the null sets in the measure space $(X, \mathfrak{A}, \mu)$ on which addition $f+g$, or multiplication fg, or scalar multiplication $0f$ is not defined, let us define $f+g$, or fg, or $0f$ to be equal to 0 so that $f+g$, or fg, or $0f$ is defined and $\mathfrak{A}$-measurable on D.*

Lemma 16.7. *For $\alpha, \beta \in \mathbb{C}$, we have*

(1) $$|\alpha+\beta|^2 \le 2\left\{|\alpha|^2+|\beta|^2\right\}.$$

More generally for any $p \in (0,\infty)$, we have

(2) $$|\alpha+\beta|^p \le 2^p\left\{|\alpha|^p+|\beta|^p\right\}.$$

Proof. To prove (1), let us note that $|\alpha+\beta| \le |\alpha|+|\beta|$ so that

(3) $$|\alpha+\beta|^2 \le \left\{|\alpha|+|\beta|\right\}^2 = |\alpha|^2+2|\alpha||\beta|+|\beta|^2.$$

Now $|\alpha|^2-2|\alpha||\beta|+|\beta|^2 = \left\{|\alpha|-|\beta|\right\}^2 \ge 0$ so that $2|\alpha||\beta| \le |\alpha|^2+|\beta|^2$. Using this in (3), we have (1). To prove (2), note that $|\alpha+\beta| \le |\alpha|+|\beta| \le 2\max\left\{|\alpha|,|\beta|\right\}$. From this we have $|\alpha+\beta|^p \le 2^p\left[\max\left\{|\alpha|,|\beta|\right\}\right]^p = 2^p\max\left\{|\alpha|^p,|\beta|^p\right\} \le 2^p\left\{|\alpha|^p+|\beta|^p\right\}$. ∎

Definition 16.8. *Given a measure space $(X, \mathfrak{A}, \mu)$. Let $p \in (0,\infty)$. For an extended complex-valued $\mathfrak{A}$-measurable function f on X, we define*

$$\|f\|_p = \left[\int_X |f|^p\,d\mu\right]^{1/p} \in [0,\infty].$$

We define $\mathcal{L}^p(X, \mathfrak{A}, \mu)$ as the collection of all extended complex-valued $\mathfrak{A}$-measurable functions f on X with $\|f\|_p < \infty$.

Proposition 16.9. *Let $(X, \mathfrak{A}, \mu)$ be a measure space. Let $p \in (0,\infty)$. Then $\mathcal{L}^p(X, \mathfrak{A}, \mu)$ is a linear space over $\mathbb{C}$.*

Proof. The addition and scalar multiplication in $\mathcal{L}^p(X, \mathfrak{A}, \mu)$ are as in Definition 16.6. Let $f, g \in \mathcal{L}^p(X, \mathfrak{A}, \mu)$. Then $\|f\|_p, \|g\|_p < \infty$ so that $|f|, |g| < \infty$ a.e. on X, that is, f and g are complex-valued a.e. on X. Thus we have $|f+g|^p \le 2^p\{|f|^p+|g|^p\}$ a.e. on X by (2) of Lemma 16.7. Then integrating over X, we have

$$\int_X |f+g|^p\,d\mu \le 2^p\left\{\int_X |f|^p\,d\mu + \int_X |g|^p\,d\mu\right\} < \infty.$$

This shows that $f+g \in \mathcal{L}^p(X, \mathfrak{A}, \mu)$.

If $f \in \mathcal{L}^p(X, \mathfrak{A}, \mu)$ and $c \in \mathbb{C}$, then $|cf| = |c||f|$ a.e. on X. Then integrating over X, we have $\int_X |cf|^p\,d\mu = |c|^p \int_X |f|^p\,d\mu < \infty$. This shows that $cf \in \mathcal{L}^p(X, \mathfrak{A}, \mu)$. ∎

Remark 16.10. Let us observe that $\|\cdot\|_p$ is not a norm on the linear space $\mathcal{L}^p(X, \mathfrak{A}, \mu)$. This is from the fact that while the origin of the linear space $\mathcal{L}^p(X, \mathfrak{A}, \mu)$ is the identically vanishing function on X, $\|f\|_p = 0$ implies $|f|^p = 0$ a.e. on X, that is, $f = 0$ a.e. on X but f need not be the identically vanishing function on X.

[II] The Linear Spaces $\mathcal{L}^p$ for $p \in [1, \infty)$

Definition 16.11. *Two real numbers $p, q \in (1, \infty)$ are called conjugates if they satisfy the equation $1/p + 1/q = 1$. The pair $p = 1$ and $q = \infty$ is also called conjugates.*

Let us note that if $p, q \in (1, \infty)$ are conjugates, then $q = p/(p-1)$. Thus $\lim_{p\downarrow 1} q = \infty$ and $\lim_{p\to\infty} q = 1$. Note also that if $p, q \in (1, \infty)$ are conjugates and $p = q$, then $p = q = 2$.

Lemma 16.12. (Young's Inequality) *Let φ be a real-valued, continuous, and strictly increasing function on $[0, \infty)$ with $\varphi(0) = 0$ and $\lim_{u\to\infty} \varphi(u) = \infty$. Let $\psi = \varphi^{-1}$ be the inverse function of φ defined on $[0, \infty)$. Then for any $a, b \in [0, \infty)$, we have*

$$ab \le \int_0^a \varphi(u)\, du + \int_0^b \psi(v)\, dv,$$

where the equality holds if and only if $b = \varphi(a)$.

Proof. Note that the inverse function ψ is also real-valued, continuous, and strictly increasing on $[0, \infty)$ with $\psi(0) = 0$ and $\lim_{v\to\infty} \psi(v) = \infty$. The Lemma follows immediately if we interpret ab as the area of the rectangle in the uv-plane with sides $[0, a]$ and $[0, b]$, $\int_0^a \varphi(u)\, du$ as the area under the curve $v = \varphi(u)$ for $u \in [0, a]$, and $\int_0^b \psi(v)\, dv$ as the area under the curve $u = \psi(v)$ for $v \in [0, b]$. ∎

Lemma 16.13. *Let $p, q \in (1, \infty)$ be conjugates. Then for any $\alpha, \beta \in \mathbb{C}$, we have*

$$|\alpha\beta| \le \frac{|\alpha|^p}{p} + \frac{|\beta|^q}{q},$$

and the equality holds if and only if $|\alpha|^p = |\beta|^q$.

Proof. Since the complex numbers α and β enter in the inequality only as nonnegative numbers $|\alpha|$ and $|\beta|$, it suffices to show that for any $a, b \in [0, \infty)$ we have

$$(1) \qquad ab \le \frac{a^p}{p} + \frac{b^q}{q},$$

and the equality holds if and only if $a^p = b^q$.

With $r := p - 1 > 0$, consider the function $\varphi(u) = u^r$ for $u \in [0, \infty)$. Now φ is real-valued, continuous, and strictly increasing on $[0, \infty)$ with $\varphi(0) = 0$ and $\lim_{u\to\infty} \varphi(u) = \infty$.

Its inverse function is given by $\psi(v) = v^{\frac{1}{r}}$ for $v \in [0, \infty)$. Thus by Lemma 16.13, we have

(2)
$$ab \le \int_0^a u^r\,du + \int_0^b v^{\frac{1}{r}}\,dv,$$

and the equality holds if and only if $b = a^r$. From $r = p - 1$ and $\frac{1}{p} + \frac{1}{q} = 1$, we have $\frac{1}{r} + 1 = q$ and $r = \frac{p}{q}$. Thus we have

(3)
$$\int_0^a u^r\,du + \int_0^b v^{\frac{1}{r}}\,dv = \left[\frac{u^{r+1}}{r+1}\right]_0^a + \left[\frac{v^{\frac{1}{r}+1}}{\frac{1}{r}+1}\right]_0^b = \frac{a^p}{p} + \frac{b^q}{q}.$$

By (2) and (3) we have (1). Also $b = a^r = a^{\frac{p}{q}}$ if and only if $b^q = a^p$. ∎

Theorem 16.14. (Hölder's Inequality for $p, q \in (1, \infty)$) *Given a measure space $(X, \mathfrak{A}, \mu)$. Let f and g be two extended complex-valued $\mathfrak{A}$-measurable functions on X such that $|f|, |g| < \infty$ a.e. on X.*
(a) *For any $p, q \in (1, \infty)$ such that $1/p + 1/q = 1$, we have*

(1)
$$\|fg\|_1 \le \|f\|_p \|g\|_q,$$

provided the product of the two nonnegative extended real numbers $\|f\|_p$ and $\|g\|_q$ exists.
(b) *If $0 < \|f\|_p < \infty$ and $0 < \|g\|_q < \infty$, then the equality in (1) holds if and only if*

(2)
$$A|f|^p = B|g|^q \quad \text{a.e. on } X \text{ for some } A, B > 0.$$

Proof. 1. We are to prove

(3)
$$\int_X |fg|\,d\mu \le \left[\int_X |f|^p\,d\mu\right]^{1/p} \left[\int_X |g|^q\,d\mu\right]^{1/q}.$$

If $\mu(X) = 0$, then the equality in (3) is trivially true. Thus assume that $\mu(X) > 0$. If at least one of f and g is equal to 0 a.e. on X, then $fg = 0$ a.e. on X and the equality in (3) holds trivially provided that the product on the right side of (3) exists. Thus assume that $\mu\{X : f \neq 0\} > 0$ and $\mu\{X : g \neq 0\} > 0$. Then $\int_X |f|^p\,d\mu = \int_{\{X:f\neq 0\}} |f|^p\,d\mu > 0$ by (d) of Lemma 8.2 and similarly $\int_X |g|^q\,d\mu > 0$. For brevity let

(4)
$$I := \int_X |f|^p\,d\mu > 0 \quad \text{and} \quad J := \int_X |g|^q\,d\mu > 0.$$

If at least one of I and J is equal to ∞, then (3) holds trivially provided that the product on the right side of (3) exists. Therefore assume that $I, J \in (0, \infty)$. Now since $|f|, |g| \in \mathbb{R}$ a.e. on X, we have by Lemma 16.13,

(5)
$$\frac{|f|}{I^{1/p}}\frac{|g|}{J^{1/q}} \le \frac{1}{p}\frac{|f|^p}{I} + \frac{1}{q}\frac{|g|^q}{J} \quad \text{a.e. on } X.$$

Integrating over X and recalling (4), we have

$$(6) \qquad \frac{1}{I^{1/p}J^{1/q}}\int_X |fg|\,d\mu \le \frac{1}{pI}\int_X |f|^p\,d\mu + \frac{1}{qJ}\int_X |g|^q\,d\mu = \frac{1}{p}+\frac{1}{q} = 1.$$

Then multiplying by $I^{1/p}J^{1/q}$ we have (3).

2. Under the assumption that $I, J \in (0,\infty)$, observe that the equality in (3) holds if and only if the equality in (6) holds. Now the equality in (6) holds if and only if the equality in (5) holds a.e. on X. But the equality in (5) holds a.e. on X if and only if $|f|^p/I = |g|^q/J$ a.e. on X according to Lemma 16.13. Thus it remains to show

$$(7) \qquad \frac{|f|^p}{I} = \frac{|g|^q}{J} \quad \text{a.e. on } X \Leftrightarrow A|f|^p = B|g|^q \quad \text{a.e. on } X \text{ for some } A, B > 0.$$

Now if the left side of (7) holds, then the right side of (7) holds with $A = 1/I \in (0,\infty)$ and $B = 1/J \in (0,\infty)$. Conversely if the right side of (7) holds, then integrating we have $AI = BJ$. Dividing the right side of (7) by this equality we have the left side of (7). This proves (7). ∎

Corollary 16.15. (Schwarz's Inequality) *Given a measure space $(X, \mathfrak{A}, \mu)$. Let f and g be two extended complex-valued $\mathfrak{A}$-measurable functions on X such that $|f|, |g| < \infty$ a.e. on X. Then we have*

$$(1) \qquad \|fg\|_1 \le \|f\|_2\|g\|_2,$$

provided the product of the two nonnegative extended real numbers $\|f\|_2$ and $\|g\|_2$ exists. If $0 < \|f\|_2 < \infty$ and $0 < \|g\|_2 < \infty$, then the equality in (1) holds if and only if

$$(2) \qquad a|f| = b|g| \quad \text{a.e. on } X \text{ for some } a, b > 0.$$

Proof. This is a particular case of Theorem 16.14 for the conjugates $p = q = 2$. Note that condition (2) in Theorem 16.14 in this case becomes $A|f|^2 = B|g|^2$ and taking square roots we have $a|f| = b|g|$ where $a = \sqrt{A}$ and $b = \sqrt{B}$. ∎

Corollary 16.16. *Given a measure space $(X, \mathfrak{A}, \mu)$. Let f be an extended complex-valued $\mathfrak{A}$-measurable functions on X such that $|f| < \infty$ a.e. on X. Then for any $p, q \in (1,\infty)$ such that $1/p + 1/q = 1$ we have*

$$\|f\|_1 \le \|f\|_p\|1\|_q,$$

that is,

$$\int_X |f|\,d\mu \le \mu(X)^{1/q}\left[\int_X |f|^p\,d\mu\right]^{1/p},$$

provided that the product on the right side exists.

Proof. Writing $f = f\mathbf{1}_X$ and applying Theorem 16.14, we have

$$\int_X |f|\,d\mu \le \left[\int_X |f|^p\,d\mu\right]^{1/p} \left[\int_X |\mathbf{1}_X|^q\,d\mu\right]^{1/q} = \mu(X)^{1/q} \left[\int_X |f|^p\,d\mu\right]^{1/p}. \;\blacksquare$$

Theorem 16.17. **(Minkowski's Inequality for** $p \in [1, \infty)$**)** *Given a measure space* $(X, \mathfrak{A}, \mu)$. *Let* f *and* g *be two extended complex-valued* $\mathfrak{A}$*-measurable functions on* X *such that* $|f|, |g| < \infty$ *a.e. on* X. *Then for every* $p \in [1, \infty)$, *we have*

$$\|f + g\|_p \le \|f\|_p + \|g\|_p.$$

Proof. If at least one of $\|f\|_p$ and $\|g\|_p$ is equal to ∞ then $\|f + g\|_p \le \|f\|_p + \|g\|_p$ holds trivially. Thus we consider the case that $\|f\|_p < \infty$ and $\|g\|_p < \infty$. In this case by applying Lemma 16.7 we have $|f + g|^p \le 2^p\{|f|^p + |g|^p\}$. Then integrating over X

(1) $$\int_X |f + g|^p\,d\mu \le 2^p\{\|f\|_p^p + \|g\|_p^p\} < \infty.$$

We are to prove

(2) $$\left[\int_X |f + g|^p\,d\mu\right]^{1/p} \le \left[\int_X |f|^p\,d\mu\right]^{1/p} + \left[\int_X |g|^p\,d\mu\right]^{1/p}.$$

Let us note that for $p = 1$, (2) is immediate since $|f + g| \le |f| + |g|$ on X and integrating over X, we have

$$\int_X |f + g|\,d\mu \le \int_X |f|\,d\mu + \int_X |g|\,d\mu.$$

Let us consider the case $p \in (1, \infty)$ and let $q \in (1, \infty)$ be its conjugate, that is, $1/p + 1/q = 1$. Then $q = p/(p-1)$ and $p - 1 = p/q$. Now

(3) $$|f + g|^p = |f + g||f + g|^{p-1} \le \{|f| + |g|\}|f + g|^{p/q}.$$

Integrating over X, we have

(4) $$\int_X |f + g|^p\,d\mu \le \int_X |f||f + g|^{p/q}\,d\mu + \int_X |g||f + g|^{p/q}\,d\mu.$$

By (3) in the Proof of Theorem 16.14, we have

(5) $$\int_X |f||f + g|^{p/q}\,d\mu \le \left[\int_X |f|^p\,d\mu\right]^{1/p} \left[\int_X |f + g|^p\,d\mu\right]^{1/q},$$

and similarly

(6) $$\int_X |g||f + g|^{p/q}\,d\mu \le \left[\int_X |g|^p\,d\mu\right]^{1/p} \left[\int_X |f + g|^p\,d\mu\right]^{1/q}.$$

Using (5) and (6) in (4), we have

$$(7)\quad \int_X |f+g|^p\,d\mu \le \left\{\left[\int_X |f|^p\,d\mu\right]^{1/p} + \left[\int_X |g|^p\,d\mu\right]^{1/p}\right\}\left[\int_X |f+g|^p\,d\mu\right]^{1/q}.$$

If $\int_X |f+g|^p\,d\mu = 0$, then (2) holds trivially. If $\int_X |f+g|^p\,d\mu > 0$, then dividing (7) by $\left[\int_X |f+g|^p\,d\mu\right]^{1/q} \in (0,\infty)$ by (1), we have

$$\left[\int_X |f+g|^p\,d\mu\right]^{1-1/q} \le \left[\int_X |f|^p\,d\mu\right]^{1/p} + \left[\int_X |g|^p\,d\mu\right]^{1/p}.$$

Since $1 - 1/q = 1/p$, we have (2). ∎

Let us consider the possibility of the equality in Minkowski's Inequality.

Observation 16.18. For $\zeta = \xi + i\eta \in \mathbb{C}$, $\zeta \neq 0$, let $\arg\zeta$ be the principal value of the argument of ζ, that is, the principal value of $\arctan \eta/\xi$. Thus $\arg\zeta \in [0, 2\pi)$. For $\zeta = 0$, let us set $\arg\zeta = 0$. For $\zeta \in \overline{\mathbb{C}} - \mathbb{C}$, $\arg\zeta$ remains undefined. Note that any $\zeta_1, \zeta_2 \in \mathbb{C}$, we have

$$|\zeta_1 + \zeta_2| = |\zeta_1| + |\zeta_2| \Leftrightarrow \arg\zeta_1 = \arg\zeta_2.$$

Let f and g be two extended complex-valued $\mathfrak{A}$-measurable functions on X in a measure space $(X, \mathfrak{A}, \mu)$. Let $E = \{x \in X : f(x) \in \mathbb{C}\} \cap \{x \in X : g(x) \in \mathbb{C}\}$. If $|f|, |g| < \infty$ a.e. on X, then $\mu(E^c) = 0$ so the condition that $\arg f = \arg g$ a.e. on E is equivalent to the condition that $\arg f = \arg g$ a.e. on X.

Proposition 16.19. *Given a measure space $(X, \mathfrak{A}, \mu)$. Let f and g be two extended complex-valued $\mathfrak{A}$-measurable functions on X such that $|f|, |g| < \infty$ a.e. on X. Then $\|f+g\|_1 = \|f\|_1 + \|g\|_1$ if and only if $\arg f = \arg g$ a.e. on X.*

Proof. Let $E = \{x \in X : f(x) \in \mathbb{C}\} \cap \{x \in X : g(x) \in \mathbb{C}\}$. Since $|f|, |g| < \infty$ a.e. on X, we have $\mu(E^c) = 0$. Since $0 \le |f+g| \le |f| + |g|$ on X, we have

$$\begin{aligned}\int_X |f+g|\,d\mu = \int_X |f|\,d\mu + \int_X |g|\,d\mu \\ \Leftrightarrow |f+g| = |f| + |g| \quad \text{a.e. on } X \\ \Leftrightarrow \arg f = \arg g \quad \text{a.e. on } X,\end{aligned}$$

by Observation 16.18. ∎

Proposition 16.20. *Given a measure space $(X, \mathfrak{A}, \mu)$. Let $p \in (1, \infty)$. Let f and g be two extended complex-valued $\mathfrak{A}$-measurable functions on X such that $|f|, |g| < \infty$ a.e. on X. Assume that $0 < \|f\|_p, \|g\|_p < \infty$. Then $\|f+g\|_p = \|f\|_p + \|g\|_p$ if and only if*

1° $\arg f = \arg g$ *a.e. on X,*

2° $a|f| = b|g|$ *a.e. on X for some $a, b > 0$.*

Proof. Recall that in proving Minkowski's Inequality, inequality occurred in (2), (4), and (5) in the Proof of Theorem 16.17 so that $\|f+g\|_p = \|f\|_p + \|g\|_p$ if and only if equality in (2) holds a.e. on X and equality in (4) and (5) holds.

Now equality in (2) holds a.e. on X if and only if $|f+g| = |f| + |g|$ a.e. on X and this condition is equivalent to the condition $\arg f = \arg g$ a.e. on X by Observation 16.14.

Equality in (4) and (5) holds if and only if

$$(1) \qquad A_1|f|^p = B_1|f+g|^p \quad \text{a.e. on } X \text{ for some } A_1, B_1 > 0,$$

and

$$(2) \qquad A_2|g|^p = B_2|f+g|^p \quad \text{a.e. on } X \text{ for some } A_2, B_2 > 0,$$

according to (2) of Theorem 16.14 (Hölder's Inequality). It is easily verified that (1) and (2) imply 2°. Conversely condition 1° (which is equivalent to the condition $|f+g| = |f|+|g|$ a.e. on X) and condition 2° imply (1) and (2). ∎

Remark 16.21. For $p \in (0, 1)$, the function $\|\cdot\|_p$ on $\mathcal{L}^p(X, \mathfrak{A}, \mu)$ does not satisfy the triangle inequality and therefore it will not be a norm on $\mathcal{L}^p(X, \mathfrak{A}, \mu)$. This is shown by the following example.

Example. Let $(X, \mathfrak{A}, \mu) = \big([0, 1), \mathfrak{M}_L \cap [0, 1), \mu_L\big)$ where $\mathfrak{M}_L \cap [0, 1)$ is a σ-algebra of subsets of $[0, 1)$ defined by $\mathfrak{M}_L \cap [0, 1) = \{E \cap [0, 1) : E \in \mathfrak{M}_L\}$. Let $f = \mathbf{1}_{[0,\frac{1}{2})}$ and $g = \mathbf{1}_{[\frac{1}{2},1)}$. Then $f + g = \mathbf{1}_{[0,1)}$ and $\int_X |f+g|^p \, d\mu_L = \mu_L([0, 1)) = 1$ so that $\|f+g\|_p = 1$. On the other hand $\int_X |f|^p \, d\mu_L = \mu_L\big([0, \frac{1}{2})\big) = \frac{1}{2}$ so that $\|f\|_p = 2^{-\frac{1}{p}}$ and similarly $\|g\|_p = 2^{-\frac{1}{p}}$. Then $\|f\|_p + \|g\|_p = 2 \cdot 2^{-\frac{1}{p}} = 2^{1-\frac{1}{p}} < 1$ when $p \in (0, 1)$. Thus $\|f+g\|_p > \|f\|_p + \|g\|_p$.

[III] The L^p Spaces for $p \in [1, \infty)$

Given a measure space $(X, \mathfrak{A}, \mu)$. Let $p \in (0, \infty)$ and consider the linear space of functions $\mathcal{L}^p(X, \mathfrak{A}, \mu)$. The origin, that is the identity of addition, of this linear space is the identically vanishing function 0 on X and for this function 0 we have $\|0\|_p = 0$. On the other hand if $f \in \mathcal{L}^p(X, \mathfrak{A}, \mu)$ and $\|f\|_p = 0$ then $f(x) = 0$ for a.e. $x \in X$ and f need not be the identically vanishing function on X. For this reason $\|\cdot\|_p$ is not a norm on the linear space $\mathcal{L}^p(X, \mathfrak{A}, \mu)$.

Let us introduce a relation $\sim$ among the members of $\mathcal{L}^p(X, \mathfrak{A}, \mu)$ by declaring $f \sim g$ for $f, g \in \mathcal{L}^p(X, \mathfrak{A}, \mu)$ if $f = g$ a.e. on X. Thus defined, $\sim$ satisfies the defining conditions for an equivalence relation:

1° $f \sim f$ for $f \in \mathcal{L}^p(X, \mathfrak{A}, \mu)$,
2° $f \sim g \Rightarrow g \sim f$ for $f, g \in \mathcal{L}^p(X, \mathfrak{A}, \mu)$,
3° $f \sim g, g \sim h \Rightarrow f \sim h$ for $f, g, h \in \mathcal{L}^p(X, \mathfrak{A}, \mu)$.

Therefore $\sim$ is an equivalence relation.

Definition 16.22. *Given a measure space $(X, \mathfrak{A}, \mu)$. For $p \in (0, \infty)$, let $L^p(X, \mathfrak{A}, \mu)$ be the collection of all equivalence classes of members of $\mathcal{L}^p(X, \mathfrak{A}, \mu)$ with respect to the equivalence relation of a.e. equality on X.*

Let $\hat{f}$ be the element in $L^p(X, \mathfrak{A}, \mu)$ which is the equivalence class to which a function $f \in \mathcal{L}^p(X, \mathfrak{A}, \mu)$ belongs. Thus if $\hat{f}, \hat{g} \in L^p(X, \mathfrak{A}, \mu)$, then $\hat{f} = \hat{g}$ if and only if $f = g$ a.e. on X. If $f, g \in \mathcal{L}^p(X, \mathfrak{A}, \mu)$ and $g \in \hat{f}$, then $f = g$ a.e. on X so that $f \in \hat{g}$.

In $L^p(X, \mathfrak{A}, \mu)$, we define scalar multiplication and addition by setting

$$\alpha \hat{f} = (\alpha f)\hat{} \qquad \text{for } \hat{f} \in L^p(X, \mathfrak{A}, \mu) \text{ and } \alpha \in \mathbb{C},$$

$$\hat{f} + \hat{g} = (f + g)\hat{} \quad \text{for } \hat{f}, \hat{g} \in L^p(X, \mathfrak{A}, \mu).$$

It is clear that $\alpha \hat{f}$ and $\hat{f} + \hat{g}$ defined above do not depend on the representatives f and g of the equivalence classes $\hat{f}$ and $\hat{g}$ used in the definition of $\alpha \hat{f}$ and $\hat{f} + \hat{g}$. With addition and scalar multiplication as defined above, $L^p(X, \mathfrak{A}, \mu)$ is a linear space over the field of scalars $\mathbb{C}$. The origin of this linear space is the equivalence class of all extended complex-valued $\mathfrak{A}$-measurable functions which vanish a.e. on X. We define the integral of $\hat{f} \in L^p(X, \mathfrak{A}, \mu)$ by setting $\int_X \hat{f}\, d\mu = \int_X f\, d\mu$. This definition does not depend on the representative f of $\hat{f}$ since any two representatives of $\hat{f}$ are equal a.e. on X.

It is customary to write f for an element $\hat{f}$ of $L^p(X, \mathfrak{A}, \mu)$. Thus we write f for an individual function in the space $\mathcal{L}^p(X, \mathfrak{A}, \mu)$ as well as for an element in $L^p(X, \mathfrak{A}, \mu)$ which is an equivalence class of functions. Which one is meant by f, a function or an equivalence class of functions, should be clear from the context.

Theorem 16.23. *Given a measure space $(X, \mathfrak{A}, \mu)$. Let $p \in [1, \infty)$. Then $\|\cdot\|_p$ is a norm on the linear space $L^p(X, \mathfrak{A}, \mu)$ over the field of scalars $\mathbb{C}$ and $L^p(X, \mathfrak{A}, \mu)$ is a Banach space with respect to the norm $\|\cdot\|_p$. We call $\|\cdot\|_p$ the L^p norm on $L^p(X, \mathfrak{A}, \mu)$.*

Proof. 1. It is easily verified that the function $\|\cdot\|_p$ on $L^p(X, \mathfrak{A}, \mu)$ satisfies the defining conditions of a norm:

1° $\|f\|_p \in [0, \infty)$ for $f \in L^p(X, \mathfrak{A}, \mu)$,
2° $\|f\|_p = 0$ if and only if $f = 0 \in L^p(X, \mathfrak{A}, \mu)$,
3° $\|\alpha f\|_p = |\alpha| \|f\|_p$ for $f \in L^p(X, \mathfrak{A}, \mu)$ and $\alpha \in \mathbb{C}$,
4° $\|f + g\|_p \le \|f\|_p + \|g\|_p$ for $f, g \in L^p(X, \mathfrak{A}, \mu)$.

In particular the fact that $\|f\|_p < \infty$ for every $f \in L^p(X, \mathfrak{A}, \mu)$ is from the definition of $\mathcal{L}^p(X, \mathfrak{A}, \mu)$ as the collection of all extended complex-valued $\mathfrak{A}$-measurable functions f on X with $\int_X |f|^p\, d\mu < \infty$. Regarding 2°, note that if $\|f\|_p = 0$ for some $f \in L^p(X, \mathfrak{A}, \mu)$ then f is the equivalence class of all extended complex-valued $\mathfrak{A}$-measurable functions which vanish a.e. on X, that is, $f = 0 \in L^p(X, \mathfrak{A}, \mu)$. Condition 4° is from Theorem 16.17 (Minkowski's Inequality).

2. Let us show that $L^p(X, \mathfrak{A}, \mu)$ is complete with respect to the norm $\|\cdot\|_p$. According to Theorem 15.18, it suffices to show that for every sequence $(f_n : n \in \mathbb{N})$ in $L^p(X, \mathfrak{A}, \mu)$ with $\sum_{n\in\mathbb{N}} \|f_n\|_p < \infty$, the series $\sum_{n\in\mathbb{N}} f_n$ converges in the norm, that is, there exists $g \in L^p(X, \mathfrak{A}, \mu)$ such that $\lim_{n\to\infty} \|g_n - g\|_p = 0$ where $g_n = \sum_{k=1}^n f_k$ for $n \in \mathbb{N}$.

Let $(f_n : n \in \mathbb{N})$ be a sequence in $L^p(X, \mathfrak{A}, \mu)$ with $B := \sum_{n\in\mathbb{N}} \|f_n\|_p < \infty$. Note that $f_n \in L^p(X, \mathfrak{A}, \mu)$ implies that $|f_n| \in L^p(X, \mathfrak{A}, \mu)$ and $\big\||f_n|\big\|_p = \|f_n\|_p$. If we let $h_n = \sum_{k=1}^n |f_k|$ for $n \in \mathbb{N}$, then $(h_n : n \in \mathbb{N})$ is a sequence in the linear space $L^p(X, \mathfrak{A}, \mu)$. By the triangle inequality of the norm $\|\cdot\|_p$, we have

$$(1) \qquad \|h_n\|_p \le \sum_{k=1}^n \big\||f_k|\big\|_p = \sum_{k=1}^n \|f_k\|_p \le B.$$

Let us define a function h on X by setting $h = \sum_{n\in\mathbb{N}} |f_n|$. Let us show $h \in L^p(X, \mathfrak{A}, \mu)$. Now $h_n \uparrow h$ and hence $h_n^p \uparrow h^p$ on X. Thus by Theorem 8.5 (Monotone Convergence Theorem) and then by (1), we have

$$(2) \qquad \int_X h^p \, d\mu = \lim_{n\to\infty} \int_X h_n^p \, d\mu = \lim_{n\to\infty} \|h_n\|_p^p \le B^p < \infty.$$

This shows that $h \in L^p(X, \mathfrak{A}, \mu)$. It also implies that $0 \le h^p < \infty$ a.e. on X and then $0 \le h = \sum_{n\in\mathbb{N}} |f_n| < \infty$ a.e. on X. Thus $\sum_{n\in\mathbb{N}} |f_n|$ converges a.e. on X. Since absolute convergence of a series implies convergence of the series, $\sum_{n\in\mathbb{N}} f_n$ converges a.e. on X. Let $g_n = \sum_{k=1}^n f_k$ for $n \in \mathbb{N}$ and $g = \sum_{n\in\mathbb{N}} f_n$. Then the sequence $(g_n : n \in \mathbb{N})$ converges to g a.e. on X. It remains to show that $g \in L^p(X, \mathfrak{A}, \mu)$ and $\lim_{n\to\infty} \|g_n - g\|_p = 0$. Now $|g| = |\sum_{n\in\mathbb{N}} f_n| \le \sum_{n\in\mathbb{N}} |f_n| = h$ a.e. on X. Then $\int_X |g|^p \, d\mu \le \int_X h^p \, d\mu < \infty$ by (2). This shows that $g \in L^p(X, \mathfrak{A}, \mu)$. Now $|g - g_n| = |g - \sum_{k=1}^n f_k| \le |g| + \sum_{k=1}^n |f_k| \le 2h$ a.e. on X and then $|g - g_n|^p \le 2^p h^p$ a.e. on X. Since $h \in L^p(X, \mathfrak{A}, \mu)$, h^p is μ-integrable on X. Thus by Theorem 9.20 (Dominated Convergence Theorem), we have

$$\lim_{n\to\infty} \int_X |g - g_n|^p \, d\mu = \int_X \lim_{n\to\infty} |g - g_n|^p \, d\mu = 0,$$

that is, $\lim_{n\to\infty} \|g - g_n\|_p^p = 0$ and then $\lim_{n\to\infty} \|g - g_n\|_p = 0$. ∎

For $z \in \mathbb{C}$, let $\overline{z}$ be the complex conjugate of z, that is, $\overline{z} = \Re z - i\Im z$. For an extended complex-valued function g, let $\overline{g} = \Re g - i\Im g$.

Theorem 16.24. *Let $(X, \mathfrak{A}, \mu)$ be an arbitrary measure space. On $L^2(X, \mathfrak{A}, \mu)$, define*

$$\langle f, g\rangle = \int_X f\overline{g} \, d\mu \quad \textit{for } f, g \in L^2(X, \mathfrak{A}, \mu).$$

Then $\langle \cdot, \cdot \rangle$ is an inner product on $L^2(X, \mathfrak{A}, \mu)$. For the norm $\|\cdot\|_h$ derived from $\langle \cdot, \cdot \rangle$, we have $\|f\|_h = \|f\|_2$ for every $f \in L^2(X, \mathfrak{A}, \mu)$. Furthermore $L^2(X, \mathfrak{A}, \mu)$ is a Hilbert space with respect to the inner product $\langle \cdot, \cdot \rangle$.

Proof. If $f \in L^2(X, \mathfrak{A}, \mu)$, then $\int_X |\overline{f}|^2 \, d\mu = \int_X |f|^2 \, d\mu < \infty$ so that $\overline{f} \in L^2(X, \mathfrak{A}, \mu)$. If $f, g \in L^2(X, \mathfrak{A}, \mu)$, then

$$|\langle f, g\rangle| = \left|\int_X f\overline{g} \, d\mu\right| \le \int_X |f\overline{g}| \, d\mu \le \|f\|_2 \|\overline{g}\|_2 < \infty.$$

This shows that $\langle f, g\rangle \in \mathbb{C}$, verifying condition 1° of Definition 15.19. Conditions 2°, 3°, 4°, and 5° are verified readily. Thus $\langle\cdot,\cdot\rangle$ is an inner product on $L^2(X, \mathfrak{A}, \mu)$. For the norm $\|\cdot\|_h$ on $L^2(X, \mathfrak{A}, \mu)$ derived from the inner product $\langle\cdot,\cdot\rangle$, we have for every $f \in L^2(X, \mathfrak{A}, \mu)$

$$\|f\|_2 = \{\langle f, f\rangle\}^{1/2} = \left\{\int_X f\overline{f}\, d\mu\right\}^{1/2} = \left\{\int_X |f|^2\, d\mu\right\}^{1/2} = \|f\|_2.$$

This shows that $\|\cdot\|_h = \|\cdot\|_2$.

Since $L^2(X, \mathfrak{A}, \mu)$ is complete with respect to the metric derived from the norm $\|\cdot\|_2$ and since the norm $\|\cdot\|_h$ is equal to the norm $\|\cdot\|_2$, $L^2(X, \mathfrak{A}, \mu)$ is complete with respect to the metric derived from the norm $\|\cdot\|_h$. Thus $L^2(X, \mathfrak{A}, \mu)$ is a Hilbert space. ∎

Let us compare the various modes of convergence of a sequence in $L^p(X, \mathfrak{A}, \mu)$. These are convergence a.e. on X, convergence in measure, convergence of the L^p norms, convergence in the L^p norm and weak convergence in $L^p(X, \mathfrak{A}, \mu)$.

Theorem 16.25. *Given a measure space $(X, \mathfrak{A}, \mu)$. Consider a sequence $(f_n : n \in \mathbb{N})$ and an element f in $L^p(X, \mathfrak{A}, \mu)$ where $p \in [1, \infty)$. If the sequence converges to f in the L^p norm, that is, $\lim_{n\to\infty} \|f_n - f\|_p = 0$, then*

1° $\lim_{n\to\infty} \|f_n\|_p = \|f\|_p$ *and hence* $\lim_{n\to\infty} \int_X |f_n|^p\, d\mu = \int_X |f|^p\, d\mu$,

2° *$(f_n : n \in \mathbb{N})$ converges to f in measure μ on X,*

3° *there exists a subsequence $(f_{n_k} : n \in \mathbb{N})$ such that $\lim_{k\to\infty} f_{n_k} = f$ a.e. on X.*

(Note that 2° and 3° assert that if $f_n, n \in \mathbb{N}$, and f are arbitrary representative functions of the equivalence classes, then the sequence of functions $(f_n : n \in \mathbb{N})$ converges to the function f in measure μ on X and there exists a subsequence $(f_{n_k} : n \in \mathbb{N})$ such that $\lim_{k\to\infty} f_{n_k} = f$ a.e. on X.)

Proof. 1. To prove 1°, note that since $\|\cdot\|_p$ is a norm we have $\big|\|f_n\|_p - \|f\|_p\big| \le \|f_n - f\|_p$ by Observation 15.6. Then $\lim_{n\to\infty} \|f_n - f\|_p = 0$ implies $\lim_{n\to\infty} \|f_n\|_p = \|f\|_p$.

2. To prove 2°, let $\varepsilon > 0$ be arbitrarily given. We have

$$\|f_n - f\|_p^p \ge \int_{\{X:|f_n - f|\ge\varepsilon\}} |f_n - f|^p\, d\mu \ge \varepsilon^p \mu\{X : |f_n - f| \ge \varepsilon\}.$$

Then $\lim_{n\to\infty} \|f_n - f\|_p = 0$ implies $\lim_{n\to\infty} \mu\{X : |f_n - f| \ge \varepsilon\} = 0$, that is, $(f_n : n \in \mathbb{N})$ converges to f in measure μ on X.

3. The convergence of $(f_n : n \in \mathbb{N})$ to f in measure implies the existence of a subsequence which converges to f a.e. on X according to Theorem 6.23 (Riesz). ∎

Observation 16.26. For $p = 1$, $\lim_{n\to\infty} \|f_n - f\|_1 = 0$ implies not only $\lim_{n\to\infty} \|f_n\|_1 = \|f\|_1$, that is, $\lim_{n\to\infty} \int_X |f_n|\, d\mu = \int_X |f|\, d\mu$, but also $\lim_{n\to\infty} \int_X f_n\, d\mu = \int_X f\, d\mu$. This is from the fact that $\left|\int_X f_n\, d\mu - \int_X f\, d\mu\right| \le \int_X |f_n - f|\, d\mu = \|f_n - f\|_1$.

Remark 16.27. (The notation is as in Theorem 16.25.) Let $(f_n : n \in \mathbb{N})$ be a sequence and f be an element in $L^p(X, \mathfrak{A}, \mu)$ where $p \in [1, \infty)$. Then

(a) $\lim_{n\to\infty} \|f_n - f\|_p = 0$ does not imply $\lim_{n\to\infty} f_n = f$ a.e. on X,

(b) $\lim_{n\to\infty} f_n = f$ a.e. on X does not imply $\lim_{n\to\infty} \|f_n - f\|_p = 0$.

See the following examples.

Example 1. Let $(X, \mathfrak{A}, \mu) = \big([0,1), \mathfrak{M}_L \cap [0,1), \mu_L\big)$ where $\mathfrak{M}_L \cap [0,1)$ is the σ-algebra of subsets of $[0,1)$ consisting of all members of $\mathfrak{M}_L$ contained in $[0,1)$. Let $\alpha_n = \big(1 + \frac{1}{2} + \cdots + \frac{1}{n}\big)_{mod 1}$ for $n \in \mathbb{N}$ and let

$$D_n = \begin{cases} [\alpha_n, \alpha_{n+1}) & \text{when } \alpha_n < \alpha_{n+1}, \\ [\alpha_n, 1) \cup [0, \alpha_{n+1}) & \text{when } \alpha_n > \alpha_{n+1}, \end{cases}$$

with $\mu_L(D_n) = 1/n$ for $n \in \mathbb{N}$. Define a sequence of functions $(f_n : n \in \mathbb{N})$ on $[0,1)$ by setting $f_n = \mathbf{1}_{D_n}$ for $n \in \mathbb{N}$ and let $f = 0$ on $[0,1)$. We showed in §6 that $(f_n : n \in \mathbb{N})$ converges to f on $[0,1)$ in measure μ_L but diverges at every point in $[0,1)$. For an arbitrary $p \in [1, \infty)$ we have

$$\|f_n - f\|_p = \Big[\int_{[0,1)} |f_n|^p \, d\mu_L\Big]^{1/p} = \Big(\frac{1^p}{n}\Big)^{1/p} = \Big(\frac{1}{n}\Big)^{1/p},$$

so that $\lim_{n\to\infty} \|f_n - f\|_p = 0$, but $\lim_{n\to\infty} f_n = f$ does not hold at any point in $[0,1)$.

Example 2. In $\big(\mathbb{R}, \mathfrak{M}_L, \mu_L\big)$, let $f_n = \mathbf{1}_{[n-1,n)}$ for $n \in \mathbb{N}$ and let $f = 0$ on $\mathbb{R}$. For an arbitrary $p \in [1, \infty)$, $\int_{\mathbb{R}} |f_n|^p \, d\mu_L = 1$ so that $f_n \in L^p\big(\mathbb{R}, \mathfrak{M}_L, \mu_L\big)$ for $n \in \mathbb{N}$ and $\int_{\mathbb{R}} |f|^p \, d\mu_L = 0$ so that $f \in L^p\big(\mathbb{R}, \mathfrak{M}_L, \mu_L\big)$. Now $\lim_{n\to\infty} f_n = f$ on $\mathbb{R}$. But

$$\|f_n - f\|_p = \Big[\int_{\mathbb{R}} |f_n|^p \, d\mu_L\Big]^{1/p} = 1$$

for every $n \in \mathbb{N}$ so that $\lim_{n\to\infty} \|f_n - f\|_p = 1 \neq 0$.

Theorem 16.28. *Given a measure space $(X, \mathfrak{A}, \mu)$. Let $p \in (0, \infty)$ and consider the linear space $L^p(X, \mathfrak{A}, \mu)$. Let $(f_n : n \in \mathbb{N})$ be a sequence and f be an element in $L^p(X, \mathfrak{A}, \mu)$. If $\lim_{n\to\infty} f_n = f$ a.e. on X and $\lim_{n\to\infty} \|f_n\|_p = \|f\|_p$, then $\lim_{n\to\infty} \|f_n - f\|_p = 0$.*

Proof. For $p \in (0, \infty)$ and $\alpha, \beta \in \mathbb{C}$, we have $|\alpha + \beta|^p \leq 2^p\{|\alpha|^p + |\beta|^p\}$ by Lemma 16.7. Since $|f_n|, |f| < \infty$ a.e. on X, we have $2^p\{|f_n|^p + |f|^p\} - |f_n - f|^p \geq 0$ a.e. on X for $n \in \mathbb{N}$. Then since $\lim_{n\to\infty} f_n = f$ a.e. on X we have

$$\lim_{n\to\infty} \big[2^p\{|f_n|^p + |f|^p\} - |f_n - f|^p\big] = 2^{p+1}|f|^p \quad \text{a.e. on } X.$$

Applying Fatou's Lemma (Theorem 8.13), we have

$$\begin{aligned}\int_X 2^{p+1}|f|^p\,d\mu &\le \liminf_{n\to\infty}\int_X \left[2^p\left\{|f_n|^p+|f|^p\right\}-|f_n-f|^p\right]d\mu\\ &=\liminf_{n\to\infty}\left\{\int_X 2^p|f_n|^p\,d\mu+\int_X 2^p|f|^p\,d\mu-\int_X|f_n-f|^p\,d\mu\right\}.\end{aligned}$$

Now for a convergent sequence of real numbers $(a_n : n\in\mathbb{N})$ and an arbitrary sequence of real numbers $(b_n : n\in\mathbb{N})$, we have $\liminf_{n\to\infty}\{a_n+b_n\}=\lim_{n\to\infty}a_n+\liminf_{n\to\infty}b_n$ and $\liminf_{n\to\infty}\{-b_n\}=-\limsup_{n\to\infty}b_n$. We have $\lim_{n\to\infty}\int_X|f_n|^p\,d\mu=\int_X|f|^p\,d\mu$. Thus

$$\int_X 2^{p+1}|f|^p\,d\mu\le\int_X 2^{p+1}|f|^p\,d\mu-\limsup_{n\to\infty}\int_X|f_n-f|^p\,d\mu.$$

Subtracting $\int_X 2^{p+1}|f|^p\,d\mu$ from both sides we have $\limsup_{n\to\infty}\int_X|f_n-f|^p\,d\mu\le 0$. Since $\limsup_{n\to\infty}\int_X|f_n-f|^p\,d\mu\ge 0$, we have $\limsup_{n\to\infty}\int_X|f_n-f|^p\,d\mu=0$. This then implies $\liminf_{n\to\infty}\int_X|f_n-f|^p\,d\mu=0$ also. Therefore $\lim_{n\to\infty}\|f_n-f\|_p=0$. ■

Remark 16.29. For the particular case that $p=1$ and $(f_n : n\in\mathbb{N})$ and f are nonnegative functions in $L^1(X,\mathfrak{A},\mu)$, Theorem 16.28 has a simple proof given below.

Proof. Assume that $\lim_{n\to\infty}f_n=f$ a.e. on X and $\lim_{n\to\infty}\|f_n\|_1=\|f\|_1$. Note that since f_n and f are nonnegative functions the last condition is equivalent to

(1) $$\lim_{n\to\infty}\int_X f_n\,d\mu=\int_X f\,d\mu.$$

For any $a,b\in\mathbb{R}$, $a\ge 0$ and $b\ge 0$, we have

(2) $$|a-b|=a+b-2\min\{a,b\}.$$

Let

$$g_n=\min\{f_n,f\}\quad\text{for } n\in\mathbb{N}.$$

The μ-integrability of f_n and f implies that $0\le f_n<\infty$ and $0\le f<\infty$ a.e. on X. Thus by (2) we have

$$|f_n-f|=f_n+f-2g_n\quad\text{a.e. on } X.$$

Integrating over X we have

(3) $$\int_X|f_n-f|\,d\mu=\int_X f_n\,d\mu+\int_X f\,d\mu-2\int_X g_n\,d\mu.$$

If $(c_n : n\in\mathbb{N})$ is a sequence in $\mathbb{R}$ converging to $c\in\mathbb{R}$ and if we let $\gamma_n=\min\{c_n,c\}$ then $\lim_{n\to\infty}\gamma_n=c$. (This follows from the fact that since γ_n is equal to either c_n or c we have $|\gamma_n-c|\le|c_n-c|$. Then $\lim_{n\to\infty}c_n=c$ implies $\lim_{n\to\infty}\gamma_n=c$.) Then since $\lim_{n\to\infty}f_n=f$ a.e.

on X we have $\lim_{n\to\infty} g_n = f$ a.e. on X. Since $0 \le g_n \le f$ on X and since f is μ-integrable on X, the Dominated Convergence Theorem (Theorem 9.20) applies so that

$$\lim_{n\to\infty}\int_X g_n\,d\mu = \int_X f\,d\mu. \tag{4}$$

If we let $n \to \infty$ in (3) then by (1) and (4) we have

$$\lim_{n\to\infty}\int_X |f_n - f|\,d\mu = 2\int_X f\,d\mu - 2\int_X f\,d\mu = 0. \ ■$$

We show next that the condition $\lim_{n\to\infty} f_n = f$ a.e. on X in Theorem 16.28 may be replaced by the condition that $(f_n : n \in \mathbb{N})$ converges to f in measure μ on X.

Theorem 16.30. *Let $(X, \mathfrak{A}, \mu)$ be a measure space. Let $p \in (0,\infty)$. Consider the linear space $L^p(X, \mathfrak{A}, \mu)$. Let $(f_n : n \in \mathbb{N})$ be a sequence and f be an element in $L^p(X, \mathfrak{A}, \mu)$. If $(f_n : n \in \mathbb{N})$ converges to f in measure μ on X and $\lim_{n\to\infty}\|f_n\|_p = \|f\|_p$, then $\lim_{n\to\infty}\|f_n - f\|_p = 0$.*

Proof. To prove $\lim_{n\to\infty}\|f_n - f\|_p = 0$, consider the sequence of real numbers $(a_n : n \in \mathbb{N})$ where $a_n = \|f_n - f\|_p$ for $n \in \mathbb{N}$. To show that $\lim_{n\to\infty} a_n = 0$, it suffices to show that an arbitrary subsequence $(a_{n_k} : k \in \mathbb{N})$ has a subsequence $(a_{n_{k_\ell}} : \ell \in \mathbb{N})$ such that $\lim_{\ell\to\infty} a_{n_{k_\ell}} = 0$. Let $(a_{n_k} : k \in \mathbb{N})$ be an arbitrary subsequence of $(a_n : n \in \mathbb{N})$. Since $(f_n : n \in \mathbb{N})$ converges to f in measure on X, the subsequence $(f_{n_k} : n \in \mathbb{N})$ converges to f in measure on X. Then by Theorem 6.23 (Riesz), there exists a subsequence $(f_{n_{k_\ell}} : \ell \in \mathbb{N})$ such that $\lim_{\ell\to\infty} f_{n_{k_\ell}} = f$ a.e. on X. Also $\lim_{n\to\infty}\|f_n\|_p = \|f\|_p$ implies that $\lim_{\ell\to\infty}\|f_{n_{k_\ell}}\|_p = \|f\|_p$. Thus the sequence $(f_{n_{k_\ell}} : \ell \in \mathbb{N})$ satisfies the conditions in Theorem 16.28 and therefore $\lim_{\ell\to\infty}\|f_{n_{k_\ell}} - f\|_p = 0$, that is, $\lim_{\ell\to\infty} a_{n_{k_\ell}} = 0$. ■

Definition 16.31. *Let $(X, \mathfrak{A}, \mu)$ be a measure space. Consider $L^p(X, \mathfrak{A}, \mu)$ where $p \in (1,\infty)$. Let $q \in (1,\infty)$ be the conjugate of p. We say that a sequence $(f_n : n \in \mathbb{N})$ in $L^p(X, \mathfrak{A}, \mu)$ converges weakly to an element f in $L^p(X, \mathfrak{A}, \mu)$ if*

$$\lim_{n\to\infty}\int_X f_n g\,d\mu = \int_X fg\,d\mu \quad \textit{for every } g \in L^q(X, \mathfrak{A}, \mu).$$

Theorem 16.32. *Let $(X, \mathfrak{A}, \mu)$ be a measure space. Consider $L^p(X, \mathfrak{A}, \mu)$ where $p \in (1,\infty)$. Let $q \in (1,\infty)$ be the conjugate of p. If a sequence $(f_n : n \in \mathbb{N})$ in $L^p(X, \mathfrak{A}, \mu)$ converges to an element f in $L^p(X, \mathfrak{A}, \mu)$ in the norm then $(f_n : n \in \mathbb{N})$ converges weakly to f.*

Proof. By Hölder's Inequality (Theorem 16.14) we have

$$\left|\int_X f_n g\,d\mu - \int_X fg\,d\mu\right| \le \int_X |f_n - f||g|\,d\mu \le \|f_n - f\|_p\|g\|_q.$$

Then $\lim_{n\to\infty} \|f_n - f\|_p = 0$ implies $\lim_{n\to\infty} \left| \int_X f_n g \, d\mu - \int_X fg \, d\mu \right| = 0$ and therefore we have $\lim_{n\to\infty} \int_X f_n g \, d\mu = \int_X fg \, d\mu$. ∎

Theorem 16.33. *Let $(X, \mathfrak{A}, \mu)$ be a σ-finite measure space. Consider $L^p(X, \mathfrak{A}, \mu)$ where $p \in (1, \infty)$. Let $q \in (1, \infty)$ be the conjugate of p. Let $(f_n : n \in \mathbb{N})$ be a sequence in $L^p(X, \mathfrak{A}, \mu)$ and let f be an element in $L^p(X, \mathfrak{A}, \mu)$. Suppose $(f_n : n \in \mathbb{N})$ satisfies the following conditions:*
1° $\lim_{n\to\infty} f_n = f$ *a.e. on* X.
2° $\|f_n\|_p \le B$ *for* $n \in \mathbb{N}$ *where* $B > 0$.
Then $(f_n : n \in \mathbb{N})$ converges weakly to f in $L^p(X, \mathfrak{A}, \mu)$.

Proof. We are to show that for every $g \in L^q(X, \mathfrak{A}, \mu)$ we have

(1)
$$\lim_{n\to\infty} \int_X f_n g \, d\mu = \int_X fg \, d\mu.$$

Let us show first that $\|f\|_p \le B$ also. Now by Fatou's Lemma (Theorem 8.13), we have

$$\begin{aligned} \|f\|_p^p &= \int_X |f|^p \, d\mu = \int_X \lim_{n\to\infty} |f_n|^p \, d\mu \\ &\le \liminf_{n\to\infty} \int_X |f_n|^p \, d\mu \le B^p. \end{aligned}$$

Thus $\|f\|_p \le B$. Then for any $A \in \mathfrak{A}$ we have

(2)
$$\begin{aligned} \left\{ \int_A |f_n - f|^p \, d\mu \right\}^{1/p} &\le \left\{ \int_X |f_n - f|^p \, d\mu \right\}^{1/p} = \|f_n - f\|_p \\ &\le \|f_n\|_p + \|f\|_p \le 2B. \end{aligned}$$

Let $g \in L^q(X, \mathfrak{A}, \mu)$ be arbitrarily chosen and let $\varepsilon > 0$. Since $\int_X |g|^q \, d\mu < \infty$, by Theorem 8.20 there exists $\delta > 0$ such that for every $E \in \mathfrak{A}$ with $\mu(E) < \delta$ we have

(3)
$$\int_E |g|^q \, d\mu < \left(\frac{\varepsilon}{6B}\right)^q.$$

Since $(X, \mathfrak{A}, \mu)$ is a σ-finite measure space there exists an increasing sequence $(F_k : k \in \mathbb{N})$ in $\mathfrak{A}$ such that $\lim_{k\to\infty} F_k = \bigcup_{k\in\mathbb{N}} F_k = X$ and $\mu(F_k) < \infty$ for $k \in \mathbb{N}$. By the Monotone Convergence Theorem (Theorem 8.5) we have

$$\lim_{k\to\infty} \int_{F_k} |g|^q \, d\mu = \lim_{k\to\infty} \int_X \mathbf{1}_{F_k} |g|^q \, d\mu = \int_X |g|^q \, d\mu < \infty.$$

Thus for sufficiently large $k_0 \in \mathbb{N}$ we have

$$\int_X |g|^q \, d\mu - \int_{F_{k_0}} |g|^q \, d\mu < \left(\frac{\varepsilon}{6B}\right)^q.$$

Let $F = F_{k_0}$. Then $\mu(F) < \infty$ and

(4) $$\int_{F^c} |g|^q \, d\mu < \left(\frac{\varepsilon}{6B}\right)^q.$$

Now since $\lim_{n\to\infty} f_n = f$ a.e. on X, we have $\lim_{n\to\infty} f_n = f$ a.e. on F. Since $\mu(F) < \infty$, by Egoroff's Theorem (Theorem 6.12) there exists $G \in \mathfrak{A}$ such that $G \subset F$, $\mu(F \setminus G) < \delta$ and $\lim_{n\to\infty} f_n = f$ uniformly on G. Now if $\|g\|_q = 0$ then $g = 0$ a.e. on X so that (1) holds trivially. Thus let us consider the case $\|g\|_q > 0$. Then since $\lim_{n\to\infty} f_n = f$ uniformly on G there exists $N \in \mathbb{N}$ such that for $n \geq N$ we have

$$|f_n(x) - f(x)| < \frac{1}{\mu(G)^{1/p}\|g\|_q}\frac{\varepsilon}{3} \quad \text{for all } x \in G.$$

Then for $n \geq N$ we have

(5) $$\left\{\int_G |f_n - f|^p \, d\mu\right\}^{1/p} < \left\{\frac{1}{\mu(G)\|g\|_q^p}\left(\frac{\varepsilon}{3}\right)^p \mu(G)\right\}^{1/p} = \frac{1}{\|g\|_q}\frac{\varepsilon}{3}.$$

Now $F \setminus G$, G, and F^c are disjoint and their union is equal to X. Thus for $n \geq N$ we have

(6) $$\begin{aligned}\left|\int_X f_n g \, d\mu - \int_X fg \, d\mu\right| &\leq \int_X |f_n - f||g| \, d\mu = \int_{F\setminus G} |f_n - f||g| \, d\mu \\ &\quad + \int_G |f_n - f||g| \, d\mu + \int_{F^c} |f_n - f||g| \, d\mu.\end{aligned}$$

Let us estimate the last three integrals by applying Hölder's Inequality (Theorem 16.14). For the first of the three integrals we have

(7) $$\begin{aligned}\int_{F\setminus G} |f_n - f||g| \, d\mu &\leq \left\{\int_{F\setminus G} |f_n - f|^p \, d\mu\right\}^{1/p}\left\{\int_{F\setminus G} |g|^q \, d\mu\right\}^{1/q} \\ &\leq 2B\left\{\int_{F\setminus G} |g|^q \, d\mu\right\}^{1/q} < 2B\frac{\varepsilon}{6B} = \frac{\varepsilon}{3},\end{aligned}$$

where the second inequality is by (2) and the third inequality is by (3). For the second integral we have

(8) $$\begin{aligned}\int_G |f_n - f||g| \, d\mu &\leq \left\{\int_G |f_n - f|^p \, d\mu\right\}^{1/p}\left\{\int_G |g|^q \, d\mu\right\}^{1/q} \\ &< \frac{1}{\|g\|_q}\frac{\varepsilon}{3}\left\{\int_G |g|^q \, d\mu\right\}^{1/q} \leq \frac{1}{\|g\|_q}\frac{\varepsilon}{3}\|g\|_q \\ &= \frac{\varepsilon}{3},\end{aligned}$$

where the second inequality is by (5). Finally for the third integral we have

(9) $$\begin{aligned}\int_{F^c} |f_n - f||g| \, d\mu &\leq \left\{\int_{F^c} |f_n - f|^p \, d\mu\right\}^{1/p}\left\{\int_{F^c} |g|^q \, d\mu\right\}^{1/q} \\ &\leq 2B\left\{\int_{F^c} |g|^q \, d\mu\right\}^{1/q} < 2B\frac{\varepsilon}{6B} = \frac{\varepsilon}{3},\end{aligned}$$

where the second inequality is by (2) and the third inequality is by (4). Substituting (7), (8), and (9) in (6), we have for $n \geq N$

$$\left| \int_X f_n g \, d\mu - \int_X f g \, d\mu \right| < \frac{1}{3} + \frac{1}{3} + \frac{1}{3} = \varepsilon.$$

This proves (1). ∎

Corollary 16.34. *The condition $\lim_{n \to \infty} f_n = f$ a.e. on X in Theorem 16.33 can be replaced by the condition that $(f_n : n \in \mathbb{N})$ converges to f in measure μ on X.*

Proof. Suppose there exists $g \in L^q(X, \mathfrak{A}, \mu)$ such that the sequence $\left(\int_X f_n g \, d\mu : n \in \mathbb{N} \right)$ does not converge to $\int_X f g \, d\mu$. Then we have

$$\limsup_{n \to \infty} \left| \int_X f_n g \, d\mu - \int_X f g \, d\mu \right| = \alpha > 0.$$

Then there exists a subsequence $(n_k : k \in \mathbb{N})$ of $(n : n \in \mathbb{N})$ such that

$$(1) \qquad \lim_{k \to \infty} \left| \int_X f_{n_k} g \, d\mu - \int_X f g \, d\mu \right| = \alpha.$$

Since $(f_n : n \in \mathbb{N})$ converges to f in measure on X, the subsequence $(f_{n_k} : n \in \mathbb{N})$ converges to f in measure on X. Then by Theorem 6.24 (Riesz) there exists a subsequence $(f_{n_{k_\ell}} : n \in \mathbb{N})$ of $(f_{n_k} : n \in \mathbb{N})$ which converges to f a.e. on X. Then by Theorem 16.33 we have

$$\lim_{\ell \to \infty} \int_X f_{n_{k_\ell}} g \, d\mu = \int_X f g \, d\mu,$$

which contradicts (1). ∎

[IV] The Space L^∞

Definition 16.35. *Given a measure space $(X, \mathfrak{A}, \mu)$. Let f be an extended complex-valued $\mathfrak{A}$-measurable function on X. We say that f is essentially bounded on X if there exists a constant $M \in [0, \infty)$ such that $\mu\{x \in X : |f(x)| > M\} = 0$, that is, $|f| \leq M$ a.e. on X; in other words, there exists a null set E in $(X, \mathfrak{A}, \mu)$ such that $|f| \leq M$ on $X \setminus E$. Such a constant M is called an essential bound for f on X.*

Definition 16.36. *Given a measure space $(X, \mathfrak{A}, \mu)$. Let f be an extended complex-valued $\mathfrak{A}$-measurable function on X. We define the essential supremum of f on X, $\|f\|_\infty$, as the infimum of the set of all essential bounds for f on X, that is,*

$$\|f\|_\infty = \inf \left\{ M \in [0, \infty) : \mu\{x \in X : |f(x)| > M\} = 0 \right\}.$$

If f has no essential bound on X, that is, if the set of all essential bounds of f on X is an empty set then we set $\|f\|_\infty = \infty$. (This is consistent with the convention that the infimum of an empty subset of $\mathbb{R}$ is equal to ∞. Recall that the smaller the set the greater

the infimum over the set.)
Thus $\|f\|_\infty$ *is defined for every extended complex-valued* $\mathfrak{A}$*-measurable function* f *on* X*;* $\|f\|_\infty \in [0, \infty]$*; and* $\|f\|_\infty = \infty$ *if and only if* f *does not have an essential bound. We call* $\|f\|_\infty$ *the essential supremum of* f*. An alternate notation for* $\|f\|_\infty$ *is* $\text{ess.}\sup_{x \in X} |f(x)|$.

Lemma 16.37. *Given a measure space* $(X, \mathfrak{A}, \mu)$*. Let* f *be an extended complex-valued* $\mathfrak{A}$*-measurable function on* X*.*
(a) f *is essentially bounded on* X *if and only if* $\|f\|_\infty < \infty$*.*
(b) $|f| \le \|f\|_\infty$ *a.e. on* X*.*
(c) *If* $\|f\|_\infty < \infty$*, then* f *is essentially bounded on* X *and* $\|f\|_\infty$ *is an essential bound for* f *on* X*.*

Proof. 1. To prove (a), note that if f is essentially bounded on X, then the set of all its essential bounds is a non-empty set of nonnegative real numbers and thus its infimum $\|f\|_\infty$ is finite. Conversely if f is not essentially bounded on X, then $\|f\|_\infty = \infty$ by definition.

2. To prove (b), note that if $\|f\|_\infty = \infty$ then $|f| \le \|f\|_\infty$ on X trivially. If $\|f\|_\infty < \infty$, then f is essentially bounded on X by (a) so that the set of essential bounds of f is a non-empty subset of $[0, \infty)$ and $\|f\|_\infty$ is the infimum of this set. Then for every $k \in \mathbb{N}$, there exists an essential bound M_k of f such that $M_k < \|f\|_\infty + \frac{1}{k}$. Since M_k is an essential bound for f, we have $\mu\{x \in X : |f(x)| > M_k\} = 0$. Since

$$\{x \in X : |f(x)| > \|f\|_\infty\} = \bigcup_{k \in \mathbb{N}} \left\{x \in X : |f(x)| > \|f\|_\infty + \tfrac{1}{k}\right\},$$

we have

$$\mu\{x \in X : |f(x)| > \|f\|_\infty\} \le \sum_{k \in \mathbb{N}} \mu \left\{x \in X : |f(x)| > \|f\|_\infty + \tfrac{1}{k}\right\}.$$

But $\|f\|_\infty + \frac{1}{k} > M_k$ implies that $\left\{x \in X : |f(x)| > \|f\|_\infty + \frac{1}{k}\right\} \subset \left\{x \in X : |f(x)| > M_k\right\}$ so that

$$\mu \left\{x \in X : |f(x)| > \|f\|_\infty + \tfrac{1}{k}\right\} \le \mu\{x \in X : |f(x)| > M_k\} = 0.$$

Thus $\mu\{x \in X : |f(x)| > \|f\|_\infty\} = 0$. This shows that $|f| \le \|f\|_\infty$ a.e. on X.

3. If $\|f\|_\infty < \infty$, then $\|f\|_\infty$ is an essential bound for f on X by (b). ∎

Definition 16.38. *Given a measure space* $(X, \mathfrak{A}, \mu)$*. Let* $\mathcal{L}^\infty(X, \mathfrak{A}, \mu)$ *be the collection of all essentially bounded extended complex-valued* $\mathfrak{A}$*-measurable functions on* X*, that is,* $\mathcal{L}^\infty(X, \mathfrak{A}, \mu)$ *is the collection of all extended complex-valued* $\mathfrak{A}$*-measurable functions* f *on* X *with* $\|f\|_\infty < \infty$*.*

Proposition 16.39. $\mathcal{L}^\infty(X, \mathfrak{A}, \mu)$ *is a linear space over the field of scalars* $\mathbb{C}$*.*

Proof. It is immediate that if f and g are essentially bounded extended complex-valued $\mathfrak{A}$-measurable functions on X, then so are $f + g$ and αf for $\alpha \in \mathbb{C}$. ∎

Theorem 16.40. (Hölder's Inequality for $p = 1$ and $q = \infty$) *Given a measure space $(X, \mathfrak{A}, \mu)$. Let f and g be two extended complex-valued $\mathfrak{A}$-measurable functions on X such that $|f|, |g| < \infty$ a.e. on X.*

(a) *Excepting the case that one of $\|f\|_1$ and $\|g\|_\infty$ is equal to 0 and the other is equal to ∞, we have*

$$(1) \qquad \|fg\|_1 \le \|f\|_1 \|g\|_\infty.$$

(b) *When $\|f\|_1, \|g\|_\infty < \infty$, then the equality in (1) holds if and only if*

$$(2) \qquad |g| = \|g\|_\infty \text{ a.e. on } \{x \in X : f(x) \neq 0\}.$$

Proof. If one of $\|f\|_1$ and $\|g\|_\infty$ is equal to 0 and the other is equal to ∞, then $\|f\|_1\|g\|_\infty$ is undefined. Thus let us exclude this case from our consideration.

1. If $\|g\|_\infty = \infty$ then $\|f\|_1 > 0$ so that $\|f\|_1\|g\|_\infty = \infty$ and (1) holds trivially. Consider the case $\|g\|_\infty < \infty$. Then by Lemma 16.37, g is essentially bounded on X and $\|g\|_\infty$ is an essential bound for f on X, that is, $|g| \le \|g\|_\infty$ a.e. on X. This implies that $|fg| = |f||g| \le \|g\|_\infty |f|$ a.e. on X and then $\int_X |fg|\, d\mu \le \|g\|_\infty \int_X |f|\, d\mu$, probing (1).

2. To prove (b) let us assume that $\|f\|_1, \|g\|_\infty < \infty$. Let $E = \{x \in X : f(x) \neq 0\}$. Now if $|g| = \|g\|_\infty$ a.e. on E, then $|fg| = \|g\|_\infty |f|$ a.e. on E and thus

$$\int_X |fg|\, d\mu = \int_E |fg|\, d\mu = \|g\|_\infty \int_E |f|\, d\mu = \|g\|_\infty \int_X |f|\, d\mu,$$

that is, $\|fg\|_1 = \|f\|_1\|g\|_\infty$. Conversely suppose $\|fg\|_1 = \|f\|_1\|g\|_\infty$. Then

$$\int_X |fg|\, d\mu = \|f\|_1\|g\|_\infty = \|g\|_\infty \int_X |f|\, d\mu = \int_X \|g\|_\infty |f|\, d\mu.$$

Since $0 \le |fg| \le \|g\|_\infty |f|$ a.e. on X, the equality $\int_X |fg|\, d\mu = \int_X \|g\|_\infty |f|\, d\mu < \infty$ implies that $|fg| = \|g\|_\infty |f|$ a.e. on X. On E, dividing the last equality by $|f| \neq 0$ we have $|g| = \|g\|_\infty$ a.e. on E. ∎

Theorem 16.41. (Minkowski's Inequality for $p = \infty$) *Let $(X, \mathfrak{A}, \mu)$ be a measure space. Let f and g be extended complex-valued $\mathfrak{A}$-measurable functions on X such that $|f|, |g| < \infty$ a.e. on X. Then $\|f + g\|_\infty \le \|f\|_\infty + \|g\|_\infty$.*

Proof. If at least one of $\|f\|_\infty$ and $\|g\|_\infty$ is equal to ∞, then Minkowski's inequality holds trivially. Thus assume that $\|f\|_\infty, \|g\|_\infty < \infty$. Then by (b) of Lemma 16.37 we have $|f + g| \le |f| + |g| \le \|f\|_\infty + \|g\|_\infty$ a.e. on X. Since $\|f\|_\infty + \|g\|_\infty < \infty$, $\|f\|_\infty + \|g\|_\infty$ is an essential bound for $f + g$. Then since $\|f + g\|_\infty$ is the infimum of the set of all essential bounds of $f + g$, we have $\|f + g\|_\infty \le \|f\|_\infty + \|g\|_\infty$. ∎

Definition 16.42. *Given a measure space $(X, \mathfrak{A}, \mu)$. Let $L^\infty(X, \mathfrak{A}, \mu)$ be the collection of the equivalence classes of members of $\mathcal{L}^\infty(X, \mathfrak{A}, \mu)$ with respect to the equivalence relation of a.e. equality on X.*

Theorem 16.43. *The essential supremum $\|\cdot\|_\infty$ is a norm on the linear space $L^\infty(X, \mathfrak{A}, \mu)$ and $L^\infty(X, \mathfrak{A}, \mu)$ is a Banach space with respect to this norm. Indeed if $(f_n : n \in \mathbb{N})$ is a Cauchy sequence in $L^\infty(X, \mathfrak{A}, \mu)$ with respect to the norm $\|\cdot\|_\infty$, then there exists $f \in L^\infty(X, \mathfrak{A}, \mu)$ such that $\lim_{n\to\infty} \|f_n - f\|_\infty = 0$ and furthermore $\lim_{n\to\infty} f_n = f$ uniformly on $X \setminus E$ where E is a null set in $(X, \mathfrak{A}, \mu)$.*

Proof. The fact that $\|\cdot\|_\infty$ is a norm on $L^\infty(X, \mathfrak{A}, \mu)$ is immediate. In particular, $\|\cdot\|_\infty$ satisfies the triangle inequality by Theorem 16.41.

To show that $L^\infty(X, \mathfrak{A}, \mu)$ is a complete normed linear space with respect to the norm $\|\cdot\|_\infty$, let $(f_n : n \in \mathbb{N})$ be a Cauchy sequence in $L^\infty(X, \mathfrak{A}, \mu)$ with respect to this norm. Now for any $m, n \in \mathbb{N}$, $f_m - f_n$ is an element in $L^\infty(X, \mathfrak{A}, \mu)$ with $\|f_m - f_n\|_\infty$ as an essential bound, that is, $|f_m - f_n| \leq \|f_m - f_n\|_\infty$ a.e. on X. Thus there exists a null set $E_{m,n}$ in $(X, \mathfrak{A}, \mu)$ such that $|f_m - f_n| \leq \|f_m - f_n\|_\infty$ on $X \setminus E_{m,n}$. If we let $E = \bigcup_{m,n\in\mathbb{N}} E_{m,n}$, then E is a null set in $(X, \mathfrak{A}, \mu)$ and

(1) $$|f_m - f_n| \leq \|f_m - f_n\|_\infty \quad \text{on} X \setminus E \text{ for all } m, n \in \mathbb{N}.$$

Since $(f_n : n \in \mathbb{N})$ is a Cauchy sequence, for every $k \in \mathbb{N}$, there exists $N_k \in \mathbb{N}$ such that

(2) $$\|f_m - f_n\|_\infty < \tfrac{1}{k} \quad \text{for } m, n \geq N_k.$$

By (1) and (2), we have

(3) $$|f_m - f_n| < \tfrac{1}{k} \quad \text{on} X \setminus E \text{ for } m, n \geq N_k.$$

This shows that $(f_n : n \in \mathbb{N})$ converges uniformly on $X \setminus E$. Let

$$f(x) = \begin{cases} \lim_{n\to\infty} f_n(x) & \text{for } x \in X \setminus E, \\ 0 & \text{for } x \in E. \end{cases}$$

It remains to show that $f \in L^\infty(X, \mathfrak{A}, \mu)$ and $\lim_{n\to\infty} \|f_n - f\|_\infty = 0$. Now by (3) and by the fact that $\lim_{m\to\infty} f_m = f$ on $X \setminus E$, we have

(4) $$|f - f_n| \leq \tfrac{1}{k} \quad \text{on} X \setminus E \text{ for } n \geq N_k.$$

From (4) we have $|f| \leq |f_n| + \frac{1}{k}$ on $X \setminus E$ for $n \geq N_k$. Thus $|f| \leq \|f_n\|_\infty + \frac{1}{k}$ a.e. on X for $n \geq N_k$. This shows that f is essentially bounded on X and thus $f \in L^\infty(X, \mathfrak{A}, \mu)$. From (4), we have also $\|f - f_n\|_\infty \leq \frac{1}{k}$ for $n \geq N_k$. This shows that $\lim_{n\to\infty} \|f_n - f\|_\infty = 0$. ∎

Theorem 16.44. *Given a measure space $(X, \mathfrak{A}, \mu)$. Let $(f_n : n \in \mathbb{N})$ be a sequence and f be an element in $L^\infty(X, \mathfrak{A}, \mu)$. If $\lim_{n\to\infty} \|f_n - f\|_\infty = 0$, then*

1° $\lim_{n\to\infty} \|f_n\|_\infty = \|f\|_\infty$,

2° $\lim_{n\to\infty} f_n = f$ *uniformly on $X \setminus E$ where E is a null set in $(X, \mathfrak{A}, \mu)$,*

3° *$(f_n : n \in \mathbb{N})$ converges to f on X in measure μ.*

Proof. Since $\|\cdot\|_\infty$ is a norm we have $\big|\|f_n\|_\infty - \|f\|_\infty\big| \le \|f_n - f\|_\infty$ by Observation 15.6. Then $\lim_{n\to\infty} \|f_n - f\|_\infty = 0$ implies $\lim_{n\to\infty} \|f_n\|_\infty = \|f\|_\infty$. This proves 1°. 2° follows from Theorem 16.43. To prove 3°, note that by 2° for every $\varepsilon > 0$ there exists $N \in \mathbb{N}$ such that $|f_n - f| < \varepsilon$ on $X \setminus E$ for $n \ge N$. Then $\mu\{X : |f_n - f| \ge \varepsilon\} = 0$ for $n \ge N$. This shows that $\lim_{n\to\infty} \mu\{X : |f_n - f| \ge \varepsilon\} = 0$ for every $\varepsilon > 0$, that is, $(f_n : n \in \mathbb{N})$ converges to f on X in measure μ. ■

Remark 16.45. Let $(f_n : n \in \mathbb{N})$ be a sequence and f be an element in $f \in L^\infty(X, \mathfrak{A}, \mu)$. The condition $\lim_{n\to\infty} f_n = f$ a.e. on X and the condition $\lim_{n\to\infty} \|f_n\|_\infty = \|f\|_\infty$ together do not imply $\lim_{n\to\infty} \|f_n - f\|_\infty = 0$, that is, Theorem 16.28 does not hold for $L^\infty(X, \mathfrak{A}, \mu)$.

Example. In $(\mathbb{R}, \mathfrak{M}_L, \mu_L)$, let

$$f_n(x) = \begin{cases} 0 & \text{for } x \in (-\infty, 0), \\ nx & \text{for } x \in \left[0, \frac{1}{n}\right], \\ 1 & \text{for } x \in \left(\frac{1}{n}, \infty\right), \end{cases}$$

for $n \in \mathbb{N}$ and let $f = \mathbf{1}_{[0,\infty)}$. Then $\lim_{n\to\infty} f_n = f$ on $\mathbb{R}$, $\|f_n\|_\infty = 1$ for $n \in \mathbb{N}$, $\|f\|_\infty = 1$ so that $\lim_{n\to\infty} \|f_n\|_\infty = \|f\|_\infty$, but $\|f_n - f\|_\infty = 1$ for every $n \in \mathbb{N}$ and thus $\lim_{n\to\infty} \|f_n - f\|_\infty = 1 \ne 0$.

Definition 16.46. *Let $(X, \mathfrak{A}, \mu)$ be an arbitrary measure space.*
(a) *Let $(f_n : n \in \mathbb{N})$ be a sequence and f be an element in $L^1(X, \mathfrak{A}, \mu)$. We say that the sequence $(f_n : n \in \mathbb{N})$ converges to f weakly in $L^1(X, \mathfrak{A}, \mu)$ if we have*

$$\lim_{n\to\infty} \int_X f_n g \, d\mu = \int_X fg \, d\mu \quad \text{for every } g \in L^\infty(X, \mathfrak{A}, \mu).$$

(b) *Let $(f_n : n \in \mathbb{N})$ be a sequence and f be an element in $L^\infty(X, \mathfrak{A}, \mu)$. We say that the sequence $(f_n : n \in \mathbb{N})$ converges to f weakly in $L^\infty(X, \mathfrak{A}, \mu)$ if we have*

$$\lim_{n\to\infty} \int_X f_n g \, d\mu = \int_X fg \, d\mu \quad \text{for every } g \in L^1(X, \mathfrak{A}, \mu).$$

Theorem 16.47. *Let $(X, \mathfrak{A}, \mu)$ be an arbitrary measure space.*
(a) *Let $(f_n : n \in \mathbb{N})$ be a sequence and f be an element in $L^1(X, \mathfrak{A}, \mu)$. If $\lim_{n\to\infty} \|f_n - f\|_1 = 0$, then $(f_n : n \in \mathbb{N})$ converges to f weakly in $L^1(X, \mathfrak{A}, \mu)$.*
(b) *Let $(f_n : n \in \mathbb{N})$ be a sequence and f be an element in $L^\infty(X, \mathfrak{A}, \mu)$. If $\lim_{n\to\infty} \|f_n - f\|_\infty = 0$, then $(f_n : n \in \mathbb{N})$ converges to f weakly in $L^\infty(X, \mathfrak{A}, \mu)$.*

Proof. To prove (a) suppose $\lim_{n\to\infty} \|f_n - f\|_1 = 0$. Let $g \in L^\infty(X, \mathfrak{A}, \mu)$. Now we have

$$\left| \int_X f_n g \, d\mu - \int_X fg \, d\mu \right| \le \int_X |(f_n - f)g| \, d\mu = \|(f_n - f)g\|_1 \le \|f_n - f\|_1 \|g\|_\infty$$

by Hölder's Inequality (Theorem 16.40). Thus the assumption $\lim_{n\to\infty} \|f_n - f\|_1 = 0$ implies

$$\lim_{n\to\infty} \left| \int_X f_n g \, d\mu - \int_X f g \, d\mu \right| = 0, \text{ that is, } \lim_{n\to\infty} \int_X f_n g \, d\mu = \int_X f g \, d\mu.$$

This proves (a). (b) is proved in the same way using Theorem 16.40 again. ∎

Remark 16.48. Theorem 16.33 which is valid for $p \in (1, \infty)$ does not hold for $p = 1$, that is, if $(f_n : n \in \mathbb{N})$ is a sequence and f is an element in $L^1(X, \mathfrak{A}, \mu)$ such that $\lim_{n\to\infty} f_n = f$ a.e. on X and $\|f_n\|_1 \le B$ for all $n \in \mathbb{N}$ for some $B > 0$, the sequence $(f_n : n \in \mathbb{N})$ need not converge to f weakly in $L^1(X, \mathfrak{A}, \mu)$. See the Example below.

Example. Consider $L^1\big((0,1), \mathfrak{M}_L \cap (0,1), \mu_L\big)$ and $L^\infty\big((0,1), \mathfrak{M}_L \cap (0,1), \mu_L\big)$. Let $f_n = n\mathbf{1}_{(0,1/n)}$ for $n \in \mathbb{N}$ and $f = 0$ on $(0,1)$. Now $\|f_n\|_1 = \int_{(0,1)} n\mathbf{1}_{(0,1/n)} \, d\mu_L = 1$ for every $n \in \mathbb{N}$ so that $(f_n : n \in \mathbb{N})$ is a sequence in $L^1\big((0,1), \mathfrak{M}_L \cap (0,1), \mu_L\big)$ and $\|f_n\|_1 = 1$ for every $n \in \mathbb{N}$. Also $f \in L^1\big((0,1), \mathfrak{M}_L \cap (0,1), \mu_L\big)$ and $\lim_{n\to\infty} f_n = f$ everywhere on $(0,1)$.

Let $g = 1$ on $(0,1)$. Then $g \in L^\infty\big((0,1), \mathfrak{M}_L \cap (0,1), \mu_L\big)$. Now for every $n \in \mathbb{N}$ we have $\int_{(0,1)} f_n g \, d\mu_L = \int_{(0,1)} f_n \, d\mu_L = 1$ while $\int_{(0,1)} f g \, d\mu_L = \int_{(0,1)} 0 \, d\mu_L = 0$ so that $\lim_{n\to\infty} \int_{(0,1)} f_n g \, d\mu_L = \int_{(0,1)} f g \, d\mu_L$ does not hold. Thus $(f_n : n \in \mathbb{N})$ does not converge to f weakly in $L^1(X, \mathfrak{A}, \mu)$.

Theorem 16.49. *Let $(X, \mathfrak{A}, \mu)$ be an arbitrary measure space. Let $(f_n : n \in \mathbb{N})$ be a sequence and f be an element in $L^\infty(X, \mathfrak{A}, \mu)$. If $\lim_{n\to\infty} f_n = f$ a.e. on X and if $\|f_n\|_\infty \le B$ for all $n \in \mathbb{N}$ for some $B \ge 0$ then the sequence $(f_n : n \in \mathbb{N})$ converges to f weakly in $L^\infty(X, \mathfrak{A}, \mu)$.*

Proof. Let $g \in L^1(X, \mathfrak{A}, \mu)$. Then $\int_X B|g| \, d\mu < \infty$, that is, $B|g|$ is μ-integrable on X. Now $\lim_{n\to\infty} f_n g = fg$ a.e. on X and $|f_n g| \le B|g|$ a.e. on X for $n \in \mathbb{N}$. Thus by the Dominated Convergence Theorem (Theorem 9.20) we have

$$\lim_{n\to\infty} \int_X f_n g \, d\mu = \int_X f g \, d\mu.$$

This shows that the sequence $(f_n : n \in \mathbb{N})$ converges to f weakly in $L^\infty(X, \mathfrak{A}, \mu)$. ∎

Given a measure space $(X, \mathfrak{A}, \mu)$. If $\mu(X) = \infty$ then for some $f \in L^\infty(X, \mathfrak{A}, \mu)$ we may have $\|f\|_p = \infty$ for every $p \in (0, \infty)$. An example of such a function is given by $f = 1$ on X. We show below that for an arbitrary measure space $(X, \mathfrak{A}, \mu)$, if $f \in L^\infty(X, \mathfrak{A}, \mu)$ and if $\|f\|_{p_0} < \infty$ for some $p_0 \in (0, \infty)$ then $\|f\|_p < \infty$ for every $p \ge p_0$ and $\lim_{p\to\infty} \|f\|_p = \|f\|_\infty$.

Theorem 16.50. *Let $(X, \mathfrak{A}, \mu)$ be an arbitrary measure space. Let $f \in L^\infty(X, \mathfrak{A}, \mu)$ be such that $f \in L^{p_0}(X, \mathfrak{A}, \mu)$ for some $p_0 \in (0, \infty)$. Then $f \in L^p(X, \mathfrak{A}, \mu)$ for every $p \ge p_0$ and moreover $\lim_{p\to\infty} \|f\|_p = \|f\|_\infty$.*

Proof. 1. For $p > p_0$ we have $p - p_0 > 0$. Then $|f|^p = |f|^{p-p_0}|f|^{p_0} \le \|f\|_\infty^{p-p_0}|f|^{p_0}$ a.e. on X so that

$$\int_X |f|^p \, d\mu \le \|f\|_\infty^{p-p_0} \int_X |f|^{p_0} \, d\mu < \infty$$

since $\|f\|_\infty$ and $\|f\|_{p_0} < \infty$. This shows that $f \in L^p(X, \mathfrak{A}, \mu)$ for $p > p_0$.

2. Let $A = \{X : |f| \in [1, \infty]\}$, $B = \{X : |f| \in (0, 1)\}$, and $C = \{X : |f| = 0\}$. Then $\{A, B, C\}$ is a disjoint collection in $\mathfrak{A}$ and $A \cup B \cup C = X$. Then we have

$$\mu(A) = 1^{p_0}\mu(A) \le \int_A |f|^{p_0} \, d\mu \le \int_X |f|^{p_0} \, d\mu < \infty. \tag{1}$$

On B we have $|f| \in (0, 1)$. Thus for $k \in \mathbb{N}$, $k \ge 2$ let $B_k = \{B : |f| \in [\frac{1}{k}, 1)\}$. Then $(B_k : k \ge 2)$ is an increasing sequence in $\mathfrak{A}$ with $\lim_{k\to\infty} B_k = \bigcup_{k\ge 2} B_k = B$. We have

$$\frac{1}{k^{p_0}}\mu(B_k) \le \int_{B_k} |f|^{p_0} \, d\mu \le \int_X |f|^{p_0} \, d\mu < \infty.$$

Thus we have

$$\mu(B_k) < \infty \quad \text{for } k \ge 2. \tag{2}$$

Since $(B_k : k \ge 2)$ is an increasing sequence with $\lim_{k\to\infty} B_k = B$ we have $\lim_{k\to\infty} \mathbf{1}_{B_k} = \mathbf{1}_B$. Thus by the Monotone Convergence Theorem (Theorem 8.5) we have

$$\begin{aligned}\lim_{k\to\infty} \int_{B_k} |f|^{p_0} \, d\mu &= \lim_{k\to\infty} \int_X \mathbf{1}_{B_k} \cdot |f|^{p_0} \, d\mu = \int_X \lim_{k\to\infty} \mathbf{1}_{B_k} \cdot |f|^{p_0} \, d\mu \\ &= \int_X \mathbf{1}_B \cdot |f|^{p_0} \, d\mu = \int_B |f|^{p_0} \, d\mu.\end{aligned}$$

Let $\varepsilon > 0$ be arbitrarily given. Then the convergence above implies that there exists $k_0 \in \mathbb{N}$ such that

$$\int_{B \setminus B_{k_0}} |f|^{p_0} \, d\mu = \int_B |f|^{p_0} \, d\mu - \int_{B_{k_0}} |f|^{p_0} \, d\mu < \varepsilon. \tag{3}$$

For brevity let us write $B^* = B_{k_0}$. Then B^* and $B \setminus B^*$ are disjoint members of $\mathfrak{A}$ and $B^* \cup (B \setminus B^*) = B$. Let $p > p_0$. Since X is the disjoint union of A, B^*, $B \setminus B^*$, and C,

$$\begin{aligned}\int_X |f|^p \, d\mu &= \int_A |f|^p \, d\mu + \int_{B^*} |f|^p \, d\mu + \int_{B\setminus B^*} |f|^p \, d\mu + \int_C |f|^p \, d\mu \\ &= \int_{A\cup B^*} |f|^p \, d\mu + \int_{B\setminus B^*} |f|^p \, d\mu.\end{aligned} \tag{4}$$

For the first integral on the right side of (4) we have

$$\int_{A\cup B^*} |f|^p \, d\mu \le \|f\|_\infty^p \mu(A \cup B^*). \tag{5}$$

For the second integral on the right side of (4), recall that $|f| \in (0,1)$ on $B \setminus B^*$. Thus for $p > p_0$ we have $|f|^p < |f|^{p_0}$ on $B \setminus B^*$ so that by (3) we have

$$\int_{B\setminus B^*} |f|^p \, d\mu \le \int_{B\setminus B^*} |f|^{p_0} \, d\mu < \varepsilon. \tag{6}$$

Using (5) and (6) in (4) we have $\int_X |f|^p \, d\mu \le \|f\|_\infty^p \mu(A \cup B^*) + \varepsilon$ and therefore

$$\|f\|_p = \left\{\int_X |f|^p \, d\mu\right\}^{1/p} \le \left\{\|f\|_\infty^p \mu(A \cup B^*) + \varepsilon\right\}^{1/p}. \tag{7}$$

Since $\psi(t) = t^{1/p}$ for $t \in (0,\infty)$ is a continuous function, letting $\varepsilon \to 0$ in (7), we have

$$\|f\|_p = \left\{\int_X |f|^p \, d\mu\right\}^{1/p} \le \|f\|_\infty \big(\mu(A \cup B^*)\big)^{1/p}. \tag{8}$$

Since $\mu(A) < \infty$ by (1) and $\mu(B^*) < \infty$ by (2), we have $\mu(A \cup B^*) < \infty$. Then we have $\lim_{p\to\infty} \big(\mu(A \cup B^*)\big)^{1/p} = 1$. Thus we have

$$\limsup_{p\to\infty} \|f\|_p \le \|f\|_\infty \cdot \lim_{p\to\infty} \big(\mu(A \cup B^*)\big)^{1/p} = \|f\|_\infty. \tag{9}$$

Now $\|f\|_\infty$ is the infimum of all essential bounds of f on X. Then $\|f\|_\infty - \eta$ for an arbitrary $\eta > 0$ is not an essential bound for f on X. This implies that if we define a set $E_\eta = \big\{X : |f| > \|f\|_\infty - \eta\big\}$ then $\mu(E_\eta) > 0$. Now

$$\int_X |f|^p \, d\mu \ge \int_{E_\eta} |f|^p \, d\mu \ge \{\|f\|_\infty - \eta\}^p \mu(E_\eta)$$

and

$$\|f\|_p = \left\{\int_X |f|^p \, d\mu\right\}^{1/p} \ge \{\|f\|_\infty - \eta\}\big(\mu(E_\eta)\big)^{1/p}.$$

Since $\mu(E_\eta) > 0$ we have $\lim_{p\to\infty} \big(\mu(E_\eta)\big)^{1/p} = 1$. Thus we have

$$\liminf_{p\to\infty} \|f\|_p \ge \big\{\|f\|_\infty - \eta\big\} \lim_{p\to\infty} \big(\mu(E_\eta)\big)^{1/p} = \|f\|_\infty - \eta.$$

By the arbitrariness of $\eta > 0$ we have

$$\liminf_{p\to\infty} \|f\|_p \ge \|f\|_\infty. \tag{10}$$

By (9) and (10) we have $\|f\|_\infty \le \liminf_{p\to\infty} \|f\|_p \le \limsup_{p\to\infty} \|f\|_p \le \|f\|_\infty$ and therefore we have $\lim_{p\to\infty} \|f\|_p = \|f\|_\infty$. ∎

[V] The L^p Spaces for $p \in (0, 1)$

Let $(X, \mathfrak{A}, \mu)$ be a measure space and let $p \in (0, 1)$. As in Definition 16.8 let $\mathcal{L}^p(X, \mathfrak{A}, \mu)$ be the collection of all extended complex-valued $\mathfrak{A}$-measurable functions f on X with $\|f\|_p = \left[\int_X |f|^p \, d\mu\right]^p < \infty$. We showed in Proposition 16.8 that $\mathcal{L}^p(X, \mathfrak{A}, \mu)$ is a linear space over $\mathbb{C}$. Let $L^p(X, \mathfrak{A}, \mu)$ be the collection of all equivalence classes of members of $\mathcal{L}^p(X, \mathfrak{A}, \mu)$ with respect to the equivalent relation of a.e. equality on X as in Definition 16.22. Then $L^p(X, \mathfrak{A}, \mu)$ is a linear space over $\mathbb{C}$. For $p \in (0, 1)$ the function $\|\cdot\|_p$ on $L^p(X, \mathfrak{A}, \mu)$ does not satisfy the triangular inequality and is thus not a norm on $L^p(X, \mathfrak{A}, \mu)$ as we showed in Remark 16.21. Thus the function

$$\|f - g\|_p = \left[\int_X |f - g|^p \, d\mu\right]^{1/p} \quad \text{for } f, g \in L^p(X, \mathfrak{A}, \mu)$$

is not a metric on $L^p(X, \mathfrak{A}, \mu)$. Let us define a function ρ on $L^p(X, \mathfrak{A}, \mu) \times L^p(X, \mathfrak{A}, \mu)$ by setting

$$\rho(f, g) = \int_X |f - g|^p \, d\mu \quad \text{for } f, g \in L^p(X, \mathfrak{A}, \mu).$$

We show below that ρ is a metric on $L^p(X, \mathfrak{A}, \mu)$ and moreover $L^p(X, \mathfrak{A}, \mu)$ is complete with respect to this metric. Note that while $\rho(f, g) = \|f - g\|_p^p$ for $f, g \in L^p(X, \mathfrak{A}, \mu)$ is a metric on $L^p(X, \mathfrak{A}, \mu)$, $\|f\|_p^p$ is not a norm on $L^p(X, \mathfrak{A}, \mu)$ since $\|\alpha f\|_p^p = |\alpha|^p \|f\|_p^p$ for $\alpha \in \mathbb{C}$ rather than $\|\alpha f\|_p^p = |\alpha| \|f\|_p^p$ which is condition 3° in Definition 15.5 for a norm.

Lemma 16.51. *Let $p \in (0, 1)$. Then for $a \geq 0$ and $b \geq 0$ we have*

(1) $$(a + b)^p \leq a^p + b^p.$$

The equality in (1) holds if and only if at least one of a and b is equal to 0. *More generally for nonnegative real numbers $a_1, \ldots, a_k$ we have*

(2) $$\left\{\sum_{i=1}^k a_i\right\} \leq \sum_{i=1}^k a_i^p.$$

Proof. Consider the function

(3) $$\varphi(t) = (1 + t)^p - 1 - t^p \quad \text{for } t \in [0, \infty).$$

We have $\varphi(0) = 0$ and $\varphi'(t) = p\{(1 + t)^{p-1} - t^{p-1}\} < 0$ for $t \in (0, \infty)$. Thus $\varphi(t) < 0$ for $t \in (0, \infty)$. Let $a > 0$ and $b > 0$. Then $\frac{b}{a} > 0$ so that $\varphi\left(\frac{b}{a}\right) < 0$. Thus by (3) we have

$$\left(1 + \frac{b}{a}\right)^p - 1 - \left(\frac{b}{a}\right)^p = \varphi\left(\frac{b}{a}\right) < 0.$$

Transposing the second and third terms on the left side of the last inequality and multiplying both sides by a^p we obtain

$$(a + b)^p < a^p + b^p.$$

If $a \geq 0$ and $b \geq 0$ and if at least one of a and b is equal to 0, then $(a+b)^p = a^p + b^p$ holds trivially. This proves (1) and the fact that the equality in (1) holds if and only if at least one of a and b is equal to 0.

If $a_1, \ldots, a_k$ are nonnegative real numbers then by iterated application of (1) we have

$$\begin{aligned}(a_1 + \cdots + a_k)^p &\leq (a_1 + \cdots + a_{k-1})^p + a_k^p \\ &\leq (a_1 + \cdots + a_{k-2})^p + a_{k-1}^p + a_k^p \\ &\vdots \\ &\leq a_1^p + \cdots + a_k^p.\end{aligned}$$

This proves (2). ∎

Theorem 16.52. *Let $(X, \mathfrak{A}, \mu)$ be an arbitrary measure space and let $p \in (0,1)$. Let us define a function ρ on $L^p(X, \mathfrak{A}, \mu) \times L^p(X, \mathfrak{A}, \mu)$ by setting*

$$\rho(f, g) = \int_X |f - g|^p \, d\mu \quad \text{for } f, g \in L^p(X, \mathfrak{A}, \mu).$$

Then ρ is a metric on $L^p(X, \mathfrak{A}, \mu)$ and moreover $L^p(X, \mathfrak{A}, \mu)$ is complete with respect to the metric ρ.

Proof. 1. Let us show that ρ is a metric on $L^p(X, \mathfrak{A}, \mu)$. Since $L^p(X, \mathfrak{A}, \mu)$ is a linear space, if $f, g \in L^p(X, \mathfrak{A}, \mu)$ then $f - g \in L^p(X, \mathfrak{A}, \mu)$. Thus $\rho(f, g) = \int_X |f - g|^p \, d\mu \in [0, \infty)$. If $f = g$ then $\rho(f, g) = \int_X 0 \, d\mu = 0$. On the other hand if $\rho(f, g) = 0$ then $\int_X |f - g|^p \, d\mu = 0$ so that $|f - g|^p = 0$ a.e. on X and then $f = g$ a.e. on X, that is, $f = g$ as elements of $L^p(X, \mathfrak{A}, \mu)$. Thus $\rho(f, g) = 0$ if and only if $f = g$. From the definition of ρ it is clear that $\rho(f, g) = \rho(g, f)$. Let us verify the triangle inequality for ρ. Let $f, g, h \in L^p(X, \mathfrak{A}, \mu)$. Now $|f - g| \leq |f - h| + |h - g|$. Then by Lemma 16.51 we have

$$|f - g|^p \leq \big\{|f - h| + |h - g|\big\}^p \leq |f - h|^p + |h - g|^p.$$

Integrating over X we have

$$\int_X |f - g|^p \, d\mu \leq \int_X |f - h|^p \, d\mu + \int_X |h - g|^p \, d\mu,$$

that is, $\rho(f, g) \leq \rho(f, h) + \rho(h, g)$. This proves the triangle inequality for ρ and completes the proof that ρ is a metric on $L^p(X, \mathfrak{A}, \mu)$.

2. Let us show that $L^p(X, \mathfrak{A}, \mu)$ is complete with respect to the metric ρ. For an arbitrary Cauchy sequence $(f_n : n \in \mathbb{N})$ in $L^p(X, \mathfrak{A}, \mu)$ with respect to the metric ρ, we construct an element f in $L^p(X, \mathfrak{A}, \mu)$ such that $\lim_{n \to \infty} \rho(f_n, f) = 0$ as follows.

Since $(f_n : n \in \mathbb{N})$ is a Cauchy sequence in $L^p(X, \mathfrak{A}, \mu)$ with respect to the metric ρ, for every $\varepsilon > 0$ there exists $N \in \mathbb{N}$ such that $\rho(f_m, f_n) < \varepsilon$ for $m, n \geq N$. Then we can select a subsequence $(f_{n_i} : i \in \mathbb{Z}_+)$ of $(f_n : n \in \mathbb{N})$ such that

(1) $$\rho\big(f_{n_{i-1}}, f_{n_i}\big) < \frac{1}{2^i} \quad \text{for } i \in \mathbb{N}.$$

Let us show that the series $\sum_{i\in\mathbb{N}}\{f_{n_i}(x)-f_{n_{i-1}}(x)\}$ converges for a.e. $x\in X$. Since absolute convergence of a series implies convergence of the series, it suffices to show that the series $\sum_{i\in\mathbb{N}}|f_{n_i}(x)-f_{n_{i-1}}(x)|$ converges for a.e. $x\in X$. Note that since $f_{n_i}\in L^p(X,\mathfrak{A},\mu)$, $f_{n_i}(x)\in\mathbb{C}$ for a.e. $x\in X$. Then since a countable union of null sets in $(X,\mathfrak{A},\mu)$ is a null set in $(X,\mathfrak{A},\mu)$, the series $\sum_{i\in\mathbb{N}}|f_{n_i}(x)-f_{n_{i-1}}(x)|$ is defined for a.e. $x\in X$. Let $(g_k : k\in\mathbb{N})$ be the sequence of partial sums of the series $\sum_{i\in\mathbb{N}}|f_{n_i}-f_{n_{i-1}}|$, that is, $g_k=\sum_{i=1}^k|f_{n_i}-f_{n_{i-1}}|$ for $k\in\mathbb{N}$ defined a.e. on X. Applying (2) of Lemma 16.51 we have $|g_k|^p\le\sum_{i=1}^k|f_{n_i}-f_{n_{i-1}}|^p$. Integrating over X and applying (1) we have

$$(2)\qquad \int_X|g_k|^p\,d\mu\le\sum_{i=1}^k\int_X|f_{n_i}-f_{n_{i-1}}|^p\,d\mu=\sum_{i=1}^k\rho\big(f_{n_{i-1}},f_{n_i}\big)\le\sum_{i\in\mathbb{N}}\frac{1}{2^i}=1.$$

Now $(g_k : k\in\mathbb{N})$ is an increasing sequence and thus $\lim_{k\to\infty}g_k$ exists a.e. on X. If we let $g=\lim_{n\to\infty}g_k$ defined a.e. on X, then $|g|^p=\lim_{n\to\infty}|g_k|^p$ a.e. on X. By Fatou's Lemma (Theorem 8.13) and by (2) we have

$$(3)\qquad \int_X|g|^p\,d\mu\le\liminf_{k\to\infty}\int_X|g|_k^p\,d\mu\le 1.$$

Thus $g\in L^p(X,\mathfrak{A},\mu)$ and hence $g<\infty$ a.e. on X. This shows that the sequence of partial sums $(g_k : k\in\mathbb{N})$ of the series $\sum_{i\in\mathbb{N}}|f_{n_i}-f_{n_{i-1}}|$ converges a.e. on X, that is, the series $\sum_{i\in\mathbb{N}}|f_{n_i}-f_{n_{i-1}}|$ converges a.e. on X. This then implies that the series $\sum_{i\in\mathbb{N}}\{f_{n_i}-f_{n_{i-1}}\}$ converges a.e. on X.

Let $(h_k : k\in\mathbb{N})$ be the sequence of partial sums of the series $\sum_{i\in\mathbb{N}}\{f_{n_i}-f_{n_{i-1}}\}$, that is, $h_k=\sum_{i=1}^k\{f_{n_i}-f_{n_{i-1}}\}$ for $k\in\mathbb{N}$ defined a.e. on X. Since the series $\sum_{i\in\mathbb{N}}\{f_{n_i}-f_{n_{i-1}}\}$ converges a.e. on X, the sequence $(h_k : k\in\mathbb{N})$ converges a.e. on X. Now we have $h_k=\sum_{i=1}^k\{f_{n_i}-f_{n_{i-1}}\}=f_{n_k}-f_{n_0}$ so that $f_{n_k}=h_k+f_{n_0}$. Let us define

$$(4)\qquad f(x)=\lim_{k\to\infty}f_{n_k}(x)=\lim_{k\to\infty}h_k(x)+f_{n_0}(x)\in\mathbb{C}\quad\text{for a.e. } x\in X.$$

Let $f=0$ on the null set on which f is not defined by (4).

It remains to show that $f\in L^p(X,\mathfrak{A},\mu)$ and $\lim_{n\to\infty}\rho(f_n,f)=0$. Let $\varepsilon>0$. Since $(f_n : n\in\mathbb{N})$ is a Cauchy sequence in $L^p(X,\mathfrak{A},\mu)$ with respect to the metric ρ there exists $N\in\mathbb{N}$ such that $\rho(f_m,f_n)<\varepsilon$ for $m,n\ge N$. Let $m\ge N$. Now since $f=\lim_{k\to\infty}f_{n_k}$ a.e. on X, by Fatou's Lemma (Theorem 8.13) we have

$$(5)\qquad \int_X|f-f_m|^p\,d\mu\le\liminf_{k\to\infty}\int_X|f_{n_k}-f_m|^p\,d\mu=\liminf_{k\to\infty}\rho\big(f_{n_k},f_m\big)\le\varepsilon.$$

Now $f=f-f_m+f_m$ implies $|f|\le|f-f_m|+|f_m|$. Applying Lemma 16.51 we have

$$|f|^p\le\{|f-f_m|+|f_m|\}^p\le|f-f_m|^p+|f_m|^p.$$

Then by (5) we have

$$\int_X|f|^p\,d\mu\le\int_X|f-f_m|^p\,d\mu+\int_X|f_m|^p\,d\mu<\infty.$$

This shows that $f \in L^p(X, \mathfrak{A}, \mu)$. Also according to (5) we have $\rho(f, f_m) \le \varepsilon$ for $m \ge N$. Thus $\limsup_{m\to\infty} \rho(f, f_m) \le \varepsilon$. By the arbitrariness of $\varepsilon > 0$ we have $\limsup_{m\to\infty} \rho(f, f_m) = 0$. This implies $\liminf_{m\to\infty} \rho(f, f_m) = 0$ and then $\lim_{m\to\infty} \rho(f, f_m) = 0$. ■

For $p \in (0, 1)$, consider its conjugate q, that is, the real number q defined by the equality $1/p + 1/q = 1$. Then $q = p/(p-1) \in (-\infty, 0)$, $\lim_{p\downarrow 0} q = 0$ and $\lim_{p\uparrow 1} q = -\infty$.

Conventions. Let g be an extended complex-valued $\mathfrak{A}$-measurable function on X in a measure space $(X, \mathfrak{A}, \mu)$ such that $0 < |g| < \infty$ a.e. on X. On the null set in $(X, \mathfrak{A}, \mu)$ where $|g| = 0$, let $|g|^{-1} = \infty$ and on the null set where $|g| = \infty$, let $|g|^{-1} = 0$ so that $|g|^{-1}$ is a nonnegative extended real-valued $\mathfrak{A}$-measurable function on X such that $0 < |g|^{-1} < \infty$ a.e. on X. Then for any $q \in (-\infty, 0)$, the function $|g|^q = |g|^{-|q|}$ is a nonnegative extended real-valued $\mathfrak{A}$-measurable function on X such that $0 < |g|^q < \infty$ a.e. on X. This implies that $\int_X |g|^q \, d\mu > 0$ if $\mu(X) > 0$ and $\int_X |g|^q \, d\mu = 0$ if $\mu(X) = 0$.

Definition 16.53. *Let g be an extended complex-valued $\mathfrak{A}$-measurable function on X in a measure space $(X, \mathfrak{A}, \mu)$ such that $0 < |g| < \infty$ a.e. on X. For $q \in (-\infty, 0)$, we define*

$$\|g\|_q = \left[\int_X |g|^q \, d\mu\right]^{1/q} \in [0, \infty]$$

with the understanding that if $\int_X |g|^q \, d\mu = \infty$ then $\|g\|_q = 0$ and if $\int_X |g|^q \, d\mu = 0$ then $\|g\|_q = \infty$.

Theorem 16.54. (Hölder's Inequality for $p \in (0, 1)$ and $q \in (-\infty, 0)$) *Given a measure space $(X, \mathfrak{A}, \mu)$. Let $p \in (0, 1)$ and $q \in (-\infty, 0)$ be such that $1/p + 1/q = 1$. Let f and g be two extended complex-valued $\mathfrak{A}$-measurable functions on X such that $|f| < \infty$ a.e. on X and $0 < |g| < \infty$ a.e. on X with $0 < \|g\|_q < \infty$. Then*

(1) $$\|fg\|_1 \ge \|f\|_p \|g\|_q.$$

If $0 < \|fg\|_1 < \infty$ and $0 < \|g\|_q < \infty$, then the equality in (1) holds if and only if

(2) $$A|f|^p = B|g|^q \quad \text{a.e. on } X \text{ for some } A, B > 0.$$

Proof. 1. Note that since $|g| > 0$ a.e. on X, any negative power of $|g|$ is finite a.e. on X and in particular $|g|^{-p}, |g|^q < \infty$ a.e. on X. Now since $p \in (0, 1)$, we have $1/p \in (1, \infty)$. Let $r \in (1, \infty)$ be the conjugate of $1/p$, that is, $p + 1/r = 1$. Applying Theorem 16.14 to the pair of functions $|f|^p |g|^p$ and $|g|^{-p}$ with conjugate exponents $1/p$ and r, we have

(3) $$\begin{aligned}\int_X |f|^p \, d\mu &= \int_X \left(|f|^p |g|^p\right) |g|^{-p} \, d\mu \\ &\le \left[\int_X |f||g| \, d\mu\right]^p \left[\int_X |g|^{-pr} \, d\mu\right]^{1/r},\end{aligned}$$

and then

$$\left[\int_X |f|^p\,d\mu\right]^{1/p} \le \left[\int_X |f||g|\,d\mu\right]\left[\int_X |g|^{-pr}\,d\mu\right]^{1/pr}.$$

From $p + 1/r = 1$ and $q = p/(p-1)$, we have $r = 1/(1-p)$ and $pr = -q$. Thus

$$\left[\int_X |f|^p\,d\mu\right]^{1/p} \le \left[\int_X |f||g|\,d\mu\right]\left[\int_X |g|^q\,d\mu\right]^{-1/q},$$

that is, $\|f\|_p \le \|fg\|_1\|g\|_q^{-1}$. Since $0 < \|g\|_q < \infty$, multiplying by $\|g\|_q$ we have $\|f\|_p\|g\|_q \le \|fg\|_1$. This proves (1).

2. By (b) of Theorem 16.14, if $0 < \|fg\|_1 < \infty$ and $0 < \|g\|_q < \infty$, then the equality in (3) holds if and only if $a\left(|f|^p|g|^p\right)^{1/p} = b\left(|g|^{-p}\right)^r$, that is, $a|fg| = b|g|^q$, a.e. on X for some $a, b > 0$. Since $0 < |g| < \infty$ a.e. on X, this condition is equivalent to $a|f| = b|g|^{q-1}$, or equivalently, $a^p|f|^p = b^p|g|^{p(q-1)} = b^p|g|^q$, a.e. on X for some $a, b > 0$, that is, $A|f|^p = B|g|^q$ a.e. on X for some $A, B > 0$. ■

Theorem 16.55. (Minkowski's Inequality for $p \in (0,1)$) *Given a measure space $(X, \mathfrak{A}, \mu)$. Let $p \in (0,1)$. Let f and g be two extended complex-valued $\mathfrak{A}$-measurable functions on X such that $|f| < \infty$ a.e. on X and $0 < |f+g| < \infty$ a.e. on X with $0 < \|f+g\|_p < \infty$. Assume that* $\arg f = \arg g$ *a.e. on X. Then $\|f+g\|_p \ge \|f\|_p + \|g\|_p$.*

Proof. Since $\arg f = \arg g$ a.e. on X, we have $|f+g| = |f| + |g|$ a.e. on X. Let $1/p + 1/q = 1$ so that $p - 1 = p/q$. Then

$$|f+g|^p = |f+g||f+g|^{p-1} = \{|f| + |g|\}|f+g|^{p/q} \quad \text{a.e. on } X$$

so that

$$(1) \qquad \int_X |f+g|^p\,d\mu = \int_X |f||f+g|^{p/q}\,d\mu + \int_X |g||f+g|^{p/q}\,d\mu.$$

Applying Theorem 16.54 to the pair of functions $|f|$ and $|f+g|^{p/q}$ with conjugate exponents $p \in (0,1)$ and $q \in (-\infty, 0)$, we have

$$\int_X |f||f+g|^{p/q}\,d\mu \ge \left[\int_X |f|^p\,d\mu\right]^{1/p}\left[\int_X |f+g|^p\,d\mu\right]^{1/q},$$

and similarly

$$\int_X |f||f+g|^{p/q}\,d\mu \ge \left[\int_X |g|^p\,d\mu\right]^{1/p}\left[\int_X |f+g|^p\,d\mu\right]^{1/q}.$$

Using these two inequalities in (1), we have $\|f+g\|_p^p \ge \{\|f\|_p + \|g\|_p\}\|f+g\|_p^{p/q}$. Dividing by $\|f+g\|_p^{p/q}$ and noting $p - p/q = 1$, we have $\|f+g\|_p \ge \|f\|_p + \|g\|_p$. ■

[VI] Extensions of Hölder's Inequality

Let us extend Hölder's Inequality to cases where there are more than two factors.

Theorem 16.56. *Let $(X, \mathfrak{A}, \mu)$ be an arbitrary measure space. Let $f_1, \ldots, f_n$ be extended complex-valued $\mathfrak{A}$-measurable functions on X such that $|f_1|, \ldots, |f_n| < \infty$ a.e. on X. Let $p_1, \cdots, p_n$ be real numbers such that*

(1) $$p_1, \ldots, p_n \in (1, \infty) \quad \text{and} \quad \frac{1}{p_1} + \cdots + \frac{1}{p_n} = 1.$$

Then we have

(2) $$\|f_1 \cdots f_n\|_1 \leq \|f_1\|_{p_1} \cdots \|f_n\|_{p_n}.$$

Proof. We prove (2) by induction on n. For $n = 2$, (2) holds by Theorem 16.14. Suppose (2) holds for some $n \geq 2$. Let us show that this hypothesis implies that (2) holds for $n + 1$. Let $p_1, \cdots, p_{n+1}$ be real numbers such that

(3) $$p_1, \ldots, p_{n+1} \in (1, \infty) \quad \text{and} \quad \frac{1}{p_1} + \cdots + \frac{1}{p_{n+1}} = 1.$$

Let $q \in \mathbb{R}$ be defined by

(4) $$q = \left\{\frac{1}{p_1} + \cdots + \frac{1}{p_n}\right\}^{-1}, \quad \text{that is,} \quad \frac{1}{q} = \frac{1}{p_1} + \cdots + \frac{1}{p_n}.$$

Now (3) implies that $\frac{1}{p_1} + \cdots + \frac{1}{p_n} \in (0, 1)$. Then (4) implies that $\frac{1}{q} \in (0, 1)$ and then $q \in (1, \infty)$. Thus we have

(5) $$q \in (1, \infty), p_{n+1} \in (1, \infty) \quad \text{and} \quad \frac{1}{q} + \frac{1}{p_{n+1}} = 1.$$

Then by Theorem 16.14, we have

(6) $$\int_X |f_1 \cdots f_{n+1}| \, d\mu \leq \left\{\int_X |f_1 \cdots f_n|^q \, d\mu\right\}^{1/q} \left\{\int_X |f_{n+1}|^{p_{n+1}} \, d\mu\right\}^{1/(p_{n+1})}.$$

On the other hand, by the fact that $p_1, \cdots, p_n \in (1, \infty)$ and $q \in (1, \infty)$ and by (4) we have

(7) $$\frac{p_1}{q}, \cdots, \frac{p_n}{q} \in (1, \infty) \quad \text{and} \quad \frac{1}{p_1/q} + \cdots + \frac{1}{p_n/q} = 1.$$

Thus by our induction hypothesis that (2) holds for n, we have

(8) $$\begin{aligned} \int_X |f_1 \cdots f_n|^q \, d\mu &\leq \left\{\int_X |f_1|^{q\frac{p_1}{q}} \, d\mu\right\}^{q/p_1} \cdots \left\{\int_X |f_n|^{q\frac{p_n}{q}} \, d\mu\right\}^{q/p_n} \\ &= \left\{\int_X |f_1|^{p_1} \, d\mu\right\}^{q/p_1} \cdots \left\{\int_X |f_n|^{p_n} \, d\mu\right\}^{q/p_n}. \end{aligned}$$

Substituting the q-th root of (8) in (6), we have

$$\int_X |f_1 \cdots f_{n+1}|\, d\mu \leq \left\{\int_X |f_1|^{p_1}\, d\mu\right\}^{1/p_1} \cdots \left\{\int_X |f_{n+1}|^{p_{n+1}}\, d\mu\right\}^{1/p_{n+1}}.$$

This shows that (2) holds for $n+1$. Then by induction (2) holds for all $n \geq 2$. ∎

Theorem 16.57. *Let $(X, \mathfrak{A}, \mu)$ be an arbitrary measure space. Let $f_1, \ldots, f_n$ be extended complex-valued $\mathfrak{A}$-measurable functions on X such that $|f_1|, \ldots, |f_n| < \infty$ a.e. on X. Let $p_1, \ldots, p_n$ and r be real numbers such that*

(1) $$p_1, \ldots, p_n \in (1, \infty), r \in (1, \infty) \quad \text{and} \quad \frac{1}{p_1} + \cdots + \frac{1}{p_n} = \frac{1}{r}.$$

Then we have

(2) $$\|f_1 \cdots f_n\|_r \leq \|f_1\|_{p_1} \cdots \|f_n\|_{p_n}.$$

Proof. From (1) we have

$$\frac{1}{p_1/r} + \cdots + \frac{1}{p_n/r} = 1.$$

Thus by (2) of Theorem 16.56 we have

(3) $$\big\||f_1 \cdots f_n|^r\big\|_1 \leq \big\||f_1|^r\big\|_{p_1/r} \cdots \big\||f_n|^r\big\|_{p_n/r}.$$

Now we have

(4) $$\big\||f_1 \cdots f_n|^r\big\|_1 = \int_X |f_1 \cdots f_n|^r\, d\mu = \|f_1 \cdots f_n\|_r^r,$$

and for $i = 1, \ldots, n$ we have

(5) $$\big\||f_i|^r\big\|_{p_i/r} = \left\{\int_X |f_i|^{r\frac{p_i}{r}}\, d\mu\right\}^{r/p_i} = \left\{\int_X |f_i|^{p_i}\, d\mu\right\}^{r/p_i} = \|f_i\|_{p_i}^r.$$

Substituting (4) and (5) in (3), we have

(6) $$\|f_1 \cdots f_n\|_r^r \leq \|f_1\|_{p_1}^r \cdots \|f_n\|_{p_n}^r.$$

Taking the r-th roots of (6) we have (2). ∎

Problems

Prob. 16.1. Let f be a Lebesgue measurable function on $[0, 1]$ and let $0 < f(x) < \infty$ for $x \in [0, 1]$. Show that

$$\left\{\int_{[0,1]} f\, d\mu\right\}\left\{\int_{[0,1]} \frac{1}{f}\, d\mu\right\} \geq 1.$$

Prob. 16.2. Let $(X, \mathfrak{A}, \mu)$ be a finite measure space. Let f be an extended complex-valued $\mathfrak{A}$-measurable function on X such that $|f| < \infty$ a.e. on X. Let $0 < p < q < \infty$. Show that

$$\|f\|_p \le \|f\|_q \{\mu(X)\}^{\frac{1}{p}-\frac{1}{q}}.$$

Prob. 16.3. Let $(X, \mathfrak{A}, \mu)$ be a measure space. Let $\vartheta \in (0, 1)$ and let $p, q, r \ge 1, p, q \ge r$, be related by

$$\tfrac{1}{r} = \tfrac{\vartheta}{p} + \tfrac{1-\vartheta}{q}.$$

Show that for every extended complex-valued $\mathfrak{A}$-measurable function f on X we have

$$\|f\|_r \le \|f\|_p^{\vartheta} \|f\|_q^{1-\vartheta}.$$

Prob. 16.4. Let $(X, \mathfrak{A}, \mu)$ be a measure space and let $p, q \in [1, \infty]$ be conjugates, that is, $\frac{1}{p} + \frac{1}{q} = 1$. Let $(f_n : n \in \mathbb{N}) \subset L^p(X, \mathfrak{A}, \mu)$ and $f \in L^p(X, \mathfrak{A}, \mu)$ and similarly $(g_n : n \in \mathbb{N}) \subset L^q(X, \mathfrak{A}, \mu)$ and $g \in L^q(X, \mathfrak{A}, \mu)$. Show that if $\lim_{n\to\infty} \|f_n - f\|_p = 0$ and $\lim_{n\to\infty} \|g_n - g\|_q = 0$ then $\lim_{n\to\infty} \|f_n g_n - fg\|_1 = 0$.

Prob. 16.5. Let $(X, \mathfrak{A}, \mu)$ be a measure space and let $p \in [1, \infty)$. Let $(f_n : n \in \mathbb{N})$ be a sequence in $L^p(X, \mathfrak{A}, \mu)$ and $f \in L^p(X, \mathfrak{A}, \mu)$ be such that $\lim_{n\to\infty} \|f_n - f\|_p = 0$. Let $(g_n : n \in \mathbb{N})$ be a sequence of complex-valued $\mathfrak{A}$-measurable functions on X such that $|g_n| \le M$ for every $n \in \mathbb{N}$ and let g be complex-valued $\mathfrak{A}$-measurable function on X such that $\lim_{n\to\infty} g_n = g$ a.e. on X. Show that $\lim_{n\to\infty} \|g_n f_n - gf\|_p = 0$.

Prob. 16.6. Let f be an extended real-valued $\mathfrak{M}_L$-measurable function on $[a, b]$ such that $\int_{[a,b]} |f|^p \, d\mu_L < \infty$ for some $p \in (1, \infty)$.
(a) Show that f is μ_L-integrable on $[a, b]$.
Let F be an indefinite integral of f on $[a, b]$ with the additive constant set to be 0 and let q be the conjugate of p.
(b) Show that there exists a nonnegative constant C such that for every $x \in [a, b]$, we have

$$F(x) \le C|x - a|^{1/q}.$$

(c) Show that for every $x \in [a, b]$, we have

$$\lim_{h\to 0} \frac{F(x+h) - F(x)}{h^{1/q}} = 0.$$

Prob. 16.7. Let $f \in L^p(\mathbb{R}, \mathfrak{M}_L, \mu_L)$ where $p \in [1, \infty)$. With $h \in \mathbb{R}$, let us define a function f_h on $\mathbb{R}$ by setting $f_h(x) = f(x+h)$ for $x \in \mathbb{R}$. Show that $f_h \in L^p(\mathbb{R}, \mathfrak{M}_L, \mu_L)$ for every $h \in \mathbb{R}$ and $\lim_{h\to 0} \|f - f_h\|_p = 0$.

Prob. 16.8. Let $f \in L^p(\mathbb{R}, \mathfrak{M}_L, \mu_L)$ and $g \in L^q(\mathbb{R}, \mathfrak{M}_L, \mu_L)$ where $p, q \in (1, \infty)$ and are conjugates. Show that the function F defined on $\mathbb{R}$ by setting

$$F(h) = \int_{\mathbb{R}} f(x+h) g(x) \mu_L(dx) \text{ for } h \in \mathbb{R},$$

is a continuous function on $\mathbb{R}$.

Prob. 16.9. Consider $(\mathbb{R}, \mathfrak{M}_L, \mu_L)$. Let $E \in \mathfrak{M}_L$ with $\mu_L(E) < \infty$. Show that the function defined by $f(x) = \mu_L(E + x \cap E)$ for $x \in \mathbb{R}$ is a real-valued continuous function on $\mathbb{R}$.

Prob. 16.10. Given a measure space $(X, \mathfrak{A}, \mu)$. Let f be an extended real-valued $\mathfrak{A}$-measurable function on X such that $\int_X |f|^p \, d\mu < \infty$ for some $p \in (0, \infty)$. Show that for every $\varepsilon > 0$ there exists $M > 0$ such that for the truncation of f at level M, that is, $f^{[M]} = (f \wedge M) \vee (-M)$, we have $\int_X |f - f^{[M]}|^p \, d\mu < \varepsilon$.

Prob. 16.11. Let f be an extended real-valued $\mathfrak{M}_L$-measurable function on $[0, 1]$ such that $\int_{[0,1]} |f|^p \, d\mu_L < \infty$ for some $p \in [1, \infty)$. Let $q \in (1, \infty]$ be the conjugate of p, that is, $\frac{1}{p} + \frac{1}{q} = 1$. Let $a \in (0, 1]$. Show that

$$\lim_{a \to 0} a^{-\frac{1}{q}} \int_{[0,t]} |f(s)| \, \mu_L(ds) = 0.$$

Prob. 16.12. Let $(X, \mathfrak{A}, \mu)$ be a measure space and let $p_0 \in (0, \infty)$. Let $\{f_\alpha : \alpha \in A\}$ be a collection in $L^{p_0}(X, \mathfrak{A}, \mu)$ such that $\sup_{\alpha \in A} \|f_\alpha\|_{p_0} < \infty$. Show that for every $p \in (0, p_0)$ the following hold:

(a)

$$\lim_{\lambda \to \infty} \sup_{\alpha \in A} \int_{\{X : |f_\alpha|^p > \lambda\}} |f_\alpha|^p \, d\mu = 0.$$

(b) For every $\varepsilon > 0$ there exists $\delta > 0$ such that for every $E \in \mathfrak{A}$ with $\mu(E) < \delta$ we have

$$\int_E |f_\alpha|^p \, d\mu < \varepsilon \quad \text{for all } \alpha \in A.$$

(Hint for (a): Let $p \in (0, p_0)$. Then for $0 < \eta < \xi$ we have $\xi^p = \xi^{p-p_0}\xi^{p_0} \le \eta^{p-p_0}\xi^{p_0}$.)

Prob. 16.13. Let $(X, \mathfrak{A}, \mu)$ be a finite measure space and let $p_0 \in (0, \infty)$. Let $(f_n : n \in \mathbb{N})$ be a sequence and f be an element in $L^{p_0}(X, \mathfrak{A}, \mu)$ such that

1° $\lim_{n \to \infty} f_n = f$, μ-a.e. on X.

2° $\|f_n\|_{p_0} \le M$ for all $n \in \mathbb{N}$.

Show that for every $p \in (0, p_0)$ we have $\lim_{n \to \infty} \|f_n - f\|_p = 0$.

(Hint: Apply Egoroff's Theorem and Prob. 16.12.)

Prob. 16.14. Let $(X, \mathfrak{A}, \mu)$ be a finite measure space and let $p, q \in (1, \infty)$ be conjugates. Let $(f_n : n \in \mathbb{N})$ be a sequence and f be an element in $L^p(X, \mathfrak{A}, \mu)$ such that

1° $\lim_{n \to \infty} f_n = f$, μ-a.e. on X.

2° $\|f_n\|_p \le M$ for all $n \in \mathbb{N}$.

Show that

(a) $\|f\|_p \le M$.

(b) $\lim_{n \to \infty} \int_X f_n g \, d\mu = \int_X f g \, d\mu$ for every $g \in L^q(X, \mathfrak{A}, \mu)$.

(c) $\lim_{n \to \infty} \int_E f_n \, d\mu = \int_E f \, d\mu$ for every $E \in \mathfrak{A}$.

(Hint: Egroroff's Theorem)

Prob. 16.15. Let $(X, \mathfrak{A}, \mu)$ be a measure space and let $1 \le p_0 < p$. Let $f \in L^p(X, \mathfrak{A}, \mu)$ and $(f_n : n \in \mathbb{N}) \subset L^p(X, \mathfrak{A}, \mu)$ be such that

1° $\lim_{n \to \infty} f_n = f$, μ-a.e. on X,

2° $\lim_{n \to \infty} \|f_n - f\|_p = 0$.

(a) Show that for every $\varepsilon > 0$ there exists $\delta > 0$ such that for $E \in \mathfrak{A}$ with $\mu(E) < \delta$ we have

$$\int_E |f_n|^p \, d\mu < \varepsilon \text{ for all } n \in \mathbb{N}.$$

(b) Show that if $(X, \mathfrak{A}, \mu)$ is a σ-finite measure space, then for every $\varepsilon > 0$ there exists $A \in \mathfrak{A}$ with $\mu(A) < \infty$ such that

$$\int_{X \setminus A} |f_n|^p \, d\mu < \varepsilon \text{ for all } n \in \mathbb{N}.$$

Prob. 16.16. Let $(X, \mathfrak{A}, \mu)$ be a measure space and let $1 \leq p_0 < p$. Let $(f_n : n \in \mathbb{N})$ be a sequence in $L^p(X, \mathfrak{A}, \mu)$. Let f be an extended complex-valued $\mathfrak{A}$-measurable function on X such that $|f| < \infty$, μ-a.e. on X and $\lim_{n\to\infty} f_n = f$, μ-a.e. on X. Suppose

1° for every $\varepsilon > 0$ there exists $\delta > 0$ such that for every $E \in \mathfrak{A}$ with $\mu(E) < \delta$ we have

$$\int_E |f_n|^p \, d\mu < \varepsilon \text{ for all } n \in \mathbb{N}.$$

2° for every $\varepsilon > 0$ there exists $A \in \mathfrak{A}$ with $\mu(A) < \infty$ such that

$$\int_{X \setminus A} |f_n|^p \, d\mu < \varepsilon \text{ for all } n \in \mathbb{N}.$$

(a) Show that $f \in L^p(X, \mathfrak{A}, \mu)$.
(b) Show that if $(X, \mathfrak{A}, \mu)$ is a σ-finite measure space, then $\lim_{n\to\infty} \|f_n - f\|_p = 0$.

Prob. 16.17. Let $(f_n : n \in \mathbb{N})$ be a sequence of real-valued differentiable functions with continuous derivatives on $[0, 1]$ such that $\sup_{n\in\mathbb{N}} \int_{[0,1]} |f_n'|^2 \, d\mu_L \leq 1$. Show that $(f_n : n \in \mathbb{N})$ is equicontinuous on $[0, 1]$, that is, for every $\varepsilon > 0$ there exists $\delta > 0$ such that for every $n \in \mathbb{N}$ we have $|f_n(x') - f_n(x'')| < \varepsilon$ for $x', x'' \in [0, 1]$ such that $|x' - x''| < \delta$.

Prob. 16.18. Given a measure space $(X, \mathfrak{A}, \mu)$. Let $D, E \in \mathfrak{A}$ and $E \subset D$. Let $(f_n : n \in \mathbb{N})$ and f be extended real-valued $\mathfrak{A}$-measurable and μ-integrable functions on D. Show that

1° $\lim_{n\to\infty} f_n = f$ a.e. on D,

2° $\lim_{n\to\infty} \int_D |f_n| \, d\mu = \int_D |f| \, d\mu$,

imply

(1) $$\lim_{n\to\infty} \int_E |f_n| \, d\mu = \int_E |f| \, d\mu,$$

(2) $$\lim_{n\to\infty} \int_E f_n \, d\mu = \int_E f \, d\mu.$$

Prob. 16.19. If condition 2° in Prob. 16.18 is replaced by

3° $\lim_{n\to\infty} \int_D f_n \, d\mu = \int_D f \, d\mu$,

then (1) and (2) do not hold. Construct an example to show this.

Prob. 16.20. Given a measure space $(X, \mathfrak{A}, \mu)$. Let $(f_n : n \in \mathbb{N})$ and f be extended real-valued $\mathfrak{A}$-measurable and μ-integrable functions on X. Show that conditions

1° $\lim_{n\to\infty} f_n = f$ a.e. on X,

2° $\lim_{n\to\infty} \|f_n\|_\infty = \|f\|_\infty$,

do not imply $\lim_{n\to\infty} \|f_n - f\|_\infty = 0$.

Prob. 16.21. Consider the function

$$f(x) = \frac{1}{\sqrt{2\pi}} e^{-\frac{x^2}{2}} \text{ for } x \in \mathbb{R}.$$

Show that $f \in L^p(\mathbb{R}, \mathfrak{M}_L, \mu_L)$ for every $p \in (0, \infty]$ and $\lim_{p\to\infty} \|f\|_p = \|f\|_\infty$.

Prob. 16.22. Construct a finite measure space $(X, \mathfrak{A}, \mu)$ and an extended real-valued $\mathfrak{A}$-measurable function f on X such that $\|f\|_p < \infty$ for every $p \in \mathbb{N}$ and $f \notin L_\infty(X, \mathfrak{A}, \mu)$.

Prob. 16.23. Let $(X, \mathfrak{A}, \mu)$ be an arbitrary measure space. Let g be an extended real-valued $\mathfrak{A}$-measurable function on X. Suppose $\int_X fg\, d\mu \in \mathbb{R}$ for every extended real-valued $\mathfrak{A}$-measurable $f \in L^1(X, \mathfrak{A}, \mu)$. Show that $g \in L^\infty(X, \mathfrak{A}, \mu)$.

Prob. 16.24. (An Extension of Hölder's Inequality) Let $(X, \mathfrak{A}, \mu)$ be a measure space. For an arbitrary $n \in \mathbb{N}$, let $f_1, \ldots, f_n$ be n extended complex-valued $\mathfrak{A}$-measurable functions on X such that $|f_1|, \ldots, |f_n| < \infty$ a.e. on X. Let $p_1, \ldots, p_n \in (1, \infty)$ be such that $1/p_1 + \cdots + 1/p_n = 1$. Show that

(1) $$\|f_1 \cdots f_n\|_1 \le \|f_1\|_{p_1} \cdots \|f_n\|_{p_n},$$

where the equality holds if and only if there exist $A_1, \ldots, A_n > 0$ such that

(2) $$A_1|f_1|^{p_1} = \cdots = A_n|f_1|^{p_n} \quad \text{a. e. on } X.$$

§17 Relation among the L^p Spaces

[I] The Modified L^p Norms for L^p Spaces with $p \in [1, \infty]$

Let $(X, \mathfrak{A}, \mu)$ be an arbitrary measure space. Let $0 < p_1 < p_2 \leq \infty$. For an extended complex-valued $\mathfrak{A}$-measurable function f on X, $\int_X |f|^{p_1}\, d\mu < \infty$ does not imply $\int_X |f|^{p_2}\, d\mu < \infty$ and $\int_X |f|^{p_2}\, d\mu < \infty$ does not imply $\int_X |f|^{p_1}\, d\mu < \infty$. Thus we have neither $L^{p_1}(X, \mathfrak{A}, \mu) \subset L^{p_2}(X, \mathfrak{A}, \mu)$ nor $L^{p_1}(X, \mathfrak{A}, \mu) \supset L^{p_2}(X, \mathfrak{A}, \mu)$. We show below that if $\mu(X) < \infty$, then $L^{p_1}(X, \mathfrak{A}, \mu) \supset L^{p_2}(X, \mathfrak{A}, \mu)$ holds for $0 < p_1 < p_2 \leq \infty$.

Definition 17.1. *Given a measure space $(X, \mathfrak{A}, \mu)$ with $\mu(X) \in (0, \infty)$. For an extended complex-valued $\mathfrak{A}$-measurable function f on X and for $p \in (0, \infty)$, we define*

$$M_p(f) = \left[\frac{1}{\mu(X)} \int_X |f|^p\, d\mu\right]^{1/p} = \frac{\|f\|_p}{\mu(X)^{1/p}} \in [0, \infty],$$

and

$$M_\infty(f) = \frac{\|f\|_\infty}{\mu(X)^{1/\infty}} = \|f\|_\infty \in [0, \infty],$$

with the convention that $1/\infty = 0$ so that $\mu(X)^{1/\infty} = 1$.

Proposition 17.2. *Given a measure space $(X, \mathfrak{A}, \mu)$ with $\mu(X) \in (0, \infty)$. Let f and g be two extended complex-valued $\mathfrak{A}$-measurable functions f and g on X such that $|f|, |g| < \infty$ a.e. on X.*

(a) *For $p, q \in [1, \infty]$ such that $1/p + 1/q = 1$, we have Hölder's Inequality*

(1) $$M_1(fg) \leq M_p(f) M_q(g)$$

provided that the product $M_p(f)M_q(g)$ is defined.

(b) *For $p \in [1, \infty]$, we have Minkowski's Inequality*

(2) $$M_p(f + g) \leq M_p(f) + M_p(g).$$

Proof. 1. By Theorem 16.14 for the case $p, q \in (1, \infty)$ and by Theorem 16.40 for the case $p = 1$ and $q = \infty$, we have $\|fg\|_1 \leq \|f\|_p \|g\|_q$. Dividing both sides of the inequality by $\mu(X) = \mu(X)^{1/p} \mu(X)^{1/q}$, we have (1).

2. By Theorem 16.17 for the case $p \in [1, \infty)$ and by Theorem 16.41 for the case $p = \infty$, we have $\|f + g\|_p \leq \|f\|_p + \|g\|_p$. Dividing both sides of the inequality by $\mu(X)^{1/p}$, we have (2). ∎

Proposition 17.3. *Given a measure space $(X, \mathfrak{A}, \mu)$ with $\mu(X) \in (0, \infty)$. For $p \in [1, \infty]$, M_p and $\|\cdot\|_p$ are two equivalent norms on the linear space $L^p(X, \mathfrak{A}, \mu)$ and thus $L^p(X, \mathfrak{A}, \mu)$ is a Banach space with respect to the norm M_p.*

Proof. Since $M_p = \left(\mu(X)^{1/p}\right)^{-1} \|\cdot\|_p$ and since $\mu(X)^{1/p} \in (0, \infty)$, the fact that M_p is a norm follows immediately from the fact that $\|\cdot\|_p$ is a norm. The two norms are equivalent since $M_p = \left(\mu(X)^{1/p}\right)^{-1} \|\cdot\|_p$ and $\|\cdot\|_p = \mu(X)^{1/p} M_p$. Then since $L^p(X, \mathfrak{A}, \mu)$ is a Banach space with respect to $\|\cdot\|_p$ by Theorem 16.23 for $p \in [1, \infty)$ and by Theorem 16.43 for $p = \infty$, $L^p(X, \mathfrak{A}, \mu)$ is a Banach space with respect to M_p by Proposition 15.12. ∎

Theorem 17.4. *Given a measure space $(X, \mathfrak{A}, \mu)$ with $\mu(X) \in (0, \infty)$. Then for $1 \le p_1 < p_2 \le \infty$, we have*

(1)
$$L^{p_1}(X, \mathfrak{A}, \mu) \supset L^{p_2}(X, \mathfrak{A}, \mu)$$

and

(2)
$$M_{p_1}(f) \le M_{p_2}(f) \quad \text{for } f \in L^{p_2}(X, \mathfrak{A}, \mu).$$

Proof. 1. Let us prove (1) and (2) for the case $p_2 = \infty$. Let $f \in L^\infty(X, \mathfrak{A}, \mu)$. We have $\|f\|_\infty < \infty$. Since $|f| \le \|f\|_\infty$ a.e. on X, we have $\int_X |f|^{p_1}\, d\mu \le \|f\|_\infty^{p_1} \mu(X) < \infty$ so that $f \in L^{p_1}(X, \mathfrak{A}, \mu)$. This proves (1). To prove (2), note that for $f \in L^\infty(X, \mathfrak{A}, \mu)$, we have

$$M_{p_1}(f) = \left[\frac{1}{\mu(X)} \int_X |f|^{p_1}\, d\mu\right]^{1/p_1} \le \left[\frac{1}{\mu(X)} \|f\|_\infty^{p_1} \mu(X)\right]^{1/p_1} = \|f\|_\infty = M_\infty(f).$$

2. Now consider the case $p_2 < \infty$. Let $f \in L^{p_2}(X, \mathfrak{A}, \mu)$. Since $1 \le p_1 < p_2 < \infty$, $\alpha := p_2/p_1 \in (1, \infty)$. Let $\beta \in (1, \infty)$ be the conjugate of α, that is, $1/\alpha + 1/\beta = 1$. By Theorem 16.14 for our conjugates $\alpha, \beta \in (1, \infty)$, we have

(3)
$$\begin{aligned} \int_X |f|^{p_1}\, d\mu &= \int_X \left(|f|^{p_2}\right)^{1/\alpha} \mathbf{1}_X\, d\mu \\ &\le \left[\int_X |f|^{p_2}\, d\mu\right]^{1/\alpha} \left[\int_X |\mathbf{1}_X|^\beta\, d\mu\right]^{1/\beta} \\ &= \|f\|_{p_2}^{p_2/\alpha} \mu(X)^{1/\beta} < \infty, \end{aligned}$$

since $\|f\|_{p_2} < \infty$. Thus $f \in L^{p_1}(X, \mathfrak{A}, \mu)$. This proves (1).

To prove (2), let $f \in L^{p_2}(X, \mathfrak{A}, \mu) \subset L^{p_1}(X, \mathfrak{A}, \mu)$. With α and β as above, we have

$$\begin{aligned} M_{p_1}(f) &= \left[\frac{1}{\mu(X)} \int_X |f|^{p_1}\, d\mu\right]^{1/p_1} \le \left[\frac{1}{\mu(X)} \|f\|_{p_2}^{p_2/\alpha} \mu(X)^{1/\beta}\right]^{1/p_1} \quad \text{by (3)} \\ &= \left[\frac{1}{\mu(X)} \left[\int_X |f|^{p_2}\, d\mu\right]^{1/\alpha} \frac{1}{\mu(X)^{-1/\beta}}\right]^{1/p_1} = \left[\frac{1}{\mu(X)^{1/\alpha}} \left[\int_X |f|^{p_2}\, d\mu\right]^{1/\alpha}\right]^{1/p_1} \\ &= \left[\frac{1}{\mu(X)} \int_X |f|^{p_2}\, d\mu\right]^{1/\alpha p_1} = M_{p_2}(f) \quad \text{since } \alpha p_1 = p_2. \end{aligned}$$ ∎

Remark 17.5. If $\mu(X) \in (0, \infty)$, then $L^p(X, \mathfrak{A}, \mu)$ decreases as p increases from 1 to ∞ and $L^p(X, \mathfrak{A}, \mu) \supset L^\infty(X, \mathfrak{A}, \mu)$ for $p \in [1, \infty)$ by Theorem 17.4. Thus we have

$L^\infty(X, \mathfrak{A}, \mu) \subset \bigcap_{p\in[1,\infty)} L^p(X, \mathfrak{A}, \mu)$. But $L^\infty(X, \mathfrak{A}, \mu) \neq \bigcap_{p\in[1,\infty)} L^p(X, \mathfrak{A}, \mu)$, that is, if $\int_X |f|^p \, d\mu < \infty$ for every $p \in [1, \infty)$, f need not be essentially bounded on X.

Example. Let $(I_n : n \in \mathbb{N})$ be a sequence of disjoint intervals in $\mathbb{R}$ with $\mu_L(I_n) = e^{-2n}$ for $n \in \mathbb{N}$. Let $X = \bigcup_{n\in\mathbb{N}} I_n$ and consider the measure space $(X, \mathfrak{M}_L \cap X, \mu_L)$ where $\mathfrak{M}_L \cap X$ is the σ-algebra of subsets of X consisting of the intersections of members of $\mathfrak{M}_L$ with X. We have $\mu_L(X) = \sum_{n\in\mathbb{N}} \mu_L(I_n) = \sum_{n\in\mathbb{N}} e^{-2n} < \infty$. Let f be a real-valued function on X defined by $f = n$ on I_n for $n \in \mathbb{N}$. Then $f \notin L^\infty(X, \mathfrak{M}_L \cap X, \mu_L)$ since for every $n \in \mathbb{N}$ we have $\mu_L\{x \in X : f(x) \geq n\} \geq \mu_L(I_n) = e^{-2n} > 0$. To show that $f \in L^p(X, \mathfrak{M}_L \cap X, \mu_L)$ for every $p \in [1, \infty)$, note that $\int_X |f|^p \, d\mu_L = \sum_{n\in\mathbb{N}} n^p/e^{2n}$. To show that this series converges, note that by l'Hôpital's rule $\lim_{n\to\infty} n^p/e^n = 0$ so that there exists $N \in \mathbb{N}$ such that $n^p/e^n \leq 1$ for $n \geq N$. Then $\sum_{n\geq N} n^p/e^{2n} \leq \sum_{n\geq N} 1/e^n < \infty$. This shows that $\int_X |f|^p \, d\mu_L < \infty$ and therefore $f \in L^p(X, \mathfrak{M}_L \cap X, \mu_L)$.

Remark 17.6. For $p \in [1, \infty]$, as we showed in Theorem 17.4, $M_p(f)$ increases as p increases. $\|f\|_p$ however does not have this property.

Example. Consider $\big([0, 2), \mathfrak{M}_L \cap [0, 2), \mu_L\big)$. Let $f = 1$ on $[0, 2)$. Then $\|f\|_1 = 2$ and $\|f\|_2 = \sqrt{2}$. Thus $\|f\|_1 > \|f\|_2$.

[II] Approximation by Continuous Functions

Let us consider approximation of functions in $L^p(X, \mathfrak{A}, \mu)$ by continuous functions on X. The notion of a continuous function on X is predicated on a topology on X. Here we consider the case $X = \mathbb{R}$ with its Euclidean metric topology and $(X, \mathfrak{A}, \mu) = \big(\mathbb{R}, \mathfrak{M}_L, \mu_L\big)$. In §19, we consider a locally compact Hausdorff space X with a Radon measure μ on the Borel σ-algebra $\mathfrak{B}_X$ of subsets of X.

Definition 17.7. *Let $C_c(\mathbb{R})$ be the collection of all complex-valued continuous functions vanishing outside of a compact set in $\mathbb{R}$ which depends on the individual function.*

Since every compact set in $\mathbb{R}$ is contained in a finite closed interval and a finite closed interval is a compact set, a complex-valued continuous function on $\mathbb{R}$ is a member of $C_c(\mathbb{R})$ if and only if it vanishes outside of a finite closed interval. Thus if $f \in C_c(\mathbb{R})$, then there exists $B > 0$ such that f vanishes outside of the interval $[-B, B]$. Then we have $\sup_{\mathbb{R}} |f| = \max_{[-B,B]} |f| < \infty$. Thus $f \in L^\infty\big(\mathbb{R}, \mathfrak{M}_L, \mu_L\big)$. Also for every $p \in [1, \infty)$, we have $\int_{\mathbb{R}} |f|^p \, d\mu_L \leq \big(\sup_{\mathbb{R}} |f|\big)^p \, 2B < \infty$ so that $f \in L^p\big(\mathbb{R}, \mathfrak{M}_L, \mu_L\big)$. This shows $C_c(\mathbb{R}) \subset L^p\big(\mathbb{R}, \mathfrak{M}_L, \mu_L\big)$ for $p \in [1, \infty]$.

Lemma 17.8. *Let $f \in L^p\big(\mathbb{R}, \mathfrak{M}_L, \mu_L\big)$ with $p \in [1, \infty]$. For every $n \in \mathbb{N}$, let a truncation of f be defined by*

$$f_n(x) = \begin{cases} f(x) & \text{for } x \in \mathbb{R} \text{ such that } |x| \leq n \text{ and } |f(x)| \leq n, \\ 0 & \text{otherwise.} \end{cases}$$

Then the sequence of functions $(f_n : n \in \mathbb{N})$ *has the following properties:*

1° $\lim_{n\to\infty} f_n = f$ *a.e. on* $\mathbb{R}$,

2° $f_n \in L^p(\mathbb{R}, \mathfrak{M}_L, \mu_L)$ *for every* $n \in \mathbb{N}$,

3° $\|f_n\|_p \le \|f\|_p$ *for every* $n \in \mathbb{N}$,

4° $\lim_{n\to\infty} \|f_n - f\|_p = 0$ *for* $p \in [1, \infty)$.

Proof. Note that 1° holds since $|f| < \infty$ a.e. on $\mathbb{R}$. 2° and 3° are immediate consequences of the definition of f_n. To prove 4°, note that from $|f_n - f| \le |f_n| + |f| \le 2|f|$ we have $|f_n - f|^p = 2^p|f|^p$ for $p \in [1, \infty)$. Since $f \in L^p(\mathbb{R}, \mathfrak{M}_L, \mu_L)$, $|f|^p$ is μ_L-integrable on $\mathbb{R}$ and so is $2^p|f|^p$. Then by the Dominated Convergence Theorem (Theorem 9.20), we have $\lim_{n\to\infty} \int_{\mathbb{R}} |f_n - f|^p \, d\mu_L = \int_{\mathbb{R}} \lim_{n\to\infty} |f_n - f|^p \, d\mu_L = 0$. This proves 4°. ∎

Remark 17.9. 4° in Lemma 17.8 does not hold for $p = \infty$. As a counter example, let $f = 1$ on $\mathbb{R}$. Then $f \in L^\infty(\mathbb{R}, \mathfrak{M}_L, \mu_L)$ but $\|f_n - f\|_\infty = 1$ for every $n \in \mathbb{N}$.

Theorem 17.10. $C_c(\mathbb{R})$ *is dense in* $L^p(\mathbb{R}, \mathfrak{M}_L, \mu_L)$ *for every* $p \in [1, \infty)$, *that is, every open set in the metric space* $L^p(\mathbb{R}, \mathfrak{M}_L, \mu_L)$ *contains an element of* $C_c(\mathbb{R})$, *or equivalently, for every* $f \in L^p(\mathbb{R}, \mathfrak{M}_L, \mu_L)$ *and* $\varepsilon > 0$, *there exists* $g \in C_c(\mathbb{R})$ *such that* $\|f - g\|_p < \varepsilon$.

Proof. 1. Let $f \in L^p(\mathbb{R}, \mathfrak{M}_L, \mu_L)$. Let us consider first the particular case where f is extended real-valued, that is, $\Im f = 0$. Let $(f_n : n \in \mathbb{N})$ be the sequence of truncated functions as defined in Lemma 17.8. Since f is extended real-valued, f_n is real-valued on $\mathbb{R}$. Let $\varepsilon > 0$. By 4° of Lemma 17.8, $\lim_{n\to\infty} \|f_n - f\|_p = 0$ and thus there exists $N \in \mathbb{N}$ such that

(1)
$$\|f - f_N\|_p < \frac{\varepsilon}{2}.$$

By definition, $|f_N| \le N$ on $\mathbb{R}$ and f_N vanishes outside of the interval $[-N, N]$. Let us approximate f_N by a real-valued continuous function which is bounded by N on $\mathbb{R}$ and vanishes outside of a finite closed interval as follows. According to Theorem 6.36, there exists a real-valued continuous function g_0 on $[-N, N]$ such that $|g_0| \le N$ on $[-N, N]$ and

(2)
$$|f_N - g_0| \le \left(\frac{1}{8N}\right)^{1/p} \frac{\varepsilon}{2} \quad \text{on } [-N, N] \setminus E,$$

where E is a $\mathfrak{M}_L$-measurable subset of $[-N, N]$ with

(3)
$$\mu_L(E) < \frac{1}{4(2N)^p}\left(\frac{\varepsilon}{2}\right)^p.$$

We have then

$$\begin{aligned}\int_{[-N,N]} |f_N - g_0|^p \, d\mu_L &= \int_{[-N,N]\setminus E} |f_N - g_0|^p \, d\mu_L + \int_E |f_N - g_0|^p \, d\mu_L \\ &\le \frac{1}{8N}\left(\frac{\varepsilon}{2}\right)^p 2N + (2N)^p \frac{1}{4(2N)^p}\left(\frac{\varepsilon}{2}\right)^p = \frac{1}{2}\left(\frac{\varepsilon}{2}\right)^p.\end{aligned}$$

To extend g_0 to a real-valued continuous function g bounded by N on $\mathbb{R}$ and vanishing outside of a finite closed interval, consider two intervals $I_1 = \left[-N - \frac{1}{4N^p}\left(\frac{\varepsilon}{2}\right)^p, -N\right]$ and $I_2 = \left[N, N + \frac{1}{4N^p}\left(\frac{\varepsilon}{2}\right)^p\right]$ and define g by setting

$$g(x) = \begin{cases} g_0(x) & \text{for } x \in [-N, N], \\ 0 & \text{for } x \in (I_1 \cup [-N, N] \cup I_2)^c, \\ \text{linear} & \text{on } I_1 \text{ and on } I_2. \end{cases}$$

Then

$$\int_{I_1} |g|^p \, d\mu_L \leq \frac{N^p}{4N^p}\left(\frac{\varepsilon}{2}\right)^p = \frac{1}{4}\left(\frac{\varepsilon}{2}\right)^p$$

and similarly we have $\int_{I_2} |g|^p \, d\mu_L \leq \frac{1}{4}\left(\frac{\varepsilon}{2}\right)^p$. Thus we have

$$\begin{aligned} \int_{\mathbb{R}} |f_N - g|^p \, d\mu_L &= \int_{[-N,N]} |f_N - g_0|^p \, d\mu_L + \int_{I_1} |g|^p \, d\mu_L + \int_{I_2} |g|^p \, d\mu_L \\ &\leq \frac{1}{2}\left(\frac{\varepsilon}{2}\right)^p + \frac{1}{4}\left(\frac{\varepsilon}{2}\right)^p + \frac{1}{4}\left(\frac{\varepsilon}{2}\right)^p = \left(\frac{\varepsilon}{2}\right)^p, \end{aligned}$$

and therefore $\|f_N - g\|_p < \frac{\varepsilon}{2}$. With (1) we have

$$\|f - g\|_p \leq \|f - f_N\|_p + \|f_N - g\|_p < \frac{\varepsilon}{2} + \frac{\varepsilon}{2} = \varepsilon.$$

2. Now consider an extended complex-valued $f \in L^p(\mathbb{R}, \mathfrak{M}_L, \mu_L)$. For an arbitrary $\varepsilon > 0$, according to **1** there exist two real-valued continuous functions g_1 and g_2 on $\mathbb{R}$ vanishing outside of a finite closed interval such that $\|\Re f - g_1\|_p < \frac{\varepsilon}{2}$ and $\|\Im f - g_1\|_p < \frac{\varepsilon}{2}$. Let $g = g_1 + ig_2$. Then g is a complex-valued continuous function on $\mathbb{R}$ vanishing outside of a finite closed interval and

$$\begin{aligned} \|f - g\|_p &= \left\|\{\Re f + i\Im f\} - \{g_1 + ig_2\}\right\|_p \\ &\leq \|\Re f - g_1\|_p + \|\Im f - g_1\|_p < \frac{\varepsilon}{2} + \frac{\varepsilon}{2} = \varepsilon. \end{aligned}$$

This completes the proof. ∎

Remark 17.11. $C_c(\mathbb{R})$ is contained but is not dense in $L^\infty(\mathbb{R}, \mathfrak{M}_L, \mu_L)$.

Example. Consider $f \in L^\infty(\mathbb{R}, \mathfrak{M}_L, \mu_L)$ defined by $f(x) = -1$ for $x \in (-\infty, 0)$ and $f(x) = 1$ for $x \in [0, \infty)$. Let us show that for $\varepsilon = \frac{1}{2}$, there does not exist $g \in C_c(\mathbb{R})$ such that $\|f - g\|_\infty < \frac{1}{2}$. Suppose there exists such $g \in C_c(\mathbb{R})$. Then $|f - g| < \frac{1}{2}$ a.e. on $\mathbb{R}$ and this implies that $g < -\frac{1}{2}$ a.e. on $(-\infty, 0)$ and $g > \frac{1}{2}$ a.e. on $[0, \infty)$ and consequently $g(x) \in \left[-\frac{1}{2}, \frac{1}{2}\right]^c$ for a.e. $x \in \mathbb{R}$. Then since g is continuous on $\mathbb{R}$ and assumes a value $a < -\frac{1}{2}$ and a value $b > \frac{1}{2}$, there exists $x_0 \in \mathbb{R}$ such that $g(x_0) = 0$ by the Intermediate Value Theorem. The continuity of g implies then that there exists $\delta > 0$ such that $g(x) \in \left(-\frac{1}{2}, \frac{1}{2}\right)$ for $x \in (x_0 - \delta, x_0 + \delta)$. This contradicts the fact that $g(x) \in \left[-\frac{1}{2}, \frac{1}{2}\right]^c$ for a.e. $x \in \mathbb{R}$.

Theorem 17.12. *Let $L^p([a, b]) = L^p([a, b], \mathfrak{M}_L \cap [a, b], \mu_L)$ where $\mathfrak{M}_L \cap [a, b]$ is the σ-algebra of subsets of $[a, b]$ consisting of all members of $\mathfrak{M}_L$ contained in $[a, b]$ and $p = [1, \infty]$. Let $C([a, b])$ be the collection of all complex-valued continuous functions on $[a, b]$. Then*

$$C([a,b]) \subset L^\infty([a,b]) \subset L^{p_2}([a,b]) \subset L^{p_1}([a,b]) \quad \text{for } 1 \le p_1 < p_2 < \infty.$$

Furthermore each of $C([a, b])$, $L^\infty([a, b])$, and $L^{p_2}([a, b])$ is dense in $L^{p_1}([a, b])$.

Proof. Every complex-valued continuous function on $[a, b]$ is bounded on $[a, b]$. Thus $C([a, b]) \subset L^\infty([a, b])$. If $f \in L^\infty([a, b])$, then $\int_{[a,b]} |f|^p \, d\mu_L \le \|f\|_\infty^p (b - a) < \infty$ and thus $L^\infty([a, b]) \subset L^p([a, b])$ for any $p \in [1, \infty)$. Finally $L^{p_2}([a, b]) \subset L^{p_1}([a, b])$ by Theorem 17.4.

Using the fact that every complex-valued continuous function on $[a, b]$ is bounded on $[a, b]$, we prove the denseness of $C([a, b])$ in $L^p([a, b])$ for any $p \in [1, \infty)$ by the same argument as in Theorem 17.10. Thus every open set in $L^p([a, b])$ contains an element of $C([a, b])$. Then since $C([a, b]) \subset L^\infty([a, b])$, every open set in $L^p([a, b])$ contains an element of $L^\infty([a, b])$ so that $L^\infty([a, b])$ is dense in $L^p([a, b])$. By the same reason $L^{p_2}([a, b])$ is dense in $L^{p_1}([a, b])$. ∎

[III] L^p Spaces with $p \in (0, 1]$

Let $(X, \mathfrak{A}, \mu)$ be a measure space with $\mu(X) \in (0, \infty)$. We show next that for $p \in (0, 1]$, the space $L^p(X, \mathfrak{A}, \mu)$ decreases as p increases. We showed in Remark 16.21 that $\|\cdot\|_p$ is not a norm when $p \in (0, 1)$ and hence M_p in Definition 17.1 is not a norm when $p \in (0, 1)$. Nevertheless as we show below for a fixed $f \in L^1(X, \mathfrak{A}, \mu)$, $M_p(f)$ is an increasing function of $p \in (0, 1]$.

Theorem 17.13. *Given a measure space $(X, \mathfrak{A}, \mu)$ with $\mu(X) \in (0, \infty)$. Then for $0 < p_1 < p_2 \le 1$ we have*

(1) $$L^{p_1}(X, \mathfrak{A}, \mu) \supset L^{p_2}(X, \mathfrak{A}, \mu)$$

and

(2) $$M_{p_1}(f) \le M_{p_2}(f) \quad \text{for } f \in L^{p_2}(X, \mathfrak{A}, \mu).$$

Proof. 1. Let us prove (1) and (2) for the special case $p_2 = 1$ first. For our $p_1 \in (0, 1)$, let $q \in (-\infty, 0)$ be its conjugate, that is, $1/p_1 + 1/q = 1$. Let $f \in L^1(X, \mathfrak{A}, \mu)$. Applying Theorem 16.54 to the pair of functions $|f|$ and $\mathbf{1}_X$ with conjugate exponents p_1 and q, we have

(3) $$\int_X |f| \, d\mu = \int_X |f| \mathbf{1}_X \, d\mu \ge \left[\int_X |f|^{p_1} \, d\mu\right]^{1/p_1} \mu(X)^{1/q}.$$

Since $f \in L^1(X, \mathfrak{A}, \mu)$, we have $\int_X |f|\,d\mu < \infty$. Then (3) implies that $\int_X |f|^{p_1}\,d\mu < \infty$. This shows that $f \in L^{p_1}(X, \mathfrak{A}, \mu)$ and proves (1). To prove (2), let $f \in L^1(X, \mathfrak{A}, \mu)$. Then by (3), we have

$$M_1(f) = \frac{1}{\mu(X)} \int_X |f|\,d\mu \geq \frac{1}{\mu(X)} \left[\int_X |f|^{p_1}\,d\mu\right]^{1/p_1} \mu(X)^{1/q}$$
$$= \frac{1}{\mu(X)^{1/p_1}} \left[\int_X |f|^{p_1}\,d\mu\right]^{1/p_1} = M_{p_1}(f).$$

2. Consider the case $0 < p_1 < p_2 < 1$. Let $\alpha = p_1/p_2 \in (0, 1)$. Let $\beta \in (-\infty, 0)$ be the conjugate of α, that is, $1/\alpha + 1/\beta = 1$. Let $f \in L^{p_2}(X, \mathfrak{A}, \mu)$. Applying Theorem 16.54 to the pair of functions $|f|^{p_1/\alpha}$ and $\mathbf{1}_X$ with conjugate exponents α and β, we have

$$(4) \qquad \int_X |f|^{p_2}\,d\mu = \int_X |f|^{p_1/\alpha} \mathbf{1}_X\,d\mu \geq \left[\int_X |f|^{p_1}\,d\mu\right]^{1/\alpha} \mu(X)^{1/\beta}.$$

Since $f \in L^{p_2}(X, \mathfrak{A}, \mu)$, we have $\int_X |f|^{p_2}\,d\mu < \infty$. Then $\int_X |f|^{p_1}\,d\mu < \infty$ by (4) so that $f \in L^{p_1}(X, \mathfrak{A}, \mu)$. This proves (1). To prove (2), note that for $f \in L^{p_2}(X, \mathfrak{A}, \mu)$ we have

$$M_{p_2}(f) = \left[\frac{1}{\mu(X)} \int_X |f|^{p_2}\,d\mu\right]^{1/p_2} \geq \left[\frac{1}{\mu(X)} \left[\int_X |f|^{p_1}\,d\mu\right]^{1/\alpha} \mu(X)^{1/\beta}\right]^{1/p_2}$$
$$= \left[\frac{1}{\mu(X)^{1/\alpha}} \left[\int_X |f|^{p_1}\,d\mu\right]^{1/\alpha}\right]^{1/p_2} = \left[\frac{1}{\mu(X)} \int_X |f|^{p_1}\,d\mu\right]^{1/\alpha p_2} = M_{p_1}(f),$$

where the inequality is by (4) and the last equality is by $\alpha p_2 = p_1$. ■

As we noted above when $p \in (0, 1)$, $\|f\|_p = \left\{\int_X |f|^p\,d\mu\right\}^{1/p}$ for $f \in L^p(X, \mathfrak{A}, \mu)$ fails to satisfy the triangle inequality and hence $\|\cdot\|_p$ is not a norm on $L^p(X, \mathfrak{A}, \mu)$. However if we define $\rho_p(f, g) = \int_X |f - g|^p\,d\mu$ for $f, g \in L^p(X, \mathfrak{A}, \mu)$, then ρ_p satisfies the triangle inequality and indeed ρ_p is a metric on $L^p(X, \mathfrak{A}, \mu)$ and moreover $L^p(X, \mathfrak{A}, \mu)$ is a complete metric space with respect to the metric ρ_p. To prove this we need the following:

Lemma 17.14. *Let $p \in (0, 1)$. Then for any $\alpha, \beta \in \mathbb{C}$, we have*

$$|\alpha + \beta|^p \leq |\alpha|^p + |\beta|^p$$

where the equality holds if and only if at least one of α and β is equal to 0.

Proof. It is obvious that if at least one of α and β is equal to 0 then the equality holds. It remains to show that if neither of α and β is equal to 0 then the strict inequality holds.

Let φ be a real-valued function on $[0, \infty)$ defined by

$$\varphi(t) = (1 + t)^p - 1 - t^p \quad \text{for } t \in [0, \infty).$$

We have $\varphi(0) = 0$ and $\varphi'(t) = p\{(1+t)^{p-1} - t^{p-1}\}$ for $t \in (0, \infty)$. Since $p \in (0, 1)$, $\varphi'(t) < 0$ for $t \in [0, \infty)$. Thus φ is strictly decreasing on $[0, \infty)$ and consequently $\varphi(t) < 0$ for $t \in (0, \infty)$. Let $a, b \in \mathbb{R}$, $a > 0$ and $b > 0$. Then $\frac{b}{a} > 0$ so that

$$\left(1 + \tfrac{b}{a}\right)^p - 1 - \left(\tfrac{b}{a}\right)^p = \varphi\left(\tfrac{b}{a}\right) < 0.$$

Multiplying the inequality by a^p, we obtain

$$(1) \qquad (a+b)^p < a^p + b^p \quad \text{for } a > 0 \text{ and } b > 0.$$

Let $\alpha, \beta \in \mathbb{C}$, $\alpha \neq 0$ and $\beta \neq 0$. Then $|\alpha| > 0$ and $|\beta| > 0$ so that by (1) we have

$$|\alpha + \beta|^p \leq \{|\alpha| + |\beta|\}^p < |\alpha|^p + |\beta|^p.$$

This completes the proof. ∎

Theorem 17.15. *Let $(X, \mathfrak{A}, \mu)$ be a measure space. For $p \in (0, 1)$ consider the linear space $L^p(X, \mathfrak{A}, \mu)$. Let*

$$\rho_p(f, g) = \int_X |f - g|^p \, d\mu \quad \text{for } f, g \in L^p(X, \mathfrak{A}, \mu).$$

Then ρ_p is a metric on $L^p(X, \mathfrak{A}, \mu)$ and $L^p(X, \mathfrak{A}, \mu)$ is a complete metric space with respect to the metric ρ_p.

Proof. 1. $L^p(X, \mathfrak{A}, \mu)$ is the linear space of the equivalence classes, the equivalence relation being that of μ-a.e. equality on $(X, \mathfrak{A})$, of extended complex-valued $\mathfrak{A}$-measurable functions f on X with $\int_X |f|^p \, d\mu < \infty$. Now $\int_X |f|^p \, d\mu < \infty$ implies that $|f| < \infty$ μ-a.e. on $(X, \mathfrak{A})$ so that f is complex-valued μ-a.e. on $(X, \mathfrak{A})$. On the null set in $(X, \mathfrak{A}, \mu)$ on which $|f| = \infty$, let us redefine f to be equal to 0 so that f is complex-valued and $\mathfrak{A}$-measurable on X. This then is a complex-valued (not extended complex-valued) representative of $f \in L^p(X, \mathfrak{A}, \mu)$. In what follows we always take a complex-valued representative of every $f \in L^p(X, \mathfrak{A}, \mu)$. Then in particular for every $f, g \in L^p(X, \mathfrak{A}, \mu)$, $f - g$ is defined and complex-valued on X.

Let us show that ρ_p is a metric on $L^p(X, \mathfrak{A}, \mu)$. For $f, g \in L^p(X, \mathfrak{A}, \mu)$, we have

$$\rho_p(f, g) = \int_X |f - g|^p \, d\mu \leq \int_X |f|^p \, d\mu + \int_X |g|^p \, d\mu < \infty$$

by Lemma 17.14. Thus $\rho_p(f, g) \in [0, \infty)$ for every $f, g \in L^p(X, \mathfrak{A}, \mu)$. If $f = g$, then clearly $\rho_p(f, g) = 0$. Conversely if $\rho_p(f, g) = 0$ then $|f - g|^p = 0$, that is, $|f - g| = 0$, μ-a.e. on $(X, \mathfrak{A})$ so that $f = g$ μ-a.e. on $(X, \mathfrak{A})$, that is, $f = g$ as elements of $L^p(X, \mathfrak{A}, \mu)$. Clearly $\rho_p(f, g) = \rho_p(g, f)$ for every $f, g \in L^p(X, \mathfrak{A}, \mu)$. To verify the triangle inequality for ρ_p, let $f, g, h \in L^p(X, \mathfrak{A}, \mu)$. Then we have

$$\begin{aligned}
\rho_p(f, g) &= \int_X |f - g|^p \, d\mu = \int_X |(f - h) + (h - g)|^p \, d\mu \\
&\leq \int_X |f - h|^p \, d\mu + \int_X |h - g|^p \, d\mu \\
&= \rho_p(f, h) + \rho_p(h, g)
\end{aligned}$$

where the inequality is by Lemma 17.14. This shows that ρ_p is a metric on $L^p(X, \mathfrak{A}, \mu)$.

2. To show that $L^p(X, \mathfrak{A}, \mu)$ is complete with respect to the metric ρ_p, let $(f_n : n \in \mathbb{N})$ be a Cauchy sequence in $L^p(X, \mathfrak{A}, \mu)$ with respect to ρ_p, that is, the sequence $(f_n : n \in \mathbb{N})$ is such that for every $\varepsilon > 0$ there exists $N \in \mathbb{N}$ such that

$$(1) \qquad \rho_p(f_m, f_n) < \varepsilon \quad \text{for } m, n \geq N.$$

Let us show that there exists $f \in L^p(X, \mathfrak{A}, \mu)$ such that $\lim_{n \to \infty} \rho_p(f_n, f) = 0$. Now by (1) for every $k \in \mathbb{N}$ there exists $n_k \in \mathbb{N}$ such that

$$\rho_p(f_m, f_n) < \frac{1}{2^k} \quad \text{for } m, n \geq n_k.$$

We can select $n_k \in \mathbb{N}$ so that $n_k < n_{k+1}$ for every $k \in \mathbb{N}$. Consider the subsequence $(f_{n_k} : i \in \mathbb{N})$ of $(f_n : n \in \mathbb{N})$. We have

$$(2) \qquad \rho_p(f_{n_{k+1}}, f_{n_k}) < \frac{1}{2^k} \quad \text{for } k \in \mathbb{N}.$$

For each $\ell \in \mathbb{N}$, let us define a nonnegative real-valued $\mathfrak{A}$-measurable function on X by

$$(3) \qquad g_\ell = \sum_{k=1}^{\ell} |f_{n_{k+1}} - f_{n_k}|.$$

Let us show that $g_\ell \in L^p(X, \mathfrak{A}, \mu)$. Now $|\alpha_1 + \cdots + \alpha_\ell|^p = |\alpha_1|^p + \cdots + |\alpha_\ell|^p$ for $\alpha_1, \ldots, \alpha_\ell \in \mathbb{R}$ by iterated application of Lemma 17.14. Thus $|g_\ell|^p \leq \sum_{k=1}^{\ell} |f_{n_{k+1}} - f_{n_k}|^p$. Integrating with respect to μ on X and recalling (2) we have

$$(4) \qquad \begin{aligned} \int_X |g_\ell|^p \, d\mu &\leq \sum_{k=1}^{\ell} \int_X |f_{n_{k+1}} - f_{n_k}|^p \, d\mu = \sum_{k=1}^{\ell} \rho_p(f_{n_{k+1}}, f_{n_k}) \\ &< \sum_{k=1}^{\ell} \frac{1}{2^k} < \sum_{k \in \mathbb{N}} \frac{1}{2^k} = 1 < \infty. \end{aligned}$$

This shows that $g_\ell \in L^p(X, \mathfrak{A}, \mu)$. Now for every $x \in X$, $(g_\ell(x) : \ell \in \mathbb{N})$ is an increasing sequence of nonnegative real numbers so that $\lim_{\ell \to \infty} g_\ell(x)$ exists. Let us define a nonnegative extended real-valued $\mathfrak{A}$-measurable function on X by setting

$$g = \lim_{\ell \to \infty} g_\ell = \lim_{\ell \to \infty} \sum_{k=1}^{\ell} |f_{n_{k+1}} - f_{n_k}|.$$

Now $|g|^p = \lim_{\ell \to \infty} |g_\ell|^p$ on X. Thus by Fatou's Lemma (Theorem 8.13) and (4), we have

$$\int_X |g|^p \, d\mu \leq \liminf_{\ell \to \infty} \int_X |g_\ell|^p \, d\mu \leq 1.$$

Thus $g < \infty$ μ-a.e. on $(X, \mathfrak{A})$. Then $\lim_{\ell\to\infty} \sum_{k=1}^{\ell} |f_{n_{k+1}} - f_{n_k}|$ converges μ-a.e. on $(X, \mathfrak{A})$, that is, the series $\sum_{k\in\mathbb{N}} |f_{n_{k+1}} - f_{n_k}|$ converges μ-a.e. on $(X, \mathfrak{A})$. Since the absolute convergence of a series of complex numbers implies convergence of the series, $\sum_{k\in\mathbb{N}} \{f_{n_{k+1}} - f_{n_k}\}$ converges μ-a.e. on $(X, \mathfrak{A})$. Let

$$f(x) = f_{n_1}(x) + \sum_{k\in\mathbb{N}} \{f_{n_{k+1}}(x) - f_{n_k}(x)\}$$

at any $x \in X$ at which $\sum_{k\in\mathbb{N}} \{f_{n_{k+1}} - f_{n_k}\}$ converges and let $f = 0$ on the null set in $(X, \mathfrak{A}, \mu)$ on which $\sum_{k\in\mathbb{N}} \{f_{n_{k+1}} - f_{n_k}\}$ does not converge. Thus defined, f is a complex-valued $\mathfrak{A}$-measurable function on X. Now

$$(5) \qquad f(x) = f_{n_1}(x) + \lim_{\ell\to\infty} \sum_{k=1}^{\ell} \{f_{n_{k+1}}(x) - f_{n_k}(x)\} = \lim_{k\to\infty} f_{n_{k+1}}(x) \quad \text{for } \mu\text{-a.e. } x \text{ in } (X, \mathfrak{A}).$$

Let us show that $f \in L^p(X, \mathfrak{A}, \mu)$. Let $\varepsilon > 0$ be arbitrarily given. Let $m, n_k \geq N$ as in (1) so that $\rho_p(f_{n_k}, f_m) < \varepsilon$. Then by (5) and Fatou's Lemma we have for $m \geq N$

$$(6) \qquad \int_X |f - f_m|^p \, d\mu \leq \liminf_{k\to\infty} \int_X |f_{n_k} - f_m|^p \, d\mu = \liminf_{k\to\infty} \rho_p(f_{n_k}, f_m) \leq \varepsilon.$$

Now since $|f|^p = |(f - f_m) + f_m|^p \leq |f - f_m|^p + |f_m|^p$ by Lemma 17.14, we have

$$\int_X |f|^p \, d\mu \leq \int_X |f - f_m|^p \, d\mu + \int_X |f_m|^p \, d\mu \leq \varepsilon + \int_X |f_m|^p \, d\mu < \infty.$$

This shows that $f \in L^p(X, \mathfrak{A}, \mu)$.

By (6) for every $\varepsilon > 0$ there exists $N \in \mathbb{N}$ such that $\rho_p(f_m, f) \leq \varepsilon$ for $m \geq N$. Then $\limsup_{m\to\infty} \rho_p(f_m, f) \leq \varepsilon$. Since this holds for every $\varepsilon > 0$, we have $\limsup_{m\to\infty} \rho_p(f_m, f) = 0$. This implies $\liminf_{m\to\infty} \rho_p(f_m, f) = 0$ and then $\lim_{m\to\infty} \rho_p(f_m, f) = 0$. ∎

Remark 17.16. Regarding the metric ρ_p in Theorem 17.15, let us observe that if we define

$$\|f\| = \int_X |f|^p \, d\mu \quad \text{for } f \in L^p(X, \mathfrak{A}, \mu),$$

then we have $\|f - g\| = \int_X |f - g|^p \, d\mu = \rho_p(f, g)$ for $f, g \in L^p(X, \mathfrak{A}, \mu)$. Nevertheless the function $\|\cdot\|$ is not a norm on $L^p(X, \mathfrak{A}, \mu)$ since the equality $\|\alpha f\| = |\alpha| \|f\|$ does not hold for some $f \in L^p(X, \mathfrak{A}, \mu)$ and $\alpha \in \mathbb{C}$. Indeed if $f \in L^p(X, \mathfrak{A}, \mu)$ is such that $\|f\| = \int_X |f|^p \, d\mu > 0$ and $\alpha \in \mathbb{C}$ is such that $\alpha \neq 0, 1$ so that $|\alpha|^p \neq |\alpha|$ for our $p \in (0, 1)$, then

$$\|\alpha f\| = \int_X |\alpha f|^p = |\alpha|^p \int_X |f|^p \neq |\alpha| \int_X |f|^p = |\alpha| \|f\|.$$

Thus for $p \in (0, 1)$, the linear space $L^p(X, \mathfrak{A}, \mu)$ is a complete metric space with respect to the metric ρ_p but it is not a Banach space.

[IV] The ℓ^p Spaces

Definition 17.17. *The measure space $\big(\mathbb{N}, \mathfrak{P}(\mathbb{N}), \nu\big)$ where $\mathfrak{P}(\mathbb{N})$ is the σ-algebra of all subsets of $\mathbb{N}$ and $\nu(E)$ is equal to the number of elements in E, and in particular $\nu(E) = \infty$ when E is an infinite set, for $E \in \mathfrak{P}(\mathbb{N})$ is called the counting measure space. For $k \in \mathbb{N}$, let $\mathbb{N}_k = \{1, \ldots, k\}$. We call $\big(\mathbb{N}_k, \mathfrak{P}(\mathbb{N}_k), \nu\big)$ a finite counting measure space.*

More generally, for an arbitrary set X a set function ν defined on the σ-algebra $\mathfrak{P}(X)$ consisting of all subsets of X by setting $\nu(E)$ for $E \in \mathfrak{P}(X)$ to be equal to the number of elements in E, and in particular $\nu(E) = \infty$ when E is an infinite set, is called a counting measure on X. (The fact that ν is indeed a measure is verified immediately.)

Note that $\big(\mathbb{N}, \mathfrak{P}(\mathbb{N}), \nu\big)$ is a σ-finite measure space and $\big(\mathbb{N}_k, \mathfrak{P}(\mathbb{N}_k), \nu\big)$ is a finite measure space. The only null sets in $\big(\mathbb{N}, \mathfrak{P}(\mathbb{N}), \nu\big)$ is the empty set $\emptyset$. Thus a.e. equality in $\big(\mathbb{N}, \mathfrak{P}(\mathbb{N}), \nu\big)$ is equivalent to equality everywhere. Since the σ-algebra $\mathfrak{P}(\mathbb{N})$ consists of all subsets of $\mathbb{N}$, every extended real-valued function on $\mathbb{N}$ is $\mathfrak{P}(\mathbb{N})$-measurable.

Every real-valued function f on $\mathbb{N}$ is given by $\big(f(n) : n \in \mathbb{N}\big)$ where $f(n) \in \mathbb{R}$ and is thus a sequence of real numbers. Conversely every sequence of real numbers $(a_n : n \in \mathbb{N})$ can be interpreted as a real-valued function on $\mathbb{N}$. Thus we identify the collection of all real-valued functions on $\mathbb{N}$ with the collection of all sequences of real numbers. For a real-valued function f on $\mathbb{N}$, we have $f^+ = \max\{f, 0\}$ and thus

$$\int_{\mathbb{N}} f^+ \, d\nu = \sum_{n \in \mathbb{N}} \max\{f(n), 0\} \nu\big(\{n\}\big) = \sum_{n \in \mathbb{N}} \max\{f(n), 0\} \in [0, \infty],$$

and similarly $f^- = -\min\{f, 0\}$ so that

$$\int_{\mathbb{N}} f^- \, d\nu = \sum_{n \in \mathbb{N}} -\min\{f(n), 0\} \in [0, \infty].$$

A real-valued function f on $\mathbb{N}$ is ν semi-integrable if and only if at least one of

$\int_{\mathbb{N}} f^+ \, d\nu$ and $\int_{\mathbb{N}} f^- \, d\nu$ is finite, and in this case

$$\begin{aligned}\int_{\mathbb{N}} f \, d\nu &= \int_{\mathbb{N}} f^+ \, d\nu - \int_{\mathbb{N}} f^- \, d\nu \\ &= \sum_{n\in\mathbb{N}} \max\{f(n), 0\} + \sum_{n\in\mathbb{N}} \min\{f(n), 0\} \\ &= \sum_{n\in\mathbb{N}} \big(\max\{f(n), 0\} + \min\{f(n), 0\}\big) \\ &= \sum_{n\in\mathbb{N}} f(n),\end{aligned}$$

where the third equality is by the fact that if $(a_n : n \in \mathbb{N})$ and $(b_n : n \in \mathbb{N})$ are two sequences of real numbers such that $\sum_{n\in\mathbb{N}} a_n$ and $\sum_{n\in\mathbb{N}} b_n$ exist in $\overline{\mathbb{R}}$ and at least one is in $\mathbb{R}$, then $\sum_{n\in\mathbb{N}}\{a_n + b_n\}$ exists in $\overline{\mathbb{R}}$ and $\sum_{n\in\mathbb{N}}\{a_n + b_n\} = \sum_{n\in\mathbb{N}} a_n + \sum_{n\in\mathbb{N}} b_n$.

The function f is ν-integrable if and only if both $\int_{\mathbb{N}} f^+ \, d\nu$ and $\int_{\mathbb{N}} f^- \, d\nu$ are finite. In this case we have $\int_{\mathbb{N}} f \, d\nu = \sum_{n\in\mathbb{N}} f(n) \in \mathbb{R}$. But f is ν-integrable if and only if $|f|$ is ν-integrable according to Observation 9.2. Thus f is ν-integrable if and only if $\sum_{n\in\mathbb{N}} |f(n)| < \infty$. (From the view point of convergence of series of real numbers, the ν-integrability of $f = \big(f(n) : n \in \mathbb{N}\big)$ is not only the convergence of the series $\sum_{n\in\mathbb{N}} f(n)$ but also its absolute convergence.)

The collection of all complex-valued functions on $\mathbb{N}$ is identified with the collection of all sequences of complex numbers. The ν semi-integrability of such a function is defined as the ν semi-integrability of both its real and imaginary parts. Similarly for the ν-integrability.

Observation 17.18. Let f be a complex-valued function on $\mathbb{N}$, that is, $f = \big(f(n) : n \in \mathbb{N}\big)$ where $f(n) \in \mathbb{C}$ for $n \in \mathbb{N}$. For $p \in (0, \infty)$, we have according to Definition 16.8

(1) $$\|f\|_p = \left[\int_{\mathbb{N}} |f|^p \, d\nu\right]^{1/p} = \left[\sum_{n\in\mathbb{N}} |f(n)|^p\right]^{1/p} \in [0, \infty],$$

and according to Definition 16.36, we have

(2) $$\begin{aligned}\|f\|_\infty &= \inf\big\{M \geq 0 : \nu\{n \in \mathbb{N} : |f(n)| > M\} = 0\big\} \\ &= \inf\big\{M \geq 0 : |f(n)| \leq M \text{ for every } n \in \mathbb{N}\big\} \\ &= \sup\big\{|f(n)| : n \in \mathbb{N}\big\} \in [0, \infty],\end{aligned}$$

where the second equality is from the fact that $\emptyset$ is the only null set in the measure space $(\mathbb{N}, \mathfrak{P}(\mathbb{N}), \nu)$.

The spaces $L^p\big(\mathbb{N}, \mathfrak{P}(\mathbb{N}), \nu\big)$ for $p \in (0, \infty]$ are particular cases of $L^p(X, \mathfrak{A}, \mu)$ as in Definition 16.22 for $p \in (0, \infty)$ and in Definition 16.42 for $p = \infty$. Thus for $p \in (0, \infty)$, $L^p\big(\mathbb{N}, \mathfrak{P}(\mathbb{N}), \nu\big)$ is the linear space of all complex-valued functions f on $\mathbb{N}$ with $\|f\|_p <$

∞ and $L^\infty(\mathbb{N}, \mathfrak{P}(\mathbb{N}), \nu)$ is the linear space of all complex-valued functions f on $\mathbb{N}$ with $\|f\|_\infty < \infty$.

Definition 17.19. *We write ℓ^p for $L^p(\mathbb{N}, \mathfrak{P}(\mathbb{N}), \nu)$ for $p \in (0, \infty]$.*

Proposition 17.20. *For $p \in [1, \infty)$, the linear space ℓ^p of all sequences of complex numbers $f = (f(n) : n \in \mathbb{N})$ satisfying the condition $\sum_{n\in\mathbb{N}} |f(n)|^p < \infty$ is a Banach space with respect to the norm $\|f\|_p = \left[\sum_{n\in\mathbb{N}} |f(n)|^p\right]^{1/p}$. The linear space ℓ^∞ of all bounded sequences of complex numbers $f = (f(n) : n \in \mathbb{N})$ is a Banach space with respect to the norm $\|f\|_\infty = \sup_{n\in\mathbb{N}} |f(n)|$.*

Proof. Since $\ell^p = L^p(\mathbb{N}, \mathfrak{P}(\mathbb{N}), \nu)$ is a particular case of $L^p(X, \mathfrak{A}, \mu)$ for $p \in [1, \infty]$, ℓ^p is a Banach space by Theorem 16.23 for $p \in [1, \infty)$ and by Theorem 16.43 for $p = \infty$. ∎

In Theorems 17.4 and 17.13, we showed that $L^{p_1}(X, \mathfrak{A}, \mu) \supset L^{p_2}(X, \mathfrak{A}, \mu)$ for $0 < p_1 < p_2 \le \infty$ provided $\mu(X) < \infty$. This result does not apply to $\ell^p = L^p(\mathbb{N}, \mathfrak{P}(\mathbb{N}), \nu)$ since $\nu(\mathbb{N}) = \infty$. On the other hand, the ℓ^p spaces have the following reverse inclusion property.

Theorem 17.21. *For $0 < p_1 < p_2 \le \infty$, we have $\ell^{p_1} \subset \ell^{p_2}$.*

Proof. Let $f \in \ell^{p_1}$ be given by $f = (\alpha_n : n \in \mathbb{N})$ where $\alpha_n \in \mathbb{C}$ for $n \in \mathbb{N}$. Since $f \in \ell^{p_1}$, we have $\|f\|_{p_1} < \infty$, that is, $\sum_{n\in\mathbb{N}} |\alpha_n|^{p_1} < \infty$. This implies $\lim_{n\to\infty} |\alpha_n|^{p_1} = 0$ and then $\lim_{n\to\infty} |\alpha_n| = 0$. Thus there exists $M > 0$ such that $|\alpha_n| \le M$ for all $n \in \mathbb{N}$. This shows that $\|f\|_\infty < \infty$ and $f \in \ell^\infty$. Now $\lim_{n\to\infty} |\alpha_n| = 0$ also implies that there exists $N \in \mathbb{N}$ such that $|\alpha_n| < 1$ for $n \ge N$. If $p_1 < p_2 < \infty$, then $|\alpha_n|^{p_2} \le |\alpha_n|^{p_1}$ for $n \ge N$ so that $\sum_{n\ge N} |\alpha_n|^{p_2} \le \sum_{n\ge N} |\alpha_n|^{p_1} < \infty$ and then $\sum_{n\in\mathbb{N}} |\alpha_n|^{p_2} < \infty$, that is, $\|f\|_{p_2} < \infty$ and therefore $f \in \ell^{p_2}$. ∎

Since the ℓ^p spaces are particular cases of the L^p spaces, Hölder's Inequality and Minkowski's Inequality that we proved for the L^p spaces hold for the ℓ^p spaces. Thus we have for instance the following inequalities.

Proposition 17.22. *Let $(\alpha_n : n \in \mathbb{N})$ and $(\beta_n : n \in \mathbb{N})$ be two sequences of complex numbers. Let $p, q \in (1, \infty)$ be conjugates, that is, $1/p + 1/q = 1$. Then we have Hölder's Inequality for series, that is,*

(1)
$$\sum_{n\in\mathbb{N}} |\alpha_n \beta_n| \le \left[\sum_{n\in\mathbb{N}} |\alpha_n|^p\right]^{1/p} \left[\sum_{n\in\mathbb{N}} |\beta_n|^q\right]^{1/q}.$$

For $p \in [1, \infty)$, we have Minkowski's Inequality for series, that is,

(2)
$$\left[\sum_{n\in\mathbb{N}} |\alpha_n + \beta_n|^p\right]^{1/p} \le \left[\sum_{n\in\mathbb{N}} |\alpha_n|^p\right]^{1/p} + \left[\sum_{n\in\mathbb{N}} |\beta_n|^p\right]^{1/p}.$$

Proof. Consider the complex-valued functions $f = (\alpha_n : n \in \mathbb{N})$ and $g = (\beta_n : n \in \mathbb{N})$ on the measure space $(\mathbb{N}, \mathfrak{P}(\mathbb{N}), \nu)$. By (1) of Observation 17.18, we have $\|fg\|_1 = \sum_{n\in\mathbb{N}} |\alpha_n\beta_n|$, $\|f\|_p = \left[\sum_{n\in\mathbb{N}} |\alpha_n|^p\right]^{1/p}$, and $\|g\|_q = \left[\sum_{n\in\mathbb{N}} |\beta_n|^q\right]^{1/q}$. Then (1) is a restatement of (1) of Theorem 16.14 and (2) is a restatement of Theorem 16.17. ∎

Observation 17.23. Consider a finite counting measure space $(\mathbb{N}_k, \mathfrak{P}(\mathbb{N}_k), \nu)$. A complex valued function f on $\mathbb{N}_k$ is a finite sequence of complex numbers $f = (f(1), \dots, f(k))$. By Definition 16.8, we have $\|f\|_p = \left[\int_{\mathbb{N}_k} |f|^p \, d\nu\right]^{1/p} = \left[\sum_{i=1}^k |f(i)|^p\right]^{1/p} \in [0, \infty)$ for $p \in (0, \infty)$. By Definition 16.36, we have $\|f\|_\infty = \max_{i=1,\dots,k} |f(i)| \in [0, \infty)$. Let Φ be the collection of all complex-valued functions on $\mathbb{N}_k$. By Definition 16.22, $L^p(\mathbb{N}_k, \mathfrak{P}(\mathbb{N}_k), \nu)$ is the collection of all complex-valued functions f on $\mathbb{N}_k$ such that $\|f\|_p < \infty$ for $p \in (0, \infty)$. Since this condition is satisfied by every $f \in \Phi$, we have $L^p(\mathbb{N}_k, \mathfrak{P}(\mathbb{N}_k), \nu) = \Phi$. By Definition 16.42, $L^\infty(\mathbb{N}_k, \mathfrak{P}(\mathbb{N}_k), \nu)$ is the collection of all complex-valued functions f on $\mathbb{N}_k$ such that $\|f\|_\infty < \infty$. Since every $f \in \Phi$ satisfies this condition, we have $L^\infty(\mathbb{N}_k, \mathfrak{P}(\mathbb{N}_k), \nu) = \Phi$. Thus we have $L^p(\mathbb{N}_k, \mathfrak{P}(\mathbb{N}_k), \nu) = \Phi$ for all $p \in (0, \infty]$. Since $\|\cdot\|_p$ is a norm on $L^p(\mathbb{N}_k, \mathfrak{P}(\mathbb{N}_k), \nu)$ and $L^p(\mathbb{N}_k, \mathfrak{P}(\mathbb{N}_k), \nu)$ is a Banach space with respect to the norm $\|\cdot\|_p$ by Theorem 16.23 for $p \in [1, \infty)$ and by Theorem 16.43 for $p = \infty$, Φ is a Banach space with respect to $\|\cdot\|_p$ for $p \in [1, \infty]$.

Proposition 17.24. *Let $(\alpha_1, \dots, \alpha_k)$ and $(\beta_1, \dots, \beta_k)$ be two finite sequences of complex numbers. Then for $p, q \in (1, \infty)$ such that $1/p + 1/q = 1$, we have Hölder's Inequality*

(1) $$\sum_{i=1}^k |\alpha_i\beta_i| \le \left[\sum_{i=1}^k |\alpha_i|^p\right]^{1/p} \left[\sum_{i=1}^k |\beta_i|^q\right]^{1/q}.$$

For $p \in [1, \infty)$, we have Minkowski's Inequality

(2) $$\left[\sum_{i=1}^k |\alpha_i + \beta_i|^p\right]^{1/p} \le \left[\sum_{i=1}^k |\alpha_i|^p\right]^{1/p} + \left[\sum_{i=1}^k |\beta_i|^p\right]^{1/p}.$$

For $1 \le p_1 < p_2 \le \infty$, we have

(3) $$\frac{1}{k}\sum_{i=1}^k |\alpha_i| \le \frac{1}{k^{1/p_1}}\left[\sum_{i=1}^k |\alpha_i|^{p_1}\right]^{1/p_1} \le \frac{1}{k^{1/p_2}}\left[\sum_{i=1}^k |\alpha_i|^{p_2}\right]^{1/p_2} \le \max_{i=1,\dots,k} |\alpha_i|.$$

Proof. (1) and (2) are particular cases of Theorems 16.14 and 16.17. To prove (3), regard $(\alpha_1, \dots, \alpha_k)$ and $(\beta_1, \dots, \beta_k)$ as two complex-valued functions on the finite counting measure space $(\mathbb{N}_k, \mathfrak{P}(\mathbb{N}_k), \nu)$. Since $\nu(\mathbb{N}_k) = k < \infty$, we have $M_p(f) = k^{-1/p}\|f\|_p$ for $p \in (0, \infty)$ and $M_\infty = \|f\|_\infty$ by Definition 17.1. By Theorem 17.4, we have $M_1(f) \le M_{p_1}(f) \le M_{p_2}(f) \le M_\infty(f)$ for $1 \le p_1 < p_2 \le \infty$. Since $\|f\|_p = \left[\sum_{i=1}^k |\alpha_i|^p\right]^{1/p}$ for $p \in [1, \infty)$ and $\|f\|_\infty = \max_{i=1,\dots,k} |\alpha_i|$, (3) follows. ∎

Proposition 17.25. *On $\mathbb{C}^k$, let us define $\|x\|_p = \left\{\sum_{i=1}^k |x_i|^p\right\}^{1/p}$ for $p \in [1, \infty)$ and $\|x\|_\infty = \max_{i=1,\dots,k} |x_i|$ for $x = (x_1, \dots, x_k) \in \mathbb{C}^k$.*

(a) *For every $p \in [1, \infty]$, $\|\cdot\|_p$ is a norm on $\mathbb{C}^k$ and $\mathbb{C}^k$ is a Banach space with respect to the norm $\|\cdot\|_p$.*

(b) *For $p \in [1, \infty)$, we have $\|\cdot\|_p \geq \|\cdot\|_\infty$ and for $1 \leq p_1 < p_2 \leq \infty$ we have*

$$(1) \qquad \frac{\|\cdot\|_1}{k} \leq \frac{\|\cdot\|_{p_1}}{k^{1/p_1}} \leq \frac{\|\cdot\|_{p_2}}{k^{1/p_2}} \leq \|\cdot\|_\infty.$$

Thus any two of the norms $\|\cdot\|_p$ for $p \in [1, \infty]$ is a pair of equivalent norms.

Proof. Every $x = (x_1, \dots, x_k) \in \mathbb{C}^k$ is represented as a sequence of k complex numbers. Thus $\mathbb{C}^k$ is identified with the collection of all complex-valued functions on $\mathbb{N}_k$. Then by Observation 17.23, $\|\cdot\|_p$ is a norm on $\mathbb{C}^k$ and $\mathbb{C}^k$ is a Banach space with respect to the norm $\|\cdot\|_p$ for $p \in [1, \infty]$.

For every $p \in [1, \infty)$, we have $\|x\|_p^p = \sum_{i=1}^k |x_i|^p \geq \left[\max_{i=1,\dots,k} |x_i|\right]^p = \|x\|_\infty^p$ and thus $\|x\|_p \geq \|x\|_\infty$. The inequality (1) is from (3) of Proposition 17.24. Thus for every $p \in [1, \infty)$ we have $\|x\|_p \geq \|x\|_\infty$ and $k^{-1/p}\|x\|_p \leq \|x\|_\infty$. This shows that $\|\cdot\|_p$ and $\|\cdot\|_\infty$ are equivalent norms. This implies that any two norms in $\left\{\|\cdot\|_p : p \in [1, \infty]\right\}$ are equivalent. ∎

Problems

Prob. 17.1. Prove the following statements about the L^p spaces on the Lebesgue measure space $(\mathbb{R}, \mathfrak{M}_L, \mu_L)$ by constructing counter examples.

(a) For $0 < p < q < \infty$, we have

$$(1) \qquad L^p(\mathbb{R}, \mathfrak{M}_L, \mu_L) \not\supset L^q(\mathbb{R}, \mathfrak{M}_L, \mu_L),$$

and

$$(2) \qquad L^p(\mathbb{R}, \mathfrak{M}_L, \mu_L) \not\subset L^q(\mathbb{R}, \mathfrak{M}_L, \mu_L).$$

(b) For $0 < p < \infty$, we have

$$(3) \qquad L^p(\mathbb{R}, \mathfrak{M}_L, \mu_L) \not\supset L^\infty(\mathbb{R}, \mathfrak{M}_L, \mu_L),$$

and

$$(4) \qquad L^p(\mathbb{R}, \mathfrak{M}_L, \mu_L) \not\subset L^\infty(\mathbb{R}, \mathfrak{M}_L, \mu_L).$$

Prob. 17.2. Given a measure space $(X, \mathfrak{A}, \mu)$. Let $f \in L^p(X, \mathfrak{A}, \mu)$ for some $p \in (0, \infty)$ and $g \in L^\infty(X, \mathfrak{A}, \mu)$. Show that $fg \in L^p(X, \mathfrak{A}, \mu)$ and $\|fg\|_p \leq \|f\|_p \|g\|_\infty$.

Prob. 17.3. Let $(X, \mathfrak{A}, \mu)$ be an arbitrary measure space. Let f be an extended real-valued $\mathfrak{A}$-measurable function on X such that $f \in L^1(X, \mathfrak{A}, \mu) \cap L^\infty(X, \mathfrak{A}, \mu)$. Show that $f \in L^p(X, \mathfrak{A}, \mu)$ for every $p \in [1, \infty]$.

Prob. 17.4. Let $(X, \mathfrak{A}, \mu)$ be a finite measure space and let $1 \leq p_0 < p$. Let $f \in L^p(X, \mathfrak{A}, \mu)$ and $(f_n : n \in \mathbb{N}) \subset L^p(X, \mathfrak{A}, \mu)$ be such that

1° $\lim_{n \to \infty} f_n = f$, μ-a.e. on X,

2° $\|f_n\|_p \leq C$ for all $n \in \mathbb{N}$ for some $C > 0$.

Show that $\lim_{n \to \infty} \|f_n - f\|_{p_0} = 0$.

Prob. 17.5. Let $(X, \mathfrak{A}, \mu)$ be a measure space and let $0 < p_1 < p < p_2 \le \infty$. Show that
$$L^p(X, \mathfrak{A}, \mu) \subset L^{p_1}(X, \mathfrak{A}, \mu) + L^{p_2}(X, \mathfrak{A}, \mu),$$
that is, if $f \in L^p(X, \mathfrak{A}, \mu)$, then $f = g + h$ for some $g \in L^{p_1}(X, \mathfrak{A}, \mu)$ and some $h \in L^{p_2}(X, \mathfrak{A}, \mu)$.
(Hint: Let $D_1 = \{x \in X : |f(x)| > 1\}$ and $D_2 = \{x \in X : |f(x)| \le 1\}$ and then let $g = f\mathbf{1}_{D_1}$ and $h = f\mathbf{1}_{D_2}$.)

Prob. 17.6. Let $(X, \mathfrak{A}, \mu)$ be a measure space and let $0 < p_1 < p < p_2 \le \infty$. Show that
$$L^{p_1}(X, \mathfrak{A}, \mu) \cap L^{p_2}(X, \mathfrak{A}, \mu) \subset L^p(X, \mathfrak{A}, \mu),$$
and in fact for every $f \in L^{p_1}(X, \mathfrak{A}, \mu) \cap L^{p_2}(X, \mathfrak{A}, \mu)$ we have
$$\|f\|_p \le \|f\|_{p_1}^{\lambda} \|f\|_{p_2}^{1-\lambda},$$
where $\lambda \in (0, 1)$ is defined by
$$\frac{1}{p_1(\lambda p)^{-1}} + \frac{1}{p_2[(1-\lambda)p]^{-1}} = 1.$$
(Hint: For the case $p_2 < \infty$, write $|f|^p = |f|^{\lambda p}|f|^{(1-\lambda)p}$ and apply Hölder's Inequality with conjugates $p_1(\lambda p)^{-1}$ and $p_2[(1-\lambda)p]^{-1}$ which are in $(1, \infty)$.)
This Problem states that if $0 < p_1 < p < p_2 \le \infty$ and if $\|f\|_{p_1} < \infty$ and $\|f\|_{p_2} < \infty$, then $\|f\|_p < \infty$ and in fact $\|f\|_p \le \|f\|_{p_1}^{\lambda} \|f\|_{p_2}^{1-\lambda}$. The result is interesting in that while $L^{p_1}(X, \mathfrak{A}, \mu)$ may not contain $L^{p_2}(X, \mathfrak{A}, \mu)$ and conversely $L^{p_2}(X, \mathfrak{A}, \mu)$ may not contain $L^{p_1}(X, \mathfrak{A}, \mu)$, their intersection is always contained in $L^p(X, \mathfrak{A}, \mu)$ for any $p \in (p_1, p_2)$.

Prob. 17.7. Let $(X, \mathfrak{A}, \mu)$ be an arbitrary measure space. Let $1 \le p < q \le \infty$.
(a) Show that $L^p(X, \mathfrak{A}, \mu) \cap L^q(X, \mathfrak{A}, \mu)$ is a linear space.
(b) For $f \in L^p(X, \mathfrak{A}, \mu) \cap L^q(X, \mathfrak{A}, \mu)$, let us define $\|f\| = \|f\|_p + \|f\|_q$. Show that $\|\cdot\|$ is a norm on the linear space $L^p(X, \mathfrak{A}, \mu) \cap L^q(X, \mathfrak{A}, \mu)$.
(c) Show that $L^p(X, \mathfrak{A}, \mu) \cap L^q(X, \mathfrak{A}, \mu)$ is a Banach space with respect to the norm $\|\cdot\|$ defined above.

Prob. 17.8. Given a measure space $(X, \mathfrak{A}, \mu)$ with $\mu(X) \in (0, \infty)$. Let f be an arbitrary extended complex-valued $\mathfrak{A}$-measurable function on X. Let $M_p(f) = \|f\|_p / \mu(X)^{1/p}$ for $p \in (0, \infty)$ and $M_\infty(f) = \|f\|_\infty$. Show that for $p \in (0, \infty)$, we have
(a) $M_p(f) \uparrow M_\infty(f)$ as $p \uparrow \infty$.
(b) $\lim_{p\to\infty} \|f\|_p = \|f\|_\infty$, and in particular $\|f\|_p \uparrow \|f\|_\infty$ as $p \uparrow \infty$ when $\mu(X) \in (0, 1]$.

Prob. 17.9. Given a measure space $(X, \mathfrak{A}, \mu)$ with $\mu(X) \in (0, \infty)$. Let $f \in L^\infty(X, \mathfrak{A}, \mu)$ and let $\alpha_n = \int_X |f|^n \, d\mu$ for $n \in \mathbb{N}$. Show that $\lim_{n\to\infty} \dfrac{\alpha_{n+1}}{\alpha_n} = \|f\|_\infty$.

Prob. 17.10. Consider a partition of the open interval $(0, 1)$ into $D_1 = (0, 2^{-1}]$ and $D_k = \left(\sum_{i=1}^{k-1} 2^{-i}, \sum_{i=1}^{k} 2^{-i}\right]$ for $k = 2, 3, \cdots$. Let f be a function on $(0, 1)$ defined by
$$f(t) = \sqrt{2^k} \text{ for } t \in D_k \text{ for } k \in \mathbb{N}.$$
Determine the range of $p \in (0, \infty)$ for which f belongs to $L_p((0, 1))$.

Prob. 17.11. Consider the $([0, 1], \mathfrak{M}_L, \mu_L)$. For an extended real-valued $\mathfrak{M}_L$-measurable

function f on $[0, 1]$, we define $\|f\|_1 = \int_{[0,1]} |f| \, d\mu_L$ and $\|f\|_2 = \left\{ \int_{[0,1]} |f|^2 \, d\mu_L \right\}^{1/2}$.
(a) Show that $\|\cdot\|_2 \geq \|\cdot\|_1$ and thus $L^2([0, 1], \mathfrak{M}_L, \mu_L) \subset L^1([0, 1], \mathfrak{M}_L, \mu_L)$.
(b) Show that $L^2([0, 1], \mathfrak{M}_L, \mu_L)$ is not complete with respect to the norm $\|\cdot\|_1$ on $L^1([0, 1], \mathfrak{M}_L, \mu_L)$.

Prob. 17.12. Given a measure space $(X, \mathfrak{A}, \mu)$.
(a) Suppose $(X, \mathfrak{A}, \mu)$ has a $\mathfrak{A}$-measurable set of arbitrarily small positive measure, that is, for every $\varepsilon > 0$ there exists $E \in \mathfrak{A}$ such that $\mu(E) \in (0, \varepsilon]$. Show that there exists a disjoint sequence $(E_n : n \in \mathbb{N})$ in $\mathfrak{A}$ with $\mu(E_n) > 0$ for every $n \in \mathbb{N}$ such that $\lim_{n\to\infty} \mu(E_n) = 0$.
(b) Suppose $(X, \mathfrak{A}, \mu)$ has a $\mathfrak{A}$-measurable set of arbitrarily large finite measure, that is, for every $M > 0$ there exists $E \in \mathfrak{A}$ such that $\mu(E) \in [M, \infty)$. Show that there exists a disjoint sequence $(E_n : n \in \mathbb{N})$ in $\mathfrak{A}$ with $\mu(E_n) \in [1, \infty)$ for every $n \in \mathbb{N}$.
(This problem is a preparation for the next.)

Prob. 17.13. Given a measure space $(X, \mathfrak{A}, \mu)$. Let $0 < p < q \leq \infty$. Prove the following statements.
(a) $L^p(X, \mathfrak{A}, \mu) \not\subset L^q(X, \mathfrak{A}, \mu)$ if and only if $(X, \mathfrak{A}, \mu)$ has a $\mathfrak{A}$-measurable set of arbitrarily small positive measure.
(b) $L^p(X, \mathfrak{A}, \mu) \not\supset L^q(X, \mathfrak{A}, \mu)$ if and only if $(X, \mathfrak{A}, \mu)$ has a $\mathfrak{A}$-measurable set of arbitrarily large finite measure.

Prob. 17.14. According to Theorem 17.21, we have $\ell^p \subset \ell^q$ for $0 < p < q \leq \infty$. Show that $\ell^p \neq \ell^q$.

Prob. 17.15. Let $0 < p_1 < p_2$. Show that $\ell^{p_1} \supset \ell^{p_2}$ does not hold.

Prob. 17.16. Let $(\varepsilon_n : n \in \mathbb{N})$ be a sequence of positive numbers. Let $p \in (0, 1)$. Prove

$$\sum_{n\in\mathbb{N}} \varepsilon_n < \infty \Rightarrow \sum_{n\in\mathbb{N}} \varepsilon_n^p < \infty.$$

§18 Bounded Linear Functionals on the L^p Spaces

[I] Bounded Linear Functionals Arising from Integration

Let $(X, \mathfrak{A}, \mu)$ be an arbitrary measure space and consider the Banach space $L^p(X, \mathfrak{A}, \mu)$ with $p \in [1, \infty)$ consisting of the equivalence classes of extended complex-valued $\mathfrak{A}$-measurable functions f on X such that $\int_X |f|^p \, d\mu < \infty$ with the norm on the space given by $\|f\|_p = \left[\int_X |f|^p \, d\mu\right]^{1/p}$. For such a function f we have $|f| < \infty$ a.e. on X. For $p = \infty$, the Banach space $L^\infty(X, \mathfrak{A}, \mu)$ consists of the equivalence classes of essentially bounded extended complex-valued $\mathfrak{A}$-measurable functions f on X, the norm on the space $\|f\|_\infty$ being the essential supremum of f on X. We shall show that for $p \in [1, \infty]$ if $g \in L^q(X, \mathfrak{A}, \mu)$ where q is the conjugate of p and if we define a complex-valued function L_g on $L^p(X, \mathfrak{A}, \mu)$ by setting $L_g(f) = \int_X fg \, d\mu$ for $f \in L^p(X, \mathfrak{A}, \mu)$, then L_g is a bounded linear functional on $L^p(X, \mathfrak{A}, \mu)$.

For $\zeta \in \mathbb{C}$, let

$$sgn\,\zeta = \begin{cases} \zeta \cdot |\zeta|^{-1} & \text{for } \zeta \neq 0, \\ 0 & \text{for } \zeta = 0. \end{cases}$$

Then

$$|sgn\,\zeta| = \begin{cases} 1 & \text{for } \zeta \neq 0, \\ 0 & \text{for } \zeta = 0, \end{cases}$$

and

$$\zeta \cdot \overline{sgn\,\zeta} = |\zeta| \quad \text{for every } \zeta \in \mathbb{C}.$$

Theorem 18.1. *For an arbitrary measure space $(X, \mathfrak{A}, \mu)$, consider $L^p(X, \mathfrak{A}, \mu)$ where $p \in [1, \infty]$. Let $g \in L^q(X, \mathfrak{A}, \mu)$ where q is the conjugate of p, that is, $1/p + 1/q = 1$.*
(a) *A complex-valued function L_g on $L^p(X, \mathfrak{A}, \mu)$ defined by*

(1) $$L_g(f) = \int_X fg \, d\mu \quad \text{for } f \in L^p(X, \mathfrak{A}, \mu)$$

is a bounded linear functional on $L^p(X, \mathfrak{A}, \mu)$.
(b) *If $p \in (1, \infty]$ and $q \in [1, \infty)$, then*

(2) $$\|L_g\|_* = \|g\|_q.$$

If $(X, \mathfrak{A}, \mu)$ is σ-finite, then (2) holds for the case $p = 1$ and $q = \infty$ also.

Proof. 1. By Hölder's Inequality (Theorem 16.14 for the case $p \in (1, \infty)$ and Theorem 16.40 for the cases $p = 1$ and $q = \infty$), we have

(3) $$|L_g(f)| = \left|\int_X fg \, d\mu\right| \leq \int_X |fg| \, d\mu \leq \|f\|_p \|g\|_q \quad \text{for } f \in L^p(X, \mathfrak{A}, \mu).$$

Since $\|f\|_p < \infty$ and $\|g\|_q < \infty$ we have $|L_g(f)| < \infty$, that is, $L_g(f) \in \mathbb{C}$. The linearity of L_g follows from the linearity of the integral with respect to the integrand. The inequality (3) shows that $\|g\|_q$ is a bound for the linear functional L_g and L_g is a bounded linear functional on $L^p(X, \mathfrak{A}, \mu)$ with $\|L_g\|_* \le \|g\|_q$. To prove (2), it remains to show the reverse inequality $\|L_g\|_* \ge \|g\|_q$.

2. Let $S(0,1) = \{f \in L^p(X, \mathfrak{A}, \mu) : \|f\|_p = 1\}$. Then $\|L_g\|_* = \sup_{f \in S(0,1)} |L_g(f)|$ by Lemma 15.36. We construct $f_0 \in S(0,1)$ such that $|L_g(f_0)| = \|g\|_q$. With the existence of such $f_0 \in S(0,1)$, we have $\|L_g\|_* = \sup_{f \in S(0,1)} |L_g(f)| \ge |L_g(f_0)| = \|g\|_q$. The construction of f_0 is divided into three cases depending on the value of $p \in [1, \infty]$. If $g = 0 \in L^q(X, \mathfrak{A}, \mu)$, then $L_g(f) = 0$ for every $f \in L^p(X, \mathfrak{A}, \mu)$ so that L_g is the identically vanishing linear functional on $L^p(X, \mathfrak{A}, \mu)$ with $\|L_g\|_* = 0 = \|g\|_q$ and (2) holds trivially. In what follows we assume $g \neq 0 \in L^p(X, \mathfrak{A}, \mu)$ so that $\|g\|_q > 0$.

2a. If $p \in (1, \infty)$ so that $q \in (1, \infty)$ also, let

$$f_0 = \|g\|_q^{1-q}\, |g|^{q-1}\, \overline{sgn\, g}.$$

Since $p = q/(q-1)$ and $p(q-1) = q$, we have

$$\begin{aligned}
\|f_0\|_p^p &= \int_X \|g\|_q^{(1-q)p}\, |g|^{(q-1)p}\, |\overline{sgn\, g}|^p \, d\mu \\
&= \|g\|_q^{-q} \int_X |g|^q\, |\overline{sgn\, g}|^p \, d\mu \\
&= \|g\|_q^{-q} \left\{ \int_{\{X:g=0\}} 0 \, d\mu + \int_{\{X:g\neq 0\}} |g|^q \cdot 1 \, d\mu \right\} \\
&= \|g\|_q^{-q} \int_X |g|^q \, d\mu = \|g\|_q^{-q} \|g\|_q^q = 1.
\end{aligned}$$

Thus $\|f_0\|_p = 1$ so that $f_0 \in S(0,1) \subset L^p(X, \mathfrak{A}, \mu)$. By Lemma 15.36, we have

$$\begin{aligned}
\|L_g\|_* &= \sup_{f \in S(0,1)} |L_g(f)| \ge |L_g(f_0)| \\
&= \left| \int_X f_0\, g \, d\mu \right| = \|g\|_q^{1-q} \left| \int_X |g|^{q-1}\, \overline{sgn\, g} \cdot g \, d\mu \right| \\
&= \|g\|_q^{1-q} \int_X |g|^{q-1}\, |g| \, d\mu = \|g\|_q^{1-q} \int_X |g|^q \, d\mu \\
&= \|g\|_q^{1-q} \|g\|_q^q = \|g\|_q.
\end{aligned}$$

Thus we have $\|L_g\|_* \ge \|g\|_q$ in this case.

2b. If $p = \infty$ and $q = 1$, let $f_0 = \overline{sgn\, g}$. Then $\|f_0\|_\infty = 1$ so that $f_0 \in S(0,1) \subset L^\infty(X, \mathfrak{A}, \mu)$. By Lemma 15.36, we have

$$\begin{aligned}
\|L_g\|_* &= \sup_{f \in S(0,1)} |L_g(f)| \ge |L_g(f_0)| \\
&= \left| \int_X f_0\, g \, d\mu \right| = \left| \int_X \overline{sgn\, g} \cdot g \, d\mu \right| \\
&= \int_X |g| \, d\mu = \|g\|_1.
\end{aligned}$$

Thus we have $\|L_g\|_* \geq \|g\|_1$ in this case.

2c. Consider the case $p = 1$ and $q = \infty$. For an arbitrary $\varepsilon > 0$, let $A_\varepsilon = \{x \in X : |g(x)| > \|g\|_\infty - \varepsilon\}$. Assume that $\varepsilon > 0$ is so small that $\|g\|_\infty - \varepsilon > 0$. Since $\|g\|_\infty$ is the infimum of all essential bounds of g, $\|g\|_\infty - \varepsilon$ is not an essential bound of g and therefore $\mu(A_\varepsilon) > 0$. If we assume that $(X, \mathfrak{A}, \mu)$ is σ-finite, then there exists a $\mathfrak{A}$-measurable subset A of A_ε such that $\mu(A) \in (0, \infty)$. Let

$$f_0 = \frac{1}{\mu(A)} \mathbf{1}_A \overline{sgn\, g}.$$

Then

$$\|f_0\|_1 = \frac{1}{\mu(A)} \int_A |\overline{sgn\, g}|\, d\mu = \frac{1}{\mu(A)} \int_A d\mu = 1.$$

Thus $f_0 \in S(0,1) \subset L^1(X, \mathfrak{A}, \mu)$. Then by Lemma 15.36, we have

$$\begin{aligned}
\|L_g\|_* &= \sup_{f \in S(0,1)} |L_g(f)| \geq |L_g(f_0)| \\
&= \left| \int_X f_0\, g\, d\mu \right| = \frac{1}{\mu(A)} \left| \int_A \overline{sgn\, g} \cdot g\, d\mu \right| \\
&= \frac{1}{\mu(A)} \int_A |g|\, d\mu \geq \frac{1}{\mu(A)} \{\|g\|_\infty - \varepsilon\} \mu(A) \\
&= \|g\|_\infty - \varepsilon.
\end{aligned}$$

By the arbitrariness of $\varepsilon > 0$, we have $\|L_g\|_* \geq \|g\|_\infty$. ∎

Let us note that when $p = 1$ and $q = \infty$, the equality $\|L_g\|_* = \|g\|_\infty$ may not hold without the assumption that $(X, \mathfrak{A}, \mu)$ is σ-finite.

Example. Let $(X, \mathfrak{A}, \mu)$ be a measure space which has an atom with infinite measure, that is, there exists $A \in \mathfrak{A}$ with $\mu(A) = \infty$ such that for every $\mathfrak{A}$-measurable subset A_0 of A we have either $\mu(A_0) = 0$ or $\mu(A_0) = \mu(A)$. With such an atom A, let $g = \mathbf{1}_A$. Then $g \in L^\infty(X, \mathfrak{A}, \mu)$ with $\|g\|_\infty = 1$. Consider the bounded linear functional L_g on $L^1(X, \mathfrak{A}, \mu)$ defined by $L_g(f) = \int_X fg\, d\mu$ for $f \in L^1(X, \mathfrak{A}, \mu)$. For an arbitrary $f \in L^1(X, \mathfrak{A}, \mu)$, we have $\int_X |f|\, d\mu < \infty$. This implies that the set $E = \{x \in X : |f(x)| \neq 0\}$ is a σ-finite set according to (f) of Observation 9.2. Thus there exists a sequence $(E_n : n \in \mathbb{N})$ in $\mathfrak{A}$ such that $\bigcup_{n \in \mathbb{N}} E_n = E$ and $\mu(E_n) < \infty$ for every $n \in \mathbb{N}$. Now $E_n \cap A$ is a $\mathfrak{A}$-measurable subset of A with $\mu(E_n \cap A) \leq \mu(E_n) < \infty$ and thus $\mu(E_n \cap A) \neq \mu(A) = \infty$. Since A is an atom, this implies that $\mu(E_n \cap A) = 0$. Since this holds for every $n \in \mathbb{N}$, we have $\mu(E \cap A) = \sum_{n \in \mathbb{N}} \mu(E_n \cap A) = 0$. This fact and the fact that $f = 0$ on E^c imply

$$L_g(f) = \int_X fg\, d\mu = \int_X f \mathbf{1}_A\, d\mu = \int_{E \cap A} f\, d\mu = 0.$$

Since this holds for every $f \in L^1(X, \mathfrak{A}, \mu)$, we have $L_g = 0$ and $\|L_g\|_* = 0 \neq 1 = \|g\|_\infty$.

[II] Approximation by Simple Functions

In Lemma 8.6 we showed that for an arbitrary nonnegative extended real-valued $\mathfrak{A}$-measurable function f on a measure space $(X, \mathfrak{A}, \mu)$ there exists an increasing sequence $(\varphi_n : n \in \mathbb{N})$ of simple functions on X such that $\varphi_n \uparrow f$ on X and moreover when $(X, \mathfrak{A}, \mu)$ is a σ-finite measure space φ_n can be chosen to be μ-integrable for all $n \in \mathbb{N}$. Let us extend this result to extended complex-valued $\mathfrak{A}$-measurable functions.

Let $(X, \mathfrak{A}, \mu)$ be an arbitrary measure space. Let φ be a complex-valued simple function on X. Then φ is μ-integrable on X if and only if $|\varphi|$ is μ-integrable on X according to Observation 16.3 and the real-valued simple function $|\varphi|$ is μ-integrable on X if and only if $\mu\{X : |\varphi| \neq 0\} < \infty$ according to 6° of Observation 7.5. Thus a complex-valued simple function φ on X is μ-integrable on X if and only if $\mu\{X : \varphi \neq 0\} < \infty$.

Lemma 18.2. *For an arbitrary measure space $(X, \mathfrak{A}, \mu)$, let $\mathcal{S}(X, \mathfrak{A}, \mu)$ be the collection of all complex-valued simple functions on X and let $\mathcal{S}_0(X, \mathfrak{A}, \mu)$ be the subcollection consisting of μ-integrable members of $\mathcal{S}(X, \mathfrak{A}, \mu)$, that is, $\mathcal{S}_0(X, \mathfrak{A}, \mu) = \mathcal{S}(X, \mathfrak{A}, \mu) \cap \mathcal{L}^1(X, \mathfrak{A}, \mu)$.*

(a) *For every extended complex-valued $\mathfrak{A}$-measurable function f on X, there exists a sequence $(\varphi_n : n \in \mathbb{N})$ in $\mathcal{S}(X, \mathfrak{A}, \mu)$ such that $\lim_{n\to\infty} \varphi_n = f$ on X and $|\varphi_n| \leq |f|$ on X*

for every $n \in \mathbb{N}$. Moreover $|\varphi_n| \uparrow |f|$ on X as $n \to \infty$.

(b) *If $(X, \mathfrak{A}, \mu)$ is a σ-finite measure space, then the sequence $(\varphi_n : n \in \mathbb{N})$ in (a) can be chosen from $\mathcal{S}_0(X, \mathfrak{A}, \mu)$.*

(c) *If $f \in L^p(X, \mathfrak{A}, \mu)$ and $p \in (0, \infty)$, then the sequence $(\varphi_n : n \in \mathbb{N})$ in (a) can be chosen from $\mathcal{S}_0(X, \mathfrak{A}, \mu)$ (without assuming σ-finiteness of $(X, \mathfrak{A}, \mu)$).*

Proof. 1. Let f be an extended complex-valued $\mathfrak{A}$-measurable function on X. If we write

$$f = \Re f + i\Im f = \{(\Re f)^+ - (\Re f)^-\} + i\{(\Im f)^+ - (\Im f)^-\},$$

then each of $(\Re f)^+$, $(\Re f)^-$, $(\Im f)^+$, and $(\Im f)^-$ is a nonnegative extended real-valued $\mathfrak{A}$-measurable functions on X and thus by Lemma 8.6 there exist four increasing sequences of nonnegative simple functions $(\varphi_{j,n} : n \in \mathbb{N})$, $j = 1, 2, 3, 4$, such that $\varphi_{1,n} \uparrow (\Re f)^+$, $\varphi_{2,n} \uparrow (\Re f)^-$, $\varphi_{3,n} \uparrow (\Im f)^+$, and $\varphi_{4,n} \uparrow (\Im f)^-$ on X. If we let $\varphi_n = \{\varphi_{1,n} - \varphi_{2,n}\} + i\{\varphi_{3,n} - \varphi_{4,n}\}$ for $n \in \mathbb{N}$, then $(\varphi_n : n \in \mathbb{N})$ is a sequence in $\mathcal{S}(X, \mathfrak{A}, \mu)$ and $\lim_{n\to\infty} \varphi_n = f$ on X. Let us show that $|\varphi_n| \leq |f|$ on X for every $n \in \mathbb{N}$. We have $0 \leq \varphi_{1,n} \leq (\Re f)^+$ and $0 \leq \varphi_{2,n} \leq (\Re f)^-$ on X for every $n \in \mathbb{N}$. At each $x \in X$, at least one of $(\Re f)^+$ and $(\Re f)^-$ is equal to 0. Thus for every $n \in \mathbb{N}$, at each $x \in X$ at least one of $\varphi_{1,n}$ and $\varphi_{2,n}$ is equal to 0. Similarly for every $n \in \mathbb{N}$, at each $x \in X$ at least one of $\varphi_{3,n}$ and $\varphi_{4,n}$ is equal to 0. Thus we have

$$\begin{aligned}
|\varphi_n|^2 &= |\varphi_{1,n} - \varphi_{2,n}|^2 + |\varphi_{3,n} - \varphi_{4,n}|^2 = \{\varphi_{1,n}^2 + \varphi_{2,n}^2\} + \{\varphi_{3,n}^2 + \varphi_{4,n}^2\} \\
&\leq \{(\Re f)^+\}^2 + \{(\Re f)^-\}^2 + \{(\Im f)^+\}^2 + \{(\Im f)^-\}^2 \\
&= |\Re f|^2 + |\Im f|^2 = |f|^2
\end{aligned}$$

so that $|\varphi_n| \le |f|$ on X for $n \in \mathbb{N}$. Now $\lim_{n\to\infty} \varphi_n = f$ on X implies $\lim_{n\to\infty} |\varphi_n| = |f|$ on X. Since $\varphi_{i,n} \uparrow$ we have $\varphi_{i,n}^2 \uparrow$ for $i = 1, \ldots, 4$ and then $|\varphi_n|^2 \uparrow$ so that $|\varphi_n| \uparrow |f|$ on X as $n \to \infty$.

2. Suppose $(X, \mathfrak{A}, \mu)$ is a σ-finite measure space. Then there exists an increasing sequence $(E_n : n \in \mathbb{N})$ in $\mathfrak{A}$ such that $\bigcup_{n\in\mathbb{N}} E_n = X$ and $\mu(E_n) < \infty$ for every $n \in \mathbb{N}$. By (a) there exists a sequence $(\varphi_n : n \in \mathbb{N})$ in $\mathcal{S}(X, \mathfrak{A}, \mu)$ such that $\lim_{n\to\infty} \varphi_n = f$ on X and $|\varphi_n| \le |f|$ on X for every $n \in \mathbb{N}$ and $|\varphi_n| \uparrow |f|$ on X as $n \to \infty$. For every $n \in \mathbb{N}$, let $\psi_n = \varphi_n \mathbf{1}_{E_n}$. Then $\psi_n \in \mathcal{S}_0(X, \mathfrak{A}, \mu)$. Also $|\psi_n| \le |\varphi_n| \le |f|$ on X for $n \in \mathbb{N}$. For every $x \in X$, we have $x \in E_n$ and thus $\psi_n(x) = \varphi_n(x)$ for sufficiently large $n \in \mathbb{N}$. Thus $\lim_{n\to\infty} \psi_n(x) = \lim_{n\to\infty} \varphi_n(x) = f(x)$. Also $|\psi_n| \le |\varphi_n| \le |f|$ on X for $n \in \mathbb{N}$. Since $(E_n : n \in \mathbb{N})$ is an increasing sequence we have $\mathbf{1}_{E_n} \uparrow$ on X as $n \to \infty$. Then since $|\varphi_n| \uparrow$ on X as $n \to \infty$, we have $|\psi_n| \uparrow |f|$ on X as $n \to \infty$.

3. Suppose $f \in L^p(X, \mathfrak{A}, \mu)$ where $p \in (0, \infty)$. Let $E = \{x \in X : |f(x)|^p \neq 0\}$. Let us show that E is a σ-finite set. For $n \in \mathbb{N}$, let $E_n = \{x \in X : |f(x)|^p \ge \frac{1}{n}\}$. Then $(E_n : n \in \mathbb{N})$ is an increasing sequence in $\mathfrak{A}$ and $\bigcup_{n\in\mathbb{N}} E_n = E$. Now we have

$$\frac{1}{n}\mu(E_n) \le \int_{E_n} |f|^p \, d\mu \le \int_X |f|^p \, d\mu < \infty,$$

and therefore $\mu(E_n) \le n \int_X |f|^p \, d\mu < \infty$ for every $n \in \mathbb{N}$. This shows that E is a σ-finite set. By (a) there exists a sequence $(\varphi_n : n \in \mathbb{N})$ in $\mathcal{S}(X, \mathfrak{A}, \mu)$ such that $\lim_{n\to\infty} \varphi_n = f$ on X and $|\varphi_n| \le |f|$ on X for every $n \in \mathbb{N}$ and $|\varphi_n| \uparrow |f|$ on X as $n \to \infty$. If we let $\psi_n = \varphi_n \mathbf{1}_{E_n}$ then $\psi_n \in \mathcal{S}_0(X, \mathfrak{A}, \mu)$ for $n \in \mathbb{N}$. Also $|\psi_n| \le |\varphi_n| \le |f|$ on X for $n \in \mathbb{N}$. On E^c, we have $f = 0$ and also $\psi_n = 0$ for every $n \in \mathbb{N}$ and thus $\lim_{n\to\infty} \psi_n = f$ trivially. On E, if $x \in E$ then $x \in E_n$ and hence $\psi_n(x) = \phi_n(x)$ for sufficiently large $n \in \mathbb{N}$ so that $\lim_{n\to\infty} \psi_n(x) = \lim_{n\to\infty} \varphi_n(x) = f(x)$. Thus $\lim_{n\to\infty} \psi_n = f$ on X. Since $|\varphi_n| \uparrow$ and $\mathbf{1}_{E_n} \uparrow$ we have $|\psi_n| \uparrow |f|$ on X as $n \to \infty$. ∎

Theorem 18.3. *Let $(X, \mathfrak{A}, \mu)$ be an arbitrary measure space and let $S_0(X, \mathfrak{A}, \mu)$ be the collection of all the equivalence classes of μ-integrable complex-valued simple functions on $(X, \mathfrak{A}, \mu)$, the equivalence relation being that of μ-a.e. equality on X. Then we have*
(a) *$S_0(X, \mathfrak{A}, \mu) \subset L^p(X, \mathfrak{A}, \mu)$ for $p \in (0, \infty]$.*
(b) *$S_0(X, \mathfrak{A}, \mu)$ is dense in $L^p(X, \mathfrak{A}, \mu)$ for $p \in (0, \infty)$.*
(c) *$S_0(X, \mathfrak{A}, \mu)$ is not dense in $L^\infty(X, \mathfrak{A}, \mu)$.*

Proof. 1. If $\varphi \in S_0(X, \mathfrak{A}, \mu)$ then φ assumes only finitely many values in $\mathbb{C}$ so that φ is essentially bounded on X and hence $\varphi \in L^\infty(X, \mathfrak{A}, \mu)$. Thus $S_0(X, \mathfrak{A}, \mu) \subset L^\infty(X, \mathfrak{A}, \mu)$.

Let $p \in (0, \infty)$. To show $S_0(X, \mathfrak{A}, \mu) \subset L^p(X, \mathfrak{A}, \mu)$, let $\varphi \in S_0(X, \mathfrak{A}, \mu)$. Then $\varphi = \sum_{i=1}^k c_i \mathbf{1}_{D_i}$ where $c_i \in \mathbb{C}$ for $i = 1, \ldots, k$ and $\{D_i : i = 1, \ldots, k\}$ is a disjoint collection in $\mathfrak{A}$. Now the μ-integrability of φ is equivalent to the μ-integrability of $|\varphi|$. Thus we have $\int_X |\varphi| \, d\mu = \sum_{i=1}^k |c_i| \mu(D_i) < \infty$. This implies that $\mu(D_i) < \infty$ for $i = 1, \ldots, k$. Then $\int_X |\varphi|^p \, d\mu = \sum_{i=1}^k |c_i|^p \mu(D_i) < \infty$ so that $\varphi \in L^p(X, \mathfrak{A}, \mu)$. This proves $S_0(X, \mathfrak{A}, \mu) \subset L^p(X, \mathfrak{A}, \mu)$.

2. To show the denseness of $S_0(X, \mathfrak{A}, \mu)$ in $L^p(X, \mathfrak{A}, \mu)$ for $p \in (0, \infty)$, let $f \in L^p(X, \mathfrak{A}, \mu)$ be arbitrarily chosen. According to (c) of Lemma 18.2, there exists a sequence $(\varphi_n : n \in \mathbb{N})$ in $S_0(X, \mathfrak{A}, \mu)$ such that $\lim_{n\to\infty} \varphi_n = f$ on X and $|\varphi_n| \leq |f|$ on X for every $n \in \mathbb{N}$. Then we have $|\varphi_n - f|^p \leq \{|\varphi_n| + |f|\}^p \leq \{2|f|\}^p = 2^p|f|^p$. Since $f \in L^p(X, \mathfrak{A}, \mu)$, $|f|^p$ is μ-integrable on X. Thus by the Dominated Convergence Theorem (Theorem 9.20) we have $\lim_{n\to\infty} \int_X |\varphi_n - f|^p \, d\mu = \int_X \lim_{n\to\infty} |\varphi_n - f|^p \, d\mu = 0$. This shows that $S_0(X, \mathfrak{A}, \mu)$ is dense in $L^p(X, \mathfrak{A}, \mu)$. ∎

3. Let us show that $S_0(X, \mathfrak{A}, \mu)$ is not dense in $L^\infty(X, \mathfrak{A}, \mu)$ by constructing an example. Consider $(\mathbb{R}, \mathfrak{M}_L, \mu_L)$. Let $f = 1$ on $\mathbb{R}$. Then $\|f\|_\infty = 1$ and $f \in L^\infty(\mathbb{R}, \mathfrak{M}_L, \mu_L)$. If $\varphi \in S_0(\mathbb{R}, \mathfrak{M}_L, \mu_L)$, then φ is a μ_L-integrable complex-valued simple function on $(\mathbb{R}, \mathfrak{M}_L, \mu_L)$ so that $\mu_L\{\mathbb{R} : \varphi \neq 0\} < \infty$. Then $\mu_L\{\mathbb{R} : \varphi = 0\} > 0$. Now $|f - \varphi| = 1$ on $\{\mathbb{R} : \varphi = 0\}$ so that $\|f - \varphi\|_\infty \geq 1$. Thus the open ball in $L^\infty(\mathbb{R}, \mathfrak{M}_L, \mu_L)$ with center at f and with radius $\frac{1}{2}$ contains no elements in $S_0(\mathbb{R}, \mathfrak{M}_L, \mu_L)$. This shows that $S_0(\mathbb{R}, \mathfrak{M}_L, \mu_L)$ is not dense in $L^\infty(\mathbb{R}, \mathfrak{M}_L, \mu_L)$.

[III] A Converse of Hölder's Inequality

If g is an extended complex-valued $\mathfrak{A}$-measurable function on X in a measure space $(X, \mathfrak{A}, \mu)$, then $\|g\|_q$ is defined by $\|g\|_q = \left\{\int_X |f|^q \, d\mu\right\}^{1/q}$ if $q \in [1, \infty)$ and $\|g\|_q$ is the essential supremum of g on X for $q = \infty$. In the next theorem, we show that if $(X, \mathfrak{A}, \mu)$ is a σ-finite measure space, then $\|g\|_q$ is equal to the supremum of $\left|\int_X fg \, d\mu\right|$ on the collection of all μ-integrable complex-valued simple functions f with $\|f\|_p = 1$ where p is the conjugate of q.

Theorem 18.4. *Let $(X, \mathfrak{A}, \mu)$ be a σ-finite measure space and let $S_0(X, \mathfrak{A}, \mu)$ be the collection of all equivalence classes of μ-integrable complex-valued simple functions on $(X, \mathfrak{A}, \mu)$. Let g be an extended complex-valued $\mathfrak{A}$-measurable function on X such that $|g| < \infty$ a.e. on X. Let $p, q \in [1, \infty]$ be conjugates, that is, $1/p + 1/q = 1$. Define*

$$M_q(g) = \sup \left\{\left|\int_X fg \, d\mu\right| : f \in S_0(X, \mathfrak{A}, \mu) \text{ and } \|f\|_p = 1\right\}.$$

Then $M_q(g) = \|g\|_q$. In particular $g \in L^q(X, \mathfrak{A}, \mu)$ if and only if $M_q(g) < \infty$.

Proof. 1. We have $S_0(X, \mathfrak{A}, \mu) \subset L^p(X, \mathfrak{A}, \mu)$ for every $p \in (0, \infty]$ by Theorem 18.3. If $f \in S_0(X, \mathfrak{A}, \mu)$ and $\|f\|_p = 1$, then by Hölder's Inequality (Theorem 16.14 for $p, q \in (1, \infty)$ and Theorem 16.40 for $p = 1$ and $q = \infty$ or $p = \infty$ and $q = 1$), we have

$$\left|\int_X fg \, d\mu\right| \leq \int_X |fg| \, d\mu \leq \|f\|_p \|g\|_q = \|g\|_q.$$

Thus $M_q(g) \leq \|g\|_q$. It remains to show that $M_q(g) \geq \|g\|_q$.

2. To prove $M_q(g) \geq \|g\|_q$ we consider three cases for the value of p.

2a. Consider the case $p, q \in (1, \infty)$. Now since $(X, \mathfrak{A}, \mu)$ is a σ-finite measure space, by (b) of Lemma 18.2 there exists a sequence $(g_n : n \in \mathbb{N})$ in $S_0(X, \mathfrak{A}, \mu)$ such that $\lim_{n\to\infty} g_n = g$ on X, $|g_n| \leq |g|$ for $n \in \mathbb{N}$ and $|g_n| \uparrow |g|$ on X as $n \to \infty$. By

Theorem 8.5 (Monotone Convergence Theorem) we have $\lim_{n\to\infty}\int_X |g_n|^q\,d\mu = \int_X |g|^q\,d\mu$ and then $\lim_{n\to\infty}\|g_n\|_q = \|g\|_q$. If $g = 0$, that is, $g(x) = 0$ for a.e. $x \in X$, then $\|g\|_q = 0$ so that $M_q(g) \geq \|g\|_q$ holds trivially. Thus consider $g \neq 0$. Then $\|g\|_q > 0$. Since $\lim_{n\to\infty}\|g_n\|_q = \|g\|_q$ there exists $N \in \mathbb{N}$ such that $\|g_n\|_q > 0$ for $n \geq N$. In what follows we consider $n \geq N$ only. Let

$$f_n = \|g_n\|_q^{1-q}\,|g_n|^{q-1}\,\overline{\mathrm{sgn}\, g}.$$

Since $g_n \in S_0(X, \mathfrak{A}, \mu)$, we have $f_n \in S_0(X, \mathfrak{A}, \mu)$. Moreover

$$\begin{aligned}
\|f_n\|_p^p &= \int_X \|g_n\|_q^{(1-q)p}\,|g_n|^{(q-1)p}\,|\overline{\mathrm{sgn}\, g}|^p\,d\mu \\
&= \|g_n\|_q^{-q}\int_X |g_n|^q\,|\overline{\mathrm{sgn}\, g}|^p\,d\mu \\
&= \|g_n\|_q^{-q}\left\{\int_{\{X:g=0\}} 0\,d\mu + \int_{\{X:g\neq 0\}} |g_n|^q\cdot 1\,d\mu\right\} \\
&= \|g_n\|_q^{-q}\int_X |g_n|^q\,d\mu = \|g_n\|_q^{-q}\|g_n\|_q^q = 1,
\end{aligned}$$

so that $\|f_n\|_p = 1$. From the definition of f_n we have

$$|f_n|^p = \|g_n\|_q^{(1-q)p}|g_n|^{(q-1)p}|\overline{\mathrm{sgn}\, g}|^p = \|g_n\|_q^{-q}|g_n|^q|\overline{\mathrm{sgn}\, g}|^p.$$

At $x \in X$ where $g(x) \neq 0$, we have $|\overline{\mathrm{sgn}\, g(x)}| = 1$ so that $|f_n(x)|^p = \|g_n\|_q^{-q}|g_n(x)|^q$. At $x \in X$ where $g(x) = 0$, we have $\overline{\mathrm{sgn}\, g(x)} = 0$ and hence $|f_n(x)|^p = 0$. Also $|g_n(x)| \leq |g(x)|$ implies $g_n(x) = 0$. Thus $|f_n(x)|^p = \|g_n\|_q^{-q}|g_n(x)|^q$ holds trivially. Therefore $|f_n|^p = \|g_n\|_q^{-q}|g_n|^q$ holds on X. Since $\|f_n\|_p, \|g_n\|_q \in (0, \infty)$ the last equality on X implies $\|f_n g_n\|_1 = \|f_n\|_p\|g_n\|_q$ according to (b) of Theorem 16.14. Thus we have

(1) $$\int_X |f_n g_n|\,d\mu = \|f_n\|_p\|g_n\|_q = \|g_n\|_q.$$

Now since $\lim_{n\to\infty}|g_n| = |g|$ and $\lim_{n\to\infty}|g_n|^q = |g|^q$ on X, Fatou's Lemma (Theorem 8.13) implies

(2) $$\begin{aligned}
\int_X |g|^q\,d\mu &\leq \liminf_{n\to\infty}\int_X |g_n|^q\,d\mu \\
&= \liminf_{n\to\infty}\left\{\int_X |f_n g_n|\,d\mu\right\}^q \quad \text{by (1)} \\
&\leq \liminf_{n\to\infty}\left\{\int_X |f_n g|\,d\mu\right\}^q \quad \text{since } |g_n| \leq |g|.
\end{aligned}$$

But $f_n g = \|g_n\|_q^{1-q}\,|g_n|^{q-1}\,\overline{\mathrm{sgn}\, g}\cdot g = \|g_n\|_q^{1-q}\,|g_n|^{q-1}\,|g| \geq 0$ so that $|f_n g| = f_n g$. Then

(3) $$\|g\|_q^q \leq \liminf_{n\to\infty}\left|\int_X f_n g\,d\mu\right|^q \leq \{M_q(g)\}^q.$$

This proves $\|g\|_q \le M_q(g)$.

2b. Next consider the case $p = \infty$ and $q = 1$. If we let $f = \overline{sgn\, g}$, then we have $f \in S_0(X, \mathfrak{A}, \mu)$ and $\|f\|_1 = 1$ so that

$$\|g\|_1 = \int_X |g|\, d\mu = \int_X \overline{sgn\, g} \cdot g\, d\mu = \int_X fg\, d\mu \le M_1(g),$$

by the definition of $M_1(g)$.

2c. Finally consider the case $p = 1$ and $q = \infty$. For an arbitrary $\varepsilon > 0$, let us define $A_\varepsilon = \{x \in X : |g(x)| \ge M_\infty(g) + \varepsilon\}$. To show $\mu(A_\varepsilon) = 0$, assume the contrary, that is, $\mu(A_\varepsilon) > 0$. By the σ-finiteness of $(X, \mathfrak{A}, \mu)$, there exists a $\mathfrak{A}$-measurable subset A of A_ε with $\mu(A_\varepsilon) \in (0, \infty)$. Let

$$f = \frac{1}{\mu(A)} \overline{sgn\, g}\, \mathbf{1}_A.$$

Then $f \in S_0(X, \mathfrak{A}, \mu)$ and

$$\int_X |f|\, d\mu = \frac{1}{\mu(A)} \int_X |\overline{sgn\, g}| \mathbf{1}_A\, d\mu = \frac{1}{\mu(A)} \int_A d\mu = 1,$$

so that $\|f\|_1 = 1$. But

$$\begin{aligned}\int_X fg\, d\mu &= \frac{1}{\mu(A)} \int_X \overline{sgn\, g} \cdot g \mathbf{1}_A\, d\mu = \frac{1}{\mu(A)} \int_A |g|\, d\mu \\ &\ge \frac{1}{\mu(A)} \{M_\infty(g) + \varepsilon\} \mu(A) = M_\infty(g) + \varepsilon,\end{aligned}$$

contradicting the definition of $M_\infty(g)$. Thus $\mu(A_\varepsilon) = 0$ and this implies that $M_\infty(g) + \varepsilon$ is an essential bound of g so that $M_\infty(g) + \varepsilon \ge \|g\|_\infty$. By the arbitrariness of $\varepsilon > 0$, we have $M_\infty(g) \ge \|g\|_\infty$. ∎

Let $p, q \in [1, \infty]$ be conjugates. If $g \in L^q(X, \mathfrak{A}, \mu)$, then there exists a constant $M \ge 0$ such that $\left|\int_X fg\, d\mu\right| \le M\|f\|_p$ for every $f \in L^p(X, \mathfrak{A}, \mu)$. In fact by Hölder's Inequality (Theorem 16.14 for $q \in (1, \infty)$ and Theorem 16.40 for $q = 1$ and $q = \infty$), we have $\left|\int_X fg\, d\mu\right| \le \int_X |fg|\, d\mu \le \|g\|_q \|f\|_p$ for every $f \in L^p(X, \mathfrak{A}, \mu)$ so that $\|g\|_q$ will do as M. A converse of this fact, that is, the existence of such constant M implies that $g \in L^q(X, \mathfrak{A}, \mu)$, is given next. It is essentially a rephrasing of Theorem 18.4.

Theorem 18.5. (Converse to Hölder's Inequality) *Let $(X, \mathfrak{A}, \mu)$ be a σ-finite measure space. Let $p, q \in [1, \infty]$ be conjugates. Let g be an extended complex-valued $\mathfrak{A}$-measurable function on X such that $|g| < \infty$ a.e. on X. If there exists a constant $M \ge 0$ such that $\left|\int_X fg\, d\mu\right| \le M\|f\|_p$ for every $f \in L^p(X, \mathfrak{A}, \mu)$, then $g \in L^q(X, \mathfrak{A}, \mu)$ and $\|g\|_q \le M$.*

Proof. Suppose such constant $M \ge 0$ exists. Then we have

$$\sup\left\{\left|\int_X fg\, d\mu\right| : \|f\|_p = 1\right\} \le M.$$

Thus for $M_q(g)$ defined in Theorem 18.4, we have

$$M_q(g) = \sup\left\{\left|\int_X fg\,d\mu\right| : f \in S_0(X, \mathfrak{A}, \mu) \text{ and } \|f\|_p = 1\right\}$$
$$\leq \sup\left\{\left|\int_X fg\,d\mu\right| : \|f\|_p = 1\right\} \leq M < \infty.$$

By Theorem 18.4, this implies that $g \in L^g(X, \mathfrak{A}, \mu)$ and $\|g\|_q = M_q(g) \leq M$. ∎

[IV] Riesz Representation Theorem on the L^p Spaces

Let $(X, \mathfrak{A}, \mu)$ be an arbitrary measure space. In Theorem 18.1, we showed that if we define a function L_g on $L^p(X, \mathfrak{A}, \mu)$ with $p \in [1, \infty]$ and $g \in L^q(X, \mathfrak{A}, \mu)$ where q is the conjugate of p by setting $L_g(f) = \int_X fg\,d\mu$ for $f \in L^p(X, \mathfrak{A}, \mu)$, then L_g is a bounded linear functional on $L^p(X, \mathfrak{A}, \mu)$ and moreover $\|L_g\|_* = \|g\|_q$ when $p \in (1, \infty]$ and the equality holds when $p = 1$ provided that the measure space $(X, \mathfrak{A}, \mu)$ is σ-finite. The Riesz Representation Theorem is a converse of this fact for $p \in [1, \infty)$.

Theorem 18.6. (Riesz Representation Theorem on L^p Spaces with $p \in [1, \infty)$) *Let $(X, \mathfrak{A}, \mu)$ be a σ-finite measure space. Let $p \in [1, \infty)$ and $q \in (1, \infty]$ be conjugates. Then for every bounded linear functional L on $L^p(X, \mathfrak{A}, \mu)$, there exists a unique g in $L^q(X, \mathfrak{A}, \mu)$ such that $\|g\|_q = \|L\|_*$ and $L(f) = \int_X fg\,d\mu$ for every $f \in L^p(X, \mathfrak{A}, \mu)$.*

Proof. 1. Consider first the case where $(X, \mathfrak{A}, \mu)$ is a finite measure space. In this case, for every $E \in \mathfrak{A}$ we have $\int_X |\mathbf{1}_E|^p\,d\mu = \int_X \mathbf{1}_E\,d\mu = \mu(E) < \infty$ so that $\mathbf{1}_E \in L^p(X, \mathfrak{A}, \mu)$. Let us define a complex-valued set function λ on $\mathfrak{A}$ by setting

$$(1) \qquad \lambda(E) = L(\mathbf{1}_E) \quad \text{for } E \in \mathfrak{A}.$$

We show that $\Re\lambda$ and $\Im\lambda$ are finite signed measures on $\mathfrak{A}$ which are absolutely continuous with respect to μ on $(X, \mathfrak{A})$.

Now $\lambda(\emptyset) = L(\mathbf{1}_\emptyset) = L(0) = 0$ so that $(\Re\lambda)(\emptyset) = 0$ and $(\Im\lambda)(\emptyset) = 0$. To show the countable additivity of $\Re\lambda$ and $\Im\lambda$, let $(E_n : n \in \mathbb{N})$ be a disjoint sequence in $\mathfrak{A}$ and let $E = \bigcup_{n\in\mathbb{N}} E_n$. The disjointness of $(E_n : n \in \mathbb{N})$ implies that the sequence of partial sums $\left(\sum_{k=1}^n \mathbf{1}_{E_k} : n \in \mathbb{N}\right)$ converges pointwise on X and $\sum_{n\in\mathbb{N}} \mathbf{1}_{E_n} = \mathbf{1}_E$. Let us show that the sequence $\left(\sum_{k=1}^n \mathbf{1}_{E_k} : n \in \mathbb{N}\right)$ converges to $\mathbf{1}_E$ in the L^p norm, that is, $\lim_{n\to\infty} \left\|\sum_{k=1}^n \mathbf{1}_{E_k} - \mathbf{1}_E\right\|_p = 0$. Now since $\mu(X) < \infty$ and since $\left|\sum_{k=1}^n \mathbf{1}_{E_k} - \mathbf{1}_E\right|^p \leq 1$, the Bounded Convergence Theorem (Theorem 7.16) implies that

$$\lim_{n\to\infty}\left\|\sum_{k=1}^n \mathbf{1}_{E_k} - \mathbf{1}_E\right\|_p^p = \lim_{n\to\infty}\int_X \left|\sum_{k=1}^n \mathbf{1}_{E_k} - \mathbf{1}_E\right|^p d\mu$$
$$= \int_X \lim_{n\to\infty}\left|\sum_{k=1}^n \mathbf{1}_{E_k} - \mathbf{1}_E\right|^p d\mu = 0.$$

(Let us remark that this does not hold in $L^\infty(X, \mathfrak{A}, \mu)$. Now the only possible values of $\sum_{k=1}^n \mathbf{1}_{E_k} - \mathbf{1}_E$ on X are 0 and 1. Suppose $\mu(E_n) > 0$ for every $n \in \mathbb{N}$. Then for every $n \in$

$\mathbb{N}$, $\sum_{k=1}^{n} \mathbf{1}_{E_k} - \mathbf{1}_E = 1$ on E_{n+1}. Since $\mu(E_{n+1}) \neq 0$, we have $\left\| \sum_{k=1}^{n} \mathbf{1}_{E_k} - \mathbf{1}_E \right\|_\infty = 1$. Thus we do not have $\lim_{n\to\infty} \left\| \sum_{k=1}^{n} \mathbf{1}_{E_k} - \mathbf{1}_E \right\|_\infty = 0$.)

Now as a bounded linear functional on $L^p(X, \mathfrak{A}, \mu)$, L is continuous on $L^p(X, \mathfrak{A}, \mu)$ according to Theorem 15.32. Thus the convergence $\lim_{n\to\infty} \sum_{k=1}^{n} \mathbf{1}_{E_k} = \mathbf{1}_E$ in L^p norm implies

$$L(\mathbf{1}_E) = L\Big(\lim_{n\to\infty} \sum_{k=1}^{n} \mathbf{1}_{E_k}\Big) = \lim_{n\to\infty} L\Big(\sum_{k=1}^{n} \mathbf{1}_{E_k}\Big)$$
$$= \lim_{n\to\infty} \sum_{k=1}^{n} L(\mathbf{1}_{E_k}) = \sum_{n\in\mathbb{N}} L(\mathbf{1}_{E_n}),$$

that is, $\lambda(E) = \sum_{n\in\mathbb{N}} \lambda(E_n)$. Thus for the real and the imaginary part of λ, we have $(\Re\lambda)(E) = \sum_{n\in\mathbb{N}} (\Re\lambda)(E_n)$ and $(\Im\lambda)(E) = \sum_{n\in\mathbb{N}} (\Im\lambda)(E_n)$. This proves the countable additivity of $\Re\lambda$ and $\Im\lambda$. To show that the two signed measures $\Re\lambda$ and $\Im\lambda$ are finite, let us note that from $\lambda(X) = L(\mathbf{1}_X) \in \mathbb{C}$, we have $(\Re\lambda)(X), (\Im\lambda)(X) \in \mathbb{R}$.

To show the absolute continuity of $(\Re\lambda)$ and $(\Im\lambda)$ with respect to μ, let $E \in \mathfrak{A}$ be such that $\mu(E) = 0$. Then $\mathbf{1}_E = 0 \in L^p(X, \mathfrak{A}, \mu)$ so that $\lambda(E) = L(\mathbf{1}_E) = L(0) = 0$ and then $(\Re\lambda)(E) = 0$ and $(\Im\lambda)(E) = 0$. This proves the absolute continuity of $\Re\lambda$ and $\Im\lambda$ with respect to μ. Since μ is a finite measure and $(\Re\lambda)$ and $(\Im\lambda)$ are finite signed measures, the Radon-Nikodym derivatives $d\Re\lambda/d\mu$ and $d\Im\lambda/d\mu$ exist by Theorem 11.14. The finiteness of $\Re\lambda$ and $\Im\lambda$ imply that $d\Re\lambda/d\mu$ and $d\Im\lambda/d\mu$ are μ-integrable on X by (c) of Observation 11.19. Let

(2) $$g = \frac{d\Re\lambda}{d\mu} + i\frac{d\Im\lambda}{d\mu}.$$

Note that the μ-integrability of $d\Re\lambda/d\mu$ and $d\Im\lambda/d\mu$ on X implies that $|d\Re\lambda/d\mu| < \infty$ and $|d\Im\lambda/d\mu| < \infty$ a.e. on X and therefore $|g| < \infty$ a.e. on X. Now for every $E \in \mathfrak{A}$ we have

(3) $$L(\mathbf{1}_E) = \lambda(E) = (\Re\lambda)(E) + i(\Im\lambda)(E)$$
$$= \int_E \frac{d\Re\lambda}{d\mu}\,d\mu + i\int_E \frac{d\Im\lambda}{d\mu}\,d\mu$$
$$= \int_E g\,d\mu = \int_X \mathbf{1}_E \cdot g\,d\mu.$$

Let $\varphi \in S_0(X, \mathfrak{A}, \mu) \subset L^p(X, \mathfrak{A}, \mu)$. By (3) and by the linearity of L and of the integral, we have

(4) $$L(\varphi) = \int_X \varphi g\,d\mu.$$

Since L is a bounded linear functional on $L^p(X, \mathfrak{A}, \mu)$, (4) implies that

(5) $$\left|\int_X \varphi g\,d\mu\right| = |L(\varphi)| \leq \|L\|_*\|\varphi\|_p.$$

Therefore we have

$$\sup\left\{\left|\int_X \varphi g\,d\mu\right| : \varphi \in S_0(X,\mathfrak{A},\mu) \text{ and } \|\varphi\|_p = 1\right\} \le \|L\|_* < \infty$$

so that $g \in L^q(X,\mathfrak{A},\mu)$ by Theorem 18.4.

Now let $f \in L^p(X,\mathfrak{A},\mu)$. Since $S_0(X,\mathfrak{A},\mu)$ is dense in $L^p(X,\mathfrak{A},\mu)$ by Theorem 18.3, there exists a sequence $(\varphi_n : n \in \mathbb{N})$ in $S_0(X,\mathfrak{A},\mu)$ such that $\lim_{n\to\infty} \|\varphi_n - f\|_p = 0$. Then by the continuity of L on $L^p(X,\mathfrak{A},\mu)$, we have

$$(6) \qquad L(f) = \lim_{n\to\infty} L(\varphi_n) = \lim_{n\to\infty} \int_X \varphi_n g\,d\mu$$

by (4). By Hölder's Inequality (Theorems 16.14 for $p \in (1,\infty)$ and Theorem 16.40 for $p = 1$), we have

$$\left|\int_X \varphi_n g\,d\mu - \int_X fg\,d\mu\right| \le \int_X |\varphi_n - f|\,|g|\,d\mu \le \|\varphi_n - f\|_p \|g\|_q.$$

Then since $\lim_{n\to\infty} \|\varphi_n - f\|_p = 0$ we have

$$\lim_{n\to\infty} \int_X \varphi_n g\,d\mu = \int_X fg\,d\mu.$$

Substituting this in (6) we have

$$L(f) = \int_X fg\,d\mu \quad \text{for } f \in L^p(X,\mathfrak{A},\mu).$$

Then $\|L\|_* = \|g\|_q$ by Theorem 18.1.

2. Let $(X,\mathfrak{A},\mu)$ be a σ-finite measure space. Then there exists an increasing sequence $(E_n : n \in \mathbb{N})$ in $\mathfrak{A}$ such that $\bigcup_{n\in\mathbb{N}} E_n = X$ and $\mu(E_n) < \infty$ for every $n \in \mathbb{N}$. For each $n \in \mathbb{N}$ if we let $\mathfrak{A}_n = \mathfrak{A} \cap E_n$ then $\mathfrak{A}_n$ is a σ-algebra of subsets of E_n and we have a finite measure space $(E_n, \mathfrak{A}_n, \mu)$. Note also that $(\mathfrak{A}_n : n \in \mathbb{N})$ is an increasing sequence of collections of subsets of X. (Indeed if $E \in \mathfrak{A}_n = \mathfrak{A} \cap E_n$ then $E = A \cap E_n$ for some $A \in \mathfrak{A}$. Then since $E_n \subset E_{n+1}$ we have $E = A \cap E_n = A \cap E_n \cap E_{n+1} \in \mathfrak{A} \cap E_{n+1} = \mathfrak{A}_{n+1}$. Thus $\mathfrak{A}_n \subset \mathfrak{A}_{n+1}$.) For each $n \in \mathbb{N}$ consider the finite measure space $(E_n, \mathfrak{A}_n, \mu)$ and the Banach space $L^p(E_n, \mathfrak{A}_n, \mu)$. For every $f_n \in L^p(E_n, \mathfrak{A}_n, \mu)$, let $\widetilde{f_n}$ be a function on X defined by setting $\widetilde{f_n} = f_n$ on E_n and $\widetilde{f_n} = 0$ on E_n^c. Then $\int_X |\widetilde{f_n}|^p\,d\mu = \int_{E_n} |f_n|^p\,d\mu < \infty$ so that $\widetilde{f_n} \in L^p(X,\mathfrak{A},\mu)$.

Let L be an arbitrary bounded linear functional on $L^p(X,\mathfrak{A},\mu)$. Let us define a complex-valued function L_n on $L^p(E_n,\mathfrak{A}_n,\mu)$ by setting

$$(7) \qquad L_n(f_n) = L(\widetilde{f_n}) \quad \text{for } f_n \in L^p(E_n,\mathfrak{A}_n,\mu).$$

The linearity of L_n on $L^p(E_n,\mathfrak{A}_n,\mu)$ follows from the linearity of L on $L^p(X,\mathfrak{A},\mu)$. Also $|L_n(f_n)| = |L(\widetilde{f_n})| \le \|L\|_* \|\widetilde{f_n}\|_p = \|L\|_* \|f_n\|_p$ for every $f_n \in L^p(E_n,\mathfrak{A}_n,\mu)$ so that $\|L\|_*$ is a bound for the linear functional L_n and thus L_n is a bounded linear functional on $L^p(E_n, \mathfrak{A}_n, \mu)$ with $\|L_n\|_* \le \|L\|_*$. Since $(E_n, \mathfrak{A}_n, \mu)$ is a finite measure space, by our results in **1** there exists $g_n \in L^q(E_n, \mathfrak{A}_n, \mu)$ such that $\|g_n\|_q = \|L_n\|_*$ and

(8) $$L_n(f_n) = \int_{E_n} f_n g_n \, d\mu \quad \text{for } f_n \in L^p(E_n, \mathfrak{A}_n, \mu).$$

Let us consider the functions g_n for $n \in \mathbb{N}$. Each g_n is defined on E_n. Now for $n < m$ we have $E_n \subset E_m$ and $\mathfrak{A}_n \subset \mathfrak{A}_m$. Let us show that $g_n = g_m$ a.e. on $(E_n, \mathfrak{A}_n, \mu)$. Let $E \in \mathfrak{A}_n$ be arbitrarily chosen. Then

$$\begin{aligned}
\textstyle\int_E g_n \, d\mu &= \textstyle\int_{E_n} \mathbf{1}_E \cdot g_n \, d\mu \quad \text{since } E \subset E_n \\
&= L_n(\mathbf{1}_E) \quad \text{by (8) since } \mathbf{1}_E \in L^p(E_n, \mathfrak{A}_n, \mu) \\
&= L(\widetilde{\mathbf{1}}_E) \quad \text{by (7)} \\
&= L_m(\mathbf{1}_E) \quad \text{by (7) since } \mathbf{1}_E \in L^p(E_m, \mathfrak{A}_m, \mu) \\
&= \textstyle\int_{E_m} \mathbf{1}_E \cdot g_m \, d\mu \quad \text{by (8)} \\
&= \textstyle\int_E g_m \, d\mu \quad \text{since } E \subset E_m.
\end{aligned}$$

Thus $\int_E g_n \, d\mu = \int_E g_m \, d\mu$ for every $E \in \mathfrak{A}_n$. By (a) of Proposition 9.16 this implies that $g_n = g_m$ a.e. on $(E_n, \mathfrak{A}_n, \mu)$ for $n < m$.

Let us define a function g on X as follows. We have $g_1 = g_2$ a.e. on $(E_1, \mathfrak{A}_1, \mu)$. Redefine g_2 on a null set in $(E_1, \mathfrak{A}_1, \mu)$ if necessary so that $g_2 = g_1$ on E_1. For each $n \in \mathbb{N}$ we have $g_n = g_{n+1}$ a.e. on $(E_n, \mathfrak{A}_n, \mu)$. Redefine g_{n+1} on a null set in $(E_n, \mathfrak{A}_n, \mu)$ if necessary so that $g_{n+1} = g_n$ on E_n. Thus proceeding we have $g_{n+1} = g_n$ on E_n for every $n \in \mathbb{N}$. Let g be a function defined on X by setting $g = g_n$ on E_n for $n \in \mathbb{N}$. For each $n \in \mathbb{N}$, let $\widetilde{g}_n$ be a function on X defined by setting $\widetilde{g}_n = g_n$ on E_n and $\widetilde{g}_n = 0$ on E_n^c. Then we have $\lim_{n\to\infty} \widetilde{g}_n = g$ and $|\widetilde{g}_n| \uparrow |g|$ on X as $n \to \infty$. Since $\widetilde{g}_n$ is $\mathfrak{A}$-measurable on X, the function $g = \lim_{n\to\infty} \widetilde{g}_n$ is $\mathfrak{A}$-measurable on X.

Let us show that $g \in L^q(X, \mathfrak{A}, \mu)$. For the case $p \in (1, \infty)$ and $q \in (1, \infty)$, we have $\int_X |g|^q \, d\mu = \lim_{n\to\infty} \int_X |\widetilde{g}_n|^q \, d\mu = \lim_{n\to\infty} \|\widetilde{g}_n\|_q^q$ by the Monotone Convergence Theorem (Theorem 8.5), and consequently we have

$$\|g\|_q = \lim_{n\to\infty} \|\widetilde{g}_n\|_q = \lim_{n\to\infty} \|g_n\|_q = \lim_{n\to\infty} \|L_n\|_* \le \|L\|_* < \infty$$

so that $g \in L^q(X, \mathfrak{A}, \mu)$. On the other hand for the case $p = 1$ and $q = \infty$, we have $\|\widetilde{g}_n\|_\infty = \|g_n\|_\infty = \|L_n\|_* \le \|L\|_* < \infty$ so that $|\widetilde{g}_n| \le \|L\|_* < \infty$ a.e. on X. Then $|\widetilde{g}_n| \uparrow |g|$ on X implies that $|g| \le \|L\|_* < \infty$ a.e. on X and thus $g \in L^\infty(X, \mathfrak{A}, \mu)$.

Let us show that $L(f) = \int_X fg \, d\mu$ for every $f \in L^p(X, \mathfrak{A}, \mu)$. Let $f \in L^p(X, \mathfrak{A}, \mu)$. Since $\lim_{n\to\infty} f \cdot \mathbf{1}_{E_n} = f$ pointwise on X and $|f \cdot \mathbf{1}_{E_n} - f|^p \le \{|f \cdot \mathbf{1}_{E_n}| + |f|\}^p \le 2^p |f|^p$, we have $\lim_{n\to\infty} \|f \cdot \mathbf{1}_{E_n} - f\|_p^p = \lim_{n\to\infty} \int_X |f \cdot \mathbf{1}_{E_n} - f|^p \, d\mu = 0$ by the Dominated Convergence Theorem (Theorem 9.20) and thus $\lim_{n\to\infty} \|f \cdot \mathbf{1}_{E_n} - f\|_p = 0$. Then by the continuity of L on $L^p(X, \mathfrak{A}, \mu)$ and by (7) and (8) we have

$$L(f) = \lim_{n\to\infty} L(f \cdot \mathbf{1}_{E_n}) = \lim_{n\to\infty} L_n(f \cdot \mathbf{1}_{E_n})$$
$$= \lim_{n\to\infty} \int_{E_n} f g_n \, d\mu = \lim_{n\to\infty} \int_X f \widetilde{g}_n \, d\mu.$$

Since $|f\widetilde{g}_n| \leq |fg|$ and since $\int_X |fg| \, d\mu \leq \|f\|_p \|g\|_q < \infty$, the Dominated Convergence Theorem again implies that $\lim_{n\to\infty} \int_X f\widetilde{g}_n \, d\mu = \int_X \lim_{n\to\infty} f\widetilde{g}_n \, d\mu = \int_X fg \, d\mu$. Thus we have $L(f) = \int_X fg \, d\mu$ for every $f \in L^p(X, \mathfrak{A}, \mu)$. Then $\|L\|_* = \|g\|_q$ by Theorem 18.1.

3. To show the uniqueness of g, suppose there exist $g_1, g_2 \in L^q(X, \mathfrak{A}, \mu)$ such that $L(f) = \int_X fg_j \, d\mu$ for every $f \in L^p(X, \mathfrak{A}, \mu)$ for $j = 1$ and 2. Let $(E_n : n \in \mathbb{N})$ be an increasing sequence in $\mathfrak{A}$ such that $\bigcup_{n\in\mathbb{N}} E_n = X$ and $\mu(E_n) < \infty$ for $n \in \mathbb{N}$. With $f = \mathbf{1}_{E_n} \in L^p(X, \mathfrak{A}, \mu)$, we have $\int_{E_n} g_j \, d\mu = \int_X fg_j \, d\mu = L(f) \in \mathbb{C}$ and thus g_j is μ-integrable on E_n. If we let $f = \mathbf{1}_E$ for an arbitrary $\mathfrak{A}$-measurable subset E of E_n, then $f \in L^p(X, \mathfrak{A}, \mu)$ so that $\int_E g_j \, d\mu = \int_X fg_j \, d\mu = L(f)$ for $j = 1$ and 2. Thus $\int_E g_1 \, d\mu = \int_E g_2 \, d\mu$ for every $\mathfrak{A}$-measurable subset E of E_n. This implies that $g_1 = g_2$ a.e. on E_n by (a) of Proposition 9.16. Since this holds for every $n \in \mathbb{N}$, we have $g_1 = g_2$ a.e. on X. ∎

The Riesz Representation Theorem on $L^p(X, \mathfrak{A}, \mu)$ holds for $p, q \in (1, \infty)$ without assumption of σ-finiteness of the measure space $(X, \mathfrak{A}, \mu)$.

Theorem 18.7. (Riesz Representation Theorem on L^p Spaces for $p \in (1, \infty)$) *Let $(X, \mathfrak{A}, \mu)$ be an arbitrary measure space. Let $p, q \in (1, \infty)$ be conjugates. Then for every bounded linear functional L on $L^p(X, \mathfrak{A}, \mu)$, there exists $g \in L^q(X, \mathfrak{A}, \mu)$ such that $\|g\|_q$ and $L(f) = \int_X fg \, d\mu$ for every $f \in L^p(X, \mathfrak{A}, \mu)$.*

Proof. Let $\mathfrak{S}$ be the collection of all σ-finite sets in $(X, \mathfrak{A}, \mu)$. Then $\mathfrak{S}$ is closed under countable unions. To show this, let $(E_n : n \in \mathbb{N})$ be a sequence in $\mathfrak{S}$. Then for each $n \in \mathbb{N}$, there exists a sequence $(E_{n,k} : k \in \mathbb{N})$ in $\mathfrak{A}$ such that $\bigcup_{k\in\mathbb{N}} E_{n,k} = E_n$ and $\mu(E_{n,k}) < \infty$ for $k \in \mathbb{N}$. Let $F_N = \bigcup_{n=1}^N \left[\bigcup_{k=1}^N E_{n,k}\right]$ for $N \in \mathbb{N}$. Then $F_N \in \mathfrak{A}$, $\mu(F_N) < \infty$, and $\bigcup_{N\in\mathbb{N}} F_N = \bigcup_{n\in\mathbb{N}} E_n$ so that $\bigcup_{n\in\mathbb{N}} E_n$ is a σ-finite set.

For an arbitrary $E \in \mathfrak{S}$, consider the σ-finite measure space $(E, \mathfrak{A} \cap E, \mu)$ and $L^p(E, \mathfrak{A} \cap E, \mu)$. For an arbitrary $f_E \in L^p(E, \mathfrak{A} \cap E, \mu)$, let

$$(1) \qquad \widetilde{f}_E = \begin{cases} f_E & \text{on } E, \\ 0 & \text{on } E^c. \end{cases}$$

Then $\|\widetilde{f}_E\|_p^p = \int_X |\widetilde{f}_E|^p \, d\mu = \int_E |f_E|^p \, d\mu = \|f_E\|_p^p < \infty$ so that $\widetilde{f}_E \in L^p(X, \mathfrak{A}, \mu)$. Let L be a bounded linear functional on $L^p(X, \mathfrak{A}, \mu)$. Let us define a function L_E on $L^p(E, \mathfrak{A} \cap E, \mu)$ by

$$(2) \qquad L_E(f_E) = L(\widetilde{f}_E) \quad \text{for } f_E \in L^p(E, \mathfrak{A} \cap E, \mu).$$

Then $|L_E(f_E)| = |L(\widetilde{f}_E)| \leq \|L\|_* \|\widetilde{f}_E\|_p = \|L\|_* \|f_E\|_p$ for every $f_E \in L^p(E, \mathfrak{A}\cap E, \mu)$. Thus $\|L\|_*$ is a bound for the linear functional L_E on $L^p(E, \mathfrak{A} \cap E, \mu)$ so that L_E is a bounded linear functional on $L^p(E, \mathfrak{A} \cap E, \mu)$ with $\|L_E\|_* \le \|L\|_*$. Since $(E, \mathfrak{A} \cap E, \mu)$ is a σ-finite measure space, by Theorem 18.6 there exists a unique $g_E \in L^q(E, \mathfrak{A} \cap E, \mu)$ with $\|g_E\|_q = \|L_E\|_*$ such that

$$(3) \qquad L_E(f_E) = \int_E f_E g_E \, d\mu \quad \text{for every } f_E \in L^p(E, \mathfrak{A} \cap E, \mu).$$

Consider the collection $\{g_E : E \in \mathfrak{S}\}$. Let us show that if $E_1, E_2 \in \mathfrak{S}$ and $E_1 \subset E_2$ then $g_{E_1} = g_{E_2}$ a.e. on E_1. Let $h_1 \in L^p(E_1, \mathfrak{A} \cap E_1, \mu)$ and let $h_2 = h_1$ on E_1 and $h_2 = 0$ on $E_2 \setminus E_1$. Then $h_2 \in L^p(E_2, \mathfrak{A} \cap E_2, \mu)$ and $\widetilde{h}_1 = \widetilde{h}_2$. Now $L(\widetilde{h}_1) = L_{E_1}(h_1) = \int_{E_1} h_1 g_{E_1} \, d\mu$ and since $\widetilde{h}_1 = \widetilde{h}_2$ we have

$$L(\widetilde{h}_1) = L(\widetilde{h}_2) = L_{E_2}(h_2) = \int_{E_2} h_2 g_{E_2} \, d\mu = \int_{E_1} h_1 g_{E_2} \, d\mu.$$

Thus we have $L(\widetilde{h}_1) = \int_{E_1} h_1 g_{E_1} \, d\mu$ and $L(\widetilde{h}_1) = \int_{E_1} h_1 g_{E_2} \, d\mu$ also for every $h_1 \in L^p(E_1, \mathfrak{A} \cap E_1, \mu)$. Then by the uniqueness of representation, we have $g_{E_1} = g_{E_2}$ a.e. on E_1. Let

$$M = \sup\{\|g_E\|_q : E \in \mathfrak{S}\}.$$

Since $\|g_E\|_q = \|L_E\|_* \le \|L\|_*$ for every $E \in \mathfrak{S}$, we have $M \le \|L\|_*$. Let $(E_n : n \in \mathbb{N})$ be a sequence in $\mathfrak{S}$ such that $\lim_{n\to\infty} \|g_{E_n}\|_q = M$. Let $E_0 = \bigcup_{n\in\mathbb{N}} E_n$. Then $E_0 \in \mathfrak{S}$ since $\mathfrak{S}$ is closed under countable unions.

Now for every $n \in \mathbb{Z}_+$, $(E_n, \mathfrak{A} \cap E_n, \mu)$ is a σ-finite measure space. Thus for the bounded linear functional L_{E_n} defined on $L^p(E_n, \mathfrak{A} \cap E_n, \mu)$ by (2), there exists $g_{E_n} \in L^q(E_n, \mathfrak{A} \cap E_n, \mu)$ such that $\|g_{E_n}\|_q = \|L_{E_n}\|_* \le \|L\|_*$ and

$$L_{E_n}(f_{E_n}) = \int_{E_n} f_{E_n} g_{E_n} \, d\mu \quad \text{for every } f_{E_n} \in L^p(E_n, \mathfrak{A} \cap E_n, \mu).$$

According to Theorem 18.4, for $n \in \mathbb{Z}_+$ we have

$$\|g_{E_n}\|_q = \sup\left\{\left|\int_{E_n} f_{E_n} g_{E_n} \, d\mu\right| : f_{E_n} \in S_0(E_n, \mathfrak{A} \cap E_n, \mu) \text{ and } \|f_{E_n}\|_p = 1\right\}.$$

Since $E_n \subset E_0$, we have $g_{E_n} = g_{E_0}$ a.e. on E_n and this implies $\|g_{E_n}\|_q \le \|g_{E_0}\|_q$ for $n \in \mathbb{N}$. Thus $M = \lim_{n\to\infty} \|g_{E_n}\|_q \le \|g_{E_0}\|_q$. On the other hand since $E_0 \in \mathfrak{S}$, we have $\|g_{E_0}\|_q \le M$. Therefore we have $\|g_{E_0}\|_q = M$.

Let $\widetilde{g}_{E_0} = g_{E_0}$ on E_0 and $\widetilde{g}_{E_0} = 0$ on E_0^c. Let us show that

$$L(f) = \int_X f \widetilde{g}_{E_0} \, d\mu \quad \text{for every } f \in L^p(X, \mathfrak{A}, \mu).$$

Let $f \in L^p(X, \mathfrak{A}, \mu)$ be arbitrarily fixed and let $F = \{x \in X : |f|^p \neq 0\}$. By (f) of Proposition 9.2, $F \in \mathfrak{S}$. Since $\mathfrak{S}$ is closed under unions, $G := E_0 \cup F \in \mathfrak{S}$. Then $G \setminus E_0 \in \mathfrak{S}$ also. Since $E_0 \subset G$ and $G \setminus E_0 \subset G$, we have $g_{E_0} = g_G$ a.e. on E_0 and $g_{G\setminus E_0} = g_G$ a.e. on $G \setminus E_0$. Let $\widetilde{g}_{E_0} = g_{E_0}$ on E_0 and $\widetilde{g}_{E_0} = 0$ on E_0^c. Define $\widetilde{g}_{G\setminus E_0}$ and $\widetilde{g}_G$ similarly. Since E_0 and $G \setminus E_0$ are disjoint and their union is equal to G, we have $\widetilde{g}_{E_0} + \widetilde{g}_{G\setminus E_0} = \widetilde{g}_G$ a.e. on X. The disjointness of E_0 and $G \setminus E_0$ also implies that at every $x \in X$, at least one of $\widetilde{g}_{E_0}(x)$ and $\widetilde{g}_{G\setminus E_0}(x)$ is equal to 0. Thus we have $|\widetilde{g}_{E_0}|^q + |\widetilde{g}_{G\setminus E_0}|^q = |\widetilde{g}_G|^q$ a.e. on X. Then $\int_X |\widetilde{g}_{E_0}|^q \, d\mu + \int_X |\widetilde{g}_{G\setminus E_0}|^q \, d\mu = \int_X |\widetilde{g}_G|^q \, d\mu$. But $\int_X |\widetilde{g}_G|^q \, d\mu = \|\widetilde{g}_G\|_q^q \le M = \int_X |\widetilde{g}_{E_0}|^q \, d\mu$. Thus we have $\int_X |\widetilde{g}_{G\setminus E_0}|^q \, d\mu = 0$ and then $\widetilde{g}_{G\setminus E_0} = 0$ a.e. on X. Therefore $\widetilde{g}_{E_0} = \widetilde{g}_G$, that is, $\widetilde{g}_{E_0} = \widetilde{g}_{E_0 \cup F}$ a.e. on X. Since $f = 0$ on F^c, we have $f = 0$ on $(E_0 \cup F)^c$ and thus $f = f \cdot \mathbf{1}_{E_0 \cup F}$. Then for our $f \in L^p(X, \mathfrak{A}, \mu)$,

$$L(f) = L_{E_0 \cup F}(f \cdot \mathbf{1}_{E_0 \cup F}) = \int_{E_0 \cup F} f \cdot \mathbf{1}_{E_0 \cup F} \cdot g_{E_0 \cup F} \, d\mu$$
$$= \int_X f \widetilde{g}_{E_0 \cup F} \, d\mu = \int_X f \widetilde{g}_{E_0} \, d\mu$$

by (2) and (3). ∎

From the uniqueness in the Riesz Representation Theorem we derive the following sufficient condition for a function to be in $L^q(X, \mathfrak{A}, \mu)$ where $q \in [1, \infty]$.

Theorem 18.8. *Given a σ-finite measure space $(X, \mathfrak{A}, \mu)$. Consider $L^p(X, \mathfrak{A}, \mu)$ for $p \in [1, \infty]$ and let $q \in [1, \infty]$ be its conjugate. Let g be an extended complex-valued $\mathfrak{A}$-measurable function on X such that $|g| < \infty$ a.e. on X. If $\int_X fg \, d\mu$ exists in $\mathbb{C}$ for every $f \in L^p(X, \mathfrak{A}, \mu)$, then $g \in L^q(X, \mathfrak{A}, \mu)$.*

Proof. 1. According to Lemma 18.2, there exists a sequence of complex-valued simple functions $(\varphi_n : n \in \mathbb{N})$ each vanishing outside of a $\mathfrak{A}$-measurable set with finite measure such that $\lim_{n\to\infty} \varphi_n = g$ on X and $|\varphi_n| \le |g|$ on X for every $n \in \mathbb{N}$. Let $f \in L^p(X, \mathfrak{A}, \mu)$. Since $\lim_{n\to\infty} \varphi_n = g$ on X, we have $\lim_{n\to\infty} f\varphi_n = fg$ on X. By our assumption, $\int_X fg \, d\mu$ exists in $\mathfrak{C}$, that is, fg is μ-integrable on X. Since $|f\varphi_n| \le |fg|$ on X, by the Dominated Convergence Theorem (Theorem 9.20), we have $\lim_{n\to\infty} \int_X f\varphi_n \, d\mu = \int_X fg \, d\mu$. Let us define functions L_n, $n \in \mathbb{N}$, and L on $L^p(X, \mathfrak{A}, \mu)$ by setting $L_n(f) = \int_X f\varphi_n \, d\mu$, $n \in \mathbb{N}$, and $L(f) = \int_X fg \, d\mu$ for $f \in L^p(X, \mathfrak{A}, \mu)$. Since φ_n is a simple function vanishing outside of a $\mathfrak{A}$-measurable set with finite measure, $\varphi_n \in L^q(X, \mathfrak{A}, \mu)$. Thus L_n is a bounded linear functional on $L^p(X, \mathfrak{A}, \mu)$ with $\|L_n\|_* = \|\varphi_n\|_q$ by Theorem 18.1. Since $L(f) = \lim_{n\to\infty} L_n(f)$ for every $f \in L^p(X, \mathfrak{A}, \mu)$, L is a bounded linear functional on $L^p(X, \mathfrak{A}, \mu)$ by Theorem 15.89.

2. Suppose $p \in [1, \infty)$. Since L is a bounded linear functional on $L^p(X, \mathfrak{A}, \mu)$, there exists a unique $h \in L^q(X, \mathfrak{A}, \mu)$ such that $L(f) = \int_X fh \, d\mu$ for $f \in L^p(X, \mathfrak{A}, \mu)$ by Theorem 18.6. But $L(f) = \int_X fg \, d\mu$ for $f \in L^p(X, \mathfrak{A}, \mu)$. Thus by the uniqueness, we have $g = h \in L^q(X, \mathfrak{A}, \mu)$.

3. For the case $p = \infty$, since L is a bounded linear functional on $L^\infty(X, \mathfrak{A}, \mu)$, we have $\left|\int_X fg \, d\mu\right| = |L(f)| \le \|L\|_* \|f\|_\infty$. Let

$$f_0 = \begin{cases} \overline{g} \cdot |g|^{-1} & \text{where } g \ne 0, \\ 0 & \text{where } g = 0. \end{cases}$$

If $g = 0$ a.e. on X, then $g \in L^1(X, \mathfrak{A}, \mu)$. If $\mu\{x \in X : g(x) \neq 0\} > 0$, then we have $\mu\{x \in X : |f_0(x)| = 1\} > 0$. Since $|f_0| \leq 1$ on X, we have $\|f_0\|_\infty = 1$ and $f_0 \in L^\infty(X, \mathfrak{A}, \mu)$. This shows that $g \in L^1(X, \mathfrak{A}, \mu)$. ∎

Problems

Prob. 18.1. Let $(X, \mathfrak{A}, \mu)$ be an arbitrary measure space and let $p \in (1, \infty)$. Consider $(f_n : n \in \mathbb{N}) \subset L^p(X, \mathfrak{A}, \mu)$ and $f \in L^p(X, \mathfrak{A}, \mu)$. Show that the sequence $(f_n : n \in \mathbb{N})$ converges weakly to f in $L^p(X, \mathfrak{A}, \mu)$ if and only if

1° $\|f_n\|_p \leq M$ for all $n \in \mathbb{N}$ for some $M > 0$,

2° $\lim_{n\to\infty} \int_E f_n \, d\mu = \int_E f \, d\mu$ for every $E \in \mathfrak{A}$ with $\mu(E) < \infty$.

Prob. 18.2. Let $(X, \mathfrak{A}, \mu)$ be a σ-finite measure space. Let $p \in [1, \infty)$ and let $q \in (1, \infty]$ be the conjugate, that is, $1/p + 1/q = 1$. Let $g_1, g_2 \in L^q(X, \mathfrak{A}, \mu)$. Suppose

$$\int_X f g_1 \, d\mu = \int_X f g_2 \, d\mu \text{ for every } f \in L^p(X, \mathfrak{A}, \mu).$$

Show that $g_1 = g_2$, μ-a.e. on X.

§19 Integration on Locally Compact Hausdorff Space

[I] Continuous Functions on a Locally Compact Hausdorff Space

Let us review some topological notions in order to fix the terminology. By a neighborhood of a point x in a topological space X we mean an open set V containing x. A topological space X is called a Hausdorff space if for every pair of distinct points x and y in X there exists a pair of disjoint neighborhoods U and V of x and y respectively. A set K in a topological space X is called a compact set if for an arbitrary collection $\mathfrak{V}$ of open sets in X such that $K \subset \bigcup_{V \in \mathfrak{V}} V$ there exists a finite subcollection $\{V_1, \ldots, V_N\}$ of $\mathfrak{V}$ such that $K \subset \bigcup_{j=1}^{N} V_j$. We refer to a collection of open sets whose union contains a set as an open cover of the set. Thus a set is a compact set if every open cover of it has a finite subcover. It follows then that an arbitrary intersection of compact sets is a compact set and a finite union of compact set is a compact set. A set which is a countable union of compact sets is called a σ-compact set.

Proposition 19.1. *A compact set K in a Hausdorff space X is always a closed set.*

Proof. If $K = X$ or $K = \emptyset$, then K is a closed set. Suppose $K \neq X$ and $K \neq \emptyset$. To show that K is closed, we show that K^c is open. We have $K^c \neq \emptyset$. For every $x \in K$ and $y \in K^c$, there exist disjoint neighborhoods $U_{x,y}$ and $V_{x,y}$ of x and y respectively by the Hausdorff property of X. Let $y \in K^c$ be fixed. The collection $\{U_{x,y} : x \in K\}$ is an open cover of K. Since K is a compact set a finite subcollection covers K. Thus there exist $x_1, \ldots, x_N \in K$ such that $K \subset \bigcup_{j=1}^{N} U_{x_j,y}$. Let $U = \bigcup_{j=1}^{N} U_{x_j,y}$ and $V = \bigcap_{j=1}^{N} V_{x_j,y}$. Since $U_{x_j,y} \cap V_{x_j,y} = \emptyset$ for $j = 1, \ldots, N$, we have $U_{x_j,y} \cap V = \emptyset$ for $j = 1, \ldots, N$ and then $U \cap V = \emptyset$. Since $K \subset U$, we have $K \cap V = \emptyset$ and then $V \subset K^c$. Thus for an arbitrary $y \in K^c$ there exists an open set V containing y and contained in K^c. This shows that K^c is an open set. ∎

Proposition 19.2. *If K is a compact set and C is a closed set in a topological space X, then $K \cap C$ is a compact set.*

Proof. Let $\mathfrak{V}$ be an arbitrary collection of open sets such that $K \cap C \subset \bigcup_{V \in \mathfrak{V}} V$. Now C^c is an open set and $K \setminus C = K \cap C^c \subset C^c$. Thus $K \subset \left(\bigcup_{V \in \mathfrak{V}} V\right) \cup C^c$. Since K is a compact set, there exists a finite subcollection $\{V_1, \ldots, V_N\}$ of $\mathfrak{V}$ such that $K \subset V_1 \cup \cdots \cup V_N \cup C^c$. Then $K \cap C \subset (V_1 \cup \cdots \cup V_N \cup C^c) \cap C \subset (V_1 \cup \cdots \cup V_N) \cup (C^c \cap C) = V_1 \cup \cdots \cup V_N$. This shows that an arbitrary open cover of $K \cap C$ has a finite subcover. Thus $K \cap C$ is a compact set. ∎

Observation 19.3. Let $\mathfrak{O}$ be the collection of all open sets and $\mathfrak{K}$ be the collection of all compact sets in a Hausdorff space X. Then
(a) $\sigma(\mathfrak{K}) \subset \mathfrak{B}_X$,
(b) $\sigma(\mathfrak{K}) = \mathfrak{B}_X$ if X is a σ-compact set.

Proof. A compact set in a Hausdorff space X is a closed set by Proposition 19.1 and is a member of $\mathfrak{B}_X$. Thus $\mathfrak{K} \subset \mathfrak{B}_X$ and then $\sigma(\mathfrak{K}) \subset \mathfrak{B}_X$. This proves (a). Suppose X is a σ-compact set. Then there exists a sequence $(K_j : j \in \mathbb{N})$ in $\mathfrak{K}$ such that $X = \bigcup_{j\in\mathbb{N}} K_j$. Let $O \in \mathfrak{O}$. Then $O^c = O^c \cap X = \bigcup_{j\in\mathbb{N}}(O^c \cap K_j)$. Since O^c is a closed set and $K_j \in \mathfrak{K}$, we have $O^c \cap K_j \in \mathfrak{K}$. Then $O^c = \bigcup_{j\in\mathbb{N}}(O^c \cap K_j) \in \sigma(\mathfrak{K})$ and hence $O \in \sigma(\mathfrak{K})$. Thus $\mathfrak{O} \subset \sigma(\mathfrak{K})$ and then $\sigma(\mathfrak{O}) \subset \sigma(\mathfrak{K})$, that is, $\mathfrak{B}_X \subset \sigma(\mathfrak{K})$. Then by (a), we have $\sigma(\mathfrak{K}) = \mathfrak{B}_X$. ∎

Definition 19.4. *A topological space X is called a locally compact space if every point x in X has a neighborhood V such that $\overline{V}$ is a compact set.*

According to the Heine-Borel Theorem, a set in $\mathbb{R}^n$ is a compact set if and only if it is a bounded and closed set. It follows then that $\mathbb{R}^n$ is a σ-compact set and a locally compact Hausdorff space.

Proposition 19.5. *Let X be a locally compact Hausdorff space and let O be an open set in X. Then for every $x \in O$, there exists a neighborhood V of x such that $\overline{V} \subset O$ and $\overline{V}$ is compact.*

Proof. Let $x \in O$ be arbitrarily fixed. Let W be a neighborhood of x such that $\overline{W}$ is a compact set and let $O_0 = O \cap W$. Now $O_0 \subset W$ implies $\overline{O_0} \subset \overline{W}$. Thus $\overline{O_0}$ is a closed set contained in a compact set $\overline{W}$ so that it is a compact set by Proposition 19.2. Then $K := \overline{O_0} \setminus O_0 = \overline{O_0} \cap O_0^c$, being the intersection of a compact set and a closed set, is a compact set.

Now if $K = \emptyset$, then $\overline{O_0} = O_0 \subset O$ so that if we let $V = O_0$ then V is a neighborhood of x such that $\overline{V} \subset O$ and $\overline{V}$ is a compact set.

Consider the case $K \neq \emptyset$. Since $x \in O_0$, $x \notin K$. For every $y \in K$, there exist disjoint neighborhoods U_y and V_y of y and x respectively. We may assume that $V_y \subset O_0$ by intersecting V_y with O_0 if necessary. Now since $\{U_y : y \in K\}$ is an open cover of the compact set K, there exist $y_1, \ldots, y_N \in K$ such that $K \subset \bigcup_{j=1}^N U_{y_j}$. Let $U = \bigcup_{j=1}^N U_{y_j}$ and $V = \bigcap_{j=1}^N V_{y_j}$. Then $U \cap V = \emptyset$, $K \subset U$, $V \subset O_0$, and V is a neighborhood of x. Now $V \subset O_0 \setminus U \subset \overline{O_0} \setminus U = \overline{O_0} \cap U^c$. Since $\overline{O_0} \cap U^c$ is a closed set, we have

$$\overline{V} \subset \overline{O_0} \cap U^c = \overline{O_0} \setminus U \subset \overline{O_0} \setminus K = \overline{O_0} \setminus (\overline{O_0} \setminus O_0) = O_0 \subset O.$$

Since $\overline{O_0}$ is a compact set and U^c is a closed set, $\overline{O_0} \cap U^c$ is a compact set by Proposition 19.2. Then the closed set $\overline{V}$ contained in the compact set $\overline{O_0} \cap U^c$ is a compact set. ∎

Proposition 19.6. *Let X be a locally compact Hausdorff space. If K is a compact set, O is an open set and $K \subset O$, then there exists an open set V such that $\overline{V}$ is a compact set and $K \subset V \subset \overline{V} \subset O$.*

Proof. Let $x \in K$. By Proposition 19.5, there exists a neighborhood V_x of x such that $\overline{V}_x \subset O$ and $\overline{V}_x$ is a compact set. The collection $\{V_x : x \in K\}$ is an open cover of the compact set K and thus there exist $x_1, \ldots, x_N \in K$ such that $K \subset \bigcup_{j=1}^N V_{x_j}$. Let $V = \bigcup_{j=1}^N V_{x_j}$. Then $\overline{V} = \bigcup_{j=1}^N \overline{V}_x \subset O$ and $\overline{V}$, being a finite union of compact sets, is a compact set. ∎

Definition 19.7. *Let f be a real-valued function on a topological space X. The support of f is defined to be the closure of the set $\{x \in X : f(x) \neq 0\}$. We write* $\text{supp}\{f\}$ *for this set.*

Observation 19.8. For real-valued functions f and g on a topological space X and for $\lambda \in \mathbb{R}$, we have
(a) $\text{supp}\{f + g\} \subset \text{supp}\{f\} \cup \text{supp}\{g\}$.
(b) $\text{supp}\{\lambda f\} = \text{supp}\{f\}$ if $\lambda \neq 0$.
(c) $\text{supp}\{fg\} \subset \text{supp}\{f\} \cap \text{supp}\{g\}$.

Proof. 1. If $f(x) + g(x) \neq 0$ for some $x \in X$, then either $f(x) \neq 0$ or $g(x) \neq 0$ or both. Thus $\{x \in X : f(x) + g(x) \neq 0\} \subset \{x \in X : f(x) \neq 0\} \cup \{x \in X : g(x) \neq 0\}$. Then (a) follows from the fact that for any three sets A, B, and C such that $A \subset B \cup C$, we have $\overline{A} \subset \overline{B \cup C} = \overline{B} \cup \overline{C}$.

2. For $\lambda \neq 0$, we have $\{x \in X : \lambda f(x) \neq 0\} = \{x \in X : f(x) \neq 0\}$. (b) follows from this equality.

3. Note that $\{x \in X : f(x)g(x) \neq 0\} = \{x \in X : f(x) \neq 0\} \cap \{x \in X : g(x) \neq 0\}$. Since $A = B \cap C$ implies $\overline{A} = \overline{B \cap C} \subset \overline{B} \cap \overline{C}$, (c) follows. ∎

Definition 19.9. *Let X be a topological space. We write $C_c(X)$ for the collection of all continuous real-valued functions f on X such that* $\text{supp}\{f\}$ *is a compact set.*

Let us note that $C_c(X)$ is a linear space. Indeed if $f, g \in C_c(X)$, then $f + g$ is a continuous function on X and furthermore by Observation 19.8, $\text{supp}\{f+g\} \subset \text{supp}\{f\} \cup \text{supp}\{g\}$ so that $\text{supp}\{f+g\}$ being a closed set contained in a compact set $\text{supp}\{f\} \cup \text{supp}\{g\}$ is a compact set. Similarly if $f \in C_c(X)$ and $\lambda \in \mathbb{R}$, then $\lambda f \in C_c(X)$.

Remark 19.10. If X is a Hausdorff space, then the collection of all real-valued continuous functions on X that have compact supports is identical with the collection of all real-valued continuous functions on X that vanish outside of compact sets.

Proof. Let f be a real-valued continuous function on a topological space X. If $\text{supp}\{f\}$ is a compact set, then f vanishes outside of a compact set. Conversely, if X is a Hausdorff space and if f vanishes outside of a compact set K, then $E := \{x \in X : f(x) \neq 0\} \subset K$. Since X is a Hausdorff space, K is a closed set by Proposition 19.1 so that $\overline{E} \subset K$. Then $\overline{E}$ is a compact set by Proposition 19.2 so that f has a compact support. ∎

Theorem 19.11. (Urysohn's Lemma) *Let X be a locally compact Hausdorff space. Let K be a compact set and V be an open set in X such that $K \subset V$. Then there exists $f \in C_c(X)$ such that*
1° $f(x) \in [0, 1]$ *for* $x \in X$.
2° $f(x) = 1$ *for* $x \in K$.
3° $\text{supp}\{f\} \subset V$.

Proof. Let Q be the collection of all rational numbers in $[0, 1]$. Let us show that there exists a collection $\{V_r : r \in Q\}$ of open sets in X such that

(i) $\overline{V_r}$ is a compact set for every $r \in Q$.
(ii) $V_r \supset \overline{V_s}$ for $r, s \in Q$ such that $r < s$.
(iii) $V \supset \overline{V_0}$ and $V_1 \supset K$.

Applying Proposition 19.6 to the compact set K contained in the open set V, we have an open set V_0 such that $\overline{V_0}$ is a compact set and $K \subset V_0 \subset \overline{V_0} \subset V$. Applying Proposition 19.6 again to the compact set K contained in the open set V_0, we have an open set V_1 such that $\overline{V_1}$ is a compact set and $K \subset V_1 \subset \overline{V_1} \subset V_0 \subset \overline{V_0} \subset V$. Let $\{r_j : j \in \mathbb{Z}_+\}$ be an arbitrary enumeration of the elements of Q with $r_0 = 0$ and $r_1 = 1$. Now suppose for some $n \geq 1$, we have selected open sets $V_{r_0}, \ldots, V_{r_n}$ such that $\overline{V}_{r_0}, \ldots, \overline{V}_{r_n}$ are compact sets and $\overline{V}_{r_j} \subset V_{r_i}$ for $r_j > r_i$. Consider r_{n+1}. We have $0 = r_0 < r_{n+1} < r_1 = 1$. Let r_i be the greatest among those of $r_0, \ldots, r_n$ that are less than r_{n+1} and let r_j be the least among those of $r_0, \ldots, r_n$ that are greater than r_{n+1}. By Proposition 19.6, there exists an open set $V_{r_{n+1}}$ such that $\overline{V}_{r_{n+1}}$ is a compact set and $\overline{V}_{r_j} \subset V_{r_{n+1}} \subset \overline{V}_{r_{n+1}} \subset V_{r_i}$. By induction on $n \in \mathbb{Z}_+$, we have a collection of open sets $\{V_r : r \in Q\}$ satisfying conditions (i), (ii), and (iii).

With the collection of open sets $\{V_r : r \in Q\}$, let us define two collections of functions $\{f_r : r \in Q\}$ and $\{g_s : s \in Q\}$ on X by setting

$$f_r(x) = \begin{cases} r & \text{if } x \in V_r, \\ 0 & \text{otherwise,} \end{cases} \quad \text{and} \quad g_s(x) = \begin{cases} 1 & \text{if } x \in \overline{V}_s, \\ s & \text{otherwise.} \end{cases}$$

Let $f = \sup_{r \in Q} f_r$ and $g = \inf_{s \in Q} g_s$. The characteristic function of an open set is lower semicontinuous and that of a closed set is upper semicontinuous by Observation 15.81. Thus f_r is lower semicontinuous and g_s is upper semicontinuous. Then by Theorem 15.84, f is lower semicontinuous and g is upper semicontinuous. Clearly $f(x) \in [0, 1]$ for $x \in X$, $f(x) = 1$ for $x \in K$, and $\operatorname{supp}\{f\} \subset V$.

Since f is lower semicontinuous and g is upper semicontinuous on X, to show that f is continuous on X it suffices to show that $f = g$ on X. Let $x \in X$ be arbitrarily fixed. Suppose $f_r(x) > g_s(x)$ for some $r, s \in Q$. Then the definitions of f_r and g_s imply that $r > s, x \in V_r$ and $x \notin \overline{V}_s$. But $r > s$ implies $V_r \subset V_s$, contradicting the statements $x \in V_r$ and $x \notin \overline{V}_s$. Thus $f_r(x) \leq g_s(x)$ for all $r, s \in Q$. Now suppose $f(x) < g(x)$. Then there exists $r, s \in Q$ such that $f(x) < r < s < g(x)$. But $f(x) < r$ implies $f_r(x) < r$ and then $x \notin V_r$ and $g(x) > s$ implies $g_s(x) > s$ and then $x \in \overline{V}_s$. This contradicts the fact that $\overline{V}_s \subset V_r$. Thus we have $f(x) = g(x)$. This shows that $f = g$ on X. ∎

Theorem 19.12. (Partition of Unity) *Let X be a locally compact Hausdorff space. Let K be a compact set, $V_1, \ldots, V_n$ be open sets in X such that $K \subset V_1 \cup \cdots \cup V_n$. Then there exist $f_1, \ldots, f_n \in C_c(X)$ such that*

1° $f_i(x) \in [0, 1]$ *for* $x \in X$ *for* $i = 1, \ldots, n$.

2° $\operatorname{supp}\{f_i\} \subset V_i$ *for* $i = 1, \ldots, n$.

3° $f_1(x) + \cdots + f_n(x) = 1$ *for* $x \in K$.

Proof. For every $x \in X$, $\{x\}$ is a compact set. If $x \in K$, then $\{x\} \subset V_i$ for some $i = 1, \dots, n$. Thus by Proposition 19.6, there exists an open set W_x containing x such that $\overline{W}_x$ is a compact set and $\overline{W}_x \subset V_i$. Then $\{W_x : x \in K\}$ is an open cover of the compact set K so that there exist $x_1, \dots, x_N \in K$ such that $K \subset W_{x_1} \cup \dots \cup W_{x_N}$. Now each $\overline{W}_{x_j}$ is contained in some V_i. For $i = 1, \dots, n$, let C_i be the union of those $\overline{W}_{x_j}$ which are contained in C_i. Then C_i is a compact set contained in the open set V_i so that by Theorem 19.11 (Urysohn's Lemma) there exists $g_i \in C_c(X)$ such that $g_i(x) \in [0, 1]$ for $x \in X$, $g_i(x) = 1$ for $x \in C_1$, and $\text{supp}\{g_i\} \subset V_i$. Let us define

$$
\begin{aligned}
f_1 &= g_1, \\
f_2 &= \{1 - g_1\}g_2, \\
&\vdots \\
f_n &= \{1 - g_1\}\{1 - g_2\} \cdots \{1 - g_{n-1}\}g_n.
\end{aligned}
$$

Clearly f_i is a continuous function on X and $f_i(x) \in [0, 1]$ for $x \in X$. Also $\text{supp}\{f_i\} \subset \text{supp}\{g_i\} \subset V_i$ by (c) of Observation 19.8. This shows that $f_i \in C_c(X)$. By induction on n it is easily verified that

$$f_1 + \dots + f_n = 1 - \{1 - g_1\}\{1 - g_2\} \cdots \{1 - g_n\}.$$

Since $K \subset C_1 \cup \cdots \cup C_n$, each $x \in K$ is in some C_i so that $g_i(x) = 1$. Thus for each $x \in K$, we have $1 - g_i(x) = 0$ for some $i = 1, \dots, n$ and therefore $f_1(x) + \dots + f_n(x) = 1$. ∎

Observation 19.13. In $\mathbb{R}^n$ every open set is a σ-compact set.

Proof. Let $\mathfrak{Q}$ be the collection of all subsets of $\mathbb{R}^n$ of the form $[a_1, b_1] \times \cdot \times [a_n, b_n]$ where $a_1, \dots, a_n, b_1, \dots, b_n$ are rational numbers. Such a set is a compact set in $\mathbb{R}^n$. Since the set of all rational numbers is a countable set, it follows that $\mathfrak{Q}$ is a countable collection of compact sets. Let O be a non-empty open set in $\mathbb{R}^n$. Then for every $x \in O$ there exist rational numbers $a_1, \dots, a_n, b_1, \dots, b_n$ such that $x \in [a_1, b_1] \times \cdot \times [a_n, b_n]$ and $[a_1, b_1] \times \cdot \times [a_n, b_n] \subset O$. Thus for every $x \in O$ there exists $Q_x \in \mathfrak{Q}$ such that $x \in Q_x$ and $Q_x \subset O$. We have $\bigcup_{x \in O} Q_x \subset O$. On the other hand $O = \bigcup_{x \in O}\{x\} \subset \bigcup_{x \in O} Q_x$. Thus we have $O = \bigcup_{x \in O} Q_x$. Now $\{Q_x : x \in O\}$, being a subcollection of the countable collection $\mathfrak{Q}$, is a countable collection. Thus $O = \bigcup_{x \in O} Q_x$ is a countable union of compact sets. This shows that O is a σ-compact set. ∎

Proposition 19.14. *Let X be a locally compact Hausdorff space. Let O be an open and σ-compact set in X. Then there exists an increasing sequence $(f_n : n \in \mathbb{N})$ in $C_c(X)$ such that* $\lim_{n \to \infty} f_n = \mathbf{1}_O$.

Proof. Since O is a σ-compact set in X there exists a sequence of compact sets in X, $(K_n : n \in \mathbb{N})$, such that $\bigcup_{n \in \mathbb{N}} K_n = O$. Let us define a sequence $(f_n : n \in \mathbb{N})$ in $C_c(X)$ inductively as follows. Now since K_1 is a compact set contained in an open set O in a locally compact Hausdorff space X, there exists $f_1 \in C_c(X)$ such that $0 \le f_1 \le 1$, $\text{supp}\{f_1\} \subset O$, and $f_1 = 1$ on K_1 by Theorem 19.11 (Urysohn). For $n > 1$, there exists

$f_n \in C_c(X)$ such that $0 \le f_n \le 1$, $\text{supp}\{f_n\} \subset O$, and $f_1 = 1$ on the compact set $\left(\bigcup_{j=1}^{n} K_j\right) \cup \left(\bigcup_{j=1}^{n1} \text{supp}\{f_j\}\right)$ by Theorem 19.11. Thus defined $(f_n : n \in \mathbb{N})$ is an increasing sequence in $C_c(X)$ such that $f_n(x) \uparrow 1$ for every $x \in O$. Therefore we have $\lim_{n\to\infty} f_n = \mathbf{1}_O$. ∎

[II] Borel and Radon Measures

Notations. Unless otherwise stated, X is a Hausdorff space, $\mathfrak{O}_X$ is the collection of all open sets in X, $\mathfrak{K}_X$ is the collection of all compact sets in X, and $\mathfrak{B}_X$ be the Borel σ-algebra on X, that is, $\mathfrak{B}_X = \sigma(\mathfrak{O}_X)$.

Definition 19.15. *Let X be a Hausdorff space. Let $\mathfrak{O}_X$, $\mathfrak{K}_X$, and $\mathfrak{B}_X$ be respectively the collection of all open sets in X, the collection of all compact sets in X, and the Borel σ-algebra of subsets of X. We call a measure on $(X, \mathfrak{B}_X)$ a Borel measure on X.*
(a) *Let μ be a Borel measure on X and let $E \in \mathfrak{B}_X$.*
1° *We say that μ is outer regular for E if*

(1) $$\mu(E) = \inf\left\{\mu(O) : O \supset E, O \in \mathfrak{O}_X\right\}.$$

2° *We say that μ is inner regular for E if*

(2) $$\mu(E) = \sup\left\{\mu(K) : K \subset E, K \in \mathfrak{K}_X\right\}.$$

3° *We say that μ is regular for E if μ is both outer and inner regular for E.*
(b) *Let $\mathfrak{E}$ be an arbitrary subcollection of $\mathfrak{B}_X$. We say that μ is outer regular for $\mathfrak{E}$, inner regular for $\mathfrak{E}$, or regular for $\mathfrak{E}$ according as μ is outer regular, inner regular, or regular for every $E \in \mathfrak{E}$.*
(c) *We say that μ is outer regular, inner regular, or regular according as μ is outer regular, inner regular, or regular for $\mathfrak{B}_X$.*

Let $(X, \mathfrak{A}, \mu)$ be an arbitrary measure space. Let $A, B \in \mathfrak{A}$, $A \subset B$, and assume that $\mu(A) < \infty$. Then for an arbitrary $\varepsilon > 0$, we have $\mu(B \setminus A) < \varepsilon$ if and only if $\mu(B) - \mu(A) < \varepsilon$. The following equivalent conditions for outer regularity and inner regularity of a Borel measure μ for $E \in \mathfrak{B}_X$ is based on this basic fact.

Lemma 19.16. *Let μ be a Borel measure on a Hausdorff space X.*
(a) *Let $E \in \mathfrak{B}_X$ and $\mu(E) < \infty$. Then*

(1) *μ is outer regular for E $\Leftrightarrow$ $\forall \varepsilon > 0, \exists O \in \mathfrak{O}_X$, $O \supset E$ and $\mu(O \setminus E) < \varepsilon$.*
(2) *μ is inner regular for E $\Leftrightarrow$ $\forall \varepsilon > 0, \exists K \in \mathfrak{K}_X$, $K \subset E$ and $\mu(E \setminus K) < \varepsilon$.*

(b) *Let $E \in \mathfrak{B}_X$ and $\mu(E) = \infty$. Then*

(3) *μ is outer regular for E.*
(4) *μ is inner regular for E $\Leftrightarrow$ $\forall M \ge 0, \exists K \in \mathfrak{K}_X$, $K \subset E$ and $\mu(K) \ge M$.*

Proof. 1. Let us prove (1). Let $E \in \mathfrak{B}_X$ and $\mu(E) < \infty$. Suppose μ is outer regular for E, that is, we have $\mu(E) = \inf\{\mu(O) : O \supset E, O \in \mathfrak{O}_X\}$. Then for every $\varepsilon > 0$ there exists $O \in \mathfrak{O}_X$ such that $O \supset E$ and $\mu(O) < \mu(E) + \varepsilon$. Then since $\mu(E) < \infty$, we have $\mu(O \setminus E) = \mu(O) - \mu(E) < \varepsilon$.

Conversely suppose for every $\varepsilon > 0$ there exists $O \in \mathfrak{O}_X$ such that $O \supset E$ and $\mu(O \setminus E) < \varepsilon$. Then since $\mu(E) < \infty$, we have $\mu(O \setminus E) = \mu(O) - \mu(E)$ and thus $\mu(O) - \mu(E), \varepsilon$. This shows that for every $\varepsilon > 0$ there exists $O \in \mathfrak{O}_X$ such that $O \supset E$ and $\mu(O) - \mu(E) < \varepsilon$, that is, $\mu(E) = \inf\{\mu(O) : O \supset E, O \in \mathfrak{O}_X\}$.

2. (2) is proved by the same argument as above.

3. Let us prove (3). If $E \in \mathfrak{B}_X$ and $\mu(E) = \infty$, then for every $O \in \mathfrak{O}_X$ such that $O \supset E$ we have $\mu(O) \geq \mu(E) = \infty$. Then $\inf\{\mu(O) : O \supset E, O \in \mathfrak{O}_X\} = \infty = \mu(E)$. This shows that μ is outer regular for E.

4. Let us prove (4). Let $E \in \mathfrak{B}_X$ and $\mu(E) = \infty$. Suppose μ is inner regular for E. Then we have $\sup\{\mu(K) : K \subset E, K \in \mathfrak{K}_X\} = \mu(E) = \infty$. Then for every $M \geq 0$ there exists $K \in \mathfrak{K}_X$ such that $K \subset E$ and $\mu(K) \geq M$.

Conversely suppose for every $M \geq 0$ there exists $K \in \mathfrak{K}_X$ such that $K \subset E$ and $\mu(K) \geq M$. Then we have $\sup\{\mu(K) : K \subset E, K \in \mathfrak{K}_X\} = \infty = \mu(E)$. this shows that μ is inner regular for E. ∎

We show next that if a Hausdorff space X is such that every open set in X is a σ-compact set, that is, a countable union of compact sets, then every Borel measure μ on X is inner regular for $\mathfrak{O}_X$. Note that $\mathbb{R}^n$ is such a space by Observation 19.13.

Observation 19.17. Let X be a Hausdorff space such that every open set in X is a σ-compact set. Then every Borel measure μ on X is inner regular for $\mathfrak{O}_X$.

Proof. Let $O \in \mathfrak{O}_X$. Since O is a σ-compact set, there exists an increasing sequence $(K_n : n \in \mathbb{N})$ in $\mathfrak{K}_X$ such that $\lim_{n\to\infty} K_n = \bigcup_{n\in\mathbb{N}} K_n = O$ and then $\lim_{n\to\infty} \mu(K_n) = \mu(O)$. Now if $\mu(O) < \infty$, then for every $\varepsilon > 0$ there exists $n \in \mathbb{N}$ such that $\mu(K_n) > \mu(O) - \varepsilon$, that is, $\mu(O \setminus K_n) < \varepsilon$, so that μ is inner regular for$O$ by (2) of Lemma 19.16. On the other hand if $\mu(O) = \infty$, then for any $M \geq 0$ there exists $n \in \mathbb{N}$ such that $\mu(K_n) \geq M$. Thus μ is inner regular for O by (4) of Lemma 19.16. ∎

Outer regularity for $\mathfrak{B}_X$ or inner regularity for $\mathfrak{B}_X$ can be used to prove the identity of two Borel measures that are equal on $\mathfrak{O}_X$ or equal on $\mathfrak{K}_X$ respectively.

Proposition 19.18. *Let μ_1 and μ_2 be two Borel measures on a Hausdorff space X.*
(a) *If μ_1 and μ_2 are outer regular for $\mathfrak{B}_X$ and $\mu_1 = \mu_2$ on $\mathfrak{O}_X$, then $\mu_1 = \mu_2$ on $\mathfrak{B}_X$.*
(b) *If μ_1 and μ_2 are inner regular for $\mathfrak{B}_X$ and $\mu_1 = \mu_2$ on $\mathfrak{K}_X$, then $\mu_1 = \mu_2$ on $\mathfrak{B}_X$.*

Proof. Suppose μ_1 and μ_2 are outer regular for $\mathfrak{B}_X$ and $\mu_1 = \mu_2$ on $\mathfrak{O}_X$. Then for every

$E \in \mathfrak{B}_X$ we have by (1) of Definition 19.15

$$\begin{aligned}\mu_1(E) &= \inf\{\mu_1(O) : O \supset E, O \in \mathfrak{O}_X\}\\ &= \inf\{\mu_2(O) : O \supset E, O \in \mathfrak{O}_X\}\\ &= \mu_2(E).\end{aligned}$$

Thus $\mu_1 = \mu_2$ on $\mathfrak{B}_X$. This proves (a). Then (b) is proved likewise with (2) of Definition 19.15. ∎

Definition 19.19. *A Radon measure is a Borel measure μ on a Hausdorff space X such that μ is outer regular for $\mathfrak{B}_X$, inner regular for $\mathfrak{O}_X$, and finite on $\mathfrak{K}_X$, that is,*

1° *outer regularity for $\mathfrak{B}_X$: $\mu(E) = \inf\{\mu(O) : O \supset E, O \in \mathfrak{O}_X\}$ for every $E \in \mathfrak{B}_X$,*

2° *inner regularity for $\mathfrak{O}_X$: $\mu(O) = \sup\{\mu(K) : K \subset O, K \in \mathfrak{K}_X\}$ for every $O \in \mathfrak{O}_X$,*

3° *finiteness on $\mathfrak{K}_X$: $\mu(K) < \infty$ for every $K \in \mathfrak{K}_X$.*

A Radon measure μ on a Hausdorff space X is outer regular for $\mathfrak{B}_X$ but it is not required to be inner regular for $\mathfrak{B}_X$. We show next that if a Radon measure μ on a Hausdorff space X is a σ-finite measure then μ is inner regular for $\mathfrak{B}_X$ and hence regular for $\mathfrak{B}_X$.

Proposition 19.20. (a) *Every Radon measure μ on a Hausdorff space X is inner regular for every $E \in \mathfrak{B}_X$ with $\mu(E) < \infty$.*
(b) *Every σ-finite Radon measure μ on a Hausdorff space X is inner regular for $\mathfrak{B}_X$ and hence regular for $\mathfrak{B}_X$. In particular every finite Radon measure is regular for $\mathfrak{B}_X$.*

Proof. 1. Let us show first the μ is inner regular for every $E \in \mathfrak{B}_X$ with $\mu(E) < \infty$. By the outer regularity of μ for E, for every $\varepsilon > 0$ there exists $O \in \mathfrak{O}_X$ such that $O \supset E$ and $\mu(O \setminus E) < \frac{\varepsilon}{3}$ by (1) of Lemma 19.16. Then by the inner regularity of μ for our $O \in \mathfrak{O}_X$ with $\mu(O) < \infty$, there exists $C \in \mathfrak{K}_X$ such that $C \subset O$ and $\mu(C) > \mu(O) - \frac{\varepsilon}{3}$ by (2) of Lemma 19.16. Since μ is outer regular for $O \setminus E \in \mathfrak{B}_X$ and since $\mu(O \setminus E) < \infty$, there exists $V \in \mathfrak{O}_X$ such that $V \supset O \setminus E$ and $\mu(V) < \mu(O \setminus E) + \frac{\varepsilon}{6} < \frac{\varepsilon}{2}$ by (1) of Lemma 19.16. Let $K = C \setminus V = C \cap V^c \in \mathfrak{K}_X$. Then $K \subset O \setminus (O \setminus E) = E$ and

$$\begin{aligned}\mu(K) &= \mu(C \setminus V) = \mu(C) - \mu(C \cap V)\\ &> \mu(O) - \tfrac{\varepsilon}{3} - \mu(V) > \mu(E) - \varepsilon.\end{aligned}$$

This shows that μ is inner regular for E by (2) of Lemma 19.16.

2. Let $E \in \mathfrak{B}_X$. If μ is a σ-finite measure on $\mathfrak{B}_X$, then every $E \in \mathfrak{B}_X$ is a σ-finite set. If $\mu(E) < \infty$, then μ is inner regular for E by our result in **1**. If $\mu(E) = \infty$ then the σ-finiteness of E implies that there exists an increasing sequence $(E_n : n \in \mathbb{N})$ in $\mathfrak{B}_X$ such that $\mu(E_n) < \infty$ for $n \in \mathbb{N}$ and $\bigcup_{n \in \mathbb{N}} E_n = E$. Let $M \geq 0$ be arbitrarily given.

Since $\lim_{n\to\infty} \mu(E_n) = \mu(E) = \infty$, there exists $N \in \mathbb{N}$ such that $\mu(E_N) \geq M + 1$. Since $E_N \in \mathfrak{B}_X$ and $\mu(E_N) < \infty$, μ is inner regular for E_N by our result in **1**. Thus there exists $K \in \mathfrak{K}_X$ such that $K \subset E_N$ and $\mu(E_N) - 1 < \mu(K)$ by (2) of Lemma 19.16. Thus for an arbitrary $M \geq 0$, we have $K \in \mathfrak{K}_X$ such that $K \subset E_N \subset E$ and $\mu(K) > \mu(E_N) - 1 \geq M$. This shows that μ is inner regular for E by (4) of Lemma 19.16. This shows that μ is inner regular for $\mathfrak{B}_X$. ∎

Corollary 19.21. *Every Radon measure μ on a compact Hausdorff space X is a finite Radon measure and hence regular for $\mathfrak{B}_X$.*

Proof. Since X is a compact Hausdorff space we have $X \in \mathfrak{K}_X$. Then since μ is a Radon measure we have $\mu(K) < \infty$ for every $K \in \mathfrak{K}_X$ and thus in particular we have $\mu(X) < \infty$. Thus μ is a finite Radon measure and then μ is regular for $\mathfrak{B}_X$ by Proposition 19.20. ∎

Proposition 19.22. *Let X be a σ-compact Hausdorff space. Then every Radon measure μ on X is inner regular for $\mathfrak{B}_X$ and hence regular for $\mathfrak{B}_X$.*

Proof. σ-compactness of X implies that there exist a sequence $(K_n : n \in \mathbb{N})$ in $\mathfrak{K}_X$ such that $X = \bigcup_{n\in\mathbb{N}} K_n$. Since μ is a Radon measure on X, we have $\mu(K_n) < \infty$ for $n \in \mathbb{N}$. Then μ is a σ-finite Radon measure on X and thus by Proposition 19.20, μ is inner regular for $\mathfrak{B}_X$. ∎

As we showed above, a σ-finite Radon measure μ on a Hausdorff space X is regular for $\mathfrak{B}_X$, that is, both outer and inner regular for $\mathfrak{B}_X$. This has the following consequence.

Theorem 19.23. *Let μ be a σ-finite Radon measure on a Hausdorff space X. We have:*
(a) *For every $E \in \mathfrak{B}_X$ and $\varepsilon > 0$, there exist an open set O and a closed set C such that $C \subset E \subset O$ and $\mu(O \setminus C) < \varepsilon$.*
(b) *For every $E \in \mathfrak{B}_X$ there exist a G_δ-set G and a F_σ-set F such that $F \subset E \subset G$ and $\mu(G \setminus F) = 0$.*

Proof. 1. σ-finiteness of the measure μ on $(X, \mathfrak{B}_X)$ implies that there exists an increasing sequence $(A_n : n \in \mathbb{N})$ in $\mathfrak{B}_X$ such that $\mu(A_n) < \infty$ for $n \in \mathbb{N}$ and $\bigcup_{n\in\mathbb{N}} A_n = X$. For $E \in \mathfrak{B}_X$, if we let $E_n = E \cap A_n$ for $n \in \mathbb{N}$, then $(E_n : n \in \mathbb{N})$ is an increasing sequence in $\mathfrak{B}_X$ such that $\mu(E_n) < \infty$ for $n \in \mathbb{N}$ and $\bigcup_{n\in\mathbb{N}} E_n = E$. Since μ is outer regular for $\mathfrak{B}_X$ and since $\mu(E_n) < \infty$, for every $\varepsilon > 0$ there exists an open set $O_n \supset E_n$ such that $\mu(O_n \setminus E_n) < \frac{\varepsilon}{2^{n+1}}$ by (1) of Lemma 19.16. If we let $O = \bigcup_{n\in\mathbb{N}} O_n$, then O is an open set, $O \supset E$, and

$$\begin{aligned} O \setminus E &= O \cap E^c = O \cap \Big(\bigcup_{n\in\mathbb{N}} E_n\Big)^c = O \cap \Big(\bigcap_{n\in\mathbb{N}} E_n^c\Big) \\ &= \bigcup_{n\in\mathbb{N}} \Big[O_n \cap \Big(\bigcap_{n\in\mathbb{N}} E_n^c\Big)\Big] \subset \bigcup_{n\in\mathbb{N}} (O_n \cap E_n^c) \\ &= \bigcup_{n\in\mathbb{N}} (O_n \setminus E_n) \end{aligned}$$

so that

$$\mu(O \setminus E) \leq \sum_{n \in \mathbb{N}} \mu(O_n \setminus E_n) < \sum_{n \in \mathbb{N}} \frac{\varepsilon}{2^{n+1}} = \frac{\varepsilon}{2}.$$

Thus we have shown that for every $E \in \mathfrak{B}_X$ and $\varepsilon > 0$, there exists an open set O such that $O \supset E$ and $\mu(O \setminus E) < \frac{\varepsilon}{2}$. Applying this to $E^c \in \mathfrak{B}_X$, we have an open set V such that $V \supset E^c$ and $\mu(V \setminus E^c) < \frac{\varepsilon}{2}$. If we let $C = V^c$, then C is a closed set, $C \subset E$. Thus $C \subset E \subset O$, $(E \setminus C) \cap (O \setminus E) = \emptyset$, and $(E \setminus C) \cup (O \setminus E) = O \setminus C$. Then $\mu(O \setminus C) = \mu(O \setminus E) + \mu(E \setminus C)$. Now $E \setminus C = E \cap C^c = E \cap V = V \setminus E^c$. Therefore we have $\mu(O \setminus C) = \mu(O \setminus E) + \mu(V \setminus E^c) < \frac{\varepsilon}{2} + \frac{\varepsilon}{2} = \varepsilon$. This proves (a).

2. Let $E \in \mathfrak{B}_X$. By (a), for every $n \in \mathbb{N}$ there exist an open set O_n and a closed set C_n such that $C_n \subset E \subset O_n$ and $\mu(O_n \setminus C_n) < \frac{1}{n}$. If we let $G = \bigcap_{n \in \mathbb{N}} O_n$ and $F = \bigcup_{n \in \mathbb{N}} E_n C_n$, then G is a G_δ-set set, F is a F_σ-set set and $F \subset E \subset G$. Now $G \setminus F \subset O_n \setminus C_n$ for every $n \in \mathbb{N}$ so that $\mu(G \setminus F) \leq \mu(O_n \setminus C_n) < \frac{1}{n}$. Since this holds for every $n \in \mathbb{N}$, we have $\mu(G \setminus F) = 0$. ∎

The foregoing results on the regularity of Borel measures have the following extension to a measure on a σ-algebra $\mathfrak{A}$ of subsets of a Hausdorff space X containing the Borel σ-algebra $\mathfrak{B}_X$ on X.

Theorem 19.24. *Let X be a Hausdorff space and let μ be measure on $(X, \mathfrak{A})$ where $\mathfrak{A}$ is a σ-algebra of subsets of X such that $\mathfrak{A} \supset \mathfrak{B}_X$. Suppose μ satisfies the following conditions:*

1° *μ is σ-finite.*
2° *μ is finite on $\mathfrak{K}_X$.*
3° *μ is outer regular for $\mathfrak{A}$, that is, $\mu(E) = \inf\{\mu(O) : O \supset E, O \in \mathfrak{O}_X\}$ for $E \in \mathfrak{A}$.*
4° *μ is inner regular for $\mathfrak{O}_X$ that is, $\mu(O) = \sup\{\mu(K) : K \subset O, K \in \mathfrak{K}_X\}$ for $O \in \mathfrak{O}_X$.*

Then we have:

5° *μ is inner regular for $\mathfrak{A}$, that is, $\mu(E) = \sup\{\mu(K) : K \subset E, K \in \mathfrak{K}_X\}$ for $E \in \mathfrak{A}$.*
6° *For every $E \in \mathfrak{A}$ and $\varepsilon > 0$ there exist an open set O and a closed set C such that $C \subset E \subset O$ and $\mu(O \setminus C) < \varepsilon$.*
7° *For every $E \in \mathfrak{A}$ there exist a G_δ-set G and a F_σ-set F such that $F \subset E \subset G$ and $\mu(G \setminus F) = 0$.*

Proof. 5° is proved by the same argument as in the Proof of Proposition 19.20. 6° and 7° follow by the same argument as in Theorem 19.23. ∎

Remark 19.25. Consider the Lebesgue measure space $(\mathbb{R}, \mathfrak{M}_L, \mu_L)$. By Theorem 3.22, for every $E \in \mathfrak{M}_L$ and $\varepsilon > 0$ there exist an open set O and a closed set C such that $C \subset E \subset O$ and $\mu_L(O \setminus C) < \varepsilon$. Actually this is a particular case of Theorem 19.24. This can be shown as follows. To start with, $\mathbb{R}$ is a Hausdorff space and $\mathfrak{B}_{\mathbb{R}} \subset \mathfrak{M}_L$. The measure μ_L on $\mathfrak{M}_L$ is σ-finite and finite on $\mathfrak{K}_{\mathbb{R}}$. μ_L is outer regular for $\mathfrak{M}_L$ according to Proposition 3.30. Every open set in $\mathbb{R}$ is a σ-compact set by Observation 19.13. Then μ_L is inner regular for $\mathfrak{O}_{\mathbb{R}}$ by Observation 19.17. This shows that $(\mathbb{R}, \mathfrak{M}_L, \mu_L)$ satisfies conditions 1° to 4° of Theorem 19.24. Thus by 5°, 6°, and 7° of Theorem 19.24, μ_L is inner regular for $\mathfrak{M}_L$, for every $E \in \mathfrak{M}_L$ and $\varepsilon > 0$ there exist an open set O and a closed set C such that $C \subset E \subset O$ and $\mu_L(O \setminus C) < \varepsilon$, and for every $E \in \mathfrak{M}_L$ there exist a G_δ-set G and an F_σ-set F such that $F \subset E \subset G$ and $\mu_L(G \setminus F) = 0$.

[III] Positive Linear Functionals on $C_c(X)$

Definition 19.26. *Let X be a locally compact Hausdorff space and let $C_c(X)$ be the linear space of real-valued continuous functions with compact support on X. Let I be a real-valued linear functional on $C_c(X)$, that is, I is real-valued, $I(f+g) = I(f) + I(g)$ for $f, g \in C_c(X)$ and $I(\lambda f) = \lambda I(f)$ for $f \in C_c(X)$ and $\lambda \in \mathbb{R}$.*
We say that a linear functional I on $C_c(X)$ is positive if $I(f) \geq 0$ for every $f \in C_c(X)$ such that $f \geq 0$ on X.

We shall show that if I is a positive linear functional on $C_c(X)$, then there exists a unique Radon measure μ on $\mathfrak{B}_X$ such that $I(f) = \int_X f \, d\mu$ for every $f \in C_c(X)$. The construction of such a measure μ is given in the next propositions.

Definition 19.27. *Let X be a locally compact Hausdorff space and let $C_c(X)$ be the linear space of real-valued continuous functions with compact support on X.*
(a) *For $O \in \mathfrak{O}_X$ and $O \neq \emptyset$ and for $f \in C_c(x)$ such that $0 \leq f \leq 1$ on X, we write $f \prec O$ if $\operatorname{supp}\{f\} \subset O$.*
(b) *For $K \in \mathfrak{K}_X$ and $f \in C_c(X)$ such that $0 \leq f \leq 1$ on X, we write $f \succ K$ if $f \geq \mathbf{1}_K$, that is, $f = 1$ on K.*

Definition 19.28. *Let X be a locally compact Hausdorff space. Let I be a positive linear functional on $C_c(X)$. Let us define a set function γ on $\mathfrak{O}_X$ by setting $\gamma(\emptyset) = 0$ and*

$$(1) \qquad \gamma(O) = \sup\left\{I(f) : f \in C_c(X), f \prec O\right\} \quad \text{for } O \in \mathfrak{O}_X, O \neq \emptyset.$$

Let us define a set function μ^* on $\mathfrak{P}(X)$ by setting

$$(2) \qquad \mu^*(E) = \inf\left\{\gamma(O) : O \supset E, O \in \mathfrak{O}_X\right\} \quad \text{for } E \in \mathfrak{P}(X).$$

Proposition 19.29. *Let X be a locally compact Hausdorff space. Let I be a positive linear functional on $C_c(X)$. For the set functions γ and μ^* in Definition 19.28 we have:*
(a) *γ is monotone and subadditive on $\mathfrak{O}_X$.*
(b) *$\mu^* = \gamma$ on $\mathfrak{O}_X$.*
(c) *μ^* is an outer measure on X.*

Proof. 1. Let $O_1, O_2 \in \mathfrak{O}_X$ and $O_1 \subset O_2$. Let $f \in C_c(X)$. If $f \prec O_1$, then $f \prec O_2$ also. Thus $\left\{f \in C_c(X) : f \prec O_1\right\} \subset \left\{f \in C_c(X) : f \prec O_2\right\}$ and hence $\gamma(O_1) \leq \gamma(O_2)$. This shows that γ is monotone on $\mathfrak{O}_X$.

Next let us show that γ is subadditive on $\mathfrak{O}_X$, that is, for $O_1, O_2 \in \mathfrak{O}_X$, we have

$$(1) \qquad \gamma(O_1 \cup O_2) \leq \gamma(O_1) + \gamma(O_2).$$

Let f be an arbitrary member of $C_c(X)$ such that $f \prec O_1 \cup O_2$. Such function f exists by Theorem 19.11 (Urysohn's Lemma). Then the compact set $\text{supp}\{f\}$ is contained in the open set $O_1 \cup O_2$ so that by Theorem 19.12 (Partition of Unity), there exist $h_1, h_2 \in C_c(X)$ such that $h_1 \prec O_1$, $h_2 \prec O_2$, and $h_1 + h_2 = 1$ on $\text{supp}\{f\}$. Then $fh_1, fh_2 \in C_c(X)$ and $fh_1 \prec O_1$ and $fh_2 \prec O_2$. Furthermore $fh_1 + fh_2 = f\{h_1 + h_2\} = f$ on $\text{supp}\{f\}$. Since $f = 0$ outside $\text{supp}\{f\}$, we have $fh_1 + fh_2 = f$ on X. Thus we have

$$(2) \qquad I(f) = I(fh_1 + fh_2) = I(fh_1) + I(fh_2) \le \gamma(O_1) + \gamma(O_2),$$

where the inequality is by the definition of γ by (1) in Definition 19.28. Then by (1) in Definition 19.28 and by (2), we have

$$\gamma(O_1 \cup O_2) = \sup\left\{I(f) : f \in C_c(X), f \prec O_1 \cup O_2\right\} \le \gamma(O_1) + \gamma(O_2).$$

This proves (1).

2. The fact that $\mu^* = \gamma$ on $\mathfrak{O}_X$ is immediate from (2) of Definition 19.28 and the monotonicity of γ on $\mathfrak{O}_X$. Indeed if $O, O_1 \in \mathfrak{O}_X$ and $O \subset O_1$ then $\gamma(O) \le \gamma(O_1)$ so that we have $\mu^*(O) = \inf\left\{\gamma(O_1) : O_1 \supset O, O_1 \in \mathfrak{O}_X\right\} = \gamma(O)$.

3. Let us show that μ^* is an outer measure on X. Clearly $\mu^*(E) \in [0, \infty]$ for every $E \in \mathfrak{P}(X)$ and $\mu^*(\emptyset) = 0$. To verify the monotonicity of μ^* on $\mathfrak{P}(X)$, let $E_1, E_2 \in \mathfrak{P}(X)$ and $E_1 \subset E_2$. If $E_2 \subset O$ for some $O \in \mathfrak{O}_X$, then $E_1 \subset O$ also. It follows from (2) of Definition 19.28 that $\mu^*(E_1) \le \mu^*(E_2)$.

Let us prove the countable subadditivity of μ^* on $\mathfrak{P}(X)$, that is, for every sequence $(E_j : j \in \mathbb{N})$ in $\mathfrak{P}(X)$, we have

$$(3) \qquad \mu^*\Big(\bigcup_{j\in\mathbb{N}} E_j\Big) \le \sum_{j\in\mathbb{N}} \mu^*(E_j).$$

Let $\varepsilon > 0$ be arbitrarily given. By (2) of Definition 19.28, there exists $O_j \in \mathfrak{O}_X$ such that $O_j \supset E_j$ and $\gamma(O_j) \le \mu^*(E_j) + \frac{\varepsilon}{2^j}$. Consider the open set $\bigcup_{j\in\mathbb{N}} O_j$. Let $f \in C_c(X)$ be such that $f \prec \bigcup_{j\in\mathbb{N}} O_j$. Since $\{O_j : j \in \mathbb{N}\}$ is an open cover of the compact set $\text{supp}\{f\}$, there exists $N \in \mathbb{N}$ such that $\text{supp}\{f\} \subset \bigcup_{j=1}^N O_j$. Then $f \prec \bigcup_{j=1}^N O_j$ so that by (1) of Definition 19.28 we have

$$(4) \qquad I(f) \le \gamma\Big(\bigcup_{j=1}^N O_j\Big) \le \sum_{j=1}^N \gamma(O_j) \le \sum_{j\in\mathbb{N}} \gamma(O_j) \le \sum_{j\in\mathbb{N}} \mu^*(E_j) + \varepsilon,$$

where the second inequality is by $N - 1$ iterated applications of (1). Since (4) holds for every $f \in C_c(X)$ such that $f \prec \bigcup_{j\in\mathbb{N}} O_j$, definition (1) in Definition 19.28 implies $\gamma\left(\bigcup_{j\in\mathbb{N}} O_j\right) \le \sum_{j\in\mathbb{N}} \mu^*(E_j) + \varepsilon$. Since $\bigcup_{j\in\mathbb{N}} E_j \subset \bigcup_{j\in\mathbb{N}} O_j$, the monotonicity of μ^* implies

$$\mu^*\Big(\bigcup_{j\in\mathbb{N}} E_j\Big) \le \mu^*\Big(\bigcup_{j\in\mathbb{N}} O_j\Big) = \gamma\Big(\bigcup_{j\in\mathbb{N}} O_j\Big) \le \sum_{j\in\mathbb{N}} \mu^*(E_j) + \varepsilon,$$

where the equality is by (b). Then by the arbitrariness of $\varepsilon > 0$, we have (3). ∎

Proposition 19.30. *Let X be a locally compact Hausdorff space. Let I be a positive linear functional on $C_c(X)$. For the outer measure μ^* on X defined by Definition 19.28, we have $\mathfrak{B}_X \subset \mathfrak{M}(\mu^*)$, the σ-algebra of μ^*-measurable subsets of X.*

Proof. If we show that $\mathfrak{O}_X \subset \mathfrak{M}(\mu^*)$, then we have $\mathfrak{B}_X = \sigma(\mathfrak{O}_X) \subset \mathfrak{M}(\mu^*)$.

Let $O \in \mathfrak{O}_X$. To show that $O \in \mathfrak{M}(\mu^*)$, it suffices, according to Observation 2.3, to show that for every $A \in \mathfrak{P}(X)$ we have

$$\mu^*(A) \geq \mu^*(A \cap O) + \mu^*(A \cap O^c). \tag{1}$$

Let $A \in \mathfrak{P}(X)$. For an arbitrary $\varepsilon > 0$, by the definition of μ^* by (2) of Definition 19.28, there exists an open set $G \supset A$ such that $\gamma(G) \leq \mu^*(A) + \varepsilon$. Now if $A \cap O = \emptyset$, then (1) reduces to $\mu^*(A) \geq 0 + \mu^*(A)$ which is trivially true. Thus we consider the case $A \cap O \neq \emptyset$. Then since $G \supset A$, we have $G \cap O \neq \emptyset$. Take an arbitrary $f \in C_c(X)$ such that $f \prec G \cap O$. Then since the compact set $\operatorname{supp}\{f\}$ is contained in $G \cap O \subset O$, by Proposition 19.6 there exists an open set V such that $\overline{V}$ is a compact set and $\operatorname{supp}\{f\} \subset V \subset \overline{V} \subset O$. Now if $O \supset A$, then (1) reduces to $\mu^*(A) \geq \mu^*(A) + 0$ which is trivially true. Thus we consider the case $O \not\supset A$. Then $O \not\supset G$. This implies $\overline{V} \not\supset G$ so that $G \setminus \overline{V} \neq \emptyset$. Take an arbitrary $g \in C_c(X)$ such that $g \prec G \setminus \overline{V}$. Since $\operatorname{supp}\{f\} \subset \overline{V}$ and $\operatorname{supp}\{g\} \subset G \setminus \overline{V}$, we have $0 \leq f + g \leq 1$ on X and $\operatorname{supp}\{f + g\} = \operatorname{supp}\{f\} \cup \operatorname{supp}\{g\}$. Since $\operatorname{supp}\{f\} \subset G \cap O$ and $\operatorname{supp}\{g\} \subset O \setminus \overline{V}$, we have $\operatorname{supp}\{f + g\} \subset G$. Thus $f + g \prec G$. This implies, by the definition of γ by (1) of Definition 19.28

$$I(f) + I(g) = I(f + g) \leq \gamma(G) \leq \mu^*(A) + \varepsilon. \tag{2}$$

Now since $f \prec G \cap O$, we have $I(f) \leq \gamma(G \cap O)$, and similarly from $g \prec G \setminus \overline{V}$, we have $I(g) \leq \gamma(G \setminus \overline{V})$. With f fixed, taking supremum on g, and then taking supremum on f in (2), we have

$$\gamma(G \cap O) + \gamma(G \setminus \overline{V}) \leq \mu^*(A) + \varepsilon. \tag{3}$$

Since $G \supset A$, we have $\mu^*(A \cap O) \leq \mu^*(G \cap O) = \gamma(G \cap O)$. Since $\overline{V} \subset O$, we have $\mu^*(A \setminus O) \leq \mu^*(G \setminus \overline{V}) = \gamma(G \setminus \overline{V})$. Using these in (3), we have $\mu^*(A \cap O) + \mu^*(A \cap O^c) \leq \mu^*(A) + \varepsilon$. By the arbitrariness of $\varepsilon > 0$, we have (1). This shows that $O \in \mathfrak{M}(\mu^*)$. Therefore $\mathfrak{O}_X \subset \mathfrak{M}(\mu^*)$ and consequently $\mathfrak{B}_X \subset \mathfrak{M}(\mu^*)$. ∎

Lemma 19.31. *Let X be a locally compact Hausdorff space. Let I be a positive linear functional on $C_c(X)$. For the outer measure μ^* on X defined by Definition 19.28, we have*

$$\mu^*(K) = \inf\big\{I(f) : f \in C_c(X), f \succ K\big\} \quad \textit{for every } K \in \mathfrak{K}_X. \tag{1}$$

In particular μ^* is finite on $\mathfrak{K}_X$.

Proof. Let $K \in \mathfrak{K}_X$. Let $f \in C_c(X)$ be such that $f \succ K$, that is, $0 \leq f \leq 1$ on X and $f \geq \mathbf{1}_K$ on K. Let us show that $\mu^*(K) \leq I(f)$. For an arbitrary $\varepsilon \in (0, 1)$, let $O_\varepsilon = \big\{x \in X : f(x) > 1 - \varepsilon\big\}$. The continuity of f on X implies that $O_\varepsilon \in \mathfrak{O}_X$. Note also that $K \subset \big\{x \in X : f(x) \geq 1\big\} \subset O_\varepsilon$. Let $g \in C_c(X)$ be such that $g \prec O_\varepsilon$. Since

$\frac{1}{1-\varepsilon}f > 1$ on O_ε and since $\text{supp}\{g\} \subset O_\varepsilon$ and $0 \le g \le 1$ on X, we have $\frac{1}{1-\varepsilon}f \ge g$ on X. By the positivity of the linear functional I on $C_c(X)$, we have $I(g) \le \frac{1}{1-\varepsilon}I(f)$. Since this holds for an arbitrary $g \in C_c(X)$ such that $g \prec O_\varepsilon$, we have $\gamma(O_\varepsilon) \le \frac{1}{1-\varepsilon}I(f)$ by the definition of γ by (1) of Definition 19.28. Then $K \subset O_\varepsilon$ implies $\mu^*(K) \le \mu^*(O_\varepsilon) = \gamma(O_\varepsilon) \le \frac{1}{1-\varepsilon}I(f)$. By the arbitrariness of $\varepsilon \in (0, 1)$, we have $\mu^*(K) \le I(f)$. Since this holds for an arbitrary $f \in C_c(X)$ such that $f \succ K$, we have

$$\mu^*(K) \le \inf\{I(f) : f \in C_c(X), f \succ K\}.$$

Let us prove the reverse inequality. Now $\mu^*(K) = \inf\{\gamma(O) : O \supset K, O \in \mathfrak{O}_X\}$ by (2) of Definition 19.28. Thus for an arbitrary $\varepsilon > 0$, there exists $O \in \mathfrak{O}_X$ such that $O \supset K$ and $\gamma(O) \le \mu^*(K)+\varepsilon$. Then by Urysohn's Lemma (Theorem 19.11), there exists $f \in C_c(X)$ such that $f \prec O$ and $f \succ K$. By the definition of γ by (1) of Definition 19.28, we have $I(f) \le \gamma(O) \le \mu^*(K)+\varepsilon$. Thus $\inf\{I(f) : f \in C_c(X), f \succ K\} \le \mu^*(K)+\varepsilon$. Then by the arbitrariness of $\varepsilon > 0$, we have $\inf\{I(f) : f \in C_c(X), f \succ K\} \le \mu^*(K)$. This proves the reverse inequality and completes the proof of (1). Since $I(f) < \infty$ for every $f \in C_c(X)$, (1) implies that $\mu^*(K) < \infty$ for every $K \in \mathfrak{K}_X$. ∎

Proposition 19.32. *Let X be a locally compact Hausdorff space. Let I be a positive linear functional on $C_c(X)$. For the outer measure μ^* on X defined by Definition 19.28, we have*

(1) $$\mu^*(O) = \sup\{\mu^*(K) : K \subset O, K \in \mathfrak{K}_X\} \quad \textit{for every } O \in \mathfrak{O}_X.$$

Proof. Let $O \in \mathfrak{O}_X$. If $\mu^*(O) = 0$, then $\mu^*(K) = 0$ for every $K \in \mathfrak{K}_X$ such that $K \subset O$ so that (1) holds trivially.

Next consider the case $\mu^*(O) \in (0, \infty)$. For an arbitrary $\varepsilon \in (0, \mu^*(O))$, we have $\mu^*(O) - \varepsilon > 0$. Since $\mu^*(O) = \gamma(O) = \sup\{I(f) : f \in C_c(X), f \prec O\}$, there exists $f \in C_c(X)$ such that $f \prec O$ and $I(f) \ge \mu^*(O) - \varepsilon$. (Note that if $\mu^*(O) = \infty$, such function f does not exist since $I(f) < \infty$ for every $f \in C_c(X)$.) Let $K = \text{supp}\{f\}$. We have $K \in \mathfrak{K}_X$ and $K \subset O$. Let $g \in C_c(X)$ be such that $g \succ K$. Then $g \ge f$ on X so that by the positivity of the linear functional I we have $I(g) \ge I(f) \ge \mu^*(O) - \varepsilon$. Then by Lemma 19.31, we have $\mu^*(K) = \inf\{I(g) : g \in C_c(X), g \succ K\} \ge \mu^*(O) - \varepsilon$. Thus we have shown that for every $\varepsilon \in (0, \mu^*(O))$ there exists $K \in \mathfrak{K}_X$ such that $K \subset O$ and $\mu^*(O) - \varepsilon \le \mu^*(K) \le \mu^*(O)$. This proves (1) for the case $\mu^*(O) \in (0, \infty)$.

Finally consider the case $\mu^*(O) = \infty$. Since

$$\mu^*(O) = \gamma(O) = \sup\{I(f) : f \in C_c(X), f \prec O\},$$

for every $M > 0$ there exists $f \in C_c(X)$ such that $f \prec O$ and $I(f) \ge M$. Let $K = \text{supp}\{f\}$. Then $K \in \mathfrak{K}_X$ and $K \subset O$. Let $g \in C_c(X)$ be such that $g \succ K$. Then $g \ge f$ on X so that $I(g) \ge I(f) \ge M$ by the positivity of the linear functional I. Then by Lemma 19.31 we have $\mu^*(K) \ge M$. Thus we have shown that for every $M > 0$ there exists $K \in \mathfrak{K}_X$ such that $K \subset O$ and $\mu^*(K) \ge M$. Therefore $\sup\{\mu^*(K) : K \subset O, K \in \mathfrak{K}_X\} = \infty = \mu^*(O)$. This proves (1) for the case $\mu^*(O) = \infty$. ∎

Theorem 19.33. *Let X be a locally compact Hausdorff space. Let I be a positive linear functional on $C_c(X)$. Let μ^* be the outer measure on X defined by Definition 19.28. Let us write μ for the restriction of μ^* to the σ-algebra $\mathfrak{M}(\mu^*)$ of the μ^*-measurable subsets of X. Then in the measure space $(X, \mathfrak{M}(\mu^*), \mu)$, the measure μ is outer regular for $\mathfrak{M}(\mu^*)$, inner regular for the open sets $\mathfrak{O}_X$, and finite on the compact sets $\mathfrak{K}_X$, that is, we have*

1° $\mu(E) = \inf\{\mu(O) : O \supset E, O \in \mathfrak{O}_X\}$ *for every* $E \in \mathfrak{M}(\mu^*)$.

2° $\mu(O) = \sup\{\mu(K) : K \subset O, K \in \mathfrak{K}_X\}$ *for every* $O \in \mathfrak{O}_X$.

3° $\mu(K) < \infty$ *for every* $K \in \mathfrak{K}_X$.

In particular the restriction of μ to $\mathfrak{B}_X \subset \mathfrak{M}(\mu^*)$ is a Radon measure on X.

Proof. 1° is from the definition of μ^* by (2) of Definition 19.28 and the fact that $\mu^* = \gamma$ on $\mathfrak{O}_X$. 2° is from Proposition 19.32. 3° is from Lemma 19.31. If we restrict μ to $\mathfrak{B}_X$, then 1° is the outer regularity condition of μ on $\mathfrak{B}_X$ and 2° is the inner regularity condition of μ on $\mathfrak{O}_X$ so that μ is a Radon measure on X. ∎

Theorem 19.34. (Riesz-Markoff Theorem) *Let X be a locally compact Hausdorff space and let I be a positive linear functional on the linear space $C_c(X)$ of real-valued continuous functions with compact support on X. Then there exists a unique Radon measure μ on the Borel σ-algebra $\mathfrak{B}_X$ of subsets of X such that $I(f) = \int_X f\,d\mu$ for every $f \in C_c(X)$.*

Proof. 1. Let us show that for the Radon measure μ given in Theorem 19.33, we have $I(f) = \int_X f\,d\mu$ for every $f \in C_c(X)$.

Let us consider first $f \in C_c(X)$ such that $0 \le f \le 1$ and $\max_{x\in X} f(x) = 1$. Let $K_0 = \operatorname{supp}\{f\}$, a compact set. With $n \in \mathbb{N}$ arbitrarily fixed, let

(1) $$K_j = \left\{x \in X : f(x) \ge \tfrac{j}{n}\right\} \quad \text{for } j = 1, \dots, n.$$

We have $K_0 \supset K_1 \supset \cdots \supset K_n$. Since $f(X) = [0, 1]$, we have $K_j \ne \emptyset$ for $j = 0, 1, \dots, n$. The continuity of f on X implies that K_j is a closed set. Being a closed set contained in a compact set K_0, K_j is a compact set for $j = 1, \dots, n$. Let us write

$$K_0 = (K_0 \setminus K_1) \cup (K_1 \setminus K_2) \cup \cdots \cup (K_{n-1} \setminus K_n) \cup K_n.$$

Note that $f(x) \in \left[\frac{j-1}{n}, \frac{j}{n}\right)$ for $x \in K_{j-1} \setminus K_j$ for $j = 1, \dots, n$ and $f(x) = 1$ for $x \in K_n$. For $j = 1, \dots, n$, let us define a function f_j on X by

(2) $$f_j(x) = \left[\left\{f(x) - \tfrac{j-1}{n}\right\} \vee 0\right] \wedge \tfrac{1}{n} \quad \text{for } x \in X.$$

Then f_j is continuous and $0 \le f_j \le 1/n$ on X and

(3) $$f_j(x) = \begin{cases} 0 & \text{for } x \notin K_{j-1}, \\ f(x) - (j-1)/n & \text{for } x \in K_{j-1} \setminus K_j, \\ 1/n & \text{for } x \in K_j. \end{cases}$$

Thus we have

$$\operatorname{supp}\{f_j\} = K_{j-1}, \tag{4}$$

$$f_j \in C_c(X), \tag{5}$$

$$\tfrac{1}{n}\mathbf{1}_{K_j} \le f_j \le \tfrac{1}{n}\mathbf{1}_{K_{j-1}}, \tag{6}$$

$$f = \sum_{j=1}^{n} f_j. \tag{7}$$

The last equality is verified as follows. If $x \notin K_0$, then $f(x) = 0$ and $f_j(x) = 0$ for $j = 1, \dots, n$ so that (7) holds at x. Suppose $x \in K_0$. Then x is in one of the $n+1$ disjoint sets $K_0 \setminus K_1, K_1 \setminus K_2, \dots, K_{n-1} \setminus K_n, K_n$. Suppose $x \in K_{j_0-1} \setminus K_{j_0}$ for some $j_0 = 1, \dots, n$. Then

$$f_j(x) = \begin{cases} 1/n & \text{for } j = 1, \dots, j_0 - 1, \\ f(x) - (j_0 - 1)/n & \text{for } j = j_0, \\ 0 & \text{for } j = j_0 + 1, \dots, n. \end{cases}$$

This implies

$$\sum_{j=1}^{n} f_j(x) = \frac{j_0 - 1}{n} + f(x) - \frac{j_0 - 1}{n} = f(x).$$

If $x \in K_n$, then $f(x) = 1$ and $f_j(x) = 1/n$ for $j = 1, \dots, n$ so that $f(x) = \sum_{j=1}^{n} f_j(x)$. This proves (7).

Now from (6) we have

$$\frac{1}{n}\mu(K_j) \le \int_X f_j \, d\mu \le \frac{1}{n}\mu(K_{j-1})$$

for $j = 1, \dots, n$. Summing over $j = 1, \dots, n$ and recalling (7) and the linearity of the integral, we have

$$\frac{1}{n}\sum_{j=1}^{n} \mu(K_j) \le \int_X f \, d\mu \le \frac{1}{n}\sum_{j=1}^{n} \mu(K_{j-1}). \tag{8}$$

Since $nf_j \in C_c(X)$, $0 \le nf_j \le 1$, and $\operatorname{supp}\{nf_j\} = \operatorname{supp}\{f_j\} = K_{j-1}$, if V is an open set containing K_{j-1}, then $nf_j \prec V$ so that $I(nf_j) \le \mu(V)$. Thus $I(f_j) \le \frac{1}{n}\mu(V)$. Since this holds for every open set V containing K_{j-1}, the outer regularity of μ on $\mathfrak{B}_X$ implies $I(f_j) \le \frac{1}{n}\mu(K_{j-1})$. On the other hand, since $nf_j \ge \mathbf{1}_{K_j}$ by (6), we have $nf_j \succ K_j$. Then by Lemma 19.31, we have $\mu(K_j) \le I(nf_j)$ so that $\frac{1}{n}\mu(K_j) \le I(f_j)$. Therefore we have $\frac{1}{n}\mu(K_j) \le I(f_j) \le \frac{1}{n}\mu(K_{j-1})$ for $j = 1, \dots, n$. Summing over $j = 1, \dots, n$ and recalling (7) and linearity of the functional I, we have

$$\frac{1}{n}\sum_{j=1}^{n} \mu(K_j) \le I(f) \le \frac{1}{n}\sum_{j=1}^{n} \mu(K_{j-1}). \tag{9}$$

By (8) and (9), we have

$$\begin{aligned}\left|I(f)-\int_X f\,d\mu\right| &\leq \frac{1}{n}\sum_{j=1}^{n}\mu(K_{j-1})-\frac{1}{n}\sum_{j=1}^{n}\mu(K_j)\\ &=\frac{1}{n}\{\mu(K_0)-\mu(K_n)\}\\ &\leq \frac{1}{n}\mu(K_0).\end{aligned}$$

Since $\mu(K_0)<\infty$, letting $n\to\infty$ we have $\left|I(f)-\int_X f\,d\mu\right|=0$ and thus for our particular $f\in C_c(X)$ we have $I(f)=\int_X f\,d\mu$.

Next consider $f\in C_c(X)$ such that $f\geq 0$ on X. The equality $I(f)=\int_X f\,d\mu$ holds trivially if f is vanishing identically on X. If f is not identically vanishing on X, then $\alpha := \max_{x\in X} f(x) > 0$. Let $f_0=\alpha^{-1}f$. Then $f_0\in C_c(X)$, $0\leq f_0\leq 1$, and $\max_{x\in X} f_0(x)=1$. Thus by our result above, we have $I(f_0)=\int_X f_0\,d\mu$. Multiplying the equality by α , we have $I(f)=\int_X f\,d\mu$ by the linearity of I and the integral. For an arbitrary $f\in C_c(X)$, let us write $f=f^+-f^-$. Since $f^+, f^-\in C_c(X)$ and $f^+\geq 0$ and $f^-\geq 0$ on X, we have

$$\begin{aligned}I(f)=I(f^+)-I(f^-)&=\int_X f^+\,d\mu-\int_X f^-\,d\mu\\ &=\int_X\{f^+-f^-\}\,d\mu=\int_X f\,d\mu.\end{aligned}$$

2. Let us prove the uniqueness of a Radon measure ν on $\mathfrak{B}_X$ such that $I(f)=\int_X f\,d\nu$ for every $f\in C_c(X)$. Let μ be the Radon measure in **1**, for which we have shown $I(f)=\int_X f\,d\mu$ for every $f\in C_c(X)$. Let O be an arbitrary nonempty open set in X. Let $f\in C_c(X)$ be such that $f\prec O$. Then since $0\leq f\leq 1$ on X and $\operatorname{supp}\{f\}\subset O$, we have $f\leq \mathbf{1}_{\operatorname{supp}\{f\}}\leq \mathbf{1}_O$. Thus $\int_X f\,d\mu=I(f)=\int_X f\,d\nu\leq\int_X \mathbf{1}_O\,d\nu=\nu(O)$. Recalling (b) of Proposition 19.29 and Definition 19.28, we have

$$\begin{aligned}\mu(O)&=\sup\left\{I(f): f\in C_c(X)\text{ and } f\prec O\right\}\\ &=\sup\left\{\textstyle\int_X f\,d\mu : f\in C_c(X)\text{ and } f\prec O\right\}\\ &\leq \nu(O).\end{aligned}$$

Since this holds for every nonempty open set O, the outer regularity of μ and ν on $\mathfrak{B}_X$ implies $\mu(E)\leq\nu(E)$ for every $E\in\mathfrak{B}_X$.

To prove the reverse inequality, let K be an arbitrary compact set in X. Let $f\in C_c(X)$ and $f\succ K$. Then $f\geq \mathbf{1}_K$ so that $\int_X f\,d\mu=I(f)=\int_X f\,d\nu\geq\int_X \mathbf{1}_K\,d\nu=\nu(K)$. For our Radon measure μ, we have according to Lemma 19.31

$$\begin{aligned}\mu(K)&=\inf\left\{I(f): f\in C_c(X)\text{ and } f\succ K\right\}\\ &=\inf\left\{\textstyle\int_X f\,d\mu : f\in C_c(X)\text{ and } f\succ K\right\}\\ &\geq \nu(K).\end{aligned}$$

Since this holds for every compact set K, the inner regularity of μ and ν on the open sets implies that $\mu(O) \geq \nu(O)$ for every open set O. Then by the outer regularity of μ and ν on $\mathfrak{B}_X$, we have $\mu(E) \geq \nu(E)$ for every $E \in \mathfrak{B}_X$. This shows that $\mu(E) = \nu(E)$ for every $E \in \mathfrak{B}_X$. ∎

Applying the Riesz-Markoff Theorem we obtain the following regularity condition for a Borel measure.

Lemma 19.35. *Let X be a σ-compact Hausdorff space and let μ be a Radon measure on X. Let ν be a Borel measure on X that is finite on $\mathfrak{K}_X$. If $\nu = \mu$ on $\mathfrak{O}_X$, then ν is a regular Radon measure on X.*

Proof. Since X is a σ-compact Hausdorff space, a Radon measure μ on X is regular for $\mathfrak{B}_X$ by Proposition 19.22. Since ν is a Borel measure on X that is finite on $\mathfrak{K}_X$, to show that ν is a regular Radon measure it remains to show that ν is both outer and inner regular for $\mathfrak{B}_X$.

1. Let $E \in \mathfrak{B}_X$. Since μ is a Radon measure on a σ-compact Hausdorff space X, μ is a σ-finite measure. Then by Theorem 19.23, for every $\varepsilon > 0$ there exist an open set O and a closed set C such that

$$(1) \qquad C \subset E \subset O \quad \text{and} \quad \mu(O \setminus C) < \varepsilon.$$

2. Let us prove the outer regularity of ν for E. If $\nu(E) = \infty$ then ν is outer regular for E by (3) of Lemma 19.16. Suppose $\nu(E) < \infty$. Now $O \setminus C = O \cap C^c \in \mathfrak{O}_X$. Since $\nu = \mu$ on $\mathfrak{O}_X$, we have $\nu(O \setminus C) = \mu(O \setminus C) < \varepsilon$. Then $\nu(O \setminus E) \leq \nu(O \setminus C) < \varepsilon$. This shows that ν is outer regular for E by (1) of Lemma 19.16.

3. Let us prove the inner regularity of ν for E. Now since X is a σ-compact space, there exists an increasing sequence $(F_n : n \in \mathbb{N})$ in $\mathfrak{K}_X$ such that $\bigcup_{n \in \mathbb{N}} F_n = X$. If we let $K_n = F_n \cap C$ for $n \in \mathbb{N}$ then $(K_n : n \in \mathbb{N})$ is an increasing sequence in $\mathfrak{K}_X$ such that $\bigcup_{n \in \mathbb{N}} K_n = C$ and hence $\lim_{n \to \infty} \nu(K_n) = \nu(C)$.

3.1. If $\nu(C) = \infty$, then since $C \subset E$ we have $\nu(E) = \infty$ also. Since $\lim_{n \to \infty} \nu(K_n) = \nu(C) = \infty$, for every $M \geq 0$ there exists $N \in \mathbb{N}$ such that $\nu(K_N) \geq M$. Thus $\nu(E) = \infty$, $K_N \in \mathfrak{K}_X$, $K_N \subset C \subset E$, and $\nu(K_N) \geq M$. This shows the inner regularity of ν for E by (4) of Lemma 19.16.

3.2. Consider the case $\nu(C) < \infty$. Then $\nu(E) - \nu(C) = \nu(E \setminus C) \leq \nu(O \setminus C) < \varepsilon$ so that we have $\nu(E) < \infty$ also. Since $\lim_{n \to \infty} \nu(K_n) = \nu(C)$, there exists $N \in \mathbb{N}$ such that $\nu(C) - \nu(K_N) < \varepsilon$. Then $\nu(E) - \nu(K_N) < \nu(E) - \nu(C) + \varepsilon < 2\varepsilon$. This shows that ν is inner regular for E by (2) of Lemma 19.16. ∎

Theorem 19.36. *Let X be a locally compact Hausdorff space and a σ-compact space. Then every Borel measure μ on X that is finite on the compact sets is a regular Radon measure.*

Proof. Since μ is a Borel measure on X that is finite on $\mathfrak{K}_X$, to show that μ is a regular Radon measure it remains to show that μ is both outer and inner regular for $\mathfrak{B}_X$.

Let us define a function on $C_c(X)$, the linear space of real-valued continuous functions with compact support on X by setting

(1) $$I(f) = \int_X f\,d\mu \quad \text{for } f \in C_c(X).$$

We have $|\int_X f\,d\mu| \le \int_X |f|\,d\mu \le \max_{x\in X}|f(x)|\mu(\text{supp}\{f\}) < \infty$ for every $f \in C_c(X)$. Thus I is real-valued on $C_c(X)$. Its linearity, that is, $I(c_1 f_1 + c_2 f_2) = c_1 I(f_1) + c_2 I(f_2)$ for $f_1, f_2 \in C_c(X)$ and $c_1, c_2 \in \mathbb{R}$, and its positivity, that is, $I(f) \ge 0$ for $f \in C_c(X)$ such that $f \ge 0$, are immediate. Thus I is a positive linear functional on $C_c(X)$. By Theorem 19.34 (Riesz-Markoff), there exists a unique Radon measure ν on $(X, \mathfrak{B}_X)$ such that

(2) $$I(f) = \int_X f\,d\nu \quad \text{for } f \in C_c(X).$$

Since X is a σ-compact Hausdorff space, the Radon measure ν on X is inner regular for $\mathfrak{B}_X$ and hence regular for $\mathfrak{B}_X$ by Proposition 19.22. Now by (1) and (2) we have

(3) $$\int_X f\,d\mu = \int_X f\,d\nu \quad \text{for } f \in C_c(X).$$

Let us show that (3) implies $\mu = \nu$ on $\mathfrak{O}_X$. Let $O \in \mathfrak{O}_X$. Since O is an open and σ-compact set, there exists an increasing sequence $(f_n : n \in \mathbb{N})$ in $C_c(X)$ such that $\lim_{n\to\infty} f_n = \mathbf{1}_O$ by Proposition 19.14. Then by the Monotone Convergence Theorem (Theorem 8.5) and by (3), we have

$$\begin{aligned}\mu(O) &= \int_X \mathbf{1}_O\,d\mu = \lim_{n\to\infty}\int_X f_n\,d\mu \\ &= \lim_{n\to\infty}\int_X f_n\,d\nu = \int_X \mathbf{1}_O\,d\nu \\ &= \nu(O).\end{aligned}$$

This proves $\mu = \nu$ on $\mathfrak{O}_X$. The fact that ν is a Radon measure on a σ-compact Hausdorff space X and μ is a Borel measure on X that is finite on $\mathfrak{K}_X$ such that $\mu = \nu$ on $\mathfrak{O}_X$ implies that μ is a regular Radon measure on $(X, \mathfrak{B}_X)$ by Lemma 19.35. ∎

[IV] Approximation by Continuous Functions

Theorem 19.37. (Lusin) *Let X be a locally compact Hausdorff space and let μ be a Radon measure on the Borel σ-algebra $\mathfrak{B}_X$ of subsets of X. Let f be an arbitrary extended real-valued $\mathfrak{B}_X$-measurable function on X which is finite a.e. on X and vanishes outside of a set $A \in \mathfrak{B}_X$ with $\mu(A) < \infty$. Then for every $\varepsilon > 0$ there exists $g \in C_c(X)$ such that $\mu\{x \in X : f(x) \ne g(x)\} < \varepsilon$. Moreover $g \in C_c(X)$ can be so chosen that we have* $\inf_{y\in X} f(y) \le g(x) \le \sup_{y\in X} f(y)$ *for $x \in X$ and in particular* $\sup_{x\in X}|g(x)| \le \sup_{x\in X}|f(x)|$.

Proof. 1. Let us show first that if f is a real-valued $\mathfrak{B}_X$-measurable function on X such that $f(x) \in [0,1)$ for $x \in X$ and $f = 0$ on A^c where A is a compact set, then for every $\varepsilon > 0$ there exists $g \in C_c(X)$ such that $g(x) \in [0,1]$ for $x \in X$ and $\mu\{X : f \neq g\} < \varepsilon$.

With $n \in \mathbb{N}$ arbitrarily fixed, let

$$I_{n,j} = \left[\tfrac{j-1}{2^n}, \tfrac{j}{2^n}\right) \quad \text{for } j = 1, \dots, 2^n,$$

$$A_{n,j} = f^{-1}(I_{n,j}) \cap A,$$

$$\varphi_n = \sum_{j=1}^{2^n} \tfrac{j-1}{2^n} \mathbf{1}_{A_{n,j}}.$$

Then $(\varphi_n : n \in \mathbb{N})$ is an increasing sequence of nonnegative simple functions on $(X, \mathfrak{B}_X)$ vanishing on A^c and $\varphi_n \uparrow f$ on X. If we define $\psi_1 = \varphi_1$ and $\psi_n = \varphi_n - \varphi_{n-1}$ for $n \geq 2$, then $(\psi_n : n \in \mathbb{N})$ is a sequence of nonnegative simple functions vanishing on A^c and $f = \sum_{n \in \mathbb{N}} \psi_n$ on X. Now ψ_n assumes at most two values, 0 and $\frac{1}{2^n}$. Thus $2^n \psi_n = \mathbf{1}_{B_n}$ for some set $B_n \in \mathfrak{B}_X$ contained in A. Since A is a compact set, there exists an open set V such that $\overline{V}$ is a compact set and $A \subset V$ by Proposition 19.6. Now $\mu(B_n) < \infty$ since B_n is a subset of the compact set A. Then by (a) of Theorem 19.20, μ is inner regular for B_n. Thus for an arbitrary $\varepsilon > 0$, there exists a compact set K_n such that $K_n \subset B_n$ and $\mu(B_n \setminus K_n) < \frac{\varepsilon}{2^{n+1}}$. By the outer regularity of μ on $\mathfrak{B}_X$, there exists an open set V_n such that $\mu(V_n \setminus B_n) < \frac{\varepsilon}{2^{n+1}}$. Since $B_n \subset V$, we can take $V_n \subset V$ by intersecting with V if necessary. For the sets $K_n \subset B_n \subset V_n$, we have $\mu(V_n \setminus K_n) = \mu(V_n \setminus B_n) + \mu(B_n \setminus K_n) < \frac{\varepsilon}{2^{n+1}} + \frac{\varepsilon}{2^{n+1}} = \frac{\varepsilon}{2^n}$. Since $K_n \subset V_n$, by Theorem 19.11 (Urysohn's Lemma) there exists $h_n \in C_c(X)$ such that $h_n \prec V_n$ and $h_n \succ K_n$. Let us define a function g by setting $g = \sum_{n \in \mathbb{N}} \frac{1}{2^n} h_n$. Since $h_n(x) \in [0,1]$ for $x \in X$ for every $n \in \mathbb{N}$, we have $g(x) \in [0,1]$ for $x \in X$. By the uniform convergence of the series in the definition of g, g is a continuous function on X. Since $h_n \prec V_n$, we have $\operatorname{supp}\{h_n\} \subset V_n \subset V$ so that $h_n = 0$ on V^c for every $n \in \mathbb{N}$. This implies that $g = 0$ on V^c so that $\{X : g \neq 0\} \subset V$ and then $\operatorname{supp}\{g\} \subset \overline{V}$. Since $\overline{V}$ is a compact set, we have $g \in C_c(X)$.

It remains to show $\mu\{X : f \neq g\} < \varepsilon$. Recall that $g = \sum_{n \in \mathbb{N}} \frac{1}{2^n} h_n$ and $f = \sum_{n \in \mathbb{N}} \psi_n$. Let us compare $\frac{1}{2^n} h_n$ and ψ_n. By $h_n \succ K_n$, we have $h_n = 1$ on K_n and then $\frac{1}{2^n} h_n = \frac{1}{2^n}$ on K_n. On the other hand we have $2^n \psi_n = \mathbf{1}_{B_n}$ and $B_n \supset K_n$. This implies that $\psi_n = \frac{1}{2^n}$ on K_n and thus $\frac{1}{2^n} h_n = \psi_n$ on K_n. Next $h_n \prec V_n$ implies $h_n = 0$ on V_n^c and then $\frac{1}{2^n} h_n = 0$ on V_n^c. On the other hand $2^n \psi_n = \mathbf{1}_{B_n}$ implies $\psi_n = 0$ on $B_n^c \supset V_n^c$. Thus we have $\frac{1}{2^n} h_n = \psi_n$ on V_n^c also. Therefore we have $\frac{1}{2^n} h_n = \psi_n$ on $K_n \cup V_n^c$. Then $\left\{X : \frac{1}{2^n} h_n = \psi_n\right\} \supset K_n \cup V_n^c$ and consequently

$$\left\{X : \tfrac{1}{2^n} h_n \neq \psi_n\right\} \subset \left(K_n \cup V_n^c\right)^c = K_n^c \cap V_n = V_n \setminus K_n.$$

Since $\{X : g = f\} \supset \bigcap_{n \in \mathbb{N}} \left\{X : \frac{1}{2^n} h_n = \psi_n\right\}$, we have

$$\begin{aligned}\{X : g \neq f\} &\subset \Big[\bigcap_{n \in \mathbb{N}} \left\{X : \tfrac{1}{2^n} h_n = \psi_n\right\}\Big]^c = \textstyle\bigcup_{n \in \mathbb{N}} \left\{X : \tfrac{1}{2^n} h_n = \psi_n\right\}^c \\ &= \bigcup_{n \in \mathbb{N}} \left\{X : \tfrac{1}{2^n} h_n \neq \psi_n\right\} \subset \textstyle\bigcup_{n \in \mathbb{N}} V_n \setminus K_n.\end{aligned}$$

Then $\mu\{X : g \neq f\} \leq \sum_{n\in\mathbb{N}} \mu(V_n \setminus K_n) < \sum_{n\in\mathbb{N}} \varepsilon/2^n = \varepsilon$.

2. Next let f be a real-valued $\mathfrak{B}_X$-measurable function on X such that $f(x) \in [0, 1)$ for $x \in X$ and $f = 0$ on A^c where $A \in \mathfrak{B}_X$ and $\mu(A) < \infty$. Then by (a) of Proposition 19.20, μ is inner regular for A. Thus for an arbitrary $\varepsilon > 0$ there exists a compact set $K \subset A$ such that $\mu(A \setminus K) < \frac{\varepsilon}{2}$. Let $f_0 = f\mathbf{1}_K$. Then $f_0(x) \in [0, 1)$ for $x \in X$ and $f_0 = 0$ on K^c. Thus by our result in **1**, there exists $g \in C_c(X)$ such that $g(x) \in [0, 1]$ for $x \in X$ and $\mu\{X : f_0 \neq g\} < \frac{\varepsilon}{2}$. Since $X = K \cup (A \setminus K) \cup A^c$, we have

$$\{X : f \neq g\} \subset \{K : f \neq g\} \cup (A \setminus K) \cup \{A^c : f \neq g\}.$$

Since $f = f_0$ on K, we have $\{K : f \neq gv\} = \{K : f_0 \neq g\} \subset \{X : f_0 \neq g\}$. Similarly, since $f = f_0 = 0$ on A^c we have $\{A^c : f \neq g\} = \{A^c : f_0 \neq g\} \subset \{X : f_0 \neq g\}$. Thus

$$\{X : f \neq g\} \subset \{X : f_0 \neq g\} \cup (A \setminus K)$$

and therefore $\mu\{X : f \neq g\} \leq \mu\{X : f_0 \neq g\} + \mu(A \setminus K) < \frac{\varepsilon}{2} + \frac{\varepsilon}{2} = \varepsilon$.

3. Next consider a real-valued $\mathfrak{B}_X$-measurable function f on X such that $f(x) \in [0, M)$ for $x \in X$ for some $M > 0$ and $f = 0$ on A^c where $A \in \mathfrak{B}_X$ with $\mu(A) < \infty$. Let $f_0 = M^{-1}f$. Then $f_0(x) \in [0, 1)$ for $x \in X$ and $f_0 = 0$ on A^c so that by our result in **2** there exists $g_0 \in C_c(X)$ such that $g_0(x) \in [0, 1]$ for $x \in X$ and $\mu\{X: f_0 \neq g_0\} < \varepsilon$. If we let $g = Mg_0$, then $g \in C_c(X)$, $g(x) \in [0, M]$ for $x \in X$ and $\mu\{X : f \neq g\} = \mu\{X : f_0 \neq g_0\} < \varepsilon$.

4. Next suppose f is $\mathfrak{B}_X$-measurable on X, $f(x) \in (-M, M)$ for $x \in X$ for some $M > 0$ and $f = 0$ on A^c for some $A \in \mathfrak{B}_X$ with $\mu(A) < \infty$. Let $f = f^+ - f^-$. Then $f^+(x), f^-(x) \in [0, M)$ for $x \in X$ and $f^+ = f^- = 0$ on A^c. By our result in **3**, there exist $g_1, g_2 \in C_c(X)$ such that $g_1(x), g_2(x) \in [0, M]$ for $x \in X$, $\mu\{X : f^+ \neq g_1\} < \frac{\varepsilon}{2}$, and $\mu\{X : f^- \neq g_2\} < \frac{\varepsilon}{2}$. If we let $g = g_1 - g_2$, then $g \in C_c(X)$ and $g(x) \in [-M, M]$ for $x \in X$. Since $\{X : f = g\} \supset \{X : f^+ = g_1\} \cap \{X : f^- = g_2\}$, taking complements we have $\{X : f \neq g\} \subset \{X : f^+ \neq g_1\} \cup \{X : f^- \neq g_2\}$. This implies

$$\mu\{X : f \neq g\} \leq \mu\{X : f^+ \neq g_1\} + \mu\{X : f^- \neq g_2\} < \frac{\varepsilon}{2} + \frac{\varepsilon}{2} = \varepsilon.$$

5. Now let f be an extended real-valued $\mathfrak{B}_X$-measurable function on X such that f is finite a.e. on X and $f = 0$ on A^c where A is a set in $\mathfrak{B}_X$ with $\mu(A) < \infty$. Let $B_n = \{X : |f| > n\}$ for $n \in \mathbb{N}$. Then $(B_n : n \in \mathbb{N})$ is a decreasing sequence in $\mathfrak{B}_X$ and $\bigcap_{n\in\mathbb{N}} B_n = \{X : |f| = \infty\}$. Since f is finite a.e. on X, we have $\mu\left(\bigcap_{n\in\mathbb{N}} B_n\right) = 0$. Now $B_1 \cap A^c = \emptyset$ so that $B_1 \subset A$ and then $\mu(B_1) \leq \mu(A) < \infty$. Thus we have $\lim_{n\to\infty} \mu(B_n) = \mu\left(\bigcap_{n\in\mathbb{N}} B_n\right) = 0$. Then for an arbitrary $\varepsilon > 0$, there exists $N \in \mathbb{N}$ such that $\mu(B_N) < \frac{\varepsilon}{2}$, that is, $\mu\{X : |f| > N\} < \frac{\varepsilon}{2}$. Let $f^{[N]} = (f \vee -N) \wedge N$. Then $f^{[N]}(x) \in (-N-1, N+1)$ for $x \in X$ and $f^{[N]} = 0$ on A^c. By our result in **4**, there exists $g \in C_c(X)$ such that $g(x) \in [-N-1, N+1]$ for $x \in X$ and $\mu\{X : f^{[N]} \neq g\} < \frac{\varepsilon}{2}$. Now

$$\{X : f \neq g\} = \{X : |f| \leq N, f \neq g\} \cup \{X : |f| > N, f \neq g\}.$$

On $\{X : |f| \leq N\}$, we have $f = f^{[N]}$. Thus

$$\begin{aligned}\{X : |f| \leq N, f \neq g\} &= \{X : |f| \leq N\} \cap \{X : f \neq g\} \\ &\subset \{X : f = f^{[N]}\} \cap \{X : f \neq g\} \\ &\subset \{X : f^{[N]} \neq g\}.\end{aligned}$$

We also have $\{X : |f| > N, f \neq g\} \subset \{X : |f| > N\}$. Therefore we have

$$\{X : f \neq g\} \subset \{X : f^{[N]} \neq g\} \cup \{X : |f| > N\},$$

and then

$$\mu\{X : f \neq g\} \leq \mu\{X : f^{[N]} \neq g\} + \mu\{X : |f| > N\} < \frac{\varepsilon}{2} + \frac{\varepsilon}{2} = \varepsilon.$$

6. Let f be an extended real-valued $\mathfrak{B}_X$-measurable function on X which is finite a.e. on X and vanishes outside of a set $A \in \mathfrak{B}_X$ with $\mu(A) < \infty$. For an arbitrary $\varepsilon > 0$, there exists $g \in C_c(X)$ such that $\mu\{X : f \neq g\} < \varepsilon$ as we showed in **5** above. Let us define a function h on X by setting $h = (g \vee \inf_X f) \wedge \sup_X f$. Then $h \in C_c(X)$, $\inf_X f \leq h \leq \sup_X f$ and $\{X : f \neq h\} \subset \{X : f \neq g\}$ so that $\mu\{X : f \neq h\} \leq \mu\{X : f \neq g\} < \varepsilon$.

Let us note that $\inf_X f \leq h \leq \sup_X f$ implies $\sup_X |h| \leq \sup_X |f|$. Indeed if $x \in X$ and $h(x) \geq 0$ then $|h(x)| = h(x) \leq \sup_X f \leq \sup_X |f|$ and if $h(x) < 0$ then

$$|h(x)| = -h(x) \leq -\inf_X f = \sup_X(-f) \leq \sup_X |f|.$$

Therefore $\sup_X |h| \leq \sup_X |f|$. ∎

Recall that for an arbitrary measure space $(X, \mathfrak{A}, \mu)$ and $p \in [1, \infty)$, $L^p(X, \mathfrak{A}, \mu)$ is the linear space of equivalent classes of extended complex-valued $\mathfrak{A}$-measurable functions f such that $\int_X |f|^p \, d\mu < \infty$.

Theorem 19.38. *Given $L^p(X, \mathfrak{B}_X, \mu)$ where μ is a Radon measure on the Borel σ-algebra $\mathfrak{B}_X$ of subsets of a locally compact Hausdorff space X and $p \in [1, \infty)$. Let $C_c(X)$ be the collection of all complex-valued continuous functions with compact support on X. Then $C_c(X)$ is a dense subset of $L^p(X, \mathfrak{B}_X, \mu)$.*

Proof. If $f \in C_c(X)$, then $\text{supp}\{f\}$ is a compact set so that $\mu(\text{supp}\{f\}) < \infty$. Also there exists $M > 0$ such that $|f| \leq M$ on X. Thus we have

$$\int_X |f|^p \, d\mu = \int_{\text{supp}\{f\}} |f|^p \, d\mu \leq M^p \mu(\text{supp}\{f\}) < \infty$$

so that $f \in L^p(X, \mathfrak{B}_X, \mu)$. This shows that $C_c(X) \subset L^p(X, \mathfrak{B}_X, \mu)$.

Next in order to show that $C_c(X)$ is dense in $L^p(X, \mathfrak{B}_X, \mu)$, we show that for every $f \in L^p(X, \mathfrak{B}_X, \mu)$ and $\varepsilon > 0$ there exists $g \in C_c(X)$ such that $\|f - g\|_p < \varepsilon$. Let $S_0(X, \mathfrak{B}_X, \mu)$ be the collection of equivalence classes of complex-valued simple functions on $(X, \mathfrak{B}_X)$ each vanishing outside a set in $\mathfrak{B}_X$ with finite measure. According to Theorem 18.3, $S_0(X, \mathfrak{B}_X, \mu)$ is a dense subset of $L^p(X, \mathfrak{B}_X, \mu)$. Thus for an arbitrary $f \in L^p(X, \mathfrak{B}_X, \mu)$ and $\varepsilon > 0$, there exists $\varphi \in S_0(X, \mathfrak{B}_X, \mu)$ such that $\|f - \varphi\|_p < \frac{\varepsilon}{2}$. Let $\alpha = \sup_{x \in X} |\varphi(x)|$. If $\alpha = 0$, then $\varphi = 0$ so that $\varphi \in C_c(X)$ and we are done. Consider the case $\alpha > 0$. Now $\varphi = \Re\varphi + i\Im\varphi$ and $\Re\varphi$ and $\Im\varphi$ are real-valued $\mathfrak{B}_X$-measurable functions on X vanishing outside a set of finite measure so that by Theorem 19.37 (Lusin) there exist real-valued $g_1, g_2 \in C_c(X)$ such that

$$\sup |g_1| \le \sup |\Re\varphi| \le \alpha, \quad \mu\{X : \Re\varphi \ne g_1\} < \frac{\varepsilon^p}{8^p\alpha^p},$$
$$\sup |g_2| \le \sup |\Im\varphi| \le \alpha, \quad \mu\{X : \Im\varphi \ne g_2\} < \frac{\varepsilon^p}{8^p\alpha^p}.$$

If we let $g = g_1 + ig_2$, then $g \in C_c(X)$ and $\|\varphi - g\|_p \le \|\Re\varphi - g_1\|_p + \|\Im\varphi - g_2\|_p$. Now

$$\begin{aligned}\|\Re\varphi - g_1\|_p^p &= \int_X |\Re\varphi - g_1|^p \, d\mu = \int_{\{X:\Re\varphi\ne g_1\}} |\Re\varphi - g_1|^p \, d\mu \\ &\le \int_{\{X:\Re\varphi\ne g_1\}} \left\{|\Re\varphi| + |g_1|\right\}^p d\mu < (2\alpha)^p \frac{\varepsilon^p}{8^p\alpha^p} = \frac{\varepsilon^p}{4^p}\end{aligned}$$

so that $\|\Re\varphi - g_1\|_p < \frac{\varepsilon}{4}$. Similarly we have $\|\Im\varphi - g_2\|_p < \frac{\varepsilon}{4}$. Therefore we have $\|\varphi - g\|_p < \frac{\varepsilon}{4} + \frac{\varepsilon}{4} = \frac{\varepsilon}{2}$. Then $\|f - g\|_p \le \|f - \varphi\|_p + \|\varphi - g\|_p < \frac{\varepsilon}{2} + \frac{\varepsilon}{2} = \varepsilon$. ∎

[V] Signed Radon Measures

Proposition 19.39. *If μ is a Radon measure on a Hausdorff space X, then so is $c\mu$ for every $c \ge 0$.*

Proof. 1. Let $c \ge 0$. Since μ is finite on $\mathfrak{K}_X$, $c\mu$ is finite on $\mathfrak{K}_X$. By the outer regularity of μ for $\mathfrak{B}_X$, for every $E \in \mathfrak{B}_X$ we have

$$\begin{aligned}(c\mu)(E) &= c\mu(E) = c \inf\left\{\mu(O) : O \supset E, O \in \mathfrak{O}_X\right\} \\ &= \inf\left\{(c\mu)(O) : O \supset E, O \in \mathfrak{O}_X\right\}.\end{aligned}$$

This shows that $c\mu$ is outer regular for every $E \in \mathfrak{B}_X$. Similarly by the inner regularity of μ for $\mathfrak{O}_X$, for every $O \in \mathfrak{O}_X$ we have

$$\begin{aligned}(c\mu)(O) &= c\mu(O) = c \inf\left\{\mu(K) : K \subset O, K \in \mathfrak{K}_X\right\} \\ &= \inf\left\{(c\mu)(K) : K \subset O, K \in \mathfrak{K}_X\right\}.\end{aligned}$$

This shows that $c\mu$ is inner regular for every $O \in \mathfrak{O}_X$. Thus $c\mu$ is a Radon measure. ∎

Proposition 19.40. *If μ_1 and μ_2 are Radon measures on a Hausdorff space X, then so is $\mu_1 + \mu_2$.*

Proof. Since μ_1 and μ_2 are finite on $\mathfrak{K}_X$, $\mu_1 + \mu_2$ is finite on $\mathfrak{K}_X$. It remains to show that $\mu_1 + \mu_2$ is outer regular for $\mathfrak{B}_X$ and inner regular for $\mathfrak{O}_X$.

1. Let us prove the outer regularity of $\mu_1+\mu_2$ for $\mathfrak{B}_X$. Let $E \in \mathfrak{B}_X$. If $(\mu_1+\mu_2)(E) = \infty$, then $\mu_1 + \mu_2$ is outer regular for E by (3) of Lemma 19.16. If $(\mu_1 + \mu_2)(E) < \infty$, to show the outer regularity of $\mu_1 + \mu_2$ on E it suffices to show that for every $\varepsilon > 0$ there exists $O \in \mathfrak{O}_X$ such that $O \supset E$ and $(\mu_1+\mu_2)(O \setminus E) < \varepsilon$ by (1) of Lemma 19.16. Now $(\mu_1+\mu_2)(E) < \infty$ implies $\mu_i(E) < \infty$ for $i = 1$ and 2. Since μ_i is outer regular for $\mathfrak{B}_X$ and $E \in \mathfrak{B}_X$, there exists $O_i \in \mathfrak{O}_X$ such that $O_i \supset E$ and $\mu_i(O_i \setminus E) < \frac{\varepsilon}{2}$ by (1) of Lemma 19.16. Let $O = O_1 \cap O_2$. Then $O \in \mathfrak{O}_X$, $O \supset E$, and

$$\begin{aligned}(\mu_1+\mu_2)(O \setminus E) &= \mu_1(O \setminus E) + \mu_2(O \setminus E)\\ &\leq \mu_1(O_1 \setminus E) + \mu_2(O_2 \setminus E)\\ &< \tfrac{\varepsilon}{2} + \tfrac{\varepsilon}{2} < \varepsilon.\end{aligned}$$

This proves the outer regularity of $\mu_1 + \mu_2$ for E.

2. Let us prove the inner regularity of $\mu_1+\mu_2$ for $\mathfrak{O}_X$. Let $O \in \mathfrak{O}_X$. If $(\mu_1+\mu_2)(O) = \infty$, then at least one of the two equalities $\mu_1(O) = \infty$ and $\mu_2(O) = \infty$ holds. We assume without loss of generality that $\mu_1(O) = \infty$. Now since μ_1 is inner regular for $\mathfrak{O}_X$, for every $M \geq 0$ there exists $K \in \mathfrak{K}_X$ such that $K \subset O$ and $\mu_1(K) \geq M$ by (4) of Lemma 19.16. Then $(\mu_1+\mu_2)(K) \geq M$. This proves the inner regularity of $\mu_1+\mu_2$ on O by (4) of Lemma 19.16.

Consider the case $(\mu_1+\mu_2)(O) < \infty$. To show that $\mu_1+\mu_2$ is inner regular for O, it suffices to show that for every $\varepsilon > 0$ there exists $K \in \mathfrak{K}_X$ such that $K \subset O$ and $(\mu_1+\mu_2)(O \setminus K) < \varepsilon$ by (2) of Lemma 19.16. Now $(\mu_1+\mu_2)(O) < \infty$ implies $\mu_i(O) < \infty$ for $i = 1$ and 2. Since μ_i is inner regular for $\mathfrak{O}_X$, there exists $K_i \in \mathfrak{K}_X$ such that $K_i \subset O_i$ and $\mu_i(O \setminus K_i) < \frac{\varepsilon}{2}$ by (2) of Lemma 19.16. If we let $K = K_1 \cup K_2$, then $K \in \mathfrak{K}_X$, $K \subset O$, and

$$\begin{aligned}(\mu_1+\mu_2)(O \setminus K) &= \mu_1(O \setminus K) + \mu_2(O \setminus K)\\ &\leq \mu_1(O_1 \setminus K_1) + \mu_2(O_2 \setminus K_2)\\ &< \tfrac{\varepsilon}{2} + \tfrac{\varepsilon}{2} < \varepsilon.\end{aligned}$$

This proves the outer regularity of $\mu_1 + \mu_2$ for O. ∎

Definition 19.41. *Let X be a Hausdorff space and let $\mathfrak{B}_X$ be the Borel σ-algebra on X. A signed measure λ on $(X, \mathfrak{B}_X)$ is called a signed Radon measure on X if its positive part λ^+ and negative part λ^- are Radon measures on X.*

By Definition 10.29, a signed measure λ on an arbitrary measurable space $(X, \mathfrak{A})$ is called a finite signed measure if $\lambda(X) \in \mathbb{R}$. We have $\lambda(X) \in \mathbb{R}$ if and only if $\lambda(E) \in \mathbb{R}$ for every $E \in \mathfrak{A}$ by Lemma 10.5. Let λ^+ and λ^- be the positive and negative parts of λ. The total variation of λ, that is, $\|\lambda\| = \lambda^+ + \lambda^-$, is a finite measure if and only if λ is a finite signed measure. (See Definition 10.22 and Observation 10.30.)

Lemma 19.42. **(a)** *Let μ be a σ-finite Borel measure on a Hausdorff space X. Then μ is regular for every $E \in \mathfrak{B}_X$ such that $\mu(E) = \infty$.*
(b) *Let μ_1 and μ_2 be σ-finite Borel measures on a Hausdorff space X such that $\mu_1 \leq \mu_2$*

on $\mathfrak{B}_X$, and such that $\mu_1(E) = \infty$ if and only if $\mu_2(E) = \infty$ for $E \in \mathfrak{B}_X$. If μ_2 is a Radon measure, then both μ_1 and μ_2 are regular Radon measures on X.
(c) *Let μ_1 and μ_2 be finite Borel measures on a Hausdorff space X such that $\mu_1 \le \mu_2$ on $\mathfrak{B}_X$. If μ_2 is a Radon measure, then both μ_1 and μ_2 are regular Radon measures on X.*

Proof. 1. Let us prove (a). Let $E \in \mathfrak{B}_x$ and $\mu(E) = \infty$. By (3) of Lemma 19.16, μ is outer regular for E. Let us show that μ is inner regular for E. Now since μ is a σ-finite measure on $(X, \mathfrak{B}_X)$ there exists an increasing sequence $(E_n : n \in \mathbb{N})$ in $\mathfrak{B}_X$ such that $\mu(E_n) < \infty$ for $n \in \mathbb{N}$ and $\bigcup_{n\in\mathbb{N}} E_n = E$. Let $M \ge 0$ be arbitrarily given. Since we have $\lim_{n\to\infty} \mu(E_n) = \mu(E) = \infty$, there exists $N \in \mathbb{N}$ such that $\mu(E_N) \ge M + 1$. Since $\mu(E_N) < \infty$, μ is inner regular for E_N by our result above. Thus there exists $K \in \mathfrak{K}_X$ such that $K \subset E_N$ and $\mu(E_N) - 1 < \mu(K)$ by (2) of Lemma 19.16. Then $\mu(K) > \mu(E_N) - 1 \ge M$. Thus for every $M \ge 0$, there exist $K \in \mathfrak{K}_X$ such that $K \subset E_N \subset E$ and $\mu(K) \ge M$. This shows that μ is inner regular for E by (4) of Lemma 19.16.

2. Let us prove (b). Suppose μ_2 is a Radon measure. Then the σ-finiteness of μ_2 implies that μ_2 is regular, that is, both outer and inner regular, for $\mathfrak{B}_X$ by Proposition 19.20. Let us show that μ_1 is a regular Radon measure. Now since μ_2 is a Radon measure, μ_2 is finite on $\mathfrak{K}_X$. Then $\mu_1 \le \mu_2$ implies that μ_1 is finite on $\mathfrak{K}_X$. It remains to show that μ_1 is both outer and inner regular for $\mathfrak{B}_X$.

Let $E \in \mathfrak{B}_X$ be arbitrarily chosen. If $\mu(E) = \infty$, then μ is regular for E by (a). Let us consider the case $\mu_1(E) < \infty$. By the assumption in (b), we have $\mu_2(E) < \infty$. By the outer regularity of μ_2 for E, for every $\varepsilon > 0$ there exists $O \in \mathfrak{O}_X$ such that $O \supset E$ and $\mu_2(O \setminus E) < \varepsilon$ by (1) of Lemma 19.16. Then $\mu_1(O \setminus E) \le \mu_2(O \setminus E) < \varepsilon$. This shows that μ_1 is outer regular for E by (1) of Lemma 19.16. Similarly the inner regularity of μ_2 for E implies that for every $\varepsilon > 0$ there exists $K \in \mathfrak{K}_X$ such that $K \subset E$ and $\mu_2(E \setminus K) < \varepsilon$ by (2) of Lemma 19.16. Then $\mu_1(E \setminus K) \le \mu_2(E \setminus K) < \varepsilon$ so that μ_1 is inner regular for E by (2) of Lemma 19.16.

3. (c) is a particular case of (b). ∎

Proposition 19.43. *Let X be a Hausdorff space and let $\mathfrak{B}_X$ be the Borel σ-algebra on X. A finite signed measure λ on $(X, \mathfrak{B}_X)$ is a signed Radon measure on X if and only if its total variation $\|\lambda\|$ is a Radon measure on X.*

Proof. 1. Suppose λ is a Radon measure on X. Then its positive part λ^+ and negative part λ^- are Radon measures on X. Since $\|\lambda\| = \lambda^+ + \lambda^-$, $\|\lambda\|$ is a Radon measure on X by Proposition 19.40.

2. Conversely suppose $\|\lambda\|$ is a Radon measure on X. Since $\lambda^+ \le \|\lambda\|$ and $\|\lambda\|$ is a finite Radon measure on X, λ^+ is a Radon measure on X by (c) of Lemma 19.42. By the same reason λ^- is a Radon measure on X. Thus λ is a signed Radon measure on X.

Proposition 19.44. *Let μ and ν be two finite Radon measures on a Hausdorff spaceX. Then $\lambda = \mu - \nu$ is a finite signed Radon measure on X.*

Proof. By Theorem 11.27, we have $\|\lambda\| \le \mu + \nu$. Since μ and ν are finite Radon measures on X so is $\mu+\nu$ by Proposition 19.40. Then $|\lambda|$ is a finite Radon measure by (c) of Lemma 19.42. This implies that λ is a finite signed Radon measure on X by Proposition 19.43. ∎

Proposition 19.45. *Let λ be a finite signed Radon measure on a Hausdorff space X. Then $c\lambda$ is a finite signed Radon measure on X for every $c \in \mathbb{R}$.*

Proof. Let λ^+ and λ^- be the positive and negative parts of a finite signed Radon measure λ on X. Let $c \in \mathbb{R}$ and consider the finite signed measure $c\lambda$. By Theorem 10.28, we have $|c\lambda| = |c||\lambda| = |c|\{\lambda^+ + \lambda^-\}$. Since λ is a signed Radon measure on X, λ^+ and λ^- are Radon measures on X by Definition 19.41. Then by Proposition 19.40 and Proposition 19.39, $|c|\{\lambda^+ = \lambda^-\}$ is a Radon measure on X. Thus $|c\lambda|$ is a finite Radon measure on X. Then by Proposition 19.43, $c\lambda$ is a finite signed Radon measure on X. ∎

Proposition 19.46. *Let λ_1 and λ_2 be finite signed Radon measures on a Hausdorff space X. Then $\lambda_1 + \lambda_2$ is a finite signed Radon measure on X.*

Proof. Let λ_i^+ and λ_i^- be the positive and negative parts of λ_i for $i = 1, 2$. By Theorem 10.28, we have

$$(1) \qquad |\lambda_1 + \lambda_2| \leq |\lambda_1| + |\lambda_2| = \{\lambda_1^+ + \lambda_1^-\} + \{\lambda_2^+ + \lambda_2^-\}.$$

Since λ_1 and λ_2 are finite signed Radon measures on X, λ_1^+, λ_1^-, λ_2^+, and λ_2^- are finite Radon measures on X by Definition 19.41. Thus $\lambda_1^+ + \lambda_1^- + \lambda_2^+ + \lambda_2^-$ is a finite Radon measure on X by Proposition 19.40. Then the inequality (1) implies that $|\lambda_1 + \lambda_2|$ is a finite Radon measure on X by (c) of Lemma 19.42. This implies that $\lambda_1 + \lambda_2$ is a finite signed Radon measure on X by Proposition 19.43. ∎

Theorem 19.47. *Let X be a Hausdorff space and let $\mathcal{M}(X, \mathfrak{B}_X)$ be the collection of all finite signed Radon measures on $(X, \mathfrak{B}_X)$. Then $\mathcal{M}(X, \mathfrak{B}_X)$ is a linear space over the field of scalars $\mathbb{R}$. Let us define*

$$\|\lambda\|_{\mathcal{M}} = |\lambda|(X) \quad \textit{for } \lambda \in \mathcal{M}(X, \mathfrak{B}_X).$$

Then $\|\cdot\|_{\mathcal{M}}$ is a norm on $\mathcal{M}(X, \mathfrak{B}_X)$.

Proof. $\mathcal{M}(X, \mathfrak{B}_X)$ is a linear space by Proposition 19.45 and Proposition 19.46. Let us show that $\|\cdot\|_{\mathcal{M}}$ is a norm on $\mathcal{M}(X, \mathfrak{B}_X)$.

Since $\lambda \in \mathcal{M}(X, \mathfrak{B}_X)$ is a finite signed measure on $(X, \mathfrak{B}_X)$, $|\lambda|$ is a finite measure on $(X, \mathfrak{B}_X)$ so that $|\lambda|(X) < \infty$. Therefore $\|\lambda\|_{\mathcal{M}} \in [0, \infty)$. Let λ^+ and λ^- be the positive and negative parts of λ. If $\lambda = 0$, then by the uniqueness of Jordan decompostion for a signed measure (Theorem 10.21), we have $\lambda^+ = 0$ and $\lambda^- = 0$ and then $|\lambda| = \lambda^+ + \lambda^- = 0$. Then $|\lambda|(X) = 0$. Thus $\|0\|_{\mathcal{M}} = 0$. Conversely if $\|\lambda\|_{\mathcal{M}} = 0$ then $|\lambda|(X) = 0$ so that $|\lambda|(E) = 0$ for every $E \in \mathfrak{B}_X$. Then since $|\lambda| = \lambda^+ + \lambda^-$, we have $\lambda^+(E) = 0$ and $\lambda^-(E) = 0$ for every $E \in \mathfrak{B}_X$. Thus $\lambda(E) = \lambda^+(E) - \lambda^-(E) = 0$ for every $E \in \mathfrak{B}_X$ so that $\lambda = 0$.

For every $\lambda \in \mathcal{M}(X, \mathfrak{B}_X)$ and $c \in \mathbb{R}$, we have

$$\|c\lambda\|_{\mathcal{M}} = |c\lambda|(X) = |c||\lambda|(X) = |c|\|\lambda\|_{\mathcal{M}}$$

by (a) of Theorem 10.28. For $\lambda_1, \lambda_2 \in \mathcal{M}(X, \mathfrak{B}_X)$, we have

$$\|\lambda_1 + \lambda_2\|_{\mathcal{M}} = |\lambda_1 + \lambda_2|(X) \leq \{|\lambda_1| + |\lambda_2|\}(X) = \|\lambda_1\|_{\mathcal{M}} + \|\lambda_2\|_{\mathcal{M}}$$

by (b) of Theorem 10.28. This completes the proof that $\|\cdot\|_{\mathcal{M}}$ is a norm on $\mathcal{M}(X, \mathfrak{B}_X)$. ∎

[VI] The Dual Space of $C(X)$

So far we have considered the linear space $C_c(X)$ of all real-valued continuous functions with compact support on a locally compact Hausdorff space X. Let us consider now the linear space $C(X)$ of all real-valued continuous functions on a compact Hausdorff space X. Let us define some notations.

Definition 19.48. *For a compact Hausdorff space X, let $C(X)$ be the collection of all real-valued continuous functions on X. We write $C^+(X)$ for the subset of $C(X)$ consisting of all nonnegative continuous functions on X.*

Observation 19.49. Let us note that since X is a compact Hausdorff space we have $C(X) = C_c(X)$. Indeed if $f \in C(X)$ then $\text{supp}\{f\}$ is a closed set in the compact space X and hence it is a compact set and then $f \in C_c(X)$. This shows that $C(X) \subset C_c(X)$. On the other hand we have $C_c(X) \subset C(X)$. Therefore $C(X) = C_c(X)$.

Clearly $C(X)$ is a linear space over the field of scalars $\mathbb{R}$. Moreover we have

$$f \wedge g, f \vee g \in C(X) \quad \text{for } f, g \in C(X),$$
$$c \in C(X) \quad \text{for every } c \in \mathbb{R}.$$

Theorem 19.50. *Let X be a compact Hausdorff space and let $C(X)$ be the linear space of all real-valued continuous functions on X. Let us define a function $\|\cdot\|_u$ on $C(X)$ by setting*

$$\|f\|_u = \sup_{x \in X} |f(x)| \quad \text{for } f \in C(X).$$

Then $\|\cdot\|_u$ is a norm on $C(X)$ and $\big(C(X), \|\cdot\|_u\big)$ is a Banach space. We call the norm $\|\cdot\|_u$ the uniform norm.

Proof. If $f \in C(X)$, then f is a real-valued continuous function on a compact set X so that f is bounded on X and $\sup_{x \in X} |f(x)| < \infty$ and indeed $\sup_{x \in X} |f(x)| = \max_{x \in X} |f(x)|$. Thus $\|f\|_u \in [0, \infty)$ for every $f \in C(X)$. The fact that $\|f\|_u = 0$ if and only if $f = 0$, $\|\lambda f\|_u = |\lambda| \, \|f\|_u$ for every $f \in C(X)$ and $\lambda \in \mathbb{R}$, and $\|f + g\|_u \leq \|f\|_u + \|g\|_u$ for $f, g \in C(X)$ follow immediately from the definition of $\|\cdot\|_u$. Thus $\|\cdot\|_u$ is a norm on $C(X)$.

To show that $C(X)$ is a Banach space with respect to the norm $\|\cdot\|_u$, we show that $C(X)$ is complete with respect to $\|\cdot\|_u$. Let $(f_n : n \in \mathbb{N})$ be a Cauchy sequence in $C(X)$ with respect to the norm $\|\cdot\|_u$, that is, for every $\varepsilon > 0$ there exists $N \in \mathbb{N}$ such that $\|f_m - f_n\|_u < \varepsilon$ for $m, n \geq N$, that is,

(1) $$|f_m(x) - f_n(x)| < \varepsilon \quad \text{for all } x \in X \text{ when } m, n \geq N.$$

Now (1) implies that for each $x \in X$ the sequence $(f_n(x) : n \in \mathbb{N})$ is a Cauchy sequence in $\mathbb{R}$ so that there exists $f(x) \in \mathbb{R}$ such that $\lim_{n\to\infty} f_n(x) = f(x)$. Let us show that $(f_n : n \in \mathbb{N})$ converges to f uniformly on X. Now for every $\varepsilon > 0$ there exists $N \in \mathbb{N}$ such that

$$(2) \qquad |f_m(x) - f_n(x)| < \tfrac{\varepsilon}{2} \quad \text{for all } x \in X \text{ when } m, n \geq N.$$

Let $n \geq N$ and $x \in X$. By (2) we have $f_m(x) \in \big(f_n(x) - \frac{\varepsilon}{2}, f_n(x) + \frac{\varepsilon}{2}\big)$ for all $m \geq N$. Then $f(x) = \lim_{n\to\infty} f_m(x) \in \big(f_n(x) - \varepsilon, f_n(x) + \varepsilon\big)$. Thus we have

$$(3) \qquad |f(x) - f_n(x)| < \varepsilon \quad \text{for all } x \in X \text{ when } n \geq N.$$

This shows that $(f_n : n \in \mathbb{N})$ converges to f uniformly on X. The continuity of f_n on X for every $n \in \mathbb{N}$ and the uniformity of the convergence of $(f_n : n \in \mathbb{N})$ to f on X imply that f is a continuous function on X. Thus $f \in C(X)$. Moreover from (3) we have

$$\|f - f_n\|_u = \sup_{x\in X} |f(x) - f_n(x)| \leq \varepsilon \quad \text{for } n \geq N,$$

that is, $\lim_{n\to\infty} \|f - f_n\|_u = 0$. This proves the completeness of $C(X)$ with respect to the norm $\|\cdot\|_u$. ∎

Let L be a bounded linear functional on the Banach space $\big(C(X), \|\cdot\|_u\big)$, that is, L is a real-valued function on $C(X)$ such that $L(\lambda f) = \lambda L(f)$ for $f \in C(X)$ and $\lambda \in \mathbb{R}$, and $L(f+g) = L(f) + L(g)$ for $f, g \in C(X)$, and moreover there exists $M \geq 0$ such that

$$|L(f)| \leq M\|f\|_u \quad \text{for every } f \in C(X).$$

Let $C(X)^*$ be the dual space of $C(X)$, that is, the linear space of all bounded linear functionals on $C(X)$. According to Theorem 15.38, if we define a function $\|\cdot\|_*$ on $C(X)^*$ by setting for every $L \in C(X)^*$

$$\|L\|_* = \inf\{M \geq 0 : |L(f)| \leq M\|f\|_u \text{ for } f \in C(X)\},$$

then $\|\cdot\|_*$ is a norm on $C(X)^*$ and $\big(C(X)^*, \|\cdot\|_*\big)$ is a Banach space. According to Proposition 15.30 we have

$$|L(f)| \leq \|L\|_*\|f\|_u \quad \text{for every } f \in C(X).$$

Let $\overline{B}(f_0, r) = \{f \in C(X) : \|f - f_0\|_u \leq r\}$ and $S(f_0, r) = \{f \in C(X) : \|f - f_0\|_u = r\}$ for $f_0 \in C(X)$ and $r > 0$. Then according to Lemma 15.35 and Theorem 15.36, we have

$$\|L\|_* = \sup_{f\in\overline{B}(0,1)} |L(f)| = \sup_{f\in S(0,1)} |L(f)|.$$

Consider $C^+(X)$, the subset of $C(X)$ consisting of nonnegative continuous functions on X. As in Definition 19.26, a linear functional L on $C(X)$ is called a positive linear functional if $L(f) \geq 0$ for every $f \in C^+(X)$. We shall show that for every bounded linear functional L on $C(X)$ there exist two positive bounded linear functionals L^+ and L^- on $C(X)$ such that $L = L^+ - L^-$.

Observation 19.51. Let L be a positive bounded linear functional on the Banach space $\big(C(X), \|\cdot\|_u\big)$ of real-valued continuous functions on a compact Hausdorff space X. Then

(1) $$|L(f)| \le L\big(|f|\big) \quad \text{for every } f \in C(X),$$

(2) $$\|L\|_* = L(1).$$

Proof. For every $f \in C(X)$, we have $-f, |f| \in C(X)$ and $f, -f \le |f|$ so that $|f|-f \ge 0$ and $|f| + f \ge 0$, that is, $|f| - f, |f| + f \in C^+(X)$. Then since L is a positive linear functional on $C(X)$, we have $L\big(|f| - f\big) \ge 0$ and $L\big(|f| + f\big) \ge 0$, that is, $L\big(|f|\big) \ge L(f)$ and $L\big(|f|\big) \ge -L(f)$. Therefore $|L(f)| \le L\big(|f|\big)$. This proves (1).

Let $f \in S(0,1) = \{g \in C(X) : \|g\|_u = 1\}$. Then $\|f\|_u = 1$ so that $|f| \le 1$. Then $1 - |f| \ge 0$ so that $1 - |f| \in C^+(X)$. Then since L is a positive linear functional on $C(X)$, we have $L\big(1 - |f|\big) \ge 0$, that is, $L(1) \ge L\big(|f|\big)$. Thus recalling (1), we have

$$\|L\|_* = \sup_{f \in S(0,1)} |L(f)| \le \sup_{f \in S(0,1)} L\big(|f|\big) \le L(1).$$

On the other hand, since $1 \in S(0,1)$, we have $\|L\|_* \ge |L(f)| = L(1)$. Thus $\|L\|_* = L(1)$. This proves (2). ∎

Lemma 19.52. *Consider the Banach space $\big(C(X), \|\cdot\|_u\big)$ of the real-valued continuous functions on a compact Hausdorff space X and its dual space $\big(C(X)^*, \|\cdot\|_u\big)$. Let $C^+(X)$ be the subset of $C(X)$ consisting of nonnegative continuous functions on X. Corresponding to each $L \in \big(C(X)^*, \|\cdot\|_*\big)$ let us define a real-valued functional Λ_L on $C^+(X)$ by*

$$\Lambda_L(g) = \sup\big\{L(h) : h \in C^+(X), 0 \le h \le g\big\} \quad \textit{for } g \in C^+(X).$$

Then Λ_L has the following properties on $C^+(X)$:
(a) *$L(g) \le \Lambda_L(g)$ for every $g \in C^+(X)$.*
(b) *monotonicity: $\Lambda_L(g_1) \le \Lambda_L(g_2)$ for $g_1, g_2 \in C^+(X)$ such that $g_1 \le g_2$.*
(c) *nonnegativity and boundedness: $\Lambda_L(0) = 0$ and $0 \le \Lambda_L(g) \le \|L\|_* \|g\|_u < \infty$ for every $g \in C^+(X)$.*
(d) *positive homogeneity: $\Lambda_L(cg) = c\Lambda_L(g)$ for $g \in C^+(X)$ and $c \ge 0$.*
(e) *additivity: $\Lambda_L(g_1 + g_2) = \Lambda_L(g_1) + \Lambda_L(g_2)$ for $g_1, g_2 \in C^+(X)$.*

Proof. 1. (a) is immediate from the definition of Λ_L.

2. Let $g_1, g_2 \in C^+(X)$ and $g_1 \le g_2$. If $h \in C^+(X)$ and $0 \le h \le g_1$, then $0 \le h \le g_2$ also. Thus $\big\{h \in C^+(X) : 0 \le h \le g_1\big\} \subset \big\{h \in C^+(X) : 0 \le h \le g_2\big\}$ and consequently $\big\{L(h) : h \in C^+(X), 0 \le h \le g_1\big\} \subset \big\{L(h) : h \in C^+(X), 0 \le h \le g_2\big\}$. Thus we have

$$\begin{aligned}\Lambda_L(g_1) &= \sup\big\{L(h) : h \in C^+(X), 0 \le h \le g_1\big\}\\ &\le \sup\big\{L(h) : h \in C^+(X), 0 \le h \le g_2\big\}\\ &= \Lambda_L(g_2).\end{aligned}$$

3. If $h \in C^+(X)$ and $0 \le h \le 0$ then $h = 0$. Thus we have

$$\Lambda_L(0) = \sup\left\{L(h) : h \in C^+(X), 0 \le h \le 0\right\} = \sup\{L(0)\} = 0.$$

If $g \in C^+(X)$ then $0 \le g$ so that $0 = \Lambda_L(0) \le \Lambda_L(g)$ by (b). This shows that Λ_L is nonnegative on $C(X)$.

If $h \in C^+(X)$ and $0 \le h \le g$, then $|L(h)| \le \|L\|_* \|h\|_u \le \|L\|_* \|g\|_u$. Thus we have $L(h) \le \|L\|_* \|g\|_u$. Then

$$\Lambda_L(g) = \sup\left\{L(h) : h \in C^+(X), 0 \le h \le g\right\} \le \|L\|_* \|g\|_u < \infty.$$

This proves (c).

4. To prove (d), note that if $c = 0$ then for any $g \in C^+(X)$ we have $cg = 0$ so that $\Lambda_L(cg) = \Lambda_L(0) = 0 = 0 \cdot \Lambda_L(g) = c\Lambda_L(g)$ by (c). If $c > 0$ then for any $g \in C^+(X)$ we have $cg \in C^+(X)$ and

$$\begin{aligned}\Lambda_L(cg) &= \sup\left\{L(h) : h \in C^+(X), 0 \le h \le cg\right\}\\ &= \sup\left\{L(h) : h \in C^+(X), 0 \le \tfrac{1}{c}h \le g\right\}\\ &= \sup\left\{L(ch) : h \in C^+(X), 0 \le h \le g\right\}\\ &= c\Lambda_L(g).\end{aligned}$$

5. To prove (e), let $g_i \in C^+(X)$ and let $h_i \in C^+(X)$ such that $0 \le h_i \le g_i$ for $i = 1$ and 2. Then $g_1 + g_2 \in C^+(X)$, $h_1 + h_2 \in C^+(X)$ and $0 \le h_1 + h_2 \le g_1 + g_2$. Now

$$\begin{aligned}\Lambda_L(g_1 + g_2) &= \sup\left\{L(h) : h \in C^+(X), 0 \le h \le g_1 + g_2\right\}\\ &\ge L(h_1 + h_2) = L(h_1) + L(h_2).\end{aligned}$$

Since this holds for every $h_1 \in C^+(X)$ such that $0 \le h_1 \le g_1$, we have

$$\begin{aligned}\Lambda_L(g_1 + g_2) &\ge \sup\left\{L(h_1) : h_1 \in C^+(X), 0 \le h_1 \le g_1\right\} + L(h_2)\\ &= \Lambda_L(g_1) + L(h_2).\end{aligned}$$

Then since this holds for every $h_2 \in C^+(X)$ such that $0 \le h_2 \le g_2$, we have

$$\Lambda_L(g_1 + g_2) \ge \Lambda_L(g_1) + \Lambda_L(g_2). \tag{1}$$

Let us prove the reverse inequality to (1). Let $h \in C^+(X)$ be such that $0 \le h \le g_1 + g_2$. Then $h \wedge g_1 \in C^+(X)$ and $0 \le h \wedge g_1 \le g_1$. Now at each $x \in X$ we have either $h \wedge g_1 = h$ or $h \wedge g_1 = g_1$. From this fact and from the fact that $h \le g_1 + g_2$ and $g_2 \ge 0$, we have $h \le (h \wedge g_1) + g_2$. Then $0 \le h - (h \wedge g_1) \le g_2$. Now $h = (h \wedge g_1) + \left\{h - (h \wedge g_1)\right\}$. Thus we have

$$\begin{aligned}L(h) &= L(h \wedge g_1) + L\big(h - (h \wedge g_1)\big)\\ &\le \Lambda_L(h \wedge g_1) + \Lambda_L\big(h - (h \wedge g_1)\big)\\ &\le \Lambda_L(g_1) + \Lambda_L(g_2),\end{aligned}$$

where the first inequality is by (a) and the second inequality is by (b). Since this holds for every $h \in C^+(X)$ such that $0 \le h \le g_1 + g_2$, we have

(2) $$\Lambda_L(g_1 + g_2) \le \Lambda_L(g_1) + \Lambda_L(g_2).$$

With (1) and (2), (e) is proved. ∎

Proposition 19.53. *Consider the Banach space $(C(X), \|\cdot\|_u)$ of the real-valued continuous functions on a compact Hausdorff space X and its dual space $(C(X)^*, \|\cdot\|_u)$. For every $L \in (C(X)^*, \|\cdot\|_*)$ there exist two positive bounded linear functionals L^+ and L^- on $C(X)$ such that*

(1) $$L = L^+ - L^-,$$

(2) $$\|L\|_* = L^+(1) + L^-(1).$$

Proof. 1. Let $L \in (C(X)^*, \|\cdot\|_*)$ and let Λ_L be the functional on $C^+(X)$ as defined in Lemma 19.52. Let us define a functional L^+ on $C(X)$ as follows. If $f \in C(X)$, then f is a bounded real-valued continuous function on X so that there exists $\gamma \ge 0$ such that $f + \gamma \ge 0$ and $f + \gamma \in C^+(X)$. Since $f + \gamma, \gamma \in C^+(X)$, $\Lambda_L(f + \gamma)$ and $\Lambda_L(\gamma)$ are defined. Let us define a functional L^+ on $C(X)$ by setting

(3) $$L^+(f) = \Lambda_L(f + \gamma) - \Lambda_L(\gamma) \quad \text{for } f \in C(X).$$

Let us show that L^+ is well defined, that is, $L^+(f)$ defined by (3) does not depend on the nonnegative number γ such that $f + \gamma \ge 0$. Let γ_1 and γ_2 be two nonnegative numbers such that $f + \gamma_1 \ge 0$ and $f + \gamma_2 \ge 0$. Then $f + \gamma_1 + \gamma_2 \ge 0$ so that

$$\Lambda_L(f + \gamma_1 + \gamma_2) = \Lambda_L(f + \gamma_1) + \Lambda_L(\gamma_2),$$

$$\Lambda_L(f + \gamma_1 + \gamma_2) = \Lambda_L(f + \gamma_2) + \Lambda_L(\gamma_1),$$

by (e) of Lemma 19.52. From these two equalities we have

$$\Lambda_L(f + \gamma_1) - \Lambda_L(\gamma_1) = \Lambda_L(f + \gamma_2) - \Lambda_L(\gamma_2).$$

This shows the independence of the definition of $L^+(f)$ by (3) from $\gamma \ge 0$.

If $f \in C(X)$, then $|f(x)| \le \|f\|_u$ for every $x \in X$ so that $f(x) + \|f\|_u \ge 0$ for every $x \in X$, that is, $f + \|f\|_u \ge 0$. Using the nonnegative number $\|f\|_u$ as γ in the expression (3) for $L^+(f)$, we have

(4) $$L^+(f) = \Lambda_L(f + \|f\|_u) - \Lambda_L(\|f\|_u) \quad \text{for } f \in C(X).$$

If $g \in C^+(X)$, then $g + 0 \ge 0$. Then $L^+(g) = \Lambda_L(g + 0) - \Lambda_L(0) = \Lambda_L(g)$ by (3) and by the fact that $\Lambda_L(0) = 0$ by (c) of Lemma 19.52. Thus we have

(5) $$L^+ = \Lambda_L \quad \text{on } C^+(X),$$

that is, L^+ is an extension of Λ_L from $C^+(X)$ to $C(X)$.

2. Let us show that L^+ is a positive linear functional on $C(X)$, that is,

$$L^+(f_1 + f_2) = L^+(f_1) + L^+(f_2) \quad \text{for } f_1, f_2 \in C(X), \tag{6}$$

$$L^+(cf) = cL^+(f) \quad \text{for } f \in C(X) \text{ and } c \in \mathbb{R}, \tag{7}$$

$$L^+(g) \geq 0 \quad \text{for } g \in C^+(X). \tag{8}$$

To prove (6), let $f_1, f_2 \in C(X)$ and let $\gamma_1, \gamma_2 \geq 0$ be such that $f_1 + \gamma_1 \geq 0$ and $f_2 + \gamma_2 \geq 0$. Then $f_1 + f_2 + \gamma_1 + \gamma_2 \geq 0$ so that

$$\begin{aligned} L^+(f_1 + f_2) &= \Lambda_L(f_1 + f_2 + \gamma_1 + \gamma_2) - \Lambda_L(\gamma_1 + \gamma_2) \\ &= \Lambda_L(f_1 + \gamma_1) + \Lambda_L(f_2 + \gamma_2) - \Lambda_L(\gamma_1) - \Lambda_L(\gamma_2) \\ &= L^+(f_1) + L^+(f_2), \end{aligned}$$

where the second equality is by (e) of Lemma 19.52. This proves (6).

To prove (7), consider first the case $c \geq 0$. Let $f \in C(X)$ and let $\gamma \geq 0$ be such that $f + \gamma \geq 0$. Then $cf + c\gamma \geq 0$ so that

$$\begin{aligned} L^+(cf) &= \Lambda_L(cf + c\gamma) - \Lambda_L(c\gamma) \\ &= c\{\Lambda_L(f + \gamma) - \Lambda_L(\gamma)\} \\ &= cL^+(f), \end{aligned} \tag{9}$$

where the second equality is by (d) of Lemma 19.52. Next for the case $c = -1$, we have $L^+(-f) + L^+(f) = L^+(-f + f) = L^+(0) = \Lambda_L(0) = 0$ by (6), (5) and (c) of Lemma 19.52. Thus we have

$$L^+(-f) = -L^+(f). \tag{10}$$

Finally for $c < 0$, we have

$$L^+(cf) = L^+(-|c|f) = -L^+(|c|f) = -|c|L^+(f) = cL^+(f), \tag{11}$$

where the second equality is by (10) and the third equality is by (9). This proves (7). For $g \in C^+(X)$, we have $L^+(g) = \Lambda_L(g) \geq 0$ by (5) and (c) of Lemma 19.52. This proves (8).

3. Let us show that the linear function L^+ on $C(X)$ is a bounded linear functional. By (4) and by the fact that Λ_L is nonnegative valued, we have

$$-\Lambda_L(\|f\|_u) \leq L^+(f) \leq \Lambda_L(f + \|f\|_u) \quad \text{for } f \in C(X).$$

According to (c) of Lemma 19.52 we have $\Lambda_L(g) \leq \|L\|_* \|g\|_u$ for every $g \in C^+(X)$. Thus

$$-\|L\|_* \|f\|_u \leq L^+(f) \leq \|L\|_* \big\| f + \|f\|_u \big\|_u \quad \text{for } f \in C(X).$$

Now $\big\| f + \|f\|_u \big\|_u \leq \|f\|_u + \|f\|_u = 2\|f\|_u$. Thus we have

$$|L^+(f)| \leq 2\|L\|_* \|f\|_u \quad \text{for } f \in C(X).$$

This shows that L^+ is a bounded linear functional on $C(X)$ with $\|L^+\|_* \leq 2\|L\|_*$.

4. Let $L^- = L^+ - L$. Since L and L^+ are bounded linear functionals on $C(X)$, so is L^-. For $g \in C^+(X)$, we have $L^+(g) = \Lambda_L(g) \geq L(g)$ by (5) and by (a) of Lemma 19.52. Thus $L^-(g) = L^+(g) - L(g) \geq 0$. This shows that L^- is a positive linear functional.

We have $L = L^+ - L^-$. Then by Observation 19.51 we have

$$(12) \qquad \|L\|_* = \|L^+ - L^-\|_* \leq \|L^+\|_* + \|L^-\|_* = L^+(1) + L^-(1).$$

To prove the reverse inequality, let $h \in C^+(X)$ be such that $0 \leq h \leq 1$. Then $|2h - 1| \leq 1$ so that $2h - 1 \in \overline{B}(0,1) = \{f \in C(X) : \|f\|_u \leq 1\}$. Thus

$$(13) \qquad \|L\|_* = \sup_{f \in \overline{B}(0,1)} |L(f)| \geq L(2h-1) = 2L(h) - L(1).$$

By the definition of Λ_L in Lemma 19.52 and by (5) above, we have

$$(14) \qquad \sup\big\{L(h) : h \in C^+(X), 0 \leq h \leq 1\big\} = \Lambda_L(1) = L^+(1).$$

Since (13) holds for every $h \in C^+(X)$ such that $0 \leq h \leq 1$, (14) implies that we have

$$(15) \qquad \|L\|_* \geq 2L^+(1) - L(1) = 2L^+(1) - \{L^+(1) - L^-(1)\} = L^+(1) + L^-(1).$$

With (12) and (15), we have $\|L\|_* = L^+(1) + L^-(1)$. ■

Theorem 19.54. *Consider the Banach space $\big(C(X), \|\cdot\|_u\big)$ of the real-valued continuous functions on a compact Hausdorff space X and its dual space $\big(C(X)^*, \|\cdot\|_u\big)$. Then for every $L \in \big(C(X)^*, \|\cdot\|_*\big)$ there exists a unique finite signed Radon measure λ on $(X, \mathfrak{B}_X)$ such that*

$$L(f) = \int_X f\, d\lambda \quad \textit{for every } f \in C(X).$$

Moreover $\|L\|_* = |\lambda|(X)$ where $|\lambda|$ is the total variation of the signed measure λ.

Proof. 1. By Proposition 19.53, for every $L \in \big(C(X)^*, \|\cdot\|_*\big)$ there exist two positive bounded linear functionals L^+ and L^- on $C(X)$ such that

$$L = L^+ - L^- \quad \text{and} \quad \|L\|_* = L^+(1) + L^-(1).$$

By Theorem 19.34 (Riesz-Markoff), there exist unique Radon measures μ and ν on $(X, \mathfrak{B}_X)$ such that

$$L^+(f) = \int_X f\, d\mu \quad \text{for } f \in C(X),$$

$$L^-(f) = \int_X f\, d\nu \quad \text{for } f \in C(X).$$

Since μ and ν are Radon measures on $(X, \mathfrak{B}_X)$, they are finite on the compact sets in X and hence finite on the compact space X. Thus μ and ν are finite Radon measures on $(X, \mathfrak{B}_X)$. If we let $\lambda = \mu - \nu$, then λ is a finite signed Radon measure on $(X, \mathfrak{B}_X)$ by Proposition 19.44. Thus we have

$$L(f) = L^+(f) - L^-(f) = \int_X f\,d\mu - \int_X f\,d\nu = \int_X f\,d\lambda \quad \text{for } f \in C(X),$$

where the last equality is by Proposition 10.34.

Let $|\lambda|$ be the total variation of the signed measure λ. By Proposition 10.33, we have

$$|L(f)| = \left|\int_X f\,d\lambda\right| \le \int_X |f|\,d|\lambda| \le \|f\|_u |\lambda|(X) \quad \text{for } f \in C(X).$$

Thus we have $\|L\|_* \le |\lambda|(X)$. On the other hand, by Theorem 11.25 we have

$$\begin{aligned} |\lambda|(X) &\le \mu(X) + \nu(X) = \int_X 1\,d\mu + \int_X 1\,d\nu \\ &= L^+(1) + L^-(1) = \|L\|_*. \end{aligned}$$

Therefore we have $\|L\|_* = |\lambda|(X)$.

2. Suppose λ_1 and λ_2 are two finite signed Radon measures on $(X, \mathfrak{B}_X)$ such that

$$L(f) = \int_X f\,d\lambda_i \quad \text{for } f \in C(X) \text{ and } i = 1, 2.$$

If we let $\lambda = \lambda_1 - \lambda_2$, then λ is a finite signed Radon measure on $(X, \mathfrak{B}_X)$ by Proposition 19.36. Let $\lambda = \lambda^+ - \lambda^-$ be the Jordan decomposition of λ. By Definition 19.32, λ^+ and λ^- are finite Radon measures on $(X, \mathfrak{B}_X)$. If we define

$$\begin{aligned} I^+(f) &= \int_X f\,d\lambda^+ \quad \text{for } f \in C(X), \\ I^-(f) &= \int_X f\,d\lambda^- \quad \text{for } f \in C(X), \end{aligned}$$

then I^+ and I^- are positive bounded linear functionals on $C(X)$. For $f \in C(X)$, we have

$$\begin{aligned} I^+(f) - I^-(f) &= \int_X f\,d\{\lambda^+ - \lambda^-\} = \int_X f\,d\lambda \\ &= \int_X f\,d\lambda_1 - \int_X f\,d\lambda_2 \\ &= L(f) - L(f) = 0, \end{aligned}$$

where the first equality is by Proposition 10.34 and the third equality is by Theorem 10.35. Thus $I^+ = I^-$. Then we have $I^+(f) = \int_X f\,d\lambda^+$ and $I^+(f) = I^-(f) = \int_X f\,d\lambda^-$ for $f \in C(X)$. Then by the uniqueness of the Radon measure in Theorem 19.34, we have $\lambda^+ = \lambda^-$. Then $\lambda = \lambda^+ - \lambda^- = 0$, that is, $\lambda_1 = \lambda_2$. This proves the uniqueness of the finite signed Radon measure λ such that $L(f) = \int_X f\,d\lambda$ for $f \in C(X)$. ∎

Let $\mathcal{M}(X, \mathfrak{B}_X)$ be the linear space of all finite signed Radon measures on a compact Hausdorff space X. For $\lambda \in \mathcal{M}(X, \mathfrak{B}_X)$, let $|\lambda|$ be the total variation of the signed measure λ. We showed in Theorem 19.47 that $\|\lambda\|_{\mathcal{M}} = |\lambda|(X)$ for $\lambda \in \mathcal{M}(X, \mathfrak{B}_X)$ is a norm on the linear space $\mathcal{M}(X, \mathfrak{B}_X)$.

Theorem 19.55. *Consider the dual space $(C(X)^*, \|\cdot\|_*)$ of the Banach space $(C(X), \|\cdot\|_u)$ of the real-valued continuous functions on a compact Hausdorff space X, and the normed linear space $(\mathcal{M}(X, \mathfrak{B}_X), \|\cdot\|_{\mathcal{M}})$ of finite signed Radon measures on X. For $\lambda \in \mathcal{M}(X, \mathfrak{B}_X)$ let us define a bounded linear functional $I(\lambda, \cdot)$ on $C(X)$ by setting*

(1) $$I(\lambda, f) = \int_X f\, d\lambda \quad \text{for } f \in C(X).$$

Then I is a linear isometry of $\mathcal{M}(X, \mathfrak{B}_X)$ onto $C(X)^*$, that is,

(2) I *maps* $\mathcal{M}(X, \mathfrak{B}_X)$ *onto* $C(X)^*$,

(3) $I(c\lambda, \cdot) = cI(\lambda, \cdot)$ *for* $\lambda \in \mathcal{M}(X, \mathfrak{B}_X)$ *and* $c \in \mathbb{R}$,

(4) $I(\lambda_1 + \lambda_2, \cdot) = I(\lambda_1, \cdot) + I(\lambda_2, \cdot)$ *for* $\lambda_1, \lambda_2 \in \mathcal{M}(X, \mathfrak{B}_X)$,

(5) $\|I(\lambda, \cdot)\|_* = \|\lambda\|_{\mathcal{M}}$.

Proof. For every $\lambda \in \mathcal{M}(X, \mathfrak{B}_X)$, the functional $I(\lambda, \cdot)$ defined by (1) is a real-valued linear functional on $C(X)$. Moreover for every $f \in C(X)$ we have

$$\begin{aligned} |I(\lambda, f)| &= \left|\int_X f\, d\lambda\right| \le \int_X |f|\, d|\lambda| \\ &\le \|f\|_u |\lambda| = \|f\|_u \|\lambda\|_{\mathcal{M}}, \end{aligned}$$

where the first inequality is by Proposition 10.33. Thus $I(\lambda, \cdot)$ is a bounded linear functional on $C(X)$, that is, $I(\lambda, \cdot) \in C(X)^*$. This shows that I maps $\mathcal{M}(X, \mathfrak{B}_X)$ into $C(X)^*$. According to Theorem 19.54, for every $L \in C(X)^*$ there exists $\lambda \in \mathcal{M}(X, \mathfrak{B}_X)$ such that $L(f) = \int_X f\, d\lambda$ for every $f \in C(X)$ and for such $\lambda \in \mathcal{M}(X, \mathfrak{B}_X)$ we have the equality $\|L\|_* = |\lambda|(X) = \|\lambda\|_{\mathcal{M}}$. This shows that I defined by (1) maps $\mathcal{M}(X, \mathfrak{B}_X)$ onto $C(X)^*$ and proves (5) also.

To prove (3), note that for $\lambda \in \mathcal{M}(X, \mathfrak{B}_X)$, $c \in \mathbb{R}$, and $f \in C(X)$, we have

$$I(c\lambda, f) = \int_X f\, d(c\lambda) = c\int_X f\, d\lambda = cI(\lambda, f),$$

where the second equality is by Theorem 10.35.

To prove (4), note that for $\lambda_1, \lambda_2 \in \mathcal{M}(X, \mathfrak{B}_X)$ and $f \in C(X)$ we have

$$\begin{aligned} I(\lambda_1 + \lambda_2, f) &= \int_X f\, d\{\lambda_1 + \lambda_2\} = \int_X f\, d\lambda_1 + \int_X f\, d\lambda_2 \\ &= I(\lambda_1, f) + I(\lambda_2, f), \end{aligned}$$

where the second equality is by Theorem 10.36. ∎

Chapter 5

Extension of Additive Set Functions to Measures

§20 Extension of Additive Set Functions on an Algebra

[I] Additive Set Function on an Algebra

An outer measure on a set X is a nonnegative extended real-valued set function μ^* defined on the σ-algebra $\mathfrak{P}(X)$ of all subsets of X such that $\mu^*(\emptyset) = 0$, μ^* is monotone and countably subadditive on $\mathfrak{P}(X)$. (See Definition 2.1.) Let $\mathfrak{M}(\mu^*)$ be the collection of all μ^*-measurable members of $\mathfrak{P}(X)$, that is, the collection of every $E \in \mathfrak{P}(X)$ satisfying the μ^*-measurability condition

$$\mu^*(A) = \mu^*(A \cap E) + \mu^*(A \cap E^c) \quad \text{for every } A \in \mathfrak{P}(X).$$

Then $\mathfrak{M}(\mu^*)$ is a σ-algebra of subsets of X and $(X, \mathfrak{M}(\mu^*), \mu^*)$ is a complete measure space. (See Theorem 2.8 and Theorem 2.9.)

A collection $\mathfrak{V} = \{V_\alpha\}$ of subsets of a set X is called a covering class for X if $\emptyset \in \mathfrak{V}$ and there exists a sequence $(V_n : n \in \mathbb{N})$ in $\mathfrak{V}$ such that $\bigcup_{n\in\mathbb{N}} V_n = X$. Let γ be an arbitrary nonnegative extended real-valued set function on a covering class $\mathfrak{V}$ for X such that $\gamma(\emptyset) = 0$. If we define a set function μ^* on $\mathfrak{P}(X)$ by setting

$$\mu^*(E) = \inf\left\{ \textstyle\sum_{n\in\mathbb{N}} \gamma(V_n) : (V_n : n \in \mathbb{N}) \subset \mathfrak{V}, \bigcup_{n\in\mathbb{N}} V_n \supset E \right\}$$

for $E \in \mathfrak{P}(X)$, then μ^* is an outer measure on X. (See Theorem 2.21.) We refer to μ^* as the outer measure based on the set function γ.

Note that aside from the condition that $\gamma(\emptyset) = 0$, γ is an arbitrary nonnegative extended real-valued set function on $\mathfrak{V}$. For instance, γ need not be monotone on $\mathfrak{V}$. However, if there exist V_1 and V_2 in $\mathfrak{V}$ such that $V_1 \subset V_2$ and $\gamma(V_1) > \gamma(V_2)$, the omission of such a set V_1 from $\mathfrak{V}$ has no effect on the value $\mu^*(E)$ for any $E \in \mathfrak{P}(X)$ as we showed in Remark 2.22. Note also that while $\mu^*(V) \le \gamma(V)$ for every $V \in \mathfrak{V}$, the equality $\mu^*(V) = \gamma(V)$ does not hold in general. Also the inclusion $\mathfrak{V} \subset \mathfrak{M}(\mu^*)$ does not hold in general. Below is an example for these failures.

Example. The collection $\mathfrak{I}_o$ consisting of $\emptyset$ and all open intervals in $\mathbb{R}$ is a covering class for $\mathbb{R}$. The set function ℓ on $\mathfrak{I}_o$ defined to be the length of an interval in $\mathfrak{I}_o$ and $\ell(\emptyset) = 0$ is a nonnegative extended real-valued function on the covering class $\mathfrak{I}_o$. The outer measure based on this set function ℓ on $\mathfrak{I}_o$ is the Lebesgue outer measure μ_L^* on $\mathbb{R}$ and the σ-algebra $\mathfrak{M}(\mu^*)$ of all μ_L^*-measurable sets is the Lebesgue σ-algebra $\mathfrak{M}_L$. Now let E_0 be a non $\mathfrak{M}_L$-measurable subset of $(0, 1)$ in $\mathbb{R}$. Let $\mathfrak{V}$ be a covering class of $\mathbb{R}$ consisting of all members of $\mathfrak{I}_o$ and E_0. Let γ be a set function on $\mathfrak{V}$ defined by setting $\gamma(I) = \ell(I)$ for every $I \in \mathfrak{I}_0$ and $\gamma(E_0) = 3$. Then for $(0, 2) \in \mathfrak{V}$, we have $E_0 \subset (0, 2)$ and $\gamma(E_0) = 3 > 2 = \gamma((0, 2))$. Let μ^* be the outer measure on $\mathbb{R}$ based on the set function γ on $\mathfrak{V}$. Since the omission of E_0 from $\mathfrak{V}$ has no effect on μ^*, we have $\mu^* = \mu_L^*$ and $\mathfrak{M}(\mu^*) = \mathfrak{M}(\mu_L^*) = \mathfrak{M}_L$. Then since $E_0 \in \mathfrak{V}$ and $E_0 \notin \mathfrak{M}_L$, we have $\mathfrak{V} \not\subset \mathfrak{M}_L$.

We shall show that if a covering class $\mathfrak{V}$ for a set X is an algebra of subsets of X and if a nonnegative extended real-valued set function γ on $\mathfrak{V}$ with $\gamma(\emptyset) = 0$ is additive on $\mathfrak{V}$, then $\mathfrak{V} \subset \mathfrak{M}(\mu^*)$ and if γ is countably additive on $\mathfrak{V}$, then $\mu^* = \gamma$ on $\mathfrak{V}$. Regarding the assumption that $\gamma(\emptyset) = 0$, let us remark that if there exists $V \in \mathfrak{V}$ with $\gamma(V) < \infty$, then the additivity of γ on $\mathfrak{V}$ implies that $\gamma(\emptyset) = 0$. Indeed since $V = V \cup \emptyset$, $\gamma(V) = \gamma(V \cup \emptyset) = \gamma(V) + \gamma(\emptyset)$. Subtracting $\gamma(V)$ from both sides, we have $0 = \gamma(\emptyset)$.

Theorem 20.1. *Let μ be a nonnegative extended real-valued set function on an algebra $\mathfrak{A}$ of subsets of a set X with $\mu(\emptyset) = 0$. Consider the outer measure μ^* on X based on μ, that is,*

$$\mu^*(E) = \inf\left\{ \sum_{n\in\mathbb{N}} \mu(A_n) : (A_n : n \in \mathbb{N}) \subset \mathfrak{A}, \bigcup_{n\in\mathbb{N}} A_n \supset E \right\}$$

for $E \in \mathfrak{P}(X)$ and the measure space $(X, \mathfrak{M}(\mu^*), \mu^*)$.
(a) *If μ is additive on $\mathfrak{A}$, then $\mathfrak{A} \subset \mathfrak{M}(\mu^*)$.*
(b) *If μ is countably additive on $\mathfrak{A}$, then $\mu^* = \mu$ on $\mathfrak{A}$.*

Proof. 1. Suppose μ is additive on $\mathfrak{A}$. To show that $\mathfrak{A} \subset \mathfrak{M}(\mu^*)$, we show that every $A \in \mathfrak{A}$ satisfies the μ^*-measurability condition

$$(1) \qquad \mu^*(B) = \mu^*(B \cap A) + \mu^*(B \cap A^c) \quad \text{for every } B \in \mathfrak{P}(X).$$

From the definition of μ^* as an infimum, for every $B \in \mathfrak{P}(X)$ and $\varepsilon > 0$, there exists a sequence $(A_n : n \in \mathbb{N})$ in $\mathfrak{A}$ such that $B \subset \bigcup_{n\in\mathbb{N}} A_n$ and

$$(2) \qquad \sum_{n\in\mathbb{N}} \mu(A_n) \le \mu^*(B) + \varepsilon.$$

By the additivity of μ on $\mathfrak{A}$, we have

$$(3) \qquad \mu(A_n) = \mu(A_n \cap A) + \mu(A_n \cap A^c).$$

Using (3) in (2), we have

$$(4) \qquad \sum_{n\in\mathbb{N}} \mu(A_n \cap A) + \sum_{n\in\mathbb{N}} \mu(A_n \cap A^c) \le \mu^*(B) + \varepsilon.$$

By the inclusion $B \subset \bigcup_{n\in\mathbb{N}} A_n$, we have $B \cap A \subset \left(\bigcup_{n\in\mathbb{N}} A_n\right) \cap A = \bigcup_{n\in\mathbb{N}}(A_n \cap A)$ and $B \cap A^c \subset \left(\bigcup_{n\in\mathbb{N}} A_n\right) \cap A^c = \bigcup_{n\in\mathbb{N}}(A_n \cap A^c)$. Now $(A_n \cap A : n \in \mathbb{N})$ and $(A_n \cap A^c : n \in \mathbb{N})$ are sequences in $\mathfrak{A}$. The definition of μ^* implies $\mu^*(B \cap A) \le \sum_{n\in\mathbb{N}} \mu(A_n \cap A)$ and $\mu^*(B \cap A^c) \le \sum_{n\in\mathbb{N}} \mu(A_n \cap A^c)$. Using these two inequalities in (4), we have

$$\mu^*(B \cap A) + \mu^*(B \cap A^c) \le \mu^*(B) + \varepsilon.$$

By the arbitrariness of $\varepsilon > 0$, we have $\mu^*(B \cap A) + \mu^*(B \cap A^c) \le \mu^*(B)$. The reverse inequality holds by the subadditivity of the outer measure μ^*. Thus (1) holds.

2. Suppose μ is countably additive on $\mathfrak{A}$. We are to show that $\mu^*(A) = \mu(A)$ for every $A \in \mathfrak{A}$. For this, let us show first that

$$(5) \qquad A \in \mathfrak{A}, (A_n : n \in \mathbb{N}) \subset \mathfrak{A}, \bigcup_{n\in\mathbb{N}} A_n \supset A \Rightarrow \mu(A) \le \sum_{n\in\mathbb{N}} \mu(A_n).$$

(Note that $\mathfrak{A}$ is only an algebra so that $\bigcup_{n\in\mathbb{N}} A_n$ may not be in $\mathfrak{A}$ and $\mu\left(\bigcup_{n\in\mathbb{N}} A_n\right)$ may not be defined.) To prove (5), let $B_1 = A_1$ and $B_n = A_n \setminus (A_1 \cup \cdots \cup A_{n-1})$ for $n \ge 2$ so that $(B_n : n \in \mathbb{N})$ is a disjoint sequence in $\mathfrak{A}$ with $\bigcup_{n\in\mathbb{N}} B_n = \bigcup_{n\in\mathbb{N}} A_n$. From $A \subset \bigcup_{n\in\mathbb{N}} A_n$, we have $A = A \cap \left(\bigcup_{n\in\mathbb{N}} A_n\right) = A \cap \left(\bigcup_{n\in\mathbb{N}} B_n\right) = \bigcup_{n\in\mathbb{N}}(A \cap B_n)$. Since $(A \cap B_n : n \in \mathbb{N})$ is a disjoint sequence in $\mathfrak{A}$ and since $\bigcup_{n\in\mathbb{N}}(A \cap B_n) = A$, the countable additivity and the monotonicity of μ on $\mathfrak{A}$ imply $\mu(A) = \sum_{n\in\mathbb{N}} \mu(A \cap B_n) \le \sum_{n\in\mathbb{N}} \mu(B_n) \le \sum_{n\in\mathbb{N}} \mu(A_n)$. This proves (5). Now let $A \in \mathfrak{A}$. Then (5) implies by the definition of μ^* that $\mu(A) \le \mu^*(A)$. On the other hand by considering the covering of A by the sequence $(A, \emptyset, \emptyset, \ldots)$ in $\mathfrak{A}$, we have $\mu^*(A) \le \mu(A)$. Thus $\mu^*(A) = \mu(A)$. ∎

The next theorem lists equivalent conditions for countable additivity of an additive set function on an algebra of subsets of a set. Among these conditions (i) is the countable additivity condition.

Theorem 20.2. *Let μ be a nonnegative extended real-valued additive set function on an algebra $\mathfrak{A}$ of subsets of a set X with $\mu(\emptyset) = 0$. Consider the following conditions on μ:*

(i) $\quad (A_n : n \in \mathbb{N}) \subset \mathfrak{A}$, *disjoint*, $\bigcup_{n\in\mathbb{N}} A_n \in \mathfrak{A} \Rightarrow \mu\left(\bigcup_{n\in\mathbb{N}} A_n\right) = \sum_{n\in\mathbb{N}} \mu(A_n)$,

(ii) $\quad (A_n : n \in \mathbb{N}) \subset \mathfrak{A}, A_n \uparrow, \lim_{n\to\infty} A_n \in \mathfrak{A} \Rightarrow \lim_{n\to\infty} \mu(A_n) = \mu\left(\lim_{n\to\infty} A_n\right)$,

(iii) $\quad (A_n : n \in \mathbb{N}) \subset \mathfrak{A}, A_n \downarrow, \lim_{n\to\infty} A_n \in \mathfrak{A} \Rightarrow \lim_{n\to\infty} \mu(A_n) = \mu\left(\lim_{n\to\infty} A_n\right)$,

(iv) $\quad (A_n : n \in \mathbb{N}) \subset \mathfrak{A}, A_n \downarrow, \lim_{n\to\infty} A_n = \emptyset \Rightarrow \lim_{n\to\infty} \mu(A_n) = 0$.

Then (i) $\Leftrightarrow$ (ii), (iii) $\Leftrightarrow$ (iv), and if $\mu(X) < \infty$, then (ii) $\Leftrightarrow$ (iii).

Proof. 1. (i) $\Rightarrow$ (ii). Assume (i). Suppose $A_n \uparrow$ and $\lim_{n\to\infty} A_n = \bigcup_{n\in\mathbb{N}} A_n \in \mathfrak{A}$. Let $B_1 = A_1$ and $B_n = A_n \setminus (A_1 \cup \cdots \cup A_{n-1})$ for $n \ge 2$ so that $(B_n : n \in \mathbb{N})$ is a disjoint sequence in $\mathfrak{A}$, $\bigcup_{k=1}^{n} B_k = \bigcup_{k=1}^{n} A_k$ for $n \in \mathbb{N}$ and $\bigcup_{n\in\mathbb{N}} B_n = \bigcup_{n\in\mathbb{N}} A_n$. Then by (i) and by the additivity of μ on $\mathfrak{A}$, we have

$$\begin{aligned}\mu\big(\lim_{n\to\infty} A_n\big) &= \mu\Big(\bigcup_{n\in\mathbb{N}} A_n\Big) = \mu\Big(\bigcup_{n\in\mathbb{N}} B_n\Big) = \sum_{n\in\mathbb{N}} \mu(B_n)\\ &= \lim_{n\to\infty} \sum_{k=1}^{n} \mu(B_k) = \lim_{n\to\infty} \mu\Big(\bigcup_{k=1}^{n} B_k\Big)\\ &= \lim_{n\to\infty} \mu\Big(\bigcup_{k=1}^{n} A_k\Big) = \lim_{n\to\infty} \mu(A_n).\end{aligned}$$

2. (ii) $\Rightarrow$ (i). Assume (ii). Let $(A_n : n \in \mathbb{N})$ be a disjoint sequence in $\mathfrak{A}$ such that $\bigcup_{n\in\mathbb{N}} A_n \in \mathfrak{A}$. Let $B_n = \bigcup_{k=1}^{n} A_k$ for $n \in \mathbb{N}$. Then $(B_n : n \in \mathbb{N}) \subset \mathfrak{A}$, $B_n \uparrow$, and $\lim_{n\to\infty} B_n = \bigcup_{n\in\mathbb{N}} B_n = \bigcup_{n\in\mathbb{N}} A_n$. Thus by (ii) and by the additivity of μ on $\mathfrak{A}$, we have

$$\begin{aligned}\mu\big(\textstyle\bigcup_{n\in\mathbb{N}} A_n\big) &= \mu\big(\lim_{n\to\infty} B_n\big) = \lim_{n\to\infty} \mu(B_n)\\ &= \lim_{n\to\infty} \sum_{k=1}^{n} \mu(A_k) = \sum_{n\in\mathbb{N}} \mu(A_n).\end{aligned}$$

3. (iii) $\Rightarrow$ (iv). (iv) is a particular case of (iii).

4. (iv) $\Rightarrow$ (iii). Assume (iv). Suppose $A_n \downarrow$ and $\lim_{n\to\infty} A_n = \bigcap_{n\in\mathbb{N}} A_n \in \mathfrak{A}$. Let $A = \bigcap_{n\in\mathbb{N}} A_n \in \mathfrak{A}$. Then $(A_n \setminus A : n \in \mathbb{N}) \subset \mathfrak{A}$, $A_n \setminus A \downarrow$, and in fact we have $\lim_{n\to\infty} (A_n \setminus A) = \bigcap_{n\in\mathbb{N}}(A_n \setminus A) = \big(\bigcap_{n\in\mathbb{N}} A_n\big) \setminus A = \emptyset$. Thus by (iv), $\lim_{n\to\infty} \mu(A_n \setminus A) = 0$. Since $\mu(A_n \setminus A) + \mu(A) = \mu(A_n)$, by letting $n \to \infty$, we have $\mu(A) = \lim_{n\to\infty} \mu(A_n)$. This proves (iii).

5. (ii) $\Rightarrow$ (iii) when $\mu(X) < \infty$. Assume (ii) and $\mu(X) < \infty$. Suppose $A_n \downarrow$ and $\lim_{n\to\infty} A_n = \bigcap_{n\in\mathbb{N}} A_n \in \mathfrak{A}$. Then $(A_n^c : n \in \mathbb{N}) \subset \mathfrak{A}$, $A_n^c \uparrow$, and $\lim_{n\to\infty} A_n^c = \bigcup_{n\in\mathbb{N}} A_n^c = \big(\bigcap_{n\in\mathbb{N}} A_n\big)^c \in \mathfrak{A}$ so that by (ii), we have

$$(1)\qquad \lim_{n\to\infty} \mu(A_n^c) = \mu\Big(\big(\textstyle\bigcap_{n\in\mathbb{N}} A_n\big)^c\Big).$$

Now $\mu\Big(\big(\bigcap_{n\in\mathbb{N}} A_n\big)^c\Big) + \mu\big(\bigcap_{n\in\mathbb{N}} A_n\big) = \mu(X)$ and $\mu(A_n^c) + \mu(A_n) = \mu(X)$. Since $\mu(X) < \infty$, the measures of all sets involved are finite. This implies that subtraction by the measure of a $\mathfrak{A}$-measurable subset of X is always possible. Thus we have

$$(2)\qquad \begin{cases} \mu\Big(\big(\bigcap_{n\in\mathbb{N}} A_n\big)^c\Big) = \mu(X) - \mu\big(\bigcap_{n\in\mathbb{N}} A_n\big),\\ \mu(A_n^c) = \mu(X) - \mu(A_n) \quad \text{for } n \in \mathbb{N}.\end{cases}$$

By (1) and (2), we have

$$\begin{aligned}\mu(X) - \mu\big(\textstyle\bigcap_{n\in\mathbb{N}} A_n\big) &= \mu\Big(\big(\textstyle\bigcap_{n\in\mathbb{N}} A_n\big)^c\Big) = \lim_{n\to\infty} \mu(A_n^c)\\ &= \lim_{n\to\infty} \big\{\mu(X) - \mu(A_n)\big\}\\ &= \mu(X) - \lim_{n\to\infty} \mu(A_n).\end{aligned}$$

Subtracting $\mu(X) < \infty$ from both sides of the last equality and then multiplying by -1, we have $\mu\left(\bigcap_{n\in\mathbb{N}} A_n\right) = \lim_{n\to\infty} \mu(A_n)$, that is, $\mu\left(\lim_{n\to\infty} A_n\right) = \lim_{n\to\infty} \mu(A_n)$.

6. (iii) $\Rightarrow$ (ii) when $\mu(X) < \infty$. Assume (iii) and $\mu(X) < \infty$. Suppose $A_n \uparrow$ and $\lim_{n\to\infty} A_n = \bigcup_{n\in\mathbb{N}} A_n \in \mathfrak{A}$. Then $(A_n^c : n \in \mathbb{N}) \subset \mathfrak{A}$, $A_n^c \downarrow$, and $\lim_{n\to\infty} A_n^c = \bigcap_{n\in\mathbb{N}} A_n^c = \left(\bigcup_{n\in\mathbb{N}} A_n\right)^c \in \mathfrak{A}$. By (iii), we have

$$\text{(3)} \qquad \lim_{n\to\infty} \mu(A_n^c) = \mu\left(\left(\bigcup_{n\in\mathbb{N}} A_n\right)^c\right).$$

From $\mu\left(\left(\bigcup_{n\in\mathbb{N}} A_n\right)^c\right) + \mu\left(\bigcup_{n\in\mathbb{N}} A_n\right) = \mu(X)$ and $\mu(A_n^c) + \mu(A_n) = \mu(X)$ and by the fact $\left(\bigcup_{n\in\mathbb{N}} A_n\right) < \infty$ and $\mu(A_n) < \infty$ for $n \in \mathbb{N}$, we have

$$\text{(4)} \qquad \begin{cases} \mu\left(\left(\bigcup_{n\in\mathbb{N}} A_n\right)^c\right) = \mu(X) - \mu\left(\bigcup_{n\in\mathbb{N}} A_n\right), \\ \mu(A_n^c) = \mu(X) - \mu(A_n) \quad \text{for } n \in \mathbb{N}. \end{cases}$$

By (3) and (4) , we have

$$\begin{aligned} \mu(X) - \mu\left(\bigcup_{n\in\mathbb{N}} A_n\right) = \mu\left(\left(\bigcup_{n\in\mathbb{N}} A_n\right)^c\right) &= \lim_{n\to\infty} \mu(A_n^c) \\ &= \lim_{n\to\infty} \{\mu(X) - \mu(A_n)\} \\ &= \mu(X) - \lim_{n\to\infty} \mu(A_n). \end{aligned}$$

Subtracting $\mu(X) < \infty$ from both sides, we have $\mu\left(\bigcup_{n\in\mathbb{N}} A_n\right) = \lim_{n\to\infty} \mu(A_n)$, that is, $\mu\left(\lim_{n\to\infty} A_n\right) = \lim_{n\to\infty} \mu(A_n)$. ∎

Here is an example of an additive set function on an algebra of subsets of a set which is not countably additive.

Example. Let us define a set function μ on the σ-algebra $\mathfrak{P}(\mathbb{R})$ of all subsets of $\mathbb{R}$ by setting $\mu(E) = 0$ when $E \in \mathfrak{P}(\mathbb{R})$ is a finite set and $\mu(E) = \infty$ when $E \in \mathfrak{P}(\mathbb{R})$ is an infinite set. Then $\mu(E) \in [0, \infty]$ for every $E \in \mathfrak{P}(\mathbb{R})$ and $\mu(\emptyset) = 0$. To show that μ is additive on $\mathfrak{P}(\mathbb{R})$, let $E_1, E_2 \in \mathfrak{P}(\mathbb{R})$ and $E_1 \cap E_2 = \emptyset$. Now if E_1 and E_2 are both finite, then so is $E_1 \cup E_2$ and we have $\mu(E_1) + \mu(E_2) = 0 + 0 = 0 = \mu(E_1 \cup E_2)$. If E_1 is a finite set and E_2 is an infinite set, then $E_1 \cup E_2$ is an infinite set and we have $\mu(E_1) + \mu(E_2) = 0 + \infty = \infty = \mu(E_1 \cup E_2)$. If E_1 and E_2 are both infinite, then so is $E_1 \cup E_2$ and we have $\mu(E_1) + \mu(E_2) = \infty + \infty = \infty = \mu(E_1 \cup E_2)$. This verifies the additivity of $\mathfrak{P}(\mathbb{R})$. To show that μ is not countably additive on $\mathfrak{P}(\mathbb{R})$, consider the disjoint sequence $(E_n : n \in \mathbb{N})$ in $\mathfrak{P}(\mathbb{R})$ where $E_n = \{n\}$. Then $\bigcup_{n\in\mathbb{N}} E_n = \mathbb{N}$ and $\mu\left(\bigcup_{n\in\mathbb{N}} E_n\right) = \infty$. But $\mu(E_n) = 0$ for every $n \in \mathbb{N}$ and $\sum_{n\in\mathbb{N}} \mu(E_n) = 0$.

If μ^* is an outer measure based on an additive set function on an algebra, then the μ^* measurability condition can be reduced. Note that μ is not assumed to be countably additive on an algebra.

Theorem 20.3. *Let μ be a nonnegative extended real-valued additive set function on an algebra $\mathfrak{A}$ of subsets of a set X with $\mu(\emptyset) = 0$. Consider the outer measure μ^* on X based on μ, that is,*

$$\mu^*(E) = \inf\left\{\sum_{n\in\mathbb{N}} \mu(A_n) : (A_n : n \in \mathbb{N}) \subset \mathfrak{A}, \bigcup_{n\in\mathbb{N}} A_n \supset E\right\}$$

for $E \in \mathfrak{P}(X)$. Then the μ^*-measurability condition

(1) $$\mu^*(B) = \mu^*(B \cap E) + \mu^*(B \cap E^c) \quad \textit{for every } B \in \mathfrak{P}(X)$$

is equivalent to the condition

(2) $$\mu^*(A) = \mu^*(A \cap E) + \mu^*(A \cap E^c) \quad \textit{for every } A \in \mathfrak{A}.$$

Proof. Since an algebra $\mathfrak{A}$ of subsets of a set X is a covering class for X, this theorem is a particular case of Theorem 2.24. ∎

[II] Extension of an Additive Set Function on an Algebra to a Measure

Definition 20.4. *Let μ be a nonnegative extended real-valued additive set function on an algebra $\mathfrak{A}$ of subsets of a set X with $\mu(\emptyset) = 0$. We say that μ is extendible to a measure on X if there exists a measure space $(X, \mathfrak{F}, \nu)$ such that $\mathfrak{A} \subset \mathfrak{F}$ and $\mu = \nu$ on $\mathfrak{A}$.*

Theorem 20.5. (Hopf Extension Theorem) *Let μ be a nonnegative extended real-valued additive set function on an algebra $\mathfrak{A}$ of subsets of a set X with $\mu(\emptyset) = 0$. Then μ is extendible to a measure on X if and only if μ is countably additive on $\mathfrak{A}$.*

Proof. 1. Suppose μ is extendible to a measure on X, that is, there exists a measure space $(X, \mathfrak{F}, \nu)$ such that $\mathfrak{A} \subset \mathfrak{F}$ and $\mu = \nu$ on $\mathfrak{A}$. To show that μ is countably additive on $\mathfrak{A}$, let $(A_n : n \in \mathbb{N})$ be a disjoint sequence in $\mathfrak{A}$ such that $\bigcup_{n\in\mathbb{N}} A_n \in \mathfrak{A}$. Since $\mathfrak{A} \subset \mathfrak{F}$ and $\mu = \nu$ on $\mathfrak{A}$ and since ν is countably additive on $\mathfrak{F}$, we have $\mu\left(\bigcup_{n\in\mathbb{N}} A_n\right) = \nu\left(\bigcup_{n\in\mathbb{N}} A_n\right) = \sum_{n\in\mathbb{N}} \nu(A_n) = \sum_{n\in\mathbb{N}} \mu(A_n)$. This proves the countable additivity of μ on $\mathfrak{A}$.

2. Conversely suppose μ is countably additive on $\mathfrak{A}$. Then according to Theorem 20.1, if we define an outer measure μ^* on X by setting

$$\mu^*(E) = \inf\left\{\sum_{n\in\mathbb{N}} \mu(A_n) : (A_n : n \in \mathbb{N}) \subset \mathfrak{A}, \bigcup_{n\in\mathbb{N}} A_n \supset E\right\}$$

for $E \in \mathfrak{P}(X)$, then $(X, \mathfrak{M}(\mu^*), \mu^*)$ is a measure space such that $\mathfrak{A} \subset \mathfrak{M}(\mu^*)$ and $\mu = \mu^*$ on $\mathfrak{A}$. Thus μ is extendible to a measure on X. ∎

[III] Regularity of an Outer Measure Derived from a Countably Additive Set Function on an Algebra

Definition 20.6. *Let $\mathfrak{A}$ be an algebra of subsets of a set X. We define $\mathfrak{A}_\sigma$ as the collection of all countable unions of members of $\mathfrak{A}$ and define $\mathfrak{A}_{\sigma\delta}$ as the collection of all countable intersections of members of $\mathfrak{A}_\sigma$.*

Observation 20.7. For an algebra $\mathfrak{A}$ of subsets of a set X, the collections $\mathfrak{A}_\sigma$ and $\mathfrak{A}_{\sigma\delta}$ are not closed under complementation in general and thus they are not algebras in general. However $\mathfrak{A}_\sigma$ is closed under countable unions since a countable union of countable unions of members of $\mathfrak{A}$ is still a countable union of members of $\mathfrak{A}$. Also $\mathfrak{A}_\sigma$ is closed under intersections and hence closed under finite intersections. To prove this, let $A, B \in \mathfrak{A}_\sigma$. Then $A = \bigcup_{n\in\mathbb{N}} A_n$ and $B = \bigcup_{n\in\mathbb{N}} B_n$ where $(A_n : n \in \mathbb{N})$ and $(B_n : n \in \mathbb{N})$ are sequences in $\mathfrak{A}$. Then $A \cap B = \left(\bigcup_{m\in\mathbb{N}} A_m\right) \cap \left(\bigcup_{n\in\mathbb{N}} B_n\right) = \bigcup_{m,n\in\mathbb{N}}(A_m \cap B_n)$. Since $\mathfrak{A}$ is an algebra, $A_m \cap B_n \in \mathfrak{A}$ and thus $A \cap B \in \mathfrak{A}_\sigma$.

Let μ be a nonnegative extended real-valued countably additive set function on an algebra $\mathfrak{A}$ of subsets of a set X with $\mu(\emptyset) = 0$. Let μ^* be an outer measure on X based on μ, that is,

$$\mu^*(E) = \inf\left\{\textstyle\sum_{n\in\mathbb{N}} \mu(A_n) : (A_n : n \in \mathbb{N}) \subset \mathfrak{A}, \bigcup_{n\in\mathbb{N}} A_n \supset E\right\}$$

for $E \in \mathfrak{P}(X)$. Consider the σ-algebra $\mathfrak{M}(\mu^*)$ consisting of those members E of $\mathfrak{P}(X)$ which satisfy the μ^*-measurability condition

$$\mu^*(B) = \mu^*(B \cap E) + \mu^*(B \cap E^c) \quad \text{for every } B \in \mathfrak{P}(X).$$

By Theorem 20.1, $\mathfrak{A} \subset \mathfrak{M}(\mu^*)$ and thus we have the following chain of inclusions:

$$\mathfrak{A} \subset \mathfrak{A}_\sigma \subset \mathfrak{A}_{\sigma\delta} \subset \sigma(\mathfrak{A}) \subset \mathfrak{M}(\mu^*).$$

Also $\mu = \mu^*$ on $\mathfrak{A}$ according to Theorem 20.1. Let us consider approximation of a set $E \in \mathfrak{P}(X)$ by members of $\mathfrak{A}_\sigma$ and $\mathfrak{A}_{\sigma\delta}$ in terms of μ^*.

Lemma 20.8. *Let μ^* be an outer measure based on a nonnegative extended real-valued countably additive set function μ on an algebra $\mathfrak{A}$ of subsets of a set X with $\mu(\emptyset) = 0$. Then for every $E \in \mathfrak{P}(X)$ and $\varepsilon > 0$, there exists $A \in \mathfrak{A}_\sigma$ such that $E \subset A$ and $\mu^*(E) \le \mu^*(A) \le \mu^*(E) + \varepsilon$.*

Proof. Let $E \in \mathfrak{P}(X)$. By the definition of μ^* as an infimum, for every $\varepsilon > 0$ there exists a sequence $(A_n : n \in \mathbb{N})$ in $\mathfrak{A}$ such that $\bigcup_{n\in\mathbb{N}} A_n \supset E$ and $\mu^*(E) \le \sum_{n\in\mathbb{N}} \mu(A_n) \le \mu^*(E) + \varepsilon$. (Note that since $\mu^*(E)$ may be equal to ∞, a strict inequality may not be possible.) Let $A = \bigcup_{n\in\mathbb{N}} A_n \in \mathfrak{A}_\sigma$. By the monotonicity and countable subadditivity of the outer measure μ^* and by the fact that $\mu = \mu^*$ on $\mathfrak{A}$ according to Theorem 20.1, we have $\mu^*(E) \le \mu^*(A) \le \sum_{n\in\mathbb{N}} \mu^*(A_n) = \sum_{n\in\mathbb{N}} \mu(A_n) \le \mu^*(E) + \varepsilon$. ∎

Proposition 20.9. *Let μ^* be an outer measure based on a nonnegative extended real-valued countably additive set function μ on an algebra $\mathfrak{A}$ of subsets of a set X with $\mu(\emptyset) = 0$. Then μ^* is a regular outer measure, that is, for every $E \in \mathfrak{P}(X)$ there exists $A \in \mathfrak{M}(\mu^*)$ such that $E \subset A$ and $\mu^*(E) = \mu^*(A)$. In fact there exists $A \in \mathfrak{A}_{\sigma\delta} \subset \sigma(\mathfrak{A}) \subset \mathfrak{M}(\mu^*)$ such that $E \subset A$ and $\mu^*(E) = \mu^*(A)$. In particular if $\mu^*(E) = 0$, then there exists $A \in \mathfrak{A}_{\sigma\delta}$ such that $E \subset A$ and $\mu^*(A) = 0$.*

Proof. Let $E \in \mathfrak{P}(X)$. By Lemma 20.8, for every $n \in \mathbb{N}$ there exists $A_n \in \mathfrak{A}_\sigma$ such that $E \subset A_n$ and $\mu^*(E) \leq \mu^*(A_n) \leq \mu^*(E) + \frac{1}{n}$. Let $A = \bigcap_{n \in \mathbb{N}} A_n$. Then $A \in \mathfrak{A}_{\sigma\delta}$, $E \subset A$, and by the monotonicity of the outer measure μ^*, $\mu^*(E) \leq \mu^*(A) \leq \mu^*(A_n) \leq \mu^*(E) + \frac{1}{n}$ for every $n \in \mathbb{N}$. Thus $\mu^*(E) = \mu^*(A)$. ∎

Definition 20.10. *A nonnegative extended real-valued set function γ on a covering class $\mathfrak{V}$ of subsets of a set X is said to be σ-finite on $\mathfrak{V}$ if there exists a sequence $(V_n : n \in \mathbb{N})$ in $\mathfrak{V}$ such that $\bigcup_{n \in \mathbb{N}} V_n = X$ and $\gamma(V_n) < \infty$ for every $n \in \mathbb{N}$.*

In connection with the definition of a σ-finite covering class of a set X, recall that an algebra $\mathfrak{A}$ of subsets of X is a covering class for X.

Theorem 20.11. *Let μ^* be an outer measure based on a nonnegative extended real-valued countably additive set function μ on an algebra $\mathfrak{A}$ of subsets of a set X with $\mu(\emptyset) = 0$. Consider the following conditions on a set $E \in \mathfrak{P}(X)$:*
1° *for every $\varepsilon > 0$, there exists $A \in \mathfrak{A}_\sigma$ such that $E \subset A$ and $\mu^*(A \setminus E) \leq \varepsilon$,*
2° *there exists $A \in \mathfrak{A}_{\sigma\delta}$ such that $E \subset A$ and $\mu^*(A \setminus E) = 0$,*
3° *$E \in \mathfrak{M}(\mu^*)$.*
Then 1° $\Rightarrow$ 2° and 2° $\Rightarrow$ 3°. If μ is σ-finite on $\mathfrak{A}$, then 3° $\Rightarrow$ 1°.

Proof. 1. 1° $\Rightarrow$ 2°. Let $E \in \mathfrak{P}(X)$. Assume 1°. Then for every $n \in \mathbb{N}$, there exists $A_n \in \mathfrak{A}_\sigma$ such that $E \subset A_n$ and $\mu^*(A_n \setminus E) \leq \frac{1}{n}$. Let $A = \bigcap_{n \in \mathbb{N}} A_n$. Then $A \in \mathfrak{A}_{\sigma\delta}$ and $E \subset A$. Since $A \subset A_n$, we have $A \setminus E \subset A_n \setminus E$ for every $n \in \mathbb{N}$. By the monotonicity of the outer measure μ^*, we have $\mu^*(A \setminus E) \leq \mu^*(A_n \setminus E) \leq \frac{1}{n}$ for every $n \in \mathbb{N}$. Thus $\mu^*(A \setminus E) = 0$. This proves 2°.

2. 2° $\Rightarrow$ 3°. Let $E \in \mathfrak{P}(X)$. Assume 2°. Then there exists $A \in \mathfrak{A}_{\sigma\delta}$ such that $E \subset A$ and $\mu^*(A \setminus E) = 0$. Now $\mu^*(A \setminus E) = 0$ implies $A \setminus E \in \mathfrak{M}(\mu^*)$ by Lemma 2.6. Since $E \subset A$, we have $E = A \setminus (A \setminus E)$. Then since $A \in \mathfrak{A}_{\sigma\delta} \subset \mathfrak{M}(\mu^*)$ and since $\mathfrak{M}(\mu^*)$ is a σ-algebra, we have $E \in \mathfrak{M}(\mu^*)$.

3. 3° $\Rightarrow$ 1° when μ is σ-finite on $\mathfrak{A}$. Suppose μ is σ-finite on $\mathfrak{A}$. Then there exists a sequence $(X_n : n \in \mathbb{N})$ in $\mathfrak{A}$ such that $\bigcup_{n \in \mathbb{N}} X_n = X$ and $\mu(X_n) < \infty$ for every $n \in \mathbb{N}$. Let $Y_1 = X_1$ and $Y_n = X_n \setminus (X_1 \cup \cdots \cup X_{n-1})$ for $n \geq 2$. Then $(Y_n : n \in \mathbb{N})$ is a disjoint sequence in $\mathfrak{A}$ and $\bigcup_{n \in \mathbb{N}} Y_n = \bigcup_{n \in \mathbb{N}} X_n = X$. By the monotonicity of μ^*, we have $\mu^*(Y_n) \leq \mu^*(X_n) < \infty$ for $n \in \mathbb{N}$. Recall that $\mu = \mu^*$ on $\mathfrak{A}$ by Theorem 20.1. Thus we have $\mu(Y_n) \leq \mu(X_n) < \infty$ for $n \in \mathbb{N}$.

Now let $E \in \mathfrak{M}(\mu^*)$. If we let $E_n = E \cap Y_n$ for $n \in \mathbb{N}$, then $(E_n : n \in \mathbb{N})$ is a disjoint sequence in $\mathfrak{M}(\mu^*)$ with $\bigcup_{n \in \mathbb{N}} E_n = E$. Also $\mu^*(E_n) \leq \mu^*(Y_n) = \mu(Y_n) < \infty$ for $n \in \mathbb{N}$. Let $\varepsilon > 0$ be arbitrarily given. For each $n \in \mathbb{N}$, there exists $A_n \in \mathfrak{A}_\sigma$ such that $E_n \subset A_n$ and $\mu^*(E_n) \leq \mu^*(A_n) \leq \mu^*(E_n) + \frac{\varepsilon}{2^n}$ by Lemma 20.8. Since $E_n, A_n \in \mathfrak{M}(\mu^*)$, μ^* is a measure on the σ-algebra $\mathfrak{M}(\mu^*)$, and $\mu^*(E_n) < \infty$, the set inclusion $E_n \subset A_n$ implies that $\mu^*(A_n \setminus E_n) = \mu^*(A_n) - \mu^*(E_n) \leq \frac{\varepsilon}{2^n}$. Then by the countable subadditivity of μ^*, we have $\mu^*\big(\bigcup_{n \in \mathbb{N}}(A_n \setminus E_n)\big) \leq \sum_{n \in \mathbb{N}} \mu^*(A_n \setminus E_n) \leq \varepsilon$. Since $\big(\bigcup_{n \in \mathbb{N}} A_n\big) \setminus E = \big(\bigcup_{n \in \mathbb{N}} A_n\big) \setminus \big(\bigcup_{n \in \mathbb{N}} E_n\big) \subset \bigcup_{n \in \mathbb{N}}(A_n \setminus E_n)$, we have $\mu^*\big(\big(\bigcup_{n \in \mathbb{N}} A_n\big) \setminus E\big) \leq \mu^*\big(\bigcup_{n \in \mathbb{N}}(A_n \setminus E_n)\big) \leq \varepsilon$. Since $\mathfrak{A}_\sigma$ is closed under countable unions by Observation 20.7, if we let $A = \bigcup_{n \in \mathbb{N}} A_n$ then $A \in \mathfrak{A}_\sigma$ and $\mu^*(A \setminus E) \leq \varepsilon$. This proves 1°. ∎

[IV] Uniqueness of Extension of a Countably Additive Set Function on an Algebra to a Measure

If μ is a nonnegative extended real-valued countably additive set function on an algebra $\mathfrak{A}$ of subsets of a set X with $\mu(\emptyset) = 0$, then the outer measure μ^* on X based on μ is an extension of μ to a measure on the σ-algebra $\mathfrak{M}(\mu^*)$. The next theorem shows that if μ is σ-finite on $\mathfrak{A}$, then any extension of μ to a measure on a σ-algebra $\mathfrak{F}$ such that $\mathfrak{A} \subset \mathfrak{F} \subset \mathfrak{M}(\mu^*)$ is unique.

Theorem 20.12. (Uniqueness of Extension to Measure) *Let μ be a nonnegative extended real-valued countably additive set function on an algebra $\mathfrak{A}$ of subsets of a set X with $\mu(\emptyset) = 0$. Assume that μ is σ-finite on $\mathfrak{A}$. Let μ^* be the outer measure on X based on the set function μ and consider the measure space $(X, \mathfrak{M}(\mu^*), \mu^*)$. If $(X, \mathfrak{F}, \nu)$ is a measure space such that $\mathfrak{A} \subset \mathfrak{F} \subset \mathfrak{M}(\mu^*)$ and $\nu = \mu$ on $\mathfrak{A}$, then $\nu = \mu^*$ on $\mathfrak{F}$.*

Proof. We show that $\nu = \mu^*$ on $\mathfrak{A}_\sigma$, on $\mathfrak{A}_{\sigma\delta}$, and on $\mathfrak{F}$ successively.

1. Let us show that $\nu = \mu^*$ on $\mathfrak{A}_\sigma$. Let $A \in \mathfrak{A}_\sigma$. Then $A = \bigcup_{n\in\mathbb{N}} A_n$ where $(A_n : n \in \mathbb{N})$ is a sequence in $\mathfrak{A}$. If we let $B_1 = A_1$ and $B_n = A_n \setminus (A_1 \cup \cdots \cup A_{n-1})$ for $n \geq 2$, then $(B_n : n \in \mathbb{N})$ is a disjoint sequence in $\mathfrak{A}$ with $\bigcup_{n\in\mathbb{N}} B_n = \bigcup_{n\in\mathbb{N}} A_n = A$. Then

$$\begin{aligned}
\nu(A) &= \nu\Big(\bigcup_{n\in\mathbb{N}} B_n\Big) = \sum_{n\in\mathbb{N}} \nu(B_n) = \sum_{n\in\mathbb{N}} \mu(B_n) \\
&= \sum_{n\in\mathbb{N}} \mu^*(B_n) = \mu^*\Big(\bigcup_{n\in\mathbb{N}} B_n\Big) = \mu^*(A),
\end{aligned}$$

where the second equality is by the countable additivity of ν on $\mathfrak{F}$, the third equality is from the fact that $\mu = \nu$ on $\mathfrak{A}$, the fourth equality is from the fact that $\mu = \mu^*$ on $\mathfrak{A}$ and the fifth equality is from the countable additivity of μ^* on $\mathfrak{M}(\mu^*)$. This shows that $\nu = \mu^*$ on $\mathfrak{A}_\sigma$.

2. Let us show that $\nu = \mu^*$ on $\mathfrak{A}_{\sigma\delta}$. Let $A \in \mathfrak{A}_{\sigma\delta}$. Then $A = \bigcap_{n\in\mathbb{N}} A_n$ where $(A_n : n \in \mathbb{N})$ is a sequence in $\mathfrak{A}_\sigma$. The σ-finiteness of μ on the algebra $\mathfrak{A}$ implies that there exists a disjoint sequence $(X_j : j \in \mathbb{N})$ in $\mathfrak{A}$ such that $\bigcup_{j\in\mathbb{N}} X_j = X$ and $\mu(X_j) < \infty$ for $j \in \mathbb{N}$. Let $B_j = A \cap X_j$ for $j \in \mathbb{N}$. Since $A \in \mathfrak{A}_{\sigma\delta}$ and $X_j \in \mathfrak{A} \subset \mathfrak{A}_\sigma$, we have $B_j \in \mathfrak{A}_{\sigma\delta}$. Thus $(B_j : j \in \mathbb{N})$ is a disjoint sequence in $\mathfrak{A}_{\sigma\delta}$ and $\bigcup_{j\in\mathbb{N}} B_j = A$. Also

$$B_j = A \cap X_j = \Big(\bigcap_{n\in\mathbb{N}} A_n\Big) \cap X_j = \bigcap_{n\in\mathbb{N}} (A_n \cap X_j) = \lim_{n\to\infty} \bigcap_{k=1}^{n} (A_k \cap X_j).$$

Since $A_k \in \mathfrak{A}_\sigma$ and $X_j \in \mathfrak{A} \subset \mathfrak{A}_\sigma$ and since $\mathfrak{A}_\sigma$ is closed under finite intersections by Observation 20.7, $\big(\bigcap_{k=1}^{n}(A_k \cap X_j) : n \in \mathbb{N}\big)$ is a decreasing sequence in $\mathfrak{A}_\sigma$. Since every member in this sequence is a subset of X_j and since $\nu(X_j) = \mu(X_j) < \infty$, we have

$$\begin{aligned}\nu(B_j) &= \nu\Big(\lim_{n\to\infty} \bigcap_{k=1}^{n} (A_k \cap X_j)\Big) = \lim_{n\to\infty} \nu\Big(\bigcap_{k=1}^{n} (A_k \cap X_j)\Big) \\ &= \lim_{n\to\infty} \mu^*\Big(\bigcap_{k=1}^{n} (A_k \cap X_j)\Big) = \mu^*\Big(\lim_{n\to\infty} \bigcap_{k=1}^{n} (A_k \cap X_j)\Big) \\ &= \mu^*(B_j)\end{aligned}$$

where the second equality is by the fact that $\nu(X_j) = \mu^*(X_j) = \mu(X_j) < \infty$, the third equality is by the fact that $\nu = \mu^*$ on $\mathfrak{A}_\sigma$ by **1** and the fourth equality is by the fact that $\mu^*(X_j) = \mu(X_j) < \infty$. Then since ν and μ^* are measures on $\mathfrak{F}$, we have

$$\nu(A) = \nu\Big(\bigcup_{j\in\mathbb{N}} B_j\Big) = \sum_{n\in\mathbb{N}} \nu(B_j) = \sum_{n\in\mathbb{N}} \mu^*(B_j) = \mu^*\Big(\bigcup_{j\in\mathbb{N}} B_j\Big) = \mu^*(A).$$

This shows that $\nu = \mu^*$ on $\mathfrak{A}_{\sigma\delta}$.

3. To show that $\nu = \mu^*$ on $\mathfrak{F}$, let $E \in \mathfrak{F}$. Then $E \in \mathfrak{M}(\mu^*)$. Since μ is σ-finite on $\mathfrak{A}$, according to Theorem 20.11 there exists $A \in \mathfrak{A}_{\sigma\delta}$ such that $E \subset A$ and $\mu^*(A \setminus E) = 0$. If we let $B = A \setminus E$, then $\mu^*(B) = 0$ so that by Proposition 20.9 there exists $C \in \mathfrak{A}_{\sigma\delta}$ such that $B \subset C$ and $\mu^*(C) = 0$. Since $E \in \mathfrak{F}$ and $A \in \mathfrak{A}_{\sigma\delta} \subset \mathfrak{F}$, we have $B \in \mathfrak{F}$. Also $C \in \mathfrak{A}_{\sigma\delta} \subset \mathfrak{F}$. Then $\nu(B) \le \nu(C) = \mu^*(C) = 0$ since ν is a measure on $\mathfrak{F}$ and since $\nu = \mu^*$ on $\mathfrak{A}_{\sigma\delta}$ by **2**. Thus $\nu(B) = 0$. Then by the fact that ν and μ^* are measures on $\mathfrak{F}$, we have

$$\begin{aligned}\nu(E) &= \nu(A \setminus B) = \nu(A) - \nu(B) = \nu(A) = \mu^*(A) \\ &= \mu^*(A) - \mu^*(B) = \mu^*(A \setminus B) = \mu^*(E).\end{aligned}$$

This shows that $\nu = \mu^*$ on $\mathfrak{F}$. ∎

Remark 20.13. Without the assumption of σ-finiteness of μ on $\mathfrak{A}$, the uniqueness of extension in Theorem 20.12 fails.

Example. Let ν be a set function on the σ-algebra $\mathfrak{P}(\mathbb{R})$ of all subsets of $\mathbb{R}$ defined by setting $\nu(E) = N$ if E is a finite set with N elements and $\nu(E) = \infty$ if E is an infinite set. Thus defined, ν is a measure on $\mathfrak{P}(\mathbb{R})$. Let us define a sequence of measures $(\nu_n : n \in \mathbb{N})$ on $\mathfrak{P}(\mathbb{R})$ by setting $\nu_n(E) = n\nu(E)$ for $E \in \mathfrak{P}(\mathbb{R})$. Let $\mathfrak{A}$ be an algebra of subsets of $\mathbb{R}$ consisting of $\emptyset$ and all finite unions of intervals of the type $(a, b]$ where $-\infty \le a < b \le \infty$ with the understanding that $(a, \infty] = (a, \infty)$. Note that $\emptyset$ is the only set in $\mathfrak{A}$ that is a finite set and thus for every $A \in \mathfrak{A}$, $A \neq \emptyset$, we have $\nu(A) = \infty$ and $\nu_n(A) = \infty$ for every $n \in \mathbb{N}$. Note also that $\sigma(\mathfrak{A}) = \mathfrak{B}_{\mathbb{R}}$, the Borel σ-algebra of subsets of $\mathbb{R}$. Let μ be the restriction of the measures ν_n on $\mathfrak{P}(\mathbb{R})$ to $\mathfrak{A}$. Since $\nu_n(\emptyset) = 0$ and $\nu_n(A) = \infty$ for every $A \in \mathfrak{A}$, $A \neq \emptyset$, for all $n \in \mathbb{N}$, μ does not depend on $n \in \mathbb{N}$, that is, μ is common to all ν_n, $n \in \mathbb{N}$. As the restriction of the measure ν_n on $\mathfrak{P}(\mathbb{R})$ to $\mathfrak{A}$, μ is countably additive on $\mathfrak{A}$. Since $\mu(A) = \infty$ for every $A \in \mathfrak{A}$, $A \neq \emptyset$, μ is not σ-finite on $\mathfrak{A}$. Let μ_n be the restriction of ν_n to $\sigma(\mathfrak{A})$. Then μ_n is a measure on $\sigma(\mathfrak{A})$. Since $\mu_n = \mu$ on $\mathfrak{A}$, μ_n is an extension of our countably additive set function μ on the algebra $\mathfrak{A}$ to a measure on the σ-algebra $\sigma(\mathfrak{A})$. Now since $\sigma(\mathfrak{A}) = \mathfrak{B}_{\mathbb{R}}$, for every $x \in \mathbb{R}$, we have $\{x\} \in \sigma(\mathfrak{A})$. Then $\mu_n(\{x\}) = \nu_n(\{x\}) = n\nu(\{x\}) = n$ for $n \in \mathbb{N}$. This shows that μ_n, $n \in \mathbb{N}$, are distinct measures on $\sigma(\mathfrak{A})$. Thus an extension of μ to measures on $\sigma(\mathfrak{A})$ is not unique.

Theorem 20.14. *Let $\mathfrak{A}$ be an algebra of subsets of a set X and let μ_1 and μ_2 be two measures on the σ-algebra $\sigma(\mathfrak{A})$. If $\mu_1 = \mu_2$ on $\mathfrak{A}$ and if μ_1 and μ_2 are σ-finite on $\mathfrak{A}$, then $\mu_1 = \mu_2$ on $\sigma(\mathfrak{A})$.*

Proof. Since μ_1 and μ_2 are measures on $\sigma(\mathfrak{A})$, they are nonnegative extended real-valued countably additive set functions on $\mathfrak{A}$ with $\mu_1(\emptyset) = \mu_2(\emptyset) = 0$. Let μ_1^* and μ_2^* be the two outer measures on X based on the two set functions μ_1 and μ_2 on $\mathfrak{A}$, that is, for $E \in \mathfrak{P}(X)$ we have $\mu_i^*(E) = \inf\left\{\sum_{n\in\mathbb{N}} \mu_i(A_n) : (A_n : n \in \mathbb{N}) \subset \mathfrak{A}, \bigcup_{n\in\mathbb{N}} A_n \supset E\right\}$ for $i = 1$ and 2. Since $\mu_1 = \mu_2$ on $\mathfrak{A}$, we have $\mu_1^* = \mu_2^*$ on $\mathfrak{P}(X)$. This implies $\mathfrak{M}_L(\mu_1^*) = \mathfrak{M}_L(\mu_2^*)$. Since the σ-algebras $\mathfrak{M}_L(\mu_1^*)$ and $\mathfrak{M}_L(\mu_2^*)$ contain $\mathfrak{A}$ by Theorem 20.1, we have $\sigma(\mathfrak{A}) \subset \mathfrak{M}_L(\mu_1^*) = \mathfrak{M}_L(\mu_2^*)$. By Theorem 20.12, $\mu_1 = \mu_1^*$ and $\mu_2 = \mu_2^*$ on $\sigma(\mathfrak{A})$. But $\mu_1^* = \mu_2^*$ on $\mathfrak{P}(X)$. Thus $\mu_1 = \mu_2$ on $\sigma(\mathfrak{A})$. ∎

[V] Approximation to a σ-algebra Generated by an Algebra

Let $\mathfrak{A}$ be an algebra of subsets of a set X. The σ-algebra generated by $\mathfrak{A}$, $\sigma(\mathfrak{A})$, is abstractly defined as the smallest σ-algebra of subsets of X containing $\mathfrak{A}$. Thus it is of interest to know how an arbitrary member of $\sigma(\mathfrak{A})$ can be approximated by members of $\mathfrak{A}$.

Theorem 20.15. *Given a finite measure space $(X, \mathfrak{F}, \nu)$. Let $\mathfrak{A}$ be an arbitrary algebra of subsets of X such that $\sigma(\mathfrak{A}) = \mathfrak{F}$. Then for every $E \in \mathfrak{F}$ and $\varepsilon > 0$, there exists $A \in \mathfrak{A}$ such that $\nu(E \triangle A) < \varepsilon$.*

Proof. Let $\mathfrak{A}$ be an arbitrary algebra of subsets of X such that $\sigma(\mathfrak{A}) = \mathfrak{F}$ and let μ be the restriction of ν to $\mathfrak{A}$. Then μ is a nonnegative extended real-valued countably additive set function on $\mathfrak{A}$ with $\mu(\emptyset) = 0$. If we let μ^* be the outer measure on X based on the set function μ on $\mathfrak{A}$, then $\mathfrak{A} \subset \mathfrak{M}(\mu^*)$ and $\mu = \mu^*$ on $\mathfrak{A}$ by Theorem 20.1. Since $\mathfrak{M}(\mu^*)$ is a σ-algebra containing $\mathfrak{A}$, we have $\mathfrak{F} = \sigma(\mathfrak{A}) \subset \mathfrak{M}(\mu^*)$. Since ν is a finite measure on $\mathfrak{F}$, its restriction μ to $\mathfrak{A}$ is a finite set function on $\mathfrak{A}$. Thus by Theorem 20.12, we have $\nu = \mu^*$ on $\mathfrak{F}$.

Let $E \in \mathfrak{F}$ and $\varepsilon > 0$. By the definition of μ^* as an infimum, there exists a sequence $(A_n : n \in \mathbb{N})$ in $\mathfrak{A}$ such that $\bigcup_{n\in\mathbb{N}} A_n \supset E$ and $\mu^*(E) \le \sum_{n\in\mathbb{N}} \mu(A_n) \le \mu^*(E) + \frac{\varepsilon}{2}$. Since $\nu = \mu^*$ on $\mathfrak{F}$ and $\nu = \mu$ on $\mathfrak{A}$, the last inequalities are equivalent to

$$(1) \qquad \nu(E) \le \sum_{n\in\mathbb{N}} \nu(A_n) \le \nu(E) + \frac{\varepsilon}{2}.$$

Now $\left(\bigcup_{k=1}^n A_k : n \in \mathbb{N}\right)$ is an increasing sequence in $\mathfrak{A} \subset \mathfrak{F}$ and $\bigcup_{k=1}^n A_k \uparrow \bigcup_{n\in\mathbb{N}} A_n$ so that $\nu\left(\bigcup_{k=1}^n A_k\right) \uparrow \nu\left(\bigcup_{n\in\mathbb{N}} A_n\right)$. Since $\nu\left(\bigcup_{n\in\mathbb{N}} A_n\right) < \infty$, there exists $N \in \mathbb{N}$ such that

$\nu\left(\bigcup_{n\in\mathbb{N}} A_n\right) - \frac{\varepsilon}{2} < \nu\left(\bigcup_{n=1}^N A_n\right)$. Then

(2) $$\nu\Big(\bigcup_{n\in\mathbb{N}} A_n \setminus \bigcup_{n=1}^N A_n\Big) = \nu\Big(\bigcup_{n\in\mathbb{N}} A_n\Big) - \nu\Big(\bigcup_{n=1}^N A_n\Big) < \frac{\varepsilon}{2}.$$

Let $A = \bigcup_{n=1}^N A_n \in \mathfrak{A}$. Since $E \subset \bigcup_{n\in\mathbb{N}} A_n$, we have

(3) $$\nu(E \setminus A) \le \nu\Big(\bigcup_{n\in\mathbb{N}} A_n \setminus A\Big) < \frac{\varepsilon}{2}$$

by the monotonicity of ν on $\mathfrak{F}$ and by (2). Similarly since $A \subset \bigcup_{n\in\mathbb{N}} A_n$ and $E \subset \bigcup_{n\in\mathbb{N}} A_n$ also, we have

(4) $$\begin{aligned}\nu(A \setminus E) &\le \nu\Big(\bigcup_{n\in\mathbb{N}} A_n \setminus E\Big) = \nu\Big(\bigcup_{n\in\mathbb{N}} A_n\Big) - \nu(E)\\ &\le \sum_{n\in\mathbb{N}} \nu(A_n) - \nu(E) \le \frac{\varepsilon}{2}\end{aligned}$$

by (1). Thus by (3) and (4), we have $\nu(E\triangle A) = \nu(E\setminus A) + \nu(A \setminus E) < \frac{\varepsilon}{2} + \frac{\varepsilon}{2} = \varepsilon$. ■

Proposition 20.16. *Given a finite measure space $(X, \sigma(\mathfrak{A}), \mu)$ where $\mathfrak{A}$ is an algebra of subsets of X. For every $E \in \sigma(\mathfrak{A})$, there exists a sequence $(A_k : k \in \mathbb{N})$ in $\mathfrak{A}$ such that*

(1) $$\lim_{k\to\infty} \mathbf{1}_{A_k} = \mathbf{1}_E \quad \textit{a.e. on } X$$

and

(2) $$\mu\big(E \triangle \liminf_{k\to\infty} A_k\big) = 0.$$

Proof. Let $E \in \sigma(\mathfrak{A})$. For every $n \in \mathbb{N}$, there exists $A_n \in \mathfrak{A}$ such that $\mu(E\triangle A_n) < \frac{1}{n}$ by Theorem 20.15. Now $\{X : |\mathbf{1}_{A_n} - \mathbf{1}_E| > 0\} = E\triangle A_n$ so that for every $n \in \mathbb{N}$, we have $\mu\{X : |\mathbf{1}_{A_n} - \mathbf{1}_E| > 0\} = \mu(E\triangle A_n) < \frac{1}{n}$. Thus for every $\varepsilon > 0$ we have

$$\begin{aligned}\lim_{n\to\infty} \mu\big\{X : |\mathbf{1}_{A_n} - \mathbf{1}_E| \ge \varepsilon\big\} &\le \lim_{n\to\infty} \mu\big\{X : |\mathbf{1}_{A_n} - \mathbf{1}_E| > 0\big\}\\ &= \lim_{n\to\infty} \mu(E\triangle A_n) \le \lim_{n\to\infty} \frac{1}{n} = 0\end{aligned}$$

that is, the sequence $(\mathbf{1}_{A_n} : n \in \mathbb{N})$ converges in measure to $\mathbf{1}_E$. Therefore there exists a subsequence $(\mathbf{1}_{A_{n_k}} : n \in \mathbb{N})$ which converges to $\mathbf{1}_E$ a.e. on X. Let us write A_k for A_{n_k}. Then $(A_k : k \in \mathbb{N})$ is a sequence in $\mathfrak{A}$ such that $\lim_{k\to\infty} \mathbf{1}_{A_k} = \mathbf{1}_E$ a.e. on X. This proves (1). Now we have

(3) $$\lim_{k\to\infty} \mathbf{1}_{A_k} = \mathbf{1}_E \quad \text{on } E_0^c,$$

where E_0 is a null set in $(X, \sigma(\mathfrak{A}), \mu)$.

Suppose $x \in E \cap E_0^c$. By (3), we have $\lim_{k\to\infty} \mathbf{1}_{A_k}(x) = \mathbf{1}_E(x) = 1$. Thus there exists $N \in \mathbb{N}$ such that $\mathbf{1}_{A_k}(x) = 1$ for $k \geq N$, that is, $x \in A_k$ for $k \geq N$, in other words $x \in \liminf_{k\to\infty} A_k$. Thus $E \cap E_0^c \subset \liminf_{k\to\infty} A_k$ and consequently we have

(4) $$E \cap E_0^c \subset \big(\liminf_{k\to\infty} A_k\big) \cap E_0^c.$$

Conversely suppose $x \in \big(\liminf_{k\to\infty} A_k\big) \cap E_0^c$. By (3), we have $\lim_{k\to\infty} \mathbf{1}_{A_k}(x) = \mathbf{1}_E(x)$. Since $x \in \liminf_{k\to\infty} A_k$, there exists $N \in \mathbb{N}$ such that $x \in A_k$ for $k \geq N$. Then $\mathbf{1}_{A_k}(x) = 1$ for $k \geq N$ so that $\lim_{k\to\infty} \mathbf{1}_{A_k}(x) = 1$. This implies $\mathbf{1}_E(x) = 1$ so that $x \in E$. Thus we have $\big(\liminf_{k\to\infty} A_k\big) \cap E_0^c \subset E$ and consequently

(5) $$\big(\liminf_{k\to\infty} A_k\big) \cap E_0^c \subset E \cap E_0^c.$$

By (4) and (5), we have $E \cap E_0^c = \big(\liminf_{k\to\infty} A_k\big) \cap E_0^c$. Therefore $E \triangle \liminf_{k\to\infty} A_k \subset E_0$ so that $\mu\big(E \triangle \liminf_{k\to\infty} A_k\big) \leq \mu(E_0) = 0$. This proves (2). ∎

Proposition 20.17. *Let $(X, \sigma(\mathfrak{A}), \mu)$ be a finite measure space where $\mathfrak{A}$ is an algebra of subsets of X. Let $\mathcal{S}$ be the collection of all nonnegative simple functions on X based on $\mathfrak{A}$, that is, functions of the type $\psi = \sum_{i=1}^{p} c_i \mathbf{1}_{A_i}$ where $c_i \geq 0$, $A_i \in \mathfrak{A}$, for $i = 1, \ldots, p$ and $p \in \mathbb{N}$. Then for every nonnegative extended real-valued $\sigma(\mathfrak{A})$-measurable function f on X, there exists a sequence $(\psi_n : n \in \mathbb{N})$ in $\mathcal{S}$ such that $\lim_{n\to\infty} \psi_n = f$ a.e. on $(X, \sigma(\mathfrak{A}), \mu)$ and $\lim_{n\to\infty} \int_X \psi_n \, d\mu = \int_X f \, d\mu$.*

Proof. If f is a nonnegative extended real-valued $\sigma(\mathfrak{A})$-measurable function on X, then there exists an increasing sequence of nonnegative simple functions $(\varphi_n : n \in \mathbb{N})$ on $(X, \sigma(\mathfrak{A}), \mu)$ such that $\varphi_n \uparrow f$ on X by Lemma 8.6. Let φ_n be given by

(1) $$\varphi_n = \sum_{j=1}^{p_n} c_{n,j} \mathbf{1}_{E_{n,j}},$$

where $c_{n,j} \geq 0$ and $E_{n,j} \in \sigma(\mathfrak{A})$ for $j = 1, \ldots, p_n$. The Proposition is trivially true if f is an identically vanishing function on X. Thus assume that f is not identically vanishing and assume that φ_n is not identically vanishing so that $\max \varphi_n > 0$ for $n \in \mathbb{N}$. Now for an arbitrary $\varepsilon > 0$, corresponding to $E_{n,j}$ there exists $A_{n,j} \in \mathfrak{A}$ such that

(2) $$\mu(E_{n,j} \triangle A_{n,j}) < \frac{\varepsilon}{2^n p_n} \min\left\{1, \frac{1}{\max \varphi_n}\right\}$$

by Theorem 20.15. Let us define a sequence $(\psi_n : n \in \mathbb{N})$ in $\mathcal{S}$ by setting

(3) $$\psi_n = \sum_{j=1}^{p_n} c_{n,j} \mathbf{1}_{A_{n,j}}.$$

Then $\{X : |\varphi_n - \psi_n| > 0\} \subset \bigcup_{j=1}^{p_n} (E_{n,j} \triangle A_{n,j})$ so that by (2) we have

$$(4) \qquad \mu\{X : |\varphi_n - \psi_n| > 0\} \le \sum_{j=1}^{p_n} \mu\left(E_{n,j} \triangle A_{n,j}\right) < \frac{\varepsilon}{2^n} \min\left\{1, \frac{1}{\max \varphi_n}\right\}.$$

This implies $\sum_{n\in\mathbb{N}} \mu\{X : |\varphi_n - \psi_n| > 0\} < \sum_{n\in\mathbb{N}} \frac{\varepsilon}{2^n} = \varepsilon < \infty$ so that by Theorem 6.6 (Borel-Cantelli), we have

$$(5) \qquad \mu\big(\limsup_{n\to\infty}\{X : |\varphi_n - \psi_n| > 0\}\big) = 0.$$

According to Lemma 1.7, the statement (5) is equivalent to the statement

$$(6) \qquad \mu\{X : \varphi_n \neq \psi_n \text{ for infinitely many } n \in \mathbb{N}\} = 0.$$

Since $\lim_{n\to\infty} \varphi_n(x) = f(x)$ for every $x \in X$, (6) implies $\lim_{n\to\infty} \psi_n(x) = \lim_{n\to\infty} \varphi_n(x) = f(x)$ for a.e. $x \in X$. By (4) we have

$$\begin{aligned} \left|\int_X \varphi_n \, d\mu - \int_X \psi_n \, d\mu\right| &\le \int_X |\varphi_n - \psi_n| \, d\mu \\ &\le 2 \max \varphi_n \cdot \mu\big\{X : |\varphi_n - \psi_n| > 0\big\} \\ &< 2 \max \varphi_n \cdot \frac{\varepsilon}{2^n} \frac{1}{\max \varphi_n} = \frac{\varepsilon}{2^{n-1}}. \end{aligned}$$

Since $\lim_{n\to\infty} \int_X \varphi_n \, d\mu = \int_X f \, d\mu$ by Theorem 8.5 (Monotone Convergence Theorem), we have $\lim_{n\to\infty} \int_X \psi_n \, d\mu = \int_X f \, d\mu$. ■

[VI] Outer Measure Based on a Measure

Unlike a measure on a σ-algebra of subsets of a set X, an outer measure on X is in general not an additive set function on the σ-algebra $\mathfrak{P}(X)$ of all subsets of X. However it has the advantage that it is defined on $\mathfrak{P}(X)$ on which it is monotone and countably subadditive. If a measure μ is given on a σ-algebra $\mathfrak{A}$ of subsets of a set X, then since $\mathfrak{A}$ is a covering class for X and since μ is a nonnegative extended real-valued set function on the covering class $\mathfrak{A}$ with $\mu(\emptyset) = 0$, μ can be extended to be an outer measure on X as in Theorem 2.21.

Theorem 20.18. *Given a measure space $(X, \mathfrak{A}, \mu)$. Let us define a set function μ^* on $\mathfrak{P}(X)$ by setting for every $E \in \mathfrak{P}(X)$*

$$\begin{aligned} \mu^*(E) &= \inf\left\{\textstyle\sum_{n\in\mathbb{N}} \mu(A_n) : (A_n : n \in \mathbb{N}) \subset \mathfrak{A}, \bigcup_{n\in\mathbb{N}} A_n \supset E\right\} \\ &= \inf\left\{\mu(A) : A \in \mathfrak{A}, A \supset E\right\}. \end{aligned}$$

Then μ^* is an outer measure on X, $\mathfrak{A} \subset \mathfrak{M}(\mu^*)$, and $\mu^* = \mu$ on $\mathfrak{A}$. Moreover μ^* is a regular outer measure on X.

Proof. Let us show first the equality of the two infima in the definition of μ^*. For brevity let us denote the two by α and β. Clearly $\alpha \leq \beta$. To prove the reverse inequality, let $(A_n : n \in \mathbb{N})$ be an arbitrary sequence in $\mathfrak{A}$ such that $\bigcup_{n\in\mathbb{N}} A_n \supset E$. If we let $A = \bigcup_{n\in\mathbb{N}} A_n$, then $A \in \mathfrak{A}$, $A \supset E$, and $\mu(A) \leq \sum_{n\in\mathbb{N}} \mu(A_n)$ by the countable subadditivity of the measure μ on $\mathfrak{A}$. From this it follows that $\beta \leq \alpha$ and therefore $\alpha = \beta$.

Since a σ-algebra $\mathfrak{A}$ of subsets of X is a covering class for X and a measure μ on $\mathfrak{A}$ is a nonnegative extended real-valued set function on $\mathfrak{A}$ with $\mu(\emptyset) = 0$, μ^* as defined above is an outer measure on X by Theorem 2.21. The countable additivity of μ on $\mathfrak{A}$ implies that $\mathfrak{A} \subset \mathfrak{M}(\mu^*)$ and $\mu^* = \mu$ on $\mathfrak{A}$ by Theorem 20.1. To show that μ^* is a regular outer measure on X, let $E \in \mathfrak{P}(X)$. Since $\mu^*(E) = \inf\{\mu(A) : A \in \mathfrak{A}, A \supset E\}$, for every $n \in \mathbb{N}$, there exists $A_n \in \mathfrak{A}$ such that $A_n \supset E$ and $\mu(A_n) \leq \mu^*(E) + \frac{1}{n}$. Let $A = \bigcap_{n\in\mathbb{N}} A_n$. Then $A \in \mathfrak{A}$, $A \supset E$, and $\mu(A) \leq \mu(A_n) \leq \mu^*(E) + \frac{1}{n}$ for every $n \in \mathbb{N}$. Then $\mu^*(A) = \mu(A) \leq \mu^*(E)$. On the other hand, $A \supset E$ implies that $\mu^*(A) \geq \mu^*(E)$. Thus $\mu^*(A) = \mu^*(E)$. Therefore we have a set $A \in \mathfrak{A} \subset \mathfrak{M}(\mu^*)$ such that $A \supset E$ and $\mu^*(A) = \mu^*(E)$. This shows that μ^* is a regular outer measure on X. ∎

§21 Extension of Additive Set Functions on a Semialgebra

[I] Semialgebras of Sets

A semialgebra of subsets of a set is a simpler structure than an algebra of subsets of the set. Thus a countably additive set function on a semialgebra is more easily constructed than a countably additive set function on an algebra. As we shall show below if μ is a nonnegative extended real-valued countably additive set function on a semialgebra $\mathfrak{S}$ of subsets of a set X with $\mu(\emptyset) = 0$, then μ can always be extended to be a nonnegative extended real-valued countably additive set function on the smallest algebra $\mathfrak{A}$ of subsets of X containing $\mathfrak{S}$. This countably additive set function on the algebra $\mathfrak{A}$ can then be extended to be a measure on X by the Hopf Extension Theorem (Theorem 20.5).

Definition 21.1. *A collection $\mathfrak{S}$ of subsets of a set X is called a semialgebra if it satisfies the following conditions:*
1° $\emptyset, X \in \mathfrak{S}$,
2° *if $E, F \in \mathfrak{S}$, then $E \cap F \in \mathfrak{S}$,*
3° *if $E \in \mathfrak{S}$, then E^c is a finite disjoint union of members of $\mathfrak{S}$.*

Example. In $\mathbb{R}^n$ let $\mathfrak{S}$ be the collection of subsets of $\mathbb{R}^n$ consisting of $\emptyset$ and all subsets of $\mathbb{R}^n$ of the type $(a_1, b_1] \times \cdots \times (a_n, b_n]$ where $-\infty \le a_i < b_i \le \infty$ with the understanding that $(a_i, \infty] = (a_i, \infty)$ for $i = 1, \dots, n$. It is easily verified that $\mathfrak{S}$ satisfies conditions 1°, 2°, and 3° of Definition 21.1. Thus $\mathfrak{S}$ is a semialgebra of subsets of $\mathbb{R}^n$.

Observation 21.2. Recall that an algebra of subsets of a set X is a collection $\mathfrak{A}$ of subsets of X satisfying the conditions:

(1) $$X \in \mathfrak{A}$$
(2) $$A \in \mathfrak{A} \Rightarrow A^c \in \mathfrak{A}$$
(3) $$A, B \in \mathfrak{A} \Rightarrow A \cup B \in \mathfrak{A}$$

(a) An algebra $\mathfrak{A}$ of subsets of a set X is a semialgebra of subsets of X.
(b) If a semialgebra $\mathfrak{S}$ of subsets of a set X satisfies the additional condition:
4° *if $E, F \in \mathfrak{S}$, then $E \cup F \in \mathfrak{S}$, then $\mathfrak{S}$ is an algebra of subsets of X.*

Proof. 1. Let us prove (a). Let $\mathfrak{A}$ be an algebra of subsets of a set X. Then by (1) and (2) we have $\emptyset = X^c \in \mathfrak{A}$. Thus $\mathfrak{A}$ satisfies condition 1° in Definition 21.1. If $A, B \in \mathfrak{A}$ then $A \cup B = \left(A^c \cup B^c\right)^c \in \mathfrak{A}$ by (2) and (3). Thus $\mathfrak{A}$ satisfies condition 2° of Definition 21.1. If $A \in \mathfrak{A}$ then $A^c \in \mathfrak{A}$ by (2). Thus $\mathfrak{A}$ satisfies condition 3° of Definition 21.1 trivially. Therefore $\mathfrak{A}$ is a semialgebra of subsets of X.

2. Suppose a semialgebra $\mathfrak{S}$ of subsets of a set X satisfies the additional condition 4°. Then $\mathfrak{S}$ satisfies conditions (1) and (3). To show that $\mathfrak{S}$ is an algebra of subsets of X, it remains to show that $\mathfrak{S}$ satisfies (2). Now if $E \in \mathfrak{S}$ then by 3° we have $E^c = E_1 \cup \cdots \cup E_n$ where $\{E_1, \dots, E_n\}$ is a disjoint collection in $\mathfrak{S}$. Then by iterated application of 4° we have

$E_1 \cup E_2 \in \mathfrak{S}$, $E_1 \cup E_2 \cup E_3 \in \mathfrak{S}, \ldots, E_1 \cup \cdots \cup E_n \in \mathfrak{S}$ so that $E^c \in \mathfrak{S}$. This shows that $\mathfrak{S}$ satisfies (2). Therefore $\mathfrak{S}$ is an algebra of subsets of X. ∎

Unlike an algebra a semialgebra of subsets of a set is not closed under unions, or complementations, or differences. However as we shall show below, the collection of all finite unions of members of a semialgebra is an algebra.

Lemma 21.3. *Let $\mathfrak{S}$ be a semialgebra of subsets of a set X. Then every finite union of members of $\mathfrak{S}$ is equal to a finite disjoint union of members of $\mathfrak{S}$.*

Proof. Consider a finite collection $\{E_i : i = 1, \ldots, m\}$ in $\mathfrak{S}$. By 3° of Definition 21.1, for each $i = 1, \ldots, m$, there exists a finite disjoint collection $\{E_{i,j_i} : j_i = 0, \ldots, n_i\}$ in $\mathfrak{S}$ such that $E_{i,0} = E_i$ and $\bigcup_{j_i=0}^{n_i} E_{i,j_i} = X$. Then we have

$$(1) \qquad X = \Big(\bigcup_{j_1=0}^{n_1} E_{1,j_1}\Big) \cap \cdots \cap \Big(\bigcup_{j_m=0}^{n_m} E_{m,j_m}\Big) = \bigcup_{j_1=0}^{n_1} \cdots \bigcup_{j_m=0}^{n_m} (E_{1,j_1} \cap \cdots \cap E_{m,j_m}).$$

Now $\mathfrak{S}$ is closed under intersections. Then by iteration $\mathfrak{S}$ is closed under finite intersections. Thus $E_{1,j_1} \cap \cdots \cap E_{m,j_m} \in \mathfrak{S}$. Consider the following subcollection of $\mathfrak{S}$:

$$(2) \qquad \mathfrak{E} = \big\{E_{1,j_1} \cap \cdots \cap E_{m,j_m} : j_1 = 0, \ldots, n_1; \ldots; j_m = 0, \ldots, n_m\big\}.$$

Let us show that $\mathfrak{E}$ is a disjoint collection. Let F and G be two distinct members of $\mathfrak{E}$ given by $F = E_{1,s_1} \cap \cdots \cap E_{m,s_m}$ and $G = E_{1,t_1} \cap \cdots \cap E_{m,t_m}$. Then since F and G are distinct, we have $(s_1, \ldots, s_m) \neq (t_1, \ldots, t_m)$. Thus there exists at least one $i = 1, \ldots, m$ such that $s_i \neq t_i$. Then since $\{E_{i,j_i} : j_i = 0, \ldots, n_i\}$ is a disjoint collection we have $E_{i,s_i} \cap E_{i,t_i} = \emptyset$. Since F is a subset of E_{i,s_i} and G is a subset of E_{i,t_i}, we have $F \cap G \subset E_{i,s_i} \cap E_{i,t_i} = \emptyset$. This proves that $\mathfrak{E}$ is a disjoint collection.

Let us show next that each member of the collection $\{E_i : i = 1, \ldots, m\}$ is equal to a union of members of $\mathfrak{E}$. Take E_1 for instance. Recall that $\{E_{1,j_1} : j_1 = 0, \ldots, n_1\}$ is a disjoint collection in $\mathfrak{S}$ such that $E_{1,0} = E_1$. Thus we have $E_1 = E_{1,0} = E_{1,0} \cap X$ and then by (1) we have

$$\begin{aligned} E_1 &= E_{1,0} \cap \Big\{\bigcup_{j_1=0}^{n_1} \cdots \bigcup_{j_m=0}^{n_m} (E_{1,j_1} \cap \cdots \cap E_{m,j_m})\Big\} \\ &= \bigcup_{j_2=0}^{n_2} \cdots \bigcup_{j_m=0}^{n_m} (E_{1,0} \cap E_{2,j_2} \cap \cdots \cap E_{m,j_m}). \end{aligned}$$

This shows that E_1 is equal to a union of members of $\mathfrak{E}$. Similarly for $E_2, \ldots, E_m$. Then $\bigcup_{i=1}^m E_i$ is equal to a union of members of $\mathfrak{E}$. Then since $\mathfrak{E}$ is a finite disjoint collection of members of $\mathfrak{S}$, $\bigcup_{i=1}^m E_i$ is equal to a finite disjoint union of members of $\mathfrak{S}$. ∎

For an arbitrary collection $\mathfrak{C}$ of subsets of a set X, let $\alpha(\mathfrak{C})$ be the algebra of subsets of X generated by $\mathfrak{C}$, that is, the smallest algebra of subsets of X containing $\mathfrak{C}$. For any collection $\mathfrak{C}$ of subsets of X, $\alpha(\mathfrak{C})$ always exists. Indeed the collection $\mathfrak{P}(X)$ of all subsets of X is an algebra of subsets of X containing $\mathfrak{C}$. The intersection of all algebras of subsets of X containing $\mathfrak{C}$ is the smallest algebra of subsets of X containing $\mathfrak{C}$.

Let $\sigma(\mathfrak{C})$ be the σ-algebra of subsets of X generated by $\mathfrak{C}$. Then $\sigma(\mathfrak{C}) = \sigma(\alpha(\mathfrak{C}))$. This can be verified as follows. Since $\mathfrak{C} \subset \alpha(\mathfrak{C})$ we have $\sigma(\mathfrak{C}) \subset \sigma(\alpha(\mathfrak{C}))$. On the other hand a σ-algebra is also an algebra so that $\sigma(\mathfrak{C})$, as an algebra containing $\mathfrak{C}$, contains $\alpha(\mathfrak{C})$ the smallest algebra containing $\mathfrak{C}$. Thus $\alpha(\mathfrak{C}) \subset \sigma(\mathfrak{C})$ and then $\sigma(\alpha(\mathfrak{C})) \subset \sigma(\sigma(\mathfrak{C})) = \sigma(\mathfrak{C})$. Therefore $\sigma(\mathfrak{C}) = \sigma(\alpha(\mathfrak{C}))$.

For a semialgebra $\mathfrak{S}$ of subsets of X, $\alpha(\mathfrak{S})$ has the following characterization.

Theorem 21.4. *Let $\mathfrak{S}$ be a semialgebra of subsets of a set X. Then $\alpha(\mathfrak{S})$ is the collection of all finite unions of members of $\mathfrak{S}$.*

Proof. Let $\mathfrak{A}$ be the collection of all finite unions of members of $\mathfrak{S}$. Let us show that $\mathfrak{A} = \alpha(\mathfrak{S})$. Now since $\alpha(\mathfrak{S})$ is an algebra containing $\mathfrak{S}$, it contains all finite unions of members of $\mathfrak{S}$. Thus $\mathfrak{A} \subset \alpha(\mathfrak{S})$. By the definition of $\mathfrak{A}$, we have $\mathfrak{S} \subset \mathfrak{A}$. If we show that $\mathfrak{A}$ is an algebra, then it contains the smallest algebra containing $\mathfrak{S}$, that is, $\alpha(\mathfrak{S}) \subset \mathfrak{A}$, and therefore $\mathfrak{A} = \alpha(\mathfrak{S})$. Thus it remains to show that $\mathfrak{A}$ is an algebra.

To show that $\mathfrak{A}$ is an algebra of subsets of X, note first that $X \in \mathfrak{S}$ and thus $X \in \mathfrak{A}$. If $A, B \in \mathfrak{A}$, then A and B are each a finite union of members of $\mathfrak{S}$ so that $A \cup B$ is a finite union of members of $\mathfrak{S}$ and thus $A \cup B \in \mathfrak{A}$. This shows that $\mathfrak{A}$ is closed under unions. To show that $\mathfrak{A}$ is closed under complementations, let us show first that $\mathfrak{A}$ is closed under finite intersections. Let $\{A_i : i = 1, \dots, m\}$ be a finite collection in $\mathfrak{A}$. Then $A_i = \bigcup_{j_i=1}^{n_i} E_{i,j_i}$ where $\{E_{i,j_i} : j_i = 1, \dots, n_i\}$ is a collection in $\mathfrak{S}$. Then

$$\bigcap_{i=1}^{m} A_i = \bigcap_{i=1}^{m}\Big(\bigcup_{j_i=1}^{n_i} E_{i,j_i}\Big) = \bigcup_{j_1=1}^{n_1} \cdots \bigcup_{j_m=1}^{n_m} \big(E_{1,j_1} \cap \cdots \cap E_{m,j_m}\big).$$

Since $\mathfrak{S}$ is closed under finite intersections, we have $E_{1,j_1} \cap \cdots \cap E_{m,j_m} \in \mathfrak{S}$. Then $\bigcup_{j_1=1}^{n_1} \cdots \bigcup_{j_m=1}^{n_m} \big(E_{1,j_1} \cap \cdots \cap E_{m,j_m}\big)$ is a finite union of members of $\mathfrak{S}$ and is therefore in $\mathfrak{A}$. This shows that $\mathfrak{A}$ is closed under finite intersections. To prove that $\mathfrak{A}$ is closed under complementations, let $A \in \mathfrak{A}$. Then $A = \bigcup_{i=1}^{m} E_i$ where $\{E_i : i = 1, \dots, m\}$ is a collection in $\mathfrak{S}$ and $A^c = \bigcap_{i=1}^{m} E_i^c$. By 3° of Definition 21.1, E_i^c is a member of $\mathfrak{A}$. Then since $\mathfrak{A}$ is closed under finite intersections, $\bigcap_{i=1}^{m} E_i^c \in \mathfrak{A}$, that is, $A^c \in \mathfrak{A}$. Thus $\mathfrak{A}$ is closed under complementations. Finally since $X \in \mathfrak{A}$ and $\mathfrak{A}$ is closed under complementations we have $\emptyset = X^c \in \mathfrak{A}$. This completes the proof that $\mathfrak{A}$ is an algebra of subsets of X. ∎

[II] Additive Set Function on a Semialgebra

By Definition 1.19 a set function γ on a collection $\mathfrak{C}$ of subsets of a nonempty set X is said to be additive on $\mathfrak{C}$ if

$$E_1, E_2 \in \mathfrak{C},\ E_1 \cap E_2 = \emptyset,\ E_1 \cup E_2 \in \mathfrak{C} \Rightarrow \gamma(E_1 \cup E_2) = \gamma(E_1) + \gamma(E_2),$$

and finitely additive on $\mathfrak{C}$ if

$$\{E_i : i = 1, \ldots, m\} \subset \mathfrak{C}, \text{ disjoint}, \bigcup_{i=1}^{m} E_i \in \mathfrak{C} \Rightarrow \gamma\Big(\bigcup_{i=1}^{m} E_i\Big) = \sum_{i=1}^{m} \gamma(E_i).$$

If a set function γ on an algebra $\mathfrak{A}$ of subsets of X is additive on $\mathfrak{A}$, then it is finitely additive on $\mathfrak{A}$. This follows by iterated application of the additivity of γ on $\mathfrak{A}$. Indeed if $\{E_i : i = 1, \ldots, m\}$ is a disjoint collection in $\mathfrak{A}$, then by the fact that $\bigcup_{i=1}^{k} E_i \in \mathfrak{A}$ for $k = 1, \ldots, m$, we have $\gamma\big(\bigcup_{i=1}^{m} E_i\big) = \gamma\big(\bigcup_{i=1}^{m-1} E_i\big) + \gamma(E_m) = \cdots = \sum_{i=1}^{m} \gamma(E_i)$. Thus the finite additivity of γ on an algebra is derived from its additivity on the algebra.

Since a semialgebra $\mathfrak{S}$ of subsets of X is not closed under unions, the additivity of a set function γ on $\mathfrak{S}$ does not imply its finite additivity on $\mathfrak{S}$. For a disjoint collection $\{E_i : i = 1, \ldots, m\}$ in $\mathfrak{S}$ such that $\bigcup_{i=1}^{m} E_i \in \mathfrak{S}$, it is possible that $E_i \cup E_j \notin \mathfrak{S}$ for any pair in $\{E_i : i = 1, \ldots, m\}$ and then the additivity of γ on $\mathfrak{S}$ does not apply. (As an example of this, consider the semialgebra of subsets of $\mathbb{R}^2$ consisting of $\emptyset$ and all rectangles in $\mathbb{R}^2$ of the type $(a_1, b_1] \times (a_2, b_2]$. Let $E_1 = \big(0, \frac{2}{3}\big] \times \big(0, \frac{1}{3}\big]$, $E_2 = \big(\frac{2}{3}, 1\big] \times \big(0, \frac{2}{3}\big]$, $E_3 = \big(\frac{1}{3}, 1\big] \times \big(\frac{2}{3}, 1\big]$, $E_4 = \big(0, \frac{1}{3}\big] \times \big(\frac{1}{3}, 1\big]$, $E_5 = \big(\frac{1}{3}, \frac{2}{3}\big] \times \big(\frac{1}{3}, \frac{2}{3}\big]$, and $E = (0, 1] \times (0, 1]$. Then $\{E_i, i = 1, \ldots, 5\}$ is a disjoint collection in the semialgebra and the union of the five rectangles is the rectangle E but the union of any two of the five is not a rectangle and hence not in the semialgebra.) Thus the finite additivity of a set function on a semialgebra is something to be postulated, not to be derived by assuming its additivity on the semialgebra.

Lemma 21.5. *Let μ be a nonnegative extended real-valued set function on a semialgebra $\mathfrak{S}$ of subsets of set X with $\mu(\emptyset) = 0$. Suppose that μ is finitely additive on $\mathfrak{S}$, that is, if $\{E_i : i = 1, \ldots, m\}$ is a disjoint collection in $\mathfrak{S}$ such that $\bigcup_{i=1}^{m} E_i \in \mathfrak{S}$, then $\mu\big(\bigcup_{i=1}^{m} E_i\big) = \sum_{i=1}^{m} \mu(E_i)$. Let us define a set function $\widetilde{\mu}$ on $\alpha(\mathfrak{A})$ by setting*

$$\widetilde{\mu}(A) = \sum_{i=1}^{m} \mu(E_i)$$

if $A \in \alpha(\mathfrak{S})$ is given by $A = \bigcup_{i=1}^{m} E_i$ where $\{E_i : i = 1, \ldots, m\}$ is a disjoint collection in $\mathfrak{S}$. Then $\widetilde{\mu}$ is a well-defined nonnegative extended real-valued set function on $\alpha(\mathfrak{S})$ with $\widetilde{\mu}(\emptyset) = 0$ and $\widetilde{\mu} = \mu$ on $\mathfrak{S}$. We call $\widetilde{\mu}$ the extension of μ from $\mathfrak{S}$ to $\alpha(\mathfrak{S})$.

Proof. If $A \in \alpha(\mathfrak{A})$ then A is a finite union of members of $\mathfrak{S}$ by Theorem 21.4. Then A is a finite disjoint union of members of $\mathfrak{S}$ by Lemma 21.3. Let us show that $\widetilde{\mu}$ is a well-defined, that is, $\widetilde{\mu}(A)$ does not depend on the expression of A as a finite disjoint union of members of $\mathfrak{S}$. Thus suppose $A = \bigcup_{i=1}^{m} E_i$ as well as $A = \bigcup_{j=1}^{n} F_j$ where $\{E_i : i = 1, \ldots, m\}$ and $\{F_j : j = 1, \ldots, n\}$ are each a disjoint collection in $\mathfrak{S}$. Let us show that $\sum_{i=1}^{m} \mu(E_i) = \sum_{j=1}^{n} \mu(F_j)$. Let $G_{i,j} = E_i \cap F_j$. Then $\{G_{i,j} : i = 1, \ldots, m; j = 1, \ldots, n\}$ is a disjoint collection in $\mathfrak{S}$ and $E_i = \bigcup_{j=1}^{n} G_{i,j}$ for $i = 1, \ldots, m$ and $F_j = \bigcup_{i=1}^{m} G_{i,j}$ for $j = 1, \ldots, n$. The finite additivity of μ on $\mathfrak{S}$ implies $\mu(E_i) = \widetilde{\mu}(E_i) = \sum_{j=1}^{n} \mu(G_{i,j})$

and $\mu(F_j) = \widetilde{\mu}(F_j) = \sum_{i=1}^{m} \mu(G_{i,j})$. Therefore we have

$$\sum_{i=1}^{m} \mu(E_i) = \sum_{i=1}^{m} \Big\{ \sum_{j=1}^{n} \mu(G_{i,j}) \Big\} = \sum_{j=1}^{n} \Big\{ \sum_{i=1}^{m} \mu(G_{i,j}) \Big\} = \sum_{j=1}^{n} \mu(F_j).$$

This shows that $\widetilde{\mu}$ is well defined on $\alpha(\mathfrak{S})$. In particular it implies that $\widetilde{\mu}(E) = \mu(E)$ for $E \in \mathfrak{S}$. ∎

Proposition 21.6. *Let μ be a nonnegative extended real-valued finitely additive set function on a semialgebra $\mathfrak{S}$ of subsets of set X with $\mu(\emptyset) = 0$.*
(a) *The extension $\widetilde{\mu}$ of μ to $\alpha(\mathfrak{S})$ as defined in Lemma 21.5 is finitely additive on $\alpha(\mathfrak{S})$.*
(b) *Moreover $\widetilde{\mu}$ is countably additive on $\alpha(\mathfrak{S})$ if and only if μ is countably additive on $\mathfrak{S}$.*

Proof. 1. Let us show first that $\widetilde{\mu}$ is additive on $\alpha(\mathfrak{S})$. Let $A, B \in \alpha(\mathfrak{S})$ be such that $A \cap B = \emptyset$. By Theorem 21.4 and Lemma 21.3, we have $A = \bigcup_{i=1}^{m} E_i$ and $B = \bigcup_{j=1}^{n} F_j$ where $\{E_i : i = 1, \ldots, m\}$ and $\{F_j : j = 1, \ldots, n\}$ are each a disjoint collection in $\mathfrak{S}$. By the definition of $\widetilde{\mu}$ on $\alpha(\mathfrak{S})$, we have $\widetilde{\mu}(A) = \sum_{i=1}^{m} \mu(E_i)$ and $\widetilde{\mu}(B) = \sum_{j=1}^{n} \mu(F_j)$. Since A and B are disjoint, $\{E_1, \ldots, E_m, F_1, \ldots, F_n\}$ is a disjoint collection in $\mathfrak{S}$ and $\big[\bigcup_{i=1}^{m} E_i\big] \cup \big[\bigcup_{j=1}^{n} F_j\big] = A \cup B \in \alpha(\mathfrak{S})$. Thus by the definition of $\widetilde{\mu}$ on $\alpha(\mathfrak{S})$, we have $\widetilde{\mu}(A \cup B) = \sum_{i=1}^{m} \mu(E_i) + \sum_{j=1}^{n} \mu(F_j) = \widetilde{\mu}(A) + \widetilde{\mu}(B)$. This proves the additivity of $\widetilde{\mu}$ on $\alpha(\mathfrak{S})$. Since $\alpha(\mathfrak{S})$ is an algebra, the additivity of $\widetilde{\mu}$ on $\alpha(\mathfrak{S})$ implies the finite additivity of $\widetilde{\mu}$ on $\alpha(\mathfrak{S})$ by iterated application of the additivity.

2. If $\widetilde{\mu}$ is countably additive on $\alpha(\mathfrak{S})$, then since $\mathfrak{S} \subset \alpha(\mathfrak{S})$ and $\mu = \widetilde{\mu}$ on $\mathfrak{S}$, μ is countably additive on $\mathfrak{S}$. Conversely suppose μ is countably additive on $\mathfrak{S}$. Let us show that $\widetilde{\mu}$ is countably additive on $\alpha(\mathfrak{S})$. Thus let $(A_n : n \in \mathbb{N})$ be a disjoint sequence in $\alpha(\mathfrak{S})$ such that $\bigcup_{n \in \mathbb{N}} A_n \in \alpha(\mathfrak{S})$. Let $A = \bigcup_{n \in \mathbb{N}} A_n$.

For each $n \in \mathbb{N}$ we have $A_n \in \alpha(\mathfrak{S})$ so that by Theorem 21.4 and Lemma 21.3 there exists a finite disjoint collection $\{F_{n,j_n} : j_n = 1, \ldots, k_n\} \subset \mathfrak{S}$ such that

$$(1) \qquad A_n = \bigcup_{j_n=1}^{k_n} F_{n,j_n}.$$

Let

$$\mathfrak{F} = \big\{F_{n,j_n} : j_n = 1, \ldots, k_n; n \in \mathbb{N}\big\} \subset \mathfrak{S}.$$

Since $(A_n : n \in \mathbb{N})$ is a disjoint sequence, $\mathfrak{F}$ is a disjoint collection. Now $A \in \alpha(\mathfrak{S})$ implies according to Theorem 21.4 and Lemma 21.3 that there exists a finite disjoint collection $\mathfrak{E} = \{E_i : i = 1, \ldots, m\} \subset \mathfrak{S}$ such that

$$(2) \qquad A = \bigcup_{i=1}^{m} E_i.$$

Since the semialgebra $\mathfrak{S}$ is closed under intersections we have $F_{n,j_n} \cap E_i \in \mathfrak{S}$. If we let

$$\mathfrak{G} = \big\{F_{n,j_n} \cap E_i : j_n = 1, \ldots, k_n; n \in \mathbb{N}; i = 1, \ldots, m\big\},$$

then by the disjointness of the collection $\mathfrak{F}$ and the disjointness of the collection $\mathfrak{E}$, $\mathfrak{G}$ is a countable disjoint collection in $\mathfrak{S}$. Now by (1) we have

$$(3)\qquad A_n \cap E_i = \Big(\bigcup_{j_n=1}^{k_n} F_{n,j_n}\Big) \cap E_i = \bigcup_{j_n=1}^{k_n} (F_{n,j_n} \cap E_i)$$

and by the fact that $E_i \subset A$ and by (3) we have

$$(4)\qquad E_i = A \cap E_i = \Big(\bigcup_{n\in\mathbb{N}} A_n\Big) \cap E_i = \bigcup_{n\in\mathbb{N}} (A_n \cap E_i) = \bigcup_{n\in\mathbb{N}} \bigcup_{j_n=1}^{k_n} (F_{n,j_n} \cap E_i).$$

Since $A_n \in \alpha(\mathfrak{S})$, $E_i \in \mathfrak{S} \subset \alpha(\mathfrak{S})$ and $\alpha(\mathfrak{S})$ is an algebra, we have $A_n \cap E_i \in \alpha(\mathfrak{S})$. By the definition of $\widetilde{\mu}$ on $\alpha(\mathfrak{S})$, by (3) and by the disjointness of the collection $\mathfrak{G}$ we have

$$(5)\qquad \widetilde{\mu}(A_n \cap E_i) = \sum_{j_n=1}^{k_n} \mu\big(F_{n,j_n} \cap E_i\big).$$

By (4), the disjointness of the collection $\mathfrak{G} \subset \mathfrak{S}$ and the countable additivity of μ on $\mathfrak{S}$ and by (5) we have

$$(6)\qquad \mu(E_i) = \sum_{n\in\mathbb{N}} \Big[\sum_{j_n=1}^{k_n} \mu(F_{n,j_n} \cap E_i)\Big] = \sum_{n\in\mathbb{N}} \widetilde{\mu}(A_n \cap E_i).$$

Then by (2) and by the definition of $\widetilde{\mu}$ on $\alpha(\mathfrak{S})$, we have

$$(7)\qquad \widetilde{\mu}(A) = \sum_{i=1}^{m} \mu(E_i) = \sum_{i=1}^{m} \Big[\sum_{n\in\mathbb{N}} \widetilde{\mu}(A_n \cap E_i)\Big] = \sum_{n\in\mathbb{N}} \Big[\sum_{i=1}^{m} \widetilde{\mu}(A_n \cap E_i)\Big].$$

Since $A_n = A_n \cap A = A_n \cap \big(\bigcup_{i=1}^m E_i\big) = \bigcup_{i=1}^m (A_n \cap E_i)$, the additivity of $\widetilde{\mu}$ on $\alpha(\mathfrak{S})$ implies that $\sum_{i=1}^m \widetilde{\mu}(A_n \cap E_i) = \widetilde{\mu}(A_n)$. Substituting this in the last equality, we have $\widetilde{\mu}(A) = \sum_{n\in\mathbb{N}} \widetilde{\mu}(A_n)$. This proves the countable additivity of $\widetilde{\mu}$ on $\alpha(\mathfrak{S})$. ∎

Corollary 21.7. *Let μ be a nonnegative extended real-valued finitely additive set function on a semialgebra $\mathfrak{S}$ of subsets of set X with $\mu(\emptyset) = 0$ and let $\widetilde{\mu}$ be the extension of μ to $\alpha(\mathfrak{S})$ as defined in Lemma 21.5.*

(a) *$\widetilde{\mu}$ is monotone and finitely subadditive on $\alpha(\mathfrak{S})$ and μ is monotone and finitely subadditive on $\mathfrak{S}$.*

(b) *If μ is countably additive on $\mathfrak{S}$, then $\widetilde{\mu}$ is countably subadditive on $\alpha(\mathfrak{S})$ and μ is countably subadditive on $\mathfrak{S}$.*

Proof. 1. By Proposition 21.6, $\widetilde{\mu}$ is finitely additive on $\alpha(\mathfrak{S})$. Since $\alpha(\mathfrak{S})$ is an algebra, monotonicity and then finite subadditivity of $\widetilde{\mu}$ on $\alpha(\mathfrak{S})$ follows from the additivity of $\widetilde{\mu}$ on $\alpha(\mathfrak{S})$ by (a) of Lemma 1.22. Since $\mathfrak{S} \subset \alpha(\mathfrak{S})$ and $\mu = \widetilde{\mu}$ on $\mathfrak{S}$, the monotonicity and finite subadditivity of $\widetilde{\mu}$ on $\alpha(\mathfrak{S})$ imply the monotonicity and finite subadditivity of μ on $\mathfrak{S}$.

2. If μ is countably additive on $\mathfrak{S}$, then $\widetilde{\mu}$ is countably additive on $\alpha(\mathfrak{S})$ by Proposition 21.6. Since $\alpha(\mathfrak{S})$ is an algebra, the countable additivity of $\widetilde{\mu}$ on $\alpha(\mathfrak{S})$ implies its countable subadditivity on $\alpha(\mathfrak{S})$ by (b) of Lemma 1.22. Then since $\mathfrak{S} \subset \alpha(\mathfrak{S})$ and $\mu = \widetilde{\mu}$ on $\mathfrak{S}$, μ is countably subadditive on $\mathfrak{S}$. ∎

Theorem 21.8. *Let μ be a nonnegative extended real-valued finitely additive set function on a semialgebra $\mathfrak{S}$ of subsets of set X with $\mu(\emptyset) = 0$. Then μ is countably additive on $\mathfrak{S}$ if and only if μ is countably subadditive on $\mathfrak{S}$.*

Proof. If μ is countably additive on $\mathfrak{S}$, then μ is countably subadditive on $\mathfrak{S}$ by (b) of Corollary 21.7. Conversely suppose μ is countably subadditive on $\mathfrak{S}$. To show that μ is countably additive on $\mathfrak{S}$, let $(E_n : n \in \mathbb{N})$ be a disjoint sequence in $\mathfrak{S}$ such that $\bigcup_{n\in\mathbb{N}} E_n \in \mathfrak{S}$. Consider the extension $\widetilde{\mu}$ of μ to $\alpha(\mathfrak{S})$. For an arbitrary $N \in \mathbb{N}$, we have $\bigcup_{n=1}^N E_n \subset \bigcup_{n\in\mathbb{N}} E_n$. Since $\bigcup_{n=1}^N E_n \in \alpha(\mathfrak{S})$ and since $\bigcup_{n\in\mathbb{N}} E_n \in \mathfrak{S} \subset \alpha(\mathfrak{S})$, the monotonicity of $\widetilde{\mu}$ implies

$$\begin{aligned}\mu\Big(\bigcup_{n\in\mathbb{N}} E_n\Big) &= \widetilde{\mu}\Big(\bigcup_{n\in\mathbb{N}} E_n\Big) \ge \widetilde{\mu}\Big(\bigcup_{n=1}^{N} E_n\Big)\\ &= \sum_{n=1}^{N} \widetilde{\mu}(E_n) = \sum_{n=1}^{N} \mu(E_n).\end{aligned}$$

Since this holds for every $N \in \mathbb{N}$, we have $\mu\left(\bigcup_{n\in\mathbb{N}} E_n\right) \ge \sum_{n\in\mathbb{N}} \mu(E_n)$. On the other hand by the countable subadditivity of μ on $\mathfrak{S}$, we have $\mu\left(\bigcup_{n\in\mathbb{N}} E_n\right) \le \sum_{n\in\mathbb{N}} \mu(E_n)$. Thus we have $\mu\left(\bigcup_{n\in\mathbb{N}} E_n\right) = \sum_{n\in\mathbb{N}} \mu(E_n)$. This proves the countable additivity of μ on $\mathfrak{S}$. ∎

[III] Outer Measures Based on Additive Set Functions on a Semialgebra

Let μ be a nonnegative extended real-valued countably additive set function on a semialgebra $\mathfrak{S}$ of subsets of set X with $\mu(\emptyset) = 0$. The extension $\widetilde{\mu}$ of μ to the algebra $\alpha(\mathfrak{S})$ generated by $\mathfrak{S}$ is a nonnegative extended real-valued countably additive set function on $\alpha(\mathfrak{S})$ by Proposition 21.6. Thus by the Hopf Extension Theorem (Theorem 20.5), $\widetilde{\mu}$ can be extended to a measure on $\sigma(\alpha(\mathfrak{S})) = \sigma(\mathfrak{S})$ and therefore our set function μ on $\mathfrak{S}$ has an extension to a measure on $\sigma(\mathfrak{S})$. Let us address the uniqueness question of extensions of μ to a measure. Now since a semialgebra of subsets of X is a covering class for X, the set function μ^* on $\mathfrak{P}(X)$ defined by

$$\mu^*(E) = \inf\left\{\textstyle\sum_{n\in\mathbb{N}} \mu(E_n) : (E_n : n \in \mathbb{N}) \subset \mathfrak{S},\ \bigcup_{n\in\mathbb{N}} E_n \supset E\right\}$$

for $E \in \mathfrak{P}(X)$ is an outer measure on X according to Theorem 2.21. Similarly the set function $\widetilde{\mu}^*$ on $\mathfrak{P}(X)$ defined by

$$\widetilde{\mu}^*(E) = \inf\left\{\textstyle\sum_{n\in\mathbb{N}} \widetilde{\mu}(A_n) : (A_n : n \in \mathbb{N}) \subset \alpha(\mathfrak{S}),\ \bigcup_{n\in\mathbb{N}} A_n \supset E\right\}$$

is an outer measure on X. Let us compare the two measure spaces $(X, \mathfrak{M}(\mu^*), \mu^*)$ and $(X, \mathfrak{M}(\widetilde{\mu}^*), \widetilde{\mu}^*)$.

Lemma 21.9. *Let μ be a nonnegative extended real-valued countably additive set function on a semialgebra $\mathfrak{S}$ of subsets of set X with $\mu(\emptyset) = 0$ and let $\widetilde{\mu}$ be the extension of μ to the algebra $\alpha(\mathfrak{S})$. For the two outer measures μ^* and $\widetilde{\mu}^*$ on X based respectively on the set function μ on $\mathfrak{S}$ and the set function $\widetilde{\mu}$ on $\alpha(\mathfrak{S})$, we have $\mu^* = \widetilde{\mu}^*$ on $\mathfrak{P}(X)$ so that $\mathfrak{M}(\mu^*) = \mathfrak{M}(\widetilde{\mu}^*)$ and $(X, \mathfrak{M}(\mu^*), \mu^*) = (X, \mathfrak{M}(\widetilde{\mu}^*), \widetilde{\mu}^*)$.*

Proof. Let $E \in \mathfrak{P}(X)$. Since $\mathfrak{S} \subset \alpha(\mathfrak{S})$, the collection of all sequences in $\mathfrak{S}$ covering E is a subcollection of the collection of all sequences in $\alpha(\mathfrak{S})$ covering E. Then since $\widetilde{\mu} = \mu$ on $\mathfrak{S}$, we have $\widetilde{\mu}^*(E) \le \mu^*(E)$. On the other hand, every member of $\alpha(\mathfrak{S})$ is a finite union of members of $\mathfrak{S}$ so that a covering of E by a sequence in $\alpha(\mathfrak{S})$ is a covering of E by a sequence in $\mathfrak{S}$. Also for every $A \in \alpha(\mathfrak{S})$ there exists a finite disjoint collection $\{E_i : i = 1, \dots, m\}$ in $\mathfrak{S}$ such that $\widetilde{\mu}(A) = \sum_{i=1}^{m} \mu(E_i)$. Thus $\mu^*(E) \le \widetilde{\mu}^*(E)$. Therefore $\mu^* = \widetilde{\mu}^*$ on $\mathfrak{P}(X)$. This implies that $\mathfrak{M}(\mu^*) = \mathfrak{M}(\widetilde{\mu}^*)$ and $(X, \mathfrak{M}(\mu^*), \mu^*) = (X, \mathfrak{M}(\widetilde{\mu}^*), \widetilde{\mu}^*)$. ∎

Theorem 21.10. (Uniqueness of Extension to Measure) *Let μ be a nonnegative extended real-valued countably additive set function on a semialgebra $\mathfrak{S}$ of subsets of set X with $\mu(\emptyset) = 0$ and let μ^* be the outer measure on X based on the set function μ.*
(a) *μ^* is a regular outer measure on X.*
(b) *$\mathfrak{S} \subset \mathfrak{M}(\mu^*)$ and $\mu = \mu^*$ on $\mathfrak{S}$.*
(c) *Let $(X, \mathfrak{F}, \nu)$ be a measure space such that $\mathfrak{S} \subset \mathfrak{F} \subset \mathfrak{M}(\mu^*)$ and $\nu = \mu$ on $\mathfrak{S}$. If μ is σ-finite on $\mathfrak{S}$, then $\nu = \mu^*$ on $\mathfrak{F}$.*
(d) *In particular, if μ is σ-finite on $\mathfrak{S}$, then an extension of μ to a measure on $\sigma(\mathfrak{S})$ is unique.*

Proof. 1. Let $\widetilde{\mu}$ be the extension of μ to $\sigma(\mathfrak{S})$ and let $\widetilde{\mu}^*$ be the outer measure on X based on $\widetilde{\mu}$. By Lemma 21.9, $\mu^* = \widetilde{\mu}^*$ on $\mathfrak{P}(X)$. Since $\widetilde{\mu}^*$ is a regular outer measure on X by Proposition 20.9, μ^* is a regular outer measure. This proves (a).

2. We have $\mathfrak{S} \subset \alpha(\mathfrak{S})$. Now since $\widetilde{\mu}$ is additive on the algebra $\alpha(\mathfrak{S})$, we have $\alpha(\mathfrak{S}) \subset \mathfrak{M}(\widetilde{\mu}^*)$ by Theorem 20.1. Then since $\mathfrak{M}(\widetilde{\mu}^*) = \mathfrak{M}(\mu^*)$ by Lemma 21.9, we have $\mathfrak{S} \subset \mathfrak{M}(\mu^*)$. Since $\mu = \widetilde{\mu}$ on $\mathfrak{S}$ by Lemma 21.5 and $\widetilde{\mu} = \widetilde{\mu}^*$ on $\alpha(\mathfrak{S})$ by Theorem 20.1 and $\widetilde{\mu}^* = \mu^*$ on $\mathfrak{P}(X)$ by Lemma 21.9, we have $\mu = \mu^*$ on $\mathfrak{S}$. This proves (b).

3. If μ is σ-finite on $\mathfrak{S}$, then $\widetilde{\mu}$ is σ-finite on $\alpha(\mathfrak{S})$. If $\mathfrak{S} \subset \mathfrak{F} \subset \mathfrak{M}(\mu^*)$, then since $\mathfrak{M}(\mu^*) = \mathfrak{M}(\widetilde{\mu}^*)$ by Lemma 21.9 we have $\alpha(\mathfrak{S}) \subset \mathfrak{F} \subset \mathfrak{M}(\widetilde{\mu}^*)$. If $\nu = \mu$ on $\mathfrak{S}$, then $\nu = \widetilde{\mu}$ on $\alpha(\mathfrak{S})$. Then by Theorem 20.12 (Uniqueness of Extension), we have $\nu = \widetilde{\mu}^*$ on $\mathfrak{F}$. Since $\widetilde{\mu}^* = \mu^*$ on $\mathfrak{P}(X)$ by Lemma 21.9, we have $\nu = \mu^*$ on $\mathfrak{F}$. This proves (c).

4. (d) is a particular case of (c) with $\mathfrak{F} = \sigma(\mathfrak{S})$. ∎

Theorem 21.11. *Let $\mathfrak{S}$ be a semialgebra of subsets of a set X. Let μ_1 and μ_2 be two measures on the σ-algebra $\sigma(\mathfrak{S})$. If $\mu_1 = \mu_2$ on $\mathfrak{S}$ and μ_1 and μ_2 are σ-finite on $\mathfrak{S}$, then $\mu_1 = \mu_2$ on $\sigma(\mathfrak{S})$.*

Proof. If μ_1 and μ_2 are σ-finite on the semialgebra $\mathfrak{S}$, then they are σ-finite on the algebra $\alpha(\mathfrak{S})$. Since every member of $\alpha(\mathfrak{S})$ is a finite disjoint union of members of $\mathfrak{S}$, if $\mu_1 = \mu_2$ on $\mathfrak{S}$, then $\mu_1 = \mu_2$ on $\alpha(\mathfrak{S})$. Then since $\sigma(\mathfrak{S}) = \sigma(\alpha(\mathfrak{S}))$, $\mu_1 = \mu_2$ on $\sigma(\mathfrak{S})$ by Theorem 20.14. ∎

The μ^*-measurability condition for an outer measure μ^* based on a countably additive set function μ on a semialgebra $\mathfrak{S}$ of subsets of a set X can be reduced to testing by the members of $\mathfrak{S}$ only.

Theorem 21.12. *Let μ be a nonnegative extended real-valued countably additive set function on a semialgebra $\mathfrak{S}$ of subsets of set X with $\mu(\emptyset) = 0$. Then for every $E \in \mathfrak{P}(X)$, the μ^*-measurability condition*

(1) $$\mu^*(B) = \mu^*(B \cap E) + \mu^*(B \cap E^c) \quad \textit{for every } B \in \mathfrak{P}(X)$$

is equivalent to the condition

(2) $$\mu^*(F) = \mu^*(F \cap E) + \mu^*(F \cap E^c) \quad \textit{for every } F \in \mathfrak{S}.$$

Proof. Since a semialgebra $\mathfrak{S}$ of subsets of a set X is a covering class for X, this theorem is a particular case of Theorem 2.24. ∎

§22 Lebesgue-Stieltjes Measure Spaces

[I] Lebesgue-Stieltjes Outer Measures

Let g be an arbitrary real-valued increasing function on $\mathbb{R}$. For $x \in \mathbb{R}$, the left-limit $g(x-) = \lim_{\xi \uparrow x} g(\xi)$ and the right-limit $g(x+) = \lim_{\xi \downarrow x} g(\xi)$ exist and furthermore $-\infty < g(x-) \le g(x) \le g(x+) < \infty$.
Let us write $g(-\infty) = \lim_{x \to -\infty} g(x)$ and $g(\infty) = \lim_{x \to \infty} g(x)$. Then $-\infty \le g(-\infty) \le g(\infty) \le \infty$.
The function g is left-continuous at $x \in \mathbb{R}$ if and only if $g(x) = g(x-)$ and right-continuous at x if and only if $g(x) = g(x+)$. There exists a countable set $E \subset \mathbb{R}$ such that g is continuous at every $x \in E^c$.

Definition 22.1. *Let g be a real-valued increasing function on $\mathbb{R}$. We define a function g_r on $\mathbb{R}$ by $g_r(x) = g(x+)$ for $x \in \mathbb{R}$ and call it the right-continuous modification of g.*

Observation 22.2. The right-continuous modification g_r of a real-valued increasing function g on $\mathbb{R}$ has the following properties:
1° g_r is a real-valued increasing function on $\mathbb{R}$.
2° $g_r = g$ except at countably many points in $\mathbb{R}$.
3° $\lim_{\xi \uparrow x} g_r(\xi) = \lim_{\xi \uparrow x} g(\xi)$ and $\lim_{\xi \downarrow x} g_r(\xi) = \lim_{\xi \downarrow x} g(\xi)$ at every $x \in \mathbb{R}$.
4° g_r is right-continuous on $\mathbb{R}$.
5° $g_r(-\infty) = g(-\infty)$ and $g_r(\infty) = g(\infty)$.
6° g_r is continuous at $x \in \mathbb{R}$ if and only if g is continuous at x. If g is continuous at $x \in \mathbb{R}$
then $g(x) = g_r(x)$.
7° There exists a null set E in the Lebesgue measure space $(\mathbb{R}, \mathfrak{M}_L, \mu_L)$ such that the derivatives $g'(x)$ and $g_r'(x)$ exist and $g'(x) = g_r'(x)$ for every $x \in E^c$.

Proof. 1. (a) is immediate from the definition of g_r.

2. $g_r(x) \ne g(x)$, that is, $g(x+) \ne g(x)$, if and only if g is not right-continuous at x. The set of points at which g is not right-continuous is a subset of the set of points at which g is not continuous. The latter set is a countable set since g is a real-valued increasing function on $\mathbb{R}$. Thus g is not right-continuous at countably many points only. Then $g_r = g$ except at countably many points in $\mathbb{R}$.

3. Since g_r and g are both real-valued increasing functions on $\mathbb{R}$, the one-sided limits $\lim_{\xi \uparrow x} g_r(\xi)$, $\lim_{\xi \uparrow x} g(\xi)$, $\lim_{\xi \downarrow x} g_r(\xi)$, and $\lim_{\xi \downarrow x} g(\xi)$ exist in $\mathbb{R}$ at every $x \in \mathbb{R}$. Then since $g_r = g$ except at countably many points in $\mathbb{R}$ by 2°, the equalities in 3° hold. (For instance, for $x \in \mathbb{R}$ let $(\xi_n : n \in \mathbb{N})$ be a decreasing sequence such that $\xi_n > x$ for every $n \in \mathbb{N}$ and $\xi_n \downarrow x$. Then $\lim_{n \to \infty} g_r(\xi_n) = \lim_{\xi \downarrow x} g_r(\xi)$ and $\lim_{n \to \infty} g(\xi_n) = \lim_{\xi \downarrow x} g(\xi)$. Then since $g_r = g$ except at countably many points in $\mathbb{R}$ we can select $(\xi_n : n \in \mathbb{N})$ so that $g_r(\xi_n) = g(\xi_n)$ for every $n \in \mathbb{N}$. Then $\lim_{n \to \infty} g_r(\xi_n) = \lim_{n \to \infty} g(\xi_n)$ so that $\lim_{\xi \downarrow x} g_r(\xi) = \lim_{\xi \downarrow x} g(\xi)$.

4. To show that g_r is right-continuous at every $x \in \mathbb{R}$, note that by 3° and by the definition of g_r we have $\lim_{\xi\downarrow x} g_r(\xi) = \lim_{\xi\downarrow x} g(\xi) = g_r(x)$.

5. Let $(\xi_n : n \in \mathbb{N})$ be a sequence such that $\xi_n \downarrow -\infty$. Then $\lim_{n\to\infty} g_r(\xi_n) = g_r(-\infty)$ and $\lim_{n\to\infty} g(\xi_n) = g(-\infty)$. By 2° we can select $(\xi_n : n \in \mathbb{N})$ so that $g_r(\xi_n) = g(\xi_n)$ for every $n \in \mathbb{N}$. Then $\lim_{n\to\infty} g_r(\xi_n) = \lim_{n\to\infty} g(\xi_n)$. Thus $g_r(-\infty) = g(-\infty)$. We show similarly that $g_r(\infty) = g(\infty)$.

6. Since g_r and g are real-valued increasing functions on $\mathbb{R}$, g_r is continuous at $x \in \mathbb{R}$ if and only if $\lim_{\xi\uparrow x} g_r(\xi) = \lim_{\xi\downarrow x} g_r(\xi)$ and similarly g_r is continuous at x if and only if $\lim_{\xi\uparrow x} g(\xi) = \lim_{\xi\downarrow x} g(\xi)$. Now by 3° we have $\lim_{\xi\uparrow x} g_r(\xi) = \lim_{\xi\downarrow x} g_r(\xi)$ if and only if $\lim_{\xi\uparrow x} g(\xi) = \lim_{\xi\downarrow x} g(\xi)$. Thus g_r is continuous at x if and only if g is continuous at x. If g is continuous at x, then $g(x) = \lim_{\xi\downarrow x} g(\xi) = g(x+) = g_r(x)$.

7. Since g and g_r are real-valued increasing functions on $\mathbb{R}$ there exist two null sets E_1 and E_2 in the Lebesgue measure space $(\mathbb{R}, \mathfrak{M}_L, \mu_L)$ such that g is differentiable at every $x \in E_1^c$ and g_r is differentiable at every $x \in E_2^c$. Let $E = E_1 \cup E_2$. Then E is a null set in $(\mathbb{R}, \mathfrak{M}_L, \mu_L)$ and both g and g_r are differentiable at every $x \in E^c$.

Let us show that $g'(x) = g_r'(x)$ at every $x \in E^c$. Now differentiability implies continuity. Thus both g and g_r are continuous at x. Then by 6° we have $g(x) = g_r(x)$. Let $(\xi_n : n \in \mathbb{N})$ be a decreasing sequence such that $\xi_n > x$ for every $n \in \mathbb{N}$ and $\xi_n \downarrow x$. Since g and g_r are differentiable at x, we have

$$\lim_{n\to\infty} \frac{g(\xi_n) - g(x)}{\xi_n - x} = g'(x) \text{ and } \lim_{n\to\infty} \frac{g_r(\xi_n) - g_r(x)}{\xi_n - x} = g'(x).$$

According to 2°, $g = g_r$ except at countably many points in $\mathbb{R}$. Thus we can select $(\xi_n : n \in \mathbb{N})$ so that $g(\xi_n) = g_r(\xi_n)$ for every $n \in \mathbb{N}$. As we noted above, $g(x) = g_r(x)$ Thus we have

$$\lim_{n\to\infty} \frac{g(\xi_n) - g(x)}{\xi_n - x} = \lim_{n\to\infty} \frac{g_r(\xi_n) - g_r(x)}{\xi_n - x}.$$

Therefore we have $g'(x) = g_r'(x)$ ∎

Definition 22.3. *Let g be a real-valued increasing function on $\mathbb{R}$ and let g_r be its right-continuous modification. Let us define a set function ℓ_g on the semialgebra $\mathfrak{I}_{oc}$ of subsets of $\mathbb{R}$ by setting $\ell_g(\emptyset) = 0$ and for $(a, b] \in \mathfrak{I}_{oc}$*

$$\ell_g((a, b]) = g_r(b) - g_r(a),$$

with the understanding that $(a, \infty] := (a, \infty)$ and $g_r(\infty) = \lim_{x\to\infty} g_r(x)$. Let μ_g^* be the outer measure on $\mathbb{R}$ based on the set function ℓ_g on $\mathfrak{I}_{oc}$, that is, for every $E \in \mathfrak{P}(\mathbb{R})$, we define

$$\mu_g^*(E) = \inf\left\{ \textstyle\sum_{n\in\mathbb{N}} \ell_g(I_n) : (I_n : n \in \mathbb{N}) \subset \mathfrak{I}_{oc}, \bigcup_{n\in\mathbb{N}} I_n \supset E \right\}.$$

We call ℓ_g the set function on $\mathfrak{I}_{oc}$ based on g and μ_g^* the outer measure on $\mathbb{R}$ based on ℓ_g.

Theorem 22.4. *Let g be a real-valued increasing function on $\mathbb{R}$ and let ℓ_g be the set function on the semialgebra $\mathfrak{I}_{oc}$ and μ_g^* be the outer measure on $\mathbb{R}$ based on ℓ_g.*

(a) *ℓ_g is a nonnegative extended real-valued countably additive set function on the semi-*

algebra $\mathfrak{I}_{oc}$ *with* $\ell_g(\emptyset) = 0$.

(b) μ_g^* *is a regular outer measure on* $\mathbb{R}$.

(c) $\mathfrak{I}_{oc} \subset \mathfrak{M}(\mu_g^*)$ *(implying* $\mathfrak{B}_{\mathbb{R}} \subset \mathfrak{M}(\mu_g^*)$ *), and* $\ell_g = \mu_g^*$ *on* $\mathfrak{I}_{oc}$.

Proof. 1. Let us verify the countable additivity of the set function ℓ_g on the semialgebra $\mathfrak{I}_{oc}$. Thus let $(I_n : n \in \mathbb{N})$ be a disjoint sequence in $\mathfrak{I}_{oc}$ such that $\bigcup_{n\in\mathbb{N}} I_n \in \mathfrak{I}_{oc}$. Let $I_n = (a_n, b_n]$ for $n \in \mathbb{N}$ and let $I = \bigcup_{n\in\mathbb{N}} I_n = (a, b]$.

Since the right-continuous modification g_r of g is a real-valued increasing function on $\mathbb{R}$, the ordering by $\leq$ in the set $\{a, b\} \cup \{a_n, b_n : n \in \mathbb{N}\}$ is preserved by the mapping g_r. In particular we have $g_r(a_n) \leq g_r(b_n)$ for every $n \in \mathbb{N}$. Let $J_n = \big(g_r(a_n), g_r(b_n)\big]$ in case $g_r(a_n) < g_r(b_n)$ and let $J_n = \{g_r(b_n)\}$ in case $g_r(a_n) = g_r(b_n)$ for each $n \in \mathbb{N}$. Let $J = \big(g_r(a), g_r(b)\big]$. Let J_n° be the interior of J_n for $n \in \mathbb{N}$. Note that if $J_n = \{g_r(b_n)\}$ then $J_n^\circ = \emptyset$. Since g_r preserves the ordering by $\leq$, $\{J_n^\circ : n \in \mathbb{N}\}$ is a disjoint collection. (This is easily verified by a contradiction argument. Let us note also that $\{J_n : n \in \mathbb{N}\}$ may not be a disjoint collection. For instance suppose $a_n < \xi_1 < b_n < a_m < b_m < \xi_2$ and the real-valued increasing function g_r is constant on (ξ_1, ξ_2). In this case we have $J_n = \big(g_r(a_n), g_r(b_n)\big]$ and $J_m = \{g_r(b_m)\} = \{g_r(b_n)\}$ so that $J_n \cap J_m = \{g_r(b_n)\}$.)

The fact that g_r preserves the ordering by $\leq$ implies also $J = \bigcup_{n\in\mathbb{N}} J_n$. (Note that while $g_r(I_n) \subset J_n$, that is, $g_r\big((a_n, b_n]\big) \subset \big(g_r(a_n), g_r(b_n)\big]$, we may have $g_r(I_n) \neq J_n$. This is the case if g_r has a jump at some $x \in (a_n, b_n]$.)

Now consider the Lebesgue measure space $\big(\mathbb{R}, \mathfrak{M}_L, \mu_L\big)$. We have

$$\begin{aligned}\mu_L(J) &= \mu_L\Big(\bigcup_{n\in\mathbb{N}} J_n\Big) = \mu_L\Big(\bigcup_{n\in\mathbb{N}} J_n^\circ\Big)\\ &= \sum_{n\in\mathbb{N}} \mu_L(J_n^\circ) = \sum_{n\in\mathbb{N}} \mu_L(J_n),\end{aligned}$$

where the second equality is from the fact that $\bigcup_{n\in\mathbb{N}} J_n^\circ \subset \bigcup_{n\in\mathbb{N}} J_n$ and $\bigcup_{n\in\mathbb{N}} J_n \setminus \bigcup_{n\in\mathbb{N}} J_n^\circ$ is a countable set having Lebesgue measure 0 and the third equality is by the disjointness of the collection $\{J_n^\circ : n \in \mathbb{N}\}$ and by the countable additivity of μ_L. Since the Lebesgue measure of an interval is equal to its length, the last equality can be written as

$$g_r(b) - g_r(a) = \sum_{n\in\mathbb{N}} \{g_r(b_n) - g_r(a_n)\},$$

that is,

$$\ell_g(I) = \sum_{n\in\mathbb{N}} \ell_g(I_n).$$

This proves the countable additivity of ℓ_g on $\mathfrak{I}_{oc}$.

2. Since ℓ_g is a nonnegative extended real-valued countably additive set function on the semialgebra $\mathfrak{I}_{oc}$, the outer measure μ_g^* is a regular outer measure, $\mathfrak{I}_{oc} \subset \mathfrak{M}(\mu_g^*)$ and $\ell_g = \mu_g^*$ on $\mathfrak{I}_{oc}$ by Theorem 21.10. Since $\sigma(\mathfrak{I}_{oc}) = \mathfrak{B}_{\mathbb{R}}$ and since $\mathfrak{M}(\mu_g^*)$ is a σ-algebra containing $\mathfrak{I}_{oc}$, we have $\mathfrak{B}_{\mathbb{R}} \subset \mathfrak{M}(\mu_g^*)$. ∎

Definition 22.5. *Let g be a real-valued increasing function on* $\mathbb{R}$. *We call* μ_g^* *the Lebesgue-Stieltjes outer measure on* $\mathbb{R}$ *determined by g. Let* μ_g *be the restriction of* μ_g^* *to the* σ-*algebra* $\mathfrak{M}(\mu_g^*)$ *of all* μ_g^*-*measurable members of* $\mathfrak{P}(\mathbb{R})$ *and call it the Lebesgue-Stieltjes* *measure determined by g. We call* $(\mathbb{R}, \mathfrak{M}(\mu_g^*), \mu_g)$ *the Lebesgue-Stieltjes measure space on* $\mathbb{R}$ *determined by g.*

Remark 22.6. By Observation 3.2, the Lebesgue outer measure μ_L^* on $\mathbb{R}$ is also given by

$$\mu_L^*(E) = \inf\Big\{ \textstyle\sum_{n\in\mathbb{N}} \ell(I_n) : (I_n : n \in \mathbb{N}) \subset \mathfrak{I}_{oc}, \bigcup_{n\in\mathbb{N}} I_n \supset E\Big\}$$

for every $E \in \mathfrak{P}(\mathbb{R})$. The identity function ι on $\mathbb{R}$ is a real-valued increasing function on $\mathbb{R}$ and since it is continuous on $\mathbb{R}$, it is identical with its right-continuous modification ι_r. Then $\ell\big((a,b]\big) = b - a = \iota(b) - \iota(a) = \iota_r(b) - \iota_r(a) = \ell_\iota\big((a,b]\big)$ for $(a,b] \in \mathfrak{I}_{oc}$. Thus the Lebesgue measure space $(\mathbb{R}, \mathfrak{M}_L, \mu_L)$ is a particular case of the Lebesgue-Stieltjes measure space $(\mathbb{R}, \mathfrak{M}(\mu_g^*), \mu_g)$ with $g = \iota$.

Proposition 22.7. *In a Lebesgue-Stieltjes measure space* $(\mathbb{R}, \mathfrak{M}(\mu_g^*), \mu_g)$ *determined by a real-valued increasing function g on* $\mathbb{R}$*, we have*

(1) $\mu_g\big((a,b]\big) = g(b+) - g(a+)$,

(2) $\mu_g\big([a,b)\big) = g(b-) - g(a-)$,

(3) $\mu_g\big((a,b)\big) = g(b-) - g(a+)$,

(4) $\mu_g\big([a,b]\big) = g(b+) - g(a-)$,

(5) $\mu_g\big(\{c\}\big) = g(c+) - g(c-)$ *for* $c \in \mathbb{R}$.

Thus $\mu_g\big(\{c\}\big) = 0$ *if and only if* $c \in \mathbb{R}$ *is a point of continuity of g. If g is discontinuous at* $c \in \mathbb{R}$*, then* $\mu_g\big(\{c\}\big) = g(c+) - g(c-) > 0$.

Proof. Note that $(a,b]$, $[a,b)$, (a,b), $[a,b]$, and $\{c\}$ are all members of $\mathfrak{B}_{\mathbb{R}}$. By Theorem 22.4, $\mathfrak{B}_{\mathbb{R}} \subset \mathfrak{M}(\mu_g^*)$. Thus these sets are all members of $\mathfrak{M}(\mu_g^*)$. To show (1), note that by Theorem 22.4, $\ell_g = \mu_g$ on $\mathfrak{I}_{oc}$ so that

$$\mu_g\big((a,b]\big) = \ell_g\big((a,b]\big) = g_r(b) - g_r(a) = g(b+) - g(a+).$$

To prove (3), note that $\big(a, b - \frac{1}{n}\big] \uparrow (a,b)$ as $n \to \infty$. Thus

$$\begin{aligned}\mu_g\big((a,b)\big) &= \lim_{n\to\infty} \mu_g\Big(\big(a, b - \tfrac{1}{n}\big]\Big) = \lim_{n\to\infty} \Big\{g_r\big(b - \tfrac{1}{n}\big) - g_r(a)\Big\} \\ &= \lim_{n\to\infty} \Big\{g\big(b - \tfrac{1}{n}+\big) - g(a+)\Big\} = g(b-) - g(a+).\end{aligned}$$

To prove (5), note that $\{c\} = (a,c] \setminus (a,c)$. Then by (1) and (3), we have

$$\begin{aligned}\mu_g\big(\{c\}\big) &= \mu_g\big((a,c]\big) - \mu_g\big((a,c)\big) \\ &= \big\{g(c+) - g(a+)\big\} - \big\{g(c-) - g(a+)\big\} \\ &= g(c+) - g(c-).\end{aligned}$$

The equalities (2) and (4) are proved likewise. ■

Theorem 22.8. *Let $(\mathbb{R}, \mathfrak{M}(\mu_g^*), \mu_g)$ be a Lebesgue-Stieltjes measure space on $\mathbb{R}$ determined by a real-valued increasing function g on $\mathbb{R}$.*
(a) *$(\mathbb{R}, \mathfrak{M}(\mu_g^*), \mu_g)$ is a σ-finite measure space.*
(b) *$\mu_g(\mathbb{R}) = g(\infty) - g(-\infty)$.*
(c) *$(\mathbb{R}, \mathfrak{M}(\mu_g^*), \mu_g)$ is a finite measure space if and only if g is a bounded function on $\mathbb{R}$.*

Proof. $\{(n-1, n] : n \in \mathbb{Z}\}$ is a countable disjoint collection in $\mathfrak{I}_{oc} \subset \mathfrak{M}(\mu_g^*)$ with $\bigcup_{n\in\mathbb{Z}}(n-1, n] = \mathbb{R}$ and $\mu_g\big((n-1, n]\big) = g_r(n) - g_r(n-1) < \infty$. This proves (a). Since $(-n, n] \uparrow \mathbb{R}$, we have $\mu_g(\mathbb{R}) = \lim_{n\to\infty} \mu_g\big((-n, n]\big) = \lim_{n\to\infty} \{g_r(n) - g_r(-n)\} = g(\infty) - g(-\infty)$. This proves (b). By (b), $\mu_g(\mathbb{R}) < \infty$ if and only if both $g(\infty)$ and $g(-\infty)$ are finite. Since g is an increasing function on $\mathbb{R}$, this is equivalent to the existence of some $M > 0$ such that $-M \le g(-\infty) \le g(x) \le g(\infty) \le M$ for every $x \in \mathbb{R}$. ∎

Theorem 22.9. *Let g and h be two real-valued increasing functions on $\mathbb{R}$. Suppose $g = h$ on a countable dense subset D of $\mathbb{R}$. Then $(\mathbb{R}, \mathfrak{M}(\mu_g^*), \mu_g) = (\mathbb{R}, \mathfrak{M}(\mu_h^*), \mu_h)$.*

Proof. Let $x \in \mathbb{R}$. Since D is dense in $\mathbb{R}$, there exists a sequence $(x_n : n \in \mathbb{N})$ in D such that $x_n \downarrow x$ as $n \to \infty$. Since g and h are increasing functions, we have $g(x+) = \lim_{n\to\infty} g(x_n)$ and $h(x+) = \lim_{n\to\infty} h(x_n)$. Since $g(x_n) = h(x_n)$ for $n \in \mathbb{N}$, we have $g(x+) = h(x+)$ for every $x \in \mathbb{R}$, that is, $g_r = h_r$ on $\mathbb{R}$. Thus by Definition 22.3, $\ell_g = \ell_h$ on $\mathfrak{I}_{oc}$ and $\mu_g^* = \mu_h^*$ on $\mathfrak{P}(\mathbb{R})$. Then $\mathfrak{M}(\mu_g^*) = \mathfrak{M}(\mu_h^*)$ and $\mu_g = \mu_h$. ∎

Note in particular that if g_r is the right-continuous modification of a real-valued increasing function g on $\mathbb{R}$, then $g_r = g$ on $\mathbb{R}$ except at countably many points so that $(\mathbb{R}, \mathfrak{M}(\mu_{g_r}^*), \mu_{g_r}) = (\mathbb{R}, \mathfrak{M}(\mu_g^*), \mu_g)$ by Theorem 22.9.

[II] Regularity of the Lebesgue-Stieltjes Outer Measures

The Lebesgue-Stieltjes outer measure μ_g^* on $\mathbb{R}$ determined by a real-valued increasing function g on $\mathbb{R}$ is always a regular outer measure as we showed in Theorem 22.4. Beyond this, μ_g^* has the following regularity properties.

Lemma 22.10. *Let μ_g^* be the Lebesgue-Stieltjes outer measure on $\mathbb{R}$ determined by a real-valued increasing function g on $\mathbb{R}$. Let $E \in \mathfrak{P}(\mathbb{R})$.*
(a) *For every $\varepsilon > 0$, there exists an open set O in $\mathbb{R}$ such that $O \supset E$ and $\mu_g^*(E) \le \mu_g^*(O) \le \mu_g^*(E) + \varepsilon$.*
(b) *There exists a G_δ-set in $\mathbb{R}$ such that $G \supset E$ and $\mu_g^*(G) = \mu_g^*(E)$.*

Proof. 1. Let $E \in \mathfrak{P}(\mathbb{R})$. By the definition of $\mu_g^*(E)$ as an infimum, there exists a sequence $(I_n : n \in \mathbb{N})$ in $\mathfrak{I}_{oc}$ such that $\bigcup_{n\in\mathbb{N}} I_n \supset E$ and $\mu_g^*(E) \le \sum_{n\in\mathbb{N}} \ell_g(I_n) \le \mu_g^*(E) + \frac{\varepsilon}{2}$. There is no generality lost if we assume that I_n is a finite interval for every $n \in \mathbb{N}$. (If I_n is an infinite interval, we decompose it into countable disjoint intervals in the class $\mathfrak{I}_{oc}$.) Let $I_n = (a_n, b_n]$ for $n \in \mathbb{N}$. Then $\ell_g(I_n) = g_r(b_n) - g_r(a_n)$. By the right-continuity of g_r, there exists $c_n \ge b_n$ such that $g_r(c_n) - g_r(b_n) < \frac{\varepsilon}{2^{n+1}}$. Then the open set $O = \bigcup_{n\in\mathbb{N}}(a_n, c_n)$

contains E and

$$\begin{aligned}
\mu_g^*(E) &\le \mu_g^*(O) = \mu_g^*\Big(\bigcup_{n\in\mathbb{N}}(a_n,c_n)\Big) \le \sum_{n\in\mathbb{N}}\mu_g^*\big((a_n,c_n)\big)\\
&= \sum_{n\in\mathbb{N}}\{g(c_n-)-g(a_n+)\} \le \sum_{n\in\mathbb{N}}\{g_r(c_n)-g_r(a_n)\}\\
&\le \sum_{n\in\mathbb{N}}\Big\{g_r(b_n)-g_r(a_n)+\frac{\varepsilon}{2^{n+1}}\Big\} = \sum_{n\in\mathbb{N}}\ell_g(I_n)+\frac{\varepsilon}{2}\\
&\le \mu_g^*(E)+\varepsilon.
\end{aligned}$$

This proves (a).

2. To prove (b), note that by (a) for every $n \in \mathbb{N}$ there exists an open set $O_n \supset E$ such that $\mu_g^*(E) \le \mu_g^*(O_n) \le \mu_g^*(E) + \frac{1}{n}$. If we let $G = \bigcap_{n\in\mathbb{N}} O_n$, then G is a G_δ-set, $G \supset E$, and $\mu_g^*(G) \le \mu_g^*(O_n) \le \mu_g^*(E) + \frac{1}{n}$ for every $n \in \mathbb{N}$ so that $\mu_g^*(G) = \mu_g^*(E)$. ∎

Theorem 22.11. *Let $\big(\mathbb{R}, \mathfrak{M}(\mu_g^*), \mu_g\big)$ be a Lebesgue-Stieltjes measure space on $\mathbb{R}$ determined by a real-valued increasing function g on $\mathbb{R}$. For $E \in \mathfrak{P}(\mathbb{R})$, the following conditions are all equivalent :*
(i) $E \in \mathfrak{M}(\mu_g^*)$.
(ii) *For every $\varepsilon > 0$, there exists an open set $O \supset E$ with $\mu_g^*(O \setminus E) < \varepsilon$.*
(iii) *There exists a G_δ-set $G \supset E$ with $\mu_g^*(G \setminus E) = 0$.*
(iv) *For every $\varepsilon > 0$, there exists a closed set $C \subset E$ with $\mu_g^*(E \setminus C) < \varepsilon$.*
(v) *There exists an F_σ-set $F \subset E$ with $\mu_g^*(E \setminus F) = 0$.*

Proof. This theorem is proved using Lemma 22.10 exactly in the same way as Theorem 3.22 was proved by using Lemma 3.21. ∎

Theorem 22.12. *Let $\big(\mathbb{R}, \mathfrak{M}(\mu_g^*), \mu_g\big)$ be a Lebesgue-Stieltjes measure space on $\mathbb{R}$ determined by a real-valued increasing function g on $\mathbb{R}$.*
(a) *A subset E of $\mathbb{R}$ is a member of $\mathfrak{M}(\mu_g^*)$ if and only if E is of the type $E = A \cup C$ where $A \in \mathfrak{B}_{\mathbb{R}}$ and C is a subset of a set $B \in \mathfrak{B}_{\mathbb{R}}$ with $\mu_g(B) = 0$.*
(b) *The Lebesgue-Stieltjes measure space $\big(\mathbb{R}, \mathfrak{M}(\mu_g^*), \mu_g\big)$ is the completion of the measure*
space $(\mathbb{R}, \mathfrak{B}_{\mathbb{R}}, \mu_g)$.

Proof. This theorem is proved exactly in the same way as Theorem 5.7. In particular (a) is proved by using Theorem 22.11 in the same way that (a) of Theorem 5.7 is based on Theorem 3.22. ∎

[III] Absolute Continuity and Singularity of a Lebesgue-Stieltjes Measure

Let us extend the notions of absolute continuity and singularity of a real-valued function on a finite closed interval to function defined on an arbitrary interval. Then we extend the

Lebesgue decomposition for real-valued increasing functions on a finite closed interval to real-valued increasing functions defined on $\mathbb{R}$.

Definition 22.13. *Let g be a real-valued function on an arbitrary interval J in $\mathbb{R}$. We say that g is absolutely continuous resp. singular on J if g is absolutely continuous resp. singular on every finite closed interval contained in J.*

Note that since absolute continuity of a function on a finite closed interval as defined in Definition 13.1 implies its absolute continuity on any closed subinterval, Definition 22.13 is consistent with Definition 13.1. Similarly Definition 22.13 is consistent with Definition 13.10.

Theorem 22.14. *Let f be a real-valued increasing function on $\mathbb{R}$. Then $f = g + h$ where g is an absolutely continuous increasing function and h is a singular increasing function on $\mathbb{R}$. The decomposition is unique up to constants.*

Proof. For every $n \in \mathbb{N}$, let f_n be the restriction of f to $[-n, n]$. By Theorem 13.20, there exist an absolutely continuous increasing function g_n and a singular increasing function h_n on $[-n, n]$ such that $f_n = g_n + h_n$ and moreover the decomposition is unique up to constants. Let us select g_n so that $g_n(0) = 0$. Consider the sequences of functions $(g_n : n \in \mathbb{N})$ and $(h_n : n \in \mathbb{N})$. Let us show that for every $n \in \mathbb{N}$, we have $g_{n+1} = g_n$ and $h_{n+1} = h_n$ on $[-n, n]$. Now $f_{n+1} = f = f_n$ on $[-n, n]$ and thus $g_{n+1} + h_{n+1} = g_n + h_n$ on $[-n, n]$. Restricted to $[-n, n]$, $g_{n+1} + h_{n+1}$ is a Lebesgue decomposition of f_n so that by the uniqueness of decomposition up to constants, we have $g_{n+1} = g_n + c$ on $[-n, n]$. Then since $g_{n+1}(0) = 0 = g_n(0)$, we have $c = 0$ so that $g_{n+1} = g_n$ on $[-n, n]$. From this it follows that $h_{n+1} = h_n$ on $[-n, n]$ also. Let us define two functions g and h on $\mathbb{R}$ by setting $g(x) = g_n(x)$ and $h(x) = h_n(x)$ for $x \in [-n, n]$ for $n \in \mathbb{N}$. Then $f = g = h$; g is absolutely continuous on every finite closed interval in $\mathbb{R}$ so that g is absolute continuous on $\mathbb{R}$ and similarly h is singular on every finite closed interval in $\mathbb{R}$ so that h is singular on $\mathbb{R}$. The fact that the function h is an increasing function on $\mathbb{R}$ and the uniqueness of decomposition up to constants follow by the same arguments as in the Proof of Theorem 13.18. ∎

Consider two Lebesgue-Stieltjes measure spaces $(\mathbb{R}, \mathfrak{M}(\mu_g^*), \mu_g)$ and $(\mathbb{R}, \mathfrak{M}(\mu_h^*), \mu_h)$ determined by two real-valued increasing functions g and h on $\mathbb{R}$. In general the two σ-algebras $\mathfrak{M}(\mu_g^*)$ and $\mathfrak{M}(\mu_h^*)$ are distinct. However they both contain the Borel σ-algebra $\mathfrak{B}_{\mathbb{R}}$ as a sub σ-algebra by (c) of Theorem 22.4. Thus we can compare the two measures μ_g and μ_h at least on the common σ-algebra $\mathfrak{B}_{\mathbb{R}}$. In connection with this role played by $\mathfrak{B}_{\mathbb{R}}$, let us first extend and modify Lebesgue's differentiation theorem for real-valued increasing functions on a finite closed interval (Theorem 12.10) as follows.

Theorem 22.15. *Let f be a real-valued increasing function on $\mathbb{R}$. Then there exists a null set E in $(\mathbb{R}, \mathfrak{B}_{\mathbb{R}}, \mu_L)$ such that the derivative f' of f exists and is nonnegative at every point in $\mathbb{R} \setminus E$ and f' is $\mathfrak{B}_{\mathbb{R}}$-measurable on $\mathbb{R} \setminus E$.*

Proof. Let $\delta > 0$ be arbitrarily fixed. For every $n \in \mathbb{Z}$, consider the restriction of f to $[n-1-\delta, n+\delta]$. By Theorem 12.10 (Lebesgue), there exists a null set A_n in $(\mathbb{R}, \mathfrak{M}_L, \mu_L)$ contained in $[n-1-\delta, n+\delta]$ such that f' exists and is nonnegative at every point in $[n-1-\delta, n+\delta] \setminus A_n$ and f' is $\mathfrak{M}_L$-measurable on $[n-1-\delta, n+\delta] \setminus A_n$. Let $A = \bigcup_{n \in \mathbb{Z}} A_n$. Then A is a null set in $(\mathbb{R}, \mathfrak{M}_L, \mu_L)$ and f' exists and is nonnegative at every point in $\mathbb{R} \setminus A$ and f' is $\mathfrak{M}_L$-measurable on $\mathbb{R} \setminus A$. Since $(\mathbb{R}, \mathfrak{M}_L, \mu_L)$ is the completion of $(\mathbb{R}, \mathfrak{B}_{\mathbb{R}}, \mu_L)$ by Theorem 5.7, there exists a null set B in $(\mathbb{R}, \mathfrak{B}_{\mathbb{R}}, \mu_L)$ such that $B \supset A$ by Observation 5.5. Then f' exists and is nonnegative at every point in $\mathbb{R} \setminus B$ and f' is $\mathfrak{M}_L$-measurable on $\mathbb{R} \setminus B$. If we let $\varphi = f'$ on $\mathbb{R} \setminus B$ and $\varphi = 0$ on B, then φ is a nonnegative extended real-valued $\mathfrak{M}_L$-measurable function on $\mathbb{R}$. By Proposition 5.9, there exist a null set C in $(\mathbb{R}, \mathfrak{B}_{\mathbb{R}}, \mu_L)$ and an extended real-valued $\mathfrak{B}_{\mathbb{R}}$-measurable function g on $\mathbb{R}$ such that $g = \varphi$ on $\mathbb{R} \setminus C$. Let $E = B \cup C$. Then E is a null set in $(\mathbb{R}, \mathfrak{B}_{\mathbb{R}}, \mu_L)$ and $g = \varphi = f'$ on $\mathbb{R} \setminus E$. Since g is $\mathfrak{B}_{\mathbb{R}}$-measurable on $\mathbb{R} \setminus E$, f' is $\mathfrak{B}_{\mathbb{R}}$-measurable on $\mathbb{R} \setminus E$. ∎

[III.1] Absolute Continuity of Lebesgue-Stieltjes Measures

Consider two Lebesgue-Stieltjes measure spaces $(\mathbb{R}, \mathfrak{M}(\mu_g^*), \mu_g)$ and $(\mathbb{R}, \mathfrak{M}(\mu_h^*), \mu_h)$ determined by two real-valued increasing functions g and h on $\mathbb{R}$. Since $\mathfrak{M}(\mu_g^*)$ and $\mathfrak{M}(\mu_h^*)$ may not be identical, Definition 11.4 for the absolute continuity of a measure with respect to another on the same σ-algebra does not apply. Since both $\mathfrak{M}(\mu_g^*)$ and $\mathfrak{M}(\mu_h^*)$ contain $\mathfrak{B}_{\mathbb{R}}$ as a sub σ-algebra, we define the absolute continuity of a Lebesgue-Stieltjes measure with respect to another as follows.

Definition 22.16. *Let μ_g and μ_h be two Lebesgue-Stieltjes measures determined by two real-valued increasing functions g and h on $\mathbb{R}$. We say that μ_h is absolutely continuous with respect to μ_g on $\mathfrak{B}_{\mathbb{R}}$ and write $\mu_h \ll \mu_g$ on $\mathfrak{B}_{\mathbb{R}}$ if $\mu_h(E) = 0$ for every $E \in \mathfrak{B}_{\mathbb{R}}$ with $\mu_g(E) = 0$.*

Proposition 22.17. *Let μ_g and μ_h be two Lebesgue-Stieltjes measures determined by two real-valued increasing functions g and h on $\mathbb{R}$. Suppose $\mu_h \ll \mu_g$ on $\mathfrak{B}_{\mathbb{R}}$.*
(a) *Every null set in $(\mathbb{R}, \mathfrak{M}(\mu_g^*), \mu_g)$ is a null set in $(\mathbb{R}, \mathfrak{M}(\mu_h^*), \mu_h)$.*
(b) $\mathfrak{M}(\mu_g^*) \subset \mathfrak{M}(\mu_h^*)$.
(c) *$\mu_h \ll \mu_g$ on $\mathfrak{M}(\mu_g^*)$, that is, if $E \in \mathfrak{M}(\mu_g^*)$ and $\mu_g(E) = 0$ then $\mu_h(E) = 0$.*

Proof. 1. Let E be a null set in $(\mathbb{R}, \mathfrak{M}(\mu_g^*), \mu_g)$. Since the measure space $(\mathbb{R}, \mathfrak{M}(\mu_g^*), \mu_g)$ is the completion of the measure space $(\mathbb{R}, \mathfrak{B}_{\mathbb{R}}, \mu_g)$ by (b) of Theorem 22.12, E is a subset of a null set B in $(\mathbb{R}, \mathfrak{B}_{\mathbb{R}}, \mu_g)$ by Observation 5.5. Since $\mu_h \ll \mu_g$ on $\mathfrak{B}_{\mathbb{R}}$ and since $\mu_g(B) = 0$, we have $\mu_h(B) = 0$. Thus B is a null set in $(\mathbb{R}, \mathfrak{M}(\mu_h^*), \mu_h)$. Then E, being a subset of a null set B in the complete measure space $(\mathbb{R}, \mathfrak{M}(\mu_h^*), \mu_h)$, is a null set in $(\mathbb{R}, \mathfrak{M}(\mu_h^*), \mu_h)$. This proves (a).

2. If $E \in \mathfrak{M}(\mu_g^*)$, then $E = A \cup C$ where $A \in \mathfrak{B}_{\mathbb{R}}$ and C is a subset of a null set B in $(\mathbb{R}, \mathfrak{B}_{\mathbb{R}}, \mu_g)$ by (a) of Theorem 22.12. Since $\mu_h \ll \mu_g$ on $\mathfrak{B}_{\mathbb{R}}$, B is a null set in $(\mathbb{R}, \mathfrak{B}_{\mathbb{R}}, \mu_h)$. Then we have $E \in \mathfrak{M}(\mu_h^*)$ by (a) of Theorem 22.12. This shows that $\mathfrak{M}(\mu_g^*) \subset \mathfrak{M}(\mu_h^*)$. This proves (b).

3. (c) follows from (a) and (b). ■

Theorem 22.18. *If* $(\mathbb{R}, \mathfrak{M}(\mu_g^*), \mu_g)$, $(\mathbb{R}, \mathfrak{M}(\mu_h^*), \mu_h)$, *and* $(\mathbb{R}, \mathfrak{M}(\mu_{g+h}^*), \mu_{g+h})$ *are three Lebesgue-Stieltjes measure spaces determined by real-valued increasing functions* g, h, *and* $g+h$ *on* $\mathbb{R}$, *then we have*

(a) $\mu_g \ll \mu_{g+h}$ *and* $\mu_h \ll \mu_{g+h}$ *on* $\mathfrak{B}_{\mathbb{R}}$.

(b) $\mathfrak{M}(\mu_{g+h}^*) \subset \mathfrak{M}(\mu_g^*) \cap \mathfrak{M}(\mu_h^*)$.

(c) $\mu_{g+h}^* = \mu_g^* + \mu_h^*$ *on* $\mathfrak{P}(\mathbb{R})$.

Proof. 1. Consider the semialgebra $\mathfrak{I}_{oc}$ consisting of all intervals of the type $(a, b]$ in $\mathbb{R}$ and $\emptyset$. For every $I = (a, b] \in \mathfrak{I}_{oc}$, we have by (1) of Proposition 22.7

$$\mu_{g+h}(I) = (g+h)(b+) - (g+h)(a+) = \{g(b+) - g(a+)\} + \{h(b+) - h(a+)\}$$
$$= \mu_g(I) + \mu_h(I).$$

Thus the two σ-finite measures μ_{g+h} and $\mu_g + \mu_h$ on the σ-algebra $\mathfrak{B}_{\mathbb{R}}$ are equal on the semialgebra $\mathfrak{I}_{oc}$. Then since $\mathfrak{B}_{\mathbb{R}} = \sigma(\mathfrak{I}_{oc})$, we have $\mu_{g+h} = \mu_g + \mu_h$ on $\mathfrak{B}_{\mathbb{R}}$ by Theorem 21.11. Then $\mu_g(E) \leq \mu_{g+h}(E)$ for every $E \in \mathfrak{B}_{\mathbb{R}}$ so that $\mu_{g+h}(E) = 0$ implies $\mu_g(E) = 0$. This shows that $\mu_g \ll \mu_{g+h}$ on $\mathfrak{B}_{\mathbb{R}}$. Similarly $\mu_h \ll \mu_{g+h}$ on $\mathfrak{B}_{\mathbb{R}}$. Then by (b) of Proposition 22.17, we have $\mathfrak{M}(\mu_{g+h}^*) \subset \mathfrak{M}(\mu_g^*)$ and $\mathfrak{M}(\mu_{g+h}^*) \subset \mathfrak{M}(\mu_h^*)$. Therefore $\mathfrak{M}(\mu_{g+h}^*) \subset \mathfrak{M}(\mu_g^*) \cap \mathfrak{M}(\mu_h^*)$. This proves (a) and (b).

2. To prove (c), recall that by Definition 22.3 for every subset E of $\mathbb{R}$, we have $\mu_g^*(E) = \inf \sum_{n\in\mathbb{N}} \ell_g(I_n)$ where $\ell_g((a,b]) = g_r(b) - g_r(a)$, g_r is the right-continuous modification of g, and the infimum is on the collection of all sequences $(I_n : n \in \mathbb{N})$ in $\mathfrak{I}_{oc}$ such that $\bigcup_{n\in\mathbb{N}} I_n \supset E$. Similarly for $\mu_h^*(E)$ and $\mu_{g+h}^*(E)$. Note that $(g+h)_r = g_r + h_r$ and $\ell_{g+h}((a,b]) = \ell_g((a,b]) + \ell_h((a,b])$. Now for an arbitrary sequence $(I_n : n \in \mathbb{N})$ in $\mathfrak{I}_{oc}$ such that $\bigcup_{n\in\mathbb{N}} I_n \supset E$, we have

$$\sum_{n\in\mathbb{N}} \ell_{g+h}(I_n) = \sum_{n\in\mathbb{N}} \ell_g(I_n) + \sum_{n\in\mathbb{N}} \ell_h(I_n)$$
$$\geq \inf \sum_{n\in\mathbb{N}} \ell_g(I_n) + \inf \sum_{n\in\mathbb{N}} \ell_h(I_n)$$
$$= \mu_g^*(E) + \mu_h^*(E).$$

Thus we have

$$(1) \qquad \mu_{g+h}^*(E) = \inf \sum_{n\in\mathbb{N}} \ell_{g+h}(I_n) \geq \mu_g^*(E) + \mu_h^*(E).$$

To prove the reverse inequality, let $(I_n : n \in \mathbb{N})$ and $(J_n : n \in \mathbb{N})$ be two arbitrary sequences in $\mathfrak{I}_{oc}$ such that $\bigcup_{n\in\mathbb{N}} I_n \supset E$ and $\bigcup_{n\in\mathbb{N}} J_n \supset E$. Then

$$\begin{aligned}\mu^*_{g+h}(E) &\le \min\left\{\sum_{n\in\mathbb{N}} \ell_{g+h}(I_n), \sum_{n\in\mathbb{N}} \ell_{g+h}(J_n)\right\}\\ &= \min\left\{\sum_{n\in\mathbb{N}} \ell_g(I_n) + \sum_{n\in\mathbb{N}} \ell_h(I_n), \sum_{n\in\mathbb{N}} \ell_g(J_n) + \sum_{n\in\mathbb{N}} \ell_h(J_n)\right\}.\end{aligned}$$

Now for any $a_1, a_2, b_1, b_2 \in [0, \infty]$, we have $\min\{a_1+a_2, b_1+b_2\} \le \frac{1}{2}\{a_1+a_2+b_1+b_2\}$. Thus we have

$$\mu^*_{g+h}(E) \le \frac{1}{2}\left\{\sum_{n\in\mathbb{N}} \ell_g(I_n) + \sum_{n\in\mathbb{N}} \ell_h(I_n) + \sum_{n\in\mathbb{N}} \ell_g(J_n) + \sum_{n\in\mathbb{N}} \ell_h(J_n)\right\}.$$

Then by the arbitrariness of the sequences $(I_n : n \in \mathbb{N})$ and $(J_n : n \in \mathbb{N})$, we have

(2) $\qquad \mu^*_{g+h}(E) \le \frac{1}{2}\{\mu^*_g(E) + \mu^*_h(E) + \mu^*_g(E) + \mu^*_h(E)\} = \mu^*_g(E) + \mu^*_h(E).$

By (1) and (2), we have $\mu^*_{g+h}(E) = \mu^*_g(E) + \mu^*_h(E)$. This proves (c). ∎

Theorem 22.19. *Let g be a real-valued increasing function on $\mathbb{R}$ and let $c > 0$. Then for the Lebesgue-Stieltjes measure spaces $(\mathbb{R}, \mathfrak{M}(\mu^*_g), \mu_g)$ and $(\mathbb{R}, \mathfrak{M}(\mu^*_{cg}), \mu_{cg})$ determined by the two real-valued increasing functions g and cg, we have*

(a) $\mu_{cg} \ll \mu_g$ *and* $\mu_g \ll \mu_{cg}$ *on* $\mathfrak{B}_{\mathbb{R}}$.

(b) $\mathfrak{M}(\mu^*_{cg}) = \mathfrak{M}(\mu^*_g)$.

(c) $\mu^*_{cg} = c\mu^*_g$ *on* $\mathfrak{P}(\mathbb{R})$.

Proof. 1. For $I = (a, b] \in \mathfrak{I}_{oc}$, we have by (1) of Proposition 22.7

$$\mu_{cg}(I) = (cg)(b+) - (cg)(a+) = c\{g(b+) - g(a+)\} = c\mu_g(I).$$

Thus the two σ-finite measures μ_{cg} and $c\mu_g$ on the σ-algebra $\mathfrak{B}_{\mathbb{R}}$ are equal on the semi-algebra $\mathfrak{I}_{oc}$. Since $\mathfrak{B}_{\mathbb{R}} = \sigma(\mathfrak{I}_{oc})$, we have $\mu_{cg} = c\mu_g$ on $\mathfrak{B}_{\mathbb{R}}$ by Theorem 21.11. Let $E \in \mathfrak{B}_{\mathbb{R}}$. If $\mu_g(E) = 0$, then $\mu_{cg}(E) = c\mu_g(E) = 0$. Thus $\mu_{cg} \ll \mu_g$ on $\mathfrak{B}_{\mathbb{R}}$. Conversely if $\mu_{cg}(E) = 0$, then $c\mu_g(E) = 0$ and thus $\mu_g(E) = 0$. This shows that $\mu_g \ll \mu_{cg}$ on $\mathfrak{B}_{\mathbb{R}}$. Now $\mu_{cg} \ll \mu_g$ on $\mathfrak{B}_{\mathbb{R}}$ implies $\mathfrak{M}(\mu^*_g) \subset \mathfrak{M}(\mu^*_{cg})$ by (b) of Proposition 22.17 and similarly $\mu_g \ll \mu_{cg}$ on $\mathfrak{B}_{\mathbb{R}}$ implies that $\mathfrak{M}(\mu^*_{cg}) \subset \mathfrak{M}(\mu^*_g)$. Therefore $\mathfrak{M}(\mu^*_{cg}) = \mathfrak{M}(\mu^*_g)$.

2. Let E be an arbitrary subset of $\mathbb{R}$. For every sequence $(I_n : n \in \mathbb{N})$ in $\mathfrak{I}_{oc}$ such that $\bigcup_{n\in\mathbb{N}} I_n \supset E$, we have $\sum_{n\in\mathbb{N}} \ell_{cg}(I_n) = c\sum_{n\in\mathbb{N}} \ell_g(I_n)$. From this it follows that $\mu^*_{cg}(E) = c\mu^*_g(E)$. ∎

In Definition 22.3, the Lebesgue-Stieltjes outer measure $\mu^*_g(E)$ of a subset E of $\mathbb{R}$ is based on sequences $(I_n : n \in \mathbb{N})$ in the class $\mathfrak{I}_{oc}$ such that $\bigcup_{n\in\mathbb{N}} I_n \supset E$. We show next that the value of $\mu^*_g(E)$ is unchanged if we restrict to coverings of E by disjoint sequences $(J_n : n \in \mathbb{N})$ in the class $\mathfrak{I}_{oc}$.

Lemma 22.20. *For the Lebesgue-Stieltjes outer measure μ_g^* determined by a real-valued increasing function g on $\mathbb{R}$, we have*

$$\mu_g^*(E) = \inf\left\{ \sum_{n\in\mathbb{N}} \ell_g(J_n) : (I_n : n \in \mathbb{N}) \subset \mathfrak{I}_{oc}, \text{disjoint}, \bigcup_{n\in\mathbb{N}} J_n \supset E \right\}$$

for every $E \in \mathfrak{P}(\mathbb{R})$.

Proof. Since the collection of all disjoint sequences $(J_n : n \in \mathbb{N})$ in $\mathfrak{I}_{oc}$ such that $\bigcup_{n\in\mathbb{N}} J_n \supset E$ is a subcollection of all sequences $(I_n : n \in \mathbb{N})$ in $\mathfrak{I}_{oc}$ such that $\bigcup_{n\in\mathbb{N}} I_n \supset E$, we have $\mu_g^*(E) \leq \inf\{\sum_{n\in\mathbb{N}} \ell_g(J_n) : (I_n : n \in \mathbb{N}) \subset \mathfrak{I}_{oc}, \text{disjoint}, \bigcup_{n\in\mathbb{N}} J_n \supset E\}$. It remains to prove the reverse inequality.

Now $\mathfrak{I}_{oc}$ is a semialgebra of subsets of $\mathbb{R}$. Thus $\alpha(\mathfrak{I}_{oc})$, the algebra of subsets of $\mathbb{R}$ generated by $\mathfrak{I}_{oc}$, is the collection of all finite unions of members of $\mathfrak{I}_{oc}$ by Theorem 21.4. Also a finite union of members of $\mathfrak{I}_{oc}$ is always a disjoint finite union of members of $\mathfrak{I}_{oc}$ according to Lemma 21.3. Let $(I_n : n \in \mathbb{N})$ be an arbitrary sequence in $\mathfrak{I}_{oc}$ such that $\bigcup_{n\in\mathbb{N}} I_n \supset E$. If we let $A_1 = I_1$ and $A_n = I_n \setminus \bigcup_{k=1}^{n-1} I_k$ for $n \geq 2$, then $(A_n : n \in \mathbb{N})$ is a disjoint sequence in the algebra $\alpha(\mathfrak{I}_{oc})$ and $\bigcup_{n\in\mathbb{N}} A_n = \bigcup_{n\in\mathbb{N}} I_n$. Since $A_n \in \alpha(\mathfrak{I}_{oc})$, A_n is a disjoint finite union of members of $\mathfrak{I}_{oc}$. Let $A_n = \bigcup_{k_n=1}^{p_n} J_{n,k_n}$ where $\{J_{n,k_n} : k_n = 1, \ldots, p_n\}$ is a disjoint collection in $\mathfrak{I}_{oc}$. Let us rename the collection $\{J_{1,1}, \ldots, J_{1,p_1}; J_{2,1}, \ldots, J_{2,p_2}; J_{3,1}, \ldots, J_{3,p_3}; \ldots\}$ as $\{J_1, J_2, J_3, \ldots\}$. Thus we have a disjoint sequence $(J_n : n \in \mathbb{N})$ in $\mathfrak{I}_{oc}$ such that $\bigcup_{n\in\mathbb{N}} J_n = \bigcup_{n\in\mathbb{N}} A_n = \bigcup_{n\in\mathbb{N}} I_n \supset E$. Now

$$\sum_{n\in\mathbb{N}} \ell_g(J_n) = \sum_{n\in\mathbb{N}} \mu_g(J_n) = \mu_g\left(\bigcup_{n\in\mathbb{N}} J_n\right) = \mu_g\left(\bigcup_{n\in\mathbb{N}} A_n\right)$$

$$= \sum_{n\in\mathbb{N}} \mu_g(A_n) \leq \sum_{n\in\mathbb{N}} \mu_g(I_n) = \sum_{n\in\mathbb{N}} \ell_g(I_n).$$

This shows that for every sequence $(I_n : n \in \mathbb{N})$ in $\mathfrak{I}_{oc}$ such that $\bigcup_{n\in\mathbb{N}} I_n \supset E$, there exists a disjoint sequence $(J_n : n \in \mathbb{N})$ in $\mathfrak{I}_{oc}$ such that $\bigcup_{n\in\mathbb{N}} J_n \supset E$ and $\sum_{n\in\mathbb{N}} \ell_g(J_n) \leq \sum_{n\in\mathbb{N}} \ell_g(I_n)$. Therefore we have

$$\inf\left\{\sum_{n\in\mathbb{N}} \ell_g(J_n) : (I_n : n \in \mathbb{N}) \subset \mathfrak{I}_{oc}, \text{disjoint}, \bigcup_{n\in\mathbb{N}} J_n \supset E\right\}$$

$$\leq \inf\left\{\sum_{n\in\mathbb{N}} \ell_g(I_n) : (I_n : n \in \mathbb{N}) \subset \mathfrak{I}_{oc}, \bigcup_{n\in\mathbb{N}} I_n \supset E\right\} = \mu_g^*(E).$$

This completes the proof of the lemma. ∎

Theorem 22.21. *Let μ_g be the Lebesgue-Stieltjes measure determined by a real-valued increasing function g on $\mathbb{R}$. Then μ_g is absolutely continuous with respect to the Lebesgue measure μ_L on $\mathfrak{B}_\mathbb{R}$ if and only if g is absolutely continuous on $\mathbb{R}$.*

Proof. 1. Suppose g is absolutely continuous on $\mathbb{R}$. To show that μ_g is absolutely continuous with respect to μ_L on $\mathfrak{B}_\mathbb{R}$, we show that if $E \in \mathfrak{B}_\mathbb{R}$ and $\mu_L(E) = 0$, then $\mu_g(E) = 0$. Let us show that for every $N \in \mathbb{N}$, if we let $E_N = E \cap (-N, N]$ then $\mu_g(E_N) = 0$. Let $\varepsilon > 0$. The absolute continuity of g on $[-N, N]$ implies according to Lemma 13.7 that there exists $\delta > 0$ such that for any sequence of non overlapping closed intervals $\big([a_n, b_n] : n \in \mathbb{N}\big)$ contained in $[-N, N]$ with $\sum_{n\in\mathbb{N}}(b_n - a_n) < \delta$, we have $\sum_{n\in\mathbb{N}}\{g(b_n) - g(a_n)\} < \varepsilon$. Now $\mu_L(E) = 0$ implies $\mu_L(E_N) = 0$. By Lemma 22.20, valid for any Lebesgue-Stieltjes outer measure and in particular for the Lebesgue outer measure, there exists a disjoint sequence in $\mathfrak{I}_{oc}$, say $\big((a_n, b_n] : n \in \mathbb{N}\big)$, such that $\bigcup_{n\in\mathbb{N}}(a_n, b_n] \supset E_N$ and $\sum_{n\in\mathbb{N}}(b_n - a_n) < \delta$. Then $\big([a_n, b_n] : n \in \mathbb{N}\big)$ is a sequence of non overlapping closed intervals contained in $[-N, -N]$ with $\sum_{n\in\mathbb{N}}(b_n - a_n) < \delta$ so that $\sum_{n\in\mathbb{N}}\{g(b_n) - g(a_n)\} < \varepsilon$. Since g is continuous, it is identical with its right-continuous modification g_r. Thus we have $\mu_g(E_N) = \mu_g^*(E_N) \le \sum_{n\in\mathbb{N}}\{g_r(b_n) - g_r(a_n)\} = \sum_{n\in\mathbb{N}}\{g(b_n) - g(a_n)\} < \varepsilon$. By the arbitrariness of $\varepsilon > 0$, we have $\mu_g(E_N) = 0$. Then $\mu_g(E) = \mu_g\big(\lim_{N\to\infty} E_N\big) = \lim_{N\to\infty} \mu(E_N) = 0$.

2. Conversely suppose μ_g is absolutely continuous with respect to μ_L on $\mathfrak{B}_{\mathbb{R}}$. To show that g is absolutely continuous on $\mathbb{R}$, that is, g is absolutely continuous on every finite closed interval in $\mathbb{R}$, it suffices to show that g is absolutely continuous on $[-N, N]$ for an arbitrary $N \in \mathbb{N}$. Let $\mathfrak{B}_{[-N,N]} = \mathfrak{B}_{\mathbb{R}} \cap [-N, N] \subset \mathfrak{B}_{\mathbb{R}}$. By the absolute continuity of μ_g with respect to μ_L on $\mathfrak{B}_{\mathbb{R}}$, we have $\mu_g(E) = 0$ for every $E \in \mathfrak{B}_{[-N,N]}$ with $\mu_L(E) = 0$. Let us show that this implies that for every $\varepsilon > 0$ there exists $\delta > 0$ such that $\mu_g(E) < \varepsilon$ for every $E \in \mathfrak{B}_{[-N,N]}$ with $\mu_L(E) < \delta$. Assume the contrary. Then there exists $\varepsilon_0 > 0$ such that for every $n \in \mathbb{N}$ there exists $E_n \in \mathfrak{B}_{[-N,N]}$ with $\mu_L(E_n) < \frac{1}{2^n}$ and $\mu_g(E_n) \ge \varepsilon_0$. Then $\sum_{n\in\mathbb{N}} \mu_L(E_n) < \infty$ so that $\mu_L\big(\limsup_{n\to\infty} E_n\big) = 0$ by the Borel-Cantelli Lemma (Theorem 6.6). On the other hand, since $\mu_g\big([-N, N]\big) < \infty$ by (4) of Proposition 22.7, we have $\mu_g\big(\limsup_{n\to\infty} E_n\big) \ge \limsup_{n\to\infty} \mu_g(E_n) \ge \varepsilon_0$ by Theorem 1.28. Thus for our $\limsup_{n\to\infty} E_n \in \mathfrak{B}_{[-N,N]}$, we have $\mu_L\big(\limsup_{n\to\infty} E_n\big) = 0$ and $\mu_g\big(\limsup_{n\to\infty} E_n\big) > 0$. This is a contradiction.

Now for every $\varepsilon > 0$, there exists $\delta > 0$ such that $\mu_g(E) < \varepsilon$ for every $E \in \mathfrak{B}_{[-N,N]}$ with $\mu_L(E) < \delta$. Let $([a_n, b_n] : n \in \mathbb{N})$ be a sequence of non overlapping closed intervals contained in $[-N, N]$ with $\sum_{n\in\mathbb{N}}(b_n - a_n) < \delta$. If we let $E = \bigcup_{n\in\mathbb{N}}(a_n, b_n]$, then $E \in \mathfrak{B}_{[-N,N]}$ and $\mu_L(E) = \sum_{n\in\mathbb{N}}(b_n - a_n) < \delta$ so that $\mu_g(E) < \varepsilon$. Thus for the right-continuous modification g_r of g, we have $\sum_{n\in\mathbb{N}}\{g_r(b_n) - g_r(a_n)\} = \sum_{n\in\mathbb{N}} \mu_g\big((a_n, b_n)\big) = \mu_g(E) < \varepsilon$. This shows that g_r is absolutely continuous on $[-N, N]$. But the right-continuous modification of a function is continuous at a point if and only if the function itself is continuous at the point. Thus the continuity of g_r on $[-N, N]$ implies that g_r is identical with g. Therefore g is absolutely continuous on $[-N, N]$. ∎

Theorem 22.22. *Let μ_g be a Lebesgue-Stieltjes measure determined by a real-valued increasing function g on $\mathbb{R}$. Suppose g is absolutely continuous on $\mathbb{R}$. Then for the Radon-Nikodym derivative $d\mu_g/d\mu_L$ of μ_g with respect to the Lebesgue measure μ_L on $\mathfrak{B}_{\mathbb{R}}$ and the derivative g' of g, we have $d\mu_g/d\mu_L = g'$ a.e. on $\big(\mathbb{R}, \mathfrak{B}_{\mathbb{R}}, \mu_L\big)$.*

Proof. By Theorem 22.21, the absolute continuity of g on $\mathbb{R}$ implies that μ_g is absolutely continuous with respect to μ_L on $\mathfrak{B}_{\mathbb{R}}$. Since μ_L and μ_g are σ-finite measures on $\mathfrak{B}_{\mathbb{R}}$, the Radon-Nikodym derivative of μ_g with respect to μ_L on $(\mathbb{R}, \mathfrak{B}_{\mathbb{R}})$ exists by Theorem 11.14, that is, there exists a nonnegative extended real-valued $\mathfrak{B}_{\mathbb{R}}$-measurable function $d\mu_g/d\mu_L$

on $\mathbb{R}$ such that

$$\mu_g(E) = \int_E \frac{d\mu_g}{d\mu_L}\, d\mu \quad \text{for every } E \in \mathfrak{B}_{\mathbb{R}}. \tag{1}$$

Moreover the σ-finiteness of μ_L implies that $d\mu_g/d\mu_L$ is unique up to a null set in $(\mathbb{R}, \mathfrak{B}_{\mathbb{R}}, \mu_L)$ according to Proposition 11.2.

Since g is absolutely continuous on $\mathbb{R}$, it is absolutely continuous on every finite closed interval $[a, b]$ in $\mathbb{R}$. Then the derivative g' of g exists $(\mathfrak{M}_L, \mu_L)$-a.e. on $[a, b]$ by Corollary 13.3 and moreover $\int_{[a,b]} g'\, d\mu = g(b) - g(a)$ by Theorem 13.17.

Since g is a real-valued increasing function on $\mathbb{R}$, there exists a null set E_0 in $(\mathbb{R}, \mathfrak{B}_{\mathbb{R}}, \mu_L)$ such that g' is nonnegative and $\mathfrak{B}_{\mathbb{R}}$-measurable on $\mathbb{R} \setminus E_0$ according to Theorem 22.15. If we let $\varphi = g'$ on $\mathbb{R} \setminus E_0$ and $\varphi = 0$ on E_0, then φ is a nonnegative extended real-valued $\mathfrak{B}_{\mathbb{R}}$-measurable function on $\mathbb{R}$. If we define a set function ν on $\mathfrak{B}_{\mathbb{R}}$ by setting

$$\nu(E) = \int_E \varphi\, d\mu \quad \text{for every } E \in \mathfrak{B}_{\mathbb{R}}, \tag{2}$$

then ν is a measure on $\mathfrak{B}_{\mathbb{R}}$. The restrictions of the two measures μ_g and ν to the semialgebra $\mathfrak{I}_{oc}$ consisting of intervals of the type $(a, b]$ and $\emptyset$ are nonnegative extended real-valued countably additive set functions on $\mathfrak{I}_{oc}$. Since

$$\nu\big((a, b]\big) = \int_{[a,b]} \varphi\, d\mu = \int_{[a,b]} g'\, d\mu = g(b) - g(a) = \mu_g\big((a, b]\big),$$

we have $\nu = \mu_g$ on $\mathfrak{I}_{oc}$. Then by (d) of Theorem 21.10 (Uniqueness of Extension to Measure), we have $\mu_g = \nu$ on $\mathfrak{B}_{\mathbb{R}} = \sigma(\mathfrak{I}_{oc})$. Thus by (2) we have

$$\mu_g(E) = \int_E \varphi\, d\mu \quad \text{for every } E \in \mathfrak{B}_{\mathbb{R}}.$$

This implies by the uniqueness of the Radon-Nikodym derivative of μ_g with respect to μ_L that $\varphi = d\mu_g/d\mu_L$ a.e. on $(\mathbb{R}, \mathfrak{B}_{\mathbb{R}}, \mu_L)$. Then $g' = d\mu_g/d\mu_L$ a.e. on $(\mathbb{R}, \mathfrak{B}_{\mathbb{R}}, \mu_L)$. ∎

Proposition 22.23. *Let $(\mathbb{R}, \mathfrak{M}(\mu_g^*), \mu_g)$ be a Lebesgue-Stieltjes measure space on $\mathbb{R}$ determined by a real-valued increasing function g on $\mathbb{R}$. Suppose g is absolutely continuous on $\mathbb{R}$. Then for every extended real-valued $\mathfrak{M}_L$-measurable function f on $\mathbb{R}$, we have*

$$\int_{\mathbb{R}} f\, d\mu_g = \int_{\mathbb{R}} f g'\, d\mu_L$$

in the sense that the existence of one side implies that of the other and the equality of the two.

Proof. By Theorem 22.22, the absolute continuity of g on $\mathbb{R}$ implies that the derivative g' is a Radon-Nikodym derivative of μ_g with respect to μ_L. Then the Proposition is a particular case of Theorem 11.21. ∎

[III.2] Singularity of Lebesgue-Stieltjes Measures

Theorem 22.24. *Let μ_g be a Lebesgue-Stieltjes measure determined by a real-valued increasing function g on $\mathbb{R}$. Then μ_g and the Lebesgue measure μ_L are mutually singular on $(\mathbb{R}, \mathfrak{B}_{\mathbb{R}})$ if and only if g is a singular function on $\mathbb{R}$.*

Proof. 1. Suppose g is a singular function on $\mathbb{R}$. Let us show that $\mu_g \perp \mu_L$ on $(\mathbb{R}, \mathfrak{B}_{\mathbb{R}})$. Since g is singular on $\mathbb{R}$, g is singular on every finite closed interval $[a, b]$ in $\mathbb{R}$. By Definition 13.10, g' exists and $g' = 0$, $(\mathfrak{M}_L, \mu_L)$-a.e. on $[a, b]$. Now since $(\mathbb{R}, \mathfrak{M}_L, \mu_L)$ is the completion of $(\mathbb{R}, \mathfrak{B}_{\mathbb{R}}, \mu_L)$ by Theorem 5.7, every null set in $(\mathbb{R}, \mathfrak{M}_L, \mu_L)$ is contained in a null set in $(\mathbb{R}, \mathfrak{B}_{\mathbb{R}}, \mu_L)$ by Observation 5.5. Thus g' exists and $g' = 0$, $(\mathfrak{B}_{\mathbb{R}}, \mu_L)$-a.e. on $[a, b]$. Then by Theorem 11.10 (Existence of Lebesgue Decomposition) and Theorem 11.13 (Uniqueness of Lebesgue Decomposition), there exist two unique measures μ_a and μ_s on $(\mathbb{R}, \mathfrak{B}_{\mathbb{R}})$ such that

$$(1) \qquad \mu_g = \mu_a + \mu_s, \quad \mu_a \ll \mu_L \text{ and } \mu_s \perp \mu_L.$$

If $\mu_a = 0$, then $\mu_g = \mu_s$ so that $\mu_g \perp \mu_L$ and we are done. Suppose $\mu_a \neq 0$. Let us show that this leads to a contradiction. Now $\mu_a \neq 0$ implies $\mu_a(\mathbb{R}) > 0$ so that there exists an interval $[\alpha, \beta] \subset \mathbb{R}$ such that $\mu_a([\alpha, \beta]) > 0$. Let us define two real-valued increasing functions g_a and g_s on $[\alpha, \beta]$ by setting $g_a(\alpha) = g_s(\alpha) = 0$ and

$$(2) \qquad \begin{cases} g_a(x) = \mu_a((\alpha, x]) + g(\alpha) & \text{for } x \in (\alpha, \beta], \\ g_s(x) = \mu_s((\alpha, x]) & \text{for } x \in (\alpha, \beta]. \end{cases}$$

Let us assume that g is right-continuous on $\mathbb{R}$. Then by (1) of Proposition 22.7 and by (1) above we have $g(x) - g(\alpha) = \mu_g((\alpha, x]) = \mu_a((\alpha, x]) + \mu_s((\alpha, x])$ and then by (2) we have

$$g(x) = \mu_a((\alpha, x]) + g(\alpha) + \mu_s((\alpha, x]) = g_a(x) + g_s(x) \quad \text{for } x \in (\alpha, \beta].$$

Since g, g_a, and g_s are real-valued increasing functions on $[\alpha, \beta]$, we have $g' \geq 0$, $g_a' \geq 0$, and $g_s' \geq 0$, $(\mathfrak{B}_{\mathbb{R}}, \mu_L)$-a.e. on $[\alpha, \beta]$ by Theorem 22.15. We have also $g' = g_a' + g_s'$, $(\mathfrak{B}_{\mathbb{R}}, \mu_L)$-a.e. on $[\alpha, \beta]$. As we noted above, $g' = 0$, $(\mathfrak{B}_{\mathbb{R}}, \mu_L)$-a.e. on $[a, b]$. This implies that $g_a' = 0$, $(\mathfrak{B}_{\mathbb{R}}, \mu_L)$-a.e. on $[\alpha, \beta]$. By (2) and by the fact that $\mu_a \ll \mu_L$, we have

$$(3) \qquad g_a(x) - g(\alpha) = \mu_a((\alpha, x]) = \int_{[\alpha, x]} \frac{d\mu_a}{d\mu_L} \, d\mu_L.$$

Then by Theorem 13.15, we have $g_a' = d\mu_a/d\mu_L$, $(\mathfrak{B}_{\mathbb{R}}, \mu_L)$-a.e. on $[\alpha, \beta]$. Thus $d\mu_a/d\mu_L = 0$, $(\mathfrak{B}_{\mathbb{R}}, \mu_L)$-a.e. on $[\alpha, \beta]$. On the other hand, since $\mu_a([\alpha, \beta]) > 0$, (3) implies that $d\mu_a/d\mu_L > 0$ on a $\mathfrak{B}_{\mathbb{R}}$-measurable subset of $[\alpha, \beta]$ with positive Lebesgue measure. This is a contradiction. Therefore $\mu_a = 0$ so that $\mu_g \perp \mu_L$ under the assumption that g is right-continuous on $\mathbb{R}$. If g is not right-continuous on $\mathbb{R}$, consider its right-continuous modification g_r. The fact that g is a singular function on $\mathbb{R}$ implies that g_r is also a singular function on $\mathbb{R}$ by 7° of Observation 22.2. Then by our result above, we have $\mu_{g_r} \perp \mu_L$. But according to Theorem 22.9, we have $\mu_{g_r} = \mu_g$. Therefore $\mu_g \perp \mu_L$.

2. Conversely suppose $\mu_g \perp \mu_L$. Let us show that g is a singular function on $\mathbb{R}$. Now since g is a real-valued increasing function on $\mathbb{R}$, $g = g_a + g_s$ where g_a is an absolutely continuous increasing function and g_s is a singular increasing function on $\mathbb{R}$ by Theorem 22.14. Let μ_{g_a} and μ_{g_s} be the Lebesgue-Stieltjes measures determined by the two real-valued increasing functions g_a and g_s. Then $\mu_g = \mu_{g_a} + \mu_{g_s}$. Since g_a is an absolutely continuous function, we have $\mu_{g_a} \ll \mu_L$ by Theorem 22.21. Now since $\mu_g \perp \mu_L$, there exist $C_1, C_2 \in \mathfrak{B}_{\mathbb{R}}$ such that $C_1 \cap C_2 = \emptyset$, $C_1 \cup C_2 = \mathbb{R}$, $\mu_L(C_1) = 0$, and $\mu_g(C_2) = 0$ by Definition 10.16. Since $\mu_g = \mu_{g_a} + \mu_{g_s}$, we have $\mu_g(C_2) = \mu_{g_a}(C_2) + \mu_{g_s}(C_2)$. Then $\mu_g(C_2) = 0$ implies $\mu_{g_a}(C_2) = 0$. This shows that $\mu_{g_a} \perp \mu_L$. Therefore we have both $\mu_{g_a} \ll \mu_L$ and $\mu_{g_a} \perp \mu_L$ and consequently $\mu_{g_a} = 0$ by Observation 11.12. Now $\mu_{g_a} = 0$ implies that the increasing function g_a is constant on $\mathbb{R}$, say $g_a = c$ on $\mathbb{R}$ where c is a real number. Then $g = c + g_s$. Since c and g_s are both singular functions on $\mathbb{R}$, so is g. ∎

Example. Let τ be the Cantor-Lebesgue function on $[0, 1]$ as in Theorem 4.34 and let g be a real-valued function on $\mathbb{R}$ defined by

$$g(x) = \begin{cases} 0 & \text{for } x \in (-\infty, 0), \\ \tau(x) & \text{for } x \in [0, 1], \\ 1 & \text{for } x \in (1, \infty). \end{cases}$$

Then the Lebesgue-Stieltjes measure μ_g determined by g and the Lebesgue measure μ_L are mutually singular, that is, there exist $C_1, C_2 \in \mathfrak{B}_{\mathbb{R}}$ such that $C_1 \cap C_2 = \emptyset$, $C_1 \cap C_2 = \mathbb{R}$, $\mu_g(C_2) = 0$ and $\mu_L(C_1) = 0$.

Proof. Let T be the Cantor ternary set contained in $[0, 1]$ as in Theorem 4.33. Consider the open set $G = [0, 1] \setminus T$. Let $\{J_n : n \in \mathbb{N}\}$ be the countable collection of disjoint open intervals contained in $[0, 1]$ and constituting G. The Cantor-Lebesgue function τ is constant on each of these open intervals and so is g. Thus $\mu_g(J_n) = 0$ by (3) of Proposition 22.7 for every $n \in \mathbb{N}$ and thus $\mu_g(G) = 0$. Consider the four disjoint sets $(-\infty, 0)$, G, T, and $(1, \infty)$ in $\mathfrak{B}_{\mathbb{R}}$ whose union is $\mathbb{R}$. Since g is constant on $(-\infty, 0)$ and on $(1, \infty)$, we have $\mu_g\big((-\infty, 0)\big) = 0$ and $\mu_g\big((1, \infty)\big) = 0$ also. Thus $\mu_g\big((-\infty, 0) \cup G \cup (1, \infty)\big) = 0$. On the other hand we have $\mu_L(T) = 0$ by Theorem 4.33. Let $C_1 = T$ and $C_2 = (-\infty, 0) \cup G \cup (1, \infty)$. Then we have $C_1, C_2 \in \mathfrak{B}_{\mathbb{R}}$, $C_1 \cap C_2 = \emptyset$, $C_1 \cap C_2 = \mathbb{R}$, $\mu_g(C_2) = 0$ and $\mu_L(C_1) = 0$. ∎

[IV] Decomposition of an Increasing Function

Our aim is to show that if f is a real-valued increasing function on $\mathbb{R}$ then $f = h + \varphi + \psi$ where h is a real-valued, increasing and continuous function, φ is a real-valued, increasing and right-continuous singular function, and ψ is a real-valued, increasing and left-continuous singular function on $\mathbb{R}$.

For a real-valued increasing function f on $\mathbb{R}$ the only kind of discontinuity at any point $x \in \mathbb{R}$ is a jump discontinuity, that is, $f(x-) < f(x+)$, with a jump $f(x+) - f(x-) > 0$. Since there can be at most countably many jump discontinuities for a real-valued increasing function, if we let $\{\xi_n : n \in \mathbb{N}\}$ be an arbitrary enumeration of the points where a jump discontinuity occurs, then the sum $\sum_{n \in \mathbb{N}} \{f(\xi_n+) - f(\xi_n-)\}$ exists in $(0, \infty]$. If we are to sum the jumps as we move in the positive direction of $\mathbb{R}$ recording the result as real-valued increasing function on $\mathbb{R}$, then since $\{\xi_n : n \in \mathbb{N}\}$ is an infinite set in $(-\infty, \infty)$ we start summing from an arbitrarily fixed point in $\mathbb{R}$ entering a jump in the sum as a positive number as we move in the positive direction and entering a jump in the sum as a negative number as we move in the negative direction of $\mathbb{R}$.

Let f be a real-valued increasing function on $\mathbb{R}$. For $x \in \mathbb{R}$, let us call $f(x) - f(x-)$, $f(x+) - f(x)$, and $f(x+) - f(x-)$ the left jump, the right jump, and the jump respectively of f at x. Then the jump of f at x is the sum of the left and the right jump of f at x. Also
(a) f is left-continuous at x if and only if the left jump of f at x is equal to 0.
(b) f is right-continuous at x if and only if the right jump of f at x is equal to 0.
(c) f is continuous at x if and only if the jump of f at x is eual to 0.

We show first that a real-valued, increasing and right-continuous singular function on $\mathbb{R}$ with preassigned left jumps can be constructed.

Lemma 22.25. *Let $D = \{\xi_n : n \in \mathbb{N}\}$ be a countable set in $\mathbb{R}$ and let $(\alpha_n : n \in \mathbb{N})$ be a sequence of positive numbers such that for every finite interval I in $\mathbb{R}$, we have*

(1) $$\sum_{\{n \in \mathbb{N} : \xi_n \in I\}} \alpha_n < \infty.$$

Let us define a real-valued increasing function φ on $\mathbb{R}$ by

(2) $$\varphi(x) = \begin{cases} \sum_{\{n \in \mathbb{N} : \xi_n \in [0,x]\}} \alpha_n & \text{for } x \geq 0, \\ -\sum_{\{n \in \mathbb{N} : \xi_n \in (x,0)\}} \alpha_n & \text{for } x < 0. \end{cases}$$

Let μ_φ be the Lebesgue-Stieltjes measure on $\mathbb{R}$ determined by φ. Then
(a) *φ is right-continuous on $\mathbb{R}$.*
(b) *D is the set of points of left-discontinuity of φ and $\mu_\varphi(\{\xi_n\}) = \varphi(\xi_n) - \varphi(\xi_n-) = \alpha_n$.*
(c) *φ is a singular function on $\mathbb{R}$.*
(d) *$\mu_\varphi \perp \mu_L$ on $\mathfrak{B}_{\mathbb{R}}$.*
(e) *$\mu_\varphi(\mathbb{R}) = \sum_{n \in \mathbb{N}} \alpha_n$.*

Proof. 1. By (1) and (2), φ is a real-valued increasing function on $\mathbb{R}$. Let us show the right-continuity of φ on $\mathbb{R}$. Let us note first that the definition of φ by (2) implies that for any $x_0 < x$ in $\mathbb{R}$, we have

(3) $$\varphi(x) - \varphi(x_0) = \sum_{\{n \in \mathbb{N} : \xi_n \in (x_0, x]\}} \alpha_n.$$

This is easily verified by considering the three possible cases regarding the positions of x_0 and x in $\mathbb{R}$, namely, the case $0 \leq x_0 < x$, the case $x_0 < x < 0$, and the case $x_0 < 0 \leq x$.

Let us show next that for every $x_0 \in \mathbb{R}$, we have

(4) $$\lim_{x \downarrow x_0} \sum_{\{n \in \mathbb{N} : \xi_n \in (x_0, x]\}} \alpha_n = 0,$$

that is, for every $\varepsilon > 0$ there exists $\delta > 0$ such that for every $x \in (x_0, x_0 + \delta)$ we have

$$\sum_{\{n \in \mathbb{N}:\, \xi_n \in (x_0, x]\}} \alpha_n < \varepsilon.$$

Let $x_1 > x_0$ be arbitrarily fixed. We have $\sum_{\{n \in \mathbb{N}:\, \xi_n \in (x_0, x_1]\}} \alpha_n \in [0, \infty)$ by (1). This implies that for every $\varepsilon > 0$ there exists $N \in \mathbb{N}$ such that $\sum_{\{n > N:\, \xi_n \in (x_0, x_1]\}} \alpha_n \in [0, \varepsilon)$. Consequently if $\{n \leq N : \xi_n \in (x_0, x_1]\} = \emptyset$, then for any $x \in (x_0, x_1]$ we have

$$\sum_{\{n \in \mathbb{N}:\, \xi_n \in (x_0, x]\}} \alpha_n = \sum_{\{n > N:\, \xi_n \in (x_0, x]\}} \alpha_n \in [0, \varepsilon).$$

If $\{n \leq N : \xi_n \in (x_0, x_1]\} \neq \emptyset$, let

$$\delta = \min\{\xi_n - x_0 : n \leq N, \xi_n \in (x_0, x_1]\}.$$

Then for $x \in (x_0, x_0 + \delta)$, we have

$$\sum_{\{n \in \mathbb{N}:\, \xi_n \in (x_0, x]\}} \alpha_n = \sum_{\{n > N:\, \xi_n \in (x_0, x]\}} \alpha_n < \varepsilon.$$

This proves (4). By (3) and (4), we have the right-continuity of φ at x_0.

2. For arbitrary $x < x_0$ in $\mathbb{R}$, we have $\varphi(x_0) - \varphi(x) = \sum_{\{n \in \mathbb{N}:\, \xi_n \in (x, x_0]\}} \alpha_n$ by (3). By the same argument as in deriving (4), we have

(5) $$\lim_{x \uparrow x_0} \sum_{\{n \in \mathbb{N}:\, \xi_n \in (x, x_0)\}} \alpha_n = 0,$$

and then

(6) $$\lim_{x \uparrow x_0} \sum_{\{n \in \mathbb{N}:\, \xi_n \in (x, x_0]\}} \alpha_n = \begin{cases} \alpha_n & \text{if } x_0 \in D \text{ and } x_0 = \xi_n, \\ 0 & \text{if } x_0 \notin D. \end{cases}$$

Thus we have

(7) $$\lim_{x \uparrow x_0} \{\varphi(x_0) - \varphi(x)\} = \begin{cases} \alpha_n & \text{if } x_0 \in D \text{ and } x_0 = \xi_n, \\ 0 & \text{if } x_0 \notin D. \end{cases}$$

This shows that φ is discontinuous at x_0 if and only if $x_0 \in D$. If $x_0 = \xi_n$, then by (5) of Proposition 22.7, we have $\mu_\varphi(\{x_0\}) = \varphi(x_0) - \varphi(x_0-) = \alpha_n$ by (7). This proves (b).

3. To show that φ is a singular function on $\mathbb{R}$, it suffices to show that $\mu_\varphi \perp \mu_L$ on $(\mathbb{R}, \mathfrak{B}_\mathbb{R})$ according to Theorem 22.24. Let us show that for every $M \in \mathbb{N}$, we have

(8) $$\mu_\varphi\big((-M, M] \setminus D\big) = 0.$$

Now by (1) of Proposition 22.7 and by the right-continuity of φ and then by (2) and (1), we have

$$(9)\qquad \begin{aligned}\mu_\varphi\big((-M,M]\big) &= \varphi(M+0)-\varphi(-M+0)=\varphi(M)-\varphi(-M)\\ &= \sum_{\{n\in\mathbb{N}:\,\xi_n\in[0,M]\}}\alpha_n + \sum_{\{n\in\mathbb{N}:\,\xi_n\in(-M,0)\}}\alpha_n\\ &= \sum_{\{n\in\mathbb{N}:\,\xi_n\in(-M,M]\}}\alpha_n<\infty.\end{aligned}$$

Now $\mu_\varphi\big((-M,M]\big)<\infty$ implies that $\mu_\varphi\big((-M,M]\cap D\big)<\infty$ and thus

$$\begin{aligned}\mu_\varphi\big((-M,M]\setminus D\big) &= \mu_\varphi\big((-M,M]\big)-\mu_\varphi\big((-M,M]\cap D\big)\\ &= \mu_\varphi\big((-M,M]\big)-\sum_{\{n\in\mathbb{N}:\,\xi_n\in(-M,M]\}}\mu_\varphi(\{\xi_n\})\\ &= \mu_\varphi\big((-M,M]\big)-\sum_{\{n\in\mathbb{N}:\,\xi_n\in(-M,M]\}}\alpha_n=0,\end{aligned}$$

where the third equality is by (b) and the last equality is by (9). This proves (8).

Since $(-M,M]\setminus D\uparrow\mathbb{R}\setminus D$, we have $\mu_\varphi(\mathbb{R}\setminus D)=\lim_{M\to\infty}\mu_\varphi\big((-M,M]\setminus D\big)=0$ by (8). On the other hand, since D is a countable set, $D\in\mathfrak{B}_{\mathbb{R}}$ and $\mu_L(D)=0$. Thus we have two disjoint sets D and $\mathbb{R}\setminus D$ in $\mathfrak{B}_{\mathbb{R}}$ whose union is equal to $\mathbb{R}$ such that $\mu_\varphi(\mathbb{R}\setminus D)=0$ and $\mu_L(D)=0$. This shows that $\mu_\varphi\perp\mu_L$ on $\mathfrak{B}_{\mathbb{R}}$, proving (d) and (c). Finally

$$\mu_\varphi(\mathbb{R})=\mu_\varphi(D)-\mu_\varphi(\mathbb{R}\setminus D)=\mu_\varphi(D)=\sum_{n\in\mathbb{N}}\mu_\varphi(\{\xi_n\})=\sum_{n\in\mathbb{N}}\alpha_n.$$

This proves (e). ∎

We show next that for an arbitrary real-valued, increasing and right-continuous function g on $\mathbb{R}$, a real-valued, increasing and right-continuous singular function φ with the same set of points of left-discontinuity and the same left jumps as g can be constructed. Then $g-\varphi$ is a real-valued, increasing and continuous function on $\mathbb{R}$.

Proposition 22.26. *Let g be a real-valued, increasing and right-continuous function on $\mathbb{R}$. Let us define*

$$(1)\qquad \varphi(x)=\begin{cases}\sum_{t\in[0,x]}\{g(t)-g(t-)\} & \text{for } x\geq 0,\\ -\sum_{t\in(x,0)}\{g(t)-g(t-)\} & \text{for } x<0.\end{cases}$$

(a) *φ is a real-valued, increasing and right-continuous singular function on $\mathbb{R}$ and has the same set of points of left-discontinuity as g and $\varphi(\xi)-\varphi(\xi-)=g(\xi)-g(\xi-)$ at such a point ξ, that is, the left jump of φ is equal to that of g.*

(b) $\mu_\varphi \perp \mu_L$ *on* $\mathfrak{B}_\mathbb{R}$.
(c) $h = g - \varphi$ *is a real-valued, increasing and continuous function on* $\mathbb{R}$.

Proof. Let D be the set of all points of discontinuity of g in $\mathbb{R}$. Since g is a real-valued increasing function on $\mathbb{R}$, D is a countable set. If $D = \emptyset$, then g is continuous at every $x \in \mathbb{R}$ so that $g(x) - g(x-) = 0$ and therefore $\varphi = 0$ on $\mathbb{R}$ and $h = g$. Suppose $D \neq \emptyset$ and let $D = \{\xi_n : n \in \mathbb{N}\}$. (Our argument below is valid if D is a finite set.) Since g is increasing on $\mathbb{R}$ and right-continuous at every $x \in \mathbb{R}$, g is discontinuous at $x \in \mathbb{R}$ if and only if $g(x) - g(x-) > 0$. Thus $g(\xi_n) - g(\xi_n -) > 0$ for $n \in \mathbb{N}$. Let $\alpha_n = g(\xi_n) - g(\xi_n -)$ for $n \in \mathbb{N}$. Since $g(x) - g(x-) = 0$ for $x \notin D$, the definition (1) of φ can be transcribed as

$$(2) \qquad \varphi(x) = \begin{cases} \sum_{\{n\in\mathbb{N}:\, \xi_n \in [0,x]\}} \alpha_n & \text{for } x \geq 0, \\ -\sum_{\{n\in\mathbb{N}:\, \xi_n \in (x,0)\}} \alpha_n & \text{for } x < 0, \end{cases}$$

which is identical with (2) of Lemma 22.25. Consider an arbitrary finite closed interval in $\mathbb{R}$ given by $[a, b]$ with $a < b$. Then we have

$$\sum_{\{n\in\mathbb{N}:\, \xi_n\in[a,b]\}} \alpha_n = \sum_{\{n\in\mathbb{N}:\, \xi_n\in[a,b]\}} \{g(\xi_n) - g(\xi_n -)\} \leq g(b) - g(a-) < \infty.$$

Thus our set D and sequence $(\alpha_n : n \in \mathbb{N})$ satisfy conditions (1) of Lemma 22.25. Then by Lemma 22.25, φ is a real-valued, increasing, and right-continuous singular function on $\mathbb{R}$ and D is the set of all points of discontinuity of φ and by (7) of Lemma 22.25 we have $\varphi(\xi_n) - \varphi(\xi_n -) = \alpha_n = g(\xi_n) - g(\xi_n -)$. For the Lebesgue-Stieltjes measure μ_φ determined by φ, we have $\mu_\varphi \perp \mu_L$ on $\mathfrak{B}_\mathbb{R}$ by (d) of Lemma 22.25. Also by (3) of Lemma 22.25, for any $x_0 < x$ in $\mathbb{R}$, we have

$$(3) \qquad \varphi(x) - \varphi(x_0) = \sum_{\{n\in\mathbb{N}:\, \xi_n\in(x_0,x]\}} \alpha_n = \sum_{\{n\in\mathbb{N}:\, \xi_n\in(x_0,x]\}} \{g(\xi_n) - g(\xi_n -)\}.$$

Let $h = g - \varphi$. It remains to show that h is an increasing continuous function on $\mathbb{R}$. Now since both g and φ have D as the set of points of discontinuity, the set of points of discontinuity of h is a subset of D. But at every $\xi_n \in D$, we have by (5) of Proposition 22.7 the equality $g(\xi_n) - g(\xi_n -) = \alpha_n = \mu_\varphi(\{\xi_n\}) = \varphi(\xi_n) - \varphi(\xi_n -)$ and thus we have $h(\xi_n) - h(\xi_n -) = \{g(\xi_n) - g(\xi_n -)\} - \{\varphi(\xi_n) - \varphi(\xi_n -)\} = 0$ so that h is continuous at ξ_n. Therefore h is continuous everywhere on $\mathbb{R}$. To show that h is an increasing function on $\mathbb{R}$, let $x_0, x \in \mathbb{R}$ and $x_0 < x$. Then

$$h(x) - h(x_0) = \{g(x) - g(x_0)\} - \{\varphi(x) - \varphi(x_0)\}$$

$$= \{g(x) - g(x_0)\} - \sum_{\{n\in\mathbb{N}:\, \xi_n\in(x_0,x]\}} \{g(\xi_n) - g(\xi_n -)\} \geq 0,$$

by (3). ■

We show next that a real-valued, increasing and left-continuous singular function on $\mathbb{R}$ with preassigned right jumps can be constructed.

Lemma 22.27. *Let $D = \{\xi_n : n \in \mathbb{N}\}$ be a countable set in $\mathbb{R}$ and let $(\alpha_n : n \in \mathbb{N})$ be a sequence of positive numbers such that for every finite interval I in $\mathbb{R}$, we have*

(1) $$\sum_{\{n\in\mathbb{N}:\,\xi_n\in I\}} \alpha_n < \infty.$$

Let us define a real-valued increasing function ψ on $\mathbb{R}$ by

(2) $$\psi(x) = \begin{cases} \sum_{\{n\in\mathbb{N}:\,\xi_n\in[0,x)\}} \alpha_n & \text{for } x \geq 0, \\ -\sum_{\{n\in\mathbb{N}:\,\xi_n\in[x,0)\}} \alpha_n & \text{for } x < 0. \end{cases}$$

Let μ_ψ be the Lebesgue-Stieltjes measure on $\mathbb{R}$ determined by ψ. Then
(a) *ψ is left-continuous on $\mathbb{R}$.*
(b) *D is the set of points of right-discontinuity of ψ and $\mu_\psi(\{\xi_n\}) = \psi(\xi_n+) - \psi(\xi_n) = \alpha_n$.*
(c) *ψ is a singular function on $\mathbb{R}$.*
(d) *$\mu_\psi \perp \mu_L$ on $\mathfrak{B}_{\mathbb{R}}$.*
(e) *$\mu_\psi(\mathbb{R}) = \sum_{n\in\mathbb{N}} \alpha_n$.*

Proof. 1. By (1) and (2), ψ is a real-valued increasing function on $\mathbb{R}$. For any $x_0, x \in \mathbb{R}$ such that $x < x_0$, we have

(3) $$\psi(x_0) - \psi(x) = \sum_{\{n\in\mathbb{N}:\,\xi_n\in[x,x_0)\}} \alpha_n.$$

As for (3) in Lemma 22.25, this is verified by considering the three possible cases regarding the positions of x_0 and x in $\mathbb{R}$, namely, the case $0 \leq x_0 < x$, the case $x_0 < x < 0$, and the case $x_0 < 0 \leq x$. By the same argument as in proving (4) in Lemma 22.25, we have for every $x_0 \in \mathbb{R}$,

(4) $$\lim_{x\uparrow x_0} \sum_{\{n\in\mathbb{N}:\,\xi_n\in[x,x_0)\}} \alpha_n = 0.$$

By (3) and (4), we have the left-continuity of ψ at x_0.

2. By (3), we have $\psi(x) - \psi(x_0) = \sum_{\{n\in\mathbb{N}:\,\xi_n\in[x_0,x)\}} \alpha_n$ for any $x_0 < x$. By the same argument as in deriving (4), we have

(5) $$\lim_{x\downarrow x_0} \sum_{\{n\in\mathbb{N}:\,\xi_n\in(x_0,x)\}} \alpha_n = 0.$$

Then

(6) $$\lim_{x\downarrow x_0} \sum_{\{n\in\mathbb{N}:\,\xi_n\in[x_0,x)\}} \alpha_n = \begin{cases} \alpha_n & \text{if } x_0 \in D \text{ and } x_0 = \xi_n, \\ 0 & \text{if } x_0 \notin D. \end{cases}$$

Thus we have

$$\text{(7)} \qquad \lim_{x \downarrow x_0} \{\psi(x) - \psi(x_0)\} = \begin{cases} \alpha_n & \text{if } x_0 \in D \text{ and } x_0 = \xi_n, \\ 0 & \text{if } x_0 \notin D. \end{cases}$$

This shows that ψ is discontinuous at x_0 if and only if $x_0 \in D$. If $x_0 = \xi_n$, then by (5) of Proposition 22.7, we have $\mu_\psi(\{x_0\}) = \psi(x_0+) - \psi(x_0) = \alpha_n$ by (7). This proves (b). Finally, (c), (d), and (e) are proved in the same way as in Lemma 22.25. ∎

We show next that for an arbitrary real-valued increasing function f on $\mathbb{R}$, there exists a real-valued, increasing and left-continuous singular function ψ with the same set of points of right-discontinuity and the same right jumps as f. Then $f - \psi$ is a real-valued, increasing and right-continuous function on $\mathbb{R}$.

Proposition 22.28. *Let f be a real-valued increasing function on $\mathbb{R}$. Let us define*

$$\text{(1)} \qquad \psi(x) = \begin{cases} \sum_{t \in [0,x)} \{f(t+) - f(t)\} & \text{for } x \geq 0, \\ -\sum_{t \in [x,0)} \{f(t+) - f(t)\} & \text{for } x < 0. \end{cases}$$

(a) *ψ is a real-valued, increasing and left-continuous singular function on $\mathbb{R}$ and has the same set of points of right-discontinuity as f and $\psi(\xi+) - \psi(\xi) = f(\xi+) - f(\xi)$ at such a point ξ, that is, the right jump of ψ is equal to that of f.*
(b) *$\mu_\psi \perp \mu_L$ on $\mathfrak{B}_{\mathbb{R}}$.*
(c) *$g = f - \psi$ is a real-valued, increasing and right-continuous function on $\mathbb{R}$.*

Proof. Let D be the set of all points of right-discontinuity of f in $\mathbb{R}$. This set is a subset of the set of all points of discontinuity of f and is thus a countable set. If $D = \emptyset$, then f is right-continuous at every $x \in \mathbb{R}$ so that $\psi = 0$ on $\mathbb{R}$ and $g = f$. Suppose $D \neq \emptyset$. Let $D = \{\xi_n : n \in \mathbb{N}\}$. If $x \notin D$, then f is right-continuous at x so that $\{f(x+) - f(x)\} = 0$. Since f is not right-continuous at ξ_n, we have $\{f(\xi_n+) - f(\xi_n)\} > 0$. Let $\alpha_n = \{f(\xi_n+) - f(\xi_n)\}$ for $n \in \mathbb{N}$. Then (1) is transcribed as

$$\text{(1)} \qquad \psi(x) = \begin{cases} \sum_{\{n \in \mathbb{N} :\, \xi_n \in [0,x)\}} \alpha_n & \text{for } x \geq 0, \\ -\sum_{\{n \in \mathbb{N} :\, \xi_n \in [x,0)\}} \alpha_n & \text{for } x < 0, \end{cases}$$

which is (2) of Lemma 22.27. For an arbitrary finite open interval (a, b) in $\mathbb{R}$, we have $\sum_{\{n \in \mathbb{N} :\, \xi_n \in (a,b]\}} \alpha_n = \sum_{\{n \in \mathbb{N} :\, \xi_n \in (a,b)\}} \{f(\xi_n+) - f(\xi_n)\} \leq f(b) - f(a) < \infty$. Thus our set D and sequence $(\alpha_n : n \in \mathbb{N})$ satisfy condition (1) of Lemma 22.27.

Then by Lemma 22.27, ψ is a real-valued, increasing, and left-continuous singular function on $\mathbb{R}$ and D is the set of all points of discontinuity of ψ and by (7) of Lemma 22.27 we have $\psi(\xi_n+) - \psi(\xi_n) = \alpha_n = f(\xi_n+) - f(\xi_n)$. For the Lebesgue-Stieltjes measure μ_ψ determined by ψ, we have $\mu_\psi \perp \mu_L$ on $\mathfrak{B}_{\mathbb{R}}$ by (d) of Lemma 22.27. By (3) of Lemma 22.27, for any $x_0, x \in \mathbb{R}$ such that $x < x_0$, we have

$$\text{(2)} \qquad \psi(x_0) - \psi(x) = \sum_{\{n \in \mathbb{N} :\, \xi_n \in [x,x_0)\}} \alpha_n = \sum_{\{n \in \mathbb{N} :\, \xi_n \in (x_0,x]\}} \{f(\xi_n+) - f(\xi_n)\}.$$

Let $g = f - \psi$. It remains to show that g is an increasing right-continuous function on $\mathbb{R}$. Now for every $x_0 \in \mathbb{R}$, we have

$$\lim_{x \downarrow x_0} \{g(x) - g(x_0)\} = \lim_{x \downarrow x_0} \{f(x) - f(x_0)\} - \lim_{x \downarrow x_0} \{\psi(x) - \psi(x_0)\}$$
$$= \begin{cases} \alpha_n - \alpha_n = 0 & \text{if } x_0 = \xi_n, \\ 0 - 0 = 0 & \text{if } x_0 \notin D. \end{cases}$$

This proves the right-continuity of g. To show that g is an increasing function on $\mathbb{R}$, let $x_0, x \in \mathbb{R}$ be such that $x_0 < x$. Then

$$g(x) - g(x_0) = \{f(x) - f(x_0)\} - \{\psi(x) - \psi(x_0)\}$$

$$= \{f(x) - f(x_0)\} - \sum_{\{n \in \mathbb{N} : \xi_n \in [x_0, x)\}} \{f(\xi_n+) - f(\xi_n)\} \geq 0,$$

by (3). ∎

Theorem 22.29. *Let f be a real-valued increasing function on $\mathbb{R}$. Then $f = h + \varphi + \psi$, where h is a real-valued, increasing and continuous function, φ is a real-valued, increasing and right-continuous singular function, and ψ is a real-valued, increasing and left-continuous singular function on $\mathbb{R}$.*

Proof. By Proposition 22.28, we have $f = g + \psi$ where g is a real-valued, increasing and right-continuous function and ψ is a real-valued, increasing and left-continuous singular function on $\mathbb{R}$. By Proposition 22.26, we have $g = h + \varphi$ where h is a real-valued, increasing and continuous function and φ is a real-valued, increasing and right-continuous singular function on $\mathbb{R}$. Thus $f = h + \varphi + \psi$. ∎

§23 Product Measure Spaces

[I] Existence and Uniqueness of Product Measure Spaces

Definition 23.1. *Given n measure spaces* $(X_1, \mathfrak{A}_1, \mu_1), \ldots, (X_n, \mathfrak{A}_n, \mu_n)$. *Consider the product measurable space* $\big(X_1 \times \cdots \times X_n, \sigma(\mathfrak{A}_1 \times \cdots \times \mathfrak{A}_n)\big)$. *A measure* μ *on* $\sigma(\mathfrak{A}_1 \times \cdots \times \mathfrak{A}_n)$ *such that*

$$\mu(E) = \mu_1(A_1) \cdots \mu_n(A_n) \quad \text{for } E = A_1 \times \cdots \times A_n \in \mathfrak{A}_1 \times \cdots \times \mathfrak{A}_n$$

with the convention that $\infty \cdot 0 = 0 \cdot \infty = 0$ *is called a product measure of* $\mu_1, \ldots, \mu_n$ *and we write* $\mu_1 \times \cdots \times \mu_n$ *for it. The measure space* $\big(X_1 \times \cdots \times X_n, \sigma(\mathfrak{A}_1 \times \cdots \times \mathfrak{A}_n), \mu_1 \times \cdots \times \mu_n\big)$ *is called a product measure space of* $(X_1, \mathfrak{A}_1, \mu_1), \ldots, (X_n, \mathfrak{A}_n, \mu_n)$.

We shall show that for n arbitrary measure spaces $(X_1, \mathfrak{A}_1, \mu_1), \ldots, (X_n, \mathfrak{A}_n, \mu_n)$, a product measure $\mu_1 \times \cdots \times \mu_n$ of $\mu_1, \ldots, \mu_n$ always exists, and furthermore if the measure spaces $(X_1, \mathfrak{A}_1, \mu_1), \ldots, (X_n, \mathfrak{A}_n, \mu_n)$ are all σ-finite, then $\mu_1 \times \cdots \times \mu_n$ is unique on the σ-algebra $\sigma(\mathfrak{A}_1 \times \cdots \times \mathfrak{A}_n)$.

Let us note that if $(X_1, \mathfrak{A}_1)$ and $(X_2, \mathfrak{A}_2)$ are two measurable spaces then the collection $\mathfrak{A}_1 \times \mathfrak{A}_2$ of subsets of $X_1 \times X_2$ need not be closed under complementations since the complement of a product set need not be a product set. Thus $\mathfrak{A}_1 \times \mathfrak{A}_2$ need not be an algebra of subsets of $X_1 \times X_2$. We show next that a finite cartesian product of semialgebras is a semialgebra.

Lemma 23.2. *Let* $\mathfrak{S}_i$ *be a semialgebra of subsets of a set* X_i *for* $i = 1, \ldots, n$. *Then* $\mathfrak{S}_1 \times \cdots \times \mathfrak{S}_n$ *is a semialgebra of subsets of* $X_1 \times \cdots \times X_n$.

Proof. We have $\emptyset = \emptyset \times \cdots \times \emptyset \in \mathfrak{S}_1 \times \cdots \times \mathfrak{S}_n$ and $X_1 \times \cdots \times X_n \in \mathfrak{S}_1 \times \cdots \times \mathfrak{S}_n$. If $E_1 \times \cdots \times E_n \in \mathfrak{S}_1 \times \cdots \times \mathfrak{S}_n$ and $F_1 \times \cdots \times F_n \in \mathfrak{S}_1 \times \cdots \times \mathfrak{S}_n$, then we have $(E_1 \times \cdots \times E_n) \cap (F_1 \times \cdots \times F_n) = (E_1 \cap F_1) \times \cdots \times (E_n \cap F_n) \in \mathfrak{S}_1 \times \cdots \times \mathfrak{S}_n$. Thus $\mathfrak{S}_1 \times \cdots \times \mathfrak{S}_n$ is closed under intersection.

Let us show that if $E_1 \times \cdots \times E_n \in \mathfrak{S}_1 \times \cdots \times \mathfrak{S}_n$, then $(E_1 \times \cdots \times E_n)^c$ is a finite disjoint union of members of $\mathfrak{S}_1 \times \cdots \times \mathfrak{S}_n$. Now since $E_i \in \mathfrak{S}_i$ and $\mathfrak{S}_i$ is a semialgebra of subsets of X_i, E_i^c is a finite disjoint union of members of $\mathfrak{S}_i$. Thus there exists a disjoint collection $\{E_{i,k_i} : k_i = 0, \ldots, p_i\}$ in $\mathfrak{S}_i$ such that $\bigcup_{k_i=0}^{p_i} E_{i,k_i} = X_i$ and $E_{i,0} = E_i$. Then

$$X_1 \times \cdots \times X_n = \Big(\bigcup_{k_1=0}^{p_1} E_{1,k_1}\Big) \times \cdots \times \Big(\bigcup_{k_n=0}^{p_n} E_{n,k_n}\Big) = \bigcup_{k_1=0}^{p_1} \cdots \bigcup_{k_n=0}^{p_n} (E_{1,k_1} \times \cdots \times E_{n,k_n}).$$

Now $\{E_{1,k_1} \times \cdots \times E_{n,k_n} : k_1 = 0, \ldots, p_1; \ldots; k_n = 0, \ldots, p_n\}$ is a disjoint collection in $\mathfrak{S}_1 \times \cdots \times \mathfrak{S}_n$. Since $E_1 \times \cdots \times E_n = E_{1,0} \times \cdots \times E_{n,0}$ is a member of this collection and the union of this collection is the set $X_1 \times \cdots \times X_n$, $(E_1 \times \cdots \times E_n)^c$ is a finite disjoint union of members of $\mathfrak{S}_1 \times \cdots \times \mathfrak{S}_n$. This completes the proof that $\mathfrak{S}_1 \times \cdots \times \mathfrak{S}_n$ is a semialgebra of subsets of $X_1 \times \cdots \times X_n$. ■

Definition 23.3. *Let $\mathfrak{C}_i$ be a collection of subsets of a nonempty set X_i for $i = 1, \dots, n$. Let us write $\mathfrak{C}_1 \otimes \cdots \otimes \mathfrak{C}_n$ for $\sigma(\mathfrak{C}_1 \times \cdots \times \mathfrak{C}_n)$, that is, the σ-algebra of subsets of $X_1 \times \cdots \times X_n$ generated by $\mathfrak{C}_1 \times \cdots \times \mathfrak{C}_n$.*

Lemma 23.4. *Let $\mathfrak{C}_i$ be a collection of subsets of a set S_i for $i = 1, 2$. Then we have*
(a) *$\sigma(\mathfrak{C}_1) \times S_2 = \sigma(\mathfrak{C}_1 \times S_2)$,*
(b) *$S_1 \times \sigma(\mathfrak{C}_2) = \sigma(S_1 \times \mathfrak{C}_2)$.*

Proof. To prove (a), let π_1 be the projection of $S_1 \times S_2$ onto S_1, that is, $\pi_1(x_1, x_2) = x_1$ for $(x_1, x_2) \in S_1 \times S_2$. Then $\pi_1^{-1}\big(\sigma(\mathfrak{C}_1)\big) = \sigma(\mathfrak{C}_1) \times S_2$. On the other hand by Theorem 1.14, we have $\pi_1^{-1}\big(\sigma(\mathfrak{C}_1)\big) = \sigma\big(\pi_1^{-1}(\mathfrak{C}_1)\big) = \sigma(\mathfrak{C}_1 \times S_2)$. Thus $\sigma(\mathfrak{C}_1) \times S_2 = \sigma(\mathfrak{C}_1 \times S_2)$. This proves (a). Similarly for (b). ∎

Proposition 23.5. *Let $\mathfrak{C}_i$ be a collection of subsets of a set X_i for $i \in \mathbb{N}$. Then for every $n \geq 2$, we have*

$$(1) \qquad \Big(\cdots\big((\mathfrak{C}_1 \otimes \mathfrak{C}_2) \otimes \mathfrak{C}_3\big) \otimes \cdots \otimes \mathfrak{C}_{n-1}\Big) \otimes \mathfrak{C}_n = \mathfrak{C}_1 \otimes \cdots \otimes \mathfrak{C}_n,$$

and

$$(2) \qquad \mathfrak{C}_1 \otimes \Big(\mathfrak{C}_2 \otimes \cdots \otimes \big(\mathfrak{C}_{n-2} \otimes (\mathfrak{C}_{n-1} \otimes \mathfrak{C}_n)\big) \cdots\Big) = \mathfrak{C}_1 \otimes \cdots \otimes \mathfrak{C}_n.$$

Proof. We prove (1) by induction on n. For $n = 2$, (1) is obviously true. Now suppose (1) holds for $n = k$ for some $k \geq 2$, that is,

$$(3) \qquad \Big(\cdots\big((\mathfrak{C}_1 \otimes \mathfrak{C}_2) \otimes \mathfrak{C}_3\big) \otimes \cdots \otimes \mathfrak{C}_{k-1}\Big) \otimes \mathfrak{C}_k = \mathfrak{C}_1 \otimes \cdots \otimes \mathfrak{C}_k.$$

Let us show that under the assumption of (3), the equality (1) holds for $n = k + 1$. For brevity, let us write $\mathfrak{C}$ for the left side of (3). To show that (1) holds for $n = k+1$, we show that

$$(4) \qquad \mathfrak{C} \otimes \mathfrak{C}_{k+1} = \mathfrak{C}_1 \otimes \cdots \otimes \mathfrak{C}_k \otimes \mathfrak{C}_{k+1}.$$

Now by (3), we have $\mathfrak{C} = \mathfrak{C}_1 \otimes \cdots \otimes \mathfrak{C}_k = \sigma(\mathfrak{C}_1 \times \cdots \times \mathfrak{C}_k) \supset \mathfrak{C}_1 \times \cdots \times \mathfrak{C}_k$. Thus $\mathfrak{C} \times \mathfrak{C}_{k+1} \supset \mathfrak{C}_1 \times \cdots \times \mathfrak{C}_k \times \mathfrak{C}_{k+1}$ and then

$$(5) \qquad \sigma(\mathfrak{C} \times \mathfrak{C}_{k+1}) \supset \sigma(\mathfrak{C}_1 \times \cdots \times \mathfrak{C}_k \times \mathfrak{C}_{k+1}).$$

On the other hand, with an arbitrary $E_{k+1} \in \mathfrak{C}_{k+1}$, we have

$$\begin{aligned} \mathfrak{C} \times E_{k+1} &= \sigma(\mathfrak{C}_1 \times \cdots \times \mathfrak{C}_k) \times E_{k+1} = \sigma(\mathfrak{C}_1 \times \cdots \times \mathfrak{C}_k \times E_{k+1}) \\ &\subset \sigma(\mathfrak{C}_1 \times \cdots \times \mathfrak{C}_k \times \mathfrak{C}_{k+1}), \end{aligned}$$

where the first equality is by (3) and the second equality is by (a) of Lemma 23.4. Since this holds for every $E_{k+1} \in \mathfrak{C}_{k+1}$, we have $\mathfrak{C} \times \mathfrak{C}_{k+1} \subset \sigma(\mathfrak{C}_1 \times \cdots \times \mathfrak{C}_k \times \mathfrak{C}_{k+1})$. Thus

$$(6) \qquad \sigma(\mathfrak{C} \times \mathfrak{C}_{k+1}) \subset \sigma(\mathfrak{C}_1 \times \cdots \times \mathfrak{C}_k \times \mathfrak{C}_{k+1}).$$

By (5) and (6), we have $\sigma(\mathfrak{C}\times\mathfrak{C}_{k+1}) = \sigma(\mathfrak{C}_1\times\cdots\times\mathfrak{C}_k\times\mathfrak{C}_{k+1}) = \mathfrak{C}_1\otimes\cdots\otimes\mathfrak{C}_k\otimes\mathfrak{C}_{k+1}$. This proves (4). Then by induction on n, we have (1). The equality (2) is proved by similar argument. ∎

Lemma 23.6. *Given two measure spaces $(X, \mathfrak{A}, \mu)$ and $(Y, \mathfrak{B}, \nu)$. Define a set function λ on the semialgebra $\mathfrak{A}\times\mathfrak{B}$ of subsets of $X \times Y$ by setting $\lambda(E) = \mu(A)\nu(B)$ for $E = A \times B \in \mathfrak{A}\times\mathfrak{B}$ with the convention that $\infty\cdot 0 = 0\cdot\infty = 0$. Then λ is a nonnegative extended real-valued countably additive set function on the semialgebra $\mathfrak{A}\times\mathfrak{B}$ with $\lambda(\emptyset) = 0$. Moreover if $(X, \mathfrak{A}, \mu)$ and $(Y, \mathfrak{B}, \nu)$ are σ-finite measure spaces, then λ is σ-finite on $\mathfrak{A}\times\mathfrak{B}$.*

Proof. 1. Let us prove the countable additivity of λ on $\mathfrak{A}\times\mathfrak{B}$. Thus let $(E_n : n \in \mathbb{N})$ be a disjoint sequence in $\mathfrak{A}\times\mathfrak{B}$ such that $\bigcup_{n\in\mathbb{N}} E_n \in \mathfrak{A}\times\mathfrak{B}$. Let us write $E_n = A_n \times B_n$ for $n \in \mathbb{N}$ and $\bigcup_{n\in\mathbb{N}} E_n = A \times B$ where $A_n, A \in \mathfrak{A}$ and $B_n, B \in \mathfrak{B}$. We are to show that $\lambda\left(\bigcup_{n\in\mathbb{N}} E_n\right) = \sum_{n\in\mathbb{N}} \lambda(E_n)$, that is,

$$\mu(A)\nu(B) = \sum_{n\in\mathbb{N}} \mu(A_n)\nu(B_n).$$

For every $x \in X$ and $y \in Y$, we have

(1)
$$\begin{aligned}\mathbf{1}_A(x)\mathbf{1}_B(y) &= \mathbf{1}_{A\times B}(x, y) = \mathbf{1}_{\bigcup_{n\in\mathbb{N}} A_n\times B_n}(x, y)\\ &= \sum_{n\in\mathbb{N}} \mathbf{1}_{A_n\times B_n}(x, y) = \sum_{n\in\mathbb{N}} \{\mathbf{1}_{A_n}(x)\mathbf{1}_{B_n}(y)\},\end{aligned}$$

where the third equality is by the disjointness of the sequence $(A_n \times B_n : n \in \mathbb{N})$. With $y \in Y$ fixed, integrating $\mathbf{1}_A(\cdot)\mathbf{1}_B(y)$ as given by (1) on X with respect to μ and applying the Monotone Convergence Theorem (Theorem 8.5) to the sequence of partial sums of the series, we have

(2)
$$\begin{aligned}\mu(A)\mathbf{1}_B(y) &= \left[\int_X \mathbf{1}_A(x)\,\mu(dx)\right]\mathbf{1}_B(y) = \int_X \mathbf{1}_A(x)\mathbf{1}_B(y)\mu(dx)\\ &= \int_X \sum_{n\in\mathbb{N}} \{\mathbf{1}_{A_n}(x)\mathbf{1}_{B_n}(y)\}\mu(dx) = \sum_{n\in\mathbb{N}} \int_X \mathbf{1}_{A_n}(x)\mathbf{1}_{B_n}(y)\mu(dx)\\ &= \sum_{n\in\mathbb{N}} \left[\int_X \mathbf{1}_{A_n}(x)\mu(dx)\right]\mathbf{1}_{B_n}(y) = \sum_{n\in\mathbb{N}} \mu(A_n)\mathbf{1}_{B_n}(y).\end{aligned}$$

Integrating with respect to ν and applying the Monotone Convergence Theorem, we have

$$\begin{aligned}\mu(A)\nu(B) &= \mu(A)\int_Y \mathbf{1}_B(y)\nu(dy) = \int_Y \mu(A)\mathbf{1}_B(y)\nu(dy)\\ &= \int_Y \sum_{n\in\mathbb{N}} \mu(A_n)\mathbf{1}_{B_n}(y)\nu(dy) = \sum_{n\in\mathbb{N}} \int_Y \mu(A_n)\mathbf{1}_{B_n}(y)\nu(dy)\\ &= \sum_{n\in\mathbb{N}} \mu(A_n)\int_Y \mathbf{1}_{B_n}(y)\,\nu(dy) = \sum_{n\in\mathbb{N}} \mu(A_n)\nu(B_n),\end{aligned}$$

where the third equality is by (2). This proves the countable additivity of λ on $\mathfrak{A} \times \mathfrak{B}$.

2. Suppose $(X, \mathfrak{A}, \mu)$ and $(Y, \mathfrak{B}, \nu)$ are σ-finite measure spaces. Then there exists a disjoint sequence $(A_n : n \in \mathbb{N})$ in $\mathfrak{A}$ such that $\bigcup_{n\in\mathbb{N}} A_n = X$ and $\mu(A_n) < \infty$ for every $n \in \mathbb{N}$ and there exists a disjoint sequence $(B_n : n \in \mathbb{N})$ in $\mathfrak{B}$ such that $\bigcup_{n\in\mathbb{N}} B_n = Y$ and $\nu(B_n) < \infty$ for every $n \in \mathbb{N}$. Then $\{A_m \times B_n : m \in \mathbb{N} \text{ and } n \in \mathbb{N}\}$ is a countable disjoint collection in $\mathfrak{A} \times \mathfrak{B}$ with $\bigcup_{m\in\mathbb{N}} \bigcup_{n\in\mathbb{N}} (A_m \times B_n) = X \times Y$ and we have $\lambda(A_m \times B_n) = \mu(A_m)\nu(B_n) < \infty$ for $m \in \mathbb{N}$ and $n \in \mathbb{N}$. This shows that λ is σ-finite on $\mathfrak{A} \times \mathfrak{B}$. ∎

Theorem 23.7. *For arbitrary n measure spaces $(X_1, \mathfrak{A}_1, \mu_1), \dots, (X_n, \mathfrak{A}_n, \mu_n)$, a product measure space $\big(X_1 \times \cdots \times X_n, \sigma(\mathfrak{A}_1 \times \cdots \times \mathfrak{A}_n), \mu_1 \times \cdots \times \mu_n\big)$ exists. Moreover if the n measure spaces are all σ-finite, then the product measure space is unique.*

Proof. 1. Consider $(X_1, \mathfrak{A}_1, \mu_1)$ and $(X_2, \mathfrak{A}_2, \mu_2)$. According to Lemma 23.6, a set function λ on the semialgebra $\mathfrak{A}_1 \times \mathfrak{A}_2$ of subsets of $X_1 \times X_2$ defined by $\lambda(E) = \mu_1(A_1)\mu_2(A_2)$ for $E = A_1 \times A_2 \in \mathfrak{A}_1 \times \mathfrak{A}_2$ is a nonnegative extended real-valued countably additive set function with $\lambda(\emptyset) = 0$. If we let λ^* be the outer measure on $X_1 \times X_2$ based on λ, then according to Theorem 21.10, the complete measure space $\big(X_1 \times X_2, \mathfrak{M}(\lambda^*), \lambda^*\big)$ has the properties that $\mathfrak{A}_1 \times \mathfrak{A}_2 \subset \mathfrak{M}(\lambda^*)$ and $\lambda = \lambda^*$ on $\mathfrak{A}_1 \times \mathfrak{A}_2$. Then $\sigma(\mathfrak{A}_1 \times \mathfrak{A}_2) \subset \mathfrak{M}(\lambda^*)$. If we restrict λ^* to $\sigma(\mathfrak{A}_1 \times \mathfrak{A}_2)$, then the measure space $\big(X_1 \times X_2, \sigma(\mathfrak{A}_1 \times \mathfrak{A}_2), \lambda^*\big)$ is a product measure space of $(X_1, \mathfrak{A}_1, \mu_1)$ and $(X_2, \mathfrak{A}_2, \mu_2)$ since $\lambda^*(A_1 \times A_2) = \lambda(A_1 \times A_2) = \mu_1(A_1)\mu_2(A_2)$ for $A_1 \times A_2 \in \mathfrak{A}_1 \times \mathfrak{A}_2$.

2. Suppose a product measure space $\big(X_1 \times \cdots \times X_k, \sigma(\mathfrak{A}_1 \times \cdots \times \mathfrak{A}_k), \mu_1 \times \cdots \times \mu_k\big)$ of the k measure spaces $(X_1, \mathfrak{A}_1, \mu_1), \dots, (X_k, \mathfrak{A}_k, \mu_k)$ exists for some $k < n$. For brevity let us write $(Y, \mathfrak{B}, \nu)$ for this product measure space. Consider the two measure spaces $(Y, \mathfrak{B}, \nu)$ and $(X_{k+1}, \mathfrak{A}_{k+1}, \mu_{k+1})$. If we define a set function λ on the semialgebra $\mathfrak{B} \times \mathfrak{A}_{k+1}$ of subsets of $Y \times X_{k+1}$ by $\lambda(E) = \nu(B)\mu_{k+1}(A_{k+1})$ for $E = B \times A_{k+1} \in \mathfrak{B} \times \mathfrak{A}_{k+1}$, then by Lemma 23.6, λ is a nonnegative extended real-valued countably additive set function with $\lambda(\emptyset) = 0$. If we let λ^* be the outer measure on $Y \times X_{k+1}$ based on λ, then the complete measure space $\big(Y \times X_{k+1}, \mathfrak{M}(\lambda^*), \lambda^*\big)$ has the properties that $\mathfrak{B} \times \mathfrak{A}_{k+1} \subset \mathfrak{M}(\lambda^*)$ and $\lambda = \lambda^*$ on $\mathfrak{B} \times \mathfrak{A}_{k+1}$ by Theorem 21.10. If we restrict λ^* to $\sigma(\mathfrak{B} \times \mathfrak{A}_{k+1}) \subset \mathfrak{M}(\lambda^*)$, then the measure space $\big(Y \times X_{k+1}, \sigma(\mathfrak{B} \times \mathfrak{A}_{k+1}), \lambda^*\big)$ is a product measure space of $(Y, \mathfrak{B}, \nu)$ and $(X_{k+1}, \mathfrak{A}_{k+1}, \mu_{k+1})$ since $\lambda^*(B \times A_{k+1}) = \lambda(B \times A_{k+1}) = \nu(B)\mu_{k+1}(A_{k+1})$ for $B \times A_{k+1} \in \mathfrak{B} \times \mathfrak{A}_{k+1}$.

Consider a product measure space $\big(Y \times X_{k+1}, \sigma(\mathfrak{B} \times \mathfrak{A}_{k+1}), \nu \times \mu_{k+1}\big)$ whose existence we have just established. Since $Y = X_1 \times \cdots \times X_k$, $\mathfrak{B} = \sigma(\mathfrak{A}_1 \times \cdots \times \mathfrak{A}_k)$, and $\nu = \mu_1 \times \cdots \times \mu_k$, we have a measure space

$$(1) \qquad \big(X_1 \times \cdots \times X_{k+1}, \sigma\big(\sigma(\mathfrak{A}_1 \times \cdots \times \mathfrak{A}_k) \times \mathfrak{A}_{k+1}\big), \mu_1 \times \cdots \times \mu_{k+1}\big).$$

Now by (1) of Proposition 23.5 we have

$$\sigma(\mathfrak{A}_1 \times \cdots \times \mathfrak{A}_k) = \mathfrak{A}_1 \otimes \cdots \otimes \mathfrak{A}_k = \big(\cdots\big((\mathfrak{A}_1 \otimes \mathfrak{A}_2) \otimes \mathfrak{A}_3\big) \otimes \cdots \otimes \mathfrak{A}_{k-1}\big) \otimes \mathfrak{A}_k.$$

Then we have

$$\begin{aligned}
&\sigma\big(\sigma(\mathfrak{A}_1 \times \cdots \times \mathfrak{A}_k) \times \mathfrak{A}_{k+1}\big)\\
=&\Big(\Big(\cdots\big((\mathfrak{A}_1 \otimes \mathfrak{A}_2) \otimes \mathfrak{A}_3\big) \otimes \cdots \otimes \mathfrak{A}_{k-1}\Big) \otimes \mathfrak{A}_k\Big) \otimes \mathfrak{A}_{k+1}\\
=&\mathfrak{A}_1 \otimes \cdots \otimes \mathfrak{A}_k \otimes \mathfrak{A}_{k+1} = \sigma(\mathfrak{A}_1 \times \cdots \times \mathfrak{A}_{k+1}),
\end{aligned}$$

where the second equality is by (1) of Proposition 23.5. Thus the measure space (1) is identical with the measure space $\big(X_1 \times \cdots \times X_{k+1}, \sigma(\mathfrak{A}_1 \times \cdots \times \mathfrak{A}_{k+1}), \mu_1 \times \cdots \times \mu_{k+1}\big)$. This shows the existence of a product measure space of $(X_1, \mathfrak{A}_1, \mu_1), \ldots, (X_{k+1}, \mathfrak{A}_{k+1}, \mu_{k+1})$. Thus by induction a product measure space of $(X_1, \mathfrak{A}_1, \mu_1), \ldots, (X_n, \mathfrak{A}_n, \mu_n)$ exists.

3. Suppose the measure spaces $(X_1, \mathfrak{A}_1, \mu_1), \ldots, (X_n, \mathfrak{A}_n, \mu_n)$ are all σ-finite. Then the set function $\mu_1 \times \cdots \times \mu_n$ on the semialgebra $\mathfrak{A}_1 \times \cdots \times \mathfrak{A}_n$ of subsets of $X_1 \times \cdots \times X_n$ is σ-finite. Thus by (d) of Theorem 21.10, the extension of $\mu_1 \times \cdots \times \mu_n$ to a measure on $\sigma(\mathfrak{A}_1 \times \cdots \times \mathfrak{A}_n)$ is unique. Thus our measure $\mu_1 \times \cdots \times \mu_n$ on $\sigma(\mathfrak{A}_1 \times \cdots \times \mathfrak{A}_n)$ is unique. ∎

[II] Integration on Product Measure Space

Definition 23.8. *Given two sets X and Y. Let $E \subset X \times Y$ and let f be an extended real-valued function on E.*

(a) *For $x \in X$ we call the set $E(x, \cdot) := \{y \in Y : (x, y) \in E\} \subset Y$ the x-section of E. For $y \in Y$ we call the set $E(\cdot, y) := \{x \in X : (x, y) \in E\} \subset X$ the y-section of E.*

(b) *For $x \in X$ we call the function $f(x, \cdot)$ on $E(x, \cdot)$ the x-section of f. For $y \in Y$ we call the function $f(\cdot, y)$ on $E(\cdot, y)$ the y-section of f.*

Proposition 23.9. *Consider the product measurable space $(X \times Y, \sigma(\mathfrak{A} \times \mathfrak{B}))$ of two measurable spaces $(X, \mathfrak{A})$ and $(Y, \mathfrak{B})$.*

(a) *If $E \in \sigma(\mathfrak{A} \times \mathfrak{B})$, then $E(x, \cdot) \in \mathfrak{B}$ for every $x \in X$ and $E(\cdot, y) \in \mathfrak{A}$ for every $y \in Y$.*

(b) *If f is an extended real-valued $\sigma(\mathfrak{A} \times \mathfrak{B})$-measurable function on $E \in \sigma(\mathfrak{A} \times \mathfrak{B})$, then $f(x, \cdot)$ is a $\mathfrak{B}$-measurable function on $E(x, \cdot) \in \mathfrak{B}$ for every $x \in X$ and $f(\cdot, y)$ is a $\mathfrak{A}$-measurable function on $E(\cdot, y) \in \mathfrak{A}$ for every $y \in Y$.*

Proof. 1. Let $\mathfrak{F}$ be the collection of all subsets E of $X \times Y$ such that $E(x, \cdot) \in \mathfrak{B}$ for every $x \in X$ and $E(\cdot, y) \in \mathfrak{A}$ for every $y \in Y$. To prove (a), we show that $\sigma(\mathfrak{A} \times \mathfrak{B}) \subset \mathfrak{F}$. Let us show that $\mathfrak{F}$ is a σ-algebra of subsets of $X \times Y$. Clearly $X \times Y \in \mathfrak{F}$. To show that $\mathfrak{F}$ is closed under complementations, let $E \in \mathfrak{F}$. Then for every $x \in X$, we have

$$E^c(x, \cdot) = \big\{y \in Y : (x, y) \in E^c\big\} = Y \setminus \big\{y \in Y : (x, y) \in E\big\} = Y \setminus E(x, \cdot) \in \mathfrak{B},$$

since $E(x, \cdot) \in \mathfrak{B}$. Similarly $E^c(\cdot, y) \in \mathfrak{A}$ for every $y \in Y$. Thus $E^c \in \mathfrak{F}$. To show that $\mathfrak{F}$ is closed under countable unions, let $(E_n : n \in \mathbb{N})$ be a sequence in $\mathfrak{F}$. Then for every

$x \in X$, we have

$$\Big(\bigcup_{n\in\mathbb{N}} E_n\Big)(x,\cdot) = \Big\{y \in Y : (x,y) \in \bigcup_{n\in\mathbb{N}} E_n\Big\}$$
$$= \bigcup_{n\in\mathbb{N}} \{y \in Y : (x,y) \in E_n\} = \bigcup_{n\in\mathbb{N}} E_n(x,\cdot) \in \mathfrak{B},$$

since $E_n(x,\cdot) \in \mathfrak{B}$ for every $n \in \mathbb{N}$. Similarly $\big(\bigcup_{n\in\mathbb{N}} E_n\big)(\cdot,y) \in \mathfrak{A}$ for every $y \in Y$. Thus $\mathfrak{F}$ is closed under countable unions. Then $\mathfrak{F}$ is a σ-algebra of subsets of $X \times Y$.

Let us show that $\mathfrak{A} \times \mathfrak{B} \subset \mathfrak{F}$. Let $E \in \mathfrak{A} \times \mathfrak{B}$. Then $E = A \times B$ where $A \in \mathfrak{A}$ and $B \in \mathfrak{B}$. Then for every $x \in X$, $E(x,\cdot) = B$ if $x \in A$ and $E(x,\cdot) = \emptyset$ if $x \in A^c$. In any case, $E(x,\cdot) \in \mathfrak{B}$ for every $x \in X$. Similarly $E(\cdot,y) \in \mathfrak{A}$ for every $y \in Y$. Thus $E \in \mathfrak{F}$. This shows that $\mathfrak{A} \times \mathfrak{B} \subset \mathfrak{F}$. Then since $\mathfrak{F}$ is a σ-algebra containing $\mathfrak{A} \times \mathfrak{B}$, it contains the smallest σ-algebra containing $\mathfrak{A} \times \mathfrak{B}$, namely $\sigma(\mathfrak{A} \times \mathfrak{B})$. Thus $\sigma(\mathfrak{A} \times \mathfrak{B}) \subset \mathfrak{F}$.

2. Let f be an extended real-valued $\sigma(\mathfrak{A}\times\mathfrak{B})$-measurable function on $E \in \sigma(\mathfrak{A}\times\mathfrak{B})$. Then for every $\alpha \in \mathbb{R}$, we have $F := \{(x,y) \in E : f(x,y) \le \alpha\} \in \sigma(\mathfrak{A} \times \mathfrak{B})$. This implies that $F(x,\cdot) \in \mathfrak{B}$ for every $x \in X$ by (a). Then for every $x \in X$, we have $\{y \in E(x,\cdot) : f(x,y) \le \alpha\} = \{y \in E(x,\cdot) : (x,y) \in F\} = F(x,\cdot) \in \mathfrak{B}$. This proves the $\mathfrak{B}$-measurability of $f(x,\cdot)$ on $E(x,\cdot)$ for every $x \in X$. Similarly $f(\cdot,y)$ is $\mathfrak{A}$-measurable on $E(\cdot,y)$ for every $y \in Y$. ∎

Definition 23.10. *A collection $\mathfrak{C}$ of subsets of a set X is called a monotone class if for every monotone sequence $(E_n : n \in \mathbb{N})$ in $\mathfrak{C}$ we have $\lim_{n\to\infty} E_n \in \mathfrak{C}$, that is, if $(E_n : n \in \mathbb{N})$ is an increasing sequence then $\bigcup_{n\in\mathbb{N}} E_n \in \mathfrak{C}$ and if $(E_n : n \in \mathbb{N})$ is a decreasing sequence then $\bigcap_{n\in\mathbb{N}} E_n \in \mathfrak{C}$.*

Observation 23.11. (a) The collection $\mathfrak{P}(X)$ of all subsets of a set X is a monotone class.
(b) A σ-algebra of subsets of a set is a monotone class.
(c) If $\{\mathfrak{C}_\alpha : \alpha \in A\}$ is a collection of monotone classes of subsets of a set X, then $\bigcap_{\alpha\in A} \mathfrak{C}_\alpha$ is a monotone class of subsets of X.

Proof. (a) and (b) are immediate. To prove (c) let $(E_n : n \in \mathbb{N})$ be a monotone sequence in $\bigcap_{\alpha\in A} \mathfrak{C}_\alpha$. Then $(E_n : n \in \mathbb{N})$ is a monotone sequence in $\mathfrak{C}_\alpha$ for every $\alpha \in A$ so that $\lim_{n\to\infty} E_n \in \mathfrak{C}_\alpha$ for every $\alpha \in A$. Thus $\lim_{n\to\infty} E_n \in \bigcap_{\alpha\in A} \mathfrak{C}_\alpha$. ∎

Observation 23.12. If a collection $\mathfrak{C}$ of subsets of a set X is both a monotone class and an algebra, then $\mathfrak{C}$ is a σ-algebra of subsets of X.

Proof. Since $\mathfrak{C}$ is an algebra, to show that it is a σ-algebra, we need only to show that $\mathfrak{C}$ is closed under countable unions. Let $(E_n : n \in \mathbb{N})$ be an arbitrary sequence in $\mathfrak{C}$. If we let $F_n = \bigcup_{k=1}^n E_k$ for $n \in \mathbb{N}$, then since $\mathfrak{C}$ is an algebra, we have $F_n \in \mathfrak{C}$ and thus $(F_n : n \in \mathbb{N})$ is an increasing sequence in $\mathfrak{C}$. Since $\mathfrak{C}$ is a monotone class, we have $\bigcup_{n\in\mathbb{N}} F_n \in \mathfrak{C}$. But $\bigcup_{n\in\mathbb{N}} F_n = \bigcup_{n\in\mathbb{N}} E_n$. Thus $\bigcup_{n\in\mathbb{N}} E_n \in \mathfrak{C}$. This shows that $\mathfrak{C}$ is closed under countable unions. ∎

Let $\mathfrak{E}$ be a collection of subsets of a set X. Then $\mathfrak{E}$ is contained in the monotone class $\mathfrak{P}(X)$. Let $\{\mathfrak{E}_\alpha : \alpha \in A\}$ be the collection of all monotone classes containing $\mathfrak{E}$. Then $\mathfrak{E} \subset \bigcap_{\alpha\in A} \mathfrak{E}_\alpha$. By Observation 23.11, $\bigcap_{\alpha\in A} \mathfrak{E}_\alpha$ is a monotone class. Thus $\bigcap_{\alpha\in A} \mathfrak{E}_\alpha$ is the smallest monotone class of subsets of X containing $\mathfrak{E}$.

Definition 23.13. *Let $\mathfrak{E}$ be a collection of subsets of a set X. We call the smallest monotone class of subsets of X containing $\mathfrak{E}$ the monotone class generated by $\mathfrak{E}$.*

Proposition 23.14. *The monotone class generated by an algebra $\mathfrak{A}$ of subsets of a set X is equal to the σ-algebra generated by $\mathfrak{A}$.*

Proof. 1. Let us show first that the monotone class $\mathfrak{K}$ generated by an algebra $\mathfrak{A}$ of subsets of a set X is again an algebra of subsets of X.

Now $X \in \mathfrak{A} \subset \mathfrak{K}$. If we show that $E \setminus F \in \mathfrak{K}$ for every pair $E, F \in \mathfrak{K}$ then since $X \in \mathfrak{K}$ we have $E^c = X \setminus E \in \mathfrak{K}$ for every $E \in \mathfrak{K}$ so that $\mathfrak{K}$ is closed under complementations. Thus to show that $\mathfrak{K}$ is an algebra of subsets of X it remains to show

(1) $$E \setminus F \in \mathfrak{K} \quad \text{and} \quad E \cup F \in \mathfrak{K} \quad \text{for every pair } E, F \in \mathfrak{K}.$$

For every $E \in \mathfrak{K}$ let us define a subcollection $\mathfrak{K}_E$ of $\mathfrak{K}$ by setting

(2) $$\mathfrak{K}_E = \big\{F \in \mathfrak{K} : E \setminus F, F \setminus E, E \cup F \in \mathfrak{K}\big\}.$$

Note that since $\emptyset \in \mathfrak{A} \subset \mathfrak{K}$ for every $E \in \mathfrak{K}$ we have $E \setminus E = \emptyset \in \mathfrak{K}$ and $E \cup E = E \in \mathfrak{K}$ so that $E \in \mathfrak{K}_E$. Note also that for every $E \in \mathfrak{K}$ we have $E \setminus \emptyset = E \in \mathfrak{K}, \emptyset \setminus E = \emptyset \in \mathfrak{K}$, and $E \cup \emptyset = E \in \mathfrak{K}$ so that $\emptyset \in \mathfrak{K}_E$. Another immediate consequence of the definition of $\mathfrak{K}_E$ by (2) is that

(3) $$F \in \mathfrak{K}_E \Leftrightarrow E \in \mathfrak{K}_F \quad \text{for every pair } E, F \in \mathfrak{K}.$$

Let us show that $\mathfrak{K}_E$ has the following properties:

1° $\mathfrak{K}_E$ is a monotone class for every $E \in \mathfrak{K}$.

2° $\mathfrak{K} \subset \mathfrak{K}_E$ for every $E \in \mathfrak{A}$.

3° $\mathfrak{K} \subset \mathfrak{K}_F$ for every $F \in \mathfrak{K}$.

To prove 1°, let $E \in \mathfrak{K}$ and let $(F_n : n \in \mathbb{N})$ be a monotone sequence in $\mathfrak{K}_E$. Let us show that $\lim_{n\to\infty} F_n \in \mathfrak{K}_E$. Now $\lim_{n\to\infty} F_n = \bigcup_{n\in\mathbb{N}} F_n$ if $(F_n : n \in \mathbb{N})$ is an increasing sequence and $\lim_{n\to\infty} F_n = \bigcap_{n\in\mathbb{N}} F_n$ if $(F_n : n \in \mathbb{N})$ is a decreasing sequence. In any case, $(E \setminus F_n : n \in \mathbb{N})$ is a monotone sequence and $E \setminus \big(\lim_{n\to\infty} F_n\big) = \lim_{n\to\infty} (E \setminus F_n)$. Now since $F_n \in \mathfrak{K}_E$, we have $E \setminus F_n \in \mathfrak{K}$. Then since $\mathfrak{K}$ is a monotone class, we have $\lim_{n\to\infty} (E \setminus F_n) \in \mathfrak{K}$. This shows that $E \setminus \big(\lim_{n\to\infty} F_n\big) \in \mathfrak{K}$. We show similarly that $\big(\lim_{n\to\infty} F_n\big) \setminus E \in \mathfrak{K}$ and $E \cup \big(\lim_{n\to\infty} F_n\big) \in \mathfrak{K}$. This verifies $\lim_{n\to\infty} F_n \in \mathfrak{K}_E$.

To prove 2°, let $E \in \mathfrak{A} \subset \mathfrak{K}$. Then for any $F \in \mathfrak{A}$, we have $F \in \mathfrak{A} \subset \mathfrak{K}$. Also since $\mathfrak{A}$ is an algebra, we have $E \setminus F, F \setminus E, E \cup F \in \mathfrak{A} \subset \mathfrak{K}$. This shows that $F \in \mathfrak{K}_E$. Therefore $\mathfrak{A} \subset \mathfrak{K}_E$. Since $\mathfrak{K}_E$ is a monotone class by 1°, it contains the smallest monotone class containing $\mathfrak{A}$, namely $\mathfrak{K}$.

To prove 3°, let $F \in \mathfrak{K}$. Then $F \in \mathfrak{K}_E$ for every $E \in \mathfrak{A}$ by 2°. This implies that $E \in \mathfrak{K}_F$ for every $E \in \mathfrak{A}$ by (3) so that $\mathfrak{A} \subset \mathfrak{K}_F$. Then since $\mathfrak{K}_F$ is a monotone class by 1°, it contains the smallest monotone class containing $\mathfrak{A}$, that is, $\mathfrak{K}$.

Now let $E, F \in \mathfrak{K}$. By 3°, $E \in \mathfrak{K}_F$ and thus $E \setminus F, F \setminus E, E \cup F \in \mathfrak{K}$. This proves that $\mathfrak{K}$ is an algebra.

2. Let us show that the monotone class $\mathfrak{K}$ generated by an algebra $\mathfrak{A}$ of subsets of a set X is equal to $\sigma(\mathfrak{A})$. Since a σ-algebra of subsets of a set is always a monotone class, $\sigma(\mathfrak{A})$ contains the smallest monotone class containing $\mathfrak{A}$. Thus $\sigma(\mathfrak{A}) \supset \mathfrak{K}$. Now $\mathfrak{K}$ is a monotone class and an algebra of subsets of X by **1.** Then by Observation 23.12, $\mathfrak{K}$ is a σ-algebra of subsets of X. Since $\mathfrak{K}$ contains $\mathfrak{A}$, $\mathfrak{K}$ contains the smallest σ-algebra of subsets of X containing $\mathfrak{A}$. Hence $\mathfrak{K} \supset \sigma(\mathfrak{A})$. Therefore we have $\mathfrak{K} = \sigma(\mathfrak{A})$. ∎

Let $\{A_i : i = 1, \dots, n\}$ and $\{B_i : i = 1, \dots, n\}$ be collections of subsets of X and Y respectively. Consider the collection $\{A_i \times B_i : i = 1, \dots, n\}$ of subsets of $X \times Y$. If at least one of the two collections $\{A_i : i = 1, \dots, n\}$ and $\{B_i : i = 1, \dots, n\}$ is a disjoint collection then $\{A_i \times B_i : i = 1, \dots, n\}$ is a disjoint collection. The converse of this is false, that is, the disjointness of $\{A_i \times B_i : i = 1, \dots, n\}$ does not imply that at least one of the two collections $\{A_i : i = 1, \dots, n\}$ and $\{B_i : i = 1, \dots, n\}$ is a disjoint collection.

For example consider the intervals $A_1 = (0, 2]$, $A_2 = (1, 3]$, $A_3 = (3, 4]$ in $\mathbb{R}_1$ and the intervals $B_1 = (0, 1]$, $B_2 = (1, 3]$, $B_3 = (2, 4]$ in $\mathbb{R}_2$. Then $\{A_i \times B_i : i = 1, 2, 3\}$ is a disjoint collection of subsets of $\mathbb{R}_1 \times \mathbb{R}_2$ but neither $\{A_i : i = 1, 2, 3\}$ nor $\{B_i : i = 1, 2, 3\}$ is a disjoint collection in $\mathbb{R}$.

Lemma 23.15. *Let $\mathfrak{A}$ and $\mathfrak{B}$ be algebras of subsets of sets X and Y respectively. Let $\{A_i \times B_i : i = 1, \dots, n\}$ be a collection in the semialgebra $\mathfrak{A} \times \mathfrak{B}$ of subsets of $X \times Y$ and let $E = \bigcup_{i=1}^n (A_i \times B_i)$. Then we have*

(a) *There exists a collection $\{A'_j : j = 1, \dots, p\}$ in $\mathfrak{A}$ and there exists a disjoint collection $\{B'_j : j = 1, \dots, p\}$ in $\mathfrak{B}$ such that $E = \bigcup_{j=1}^p (A'_j \times B'_j)$.*

(b) *There exists a disjoint collection $\{A''_k : k = 1, \dots, q\}$ in $\mathfrak{A}$ and there exists a collection $\{B''_k : k = 1, \dots, q\}$ in $\mathfrak{B}$ such that $E = \bigcup_{k=1}^q (A''_k \times B''_k)$.*

Proof. To prove (a), for $i = 1, \dots, n$ let $B_i^0 = B_i$ and $B_i^1 = B_i^c$ so that $Y = \bigcup_{\delta_i=0}^1 B_i^{\delta_i}$. Then since $Y = Y \cap \dots \cap Y$, we have

$$Y = \left(\bigcup_{\delta_1=0}^1 B_1^{\delta_1}\right) \cap \dots \cap \left(\bigcup_{\delta_n=0}^1 B_n^{\delta_n}\right) = \bigcup_{\delta_1=0}^1 \cdots \bigcup_{\delta_n=0}^1 (B_1^{\delta_1} \cap \dots \cap B_n^{\delta_n}).$$

Now $\{B_1^{\delta_1} \cap \dots \cap B_n^{\delta_n} : \delta_1 = 0, 1; \cdots; \delta_n = 0, 1\}$ is a disjoint collection in $\mathfrak{B}$. From this collection select only those which are subsets of B_i for some $i = 1, \dots, n$ and let $\{B'_j : j = 1, \dots, p\}$ be the subcollection thus selected. For each $j = 1, \dots, p$, let us write $\bigcup_{\{i : B_i \supset B'_j\}} A_i$ for the union of such A_i that $B_i \supset B'_j$. Then we have

$$E = \bigcup_{j=1,\dots,p} \left[\left(\bigcup_{\{i : B_i \supset B'_j\}} A_i\right) \times B'_j\right].$$

Since $\mathfrak{A}$ is an algebra, we have $\bigcup_{\{i:B_i\supset B'_j\}} A_i \in \mathfrak{A}$. Then let $A'_j = \bigcup_{\{i:B_i\supset B'_j\}} A_i$ for $j = 1, \ldots, p$. Thus we have $E = \bigcup_{j=1}^{p}(A'_j \times B'_j)$ where $\{A'_j : j = 1, \ldots, p\}$ is a collection in $\mathfrak{A}$ and $\{B'_j : j = 1, \ldots, p\}$ is a disjoint collection in $\mathfrak{B}$. This proves (a). Similarly for (b). ∎

Proposition 23.16. *Let $\big(X\times Y, \sigma(\mathfrak{A}\times\mathfrak{B}), \mu\times\nu\big)$ be the product measure space of two σ-finite measure spaces $(X, \mathfrak{A}, \mu)$ and $(Y, \mathfrak{B}, \nu)$. Then for every $E \in \sigma(\mathfrak{A}\times\mathfrak{B})$, $\nu\big(E(x,\cdot)\big)$ is a $\mathfrak{A}$-measurable function of $x \in X$ and $\mu\big(E(\cdot,y)\big)$ is a $\mathfrak{B}$-measurable function of $y \in Y$. Furthermore we have*

$$(\mu\times\nu)(E) = \int_X \nu\big(E(x,\cdot)\big)\mu(dx) = \int_Y \mu\big(E(\cdot,y)\big)\nu(dy).$$

Proof. 1. Consider first the case where $(X, \mathfrak{A}, \mu)$ and $(Y, \mathfrak{B}, \nu)$ are finite measure spaces. To prove the Proposition, let $\mathfrak{F}$ be the collection of all members of $\sigma(\mathfrak{A}\times\mathfrak{B})$ for which the assertion of the Proposition is valid. By definition we have $\mathfrak{F} \subset \sigma(\mathfrak{A}\times\mathfrak{B})$. Now if we show that $\sigma(\mathfrak{A}\times\mathfrak{B}) \subset \mathfrak{F}$ then $\mathfrak{F} = \sigma(\mathfrak{A}\times\mathfrak{B})$ and we are done. Thus it remains to show $\sigma(\mathfrak{A}\times\mathfrak{B}) \subset \mathfrak{F}$.

1.1. Let us show that $\mathfrak{A}\times\mathfrak{B} \subset \mathfrak{F}$. Suppose $E = A\times B \in \mathfrak{A}\times\mathfrak{B}$. Then

$$E(x,\cdot) = \begin{cases} B & \text{for } x \in A, \\ \emptyset & \text{for } x \in A^c, \end{cases}$$

so that

$$\nu\big(E(x,\cdot)\big) = \begin{cases} \nu(B) & \text{for } x \in A, \\ 0 & \text{for } x \in A^c. \end{cases}$$

Thus $\nu\big(E(x,\cdot)\big)$ is a $\mathfrak{A}$-measurable function of $x \in X$. Furthermore we have the equality $\int_X \nu\big(E(x,\cdot)\big)\mu(dx) = \nu(B)\mu(A) = (\mu\times\nu)(A\times B)$. Similarly $\mu\big(E(\cdot,y)\big)$ is a $\mathfrak{B}$-measurable function of $y \in Y$ and $\int_Y \mu\big(E(\cdot,y)\big)\nu(dy) = (\mu\times\nu)(A\times B)$. This shows that $\mathfrak{A}\times\mathfrak{B} \subset \mathfrak{F}$.

1.2. Next we show that $\alpha(\mathfrak{A}\times\mathfrak{B}) \subset \mathfrak{F}$. Let $E \in \alpha(\mathfrak{A}\times\mathfrak{B})$. By Lemma 21.3 and Theorem 21.4, E is a finite disjoint union of members of $\mathfrak{A}\times\mathfrak{B}$, say $E = \bigcup_{i=1}^{n}(A_i\times B_i)$ where $A_i \in \mathfrak{A}$ and $B_i \in \mathfrak{B}$ for $i = 1, \ldots, n$. Now the disjointness of the collection $\{A_i \times B_i : i = 1, \ldots, n\}$ implies neither the disjointness of the collection $\{A_i : i = 1, \ldots, n\}$ nor the disjointness of the collection $\{B_i : i = 1, \ldots, n\}$. However according to Lemma 23.15, E can be expressed as $E = \bigcup_{i=1}^{p}(A'_i\times B'_i)$ where $\{A'_i : i = 1, \ldots, p\}$ is a collection in $\mathfrak{A}$ and $\{B'_i : i = 1, \ldots, n\}$ is a disjoint collection in $\mathfrak{B}$. Similarly E can be expressed as $E = \bigcup_{i=1}^{q}(A''_i\times B''_i)$ where $\{A''_i : i = 1, \ldots, q\}$ is a disjoint collection in $\mathfrak{A}$ and $\{B''_i : i = 1, \ldots, q\}$ is a collection in $\mathfrak{B}$. The disjointness of $\{B'_i : i = 1, \ldots, n\}$ implies the disjointness of $\{A'_i\times B'_i : i = 1, \ldots, p\}$. Thus $(\mu\times\nu)(E) = \sum_{i=1}^{p}(\mu\times\nu)(A'_i\times B'_i)$. Now for every $x \in X$, we have

$$E(x,\cdot) = \Big(\bigcup_{i=1}^{p}(A'_i\times B'_i)\Big)(x,\cdot) = \bigcup_{\{i:A'_i\ni x\}} B'_i.$$

Since $\{B_i' : i = 1, \ldots, n\}$ is a disjoint collection, we have

$$\nu\big(E(x,\cdot)\big) = \nu\Big(\bigcup_{\{i:A_i'\ni x\}} B_i'\Big) = \sum_{\{i:A_i'\ni x\}} \nu(B_i') = \sum_{i=1}^{p} \mathbf{1}_{A_i'}(x)\nu(B_i').$$

Thus $\nu\big(E(x,\cdot)\big)$ is a $\mathfrak{A}$-measurable function of $x \in X$ and

$$\int_X \nu\big(E(x,\cdot)\big)\mu(dx) = \int_X \sum_{i=1}^{p} \mathbf{1}_{A_i'}(x)\nu(B_i')\mu(dx)$$
$$= \sum_{i=1}^{p}\Big[\int_X \mathbf{1}_{A_i'}(x)\mu(dx)\Big]\nu(B_i') = \sum_{i=1}^{p} \mu(A_i')\nu(B_i')$$
$$= \sum_{i=1}^{p} (\mu\times\nu)(A_i'\times B_i') = (\mu\times\nu)(E).$$

Similarly $\mu\big(E(\cdot,y)\big)$ is a $\mathfrak{B}$-measurable function of $y \in Y$ and we have the equality $\int_Y \mu\big(E(\cdot,y)\big)\nu(dy) = (\mu\times\nu)(E)$. This shows that $E \in \mathfrak{F}$. Therefore $\alpha(\mathfrak{A}\times\mathfrak{B}) \subset \mathfrak{F}$.

1.3. Let us show that $\mathfrak{F}$ is a monotone class. Let $(E_n : n \in \mathbb{N})$ be a monotone sequence in $\mathfrak{F}$ and let $E = \lim_{n\to\infty} E_n$. We are to show that $E \in \mathfrak{F}$. Now $E = \bigcup_{n\in\mathbb{N}} E_n$ or $E = \bigcap_{n\in\mathbb{N}} E_n$ according as $(E_n : n \in \mathbb{N})$ is an increasing sequence or a decreasing sequence. Since $E_n \in \mathfrak{F} \subset \sigma(\mathfrak{A}\times\mathfrak{B})$, we have $E \in \sigma(\mathfrak{A}\times\mathfrak{B})$. For every $x \in X$, we have $E_n(x,\cdot) \uparrow E(x,\cdot)$ or $E_n(x,\cdot) \downarrow E(x,\cdot)$ according as $E_n \uparrow$ or $E_n \downarrow$. In any case, since ν is a finite measure, we have

$$\lim_{n\to\infty} \nu\big(E_n(x,\cdot)\big) = \nu\big(\lim_{n\to\infty} E_n(x,\cdot)\big) = \nu\big(E(x,\cdot)\big).$$

Since $E_n \in \mathfrak{F}$, $\nu\big(E_n(x,\cdot)\big)$ is a $\mathfrak{A}$-measurable function of $x \in X$. Thus the limit function $\nu\big(E(x,\cdot)\big)$ is a $\mathfrak{A}$-measurable function of $x \in X$. The sequence of $\mathfrak{A}$-measurable functions of $x \in X$, $\big(\nu\big(E_n(x,\cdot)\big) : n \in \mathbb{N}\big)$ is uniformly bounded on X by a constant since $\nu\big(E_n(x,\cdot)\big) \le \nu(Y) < \infty$ for all $x \in X$ and $n \in \mathbb{N}$. Since $\mu(X) < \infty$, the Bounded Convergence Theorem (Theorem 7.16) is applicable and we have

$$\int_X \nu\big(E(x,\cdot)\big)\mu(dx) = \lim_{n\to\infty}\int_X \nu\big(E_n(x,\cdot)\big)\mu(dx)$$
$$= \lim_{n\to\infty}(\mu\times\nu)(E_n) = (\mu\times\nu)\big(\lim_{n\to\infty} E_n\big) = (\mu\times\nu)(E),$$

where the second equality is by the fact that $E_n \in \mathfrak{F}$ and the third equality is by the fact that $(E_n : n \in \mathbb{N})$ is a monotone sequence and $\mu\times\nu$ is a finite measure. Similarly $\mu\big(E(\cdot,y)\big)$ is a $\mathfrak{B}$-measurable function of $y \in Y$ and $\int_Y \mu\big(E(\cdot,y)\big)\nu(dy) = (\mu\times\nu)(E)$. This shows that $\mathfrak{F}$ is a monotone class.

1.4. We have shown above that the algebra $\alpha(\mathfrak{A}\times\mathfrak{B})$ is contained in $\mathfrak{F}$ and $\mathfrak{F}$ is a monotone class. According to Proposition 23.14, the smallest σ algebra containing an algebra $\alpha(\mathfrak{A}\times\mathfrak{B})$ is equal to the smallest monotone class containing the algebra $\alpha(\mathfrak{A}\times\mathfrak{B})$

and is hence contained in the monotone class $\mathfrak{F}$ containing $\alpha(\mathfrak{A} \times \mathfrak{B})$. Thus we have $\sigma(\alpha(\mathfrak{A} \times \mathfrak{B})) \subset \mathfrak{F}$. But $\sigma(\alpha(\mathfrak{A} \times \mathfrak{B})) = \sigma(\mathfrak{A} \times \mathfrak{B})$. Thus we have $\sigma(\mathfrak{A} \times \mathfrak{B}) \subset \mathfrak{F}$. This completes the proof of the Proposition for the case that $(X, \mathfrak{A}, \mu)$ and $(Y, \mathfrak{B}, \nu)$ are finite measure spaces.

2. Let $(X, \mathfrak{A}, \mu)$ and $(Y, \mathfrak{B}, \nu)$ be two σ-finite measure spaces. Then there exists a disjoint sequence $(A_n : n \in \mathbb{N})$ in $\mathfrak{A}$ such that $\bigcup_{n\in\mathbb{N}} A_n = X$ and $\mu(A_n) < \infty$ for $n \in \mathbb{N}$ and there exists a disjoint sequence $(B_n : n \in \mathbb{N})$ in $\mathfrak{B}$ such that $\bigcup_{n\in\mathbb{N}} B_n = Y$ and $\nu(B_n) < \infty$ for $n \in \mathbb{N}$. Then $\{A_i \times B_j : i \in \mathbb{N} \text{ and } j \in \mathbb{N}\}$ is a disjoint collection in $\mathfrak{A} \times \mathfrak{B}$ and $\bigcup_{i\in\mathbb{N}} \bigcup_{j\in\mathbb{N}} (A_i \times B_j) = X \times Y$. Let $\mathfrak{A}_i = \mathfrak{A} \cap A_i$ and $\mathfrak{B}_j = \mathfrak{B} \cap B_j$. Consider the product measure space $\left(A_i \times B_j, \sigma(\mathfrak{A}_i \times \mathfrak{B}_j), \mu \times \nu\right)$ of the two finite measure spaces $(A_i, \mathfrak{A}_i, \mu)$ and $(B_j, \mathfrak{B}_j, \nu)$. Now $\mathfrak{A}_i \times \mathfrak{B}_j = (\mathfrak{A} \cap A_i) \times (\mathfrak{B} \cap B_j) = (\mathfrak{A} \times \mathfrak{B}) \cap (A_i \times B_j)$ and

$$\sigma(\mathfrak{A}_i \times \mathfrak{B}_j) = \sigma\left((\mathfrak{A} \times \mathfrak{B}) \cap (A_i \times B_j)\right) = \sigma(\mathfrak{A} \times \mathfrak{B}) \cap (A_i \times B_j)$$

by Theorem 1.15. Thus we have

$$\left(A_i \times B_j, \sigma(\mathfrak{A}_i \times \mathfrak{B}_j), \mu \times \nu\right) = \left(A_i \times B_j, \sigma(\mathfrak{A} \times \mathfrak{B}) \cap (A_i \times B_j), \mu \times \nu\right).$$

Let $E \in \sigma(\mathfrak{A} \times \mathfrak{B})$ and let $E_{i,j} = E \cap (A_i \times B_j) \in \sigma(\mathfrak{A} \times \mathfrak{B}) \cap (A_i \times B_j) = \sigma(\mathfrak{A}_i \times \mathfrak{B}_j)$. Then by our result in **1**, $\nu\left(E_{i,j}(x, \cdot)\right)$ is a $\mathfrak{A}_i$-measurable function on A_i, $\mu\left(E_{i,j}(\cdot, y)\right)$ is a $\mathfrak{B}_j$-measurable function on B_j, and

$$(\mu \times \nu)(E_{i,j}) = \int_{A_i} \nu\left(E_{i,j}(x, \cdot)\right) \mu(dx) = \int_{B_j} \mu\left(E_{i,j}(\cdot, y)\right) \nu(dy).$$

Let $x \in X$. Then $x \in A_i$ for some $i \in \mathbb{N}$. Thus $\nu\left(E(x, \cdot)\right) = \sum_{j\in\mathbb{N}} \nu\left(E_{i,j}(x, \cdot)\right)$. Since $\nu\left(E_{i,j}(x, \cdot)\right)$ is $\mathfrak{A}_i$-measurable and hence $\mathfrak{A}$-measurable, the sum $\nu\left(E(x, \cdot)\right)$ is $\mathfrak{A}$-measurable. Also

$$\begin{aligned}
(\mu \times \nu)(E) &= \sum_{i\in\mathbb{N}} \sum_{j\in\mathbb{N}} (\mu \times \nu)(E_{i,j}) = \sum_{i\in\mathbb{N}} \sum_{j\in\mathbb{N}} \int_{A_i} \nu\left(E_{i,j}(x, \cdot)\right) \mu(dx) \\
&= \sum_{i\in\mathbb{N}} \int_{A_i} \sum_{j\in\mathbb{N}} \nu\left(E_{i,j}(x, \cdot)\right) \mu(dx) = \sum_{i\in\mathbb{N}} \int_{A_i} \nu\left(E(x, \cdot)\right) \mu(dx) \\
&= \int_X \nu\left(E(x, \cdot)\right) \mu(dx).
\end{aligned}$$

Similarly $\mu\left(E(\cdot, y)\right)$ is $\mathfrak{B}$-measurable and $(\mu \times \nu)(E) = \int_Y \mu\left(E(\cdot, y)\right) \nu(dy)$. ∎

Theorem 23.17. (Tonelli's Theorem) *Let $\left(X \times Y, \sigma(\mathfrak{A} \times \mathfrak{B}), \mu \times \nu\right)$ be the product measure space of two σ-finite measure spaces $(X, \mathfrak{A}, \mu)$ and $(Y, \mathfrak{B}, \nu)$. Let f be a nonnegative extended real-valued $\sigma(\mathfrak{A} \times \mathfrak{B})$-measurable function on $X \times Y$. Then*

(a) *$F^1(x) := \int_Y f(x, \cdot)\, d\nu$ is a $\mathfrak{A}$-measurable function of $x \in X$.*

(b) *$F^2(y) := \int_X f(\cdot, y)\, d\mu$ is a $\mathfrak{B}$-measurable function of $y \in Y$.*

(c) *$\int_{X\times Y} f\, d(\mu \times \nu) = \int_X F^1\, d\mu = \int_Y F^2\, d\nu$, that is,*

$$\int_{X\times Y} f\, d(\mu \times \nu) = \int_X \left[\int_Y f(x, \cdot)\, d\nu\right] \mu(dx) = \int_Y \left[\int_X f(\cdot, y)\, d\mu\right] \nu(dy).$$

Proof. Note that by Proposition 23.9, the $\sigma(\mathfrak{A} \times \mathfrak{B})$-measurability of the function f on $X \times Y$ implies the $\mathfrak{B}$-measurability of the function $f(x, \cdot)$ on Y for every $x \in X$ and the $\mathfrak{A}$-measurability of the function $f(\cdot, y)$ on X for every $y \in Y$.

1. Consider first the case $f = \mathbf{1}_E$ where $E \in \sigma(\mathfrak{A} \times \mathfrak{B})$. In this case, $f(x, \cdot) = \mathbf{1}_{E(x,\cdot)}$ and $F^1(x) = \int_Y \mathbf{1}_{E(x,\cdot)}\, d\nu = \nu\big(E(x, \cdot)\big)$, which is a $\mathfrak{A}$-measurable function of $x \in X$ by Proposition 23.16. Similarly $F^2(y)$ is a $\mathfrak{B}$-measurable function of $y \in Y$. This verifies (a) and (b). Now

$$(1) \qquad \int_{X\times Y} f\, d(\mu \times \nu) = \int_{X\times Y} \mathbf{1}_E\, d(\mu \times \nu) = (\mu \times \nu)(E).$$

On the other hand, we have

$$(2) \qquad \int_X \left[\int_Y f(x, \cdot)\, d\nu\right] \mu(dx) = \int_X \left[\int_Y \mathbf{1}_{E(x,\cdot)}\, d\nu\right] \mu(dx) = \int_X \nu\big(E(x, \cdot)\big)\mu(dx)$$

and

$$(3) \qquad \int_Y \left[\int_X f(\cdot, y)\, d\mu\right] \nu(dy) = \int_Y \left[\int_X \mathbf{1}_{E(\cdot,y)}\, d\mu\right] \nu(dy) = \int_Y \mu\big(E(\cdot, y)\big)\nu(dy).$$

According to Proposition 23.16, the right sides of (1), (2), and (3) are all equal. Thus the left sides of (1), (2), and (3) are all equal. This verifies (c) for this case.

2. If f is a nonnegative simple function on $X \times Y$, then $f = \sum_{k=1}^n a_k \mathbf{1}_{E_k}$ where $\{E_k : k = 1, \dots, n\}$ is a disjoint collection in $\sigma(\mathfrak{A} \times \mathfrak{B})$ and $a_k \geq 0$ for $k = 1, \dots, n$. Then (a), (b), and (c) hold for f by our result in **1** and by the linearity of the integrals.

3. If f is a nonnegative extended real-valued $\sigma(\mathfrak{A} \times \mathfrak{B})$-measurable function on $X \times Y$, then there exists an increasing sequence $(f_n : n \in \mathbb{N})$ of nonnegative simple functions on $(X \times Y, \sigma(\mathfrak{A} \times \mathfrak{B}))$ such that $f_n \uparrow f$ on $X \times Y$ by Lemma 8.6. By our results in **2**, (a), (b), and (c) hold for f_n for each $n \in \mathbb{N}$ so that $F_n^1(x) := \int_Y f_n(x, \cdot)\, d\nu$ is a $\mathfrak{A}$-measurable function of $x \in X$, $F_n^2(y) := \int_X f_n(\cdot, y)\, d\mu$ is a $\mathfrak{B}$-measurable function of $y \in Y$, and

$$(4) \qquad \int_{X\times Y} f_n\, d(\mu \times \nu) = \int_X \left[\int_Y f_n(x, \cdot)\, d\nu\right] \mu(dx) = \int_Y \left[\int_X f_n(\cdot, y)\, d\mu\right] \nu(dy).$$

For each $x \in X$, $(f_n(x, \cdot) : n \in \mathbb{N})$ is an increasing sequence of nonnegative simple functions with $f_n(x, \cdot) \uparrow f(x, \cdot)$ on Y. Thus by Theorem 8.5 (Monotone Convergence Theorem), we have

$$(5) \qquad F^1(x) = \int_Y f(x, \cdot)\, d\nu = \lim_{n\to\infty} \int_Y f_n(x, \cdot)\, d\nu.$$

Since $\int_Y f_n(x, \cdot)\, d\nu$ is a $\mathfrak{A}$-measurable function of $x \in X$ for every $n \in \mathbb{N}$ by our result in **2** and since $\left(\int_Y f_n(x, \cdot)\, d\nu : n \in \mathbb{N}\right)$ is an increasing sequence, $F^1(x)$ is a $\mathfrak{A}$-measurable function of $x \in X$ by Theorem 4.21. This proves (a). Similarly for (b).

To prove (c), note that by Theorem 8.5, we have

$$(6) \qquad \lim_{n\to\infty} \int_{X\times Y} f_n\, d(\mu \times \nu) = \int_{X\times Y} f\, d(\mu \times \nu).$$

Since $\left(\int_Y f_n(x,\cdot)\,d\nu : n \in \mathbb{N}\right)$ is an increasing sequence of nonnegative extended real-valued $\mathfrak{A}$-measurable functions of $x \in X$, applying Theorem 8.5 we have

$$(7)\qquad \lim_{n\to\infty}\int_X\left[\int_Y f_n(x,\cdot)\,d\nu\right]\mu(dx)=\int_X\lim_{n\to\infty}\left[\int_Y f_n(x,\cdot)\,d\nu\right]\mu(dx)$$
$$=\int_X\left[\int_Y f(x,\cdot)\,d\nu\right]\mu(dx)$$

by (5). Applying (6) and (7) to (4), we have $\int_{X\times Y} f\,d(\mu\times\nu)=\int_X\left[\int_Y f(x,\cdot)\,d\nu\right]\mu(dx)$. Similarly $\int_{X\times Y} f\,d(\mu\times\nu)=\int_Y\left[\int_X f(\cdot,y)\,d\mu\right]\nu(dy)$. ∎

Theorem 23.18. (Fubini's Theorem) *Let $\left(X\times Y, \sigma(\mathfrak{A}\times\mathfrak{B}), \mu\times\nu\right)$ be the product measure space of two σ-finite measure spaces $(X,\mathfrak{A},\mu)$ and $(Y,\mathfrak{B},\nu)$. Let f be a $\mu\times\nu$-integrable extended real-valued $\sigma(\mathfrak{A}\times\mathfrak{B})$-measurable function on $X\times Y$. Then*

(a) *The $\mathfrak{B}$-measurable function $f(x,\cdot)$ is ν-integrable on Y for μ-a.e. $x\in X$ and the $\mathfrak{A}$-measurable function $f(\cdot,y)$ is μ-integrable on X for ν-a.e. $y\in Y$.*

(b) *The function $F^1(x) := \int_Y f(x,\cdot)\,d\nu$ is defined for μ-a.e. $x\in X$, $\mathfrak{A}$-measurable and μ-integrable on X.*
The function $F^2(y) := \int_X f(\cdot,y)\,d\mu$ is defined for ν-a.e. $y\in Y$, $\mathfrak{B}$-measurable and ν-integrable on Y.

(c) *We have the equalities: $\int_{X\times Y} f\,d(\mu\times\nu)=\int_X F^1\,d\mu=\int_Y F^2\,d\nu$, that is,*

$$\int_{X\times Y} f\,d(\mu\times\nu)=\int_X\left[\int_Y f(x,\cdot)\,d\nu\right]\mu(dx)=\int_Y\left[\int_X f(\cdot,y)\,d\mu\right]\nu(dy).$$

Proof. 1. Let $f = f^+ - f^-$. Then f^+ and f^- are nonnegative extended real-valued $\sigma(\mathfrak{A}\times\mathfrak{B})$-measurable functions. The $\mu\times\nu$-integrability of f on $X\times Y$ implies that f^+ and f^- are $\mu\times\nu$-integrable on $X\times Y$. Applying (c) of Theorem 23.17 to f^+ and f^-, we have

$$(1)\qquad \int_{X\times Y} f^+\,d(\mu\times\nu)=\int_X\left[\int_Y f^+(x,\cdot)\,d\nu\right]\mu(dx)$$
$$=\int_Y\left[\int_X f^+(\cdot,y)\,d\mu\right]\nu(dy)<\infty$$

and similarly

$$(2)\qquad \int_{X\times Y} f^-\,d(\mu\times\nu)=\int_X\left[\int_Y f^-(x,\cdot)\,d\nu\right]\mu(dx)$$
$$=\int_Y\left[\int_X f^-(\cdot,y)\,d\mu\right]\nu(dy)<\infty.$$

Since $\int_X\left[\int_Y f^+(x,\cdot)\,d\nu\right]\mu(dx)<\infty$, we have $\int_Y f^+(x,\cdot)\,d\nu<\infty$ for μ-a.e. $x\in X$ by (a) of Lemma 8.2, that is, $f^+(x,\cdot)$ is ν-integrable on Y for μ-a.e. $x\in X$. Likewise

$\int_X \left[\int_Y f^-(x,\cdot)\,d\nu\right]\mu(dx) < \infty$ implies that $f^-(x,\cdot)$ is ν-integrable on Y for μ-a.e. $x \in X$. Thus $f(x,\cdot) = f^+(x,\cdot) - f^-(x,\cdot)$ is ν-integrable on Y for μ-a.e. $x \in X$. Similarly $f(\cdot,y)$ is μ-integrable on X for ν-a.e. $y \in Y$. This proves (a).

2. $\int_Y f^+(x,\cdot)\,d\nu$ and $\int_Y f^-(x,\cdot)\,d\nu$ are $\mathfrak{A}$-measurable functions of $x \in X$ by (a) of Theorem 23.17. Since $\int_Y f^+(x,\cdot)\,d\nu < \infty$ and $\int_Y f^-(x,\cdot)\,d\nu < \infty$ for μ-a.e. $x \in X$, the function $F^1(x) = \int_Y f(x,\cdot)\,d\nu = \int_Y f^+(x,\cdot)\,d\nu - \int_Y f^-(x,\cdot)\,d\nu$ is defined for μ-a.e. $x \in X$ and $\mathfrak{A}$-measurable on X. Therefore, recalling Definition 9.11, we have

$$\text{(3)}\qquad \int_X F^1(x)\mu(dx) = \int_X \left[\int_Y f^+(x,\cdot)\,d\nu - \int_Y f^-(x,\cdot)\,d\nu\right]\mu(dx)$$
$$= \int_X \left[\int_Y f^+(x,\cdot)\,d\nu\right]\mu(dx) - \int_X \left[\int_Y f^-(x,\cdot)\,d\nu\right]\mu(dx) \in \mathbb{R},$$

by (1) and (2). Thus F^1 is μ-integrable on X. Similarly F^2 is $\mathfrak{B}$-measurable and ν-integrable on Y. This proves (b).

3. Finally since f is $\mu \times \nu$-integrable on $X \times Y$, we have

$$\int_{X\times Y} f\,d(\mu\times\nu) = \int_{X\times Y} f^+\,d(\mu\times\nu) - \int_{X\times Y} f^-\,d(\mu\times\nu)$$
$$= \int_X \left[\int_Y f^+(x,\cdot)\,d\nu\right]\mu(dx) - \int_X \left[\int_Y f^-(x,\cdot)\,d\nu\right]\mu(dx)$$
$$= \int_X F^1(x)\mu(dx) = \int_X \left[\int_Y f(x,\cdot)\,d\nu\right]\mu(dx),$$

where the second equality is by (1) and (2) and the third equality is by (3). Similarly we have $\int_{X\times Y} f\,d(\mu\times\nu) = \int_Y \left[\int_X f(\cdot,y)\,d\mu\right]\nu(dy)$. This proves (c). ∎

Combining Theorem 23.17 and Theorem 23.18, we have a particularly useful theorem.

Theorem 23.19. (Fubini-Tonelli) *Let $(X \times Y, \sigma(\mathfrak{A} \times \mathfrak{B}), \mu \times \nu)$ be the product measure space of two σ-finite measure spaces $(X, \mathfrak{A}, \mu)$ and $(Y, \mathfrak{B}, \nu)$. Let f be an extended real-valued $\sigma(\mathfrak{A} \times \mathfrak{B})$-measurable function on $X \times Y$. If either $\int_X \left[\int_Y |f|\,d\nu\right]d\mu < \infty$ or $\int_Y \left[\int_X |f|\,d\mu\right]d\nu < \infty$, then f is $\mu \times \nu$-integrable on $X \times Y$ and furthermore*

$$\int_{X\times Y} f\,d(\mu\times\nu) = \int_X \left[\int_Y f\,d\nu\right]d\mu = \int_Y \left[\int_X f\,d\mu\right]d\nu.$$

Proof. $|f|$ is a nonnegative extended real-valued $\sigma(\mathfrak{A}\times\mathfrak{B})$-measurable function on $X\times Y$. Thus by Theorem 23.17 (Tonelli), we have

$$\int_{X\times Y} |f|\,d(\mu\times\nu) = \int_X \left[\int_Y |f|\,d\nu\right]d\mu = \int_Y \left[\int_X |f|\,d\mu\right]d\nu.$$

Thus if either $\int_X \left[\int_Y |f|\,d\nu\right]d\mu < \infty$ or $\int_Y \left[\int_X |f|\,d\mu\right]d\nu < \infty$, then $|f|$ is $\mu \times \nu$-integrable on $X \times Y$. This implies that f is $\mu \times \nu$-integrable on $X \times Y$. Thus by Theorem 23.18 (Fubini), we have $\int_{X\times Y} f\,d(\mu\times\nu) = \int_X \left[\int_Y f\,d\nu\right]d\mu = \int_Y \left[\int_X f\,d\mu\right]d\nu$. ∎

According to Theorem 23.18 (Fubini), the integrability of f on $X \times Y$, that is existence and finiteness of the integral $\int_{X\times Y} f\,d(\mu\times\nu)$, implies that the integral $\int_{X\times Y} f\,d(\mu\times\nu)$ is equal to both of the two repeated integrals $\int_X \big[\int_Y f\,d\nu\big]\,d\mu$ and $\int_Y \big[\int_X f\,d\mu\big]\,d\nu$. We show in the next theorem that actually the semi-integrability of f on $X \times Y$, that is, the existence of $\int_{X\times Y} f\,d(\mu\times\nu)$ as an extended real number, is sufficient for this. This is an important point. The possibility of calculating the integral $\int_{X\times Y} f\,d(\mu\times\nu)$ as a repeated integral $\int_X \big[\int_Y f\,d\nu\big]\,d\mu$ or $\int_Y \big[\int_X f\,d\mu\big]\,d\nu$ without knowing that $\int_{X\times Y} f\,d(\mu\times\nu)$ is finite presents a way to determine whether or not $\int_{X\times Y} f\,d(\mu\times\nu)$ is finite.

Theorem 23.20. (Fubini) *Let $\big(X\times Y, \sigma(\mathfrak{A}\times\mathfrak{B}), \mu\times\nu\big)$ be the product measure space of two σ-finite measure spaces $(X,\mathfrak{A},\mu)$ and $(Y,\mathfrak{B},\nu)$. Let f be a $\mu\times\nu$ semi-integrable extended real-valued $\sigma(\mathfrak{A}\times\mathfrak{B})$-measurable function on $X\times Y$, that is, $\int_{X\times Y} f\,d(\mu\times\nu)$ exists in $\overline{\mathbb{R}}$, then*

$$\int_{X\times Y} f\,d(\mu\times\nu) = \int_X \left[\int_Y f\,d\nu\right] d\mu = \int_Y \left[\int_X f\,d\mu\right] d\nu.$$

Proof. If f is $\mu\times\nu$ semi-integrable on $X\times Y$, then at least one of $\int_{X\times Y} f^+\,d(\mu\times\nu)$ and $\int_{X\times Y} f^-\,d(\mu\times\nu)$ is finite. Let us consider the case $\int_{X\times Y} f^+\,d(\mu\times\nu) < \infty$. By (c) of Theorem 23.17 (Tonelli), we have

$$(1)\qquad \int_X \left[\int_Y f^+\,d\nu\right] d\mu = \int_Y \left[\int_X f^+\,d\mu\right] d\nu = \int_{X\times Y} f^+\,d(\mu\times\nu) \in [0,\infty)$$

and

$$(2)\qquad \int_X \left[\int_Y f^-\,d\nu\right] d\mu = \int_Y \left[\int_X f^-\,d\mu\right] d\nu = \int_{X\times Y} f^-\,d(\mu\times\nu) \in [0,\infty].$$

Now $\int_X \big[\int_Y f^+\,d\nu\big]\,d\mu < \infty$ implies that $\int_Y f^+(x,\cdot)\,d\nu < \infty$ for μ-a.e. $x \in X$. Thus the difference $\int_Y f^+(x,\cdot)\,d\nu - \int_Y f^-(x,\cdot)\,d\nu$ exists in $\overline{\mathbb{R}}$ for μ-a.e. $x\in X$. This implies according to (c) of Corollary 9.15 that

$$(3)\qquad \int_Y f(x,\cdot)\,d\nu = \int_Y f^+(x,\cdot)\,d\nu - \int_Y f^-(x,\cdot)\,d\nu \in \overline{\mathbb{R}} \quad \text{for } \mu\text{-a.e. } x\in X.$$

Now the difference $\int_X \big[\int_Y f^+\,d\nu\big]\,d\mu - \int_X \big[\int_Y f^-\,d\nu\big]\,d\mu$ exists in $\overline{\mathbb{R}}$ since according to (1) we have $\int_X \big[\int_Y f^+\,d\nu\big]\,d\mu < \infty$. Thus

$$\begin{aligned}
\int_X \left[\int_Y f(x,\cdot)\,d\nu\right] d\mu &= \int_X \left[\int_Y f^+(x,\cdot)\,d\nu - \int_Y f^-(x,\cdot)\,d\nu\right] d\mu \\
&= \int_X \left[\int_Y f^+(x,\cdot)\,d\nu\right] d\mu - \int_X \left[\int_Y f^-(x,\cdot)\,d\nu\right] d\mu \\
&= \int_{X\times Y} f^+\,d(\mu\times\nu) - \int_{X\times Y} f^-\,d(\mu\times\nu) \\
&= \int_{X\times Y} f\,d(\mu\times\nu),
\end{aligned}$$

where the first equality is by (3), the second equality is by (c) of Corollary 9.15, and the third equality is by (1) and (2). We show similarly that $\int_X \left[\int_Y f(\cdot, y)\, d\nu\right] d\mu = \int_{X\times Y} f\, d(\mu \times \nu)$. The case $\int_{X\times Y} f^-\, d(\mu \times \nu) < \infty$ is handled similarly. ∎

Reduction of integrals on a product measure space of two measure spaces to repeated integrals can be extended to integrals on a product measure space of any finite number of measure spaces.

Theorem 23.21. *Let $\left(X_1 \times \cdots \times X_n, \sigma(\mathfrak{A}_1 \times \cdots \times \mathfrak{A}_n), \mu_1 \times \cdots \times \mu_n\right)$ be the product measure space of σ-finite measure spaces $(X_i, \mathfrak{A}_i, \mu_i)$ for $i = 1, \ldots, n$. If f is a nonnegative extended real-valued $\sigma(\mathfrak{A}_1 \times \cdots \times \mathfrak{A}_n)$-measurable function on $X_1 \times \cdots \times X_n$, then*

$$\begin{aligned}\int_{X_1\times\cdots\times X_n} f\, d(\mu_1 \times \cdots \times \mu_n) &= \int_{X_n} \cdots \left[\int_{X_2} \left[\int_{X_1} f\, d\mu_1\right] d\mu_2\right] \cdots d\mu_n \\ &= \int_{X_1} \cdots \left[\int_{X_{n-1}} \left[\int_{X_n} f\, d\mu_n\right] d\mu_{n-1}\right] \cdots d\mu_1.\end{aligned}$$

The equalities hold also if f is a $\mu_1 \times \cdots \times \mu_n$ semi-integrable extended real-valued $\sigma(\mathfrak{A}_1 \times \cdots \times \mathfrak{A}_n)$-measurable function on $X_1 \times \cdots \times X_n$.

Proof. By Proposition 23.5, we have

$$\begin{aligned}\sigma(\mathfrak{A}_1 \times \cdots \times \mathfrak{A}_n) &= \mathfrak{A}_1 \otimes \cdots \otimes \mathfrak{A}_n \\ &= \left(\cdots\left((\mathfrak{A}_1 \otimes \mathfrak{A}_2) \otimes \mathfrak{A}_3\right) \otimes \cdots \otimes \mathfrak{A}_{n-1}\right) \otimes \mathfrak{A}_n \\ &= \mathfrak{A}_1 \otimes \left(\mathfrak{A}_2 \otimes \cdots \otimes \left(\mathfrak{A}_{n-2} \otimes (\mathfrak{A}_{n-1} \otimes \mathfrak{A}_n)\right) \cdots\right).\end{aligned}$$

By the uniqueness of the product measure space (Theorem 23.7), we have

$$\begin{aligned}&\left(X_1 \times \cdots \times X_n, \sigma(\mathfrak{A}_1 \times \cdots \times \mathfrak{A}_n), \mu_1 \times \cdots \times \mu_n\right) \\ =&\left(X_1 \times \cdots \times X_n, \left(\cdots\left((\mathfrak{A}_1 \otimes \mathfrak{A}_2) \otimes \mathfrak{A}_3\right) \otimes \cdots \otimes \mathfrak{A}_{n-1}\right) \otimes \mathfrak{A}_n, \mu_1 \times \cdots \times \mu_n\right) \\ =&\left(X_1 \times \cdots \times X_n, \mathfrak{A}_1 \otimes \left(\mathfrak{A}_2 \otimes \cdots \otimes \left(\mathfrak{A}_{n-2} \otimes (\mathfrak{A}_{n-1} \otimes \mathfrak{A}_n)\right) \cdots\right), \mu_1 \times \cdots \times \mu_n\right).\end{aligned}$$

If f is nonnegative, then by repeated application of Theorem 23.17 (Tonelli), we have the equalities for the integrals. If f is $\mu_1 \times \cdots \times \mu_n$ semi-integrable, then by repeated application of Theorem 23.20, we have the equalities for the integrals. ∎

Theorem 23.22. *Let $\left(X_1 \times \cdots \times X_n, \sigma(\mathfrak{A}_1 \times \cdots \times \mathfrak{A}_n), \mu_1 \times \cdots \times \mu_n\right)$ be the product measure space of σ-finite measure spaces $(X_i, \mathfrak{A}_i, \mu_i)$ for $i = 1, \ldots, n$. Let f be an extended real-valued $\sigma(\mathfrak{A}_1 \times \cdots \times \mathfrak{A}_n)$-measurable function on $X_1 \times \cdots \times X_n$. If either*

(1) $$\int_{X_n} \cdots \left[\int_{X_2} \left[\int_{X_1} |f|\, d\mu_1\right] d\mu_2\right] \cdots d\mu_n < \infty$$

or

$$(2) \qquad \int_{X_1} \cdots \left[\int_{X_{n-1}} \left[\int_{X_n} |f| \, d\mu_n \right] d\mu_{n-1} \right] \cdots d\mu_1 < \infty,$$

then f is $\mu_1 \times \cdots \times \mu_n$-integrable on $X_1 \times \cdots \times X_n$ and thus

$$(3) \qquad \int_{X_1 \times \cdots \times X_n} f \, d(\mu_1 \times \cdots \times \mu_n) = \int_{X_n} \cdots \left[\int_{X_2} \left[\int_{X_1} f \, d\mu_1 \right] d\mu_2 \right] \cdots d\mu_n$$
$$= \int_{X_1} \cdots \left[\int_{X_{n-1}} \left[\int_{X_n} f \, d\mu_n \right] d\mu_{n-1} \right] \cdots d\mu_1.$$

Proof. If one of (1) and (2) holds, then by the equalities in Theorem 23.21, then $|f|$ is $\mu_1 \times \cdots \times \mu_n$-integrable on $X_1 \times \cdots \times X_n$ and thus f is $\mu_1 \times \cdots \times \mu_n$-integrable on $X_1 \times \cdots \times X_n$. The integrability of f then implies (3) by Theorem 23.21. ∎

[III] Completion of Product Measure Space

Let $(X, \mathfrak{A}, \mu)$ and $(Y, \mathfrak{B}, \nu)$ be two measure spaces. As we have shown in Lemma 23.6, the set function λ on the semialgebra $\mathfrak{A} \times \mathfrak{B}$ of subsets of $X \times Y$ defined by setting $\lambda(E) = \mu(A)\nu(B)$ for $E = A \times B \in \mathfrak{A} \times \mathfrak{B}$ is a countably additive set function on the semialgebra. Let λ^* be the outer measure on $X \times Y$ based on λ. With the σ-algebra $\mathfrak{M}(\lambda^*)$ of λ^*-measurable subsets of $X \times Y$, we have a complete measure space $(X \times Y, \mathfrak{M}(\lambda^*), \lambda^*)$ by Theorem 2.9. Restricting λ^* to $\sigma(\mathfrak{A} \times \mathfrak{B}) \subset \mathfrak{M}(\lambda^*)$, we have a product measure space $(X \times Y, \sigma(\mathfrak{A} \times \mathfrak{B}), \mu \times \nu)$ of $(X, \mathfrak{A}, \mu)$ and $(Y, \mathfrak{B}, \nu)$. Unlike $(X \times Y, \mathfrak{M}(\lambda^*), \lambda^*)$, the measure space $(X \times Y, \sigma(\mathfrak{A} \times \mathfrak{B}), \mu \times \nu)$ may not be complete even when $(X, \mathfrak{A}, \mu)$ and $(Y, \mathfrak{B}, \nu)$ are. Below is such an example. Now if we let $\left(X \times Y, \overline{\sigma(\mathfrak{A} \times \mathfrak{B})}, \mu \times \nu\right)$ be the completion of $(X \times Y, \sigma(\mathfrak{A} \times \mathfrak{B}), \mu \times \nu)$, then since $\left(X \times Y, \overline{\sigma(\mathfrak{A} \times \mathfrak{B})}, \mu \times \nu\right)$ is the smallest extension to a complete measure space, we have $\overline{\sigma(\mathfrak{A} \times \mathfrak{B})} \subset \mathfrak{M}(\lambda^*)$.

Example. Let $(X, \mathfrak{A}, \mu) = \left(\mathbb{R}, \mathfrak{M}_L, \mu_L\right)$ and $(Y, \mathfrak{B}, \nu) = \left(\mathbb{R}, \mathfrak{M}_L, \mu_L\right)$ also. Consider the product measure space $\left(X \times Y, \sigma(\mathfrak{A} \times \mathfrak{B}), \mu \times \nu\right) = \left(\mathbb{R} \times \mathbb{R}, \sigma(\mathfrak{M}_L \times \mathfrak{M}_L), \mu_L \times \mu_L\right)$. Let E be a non $\mathfrak{M}_L$-measurable subset of $\mathbb{R}$ and let F be a $\mathfrak{M}_L$-measurable subset of $\mathbb{R}$ with $\mu_L(F) = 0$. Then $\mathbb{R} \times F \in \sigma(\mathfrak{M}_L \times \mathfrak{M}_L)$ and $(\mu_L \times \mu_L)(\mathbb{R} \times F) = 0$. Thus $\mathbb{R} \times F$ is a null set in $\left(\mathbb{R} \times \mathbb{R}, \sigma(\mathfrak{M}_L \times \mathfrak{M}_L), \mu_L \times \mu_L\right)$. Consider $E \times F \subset \mathbb{R} \times F$. Let $y \in F$. Then the y-section of $E \times F$ is E which is not in $\mathfrak{M}_L$. Then by (a) of Proposition 23.9, $E \times F$ is not in $\sigma(\mathfrak{M}_L \times \mathfrak{M}_L)$. This shows that $\left(\mathbb{R} \times \mathbb{R}, \sigma(\mathfrak{M}_L \times \mathfrak{M}_L), \mu_L \times \mu_L\right)$ is not a complete measure space.

Theorem 23.23. *Consider the product measure space $\left(\times_{i=1}^n X_i, \sigma(\times_{i=1}^n \mathfrak{A}_i), \times_{i=1}^n \mu_i\right)$ of n measure spaces $(X_i, \mathfrak{A}_i, \mu_i)$ for $i = 1, \ldots, n$ and the product measure space $\left(\times_{i=1}^n X_i, \sigma(\times_{i=1}^n \overline{\mathfrak{A}}_i), \times_{i=1}^n \mu_i\right)$ of the completions $(X_i, \overline{\mathfrak{A}}_i, \mu_i)$ of $(X_i, \mathfrak{A}_i, \mu_i)$ for $i = 1, \ldots, n$. The completions of these two product measure spaces are identical, that is,*

$$\left(\times_{i=1}^n X_i, \overline{\sigma(\times_{i=1}^n \mathfrak{A}_i)}, \times_{i=1}^n \mu_i\right) = \left(\times_{i=1}^n X_i, \overline{\sigma(\times_{i=1}^n \overline{\mathfrak{A}}_i)}, \times_{i=1}^n \mu_i\right).$$

Proof. Recall that by Theorem 5.4 and Definition 5.2, if $(X, \overline{\mathfrak{A}}, \mu)$ is the completion of a measure space $(X, \mathfrak{A}, \mu)$, then $\overline{\mathfrak{A}}$ is the collection of all subsets E of X of the type $E = A \cup C$ where $A \in \mathfrak{A}$ and C is a subset of a null set in $(X, \mathfrak{A}, \mu)$.

To prove the Proposition, we show that

$$(1) \qquad \overline{\sigma\big(\times_{i=1}^{n}\mathfrak{A}_i\big)} = \overline{\sigma\big(\times_{i=1}^{n}\overline{\mathfrak{A}}_i\big)}.$$

Since $\mathfrak{A}_i \subset \overline{\mathfrak{A}}_i$ for $i = 1, \ldots, n$, we have $\sigma\big(\times_{i=1}^{n}\mathfrak{A}_i\big) \subset \sigma\big(\times_{i=1}^{n}\overline{\mathfrak{A}}_i\big)$. This implies then $\overline{\sigma\big(\times_{i=1}^{n}\mathfrak{A}_i\big)} \subset \overline{\sigma\big(\times_{i=1}^{n}\overline{\mathfrak{A}}_i\big)}$. Thus it remains to show that

$$(2) \qquad \overline{\sigma\big(\times_{i=1}^{n}\overline{\mathfrak{A}}_i\big)} \subset \overline{\sigma\big(\times_{i=1}^{n}\mathfrak{A}_i\big)}.$$

If we show that

$$(3) \qquad \sigma\big(\times_{i=1}^{n}\overline{\mathfrak{A}}_i\big) \subset \overline{\sigma\big(\times_{i=1}^{n}\mathfrak{A}_i\big)},$$

then since $\overline{\sigma\big(\times_{i=1}^{n}\mathfrak{A}_i\big)}$ is already complete with respect to $\times_{i=1}^{n}\mu_i$, its completion with respect to $\times_{i=1}^{n}\mu_i$ is identical with itself. Hence completing both sides of (3) with respect to $\times_{i=1}^{n}\mu_i$ we have (2). To prove (3), since $\overline{\sigma\big(\times_{i=1}^{n}\mathfrak{A}_i\big)}$ is a σ-algebra, it suffices to show

$$(4) \qquad \times_{i=1}^{n}\overline{\mathfrak{A}}_i \subset \overline{\sigma\big(\times_{i=1}^{n}\mathfrak{A}_i\big)}.$$

Let $\times_{i=1}^{n}E_i \in \times_{i=1}^{n}\overline{\mathfrak{A}}_i$. Since $E_i \in \overline{\mathfrak{A}}_i$, we have $E_i = A_i \cup C_i$ where $A_i \in \mathfrak{A}_i$ and C_i is a subset of a null set B_i in $(X_i, \mathfrak{A}_i, \mu_i)$. For $i = 1, \ldots, n$, let

$$P_i = F_{i,1} \times \cdots \times F_{i,n} \text{ where } F_{i,j} = \begin{cases} A_j \cup C_j & \text{for } j \neq i, \\ C_i & \text{for } j = i, \end{cases}$$

and

$$Q_i = G_{i,1} \times \cdots \times G_{i,n} \text{ where } G_{i,j} = \begin{cases} A_j \cup B_j & \text{for } j \neq i, \\ B_i & \text{for } j = i. \end{cases}$$

Then

$$(5) \qquad \times_{i=1}^{n}E_i = \times_{i=1}^{n}(A_i \cup C_i) = \big(\times_{i=1}^{n}A_i\big) \cup \big(\textstyle\bigcup_{i=1}^{n} P_i\big).$$

Now $\times_{i=1}^{n}A_i \in \times_{i=1}^{n}\mathfrak{A}_i$. We have $P_i \subset Q_i$ and since $\mu_i(G_{i,i}) = \mu_i(B_i) = 0$ we have $\big(\times_{i=1}^{n}\mu_i\big)(Q_i) = \mu_1(G_{i,1}) \cdot \cdots \cdot \mu_n(G_{i,n}) = 0$. Thus P_i is a subset of a null set Q_i in the measure space $\big(\times_{i=1}^{n}X_i, \sigma\big(\times_{i=1}^{n}\mathfrak{A}_i\big), \times_{i=1}^{n}\mu_i\big)$ and then $\bigcup_{1=1}^{n} P_i$ is a subset of a null set $\bigcup_{1=1}^{n} Q_i$ in the measure space. Thus according to (5), $\times_{i=1}^{n}E_i$ is the union of a set $\times_{i=1}^{n}A_i$ in $\sigma\big(\times_{i=1}^{n}\mathfrak{A}_i\big)$ and a subset $\bigcup_{1=1}^{n} P_i$ of a null set $\bigcup_{1=1}^{n} Q_i$ in the measure space $\big(\times_{i=1}^{n}X_i, \sigma\big(\times_{i=1}^{n}\mathfrak{A}_i\big), \times_{i=1}^{n}\mu_i\big)$. Therefore $\times_{i=1}^{n}E_i$ is in the completion $\overline{\sigma\big(\times_{i=1}^{n}\mathfrak{A}_i\big)}$ of $\sigma\big(\times_{i=1}^{n}\mathfrak{A}_i\big)$ with respect to $\times_{i=1}^{n}\mu_i$. This proves (4). ∎

Observation 23.24. *Consider the product measure space $\big(X \times Y, \sigma(\mathfrak{A} \times \mathfrak{B}), \mu \times \nu\big)$ of two measure spaces $(X, \mathfrak{A}, \mu)$ and $(Y, \mathfrak{B}, \nu)$. Let E be a null set in the product measure space. Then*

1° *$E(x, \cdot)$ is a null set in $(Y, \mathfrak{B}, \nu)$ for μ-a.e. $x \in X$,*

2° *$E(\cdot, y)$ is a null set in $(X, \mathfrak{A}, \mu)$ for ν-a.e. $y \in Y$.*

Proof. We have $\int_X \nu\big(E(x, \cdot)\big)\mu(dx) = \int_Y \mu\big(E(\cdot, y)\big)\nu(dy) = (\mu \times \nu)(E) = 0$ by Proposition 23.16. The fact that $\nu\big(E(x, \cdot)\big) \geq 0$ and $\int_X \nu\big(E(x, \cdot)\big)\mu(dx) = 0$ implies that $\nu\big(E(x, \cdot)\big) = 0$ for μ-a.e. $x \in X$. Similarly $\mu\big(E(\cdot, y)\big) = 0$ for ν-a.e. $y \in Y$. ∎

Regarding Observation 23.24 let us note that for a null set E in the product measure space $\big(X \times Y, \sigma(\mathfrak{A} \times \mathfrak{B}), \mu \times \nu\big)$ there may be $x \in X$ such that $\nu\big(E(x, \cdot)\big) > 0$ and there may be $y \in Y$ such that $\mu\big(E(\cdot, y)\big) > 0$. For instance consider the product measure space $\big(\mathbb{R} \times \mathbb{R}, \sigma(\mathfrak{M}_L \times \mathfrak{M}_L), \mu_L \times \mu_L\big)$. Let $E = \{c\} \times [a, b]$ where $c \in \mathbb{R}$ and $[a, b] \subset \mathbb{R}$. We have $E \in \mathfrak{M}_L \times \mathfrak{M}_L \subset \sigma(\mathfrak{M}_L \times \mathfrak{M}_L)$ and $(\mu_L \times \mu_L)(E) = \mu_L\big(\{c\}\big)\mu_L\big([a, b]\big) = 0 \cdot (b - a) = 0$ so that E is a null set in the product measure space. Now $E(c, \cdot) = [a, b]$ and $\mu_L\big(E(c, \cdot)\big) = \mu_L\big([a, b]\big) = b - a > 0$.

Proposition 23.25. *Let $\big(X \times Y, \overline{\sigma(\mathfrak{A} \times \mathfrak{B})}, \overline{\mu \times \nu}\big)$ be the completion of the product measure space $\big(X \times Y, \sigma(\mathfrak{A} \times \mathfrak{B}), \mu \times \nu\big)$ of two complete measure spaces $(X, \mathfrak{A}, \mu)$ and $(Y, \mathfrak{B}, \nu)$.*

(a) *If $E \in \overline{\sigma(\mathfrak{A} \times \mathfrak{B})}$, then $E(x, \cdot) \in \mathfrak{B}$ for μ-a.e. $x \in X$ and $E(\cdot, y) \in \mathfrak{A}$ for ν-a.e. $y \in Y$.*

(b) *If f is an extended real-valued $\overline{\sigma(\mathfrak{A} \times \mathfrak{B})}$-measurable function on $X \times Y$, then $f(x, \cdot)$ is a $\mathfrak{B}$-measurable function on Y for μ-a.e. $x \in X$ and $f(\cdot, y)$ is a $\mathfrak{A}$-measurable function on X for ν-a.e. $y \in Y$.*

Proof. Let us prove (b) first. Now if f is an extended real-valued $\overline{\sigma(\mathfrak{A} \times \mathfrak{B})}$-measurable function on $X \times Y$, then by Theorem 5.6 there exist a null set E in $\big(X \times Y, \sigma(\mathfrak{A} \times \mathfrak{B}), \mu \times \nu\big)$ and an extended real-valued $\sigma(\mathfrak{A} \times \mathfrak{B})$-measurable function g on $X \times Y$ such that $f = g$ on E^c. Then we have

1° $f(x, \cdot) = g(x, \cdot)$ on $E(x, \cdot)^c$ for $x \in X$,

2° $f(\cdot, y) = g(\cdot, y)$ on $E(\cdot, y)^c$ for $y \in Y$.

Now $g(x, \cdot)$ is $\mathfrak{B}$-measurable on Y for every $x \in X$ by Proposition 23.9. By Observation 23.24, $E(x, \cdot)$ is a null set in $(Y, \mathfrak{B}, \nu)$ for μ-a.e. $x \in X$. The fact that $g(x, \cdot)$ is $\mathfrak{B}$-measurable on Y and $f(x, \cdot) = g(x, \cdot)$ on $E(x, \cdot)^c$ where $E(x, \cdot)$ is a null set in a complete measure space $(Y, \mathfrak{B}, \nu)$ implies that $f(x, \cdot)$ is $\mathfrak{B}$-measurable on Y by Observation 4.20. Thus $f(x, \cdot)$ is a $\mathfrak{B}$-measurable function on Y for μ-a.e. $x \in X$. Similarly $f(\cdot, y)$ is a $\mathfrak{A}$-measurable function on X for ν-a.e. $y \in Y$.

To prove (a), let $E \in \overline{\sigma(\mathfrak{A} \times \mathfrak{B})}$. The $\overline{\sigma(\mathfrak{A} \times \mathfrak{B})}$-measurability of E is equivalent to the $\overline{\sigma(\mathfrak{A} \times \mathfrak{B})}$-measurability of $\mathbf{1}_E$ by Observation 4.3. Thus $\mathbf{1}_E$ is a $\overline{\sigma(\mathfrak{A} \times \mathfrak{B})}$-measurable function on $X \times Y$. Then by (a), $\mathbf{1}_E(x, \cdot)$ is a $\mathfrak{B}$-measurable function on Y for μ-a.e. $x \in X$. Since the $\mathfrak{B}$-measurability of $\mathbf{1}_E(x, \cdot)$ is equivalent to the $\mathfrak{B}$-measurability of E by Observation 4.3, we have $E(x, \cdot) \in \mathfrak{B}$ for μ-a.e. $x \in X$. Similarly $E(\cdot, y) \in \mathfrak{A}$ for ν-a.e. $y \in Y$. ∎

Theorem 23.26. (Tonelli's Theorem) *Let $\left(X \times Y, \overline{\sigma(\mathfrak{A} \times \mathfrak{B})}, \mu \times \nu\right)$ be the completion of the product measure space $\left(X \times Y, \sigma(\mathfrak{A} \times \mathfrak{B}), \mu \times \nu\right)$ of two complete and σ-finite measure spaces $(X, \mathfrak{A}, \mu)$ and $(Y, \mathfrak{B}, \nu)$. Let f be a nonnegative extended real-valued $\overline{\sigma(\mathfrak{A} \times \mathfrak{B})}$-measurable function on $X \times Y$. Then*

(a) *$\int_Y f(x, \cdot)\, d\nu$ is defined for μ-a.e. $x \in X$ and $\mathfrak{A}$-measurable.*

(b) *$\int_X f(\cdot, y)\, d\mu$ is defined for ν-a.e. $y \in Y$ and $\mathfrak{B}$-measurable.*

(c) *$\int_{X \times Y} f\, d(\mu \times \nu) = \int_X \left[\int_Y f\, d\nu\right] d\mu = \int_Y \left[\int_X f\, d\mu\right] d\nu$.*

Proof. Since f is an extended real-valued $\overline{\sigma(\mathfrak{A} \times \mathfrak{B})}$-measurable function on $X \times Y$, by Theorem 5.6 there exist a null set E in $\left(X \times Y, \sigma(\mathfrak{A} \times \mathfrak{B}), \mu \times \nu\right)$ and an extended real-valued $\sigma(\mathfrak{A} \times \mathfrak{B})$-measurable function g on $X \times Y$ such that $f = g$ on E^c. Redefine g on E by setting $g = 0$ on E. We still have $f = g$ on E^c but g is now a nonnegative extended real-valued $\sigma(\mathfrak{A} \times \mathfrak{B})$-measurable function on $X \times Y$. Thus by Theorem 23.17 (Tonelli), $\int_Y g(x, \cdot)\, d\nu$ is a $\mathfrak{A}$-measurable function of $x \in X$. By 1° of Observation 23.24 and 1° in the Proof of Proposition 23.25, we have $\int_Y f(x, \cdot)\, d\nu = \int_Y g(x, \cdot)\, d\nu$ for μ-a.e. $x \in X$. Then by the completeness of the measure space $(X, \mathfrak{A}, \mu)$, $\int_Y f(x, \cdot)\, d\nu$ is a $\mathfrak{A}$-measurable function of $x \in X$. Similarly $\int_X f(\cdot, y)\, d\mu$ is a $\mathfrak{B}$-measurable function of $y \in Y$.

Finally since $f = g$ on E^c and E is a null set in $\left(X \times Y, \sigma(\mathfrak{A} \times \mathfrak{B}), \mu \times \nu\right)$, we have

$$\int_{X \times Y} f\, d(\mu \times \nu) = \int_{X \times Y} g\, d(\mu \times \nu) = \int_X \left[\int_Y g(x, \cdot)\, d\nu\right] d\mu = \int_X \left[\int_Y f(x, \cdot)\, d\nu\right] d\mu,$$

where the second equality is by Theorem 23.17. Similarly we have

$$\int_{X \times Y} f\, d(\mu \times \nu) = \int_X \left[\int_Y f(x, \cdot)\, d\nu\right] d\mu. \ \blacksquare$$

Theorem 23.27. (Fubini's Theorem) *Let $\left(X \times Y, \overline{\sigma(\mathfrak{A} \times \mathfrak{B})}, \mu \times \nu\right)$ be the completion of the product measure space $(X \times Y, \sigma(\mathfrak{A} \times \mathfrak{B}), \mu \times \nu)$ of two complete and σ-finite measure spaces $(X, \mathfrak{A}, \mu)$ and $(Y, \mathfrak{B}, \nu)$. Let f be an $\mu \times \nu$-integrable extended real-valued $\overline{\sigma(\mathfrak{A} \times \mathfrak{B})}$-measurable function on $X \times Y$. Then*

(a) *$f(x, \cdot)$ is ν-integrable on Y for μ-a.e. $x \in X$.*

$f(\cdot, y)$ is μ-integrable on X for ν-a.e. $y \in Y$.

(b) *$\int_Y f(x, \cdot)\, d\nu$ is defined for μ-a.e. $x \in X$, $\mathfrak{A}$-measurable and μ-integrable on X.*

$\int_X f(\cdot, y)\, d\mu$ is defined for ν-a.e. $y \in Y$, $\mathfrak{B}$-measurable and ν-integrable on Y.

(c) *$\int_{X \times Y} f\, d(\mu \times \nu) = \int_X \left[\int_Y f\, d\nu\right] d\mu = \int_Y \left[\int_X f\, d\mu\right] d\nu$.*

Proof. Since f is an extended real-valued $\overline{\sigma(\mathfrak{A} \times \mathfrak{B})}$-measurable function on $X \times Y$, Theorem 5.6 implies that there exist a null set E in $\left(X \times Y, \sigma(\mathfrak{A} \times \mathfrak{B}), \mu \times \nu\right)$ and an extended real-valued $\sigma(\mathfrak{A} \times \mathfrak{B})$-measurable function g on $X \times Y$ such that $f = g$ on E^c. Being a null set in $\left(X \times Y, \sigma(\mathfrak{A} \times \mathfrak{B}), \mu \times \nu\right)$, E is a null set in $\left(X \times Y, \overline{\sigma(\mathfrak{A} \times \mathfrak{B})}, \mu \times \nu\right)$ also. The function g is $\sigma(\mathfrak{A} \times \mathfrak{B})$-measure and hence $\overline{\sigma(\mathfrak{A} \times \mathfrak{B})}$-measurable. Since $g = f$

on E^c and f is $\mu \times \nu$-integrable on $X \times Y$, we have $\int_{X\times Y} |g|\,d(\mu\times\nu) = \int_{X\times Y} |f|\,d(\mu \times \nu) < \infty$ so that g is $\mu \times \nu$-integrable on $X \times Y$. Thus by Theorem 23.18 (Fubini), $g(x,\cdot)$ is ν-integrable on Y for μ-a.e. $x \in X$. Since $f(x,\cdot) = g(x,\cdot)$ on $E(x,\cdot)^c$ for every $x \in X$ and since $E(x,\cdot)$ is a null set in $(Y, \mathfrak{B}, \nu)$ for μ-a.e. $x \in X$, $f(x,\cdot)$ is ν-integrable on Y for μ-a.e. $x \in X$. Similarly $f(\cdot, y)$ is μ-integrable on X for ν-a.e. $y \in Y$. This proves (a). Then (b) and (c) are proved by the same argument as in the Proof of Theorem 23.26. ∎

Combining Theorem 23.26 and Theorem 23.27, we have

Theorem 23.28. (Fubini-Tonelli Theorem) *Let $\big(X\times Y, \overline{\sigma(\mathfrak{A}\times\mathfrak{B})}, \mu\times\nu\big)$ be the completion of the product measure space $\big(X\times Y, \sigma(\mathfrak{A}\times\mathfrak{B}), \mu\times\nu\big)$ of two complete and σ-finite measure spaces $(X, \mathfrak{A}, \mu)$ and $(Y, \mathfrak{B}, \nu)$. Let f be an extended real-valued $\overline{\sigma(\mathfrak{A}\times\mathfrak{B})}$-measurable function on $X \times Y$. If either $\int_X \big[\int_Y |f|\,d\nu\big]\,d\mu < \infty$ or $\int_Y \big[\int_X |f|\,d\mu\big]\,d\nu < \infty$, then f is $\mu\times\nu$-integrable on $X \times Y$ and furthermore we have*

$$\int_{X\times Y} f\,d(\mu\times\nu) = \int_X \left[\int_Y f\,d\nu\right] d\mu = \int_Y \left[\int_X f\,d\mu\right] d\nu.$$

[IV] Convolution of Functions

Consider two sequences of real numbers $(a_n : n \in \mathbb{Z}_+)$ and $(b_n : n \in \mathbb{Z}_+)$ and the sequence $(a_n b_n : n \in \mathbb{Z}_+)$. Then $\sum_{n\in\mathbb{Z}_+} |a_n b_n| \le \big[\sum_{n\in\mathbb{Z}_+} |a_n|^2\big]^{1/2}\big[\sum_{n\in\mathbb{Z}_+} |b_n|^2\big]^{1/2}$ by Proposition 17.22. If the two series $\sum_{n\in\mathbb{Z}_+} a_n$ and $\sum_{n\in\mathbb{Z}_+} b_n$ are absolutely convergent, that is, $\sum_{n\in\mathbb{Z}_+} |a_n| < \infty$ and $\sum_{n\in\mathbb{Z}_+} |b_n| < \infty$, then $\sum_{n\in\mathbb{Z}_+} |a_n|^2 < \infty$ and $\sum_{n\in\mathbb{Z}_+} |b_n|^2 < \infty$ by Theorem 17.21 so that the series $\sum_{n\in\mathbb{Z}_+} a_n b_n$ is absolutely convergent.

The Cauchy product of $(a_n : n \in \mathbb{Z}_+)$ and $(b_n : n \in \mathbb{Z}_+)$ is the sequence $(c_n : n \in \mathbb{Z}_+)$ where $c_n = \sum_{k=0}^n a_{n-k} b_k$, that is, $c_n = a_0 b_n + a_1 b_{n-1} + a_2 b_{n-2} + \cdots + a_n b_0$, for $n \in \mathbb{Z}_+$. If $\sum_{n\in\mathbb{Z}_+} a_n$ and $\sum_{n\in\mathbb{Z}_+} b_n$ are absolutely convergent, then so is $\sum_{n\in\mathbb{Z}_+} c_n$ and furthermore $\sum_{n\in\mathbb{Z}_+} c_n = \big[\sum_{n\in\mathbb{Z}_+} a_n\big]\big[\sum_{n\in\mathbb{Z}_+} b_n\big]$. (For a proof, see R. C. Buck [5].)

The convolution $f * g$ of two extended real-valued $\mathfrak{M}_L$-measurable functions f and g on $\mathbb{R}$ is defined by $(f * g)(x) = \int_{\mathbb{R}} f(x-y) g(y)\,\mu_L(dy)$ for $x \in \mathbb{R}$ in analogy to the Cauchy product of two sequences. We shall show that if f and g are μ_L-integrable on $\mathbb{R}$, then $(f * g)(x)$ is defined for a.e. $x \in \mathbb{R}$, $f * g$ is $\mathfrak{M}_L$-measurable and furthermore we have the equality $\int_{\mathbb{R}} f * g\,d\mu_L = \big[\int_{\mathbb{R}} f\,d\mu_L\big]\big[\int_{\mathbb{R}} g\,d\mu_L\big]$.

Convolution of functions with higher order integrability will be treated. For this we define the following notations.

Definition 23.29. *For $p \in [1, \infty)$, we write $L_r^p(\mathbb{R}, \mathfrak{M}_L, \mu_L)$ for the linear space of equivalence classes of extended real-valued $\mathfrak{M}_L$-measurable functions f on $\mathbb{R}$ with $\|f\|_p = \big\{\int_{\mathbb{R}} |f|^p\,d\mu_L\big\}^{1/p} < \infty$, the equivalence relation being that of μ_L-a.e. equality on $\mathbb{R}$. We write $L_r^\infty(\mathbb{R}, \mathfrak{M}_L, \mu_L)$ for the linear space of equivalence classes of extended real-*

valued $\mathfrak{M}_L$-measurable functions f on $\mathbb{R}$ with essential supremum $\|f\|_\infty < \infty$, the equivalence relation being that of μ_L-a.e. equality on $\mathbb{R}$.

[IV.1] Convolution of Integrable Functions

Definition 23.30. *Let f and g be two extended real-valued $\mathfrak{M}_L$-measurable functions on $\mathbb{R}$. We define the convolution $f * g$ of f and g by setting*

$$(f * g)(x) = \int_{\mathbb{R}} f(x-y)g(y)\,\mu_L(dy)$$

for $x \in \mathbb{R}$ for which the integral exists in $\overline{\mathbb{R}}$.

Proposition 23.31. *If f and g are two extended real-valued $\mathfrak{M}_L$-measurable functions on $\mathbb{R}$ and if $(f*g)(x)$ exists for some $x \in \mathbb{R}$, then $(g*f)(x)$ exists and $(f*g)(x) = (g*f)(x)$.*

Proof. By definition

$$\begin{aligned}(f * g)(x) &= \int_{\mathbb{R}} f(x-y)g(y)\,\mu_L(dy) = \int_{\mathbb{R}} f\big(x-(y+x)\big)g(y+x)\,\mu_L(dy)\\ &= \int_{\mathbb{R}} f(-y)g(y+x)\,\mu_L(dy) = \int_{\mathbb{R}} f(y)g(-y+x)\,\mu_L(dy)\\ &= (g * f)(x)\end{aligned}$$

where the second equality is by Theorem 9.31 (Translation Invariance) replacing y with $y+x$ and the fourth equality is by Theorem 9.32 (Linear Transformation) replacing y with $-y$. ■

Lemma 23.32. *Let $(\mathbb{R}\times\mathbb{R}, \mathfrak{M}_L^2, \mu_L \times \mu_L)$ be the completion of the product measure space $(\mathbb{R}\times\mathbb{R}, \sigma(\mathfrak{M}_L \times \mathfrak{M}_L), \mu_L \times \mu_L)$, that is, $\mathfrak{M}_L^2 = \overline{\sigma(\mathfrak{M}_L \times \mathfrak{M}_L)}$, the completion of the σ-algebra $\sigma(\mathfrak{M}_L \times \mathfrak{M}_L)$ with respect to the measure $\mu_L \times \mu_L$. Let f be a real-valued $\mathfrak{M}_L$-measurable function on $\mathbb{R}$, that is, f is a $\mathfrak{M}_L/\mathfrak{B}_{\mathbb{R}}$-measurable mapping of $\mathbb{R}$ into $\mathbb{R}$, and define a real-valued function h on $\mathbb{R}\times\mathbb{R}$ by setting*

$$(1) \qquad h(x,y) = f(x-y) \quad \text{for } (x,y) \in \mathbb{R}\times\mathbb{R}.$$

Then h is a real-valued $\mathfrak{M}_L^2$-measurable function on $\mathbb{R}\times\mathbb{R}$, that is, a $\mathfrak{M}_L^2/\mathfrak{B}_{\mathbb{R}}$-measurable mapping of $\mathbb{R}\times\mathbb{R}$ into $\mathbb{R}$.

Proof. According to Theorem 24.31, for every non singular linear mapping S of $\mathbb{R}\times\mathbb{R}$ onto $\mathbb{R}\times\mathbb{R}$ we have $S(E) \in \mathfrak{M}_L^2$ for every $E \in \mathfrak{M}_L^2$. Then since S^{-1} is also a non singular linear mapping of $\mathbb{R}\times\mathbb{R}$ onto $\mathbb{R}\times\mathbb{R}$, we have $S^{-1}(E) \in \mathfrak{M}_L^2$ for every $E \in \mathfrak{M}_L^2$. Thus S is a $\mathfrak{M}_L^2/\mathfrak{M}_L^2$-measurable mapping of $\mathbb{R}\times\mathbb{R}$ onto $\mathbb{R}\times\mathbb{R}$.

Let π_1 be the projection of $\mathbb{R}_1 \times \mathbb{R}_2$ onto $\mathbb{R}_1$. Then for every $F \in \mathfrak{M}_L$ we have

$$\pi_1^{-1}(F) = F \times \mathbb{R}_2 \in \mathfrak{M}_L \times \mathfrak{M}_L \subset \mathfrak{M}_L^2.$$

Thus π_1 is a $\mathfrak{M}_L^2/\mathfrak{M}_L$-measurable mapping of $\mathbb{R}_1 \times \mathbb{R}_2$ onto $\mathbb{R}$.

Now let f be a real-valued $\mathfrak{M}_L$-measurable function on $\mathbb{R}$ and consider the real-valued function h on $\mathbb{R} \times \mathbb{R}$ defined by (1). Let S be a non singular linear mapping of $\mathbb{R} \times \mathbb{R}$ onto $\mathbb{R} \times \mathbb{R}$ defined by

$$S(x, y) = (x - y, x + y) \quad \text{for } (x, y) \in \mathbb{R} \times \mathbb{R}.$$

(This non singular linear mapping is represented by rotating the xy-plane by an angle $\frac{\pi}{4}$ and then multiplying by a scalar $\sqrt{2}$.) In terms of the mappings S and π_1, the function h is expressed as

$$h(x, y) = \big(f \circ \pi_1 \circ S\big)(x, y) \quad \text{for } (x, y) \in \mathbb{R} \times \mathbb{R}.$$

Since S is a $\mathfrak{M}_L^2/\mathfrak{M}_L^2$-measurable mapping of $\mathbb{R} \times \mathbb{R}$ onto $\mathbb{R} \times \mathbb{R}$, π_1 is a $\mathfrak{M}_L^2/\mathfrak{M}_L$-measurable mapping of $\mathbb{R}_1 \times \mathbb{R}_2$ onto $\mathbb{R}$, and f is a $\mathfrak{M}_L/\mathfrak{B}_{\mathbb{R}}$-measurable mapping of $\mathbb{R}$ into $\mathbb{R}$, Theorem 1.40 (Chain Rule for Measurable Mappings) implies that $h = f \circ \pi_1 \circ S$ is a $\mathfrak{M}_L^2/\mathfrak{B}_{\mathbb{R}}$-measurable mapping of $\mathbb{R} \times \mathbb{R}$ into $\mathbb{R}$. ■

Theorem 23.33. *Let $f, g \in L_r^1(\mathbb{R}, \mathfrak{M}_L, \mu_L)$. Let us define a function $f * g$ on $\mathbb{R}$ by setting*

(1) $$(f * g)(x) = \int_{\mathbb{R}} f(x - y) g(y)\, \mu_L(dy)$$

for $x \in \mathbb{R}$ for which the integral exists in $\overline{\mathbb{R}}$. Then $(f * g)(x)$ is defined and real-valued for μ_L-a.e. $x \in \mathbb{R}$ and $f * g$ is $\mathfrak{M}_L$-measurable. For $x \in \mathbb{R}$ for which the integral in (1) does not exist, we set $(f * g)(x) = 0$. Then the function $f * g$ is $\mathfrak{M}_L$-measurable and μ_L-integrable on $\mathbb{R}$ and we have

(2) $$\int_{\mathbb{R}} |f * g|\, d\mu_L \le \left[\int_{\mathbb{R}} |f|\, d\mu_L\right]\left[\int_{\mathbb{R}} |g|\, d\mu_L\right].$$

Thus $f * g \in L_r^1(\mathbb{R}, \mathfrak{M}_L, \mu_L)$ with $\|f * g\|_1 \le \|f\|_1 \|g\|_1$. Furthermore we have

(3) $$\int_{\mathbb{R}} f * g\, d\mu_L = \left[\int_{\mathbb{R}} f\, d\mu_L\right]\left[\int_{\mathbb{R}} g\, d\mu_L\right].$$

Proof. Since $f, g \in L_r^1(\mathbb{R}, \mathfrak{M}_L, \mu_L)$, f and g are real-valued μ_L-a.e. on $\mathbb{R}$. On the null set of $(\mathbb{R}, \mathfrak{M}_L, \mu_L)$ on which f is not real-valued redefine f to be equal to 0. Thus redefined, f is real-valued and $\mathfrak{M}_L$-measurable on $\mathbb{R}$. Redefine g similarly. Then the product $f(x - y)g(y)$ is defined and real-valued for every $(x, y) \in \mathbb{R} \times \mathbb{R}$.

Since f is a real-valued $\mathfrak{M}_L$-measurable function on $\mathbb{R}$, the function $f(x - y)$ for $(x, y) \in \mathbb{R} \times \mathbb{R}$ is a real-valued $\mathfrak{M}_L^2$-measurable function on $\mathbb{R} \times \mathbb{R}$ by Lemma 23.32. Since g is a real-valued $\mathfrak{M}_L$-measurable function on $\mathbb{R}$, the function g is trivially a real-valued $\mathfrak{M}_L^2$-measurable function on $\mathbb{R} \times \mathbb{R}$. Thus $f(x - y)g(y)$ for $(x, y) \in \mathbb{R} \times \mathbb{R}$ is a real-valued $\mathfrak{M}_L^2$-measurable function on $\mathbb{R} \times \mathbb{R}$. Now

$$
\begin{aligned}
(4) \qquad & \int_{\mathbb{R}} \left[\int_{\mathbb{R}} |f(x-y)g(y)| \, \mu_L(dx) \right] \mu_L(dy) \\
&= \int_{\mathbb{R}} |g(y)| \left[\int_{\mathbb{R}} |f(x-y)| \, \mu_L(dx) \right] \mu_L(dy) \\
&= \int_{\mathbb{R}} |g(y)| \left[\int_{\mathbb{R}} |f(x)| \, \mu_L(dx) \right] \mu_L(dy) \\
&= \left[\int_{\mathbb{R}} |f(x)| \, \mu_L(dx) \right] \left[\int_{\mathbb{R}} |g(x)| \, \mu_L(dx) \right] \\
&= \|f\|_1 \|g\|_1 < \infty,
\end{aligned}
$$

where the second equality is by Theorem 9.31 (Translation Invariance). Thus by Theorem 23.28 (Fubini-Tonelli), the function $f(x-y)g(y)$ for $(x, y) \in \mathbb{R} \times \mathbb{R}$ is $\mu_L \times \mu_L$-integrable on $\mathbb{R} \times \mathbb{R}$. Then by Theorem 23.27 (Fubini), $f(x-y)g(y)$ is $\mu_L(dy)$-integrable on $\mathbb{R}$ for μ_L-a.e. $x \in \mathbb{R}$ and the function $(f * g)(x)$ is $\mathfrak{M}_L$-measurable and $\mu_L(dx)$-integrable on $\mathbb{R}$. The μ_L-integrability of $f * g$ on $\mathbb{R}$ implies that $(f * g)(x) \in \mathbb{R}$ for μ_L-a.e. $x \in \mathbb{R}$.

To prove (2), note that

$$
\begin{aligned}
\int_{\mathbb{R}} |f * g| \, d\mu_L &= \int_{\mathbb{R}} \left| \int_{\mathbb{R}} f(x-y)g(y) \, \mu_L(dy) \right| \mu_L(dx) \\
&\leq \int_{\mathbb{R}} \left[\int_{\mathbb{R}} |f(x-y)g(y)| \, \mu_L(dy) \right] \mu_L(dx) \\
&= \int_{\mathbb{R}} \left[\int_{\mathbb{R}} |f(x-y)g(y)| \, \mu_L(dx) \right] \mu_L(dy) \\
&= \left[\int_{\mathbb{R}} |f(x)| \, \mu_L(dx) \right] \left[\int_{\mathbb{R}} |g(x)| \, \mu_L(dx) \right]
\end{aligned}
$$

where the second equality is by Theorem 23.26 (Tonelli) and the last equality is by (4).

Similarly we have

$$
\begin{aligned}
\int_{\mathbb{R}} (f * g)(x) \, \mu_L(dx) &= \int_{\mathbb{R}} \left[\int_{\mathbb{R}} f(x-y)g(y) \, \mu_L(dy) \right] \mu_L(dx) \\
&= \int_{\mathbb{R}} \left[\int_{\mathbb{R}} f(x-y)g(y) \, \mu_L(dx) \right] \mu_L(dy) \\
&= \int_{\mathbb{R}} g(y) \left[\int_{\mathbb{R}} f(x-y) \, \mu_L(dx) \right] \mu_L(dy) \\
&= \int_{\mathbb{R}} g(y) \left[\int_{\mathbb{R}} f(x) \, \mu_L(dx) \right] \mu_L(dy) \\
&= \left[\int_{\mathbb{R}} f(x) \, \mu_L(dx) \right] \left[\int_{\mathbb{R}} g(x) \, \mu_L(dx) \right]
\end{aligned}
$$

where the second equality is by Theorem 23.27 (Fubini) and the fourth equality is by Theorem 9.31 (Translation Invariance). This proves (3). ∎

Observation 23.34. Let $f \in L_r^1(\mathbb{R}, \mathfrak{M}_L, \mu_L)$ be such that $f \notin L_r^2(\mathbb{R}, \mathfrak{M}_L, \mu_L)$, that is, $\int_{\mathbb{R}} |f|^2 \, d\mu_L = \infty$. Assume further that f is an even function, that is, $f(-x) = f(x)$ for $x \in \mathbb{R}$. By Theorem 23.33, $f \in L_r^1(\mathbb{R}, \mathfrak{M}_L, \mu_L)$ implies that $(f * f)(x) \in \mathbb{R}$ for μ_L-a.e. $x \in \mathbb{R}$. However, we have

$$(f * f)(0) = \int_{\mathbb{R}} f(-y) f(y) \, \mu_L(dy) = \int_{\mathbb{R}} |f(y)|^2 \, \mu_L(dy) = \infty.$$

This shows that $f * f$ is not real-valued at $0 \in \mathbb{R}$ and hence $f * f$ is not continuous at $0 \in \mathbb{R}$.

Theorem 23.35. *Let $f, g \in L_r^1(\mathbb{R}, \mathfrak{M}_L, \mu_L)$. Then for any $p \in [1, \infty]$ we have*

$$\|f * g\|_p \leq \|f\|_p \|g\|_1. \tag{1}$$

Proof. 1. For $p = 1$ the inequality (1) was proved in Theorem 23.33.

2. According to Theorem 23.33, for μ_L-a.e. $x \in \mathbb{R}$ we have

$$\begin{aligned} |(f * g)(x)| &= \left| \int_{\mathbb{R}} f(x-y) g(y) \, \mu_L(dy) \right| \leq \int_{\mathbb{R}} |f(x-y)| \, |g(y)| \, \mu_L(dy) \\ &\leq \int_{\mathbb{R}} \|f\|_\infty |g(y)| \, \mu_L(dy) = \|f\|_\infty \|g\|_1. \end{aligned}$$

It follows from this that $\|f * g\|_\infty \leq \|f\|_\infty \|g\|_1$.

3. Consider the case $p \in (1, \infty)$. Let q be the conjugate of p, that is, $\frac{1}{p} + \frac{1}{q} = 1$. Let us assume that f and g are nonnegative so that $f * g$ is nonnegative. Now

$$\begin{aligned} \|f * g\|_p^p &= \int_{\mathbb{R}} \left|(f * g)(x)\right|^p \mu_L(dx) \\ &= \int_{\mathbb{R}} \left| \int_{\mathbb{R}} f(x-y) g(y) \mu_L(dy) \right|^p \mu_L(dx) \\ &= \int_{\mathbb{R}} \left| \int_{\mathbb{R}} f(x-y) g(y)^{1/p} g(y)^{1/q} \mu_L(dy) \right|^p \mu_L(dx). \end{aligned} \tag{2}$$

For the inner integral above, we have by the non-negativity of f and g and by Theorem 16.14 (Hölder)

$$\begin{aligned} &\left| \int_{\mathbb{R}} f(x-y) g(y)^{1/p} g(y)^{1/q} \mu_L(dy) \right| \\ &= \int_{\mathbb{R}} f(x-y) g(y)^{1/p} g(y)^{1/q} \mu_L(dy) \\ &\leq \left\{ \int_{\mathbb{R}} f(x-y)^p g(y) \mu_L(dy) \right\}^{1/p} \left\{ \int_{\mathbb{R}} g(y) \mu_L(dy) \right\}^{1/q}. \end{aligned} \tag{3}$$

Substituting (3) in (2) we have

$$\begin{aligned}\|f*g\|_p^p &\le \int_{\mathbb{R}}\Big\{\int_{\mathbb{R}} f(x-y)^p g(y)\mu_L(dy)\Big\}\Big\{\int_{\mathbb{R}} g(y)\mu_L(dy)\Big\}^{p/q}\mu_L(dx)\\ &= \Big\{\int_{\mathbb{R}} g(y)\mu_L(dy)\Big\}^{p/q}\int_{\mathbb{R}}\Big\{\int_{\mathbb{R}} f(x-y)^p g(y)\mu_L(dy)\Big\}\mu_L(dx)\\ &= \|g\|_1^{p/q}\int_{\mathbb{R}}\Big\{\int_{\mathbb{R}} f(x-y)^p\mu_L(dx)\Big\}g(y)\,\mu_L(dy)\\ &= \|g\|_1^{p/q}\|f\|_p^p\|g\|_1 = \|f\|_p^p\|g\|_1^p\end{aligned}$$

where the second equality is by Theorem 23.17 (Tonelli), the third equality is by Theorem 9.31 (Translation Invariance), and the last equality is by $\frac{p}{q}+1=p$. Taking the p-th roots we have (1) for $f,g\in L_r^1(\mathbb{R},\mathfrak{M}_L,\mu_L)$ such that $f\ge 0$ and $g\ge 0$.

For $f,g\in L_r^1(\mathbb{R},\mathfrak{M}_L,\mu_L)$, we have

$$(4)\qquad \begin{aligned}|(f*g)(x)| &= \Big|\int_{\mathbb{R}} f(x-y)g(y)\,\mu_L(dy)\Big|\\ &\le \int_{\mathbb{R}} |f(x-y)|\,|g(y)|\,\mu_L(dy) = \big(|f|*|g|\big)(x).\end{aligned}$$

Now $|f|,|g|\in L_r^1(\mathbb{R},\mathfrak{M}_L,\mu_L)$ and $|f|\ge 0$ and $|g|\ge 0$. Thus we have

$$\|f*g\|_p \le \big\||f|*|g|\big\|_p \le \big\||f|\big\|_p\big\||g|\big\|_1 = \|f\|_p\|g\|_1,$$

where the second inequality is by our result above for $f,g\in L_r^1(\mathbb{R},\mathfrak{M}_L,\mu_L)$ such that $f\ge 0$ and $g\ge 0$. This completes the proof. ∎

Consider the Banach space $L_r^1(\mathbb{R},\mathfrak{M}_L,\mu_L)$ of extended real-valued $\mathfrak{M}_L$-measurable and μ_L-integrable functions on $\mathbb{R}$. By Theorem 23.33, for any $f,g\in L_r^1(\mathbb{R},\mathfrak{M}_L,\mu_L)$ we have $f*g\in L_r^1(\mathbb{R},\mathfrak{M}_L,\mu_L)$. Thus we regard $*$ as a multiplication defined on $L_r^1(\mathbb{R},\mathfrak{M}_L,\mu_L)$. We show next that commutative law, associative law, and distributive law hold for this multiplication.

Theorem 23.36. *Let $f,g,h\in L_r^1(\mathbb{R},\mathfrak{M}_L,\mu_L)$ and $\alpha,\beta\in\mathbb{R}$. Then we have*

(1) *Distributive Law:* $(\alpha f+\beta g)*h=\alpha(f*h)+\beta(g*h)$,

(2) *Commutative Law:* $f*g=g*f$,

(3) *Associative Law:* $(f*g)*h=f*(g*h)$.

Proof. (1) follows from the linearity of the integral. (2) follows from Proposition 23.31.

Let us prove (3). According to Theorem 23.33 if $f,g\in L_r^1(\mathbb{R},\mathfrak{M}_L,\mu_L)$, then we have $(f*g)(x)=\int_{\mathbb{R}} f(x-y)g(y)\,\mu_L(dy)\in\mathbb{R}$ for μ_L-a.e. $x\in\mathbb{R}$ and $f*g\in L_r^1(\mathbb{R},\mathfrak{M}_L,\mu_L)$.

Thus if $f, g, h \in L_r^1(\mathbb{R}, \mathfrak{M}_L, \mu_L)$, then $f * g \in L_r^1(\mathbb{R}, \mathfrak{M}_L, \mu_L)$ so that for μ_L-a.e. $x \in \mathbb{R}$ we have

$$(4) \qquad [(f * g) * h](x) = \int_{\mathbb{R}} (f * g)(x-y)h(y)\,\mu_L(dy)$$
$$= \int_{\mathbb{R}} \left[\int_{\mathbb{R}} f(x-y-z)g(z)\,\mu_L(dz) \right] h(y)\,\mu_L(dy)$$
$$= \int_{\mathbb{R}} \left[\int_{\mathbb{R}} f(x-z)g(z-y)\,\mu_L(dz) \right] h(y)\,\mu_L(dy)$$

where the last equality is by Theorem 9.31 (Translation Invariance) replacing the variable of integration z with $z - y$. Similarly for μ_L-a.e. $x \in \mathbb{R}$ we have

$$(5) \qquad [f * (g * h)](x) = \int_{\mathbb{R}} f(x-y)(g * h)(y)\,\mu_L(dy)$$
$$= \int_{\mathbb{R}} f(x-y) \left[\int_{\mathbb{R}} g(y-z)h(z)\,\mu_L(dz) \right] \mu_L(dy)$$
$$= \int_{\mathbb{R}} f(x-z) \left[\int_{\mathbb{R}} g(z-y)h(y)\,\mu_L(dy) \right] \mu_L(dz)$$

where the last equality is obtained by relabeling the variable y as z and the variable z as y. To prove (3), it remains to show that the last member of (4) is equal to that of (5) for μ_L-a.e. $x \in \mathbb{R}$. This is done by applying Theorem 23.28 (Fubini-Tonelli). Indeed we have

$$\int_{\mathbb{R}} \left[\int_{\mathbb{R}} |f(x-z)g(z-y)h(y)|\,\mu_L(dy) \right] \mu_L(dz)$$
$$= \int_{\mathbb{R}} \left[\int_{\mathbb{R}} |g(z-y)|\,|h(y)|\,\mu_L(dy) \right] |f(x-z)|\,\mu_L(dz)$$
$$= \int_{\mathbb{R}} \big(|g| * |h|\big)(z)|f(x-z)|\,\mu_L(dz)$$
$$= \Big[|f| * \big(|g| * |h|\big)\Big](x) \in \mathbb{R} \quad \text{for } \mu_L\text{-a.e. } x \in \mathbb{R}.$$

Thus by Theorem 23.28 (Fubini-Tonelli), the last member of (4) is equal to that of (5) for μ_L-a.e. $x \in \mathbb{R}$. Therefore $(f * g) * h = f * (g * h)$. ∎

Differentiability of one of the two factor functions f and g implies differentiability of $f * g$. To show this, let us consider differentiation under the integral.

Proposition 23.37. *Let I and J be two open intervals in $\mathbb{R}$. Let h be a real-valued function of $(x, y) \in I \times J$ such that the partial derivative $\frac{\partial h}{\partial x}$ exists on $I \times J$. Suppose*

1° *$h(x, \cdot)$ is $\mathfrak{M}_L$-measurable on J for every $x \in I$,*
2° *$h(x_0, \cdot)$ is μ_L-integrable on J for some $x_0 \in I$,*
3° *there exists a nonnegative extended real-valued $\mathfrak{M}_L$-measurable and μ_L-integrable function ψ on J such that $\left|\frac{\partial h}{\partial x}(x, y)\right| \leq \psi(y)$ for $(x, y) \in I \times J$.*

Then we have

(a) *$h(x, \cdot)$ is μ_L-integrable on J for every $x \in I$,*
(b) *$\frac{\partial h}{\partial x}(x, \cdot)$ is $\mathfrak{M}_L$-measurable and μ_L-integrable on J for every $x \in I$,*
(c) *the real-valued function H on I, defined by $H(x) = \int_J h(x, \cdot)\, d\mu_L$ for $x \in I$, is differentiable on I and $\frac{dH}{dx}(x) = \int_J \frac{\partial h}{\partial x}(x, \cdot)\, d\mu_L$.*

Proof. 1. To prove (a), let $x \in I$ be arbitrarily chosen. To prove the μ_L-integrability of the function $h(x, \cdot)$ on J let us note that by the Mean Value Theorem applied to the function $h(\cdot, y)$ on I for each $y \in J$ we have $h(x, y) - h(x_0, y) = \frac{\partial h}{\partial x}(\vartheta(y), y)(x - x_0)$ where $\vartheta(y)$ is a point in I intermediate to x_0 and x. Thus we have

$$\begin{aligned} |h(x, y)| &\le |h(x_0, y)| + \left| \frac{\partial h}{\partial x}(\vartheta(y), y) \right| |x - x_0| \\ &\le |h(x_0, y)| + \psi(y)|x - x_0| \end{aligned}$$

by 3°. Then since $h(x_0, \cdot)$ and ψ are μ_L-integrable on J, $h(x, \cdot)$ is μ_L-integrable on J.

2. Since $\frac{\partial h}{\partial x}(x, y)$ exists at every $(x, y) \in I \times J$, we have

$$\begin{aligned} \frac{\partial h}{\partial x}(x, y) &= \lim_{\Delta x \to 0} \frac{h(x + \Delta x, y) - h(x, y)}{\Delta x} \\ &= \lim_{n \to \infty} \frac{h\left(x + \frac{1}{n}, y\right) - h(x, y)}{\frac{1}{n}} \\ &= \lim_{n \to \infty} n \left\{ h\left(x + \tfrac{1}{n}, y\right) - h(x, y) \right\}. \end{aligned}$$

Since $h(x, \cdot)$ is $\mathfrak{M}_L$-measurable on J for every $x \in I$, $h\left(x + \frac{1}{n}, \cdot\right)$ is $\mathfrak{M}_L$-measurable on J for $n \in \mathbb{N}$ so large that $x + \frac{1}{n} \in I$. Thus $n\left\{h\left(x + \frac{1}{n}, \cdot\right) - h(x, \cdot)\right\}$ is $\mathfrak{M}_L$-measurable on J for sufficiently large $n \in \mathbb{N}$. Then $\lim_{n \to \infty} n\left\{h\left(x + \frac{1}{n}, \cdot\right) - h(x, \cdot)\right\}$ is $\mathfrak{M}_L$-measurable on J. This shows that $\frac{\partial h}{\partial x}(x, \cdot)$ is $\mathfrak{M}_L$-measurable on J for every $x \in I$. Then by 3°, $\frac{\partial h}{\partial x}(x, \cdot)$ is μ_L-integrable on J for every $x \in I$.

3. To show that H is differentiable on I, we show that $\lim_{\Delta x \to 0} \frac{H(x+\Delta x) - H(x)}{\Delta x}$ exists in $\mathbb{R}$ for every $x \in I$. It suffices to show that for each fixed $x \in I$ there exists a real number λ such that for every sequence $(x_n : n \in \mathbb{N})$ in I such that $\lim_{n \to \infty} x_n = x$ and $x_n \neq x$ for $n \in \mathbb{N}$ we have $\lim_{n \to \infty} \frac{H(x_n) - H(x)}{x_n - x} = \lambda$. Let us show that $\int_J \frac{\partial h}{\partial x}(x, \cdot)\, d\mu_L$ is such a real number. This then shows not only the differentiability of H at $x \in I$ but also $\frac{dH}{dx}(x) = \int_J \frac{\partial h}{\partial x}(x, \cdot)\, d\mu_L$. Now

$$\frac{H(x_n) - H(x)}{x_n - x} = \int_J \frac{h(x_n, \cdot) - h(x, \cdot)}{x_n - x}\, d\mu_L.$$

By the Mean Value Theorem and by 3°, we have

$$\left| \frac{h(x_n, y) - h(x, y)}{x_n - x} \right| = \left| \frac{\partial h}{\partial x}(\vartheta_n(y), y) \right| \le \psi(y)$$

where $\vartheta_n(y)$ is a point in I intermediate to x and x_n. Since ψ is μ_L-integrable on J

$$\lim_{n \to \infty} \frac{H(x_n) - H(x)}{x_n - x} = \int_J \lim_{n \to \infty} \frac{h(x_n, \cdot) - h(x, \cdot)}{x_n - x}\, d\mu_L = \int_J \frac{\partial h}{\partial x}(x, \cdot)\, d\mu_L$$

by the Dominated Convergence Theorem (Theorem 9.20). This completes the proof. ∎

Definition 23.38. **(a)** *Let $C(\mathbb{R})$ be the set of all real-valued continuous functions f on $\mathbb{R}$. For $n \in \mathbb{N}$, we write $C^n(\mathbb{R})$ for the subset of $C_c(\mathbb{R})$ consisting of every member f of $C(\mathbb{R})$ such that the derivative $D^k f$ exists and is continuous on $\mathbb{R}$ for $k = 1, \dots, n$.*
We write $C^\infty(\mathbb{R})$ for the subset of $C_c(\mathbb{R})$ consisting of every member f of $C(\mathbb{R})$ such that the derivative $D^n f$ exists and is continuous on $\mathbb{R}$ for every $n \in \mathbb{N}$. Thus we have

$$C^\infty(\mathbb{R}) = \bigcap_{n \in \mathbb{N}} C^n(\mathbb{R}).$$

(b) *We write $C_c(\mathbb{R})$ for the subset of $C(\mathbb{R})$ consisting of those members f of $C(\mathbb{R})$ whose supports* supp$\{f\}$ *are compact sets in $\mathbb{R}$. We define*

$$C_c^n(\mathbb{R}) := C^n(\mathbb{R}) \cap C_c(\mathbb{R}) \quad \text{for } n \in \mathbb{N},$$

$$C_c^\infty(\mathbb{R}) := C^\infty(\mathbb{R}) \cap C_c(\mathbb{R}).$$

Thus we have

$$C_c^\infty(\mathbb{R}) = \bigcap_{n \in \mathbb{N}} C_c^n(\mathbb{R}).$$

The set $C_c(\mathbb{R})$ as defined above is equal to the set of every real-valued continuous function f on $\mathbb{R}$ that vanishes outside of a finite closed interval. This can be shown as follows. Suppose $f \in C_c(\mathbb{R})$. Then supp$\{f\}$ is a compact set in $\mathbb{R}$ and thus it is a bounded closed set in $\mathbb{R}$. Then there exists a finite closed interval I such that $I \supset \text{supp}\{f\}$ and then $I^c \subset \text{supp}\{f\}^c$. Then since f vanishes on supp$\{f\}^c$, f vanishes on I^c. Conversely suppose f is a real-valued continuous function on $\mathbb{R}$ and f vanishes outside of a finite closed interval I. Then $\text{supp}\{f\} = \overline{\{x \in \mathbb{R} : f(x) \neq 0\}} \subset I$ so that supp$\{f\}$ is a bounded closed set in $\mathbb{R}$, that is, supp$\{f\}$ is a compact set in $\mathbb{R}$. This shows that $f \in C_c(\mathbb{R})$.

Theorem 23.39. *Consider the class $C_c^n(\mathbb{R})$ of real-valued continuous functions with compact support and having continuous derivatives $D^k f$ of order $k = 1, \cdots, n$ on $\mathbb{R}$.*
(a) *If $f \in C_c^n(\mathbb{R})$ and $g \in L_r^1(\mathbb{R}, \mathfrak{M}_L, \mu_L)$, then $D^k(f * g)$ exists on $\mathbb{R}$ and*

(1) $$D^k(f * g) = D^k f * g \quad \text{for } k \in \{1, \dots, n\}.$$

(b) *If $f \in L_r^1(\mathbb{R}, \mathfrak{M}_L, \mu_L)$ and $g \in C_c^n(\mathbb{R})$, then $D^k(f * g)$ exists on $\mathbb{R}$ and*

(2) $$D^k(f * g) = f * D^k g \quad \text{for } k \in \{1, \dots, n\}.$$

(c) *If $f \in C_c^m(\mathbb{R})$ and $g \in C_c^n(\mathbb{R})$, then $D^{k+\ell}(f * g)$ exists on $\mathbb{R}$ and*

(3) $$D^{k+\ell}(f * g) = (D^k f) * (D^\ell g) \quad \text{for } k \in \{1, \dots, m\}, \ell \in \{1, \dots, n\}.$$

Proof. Suppose $f \in C_c^n(\mathbb{R})$ and $g \in L_r^1(\mathbb{R}, \mathfrak{M}_L, \mu_L)$. Then $\int_{\mathbb{R}} |g|\, d\mu_L < \infty$ and this implies that $g(x) \in \mathbb{R}$ for μ_L-a. e. $x \in \mathbb{R}$. Let us redefine g to be equal to 0 on the null set in $(\mathbb{R}, \mathfrak{M}_L, \mu_L)$ on which g is not real-valued so that g is real-valued and $\mathfrak{M}_L$-measurable on $\mathbb{R}$. Consider a real-valued function h on $\mathbb{R} \times \mathbb{R}$ defined by $h(x, y) = f(x - y)g(y)$ for $(x, y) \in \mathbb{R} \times \mathbb{R}$. Let us verify that h satisfies conditions 1°, 2°, and 3° of Proposition 23.37. Now since $f(y)$ for $y \in \mathbb{R}$ is a $\mathfrak{M}_L$-measurable function of $y \in \mathbb{R}$, $f(-y)$ for $y \in \mathbb{R}$ is a $\mathfrak{M}_L$-measurable function of $y \in \mathbb{R}$ by Theorem 9.32 (Linear Transformation). Then for every $x \in \mathbb{R}$, $f(-y+x)$ is a $\mathfrak{M}_L$-measurable function of $y \in \mathbb{R}$ by Theorem 9.31 (Translation Invariance). Thus $h(x, y) = f(-y+x)g(y)$ is a $\mathfrak{M}_L$-measurable function of $y \in \mathbb{R}$ for each fixed $x \in \mathbb{R}$. This verifies condition 1° of Proposition 23.37 for h.

Next by (4) in the Proof of Theorem 23.33, we have

$$\int_{\mathbb{R}} \left[\int_{\mathbb{R}} |h(x, y)|\, \mu_L(dy) \right] \mu_L(dx) = \int_{\mathbb{R}} \left[\int_{\mathbb{R}} |f(x - y)g(y)|\, \mu_L(dy) \right] \mu_L(dx) < \infty.$$

This implies that $\int_{\mathbb{R}} |h(x, y)|\, \mu_L(dy) < \infty$ for μ_L-a.e. $x \in \mathbb{R}$. This verifies condition 2° of Proposition 23.37 for h.

Since $h(x, y) = f(x - y)g(y)$ and f is differentiable on $\mathbb{R}$, $\frac{\partial h}{\partial x}(x, y) = (Df)(x - y)g(y)$ for $(x, y) \in \mathbb{R} \times \mathbb{R}$. Since $f \in C_c^n(\mathbb{R})$, Df is a bounded real-valued function on $\mathbb{R}$. If we let $M > 0$ be a bound of Df on $\mathbb{R}$, then $\left|\frac{\partial h}{\partial x}(x, y)\right| \le M|g(y)|$ for $(x, y) \in \mathbb{R} \times \mathbb{R}$. If we let $\psi = M|g|$, then ψ is a μ_L-integrable function on $\mathbb{R}$. This shows that h satisfies condition 3° of Proposition 23.37. Thus by Proposition 23.37, $H(x) = \int_{\mathbb{R}} h(x, \cdot)\, d\mu_L$ is a real-valued function of $x \in \mathbb{R}$, differentiable with derivative $\frac{dH}{dx}(x) = \int_{\mathbb{R}} \frac{\partial h}{\partial x}(x, \cdot)\, d\mu_L$. Now

$$(4) \qquad H(x) = \int_{\mathbb{R}} h(x, \cdot)\, d\mu_L = \int_{\mathbb{R}} f(x - y)g(y)\, \mu_L(dy) = (f * g)(x)$$

and

$$(5) \qquad \frac{dH}{dx}(x) = \int_{\mathbb{R}} \frac{\partial h}{\partial x}(x, \cdot)\, d\mu_L = \int_{\mathbb{R}} (Df)(x - y)g(y)\, \mu_L(dy) = (Df * g)(x)$$

since $Df \in L_r^1(\mathbb{R}, \mathfrak{M}_L, \mu_L)$. By (4) and (5), we have $D(f * g) = Df * g$. Repeating the argument we have $D^2(f * g) = D(Df * g) = D^2 * g$ and so on. This proves (1).

We have $D^k(f * g) = D^k(g * f) = D^k g * f = f * D^k g$ by (2) of Theorem 23.36 and (1) above. This proves (2). The equality (3) follows from (1) and (2). ∎

[IV.2] Convolution of L^p-Functions

Lemma 23.40. *With $y \in \mathbb{R}$ let us define a transformation T_y of an extended real-valued function f on $\mathbb{R}$ into an extended real-valued function on $\mathbb{R}$ by setting*

$$(1) \qquad T_y f(x) = f(x + y) \quad \text{for } x \in \mathbb{R}.$$

Let us define a transformation R of an extended real-valued function f on $\mathbb{R}$ into an extended real-valued function on $\mathbb{R}$ by setting

$$(2) \qquad Rf(x) = f(-x) \quad \text{for } x \in \mathbb{R}.$$

Let $p \in [1, \infty]$. *Then for any extended real-valued function* $f \in L_r^p(\mathbb{R}, \mathfrak{M}_L, \mu_L)$, *we have*

$$\|T_y f\|_p = \|Rf\|_p = \|(T_y \circ R)f\|_p = \|f\|_p < \infty, \tag{3}$$

and therefore

$$T_y f,\ Rf,\ (T_y \circ R)f \in L_r^p(\mathbb{R}, \mathfrak{M}_L, \mu_L). \tag{4}$$

Proof. 1. It follows immediately from the definition of $\|\cdot\|_\infty$ in Definition 16.31 that for any extended real-valued $\mathfrak{M}_L$-measurable function f on $\mathbb{R}$, $\|T_y f\|_\infty$, $\|Rf\|_\infty$, and $\|(T_y \circ R)f\|_\infty$ are all equal to $\|f\|_\infty$. If $f \in L_r^\infty(\mathbb{R}, \mathfrak{M}_L, \mu_L)$ then $\|f\|_\infty < \infty$ so that we have $\|T_y f\|_\infty, \|Rf\|_\infty, \|(T_y \circ R)f\|_\infty < \infty$ and consequently $T_y f, Rf, (T_y \circ R)f \in L_r^\infty(\mathbb{R}, \mathfrak{M}_L, \mu_L)$.

2. For $p \in [1, \infty)$, we have

$$\int_{\mathbb{R}} |T_y f(x)|^p \,\mu_L(dx) = \int_{\mathbb{R}} |f(x+y)|^p \,\mu_L(dx) = \int_{\mathbb{R}} |f(x)|^p \,\mu_L(dx) \tag{5}$$

by Theorem 9.31 (Translation Invariance) and

$$\int_{\mathbb{R}} |Rf(x)|^p \,\mu_L(dx) = \int_{\mathbb{R}} |f(-x)|^p \,\mu_L(dx) = \int_{\mathbb{R}} |f(x)|^p \,\mu_L(dx) \tag{6}$$

by Theorem 9.32 (Linear transformation), and then by (5) and (6) we have

$$\int_{\mathbb{R}} |(T_y \circ R)f(x)|^p \,\mu_L(dx) = \int_{\mathbb{R}} |Rf(x)|^p \,\mu_L(dx) = \int_{\mathbb{R}} |f(x)|^p \,\mu_L(dx). \tag{7}$$

Taking the p-th roots of (5), (6), and (7) we have the equalities (3). The finiteness of $\|T_y f\|_p$, $\|Rf\|_p$, and $\|(T_y \circ R)f\|_p$ implies that $T_y f$, Rf, and $(T_y \circ R)f$ are elements of $L_r^p(\mathbb{R}, \mathfrak{M}_L, \mu_L)$. ∎

Lemma 23.41. *Let* $p \in [1, \infty)$ *and* $f \in L_r^p(\mathbb{R}, \mathfrak{M}_L, \mu_L)$. *For* $y \in \mathbb{R}$, *consider the function* $T_y f(x) = f(x+y)$ *for* $x \in \mathbb{R}$. *Then* $T_y f \in L_r^p(\mathbb{R}, \mathfrak{M}_L, \mu_L)$ *with* $\|T_y f\|_p = \|f\|_p$ *and*

$$\lim_{y \to 0} \|T_y f - f\|_p = 0. \tag{1}$$

Proof. By Lemma 23.40, we have $T_y f \in L_r^p(\mathbb{R}, \mathfrak{M}_L, \mu_L)$ with $\|T_y f\|_p = \|f\|_p$. Thus it remains to prove (1).

1. Consider first the case $f \in C_c(\mathbb{R})$, that is, f is a real-valued continuous function on $\mathbb{R}$ and $\text{supp}\{f\}$ is a compact set in $\mathbb{R}$. Since every compact set in $\mathbb{R}$ is a bounded closed set in $\mathbb{R}$, we have $\text{supp}\{f\} \subset [-a, a]$ for sufficiently large $a > 0$. Since $f = 0$ on $\text{supp}\{f\}^c$ and since $\text{supp}\{f\}^c \supset [-a, a]^c$, $f = 0$ on $[-a, a]^c$. Thus f is uniformly continuous on $[-a, a]^c$. Now $[-a-1, a+1]$ is a compact set in $\mathbb{R}$. Then the continuity of f on $[-a-1, a+1]$ implies uniform continuity of f on $[-a-1, a+1]$. Then the uniform continuity of f on $[-a, a]^c$ and the uniform continuity of f on $[-a-1, a+1]$ imply the uniform continuity of f on $[-a-1, a+1] \cup [-a, a]^c = \mathbb{R}$. Now that f uniformly continuous on $\mathbb{R}$, for every $\varepsilon > 0$ there exists $\delta > 0$ such that

$$x, y \in \mathbb{R} \text{ and } |y| < \delta \Rightarrow |f(x+y) - f(x)| < \varepsilon.$$

We may assume without loss of generality that $\delta \in (0, 1)$. Now

$$x \in [-a-\delta, a+\delta]^c \text{ and } |y| < \delta \Rightarrow x - y \in [-a, a]^c.$$

Since $f = 0$ on $[-a, a]^c$, we have $f(x-y) = 0$ for $x - y \in [-a, a]^c$ and $f(x) = 0$ for $x \in [-a-\delta, a+\delta]^c \subset [-a, a]^c$. Thus for $y \in \mathbb{R}$ such that $|y| < \delta$, we have

$$\begin{aligned} \|T_y f - f\|_p^p &= \int_{\mathbb{R}} |f(x+y) - f(x)|^p \, \mu_L(dx) \\ &= \int_{[-a-\delta, a+\delta]} |f(x+y) - f(x)|^p \, \mu_L(dx) \\ &\leq \int_{[-a-\delta, a+\delta]} \varepsilon^p \, \mu_L(dx) \leq 2(a+1)\varepsilon^p. \end{aligned}$$

Therefore for every $\varepsilon > 0$ there exists $\delta > 0$ such that for $y \in \mathbb{R}$ where $|y| < \delta$ we have

$$\|T_y f - f\|_p \leq \big(2(a+1)\big)^{\frac{1}{p}} \varepsilon.$$

This proves $\lim_{y \to 0} \|T_y f - f\|_p = 0$.

2. Now consider the general case $f \in L_r^p(\mathbb{R}, \mathfrak{M}_L, \mu_L)$ where $p \in [1, \infty)$. According to Theorem 17.10, $C_c(\mathbb{R})$ is dense in $L_r^p(\mathbb{R}, \mathfrak{M}_L, \mu_L)$ when $p \in [1, \infty)$. Thus for any $f \in L_r^p(\mathbb{R}, \mathfrak{M}_L, \mu_L)$ and $\varepsilon > 0$ there exists $g \in C_c(\mathbb{R})$ such that $\|f - g\|_p < \varepsilon$. Now

$$\|f - T_y f\|_p \leq \|f - g\|_p + \|g - T_y g\|_p + \|T_y g - T_y f\|_p.$$

By Lemma 23.40 we have $\|T_y g - T_y f\|_p = \|T_y(g - f)\|_p = \|g - f\|_p$. Thus we have

$$\|f - T_y f\|_p \leq 2\|f - g\|_p + \|g - T_y g\|_p < 2\varepsilon + \|g - T_y g\|_p.$$

For $g \in C_c(\mathbb{R})$ we have $\lim_{y \to 0} \|g - T_y g\|_p = 0$ by our result in **1**. Thus we have

$$\limsup_{y \to 0} \|f - T_y f\|_p \leq 2\varepsilon + \lim_{y \to 0} \|g - T_y g\|_p = 2\varepsilon.$$

By the arbitrariness of $\varepsilon > 0$ we have $\limsup_{y \to 0} \|f - T_y f\|_p = 0$. This then implies $\liminf_{y \to 0} \|f - T_y f\|_p = 0$ and then $\lim_{y \to 0} \|f - T_y f\|_p = 0$. ■

Theorem 23.42. *Let $p \in [1, \infty]$. Let $f \in L_r^p(\mathbb{R}, \mathfrak{M}_L, \mu_L)$ and $g \in L_r^1(\mathbb{R}, \mathfrak{M}_L, \mu_L)$. Let us define a function $f * g$ on $\mathbb{R}$ by setting*

$$(1) \qquad (f * g)(x) = \int_{\mathbb{R}} f(x - y) g(y) \, \mu_L(dy),$$

for $x \in \mathbb{R}$ such that the integral exists in $\overline{\mathbb{R}}$. Then $(f * g)(x)$ is defined and real-valued for μ_L-a.e. $x \in \mathbb{R}$ and $f * g$ is $\mathfrak{M}_L$-measurable. For $x \in \mathbb{R}$ such that the integral in (1) does not exist, we set $(f * g)(x) = 0$. Then $f * g \in L_r^p(\mathbb{R}, \mathfrak{M}_L, \mu_L)$ and

(2) $$\|f * g\|_p \leq \|f\|_p \|g\|_1.$$

Proof. 1. For the case $p = 1$ the theorem has been proved as Theorem 23.33. Thus consider the case $p \in (1, \infty]$.

If $f \in L_r^p(\mathbb{R}, \mathfrak{M}_L, \mu_L)$ for some $p \in (1, \infty]$ and $g \in L_r^1(\mathbb{R}, \mathfrak{M}_L, \mu_L)$, f and g are real-valued μ_L-a.e. on $\mathbb{R}$. On the null set of $(\mathbb{R}, \mathfrak{M}_L, \mu_L)$ on which f is not real-valued, redefine f to be equal to 0 so that f is real-valued and $\mathfrak{M}_L$-measurable on $\mathbb{R}$. Redefine g similarly so that g is real-valued and $\mathfrak{M}_L$-measurable on $\mathbb{R}$. By Lemma 23.32, $f(x - y)$ for $(x, y) \in \mathbb{R} \times \mathbb{R}$ is real-valued and $\mathfrak{M}_L^2$-measurable on $\mathbb{R} \times \mathbb{R}$. Let $q \in [1, \infty)$ be the conjugate of $p \in (1, \infty]$, that is, $\frac{1}{p} + \frac{1}{q} = 1$. Let $h \in L_r^q(\mathbb{R}, \mathfrak{M}_L, \mu_L)$ be such that $h(x) \in (0, \infty)$ for every $x \in \mathbb{R}$. Then the function $h(x)f(x - y)g(y)$ for $(x, y) \in \mathbb{R} \times \mathbb{R}$ is a real-valued $\mathfrak{M}_L^2$-measurable function on $\mathbb{R} \times \mathbb{R}$. Now

(3) $$\begin{aligned}
&\int_{\mathbb{R}} |h(x)| \Big\{ \int_{\mathbb{R}} |f(x - y)g(y)| \, \mu_L(dy) \Big\} \, \mu_L(dx) \\
&= \int_{\mathbb{R} \times \mathbb{R}} |h(x)f(x - y)g(y)| \, (\mu_L \times \mu_L)(d(x, y)) \\
&= \int_{\mathbb{R}} |g(y)| \Big\{ \int_{\mathbb{R}} |f(x - y)h(x)| \, \mu_L(dx) \Big\} \, \mu_L(dy) \\
&= \int_{\mathbb{R}} |g(y)| \Big\{ \int_{\mathbb{R}} |T_{-y} f(x)h(x)| \, \mu_L(dx) \Big\} \, \mu_L(dy) \\
&\leq \int_{\mathbb{R}} |g(y)| \|T_{-y} f\|_p \|h\|_q \, \mu_L(dy) \\
&= \|g\|_1 \|f\|_p \|h\|_q < \infty
\end{aligned}$$

where the first and the second equalities are by Theorem 23.17 (Tonelli), the inequality is by Theorem 16.14 (Hölder) for $p \in (1, \infty)$ and by Theorem 16.40 (Hölder) for $p = \infty$, and the last equality is by (3) of Lemma 23.40. The finiteness of the integral in (3) implies that we have for μ_L-a.e. $x \in \mathbb{R}$

$$|h(x)| \Big\{ \int_{\mathbb{R}} |f(x - y)g(y)| \, \mu_L(dy) \Big\} < \infty.$$

Since $h(x) \in (0, \infty)$ for every $x \in \mathbb{R}$, we have for μ_L-a.e. $x \in \mathbb{R}$

$$\int_{\mathbb{R}} |f(x - y)g(y)| \, \mu_L(dy) < \infty.$$

Thus we have for μ_L-a.e. $x \in \mathbb{R}$

$$(f * g)(x) = \int_{\mathbb{R}} f(x - y)g(y) \, \mu_L(dy) \in \mathbb{R}.$$

The $\mathfrak{M}_L$-measurability of $f * g$ on $\mathbb{R}$ follows from (a) of Theorem 23.17 (Tonelli).

2. Let us show that $f * g \in L_r^p(\mathbb{R}, \mathfrak{M}_L, \mu_L)$. Let $q \in [1, \infty)$ be the conjugate of our $p \in (1, \infty]$, that is, $\frac{1}{p} + \frac{1}{q} = 1$. With the extended real-valued $\mathfrak{M}_L$-measurable function $f * g$ on $\mathbb{R}$, let us define a functional L_{f*g} on $h \in L_r^q(\mathbb{R}, \mathfrak{M}_L, \mu_L)$ by setting

$$(4) \qquad L_{f*g}(h) = \int_{\mathbb{R}} h(x)(f * g)(x)\, \mu_L(dx).$$

Clearly L_{f*g} is a linear functional on $L_r^q(\mathbb{R}, \mathfrak{M}_L, \mu_L)$. By (4), (1), and (3) we have

$$(5) \qquad \begin{aligned} |L_{f*g}(h)| &\le \int_{\mathbb{R}} |h(x)|\big|(f * g)(x)\big|\, \mu_L(dx) \\ &\le \int_{\mathbb{R}} |h(x)| \Big\{ \int_{\mathbb{R}} |f(x-y)g(y)|\, \mu_L(dy) \Big\}\, \mu_L(dx) \\ &\le \|f\|_p \|g\|_1 \|h\|_q. \end{aligned}$$

Thus L_{f*g} is a bound linear functional on $L_r^q(\mathbb{R}, \mathfrak{M}_L, \mu_L)$ with norm

$$(6) \qquad \|L_{f*g}\|_* \le \|f\|_p \|g\|_1.$$

Now L_{f*g} is a bounded linear functional on $L_r^q(\mathbb{R}, \mathfrak{M}_L, \mu_L)$ and $q \in [1, \infty)$. Then by Theorem 18.6 (Riesz Representation Theorem), there exists a unique $\varphi \in L_r^p(\mathbb{R}, \mathfrak{M}_L, \mu_L)$ such that

$$(7) \qquad L_{f*g}(h) = \int_{\mathbb{R}} h\varphi\, d\mu_L$$

and

$$(8) \qquad \|\varphi\|_p = \|L_{f*g}\|_* \le \|f\|_p \|g\|_1.$$

Thus, to show that $f * g \in L_r^p(\mathbb{R}, \mathfrak{M}_L, \mu_L)$ and $\|f * g\|_p \le \|f\|_p \|g\|_1$, it suffices to show that $f * g = \varphi$, that is, $(f * g)(x) = \varphi(x)$ for μ_L-a.e. $x \in \mathbb{R}$. Since both $f * g$ and $\varphi \in L_r^p(\mathbb{R}, \mathfrak{M}_L, \mu_L)$ are real-valued μ_L-a.e. on $\mathbb{R}$, $f * g - \varphi$ is defined μ_L-a.e. on $\mathbb{R}$. Thus to show that $f * g = \varphi$, μ_L-a.e. on $\mathbb{R}$, it suffices to show that $\mu_L(D) = 0$ where $D := \{x \in \mathbb{R} : (f * g)(x) - \varphi(x) \neq 0\} \in \mathfrak{M}_L$. Suppose $\mu_L(D) > 0$. Now D is the union of two disjoint $\mathfrak{M}_L$-measurable sets $D_1 = \{x \in \mathbb{R} : (f * g)(x) - \varphi(x) > 0\}$ and $D_2 = \{x \in \mathbb{R} : (f * g)(x) - \varphi(x) < 0\}$. If $\mu_L(D) > 0$, then at least one of $\mu_L(D_1)$ and $\mu_L(D_2)$ is positive. Consider the case $\mu_L(D_1) > 0$. Since $(\mathbb{R}, \mathfrak{M}_L, \mu_L)$ is a σ-finite measure space, there exists a $\mathfrak{M}_L$-measurable subset E of D_1 such that $\mu_L(E) \in (0, \infty)$. Let $h_0 = \mathbf{1}_E$. We have $\|h\|_q^q = \int_{\mathbb{R}} |\mathbf{1}_E|^q\, d\mu = \int_E 1\, d\mu < \infty$ so that $h_0 \in L_r^q(\mathbb{R}, \mathfrak{M}_L, \mu_L)$. Now by the definition of L_{f*g} by (4), we have

$$L_{f*g}(h_0) - \int_{\mathbb{R}} h_0\varphi\, d\mu_L = \int_{\mathbb{R}} h_0\{(f * g) - \varphi\}\, d\mu_L = \int_E \{(f * g) - \varphi\}\, d\mu_L > 0$$

since $(f * g) - \varphi > 0$ on $E \subset D_1$ and $\mu_L(E) > 0$. This contradicts (7). Similarly the assumption that $\mu_L(D_2) > 0$ leads to a contradiction of (7). This shows that $\mu_L(D) = 0$. Therefore $f * g = \varphi$, μ_L-a.e. on $\mathbb{R}$. ∎

Lemma 23.43. *Let $p, q, r \in (1, \infty)$ be such that $\frac{1}{p} + \frac{1}{q} = 1 + \frac{1}{r}$. Let $s \in (1, \infty)$ be the conjugate of r, that is, $\frac{1}{r} + \frac{1}{s} = 1$. Then there exist $\xi, \eta, \zeta \in (1, \infty)$ such that*

(1)
$$\frac{1}{\xi} + \frac{1}{\eta} = \frac{1}{p},$$

(2)
$$\frac{1}{\eta} + \frac{1}{\zeta} = \frac{1}{q},$$

(3)
$$\frac{1}{\xi} + \frac{1}{\zeta} = \frac{1}{s},$$

(4)
$$\frac{1}{\xi} + \frac{1}{\eta} + \frac{1}{\zeta} = 1.$$

Proof. Equations (1), (2), and (3) constitute a system of linear equations in three unknowns $\frac{1}{\xi}$, $\frac{1}{\eta}$, and $\frac{1}{\zeta}$. Solving this system we have

$$\frac{1}{\xi} = 1 - \frac{1}{q} \in (0, 1),$$
$$\frac{1}{\eta} = \frac{1}{r} \in (0, 1),$$
$$\frac{1}{\zeta} = 1 - \frac{1}{p} \in (0, 1).$$

Thus $\xi, \eta, \zeta \in (1, \infty)$. Also

$$\frac{1}{\xi} + \frac{1}{\eta} + \frac{1}{\zeta} = 1 - \frac{1}{q} + \frac{1}{r} + 1 - \frac{1}{p} = 1 - \left\{1 + \frac{1}{r}\right\} + \frac{1}{r} + 1 = 1$$

so that (4) is satisfied. ∎

Theorem 23.44. (Young's Convolution Theorem) *Let $p, q, r \in (1, \infty)$ be such that*

(1)
$$\frac{1}{p} + \frac{1}{q} = 1 + \frac{1}{r}.$$

Let $f \in L_r^p(\mathbb{R}, \mathfrak{M}_L, \mu_L)$ and $g \in L_r^q(\mathbb{R}, \mathfrak{M}_L, \mu_L)$. Let us define a function $f * g$ on $\mathbb{R}$ by setting

(2)
$$(f * g)(x) = \int_{\mathbb{R}} f(x - y) g(y)\, \mu_L(dy)$$

for $x \in \mathbb{R}$ for which the integral exists in $\overline{\mathbb{R}}$. Then $(f * g)(x)$ is defined and real-valued for μ_L-a.e. $x \in \mathbb{R}$ and $f * g$ is $\mathfrak{M}_L$-measurable. For every $x \in \mathbb{R}$ for which the integral in (2) does not exist, we set $(f * g)(x) = 0$. Then $f * g \in L_r^r(\mathbb{R}, \mathfrak{M}_L, \mu_L)$ and

(3)
$$\|f * g\|_r \leq \|f\|_p \|g\|_q.$$

Proof. 1. If $f \in L_r^p(\mathbb{R}, \mathfrak{M}_L, \mu_L)$ and $g \in L_r^q(\mathbb{R}, \mathfrak{M}_L, \mu_L)$, then f and g are real-valued μ_L-a.e. on $\mathbb{R}$. On the null set in $(\mathbb{R}, \mathfrak{M}_L, \mu_L)$ on which f is not real-valued, redefine f to be equal to 0 so that f is real-valued and $\mathfrak{M}_L$-measurable on $\mathbb{R}$. Redefine g similarly. Then by Lemma 23.32, $f(x-y)g(y)$ for $(x, y) \in \mathbb{R} \times \mathbb{R}$ is real-valued and $\mathfrak{M}_L^2$-measurable on $\mathbb{R} \times \mathbb{R}$.

Let $s \in (1, \infty)$ be the conjugate of r, that is, $\frac{1}{r} + \frac{1}{s} = 1$. Let $h \in L_r^s(\mathbb{R}, \mathfrak{M}_L, \mu_L)$ be such that $h(x) \in (0, \infty)$ for every $x \in \mathbb{R}$. Then the function $h(x)f(x-y)g(y)$ for $(x, y) \in \mathbb{R} \times \mathbb{R}$ is real-valued and $\mathfrak{M}_L^2$-measurable on $\mathbb{R} \times \mathbb{R}$. Let us show that

$$(4) \qquad \int_{\mathbb{R}\times\mathbb{R}} |h(x)f(x-y)g(y)| \, (\mu_L \times \mu_L)(d(x,y)) \leq \|f\|_p \|g\|_q \|h\|_s.$$

Now $p, q, r \in (1, \infty)$, $\frac{1}{p} + \frac{1}{q} = 1 + \frac{1}{r}$ and $\frac{1}{r} + \frac{1}{s} = 1$. Then by Lemma 23.43, there exist $\xi, \eta, \zeta \in (1, \infty)$ such that

$$\frac{1}{\xi} + \frac{1}{\eta} = \frac{1}{p}, \quad \frac{1}{\eta} + \frac{1}{\zeta} = \frac{1}{q}, \quad \frac{1}{\xi} + \frac{1}{\zeta} = \frac{1}{s}, \quad \frac{1}{\xi} + \frac{1}{\eta} + \frac{1}{\zeta} = 1.$$

Let $a_1, a_2, b_1, b_2, c_1, c_2 > 0$ be defined by

$$(5) \qquad a_1 = \frac{s}{\xi}, \quad a_2 = \frac{s}{\zeta}, \quad b_1 = \frac{p}{\xi}, \quad b_2 = \frac{p}{\eta}, \quad c_1 = \frac{q}{\eta}, \quad c_2 = \frac{q}{\zeta}.$$

Then we have

$$a_1 + a_2 = \left\{\frac{1}{\xi} + \frac{1}{\zeta}\right\}s = 1, \quad b_1 + b_2 = \left\{\frac{1}{\xi} + \frac{1}{\eta}\right\}p = 1, \quad c_1 + c_2 = \left\{\frac{1}{\eta} + \frac{1}{\zeta}\right\}q = 1.$$

Thus the integrand in (4) can be written as

$$|h(x)f(x-y)g(y)| = \left\{|h(x)|^{a_1}|f(x-y)|^{b_1}\right\}\left\{|f(x-y)|^{b_2}|g(y)|^{c_1}\right\}\left\{|g(y)|^{c_2}|h(x)|^{a_2}\right\}.$$

Now $\xi, \eta, \zeta \in (1, \infty)$ and $\frac{1}{\xi} + \frac{1}{\eta} + \frac{1}{\zeta} = 1$. Then applying Theorem 16.56, an extension of Hölder's inequality, we have

$$(6) \qquad \begin{aligned} &\int_{\mathbb{R}\times\mathbb{R}} |h(x)f(x-y)g(y)| \, (\mu_L \times \mu_L)(d(x,y)) \\ &\leq \left\{\int_{\mathbb{R}\times\mathbb{R}} |h(x)|^{a_1\xi} |f(x-y)|^{b_1\xi} \, (\mu_L \times \mu_L)(d(x,y))\right\}^{1/\xi} \\ &\cdot \left\{\int_{\mathbb{R}\times\mathbb{R}} |f(x-y)|^{b_2\eta} |g(y)|^{c_1\eta} \, (\mu_L \times \mu_L)(d(x,y))\right\}^{1/\eta} \\ &\cdot \left\{\int_{\mathbb{R}\times\mathbb{R}} |g(y)|^{c_2\zeta} |h(x)|^{a_2\zeta} \, (\mu_L \times \mu_L)(d(x,y))\right\}^{1/\zeta}. \end{aligned}$$

Substituting (5) in (6) we have

$$
\begin{aligned}
(7) \qquad & \int_{\mathbb{R}\times\mathbb{R}} |h(x)f(x-y)g(y)|\,(\mu_L\times\mu_L)\big(d(x,y)\big) \\
\le & \Big\{\int_{\mathbb{R}\times\mathbb{R}} |h(x)|^s|f(x-y)|^p\,(\mu_L\times\mu_L)\big(d(x,y)\big)\Big\}^{1/\xi} \\
\cdot & \Big\{\int_{\mathbb{R}\times\mathbb{R}} |f(x-y)|^p|g(y)|^q\,(\mu_L\times\mu_L)\big(d(x,y)\big)\Big\}^{1/\eta} \\
\cdot & \Big\{\int_{\mathbb{R}\times\mathbb{R}} |g(y)|^q|h(x)|^s\,(\mu_L\times\mu_L)\big(d(x,y)\big)\Big\}^{1/\zeta}.
\end{aligned}
$$

For the first of the three integrals on the right side of (7), we have

$$
\begin{aligned}
& \int_{\mathbb{R}\times\mathbb{R}} |h(x)|^s|f(x-y)|^p\,(\mu_L\times\mu_L)\big(d(x,y)\big) \\
= & \int_{\mathbb{R}} |h(x)|^s\Big\{\int_{\mathbb{R}} |f(x-y)|^p\,\mu_L(dy)\Big\}\,\mu_L(dx) \\
= & \int_{\mathbb{R}} |h(x)|^s\,\|f\|_p^p\,\mu_L(dx) = \|h\|_s^s\|f\|_p^p,
\end{aligned}
$$

where the first equality is by Theorem 23.17 (Tonelli) and the second equality is by Theorem 9.31 (Translation Invariance) and Theorem 9.32 (Linear Transformation). By similar computation the second integral on the right side of (7) is equal to $\|f\|_p^p\|g\|_q^q$ and the third integral is equal to $\|g\|_q^q\|h\|_s^s$. Substituting these results in (7) we have

$$
\begin{aligned}
& \int_{\mathbb{R}\times bbr} |h(x)f(x-y)g(y)|\,(\mu_L\times\mu_L)\big(d(x,y)\big) \\
\le & \|h\|_s^{\frac{s}{\xi}}\|f\|_p^{\frac{p}{\xi}}\|f\|_p^{\frac{p}{\eta}}\|g\|_q^{\frac{q}{\eta}}\|g\|_q^{\frac{q}{\zeta}}\|h\|_s^{\frac{s}{\zeta}} = \|h\|_s^{s\{\frac{1}{\xi}+\frac{1}{\zeta}\}}\|f\|_p^{p\{\frac{1}{\xi}+\frac{1}{\eta}\}}\|g\|_q^{q\{\frac{1}{\eta}+\frac{1}{\zeta}\}} \\
= & \|h\|_s\|f\|_p\|g\|_q.
\end{aligned}
$$

This completes the proof of (4).

Now for $f \in L_r^p(\mathbb{R}, \mathfrak{M}_L, \mu_L)$, $g \in L_r^q(\mathbb{R}, \mathfrak{M}_L, \mu_L)$, and $h \in L_r^s(\mathbb{R}, \mathfrak{M}_L, \mu_L)$, we have

$$
\begin{aligned}
& \int_{\mathbb{R}} |h(x)|\Big\{\int_{\mathbb{R}} |f(x-y)g(y)|\,\mu_L(dy)\Big\}\,\mu_L(dx) \\
= & \int_{\mathbb{R}\times\mathbb{R}} |h(x)f(x-y)g(y)|\,(\mu_L\times\mu_L)\big(d(x,y)\big) \\
\le & \|h\|_s\|f\|_p\|g\|_q < \infty,
\end{aligned}
$$

where the equality is by Theorem 23.17 (Tonelli) and the first inequality is by (4). This implies that $|h(x)|\{\int_{\mathbb{R}} |f(x-y)g(y)|\,\mu_L(dy)\} < \infty$ for μ_L-a.e. $x \in \mathbb{R}$. Since $h(x) \in (0,\infty)$ for every $x \in \mathbb{R}$, we have $\int_{\mathbb{R}} |f(x-y)g(y)|\,\mu_L(dy) < \infty$ for μ_L-a.e. $x \in \mathbb{R}$. Thus $(f*g)(x) = \int_{\mathbb{R}} f(x-y)g(y)\,\mu_L(dy) \in \mathbb{R}$ for μ_L-a.e. $x \in \mathbb{R}$. The $\mathfrak{M}_L$-measurability of $f*g$ on $\mathbb{R}$ follows from (a) of Theorem 23.17 (Tonelli).

2. Let us show that $f * g \in L_r^r(\mathbb{R}, \mathfrak{M}_L, \mu_L)$. With our extended real-valued $\mathfrak{M}_L$-measurable function $f * g$ on $\mathbb{R}$, let us define a functional L_{f*g} on $L_r^s(\mathbb{R}, \mathfrak{M}_L, \mu_L)$ by setting for $h \in L_r^s(\mathbb{R}, \mathfrak{M}_L, \mu_L)$

$$L_{f*g}(h) = \int_{\mathbb{R}} h(x)(f * g)(x)\, \mu_L(dx). \tag{8}$$

Then L_{f*g} is a linear functional on $L_r^s(\mathbb{R}, \mathfrak{M}_L, \mu_L)$ and moreover by (4) we have

$$\begin{aligned} |L_{f*g}(h)| &\leq \int_{\mathbb{R}} |h(x)| \big|(f * g)(x)\big|\, \mu_L(dx) \\ &\leq \int_{\mathbb{R}} |h(x)| \Big\{ \int_{\mathbb{R}} |f(x-y)g(y)|\, \mu_L(dy) \Big\}\, \mu_L(dx) \\ &= \int_{\mathbb{R}\times\mathbb{R}} |h(x)f(x-y)g(y)|\, (\mu_L \times \mu_L)\big(d(x,y)\big) \\ &\leq \|f\|_p \|g\|_q \|h\|_s. \end{aligned} \tag{9}$$

Thus L_{f*g} is a bounded linear functional on $L_r^s(\mathbb{R}, \mathfrak{M}_L, \mu_L)$ with norm

$$\|L_{f*g}\|_* \leq \|f\|_p \|g\|_q. \tag{10}$$

Now $s \in (1, \infty)$ is the conjugate of $r \in (1, \infty)$. Thus by Theorem 18.6 (Riesz Representation Theorem), there exists a unique $\varphi \in L_r^r(\mathbb{R}, \mathfrak{M}_L, \mu_L)$ such that for every $h \in L_r^s(\mathbb{R}, \mathfrak{M}_L, \mu_L)$ we have

$$L_{f*g}(h) = \int_{\mathbb{R}} h\varphi\, d\mu_L \tag{11}$$

and

$$\|\varphi\|_r = \|L_{f*g}\|_* \leq \|f\|_p \|g\|_q. \tag{12}$$

By (8) and (11) we have for all $h \in L_r^s(\mathbb{R}, \mathfrak{M}_L, \mu_L)$

$$\int_{\mathbb{R}} h(x)(f * g)(x)\, \mu_L(dx) = \int_{\mathbb{R}} h\varphi\, d\mu_L. \tag{13}$$

By the same argument as in the last paragraph in the Proof of Theorem 23.42, we show that the equality (13) implies that $f * g = \varphi$, μ_L-a.e. on $\mathbb{R}$ so that $f * g \in L_r^r(\mathbb{R}, \mathfrak{M}_L, \mu_L)$ and $\|f * g\|_r = \|\varphi\|_r \leq \|f\|_p \|g\|_q$. ∎

Definition 23.45. (a) *For an extended real-valued function f on $\mathbb{R}$, we define*

$$\|f\|_u = \sup_{x\in\mathbb{R}} |f(x)| \in [0, \infty].$$

Let $C_b(\mathbb{R})$ be the linear space of all bounded real-valued continuous functions on $\mathbb{R}$. Then $\|f\|_u \in [0, \infty)$ for every $f \in C_b(\mathbb{R})$ and indeed $\|\cdot\|_u$ is a norm on the linear space $C_b(\mathbb{R})$.

We call this norm the uniform norm.
(b) *Let $C_0(\mathbb{R})$ be the collection of all real-valued continuous functions f on $\mathbb{R}$ satisfying the condition $\lim_{|x|\to\infty} f(x) = 0$. We have $C_0(\mathbb{R}) \subset C_b(\mathbb{R})$.*

Theorem 23.46. *Let $p \in [1,\infty]$ and $q \in [1,\infty]$ be conjugates, that is, $\frac{1}{p}+\frac{1}{q}=1$. Let $f \in L_r^p(\mathbb{R}, \mathfrak{M}_L, \mu_L)$ and $g \in L_r^q(\mathbb{R}, \mathfrak{M}_L, \mu_L)$. Define a function $f * g$ on $\mathbb{R}$ by setting*

$$(1) \qquad (f*g)(x) = \int_{\mathbb{R}} f(x-y)g(y)\,\mu_L(dy),$$

for $x \in \mathbb{R}$ for which the integral exists in $\overline{\mathbb{R}}$. Then we have:
(a) *The integral $\int_{\mathbb{R}} f(x-y)g(y)\,\mu_L(dy)$ exists in $\mathbb{R}$ for every $x \in \mathbb{R}$.*
(b) *$f*g$ is a real-valued uniformly continuous function on $\mathbb{R}$.*
(c) *For $p \in (1,\infty)$, we have $f*g \in C_0(\mathbb{R})$.*
(d) *$\|f*g\|_u \le \|f\|_p\|g\|_q$.*

Proof. 1. Let $f \in L_r^p(\mathbb{R}, \mathfrak{M}_L, \mu_L)$ and $g \in L_r^q(\mathbb{R}, \mathfrak{M}_L, \mu_L)$. Let us show first that $f*g$ is a real-valued function on $\mathbb{R}$ and is bounded by $\|f\|_p\|g\|_q$ on $\mathbb{R}$. Let $x \in \mathbb{R}$ be arbitrarily fixed. With the translation operator T_x and the reflexion operator R as defined in Lemma 23.40, we have $(T_x \circ R)f(y) = f(-y+x)$ for $y \in \mathbb{R}$. Then for $(f*g)(x)$ defined by (1) we have

$$(2) \qquad (f*g)(x) = \int_{\mathbb{R}} (T_x \circ R)f(y)g(y)\,\mu_L(dy).$$

By Lemma 23.40 we have

$$(3) \qquad (T_x \circ R)f \in L_r^p(\mathbb{R}, \mathfrak{M}_L, \mu_L) \quad \text{with} \quad \|(T_x \circ R)f\|_p = \|f\|_p.$$

With our $g \in L_r^q(\mathbb{R}, \mathfrak{M}_L, \mu_L)$, let us define a functional L_g on $h \in L_r^p(\mathbb{R}, \mathfrak{M}_L, \mu_L)$ by

$$(4) \qquad L_g(h) = \int_{\mathbb{R}} hg\,d\mu_L.$$

According to Theorem 18.1, L_g is a bounded linear functional on $L_r^p(\mathbb{R}, \mathfrak{M}_L, \mu_L)$. Thus $L_g(h) \in \mathbb{R}$ for every $h \in L_r^p(\mathbb{R}, \mathfrak{M}_L, \mu_L)$. Then by (2), (4), and (3) we have

$$(f*g)(x) = L_g\big((T_x \circ R)f\big) \in \mathbb{R}.$$

Thus we have $(f*g)(x) = \int_{\mathbb{R}} f(x-y)g(y)\,\mu_L(dy) \in \mathbb{R}$ for every $x \in \mathbb{R}$. This proves (a).

2. Let us prove the uniform continuity of $f*g$ on $\mathbb{R}$. Let us restrict ourselves first to the case $p \in [1,\infty)$ to which Lemma 23.41 is applicable. Let $x \in \mathbb{R}$ and $y \in \mathbb{R}$. Using the expression (2) for $f*g$, we have

$$(5) \qquad \begin{aligned} &|(f*g)(x+y) - (f*g)(x)| \\ &\le \int_{\mathbb{R}} \big|(T_{x+y} \circ R)f(z) - (T_x \circ R)f(z)\big||g(z)|\,\mu_L(dz) \\ &\le \|(T_{x+y} \circ R)f - (T_x \circ R)f\|_p\|g\|_q \end{aligned}$$

by Theorem 16.14 (Hölder) for $p \in (1, \infty)$ and by Theorem 16.40 (Hölder) for $p = 1$. Now $T_{x+y} = T_{y+x} = T_y \circ T_x$. Thus

$$(6) \qquad \|(T_{x+y} \circ R)f - (T_x \circ R)f\|_p = \|T_y(T_x \circ R)f - (T_x \circ R)f\|_p.$$

Since $(T_x \circ R)f \in L_r^p(\mathbb{R}, \mathfrak{M}_L, \mu_L)$, Lemma 23.41 implies

$$\lim_{y \to 0} \|T_y(T_x \circ R)f - (T_x \circ R)f\|_p = 0.$$

Thus for every $\varepsilon > 0$ there exists $\delta > 0$ such that

$$(7) \qquad \|T_y(T_x \circ R)f - (T_x \circ R)f\|_p < \varepsilon \quad \text{for } |y| < \delta.$$

By (5), (6), and (7), for every $\varepsilon > 0$ there exists $\delta > 0$ such that

$$|(f * g)(x + y) - (f * g)(x)| \leq \varepsilon \|g\|_q \quad \text{for } |y| < \delta.$$

This proves the uniform continuity of $f * g$ on $\mathbb{R}$ for the case $p \in [1, \infty)$.

Now consider the case $p = \infty$. In this case we have $q = 1$ and thus $g \in L_r^1(\mathbb{R}, \mathfrak{M}_L, \mu_L)$. Thus by our result above $g * f$ is uniformly continuous on $\mathbb{R}$. As we showed above we have $(f * g)(x) = \int_{\mathbb{R}} f(x - y)g(y)\, \mu_L(dy) \in \mathbb{R}$ for every $x \in \mathbb{R}$. Then $(g * f)(x) = (f * g)(x)$ for every $x \in \mathbb{R}$ by Proposition 23.31. Thus the uniform continuity of $g * f$ on $\mathbb{R}$ implies the uniform continuity of $f * g$ on $\mathbb{R}$.

3. Let us show that if $p \in (1, \infty)$ then $f * g \in C_0(\mathbb{R})$, that is, $\lim_{|x| \to \infty} (f * g)(x) = 0$. Now for $p \in (1, \infty)$ we have $q \in (1, \infty)$. Then $f \in L_r^p(\mathbb{R}, \mathfrak{M}_L, \mu_L)$ and $g \in L_r^q(\mathbb{R}, \mathfrak{M}_L, \mu_L)$ imply

$$(8) \qquad \int_{\mathbb{R}} |f|^p \, d\mu_L < \infty \quad \text{and} \quad \int_{\mathbb{R}} |g|^q \, d\mu_L < \infty.$$

Now (c) of Proposition 8.11 implies that $\lim_{M \to \infty} \int_{[-M,M]} |f|^p \, d\mu_L = \int_{\mathbb{R}} |f|^p \, d\mu_L$. Then (8) implies that $\lim_{M \to \infty} \int_{[-M,M]^c} |f|^p \, d\mu_L = 0$. The same holds for $|g|^q$. Thus for every $\varepsilon > 0$ there exists $M > 0$ such that

$$(9) \qquad \int_{[-M,M]^c} |f|^p \, d\mu_L < \varepsilon \quad \text{and} \quad \int_{[-M,M]^c} |g|^q \, d\mu_L < \varepsilon.$$

Now for every $x \in \mathbb{R}$ we have

$$\begin{aligned}
(10) \qquad |(f * g)(x)| &\leq \int_{[-M,M]} |f(x - y)g(y)| \, \mu_L(dy) \\
&\quad + \int_{[-M,M]^c} |f(x - y)g(y)| \, \mu_L(dy) \\
&\leq \Big\{ \int_{[-M,M]} |f(x - y)|^p \, \mu_L(dy) \Big\}^{1/p} \|g\|_q \\
&\quad + \|f\|_p \Big\{ \int_{[-M,M]^c} |g(y)|^q \, \mu_L(dy) \Big\}^{1/q}
\end{aligned}$$

by Theorem 16.14 (Hölder). Let $x \in \mathbb{R}$ be such that $|x| > 2M$. Then we have

$$(11)\qquad \int_{[-M,M]} |f(x-y)|^p \, \mu_L(dy) = \int_{[x-M,x+M]} |f(-y)|^p \, \mu_L(dy)$$

$$= \int_{[-x-M,-x+M]} |f(y)|^p \, \mu_L(dy) \leq \int_{[-M,+M]^c} |f(y)|^p \, \mu_L(dy)$$

where the first equality is by Theorem 9.31 (Translation Invariance), the second equality is by Theorem 9.32 (Linear Transformation) and the inequality is by the fact that $|x| > 2M$ implies $[-x-M, -x+M] \subset [-M, M]^c$. Substituting (11) in (10), we have for $x \in \mathbb{R}$ such that $|x| > 2M$

$$|(f * g)(x)| \leq \left\{ \int_{[-M,M]^c} |f(y)|^p \, \mu_L(dy) \right\}^{1/p} \|g\|_q$$

$$+ \|f\|_p \left\{ \int_{[-M,M]^c} |g(y)|^q \, \mu_L(dy) \right\}^{1/q}$$

$$< \varepsilon^{1/p} \|g\|_q + \varepsilon^{1/q} \|f\|_p < \{\|f\|_p + \|g\|_q\} \varepsilon.$$

Thus for an arbitrary $\varepsilon > 0$ there exists $M > 0$ such that

$$|(f * g)(x)| < \{\|f\|_p + \|g\|_q\} \varepsilon \quad \text{for } |x| > 2M.$$

This shows that $\lim_{|x| \to \infty} (f * g)(x) = 0$.

4. To prove (d), note that for every $x \in \mathbb{R}$ we have by (2)

$$|(f * g)(x)| \leq \int_{\mathbb{R}} \big|(T_x \circ R) f(y)\big| |g(y)| \, \mu_L(dy)$$

$$\leq \|(T_x \circ R) f\|_p \|g\|_q = \|f\|_p \|g\|_q$$

where the second inequality is by Theorem 16.14 (Hölder) for $p \in (1, \infty)$ and by Theorem 16.40 (Hölder) for $p = 1$ and $p = \infty$ and the last equality is by (3). Then we have

$$\|f * g\|_u = \sup_{x \in \mathbb{R}} |(f * g)(x)| \leq \|f\|_p \|g\|_q.$$

This completes the proof. ∎

Remark 23.47. In the foregoing existence theorems for the convolution product $f * g$, (that is, Theorem 23.33, Theorem 23.42, Theorem 23.44, and Theorem 23.46), for $f \in L_r^p(\mathbb{R}, \mathfrak{M}_L, \mu_L)$ and $g \in L_r^{p'}(\mathbb{R}, \mathfrak{M}_L, \mu_L)$ with various combinations of $p \in [1, \infty]$ and $p' \in [1, \infty]$, we defined $(f * g)(x) := \int_{\mathbb{R}} f(x-y) g(y) \, \mu_L(dy)$ for $x \in \mathbb{R}$ for which the integral exists in $\overline{\mathbb{R}}$ and $(f * g)(x) := 0$ for $x \in \mathbb{R}$ for which the integral does not exist. In Theorem 23.46, we showed that if $p \in [1, \infty]$, $q \in [1, \infty]$ and $\frac{1}{p} + \frac{1}{q} = 1$, then for $f \in L_r^p(\mathbb{R}, \mathfrak{M}_L, \mu_L)$ and $g \in L_r^q(\mathbb{R}, \mathfrak{M}_L, \mu_L)$ the integral $\int_{\mathbb{R}} f(x-y) g(y) \, \mu_L(dy)$

does exist in $\mathbb{R}$ for every $x \in \mathbb{R}$ and thus $(f * g)(x) = \int_{\mathbb{R}} f(x-y)g(y)\,\mu_L(dy)$ for every $x \in \mathbb{R}$. In some questions, for instance for the convergence of a sequence $\big((f * g_n)(x_0) : n \in \mathbb{N}\big)$ in $\overline{\mathbb{R}}$ with a fixed $x_0 \in \mathbb{R}$, it is pertinent to know that $(f * g_n)(x_0) = \int_{\mathbb{R}} f(x_0 - y)g_n(y)\,\mu_L(dy)$. The next theorem gives a sufficient condition for having $(f * g)(x) = \int_{\mathbb{R}} f(x-y)g(y)\,\mu_L(dy)$ for every $x \in \mathbb{R}$.

Theorem 23.48. *Let $f \in L_r^p(\mathbb{R}, \mathfrak{M}_L, \mu_L)$ for some $p \in [1, \infty]$. Let*

(1)
$$g \in \bigcap_{p' \in [1,\infty]} L_r^{p'}(\mathbb{R}, \mathfrak{M}_L, \mu_L).$$

Then we have

(2)
$$(f * g)(x) := \int_{\mathbb{R}} f(x-y)g(y)\,\mu_L(dy) \in \mathbb{R} \quad \textit{for every } x \in \mathbb{R}.$$

Proof. Let $q \in [1, \infty]$ be the conjugate of $p \in [1, \infty]$, that is, $\frac{1}{p} + \frac{1}{q} = 1$. Now (1) implies that $g \in L_r^q(\mathbb{R}, \mathfrak{M}_L, \mu_L)$. Then by Theorem 23.46, the integral $\int_{\mathbb{R}} f(x-y)g(y)\,\mu_L(dy)$ exists in $\mathbb{R}$ for every $x \in \mathbb{R}$. ∎

The following propositions show examples of functions in $\bigcap_{p' \in [1,\infty]} L_r^{p'}(\mathbb{R}, \mathfrak{M}_L, \mu_L)$.

Proposition 23.49. **(a)** *If $g \in L_r^1(\mathbb{R}, \mathfrak{M}_L, \mu_L) \cap L_r^\infty(\mathbb{R}, \mathfrak{M}_L, \mu_L)$, then*

$$g \in \bigcap_{p \in [1,\infty]} L_r^p(\mathbb{R}, \mathfrak{M}_L, \mu_L).$$

(b) *Let g be a bounded real-valued $\mathfrak{M}_L$-measurable function on $\mathbb{R}$ with a compact support, that is,* $\operatorname{supp}\{g\} = \overline{\{x \in \mathbb{R} : g(x) \neq 0\}}$ *is a compact set in $\mathbb{R}$. Then we have*

$$g \in L_r^1(\mathbb{R}, \mathfrak{M}_L, \mu_L) \cap L_r^\infty(\mathbb{R}, \mathfrak{M}_L, \mu_L).$$

Proof. 1. Let us prove (a). Let $p \in [1, \infty)$. Since $g \in L_r^\infty(\mathbb{R}, \mathfrak{M}_L, \mu_L)$, we have $\|g\|_\infty < \infty$ and $|g| \le \|g\|_\infty$ μ_L-a.e. on $\mathbb{R}$. Let $E = \{x \in \mathbb{R} : |g(x)| \ge 1\}$. Then since $g \in L_r^1(\mathbb{R}, \mathfrak{M}_L, \mu_L)$, we have $\mu_L(E) = 1 \cdot \mu_L(E) \le \int_E |g|\,d\mu_L \le \int_{\mathbb{R}} |g|\,d\mu_L = \|g\|_1 < \infty$. On E^c we have $|g| < 1$ so that $|g|^p \le |g|$ for $p \in [1, \infty)$. Then we have

$$\int_{\mathbb{R}} |g|^p\,d\mu_L = \int_E |g|^p\,d\mu_L + \int_{E^c} |g|^p\,d\mu_L \le \|g\|_\infty^p \mu_L(E) + \int_{E^c} |g|\,d\mu_L < \infty.$$

This shows that $g \in L_r^p(\mathbb{R}, \mathfrak{M}_L, \mu_L)$. Thus $g \in \bigcap_{p \in [1,\infty]} L_r^p(\mathbb{R}, \mathfrak{M}_L, \mu_L)$.

2. Let us prove (b). Since g is bounded on $\mathbb{R}$ we have $\|g\|_\infty < \infty$ so that $g \in L_r^\infty(\mathbb{R}, \mathfrak{M}_L, \mu_L)$. Since $\operatorname{supp}\{g\}$ is a compact set in $\mathbb{R}$, it is a bounded set in $\mathbb{R}$. Thus $\mu_L(\operatorname{supp}\{g\}) < \infty$. Then we have

$$\|g\|_1 = \int_{\mathbb{R}} |g|\,d\mu_L = \int_{\operatorname{supp}\{g\}} |g|\,d\mu_L \le \|g\|_\infty \mu_L(\operatorname{supp}\{g\}) < \infty.$$

Thus $g \in L_r^1(\mathbb{R}, \mathfrak{M}_L, \mu_L)$. Therefore $g \in L_r^1(\mathbb{R}, \mathfrak{M}_L, \mu_L) \cap L_r^\infty(\mathbb{R}, \mathfrak{M}_L, \mu_L)$. ∎

Proposition 23.50. **(a)** *With $c > 0$ and $\delta > 0$, let h be a nonnegative real-valued $\mathfrak{M}_L$-measurable function on $\mathbb{R}$ defined by*

$$(1) \qquad h(x) = c\{|x|^{-(1+\delta)} \wedge 1\} \quad \text{for } x \in \mathbb{R}.$$

Then

$$(2) \qquad \|h\|_\infty = c < \infty,$$

$$(3) \qquad \|h\|_p = 2^{1/p}c\{1 + \{(1+\delta)p - 1\}^{-1}\}^{1/p} < \infty \quad \text{for } p \in [1, \infty),$$

$$(4) \qquad h \in \bigcap_{p \in [1,\infty]} L_r^p(\mathbb{R}, \mathfrak{M}_L, \mu_L).$$

(b) *Let g be a real-valued $\mathfrak{M}_L$-measurable function on $\mathbb{R}$ such that $|g| \leq h$ on $\mathbb{R}$. Then*

$$(5) \qquad g \in \bigcap_{p \in [1,\infty]} L_r^p(\mathbb{R}, \mathfrak{M}_L, \mu_L).$$

Proof. 1. We have $0 \leq h \leq c$ on $\mathbb{R}$. Now we have

$$\begin{aligned} \mu_L\{x \in \mathbb{R} : h(x) = c\} &= \mu_L\{x \in \mathbb{R} : |x|^{-(1+\delta)} \wedge 1 = 1\} \\ &= \mu_L\{x \in \mathbb{R} : |x|^{-(1+\delta)} \geq 1\} \\ &= \mu_L\{x \in \mathbb{R} : |x| \leq 1\} = 2. \end{aligned}$$

Then for any $\alpha \in [0, c)$ we have $\{x \in \mathbb{R} : h(x) > \alpha\} \supset \{x \in \mathbb{R} : h(x) = c\}$ so that

$$\mu_L\{x \in \mathbb{R} : h(x) > \alpha\} \geq \mu_L\{x \in \mathbb{R} : h(x) = c\} = 2 > 0.$$

This shows that any $\alpha \in [0, c)$ is not an essential bound of h by Definition 16.35 and therefore $\|h\|_\infty$, the infimum of all essential bounds of h, is equal to c. This proves (2).

For $p \in [1, \infty)$ we have

$$\begin{aligned} \|h\|_p^p &= \int_{\mathbb{R}} h^p \, d\mu_L = 2c^p \int_{[0,\infty)} \{x^{-(1+\delta)} \wedge 1\}^p \, \mu_L(dx) \\ &= 2c^p \Big\{ \int_{[0,1)} 1 \, \mu_L(dx) + \int_{[1,\infty)} x^{-(1+\delta)p} \, \mu_L(dx) \Big\} \\ &= 2c^p \Big\{ 1 + \{(1+\delta)p - 1\}^{-1} \big[x^{-(1+\delta)p+1}\big]_\infty^1 \Big\} \\ &= 2c^p \Big\{ 1 + \{(1+\delta)p - 1\}^{-1} \Big\}. \end{aligned}$$

Taking the p-th root we have (3). By (2) and (3) we have $\|h\|_p < \infty$ for all $p \in [1, \infty]$. Thus $h \in L_r^p(\mathbb{R}, \mathfrak{M}_L, \mu_L)$ for all $p \in [1, \infty]$. This proves (4).

2. To prove (5), note that the condition $|g| \leq h$ on $\mathbb{R}$ implies $\|g\|_p \leq \|h\|_p < \infty$ for all $p \in [1, \infty]$ so that $g \in L_r^p(\mathbb{R}, \mathfrak{M}_L, \mu_L)$ for all $p \in [1, \infty]$. ∎

[IV.3] Approximate Identity in Convolution Product

Let $p \in [1, \infty]$. Let $f \in L_r^p(\mathbb{R}, \mathfrak{M}_L, \mu_L)$ and $g \in L_r^1(\mathbb{R}, \mathfrak{M}_L, \mu_L)$. We showed in Theorem 23.42 that the convolution product $f * g$ is defined and $f * g \in L_r^p(\mathbb{R}, \mathfrak{M}_L, \mu_L)$. We show next that the identity of multiplication in convolution product does not exist.

Proposition 23.51. *Let $p \in [1, \infty]$. Let $f \in L_r^p(\mathbb{R}, \mathfrak{M}_L, \mu_L)$ and $g \in L_r^1(\mathbb{R}, \mathfrak{M}_L, \mu_L)$ and consider the convolution product $f * g$. Identity of multiplication for the convolution product does not exist, that is, there does not exist $g \in L_r^1(\mathbb{R}, \mathfrak{M}_L, \mu_L)$ such that*

$$(1) \qquad f * g = f \quad \textit{for every } f \in L_r^p(\mathbb{R}, \mathfrak{M}_L, \mu_L).$$

Proof. Suppose there exists $g \in L_r^1(\mathbb{R}, \mathfrak{M}_L, \mu_L)$ such that (1) holds. Let us show that this leads to a contradiction. Now since $g \in L_r^1(\mathbb{R}, \mathfrak{M}_L, \mu_L)$ is μ_L-integrable on $\mathbb{R}$, by Theorem 9.26 there exists $\delta > 0$ such that

$$(2) \qquad \int_{[-2\delta, 2\delta]} |g(y)| \, \mu_L(dy) < \frac{1}{2}.$$

Let $f = \mathbf{1}_{[-\delta,\delta]}$. Then $f \in L_r^p(\mathbb{R}, \mathfrak{M}_L, \mu_L)$. Thus by (1) we have for μ_L-a.e. $x \in \mathbb{R}$

$$(3) \qquad \begin{aligned} f(x) = (f * g)(x) = (g * f)(x) &= \int_{\mathbb{R}} g(x-y) f(y) \, \mu_L(dy) \\ &= \int_{[-\delta,\delta]} g(x-y) \, \mu_L(dy) = \int_{[-\delta,\delta]} g(y-x) \, \mu_L(dy) \\ &= \int_{[x-\delta, x+\delta]} g(y) \, \mu_L(dy), \end{aligned}$$

where the second equality is by Proposition 23.31, the fifth equality is by Theorem 9.32 (Linear transformation) and the last equality is by Theorem 9.31 (Translation Invariance). Since the equality (3) holds for μ_L-a.e. $x \in \mathbb{R}$ and since $\mu_L([-\delta, \delta]) = 2\delta > 0$, there exists $x_0 \in [-\delta, \delta]$ for which (3) holds. For $x_0 \in [-\delta, \delta]$, we have $f(x_0) = 1$. Thus

$$(4) \qquad 1 = f(x_0) = \int_{[x_0-\delta, x_0+\delta]} g(y) \, \mu_L(dy).$$

Then by (4) and by (2) we have

$$\begin{aligned} 1 &= \left| \int_{[x_0-\delta, x_0+\delta]} g(y) \, \mu_L(dy) \right| \leq \int_{[x_0-\delta, x_0+\delta]} |g(y)| \, \mu_L(dy) \\ &\leq \int_{[-2\delta, 2\delta]} |g(y)| \, \mu_L(dy) < \frac{1}{2}. \end{aligned}$$

This is a contradiction. ∎

Definition 23.52. *Let $p \in [1, \infty]$. Let $f \in L_r^p(\mathbb{R}, \mathfrak{M}_L, \mu_L)$ and $g \in L_r^1(\mathbb{R}, \mathfrak{M}_L, \mu_L)$ and consider the convolution product $f * g$. A collection of functions $\{g_\varepsilon : \varepsilon \in (0, \infty)\} \subset L_r^1(\mathbb{R}, \mathfrak{M}_L, \mu_L)$ is called an approximate identity of the convolution product if*

(1) $$\lim_{\varepsilon \to 0} f * g = f \quad \textit{in some sense for every } f \in L_r^p(\mathbb{R}, \mathfrak{M}_L, \mu_L).$$

In particular we call $\{g_\varepsilon : \varepsilon \in (0, \infty)\}$ an approximate identity of the convolution product with respect to norm convergence if

(2) $$\lim_{\varepsilon \to 0} \| f * g - f \|_p = 0 \quad \textit{for every } f \in L_r^p(\mathbb{R}, \mathfrak{M}_L, \mu_L).$$

Examples of approximate identities of the convolution product are constructed below.

Definition 23.53. *Let φ be an extended real-valued function on $\mathbb{R}$. For $\varepsilon > 0$, the ε-dilation of φ is a function on $\mathbb{R}$ defined by*

$$\varphi_\varepsilon(x) = \frac{1}{\varepsilon}\varphi\left(\frac{x}{\varepsilon}\right) \quad \textit{for } x \in \mathbb{R}.$$

Observation 23.54. Let φ_ε be the ε-dilation of an extended real-valued function φ on $\mathbb{R}$. Then we have

(1) $$\sup |\varphi_\varepsilon| = \frac{1}{\varepsilon} \sup |\varphi|.$$

Unless φ is the identically vanishing function on $\mathbb{R}$, we have

(2) $$\lim_{\varepsilon \to 0} \sup |\varphi_\varepsilon| = \infty.$$

Regarding the supports of the functions φ and φ_ε, we have

(3) $$\operatorname{supp}\{\varphi_\varepsilon\} = \varepsilon \operatorname{supp}\{\varphi\}.$$

Thus as $\varepsilon \to 0$ the set $\operatorname{supp}\{\varphi_\varepsilon\}$ becomes smaller and moves toward $0 \in \mathbb{R}$.
If $\operatorname{supp}\{\varphi\}$ is a compact set then so is $\operatorname{supp}\{\varphi_\varepsilon\}$ and moreover we have

(4) $$\lim_{\varepsilon \to 0} \mu_L\big(\operatorname{supp}\{\varphi_\varepsilon\}\big) = 0.$$

Proof. 1. The equality (1) follows immediately from the definition of φ_ε in Definition 23.53. If φ is not identically vanishing on $\mathbb{R}$ then $\sup |\varphi| > 0$. Then (2) follows from (1).

2. Let us prove (3). Now by definition we have $\operatorname{supp}\{\varphi\} = \overline{\{x \in \mathbb{R} : \varphi(x) \neq 0\}}$ and $\operatorname{supp}\{\varphi_\varepsilon\} = \overline{\{x \in \mathbb{R} : \varphi_\varepsilon(x) \neq 0\}}$. Observe that

$$\begin{aligned}
&\{x \in \mathbb{R} : \varphi_\varepsilon(x) \neq 0\} = \left\{x \in \mathbb{R} : \frac{1}{\varepsilon}\varphi\left(\frac{x}{\varepsilon}\right) \neq 0\right\} \\
=&\left\{x \in \mathbb{R} : \varphi\left(\frac{x}{\varepsilon}\right) \neq 0\right\} = \{\varepsilon x \in \mathbb{R} : \varphi(x) \neq 0\} \\
=&\varepsilon\{x \in \mathbb{R} : \varphi(x) \neq 0\}.
\end{aligned}$$

Then

$$\operatorname{supp}\{\varphi_\varepsilon\} = \overline{\{x \in \mathbb{R} : \varphi_\varepsilon(x) \neq 0\}} = \overline{\varepsilon\{x \in \mathbb{R} : \varphi(x) \neq 0\}}$$
$$= \varepsilon\overline{\{x \in \mathbb{R} : \varphi(x) \neq 0\}} = \varepsilon \operatorname{supp}\{\varphi\}.$$

3. If $\operatorname{supp}\{\varphi\}$ is a compact set, that is, a bounded closed set in $\mathbb{R}$, then so is $\operatorname{supp}\{\varphi_\varepsilon\}$ by (3). Moreover if $\operatorname{supp}\{\varphi\}$ is a compact set then $\mu_L(\operatorname{supp}\{\varphi\}) < \infty$. Then

$$\lim_{\varepsilon \to 0} \mu_L(\operatorname{supp}\{\varphi_\varepsilon\}) = \lim_{\varepsilon \to 0} \mu_L(\varepsilon \operatorname{supp}\{\varphi\}) = \lim_{\varepsilon \to 0} \varepsilon \mu_L(\operatorname{supp}\{\varphi\}) = 0,$$

where the second equality is by Theorem 3.18 (Positive Homogeneity). This proves (4). ∎

Proposition 23.55. *Let $\varphi \in L_r^1(\mathbb{R}, \mathfrak{M}_L, \mu_L)$. For $\varepsilon > 0$ consider the ε-dilation φ_ε of φ. Then for every $D \in \mathfrak{M}_L$ we have*

(1) $$\int_D \varphi_\varepsilon(x)\, \mu_L(dx) = \int_{\frac{1}{\varepsilon}D} \varphi(x)\, \mu_L(dx),$$

and in particular we have

(2) $$\int_{\mathbb{R}} \varphi_\varepsilon(x)\, \mu_L(dx) = \int_{\mathbb{R}} \varphi(x)\, \mu_L(dx).$$

Moreover for every $M > 0$ we have

(3) $$\lim_{\varepsilon \to 0} \int_{\{x \in \mathbb{R} : |x| > M\}} \varphi_\varepsilon(x)\, \mu_L(dx) = 0.$$

Proof. 1. Let $D \in \mathfrak{M}_L$ and $\varepsilon > 0$. Then by (1) of Theorem 9.32 (Linear Transformation) we have

$$\int_D \varphi_\varepsilon(x)\, \mu_L(dx) = \frac{1}{\varepsilon} \int_D \varphi\Big(\frac{x}{\varepsilon}\Big)\, \mu_L(dx) = \int_{\frac{1}{\varepsilon}D} \varphi(x)\, \mu_L(dx).$$

This proves (1). When $D = \mathbb{R}$ we have $\frac{1}{\varepsilon}\mathbb{R} = \mathbb{R}$. Thus (2) is a particular case of (1).

2. Let us prove (3). Let $M > 0$ and let $D = \{x \in \mathbb{R} : |x| > M\}$. Then we have $\frac{1}{\varepsilon}D = \{x \in \mathbb{R} : |x| > \frac{M}{\varepsilon}\}$. By (1) we have

(4) $$\Big|\int_{\{x \in \mathbb{R} : |x| > M\}} \varphi_\varepsilon(x)\, \mu_L(dx)\Big| = \Big|\int_{\{x \in \mathbb{R} : |x| > \frac{M}{\varepsilon}\}} \varphi(x)\, \mu_L(dx)\Big|$$
$$\leq \int_{\{x \in \mathbb{R} : |x| > \frac{M}{\varepsilon}\}} |\varphi(x)|\, \mu_L(dx).$$

Now since $\int_{\mathbb{R}} |\varphi(x)|\, \mu_L(dx) < \infty$, we have $\lim_{B \to \infty} \int_{\{x \in \mathbb{R} : |x| > B\}} |\varphi(x)|\, \mu_L(dx) = 0$. With $M > 0$ fixed, if we let $\varepsilon \to 0$ then $\frac{M}{\varepsilon} \to \infty$. Thus we have

(5) $$\lim_{\varepsilon \to 0} \int_{\{x \in \mathbb{R} : |x| > \frac{M}{\varepsilon}\}} |\varphi(x)|\, \mu_L(dx) = 0.$$

Applying (5) to (4) we have $\lim_{\varepsilon\to 0}\left|\int_{\{x\in\mathbb{R}:|x|>M\}}\varphi_\varepsilon(x)\,\mu_L(dx)\right|=0$ which implies (3). ∎

Proposition 23.56. *Let φ be an extended real-valued $\mathfrak{M}_L$-measurable function on $\mathbb{R}$. For $\varepsilon>0$ let φ_ε be the ε-dilation of φ. Then we have*

(1)
$$\|\varphi_\varepsilon\|_p=\left(\frac{1}{\varepsilon}\right)^{\frac{p-1}{p}}\|\varphi\|_p \quad \textit{for } p\in[1,\infty),$$

(2)
$$\|\varphi_\varepsilon\|_\infty=\frac{1}{\varepsilon}\|\varphi\|_\infty.$$

Thus, for $p\in[1,\infty]$, if $\varphi\in L_r^p(\mathbb{R},\mathfrak{M}_L,\mu_L)$ then $\varphi_\varepsilon\in L_r^p(\mathbb{R},\mathfrak{M}_L,\mu_L)$.

Proof. 1. Let $p\in[1,\infty)$. Then we have

$$\begin{aligned}\|\varphi_\varepsilon\|_p^p&=\int_{\mathbb{R}}|\varphi_\varepsilon|^p\,d\mu_L=\int_{\mathbb{R}}\left(\frac{1}{\varepsilon}\right)^p\varphi\left(\frac{x}{\varepsilon}\right)^p\mu_L(dx)\\&=\left(\frac{1}{\varepsilon}\right)^p\varepsilon\int_{\mathbb{R}}\varphi(x)^p\,\mu_L(dx)=\left(\frac{1}{\varepsilon}\right)^{p-1}\|\varphi\|_p^p,\end{aligned}$$

where the third equality is by Theorem 9.32 (Linear Transformation). Taking the p-th roots we have (1). The equality (2) is immediate from the definition $\varphi_\varepsilon(x)=\frac{1}{\varepsilon}\varphi\left(\frac{x}{\varepsilon}\right)$ for $x\in\mathbb{R}$.

2. Let $p\in[1,\infty]$. If $\varphi\in L_r^p(\mathbb{R},\mathfrak{M}_L,\mu_L)$ then $\|\varphi\|_p<\infty$. Then by (1) and (2) we have $\|\varphi_\varepsilon\|_p<\infty$ and thus $\varphi_\varepsilon\in L_r^p(\mathbb{R},\mathfrak{M}_L,\mu_L)$. ∎

Theorem 23.57. *Let $p\in[1,\infty]$. Let $f\in L_r^p(\mathbb{R},\mathfrak{M}_L,\mu_L)$ and let $\varphi\in L_r^1(\mathbb{R},\mathfrak{M}_L,\mu_L)$ be such that $\varphi\geq 0$ on $\mathbb{R}$ and $\|\varphi\|_1=1$. Then for every ε-dilation φ_ε of φ we have $f*\varphi_\varepsilon\in L_r^p(\mathbb{R},\mathfrak{M}_L,\mu_L)$ and furthermore*

(1)
$$\lim_{\varepsilon\to 0}\|f*\varphi_\varepsilon-f\|_p=0.$$

Proof. 1. The nonnegativity of φ implies the nonnegativity of φ_ε for every $\varepsilon>0$. Then by (2) of Proposition 23.55, we have

(2)
$$\int_{\mathbb{R}}|\varphi_\varepsilon|\,d\mu_L=\int_{\mathbb{R}}\varphi_\varepsilon\,d\mu_L=\int_{\mathbb{R}}\varphi\,d\mu_L=\int_{\mathbb{R}}|\varphi|\,d\mu_L=\|\varphi\|_1=1.$$

This shows that $\varphi_\varepsilon\in L_r^1(\mathbb{R},\mathfrak{M}_L,\mu_L)$ and $\|\varphi_\varepsilon\|_1=1$. Then by Theorem 23.42, we have $f*\varphi_\varepsilon\in L_r^p(\mathbb{R},\mathfrak{M}_L,\mu_L)$ and thus $f*\varphi_\varepsilon-f\in L_r^p(\mathbb{R},\mathfrak{M}_L,\mu_L)$.

2. Let us prove (1). Let $q\in[1,\infty]$ be the conjugate of our $p\in[1,\infty]$. Let us define a linear functional L_ε on $g\in L_r^q(\mathbb{R},\mathfrak{M}_L,\mu_L)$ by setting

(3)
$$L_\varepsilon(g)=\int_{\mathbb{R}}(f*\varphi_\varepsilon-f)g\,d\mu_L.$$

According to Theorem 18.1, L_ε is a bounded linear functional on $L_r^q(\mathbb{R},\mathfrak{M}_L,\mu_L)$ with norm given by

(4)
$$\|L_\varepsilon\|_*=\|f*\varphi_\varepsilon-f\|_p.$$

Thus, to prove (1) we show that $\lim_{\varepsilon \to 0} \|L_\varepsilon\|_* = 0$. To show this convergence we show that for every $\eta > 0$ there exists $\delta > 0$ such that

$$(5) \qquad \|L_\varepsilon\|_* \le \eta \quad \text{for } \varepsilon \in (0, \delta).$$

Let us estimate $L_\varepsilon(g)$ as defined by (3). Note first that $\int_{\mathbb{R}} \varphi_\varepsilon \, d\mu_L = 1$ by (2) so that for every $x \in \mathbb{R}$ we have $f(x) = f(x) \int_{\mathbb{R}} \varphi_\varepsilon(y) \, \mu_L(dy) = \int_{\mathbb{R}} f(x)\varphi_\varepsilon(y) \, \mu_L(dy)$. Then since $(f * \varphi_\varepsilon)(x) = \int_{\mathbb{R}} f(x-y)\varphi_\varepsilon(y) \, d\mu_L(dy)$ for μ_L-a.e. $x \in \mathbb{R}$, we have

$$(6) \qquad (f * \varphi_\varepsilon)(x) - f(x) = \int_{\mathbb{R}} \{f(x-y) - f(x)\}\varphi_\varepsilon(y) \, \mu_L(dy),$$

for μ_L-a.e. $x \in \mathbb{R}$. Then we have

$$\begin{aligned}
(7) \qquad |L_\varepsilon(g)| &= \Big| \int_{\mathbb{R}} \{f * \varphi_\varepsilon)(x) - f(x)\} g(x) \, \mu_L(dx) \Big| \\
&= \Big| \int_{\mathbb{R}} \Big\{ \int_{\mathbb{R}} \{f(x-y) - f(x)\}\varphi_\varepsilon(y) \, \mu_L(dy) \Big\} g(x) \, \mu_L(dx) \Big| \\
&\le \int_{\mathbb{R}} \Big\{ \int_{\mathbb{R}} |f(x-y) - f(x)| \varphi_\varepsilon(y) \, \mu_L(dy) \Big\} |g(x)| \, \mu_L(dx) \\
&= \int_{\mathbb{R}} \varphi_\varepsilon(y) \Big\{ \int_{\mathbb{R}} |f(x-y) - f(x)| |g(x)| \, \mu_L(dx) \Big\} \, \mu_L(dy)
\end{aligned}$$

by Theorem 23.17 (Tonelli). For the inner integral in the last member of (7), we have

$$\begin{aligned}
\int_{\mathbb{R}} |f(x-y) - f(x)| |g(x)| \, \mu_L(dx) &= \int_{\mathbb{R}} \big|(T_{-y}f)(x) - f(x)\big| |g(x)| \, \mu_L(dx) \\
&\le \big\|T_{-y}f - f\big\|_p \|g\|_q
\end{aligned}$$

by Theorem 16.14 (Hölder) for $p \in (1, \infty)$ and by Theorem 16.40 (Hölder) for $p = 1$. Substituting this in (7) we have

$$(8) \qquad |L_\varepsilon(g)| \le \Big\{ \int_{\mathbb{R}} \varphi_\varepsilon(y) \big\|T_{-y}f - f\big\|_p \, \mu_L(dy) \Big\} \|g\|_q.$$

By (3) of Lemma 23.40 we have

$$(9) \qquad \big\|T_{-y}f - f\big\|_p \le \|T_{-y}f\|_p + \|f\|_p = 2\|f\|_p.$$

According to Lemma 23.41 we have $\lim_{y \to 0} \|T_y f - f\|_p = 0$. Thus for every $\eta > 0$ there exists $M > 0$ such that

$$(10) \qquad \|T_y f - f\|_p < \frac{\eta}{2} \quad \text{for } y \in (-M, M).$$

For our $M > 0$, by (3) of Proposition 23.55 there exists $\delta > 0$ such that

$$(11) \qquad \|f\|_p \int_{[-M,M]^c} \varphi_\varepsilon(y) \, \mu_L(dy) < \frac{\eta}{4} \quad \text{for } \varepsilon \in (0, \delta).$$

Then by (8) we have for $\varepsilon \in (0, \delta)$

$$(12) \qquad |L_\varepsilon(g)| \le \Big\{\int_{[-M,M]} \varphi_\varepsilon(y)\big\|T_{-y}f - f\big\|_p \mu_L(dy)\Big\}\|g\|_q + \Big\{\int_{[-M,M]^c} \varphi_\varepsilon(y)\big\|T_{-y}f - f\big\|_p \mu_L(dy)\Big\}\|g\|_q.$$

For the first integral on the right side of (12), we have

$$(13) \qquad \int_{[-M,M]} \varphi_\varepsilon(y)\big\|T_{-y}f - f\big\|_p \mu_L(dy) \le \frac{\eta}{2}\int_{[-M,M]} \varphi_\varepsilon(y)\,\mu_L(dy) \le \frac{\eta}{2},$$

where the first inequality is by (10) and the second inequality is by the fact that

$$\int_{[-M,M]} \varphi_\varepsilon(y)\,\mu_L(dy) \le \int_{\mathbb{R}} \varphi_\varepsilon(y)\,\mu_L(dy) = \int_{\mathbb{R}} \varphi(y)\,\mu_L(dy) = 1$$

by (2) of Proposition 23.55. For the second integral on the right side of (12), we have

$$(14) \qquad \int_{[-M,M]^c} \varphi_\varepsilon(y)\big\|T_{-y}f - f\big\|_p \mu_L(dy) \le 2\|f\|_p \int_{[-M,M]^c} \varphi_\varepsilon(y)\,\mu_L(dy) \le \frac{\eta}{2}$$

by (9) and (11). Substituting (13) and (14) in (12), we have for $\varepsilon \in (0, \delta)$

$$(15) \qquad |L_\varepsilon(g)| \le \eta\|g\|_q.$$

This shows that for $\varepsilon \in (0, \delta)$, L_ε is a bounded linear functional on $L_r^q(\mathbb{R}, \mathfrak{M}_L, \mu_L)$ with norm $\|L_\varepsilon\|_* \le \eta$. This proves (5). ∎

Observation 23.58. We can apply Theorem 23.57 to obtain an approximation theorem for approximating a function f in the class $L_r^p(\mathbb{R}, \mathfrak{M}_L, \mu_L)$ by functions in a more restricted class. For instance, if $f \in L_r^1(\mathbb{R}, \mathfrak{M}_L, \mu_L)$ and $\varphi \in C_c^1(\mathbb{R})$, then $\varphi_\varepsilon \in C_c^1(\mathbb{R})$ and then we have $f * \varphi_\varepsilon \in C_c^1(\mathbb{R})$ by Theorem 23.39. Then Theorem 23.57, $\lim_{\varepsilon\to 0} \|f - f * \varphi_\varepsilon\|_1 = 0$, approximates f by continuously differentiable functions $f * \varphi_\varepsilon$. The next theorem is an example of approximation theorems derived from Theorem 23.57.

Theorem 23.59. *$C_c^\infty(\mathbb{R})$ is dense in $L_r^p(\mathbb{R}, \mathfrak{M}_L, \mu_L)$ for $p \in [1, \infty)$, that is, for every $f \in L_r^p(\mathbb{R}, \mathfrak{M}_L, \mu_L)$ and $\delta > 0$ there exists $g \in C_c^\infty(\mathbb{R})$ such that $\|f - g\|_p < \delta$.*

Proof. 1. Let $p \in [1, \infty)$ and $f \in L_r^p(\mathbb{R}, \mathfrak{M}_L, \mu_L)$. Then $\int_{\mathbb{R}} |f(x)|^p \mu_L(dx) < \infty$ and this implies that for every $\eta > 0$ there exists $M > 0$ such that

$$(1) \qquad \int_{\{x\in\mathbb{R}:|x|>M\}} |f(x)|^p \mu_L(dx) \le \eta^p.$$

Let us define a function f_M on $\mathbb{R}$ by setting

$$(2) \qquad f_M(x) = \begin{cases} f(x) & \text{for } x \in [-M, M] \\ 0 & \text{for } x \in [-M, M]^c. \end{cases}$$

Since $|f_M| \le |f|$ on $\mathbb{R}$, we have $\int_{\mathbb{R}} |f_M|^p \, d\mu_L \le \int_{\mathbb{R}} |f|^p \, d\mu_L = \|f\|_p^p < \infty$, we have $f_M \in L_r^p(\mathbb{R}, \mathfrak{M}_L, \mu_L)$ and moreover we have

$$\|f - f_M\|_p^p = \int_{\mathbb{R}} |f - f_M|^p \, d\mu_L = \int_{\{x \in \mathbb{R} : |x| > M\}} |f|^p \, d\mu_L \le \eta^p.$$

Thus we have

$$\|f - f_M\|_p \le \eta. \tag{3}$$

Let us show that

$$f_M \in L_r^1(\mathbb{R}, \mathfrak{M}_L, \mu_L). \tag{4}$$

According to Theorem 17.4, if $p \ge 1$ and $(X, \mathfrak{A}, \mu)$ is a finite measure space, then for every extended real-valued $\mathfrak{A}$-measurable function g on X we have

$$\frac{1}{\mu(X)} \int_X |g| \, d\mu \le \left[\frac{1}{\mu(X)} \int_X |g|^p \, d\mu \right]^{\frac{1}{p}}. \tag{5}$$

Our function f_M on $\mathbb{R}$ vanishes outside of the finite closed interval $[-M, M]$. The restriction of f_M to $[-M, M]$ is a function on the finite measure space $\big([-M, M], \mathfrak{M}_L, \mu_L\big)$. By (5) we have

$$\frac{1}{2M} \int_{[-M,M]} |f_M| \, d\mu_L \le \left[\frac{1}{2M} \int_{[-M,M]} |f_M|^p \, d\mu_L \right]^{\frac{1}{p}}.$$

Then we have

$$\begin{aligned} \int_{\mathbb{R}} |f_M| \, d\mu_L &= \int_{[-M,M]} |f_M| \, d\mu_L \le (2M)^{1-\frac{1}{p}} \left[\int_{[-M,M]} |f_M|^p \, d\mu_L \right]^{\frac{1}{p}} \\ &\le (2M)^{1-\frac{1}{p}} \left[\int_{\mathbb{R}} |f|^p \, d\mu_L \right]^{\frac{1}{p}} = (2M)^{1-\frac{1}{p}} \|f\|_p < \infty. \end{aligned}$$

This shows that $f_M \in L_r^1(\mathbb{R}, \mathfrak{M}_L, \mu_L)$.

2. Let $\varphi \in C_c^\infty(\mathbb{R})$ be such that $\varphi \ge 0$ on $\mathbb{R}$ and $\int_{\mathbb{R}} \varphi \, d\mu_L = 1$. Now since φ is bounded on $\mathbb{R}$ and vanishes outside of a compact set in $\mathbb{R}$ we have $\varphi \in L_r^{p'}(\mathbb{R}, \mathfrak{M}_L, \mu_L)$ for every $p' \in [1, \infty]$. Thus we have

$$\varphi \in \bigcap_{p' \in [1,\infty]} L_r^{p'}(\mathbb{R}, \mathfrak{M}_L, \mu_L). \tag{6}$$

For $\varepsilon > 0$, let φ_ε be the ε-dilation of φ. Then $\varphi_\varepsilon \in C_c^\infty(\mathbb{R})$, $\varphi_\varepsilon \ge 0$ on $\mathbb{R}$, $\|\varphi_\varepsilon\|_1 = 1$ by Proposition 23.55 and moreover

$$\varphi_\varepsilon \in \bigcap_{p' \in [1,\infty]} L_r^{p'}(\mathbb{R}, \mathfrak{M}_L, \mu_L). \tag{7}$$

3. Let us consider $f_M * \varphi_\varepsilon$. Since $f_M \in L_r^p(\mathbb{R}, \mathfrak{M}_L, \mu_L)$ and $\varphi_\varepsilon \in L_r^1(\mathbb{R}, \mathfrak{M}_L, \mu_L)$, we have by Theorem 23.42

$$(8) \qquad f_M * \varphi_\varepsilon \in L_r^p(\mathbb{R}, \mathfrak{M}_L, \mu_L).$$

Now since $\varphi_\varepsilon \in C_c^\infty(\mathbb{R}) = \bigcap_{n\in\mathbb{N}} C_c^n(\mathbb{R})$, we have $\varphi_\varepsilon \in C_c^n(\mathbb{R})$ for every $n \in \mathbb{N}$. We have also $f_M \in L_r^1(\mathbb{R}, \mathfrak{M}_L, \mu_L)$ by (4). Then by Theorem 23.39, we have $f_M * \varphi_\varepsilon \in C^n(\mathbb{R})$. Since this holds for every $n \in \mathbb{N}$, we have

$$(9) \qquad f_M * \varphi_\varepsilon \in \bigcap_{n\in\mathbb{N}} C^n(\mathbb{R}) = C^\infty(\mathbb{R}).$$

Let us show that $f_M * \varphi_\varepsilon$ has a compact support. To show this, it suffices to show that $f_M * \varphi_\varepsilon$ vanishes outside of some finite closed interval in $\mathbb{R}$. Now since $f_M \in L_r^p(\mathbb{R}, \mathfrak{M}_L, \mu_L)$ and (7) holds, Theorem 23.48 is applicable to our $f_M * \varphi_\varepsilon$. Thus for every $x \in \mathbb{R}$ we have

$$(10) \qquad (f_M * \varphi_\varepsilon)(x) = \int_{\mathbb{R}} f_M(x-y)\varphi_\varepsilon(y)\,\mu_L(dy) = \frac{1}{\varepsilon}\int_{\mathbb{R}} f_M(x-y)\varphi\Big(\frac{y}{\varepsilon}\Big)\,\mu_L(dy).$$

Now since φ has a compact support, φ vanishes outside of a finite closed interval in $\mathbb{R}$. Thus for sufficiently large $a > 0$ we have $\varphi(y) = 0$ for $y \in [-a, a]^c$. This implies that $\varphi(\frac{y}{\varepsilon}) = 0$ for $y \in [-\varepsilon a, \varepsilon a]^c$. Using this fact in (10), we have for every $x \in \mathbb{R}$

$$(11) \qquad (f_M * \varphi_\varepsilon)(x) = \frac{1}{\varepsilon}\int_{[-\varepsilon a, \varepsilon a]} f_M(x-y)\varphi\Big(\frac{y}{\varepsilon}\Big)\,\mu_L(dy).$$

Consider a finite closed interval $J = [-M - \varepsilon a, M + \varepsilon a]$. If $x \in J^c$ then we have $x - y \in [-M, M]^c$ for all $y \in [-\varepsilon a, \varepsilon a]$ so that $f_M(x-y) = 0$ for all $y \in [-\varepsilon a, \varepsilon a]$. Applying this fact to (11), we have $(f_M * \varphi_\varepsilon)(x) = 0$ for $x \in J^c$. Thus $f_M * \varphi_\varepsilon$ vanishes outside of the finite closed interval J. This shows that $f_M * \varphi_\varepsilon$ has a compact support. Therefore we have

$$(12) \qquad f_M * \varphi_\varepsilon \in C_c^\infty(\mathbb{R}).$$

4. Now we have

$$\|f - f_M * \varphi_\varepsilon\|_p \le \|f - f_M\|_p + \|f_M - f_M * \varphi_\varepsilon\|_p \le \eta + \|f_M - f_M * \varphi_\varepsilon\|_p.$$

By Theorem 23.57, we have $\lim_{\varepsilon\to 0} \|f_M - f_M * \varphi_\varepsilon\|_p = 0$. Thus we have

$$\limsup_{\varepsilon\to 0} \|f - f_M * \varphi_\varepsilon\|_p \le \eta.$$

Since this holds for every $\eta > 0$, the limit superior is equal to 0. This implies that the limit inferior is also equal to 0 and then the limit exists and is equal to 0. Thus we have

$$\lim_{\varepsilon\to 0} \|f - f_M * \varphi_\varepsilon\|_p = 0.$$

Thus for every $\delta > 0$ there exists $g := f_M * \varphi_\varepsilon \in C_c^\infty(\mathbb{R})$ such that $\|f - g\|_p < \delta$. ∎

[IV.4] Approximate Identity Relative to Pointwise Convergence

Consider $L_r^p(\mathbb{R}, \mathfrak{M}_L, \mu_L)$ where $p \in [1, \infty]$. We showed in Theorem 23.57 that there exists $\varphi \in L_r^1(\mathbb{R}, \mathfrak{M}_L, \mu_L)$ such that for its ε-dilation φ_ε we have

$$\lim_{\varepsilon \to 0} \|f * \varphi_\varepsilon - f\|_p = 0 \quad \text{for every } f \in L_r^p(\mathbb{R}, \mathfrak{M}_L, \mu_L).$$

Next we construct $\varphi \in L_r^1(\mathbb{R}, \mathfrak{M}_L, \mu_L)$ so as to have pointwise convergence

$$\lim_{\varepsilon \to 0} (f * \varphi_\varepsilon)(x) = f(x) \quad \text{for } x \in A,$$

where A is a subset of $\mathbb{R}$ which may depend on f.

By our definition of convolution product, we have $(f * \varphi_\varepsilon)(x) = \int_{\mathbb{R}} f(x-y)\varphi_\varepsilon(y)\,\mu_L$ for μ_L-a.e. $x \in \mathbb{R}$ and not necessarily for every $x \in \mathbb{R}$. However, in considering the pointwise convergence $\lim_{\varepsilon \to 0} \{(f * \varphi_\varepsilon)(x) - f(x)\} = 0$, it is pertinent to know whether or not we have $(f * \varphi_\varepsilon)(x) = \int_{\mathbb{R}} f(x-y)\varphi_\varepsilon(y)\,\mu_L$ for the point $x \in \mathbb{R}$ in question. The next lemma is related to this question.

Lemma 23.60. *Let $p \in [1, \infty]$. Let $f \in L_r^p(\mathbb{R}, \mathfrak{M}_L, \mu_L)$ and let $g \in L_r^1(\mathbb{R}, \mathfrak{M}_L, \mu_L)$ be such that $g \geq 0$ on $\mathbb{R}$ and $\|g\|_1 = 1$. Then we have:*
(a) *For μ_L-a.e. $x \in \mathbb{R}$, we have*

$$(1) \qquad (f * g)(x) - f(x) = \int_{\mathbb{R}} \{f(x-y) - f(x)\} g(y)\,\mu_L(dy).$$

(b) *If $\int_{\mathbb{R}} f(x-y)g(y)\,\mu_L(dy)$ exists for every $x \in \mathbb{R}$, then (1) holds for every $x \in \mathbb{R}$.*
(c) *If g satisfies the additional condition that $g \in \bigcap_{p' \in [1,\infty]} L_r^{p'}(\mathbb{R}, \mathfrak{M}_L, \mu_L)$, then (1) holds*
for every $x \in \mathbb{R}$.

Proof. 1. Since $g \geq 0$ on $\mathbb{R}$ and $\|g\|_1 = 1$, we have $\int_{\mathbb{R}} g(y)\,\mu_L(dy) = 1$. Thus for every $x \in \mathbb{R}$, we have $f(x) = \int_{\mathbb{R}} f(x)g(y)\,\mu_L(dy)$. Now $(f * g)(x) = \int_{\mathbb{R}} f(x-y)g(y)\,\mu_L(dy)$ for μ_L-a.e. $x \in \mathbb{R}$ according to Theorem 23.42. Subtracting the first equality above from the second, we have (1). This proves (a) as well as (b).

2. If $g \in \bigcap_{p' \in [1,\infty]} L_r^{p'}(\mathbb{R}, \mathfrak{M}_L, \mu_L)$, then $(f * g)(x) = \int_{\mathbb{R}} f(x-y)g(y)\,\mu_L$ for every $x \in \mathbb{R}$ by Theorem 23.48. Then by (b), (1) holds for every $x \in \mathbb{R}$. ∎

Theorem 23.61. *Let $f \in L_r^\infty(\mathbb{R}, \mathfrak{M}_L, \mu_L)$. Let $\varphi \in L_r^1(\mathbb{R}, \mathfrak{M}_L, \mu_L)$ be such that $\varphi \geq 0$ on $\mathbb{R}$ and $\|\varphi\|_1 = 1$. For $\varepsilon > 0$ let φ_ε be the ε-dilation of φ. If f is continuous at $x_0 \in \mathbb{R}$, then $\lim_{\varepsilon \to 0} (f * \varphi_\varepsilon)(x_0) = f(x_0)$. Thus if f is continuous on $\mathbb{R}$, then $\lim_{\varepsilon \to 0} (f * \varphi_\varepsilon)(x) = f(x)$ for every $x \in \mathbb{R}$.*

Proof. 1. We have $\int_{\mathbb{R}} \varphi_\varepsilon \, d\mu_L = \int_{\mathbb{R}} \varphi \, d\mu_L = \|\varphi\|_1 = 1$ by Proposition 23.55. Note also that since $\varphi \geq 0$ on $\mathbb{R}$, we have $\varphi_\varepsilon \geq 0$ on $\mathbb{R}$. If $f \in L_r^\infty(\mathbb{R}, \mathfrak{M}_L, \mu_L)$, then for every $x \in \mathbb{R}$ we have

$$\begin{aligned}\left|\int_{\mathbb{R}} f(x-y)\varphi_\varepsilon(y)\,\mu_L(dy)\right| &\leq \int_{\mathbb{R}} |f(x-y)|\,\varphi_\varepsilon(y)\,\mu_L(dy)\\ &\leq \|f\|_\infty \int_{\mathbb{R}} \varphi_\varepsilon(y)\,\mu_L(dy)\\ &= \|f\|_\infty < \infty.\end{aligned}$$

Then by (b) of Lemma 23.60 we have for every $x \in \mathbb{R}$

$$(1) \qquad (f * \varphi_\varepsilon)(x) - f(x) = \int_{\mathbb{R}} \{f(x-y) - f(x)\}\,\varphi_\varepsilon(y)\,\mu_L(dy).$$

2. If f is continuous at $x_0 \in \mathbb{R}$, then for every $\eta > 0$ there exists $M > 0$ such that

$$(2) \qquad |f(x_0 - y) - f(x_0)| \leq \eta \quad \text{for } |y| \leq M.$$

Now (1) holds for every $x \in \mathbb{R}$ so that it holds for our $x_0 \in \mathbb{R}$. Then

$$(3) \qquad |(f * \varphi_\varepsilon)(x_0) - f(x_0)| \leq \int_{\mathbb{R}} |f(x_0 - y) - f(x_0)|\,\varphi_\varepsilon(y)\,\mu_L(dy).$$

Applying (2) we have

$$\begin{aligned}(4) \quad |(f * \varphi_\varepsilon)(x_0) - f(x_0)| &\leq \int_{\{y\in\mathbb{R}:|y|\leq M\}} |f(x_0 - y) - f(x_0)|\,\varphi_\varepsilon(y)\,\mu_L(dy)\\ &\quad + \int_{\{y\in\mathbb{R}:|y|> M\}} |f(x_0 - y) - f(x_0)|\,\varphi_\varepsilon(y)\,\mu_L(dy)\\ &\leq \eta \int_{\mathbb{R}} \varphi_\varepsilon\,d\mu_L + 2\|f\|_\infty \int_{\{y\in\mathbb{R}:|y|>M\}} \varphi_\varepsilon(y)\,\mu_L(dy).\end{aligned}$$

According to Proposition 23.55 if we let $\varepsilon \to 0$ then the last integral above converges to 0. Thus letting $\varepsilon \to 0$ in (4), we have

$$(5) \qquad \limsup_{\varepsilon\to 0} |(f * \varphi_\varepsilon)(x_0) - f(x_0)| \leq \eta.$$

Since this holds for every $\eta > 0$, we have $\limsup_{\varepsilon\to 0} |(f*\varphi_\varepsilon)(x_0) - f(x_0)| = 0$. This implies $\liminf_{\varepsilon\to 0} |(f * \varphi_\varepsilon)(x_0) - f(x_0)| = 0$ and then $\lim_{\varepsilon\to 0} |(f * \varphi_\varepsilon)(x_0) - f(x_0)| = 0$, which is equivalent to $\lim_{\varepsilon\to 0} (f * \varphi_\varepsilon)(x_0) = f(x_0)$. ■

Theorem 23.62. *Let $p \in [1, \infty]$. Let $f \in L_r^p(\mathbb{R}, \mathfrak{M}_L, \mu_L)$ and let $\varphi \in L_r^1(\mathbb{R}, \mathfrak{M}_L, \mu_L)$ be such that $\varphi \geq 0$ on $\mathbb{R}$ and $\|\varphi\|_1 = 1$. Assume further that φ is bounded and has a compact support in $\mathbb{R}$. For $\varepsilon > 0$ let φ_ε be the ε-dilation of φ. If f is continuous at $x_0 \in \mathbb{R}$, then we* *have* $\lim_{\varepsilon \to 0} (f * \varphi_\varepsilon)(x_0) = f(x_0)$. *Thus if f is continuous on $\mathbb{R}$, then* $\lim_{\varepsilon \to 0} (f * \varphi_\varepsilon)(x) = f(x)$ *for every* $x \in \mathbb{R}$.

Proof. 1. If φ is bounded and has a compact support in $\mathbb{R}$ then for every $\varepsilon > 0$, φ_ε is bounded and has a compact support in $\mathbb{R}$ by Observation 23.54. Then $\varphi_\varepsilon \in L^1_r(\mathbb{R}, \mathfrak{M}_L, \mu_L)$ and $\varphi_\varepsilon \in L^\infty_r(\mathbb{R}, \mathfrak{M}_L, \mu_L)$ also. This implies that $\varphi_\varepsilon \in \bigcap_{p' \in [1,\infty]} L^{p'}_r(\mathbb{R}, \mathfrak{M}_L, \mu_L)$ by Proposition 23.49. Then by (c) of Lemma 23.60, we have for every $x \in \mathbb{R}$

$$(1) \qquad (f * \varphi_\varepsilon)(x) - f(x) = \int_{\mathbb{R}} \{f(x-y) - f(x)\} \varphi_\varepsilon(y)\, \mu_L(dy).$$

2. If f is continuous at $x_0 \in \mathbb{R}$ then for every $\eta > 0$ there exists $M > 0$ such that

$$(2) \qquad \big|f(x_0 - y) - f(x_0)\big| \le \eta \quad \text{for } |y| \le M.$$

Now (1) holds for every $x \in \mathbb{R}$ and in particular for our $x_0 \in \mathbb{R}$. Recall that $\varphi \ge 0$ on $\mathbb{R}$ implies $\varphi_\varepsilon \ge 0$ on $\mathbb{R}$. Thus we have, applying (2),

$$(3) \qquad \big|(f * \varphi_\varepsilon)(x_0) - f(x_0)\big| \le \int_{\mathbb{R}} \big|f(x_0 - y) - f(x_0)\big| \varphi_\varepsilon(y)\, \mu_L(dy)$$
$$\le \eta \int_{\mathbb{R}} \varphi_\varepsilon(y)\, \mu_L(dy) + \int_{\{y \in \mathbb{R} : |y| > M\}} \big|f(x_0 - y) - f(x_0)\big| \varphi_\varepsilon(y)\, \mu_L(dy).$$

For the last integral above, we have

$$(4) \qquad \int_{\{y \in \mathbb{R} : |y| > M\}} \big|f(x_0 - y) - f(x_0)\big| \varphi_\varepsilon(y)\, \mu_L(dy)$$
$$= \frac{1}{\varepsilon} \int_{\varepsilon\{y \in \mathbb{R} : |y| > \frac{M}{\varepsilon}\}} \big|f(x_0 - y) - f(x_0)\big| \varphi\Big(\frac{y}{\varepsilon}\Big)\, \mu_L(dy)$$
$$= \int_{\{y \in \mathbb{R} : |y| > \frac{M}{\varepsilon}\}} \big|f(x_0 - \varepsilon y) - f(x_0)\big| \varphi(y)\, \mu_L(dy)$$

where the last equality is by Theorem 9.32 (Linear Transformation). Substituting (4) in (3),

$$(5) \qquad |(f * \varphi_\varepsilon)(x_0) - f(x_0)| \le \eta + \int_{\{y \in \mathbb{R} : |y| > \frac{M}{\varepsilon}\}} \big|f(x_0 - \varepsilon y) - f(x_0)\big| \varphi(y)\, \mu_L(dy).$$

Now since $\operatorname{supp}\{\varphi\}$ is a compact set in $\mathbb{R}$, there exists $B > 0$ such that $\operatorname{supp}\{\varphi\} \subset [-B, B]$. For our fixed $M > 0$, there exists $\delta > 0$ such that $\frac{M}{\varepsilon} > B$ for $\varepsilon \in (0, \delta)$. Then we have $\big[-\frac{M}{\varepsilon}, \frac{M}{\varepsilon}\big] \supset [-B, B] \supset \operatorname{supp}\{\varphi\}$ and then $\big\{y \in \mathbb{R} : |y| > \frac{M}{\varepsilon}\big\} = \big[-\frac{M}{\varepsilon}, \frac{M}{\varepsilon}\big]^c \subset \operatorname{supp}\{\varphi\}^c$ for $\varepsilon \in (0, \delta)$. Thus for $\varepsilon \in (0, \delta)$, we have

$$(6) \qquad \int_{\{y \in \mathbb{R} : |y| > \frac{M}{\varepsilon}\}} \big|f(x_0 - \varepsilon y) - f(x_0)\big| \varphi(y)\, \mu_L(dy)$$
$$\le \int_{\operatorname{supp}\{\varphi\}^c} \big|f(x_0 - \varepsilon y) - f(x_0)\big| \varphi(y)\, \mu_L(dy) = 0.$$

Substituting (6) into (5), we have $\left|(f * \varphi_\varepsilon)(x_0) - f(x_0)\right| \leq \eta$ for $\varepsilon \in (0, \delta)$. This shows that $\lim_{\varepsilon \to 0} (f * \varphi_\varepsilon)(x_0) = f(x_0)$. ∎

Lemma 23.63. *Let $\varphi \in L_r^1(\mathbb{R}, \mathfrak{M}_L, \mu_L)$ be such that $\varphi \geq 0$ on $\mathbb{R}$ and $\|\varphi\|_1 = 1$. Suppose φ satisfies the condition that for some $c > 0$ and $\delta > 0$ we have*

$$(1) \qquad \varphi(x) \leq c\left\{|x|^{-(1+\delta)} \wedge 1\right\} \quad \text{for } x \in \mathbb{R}.$$

For $\varepsilon > 0$, let φ_ε be the ε-dilation of φ. Then we have

$$(2) \qquad \varphi_\varepsilon \in \bigcap_{p \in [1,\infty]} L_r^p(\mathbb{R}, \mathfrak{M}_L, \mu_L).$$

Moreover for every $p \in [1, \infty)$ and $M > 0$ we have

$$(3) \qquad \lim_{\varepsilon \to 0} \int_{\{x \in \mathbb{R} : |x| > M\}} \varphi_\varepsilon(x)^p \, \mu_L(dx) = 0.$$

Proof. 1. Condition (1) implies that $\varphi \in \bigcap_{p \in [1,\infty]} L_r^p(\mathbb{R}, \mathfrak{M}_L, \mu_L)$ by (b) of Proposition 23.50. This then implies that $\varphi_\varepsilon \in \bigcap_{p \in [1,\infty]} L_r^p(\mathbb{R}, \mathfrak{M}_L, \mu_L)$ by Proposition 23.56.

2. To prove (3), let $p \in [1, \infty)$ and $M > 0$. We have

$$(4) \qquad \begin{aligned} \int_{\{x \in \mathbb{R} : |x| > M\}} \varphi_\varepsilon(x)^p \, \mu_L(dx) &= \int_{\{x \in \mathbb{R} : |x| > M\}} \left(\frac{1}{\varepsilon}\right)^p \varphi\left(\frac{x}{\varepsilon}\right)^p \mu_L(dx) \\ &= \left(\frac{1}{\varepsilon}\right)^{p-1} \int_{\{x \in \mathbb{R} : |x| > \frac{M}{\varepsilon}\}} \varphi(x)^p \, \mu_L(dx) \\ &:= I(\varepsilon), \end{aligned}$$

where the second equality is by Theorem 9.32 (Linear Transformation).

Let us estimate $I(\varepsilon)$. Let $\varepsilon > 0$ be so small that $\frac{M}{\varepsilon} > 1$. Then by (1) we have

$$(5) \qquad \begin{aligned} I(\varepsilon) &\leq \left(\frac{1}{\varepsilon}\right)^{p-1} c^p \int_{\{x \in \mathbb{R} : |x| > \frac{M}{\varepsilon}\}} \left\{|x|^{-(1+\delta)} \wedge 1\right\}^p \mu_L(dx) \\ &= \left(\frac{1}{\varepsilon}\right)^{p-1} 2c^p \int_{[\frac{M}{\varepsilon}, \infty)} x^{-(1+\delta)p} \, \mu_L(dx) \\ &= \left(\frac{1}{\varepsilon}\right)^{p-1} 2c^p \frac{1}{(1+\delta)p - 1} \left[x^{-(1+\delta)p+1}\right]_{\infty}^{\frac{M}{\varepsilon}} \\ &= \varepsilon^{\delta p} \frac{2c^p M^{-(1+\delta)p+1}}{(1+\delta)p - 1}. \end{aligned}$$

By (4) and (5) we have

$$\lim_{\varepsilon \to 0} \int_{\{x \in \mathbb{R} : |x| > M\}} \varphi_\varepsilon(x)^p \, \mu_L(dx) \leq \lim_{\varepsilon \to 0} \varepsilon^{\delta p} \frac{2c^p M^{-(1+\delta)p+1}}{(1+\delta)p - 1} = 0.$$

This proves (3). ∎

Theorem 23.64. *Let $p \in (1, \infty]$. Let $f \in L_r^p(\mathbb{R}, \mathfrak{M}_L, \mu_L)$. Let $\varphi \in L_r^1(\mathbb{R}, \mathfrak{M}_L, \mu_L)$ be such that $\varphi \geq 0$ on $\mathbb{R}$ and $\|\varphi\|_1 = 1$. Assume further that φ satisfies the condition that for some $c > 0$ and $\delta > 0$*

$$(1) \qquad \varphi(x) \leq c\{|x|^{-(1+\delta)} \wedge 1\} \quad \text{for } x \in \mathbb{R}.$$

For $\varepsilon > 0$ let φ_ε be the ε-dilation of φ. If f is continuous at $x_0 \in \mathbb{R}$, then we have $\lim_{\varepsilon \to 0}(f * \varphi_\varepsilon)(x_0) = f(x_0)$. Thus if f is continuous on $\mathbb{R}$, then $\lim_{\varepsilon \to 0}(f * \varphi_\varepsilon)(x) = f(x)$ for every $x \in \mathbb{R}$.

Proof. 1. If φ satisfies (1), then $\varphi \in \bigcap_{p' \in [1,\infty]} L_r^{p'}(\mathbb{R}, \mathfrak{M}_L, \mu_L)$ by (b) of Proposition 23.50. Then for every $\varepsilon > 0$, we have $\varphi_\varepsilon \in \bigcap_{p' \in [1,\infty]} L_r^{p'}(\mathbb{R}, \mathfrak{M}_L, \mu_L)$ by Proposition 23.56. This then implies by (c) of Lemma 23.60 that for every $x \in \mathbb{R}$ we have

$$(2) \qquad (f * \varphi_\varepsilon)(x) - f(x) = \int_{\mathbb{R}} \{f(x-y) - f(x)\} \varphi_\varepsilon(y)\, \mu_L(dy).$$

2. If f is continuous at $x_0 \in \mathbb{R}$ then for every $\eta > 0$ there exists $M > 0$ such that

$$(3) \qquad \big|f(x_0 - y) - f(x_0)\big| \leq \eta \quad \text{for } |y| \leq M.$$

By (2) and (3) we have

$$\begin{aligned}
&\big|(f * \varphi_\varepsilon(x_0) - f(x_0)\big| \leq \int_{\mathbb{R}} \big|f(x_0 - y) - f(x_0)\big| \varphi_\varepsilon(y)\, \mu_L(dy) \\
&\leq \eta \int_{\{y \in \mathbb{R}: |y| \leq M\}} \varphi_\varepsilon(y)\, \mu_L(dy) + \int_{\{y \in \mathbb{R}: |y| > M\}} \big|f(x_0 - y) - f(x_0)\big| \varphi_\varepsilon(y)\, \mu_L(dy) \\
&\leq \eta + |f(x_0)| \int_{\{y \in \mathbb{R}: |y| > M\}} \varphi_\varepsilon(y)\, \mu_L(dy) + \int_{\{y \in \mathbb{R}: |y| > M\}} |f(x_0 - y)|\, \varphi_\varepsilon(y)\, \mu_L(dy).
\end{aligned}$$

Let us estimate the last integral above. For the case $p \in (1, \infty)$, let $q \in (1, \infty)$ be the conjugate of p. In this case we have

$$\begin{aligned}
(4) \qquad &\int_{\{y \in \mathbb{R}: |y| > M\}} |f(x_0 - y)|\, \varphi_\varepsilon(y)\, \mu_L(dy) \\
&\leq \Big\{ \int_{\{y \in \mathbb{R}: |y| > M\}} |f(x_0 - y)|^p\, \mu_L(dy) \Big\}^{1/p} \Big\{ \int_{\{y \in \mathbb{R}: |y| > M\}} \varphi_\varepsilon(y)^q\, \mu_L(dy) \Big\}^{1/q} \\
&= \Big\{ \int_{\{y \in \mathbb{R}: |y| > M\}} |f(y)|^p\, \mu_L(dy) \Big\}^{1/p} \Big\{ \int_{\{y \in \mathbb{R}: |y| > M\}} \varphi_\varepsilon(y)^q\, \mu_L(dy) \Big\}^{1/q} \\
&\leq \|f\|_p \Big\{ \int_{\{y \in \mathbb{R}: |y| > M\}} \varphi_\varepsilon(y)^q\, \mu_L(dy) \Big\}^{1/q},
\end{aligned}$$

where the first inequality is by Theorem 16.14 (Hölder) and the equality is by Theorem 9.31 and Theorem 9.32. For the case $p = \infty$, we have

$$(5) \qquad \int_{\{y\in\mathbb{R}:|y|>M\}} |f(x_0-y)|\,\varphi_\varepsilon(y)\,\mu_L(dy) \le \|f\|_\infty \int_{\{y\in\mathbb{R}:|y|>M\}} \varphi_\varepsilon(y)\,\mu_L(dy).$$

Substituting (5) in (4), we have for the case $p \in (1,\infty)$,

$$(6) \qquad \big|(f*\varphi_\varepsilon(x_0) - f(x_0)\big| \le \eta + |f(x_0)| \int_{\{y\in\mathbb{R}:|y|>M\}} \varphi_\varepsilon(y)\,\mu_L(dy) + \|f\|_p \Big\{\int_{\{y\in\mathbb{R}:|y|>M\}} \varphi_\varepsilon(y)^q\,\mu_L(dy)\Big\}^{1/q}.$$

Similarly substituting (6) in (4), we have for the case $p = \infty$,

$$(7) \qquad \big|(f*\varphi_\varepsilon(x_0) - f(x_0)\big| \le \eta + \big\{|f(x_0)| + \|f\|_\infty\big\} \int_{\{y\in\mathbb{R}:|y|>M\}} \varphi_\varepsilon(y)\,\mu_L(dy).$$

By Proposition 23.55 we have $\lim\limits_{\varepsilon\to0} \int_{\{y\in\mathbb{R}:|y|>M\}} \varphi_\varepsilon(y)\,\mu_L(dy) = 0$ and by Lemma 23.63 we have $\lim\limits_{\varepsilon\to0} \int_{\{y\in\mathbb{R}:|y|>M\}} \varphi_\varepsilon(y)^q\,\mu_L(dy) = 0$. Thus for $p \in (1,\infty]$, letting $\varepsilon \to 0$ in (7) and (8) we have $\limsup\limits_{\varepsilon\to0} \big|(f*\varphi_\varepsilon(x_0) - f(x_0)\big| \le \eta$. By the arbitrariness of $\eta > 0$, we have $\limsup\limits_{\varepsilon\to0} \big|(f*\varphi_\varepsilon(x_0) - f(x_0)\big| = 0$. This then implies $\lim\limits_{\varepsilon\to0} (f*\varphi_\varepsilon)(x_0) = f(x_0)$. ∎

Lemma 23.65. *Let ψ be a positive valued $\mathfrak{M}_L$-measurable function on $\mathbb{R}$ such that*

$$(1) \qquad \lim_{|x|\to\infty} \psi(x) = 0.$$

Let $\varphi \in L_r^1(\mathbb{R}, \mathfrak{M}_L, \mu_L)$ be such that $\varphi \ge 0$ on $\mathbb{R}$ with $\|\varphi\|_1 = 1$ and satisfies the condition that with some $c > 0$ we have

$$(2) \qquad \varphi(x) \le c\big\{\psi(x)|x|^{-1} \wedge 1\big\} \quad \text{for } x \in \mathbb{R}.$$

For $\varepsilon > 0$, let φ_ε be the ε-dilation of φ. Then we have

$$(3) \qquad \varphi_\varepsilon \in \bigcap_{p\in[1,\infty]} L_r^p(\mathbb{R}, \mathfrak{M}_L, \mu_L).$$

Moreover for every $M > 0$ we have

$$(4) \qquad \lim_{\varepsilon\to0} \Big[\sup_{\{x\in\mathbb{R}:|x|>\frac{M}{\varepsilon}\}} \psi(x)\Big] = 0,$$

and for every $p \in (1,\infty)$ and $M > 0$ we have

$$(5) \qquad \lim_{\varepsilon\to0} \int_{\{x\in\mathbb{R}:|x|>M\}} \varphi_\varepsilon(x)^p\,\mu_L(dx) = 0.$$

Proof. 1. To prove (3), note that since φ is bounded on $\mathbb{R}$ we have $\varphi \in \varphi \in L_r^\infty(\mathbb{R}, \mathfrak{M}_L, \mu_L)$. Thus we have $\varphi \in L_r^1(\mathbb{R}, \mathfrak{M}_L, \mu_L) \cap L_r^\infty(\mathbb{R}, \mathfrak{M}_L, \mu_L)$ and then by Proposition 23.49 we have $\varphi \in \bigcap_{p\in[1,\infty]} L_r^p(\mathbb{R}, \mathfrak{M}_L, \mu_L)$. Then we have $\varphi_\varepsilon \in \bigcap_{p\in[1,\infty]} L_r^p(\mathbb{R}, \mathfrak{M}_L, \mu_L)$ by Proposition 23.56.

2. To prove (4), let $M > 0$ be arbitrarily given. Since $\lim_{|x|\to\infty} \psi(x) = 0$, for every $\eta > 0$ there exists $B > 0$ such that $\psi(x) \leq \eta$ for $|x| > B$. Let $\varepsilon > 0$ be so small that $\frac{M}{\varepsilon} > B$. Then $\{x \in \mathbb{R} : |x| > \frac{M}{\varepsilon}\} \subset \{x \in \mathbb{R} : |x| > B\}$ and hence

$$\sup_{\{x\in\mathbb{R}:|x|>\frac{M}{\varepsilon}\}} \psi(x) \leq \sup_{\{x\in\mathbb{R}:|x|>B\}} \psi(x) \leq \eta.$$

Since this holds for sufficiently small $\varepsilon > 0$, we have

$$\limsup_{\varepsilon\to 0} \Big[\sup_{\{x\in\mathbb{R}:|x|>\frac{M}{\varepsilon}\}} \psi(x) \Big] \leq \eta.$$

By the arbitrariness of $\eta > 0$, the limit superior is equal to 0. This implies that the limit inferior is also equal to 0 and then the limit exists and is equal to 0. This proves (4).

3. Let us prove (5). Let $p \in (1, \infty)$ and $M > 0$. Let us write $p = 1 + \delta$ where $\delta > 0$. Now

$$(6) \qquad \int_{\{x\in\mathbb{R}:|x|>M\}} \varphi_\varepsilon(x)^p \, \mu_L(dx)$$
$$= \int_{\{x\in\mathbb{R}:|x|>M\}} \Big(\frac{1}{\varepsilon}\Big)^p \varphi\Big(\frac{x}{\varepsilon}\Big)^p \, \mu_L(dx)$$
$$= \Big(\frac{1}{\varepsilon}\Big)^{p-1} \int_{\{x\in\mathbb{R}:|x|>\frac{M}{\varepsilon}\}} \varphi(x)^p \, \mu_L(dx)$$
$$\leq \Big(\frac{1}{\varepsilon}\Big)^{p-1} \int_{\{x\in\mathbb{R}:|x|>\frac{M}{\varepsilon}\}} c^p \psi(x)^p |x|^{-p} \, \mu_L(dx)$$
$$= \Big(\frac{1}{\varepsilon}\Big)^{\delta} c^{1+\delta} \int_{\{x\in\mathbb{R}:|x|>\frac{M}{\varepsilon}\}} \psi(x)^{1+\delta} |x|^{-(1+\delta)} \, \mu_L(dx),$$

where the second equality is by Theorem 9.32 (Linear Transformation). In view of (6), to prove (5) it suffices to show

$$(7) \qquad \lim_{\varepsilon\to 0} \Big(\frac{1}{\varepsilon}\Big)^{\delta} \int_{\{x\in\mathbb{R}:|x|>\frac{M}{\varepsilon}\}} \psi(x)^{1+\delta} |x|^{-(1+\delta)} \, \mu_L(dx) = 0.$$

Let $\eta > 0$ be arbitrarily given. Since $\lim_{|x|\to\infty} \psi(x) = 0$, there exists $B > 0$ such that

$\psi(x) < \eta$ for $|x| > B$. Let $\varepsilon > 0$ be so small that $\frac{M}{\varepsilon} > B$. Then

$$\begin{aligned}\left(\frac{1}{\varepsilon}\right)^{\delta} \int_{\{x\in\mathbb{R}:|x|>\frac{M}{\varepsilon}\}} \psi(x)^{1+\delta}|x|^{-(1+\delta)}\,\mu_L(dx) &\le \left(\frac{1}{\varepsilon}\right)^{\delta} \int_{\{x\in\mathbb{R}:|x|>\frac{M}{\varepsilon}\}} \eta^{1+\delta}|x|^{-(1+\delta)}\,\mu_L(dx) \\ &= 2\left(\frac{1}{\varepsilon}\right)^{\delta} \eta^{1+\delta} \int_{[\frac{M}{\varepsilon},\infty)} |x|^{-(1+\delta)}\,\mu_L(dx) \\ &= 2\left(\frac{1}{\varepsilon}\right)^{\delta} \eta^{1+\delta} \left[\frac{x^{-\delta}}{\delta}\right]_{\infty}^{\frac{M}{\varepsilon}} = 2\eta^{1+\delta}\delta^{-1}M^{-\delta}.\end{aligned}$$

Since this holds for sufficiently small $\varepsilon > 0$, we have

$$\limsup_{\varepsilon\to 0} \left(\frac{1}{\varepsilon}\right)^{\delta} \int_{\{x\in\mathbb{R}:|x|>\frac{M}{\varepsilon}\}} \psi(x)^{1+\delta}|x|^{-(1+\delta)}\,\mu_L(dx) \le 2\eta^{1+\delta}\delta^{-1}M^{-\delta}.$$

Then since this holds for every $\eta > 0$, we have

$$\limsup_{\varepsilon\to 0} \left(\frac{1}{\varepsilon}\right)^{\delta} \int_{\{x\in\mathbb{R}:|x|>\frac{M}{\varepsilon}\}} \psi(x)^{1+\delta}|x|^{-(1+\delta)}\,\mu_L(dx) = 0.$$

This implies that the limit inferior is also equal to 0 and then the limit exists and is equal to 0. This proves (7). ∎

Theorem 23.66. *Let $p \in [1,\infty)$. Let $f \in L_r^p(\mathbb{R}, \mathfrak{M}_L, \mu_L)$. Let $\varphi \in L_r^1(\mathbb{R}, \mathfrak{M}_L, \mu_L)$ be such that $\varphi \ge 0$ on $\mathbb{R}$ and $\|\varphi\|_1 = 1$. Assume further that φ satisfies the following condition*

(1) $$\varphi(x) \le c\{\psi(x)|x|^{-1} \wedge 1\} \quad \text{for } x \in \mathbb{R}$$

where $c > 0$ and ψ is a positive valued $\mathfrak{M}_L$-measurable function on $\mathbb{R}$ and $\lim_{|x|\to\infty} \psi(x) = 0$. For $\varepsilon > 0$, let φ_ε be the ε-dilation of φ. If f is continuous at $x_0 \in \mathbb{R}$ then

(2) $$\lim_{\varepsilon\to 0} (f * \varphi_\varepsilon)(x_0) = f(x_0).$$

Thus if f is continuous on $\mathbb{R}$ then $\lim_{\varepsilon\to 0} (f * \varphi_\varepsilon)(x) = f(x)$ for every $x \in \mathbb{R}$.

Proof. We have $\varphi_\varepsilon \in L_r^{p'}(\mathbb{R}, \mathfrak{M}_L, \mu_L)$ for all $p' \in [1,\infty]$ by Lemma 23.65. Then by (c) of Lemma 23.60 we have for every $x \in \mathbb{R}$

(3) $$(f * \varphi_\varepsilon)(x) - f(x) = \int_{\mathbb{R}} \{f(x-y) - f(x)\}\varphi_\varepsilon(y)\,\mu_L(dy).$$

If f is continuous at $x_0 \in \mathbb{R}$ then for every $\eta > 0$ there exists $M > 0$ such that

(4) $$|f(x_0 - y) - f(x_0)| \le \eta \quad \text{for } |y| \le M.$$

By (3) and (4) we have

$$(5) \qquad |(f*\varphi_\varepsilon)(x_0)-f(x_0)| \le \int_{\mathbb{R}} |f(x_0-y)-f(x_0)|\varphi_\varepsilon(y)\,\mu_L(dy)$$
$$\le \eta\int_{\mathbb{R}}\varphi_\varepsilon\,d\mu_L + \int_{\{y\in\mathbb{R}:|y|>M\}} |f(x_0-y)|\varphi_\varepsilon(y)\,\mu_L(dy)$$
$$+|f(x_0)|\int_{\{y\in\mathbb{R}:|y|>M\}}\varphi_\varepsilon(y)\,\mu_L(dy)$$
$$:=\eta + I_1(\varepsilon)+|f(x_0)|I_2(\varepsilon).$$

By Proposition 23.55 we have

$$(6) \qquad \lim_{\varepsilon\to 0} I_2(\varepsilon)=\lim_{\varepsilon\to 0}\int_{\{y\in\mathbb{R}:|y|>M\}}\varphi_\varepsilon(y)\,\mu_L(dy)=0.$$

Let us show that

$$(7) \qquad \lim_{\varepsilon\to 0} I_1(\varepsilon)=\lim_{\varepsilon\to 0}\int_{\{y\in\mathbb{R}:|y|>M\}}|f(x_0-y)|\varphi_\varepsilon(y)\,\mu_L(dy)=0.$$

Now $f\in L_r^p(\mathbb{R},\mathfrak{M}_L,\mu_L)$ where $p\in[1,\infty)$. We argue for (7) separately for the case $p=1$ and for the case $p\in(1,\infty)$. For the case $p=1$ we have

$$(8) \qquad I_1(\varepsilon)=\int_{\{y\in\mathbb{R}:|y|>M\}}|f(x_0-y)|\varphi_\varepsilon(y)\,\mu_L(dy)$$
$$\le\Big\{\sup_{\{y\in\mathbb{R}:|y|>M\}}\varphi_\varepsilon(y)\Big\}\int_{\{y\in\mathbb{R}:|y|>M\}}|f(x_0-y)|\,\mu_L(dy)$$
$$\le\Big\{\sup_{\{y\in\mathbb{R}:|y|>M\}}\varphi_\varepsilon(y)\Big\}\|f\|_1.$$

By (1) we have

$$\varphi_\varepsilon(y)=\frac{1}{\varepsilon}\varphi\Big(\frac{y}{\varepsilon}\Big)\le\frac{1}{\varepsilon}c\Big\{\psi\Big(\frac{y}{\varepsilon}\Big)\Big|\frac{y}{\varepsilon}\Big|^{-1}\wedge 1\Big\}=c\psi\Big(\frac{y}{\varepsilon}\Big)|y|^{-1}.$$

Then we have

$$\sup_{\{y\in\mathbb{R}:|y|>M\}}\varphi_\varepsilon(y)\le\sup_{\{y\in\mathbb{R}:|y|>M\}}\psi\Big(\frac{y}{\varepsilon}\Big)|y|^{-1}$$
$$\le\frac{1}{M}\sup_{\{y\in\mathbb{R}:|y|>M\}}\psi\Big(\frac{y}{\varepsilon}\Big)=\frac{1}{M}\sup_{\{x\in\mathbb{R}:|x|>\frac{M}{\varepsilon}\}}\psi(x).$$

Substituting this in (8) we have

$$I_1(\varepsilon)\le\frac{1}{M}\Big\{\sup_{\{x\in\mathbb{R}:|x|>\frac{M}{\varepsilon}\}}\psi(x)\Big\}\|f\|_1.$$

Then by (5) of Lemma 23.65, we have $\lim_{\varepsilon \to 0} I_1(\varepsilon) = 0$. This proves (7) for the case $p = 1$.

For the case $p \in (1, \infty)$, let $q \in (1, \infty)$ be the conjugate of p. Then

$$(9)\quad I_1(\varepsilon) = \int_{\{y \in \mathbb{R}: |y| > M\}} |f(x_0 - y)| \varphi_\varepsilon(y)\, \mu_L(dy)$$
$$\leq \Big\{ \int_{\{y \in \mathbb{R}: |y| > M\}} |f(x_0 - y)|^p \, \mu_L(dy) \Big\}^{1/p} \Big\{ \int_{\{y \in \mathbb{R}: |y| > M\}} \varphi_\varepsilon(y)^q \, \mu_L(dy) \Big\}^{1/q}$$
$$\leq \|f\|_p \Big\{ \int_{\{y \in \mathbb{R}: |y| > M\}} \varphi_\varepsilon(y)^q \, \mu_L(dy) \Big\}^{1/q}.$$

By Lemma 23.65 we have $\lim_{\varepsilon \to 0} \int_{\{y \in \mathbb{R}: |y| > M\}} \varphi_\varepsilon(y)^q \, \mu_L(dy) = 0$. Applying this to (9) we have $\lim_{\varepsilon \to 0} I_1(\varepsilon) = 0$. This completes the proof of (7).

Letting $\varepsilon \to 0$ in (5) and applying (6) and (7) we have $\limsup_{\varepsilon \to 0} |(f * \varphi_\varepsilon)(x_0) - f(x_0)| \leq \eta$. Since this holds for every $\eta > 0$ we have $\limsup_{\varepsilon \to 0} |(f * \varphi_\varepsilon)(x_0) - f(x_0)| = 0$. This then implies $\liminf_{\varepsilon \to 0} |(f * \varphi_\varepsilon)(x_0) - f(x_0)| = 0$ and hence $\lim_{\varepsilon \to 0} |(f * \varphi_\varepsilon)(x_0) - f(x_0)| = 0$, that is, $\lim_{\varepsilon \to 0} (f * \varphi_\varepsilon)(x_0) = f(x_0)$. ∎

[V] Some Related Theorems

We include here several theorems related to product measure spaces.

[V.1] Cavalieri's Formula and Extensions

Theorem 23.67. (Cavalieri's Formula) *Let $(X, \mathfrak{A}, \mu)$ be a σ-finite measure space.*
(a) *If f is a nonnegative extended real-valued $\mathfrak{A}$-measurable function on X, then we have*

$$(1)\qquad \int_X f(x)\, \mu(dx) = \int_{[0,\infty)} \mu\{x \in X : f(x) \geq t\}\, \mu_L(dt).$$

(b) *Let f be a nonnegative real-valued $\mathfrak{A}$-measurable function on X. Let ν be a σ-finite measure on the measurable space $\big([0,\infty), \mathfrak{B}_{[0,\infty)}\big)$ where $\mathfrak{B}_{[0,\infty)}$ is the σ-algebra of Borel sets in $[0,\infty)$. Let φ be a function on $[0,\infty)$ defined by setting $\varphi(t) = \nu\big([0,t]\big)$ for $t \in [0,\infty)$. Then we have*

$$(2)\qquad \int_X \varphi\big(f(x)\big)\, \mu(dx) = \int_{[0,\infty)} \mu\{x \in X : f(x) \geq t\}\, \nu(dt).$$

Proof. 1. Let us prove (2) and derive (1) from (2). Thus let ν be a σ-finite measure on the measurable space $\big([0,\infty), \mathfrak{B}_{[0,\infty)}\big)$. Then $\varphi(t) = \nu\big([0,t]\big)$ for $t \in [0,\infty)$ is real-valued increasing function on $[0,\infty)$ and is therefore a $\mathfrak{B}_{[0,\infty)}$-measurable function on $[0,\infty)$. Consider the product measure space $\big(X \times [0,\infty), \sigma(\mathfrak{A} \times \mathfrak{B}_{[0,\infty)}), \mu \times \nu\big)$ and a subset of $X \times [0,\infty)$ defined by

$$(3)\qquad E = \big\{(x,t) \in X \times [0,\infty) : f(x) \geq t\big\}.$$

The $\mathfrak{A}$-measurable function f on X may be regarded as a $\sigma(\mathfrak{A} \times \mathfrak{B}_{[0,\infty)})$-measurable function on $X \times [0, \infty)$. Now $g(t) = t$ for $t \in [0, \infty)$ is a $\mathfrak{B}_{[0,\infty)}$-measurable function on $[0, \infty)$ and thus it can be regarded as a $\sigma(\mathfrak{A} \times \mathfrak{B}_{[0,\infty)})$-measurable function on $X \times [0, \infty)$. Then $E = \{(x, t) \in X \times [0, \infty) : f(x) \geq g(t)\} \in \sigma(\mathfrak{A} \times \mathfrak{B}_{[0,\infty)})$ by Theorem 4.16. Now

$$(4) \qquad (\mu \times \nu)(E) = \int_{[0,\infty)} \mu(E(\cdot, t))\, \nu(dt) = \int_{[0,\infty)} \mu\{x \in X : f(x) \geq t\}\, \nu(dt)$$

by Proposition 23.16 and similarly

$$(5) \qquad (\mu \times \nu)(E) = \int_X \nu(E(x, \cdot))\, \mu(dx) = \int_X \nu\{t \in [0, \infty) : f(x) \geq t\}\, \mu(dx)$$
$$= \int_X \nu([0, f(x)])\, \mu(dx) = \int_X \varphi(f(x))\, \mu(dx).$$

By (4) and (5) we have (2).

2. Let us prove (1). Consider first the case that f is a nonnegative real-valued $\mathfrak{A}$-measurable function on X. Let $\nu = \mu_L$. Then we have $\varphi(t) = \mu_L([0, t]) = t$ for $t \in [0, \infty)$ and hence $\varphi(f(x)) = f(x)$. Thus (2) reduces to (1).

Consider the general case that f is a nonnegative extended real-valued $\mathfrak{A}$-measurable function on X. Let $A = \{x \in X : f(x) = \infty\}$. Consider first the case that $\mu(A) = 0$. Let us define a function h on X by setting $h(x) = f(x)$ for $x \in A^c$ and $h(x) = 0$ for $x \in A$. Then h is a nonnegative real-valued $\mathfrak{A}$-measurable function on A and $h = f$ a.e. on X so that $\int_X f\, d\mu = \int_X h\, d\mu$. Since h is nonnegative real-valued, (1) holds for h as we pointed above. Thus

$$(6) \qquad \int_X f(x)\, \mu(dx) = \int_X h(x)\, \mu(dx) = \int_{[0,\infty)} \mu\{x \in X : h(x) \geq t\}\, \mu_L(dt).$$

The fact that $h = f$ a.e. on X implies also

$$(7) \qquad \mu\{x \in X : h(x) \geq t\} = \mu\{x \in X : f(x) \geq t\}$$

for every $t \in [0, \infty)$. By (6) and (7), we have (1). Finally for the case $\mu(A) = a > 0$, we have $\int_X f\, d\mu = \infty$. Also for any $t \in [0, \infty)$, we have $\{x \in X : f(t) \geq t\} \supset A$ so that $\mu\{x \in X : f(t) \geq t\} \geq \mu(A) = a$ and then

$$\int_{[0,\infty)} \mu\{x \in X : f(x) \geq t\}\, \mu_L(dt) \geq \int_{[0,\infty)} a\, \mu_L(dt) = \infty.$$

Thus (1) holds trivially for this case. ∎

Theorem 23.68. *Let $(X, \mathfrak{A}, \mu)$ be a σ-finite measure space and let f be a nonnegative real-valued $\mathfrak{A}$-measurable function on X. Let φ be a nonnegative real-valued increasing and absolutely continuous function on $[0, \infty)$ such that $\varphi(0) = 0$. Then we have*

$$(1) \qquad \int_X \varphi(f(x))\, \mu(dx) = \int_{[0,\infty)} \mu\{x \in X : f(x) \geq t\}\, \varphi'(t)\, \mu_L(dt).$$

In particular for $p \geq 1$, we have

$$(2) \qquad \int_X \big(f(x)\big)^p\,\mu(dx) = p\int_{[0,\infty)} t^{p-1}\mu\{x \in X : f(x) \geq t\}\,\mu_L(dt).$$

Proof. 1. Let φ be a nonnegative real-valued increasing and absolutely continuous function on $[0,\infty)$ such that $\varphi(0) = 0$. Let us extend the domain of definition of φ to $\mathbb{R}$ by setting $\varphi(t) = 0$ for $t \in (-\infty, 0)$. Then φ is a real-valued increasing and absolutely continuous function on $\mathbb{R}$ with $\varphi(0) = 0$. Let ν be the Lebesgue-Stieltjes measure on $(\mathbb{R}, \mathfrak{B}_{\mathbb{R}})$ determined by the increasing function φ on $\mathbb{R}$. By Theorem 22.8, ν is a σ-finite measure on $(\mathbb{R}, \mathfrak{B}_{\mathbb{R}})$ and by Theorem 22.22, ν is absolute continuous with respect to μ_L with Radon-Nikodym derivative $\frac{d\nu}{d\mu_L} = \varphi'$ a.e. on $(\mathbb{R}, \mathfrak{B}_{\mathbb{R}}, \mu_L)$. Also by Proposition 22.7, we have $\nu([0,t]) = \varphi(t) - \varphi(0) = \varphi(t)$ for $t \in [0,\infty)$. Let us restrict ν to the measurable space $([0,\infty), \mathfrak{B}_{[0,\infty)})$. Then ν and φ satisfy the conditions in (b) of Theorem 23.68 and consequently we have

$$\begin{aligned}
\int_X \varphi\big(f(x)\big)\,\mu(dx) &= \int_{[0,\infty)} \mu\{x \in X : f(x) \geq t\}\,\nu(dt) \\
&= \int_{[0,\infty)} \mu\{x \in X : f(x) \geq t\}\Big(\frac{d\nu}{d\mu_L}\Big)(t)\,\mu_L(dt) \\
&= \int_{[0,\infty)} \mu\{x \in X : f(x) \geq t\}\,\varphi'(t)\,\mu_L(dt)
\end{aligned}$$

where the second equality is by Theorem 11.18. This proves (1).

2. For $p \geq 1$ the function $\varphi(t) = t^p$ for $t \in [0,\infty)$ is a nonnegative real-valued increasing and absolutely continuous function on $[0,\infty)$ such that $\varphi(0) = 0$. We have $\varphi'(t) = pt^{p-1}$ for $t \in [0,\infty)$. Thus (2) follows from (1). ∎

[V.2] Minkowski's Inequality for Integrals

Theorem 23.69. (Minkowski's Inequality for Integrals) *Let $(X \times Y, \sigma(\mathfrak{A} \times \mathfrak{B}), \mu \times \nu)$ be the product measure space of two σ-finite measure spaces $(X, \mathfrak{A}, \mu)$ and $(Y, \mathfrak{B}, \nu)$. Let f be a nonnegative extended real-valued $\sigma(\mathfrak{A} \times \mathfrak{B})$-measurable function on $X \times Y$. Then for $p \in [1,\infty)$ we have*

$$\left[\int_X \left\{\int_Y f(x,y)\,\nu(dy)\right\}^p \mu(dx)\right]^{1/p} \leq \int_Y \left\{\int_X f(x,y)^p\,\mu(dx)\right\}^{1/p} \nu(dy).$$

Proof. For $p = 1$ the equality holds by Theorem 23.17 (Tonelli). Consider the case $p > 1$. Let q be its conjugate, that is, $\frac{1}{p} + \frac{1}{q} = 1$. Let us define a function of $x \in \mathbb{R}$ by setting

$$J(x) = \int_Y f(x,y)\,\nu(dy).$$

By Theorem 23.17, J is a $\mathfrak{A}$-measurable function on X. Now

$$(1)\quad \int_X \Big\{\int_Y f(x,y)\,\nu(dy)\Big\}^p \mu(dx) = \int_X \Big\{\int_Y f(x,y)\,\nu(dy)\Big\} J(x)^{p-1}\,\mu(dx)$$
$$= \int_Y \Big\{\int_X f(x,y) J(x)^{p-1}\,\mu(dx)\Big\}\nu(dy)$$

by Theorem 23.17. Now $p-1 = \frac{p}{q}$ so that by Theorem 16.14 (Hölder) we have

$$(2)\quad \int_X f(x,y)J(x)^{p-1}\,\mu(dx) = \int_X f(x,y)J(x)^{p/q}\,\mu(dx)$$
$$\leq \Big\{\int_X f(x,y)^p\,\mu(dx)\Big\}^{1/p}\Big\{\int_X J(x)^p\,\mu(dx)\Big\}^{1/q}.$$

Substituting (2) into (1) we have

$$\int_X \Big\{\int_Y f(x,y)\,\nu(dy)\Big\}^p \mu(dx) \leq \Big\{\int_X J(x)^p\,\mu(dx)\Big\}^{1/q} \int_Y \Big\{\int_X f(x,y)^p\,\mu(dx)\Big\}^{1/p} \nu(dy)$$
$$= \Big[\int_X \Big\{\int_Y f(x,y)\,\nu(dy)\Big\}^p \mu(dx)\Big]^{1/q} \int_Y \Big\{\int_X f(x,y)^p\,\mu(dx)\Big\}^{1/p} \nu(dy).$$

Dividing both sides by the first factor in the left side and recalling $1 - \frac{1}{q} = \frac{1}{p}$ we complete the proof. ∎

For $p \in [1,\infty)$, let $\|f\|_p = \big\{\int_X |f|^p\,d\mu\big\}^{1/p}$ for an extended complex-valued $\mathfrak{A}$-measurable function f on a measure space $(X, \mathfrak{A}, \mu)$. In Theorem 16.17 we proved Minkowski's Inequality for the function $\|\cdot\|_p$, that is, $\|f+g\|_p \leq \|f\|_p + \|g\|_p$. Actually this inequality is a particular case of Minkowski's Inequality for Integrals, provided that $(X, \mathfrak{A}, \mu)$ is a σ-finite measure space. We show this in the next corollary to Theorem 23.69.

Corollary 23.70. *Let $(X, \mathfrak{A}, \mu)$ be a σ-finite measure space. Let f and g be two extended complex-valued $\mathfrak{A}$-measurable functions on X such that $|f|, |g| < \infty$ a.e. on $(X, \mathfrak{A}, \mu)$. Then for $p \in [1,\infty)$, we have*

$$\|f+g\|_p \leq \|f\|_p + \|g\|_p.$$

Proof. Consider a measure space $(Y, \mathfrak{B}, \nu)$ where Y is a set with two elements, that is, $Y = \{y_1, y_2\}$, $\mathfrak{B}$ is the σ-algebra of all subsets of Y, and ν is the counting measure on $\mathfrak{B}$, that is, $\nu(\emptyset) = 0$, $\nu(\{y_1\}) = 1$, $\nu(\{y_2\}) = 1$, and $\nu(\{y_1, y_2\}) = 2$.

Consider the product measure space $\big(X \times Y, \sigma(\mathfrak{A} \times \mathfrak{B}), \mu \times \nu\big)$. Let h be a nonnegative extended real-valued $\sigma(\mathfrak{A} \times \mathfrak{B})$-measurable function on $X \times Y$ defined by

$$(1)\quad \begin{cases} h(x, y_1) = |f(x)| & \text{for } x \in X,\\ h(x, y_2) = |g(x)| & \text{for } x \in X. \end{cases}$$

For $p \in [1, \infty)$, we have by Theorem 23.69

$$(2)\quad \left[\int_X \left\{\int_Y h(x,y)\,\nu(dy)\right\}^p \mu(dx)\right]^{1/p} \le \int_Y \left\{\int_X h(x,y)^p\,\mu(dx)\right\}^{1/p} \nu(dy).$$

Now by (1) we have

$$(3)\quad \left[\int_X \left\{\int_Y h(x,y)\,\nu(dy)\right\}^p \mu(dx)\right]^{1/p} = \left[\int_X \left\{|f(x)| + |g(x)|\right\}^p \mu(dx)\right]^{1/p} = \big\| |f| + |g| \big\|_p$$

and

$$(4)\quad \int_Y \left\{\int_X h(x,y)^p\,\mu(dx)\right\}^{1/p} \nu(dy) = \|f\|_p \cdot 1 + \|g\|_p \cdot 1 = \|f\|_p + \|g\|_p.$$

Substituting (3) and (4) in (2), we have

$$(5)\quad \big\| |f| + |g| \big\|_p \le \|f\|_p + \|g\|_p.$$

Since $|f|, |g| < \infty$ a.e. on $(X, \mathfrak{A}, \mu)$, $f + g$ is defined a.e. on $(X, \mathfrak{A}, \mu)$ and we have $|f + g| \le |f| + |g|$. Then

$$(6)\quad \|f+g\|_p = \left\{\int_X |f+g|^p\,d\mu\right\}^{1/p} \le \left\{\int_X \{|f| + |g|\}^p\,d\mu\right\}^{1/p} = \big\| |f| + |g| \big\|_p.$$

By (5) and (6), we have $\|f+g\|_p \le \|f\|_p + \|g\|_p$. ∎

[V.3] Approximation by Product Simple Functions

Theorem 23.71. *Let $(X, \mathfrak{A}, \mu)$ and $(Y, \mathfrak{B}, \nu)$ be two finite measure spaces. Consider the product measure space $\big(X \times Y, \sigma(\mathfrak{A} \times \mathfrak{B}), \mu \times \nu\big)$. Let* **P** *be the collection of nonnegative functions on $X \times Y$ of the type*

$$\psi(x, y) = \sum_{j=1}^{k} c_j \mathbf{1}_{A_j}(x) \mathbf{1}_{B_j}(y)$$

where $c_j > 0$, $A_j \in \mathfrak{A}$, and $B_j \in \mathfrak{B}$ for $j = 1, \ldots, k$. Then for every nonnegative extended real-valued $\sigma(\mathfrak{A} \times \mathfrak{B})$-measurable function f on $X \times Y$, there exists an increasing sequence $(\psi_n : n \in \mathbb{N})$ in **P** *such that $\psi_n \uparrow f$ a.e. on $X \times Y$.*

Proof. By Lemma 8.6, there exists an increasing sequence $(\varphi_n : n \in \mathbb{N})$ of nonnegative simple functions on $\big(X \times Y, \sigma(\mathfrak{A} \times \mathfrak{B})\big)$ such that $\varphi_n \uparrow f$ on $X \times Y$. Let φ_n be given by

$$(1)\quad \varphi_n = \sum_{j=1}^{p_n} c_{n,j} \mathbf{1}_{E_{n,j}}$$

where $c_{n,j} > 0$ and $E_{n,j} \in \sigma(\mathfrak{A} \times \mathfrak{B})$. Let $\alpha(\mathfrak{A} \times \mathfrak{B})$ be the algebra generated by the semialgebra $\mathfrak{A} \times \mathfrak{B}$. Then $\sigma\big(\alpha(\mathfrak{A} \times \mathfrak{B})\big) = \sigma(\mathfrak{A} \times \mathfrak{B})$ and $E_{n,j} \in \sigma\big(\alpha(\mathfrak{A} \times \mathfrak{B})\big)$ so that by Theorem 20.15 for an arbitrary $\varepsilon > 0$ there exists $F_{n,j} \in \alpha(\mathfrak{A} \times \mathfrak{B})$ such that $(\mu \times \nu)(E_{n,j} \triangle F_{n,j}) < \frac{\varepsilon}{2^n p_n}$. Let

$$\psi_n = \sum_{j=1}^{p_n} c_{n,j} \mathbf{1}_{F_{n,j}}. \tag{2}$$

Let us show that $\lim_{n\to\infty} \psi_n = \lim_{n\to\infty} \varphi_n = f$ a.e. on $X \times Y$. Now we have

$$\big\{X \times Y : |\varphi_n - \psi_n| > 0\big\} \subset \bigcup_{j=1}^{p_n} \big\{X \times Y : |\mathbf{1}_{E_{n,j}} - \mathbf{1}_{F_{n,j}}| > 0\big\} = \bigcup_{j=1}^{p_n} E_{n,j} \triangle F_{n,j}.$$

This implies

$$(\mu \times \nu)\big\{X \times Y : |\varphi_n - \psi_n| > 0\big\} \leq \sum_{j=1}^{p_n} (\mu \times \nu)\big(E_{n,j} \triangle F_{n,j}\big) < \frac{\varepsilon}{2^n}.$$

Then

$$\sum_{n\in\mathbb{N}} (\mu \times \nu)\big\{X \times Y : |\varphi_n - \psi_n| > 0\big\} < \sum_{n\in\mathbb{N}} \frac{\varepsilon}{2^n} < \infty$$

so that by Theorem 6.6 (Borel-Cantelli), we have

$$(\mu \times \nu)\big(\limsup_{n\to\infty} \{X \times Y : |\varphi_n - \psi_n| > 0\}\big) = 0.$$

By Lemma 1.7, this is equivalent to

$$(\mu \times \nu)\big\{X \times Y : \varphi_n \neq \psi_n \text{ for infinitely many } n \in \mathbb{N}\big\} = 0.$$

Thus $\lim_{n\to\infty} \psi_n = \lim_{n\to\infty} \varphi_n$ a.e. on $X \times Y$ and then $\lim_{n\to\infty} \psi_n = f$ a.e. on $X \times Y$.

To complete the proof, it remains to show that $\psi_n \in \mathbf{P}$ for $n \in \mathbb{N}$. By Theorem 21.4 and Lemma 21.3, every set $F \in \alpha(\mathfrak{A} \times \mathfrak{B})$ is a finite disjoint union of members of the semialgebra $\mathfrak{A} \times \mathfrak{B}$, that is, $F = (A_1 \times B_1) \cup \cdots \cup (A_k \times B_k)$ where $\{A_1 \times B_1, \ldots, A_k \times B_k\}$ is a disjoint collection in $\mathfrak{A} \times \mathfrak{B}$. Then

$$\mathbf{1}_F(x, y) = \sum_{j=1}^{k} \mathbf{1}_{A_j \times B_j}(x, y) = \sum_{j=1}^{k} \mathbf{1}_{A_j}(x) \mathbf{1}_{B_j}(y).$$

This shows that ψ_n given by (2) is a member of $\mathbf{P}$. ∎

Problems

Prob. 23.1. Consider the product measure space $(\mathbb{R}\times\mathbb{R}, \sigma(\mathfrak{B}_{\mathbb{R}}\times\mathfrak{B}_{\mathbb{R}}), \mu_L\times\mu_L)$. Let $D=\{(x,y)\in\mathbb{R}\times\mathbb{R} : x=y\}$. Show that $D\in\sigma(\mathfrak{B}_{\mathbb{R}}\times\mathfrak{B}_{\mathbb{R}})$ and $(\mu_L\times\mu_L)(D)=0$.

Prob. 23.2. Given the product measure space $(\mathbb{R}\times\mathbb{R}, \sigma(\mathfrak{B}_{\mathbb{R}}\times\mathfrak{B}_{\mathbb{R}}), \mu_L\times\mu_L)$. Let f be a real-valued continuous function on $\mathbb{R}$. Consider the graph of f which is a subset of $\mathbb{R}\times\mathbb{R}$ defined by

$$G=\{(x,y)\in\mathbb{R}\times\mathbb{R} : y=f(x) \text{ for } x\in\mathbb{R}\}.$$

Show that $G\in\sigma(\mathfrak{B}_{\mathbb{R}}\times\mathfrak{B}_{\mathbb{R}})$ and $(\mu_L\times\mu_L)(G)=0$.

Prob. 23.3. Given the product measure space $(\mathbb{R}\times\mathbb{R}, \sigma(\mathfrak{B}_{\mathbb{R}}\times\mathfrak{B}_{\mathbb{R}}), \mu_L\times\mu_L)$. Let f be a real-valued function of bounded variation on $[a,b]$. Consider the graph of f defined by

$$G=\{(x,y)\in\mathbb{R}\times\mathbb{R} : y=f(x) \text{ for } x\in[a,b]\}.$$

Show that $G\in\sigma(\mathfrak{B}_{\mathbb{R}}\times\mathfrak{B}_{\mathbb{R}})$ and $(\mu_L\times\mu_L)(G)=0$.

Prob. 23.4. Let $(X,\mathfrak{A},\mu)$ be a finite measure space and $(Y,\mathfrak{B},\nu)$ be a non σ-finite measure space given by

$$X=Y=[0,1],$$
$$\mathfrak{A}=\mathfrak{B}=\mathfrak{B}_{[0,1]}, \text{ the } \sigma\text{-algebra of the Borel sets in } [0,1],$$
$$\mu=\mu_L \text{ and } \nu \text{ is the counting measure.}$$

Consider the product measurable space $(X\times Y, \sigma(\mathfrak{A}\times\mathfrak{B}))$ and a subset in it defined by $E=\{(x,y)\in X\times Y : x=y\}$.

(a) Show that $E\in\sigma(\mathfrak{A}\times\mathfrak{B})$.

(b) Show that

$$\int_X\left\{\int_Y \mathbf{1}_E\,d\nu\right\}d\mu \neq \int_Y\left\{\int_X \mathbf{1}_E\,d\mu\right\}d\nu,$$

thereby showing that the σ-finiteness condition on $(X,\mathfrak{A},\mu)$ and $(Y,\mathfrak{B},\nu)$ in Theorem 23.17 cannot be dropped.

Prob. 23.5. Given two σ-finite measure spaces $(X,\mathfrak{A},\mu)$ and $(Y,\mathfrak{B},\nu)$ where

$$X=Y=\mathbb{Z}_+,$$
$$\mathfrak{A}=\mathfrak{B}=\mathfrak{P}(\mathbb{Z}_+), \text{ the } \sigma\text{-algebra of all subsets of } \mathbb{Z}_+,$$
$$\mu \text{ and } \nu \text{ are the counting measures.}$$

Consider the product measure space $(X\times Y, \sigma(\mathfrak{A}\times\mathfrak{B}), \mu\times\nu)$. Define a function f on $X\times Y$ by

$$f(x,y)=\begin{cases} 1+2^{-x} & \text{when } x=y,\\ -1-2^{-x} & \text{when } x=y+1,\\ 0 & \text{otherwise.}\end{cases}$$

Show that

(a) $\int_X\{\int_Y f\,d\nu\}\,d\mu \neq \int_Y\{\int_X f\,d\mu\}\,d\nu$,

(b) $\int_{X\times Y}|f|\,d(\mu\times\nu)=\infty$,

(c) $\int_{X\times Y} f\,d(\mu\times\nu)$ does not exist.

Prob. 23.6. Consider the product measure space $(X\times Y, \sigma(\mathfrak{A}\times\mathfrak{B}), \mu\times\nu)$ of two σ-finite measure spaces $(X, \mathfrak{A}, \mu)$ and $(Y, \mathfrak{B}, \nu)$. Let f be an extended real-valued $\mathfrak{A}$-measurable and μ-integrable function on X and g be an extended real-valued $\mathfrak{B}$-measurable and ν-integrable function on Y. Let

$$h(x, y) = \begin{cases} f(x)g(y) & \text{for } (x, y) \in X \times Y \text{ for which the product exists,} \\ 0 & \text{otherwise.} \end{cases}$$

(a) Show that h is $\sigma(\mathfrak{A} \times \mathfrak{B})$-measurable on $X \times Y$.
(b) Show that h is $\mu \times \nu$ integrable on $X \times Y$ and

$$\int_{X \times Y} h \, d(\mu \times \nu) = \left\{\int_X f \, d\mu\right\}\left\{\int_Y g \, d\nu\right\}.$$

Prob. 23.7. Let D be a set contained in $[a, b] \times [c, d] \subset \mathbb{R} \times \mathbb{R}$ defined by

$$D = \{(x, y) \in \mathbb{R} \times \mathbb{R} : y \in [g_1(x), g_2(x)] \text{ for } x \in [a, b]\},$$

and suppose that D can also be given as

$$D = \{(x, y) \in \mathbb{R} \times \mathbb{R} : x \in [h_1(y), h_2(y)] \text{ for } y \in [c, d]\},$$

where g_1 and g_2 are real-valued continuous functions on $[a, b]$ such that $c \leq g_1(x) \leq g_2(x) \leq d$ for $x \in [a, b]$, and h_1 and h_2 are real valued continuous function on $[c, d]$ such that $a \leq h_1(y) \leq h_2(y) \leq b$ for $y \in [c, d]$.
(a) Show that $D \in \sigma(\mathfrak{B}_{\mathbb{R}} \times \mathfrak{B}_{\mathbb{R}})$.
(b) Show that if f is a real-valued continuous function on D, then f is $\mu_L \times \mu_L$-integrable on D and furthermore

$$\begin{aligned} \int_D f \, d(\mu_L \times \mu_L) &= \int_{[a,b]} \left\{ \int_{[g_1(x), g_2(x)]} f(x, y) \, \mu_L(dy) \right\} \mu_L(dx) \\ &= \int_{[c,d]} \left\{ \int_{[h_1(y), h_2(y)]} f(x, y) \, \mu_L(dx) \right\} \mu_L(dy). \end{aligned}$$

Prob. 23.8. Evaluate the following integrals:

(a) $\int_0^2 \int_x^2 y^2 \sin xy \, dy \, dx$

(b) $\int_0^\pi \int_x^\pi \frac{\sin x}{y} \, dy \, dx$

(c) $\int_0^1 \int_{2y}^2 \cos(x^2) \, dx \, dy$

(d) $\int_0^1 \int_y^1 x^2 e^{xy} \, dx \, dy$

(e) $\int_0^8 \int_{\sqrt[3]{x}}^2 \frac{1}{y^4+1} \, dy \, dx$

(f) $\int_0^2 \int_0^{4-x^2} \frac{xe^{2y}}{4-y} \, dy \, dx.$

Prob. 23.9. Let f be a real valued $\sigma(\mathfrak{M}_L \times \mathfrak{M}_L)$-measurable function on $\mathbb{R} \times \mathbb{R}$ satisfying the condition $\int_{\mathbb{R}} \left\{ \int_{\mathbb{R}} |f(x, y)| \mu_L(dy) \right\} \mu_L(dx) < \infty$. Suppose
1° f is continuous on $\mathbb{R} \times \mathbb{R}$,
2° for every $x \in \mathbb{R}$, the improper Riemann integral $g(x) = \int_{-\infty}^{\infty} f(x, y) \, dy$ exists as a real number,
3° g is a continuous function on $\mathbb{R}$,
4° the improper Riemann integral $\int_{-\infty}^{\infty} g(x) \, dx$ exists as a real number.
Show that

$$\int_{\mathbb{R} \times \mathbb{R}} f \, d(\mu_L \times \mu_L) = \int_{-\infty}^{\infty} \left\{ \int_{-\infty}^{\infty} f(x, y) \, dy \right\} dx$$

that is, $\int_{\mathbb{R} \times \mathbb{R}} f \, d(\mu_L \times \mu_L)$ can be evaluated as an iterated improper Riemann integral.

Prob. 23.10. From the improper Riemann integral $\int_{-\infty}^{\infty} e^{-x^2} dx = \sqrt{\pi}$ we have

$$\int_{[0,\infty)} e^{-\alpha x^2}\,\mu_L(dx) = \frac{1}{2}\sqrt{\frac{\pi}{\alpha}} \quad \text{for } \alpha \in (0,\infty).$$

(a) Differentiating the integrand with respect to α n times, we have

$$\frac{d^n}{d\alpha^n} e^{-\alpha x^2} = e^{-\alpha x^2}(-x^2)^n \quad \text{for } \alpha \in (0,\infty) \text{ and } x \in [0,\infty).$$

Note that $|e^{-\alpha x^2}(-x^2)^n| \le e^{-\alpha x^2} x^{2n}$. Show that

$$\int_{[0,\infty)} e^{-\alpha x^2} x^{2n}\,\mu_L(dx) < \infty \quad \text{for } \alpha \in (0,\infty).$$

(b) Show that for $\alpha \in (0,\infty)$, we have

$$\int_{[0,\infty)} e^{-\alpha x^2}(-x^2)^n\,\mu_L(dx) = \tfrac{1}{2}\left(-\tfrac{1}{2}\right)\left(-\tfrac{3}{2}\right)\left(-\tfrac{5}{2}\right)\cdots\left(-\tfrac{2n-1}{2}\right)\sqrt{\pi}\,\alpha^{-\frac{1}{2}-n},$$

and then

$$\int_{[0,\infty)} e^{-\alpha x^2} x^{2n}\,\mu_L(dx) = 2^{-(n+1)}\cdot 1\cdot 3\cdot 5\cdots(2n-1)\sqrt{\pi}\,\alpha^{-\frac{1}{2}-n},$$

and in particular

$$\int_{[0,\infty)} e^{-x^2} x^{2n}\,\mu_L(dx) = 2^{-(n+1)}\cdot 1\cdot 3\cdot 5\cdots(2n-1)\sqrt{\pi}.$$

(Hint: Apply Proposition 23.37.)

Prob. 23.11. Let $f \in L_1(\mathbb{R}, \mathfrak{M}_L, \mu_L)$. Show that

$$\int_{\mathbb{R}} \left\{ \int_{\mathbb{R}} f(x-y) f(y)\,\mu_L(dy) \right\} \mu_L(dx) \ge 0.$$

Prob. 23.12. Let $f \in L_1(\mathbb{R}, \mathfrak{M}_L, \mu_L)$. Suppose

$$\int_{\mathbb{R}} \left\{ \int_{\mathbb{R}} f(\alpha x - y) f(\beta y)\,\mu_L(dy) \right\} \mu_L(dx) = \gamma$$

where $\alpha, \beta, \gamma \in \mathbb{R}$ and $\alpha \neq 0$, $\beta \neq 0$. Show that $\gamma \ge 0$ and $\int_{\mathbb{R}} f\,d\mu_L = \pm\sqrt{|\alpha\beta|\gamma}$.

Prob. 23.13. Let $f \in L^1(\mathbb{R}, \mathfrak{M}_L, \mu_L)$. With $h > 0$ fixed, let us define a function φ_h on $\mathbb{R}$ by setting

$$\varphi_h(x) = \frac{1}{2h}\int_{[x-h,x+h]} f(t)\,\mu_L(dt) \text{ for } x \in \mathbb{R}.$$

(a) Show that φ_h is a real-valued continuous function on $\mathbb{R}$.
(b) Show that $\varphi_h \in L^1(\mathbb{R}, \mathfrak{M}_L, \mu_L)$ and $\|\varphi_h\|_1 \le \|f\|_1$.

Prob. 23.14. Let $(X, \mathfrak{A}, \mu)$ be a σ-finite measure space. Let $D \in \mathfrak{A}$ and let f be a nonnegative real-valued $\mathfrak{A}$-measurable function on D. Consider the σ-finite measure space $(\mathbb{R}, \mathfrak{B}_{\mathbb{R}}, \mu_L)$ and the product measure space $(X \times \mathbb{R}, \sigma(\mathfrak{A} \times \mathfrak{B}_{\mathbb{R}})\mu \times \mu_L)$. Let E be a subset of $X \times \mathbb{R}$ such that

$$\begin{cases} E(x,\cdot) = \emptyset & \text{for } x \in D^c, \\ E(x,\cdot) = [0, f(x)) & \text{for } x \in D. \end{cases}$$

Show that $E \in \sigma(\mathfrak{A} \times \mathfrak{B}_{\mathbb{R}})$ and moreover $(\mu \times \mu_L)(E) = \int_D f(x)\,\mu(dx)$.

Chapter 6

Measure and Integration on the Euclidean Space

§24 Lebesgue Measure Space on the Euclidean Space

[I] Lebesgue Outer Measure on the Euclidean Space

Let $\mathbb{R}^n$ be the Euclidean n-space. The topology on $\mathbb{R}^n$ is the metric topology derived from the Euclidean metric $d(x, y) = \left\{\sum_{j=1}^n |x_j - y_j|^2\right\}^{1/2}$ for $x = (x_1, \ldots, x_n)$ and $y = (y_1, \ldots, y_n)$ in $\mathbb{R}^n$. To fix the notations, let us repeat part of Definition 3.21.

Definition 24.1. *Let $\mathfrak{I}_o$ be the collection of all open intervals and $\emptyset$ in $\mathbb{R}$. Similarly let $\mathfrak{I}_c$, $\mathfrak{I}_{oc}$, and $\mathfrak{I}_{co}$ be the collections of all closed intervals, all intervals of the type $(a, b]$, and all intervals of the type $[a, b)$ respectively, and $\emptyset$, where $-\infty \le a < b \le \infty$ with $(a, \infty] := (a, \infty)$ and $[-\infty, b] := (-\infty, b]$. Let $\mathfrak{I} = \mathfrak{I}_o \cup \mathfrak{I}_c \cup \mathfrak{I}_{oc} \cup \mathfrak{I}_{co}$, the collection of all intervals and $\emptyset$.*

Notations. (a) Given n sets $E_1, \ldots, E_n$, we write $\times_{i=1}^n E_i$ for the Cartesian product $E_1 \times \cdots \times E_n$. We write $\times_{i=1}^n E$ for the n-fold Cartesian product of a set E.
(b) For n classes of sets $\mathfrak{E}_1, \ldots, \mathfrak{E}_n$, we write $\times_{i=1}^n \mathfrak{E}_i$ for

$$\mathfrak{E}_1 \times \cdots \times \mathfrak{E}_n = \left\{E_1 \times \cdots \times E_n : E_i \in \mathfrak{E}_i, i = 1, \ldots, n\right\}.$$

We write $\times_{i=1}^n \mathfrak{E}$ for the n-fold Cartesian product of a class $\mathfrak{E}$ of sets, that is, $\mathfrak{E} \times \cdots \times \mathfrak{E} = \{E_1 \times \cdots \times E_n : E_i \in \mathfrak{E}, i = 1, \ldots, n\}$.

Definition 24.2. *We define classes of subsets of $\mathbb{R}^n$ by letting*

$$\mathfrak{I}_{oc}^n = \times_{i=1}^n \mathfrak{I}_{oc}, \quad \mathfrak{I}_{co}^n = \times_{i=1}^n \mathfrak{I}_{co}, \quad \mathfrak{I}_o^n = \times_{i=1}^n \mathfrak{I}_o, \quad \mathfrak{I}_c^n = \times_{i=1}^n \mathfrak{I}_c,$$

and

$$\mathfrak{I}^n = \mathfrak{I}_{oc}^n \cup \mathfrak{I}_{co}^n \cup \mathfrak{I}_o^n \cup \mathfrak{I}_c^n.$$

Proposition 24.3. *$\mathfrak{J}^n_{oc}$ and $\mathfrak{J}^n_{co}$ are semialgebras of subsets of $\mathbb{R}^n$.*

Proof. It is obvious that $\mathfrak{J}^n_{oc}$ satisfies conditions 1° and 2° of Definition 21.1. Let us verify 3° for $\mathfrak{J}^n_{oc}$. Take an arbitrary member of $\mathfrak{J}^n_{oc}$ given by $R = \times_{j=1}^n (a_j, b_j]$. We let $I_{j,0} = (-\infty, a_j]$, $I_{j,1} = (a_j, b_j]$, and $I_{j,2} = (b_j, \infty]$ with $(-\infty, -\infty] := \emptyset$ and $(\infty, \infty] := \emptyset$ for $j = 1, \dots, n$. Then $\{I_{1,j_1} \times \cdots \times I_{n,j_n} : j_1, \dots, j_n = 0, 1, 2\}$ is a disjoint collection of members of $\mathfrak{J}^n_{oc}$ whose union is equal to $\mathbb{R}^n$. Since $R = I_{1,1} \times \cdots \times I_{n,1}$, R^c is a finite disjoint union of members of $\mathfrak{J}^n_{oc}$. This shows that $\mathfrak{J}^n_{oc}$ is a semialgebra of subsets of $\mathbb{R}^n$. By similar argument, $\mathfrak{J}^n_{co}$ is a semialgebra of subsets of $\mathbb{R}^n$. ∎

Note that $\mathfrak{J}^n_o$ and $\mathfrak{J}^n_c$ fail to satisfy condition 3° of Definition 21.1.

Proposition 24.4. *$\sigma(\mathfrak{J}^n_{oc}) = \mathfrak{B}_{\mathbb{R}^n}$, that is, the smallest σ-algebra of subsets of $\mathbb{R}^n$ containing $\mathfrak{J}^n_{oc}$ is equal to the smallest σ-algebra of subsets of $\mathbb{R}^n$ containing the open sets in $\mathbb{R}^n$. Similarly $\sigma(\mathfrak{J}^n_{co}) = \mathfrak{B}_{\mathbb{R}^n}$, $\sigma(\mathfrak{J}^n_o) = \mathfrak{B}_{\mathbb{R}^n}$, $\sigma(\mathfrak{J}^n_c) = \mathfrak{B}_{\mathbb{R}^n}$ and $\sigma(\mathfrak{J}^n) = \mathfrak{B}_{\mathbb{R}^n}$.*

Proof. Let us prove $\sigma(\mathfrak{J}^n_{oc}) = \mathfrak{B}_{\mathbb{R}^n}$. To show that $\sigma(\mathfrak{J}^n_{oc}) \subset \mathfrak{B}_{\mathbb{R}^n}$, take an arbitrary member $R = \times_{j=1}^n (a_j, b_j]$ of $\mathfrak{J}^n_{oc}$. For $k \in \mathbb{N}$, let $O_k = \times_{j=1}^n \left(a_j, b_j + \frac{1}{k}\right)$. Then $(O_k : k \in \mathbb{N})$ is a decreasing sequence of open sets in $\mathbb{R}^n$ and $\bigcap_{k\in\mathbb{N}} O_k = R$. Thus $R \in \mathfrak{B}_{\mathbb{R}^n}$. This shows that $\mathfrak{J}^n_{oc} \subset \mathfrak{B}_{\mathbb{R}^n}$. Consequently we have $\sigma(\mathfrak{J}^n_{oc}) \subset \mathfrak{B}_{\mathbb{R}^n}$.

Let us prove the reverse inclusion $\mathfrak{B}_{\mathbb{R}^n} \subset \sigma(\mathfrak{J}^n_{oc})$. If we let $\mathfrak{R}$ be the subcollection of $\mathfrak{J}^n_{oc}$ consisting of members $\times_{j=1}^n (a_j, b_j]$ of $\mathfrak{J}^n_{oc}$ with rational a_j and b_j for $j = 1, \dots, n$, then $\mathfrak{R}$ is a countable collection. Let $\mathfrak{O}$ be the collection of all open sets in $\mathbb{R}^n$. Now if $O \in \mathfrak{O}$ and $O \neq \emptyset$, then every point in O is contained in a member of $\mathfrak{R}$ which is contained in O. Thus O is equal to a union of members of $\mathfrak{R}$. Since $\mathfrak{R}$ is a countable subcollection of $\mathfrak{J}^n_{oc}$, O is a countable union of members of $\mathfrak{J}^n_{oc}$. Thus $O \in \sigma(\mathfrak{J}^n_{oc})$. Then $\mathfrak{O} \subset \sigma(\mathfrak{J}^n_{oc})$. This implies that $\mathfrak{B}_{\mathbb{R}^n} = \sigma(\mathfrak{O}) \subset \sigma(\mathfrak{J}^n_{oc})$. Therefore $\mathfrak{B}_{\mathbb{R}^n} \subset \sigma(\mathfrak{J}^n_{oc})$.

The equalities $\sigma(\mathfrak{J}^n_{co}) = \mathfrak{B}_{\mathbb{R}^n}$, $\sigma(\mathfrak{J}^n_o) = \mathfrak{B}_{\mathbb{R}^n}$, and $\sigma(\mathfrak{J}^n_c) = \mathfrak{B}_{\mathbb{R}^n}$ are proved by similar arguments. Finally from $\mathfrak{J}^n_{oc} \subset \mathfrak{J}^n \subset \mathfrak{B}_{\mathbb{R}^n}$, we have $\mathfrak{B}_{\mathbb{R}^n} = \sigma(\mathfrak{J}^n_{oc}) \subset \sigma(\mathfrak{J}^n) \subset \mathfrak{B}_{\mathbb{R}^n}$ so that $\sigma(\mathfrak{J}^n) = \mathfrak{B}_{\mathbb{R}^n}$. ∎

Each one of the four classes $\mathfrak{J}^n_{oc}$, $\mathfrak{J}^n_{co}$, $\mathfrak{J}^n_o$ and $\mathfrak{J}^n_c$ is a covering class of subsets of $\mathbb{R}^n$. We define the Lebesgue outer measure on $\mathbb{R}^n$ as follows.

Definition 24.5. *Let us define a nonnegative extended real-valued set function v on $\mathfrak{J}^n = \mathfrak{J}^n_{oc} \cup \mathfrak{J}^n_{co} \cup \mathfrak{J}^n_o \cup \mathfrak{J}^n_c$ by setting $v(\emptyset) = 0$ and $v(E) = \prod_{j=1}^n (b_j - a_j)$ for $E = \times_{j=1}^n (a_j, b_j]$, $\times_{j=1}^n [a_j, b_j)$, $\times_{j=1}^n (a_j, b_j)$, or $\times_{j=1}^n [a_j, b_j]$. With the set function v restricted to $\mathfrak{J}^n_{oc}$, let us define an outer measure $(\mu^n_L)^*$ on $\mathbb{R}^n$ by setting for every $E \in \mathfrak{P}(\mathbb{R}^n)$*

$$(\mu^n_L)^*(E) = \inf\left\{ \sum_{k\in\mathbb{N}} v(R_k) : (R_k : k \in \mathbb{N}) \subset \mathfrak{J}^n_{oc}, \bigcup_{k\in\mathbb{N}} R_k \supset E \right\}.$$

Let us call $(\mu^n_L)^*$ the Lebesgue outer measure on $\mathbb{R}^n$. (The fact that $(\mu^n_L)^*$ is indeed an outer measure on $\mathbb{R}^n$ is by Theorem 2.21.)

We refer to a member E of $\mathfrak{J}^n = \mathfrak{J}^n_{oc} \cup \mathfrak{J}^n_{co} \cup \mathfrak{J}^n_o \cup \mathfrak{J}^n_c$ as a box in $\mathbb{R}^n$ and $v(E)$ as the volume of E. We show next that $\mathfrak{J}^n_{oc}$ in the definition of $(\mu^n_L)^*$ may be replaced by $\mathfrak{J}^n_{co}$, or $\mathfrak{J}^n_o$, or $\mathfrak{J}^n_c$.

Proposition 24.6. *For every $E \in \mathfrak{P}(\mathbb{R}^n)$, we have*

(1) $(\mu^n_L)^*(E) = \inf\left\{\sum_{k\in\mathbb{N}} v(L_k) : (L_k : k \in \mathbb{N}) \subset \mathfrak{J}^n_{co}, \bigcup_{k\in\mathbb{N}} L_k \supset E\right\}$,

(2) $(\mu^n_L)^*(E) = \inf\left\{\sum_{k\in\mathbb{N}} v(O_k) : (O_k : k \in \mathbb{N}) \subset \mathfrak{J}^n_o, \bigcup_{k\in\mathbb{N}} O_k \supset E\right\}$,

(3) $(\mu^n_L)^*(E) = \inf\left\{\sum_{k\in\mathbb{N}} v(C_k) : (C_k : k \in \mathbb{N}) \subset \mathfrak{J}^n_c, \bigcup_{k\in\mathbb{N}} C_k \supset E\right\}$.

Proof. To prove (1), let us write $\lambda(E)$ for the right hand side of (1). Let $\varepsilon > 0$ be arbitrarily given. Let $(R_k : k \in \mathbb{N})$ be an arbitrary sequence in $\mathfrak{J}^n_{oc}$ such that $\bigcup_{k\in\mathbb{N}} R_k \supset E$. Let $(L_k : k \in \mathbb{N})$ be an arbitrary sequence in $\mathfrak{J}^n_{co}$ such that $L_k \supset R_k$ and $v(L_k) \leq v(R_k) + \varepsilon/2^k$ for $k \in \mathbb{N}$. Then $\sum_{k\in\mathbb{N}} v(L_k) \leq \sum_{k\in\mathbb{N}} v(R_k) + \varepsilon$. Thus by the definition of $\lambda(E)$ as the infimum in (1), we have $\lambda(E) \leq \sum_{k\in\mathbb{N}} v(R_k) + \varepsilon$. Since this holds for an arbitrary sequence $(R_k : k \in \mathbb{N})$ in $\mathfrak{J}^n_{oc}$ such that $\bigcup_{k\in\mathbb{N}} R_k \supset E$, we have $\lambda(E) \leq (\mu^n_L)^*(E) + \varepsilon$. By the arbitrariness of $\varepsilon > 0$, we have $\lambda(E) \leq (\mu^n_L)^*(E)$. Interchanging the roles of $\mathfrak{J}^n_{oc}$ and $\mathfrak{J}^n_{co}$ in the argument above, we have $(\mu^n_L)^*(E) \leq \lambda(E)$. Thus $\lambda(E) = (\mu^n_L)^*(E)$. This proves (1). The equalities (2) and (3) are proved likewise. ∎

Let us observe that the expression (2) has the advantage that the covering $\bigcup_{k\in\mathbb{N}} O_k \supset E$ is an open cover. If E is a compact set, then the open cover has a finite subcover.

Proposition 24.7. *For every $E \in \mathfrak{J}^n = \mathfrak{J}^n_{oc} \cup \mathfrak{J}^n_{co} \cup \mathfrak{J}^n_o \cup \mathfrak{J}^n_c$, we have $(\mu^n_L)^*(E) = v(E)$.*

Proof. 1. Let us consider first the case that $E \in \mathfrak{J}^n$ is a finite box, that is, every edge of the box is a finite interval, or equivalently, $v(E) < \infty$.

If $E \in \mathfrak{J}^n_{oc}$, then (E) is a one-term covering sequence of E in $\mathfrak{J}^n_{oc}$. Thus by Definition 24.5, we have $(\mu^n_L)^*(E) \leq v(E)$. If $E \in \mathfrak{J}^n_{co}$, then (E) is a one-term covering sequence of E in $\mathfrak{J}^n_{co}$. Thus by (1) of Proposition 24.6, we have $(\mu^n_L)^*(E) \leq v(E)$. Similarly if $E \in \mathfrak{J}^n_o$ or if $E \in \mathfrak{J}^n_c$, then we have $(\mu^n_L)^*(E) \leq v(E)$ by (2) or (3) of Proposition 24.6 respectively. Thus for every finite box $E \in \mathfrak{J}^n$, we have $(\mu^n_L)^*(E) \leq v(E)$. It remains to prove the reverse inequality:

$$(1) \qquad v(E) \leq (\mu^n_L)^*(E).$$

If $E \in \mathfrak{J}^n_c$, let $(O_k : k \in \mathbb{N})$ be an arbitrary sequence in $\mathfrak{J}^n_o$ such that $\bigcup_{k\in\mathbb{N}} O_k \supset E$. Since E is a compact set in $\mathbb{R}^n$, there exists $N \in \mathbb{N}$ such that $\bigcup_{k=1}^N O_k \supset E$. Then $E = E \cap \left(\bigcup_{k=1}^N O_k\right) = \bigcup_{k=1}^N (E \cap O_k)$ and $v(E) \leq \sum_{k=1}^N v(E \cap O_k) \leq \sum_{k=1}^N v(O_k)$. Thus by the arbitrariness of the sequence $(O_k : k \in \mathbb{N})$ in $\mathfrak{J}^n_o$, we have

$$v(E) \leq \inf\left\{\sum_{k\in\mathbb{N}} v(O_k) : (O_k : k \in \mathbb{N}) \subset \mathfrak{J}^n_o, \bigcup_{k\in\mathbb{N}} O_k \supset E\right\} = (\mu^n_L)^*(E),$$

where the last equality is by (2) of Proposition 24.6. Thus (1) holds in this case.
If $E \in \mathfrak{J}^n_o$, let E be given by $E = \times_{i=1}^n (a_i, b_i)$ where $-\infty < a_i < b_i < \infty$ for

$i = 1, \ldots, n$. For $k \in \mathbb{N}$, let $C_k = \times_{i=1}^{n} \left[a_i + \frac{1}{k}, b_i - \frac{1}{k}\right]$. Then $(C_k : k \in \mathbb{N})$ is an increasing sequence in $\mathfrak{I}_c^n$ and $v(C_k) \uparrow v(E)$. Since $C_k \in \mathfrak{I}_c^n$, we have $v(C_k) = \left(\mu_L^n\right)^*(C_k)$ by our result above. Since $C_k \subset E$, we have $\left(\mu_L^n\right)^*(C_k) \leq \left(\mu_L^n\right)^*(E)$ by the monotonicity of the outer measure $\left(\mu_L^n\right)^*$. Thus we have $v(E) = \lim_{k\to\infty} v(C_k) \leq \left(\mu_L^n\right)^*(E)$, which proves (1) for this case.

If $E \in \mathfrak{I}_{oc}^n$ and E is given by $E = \times_{i=1}^{n}(a_i, b_i]$ where $-\infty < a_i < b_i < \infty$ for $i = 1, \ldots, n$, we prove (1) by the same argument as above using the same sequence $(C_k : k \in \mathbb{N})$ as above. Similarly for $E \in \mathfrak{I}_{co}^n$.

2. If $E \in \mathfrak{I}^n$ is an infinite box, then $v(E) = \infty$ and for every $k \in \mathbb{N}$ there exists a finite box $E_k \in \mathfrak{I}^n$ such that $E_k \subset E$ and $v(E_k) \geq k$. Since E_k is a finite box, we have $v(E_k) = \left(\mu_L^n\right)^*(E_k)$ by our result in **1.** Since $E_k \subset E$, we have $\left(\mu_L^n\right)^*(E_k) \leq \left(\mu_L^n\right)^*(E)$ by the monotonicity on $\left(\mu_L^n\right)^*$. Thus $k \leq \left(\mu_L^n\right)^*(E)$ for every $k \in \mathbb{N}$ and hence $\left(\mu_L^n\right)^*(E) = \infty$. Therefore $\left(\mu_L^n\right)^*(E) = v(E)$. ∎

Definition 24.8. *Let $\mathfrak{M}_L^n = \mathfrak{M}\left(\left(\mu_L^n\right)^*\right)$, the σ-algebra of subsets of $\mathbb{R}^n$ consisting of the $\left(\mu_L^n\right)^*$-measurable subsets of $\mathbb{R}^n$, and call it the Lebesgue σ-algebra of subsets of $\mathbb{R}^n$. We call the members of $\mathfrak{M}_L^n$ the Lebesgue measurable subsets of $\mathbb{R}^n$. Let μ_L^n be the restriction of $\left(\mu_L^n\right)^*$ to $\mathfrak{M}_L^n$ and call it the Lebesgue measure on $\mathbb{R}^n$. We call $\left(\mathbb{R}^n, \mathfrak{M}_L^n\right)$ the Lebesgue measurable space and $\left(\mathbb{R}^n, \mathfrak{M}_L^n, \mu_L^n\right)$ the Lebesgue measure space on $\mathbb{R}^n$.*

Thus defined, the Lebesgue measure space $\left(\mathbb{R}^n, \mathfrak{M}_L^n, \mu_L^n\right)$ is a complete measure space by Theorem 2.9. We show next that the Lebesgue outer measure $\left(\mu_L^n\right)^*$ on $\mathbb{R}^n$ is a metric outer measure. This then implies that $\mathfrak{B}_{\mathbb{R}^n} \subset \mathfrak{M}_L^n$ by Theorem 2.19, that is, every Borel set in $\mathbb{R}^n$ is a Lebesgue measurable set.

Let us note that while $\mathbb{R}^n$, $\mathfrak{I}_{oc}^n$, $\mathfrak{I}_{co}^n$, $\mathfrak{I}_o^n$ and $\mathfrak{I}_c^n$ are defined as n-fold Cartesian products of $\mathbb{R}^n$, $\mathfrak{I}_{oc}$, $\mathfrak{I}_{co}$, $\mathfrak{I}_o$ and $\mathfrak{I}_c$ respectively, $\mathfrak{M}_L^n$ is not defined as the n-fold Cartesian product of $\mathfrak{M}_L$. It will be shown that $\mathfrak{M}_L^n$ is the completion of the n-fold Cartesian product of $\mathfrak{M}_L$ with respect to the measure μ_L^n.

Theorem 24.9. *The Lebesgue outer measure $\left(\mu_L^n\right)^*$ on $\mathbb{R}^n$ is a metric outer measure. In particular the Borel σ-algebra $\mathfrak{B}_{\mathbb{R}^n}$ is contained in the Lebesgue σ-algebra $\mathfrak{M}_L^n$.*

Proof. To show that $\left(\mu_L^n\right)^*$ is a metric outer measure we show that if $E_1, E_2 \in \mathfrak{P}(\mathbb{R}^n)$ are such that $d(E_1, E_2) > 0$, then $\left(\mu_L^n\right)^*(E_1 \cup E_2) = \left(\mu_L^n\right)^*(E_1) + \left(\mu_L^n\right)^*(E_2)$. Since $\left(\mu_L^n\right)^*$ is subadditive, it suffices to show

(1) $$\left(\mu_L^n\right)^*(E_1 \cup E_2) \geq \left(\mu_L^n\right)^*(E_1) + \left(\mu_L^n\right)^*(E_2).$$

Let us show that for every sequence $(R_k : k \in \mathbb{N})$ in $\mathfrak{I}_{oc}^n$ such that $\bigcup_{k\in\mathbb{N}} R_k \supset E_1 \cup E_2$, we have

(2) $$\sum_{k\in\mathbb{N}} v(R_k) \geq \left(\mu_L^n\right)^*(E_1) + \left(\mu_L^n\right)^*(E_2).$$

Now if $v(R_k) = \infty$ for some $k \in \mathbb{N}$, then (2) holds trivially. Thus consider the case that $v(R_k) < \infty$ for every $k \in \mathbb{N}$. Let $\delta = d(E_1, E_2) > 0$. Now for each $k \in \mathbb{N}$, R_k can be decomposed into a finite collection $\{R_{k,1}, \ldots, R_{k,p_k}\}$ in $\mathfrak{I}^n_{oc}$ in such a way that

1° $\{R_{k,1}, \ldots, R_{k,p_k}\}$ is a disjoint collection and $R_k = R_{k,1} \cup \cdots \cup R_{k,p_k}$,

2° the lengths of the edges of $R_{k,1}, \ldots, R_{k,p_k}$ are all bounded above by $\delta/\left(2\sqrt{n}\right)$.

Note that 1° implies that $v(R_k) = v(R_{k,1}) + \cdots + v(R_{k,p_k})$. Let the countable collection $\mathfrak{R} = \{R_{k,j_k} : j_k = 1, \ldots, p_k;\, k \in \mathbb{N}\}$ be arbitrarily enumerated as $\mathfrak{R} = \{F_m : m \in \mathbb{N}\}$. We have $\bigcup_{m\in\mathbb{N}} F_m \supset E_1 \cup E_2$ and

$$(3) \qquad \sum_{m\in\mathbb{N}} v(F_m) = \sum_{k\in\mathbb{N}} \sum_{j_k=1}^{p_k} v(R_{k,j_k}) = \sum_{k\in\mathbb{N}} v(R_k).$$

Let us show that no member of $\mathfrak{R} = \{F_m : m \in \mathbb{N}\}$ can intersect both E_1 and E_2. Assume the contrary, that is, for some $m \in \mathbb{N}$ the set F_m intersects both E_1 and E_2. Then $F_m \cap E_1 \neq \emptyset$ and $F_m \cap E_2 \neq \emptyset$. Let $x_1 \in F_m \cap E_1$ and $x_2 \in F_m \cap E_2$. Then $\delta = d(E_1, E_2) \leq d(x_1, x_2)$. Now since $x_1, x_2 \in F_m$ and every edge of F_m has length $\leq \delta/\left(2\sqrt{n}\right)$, we have $d(x_1, x_2) \leq \left[n\left(\delta/\left(2\sqrt{n}\right)\right)^2\right]^{1/2} = \frac{\delta}{2}$. Thus we have $\delta \leq d(x_1, x_2) \leq \delta/2$, a contradiction. This shows that no member of $\mathfrak{R} = \{F_m : m \in \mathbb{N}\}$ can intersect both E_1 and E_2. Let us classify $\{F_m : m \in \mathbb{N}\}$ into three subclasses: $\{F'_m : m \in \mathbb{N}\}$ consisting of those F_m which intersect E_1, $\{F''_m : m \in \mathbb{N}\}$ consisting of those F_m which intersect E_2, and $\{F'''_m : m \in \mathbb{N}\}$ consisting of those F_m which are disjoint from both E_1 and E_2. Then we have $E_1 \subset \bigcup_{m\in\mathbb{N}} F'_m$ and $E_2 \subset \bigcup_{m\in\mathbb{N}} F''_m$ and therefore

$$(4) \qquad \begin{aligned} \sum_{m\in\mathbb{N}} v(F_m) &= \sum_{m\in\mathbb{N}} v(F'_m) + \sum_{m\in\mathbb{N}} v(F''_m) + \sum_{m\in\mathbb{N}} v(F'''_m) \\ &\geq \sum_{m\in\mathbb{N}} v(F'_m) + \sum_{m\in\mathbb{N}} v(F''_m) \\ &\geq \left(\mu^n_L\right)^*(E_1) + \left(\mu^n_L\right)^*(E_2). \end{aligned}$$

By (3) and (4), we have (2). Now since (2) holds for every sequence $(R_k : k \in \mathbb{N})$ in $\mathfrak{I}^n_{oc}$ such that $\bigcup_{k\in\mathbb{N}} R_k \supset E_1 \cup E_2$, we have $\left(\mu^n_L\right)^*(E_1 \cup E_2) \geq \left(\mu^n_L\right)^*(E_1) + \left(\mu^n_L\right)^*(E_2)$. This proves (1) and completes the proof that $\left(\mu^n_L\right)^*$ is a metric outer measure. Then by Theorem 2.19, we have $\mathfrak{B}_{\mathbb{R}^n} \subset \mathfrak{M}\left(\left(\mu^n_L\right)^*\right) = \mathfrak{M}^n_L$. ∎

The following decomposition of a nonempty open set in $\mathbb{R}^n$ into countably many disjoint binary cubes is often useful.

Proposition 24.10. *For $k \in \mathbb{N}$, let $\mathfrak{Q}_k$ be the subcollection of $\mathfrak{I}^n_{oc}$ of n-dimensional binary cubes of the form $\left(\frac{z_1}{2^k}, \frac{z_1+1}{2^k}\right] \times \cdots \times \left(\frac{z_n}{2^k}, \frac{z_n+1}{2^k}\right]$ where $z_1, \ldots, z_n \in \mathbb{Z}$. The classes $\mathfrak{Q}_k$, $k \in \mathbb{N}$, of members of $\mathfrak{I}^n_{oc}$ have the following properties:*

(a) *For each $k \in \mathbb{N}$, $\mathfrak{Q}_k$ is a countable disjoint collection of members of $\mathfrak{I}^n_{oc}$ and the union of all members of $\mathfrak{Q}_k$ is equal to $\mathbb{R}^n$.*

(b) *If $R_1 \in \mathfrak{Q}_k$ and $R_2 \in \mathfrak{Q}_\ell$ where $k, \ell \in \mathbb{N}$ and $k < \ell$, then either $R_1 \cap R_2 = \emptyset$ or*

$R_1 \supset R_2$. Also every member of $\mathfrak{Q}_k$ is the union of $(2^{\ell-k})^n$ members of $\mathfrak{Q}_\ell$.
(c) *Every nonempty open set in $\mathbb{R}^n$ is a countable disjoint union of members of $\mathfrak{I}^n_{oc}$ in $\bigcup_{k\geq N} \mathfrak{Q}_k$ where N is an arbitrarily chosen positive integer.*

Proof. (a) and (b) are immediate from the definition of $\mathfrak{Q}_k$ for $k \in \mathbb{N}$. To prove (c), let $N \in \mathbb{N}$ be arbitrarily chosen and let V be an arbitrary nonempty open set in $\mathbb{R}^n$. Then for every $x \in V$, there exists an open ball B containing x and contained in V. For sufficiently large $k \geq N$, there exists $R \in \mathfrak{Q}_k$ such that $x \in R \subset B$. Thus V is a union of members of $\bigcup_{k\geq N} \mathfrak{Q}_k$. Since $\bigcup_{k\geq N} \mathfrak{Q}_k$ is a countable collection of members of $\mathfrak{I}^n_{oc}$, V is a countable union of members of $\mathfrak{I}^n_{oc}$. In the collection of those cubes in $\bigcup_{k\geq N} \mathfrak{Q}_k$ whose union is V, select those which belong to $\mathfrak{Q}_N$ and drop those which are in $\bigcup_{k\geq N+1} \mathfrak{Q}_k$ and are contained in any of the cubes that have been selected. In the collection of the remaining cubes, select those which belong to $\mathfrak{Q}_{N+1}$ and drop those which are in $\bigcup_{k\geq N+2} \mathfrak{Q}_k$ and are contained in any of the cubes which have been selected. Thus proceeding we show that V is a disjoint union of members of $\bigcup_{k\geq N} \mathfrak{Q}_k$. ∎

[II] Regularity Properties of Lebesgue Measure Space on $\mathbb{R}^n$

Lemma 24.11. (Borel Regularity of the Lebesgue Outer Measure on $\mathbb{R}^n$)
The Lebesgue outer measure $(\mu^n_L)^*$ on $\mathbb{R}^n$ has the following properties.
(a) *For every $E \in \mathfrak{P}(\mathbb{R}^n)$ and $\varepsilon > 0$, there exists an open set O in $\mathbb{R}^n$ such that $O \supset E$ and*

$$(\mu^n_L)^*(E) \leq (\mu^n_L)^*(O) \leq (\mu^n_L)^*(E) + \varepsilon.$$

(b) *For every $E \in \mathfrak{P}(\mathbb{R}^n)$, there exists a G_δ-set G in $\mathbb{R}^n$ such that $G \supset E$ and*

$$(\mu^n_L)^*(G) = (\mu^n_L)^*(E).$$

(c) *$(\mu^n_L)^*$ is a Borel regular outer measure.*

Proof. 1. Let $E \in \mathfrak{P}(\mathbb{R}^n)$. By Proposition 24.6 we have

$$(\mu^n_L)^*(E) = \inf\left\{\sum_{k\in\mathbb{N}} v(O_k) : (O_k : k \in \mathbb{N}) \subset \mathfrak{I}^n_o, \bigcup_{k\in\mathbb{N}} O_k \supset E\right\}.$$

Thus for every $\varepsilon > 0$ there exists $(O_k : k \in \mathbb{N}) \subset \mathfrak{I}^n_o$ such that $\bigcup_{k\in\mathbb{N}} O_k \supset E$ and

$$(1) \qquad (\mu^n_L)^*(E) \leq \sum_{k\in\mathbb{N}} v(O_k) \leq (\mu^n_L)^*(E) + \varepsilon.$$

Let $O = \bigcup_{k\in\mathbb{N}} O_k$. Then O is an open set and $E \subset \bigcup_{k\in\mathbb{N}} O_k = O$. Thus by the monotonicity of the outer measure $(\mu^n_L)^*$ we have

$$(2) \qquad (\mu^n_L)^*(E) \leq (\mu^n_L)^*(O).$$

Also, by the countable subadditivity of the outer measure $(\mu^n_L)^*$ we have

$$(3) \qquad (\mu^n_L)^*(O) = (\mu^n_L)^*\Big(\bigcup_{k\in\mathbb{N}} O_k\Big) \leq \sum_{k\in\mathbb{N}} (\mu^n_L)^*(O_k)$$
$$= \sum_{k\in\mathbb{N}} v(O_k) \leq (\mu^n_L)^*(E) + \varepsilon,$$

where the second equality is by Proposition 24.7 and the last inequality is by (1). By (2) and (3) we have (a).

2. Let $E \in \mathfrak{P}(\mathbb{R}^n)$. By (a), for every $k \in \mathbb{N}$ there exists an open set O_k such that $O_k \supset E$ and $(\mu_L^n)^*(E) \leq (\mu_L^n)^*(O_k) \leq (\mu_L^n)^*(E) + \frac{1}{k}$. Let $G = \bigcap_{k\in\mathbb{N}} O_k$. Then G is a G_δ-set and $G \supset E$. Since $G \subset O_k$ for every $k \in \mathbb{N}$, we have $(\mu_L^n)^*(E) \leq (\mu_L^n)^*(G) \leq (\mu_L^n)^*(O_k) \leq (\mu_L^n)^*(E) + \frac{1}{k}$ by the monotonicity of the outer measure $(\mu_L^n)^*$. Since this holds for every $k \in \mathbb{N}$, we have $(\mu_L^n)^*(E) \leq (\mu_L^n)^*(G) \leq (\mu_L^n)^*(E)$ and therefore $(\mu_L^n)^*(E) = (\mu_L^n)^*(G)$.

3. By Theorem 24.9 and Definition 24.8, we have $\mathfrak{B}_{\mathbb{R}^n} \subset \mathfrak{M}_L^n = \mathfrak{M}((\mu_L^n)^*)$. Thus $(\mu_L^n)^*$ is a Borel outer measure on $\mathbb{R}^n$. Since a G_δ-set is a member of $\mathfrak{B}_{\mathbb{R}^n}$, we have $G \in \mathfrak{B}_{\mathbb{R}^n}$. Then by (b), $(\mu_L^n)^*$ is a Borel regular outer measure. ∎

Theorem 24.12. *For $E \in \mathfrak{P}(\mathbb{R}^n)$, the following conditions are all equivalent:*
(i) $E \in \mathfrak{M}((\mu_L^n)^*)$.
(ii) *For every $\varepsilon > 0$, there exists an open set $O \supset E$ with $(\mu_L^n)^*(O \setminus E) < \varepsilon$.*
(iii) *There exists a G_δ-set $G \supset E$ with $(\mu_L^n)^*(G \setminus E) = 0$.*
(iv) *For every $\varepsilon > 0$, there exists a closed set $C \subset E$ with $(\mu_L^n)^*(E \setminus C) < \varepsilon$.*
(v) *There exists an F_σ-set $F \subset E$ with $(\mu_L^n)^*(E \setminus F) = 0$.*

Proof. The equivalence of these conditions can be shown in the same way as in the Proof of Theorem 3.22, with Lemma 24.11 replacing Lemma 3.21. ∎

Proposition 24.13. *If $E \in \mathfrak{M}_L^n$, then $E = F \cup N$ where $F \cap N = \emptyset$, F is an F_σ-set and N is a subset of a G_δ-set G_0 with $\mu_L^n(G_0) = 0$.*

Proof. If $E \in \mathfrak{M}_L^n$, then by (iii) and (v) of Theorem 24.12, there exists a G_δ-set $G \supset E$ with $\mu_L^n(G \setminus E) = 0$ and there exists an F_σ-set $F \subset E$ with $\mu_L^n(E \setminus F) = 0$. Then $F \subset E \subset G$ and $G \setminus F = (G \setminus E) \cup (E \setminus F)$ where $(G \setminus E) \cap (E \setminus F) = \emptyset$ so that $\mu_L^n(G \setminus F) = \mu_L^n(G \setminus E) + \mu_L^n(E \setminus F) = 0$. Now $E = F \cup (E \setminus F)$ and $E \setminus F \subset G \setminus F$. Let $N = E \setminus F$ and $G_0 = G \setminus F$. Since F is an F_σ-set, F^c is a G_δ-set. Then $G_0 = G \cap F^c$ being the intersection of two G_δ-sets is a G_δ-set. ∎

Definition 24.14. *Given the Lebesgue measure space $(\mathbb{R}^n, \mathfrak{M}_L^n, \mu_L^n)$ on $\mathbb{R}^n$. Consider the restriction of μ_L^n to $\mathfrak{B}_{\mathbb{R}^n}$. Let us call $(\mathbb{R}^n, \mathfrak{B}_{\mathbb{R}^n}, \mu_L^n)$ the Borel measure space on $\mathbb{R}^n$, or simply the n-dimensional Borel measure space.*

Theorem 24.15. *The Lebesgue measure space $(\mathbb{R}^n, \mathfrak{M}_L^n, \mu_L^n)$ is the completion of the Borel measure space $(\mathbb{R}^n, \mathfrak{B}_{\mathbb{R}^n}, \mu_L^n)$.*

Proof. Let $\overline{\mathfrak{B}_{\mathbb{R}^n}}$ be the completion of $\mathfrak{B}_{\mathbb{R}^n}$ with respect to μ_L^n, that is, $\overline{\mathfrak{B}_{\mathbb{R}^n}}$ is the collection of all subsets E of $\mathbb{R}^n$ of the type $E = A \cup C$ where $A \in \mathfrak{B}_{\mathbb{R}^n}$ and C is a subset of a null set in $(\mathbb{R}^n, \mathfrak{B}_{\mathbb{R}^n}, \mu_L^n)$. Then since $(\mathbb{R}^n, \mathfrak{M}_L^n, \mu_L^n)$ is a complete measure space and since $\mathfrak{B}_{\mathbb{R}^n} \subset \mathfrak{M}_L^n$ by Theorem 24.9, we have $E \in \mathfrak{M}_L^n$. This shows that $\overline{\mathfrak{B}}_{\mathbb{R}^n} \subset \mathfrak{M}_L^n$. Conversely, if E is a member of $\mathfrak{M}_L^n$, then E is a member of $\overline{\mathfrak{B}}_{\mathbb{R}^n}$ by Proposition 24.13. This shows that $\mathfrak{M}_L^n \subset \overline{\mathfrak{B}}_{\mathbb{R}^n}$. Therefore $\mathfrak{M}_L^n = \overline{\mathfrak{B}}_{\mathbb{R}^n}$. ∎

Proposition 24.16.
(a) *Every F_σ-set and in particular a closed set in $\mathbb{R}^n$ is a σ-compact set, that is, a countable union of compact sets.*
(b) *For every $E \in \mathfrak{M}_L^n$, there exists an increasing sequence $(K_j : j \in \mathbb{N})$ of compact sets contained in E such that $\lim_{j\to\infty} \mu_L^n(K_j) = \mu_L^n(E)$.*
(c) *For every $E \in \mathfrak{M}_L^n$, we have $\mu_L^n(E) = \sup\{\mu_L^n(K) : K \subset E$ and K is a compact set$\}$.*
(d) *For every $E \in \mathfrak{M}_L^n$, we have $\mu_L^n(E) = \sup\{\mu_L^n(C) : C \subset E$ and C is a closed set$\}$.*

Proof. 1. Let C be a closed set in $\mathbb{R}^n$. Consider an increasing sequence of compact sets $(B_j : j \in \mathbb{N})$ defined by $B_j = \{x \in \mathbb{R}^n : |x| \le j\}$ for $j \in \mathbb{N}$. Since $\bigcup_{j\in\mathbb{N}} B_j = \mathbb{R}^n$, we have $C = C \cap \bigcup_{j\in\mathbb{N}} B_j = \bigcup_{j\in\mathbb{N}}(C \cap B_j)$. Since C is a closed set and B_j is a compact set, $C \cap B_j$ is a compact set for $j \in \mathbb{N}$. Thus C is a countable union of compact sets. If F is an F_σ-set then F is a countable union of closed set. Since every closed set is a countable union of compact sets, F is a countable union of compact sets. This proves (a).

2. If $E \in \mathfrak{M}_L^n$, then there exists an F_σ-set $F \subset E$ with $\mu_L^n(E \setminus F) = 0$ by Theorem 24.12. By (a) there exists a countable collection of compact sets $\{C_j : j \in \mathbb{N}\}$ such that $F = \bigcup_{j\in\mathbb{N}} C_j$. For every $j \in \mathbb{N}$, let $K_j = \bigcup_{i=1}^j C_i$. Since a finite union of compact sets is a compact set, $(K_j : j \in \mathbb{N})$ is an increasing sequence of compact sets and we have $\lim_{j\to\infty} K_j = \bigcup_{j\in\mathbb{N}} K_j = \bigcup_{j\in\mathbb{N}} C_j = F$. Thus $\mu_L^n(F) = \lim_{j\to\infty} \mu_L^n(K_j)$. Now since E is the disjoint union of F and $E \setminus F$, we have

$$\mu_L^n(E) = \mu_L^n(F) + \mu_L^n(E \setminus F) = \mu_L^n(F) = \lim_{j\to\infty} \mu_L^n(K_j).$$

This proves (b).

3. We have $\mu_L^n(E) \ge \sup\{\mu_L^n(K) : K \subset E$ and K is compact$\}$ by the monotonicity of the measure μ_L^n. Since there exists an increasing sequence $(K_j : j \in \mathbb{N})$ of compact sets contained in E such that $\lim_{j\to\infty} \mu_L^n(K_j) = \mu_L^n(E)$ by (b), the equality in (c) holds.

4. We have $\sup\{\mu_L^n(C) : C \subset E$ and C is a closed set$\} \le \mu_L^n(E)$ by the monotonicity of μ_L^n. By (c), we have

$$\begin{aligned}\mu_L^n(E) &= \sup\left\{\mu_L^n(K) : K \subset E \text{ and } K \text{ is a compact set}\right\}\\ &\le \sup\left\{\mu_L^n(C) : C \subset E \text{ and } C \text{ is a closed set}\right\}.\end{aligned}$$

Therefore (d) holds. ∎

Proposition 24.17. $(\mu_L^n)^*(E) = \inf\{\mu_L^n(O) : E \subset O, O$ *is open*$\}$ *for every* $E \in \mathfrak{P}(\mathbb{R}^n)$.

Proof. If O is an open set and $O \supset E$, then by the monotonicity of $(\mu_L^n)^*$, we have $(\mu_L^n)^*(E) \le (\mu_L^n)^*(O) = \mu_L^n(O)$. Thus $(\mu_L^n)^*(E) \le \inf\{\mu_L^n(O) : E \subset O, O$ is open$\}$.

To prove the reverse inequality, note that for every $\varepsilon > 0$ there exists an open set O such that $O \supset E$ and $\mu_L^n(O) \le (\mu_L^n)^*(E) + \varepsilon$ by Lemma 24.1. This implies that

$$\inf\{\mu_L^n(O) : E \subset O, O \text{ is open}\} \le (\mu_L^n)^*(E) + \varepsilon.$$

By the arbitrariness of $\varepsilon > 0$, we have $\inf\{\mu_L^n(O) : E \subset O, O \text{ is open}\} \le (\mu_L^n)^*(E)$. ∎

[III] Approximation by Continuous Functions

Proposition 24.18. *Let $E \in \mathfrak{M}_L^n$ with $\mu_L^n(E) < \infty$. For every $\varepsilon > 0$, there exists a disjoint finite collection $\{O_1, \dots, O_p\}$ in $\mathfrak{I}_o^n$ such that*

$$\mu_L^n\Big(E \triangle \bigcup_{j=1}^{p} O_j\Big) < \varepsilon.$$

Proof. If $E \in \mathfrak{M}_L^n$ and $\mu_L^n(E) < \infty$, then for an arbitrary $\varepsilon > 0$ there exists a sequence $(R_k : k \in \mathbb{N})$ in $\mathfrak{I}_{oc}^n$ such that $\bigcup_{n\in\mathbb{N}} R_k \supset E$ and $\sum_{k\in\mathbb{N}} \mu_L^n(R_k) - \frac{\varepsilon}{2} \le \mu_L^n(E) \le \sum_{k\in\mathbb{N}} \mu_L^n(R_k)$. Since $\mu_L^n(E) < \infty$, we have $\sum_{k\in\mathbb{N}} \mu_L^n(R_k) < \infty$ and thus there exists $N \in \mathbb{N}$ such that $\sum_{k\in\mathbb{N}} \mu_L^n(R_k) - \frac{\varepsilon}{2} < \sum_{k\le N} \mu_L^n(R_k)$, that is, $\sum_{k>N} \mu_L^n(R_k) < \frac{\varepsilon}{2}$. Consider the finite collection $\{R_1, \dots, R_N\}$ in $\mathfrak{I}_{oc}^n$. Since $\mathfrak{I}_{oc}^n$ is a semialgebra of subsets of $\mathbb{R}^n$, there exists a disjoint finite collection $\{S_1, \dots, S_p\}$ in $\mathfrak{I}_{oc}^n$ such that $\bigcup_{k=1}^{N} R_k = \bigcup_{j=1}^{p} S_j$ by Lemma 21.3. Now

$$\mu_L^n\Big(E \triangle \bigcup_{j=1}^{p} S_j\Big) = \mu_L^n\big(E \setminus \bigcup_{j=1}^{p} S_j\big) + \mu_L^n\Big(\bigcup_{j=1}^{p} S_j \setminus E\Big).$$

Since $E \setminus \bigcup_{j=1}^{p} S_j \subset \bigcup_{k\in\mathbb{N}} R_k \setminus \bigcup_{k=1}^{N} R_k \subset \bigcup_{k>N} R_k$, we have

$$\mu_L^n\Big(E \setminus \bigcup_{j=1}^{p} S_j\Big) \le \sum_{k>N} \mu_L^n(R_k) < \frac{\varepsilon}{2}.$$

We have also

$$\mu_L^n\Big(\bigcup_{j=1}^{p} S_j \setminus E\Big) = \mu_L^n\Big(\bigcup_{k=1}^{N} R_k \setminus E\Big) \le \mu_L^n\Big(\bigcup_{k\in\mathbb{N}} R_k \setminus E\Big)$$
$$= \mu_L^n\Big(\bigcup_{k\in\mathbb{N}} R_k\Big) - \mu_L^n(E) \le \sum_{n\in\mathbb{N}} \mu_L^n(R_k) - \mu_L^n(E) \le \frac{\varepsilon}{2}.$$

Thus we have

$$\text{(1)} \qquad \mu_L^n\Big(E \triangle \bigcup_{j=1}^{p} S_j\Big) < \frac{\varepsilon}{2} + \frac{\varepsilon}{2} = \varepsilon.$$

By the triangle inequality for symmetric differences, we have $A\triangle B \subset (A\triangle C)\cup(C\triangle B)$ for any three sets A, B, and C. If $A, B, C \in \mathfrak{M}^n_L$, then $\mu^n_L(A\triangle B) \leq \mu^n_L(A\triangle C) + \mu^n_L(C\triangle B)$. Let $O_j = S_j^\circ$, the interior of S_j, for $j = 1, \dots, p$. Then $O_j \in \mathfrak{I}^n_o$ and $\{O_1, \dots, O_p\}$ is a disjoint collection since $\{S_1, \dots, S_p\}$ is. Also $\mu^n_L(S_j \setminus O_j) = 0$. Now

$$\bigcup_{j=1}^p S_j \triangle \bigcup_{j=1}^p O_j = \bigcup_{j=1}^p S_j \setminus \bigcup_{j=1}^p O_j = \bigcup_{j=1}^p (S_j \setminus O_j)$$

so that

$$\mu^n_L\Big(\bigcup_{j=1}^p S_j \triangle \bigcup_{j=1}^p O_j\Big) = \sum_{j=1}^p \mu^n_L(S_j \setminus O_j) = 0. \tag{2}$$

Thus by (1) and (2) we have

$$\mu^n_L\Big(E \triangle \bigcup_{j=1}^p O_j\Big) \leq \mu^n_L\Big(E \triangle \bigcup_{j=1}^p S_j\Big) + \mu^n_L\Big(\bigcup_{j=1}^p S_j \triangle \bigcup_{j=1}^p O_j\Big) < \varepsilon.$$

This completes the proof. ∎

Theorem 24.19. *Let f be a μ^n_L-integrable extended real-valued $\mathfrak{M}^n_L$-measurable function on $\mathbb{R}^n$. Then for every $\varepsilon > 0$, there exists $\psi = \sum_{i=1}^N a_i \mathbf{1}_{O_i}$ where $a_i \in \mathbb{R}$, $O_i \in \mathfrak{I}^n_o$ with $\mu^n_L(O_i) < \infty$ for $i = 1, \dots, N$, such that $\int_{\mathbb{R}^n} |f - \psi|\, d\mu^n_L < \varepsilon$.*

Proof. **1.** Let us consider first the case that f is nonnegative. According to Lemma 8.6, there exists an increasing sequence $(\varphi_k : k \in \mathbb{N})$ of nonnegative simple functions on $(\mathbb{R}^n, \mathfrak{M}^n_L, \mu^n_L)$ such that $\varphi_k \uparrow f$ on $\mathbb{R}^n$. By Theorem 8.5 (Monotone Convergence Theorem), we have $\int_{\mathbb{R}^n} \varphi_k\, d\mu^n_L \uparrow \int_{\mathbb{R}^n} f\, d\mu^n_L$. Now $\int_{\mathbb{R}^n} f\, d\mu^n_L < \infty$ implies $\int_{\mathbb{R}^n} \varphi_k\, d\mu^n_L < \infty$ for every $k \in \mathbb{N}$. Thus for an arbitrary $\varepsilon > 0$, there exists $k_0 \in \mathbb{N}$ such that

$$\int_{\mathbb{R}^n} \{f - \varphi_{k_0}\}\, d\mu^n_L = \int_{\mathbb{R}^n} f\, d\mu^n_L - \int_{\mathbb{R}^n} \varphi_{k_0}\, d\mu^n_L < \frac{\varepsilon}{2}.$$

For brevity, let us write φ for φ_{k_0}. Let us express the nonnegative simple function φ as $\varphi = \sum_{i=1}^N a_i \mathbf{1}_{E_i}$ where $a_i > 0$ and $E_i \in \mathfrak{M}^n_L$ for $i = 1, \dots, N$. Then $\int_{\mathbb{R}^n} \varphi\, d\mu^n_L < \infty$ implies that $\mu^n_L(E_i) < \infty$ for $i = 1, \dots, N$. For each $i = 1, \dots, N$, Proposition 24.18 implies that there exists a disjoint finite collection $\{O_{i,1}, \dots, O_{i,p_i}\}$ in $\mathfrak{I}^n_o$ such that $\mu^n_L\big(E_i \triangle \bigcup_{j=1}^{p_i} O_{i,j}\big) < \varepsilon/(2a_i N)$. Let $\psi = \sum_{i=1}^N a_i \mathbf{1}_{\bigcup_{j=1}^{p_i} O_{i,j}} = \sum_{i=1}^N \big\{\sum_{j=1}^{p_i} a_i \mathbf{1}_{O_{i,j}}\big\}$ by the disjointness of the collection $\{O_{i,1}, \dots, O_{i,p_i}\}$ for each $i = 1, \dots, N$. To show that $\int_{\mathbb{R}^n} |f - \psi|\, d\mu^n_L < \varepsilon$, let us observe that for any pair of sets A and B in $\mathbb{R}^n$, we have

$$\begin{aligned}\mathbf{1}_A - \mathbf{1}_B &= \big\{\mathbf{1}_{A\cap B} + \mathbf{1}_{A\setminus B}\big\} - \big\{\mathbf{1}_{A\cap B} + \mathbf{1}_{B\setminus A}\big\} \\ &= \mathbf{1}_{A\setminus B} - \mathbf{1}_{B\setminus A} \leq \mathbf{1}_{A\setminus B} + \mathbf{1}_{B\setminus A} \\ &= \mathbf{1}_{A\triangle B}.\end{aligned}$$

Interchanging the roles of A and B, we have $\mathbf{1}_B - \mathbf{1}_A \le \mathbf{1}_{A\triangle B}$. Therefore $|\mathbf{1}_A - \mathbf{1}_B| \le \mathbf{1}_{A\triangle B}$. Applying this inequality, we have

$$\begin{aligned}\int_{\mathbb{R}^n} \left| a_i \mathbf{1}_{E_i} - a_i \mathbf{1}_{\bigcup_{j=1}^{p_i} O_{i,j}} \right| d\mu_L^n &\le a_i \int_{\mathbb{R}^n} \mathbf{1}_{E_i \triangle \bigcup_{j=1}^{p_i} O_{i,j}}\, d\mu_L^n \\ &= a_i \mu_L^n\Big(E_i \triangle \bigcup_{j=1}^{p_i} O_{i,j}\Big) < \frac{\varepsilon}{2N}\end{aligned}$$

and

$$\int_{\mathbb{R}^n} |\varphi - \psi|\, d\mu_L^n = \int_{\mathbb{R}^n} \left| \sum_{i=1}^N a_i \mathbf{1}_{E_i} - \sum_{i=1}^N a_i \mathbf{1}_{\bigcup_{j=1}^{p_i} O_{i,j}} \right| d\mu_L^n < N\frac{\varepsilon}{2N} = \frac{\varepsilon}{2}.$$

Then

$$\int_{\mathbb{R}^n} |f - \psi|\, d\mu_L^n \le \int_{\mathbb{R}^n} |f - \varphi|\, d\mu_L^n + \int_{\mathbb{R}^n} |\varphi - \psi|\, d\mu_L^n < \frac{\varepsilon}{2} + \frac{\varepsilon}{2} = \varepsilon.$$

2. Now consider the general case that f is a μ_L^n-integrable extended real-valued $\mathfrak{M}_L^n$-measurable function on $\mathbb{R}^n$. Let $f = f^+ - f^-$. By our result above for nonnegative functions, there exist $\psi_1 = \sum_{i=1}^p a_i \mathbf{1}_{V_i}$ and $\psi_2 = \sum_{j=1}^q b_i \mathbf{1}_{W_i}$ where $a_i, b_j > 0$, $V_i, W_j \in \mathfrak{I}_o^n$ with $\mu_L^n(V_i), \mu_L^n(W_i) < \infty$ such that $\int_{\mathbb{R}^n} |f^+ - \psi_1|\, d\mu_L^n < \frac{\varepsilon}{2}$ and $\int_{\mathbb{R}^n} |f^- - \psi_2|\, d\mu_L^n < \frac{\varepsilon}{2}$. If we let $\psi = \psi_1 - \psi_2 = \sum_{i=1}^p a_i \mathbf{1}_{V_i} - \sum_{j=1}^q b_i \mathbf{1}_{W_i}$, then

$$\int_{\mathbb{R}^n} |f - \psi|\, d\mu_L^n \le \int_{\mathbb{R}^n} |f^+ - \psi_1|\, d\mu_L^n + \int_{\mathbb{R}^n} |f^- - \psi_2|\, d\mu_L^n < \frac{\varepsilon}{2} + \frac{\varepsilon}{2} = \varepsilon. \ \blacksquare$$

Let (X, d) be a metric space. The distance between a point $x \in X$ and a set $A \subset X$ is a nonnegative real-valued binary function defined by $d(x, A) = \inf_{a\in A} d(x, a)$. If A is a closed set, then $d(x, A) = 0$ if and only if $x \in A$.

Observation 24.20. Let (X, d) be an arbitrary metric space. Then
(a) $|d(x, A) - d(y, A)| \le d(x, y)$ for any $x, y \in X$ and $A \subset X$.
(b) For a fixed set $A \subset X$, $d(x, A)$ is a continuous function of $x \in X$.

Proof. Let $x, y \in X$ and $A \subset X$. For $a \in A$, we have $d(x, a) \le d(x, y) + d(y, a)$. This implies $\inf_{a\in A} d(x, a) \le d(x, y) + \inf_{a\in A} d(y, a)$, that is, $d(x, A) \le d(x, y) + d(y, A)$, and thus $d(x, A) - d(y, A) \le d(x, y)$. Interchanging the roles of x and y, we have $d(y, A) - d(x, A) \le d(x, y)$. Therefore $|d(x, A) - d(y, A)| \le d(x, y)$. This proves (a). (b) follows from (a). $\blacksquare$

Proposition 24.21. *Let (X, d) be a metric space. If C_1 and C_2 are two disjoint closed sets in X, then there exists a real-valued continuous function f on X such that*
1° $f(x) \in [0, 1]$ *for* $x \in X$.

2° $f(x) = 1$ *for* $x \in C_1$.
3° $f(x) = 0$ *for* $x \in C_2$.

Proof. Given two disjoint closed sets C_1 and C_2 in a metric space (X, d), let us define a real-valued function on X by

$$f(x) = \frac{d(x, C_2)}{d(x, C_1) + d(x, C_2)} \quad \text{for } x \in X.$$

By Observation 24.20, both the numerator and the denominator of f are nonnegative real-valued continuous functions on X. Since $C_1 \cap C_2 = \emptyset$, no point $x \in X$ can be in both C_1 and C_2. Since C_1 and C_2 are closed sets, this implies that for any $x \in X$ not both $d(x, C_1) = 0$ and $d(x, C_2) = 0$ can hold. Thus the denominator of f is never equal to 0 at any point $x \in X$. Therefore f is continuous on X. Property 1° follows from the fact that $d(x, A) \geq 0$ for any $x \in X$ and $A \subset X$. Properties 2° and 3° follow from the fact that if $A \subset X$ is a closed set, then $d(x, A) = 0$ if and only if $x \in A$. ∎

Theorem 24.22. *Let f be a μ_L^n-integrable extended real-valued $\mathfrak{M}_L^n$-measurable function on $\mathbb{R}^n$. Then for every $\varepsilon > 0$, there exists a real-valued continuous function g on $\mathbb{R}^n$ vanishing outside a bounded set in $\mathbb{R}^n$ such that $\int_{\mathbb{R}^n} |f - g|\, d\mu_L^n < \varepsilon$.*

Proof. Let $\varepsilon > 0$ be arbitrarily given. By Theorem 24.19, there exists $\psi = \sum_{i=1}^N a_i \mathbf{1}_{O_i}$ where $a_i \in \mathbb{R}$, $O_i \in \mathfrak{I}_o^n$ with $\mu_L^n(O_i) < \infty$ for $i = 1, \ldots, N$, such that $\int_{\mathbb{R}^n} |f - \psi|\, d\mu_L^n < \frac{\varepsilon}{2}$. For each $O_i \in \mathfrak{I}_o^n$ with $\mu_L^n(O_i) < \infty$, there exists $C_i \in \mathfrak{I}_c^n$ such that $C_i \subset O_i$ and $\mu_L^n(O_i) - \varepsilon/(2|a_i|N) < \mu_L^n(C_i) \leq \mu_L^n(O_i)$. Now since C_i and O_i^c are two disjoint closed sets in $\mathbb{R}^n$, Proposition 24.21 implies that there exists a real-valued continuous function h_i on $\mathbb{R}^n$ such that
1° $h_i(x) \in [0, 1]$ for $x \in \mathbb{R}^n$.
2° $h_i(x) = 1$ for $x \in C_i$.
3° $h_i(x) = 0$ for $x \in O_i^c$.
Let $g = \sum_{i=1}^N a_i h_i$. Since each h_i is a real-valued continuous function on $\mathbb{R}^n$ vanishing outside the bounded set O_i, g is a real-valued continuous function on $\mathbb{R}^n$ vanishing outside the bounded set $\bigcup_{i=1}^N O_i$. From the fact that $h_i = \mathbf{1}_{O_i}$ on C_i and on O_i^c and the fact that $0 \leq h_i \leq \mathbf{1}_{O_i}$ on $\mathbb{R}^n$, we have

$$\int_{\mathbb{R}^n} |\mathbf{1}_{O_i} - h_i|\, d\mu_L^n = \int_{O_i \setminus C_i} |\mathbf{1}_{O_i} - h_i|\, d\mu_L^n \leq \mu_L^n(O_i \setminus C_i) < \frac{\varepsilon}{2|a_i|N}.$$

Then

$$\begin{aligned}\int_{\mathbb{R}^n} |\psi - g|\, d\mu_L^n &= \int_{\mathbb{R}^n} \Big| \sum_{i=1}^N a_i \mathbf{1}_{O_i} - \sum_{i=1}^N a_i h_i \Big|\, d\mu_L^n \\ &\leq \sum_{i=1}^N |a_i| \int_{\mathbb{R}^n} |\mathbf{1}_{O_i} - h_i|\, d\mu_L^n < \sum_{i=1}^N |a_i| \frac{\varepsilon}{2|a_i|N} = \frac{\varepsilon}{2}.\end{aligned}$$

Therefore

$$\int_{\mathbb{R}^n} |f - g| \, d\mu_L^n \le \int_{\mathbb{R}^n} |f - \psi| \, d\mu_L^n + \int_{\mathbb{R}^n} |\psi - g| \, d\mu_L^n < \frac{\varepsilon}{2} + \frac{\varepsilon}{2} = \varepsilon. \quad \blacksquare$$

[IV] Lebesgue Measure Space on $\mathbb{R}^n$ as the Completion of a Product Measure Space

We show that the Lebesgue measure space $(\mathbb{R}^n, \mathfrak{M}_L^n, \mu_L^n)$ on $\mathbb{R}^n$, as defined by Definition 24.5 and Definition 24.8, is the completion of the n-fold product measure space of the Lebesgue measure space $(\mathbb{R}, \mathfrak{M}_L, \mu_L)$ on $\mathbb{R}$. Let us show first that $\mathfrak{B}_{\mathbb{R}^n} = \sigma\left(\mathsf{X}_{i=1}^n \mathfrak{B}_{\mathbb{R}}\right)$. This is based on the following lemma.

Lemma 24.23. *Let $\mathfrak{C}_i$ be a collection of subsets of a set X_i for $i = 1, \ldots, n$. Then*

(a) $\quad \mathsf{X}_{i=1}^n \sigma(\mathfrak{C}_i) \subset \sigma\left(\mathsf{X}_{i=1}^n \mathfrak{C}_i\right),$

(b) $\quad \sigma\left(\mathsf{X}_{i=1}^n \sigma(\mathfrak{C}_i)\right) = \sigma\left(\mathsf{X}_{i=1}^n \mathfrak{C}_i\right).$

In particular, we have

(c) $\quad \sigma\left(\mathsf{X}_{i=1}^n \mathfrak{B}_{\mathbb{R}}\right) = \sigma(\mathfrak{I}_{oc}^n).$

Proof. 1. Let $E_2 \in \mathfrak{C}_2, \ldots, E_n \in \mathfrak{C}_n$ be arbitrarily fixed and let π_1 be the projection of $X_1 \times \mathsf{X}_{i=2}^n E_i$ onto X_1. Then by Theorem 1.14, we have $\pi_1^{-1}\left(\sigma(\mathfrak{C}_1)\right) = \sigma\left(\pi_1^{-1}(\mathfrak{C}_1)\right)$, that is, $\sigma(\mathfrak{C}_1) \times \mathsf{X}_{i=2}^n E_i = \sigma\left(\mathfrak{C}_1 \times \mathsf{X}_{i=2}^n E_i\right) \subset \sigma\left(\mathsf{X}_{i=1}^n \mathfrak{C}_i\right)$. Thus by the arbitrariness of $E_2 \in \mathfrak{C}_2, \ldots, E_n \in \mathfrak{C}_n$, we have

(1) $$\sigma(\mathfrak{C}_1) \times \mathsf{X}_{i=2}^n \mathfrak{C}_i \subset \sigma\left(\mathsf{X}_{i=1}^n \mathfrak{C}_i\right).$$

Next, let $F_1 \in \sigma(\mathfrak{C}_1), E_3 \in \mathfrak{C}_3, \ldots, E_n \in \mathfrak{C}_n$ be arbitrarily fixed. By (1) we have

(2) $$F_1 \times \mathfrak{C}_2 \times \mathsf{X}_{i=3}^n E_i \subset \sigma\left(\mathsf{X}_{i=1}^n \mathfrak{C}_i\right).$$

Let π_2 be the projection of $F_1 \times X_2 \times \mathsf{X}_{i=3}^n E_i$ onto X_2. Then by Theorem 1.14, we have $\pi_2^{-1}\left(\sigma(\mathfrak{C}_2)\right) = \sigma\left(\pi_2^{-1}(\mathfrak{C}_2)\right)$, that is,

$$F_1 \times \sigma(\mathfrak{C}_2) \times \mathsf{X}_{i=3}^n E_i = \sigma\left(F_1 \times \mathfrak{C}_2 \times \mathsf{X}_{i=3}^n E_i\right) \subset \sigma\left(\mathsf{X}_{i=1}^n \mathfrak{C}_i\right),$$

where the last set inclusion is by (2). From the arbitrariness of $F_1 \in \sigma(\mathfrak{C}_1), E_3 \in \mathfrak{C}_3, \ldots, E_n \in \mathfrak{C}_n$, we have $\sigma(\mathfrak{C}_1) \times \sigma(\mathfrak{C}_2) \times \mathsf{X}_{i=3}^n \mathfrak{C}_i \subset \sigma\left(\mathsf{X}_{i=1}^n \mathfrak{C}_i\right)$. Thus proceeding we have finally $\sigma(\mathfrak{C}_1) \times \sigma(\mathfrak{C}_2) \times \cdots \times \sigma(\mathfrak{C}_n) \subset \sigma\left(\mathsf{X}_{i=1}^n \mathfrak{C}_i\right)$. This proves (a).

2. Now (a) implies $\sigma\left(\mathsf{X}_{i=1}^n \sigma(\mathfrak{C}_i)\right) \subset \sigma\left(\mathsf{X}_{i=1}^n \mathfrak{C}_i\right)$. But $\mathsf{X}_{i=1}^n \sigma(\mathfrak{C}_i) \supset \mathsf{X}_{i=1}^n \mathfrak{C}_i$ and then $\sigma\left(\mathsf{X}_{i=1}^n \sigma(\mathfrak{C}_i)\right) \supset \sigma\left(\mathsf{X}_{i=1}^n \mathfrak{C}_i\right)$. Therefore we have $\sigma\left(\mathsf{X}_{i=1}^n \sigma(\mathfrak{C}_i)\right) = \sigma\left(\mathsf{X}_{i=1}^n \mathfrak{C}_i\right)$. This proves (b).

3. Consider the case $\mathfrak{C}_i = \mathfrak{I}_{oc}$ for $i + 1, \ldots, n$. Now $\sigma(\mathfrak{I}_{oc}) = \mathfrak{B}_{\mathbb{R}}$ by Proposition 24.4 for the case $n = 1$. Then by (b) we have $\sigma\left(\mathsf{X}_{i=1}^n \mathfrak{B}_{\mathbb{R}}\right) = \sigma\left(\mathsf{X}_{i=1}^n \mathfrak{I}_{oc}\right) = \sigma(\mathfrak{I}_{oc}^n)$. This proves (c). $\quad \blacksquare$

Theorem 24.24. *For the Borel σ-algebra of subsets of $\mathbb{R}^n$, we have* $\mathfrak{B}_{\mathbb{R}^n} = \sigma\big(\times_{i=1}^n \mathfrak{B}_{\mathbb{R}}\big)$.

Proof. We have $\sigma\big(\times_{i=1}^n \mathfrak{B}_{\mathbb{R}}\big) = \sigma(\mathfrak{I}_{oc}^n)$ by (c) of Lemma 24.23. But $\sigma(\mathfrak{I}_{oc}^n) = \mathfrak{B}_{\mathbb{R}^n}$ by Proposition 24.4. Thus $\sigma\big(\times_{i=1}^n \mathfrak{B}_{\mathbb{R}}\big) = \mathfrak{B}_{\mathbb{R}^n}$. ∎

Theorem 24.25. *Let $\big(\mathbb{R}^n, \sigma\big(\times_{i=1}^n \mathfrak{M}_L\big), \times_{i=1}^n \mu_L\big)$ be the n-fold product measure space of the Lebesgue measure space $(\mathbb{R}, \mathfrak{M}_L, \mu_L)$. Let $\overline{\sigma\big(\times_{i=1}^n \mathfrak{M}_L\big)}$ be the completion of $\sigma\big(\times_{i=1}^n \mathfrak{M}_L\big)$ with respect to the measure $\times_{i=1}^n \mu_L$. Then the Lebesgue measure space $(\mathbb{R}^n, \mathfrak{M}_L^n, \mu_L^n)$ is identical with the measure space $\big(\mathbb{R}^n, \overline{\sigma\big(\times_{i=1}^n \mathfrak{M}_L\big)}, \times_{i=1}^n \mu_L\big)$.*

Proof. Let $\overline{\mathfrak{B}}_{\mathbb{R}^n}$ be the completion of the Borel σ-algebra $\mathfrak{B}_{\mathbb{R}^n}$ with respect to the measure μ_L^n on it. According to Theorem 24.15, we have $(\mathbb{R}^n, \mathfrak{M}_L^n, \mu_L^n) = (\mathbb{R}^n, \overline{\mathfrak{B}}_{\mathbb{R}^n}, \mu_L^n)$. Thus it remains to show

$$(1) \qquad (\mathbb{R}^n, \overline{\mathfrak{B}}_{\mathbb{R}^n}, \mu_L^n) = \big(\mathbb{R}^n, \overline{\sigma\big(\times_{i=1}^n \mathfrak{M}_L\big)}, \times_{i=1}^n \mu_L\big).$$

For an arbitrary $R \in \mathfrak{I}_{oc}^n$ given by $R = \times_{i=1}^n (a_i, b_i]$, we have

$$\mu_L^n(R) = v(R) = \prod_{i=1}^n (b_i - a_i) = \big(\times_{i=1}^n \mu_L\big)(R).$$

Therefore $\mu_L^n = \times_{i=1}^n \mu_L$ on the semialgebra $\mathfrak{I}_{oc}^n$. Now $\mathfrak{B}_{\mathbb{R}^n} = \sigma(\mathfrak{I}_{oc}^n)$ by Proposition 24.4. Then by the uniqueness of extension (Theorem 21.11), we have $\mu_L^n = \times_{i=1}^n \mu_L$ on $\mathfrak{B}_{\mathbb{R}^n}$. Thus we have $(\mathbb{R}^n, \mathfrak{B}_{\mathbb{R}^n}, \mu_L^n) = \big(\mathbb{R}^n, \mathfrak{B}_{\mathbb{R}^n}, \times_{i=1}^n \mu_L\big)$ and the completion $\overline{\mathfrak{B}}_{\mathbb{R}^n}$ of $\mathfrak{B}_{\mathbb{R}^n}$ with respect to μ_L^n is identical with the completion of $\mathfrak{B}_{\mathbb{R}^n}$ with respect to $\times_{i=1}^n \mu_L$. Therefore to prove (1), it remains to show

$$\overline{\mathfrak{B}}_{\mathbb{R}^n} = \overline{\sigma\big(\times_{i=1}^n \mathfrak{M}_L\big)}.$$

Now we have $\mathfrak{B}_{\mathbb{R}^n} = \sigma\big(\times_{i=1}^n \mathfrak{B}_{\mathbb{R}}\big)$ by Theorem 24.24. Then by Theorem 23.23 we have $\overline{\mathfrak{B}}_{\mathbb{R}^n} = \overline{\sigma\big(\times_{i=1}^n \mathfrak{B}_{\mathbb{R}}\big)} = \overline{\sigma\big(\times_{i=1}^n \overline{\mathfrak{B}}_{\mathbb{R}}\big)}$. But $\overline{\mathfrak{B}}_{\mathbb{R}} = \mathfrak{M}_L$ by Theorem 5.7. Therefore we have $\overline{\mathfrak{B}}_{\mathbb{R}^n} = \overline{\sigma\big(\times_{i=1}^n \mathfrak{M}_L\big)}$. ∎

[V] Translation of the Lebesgue Integral on $\mathbb{R}^n$

Lemma 24.26. *The Lebesgue outer measure $(\mu_L^n)^*$ on $\mathbb{R}^n$ is translation invariant, that is, $(\mu_L^n)^*(E + x) = (\mu_L^n)^*(E)$ for every $E \subset \mathbb{R}^n$ and $x \in \mathbb{R}^n$.*

Proof. Let $x \in \mathbb{R}^n$ be arbitrarily fixed. Let us define a mapping T_x of the collection $\mathfrak{P}(\mathbb{R}^n)$ of all subsets of $\mathbb{R}^n$ into itself by setting $T_x(E) = E + x$ for $E \in \mathfrak{P}(\mathbb{R}^n)$. T_x maps $\mathfrak{P}(\mathbb{R}^n)$ one-to-one onto $\mathfrak{P}(\mathbb{R}^n)$ with its inverse mapping given by T_{-x}. Consider $\mathfrak{I}_{oc}^n \subset \mathfrak{P}(\mathbb{R}^n)$. Our T_x maps $\mathfrak{I}_{oc}^n$ one-to-one onto $\mathfrak{I}_{oc}^n$. Also $\mu_L^n\big(T_x(R)\big) = \mu_L^n(R)$, that is, $\mu_L^n(R + x) = \mu_L^n(R)$ for every $R \in \mathfrak{I}_{oc}^n$. Indeed if R is given by $R = \times_{i=1}^n I_i$ where

$I_i \in \mathfrak{I}_{oc}$ for $i = 1, \ldots, n$, then $R + x = \times_{i=1}^{n}(I_i + x_i)$ where $x = (x_1, \ldots, x_n)$. Then $\mu_L^n(R+x) = \prod_{i=1}^n \mu_L(I_i + x_i) = \prod_{i=1}^n \mu_L(I_i)$ by Theorem 3.16 (Translation Invariance of the Lebesgue Measure on $\mathbb{R}$). Thus $\mu_L^n(R+x) = \mu_L^n(R)$. Now let $E \in \mathfrak{P}(\mathbb{R}^n)$. Then since $(\mu_L^n)^*(E) = \inf\{\sum_{k\in\mathbb{N}} \mu_L^n(R_k) : (R_k : k \in \mathbb{N}) \subset \mathfrak{I}_{oc}^n, \bigcup_{k\in\mathbb{N}} R_k \supset E\}$, we have

$$
\begin{aligned}
(\mu_L^n)^*(E+x) &= \inf\left\{\textstyle\sum_{k\in\mathbb{N}} \mu_L^n(R_k) : (R_k : k \in \mathbb{N}) \subset \mathfrak{I}_{oc}^n, \bigcup_{k\in\mathbb{N}} R_k \supset E + x\right\} \\
&= \inf\left\{\textstyle\sum_{k\in\mathbb{N}} \mu_L^n(R_k) : (R_k : k \in \mathbb{N}) \subset \mathfrak{I}_{oc}^n, \bigcup_{k\in\mathbb{N}} (R_k - x) \supset E\right\} \\
&= \inf\left\{\textstyle\sum_{k\in\mathbb{N}} \mu_L^n(S_k + x) : (S_k : k \in \mathbb{N}) \subset \mathfrak{I}_{oc}^n, \bigcup_{k\in\mathbb{N}} S_k \supset E\right\} \\
&= \inf\left\{\textstyle\sum_{k\in\mathbb{N}} \mu_L^n(S_k) : (S_k : k \in \mathbb{N}) \subset \mathfrak{I}_{oc}^n, \bigcup_{k\in\mathbb{N}} S_k \supset E\right\} \\
&= (\mu_L^n)^*(E),
\end{aligned}
$$

where the third equality is from the fact that T_x is a one-to-one mapping of $\mathfrak{I}_{oc}^n$ onto $\mathfrak{I}_{oc}^n$. ■

Theorem 24.27. (Translation Invariance of the Lebesgue Measure Space on $\mathbb{R}^n$) *The Lebesgue measure space $(\mathbb{R}^n, \mathfrak{M}_L^n, \mu_L^n)$ is translation invariant, that is, for every $E \in \mathfrak{M}_L^n$ and $x \in \mathbb{R}^n$, we have $E + x \in \mathfrak{M}_L^n$ and $\mu_L^n(E+x) = \mu_L^n(E)$.*

Proof. The σ-algebra $\mathfrak{M}_L^n = \mathfrak{M}((\mu_L^n)^*)$ consists of all subsets E of $\mathbb{R}^n$ satisfying the Carathéodory condition $(\mu_L^n)^*(A) = (\mu_L^n)^*(A \cap E) + (\mu_L^n)^*(A \cap E^c)$ for every subset A of $\mathbb{R}^n$. (See Definition 2.2.) Suppose $E \in \mathfrak{M}((\mu_L^n)^*)$ and $x \in \mathbb{R}^n$. Then for $E + x$ we have

$$
\begin{aligned}
&(\mu_L^n)_L^*(A \cap (E+x)) + (\mu_L^n)_L^*(A \cap (E+x)^c) \\
=&(\mu_L^n)_L^*(\{A \cap (E+x)\} - x) + (\mu_L^n)_L^*(\{A \cap (E+x)^c\} - x) \\
=&(\mu_L^n)_L^*((A-x) \cap E) + (\mu_L^n)_L^*((A-x) + E^c) \\
=&(\mu_L^n)_L^*(A - x) = (\mu_L^n)_L^*(A),
\end{aligned}
$$

where the first equality is by Lemma 24.26, the third equality is by the fact that E satisfies the Carathéodory condition with $A - x$ as a testing set, and the last equality is by Lemma 24.26. This shows that if $E \in \mathfrak{M}((\mu_L^n)^*)$, then $E + x \in \mathfrak{M}((\mu_L^n)^*)$, that is, if $E \in \mathfrak{M}_L^n$, then $E + x \in \mathfrak{M}_L^n$. Then $\mu_L^n(E+x) = \mu_L^n(E)$ by Lemma 24.26. ■

Theorem 24.28. (Translation Invariance of the Lebesgue Integral on $\mathbb{R}^n$) *Consider the Lebesgue measure space $(\mathbb{R}^n, \mathfrak{M}_L^n, \mu_L^n)$. Let f be an extended real-valued function on a set $D \in \mathfrak{M}_L^n$. Let $h \in \mathbb{R}^n$ and let g be a function on $D - h$ defined by $g(x) = f(x+h)$ for $x \in D - h$.*

(a) *If f is $\mathfrak{M}_L^n$-measurable on D, then g is $\mathfrak{M}_L^n$-measurable on $D - h$.*

(b) *If f is $\mathfrak{M}_L^n$-measurable on D, then*

(1)
$$\int_D f(x)\,\mu_L^n(dx) = \int_{D-h} f(x+h)\,\mu_L^n(dx),$$

in the sense that if one of the two integrals exists then so does the other and the two are equal. In particular if f is nonnegative, then (1) always holds.
(c) *If f is defined and $\mathfrak{M}_L^n$-measurable on $\mathbb{R}^n$, then*

$$\int_{\mathbb{R}^n} f(x)\,\mu_L^n(dx) = \int_{\mathbb{R}^n} f(x+h)\,\mu_L^n(dx), \tag{2}$$

in the same sense as above.

Proof. This theorem is proved by using Theorem 24.27 in the same way Theorem 9.31 was proved using Theorem 3.16. ■

[VI] Linear Transformation of the Lebesgue Integral on $\mathbb{R}^n$

Let T be a linear transformation of $\mathbb{R}^n$ into $\mathbb{R}^n$, that is, T is a mapping of $\mathbb{R}^n$ into $\mathbb{R}^n$ satisfying the conditions:
1° $T(\alpha x) = \alpha T(x)$ for $x \in \mathbb{R}^n$ and $\alpha \in \mathbb{R}$,
2° $T(x+y) = T(x) + T(y)$ for $x, y \in \mathbb{R}^n$.
It is a well known fact in linear algebra that if T is one-to-one, then T maps $\mathbb{R}^n$ onto $\mathbb{R}^n$ so that its inverse transformation T^{-1} is defined on $\mathbb{R}^n$ and is a linear transformation of $\mathbb{R}^n$ onto $\mathbb{R}^n$. Let us call such linear transformation non-singular. Let $\{e_1, \dots, e_n\}$ be the standard basis for $\mathbb{R}^n$ and let M_T be the matrix of T with respect to the standard basis. Thus M_T is an $n \times n$ matrix whose columns are the column vectors $T(e_1), \dots, T(e_n)$. T is non-singular if and only if $\det M_T \neq 0$.

Observation 24.29. It is well known in linear algebra that every non-singular linear transformation T of $\mathbb{R}^n$ into $\mathbb{R}^n$ is a finite product of non-singular linear transformations of the following three types:
1° $T(\dots, x_j, \dots, x_k, \dots) = (\dots, x_k, \cdots, x_j, \cdots)$ for some $j, k = 1, \dots, n$, $j < k$.
2° $T(x_1, x_2, \dots, x_n) = (\alpha x_1, x_2, \dots, x_n)$ for some $\alpha \in \mathbb{R}$, $\alpha \neq 0$.
3° $T(x_1, x_2, \dots, x_n) = (x_1 + x_2, x_2, \dots, x_n)$.
For type 1°, M_T is the matrix obtained by interchanging the j-th and k-th columns of the $n \times n$ identity matrix so that $\det M_T = \pm 1$.
For type 2°, M_T is the matrix obtained by multiplying the first row of the $n \times n$ identity matrix by α so that $\det M_T = \alpha$.
For type 3°, M_T is the matrix obtained by adding the second row to the first row of the $n \times n$ identity matrix so that $\det M_T = 1$.

Proposition 24.30. *Consider the n-dimensional Borel measure space $(\mathbb{R}^n, \mathfrak{B}_{\mathbb{R}^n}, \mu_L^n)$. Let T be a non-singular linear transformation of $\mathbb{R}^n$ into $\mathbb{R}^n$. Then*

(a) $T(E) \in \mathfrak{B}_{\mathbb{R}^n}$ *for every* $E \in \mathfrak{B}_{\mathbb{R}^n}$.

(b) $\mu_L^n(T(E)) = |\det M_T| \mu_L^n(E)$ *for every* $E \in \mathfrak{B}_{\mathbb{R}^n}$.

(c) $(\mu_L^n)^*(T(E)) = |\det M_T| (\mu_L^n)^*(E)$ *for every* $E \in \mathfrak{P}(\mathbb{R}^n)$.

Proof. 1. Let $\mathfrak{F} = \{E \subset \mathbb{R}^n : T(E) \in \mathfrak{B}_{\mathbb{R}^n}\}$. To prove (a), it suffices to show that $\mathfrak{B}_{\mathbb{R}^n} \subset \mathfrak{F}$. Now $T(\mathbb{R}^n) = \mathbb{R}^n \in \mathfrak{B}_{\mathbb{R}^n}$ so that $\mathbb{R}^n \in \mathfrak{F}$. Suppose $E \in \mathfrak{F}$. Then $T(E) \in \mathfrak{B}_{\mathbb{R}^n}$. Since T maps $\mathbb{R}^n$ one-to-one onto $\mathbb{R}^n$, we have $T(E^c) = \big(T(E)\big)^c \in \mathfrak{B}_{\mathbb{R}^n}$ so that $E^c \in \mathfrak{F}$. Thus $\mathfrak{F}$ is closed under complementations. If $(E_n : n \in \mathbb{N})$ is a sequence in $\mathfrak{F}$, then $T(E_n) \in \mathfrak{B}_{\mathbb{R}^n}$ for every $n \in \mathbb{N}$ so that $T\big(\bigcup_{n\in\mathbb{N}} E_n\big) = \bigcup_{n\in\mathbb{N}} T(E_n) \in \mathfrak{B}_{\mathbb{R}^n}$. Thus $\mathfrak{F}$ is closed under countable unions. Therefore $\mathfrak{F}$ is a σ-algebra of subsets of $\mathbb{R}^n$. Let $\mathfrak{O}$ be the collection of all open sets in $\mathbb{R}^n$. Since T is a homeomorphism of $\mathbb{R}^n$, for every $O \in \mathfrak{O}$ we have $T(O) \in \mathfrak{O} \subset \mathfrak{B}_{\mathbb{R}^n}$. Thus $\mathfrak{O} \subset \mathfrak{F}$. Then since $\mathfrak{F}$ is a σ-algebra of subsets of $\mathbb{R}^n$ containing all the open sets in $\mathbb{R}^n$, we have $\mathfrak{F} \supset \mathfrak{B}_{\mathbb{R}^n}$. This shows that $T(E) \in \mathfrak{B}_{\mathbb{R}^n}$ for every $E \in \mathfrak{B}_{\mathbb{R}^n}$.

2. Let us prove (b). As the first step, let us show that if T is one of the types 1°, 2°, and 3° in Observation 24.29, then

$$(1) \qquad \mu_L^n\big(T(E)\big) = |\det M_T|\mu_L^n(E) \quad \text{for every } E \in \mathfrak{I}_{oc}^n.$$

Let $E \in \mathfrak{I}_{oc}^n$ be given by $E = \times_{i=1}^n I_i = \times_{i=1}^n (a_i, b_i]$. Now if T is type 1°, then we have $T(E) = I_1 \times \cdots \times I_k \times \cdots \times I_j \times \cdots \times I_n$ so that

$$\mu_L^n\big(T(E)\big) = \mu_L(I_1)\cdots\mu_L(I_k)\cdots\mu_L(I_j)\cdots\mu_L(I_n) = \prod_{i=1}^n \mu_L(I_i) = \mu_L^n(E).$$

Since $\det M_T = \pm 1$ for type 1°, $|\det M_T| = 1$ so that (1) holds in this case. If T is type 2°, then $T(E) = \alpha I_1 \times \times_{i=2}^n I_i$ so that

$$\mu_L^n\big(T(E)\big) = \mu_L(\alpha I_1)\prod_{i=2}^n \mu_L(I_i) = |\alpha|\mu_L(I_1)\prod_{i=2}^n \mu_L(I_i) = |\alpha|\mu_L^n(E)$$

where the equality $\mu_L(\alpha I_1) = |\alpha|\mu_L(I_1)$ is by Theorem 3.18 for the linear transformation of the Lebesgue measure space on $\mathbb{R}$. Since $\det M_T = \alpha$ for type 2°, we have (1) in this case. Finally, if T is type 3°, then $T(E) = P \times \times_{i=3}^n I_i$ where P is a parallelogram in $\mathbb{R}^2$ with the same area as the rectangle $I_1 \times I_2$, that is, $\mu_L^2(P) = \mu_L(I_1)\mu_L(I_2)$. Thus

$$\mu_L^n\big(T(E)\big) = \mu_L^2(P)\prod_{i=3}^n \mu_L(I_i) = \prod_{i=1}^n \mu_L(I_i) = \mu_L^n(E).$$

For type 3°, we have $\det M_T = 1$. Thus (1) holds in this case.

3. As the second step toward proving (b), we show that if T is one of the types 1°, 2°, and 3° in Observation 24.29, then

$$(2) \qquad \mu_L^n\big(T(E)\big) = |\det M_T|\mu_L^n(E) \quad \text{for every } E \in \mathfrak{B}_{\mathbb{R}^n}.$$

Let us define two set functions μ and ν on the σ-algebra $\mathfrak{B}_{\mathbb{R}^n}$ by setting

$$(3) \qquad \begin{cases} \mu(E) = \mu_L^n\big(T(E)\big) & \text{for } E \in \mathfrak{B}_{\mathbb{R}^n}, \\ \nu(E) = |\det M_T|\mu_L^n(E) & \text{for } E \in \mathfrak{B}_{\mathbb{R}^n}. \end{cases}$$

Being a positive constant times the measure μ_L^n on $\mathfrak{B}_{\mathbb{R}^n}$, ν is a measure on $\mathfrak{B}_{\mathbb{R}^n}$. To verify that μ is a measure on $\mathfrak{B}_{\mathbb{R}^n}$, note that $\mu(E) \in [0, \infty]$ for every $E \in \mathfrak{B}_{\mathbb{R}^n}$ and $\mu(\emptyset) = 0$ as immediate consequences of the definition of μ by (3). If $(E_n : n \in \mathbb{N})$ is a disjoint sequence in $\mathfrak{B}_{\mathbb{R}^n}$, then $(T(E_n) : n \in \mathbb{N})$ is a disjoint sequence in $\mathfrak{B}_{\mathbb{R}^n}$ by (a) and by the fact that the mapping T is one-to-one. Then

$$\mu\Big(\bigcup_{n\in\mathbb{N}} E_n\Big) = \mu_L^n\Big(T\Big(\bigcup_{n\in\mathbb{N}} E_n\Big)\Big) = \mu_L^n\Big(\bigcup_{n\in\mathbb{N}} T(E_n)\Big)$$
$$= \sum_{n\in\mathbb{N}} \mu_L^n(T(E_n)) = \sum_{n\in\mathbb{N}} \mu(E_n).$$

This shows that μ is countably additive on $\mathfrak{B}_{\mathbb{R}^n}$ and completes the verification that μ is a measure on $\mathfrak{B}_{\mathbb{R}^n}$. By (1), the two measures μ and ν on $\mathfrak{B}_{\mathbb{R}^n}$ are equal on the semialgebra $\mathfrak{I}_{oc}^n$. Then since $\mathfrak{B}_{\mathbb{R}^n} = \sigma(\mathfrak{I}_{oc}^n)$ by Proposition 24.4 and since μ and ν are σ-finite measures, we have $\mu = \nu$ on $\mathfrak{B}_{\mathbb{R}^n}$ by Theorem 21.11. This proves (2).

4. To complete the proof of (b), let T be an arbitrary non-singular linear transformation of $\mathbb{R}^n$ into $\mathbb{R}^n$. By Observation 24.29, we have $T = T_p T_{p-1} \cdots T_1$ where each T_k, $k = 1, \dots, p$, is of type 1°, 2°, or 3°. Now for every $E \in \mathfrak{B}_{\mathbb{R}^n}$, $T(E) = T_p \circ T_{p-1} \circ \cdots \circ T_1(E)$. By repeated application of (a), we have $T_k \circ T_{k-1} \circ \cdots \circ T_1(E) \in \mathfrak{B}_{\mathbb{R}^n}$ for $k = 1, \dots, p$. By repeated application of (2), we have

$$\mu_L^n(T_1(E)) = |\det M_{T_1}|\, \mu_L^n(E),$$
$$\mu_L^n(T_2 \circ T_1(E)) = |\det M_{T_2}|\, \mu_L^n(T_1(E)) = |\det M_{T_2}|\, |\det M_{T_1}|\, \mu_L^n(E),$$

and so on and finally

$$\mu_L^n(T(E)) = \prod_{k=1}^{p} |\det M_{T_k}| \mu_L^n(E) = \Big|\prod_{k=1}^{p} \det M_{T_k}\Big| \mu_L^n(E) = |\det M_T|\, \mu_L^n(E).$$

This proves (b).

5. To prove (c), let us find $(\mu_L^n)^*(T(E))$ for an arbitrary $E \in \mathfrak{P}(\mathbb{R}^n)$. According to Definition 24.5, for every $E \in \mathfrak{P}(\mathbb{R}^n)$ we have

$$(\mu_L^n)^*(E) = \inf\Big\{ \textstyle\sum_{k\in\mathbb{N}} v(R_k) : (R_k : k \in \mathbb{N}) \subset \mathfrak{I}_{oc}^n, \bigcup_{k\in\mathbb{N}} R_k \supset E \Big\}.$$

Let $T(\mathfrak{I}_{oc}^n) = \{T(R) : R \in \mathfrak{I}_{oc}^n\}$, the collection of all parallelopipeds in $\mathbb{R}^n$ that are images of members $\mathfrak{I}_{oc}^n$ by the mapping T. Let us show that for an arbitrary countable collection $\{R_K : k \in \mathbb{N}\} \subset \mathfrak{I}_{oc}^n$ and $\varepsilon > 0$ there exists a countable collection $\{P_K : k \in \mathbb{N}\} \subset T(\mathfrak{I}_{oc}^n)$ such that

$$(4) \qquad \bigcup_{k\in\mathbb{N}} R_k \subset \bigcup_{k\in\mathbb{N}} P_k \quad \text{and} \quad \sum_{k\in\mathbb{N}} v(R_k) \geq \sum_{k\in\mathbb{N}} v(P_k) - \varepsilon.$$

Now for R_k there exists a finite disjoint collection $\{P_{k,i_k} : i_k = 1, \dots, m_k\} \subset T(\mathfrak{I}_{oc}^n)$ such that

$$R_k \subset \bigcup_{i_k=1}^{m_k} P_{k,i_k} \quad \text{and} \quad \sum_{i_k=1}^{m_k} v(P_{k,i_k}) \leq v(R_k) + \frac{\varepsilon}{2^k}.$$

Then $\{P_{k,i_k} : i_k = 1, \ldots, m_k; k \in \mathbb{N}\}$ is a countable collection in $T(\mathfrak{I}^n_{oc})$ such that $\bigcup_{k\in\mathbb{N}} R_k \subset \bigcup_{k\in\mathbb{N}} \bigcup_{i_k=1}^{m_k} P_{k,i_k}$, and

$$\sum_{k\in\mathbb{N}} \sum_{i_k=1}^{m_k} v(P_{k,i_k}) \leq \sum_{k\in\mathbb{N}} \left\{ v(R_k) + \frac{\varepsilon}{2^k} \right\} = \sum_{k\in\mathbb{N}} v(R_k) + \varepsilon.$$

If we renumber $\{P_{k,i_k} : i_k = 1, \ldots, m_k; k \in \mathbb{K}\}$ as $\{P_k : k \in \mathbb{N}\}$ then we have (4). We show similarly that for an arbitrary countable collection $\{P_K : k \in \mathbb{N}\} \subset T(\mathfrak{I}^n_{oc})$ and $\varepsilon > 0$ there exists a countable collection $\{R_K : k \in \mathbb{N}\} \subset \mathfrak{I}^n_{oc}$ such that

$$(5) \qquad \bigcup_{k\in\mathbb{N}} P_k \subset \bigcup_{k\in\mathbb{N}} R_k \quad \text{and} \quad \sum_{k\in\mathbb{N}} v(P_k) \geq \sum_{k\in\mathbb{N}} v(R_k) - \varepsilon.$$

Now for an arbitrary $E \in \mathfrak{P}(\mathbb{R}^n)$, let

$$\alpha(E) = \inf \left\{ \textstyle\sum_{k\in\mathbb{N}} v(R_k) : (R_k : k \in \mathbb{N}) \subset \mathfrak{I}^n_{oc}, \bigcup_{k\in\mathbb{N}} R_k \supset E \right\}$$
$$\beta(E) = \inf \left\{ \textstyle\sum_{k\in\mathbb{N}} v(P_k) : (P_k : k \in \mathbb{N}) \subset T(\mathfrak{I}^n_{oc}), \bigcup_{k\in\mathbb{N}} P_k \supset E \right\}.$$

Let $\varepsilon > 0$ be arbitrary given. By (4) we have $\alpha(E) \geq \beta(E) - \varepsilon$. Thus $\alpha(E) \geq \beta(E)$. Similarly (5) implies that $\beta(E) \geq \alpha(E) - \varepsilon$ for every $\varepsilon > 0$ and hence $\beta(E) \geq \alpha(E)$. Thus $\alpha(E) = \beta(E)$. Therefore we have

$$(6) \qquad (\mu^n_L)^*(E) = \inf \left\{ \textstyle\sum_{k\in\mathbb{N}} v(R_k) : (R_k : k \in \mathbb{N}) \subset \mathfrak{I}^n_{oc}, \bigcup_{k\in\mathbb{N}} R_k \supset E \right\}$$
$$= \inf \left\{ \textstyle\sum_{k\in\mathbb{N}} v(P_k) : (P_k : k \in \mathbb{N}) \subset T(\mathfrak{I}^n_{oc}), \bigcup_{k\in\mathbb{N}} P_k \supset E \right\}.$$

Let $E \in \mathfrak{P}(\mathbb{R}^n)$. Let us estimate $(\mu^n_L)^*(T(E))$. By (6) for an arbitrary $\varepsilon > 0$ there exists $(P_k : k \in \mathbb{N}) \subset T(\mathfrak{I}^n_{oc})$ such that $\bigcup_{k\in\mathbb{N}} P_k \supset T(E)$ and $(\mu^n_L)^*(T(E)) + \varepsilon \geq \sum_{k\in\mathbb{N}} v(P_k)$. Let $R_k = T^{-1}(P_k)$ for $k \in \mathbb{N}$. Then $(R_k : k \in \mathbb{N}) \subset \mathfrak{I}^n_{oc}$ and $\bigcup_{k\in\mathbb{N}} R_k \supset E$. Now by (b),

$$v(P_k) = \mu^n_L(P_k) = \mu^n_L(T(R_k)) = |\det M_T| \mu^n_L(R_k).$$

Then we have

$$(\mu^n_L)^*(T(E)) + \varepsilon \geq \sum_{k\in\mathbb{N}} v(P_k) = |\det M_T| \sum_{k\in\mathbb{N}} \mu^n_L(R_k) \geq |\det M_T| (\mu^n_L)^*(E).$$

Since this holds for every $\varepsilon > 0$, we have

$$(7) \qquad (\mu^n_L)^*(T(E)) \geq |\det M_T| (\mu^n_L)^*(E).$$

Now (7) holds for every non-singular linear transformation T of $\mathbb{R}^n$ into $\mathbb{R}^n$ and every $E \in \mathfrak{P}(\mathbb{R}^n)$. Applying (7) to the non-singular linear transformation T^{-1} and the set $T(E) \in \mathfrak{P}(\mathbb{R}^n)$, we have

$$(\mu^n_L)^*(T^{-1}(T(E))) \geq |\det M_{T^{-1}}| (\mu^n_L)^*(T(E)).$$

Now $T^{-1}(T(E)) = E$ and $\det M_{T^{-1}} = (\det M_T)^{-1}$. Thus the last inequality reduces to

$$(8) \qquad |\det M_T|(\mu_L^n)^*(E) \ge (\mu_L^n)^*(T(E)).$$

By (7) and (8), we have (c). ∎

Theorem 24.31. (Linear Transformation of the Lebesgue Measure on $\mathbb{R}^n$) *Consider the n-dimensional Lebesgue measure space $(\mathbb{R}^n, \mathfrak{M}_L^n, \mu_L^n)$. Let T be a non-singular linear transformation of $\mathbb{R}^n$ into $\mathbb{R}^n$. Then*

(a) $T(E) \in \mathfrak{M}_L^n$ *for every* $E \in \mathfrak{M}_L^n$.

(b) $\mu_L^n(T(E)) = |\det M_T|\mu_L^n(E)$ *for every* $E \in \mathfrak{M}_L^n$.

Proof. By Theorem 24.15, $\mathfrak{M}_L^n$ is the completion of $\mathfrak{B}_{\mathbb{R}^n}$ with respect to μ_L^n. Thus if $E \in \mathfrak{M}_L^n$, then $E = A \cup C$ where $A \in \mathfrak{B}_{\mathbb{R}^n}$ and C is a subset of a set $B \in \mathfrak{B}_{\mathbb{R}^n}$ with $\mu_L^n(B) = 0$. Then $T(E) = T(A) \cup T(C)$ and $T(C) \subset T(B)$. Since $A \in \mathfrak{B}_{\mathbb{R}^n}$, we have $T(A) \in \mathfrak{B}_{\mathbb{R}^n}$ and $\mu_L^n(T(A)) = |\det M_T|\mu_L^n(A)$ by Proposition 24.30. Similarly $T(B) \in \mathfrak{B}_{\mathbb{R}^n}$ and $\mu_L^n(T(B)) = |\det M_T|\mu_L^n(B) = 0$. Then since $T(E) = T(A) \cup T(C)$ where $T(A) \in \mathfrak{B}_{\mathbb{R}^n}$ and $T(C)$ is a subset of $T(B) \in \mathfrak{B}_{\mathbb{R}^n}$ with $\mu_L^n(T(B)) = 0$, we have $T(E) \in \mathfrak{M}_L^n$. This proves (a). (b) follows from (c) of Proposition 24.30. ∎

Theorem 24.32. (Linear Transformation of the Lebesgue Integral on $\mathbb{R}^n$) *Consider the Lebesgue measure space $(\mathbb{R}^n, \mathfrak{M}_L^n, \mu_L^n)$. Let f be an extended real-valued function on a set $D \in \mathfrak{M}_L^n$. Let T be a non-singular linear transformation of $\mathbb{R}^n$ into $\mathbb{R}^n$.*
(a) *If f is $\mathfrak{M}_L^n$-measurable on D, then $f \circ T$ is $\mathfrak{M}_L^n$-measurable on $T^{-1}(D)$ and*

$$(1) \qquad \int_D f(x)\,\mu_L^n(dx) = |\det M_T| \int_{T^{-1}(D)} (f \circ T)(x)\,\mu_L^n(dx),$$

in the sense that if one of the two integrals exists then so does the other and the two are equal. In particular if f is nonnegative, then (1) always holds.
(b) *If f is defined and $\mathfrak{M}_L^n$-measurable on $\mathbb{R}^n$, then*

$$(2) \qquad \int_{\mathbb{R}^n} f(x)\,\mu_L^n(dx) = |\det M_T| \int_{\mathbb{R}^n} (f \circ T)(x)\,\mu_L^n(dx),$$

in the same sense as above.

Proof. If T is a non-singular linear transformation of $\mathbb{R}^n$ into $\mathbb{R}^n$, then the inverse transformation of T exists and is a non-singular linear transformation of $\mathbb{R}^n$ into $\mathbb{R}^n$. Let us write S for the inverse transformation of T. Applying (a) of Theorem 24.31 to the non-singular linear transformation S, we have $S(E) \in \mathfrak{M}_L^n$ for every $E \in \mathfrak{M}_L^n$. Thus $T^{-1}(E) = S(E) \in \mathfrak{M}_L^n$ for every $E \in \mathfrak{M}_L^n$. This shows that T is a $\mathfrak{M}_L^n/\mathfrak{M}_L^n$-measurable mapping of $\mathbb{R}^n$ into $\mathbb{R}^n$.

If f is an extended real-valued $\mathfrak{M}_L^n$-measurable function on a set $D \in \mathfrak{M}_L^n$, then for the function $f \circ T$ defined on the set $T^{-1}(D) \in \mathfrak{M}_L^n$ we have for every $c \in \mathbb{R}$,

$$(f \circ T)^{-1}([-\infty, c]) = T^{-1}(f^{-1}([-\infty, c])) \in \mathfrak{M}_L^n$$

by the fact $f^{-1}([-\infty, c]) \in \mathfrak{M}_L^n$ and the fact that T is a $\mathfrak{M}_L^n/\mathfrak{M}_L^n$-measurable mapping. This shows that $f \circ T$ is a $\mathfrak{M}_L^n$-measurable function on $T^{-1}(D)$. To prove (1), let us note first that if $D, E \in \mathfrak{M}_L^n$, then since the inverse transformation S of T is a one-to-one mapping, we have $S(D \cap E) = S(D) \cap S(E) = T^{-1}(D) \cap T^{-1}(E)$. Then by (b) of Theorem 24.31, we have $\mu_L^n(D \cap E) = \mu_L^n\big((T \circ S)(D \cap E)\big) = |\det M_T| \mu_L^n\big(S(D \cap E)\big)$. Thus for $f = \mathbf{1}_E$, we have

$$\begin{aligned}
\int_D f(x)\, \mu_L^n(dx) &= \int_D \mathbf{1}_E(x)\, \mu_L^n(dx) = \mu_L^n(D \cap E) \\
&= |\det M_T| \mu_L^n\big(T^{-1}(D) \cap T^{-1}(E)\big) \\
&= |\det M_T| \int_{T^{-1}(D)} \mathbf{1}_{T^{-1}(E)}(x)\, \mu_L^n(dx) \\
&= |\det M_T| \int_{T^{-1}(D)} \mathbf{1}_E\big(T(x)\big)\, \mu_L^n(dx) \\
&= |\det M_T| \int_{T^{-1}(D)} (f \circ T)(x)\, \mu_L^n(dx),
\end{aligned}$$

proving (1) for this case. Then by the linearity of the integrals with respect to their integrands, (1) holds for nonnegative simple function f on D. Applying Lemma 8.6 and Theorem 8.5 (Monotone Convergence Theorem), we have (1) for every nonnegative extended real-valued $\mathfrak{M}_L^n$-measurable function f on D. If f is an extended real-valued $\mathfrak{M}_L^n$-measurable function on D, then we decompose $f = f^+ - f^-$ and apply the result above to f^+ and f^-. ∎

We call a mapping T of $\mathbb{R}^n$ into $\mathbb{R}^n$ of the form $T(x) = \alpha x$ for $x \in \mathbb{R}^n$ with fixed $\alpha \in \mathbb{R}$ a dilation with factor α. For a dilation T with factor $\alpha \in \mathbb{R}$, we have

$$T(\gamma x) = \alpha(\gamma x) = \gamma(\alpha x) = \gamma T(x) \quad \text{for } x \in \mathbb{R}^n \text{ and } \gamma \in \mathbb{R},$$

$$T(x + y) = \alpha(x + y) = \alpha x + \alpha y = T(x) + T(y) \quad \text{for } x, y \in \mathbb{R}^n.$$

Thus a dilation of $\mathbb{R}^n$ is a linear transformation of $\mathbb{R}^n$ into $\mathbb{R}^n$.

Corollary 24.33. (Positive Homogeneity of the Lebesgue Measure Space on $\mathbb{R}^n$) *Let T be a dilation on $\mathbb{R}^n$ given by $T(x) = \alpha x$ for $x \in \mathbb{R}^n$ where $\alpha \in \mathbb{R}$ and $\alpha \neq 0$. Then for every $E \in \mathfrak{M}_L^n$, we have $T(E) = \alpha E \in \mathfrak{M}_L^n$ and*

(1) $$\mu_L^n(\alpha E) = |\alpha|^n \mu_L^n(E).$$

If f is a $\mathfrak{M}_L^n$-measurable extended real-valued function on a set $D \in \mathfrak{M}_L^n$, then $f \circ T$ is $\mathfrak{M}_L^n$-measurable on $\frac{1}{\alpha} D$ and

(2) $$\int_D f(x)\, \mu_L^n(dx) = |\alpha|^n \int_{\frac{1}{\alpha} D} f(\alpha x)\, \mu_L^n(dx)$$

in the sense that the existence of one of the two integrals implies that of the other and the equality of the two.

Proof. A dilation $T(x) = \alpha x$ for $x \in \mathbb{R}^n$ with $\alpha \in \mathbb{R}$, $\alpha \neq 0$, is a linear transformation of $\mathbb{R}^n$ into $\mathbb{R}^n$ whose matrix M_T is equal to αI where I is the n by n identity matrix. Since $\det M_T = \alpha^n \neq 0$, T is non-singular. Thus (1) follows from Theorem 24.31 and (2) follows from Theorem 24.32. ∎

Proposition 24.34. *For $x \in \mathbb{R}^n$ and $r > 0$, let $B(x,r) = \{y \in \mathbb{R}^n : |y - x| < r\}$.*
(a) *For $x' \in \mathbb{R}^n$, we have $B(x', r) = B(x,r) + (x' - x)$ and $\mu_L^n(B(x', r)) = \mu_L^n(B(x,r))$.*
(b) *For $\alpha > 0$, we have $\alpha B(x,r) = B(\alpha x, \alpha r)$ and $\mu_L^n(\alpha B(x,r)) = \alpha^n \mu_L^n(B(x,r))$.*
(c) *For $\alpha > 0$, we have $\mu_L^n(B(x, \alpha r)) = \alpha^n \mu_L^n(B(x,r))$.*

Proof. 1. Let us prove (a). To prove $B(x', r) = B(x,r) + (x' - x)$, we show that for any $y \in \mathbb{R}^n$, $y \in B(x', r)$ if and only if $y \in B(x,r) + (x' - x)$. Now we have

$$y \in B(x,r) + (x' - x) \Leftrightarrow y - (x' - x) \in B(x,r)$$

$$\Leftrightarrow |y - (x' - x) - x| < r \Leftrightarrow |y - x'| < r \Leftrightarrow y \in B(x', r).$$

This proves $B(x', r) = B(x,r) + (x' - x)$. Then by Theorem 24.27 (Translation Invariance), we have $\mu_L^n(B(x', r)) = \mu_L^n(B(x,r))$.

2. To prove (b), note that

$$y \in B(\alpha x, \alpha r) \Leftrightarrow |y - \alpha x| < \alpha r \Leftrightarrow \left|\tfrac{1}{\alpha} y - x\right| < r$$

and

$$y \in \alpha B(x,r) \Leftrightarrow \tfrac{1}{\alpha} y \in B(x,r) \Leftrightarrow \left|\tfrac{1}{\alpha} y - x\right| < r.$$

This shows that $\alpha B(x,r) = B(\alpha x, \alpha r)$. Then $\mu_L^n(B(\alpha x, \alpha r)) = \alpha^n \mu_L^n(B(x,r))$ by (1) of Corollary 24.33.

3. By (b), we have $B(x, \alpha r) = B\left(\alpha \tfrac{1}{\alpha} x, \alpha r\right) = \alpha B\left(\tfrac{1}{\alpha} x, r\right)$. Then we have

$$\mu_L^n(B(x, \alpha r)) = \mu_L^n\left(\alpha B\left(\tfrac{1}{\alpha} x, r\right)\right) = \alpha^n \mu_L^n\left(B\left(\tfrac{1}{\alpha} x, r\right)\right) = \alpha^n \mu_L^n(B(x,r))$$

where the second equality is by (b) and the last equality is by (a). This proves (c). ∎

Observation 24.35. It is an elementary fact in linear algebra that the following three conditions on a linear transformation T of $\mathbb{R}^n$ into $\mathbb{R}^n$ are equivalent:

(i) T preserves the inner product, that is, $\langle T(x), T(y)\rangle = \langle x, y\rangle$ for $x, y \in \mathbb{R}^n$.
(ii) If $\{e_1, \ldots, e_n\}$ is an orthonormal basis for $\mathbb{R}^n$, then so is $\{T(e_1), \ldots, T(e_n)\}$.
(iii) If M_T is the matrix of T with respect to an orthonormal basis for $\mathbb{R}^n$, its transpose M_T^t satisfies the condition $M_T^t M_T = M_T M_T^t = I$.

If T satisfies any of the three conditions above, then T is called an orthogonal transformation. If T is an orthogonal transformation, then by (iii) we have $\det M_T^t M_T = \det I = 1$ so that $(\det M_T)^2 = 1$ and thus $\det M_T = \pm 1$.

Corollary 24.36. *For the Lebesgue measure of linear subspaces of* $\mathbb{R}^n$, *we have :*
(a) *If* E *is a k-dimensional linear subspace of* $\mathbb{R}^n$ *and* $k < n$, *then* $\mu_L^n(E) = 0$.
(b) *If* S *is a singular linear transformation of* $\mathbb{R}^n$ *into* $\mathbb{R}^n$, *then* $\mu_L^n\big(S(\mathbb{R}^n)\big) = 0$.
In particular, we have $S(E) \in \mathfrak{M}_L^n$ *with* $\mu_L^n\big(S(E)\big) = 0$ *for every* $E \in \mathfrak{P}(\mathbb{R}^n)$.

Proof. 1. Let E be a k-dimensional linear subspace of $\mathbb{R}^n$ with $k < n$. If $k = 0$, then E is a singleton so that $\mu_L^n(E) = 0$. Suppose $1 \le k < n$. Let $\{f_1, \dots, f_k\}$ be an orthonormal basis for E. Extend the set to an orthonormal basis $\{f_1, \dots, f_n\}$ for $\mathbb{R}^n$. Let $\{e_1, \dots, e_n\}$ be the standard orthonormal basis for $\mathbb{R}^n$. Consider the linear transformation T of $\mathbb{R}^n$ into $\mathbb{R}^n$ determined by $T(f_i) = e_i$ for $i = 1, \dots, n$. Then T is an orthogonal transformation by condition (ii) of Observation 24.35 and $T(E) = \mathbb{R}^k$. Thus T is a nonsingular linear transformation of $\mathbb{R}^n$ into $\mathbb{R}^n$ with $\det M_T = \pm 1$. Then by Theorem 24.31, we have $\mu_L^n(E) = \mu_L^n\big(T(E)\big) = \mu_L^n\big(\mathbb{R}^k\big)$. But $\mu_L^n\big(\mathbb{R}^k\big) = \big[\mu_L(\mathbb{R})\big]^k\big[\mu_L(\{0\})\big]^{n-k} = 0$ by our convention $\infty \cdot 0 = 0$ in the definition of product measures. Thus $\mu_L^n(E) = 0$.

2. If S is a singular linear transformation of $\mathbb{R}^n$ into $\mathbb{R}^n$, then $S(\mathbb{R}^n)$ is a k-dimensional linear subspace of $\mathbb{R}^n$ with $k < n$ so that $\mu_L^n\big(S(\mathbb{R}^n)\big) = 0$ by (a). If $E \in \mathfrak{P}(\mathbb{R}^n)$, then $S(E) \subset S(\mathbb{R}^n)$. Since $S(\mathbb{R}^n)$ is a null set in the complete measure space $\big(\mathbb{R}^n, \mathfrak{M}_L^n, \mu_L^n\big)$, we have $S(E) \in \mathfrak{M}_L^n$ with $\mu_L^n\big(S(E)\big) = 0$. ■

Problems

Prob. 24.1. Let O be an open set in $\mathbb{R}^n$.
(a) Show that if $O \neq \emptyset$, then $\mu_L^n(O) > 0$.
(b) Show that if O is a bounded set, that is, it is contained in a ball with finite radius, then $\mu_L^n(O) < \infty$.

Prob. 24.2. Given $\mathbb{R}^{n+1}$ where $n \in \mathbb{N}$. Let us identify $\mathbb{R}^n$ with the subset of $\mathbb{R}^{n+1}$ given by $\mathbb{R}^n \times \{0\}$. Show that $\mu_L^{n+1}\big(\mathbb{R}^n\big) = 0$.

Prob. 24.3. Let ν be a measure on the measurable space $\big(\mathbb{R}^n, \mathfrak{B}_{\mathbb{R}^n}\big)$. Suppose ν is translation invariant, $\nu(O) > 0$ for every nonempty open set O in $\mathbb{R}^n$ and $\upsilon(K) < \infty$ for every compact set K in $\mathbb{R}^n$. Show that there exists a constant $c \in (0, \infty)$ such that $\nu = c\mu_L^n$ on $\mathfrak{B}_{\mathbb{R}^n}$, and indeed $c = \nu\big((0, 1] \times \cdots \times (0, 1]\big)$.

Prob. 24.4. By a curve in $\mathbb{R}^n$, we mean a mapping $C = (\varphi_1, \dots, \varphi_n)$ of an interval $[a, b]$ into $\mathbb{R}^n$ where φ_i is a real-valued function on $[a, b]$ for $i = 1, \dots, n$. The graph of the curve C is the image of $[a, b]$ by the mapping C, that is, $\Gamma = C\big([a, b]\big) = (\varphi_1, \dots, \varphi_n)\big([a, b]\big)$. Show that if φ_i satisfies a Lipschitz condition $|\varphi_i(s) - \varphi_i(t)| \le \alpha_i |s - t|$ for $s, t \in [a, b]$ where $\alpha_i > 0$ for $i = 1, \dots, n$, then $\Gamma \in \mathfrak{B}_{\mathbb{R}^n}$ and $\mu_L^n(\Gamma) = 0$.

§25 Differentiation on the Euclidean Space

[I] The Lebesgue Differentiation Theorem on $\mathbb{R}^n$

Notations. For $x \in \mathbb{R}^n$ and $r \in (0,\infty)$, let $B(x,r) = \{y \in \mathbb{R}^n : |y-x| < r\}$, let $\overline{B}(x,r) = \{y \in \mathbb{R}^n : |y-x| \le r\}$, and let $S(x,r) = \{y \in \mathbb{R}^n : |y-x| = r\}$ where $|\cdot|$ is the Euclidean norm on $\mathbb{R}^n$, that is, $|x| = \{\xi_1^2 + \cdots + \xi_n^2\}^{1/2}$ for $x = (\xi_1,\dots,\xi_n) \in \mathbb{R}^n$. We refer to these three sets as an open ball, a closed ball and a spherical hypersurface respectively with center at x and radius r. We write $B(x,r)^c$, $\overline{B}(x,r)^c$, and $S(x,r)^c$ for the complements of these sets. Recall that a point x in a topological space X is a boundary point of a set E in X if every open set containing x contains at least one point of E and at least one point of E^c and the boundary of E, denoted by ∂E, is the set of all boundary points of E. Thus $\partial E = (E^\circ \cup (E^c)^\circ)^c$. The spherical hypersurface $S(x,r)$ is the boundary of the open ball $B(x,r)$ as well as that of the closed ball $\overline{B}(x,r)$. Note that $B(x,r)$, $\overline{B}(x,r)$, and $S(x,r)$ are members of $\mathfrak{B}_{\mathbb{R}^n}$.

A subset E of $\mathbb{R}^n$ is called a bounded set if $E \subset B(0,r)$ for some $r \in (0,\infty)$. If $E \in \mathfrak{M}_L^n$ is a bounded set then there exists $r \in (0,\infty)$ such that $E \subset B(0,r) \subset \times_{i=1}^n[-r,r]$ so that $\mu_L^n(E) \le \mu_L^n(\times_{i=1}^n[-r,r]) = (2r)^n < \infty$.

Since $B(x,r)$ is a bounded set we have $\mu_L^n(B(x,r)) < \infty$. We postpone the computation of $\mu_L^n(B(x,r))$ to §26. By the translation invariance and the positive homogeneity of μ_L^n, we have $\mu_L^n(\overline{B}(x,r)) = \mu_L^n(\overline{B}(0,r)) = \mu_L^n(r\overline{B}(0,1)) = r^n \mu_L^n(\overline{B}(0,1))$ as in Proposition 24.34. As a rough estimate for $\mu_L^n(\overline{B}(0,1))$, consider the n-dimensional closed cube inscribed in $\overline{B}(0,1)$ given by $C_1 = \times_{i=1}^n\left[-n^{-1/2}, n^{-1/2}\right] \subset \overline{B}(0,1)$ with $\mu_L^n(C_1) = \left(2n^{-1/2}\right)^n$. For the closed n-dimensional cube $C_2 = \times_{i=1}^n[-1,1] \supset \overline{B}(0,1)$, we have $\mu_L^n(C_2) = 2^n$. Thus $\mu_L^n(\overline{B}(0,1)) \in \left[\left(2n^{-1/2}\right)^n, 2^n\right]$.

Definition 25.1. *Let f be an extended real-valued $\mathfrak{M}_L^n$-measurable function on $\mathbb{R}^n$. We say that f is locally μ_L^n-integrable if f is μ_L^n-integrable on every bounded set $E \in \mathfrak{M}_L^n$. We write $\mathcal{L}_{loc}^1(\mathbb{R}^n, \mathfrak{M}_L^n, \mu_L^n)$ for the class of all locally μ_L^n-integrable functions on $\mathbb{R}^n$.*

Observation 25.2. (a) If f is an extended real-valued $\mathfrak{M}_L^n$-measurable function on $\mathbb{R}^n$, then $f \in \mathcal{L}_{loc}^1(\mathbb{R}^n, \mathfrak{M}_L^n, \mu_L^n)$ if and only if $|f| \in \mathcal{L}_{loc}^1(\mathbb{R}^n, \mathfrak{M}_L^n, \mu_L^n)$.
(b) $\mathcal{L}^1(\mathbb{R}^n, \mathfrak{M}_L^n, \mu_L^n) \subset \mathcal{L}_{loc}^1(\mathbb{R}^n, \mathfrak{M}_L^n, \mu_L^n)$.
(c) If $f, g \in \mathcal{L}_{loc}^1(\mathbb{R}^n, \mathfrak{M}_L^n, \mu_L^n)$ and $\alpha, \beta \in \mathbb{R}$, then $\alpha f + \beta g \in \mathcal{L}_{loc}^1(\mathbb{R}^n, \mathfrak{M}_L^n, \mu_L^n)$.
(d) If f is a real-valued $\mathfrak{M}_L^n$-measurable function on $\mathbb{R}^n$ and f is bounded on every bounded
subset of $\mathbb{R}^n$, then $f \in \mathcal{L}_{loc}^1(\mathbb{R}^n, \mathfrak{M}_L^n, \mu_L^n)$.
(e) If f is a real-valued continuous function $\mathbb{R}^n$, then $f \in \mathcal{L}_{loc}^1(\mathbb{R}^n, \mathfrak{M}_L^n, \mu_L^n)$.
(f) If $f \in \mathcal{L}_{loc}^1(\mathbb{R}^n, \mathfrak{M}_L^n, \mu_L^n)$, then $\{\mathbb{R}^n : |f| = \infty\}$ is a null set in $(\mathbb{R}^n, \mathfrak{M}_L^n, \mu_L^n)$.

Proof. (a) - (e) are immediate from Definition 25.1. Let us prove (f). Thus assume that

$f \in \mathcal{L}^1_{loc}(\mathbb{R}^n, \mathfrak{M}^n_L, \mu^n_L)$. Now $\mathbb{R}^n = \bigcup_{k\in\mathbb{N}} B(0,k)$. Since $B(0,k)$ is a bounded set in $\mathfrak{M}^n_L$, f is μ^n_L-integrable on $B(0,k)$ and this implies that $|f| < \infty$ a.e. on $B(0,k)$, that is, there exists a null set E_k in $(\mathbb{R}^n, \mathfrak{M}^n_L, \mu^n_L)$ contained in $B(0,k)$ such that $|f| < \infty$ on $B(0,k) \setminus E_k$. Let $E = \bigcup_{k\in\mathbb{N}} E_k$, a null set in $(\mathbb{R}^n, \mathfrak{M}^n_L, \mu^n_L)$. Now we have

$$|f| < \infty \quad \text{on} \bigcup_{K\in\mathbb{N}} \{B(0,k) \setminus E_k\} = \bigcup_{n\in\mathbb{N}} B(0,k) \setminus \bigcup_{k\in\mathbb{N}} E_k = \mathbb{R}^n \setminus E.$$

This implies that $\{\mathbb{R}^n : |f| = \infty\} \subset E$. Then since E is a null set in the complete measure space $(\mathbb{R}^n, \mathfrak{M}^n_L, \mu^n_L)$, an arbitrary subset of E is a null set in $(\mathbb{R}^n, \mathfrak{M}^n_L, \mu^n_L)$. Thus $\{\mathbb{R}^n : |f| = \infty\}$ is a null set in $(\mathbb{R}^n, \mathfrak{M}^n_L, \mu^n_L)$. ∎

Definition 25.3. *Let $f \in \mathcal{L}^1_{loc}(\mathbb{R}^n, \mathfrak{M}^n_L, \mu^n_L)$. For each $r \in (0,\infty)$, we define a real-valued function $A_r f$ on $\mathbb{R}^n$ by setting for $x \in \mathbb{R}^n$*

(1) $$(A_r f)(x) = \frac{1}{\mu^n_L(B(x,r))} \int_{B(x,r)} f(y)\, \mu^n_L(dy).$$

We define a function Af on $\mathbb{R}^n$ by setting for $x \in \mathbb{R}^n$

(2) $$(Af)(x) = \lim_{r\to 0} (A_r f)(x),$$

provided the limit exists.

($(A_r f)(x)$ is the mean value of f over an open ball of radius r centered at x. $(Af)(x)$ is the limit of the average functions $(A_r f)(x)$ as $r \to 0$.)

We shall show that for every $f \in \mathcal{L}^1_{loc}(\mathbb{R}^n, \mathfrak{M}^n_L, \mu^n_L)$, $(Af)(x) = \lim_{r\to 0}(A_r f)(x)$ exists and is equal to $f(x)$ for a.e. x in $(\mathbb{R}^n, \mathfrak{M}^n_L, \mu^n_L)$. As the first step we show that this is the case at every continuity point of f.

Lemma 25.4. *Let $f \in \mathcal{L}^1_{loc}(\mathbb{R}^n, \mathfrak{M}^n_L, \mu^n_L)$. If f is continuous at $x_0 \in \mathbb{R}^n$, then $(Af)(x_0)$ exists and moreover $(Af)(x_0) = f(x_0)$.*

Proof. Since $f \in \mathcal{L}^1_{loc}(\mathbb{R}^n, \mathfrak{M}^n_L, \mu^n_L)$, f is μ^n_L-integrable on $B(x_0, r)$ for every $r \in (0,\infty)$. If f is continuous at x_0, then for every $\varepsilon > 0$ there exists $\delta > 0$ such that $|f(x) - f(x_0)| < \varepsilon$ for $x \in B(x_0, \delta)$. Thus for $r \in (0,\delta)$, we have

$$\begin{aligned} |(A_r f)(x_0) - f(x_0)| &= \left| \frac{1}{\mu^n_L(B(x_0,r))} \int_{B(x_0,r)} f(y)\, \mu^n_L(dy) - f(x_0) \right| \\ &\le \frac{1}{\mu^n_L(B(x_0,r))} \int_{B(x_0,r)} |f(y) - f(x_0)|\, \mu^n_L(dy) \\ &\le \varepsilon. \end{aligned}$$

This shows that $\lim_{r\to 0}(A_r f)(x_0) = f(x_0)$. ∎

Observation 25.5. Let $c = \mu_L^n(B(0,1))$. Then $c \in (0,\infty)$. For any $x \in \mathbb{R}^n$ and any $r \in (0,\infty)$, we have

(a) $\mu_L^n(B(x,r)) = r^n c$.

(b) $\mu_L^n(\overline{B}(x,r)) = r^n c$.

(c) $\mu_L^n(S(x,r)) = 0$.

(d) For fixed $x_0 \in \mathbb{R}^n$ and $r_0 \in (0,\infty)$, we have $\lim_{x\to x_0} \mathbf{1}_{B(x,r_0)}(y) = \mathbf{1}_{B(x_0,r_0)}(y)$ for $y \notin S(x_0,r_0)$. Thus $\lim_{x\to x_0} \mathbf{1}_{B(x,r_0)} = \mathbf{1}_{B(x_0,r_0)}$, μ_L^n-a.e. on $\mathbb{R}^n$.

(e) For fixed $r_0 \in (0,\infty)$ and $x_0 \in \mathbb{R}^n$, we have $\lim_{r\to r_0} \mathbf{1}_{B(x_0,r)}(y) = \mathbf{1}_{B(x_0,r_0)}(y)$ for $y \notin S(x_0,r_0)$. Thus $\lim_{r\to r_0} \mathbf{1}_{B(x_0,r)} = \mathbf{1}_{B(x_0,r_0)}$, μ_L^n-a.e. on $\mathbb{R}^n$.

Proof. 1. Note that we have $\times_{i=1}^n[-\varepsilon,\varepsilon] \subset B(0,1) \subset \times_{i=1}^n[-1,1]$ for sufficiently small $\varepsilon > 0$. Then we have $\mu_L^n(\times_{i=1}^n[-\varepsilon,\varepsilon]) \le \mu_L^n(B(0,1)) \le \mu_L^n(\times_{i=1}^n[-1,1])$ by the monotonicity of the measure μ_L^n. Thus $c \in [(2\varepsilon)^n, 2^n]$ and hence $c \in (0,\infty)$.

Since $B(x,r)$ is a translate of $B(0,r)$ and $B(0,r) = rB(0,1)$, we have by Theorem 24.27 and Corollary 24.33 the equalities $\mu_L^n(B(x,r)) = \mu_L^n(B(0,r)) = r^n\mu_L^n(B(0,1)) = r^n c$. This proves (a). Since $\left(B\left(x, r+\frac{1}{k}\right) : k \in \mathbb{N}\right)$ is a decreasing sequence of sets in $\mathfrak{B}_{\mathbb{R}^n}$ with finite μ_L^n measures and since $\overline{B}(x,r) = \bigcap_{n\in\mathbb{N}} B\left(x, r+\frac{1}{k}\right)$, we have

$$\mu_L^n(\overline{B}(x,r)) = \lim_{k\to\infty} \mu_L^n\left(B\left(x, r+\tfrac{1}{k}\right)\right) = \lim_{k\to\infty}\left(r+\tfrac{1}{k}\right)^n c = r^n c.$$

This proves (b). Since $B(x,r) \cup S(x,r) = \overline{B}(x,r)$ and $B(x,r) \cap S(x,r) = \emptyset$, we have

$$\mu_L^n(\overline{B}(x,r)) = \mu_L^n(B(x,r)) + \mu_L^n(S(x,r)).$$

Then since $\mu_L^n(\overline{B}(x,r)) = \mu_L^n(B(x,r)) = r^n c$ by (a) and (b), we have $\mu_L^n(S(x,r)) = 0$. This proves (c).

2. Next let us prove (d). Let $x_0 \in \mathbb{R}^n$ and $r_0 \in (0,\infty)$ be fixed. Let us show that $\lim_{x\to x_0} \mathbf{1}_{B(x,r_0)}(y) = \mathbf{1}_{B(x_0,r_0)}(y)$ for every $y \in S(x_0,r_0)^c = B(x_0,r_0) \cup \overline{B}(x_0,r_0)^c$. First, let $y \in B(x_0,r_0)$ be arbitrarily chosen and fixed. Let $\delta = r_0 - |y - x_0| \in (0,r_0)$. Then for any $x \in B(x_0,\delta)$, we have $|y-x| \le |y-x_0| + |x_0-x| < r_0$ so that $y \in B(x,r_0)$. Thus $\mathbf{1}_{B(x,r_0)}(y) = 1$ for every $x \in B(x_0,\delta)$. Then $\lim_{x\to x_0} \mathbf{1}_{B(x,r_0)}(y) = 1 = \mathbf{1}_{B(x_0,r_0)}(y)$. Next, let $y \in \overline{B}(x_0,r_0)^c$ be arbitrarily fixed. Let $\delta = |y-x_0| - r_0 \in (0,\infty)$. Then for any $x \in B(x_0,\delta)$, we have $|y-x| \ge |y-x_0| - |x_0-x| > r_0+\delta-\delta = r_0$ so that $y \notin B(x,r_0)$. Thus $\mathbf{1}_{B(x,r_0)}(y) = 0$ for every $x \in B(x_0,\delta)$. Then $\lim_{x\to x_0} \mathbf{1}_{B(x,r_0)}(y) = 0 = \mathbf{1}_{B(x_0,r_0)}(y)$. This proves (d).

3. To prove (e), first let $y \in B(x_0,r_0)$ be arbitrarily fixed. If $y = x_0$, then $y \in B(x_0,r)$ for every $r \in (0,\infty)$ so that $\mathbf{1}_{B(x_0,r)}(y) = 1$ for every $r \in (0,\infty)$ and this implies that $\lim_{r\to r_0} \mathbf{1}_{B(x_0,r)}(y) = 1 = \mathbf{1}_{B(x_0,r_0)}(y)$. If $y \in B(x_0,r_0) \setminus \{x_0\}$, let $\delta = r_0 - |y-y_0| \in (0,r_0)$. Then $y \in B(x_0,r)$ for every $r \in (r_0-\delta,\infty)$ so that $\lim_{r\to r_0} \mathbf{1}_{B(x_0,r)}(y) = 1 = \mathbf{1}_{B(x_0,r_0)}(y)$. Next, let $y \in \overline{B}(x_0,r_0)^c$ be arbitrarily fixed. Let $\delta = |y-x_0| - r_0 \in (0,\infty)$ so that $y \notin B(x_0,r)$ for every $r \in (0,r_0+\delta)$. Thus $\mathbf{1}_{B(x_0,r)}(y) = 0$ for every $r \in (0,r_0+\delta)$ and then $\lim_{r\to r_0} \mathbf{1}_{B(x_0,r)}(y) = 0 = \mathbf{1}_{B(x_0,r_0)}(y)$. This proves (e). ∎

Lemma 25.6. *For every $f \in \mathcal{L}^1_{loc}(\mathbb{R}^n, \mathfrak{M}^n_L, \mu^n_L)$, the function $(A_r f)(x)$ for $x \in \mathbb{R}^n$ and $r \in (0, \infty)$ has the following continuity properties:*
(a) *For every $r \in (0, \infty)$, $(A_r f)(x)$ is a continuous function of $x \in \mathbb{R}^n$.*
(b) *For every $x \in \mathbb{R}^n$, $(A_r f)(x)$ is a continuous function of $r \in (0, \infty)$.*

Proof. **1.** Let $r_0 \in (0, \infty)$ be fixed. To show that $(A_{r_0} f)(x)$ is a continuous function of $x \in \mathbb{R}^n$, we show that for every $x_0 \in \mathbb{R}^n$ we have $\lim_{x \to x_0} (A_{r_0} f)(x) = (A_{r_0} f)(x_0)$. Since $\mu^n_L(B(x, r_0)) = \mu^n_L(B(x_0, r_0)) > 0$ for every $x \in \mathbb{R}^n$, it suffices to show that $\lim_{x \to x_0} \int_{B(x,r_0)} f \, d\mu^n_L = \int_{B(x_0,r_0)} f \, d\mu^n_L$. Let $\delta > 0$ be arbitrarily fixed and let $R > |x_0| + \delta + r_0$. Then for every $x \in B(x_0, \delta)$, we have $B(x, r_0) \subset B(0, R)$. Since $f \in \mathcal{L}^1_{loc}(\mathbb{R}^n, \mathfrak{M}^n_L, \mu^n_L)$, f is μ^n_L-integrable on $B(0, R)$. For every $x \in B(x_0, \delta)$, we have

(1) $$\int_{B(x,r_0)} f \, d\mu^n_L = \int_{B(0,R)} \mathbf{1}_{B(x,r_0)} f \, d\mu^n_L.$$

Now $|\mathbf{1}_{B(x,r_0)} f| \leq |\mathbf{1}_{B(0,R)} f|$ which is μ^n_L-integrable on $B(0, R)$ and by (d) of Observation 25.5 we have $\lim_{x \to x_0} \mathbf{1}_{B(x,r_0)} = \mathbf{1}_{B(x_0,r_0)}$, μ^n_L-a.e. on $\mathbb{R}^n$. Thus by the Dominated Convergence Theorem (Theorem 9.20), we have

(2) $$\lim_{x \to x_0} \int_{B(x,r_0)} f \, d\mu^n_L = \lim_{x \to x_0} \int_{B(0,R)} \mathbf{1}_{B(x,r_0)} f \, d\mu^n_L$$
$$= \int_{B(0,R)} \mathbf{1}_{B(x_0,r_0)} f \, d\mu^n_L = \int_{B(x_0,r_0)} f \, d\mu^n_L.$$

2. Let $x_0 \in \mathbb{R}^n$ be fixed. To show that $(A_r f)(x_0)$ is a continuous function of $r \in (0, \infty)$, we show that for every $r_0 \in (0, \infty)$, we have $\lim_{r \to r_0} (A_r f)(x_0) = (A_{r_0} f)(x_0)$. Since $\lim_{r \to r_0} \mu^n_L(B(x_0, r)) = \lim_{r \to r_0} r^n \mu^n_L(B(x_0, 1)) = r_0^n \mu^n_L(B(x_0, 1)) = \mu^n_L(B(x_0, r_0)) > 0$, it suffices to show that $\lim_{r \to r_0} \int_{B(x_0,r)} f \, d\mu^n_L = \int_{B(x_0,r_0)} f \, d\mu^n_L$. Let $\delta \in (0, r_0)$ be arbitrarily fixed. Let $R > r_0 + \delta + |x_0|$. Then for every $r \in (r_0 - \delta, r_0 + \delta)$, we have $B(x_0, r) \subset B(0, R)$ so that

(3) $$\int_{B(x_0,r)} f \, d\mu^n_L = \int_{B(0,R)} \mathbf{1}_{B(x_0,r)} f \, d\mu^n_L.$$

Then since we have $|\mathbf{1}_{B(x_0,r)} f| \leq |\mathbf{1}_{B(0,R)} f|$ which is μ^n_L-integrable on $B(0, R)$ and since $\lim_{r \to r_0} \mathbf{1}_{B(x_0,r)} = \mathbf{1}_{B(x_0,r_0)}$, μ^n_L-a.e. on $\mathbb{R}^n$ by (e) of Observation 25.5, we have

$$\lim_{r \to r_0} \int_{B(x_0,r)} f \, d\mu^n_L = \int_{B(x_0,r_0)} f \, d\mu^n_L$$

by the Dominated Convergence Theorem. ■

We introduce the Hardy-Littlewood maximal function which will serve in estimating the family of averaging functions $\{A_r f : r \in (0, \infty)\}$.

Definition 25.7. *For $f \in \mathcal{L}^1_{loc}(\mathbb{R}^n, \mathfrak{M}^n_L, \mu^n_L)$, the Hardy-Littlewood maximal function Mf of f is a nonnegative extended real-valued function on $\mathbb{R}^n$ defined by setting for $x \in \mathbb{R}^n$*

$$(Mf)(x) = \sup_{r \in (0,\infty)} \big(A_r|f|\big)(x) = \sup_{r \in (0,\infty)} \frac{1}{\mu^n_L\big(B(x,r)\big)} \int_{B(x,r)} |f|\, d\mu^n_L.$$

Observation 25.8. Note that if $f \in \mathcal{L}^1_{loc}(\mathbb{R}^n, \mathfrak{M}^n_L, \mu^n_L)$, then $|f| \in \mathcal{L}^1_{loc}(\mathbb{R}^n, \mathfrak{M}^n_L, \mu^n_L)$ also, and Mf is defined in terms of $|f|$ rather than f itself. Thus $Mf = M|f|$. By (a) of Lemma 25.6, for every $r \in (0, \infty)$, $A_r|f|$ is a real-valued continuous function on $\mathbb{R}^n$. As the supremum of a collection of continuous functions on $\mathbb{R}^n$, Mf is a lower semicontinuous function on $\mathbb{R}^n$ by (a) of Theorem 15.84. Then for every $\alpha \in \mathbb{R}$, $\{\mathbb{R}^n : Mf > \alpha\}$ is an open set in $\mathbb{R}^n$ by (d) of Observation 15.81 and so that Mf is a $\mathfrak{B}_{\mathbb{R}^n}$-measurable function on $\mathbb{R}^n$.

Let us note also that according to (b) of Lemma 25.6, for each fixed $x \in \mathbb{R}^n$, $\big(A_r|f|\big)(x)$ is a continuous function of $r \in (0, \infty)$. Thus for an arbitrary countable dense subset D of $(0, \infty)$, we have

$$(Mf)(x) = \sup_{r \in (0,\infty)} \big(A_r|f|\big)(x) = \sup_{r \in D} \big(A_r|f|\big)(x).$$

Then Mf, as the supremum of countably many continuous and hence $\mathfrak{B}_{\mathbb{R}^n}$-measurable functions on $\mathbb{R}^n$, is a $\mathfrak{B}_{\mathbb{R}^n}$-measurable functions on $\mathbb{R}^n$ by Theorem 4.22.

Let f be an extended real-valued $\mathfrak{A}$-measurable function on a measure space $(X, \mathfrak{A}, \mu)$. For $\alpha \in (0, \infty)$ consider the set $\{X : |f| > \alpha\} \in \mathfrak{A}$. The set decreases as α increases. Thus $\mu\{X : |f| > \alpha\}$ decreases as α increases. Let us consider the behavior of $\alpha\mu\{X : |f| > \alpha\}$ as α increases. We show that if f in μ-integrable on X then $\lim_{\alpha\to\infty} \alpha\mu\{X : |f| > \alpha\} = 0$.

Observation 25.9. (a) Let $(X, \mathfrak{A}, \mu)$ be a measure space and let $f \in \mathcal{L}^1(X, \mathfrak{A}, \mu)$. Then

(1) $$\alpha\mu\{X : |f| > \alpha\} \le \int_X |f|\, d\mu < \infty \quad \text{for } \alpha \in (0, \infty)$$

(2) $$\lim_{\alpha\to\infty} \alpha\mu\{X : |f| > \alpha\} = 0.$$

(b) If $f \in \mathcal{L}^1(\mathbb{R}^n, \mathfrak{M}^n_L, \mu^n_L)$, then there exists a constant $C \ge 0$ such that

(3) $$\alpha\mu^n_L\{\mathbb{R}^n : |f| > \alpha\} \le C \quad \text{for } \alpha \in (0, \infty).$$

If $f \in \mathcal{L}^1_{loc}(\mathbb{R}^n, \mathfrak{M}^n_L, \mu^n_L)$ and $f \notin \mathcal{L}^1(\mathbb{R}^n, \mathfrak{M}^n_L, \mu^n_L)$, such a constant $C \ge 0$ may not exist.

(c) Let f be an extended real-valued $\mathfrak{M}^n_L$-measurable function on $\mathbb{R}^n$. The existence of a constant $C \ge 0$ such that $\alpha\mu^n_L\{\mathbb{R}^n : |f| > \alpha\} \le C$ for $\alpha \in (0, \infty)$ does not imply that $f \in \mathcal{L}^1_{loc}(\mathbb{R}^n, \mathfrak{M}^n_L, \mu^n_L)$.

Proof. 1. If f is a μ-integrable extended real-valued $\mathfrak{A}$-measurable function on a measure space $(X, \mathfrak{A}, \mu)$, then for every $\alpha \in (0, \infty)$ we have

$$\int_X |f|\, d\mu \ge \int_{\{X:|f|>\alpha\}} |f|\, d\mu \ge \alpha\mu\{X : |f| > \alpha\}.$$

This proves (1). To prove (2), let $F = \{X : |f| < \infty\}$. Then $F^c = \{X : |f| = \infty\}$. The μ-integrability of f on X implies that f is finite μ-a.e. on X by (f) of Observation 9.2. Thus we have $\mu(F^c) = 0$. For $\alpha \in (0, \infty)$, let $D_\alpha = \{X : |f| > \alpha\}$. Then $D_\alpha^c = \{X : |f| \leq \alpha\}$. Thus we have $F = \bigcup_{\alpha \in (0,\infty)} D_\alpha^c$ and $\lim_{\alpha \to \infty} \mathbf{1}_{D_\alpha^c} = \mathbf{1}_F$ on X. Now we have for $\alpha \in (0, \infty)$

$$(4) \qquad \int_X |f| \, d\mu = \int_{D_\alpha} |f| \, d\mu + \int_{D_\alpha^c} |f| \, d\mu.$$

For the second integral on the right side of (4), we have

$$\begin{aligned} \lim_{\alpha \to \infty} \int_{D_\alpha^c} |f| \, d\mu &= \lim_{\alpha \to \infty} \int_X \mathbf{1}_{D_\alpha^c} |f| \, d\mu = \int_X \lim_{\alpha \to \infty} \mathbf{1}_{D_\alpha^c} |f| \, d\mu \\ &= \int_X \mathbf{1}_F |f| \, d\mu = \int_F |f| \, d\mu + \int_{F^c} |f| \, d\mu \\ &= \int_X |f| \, d\mu \end{aligned}$$

where the second equality is by the Dominated Convergence Theorem and the fourth equality is by the fact that $\mu(F^c) = 0$ and $\int_{F^c} |f| \, d\mu = 0$. Thus letting $\alpha \to \infty$ on the right side of (4), we obtain

$$(5) \qquad \lim_{\alpha \to \infty} \int_{D_\alpha} |f| \, d\mu = 0.$$

Then since $\int_{D_\alpha} |f| \, d\mu \geq \alpha \mu(D_\alpha) \geq 0$, (5) implies $\lim_{\alpha \to \infty} \alpha \mu(D_\alpha) = 0$. This proves (2).

2. If $f \in \mathcal{L}^1(\mathbb{R}^n, \mathfrak{M}_L^n, \mu_L^n)$, then by (1) we have

$$\alpha \mu_L^n \{\mathbb{R}^n : |f| > \alpha\} \leq C \quad \text{for } \alpha \in (0, \infty)$$

where $C = \int_{\mathbb{R}^n} |f| \, d\mu_L^n \geq 0$. This proves (3).

Let us construct a function f on $\mathbb{R}^n$ such that $f \in \mathcal{L}_{loc}^1(\mathbb{R}^n, \mathfrak{M}_L^n, \mu_L^n)$ and $f \notin \mathcal{L}^1(\mathbb{R}^n, \mathfrak{M}_L^n, \mu_L^n)$ and (3) does not hold. Let $\gamma > 1$ and let f be a constant function $f = \gamma$ on $\mathbb{R}^n$. Then $\int_{\mathbb{R}^n} |f| \, d\mu_L^n = \gamma \, \mu_L^n(\mathbb{R}^n) = \infty$ so that $f \notin \mathcal{L}^1(\mathbb{R}^n, \mathfrak{M}_L^n, \mu_L^n)$. On the other hand since f is bounded on $\mathbb{R}^n$ so that f is locally μ_L^n-integrable and hence $f \in \mathcal{L}_{loc}^1(\mathbb{R}^n, \mathfrak{M}_L^n, \mu_L^n)$. Now consider $\alpha = \frac{1}{\gamma} \in (0, \infty)$. We have

$$\alpha \mu_L^n \{\mathbb{R}^n : |f| > \alpha\} = \tfrac{1}{\gamma} \mu_L^n \{\mathbb{R}^n : \gamma > \tfrac{1}{\gamma}\} = \tfrac{1}{\gamma} \mu_L^n(\mathbb{R}^n) = \infty.$$

Thus (3) does not hold.

3. To prove (c), we construct a function $f \notin \mathcal{L}_{loc}^1(\mathbb{R}^n, \mathfrak{M}_L^n, \mu_L^n)$ for which (3) holds. We do this for the case $n = 1$. Let f be a real-valued $\mathfrak{M}_L$-measurable function on $\mathbb{R}$ defined by

$$f(x) = \begin{cases} \frac{1}{x} & \text{for } x \in (0, 1), \\ 0 & \text{for } x \in (0, 1)^c. \end{cases}$$

We have $\int_{(0,1)} |f| \, d\mu_L = \int_{(0,1)} \frac{1}{x} \mu_L(dx) = \infty$. Thus f is not μ_L-integrable on the bounded set $(0, 1)$ and hence $f \notin \mathcal{L}^1_{loc}(\mathbb{R}^n, \mathfrak{M}^n_L, \mu^n_L)$. Let us show that f satisfies (3) with $C = 1$. Now for $\alpha \in (0, 1)$, we have $\{\mathbb{R} : |f| > \alpha\} \subset (0, 1)$ so that we have

$$\alpha \mu_L\{\mathbb{R} : |f| > \alpha\} \leq \alpha \mu_L\big((0, 1)\big) = \alpha < 1.$$

For $\alpha \in [1, \infty)$, $\{\mathbb{R} : |f| > \alpha\} = \left\{x \in (0, 1) : \frac{1}{x} > \alpha\right\} = \left\{x \in (0, 1) : x < \frac{1}{\alpha}\right\} = \left(0, \frac{1}{\alpha}\right)$ so that we have

$$\alpha \mu_L\{\mathbb{R} : |f| > \alpha\} = \alpha \mu_L\left(0, \tfrac{1}{\alpha}\right) = \alpha \tfrac{1}{\alpha} = 1.$$

Thus we have shown that $\alpha \mu_L\{\mathbb{R} : |f| > \alpha\} \leq 1$ for $\alpha \in (0, \infty)$. ∎

The Hardy-Littlewood Maximal Theorem shows that if $f \in \mathcal{L}^1(\mathbb{R}^n, \mathfrak{M}^n_L, \mu^n_L)$, then its maximal function Mf satisfies condition (3) above with the constant $C = 3^n \|f\|_1$. To prove this we need the following lemma.

Lemma 25.10. *Let $\mathfrak{C}$ be an arbitrary collection of open balls in $\mathbb{R}^n$. If $c \in \mathbb{R}$ is such that $0 < c < \mu^n_L\left(\bigcup_{B \in \mathfrak{C}} B\right)$, then there exists a finite disjoint subcollection $\{B_1, \ldots, B_p\}$ of $\mathfrak{C}$ such that $c < 3^n \sum_{k=1}^p \mu^n_L(B_k)$.*

(Thus if $\mu^n_L\left(\bigcup_{B \in \mathfrak{C}} B\right) = \infty$ then we can choose a finite disjoint subcollection such that the sum of the measures is as large as we wish and if $\mu^n_L\left(\bigcup_{B \in \mathfrak{C}} B\right) < \infty$ then for any positive number $c < \mu^n_L\left(\bigcup_{B \in \mathfrak{C}} B\right)$ we can choose a finite disjoint subcollection such that the sum of the measures is greater than $\frac{c}{3^n}$.)

Proof. Note that $\bigcup_{B \in \mathfrak{C}} B$ may be an uncountable union but it is an open set so that it is a member of $\mathfrak{B}_{\mathbb{R}^n} \subset \mathfrak{M}^n_L$. Let $c < \mu^n_L\left(\bigcup_{B \in \mathfrak{C}} B\right)$. By (b) of Proposition 24.16, there exists an increasing sequence $(K_j : j \in \mathbb{N})$ of compact sets contained in $\bigcup_{B \in \mathfrak{C}} B$ with $\lim_{j \to \infty} \mu^n_L(K_j) = \mu^n_L\left(\bigcup_{B \in \mathfrak{C}} B\right)$. Thus there exists a compact set $K \subset \bigcup_{B \in \mathfrak{C}} B$ such that $\mu^n_L(K) > c$. Since $\bigcup_{B \in \mathfrak{C}} B \supset K$, there exists a finite subcollection $\{B_1, \ldots, B_N\}$ of $\mathfrak{C}$ such that $\bigcup_{i=1}^N B_i \supset K$. Let $B_i = B(x_i, r_i)$ for $i = 1, \ldots, N$. Renumber $B_1, \ldots, B_N$ if necessary so that $r_1 \geq \cdots \geq r_N$. From the finite sequence $(B_i : i = 1, \ldots, N)$, we select a disjoint subsequence as follows. Let $i_1 = 1$. Discard all B_i that intersect B_{i_1} and let B_{i_2} be the first of the remaining ones if there are any. Discard all B_i that intersect B_{i_2} and let B_{i_3} be the first of the remaining ones if there are any, and so on. The process stops after a finite number of steps, say p steps. Consider the finite disjoint collection $\{B_{i_1}, \ldots, B_{i_p}\}$. Every discarded B_i intersects some B_{i_k} with $r_{i_k} \geq r_i$ so that $B_i = B(x_i, r_i) \subset B(x_{i_k}, 3r_{i_k})$. Thus we have $\bigcup_{i=1}^N B_i \subset \bigcup_{k=1}^p B(x_{i_k}, 3r_{i_k})$. Then

$$\begin{aligned} c < \mu^n_L(K) &\leq \mu^n_L\Big(\bigcup_{i=1}^N B_i\Big) \leq \mu^n_L\Big(\bigcup_{k=1}^p B(x_{i_k}, 3r_{i_k})\Big) \\ &\leq \sum_{k=1}^p \mu^n_L\big(B(x_{i_k}, 3r_{i_k})\big) = 3^n \sum_{k=1}^p \mu^n_L\big(B(x_{i_k}, r_{i_k})\big). \end{aligned}$$

This completes the proof of the lemma. ∎

As an estimate for the Hardy-Littlewood maximal function we have the following:

Theorem 25.11. (Hardy-Littlewood Maximal Theorem) *If $f \in \mathcal{L}^1(\mathbb{R}^n, \mathfrak{M}_L^n, \mu_L^n)$, then for the maximal function Mf of f we have*

$$\mu_L^n\{\mathbb{R}^n : Mf > \alpha\} \leq \frac{3^n}{\alpha}\int_{\mathbb{R}^n} |f|\, d\mu_L^n \quad \textit{for } \alpha \in (0, \infty).$$

Proof. For $\alpha \in (0, \infty)$, let $D_\alpha = \{x \in \mathbb{R}^n : (Mf)(x) > \alpha\} \in \mathfrak{B}_{\mathbb{R}^n}$. (Recall that Mf is nonnegative valued so that $|Mf| = Mf$ and that Mf is $\mathfrak{B}_{\mathbb{R}^n}$-measurable on $\mathbb{R}^n$ by Observation 25.8.) Since $(Mf)(x) = \sup_{r \in (0,\infty)} (A_r|f|)(x)$ for every $x \in \mathbb{R}^n$, for each $x \in D_\alpha$ there exists $r_x \in (0, \infty)$ such that

(1) $$(A_{r_x}|f|)(x) > \alpha.$$

The collection $\mathfrak{C} = \{B(x, r_x) : x \in D_\alpha\}$ of open balls covers D_α, that is, $\bigcup_{x \in D_\alpha} B(x, r_x) \supset D_\alpha$, so that $\mu_L^n(D_\alpha) \leq \mu_L^n\left(\bigcup_{x \in D_\alpha} B(x, r_x)\right)$. Thus for any $c \in \mathbb{R}$ such that $c < \mu_L^n(D_\alpha)$, we have $c < \mu_L^n\left(\bigcup_{x \in D_\alpha} B(x, r_x)\right)$. Then by Lemma 25.10, there exists a finite disjoint collection of open balls $\{B(x_k, r_k) : k = 1, \ldots, p\}$ in $\mathfrak{C}$ such that $c < 3^n \sum_{k=1}^p \mu_L^n(B(x_k, r_k))$. By Definition 25.3 for $A_{r_k}|f|$ and by (1), we have

(2) $$\int_{B(x_k,r_k)} |f|\, d\mu_L^n = \mu_L^n(B(x_k, r_k))(A_{r_k}|f|)(x_k) > \alpha \mu_L^n(B(x_k, r_k)).$$

Thus we have

$$c < 3^n \sum_{k=1}^p \mu_L^n(B(x_k, r_k)) \leq \frac{3^n}{\alpha}\sum_{k=1}^p \int_{B(x_k,r_k)} |f|\, d\mu_L^n \leq \frac{3^n}{\alpha}\int_{\mathbb{R}^n} |f|\, d\mu_L^n,$$

where the last inequality is by the disjointness of $B(x_k, r_k)$, $k = 1, \ldots, p$. Since this holds for every $c < \mu_L^n(D_\alpha)$, we have $\mu_L^n(D_\alpha) \leq \frac{3^n}{\alpha}\int_{\mathbb{R}^n} |f|\, d\mu_L^n$. ∎

Theorem 25.12. *For every $f \in \mathcal{L}^1_{loc}(\mathbb{R}^n, \mathfrak{M}_L^n, \mu_L^n)$, $Af = \lim_{r \to 0} A_r f$ exists and $Af = f$ a.e. on $(\mathbb{R}^n, \mathfrak{M}_L^n, \mu_L^n)$, that is, for a.e. x in $(\mathbb{R}^n, \mathfrak{M}_L^n, \mu_L^n)$ we have*

$$\lim_{r \to 0} \frac{1}{\mu_L^n(B(x, r))}\int_{B(x,r)} f(y)\, \mu_L^n(dy) = f(x).$$

In particular $\mathfrak{D}(Af) \in \mathfrak{M}_L^n$, $\mu_L^n(\mathfrak{D}(Af)^c) = 0$, and Af is a $\mathfrak{M}_L^n$-measurable function on $\mathfrak{D}(Af)$.

Proof. 1. Consider first the case $f \in \mathcal{L}^1(\mathbb{R}^n, \mathfrak{M}_L^n, \mu_L^n)$. Then for every $\varepsilon > 0$, there exists a μ_L^n-integrable real-valued continuous function g on $\mathbb{R}^n$ such that $\int_{\mathbb{R}^n} |f - g|\, d\mu_L^n < \varepsilon$ by Theorem 24.22. Now for every $r \in (0, \infty)$ and $x \in \mathbb{R}^n$, we have

$$\begin{aligned} |(A_r f)(x) - f(x)| &= |A_r(f - g)(x) + (A_r g)(x) - g(x) + g(x) - f(x)| \\ &\leq (A_r|f - g|)(x) + |(A_r g)(x) - g(x)| + |g(x) - f(x)|. \end{aligned}$$

The continuity of g implies $\lim_{r\to 0}(A_r g)(x) = g(x)$ by Lemma 25.4. Thus we have for $x \in \mathbb{R}^n$

(1)
$$\limsup_{r\to 0}\left|(A_r f)(x) - f(x)\right| \le \sup_{r\in(0,\infty)}\left(A_r|f-g|\right)(x) + \left|g-f\right|(x)$$
$$\le M(f-g)(x) + \left|g-f\right|(x).$$

For $\alpha \in (0,\infty)$, by (1) we have

(2)
$$E_\alpha := \left\{\mathbb{R}^n : \limsup_{r\to 0}|A_r f - f| > \alpha\right\}$$
$$\subset \left\{\mathbb{R}^n : M(f-g) + |g-f| > \alpha\right\}.$$

Now

$$\left\{\mathbb{R}^n : M(f-g) \le \tfrac{\alpha}{2}\right\} \cap \left\{\mathbb{R}^n : |g-f| \le \tfrac{\alpha}{2}\right\} \subset \left\{\mathbb{R}^n : M(f-g) + |g-f| \le \alpha\right\}.$$

Taking complements we have

$$\left\{\mathbb{R}^n : M(f-g) > \tfrac{\alpha}{2}\right\} \cup \left\{\mathbb{R}^n : |g-f| > \tfrac{\alpha}{2}\right\} \supset \left\{\mathbb{R}^n : M(f-g) + |g-f| > \alpha\right\}.$$

Thus we have

(3)
$$\mu_L^n(E_\alpha) \le \mu_L^n\left\{\mathbb{R}^n : M(f-g) + |g-f| > \alpha\right\}$$
$$\le \mu_L^n\left\{\mathbb{R}^n : M(f-g) > \tfrac{\alpha}{2}\right\} + \mu_L^n\left\{\mathbb{R}^n : |g-f| > \tfrac{\alpha}{2}\right\}.$$

By Theorem 25.11, we have

$$\mu_L^n\left\{\mathbb{R}^n : M(f-g) > \tfrac{\alpha}{2}\right\} \le \frac{2}{\alpha}3^n \int_{\mathbb{R}^n} |f-g|\, d\mu_L^n < \frac{2}{\alpha}3^n\varepsilon.$$

Also

$$\frac{\alpha}{2}\mu_L^n\left\{\mathbb{R}^n : |g-f| > \tfrac{\alpha}{2}\right\} \le \int_{\{\mathbb{R}^n : |g-f| > \frac{\alpha}{2}\}} |g-f|\, d\mu_L^n \le \int_{\mathbb{R}^n} |g-f|\, d\mu_L^n < \varepsilon.$$

Using these estimates in (3), we have $\mu_L^n(E_\alpha) < \frac{2}{\alpha}(3^n+1)\varepsilon$. By the arbitrariness of $\varepsilon > 0$, we have $\mu_L^n(E_\alpha) = 0$ for every $\alpha \in (0,\infty)$. Now since

$$\left\{\mathbb{R}^n : \limsup_{r\to 0}|A_r f - f| \ne 0\right\} = \bigcup_{k\in\mathbb{N}} E_{\frac{1}{k}},$$

we have

$$\mu_L^n\left\{\mathbb{R}^n : \limsup_{r\to 0}|A_r f - f| \ne 0\right\} \le \sum_{k\in\mathbb{N}} \mu_L^n(E_{\frac{1}{k}}) = 0.$$

Therefore $\limsup_{r\to 0}|A_r f - f| = 0$ a.e. on $\mathbb{R}^n$. This implies $\liminf_{r\to 0}|A_r f - f| = 0$ a.e. on $\mathbb{R}^n$ and thus $\lim_{r\to 0}|A_r f - f| = 0$ a.e. on $\mathbb{R}^n$. This shows that $\lim_{r\to 0} A_r f = f$ a.e. on $\mathbb{R}^n$.

2. Now let $f \in \mathcal{L}^1_{loc}(\mathbb{R}^n, \mathfrak{M}^n_L, \mu^n_L)$. Then for every $k \in \mathbb{N}$, f is μ^n_L-integrable on $B(0,k)$. If we define a function f_k on $\mathbb{R}^n$ by setting $f_k = f$ on $B(0,k+1)$ and $f_k = 0$ on $B(0,k+1)^c$, then $f_k \in \mathcal{L}^1(\mathbb{R}^n, \mathfrak{M}^n_L, \mu^n_L)$ and thus by **1** we have $\lim_{r\to 0} A_r f_k = f_k$ a.e. on $\mathbb{R}^n$. Now for every $x \in B(0,k)$, we have $B(x,r) \subset B(0,k+1)$ as long as $r \in (0,1)$. Then since $f = f_k$ on $B(0,k+1)$, we have $f = f_k$ on $B(x,r)$ for $r \in (0,1)$ so that

$$(A_r f)(x) = \frac{1}{\mu^n_L(B(x,r))}\int_{B(x,r)} f\,d\mu^n_L = \frac{1}{\mu^n_L(B(x,r))}\int_{B(x,r)} f_k\,d\mu^n_L = (A_r f_k)(x).$$

Since this equality holds for all $r \in (0,1)$, we have for a.e. $x \in B(0,k)$

$$(Af)(x) = \lim_{r\to 0}(A_r f)(x) = \lim_{r\to 0}(A_r f_k)(x) = f_k(x) = f(x).$$

Since this holds for every $k \in \mathbb{N}$ and since $\bigcup_{k\in\mathbb{N}} B(0,k) = \mathbb{R}^n$, there exists a null set E in $(\mathbb{R}^n, \mathfrak{M}^n_L, \mu^n_L)$ such that $(Af)(x) = f(x)$ for $x \in E^c$. This implies $\mathfrak{D}(Af) \supset E^c$ and then $\mathfrak{D}(Af)^c \subset E$. Since $(\mathbb{R}^n, \mathfrak{M}^n_L, \mu^n_L)$ is a complete measure space, $\mathfrak{D}(Af)^c \in \mathfrak{M}^n_L$ and then $\mathfrak{D}(Af) \in \mathfrak{M}^n_L$. Since f is $\mathfrak{M}^n_L$-measurable on $\mathbb{R}^n$ and $Af = f$ a.e. on the complete measure space $(\mathbb{R}^n, \mathfrak{M}^n_L, \mu^n_L)$, Af is $\mathfrak{M}^n_L$-measurable on $\mathfrak{D}(Af)$. ∎

According to Theorem 25.12, if $f \in \mathcal{L}^1_{loc}(\mathbb{R}^n, \mathfrak{M}^n_L, \mu^n_L)$, then $\lim_{r\to 0}(A_r f)(x) = f(x)$ for a.e. $x \in \mathbb{R}^n$ and thus for a.e. $x \in \mathbb{R}^n$ we have

$$\lim_{r\to 0}\frac{1}{\mu^n_L(B(x,r))}\int_{B(x,r)} \{f(y) - f(x)\}\,\mu^n_L(dy) = 0.$$

We show below that actually we have the stronger result that for a.e. $x \in \mathbb{R}^n$

$$\lim_{r\to 0}\frac{1}{\mu^n_L(B(x,r))}\int_{B(x,r)} |f(y) - f(x)|\,\mu^n_L(dy) = 0.$$

Definition 25.13. (Lebesgue Point and Lebesgue Set) *Let $f \in \mathcal{L}^1_{loc}(\mathbb{R}^n, \mathfrak{M}^n_L, \mu^n_L)$. A point $x \in \mathbb{R}^n$ is called a Lebesgue point of the function f if*

$$\lim_{r\to 0}\frac{1}{\mu^n_L(B(x,r))}\int_{B(x,r)} |f(y) - f(x)|\,\mu^n_L(dy) = 0.$$

We call the set of all Lebesgue points of f the Lebesgue set of f and write $\Lambda(f)$ for it.

Observation 25.14. If $f \in \mathcal{L}^1_{loc}(\mathbb{R}^n, \mathfrak{M}^n_L, \mu^n_L)$ and f is continuous at $x \in \mathbb{R}$ then $x \in \Lambda(f)$, that is, every continuity point of f is a Lebesgue point of f.

Proof. Let $f \in \mathcal{L}^1_{loc}(\mathbb{R}^n, \mathfrak{M}^n_L, \mu^n_L)$ and let $x \in \mathbb{R}^n$ be a continuity point of f. To show that x is a Lebesgue point of f we show that for every $\varepsilon > 0$ there exists $\delta > 0$ such that

$$(1) \qquad \frac{1}{\mu^n_L(B(x,r))}\int_{B(x,r)} |f(y) - f(x)|\,\mu^n_L(dy) < \varepsilon \quad \text{for } r \in (0,\delta).$$

Let $\varepsilon > 0$ be arbitrarily given. By the continuity of f at x, there exists $\delta > 0$ such that $|f(y) - f(x)| < \varepsilon$ for $y \in B(x,\delta)$. Then for $r \in (0,\delta)$ we have $B(x,r) \subset B(x,\delta)$ so that

$$\frac{1}{\mu_L^n\big(B(x,r)\big)} \int_{B(x,r)} |f(y) - f(x)|\, \mu_L^n(dy) \le \frac{1}{\mu_L^n\big(B(x,r)\big)} \int_{B(x,r)} \varepsilon\, \mu_L^n(dy) = \varepsilon.$$

This proves (1). Therefore $x \in \Lambda(f)$. ∎

We show next that for every $f \in \mathcal{L}^1_{loc}\big(\mathbb{R}^n, \mathfrak{M}_L^n, \mu_L^n\big)$, almost every point in $\mathbb{R}^n$ is a Lebesgue point of f.

Theorem 25.15. *For every $f \in \mathcal{L}^1_{loc}\big(\mathbb{R}^n, \mathfrak{M}_L^n, \mu_L^n\big)$, we have the Lebesgue set $\Lambda(f) \in \mathfrak{M}_L^n$ and moreover for $\Lambda^c(f) := \mathbb{R}^n \setminus \Lambda(f)$ we have $\mu_L^n\big(\Lambda^c(f)\big) = 0$.*

Proof. If $f \in \mathcal{L}^1_{loc}\big(\mathbb{R}^n, \mathfrak{M}_L^n, \mu_L^n\big)$, then for every $k \in \mathbb{N}$, f is μ_L^n-integrable on $B(0,k)$ so that $|f| < \infty$ a.e. on $B(0,k)$. Then since $\mathbb{R}^n = \bigcup_{k\in\mathbb{N}} B(0,k)$, $|f| < \infty$ a.e. on $\mathbb{R}^n$. Thus if we let $F = \{\mathbb{R}^n : |f| = \infty\}$, then F is a null set in $\big(\mathbb{R}^n, \mathfrak{M}_L^n, \mu_L^n\big)$. For each $\xi \in \mathbb{R}$, let us define a function g_ξ on $\mathbb{R}^n$ by setting $g_\xi(x) = |f(x) - \xi|$ for $x \in \mathbb{R}^n$. Then $g_\xi \in \mathcal{L}^1_{loc}\big(\mathbb{R}^n, \mathfrak{M}_L^n, \mu_L^n\big)$ so that by Theorem 25.12 there exists a null set E_ξ in $\big(\mathbb{R}^n, \mathfrak{M}_L^n, \mu_L^n\big)$ such that $Ag_\xi(x) = g_\xi(x)$ for $x \in E_\xi^c$, that is,

$$(1) \qquad \lim_{r\to 0} \frac{1}{\mu_L^n\big(B(x,r)\big)} \int_{B(x,r)} |f(y) - \xi|\, \mu_L^n(dy) = |f(x) - \xi|.$$

Let D be an arbitrary countable dense subset of $\mathbb{R}$. Then $E = \bigcup_{\xi\in D} E_\xi$ is a null set in $\big(\mathbb{R}^n, \mathfrak{M}_L^n, \mu_L^n\big)$. Consider the null set $F \cup E$. We have $(F\cup E)^c = F^c \cap \big(\bigcap_{\xi\in D} E_\xi^c\big)$. Now if $x \in (F \cup E)^c$, then $f(x) \in \mathbb{R}$. Thus for an arbitrary $\varepsilon > 0$, the denseness of D in $\mathbb{R}$ implies that there exists $\xi \in D$ such that $|f(x) - \xi| < \varepsilon$. Then for any $y \in \mathbb{R}^n$, we have

$$(2) \qquad |f(y) - f(x)| \le |f(y) - \xi| + |\xi - f(x)| \le |f(y) - \xi| + \varepsilon.$$

By (2) and then by (1) which is applicable since $x \in E_\xi^c$, we have

$$\begin{aligned} &\limsup_{r\to 0} \frac{1}{\mu_L^n\big(B(x,r)\big)} \int_{B(x,r)} |f(y) - f(x)|\, \mu_L^n(dy) \\ &\le \limsup_{r\to 0} \frac{1}{\mu_L^n\big(B(x,r)\big)} \int_{B(x,r)} |f(y) - \xi|\, \mu_L^n(dy) + \varepsilon \\ &\le |f(x) - \xi| + \varepsilon < 2\varepsilon. \end{aligned}$$

By the arbitrariness of $\varepsilon > 0$, we have for every $x \in (F \cup E)^c$

$$\limsup_{r\to 0} \frac{1}{\mu_L^n\big(B(x,r)\big)} \int_{B(x,r)} |f(y) - f(x)|\, \mu_L^n(dy) = 0.$$

This implies that for every $x \in (F \cup E)^c$, we have

$$\lim_{r\to 0} \frac{1}{\mu_L^n\big(B(x,r)\big)} \int_{B(x,r)} |f(y) - f(x)|\, \mu_L^n(dy) = 0.$$

Therefore $(F \cup E)^c \subset \Lambda(f)$ so that $\Lambda^c(f) \subset F \cup E$. Then since $F \cup E$ is a null set in the complete measure space $(\mathbb{R}^n, \mathfrak{M}_L^n, \mu_L^n)$, its subset $\Lambda^c(f)$ is in $\mathfrak{M}_L^n$ with $\mu_L^n(\Lambda^c(f)) = 0$. Then $\Lambda^c(f) \in \mathfrak{M}_L^n$ implies $\Lambda(f) \in \mathfrak{M}_L^n$. ∎

Definition 25.16. *A collection $\mathcal{E} = \{E_r(x) : r \in (0, \infty)\}$ in $\mathfrak{B}_{\mathbb{R}^n}$ is said to shrink nicely to $x \in \mathbb{R}^n$ if it satisfies the following conditions :*

1° *$E_r(x) \subset B(x, r)$ for every $r \in (0, \infty)$.*

2° *There exists a constant $\alpha > 0$ such that $\mu_L^n(E_r(x)) \geq \alpha \mu_L^n(B(x, r))$ for $r \in (0, \infty)$.*

(Note that $E_r(x)$ need not contain x for any $r \in (0, \infty)$. Condition 1° implies that $E_r(x)$ shrinks to $\{x\}$ as $r \to 0$ and condition 2° ensures that this shrinking is not too fast for differentiation purposes.)

Examples. (a) For every $x \in \mathbb{R}^n$, $\{B(x, r) : r \in (0, \infty)\}$ is a collection in $\mathfrak{B}_{\mathbb{R}^n}$ that shrinks nicely to x with the constant α in Definition 25.16 being equal to 1.

(b) For $x = (\xi_1, \dots, \xi_n) \in \mathbb{R}^n$ and $r > 0$, let $E_r(x)$ be an open cube inscribed in $B(x, r)$, that is, $E_r(x) = (\xi_1 - n^{-1/2}r, \xi_1 + n^{-1/2}r) \times \cdots \times (\xi_n - n^{-1/2}r, \xi_n + n^{-1/2}r)$. Then $E_r(x) \subset B(x, r)$ with $\mu_L^n(E_r(x)) = (2n^{-1/2})^n r^n$. Since $\mu_L^n(B(x, r)) = r^n c$ where $c = \mu_L^n(B(0, 1))$, if we let $\alpha = \frac{1}{c}(2n^{-1/2})^n$ then we have $\alpha \mu_L^n(B(x, r)) = \mu_L^n(E_r(x))$ for every $r \in (0, \infty)$. Therefore $\{E_r(x) : r \in (0, \infty)\}$ is a collection in $\mathfrak{B}_{\mathbb{R}^n}$ that shrinks nicely to x.

(c) Let F be an arbitrary member of $\mathfrak{B}_{\mathbb{R}^n}$ such that $F \subset B(0, 1)$ and $\mu_L^n(F) > 0$. Let $\alpha = \mu_L^n(F)/\mu_L^n(B(0, 1)) \in (0, 1]$. For an arbitrary $x \in \mathbb{R}^n$ and $r > 0$, let $E_r(x) = x + rF$. We have $E_r(x) \subset x + rB(0, 1) = B(x, r)$. By the translation invariance of μ_L^n and by Corollary 24.33, we have

$$\begin{aligned}\mu_L^n(E_r(x)) &= \mu_L^n(rF) = r^n \mu_L^n(F) = \alpha r^n \mu_L^n(B(0, 1)) \\ &= \alpha \mu_L^n(B(0, r)) = \alpha \mu_L^n(B(x, r)).\end{aligned}$$

This shows that $\{E_r(x) : r \in (0, \infty)\}$ is a collection in $\mathfrak{B}_{\mathbb{R}^n}$ that shrinks nicely to x. If we choose $F = B(0, 1) \setminus B(0, \frac{1}{2})$, then $E_r(x) = x + rF = B(x, r) \setminus B(x, \frac{r}{2})$ so that $x \notin E_r(x)$ for any $r \in (0, \infty)$.

Theorem 25.17. (Lebesgue Differentiation Theorem) *Let $f \in \mathcal{L}^1_{loc}(\mathbb{R}^n, \mathfrak{M}_L^n, \mu_L^n)$. Then for every collection $\mathcal{E} = \{E_r(x) : r \in (0, \infty)\}$ in $\mathfrak{B}_{\mathbb{R}^n}$ that shrinks nicely to a point x in the Lebesgue set $\Lambda(f)$ of f, we have*

(1) $$\lim_{r \to 0} \frac{1}{\mu_L^n(E_r(x))} \int_{E_r(x)} |f(y) - f(x)| \, \mu_L^n(dy) = 0,$$

and in particular

(2) $$\lim_{r \to 0} \frac{1}{\mu_L^n(E_r(x))} \int_{E_r(x)} f(y) \, \mu_L^n(dy) = f(x).$$

Proof. Let $x \in \Lambda(f)$ and let $\{E_r(x) : r \in (0,\infty)\}$ be an arbitrary collection in $\mathfrak{B}_{\mathbb{R}^n}$ that shrinks nicely to x. Then there exists a constant $\alpha > 0$ such that $\alpha\mu_L^n(B(x,r)) \le \mu_L^n(E_r(x)) \le \mu_L^n(B(x,r))$ for all $r \in (0,\infty)$. Thus

$$\begin{aligned}
&\frac{1}{\mu_L^n(E_r(x))}\int_{E_r(x)} |f(y)-f(x)|\,\mu_L^n(dy)\\
&\le \frac{1}{\mu_L^n(E_r(x))}\int_{B(x,r)} |f(y)-f(x)|\,\mu_L^n(dy)\\
&\le \frac{1}{\alpha\mu_L^n(B(x,r))}\int_{B(x,r)} |f(y)-f(x)|\,\mu_L^n(dy).
\end{aligned}$$

Since $x \in \Lambda(f)$, the last member of the inequalities above converges to 0 as $r \to 0$. Thus (1) holds. (2) is implied by (1). ∎

The Lebesgue differentiation theorem above is not about differentiation of real-valued functions on $\mathbb{R}$ or $\mathbb{R}^n$. Rather it is related to differentiation of indefinite integrals. This point will be discussed later.

[II] Differentiation of Set Functions with Respect to the Lebesgue Measure

Definition 25.18. *We write $(\mathfrak{B}_{\mathbb{R}^n})_b$ for the subcollection of $\mathfrak{B}_{\mathbb{R}^n}$ consisting of those members of $\mathfrak{B}_{\mathbb{R}^n}$ that are bounded sets in $\mathbb{R}^n$.*
(Note that if E and F are two members of $(\mathfrak{B}_{\mathbb{R}^n})_b$ then $E \cup F$, $E \cap F$, and $E \setminus F$ are members of $(\mathfrak{B}_{\mathbb{R}^n})_b$. However $(\mathfrak{B}_{\mathbb{R}^n})_b$ is not an algebra of subsets of $\mathbb{R}^n$ since $\mathbb{R}^n \notin (\mathfrak{B}_{\mathbb{R}^n})_b$.)

Definition 25.19. *Let $(\mathfrak{B}_{\mathbb{R}^n})_b$ be the subcollection of $\mathfrak{B}_{\mathbb{R}^n}$ consisting of all bounded Borel sets in $\mathbb{R}^n$. Let ν be an extended real-valued set function on a collection $\mathfrak{C}$ of subsets of $\mathbb{R}^n$ containing $(\mathfrak{B}_{\mathbb{R}^n})_b$. For $x \in \mathbb{R}^n$ and a collection $\mathcal{E} = \{E_r(x) : r \in (0,\infty)\}$ in $\mathfrak{B}_{\mathbb{R}^n}$ that shrinks nicely to x, we define*

$$(\mathcal{D}_{\mathcal{E}}\nu)(x) = \lim_{r\to 0}\frac{\nu(E_r(x))}{\mu_L^n(E_r(x))}$$

provided the limit exists in $\overline{\mathbb{R}}$. We call $(\mathcal{D}_{\mathcal{E}}\nu)(x)$ the derivative of ν with respect to the Lebesgue measure μ_L^n along $\mathcal{E}$ at x. In particular for the collection $\mathcal{S} = \{B(x,r) : r \in (0,\infty)\}$, the limit

$$(\mathcal{D}_{\mathcal{S}}\nu)(x) = \lim_{r\to 0}\frac{\nu(B(x,r))}{\mu_L^n(B(x,r))},$$

if it exists in $\overline{\mathbb{R}}$, is called the symmetric derivative of ν with respect to the Lebesgue measure μ_L^n at x.

Theorem 25.20. *Let ν be an extended real-valued set function defined on $(\mathfrak{B}_{\mathbb{R}^n})_b$. Let $x \in \mathbb{R}^n$. Suppose for every collection $\mathcal{E}$ in $\mathfrak{B}_{\mathbb{R}^n}$ that shrinks nicely to x the derivative $(\mathcal{D}_{\mathcal{E}}\nu)(x)$ exists. Then $(\mathcal{D}_{\mathcal{E}}\nu)(x)$ does not depend on $\mathcal{E}$, that is, if $\mathcal{E}'$ and $\mathcal{E}''$ are two collections in $\mathfrak{B}_{\mathbb{R}^n}$ that shrink nicely to x then $(\mathcal{D}_{\mathcal{E}'}\nu)(x) = (\mathcal{D}_{\mathcal{E}''}\nu)(x)$. Thus for every collection $\mathcal{E}$ in $\mathfrak{B}_{\mathbb{R}^n}$ that shrinks nicely to x we have $(\mathcal{D}_{\mathcal{E}}\nu)(x) = (\mathcal{D}_{\mathcal{S}}\nu)(x)$.*

Proof. Suppose that for every collection $\mathcal{E}$ in $\mathfrak{B}_{\mathbb{R}^n}$ that shrinks nicely to x the derivative $(\mathcal{D}_{\mathcal{E}}\nu)(x)$ exists. Let us show that for any two collections $\mathcal{E}'$ and $\mathcal{E}''$ that shrink nicely to x, we have $(\mathcal{D}_{\mathcal{E}'}\nu)(x) = (\mathcal{D}_{\mathcal{E}''}\nu)(x)$.

Assume the contrary. Then $(\mathcal{D}_{\mathcal{E}'}\nu)(x) = \gamma'$ and $(\mathcal{D}_{\mathcal{E}''}\nu)(x) = \gamma''$ where $\gamma', \gamma'' \in \overline{\mathbb{R}}$ and $\gamma' \neq \gamma''$. Let $\mathcal{E}'$ and $\mathcal{E}''$ be given respectively by $\mathcal{E}' = \{E'_r(x) : r \in (0,\infty)\}$ and $\mathcal{E}'' = \{E''_r(x) : r \in (0,\infty)\}$. By Definition 25.16, $E'_r(x) \subset B(x,r)$, $E''_r(x) \subset B(x,r)$, and there exist $\alpha' > 0$ and $\alpha'' > 0$ such that $\mu^n_L(E'_r(x)) \geq \alpha'\mu^n_L(B(x,r))$ and $\mu^n_L(E''_r(x)) \geq \alpha''\mu^n_L(B(x,r))$ for $r \in (0,\infty)$. Consider a collection $\mathcal{E} = \{E_r(x) : r \in (0,\infty)\}$ where $E_r(x)$ is defined by letting $E_r(x) = E'_r(x)$ if r is a positive rational number and $E_r(x) = E''_r(x)$ if r is a positive irrational number. Then $E_r(x) \subset B(x,r)$ for $r \in (0,\infty)$ and if we let $\alpha = \min\{\alpha', \alpha''\}$ then $\mu^n_L(E_r(x)) \geq \alpha\mu^n_L(B(x,r))$ for $r \in (0,\infty)$. Thus $\mathcal{E}$ shrink nicely to x. Let $(q_k : k \in \mathbb{N})$ be a sequence of positive rational numbers such that $q_k \downarrow 0$ and let $(p_k : k \in \mathbb{N})$ be a sequence of irrational numbers such that $p_k \downarrow 0$. Then we have

$$\lim_{k\to\infty} \frac{\nu(E_{q_k}(x))}{\mu^n_L(E_{q_k}(x))} = \lim_{k\to\infty} \frac{\nu(E'_{q_k}(x))}{\mu^n_L(E'_{q_k}(x))} = \gamma'$$

and

$$\lim_{k\to\infty} \frac{\nu(E_{p_k}(x))}{\mu^n_L(E_{p_k}(x))} = \lim_{k\to\infty} \frac{\nu(E''_{p_k}(x))}{\mu^n_L(E''_{p_k}(x))} = \gamma''.$$

Thus $(\mathcal{D}_{\mathcal{E}}\nu)(x) = \lim_{r\to 0} \frac{\nu(E_r(x))}{\mu^n_L(E_r(x))}$ does not exist. This contradicts the assumption that $(\mathcal{D}_{\mathcal{E}}\nu)(x)$ exists for every collection $\mathcal{E}$ that shrinks nicely to x. ∎

Definition 25.21. *Let ν be an extended real-valued set function defined on $(\mathfrak{B}_{\mathbb{R}^n})_b$. Let $x \in \mathbb{R}^n$. If for every collection $\mathcal{E}$ in $\mathfrak{B}_{\mathbb{R}^n}$ that shrinks nicely to x the derivative $(\mathcal{D}_{\mathcal{E}}\nu)(x)$ of ν with respect to the Lebesgue measure μ^n_L at x along $\mathcal{E}$ exists, then we say that ν is differentiable with respect to the Lebesgue measure μ^n_L at x. We call $(\mathcal{D}_{\mathcal{E}}\nu)(x)$, which does not depend on $\mathcal{E}$ according to Theorem 25.20, the derivative of ν with respect to μ^n_L at x and we write $(\mathcal{D}\nu)(x)$ for it.*

As an example of the derivative of a set function with respect to the Lebesgue measure, let $f \in \mathcal{L}^1_{loc}(\mathbb{R}^n, \mathfrak{M}^n_L, \mu^n_L)$ and consider a set function ν on $(\mathfrak{B}_{\mathbb{R}^n})_b$ be defined by setting

$$\nu(E) = \int_E f \, d\mu^n_L \quad \text{for } E \in (\mathfrak{B}_{\mathbb{R}^n})_b.$$

Let $x \in \mathbb{R}^n$ and let $\mathcal{E} = \{E_r(x) : r \in (0,\infty)\}$ an arbitrary collection in $\mathfrak{B}_{\mathbb{R}^n}$ that shrinks nicely to x. Then for every $x \in \Lambda(f)$, we have by (2) of Theorem 25.17

$$(\mathcal{D}_{\mathcal{E}}\nu)(x) = \lim_{r\to 0} \frac{\nu\big(E_r(x)\big)}{\mu_L^n\big(E_r(x)\big)} = \lim_{r\to 0} \frac{1}{\mu_L^n\big(E_r(x)\big)} \int_{E_r(x)} f(y)\,\mu_L^n(dy) = f(x).$$

Thus we have

$$(\mathcal{D}\nu)(x) = f(x) \quad \text{for } x \in \Lambda(f).$$

As further examples of differentiation of set functions, we consider below differentiation of indefinite integrals of locally integrable functions, density of measurable sets, and differentiation of Borel measures with respect to the Lebesgue measure.

[III] Differentiation of the Indefinite Integral

Let f be a μ_L-integrable extended real-valued $\mathfrak{M}_L$-measurable function on $[a, b]$. By Definition 13.12, an indefinite integral of f is a real-valued function F on $[a, b]$ defined by $F(x) = \int_{[a,x]} f\,d\mu_L + c$ for $x \in [a, x]$ where c is an arbitrary real number. To extend the notion of indefinite integral to functions defined on $\mathbb{R}^n$, let us replace the function F defined on $[a, b]$ by a set function $\Phi(E) = \int_E f\,d\mu_L$ for $E \in \mathfrak{B}_{\mathbb{R}} \cap [a, b]$.

Definition 25.22. *Let $f \in \mathcal{L}^1_{loc}(\mathbb{R}^n, \mathfrak{M}^n_L, \mu^n_L)$. Let us call the real-valued set function Φ defined on $(\mathfrak{B}_{\mathbb{R}^n})_b$ by $\Phi(E) = \int_E f\,d\mu^n_L$ for $E \in (\mathfrak{B}_{\mathbb{R}^n})_b$ the indefinite integral of f.*

Theorem 25.23. *Let Φ be the indefinite integral of a function $f \in \mathcal{L}^1_{loc}(\mathbb{R}^n, \mathfrak{M}^n_L, \mu^n_L)$. Then for every point x in the Lebesgue set $\Lambda(f)$ of f and for every collection $\mathcal{E} = \{E_r(x) : r \in (0, \infty)\}$ in $\mathfrak{B}_{\mathbb{R}^n}$ that shrinks nicely to x, we have*

$$(1) \qquad (\mathcal{D}_{\mathcal{E}}\Phi)(x) = \lim_{r\to 0} \frac{\Phi\big(E_r(x)\big)}{\mu^n_L\big(E_r(x)\big)} = f(x).$$

In particular for the symmetric derivative $\mathcal{D}_{\mathcal{S}}\Phi$ of Φ, we have

$$(2) \qquad (\mathcal{D}_{\mathcal{S}}\Phi)(x) = \lim_{r\to 0} \frac{\Phi\big(B(x,r)\big)}{\mu^n_L\big(B(x,r)\big)} = f(x).$$

Proof. Let $x \in \Lambda(f)$. Then by (2) of Theorem 25.17 (Lebesgue Differentiation Theorem), we have

$$\lim_{r\to 0} \frac{\Phi\big(E_r(x)\big)}{\mu^n_L\big(E_r(x)\big)} = \lim_{r\to 0} \frac{1}{\mu^n_L\big(E_r(x)\big)} \int_{E_r(x)} f(y)\,\mu^n_L(dy) = f(x).$$

This proves (1). Since $\mathcal{S} = \big\{B(x,r) : r \in (0, \infty)\big\}$ is a collection in $\mathfrak{B}_{\mathbb{R}^n}$ shrinking nicely to x, (1) implies (2). ∎

Let f be an extended real-valued $\mathfrak{M}_L$-measurable and μ_L-integrable function on $[a, b]$. Consider the function F defined by $F(x) = \int_{[a,b]} f\,d\mu_L$ for $x \in [a, b]$. According to the

Lebesgue Differentiation Theorem for the indefinite integral (Theorem 13.15), the derivative $F'(x)$ exists and is equal to $f(x)$ for a.e. $x \in [a, b]$. Let us show that this result is contained in our Theorem 25.23 as a particular case.

Corollary 25.24. *Let f be an extended real-valued $\mathfrak{M}_L$-measurable and μ_L-integrable function on $[a, b]$. Let a function F on $[a, b]$ be defined by*

$$F(x) = \int_{[a,x]} f(y)\,\mu_L(dy) \quad \text{for } x \in [a, b].$$

Then the derivative $F'(x)$ exists and is equal to $f(x)$ for a.e. $x \in [a, b]$.

Proof. To show that the derivative $F'(x)$ exists and is equal to $f(x)$ for a.e. $x \in [a, b]$, we show that the right-hand derivative and the left-hand derivative of F exist and are equal to $f(x)$ for a.e. x.

Let us extend the definition of f to $\mathbb{R}$ by setting $f = 0$ on $\mathbb{R} \setminus [a, b]$. Then we have $f \in \mathcal{L}^1_{loc}(\mathbb{R}, \mathfrak{M}_L, \mu_L)$. As in Definition 25.22, let us define a set function Φ on $(\mathfrak{B}_\mathbb{R})_b$ by setting

$$\Phi(E) = \int_E f(y)\,\mu_L(dy) \quad \text{for } E \in (\mathfrak{B}_\mathbb{R})_b.$$

To find the right-hand derivative of F at $x \in \mathbb{R}$, let $E_r(x) = (x, x + r)$ for $r \in (0, \infty)$. Then $E_r(x) \subset B(x, r)$ and $\mu_L\big(E_r(x)\big) = r = \frac{1}{2}\mu_L\big(B(x, r)\big)$ for $r \in (0, \infty)$ so that $\mathcal{E} = \big\{E_r(x) : r \in (0, \infty)\big\}$ is a collection in $\mathfrak{B}_\mathbb{R}$ that shrinks nicely to x. Let $x \in \Lambda(f)$. Then we have

$$\begin{aligned}\lim_{r\to 0} \frac{1}{r}\{F(x + r) - F(x)\} &= \lim_{r\to 0} \frac{1}{r} \int_{(x,x+r)} f(y)\,\mu_L(dy)\\ &= \lim_{r\to 0} \frac{\Phi\big(E_r(x)\big)}{\mu_L\big(E_r(x)\big)} = f(x)\end{aligned}$$

by Theorem 25.23. This shows that the right-hand derivative of F exists and is equal to $f(x)$ at $x \in \Lambda(f)$.

To find the left-hand derivative of F at $x \in \mathbb{R}$, let $E'_r(x) = (x - r, x)$ for $r \in (0, \infty)$. Then $\mathcal{E}' = \{E'_r(x) : r \in (0, \infty)\}$ is a collection in $\mathfrak{B}_\mathbb{R}$ that shrinks nicely to x. Now

$$\lim_{r\to 0} \frac{1}{-r}\{F(x - r) - F(x)\} = \lim_{r\to 0} \frac{1}{r}\{F(x) - F(x - r)\}$$
$$= \lim_{r\to 0} \frac{1}{r} \int_{(x-r,x)} f(y)\,\mu_L(dy) = \lim_{r\to 0} \frac{\Phi\big(E'_r(x)\big)}{\mu_L\big(E'_r(x)\big)} = f(x)$$

by Theorem 25.23. This shows that the left-hand derivative of F exists and is equal to $f(x)$ at $x \in \Lambda(f)$. Therefore the derivative of F exists and is equal to $f(x)$ at every x in the Lebesgue set $\Lambda(f)$ of f. Since $\mu_L\big(\Lambda^c(f)\big) = 0$, the derivative of F exists and is equal to $f(x)$ for a.e. $x \in [a, b]$. ∎

[IV] Density of Lebesgue Measurable Sets Relative to the Lebesgue Measure

Definition 25.25. *For $E \in \mathfrak{M}^n_L$, consider a set function ν on $(\mathfrak{B}_{\mathbb{R}^n})_b$ defined by setting $\nu(B) = \mu^n_L(E \cap B)$ for $B \in (\mathfrak{B}_{\mathbb{R}^n})_b$. For $x \in \mathbb{R}^n$ and a collection $\mathcal{E} = \{E_r(x) : r \in (0, \infty)\}$ in $\mathfrak{B}_{\mathbb{R}^n}$ that shrinks nicely to x, let us define $\delta_{\mathcal{E}}(E, x) = (\mathcal{D}_{\mathcal{E}}\nu)(x)$, that is,*

$$\delta_{\mathcal{E}}(E, x) = (\mathcal{D}_{\mathcal{E}}\nu)(x) = \lim_{r \to 0} \frac{\nu(E_r(x))}{\mu^n_L(E_r(x))} = \lim_{r \to 0} \frac{\mu^n_L(E \cap E_r(x))}{\mu^n_L(E_r(x))}$$

provided the limit exists. We call $\delta_{\mathcal{E}}(E, x)$ the density of E at x with respect to the Lebesgue measure μ^n_L along $\mathcal{E}$. In particular for the collection $\mathcal{S} = \{B(x, r) : r \in (0, \infty)\}$, the limit

$$\delta_{\mathcal{S}}(E, x) = (\mathcal{D}_{\mathcal{S}}\nu)(x) = \lim_{r \to 0} \frac{\nu(B(x, r))}{\mu^n_L(B(x, r))} = \lim_{r \to 0} \frac{\mu^n_L(E \cap B(x, r))}{\mu^n_L(B(x, r))},$$

if it exists, is called the symmetric density, or the Lebesgue density, of E at x with respect to the Lebesgue measure μ^n_L. If $\delta_{\mathcal{S}}(E, x) = 1$, we call x a density point of E. If $\delta_{\mathcal{S}}(E, x) = 0$, we call x a dispersion point of E.

Observation 25.26. Let $E \in \mathfrak{M}^n_L$. Let $x \in \mathbb{R}^n$ and let $\mathcal{E} = \{E_r(x) : r \in (0, \infty)\}$ be a collection in $\mathfrak{B}_{\mathbb{R}^n}$ that shrinks nicely to x. If $\delta_{\mathcal{E}}(E, x)$ exists, then $\delta_{\mathcal{E}}(E, x) \in [0, 1]$.

Proof. Since $E \cap E_r(x) \subset E_r(x)$, we have $\mu^n_L(E \cap E_r(x)) \leq \mu^n_L(E_r(x))$ for all $r \in (0, \infty)$. By Definition 25.16, there exists $\alpha > 0$ such that $\mu^n_L(E_r(x)) \geq \alpha \mu^n_L(B(x, r))$ for all $r \in (0, \infty)$ so that $\mu^n_L(E_r(x)) > 0$ for all $r \in (0, \infty)$. Thus we have

$$\frac{\mu^n_L(E \cap E_r(x))}{\mu^n_L(E_r(x))} \in [0, 1] \quad \text{for all } r \in (0, \infty).$$

Then if the limit of the quotient exists as $r \to 0$ then the limit is in the range $[0, 1]$, that is, if $\delta_{\mathcal{E}}(E, x)$ exists, then $\delta_{\mathcal{E}}(E, x) \in [0, 1]$. ∎

Remark 25.27. Let $E \in \mathfrak{M}^n_L$. Let $x \in \mathbb{R}^n$ and let $\mathcal{E}$ and $\mathcal{E}'$ be two collections in $\mathfrak{B}_{\mathbb{R}^n}$ that shrink nicely to x.
(a) Existence of $\delta_{\mathcal{E}}(E, x)$ does not imply that of $\delta_{\mathcal{E}'}(E, x)$. (See Example 1 below.)
(b) Existence of both $\delta_{\mathcal{E}}(E, x)$ and $\delta_{\mathcal{E}'}(E, x)$ does not imply $\delta_{\mathcal{E}}(E, x) = \delta_{\mathcal{E}'}(E, x)$. (See Example 2 below.)

Example 1. In this example, $\delta_{\mathcal{E}}(E, x)$ exists for some collection $\mathcal{E}$ in $\mathfrak{B}_{\mathbb{R}}$ that shrinks nicely to x but $\delta_{\mathcal{S}}(E, x)$ does not exist. In $\mathbb{R}$ let $I_k = \left[\frac{1}{2^{2k+1}}, \frac{1}{2^{2k}}\right]$ with $\mu_L(I_k) = \frac{1}{2^{2k+1}}$ for $k \in \mathbb{Z}_+$, that is,

$$I_0 = \left[\tfrac{1}{2^1}, \tfrac{1}{2^0}\right],\ I_1 = \left[\tfrac{1}{2^3}, \tfrac{1}{2^2}\right],\ I_2 = \left[\tfrac{1}{2^5}, \tfrac{1}{2^4}\right],\ I_3 = \left[\tfrac{1}{2^7}, \tfrac{1}{2^6}\right], \ldots$$

with

$$\mu_L(I_0) = \frac{1}{2^1},\ \mu_L(I_1) = \frac{1}{2^3},\ \mu_L(I_2) = \frac{1}{2^5},\ \mu_L(I_3) = \frac{1}{2^7}, \dots$$

Let $E = \bigcup_{k \in \mathbb{Z}_+} I_k$.

1. If we let $\mathcal{E} = \{(-r, 0) : r \in (0, \infty)\}$, then $\mathcal{E}$ is a collection in $\mathfrak{B}_{\mathbb{R}}$ that shrinks nicely to 0. Now for $\mathcal{E}$ we have

$$\delta_{\mathcal{E}}(E, 0) = \lim_{r \to 0} \frac{\mu_L(E \cap (-r, 0))}{\mu_L((-r, 0))} = \lim_{r \to 0} \frac{\mu_L(\emptyset)}{\mu_L((-r, 0))} = 0.$$

2. Next let us show that $\delta_{\mathcal{E}}(E, 0)$ does not exist. Now for every $k \in \mathbb{Z}_+$, we have $E \cap B(0, \frac{1}{2^{2k}}) = I_k \cup I_{k+1} \cup I_{k+2} \cup \cdots$ so that

$$\mu_L\big(E \cap B(0, \tfrac{1}{2^{2k}})\big) = \frac{1}{2^{2k+1}} + \frac{1}{2^{2k+3}} + \frac{1}{2^{2k+5}} + \cdots = \frac{4}{3} \frac{1}{2^{2k+1}}.$$

On the other hand $E \cap B(0, \frac{1}{2^{2k+1}}) = I_{k+1} \cup I_{k+2} \cup I_{k+3} \cup \cdots$ so that

$$\mu_L\big(E \cap B(0, \tfrac{1}{2^{2k+1}})\big) = \frac{1}{2^{2k+3}} + \frac{1}{2^{2k+5}} + \frac{1}{2^{2k+7}} + \cdots = \frac{4}{3} \frac{1}{2^{2k+3}}.$$

Thus we have

$$\frac{\mu_L\big(E \cap B(0, \frac{1}{2^{2k}})\big)}{\mu_L\big(B(0, \frac{1}{2^{2k}})\big)} = \frac{4}{3} \frac{1}{2^{2k+1}} \frac{2^{2k}}{2} = \frac{1}{3},$$

and

$$\frac{\mu_L\big(E \cap B(0, \frac{1}{2^{2k+1}})\big)}{\mu_L\big(B(0, \frac{1}{2^{2k+1}})\big)} = \frac{4}{3} \frac{1}{2^{2k+3}} \frac{2^{2k+1}}{2} = \frac{1}{6}.$$

Therefore the quotient $\mu_L(E \cap B(0, r))/\mu_L(B(0, r))$ assumes the value $\frac{1}{3}$ for $r = \frac{1}{2^{2k}}$ and the value $\frac{1}{6}$ for $r = \frac{1}{2^{2k+1}}$. This shows that $\delta_{\mathcal{E}}(E, 0) = \lim_{r \to 0} \{\mu_L(E \cap B(0, r))/\mu_L(B(0, r))\}$ does not exist. ∎

Example 2. Let $E = (0, \infty) \subset \mathbb{R}$. Consider two collections $\mathcal{E}$ and $\mathcal{E}'$ in $\mathfrak{B}_{\mathbb{R}}$ that shrink nicely to $0 \in \mathbb{R}$ given by $\mathcal{E} = \{(-r, 0) : r \in (0, \infty)\}$ and $\mathcal{E}' = \{(0, r) : r \in (0, \infty)\}$. Then

$$\delta_{\mathcal{E}}(E, 0) = \lim_{r \to 0} \frac{\mu_L(E \cap (-r, 0))}{\mu_L((-r, 0))} = \lim_{r \to 0} \frac{\mu_L(\emptyset)}{\mu_L((-r, 0))} = 0,$$

$$\delta_{\mathcal{E}'}(E, 0) = \lim_{r \to 0} \frac{\mu_L(E \cap (0, r))}{\mu_L((0, r))} = \lim_{r \to 0} \frac{\mu_L((0, r))}{\mu_L((0, r))} = 1.$$

Theorem 25.28. *Let $E \in \mathfrak{M}_L^n$ and $x \in \mathbb{R}^n$. Suppose for every collection $\mathcal{E}$ in $\mathfrak{B}_{\mathbb{R}^n}$ that shrinks nicely to x the density $\delta_{\mathcal{E}}(E, x)$ of E with respect to μ_L^n at x along $\mathcal{E}$ exists. Then $\delta_{\mathcal{E}}(E, x)$ does not depend on $\mathcal{E}$, that is, if $\mathcal{E}'$ and $\mathcal{E}''$ are two collections in $\mathfrak{B}_{\mathbb{R}^n}$ that shrink* *nicely to x then* $\delta_{\mathcal{E}'}(E,x) = \delta_{\mathcal{E}''}(E,x)$. *Thus for every collection* $\mathcal{E}$ *in* $\mathfrak{B}_{\mathbb{R}^n}$ *that shrinks nicely to x, we have* $\delta_{\mathcal{E}}(E,x) = \delta_{\mathcal{S}}(E,x)$.

Proof. By Definition 25.25, $\delta_{\mathcal{E}}(E,x) = (\mathcal{D}_{\mathcal{E}}\nu)(x)$ where ν is the set function on $(\mathfrak{B}_{\mathbb{R}^n})_b$ defined by $\nu(B) = \mu_L^n(E \cap B)$ for $B \in (\mathfrak{B}_{\mathbb{R}^n})_b$. Thus our theorem is a particular case of Theorem 25.20. ■

For every $E \in \mathfrak{M}_L^n$, we have $\mathbf{1}_E \in \mathcal{L}_{loc}^1(\mathbb{R}^n, \mathfrak{M}_L^n, \mu_L^n)$. Consider the Lebesgue set $\Lambda(\mathbf{1}_E)$ of the function $\mathbf{1}_E$. We show below that if $x \in \Lambda(\mathbf{1}_E)$ then for every collection $\mathcal{E}$ in $\mathfrak{B}_{\mathbb{R}^n}$ that shrinks nicely to x the density $\delta_{\mathcal{E}}(E,\cdot)$ exists.

Lemma 25.29. *Let* $E \in \mathfrak{M}_L^n$. *Then*
(a) $E^\circ \cup (E^c)^\circ \subset \Lambda(\mathbf{1}_E)$.
(b) $\partial E \supset \Lambda^c(\mathbf{1}_E)$.
(c) *If E is a null set in* $(\mathbb{R}^n, \mathfrak{M}_L^n, \mu_L^n)$, *then* $E \subset \Lambda^c(\mathbf{1}_E)$.

Proof. By Observation 25.14, if $f \in \mathcal{L}_{loc}^1(\mathbb{R}^n, \mathfrak{M}_L^n, \mu_L^n)$ is continuous at $x \in \mathbb{R}^n$ then $x \in \Lambda(f)$. Let $E \in \mathfrak{M}_L^n$. If $x \in E^\circ$, then $\mathbf{1}_E$ is continuous at x so that $x \in \Lambda(\mathbf{1}_E)$. Thus $E^\circ \subset \Lambda(\mathbf{1}_E)$. Similarly if $x \in (E^c)^\circ$, then $\mathbf{1}_E$ is continuous at x so that $x \in \Lambda(\mathbf{1}_E)$ and thus $(E^c)^\circ \subset \Lambda(\mathbf{1}_E)$. Therefore $E^\circ \cup (E^c)^\circ \subset \Lambda(\mathbf{1}_E)$. Since $\partial E = \big(E^\circ \cup (E^c)^\circ\big)^c$, we have $\partial E \supset \Lambda^c(\mathbf{1}_E)$.

Suppose E is a null set in $(\mathbb{R}^n, \mathfrak{M}_L^n, \mu_L^n)$. If $E = \emptyset$, then $E \subset \Lambda^c(\mathbf{1}_E)$ holds trivially. If $E \neq \emptyset$, let $x \in E$. Then $\mathbf{1}_E = 0$ a.e. on $\mathbb{R}^n$ and $\mathbf{1}_E(x) = 1$ so that

$$\begin{aligned}&\lim_{r\to 0} \frac{1}{\mu_L^n(B(x,r))} \int_{B(x,r)} |\mathbf{1}_E(y) - \mathbf{1}_E(x)|\, \mu_L^n(dy)\\ &= \lim_{r\to 0} \frac{1}{\mu_L^n(B(x,r))} \int_{B(x,r)} 1\, \mu_L^n(dy) = 1 \neq 0.\end{aligned}$$

This shows that $x \in \Lambda^c(\mathbf{1}_E)$. Therefore $E \subset \Lambda^c(\mathbf{1}_E)$. ■

Theorem 25.30. *Let* $E \in \mathfrak{M}_L^n$ *and* $x \in \Lambda(\mathbf{1}_E)$, *the Lebesgue set of the function* $\mathbf{1}_E$. *Then for every collection* $\mathcal{E} = \{E_r(x) : r \in (0,\infty)\}$ *in* $\mathfrak{B}_{\mathbb{R}^n}$ *that shrinks nicely to x, the density* $\delta_{\mathcal{E}}(E,x)$ *exists. Moreover we have*

$$\delta_{\mathcal{E}}(E,x) = \lim_{r\to 0} \frac{\mu_L^n(E \cap E_r(x))}{\mu_L^n(E_r(x))} = \begin{cases} 1 & \text{if } x \in E \cap \Lambda(\mathbf{1}_E), \\ 0 & \text{if } x \in E^c \cap \Lambda(\mathbf{1}_E). \end{cases}$$

Proof. If $E \in \mathfrak{M}_L^n$, then $\mathbf{1}_E \in \mathcal{L}_{loc}^1(\mathbb{R}^n, \mathfrak{M}_L^n, \mu_L^n)$. By (2) of Theorem 25.17, for every point x in the Lebesgue set $\Lambda(\mathbf{1}_E)$ and every collection $\{E_r(x) : r \in (0,\infty)\}$ in $\mathfrak{B}_{\mathbb{R}^n}$ that shrinks nicely to x, we have

$$\lim_{r\to 0} \frac{\mu_L^n(E \cap E_r(x))}{\mu_L^n(E_r(x))} = \lim_{r\to 0} \frac{1}{\mu_L^n(E_r(x))} \int_{E_r(x)} \mathbf{1}_E(y)\, \mu_L^n(dy) = \mathbf{1}_E(x).$$

Thus $\delta_{\mathcal{E}}(E, x)$ exists and $\delta_{\mathcal{E}}(E, x) = \mathbf{1}_E(x)$. Therefore $\delta_{\mathcal{E}}(E, x) = 1$ for $x \in E \cap \Lambda(\mathbf{1}_E)$ and $\delta_{\mathcal{E}}(E, x) = 0$ for $x \in E^c \cap \Lambda(\mathbf{1}_E)$. ∎

Corollary 25.31. *Let $E \in \mathfrak{M}_L^n$ and $x \in \mathbb{R}^n$. Then for every collection $\{E_r(x) : r \in (0, \infty)\}$ in $\mathfrak{B}_{\mathbb{R}^n}$ that shrinks nicely to x, we have*

$$\delta_{\mathcal{E}}(E, x) = \lim_{r \to 0} \frac{\mu_L^n(E \cap E_r(x))}{\mu_L^n(E_r(x))} = \begin{cases} 1 & \text{if } x \in E^\circ, \\ 0 & \text{if } x \in (E^c)^\circ. \end{cases}$$

Proof. By Lemma 25.29, if $x \in E^\circ$ then $x \in E \cap \Lambda(\mathbf{1}_E)$ and similarly if $x \in (E^c)^\circ$ then $x \in E^c \cap \Lambda(\mathbf{1}_E)$. Thus the Corollary follows from Theorem 25.30. ∎

Corollary 25.32. (Lebesgue Density Theorem) *Let $E \in \mathfrak{M}_L^n$. Then for the symmetric density $\delta_{\mathcal{S}}(E, \cdot)$ of E, we have*

$$(1) \qquad \delta_{\mathcal{S}}(E, x) = \begin{cases} 1 & \text{for } x \in E \cap \Lambda(\mathbf{1}_E), \\ 0 & \text{for } x \in E^c \cap \Lambda(\mathbf{1}_E), \end{cases}$$

and therefore

$$(2) \qquad \delta_{\mathcal{S}}(E, x) = \begin{cases} 1 & \text{for a.e. } x \in E, \\ 0 & \text{for a.e. } x \in E^c. \end{cases}$$

Proof. Since $\{B(x, r) : r \in (0, \infty)\}$ is a particular case of a collection in $\mathfrak{B}_{\mathbb{R}^n}$ that shrinks nicely to x, (1) is a particular case of Theorem 25.30. Then since $\mu_L^n(\Lambda^c(\mathbf{1}_E)) = 0$ by Theorem 25.15, (2) follows from (1). In fact if we compare E with its subset $E \cap \Lambda(\mathbf{1}_E)$, then $E \setminus (E \cap \Lambda(\mathbf{1}_E)) = E \cap (E \cap \Lambda(\mathbf{1}_E))^c = E \cap (E^c \cup \Lambda^c(\mathbf{1}_E)) = E \cap \Lambda^c(\mathbf{1}_E)$ so that $\mu_L^n(E \setminus (E \cap \Lambda(\mathbf{1}_E))) = 0$. Similarly $\mu_L^n(E^c \setminus (E^c \cap \Lambda(\mathbf{1}_E))) = 0$. ∎

Remark 25.33. Let $E \in \mathfrak{M}_L^n$ and $x \in \Lambda(\mathbf{1}_E)$. According to Theorem 25.30, for an arbitrary collection $\mathcal{E} = \{E_r(x) : r \in (0, \infty)\}$ in $\mathfrak{B}_{\mathbb{R}^n}$ that shrinks nicely to x, $\delta_{\mathcal{E}}(E, x)$ always exists and is equal to 1 or 0 according as $x \in E$ or $x \in E^c$. If $x \notin \Lambda(\mathbf{1}_E)$, $\delta_{\mathcal{E}}(E, x)$ may still exist for some particular collection $\mathcal{E} = \{E_r(x) : r \in (0, \infty)\}$ in $\mathfrak{B}_{\mathbb{R}^n}$ that shrinks nicely to x.

Example 1. (In this example $\delta_{\mathcal{E}}(E, x)$ exists for some $x \notin \Lambda(\mathbf{1}_E)$.) In $\mathbb{R}$, let $E = [0, \infty)$. Then $\mathbb{R} \setminus \{0\} = E^\circ \cup (E^c)^\circ \subset \Lambda(\mathbf{1}_E)$ by Lemma 25.29. On the other hand we have

$$\begin{aligned} &\lim_{r \to 0} \frac{1}{\mu_L(B(0, r))} \int_{B(0,r)} |\mathbf{1}_E(y) - \mathbf{1}_E(0)| \, \mu_L(dy) \\ &= \lim_{r \to 0} \frac{1}{2r} \int_{(-r,r)} \{1 - \mathbf{1}_E(y)\} \, \mu_L(dy) \\ &= \lim_{r \to 0} \frac{1}{2r} \Big\{ 2r - \int_{[0,r)} \mu_L(dy) \Big\} = \frac{1}{2} \neq 0. \end{aligned}$$

Thus $0 \notin \Lambda(\mathbf{1}_E)$ according to Definition 25.13. Nevertheless for the collection in $\mathfrak{B}_{\mathbb{R}^n}$, $\mathcal{S} = \{B(0,r) : r \in (0,\infty)\}$, that shrinks nicely to 0, we have

$$\delta_{\mathcal{S}}(E,0) = \lim_{r\to 0} \frac{\mu_L\big(E \cap B(0,r)\big)}{\mu_L\big(B(0,r)\big)} = \lim_{r\to 0} \frac{r}{2r} = \frac{1}{2}. \ \blacksquare$$

Example 2. Let $\alpha \in (0,1)$ and let $\vartheta \in (0,2\pi)$ be such that $\frac{\vartheta}{2\pi} = \alpha$. Let E be a sector of the plane $\mathbb{R}^2$ sustaining an angle ϑ at $0 \in \mathbb{R}^2$. Then we have

$$\delta_{\mathcal{S}}(E,0) = \lim_{r\to 0} \frac{\mu_L^2\big(E \cap B(0,r)\big)}{\mu_L^2\big(B(0,r)\big)} = \lim_{r\to 0} \frac{\vartheta}{2\pi} = \alpha.$$

Since α is neither 1 nor 0, we have $0 \notin \Lambda(\mathbf{1}_E)$ by (1) of Corollary 25.32. ∎

Proposition 25.34. *Let E be the Cantor ternary set in $\mathbb{R}$. Then we have $\Lambda(\mathbf{1}_E) = E^c$ and $\delta_{\mathcal{S}}(E,x)$ exists and $\delta_{\mathcal{S}}(E,x) = 0$ for every $x \in \mathbb{R}$.*

Proof. By Theorem 4.31, E is a closed set in $\mathbb{R}$ and is also a null set in $(\mathbb{R}, \mathfrak{B}_{\mathbb{R}}, \mu_L)$. Thus E^c is an open set and therefore $E^c = (E^c)^\circ \subset \Lambda(\mathbf{1}_E)$ by (a) of Lemma 25.29. Since E is a null set in $(\mathbb{R}, \mathfrak{M}_L, \mu_L)$, we have $E \subset \Lambda^c(\mathbf{1}_E)$ by (c) of Lemma 25.29. This implies $E^c \supset \Lambda(\mathbf{1}_E)$ and therefore $\Lambda(\mathbf{1}_E) = E^c$.

Since E is a null set in $(\mathbb{R}, \mathfrak{M}_L, \mu_L)$, $E \cap B(x,r)$ is a null set in $(\mathbb{R}, \mathfrak{M}_L, \mu_L)$ for every $x \in \mathbb{R}$ and $r > 0$. Thus $\mu_L\big(E \cap B(x,r)\big)/\mu_L\big(B(x,r)\big) = 0$ and consequently we have $\delta_{\mathcal{S}}(E,x) = \lim_{r\to 0} \{\mu_L\big(E \cap B(x,r)\big)/\mu_L\big(B(x,r)\big)\} = 0$. ∎

Theorem 25.35. *Let $E \in \mathfrak{M}_L^n$. Let $\mathfrak{D}\big(\delta_{\mathcal{S}}(E,\cdot)\big)$ be the domain of definition of $\delta_{\mathcal{S}}(E,\cdot)$. Then $\mathfrak{D}\big(\delta_{\mathcal{S}}(E,\cdot)\big) \in \mathfrak{M}_L^n$, $\delta_{\mathcal{S}}(E,\cdot)$ is $\mathfrak{M}_L^n$-measurable on $\mathfrak{D}\big(\delta_{\mathcal{S}}(E,\cdot)\big)$ and*

$$\int_{\mathbb{R}^n} \delta_{\mathcal{S}}(E,y)\, \mu_L^n(dy) = \mu_L^n(E).$$

Proof. 1. We have $\mathfrak{D}\big(\delta_{\mathcal{S}}(E,\cdot)\big) = \{x \in \mathbb{R}^n : \delta_{\mathcal{S}}(E,x) \text{ exists}\}$. By Theorem 25.30, $\delta_{\mathcal{S}}(E,x)$ exists for every $x \in \Lambda(\mathbf{1}_E)$. Thus we have

(1) $$\mathfrak{D}\big(\delta_{\mathcal{S}}(E,\cdot)\big) \supset \Lambda(\mathbf{1}_E).$$

Let

(2) $$A = \mathfrak{D}\big(\delta_{\mathcal{S}}(E,\cdot)\big) \cap \Lambda^c(\mathbf{1}_E).$$

Then $\Lambda(\mathbf{1}_E) \cap A = \emptyset$ and

(3) $$\mathfrak{D}\big(\delta_{\mathcal{S}}(E,\cdot)\big) = \Lambda(\mathbf{1}_E) \cup A.$$

By Theorem 25.15, $\Lambda(\mathbf{1}_E) \in \mathfrak{M}_L^n$ and $\Lambda^c(\mathbf{1}_E)$ is a null set in the complete measure space $(\mathbb{R}^n, \mathfrak{M}_L^n, \mu_L^n)$. Thus the subset $A \subset \Lambda^c(\mathbf{1}_E)$ is a $\mathfrak{M}_L^n$-measurable set. Then by (3), $\mathfrak{D}\big(\delta_{\mathcal{S}}(E,\cdot)\big)$ is a $\mathfrak{M}_L^n$-measurable set.

2. Let us show that $\delta_{\mathfrak{S}}(E,\cdot)$ is $\mathfrak{M}_L^n$-measurable on $\mathfrak{D}\big(\delta_{\mathfrak{S}}(E,\cdot)\big)$. By (3), $\mathfrak{D}\big(\delta_{\mathfrak{S}}(E,\cdot)\big)$ is the union of two disjoint $\mathfrak{M}_L^n$-measurable sets $\Lambda(\mathbf{1}_E)$ and A. Thus it suffices to show that $\mathfrak{D}\big(\delta_{\mathfrak{S}}(E,\cdot)\big)$ is $\mathfrak{M}_L^n$-measurable on each of these two sets. Now $\Lambda(\mathbf{1}_E)$ is the union of two disjoint $\mathfrak{M}_L^n$-measurable sets $E\cap\Lambda(\mathbf{1}_E)$ and $E^c\cap\Lambda(\mathbf{1}_E)$ and $\delta_{\mathfrak{S}}(E,\cdot)$ assumes the constant value 1 on the first of these two $\mathfrak{M}_L^n$-measurable sets and the constant value 0 on the second according to Theorem 25.30. Therefore $\delta_{\mathfrak{S}}(E,\cdot)$ is $\mathfrak{M}_L^n$-measurable on $\Lambda(\mathbf{1}_E)$. Since A is a null set in the complete measure space $(\mathbb{R}^n,\mathfrak{M}_L^n,\mu_L^n)$, every subset of A is a $\mathfrak{M}_L^n$-measurable set and then every extended real-valued function on A is $\mathfrak{M}_L^n$-measurable on A. In particular $\delta_{\mathfrak{S}}(E,\cdot)$ is $\mathfrak{M}_L^n$-measurable on A.

3. We have $\mathbb{R}^n=\{E\cap\Lambda(\mathbf{1}_E)\}\cup\{E^c\cap\Lambda(\mathbf{1}_E)\}\cup\Lambda^c(\mathbf{1}_E)$, a union of three disjoint $\mathfrak{M}_L^n$-measurable sets. Now $\Lambda^c(\mathbf{1}_E)$ is a null set in $(\mathbb{R}^n,\mathfrak{M}_L^n,\mu_L^n)$. Thus we have

$$
\begin{aligned}
(4)\qquad \int_{\mathbb{R}^n}\delta_{\mathfrak{S}}(E,\cdot)\,d\mu_L^n &= \int_{E\cap\Lambda(\mathbf{1}_E)}\delta_{\mathfrak{S}}(E,\cdot)\,d\mu_L^n+\int_{E^c\cap\Lambda(\mathbf{1}_E)}\delta_{\mathfrak{S}}(E,\cdot)\,d\mu_L^n\\
&= \int_{E\cap\Lambda(\mathbf{1}_E)}1\,d\mu_L^n+\int_{E^c\cap\Lambda(\mathbf{1}_E)}0\,d\mu_L^n\\
&= \mu_L^n\big(E\cap\Lambda(\mathbf{1}_E)\big).
\end{aligned}
$$

Now $E\cap\Lambda(\mathbf{1}_E)=E\setminus\Lambda^c(\mathbf{1}_E)=E\setminus\big(E\cap\Lambda^c(\mathbf{1}_E)\big)$ and since $E\cap\Lambda^c(\mathbf{1}_E)$ is a null set in $(\mathbb{R}^n,\mathfrak{M}_L^n,\mu_L^n)$ we have $\mu_L^n\big(E\cap\Lambda(\mathbf{1}_E)\big)=\mu_L^n(E)-\mu_L^n\big(E\cap\Lambda^c(\mathbf{1}_E)\big)=\mu_L^n(E)$. Substituting this in (4), we have $\int_{\mathbb{R}^n}\delta_{\mathfrak{S}}(E,y)\,\mu_L^n(dy)=\mu_L^n(E)$. ∎

[V] Signed Borel Measures on $\mathbb{R}^n$

Definition 25.36.
(a) *We call a measure on the σ-algebra $\mathfrak{B}_{\mathbb{R}^n}$ of subsets of $\mathbb{R}^n$ a Borel measure on $\mathbb{R}^n$.*
(b) *We say that a Borel measure ν on $\mathbb{R}^n$ is outer regular if for every $E\in\mathfrak{B}_{\mathbb{R}^n}$ we have $\nu(E)=\inf\{\nu(O):O\supset E$ and O is an open set$\}$.*
(c) *We say that a Borel measure ν on $\mathbb{R}^n$ is inner regular if for every $E\in\mathfrak{B}_{\mathbb{R}^n}$ we have $\nu(E)=\sup\{\nu(K):K\subset E$ and K is a compact set$\}$.*
(d) *We say that a Borel measure ν on $\mathbb{R}^n$ is regular if it is both outer and inner regular.*
(e) *By a signed Borel measure on $\mathbb{R}^n$ we mean a signed measure on $\mathfrak{B}_{\mathbb{R}^n}$.*
(f) *We say that a signed Borel measure ν on $\mathbb{R}^n$ is outer regular, inner regular, or regular if both its positive part ν^+ and its negative part ν^- are outer regular, inner regular, or regular respectively.*

If we restrict Lebesgue measure μ_L^n on $\mathfrak{M}_L^n$ to $\mathfrak{B}_{\mathbb{R}^n}$, then we have a Borel measure on $\mathbb{R}^n$ which is outer regular by Proposition 24.17 and inner regular by Proposition 24.16. Note also that μ_L^n is finite on the compact sets in $\mathbb{R}^n$ since a compact set in $\mathbb{R}^n$ is a bounded set in $\mathbb{R}^n$.

Observation 25.37. Let ν be a signed Borel measure on $\mathbb{R}^n$ which is finite on the compact sets in $\mathbb{R}^n$. Then we have:
(a) ν is σ-finite signed measure on $\mathfrak{B}_{\mathbb{R}^n}$.

(b) If $\nu \ll \mu_L^n$ on $(\mathbb{R}^n, \mathfrak{B}_{\mathbb{R}^n})$, then every version $\frac{d\nu}{d\mu_L^n}$ of the Radon-Nikodym derivative of ν with respect to μ_L^n is in $\mathcal{L}^1_{loc}(\mathbb{R}^n, \mathfrak{B}_{\mathbb{R}^n}, \mu_L^n)$.

Proof. 1. For every $k \in \mathbb{N}$, the closed ball $\overline{B}(0, k)$ is a compact set in $\mathbb{R}^n$ and thus $\nu(\overline{B}(0, k)) \in \mathbb{R}$. Then since $\mathbb{R}^n = \bigcup_{k\in\mathbb{N}} \overline{B}(0, k)$, ν is a σ-finite signed measure.

2. Suppose $\nu \ll \mu_L^n$ on $(\mathbb{R}^n, \mathfrak{B}_{\mathbb{R}^n})$. Since μ_L^n is a σ-finite positive measure, $\nu \ll \mu_L^n$ implies the existence of the Radon-Nikodym derivative of ν with respect to μ_L^n by Theorem 11.16. Take an arbitrary version of the Radon-Nikodym derivative $\frac{d\nu}{d\mu_L^n}$. Then for every $k \in \mathbb{N}$, we have $\int_{\overline{B}(0,k)} \frac{d\nu}{d\mu_L^n}\, d\mu_L^n = \nu(\overline{B}(0, k)) \in \mathbb{R}$ since $\overline{B}(0, k)$ is a compact set in $\mathbb{R}^n$. Since every bounded Borel set E in $\mathbb{R}^n$ is contained in some $\overline{B}(0, k)$ with sufficiently large k and since $\nu(\overline{B}(0, k)) \in \mathbb{R}$, we have $\int_E \frac{d\nu}{d\mu_L^n}\, d\mu_L^n = \nu(E) \in \mathbb{R}$ by (a) of Lemma 10.5. This shows that $\frac{d\nu}{d\mu_L^n} \in \mathcal{L}^1_{loc}(\mathbb{R}^n, \mathfrak{B}_{\mathbb{R}^n}, \mu_L^n)$. ∎

Definition 25.38. *Let ν be a signed Borel measure on $\mathbb{R}^n$ which is finite on the compact sets in $\mathbb{R}^n$. We define the maximal function $M\nu$ of ν by setting for $x \in \mathbb{R}^n$*

$$(M\nu)(x) = \sup_{r\in(0,\infty)} \frac{\|\nu\|(B(x,r))}{\mu_L^n(B(x,r))}$$

where $\|\nu\|$ is the total variation of ν, that is, $\|\nu\| = \nu^+ + \nu^-$ and ν^+ and ν^- are the positive and negative parts of ν.

Proposition 25.39. *The maximal function $M\nu$ of a signed Borel measure ν on $\mathbb{R}^n$ which is finite on the compact sets in $\mathbb{R}^n$ is a lower semicontinuous function on $\mathbb{R}^n$ and in particular it is a $\mathfrak{B}_{\mathbb{R}^n}$-measurable function on $\mathbb{R}^n$.*

Proof. Since $M\nu$ is a nonnegative extended real-valued function on $\mathbb{R}^n$, to show that it is lower semicontinuous on $\mathbb{R}^n$ it suffices to show that the set $D_\alpha = \{\mathbb{R}^n : M\nu > \alpha\}$ is an open set in $\mathbb{R}^n$ for every $\alpha > 0$. To show that D_α is an open set, we show that for every $x \in D_\alpha$ there exists $\delta > 0$ such that $B(x, \delta) \subset D_\alpha$. Now if $x \in D_\alpha$, then $(M\nu)(x) > \alpha$ so that there exists $\rho > 0$ such that $\|\nu\|(B(x, \rho)) = \beta\mu_L^n(B(x, \rho))$ for some $\beta > \alpha$. Let $\delta > 0$ be so small that $\left(1 + \frac{\delta}{\rho}\right) < \left(\frac{\beta}{\alpha}\right)^{1/n}$, or equivalently, $(\rho + \delta)^n < \rho^n \frac{\beta}{\alpha}$. Consider $B(x, \delta)$. To show $B(x, \delta) \subset D_\alpha$, let $y \in B(x, \delta)$ be arbitrarily chosen. Then $B(y, \rho + \delta) \supset B(x, \rho)$ so that by the translation invariance of μ_L^n and by Proposition 24.34 we have

$$\begin{aligned}\|\nu\|(B(y, \rho + \delta)) &\geq \|\nu\|(B(x, \rho)) = \beta\mu_L^n(B(x, \rho)) \\ &= \beta\left(\frac{\rho}{\rho + \delta}\right)^n \mu_L^n(B(y, \rho + \delta)) \\ &> \alpha\mu_L^n(B(y, \rho + \delta)),\end{aligned}$$

and therefore

$$(M\nu)(y) = \sup_{r\in(0,\infty)} \frac{\|\nu\|(B(y,r))}{\mu_L^n(B(y,r))} > \alpha.$$

This shows that $y \in D_\alpha$. Thus $B(x, \delta) \subset D_\alpha$. $\blacksquare$

Theorem 25.40. (Maximal Theorem for a Signed Borel Measure) *For the maximal function $M\nu$ of a signed Borel measure ν on $\mathbb{R}^n$ which is finite on the compact sets in $\mathbb{R}^n$, we have*

$$\mu_L^n\{\mathbb{R}^n : M\nu > \alpha\} \leq \frac{3^n}{\alpha}|\nu|(\mathbb{R}^n) \quad \text{for } \alpha \in (0, \infty).$$

Proof. For $\alpha \in (0, \infty)$, let $D_\alpha = \{\mathbb{R}^n : M\nu > \alpha\}$. Then for every $x \in D_\alpha$, we have

$$(M\nu)(x) = \sup_{r \in (0, \infty)} \frac{|\nu|(B(x, r))}{\mu_L^n(B(x, r))} > \alpha$$

so that there exists $r_x \in (0, \infty)$ such that $|\nu|(B(x, r_x)) > \alpha \mu_L^n(B(x, r_x))$. Consider the collection of open balls $\mathfrak{C} = \{B(x, r_x) : x \in D_\alpha\}$. Since $\bigcup_{x \in D_\alpha} B(x, r_x)$ is an open set, it is a $\mathfrak{B}_{\mathbb{R}^n}$-measurable set. By Lemma 25.10, for every real number c such that $c < \mu_L^n(\bigcup_{x \in D_\alpha} B(x, r_x))$, there exists a finite disjoint subcollection $\{B_1, \ldots, B_p\}$ of $\mathfrak{C}$ such that

$$c < 3^n \sum_{k=1}^p \mu_L^n(B_k) < \frac{3^n}{\alpha} \sum_{k=1}^p |\nu|(B_k) = \frac{3^n}{\alpha}|\nu|\left(\bigcup_{k=1}^p B_k\right) \leq \frac{3^n}{\alpha}|\nu|(\mathbb{R}^n).$$

Since this holds for every $c \in \mathbb{R}$ such that $c < \mu_L^n(\bigcup_{x \in D_\alpha} B(x, r_x))$, the inequality above implies $\mu_L^n(\bigcup_{x \in D_\alpha} B(x, r_x)) \leq \frac{3^n}{\alpha}|\nu|(\mathbb{R}^n)$. Then since $D_\alpha \subset \bigcup_{x \in D_\alpha} B(x, r_x)$, we have $\mu_L^n(D_\alpha) \leq \frac{3^n}{\alpha}|\nu|(\mathbb{R}^n)$. $\blacksquare$

[VI] Differentiation of Borel Measures with Respect to the Lebesgue Measure

Let ν be an arbitrary σ-finite signed measure on $(\mathbb{R}^n, \mathfrak{B}_{\mathbb{R}^n})$. If $\nu \ll \mu_L^n$ is on $(\mathbb{R}^n, \mathfrak{B}_{\mathbb{R}^n})$, that is, $\nu(E) = 0$ for every $E \in \mathfrak{B}_{\mathbb{R}^n}$ such that $\mu_L^n(E) = 0$, then the Radon-Nikodym derivative of ν with respect to μ_L^n exists, that is, there exists an extended real-valued $\mathfrak{B}_{\mathbb{R}^n}$-measurable function on $\mathbb{R}^n$, denoted by $\frac{d\nu}{d\mu_L^n}$, such that $\nu(E) = \int_E \frac{d\nu}{d\mu_L^n} d\mu_L^n$ for every $E \in \mathfrak{B}_{\mathbb{R}^n}$. (See Definition 11.1.) Since $\int_E \frac{d\nu}{d\mu_L^n} d\mu_L^n = \nu(E) \in \mathbb{R}$ for every $E \in \mathfrak{B}_{\mathbb{R}^n}$, $\frac{d\nu}{d\mu_L^n}$ is a μ_L^n semi-integrable function on $\mathbb{R}^n$. Since the Radon-Nikodym derivative is unique up to a null set in $(\mathbb{R}^n, \mathfrak{B}_{\mathbb{R}^n}, \mu_L^n)$, we speak of versions of the Radon-Nikodym derivative.

The Radon-Nikodym derivative $\frac{d\nu}{d\mu_L^n}$ of ν with respect to μ_L^n was not defined as the derivative of a set function (and in particular a signed measure) on $\mathfrak{B}_{\mathbb{R}}^n$ with respect to the Lebesgue measure μ_L^n in the sense of Definition 25.19. In the next theorem we show that if ν is a signed Borel measure which is finite on the compact sets in $\mathbb{R}^n$ and if $\nu \ll \mu_L^n$ on $(\mathbb{R}^n, \mathfrak{B}_{\mathbb{R}^n})$, then the symmetric derivative $\mathcal{D}_S\nu$ of ν with respect to μ_L^n exists a.e. on $(\mathbb{R}^n, \mathfrak{B}_{\mathbb{R}^n}, \mu_L^n)$ and moreover $\mathcal{D}_S\nu = \frac{d\nu}{d\mu_L^n}$ a.e. on $(\mathbb{R}^n, \mathfrak{B}_{\mathbb{R}^n}, \mu_L^n)$ for an arbitrary version of the Radon-Nikodym derivative.

Theorem 25.41. *Let ν be a signed Borel measure on $\mathbb{R}^n$ which is finite on the compact sets in $\mathbb{R}^n$. Suppose $\nu \ll \mu_L^n$ on $(\mathbb{R}^n, \mathfrak{B}_{\mathbb{R}^n})$ and let $\frac{d\nu}{d\mu_L^n}$ be an arbitrary version of the Radon-Nikodym derivative of ν with respect to μ_L^n on $(\mathbb{R}^n, \mathfrak{B}_{\mathbb{R}^n})$. Then for every point x in the Lebesgue set $\Lambda\left(\frac{d\nu}{d\mu_L^n}\right)$ and for every collection $\mathcal{E} = \{E_r(x) : r \in (0, \infty)\}$ in $\mathfrak{B}_{\mathbb{R}^n}$ that shrinks nicely to x, we have*

(1)
$$(\mathcal{D}_{\mathcal{E}}\nu)(x) = \lim_{r\to 0} \frac{\nu(E_r(x))}{\mu_L^n(E_r(x))} = \frac{d\nu}{d\mu_L^n}(x).$$

In particular for every $x \in \Lambda\left(\frac{d\nu}{d\mu_L^n}\right)$, we have

(2)
$$(\mathcal{D}_{\mathcal{S}}\nu)(x) = \lim_{r\to 0} \frac{\nu(B(x,r))}{\mu_L^n(B(x,r))} = \frac{d\nu}{d\mu_L^n}(x),$$

and therefore

(3)
$$\mathcal{D}_{\mathcal{S}}\nu = \frac{d\nu}{d\mu_L^n} \quad \text{a.e. on } (\mathbb{R}^n, \mathfrak{B}_{\mathbb{R}^n}, \mu_L^n).$$

Proof. If ν is a signed Borel measure on $\mathbb{R}^n$ which is finite on the compact sets in $\mathbb{R}^n$ and if $\nu \ll \mu_L^n$ on $(\mathbb{R}^n, \mathfrak{B}_{\mathbb{R}^n})$, then by Observation 25.37, ν is a σ-finite signed measure on $(\mathbb{R}^n, \mathfrak{B}_{\mathbb{R}^n})$, and $\frac{d\nu}{d\mu_L^n} \in \mathcal{L}^1_{loc}(\mathbb{R}^n, \mathfrak{B}_{\mathbb{R}^n}, \mu_L^n)$. This implies according to Theorem 25.17 that for every $x \in \Lambda\left(\frac{d\nu}{d\mu_L^n}\right)$ and every collection $\mathcal{E} = \{E_r(x) : r \in (0, \infty)\}$ in $\mathfrak{B}_{\mathbb{R}^n}$ that shrinks nicely to x, we have

$$\frac{d\nu}{d\mu_L^n}(x) = \lim_{r\to 0} \frac{1}{\mu_L^n(E_r(x))} \int_{E_r(x)} \frac{d\nu}{d\mu_L^n}(y)\, \mu_L^n(dy) = \lim_{r\to 0} \frac{\nu(E_r(x))}{\mu_L^n(E_r(x))}.$$

This proves (1).

With $\mathcal{S} = \{B(x,r) : r \in (0, \infty)\}$ as a collection in $\mathfrak{B}_{\mathbb{R}^n}$ that shrinks nicely to x, we have (2) from (1). By Theorem 25.15, $\Lambda^c\left(\frac{d\nu}{d\mu_L^n}\right)$ is a null set in $(\mathbb{R}^n, \mathfrak{M}_L^n, \mu_L^n)$. By Lemma 24.11, there exists a G_δ-set $G \supset \Lambda^c\left(\frac{d\nu}{d\mu_L^n}\right)$ such that $\mu_L^n(G) = 0$. Now $G^c \subset \Lambda\left(\frac{d\nu}{d\mu_L^n}\right)$. Thus by (2) we have $(\mathcal{D}_{\mathcal{S}}\nu)(x) = \frac{d\nu}{d\mu_L^n}(x)$ for $x \in G^c$. Since $G \in \mathfrak{B}_{\mathbb{R}^n}$, (3) holds. ∎

Theorem 25.42. *Let ν be an outer regular signed Borel measure on $\mathbb{R}^n$ which is finite on the compact sets in $\mathbb{R}^n$. If $\nu \perp \mu_L^n$ on $(\mathbb{R}^n, \mathfrak{B}_{\mathbb{R}^n})$, then for a.e. x in $(\mathbb{R}^n, \mathfrak{M}_L^n, \mu_L^n)$ and for every collection $\mathcal{E} = \{E_r(x) : r \in (0, \infty)\}$ in $\mathfrak{B}_{\mathbb{R}^n}$ that shrinks nicely to x, we have*

(1)
$$(\mathcal{D}_{\mathcal{E}}\nu)(x) = \lim_{r\to 0} \frac{\nu(E_r(x))}{\mu_L^n(E_r(x))} = 0.$$

In particular for a.e. x in $(\mathbb{R}^n, \mathfrak{M}_L^n, \mu_L^n)$, we have

(2)
$$(\mathcal{D}_{\mathcal{S}}\nu)(x) = \lim_{r\to 0} \frac{\nu(B(x,r))}{\mu_L^n(B(x,r))} = 0,$$

and therefore

(3) $$\mathcal{D}_{\mathcal{S}}\nu = 0 \quad a.e. \text{ on } \left(\mathbb{R}^n, \mathfrak{B}_{\mathbb{R}^n}, \mu_L^n\right).$$

Proof. Let $|\nu|$ be the total variation of ν, that is, $|\nu| = \nu^+ + \nu^-$ where ν^+ and ν^- are the positive and negative parts of ν. Now $\nu \perp \mu_L^n$ on $\left(\mathbb{R}^n, \mathfrak{B}_{\mathbb{R}^n}\right)$ implies $|\nu| \perp \mu_L^n$ on $(\mathbb{R}^n, \mathfrak{B}_{\mathbb{R}^n})$ by Proposition 10.27. Thus there exists a set $A \in \mathfrak{B}_{\mathbb{R}^n}$ such that $|\nu|(A) = \mu_L^n(A^c) = 0$. Since ν is an outer regular signed Borel measure on $\mathfrak{B}_{\mathbb{R}^n}$, ν^+ and ν^- are outer regular positive Borel measures on $\mathfrak{B}_{\mathbb{R}^n}$ and then so is $|\nu|$.

For brevity, let us write

$$(\mathcal{Q}_r|\nu|)(x) = \frac{|\nu|\left(B(x,r)\right)}{\mu_L^n\left(B(x,r)\right)}$$

so that

$$\left(\mathcal{D}_{\mathcal{S}}|\nu|\right)(x) = \lim_{r\to 0}\frac{|\nu|\left(B(x,r)\right)}{\mu_L^n\left(B(x,r)\right)} = \lim_{r\to 0}\left(\mathcal{Q}_r|\nu|\right)(x).$$

To prove (1), let us show first that for a.e. x in $\left(\mathbb{R}^n, \mathfrak{M}_L^n, \mu_L^n\right)$, we have

(4) $$\mathcal{D}_{\mathcal{S}}\left(|\nu|\right)(x) = \lim_{r\to 0}\left(\mathcal{Q}_r|\nu|\right)(x) = 0.$$

To prove (4), it suffices to show that for a.e. x in $\left(\mathbb{R}^n, \mathfrak{M}_L^n, \mu_L^n\right)$, we have

(5) $$\limsup_{r\to 0}\left(\mathcal{Q}_r|\nu|\right)(x) = \lim_{m\to\infty}\sup_{0<r<1/m}\left(\mathcal{Q}_r|\nu|\right)(x) = 0.$$

Note that $\sup_{0<r<1/m}\mathcal{Q}_r|\nu|$ is a lower semicontinuous function on $\mathbb{R}^n$ so that it is a $\mathfrak{B}_{\mathbb{R}^n}$-measurable function by Proposition 25.39 for every $m \in \mathbb{N}$. Thus $\limsup_{r\to 0}\mathcal{Q}_r|\nu|$ is a $\mathfrak{B}_{\mathbb{R}^n}$-measurable function on $\mathbb{R}^n$. Now since $\mu_L^n(A^c) = 0$, to prove (5) it suffices to show

(6) $$\mu_L^n\{x \in A : \limsup_{r\to 0}\left(\mathcal{Q}_r|\nu|\right)(x) \neq 0\} = 0.$$

For $k \in \mathbb{N}$, let $F_k = \left\{x \in A : \limsup_{r\to 0}\left(\mathcal{Q}_r|\nu|\right)(x) > \frac{1}{k}\right\}$. Since the set in (6) is equal to the union $\bigcup_{k\in\mathbb{N}} F_k$, to prove (6) it suffices to show that $\mu_L^n(F_k) = 0$ for every $k \in \mathbb{N}$. Now since $|\nu|(A) = 0$ and since $|\nu|$ is an outer regular Borel measure on $\mathfrak{B}_{\mathbb{R}^n}$, there exists an open set $O_\varepsilon \supset A$ such that $|\nu|(O_\varepsilon) < \varepsilon$. For each $x \in F_k \subset A \subset O_\varepsilon$, there exists an open ball B_x with center at x such that $B_x \subset O_\varepsilon$ and $|\nu|(B_x) > \frac{1}{k}\mu_L^n(B_x)$ by the definition of the set F_k. Consider the collection $\mathfrak{C} = \{B_x : x \in F_k\}$. Since $\bigcup_{x\in F_k} B_x$ is an open set, $\bigcup_{x\in F_k} B_x \in \mathfrak{B}_{\mathbb{R}^n}$. If $c \in \mathbb{R}$ and $c < \mu_L^n\left(\bigcup_{x\in F_k} B_x\right)$, then by Lemma 25.10, there exists a disjoint subcollection $\{B_{x_1}, \ldots, B_{x_p}\}$ such that

$$\begin{aligned} c &< 3^n\sum_{j=1}^{p}\mu_L^n(B_{x_j}) < 3^n k\sum_{j=1}^{p}|\nu|(B_{x_j}) = 3^n k|\nu|\Big(\bigcup_{j=1}^{p} B_{x_j}\Big) \\ &\le 3^n k|\nu|\Big(\bigcup_{x\in F_k} B_x\Big) \le 3^n k|\nu|(O_\varepsilon) < 3^n k\varepsilon. \end{aligned}$$

Since this holds for every $c < \mu_L^n\big(\bigcup_{x\in F_k} B_x\big)$, we have $\mu_L^n\big(\bigcup_{x\in F_k} B_x\big) \le 3^n k\varepsilon$. Then by the arbitrariness of $\varepsilon > 0$, we have $\mu_L^n\big(\bigcup_{x\in F_k} B_x\big) = 0$. Since $F_k \subset \bigcup_{x\in F_k} B_x$, and since $(\mathbb{R}^n, \mathfrak{M}_L^n, \mu_L^n)$ is a complete measure space, $F_k \in \mathfrak{M}_L^n$ with $\mu_L^n(F_k) = 0$. This proves (6) and completes the proof of (4).

By (4), there exists a null set E in $(\mathbb{R}^n, \mathfrak{M}_L^n, \mu_L^n)$ such that $\lim_{r\to 0}\big(Q_r\|\nu\|\big)(x) = 0$ for $x \in E^c$. Let $\mathcal{E} = \big\{E_r(x) : r \in (0,\infty)\big\}$ be a collection in $\mathfrak{B}_{\mathbb{R}^n}$ that shrinks nicely to x. By the fact that $E_r(x) \subset B(x,r)$ and the fact that there exists a constant $\alpha > 0$ such that $\alpha\mu_L^n\big(B(x,r)\big) \le \mu_L^n\big(E_r(x)\big) \le \mu_L^n\big(B(x,r)\big)$ for $r \in (0,\infty)$, we have

$$\frac{\|\nu\|\big(E_r(x)\big)}{\mu_L^n\big(E_r(x)\big)} \le \frac{\|\nu\|\big(B(x,r)\big)}{\mu_L^n\big(E_r(x)\big)} \le \frac{1}{\alpha}\frac{\|\nu\|\big(B(x,r)\big)}{\mu_L^n\big(B(x,r)\big)} = \frac{1}{\alpha}\big(Q_r\|\nu\|\big)(x).$$

Thus for $x \in E^c$, we have

$$\lim_{r\to 0}\frac{\|\nu\|\big(E_r(x)\big)}{\mu_L^n\big(E_r(x)\big)} = 0. \tag{7}$$

Since $|\nu(E_r(x))| \le \|\nu\|\big(E_r(x)\big)$ for every $x \in \mathbb{R}^n$ and every $r \in (0,\infty)$, (7) implies that $\lim_{r\to 0}\nu\big(E_r(x)\big)/\mu_L^n\big(E_r(x)\big) = 0$ for $x \in E^c$. This proves (1). In particular we have $\lim_{r\to 0}\nu\big(B(x,r)\big)/\mu_L^n\big(B(x,r)\big) = 0$, that is, $(\mathcal{D}_{\mathcal{S}}\nu)(x) = 0$ for $x \in E^c$. This proves (2). Since E is a null set in $(\mathbb{R}^n, \mathfrak{M}_L^n, \mu_L^n)$, there exists a null set G in $(\mathbb{R}^n, \mathfrak{B}_{\mathbb{R}^n}, \mu_L^n)$ such that $G \supset E$. Then we have $(\mathcal{D}_{\mathcal{S}}\nu)(x) = 0$ for $x \in G^c \subset E^c$. This proves (3). ∎

Combining Theorem 25.41 and Theorem 25.42, we have the following theorem for the symmetric derivative $\mathcal{D}_{\mathcal{S}}\nu$ of a signed Borel measure ν on $\mathfrak{B}_{\mathbb{R}^n}$ with respect to the Lebesgue measure μ_L^n.

Theorem 25.43. *Let ν be an outer regular signed Borel measure on $\mathbb{R}^n$ which is finite on the compact sets in $\mathbb{R}^n$. Let $\nu = \nu_a + \nu_s$ be the Lebesgue decomposition of ν with respect to $(\mathbb{R}^n, \mathfrak{B}_{\mathbb{R}^n}, \mu_L^n)$ with $\nu_a \ll \mu_L^n$ and $\nu_s \perp \mu_L^n$ on $(\mathbb{R}^n, \mathfrak{B}_{\mathbb{R}^n})$. Then for a.e. x in the measure space $(\mathbb{R}^n, \mathfrak{M}_L^n, \mu_L^n)$ and for every collection $\mathcal{E} = \big\{E_r(x) : r \in (0,\infty)\big\}$ in $\mathfrak{B}_{\mathbb{R}^n}$ that shrinks nicely to x, we have*

$$(\mathcal{D}_{\mathcal{E}}\nu)(x) = \lim_{r\to 0}\frac{\nu\big(E_r(x)\big)}{\mu_L^n\big(E_r(x)\big)} = \frac{d\nu_a}{d\mu_L^n}(x). \tag{1}$$

In particular for a.e. x in $(\mathbb{R}^n, \mathfrak{M}_L^n, \mu_L^n)$, we have

$$(\mathcal{D}_{\mathcal{S}}\nu)(x) = \frac{d\nu_a}{d\mu_L^n}(x), \tag{2}$$

and therefore

$$\mathcal{D}_{\mathcal{S}}\nu = \frac{d\nu_a}{d\mu_L^n} \quad \text{a.e. on } (\mathbb{R}^n, \mathfrak{B}_{\mathbb{R}^n}, \mu_L^n). \tag{3}$$

Proof. Since $\nu = \nu_a + \nu_s$, (1) follows from (1) of Theorem 25.41 and (1) of Theorem 25.42. Since $\{B(x,r) : r \in (0,\infty)\}$ is a collection in $\mathfrak{B}_{\mathbb{R}^n}$ shrinking nicely to x, (1) implies (2). Since $\frac{d\nu_a}{d\mu_L^n}$ is a $\mathfrak{B}_{\mathbb{R}^n}$-measurable function on $\mathbb{R}^n$ and since every null set in $(\mathbb{R}^n, \mathfrak{M}_L^n, \mu_L^n)$ is contained in a null set in $(\mathbb{R}^n, \mathfrak{B}_{\mathbb{R}^n}, \mu_L^n)$, (3) follows from (2). ■

Problems

Prob. 25.1. **(a)** Show that if f is a real-valued $\mathfrak{M}_L^n$-measurable function on $\mathbb{R}^n$ and f is bounded on every bounded subset of $\mathbb{R}^n$, then $f \in \mathcal{L}^1_{loc}(\mathbb{R}^n, \mathfrak{M}_L^n, \mu_L^n)$.
(b) Show that if f is a continuous real-valued $\mathfrak{M}_L^n$-measurable function on $\mathbb{R}^n$, then $f \in \mathcal{L}^1_{loc}(\mathbb{R}^n, \mathfrak{M}_L^n, \mu_L^n)$.

Prob. 25.2. Let f be a real-valued function on $\mathbb{R}$ defined by $f(x) = 0$ if $x \in \mathbb{R}$ is rational and $f(x) = 1$ if $x \in \mathbb{R}$ is irrational.
(a) Show that $f \in \mathcal{L}^1_{loc}(\mathbb{R}, \mathfrak{M}_L, \mu_L)$.
(b) Find Af.
(c) Show that $Af = f$ a.e. on $(\mathbb{R}, \mathfrak{M}_L, \mu_L)$.

Prob. 25.3. **(a)** Show that if $f \in \mathcal{L}^1(\mathbb{R}^n, \mathfrak{M}_L^n, \mu_L^n)$, then for each $r \in (0,\infty)$ the function $A_r f$ is uniformly continuous on $\mathbb{R}^n$.
(b) Show that if $f \in \mathcal{L}^1_{loc}(\mathbb{R}^n, \mathfrak{M}_L^n, \mu_L^n)$, then for each $r \in (0,\infty)$ the function $A_r f$ is uniformly continuous on every bounded subset of $\mathbb{R}^n$.

Prob. 25.4. For $f \in \mathcal{L}^1_{loc}(\mathbb{R}^n, \mathfrak{M}_L^n, \mu_L^n)$, the Hardy-Littlewood maximal function Mf of f is a nonnegative extended real-valued function on $\mathbb{R}^n$ defined by setting for $x \in \mathbb{R}^n$

$$(Mf)(x) = \sup_{r \in (0,\infty)} \frac{1}{\mu_L^n\big(B(x,r)\big)} \int_{B(x,r)} |f| \, d\mu_L^n.$$

For $x \in \mathbb{R}^n$, let $\mathfrak{C}_x$ be the collection of all open balls in $\mathbb{R}^n$ that contain x. Let us define a nonnegative extended real-valued function M^*f on $\mathbb{R}^n$ by setting for $x \in \mathbb{R}^n$

$$(M^*f)(x) = \sup_{C \in \mathfrak{C}_x} \frac{1}{\mu_L^n(C)} \int_C |f| \, d\mu_L^n.$$

Show that $(Mf)(x) \le (M^*f)(x) \le 2^n (Mf)(x)$ for $x \in \mathbb{R}^n$.

Prob. 25.5. Let $f \in \mathcal{L}^1_{loc}(\mathbb{R}^n, \mathfrak{M}_L^n, \mu_L^n)$. Show that if f is continuous at $x \in \mathbb{R}^n$, then $x \in \Lambda(f)$.

Prob. 25.6. Let $E \in \mathfrak{M}_L^n$. Show directly from the definition of the symmetric density $\delta_{\mathbb{S}}(E, \cdot)$ of E, without recourse to the Lebesgue set $\Lambda(\mathbf{1}_E)$, that

$$\delta_{\mathbb{S}}(E,x) = \lim_{r \to 0} \frac{\mu_L^n\big(E \cap B(x,r)\big)}{\mu_L^n\big(B(x,r)\big)} = \begin{cases} 1 & \text{for a.e. } x \in E, \\ 0 & \text{for a.e. } x \in E^c. \end{cases}$$

Prob. 25.7. In $\mathbb{R}^2$ let $E = \{(x_1,x_2) \in \mathbb{R}^2 : x_1 \ge 0, x_2 \ge 0\}$.
(a) Show that $\Lambda(\mathbf{1}_E) = E^\circ \cup (E^c)^\circ$ and thus $\Lambda^c(\mathbf{1}_E) = \partial E$, the boundary of E.
(b) Let $x = (1,0) \notin \Lambda(\mathbf{1}_E)$. Let $\{E_r(x) : r \in (0,\infty)\}$ be a collection in $\mathfrak{B}_{\mathbb{R}^n}$ shrinking nicely to x and consider

(1) $$\lim_{r\to 0} \frac{\mu_L^n\big(E\cap E_r(x)\big)}{\mu_L^n\big(E_r(x)\big)}$$

(b.1) Find a collection $\{E_r(x) : r \in (0,\infty)\}$ for which the limit in (1) is equal to 1.
(b.2) Find a collection $\{E_r(x) : r \in (0,\infty)\}$ for which the limit in (1) is equal to 0.
(b.3) Find a collection $\{E_r(x) : r \in (0,\infty)\}$ for which the limit in (1) is equal to $\frac{1}{2}$.
(b.4) Find a collection $\{E_r(x) : r \in (0,\infty)\}$ for which the limit in (1) does not exist.
(c) Show that for our $x = (1,0) \notin \Lambda(\mathbf{1}_E)$, we have nevertheless

$$\delta_{\mathfrak{S}}(E,x) = \lim_{r\to 0} \frac{\mu_L^n\big(E\cap B(x,r)\big)}{\mu_L^n\big(B(x,r)\big)} = \frac{1}{2}.$$

Show similarly that for $x = (0,0) \notin \Lambda(\mathbf{1}_E)$, we have $\delta_{\mathfrak{S}}(E,x) = \frac{1}{4}$.

Prob. 25.8. Let $\alpha \in (0,1)$ and let $\vartheta \in (0,2\pi)$ be such that $\frac{\vartheta}{2\pi} = \alpha$. Let E be a sector of the plane $\mathbb{R}^2$ sustaining an angle ϑ at $0 \in \mathbb{R}^2$. Find $\Lambda(\mathbf{1}_E)$. Determine $\delta_{\mathfrak{S}}(E,\cdot)$.

Prob. 25.9. Let $E = \{x \in \mathbb{R}^2 : |x| \le a\}$ where $a > 0$. Find $\Lambda(\mathbf{1}_E)$. Determine $\delta_{\mathfrak{S}}(E,\cdot)$.

Prob. 25.10. **(a)** Let E be the set of rational numbers in $\mathbb{R}$. Find $\Lambda(\mathbf{1}_E)$ and $\delta_{\mathfrak{S}}(E,\cdot)$.
(b) Let E be the set of irrational numbers in $\mathbb{R}$. Find $\Lambda(\mathbf{1}_E)$ and $\delta_{\mathfrak{S}}(E,\cdot)$.

§26 Change of Variable of Integration on the Euclidean Space

[I] Change of Variable of Integration by Differentiable Transformations

Consider a mapping of $\mathbb{R}^n$ into $\mathbb{R}^n$ given by $y = T(x) \in \mathbb{R}^n$ for $x \in \mathbb{R}^n$. If T is a translation, that is, $T(x) = x + h$ for $x \in \mathbb{R}^n$ where $h \in \mathbb{R}^n$ is fixed, then by Theorem 24.28, for every nonnegative extended real-valued $\mathfrak{M}_L^n$-measurable function f on a set $D \in \mathfrak{M}_L^n$, we have $\int_D f(y)\,\mu_L^n(dy) = \int_{T^{-1}(D)} (f \circ T)(x)\,\mu_L^n(dx)$. Now since not only $T^{-1}(D) \in \mathfrak{M}_L^n$ but also $T(D) \in \mathfrak{M}_L^n$ for every $D \in \mathfrak{M}_L^n$ by Theorem 24.27, we have

$$\int_{T(E)} f(x)\,\mu_L^n(dx) = \int_E (f \circ T)(y)\,\mu_L^n(dy)$$

for every $E \in \mathfrak{M}_L^n$.

If T is a non-singular linear transformation of $\mathbb{R}^n$ into $\mathbb{R}^n$, then $T(D) \in \mathfrak{M}_L^n$ for every $D \in \mathfrak{M}_L^n$ by Theorem 24.31 so that Theorem 24.32 can be restated as

$$\int_{T(E)} f(x)\,\mu_L^n(dx) = |\det M_T| \int_E (f \circ T)(y)\,\mu_L^n(dy)$$

for every $E \in \mathfrak{M}_L^n$.

These are examples of change of variable of integration. We consider next a change of variable by a nonlinear but differentiable transformation. Let Ω be an open set in $\mathbb{R}^n$ and let T be a one-to-one mapping of Ω into $\mathbb{R}^n$ whose component functions have continuous first order partial derivatives on Ω. We shall show that for an extended real-valued $\mathfrak{M}_L^n$-measurable function f on Ω, we have

$$\int_{T(\Omega)} f(x)\,\mu_L^n(dx) = \int_\Omega (f \circ T)(y)|\det J_T(y)|\,\mu_L^n(dy)$$

where J_T is the Jacobian matrix of the transformation T. In preparation for this change of variable theorem, let us review some facts in the calculus of several variables.

Differential of a Mapping from $\mathbb{R}^n$ into $\mathbb{R}^m$. Let T be a mapping of an open set Ω in $\mathbb{R}^n$ into $\mathbb{R}^m$. We say that T is differentiable at a point $p \in \Omega$ if there exists a linear transformation L of $\mathbb{R}^n$ into $\mathbb{R}^m$ such that if we let

(1) $$R(p; h) = T(p + h) - T(p) - L(h)$$

for $h \in \mathbb{R}^n$, $h \neq 0$, such that $p + h \in \Omega$, then

(2) $$\lim_{h \to 0} \frac{R(p; h)}{|h|} = 0 \in \mathbb{R}^m.$$

We call the linear transformation L the differential of T at p and write $dT(p; \cdot)$ for it. If $dT(p; \cdot)$ exists, then for every $u \in \mathbb{R}^n$ with $|u| = 1$, we have

(3) $$dT(p; u) = \lim_{\lambda \downarrow 0} \frac{T(p + \lambda u) - T(p)}{\lambda}.$$

We call $dT(p; u)$ the directional derivative of T at p in the direction u. If T is differentiable at p, then T is continuous at p.

Uniqueness of the Differential. If the linear transformation $dT(p; \cdot)$ satisfying conditions (1) and (2) exists for some $p \in \Omega$, then it is unique. In particular, if T itself is a linear transformation of $\mathbb{R}^n$ into $\mathbb{R}^m$, then for every $p \in \mathbb{R}^n$ the linear transformation $dT(p; \cdot)$ exists and $dT(p; \cdot) = T$.

Existence of the Differential. For $\mathbb{R}^m = \mathbb{R}_1 \times \cdots \times \mathbb{R}_m$, let π_i be the projection of $\mathbb{R}^m$ onto $\mathbb{R}_i$, that is, $\pi_i(q) = q_i$ for $q = (q_1, \ldots, q_m) \in \mathbb{R}^m$. For a mapping T of an open set Ω in $\mathbb{R}^n$ into $\mathbb{R}^m$, let $(g_1, \ldots, g_m)$ be its component functions, that is, g_i is a real-valued function on Ω defined by $g_i = \pi_i \circ T$ on Ω. If all the partial derivatives $\frac{\partial g_i}{\partial x_j}$ for $i = 1, \ldots, m$ and $j = 1, \ldots, n$ exist at $p \in \Omega$, then we call the $m \times n$ matrix

$$J_T(p) = \begin{pmatrix} \frac{\partial g_1}{\partial x_1}(p) & \cdots & \frac{\partial g_1}{\partial x_n}(p) \\ \vdots & & \vdots \\ \frac{\partial g_m}{\partial x_1}(p) & \cdots & \frac{\partial g_m}{\partial x_n}(p) \end{pmatrix} \tag{4}$$

the Jacobian matrix of the mapping T at p.

We say that T is a mapping of class $C^1(\Omega)$ if all the partial derivatives $\frac{\partial g_i}{\partial x_j}$ for $i = 1, \ldots, m$ and $j = 1, \ldots, n$ exist and are continuous on Ω.

If T is a mapping of class $C^1(\Omega)$, then T is differentiable at every $p \in \Omega$ and furthermore the Jacobian matrix $J_T(p)$ is the matrix of the linear transformation $dT(p; \cdot)$.

Chain Rule for Differentials. Let T be a mapping whose domain of definition $\mathfrak{D}(T)$ is an open set in $\mathbb{R}^n$ and whose range $\mathfrak{R}(T)$ is a subset of $\mathbb{R}^m$. Let S be a mapping whose domain of definition $\mathfrak{D}(S)$ is an open set in $\mathbb{R}^m$ containing $\mathfrak{R}(T)$ and whose range $\mathfrak{R}(S)$ is a subset of $\mathbb{R}^\ell$. Consider the mapping $S \circ T$ of $\mathfrak{D}(T)$ into $\mathbb{R}^\ell$. If T is differentiable at some $p \in \mathfrak{D}(T)$ and S is differentiable at $T(p) \in \mathfrak{D}(S)$, then $S \circ T$ is differentiable at p and its differential at p is given by

$$d(S \circ T)(p; \cdot) = dS\big(T(p); \cdot\big) \circ dT(p; \cdot) = dS\big(T(p); dT(p; \cdot)\big), \tag{5}$$

and, assuming the existence of the first order partial derivatives of the component functions of T at p and those of S at $T(p)$, the Jacobian matrix of $S \circ T$ at p is a product of two matrices given by

$$J_{S \circ T}(p) = J_S\big(T(p)\big) J_T(p). \tag{6}$$

Proofs of these statements can be found for instance in R. C. Buck [5].

To show that if T is a mapping of an open set Ω in $\mathbb{R}^n$ into $\mathbb{R}^n$ of class $C^1(\Omega)$, then $T(E)$ is a null set in $(\mathbb{R}^n, \mathfrak{M}_L^n, \mu_L^n)$ for every null set E in $(\mathbb{R}^n, \mathfrak{M}_L^n, \mu_L^n)$ contained in Ω, we prepare a covering theorem for a null set in $(\mathbb{R}^n, \mathfrak{M}_L^n, \mu_L^n)$ in the following Lemma.

Lemma 26.1. *Let E be a null set in $(\mathbb{R}^n, \mathfrak{M}_L^n, \mu_L^n)$. Then for every $\varepsilon > 0$ and $\eta > 0$, there exists a countable collection of open balls $\{B(x_j, r_j) : j \in \mathbb{N}\}$ such that $x_j \in E$, $r_j < \eta$ for $j \in \mathbb{N}$, $E \subset \bigcup_{j \in \mathbb{N}} B(x_j, r_j)$ and $\sum_{j \in \mathbb{N}} \mu_L^n\big(B(x_j, r_j)\big) < \varepsilon$.*

Proof. Since $\mu_L^n(E) = 0$, by Lemma 24.11 for every $\varepsilon > 0$ there exists an open set $O \supset E$ with $\mu_L^n(O) < 2^{-n}n^{-n/2}\varepsilon$. According to (c) of Proposition 24.10, for an arbitrary $\eta > 0$, O is a countable disjoint union of n-dimensional cubes with lengths of the edges less than $n^{-1/2}\eta$. Drop from this collection of cubes those which are disjoint from E. Let the remaining disjoint countable collection be denoted by $\{Q_j : j \in \mathbb{N}\}$. We have $\bigcup_{j\in\mathbb{N}} Q_j \supset E$. Let a_j be the length of the edges of Q_j for $j \in \mathbb{N}$. We have $a_j < n^{-1/2}\eta$. The length of the diagonal of Q_j is equal to $n^{1/2}a_j < \eta$. Let $x_j \in Q_j \cap E$ be arbitrarily chosen. Let us replace the cube Q_j with an open ball $B_j = B(x_j, r_j)$ where $r_j = n^{1/2}a_j < \eta$. We have $Q_j \subset B_j$ for $j \in \mathbb{N}$ and thus $\bigcup_{n\in\mathbb{N}} B_j \supset \bigcup_{n\in\mathbb{N}} Q_j \supset E$. Now B_j is contained in a cube with length of the edge equal to twice the radius of B_j, that is, $2r_j = 2n^{1/2}a_j$. Thus we have

$$\mu_L^n(B_j) \le \left(2n^{1/2}a_j\right)^n = 2^n n^{n/2} a_j^n = 2^n n^{n/2}\mu_L^n(Q_j)$$

and then by the disjointness of $\{Q_j : j \in \mathbb{N}\}$, we have

$$\sum_{j\in\mathbb{N}} \mu_L^n(B_j) \le 2^n n^{n/2} \sum_{j\in\mathbb{N}} \mu_L^n(Q_j) = 2^n n^{n/2} \mu_L^n\Big(\bigcup_{j\in\mathbb{N}} Q_j\Big) \le 2^n n^{n/2}\mu_L^n(O) < \varepsilon. \quad \blacksquare$$

If a mapping T of a set $E \in \mathfrak{M}_L^n$ into $\mathbb{R}^n$ satisfies the Lipschitz condition that for some constant $C \ge 0$ we have $|T(x) - T(y)| \le C|x - y|$ for every $x, y \in E$, then $\limsup_{y\to x,\, y\in E} \frac{|T(y)-T(x)|}{|y-x|} \le C$ for every $x \in E$. We show next that if T satisfies the weaker condition that $\limsup_{y\to x,\, y\in E} \frac{|T(y)-T(x)|}{|y-x|} < \infty$ for every $x \in E$ and if E is a null set in $(\mathbb{R}^n, \mathfrak{M}_L^n, \mu_L^n)$, then $T(E)$ is a null set in $(\mathbb{R}^n, \mathfrak{M}_L^n, \mu_L^n)$.

Proposition 26.2. *Let $E \in \mathfrak{M}_L$ and let T be a mapping of E into $\mathbb{R}^n$ satisfying the condition that for every $x \in E$*

$$\text{(1)} \qquad \limsup_{y\to x,\, y\in E} \frac{|T(y) - T(x)|}{|y - x|} < \infty.$$

(a) *If E is a null set in $(\mathbb{R}^n, \mathfrak{M}_L^n, \mu_L^n)$, then so is $T(E)$.*
(b) *If T is continuous on E, then $T(E) \in \mathfrak{M}_L^n$.*

Proof. 1. Suppose E is a null set in $(\mathbb{R}^n, \mathfrak{M}_L^n, \mu_L^n)$. If T satisfies condition (1) at every $x \in E$, then for every $x \in E$ there exist $m, k \in \mathbb{N}$ depending on x such that $|T(y) - T(x)| \le k|y - x|$ for every $y \in B\left(x, \frac{1}{m}\right) \cap E$. For every $m, k \in \mathbb{N}$, consider subsets of E defined by

$$\text{(2)} \qquad E_{m,k} = \left\{x \in E : |T(y) - T(x)| \le k|y - x| \text{ for } y \in B\left(x, \tfrac{1}{m}\right) \cap E\right\}.$$

Since $E_{m,k}$ is a subset of the null set E in the complete measure space $(\mathbb{R}^n, \mathfrak{M}_L^n, \mu_L^n)$, it is a null set in $(\mathbb{R}^n, \mathfrak{M}_L^n, \mu_L^n)$. Now $E = \bigcup_{k\in\mathbb{N}} \bigcup_{m\in\mathbb{N}} E_{m,k}$ and

$$\text{(3)} \qquad T(E) = \bigcup_{k\in\mathbb{N}} \bigcup_{m\in\mathbb{N}} T(E_{m,k}).$$

Let us show that $T(E_{m,k})$ is a null set in $(\mathbb{R}^n, \mathfrak{M}_L^n, \mu_L^n)$ for every $m, k \in \mathbb{N}$. Now since $E_{m,k}$ is a null set in $(\mathbb{R}^n, \mathfrak{M}_L^n, \mu_L^n)$, by Lemma 26.1 for an arbitrary $\varepsilon > 0$ there exists a countable collection of open balls $\{B(x_j, r_j) : j \in \mathbb{N}\}$ with $x_j \in E_{m,k}$ and $r_j \in (0, \frac{1}{m})$ for $j \in \mathbb{N}$ such that $E_{m,k} \subset \bigcup_{j\in\mathbb{N}} B(x_j, r_j)$ and $\sum_{j\in\mathbb{N}} \mu_L^n(B(x_j, r_j)) < \varepsilon$. Thus we have

$$E_{m,k} = \bigcup_{j\in\mathbb{N}} \big(E_{m,k} \cap B(x_j, r_j)\big).$$

Now if $x \in E_{m,k} \cap B(x_j, r_j)$ then $|x - x_j| < r_j$. Since $x_j \in E_{m,k}$, (2) implies that $|T(x) - T(x_j)| \le k|x - x_j| < kr_j$. Thus $T(E_{m,k} \cap B(x_j, r_j)) \subset B(T(x_j), kr_j)$. Therefore

$$T(E_{m,k}) = \bigcup_{j\in\mathbb{N}} T\big(E_{m,k} \cap B(x_j, r_j)\big) \subset \bigcup_{j\in\mathbb{N}} B\big(T(x_j), kr_j\big).$$

By the monotonicity and countable subadditivity of the Lebesgue outer measure $(\mu_L^n)^*$ and by the translation invariance of μ_L^n, we have

$$\begin{aligned}(\mu_L^n)^*(T(E_{m,k})) &\le (\mu_L^n)^*\Big(\bigcup_{j\in\mathbb{N}} (T(x_j), kr_j)\Big) \le \sum_{j\in\mathbb{N}} \mu_L^n(B(T(x_j), kr_j)) \\ &= \sum_{j\in\mathbb{N}} \mu_L^n(B(x_j, kr_j)) = k^n \sum_{j\in\mathbb{N}} \mu_L^n(B(x_j, r_j)) < k^n \varepsilon,\end{aligned}$$

where the second equality is by Proposition 24.34. By the arbitrariness of $\varepsilon > 0$, we have $(\mu_L^n)^*(T(E_{m,k})) = 0$. Thus $T(E_{m,k}) \in \mathfrak{M}_L^n$ and $\mu_L^n(T(E_{m,k})) = 0$. Then by (3), $T(E) \in \mathfrak{M}_L^n$ and $\mu_L^n(T(E)) = 0$.

2. Suppose $E \in \mathfrak{M}_L^n$ and T is continuous on E. Now $E = F \cup N$ where F is an F_σ-set and N is a null set in $(\mathbb{R}^n, \mathfrak{M}_L^n, \mu_L^n)$ by Proposition 24.13. By Proposition 24.16, $F = \bigcup_{j\in\mathbb{N}} K_j$ where K_j is a compact set for $j \in \mathbb{N}$. The continuity of T implies that $T(K_j)$ is a compact set and thus $T(K_j) \in \mathfrak{B}_{\mathbb{R}^n}$. Then $T(F) = \bigcup_{j\in\mathbb{N}} T(K_j) \in \mathfrak{B}_{\mathbb{R}^n}$. On the other hand since N is a null set in $(\mathbb{R}^n, \mathfrak{M}_L^n, \mu_L^n)$, so is $T(N)$ by (a). This shows $T(E) = T(F) \cup T(N) \in \mathfrak{M}_L^n$. ∎

Proposition 26.3. *Let Ω be an open set in $\mathbb{R}^n$ and let T be a mapping of Ω of class $C^1(\Omega)$ into $\mathbb{R}^n$. Then we have:*
(a) *For every null set E in $(\mathbb{R}^n, \mathfrak{M}_L^n, \mu_L^n)$, $T(E \cap \Omega)$ is a null set in $(\mathbb{R}^n, \mathfrak{M}_L^n, \mu_L^n)$.*
(b) *For every $E \in \mathfrak{M}_L^n$, we have $T(E \cap \Omega) \in \mathfrak{M}_L^n$.*

Proof. Let us show that T satisfies condition (1) of Proposition 26.2. Now if $E \in \mathfrak{M}_L^n$, then $E \cap \Omega \in \mathfrak{M}_L^n$. Since T is differentiable at every $x \in E \cap \Omega$, if we let

$$R(x; y - x) = T(y) - T(x) - dT(x; y - x),$$

then

$$\lim_{y \to x} \frac{R(x; y - x)}{|y - x|} = 0.$$

Since

$$\frac{|T(y)-T(x)|}{|y-x|} \le \frac{|R(x;y-x)|}{|y-x|} + \left|dT\left(x;\frac{y-x}{|y-x|}\right)\right|,$$

we have

$$\limsup_{y\to x}\frac{|T(y)-T(x)|}{|y-x|} \le \limsup_{y\to x}\left|dT\left(x;\frac{y-x}{|y-x|}\right)\right|.$$

This implies that for the compact set $\overline{B}(0,1)$ in $\mathbb{R}^n$, its image $dT\big(x;\overline{B}(0,1)\big)$ is a compact set in $\mathbb{R}^n$ and therefore $dT\big(x;\overline{B}(0,1)\big) \subset B(0,C)$ for some constant $C>0$ depending on x. This implies that $\left|dT\big(x;\frac{y-x}{|y-x|}\big)\right| \le C$ for every $y \in E\cap\Omega$, $y \ne x$. Thus we have $\limsup_{y\to x}\left(\frac{|T(y)-T(x)|}{|y-x|}\right) \le C$. This shows that T satisfies condition (1) of Proposition 26.2. Thus by (a) of Proposition 26.2, $T(E\cap\Omega)$ is a null set in $\big(\mathbb{R}^n, \mathfrak{M}^n_L, \mu^n_L\big)$. The continuity of T on $E\cap\Omega$ implies $T(E\cap\Omega)\in\mathfrak{M}^n_L$ by (b) of Proposition 26.2. ∎

According to Proposition 26.3, if T is a mapping of class $C^1(\Omega)$ of an open set $\Omega\subset\mathbb{R}^n$ into $\mathbb{R}^n$, then $T(E\cap\Omega)\in\mathfrak{M}^n_L$ for every $E\in\mathfrak{M}^n_L$. Let us define a set function ν on $\mathfrak{M}^n_L$ by setting $\nu(E) = \mathfrak{M}^n_L\big(T(E\cap\Omega)\big)$ for $E\in\mathfrak{M}^n_L$. In Theorem 26.5 below, we show that the symmetric derivative $\mathcal{D}_{\mathcal{S}}\nu$ of ν with respect to μ^n_L exists and moreover $\mathcal{D}_{\mathcal{S}}\nu(x) = |\det J_T(x)|$ for every $x\in\Omega$. In Proposition 26.5 we show that if T is a one-to-one mapping then ν is a measure on $\mathfrak{M}^n_L$ and moreover $\nu \ll \mu^n_L$ on $\big(\mathbb{R}^n,\mathfrak{M}^n_L\big)$, and if ν is finite on the compact sets in $\mathbb{R}^n$ then $\frac{d\nu}{d\mu^n_L} = \mathcal{D}_{\mathcal{S}}\nu$ a.e. on $\big(\mathbb{R}^n, \mathfrak{B}_{\mathbb{R}^n}, \mu^n_L\big)$.

Lemma 26.4. *Let O be an open set in $\mathbb{R}^n$ containing $0\in\mathbb{R}^n$. Let T be a mapping of O into $\mathbb{R}^n$. Suppose T is of class $C^1(O)$ and $\det J_T(x)\ne 0$ for every $x\in O$. Let $r\in(0,\infty)$ be so small that $\overline{B}(0,r)\subset O$. If there exists a positive constant $\varepsilon\in\big(0,\frac{1}{2}\big)$ such that for every $x\in\overline{B}(0,r)$, we have*

(1) $$|T(x)-x| \le \varepsilon r,$$

then

(2) $$T\big(B(0,r)\big) \supset B\big(0,(1-2\varepsilon)r\big).$$

Proof. To prove (2), we show that for every $q\in B\big(0,(1-2\varepsilon)r\big)$, there exists some $p\in B(0,r)$ such that $T(p)=q$.

Let us define a real-valued continuous function f on O by setting $f(x)=|T(x)-q|$ for $x\in O$. The continuity of f and the fact that $\overline{B}(0,r)$ is a compact set imply that f has a minimum on $\overline{B}(0,r)$. Let $m=\min\big\{f(x): x\in\overline{B}(0,r)\big\}$. Note that by (1), we have $f(q)=|T(q)-q|\le\varepsilon r$. This implies that $m\le\varepsilon r$. Let us show that the minimum m cannot be attained by f at any point in $S(0,r)=\{x\in\mathbb{R}^n: |x|=r\}$. Indeed if $x\in S(0,r)$, then

$$f(x)=|T(x)-x+x-q| \ge |x-q|-|T(x)-x| \ge |x|-|q|-\varepsilon r.$$

But $|q| < (1-2\varepsilon)r$ and $|x| = r$ so that $f(x) > r - (1-2\varepsilon)r - \varepsilon r = \varepsilon r \geq m$. Thus the minimum m is not attained at any $x \in S(0,r)$. Therefore the minimum m is attained at some point $p \in B(0,r)$. Let us show that $f(p) = q$. Let $T = (g_1, \dots, g_n)$ and $q = (a_1, \dots, a_n)$. Then

$$f(x)^2 = \{g_1(x) - a_1\}^2 + \cdots + \{g_n(x) - a_n\}^2.$$

Since the function f^2 attains a minimum m^2 at $p \in O$, the partial derivatives of f^2 which exist and are continuous on O must vanish at p. Thus we have

$$\begin{cases} \{g_1(p) - a_1\}\frac{\partial g_1}{\partial x_1}(p) + \cdots + \{g_n(p) - a_n\}\frac{\partial g_n}{\partial x_1}(p) = 0, \\ \vdots \\ \{g_1(p) - a_1\}\frac{\partial g_1}{\partial x_n}(p) + \cdots + \{g_n(p) - a_n\}\frac{\partial g_n}{\partial x_n}(p) = 0. \end{cases}$$

The coefficient matrix M of this system of linear equations is the transpose of $J_T(p)$. Thus $\det M = \det J_T(p) \neq 0$. This implies that the solution of the system is unique. Since $(g_1(p) - a_1, \dots, g_n(p) - a_n)$ and $(0, \dots, 0)$ are solutions of the system, we have $(g_1(p) - a_1, \dots, g_n(p) - a_n) = (0, \dots, 0)$ by the uniqueness. Thus we have the equality $(g_1(p), \dots, g_n(p)) = (a_1, \dots, a_n)$, that is, $T(p) = q$. ∎

Theorem 26.5. *Let T be a mapping of class $C^1(\Omega)$ of an open set Ω in $\mathbb{R}^n$ into $\mathbb{R}^n$. Define a set function ν on $\mathfrak{M}_L^n$ by setting*

$$\nu(E) = (\mu_L^n \circ T)(E \cap \Omega) \quad \text{for } E \in \mathfrak{M}_L^n.$$

Then $\mathfrak{D}_S \nu$, the symmetric derivative of ν with respect to μ_L^n, exists on Ω and furthermore for every $x \in \Omega$ and $r \in (0, \infty)$ so small that $B(x,r) \subset \Omega$, we have

$$(\mathfrak{D}_S \nu)(x) = \lim_{r \to 0} \frac{\nu(B(x,r))}{\mu_L^n(B(x,r))} = |\det J_T(x)|,$$

that is,

$$\text{(1)} \qquad \lim_{r \to 0} \frac{\mu_L^n(T(B(x,r)))}{\mu_L^n(B(x,r))} = |\det J_T(x)|.$$

Proof. 1. By (b) of Proposition 26.3, $T(E \cap \Omega) \in \mathfrak{M}_L^n$ so that $\nu(E) = (\mu_L^n \circ T)(E \cap \Omega)$ is defined. Consider first the case $0 \in \Omega$, $T(0) = 0$ and $\det J_T(0) \neq 0$. Let us prove

$$\text{(2)} \qquad \lim_{r \to 0} \frac{\mu_L^n(T(B(0,r)))}{\mu_L^n(B(0,r))} = |\det J_T(0)|.$$

The fact that T is of class $C^1(\Omega)$ implies that $\det J_T(x)$ is a continuous function of $x \in \mathbb{R}^n$. Thus the condition $\det J_T(0) \neq 0$ implies that there exists an open set O in $\mathbb{R}^n$ such that $0 \in O \subset \Omega$ and $\det J_T(x) \neq 0$ for every $x \in O$. Since the Jacobian matrix $J_T(0)$ is the matrix of the differential $dT(0;\cdot)$, the condition $\det J_T(0) \neq 0$ implies that $dT(0;\cdot)$ is a non-singular linear transformation of $\mathbb{R}^n$ into $\mathbb{R}^n$. For brevity let us write L for $dT(0;\cdot)$. Consider the inverse transformation L^{-1} of the non-singular linear transformation L. Let us define a mapping A of O into $\mathbb{R}^n$ by

$$A = L^{-1} \circ T.$$

Let us show that A satisfies the conditions of Lemma 26.4 on O. The fact that A is of class $C^1(O)$ follows from the fact that T is of class $C^1(O)$ and L^{-1} is a linear transformation. By the Chain Rule for the differentials we have

$$dA(x;\cdot) = L^{-1} \circ dT(x;\cdot) \quad \text{for } x \in O,$$

and $J_A(x) = M_{L^{-1}} J_T(x)$ where $M_{L^{-1}}$ is the matrix of the linear transformation L^{-1}. Thus $\det J_A(x) = \det M_{L^{-1}} \det J_T(x) \neq 0$ for $x \in O$. Writing ι for the identity mapping, we have $dA(0;\cdot) = L^{-1} \circ dT(0;\cdot) = L^{-1} \circ L = \iota$. By the definition of the differential $dA(0;\cdot)$, if we let

$$R(0;x) = A(x) - A(0) - dA(0;x) \quad \text{for } x \in O,$$

then

$$\lim_{x \to 0} \frac{R(0;x)}{|x|} = 0.$$

Since $A(0) = (L^{-1} \circ T)(0) = L^{-1}(0) = 0$ and since $dA(0;x) = \iota(x) = x$, we have $R(0;x) = A(x) - x$. Thus

$$\lim_{x \to 0} \frac{A(x) - x}{|x|} = 0.$$

This implies that for an arbitrary $\varepsilon \in \left(0, \frac{1}{2}\right)$, there exists $\delta \in (0, \infty)$ such that $\overline{B}(0,\delta) \subset O$ and $|A(x) - x| < \varepsilon |x|$ for $x \in \overline{B}(0,\delta)$. Then for $r \in (0,\delta)$ we have

(3) $$|A(x) - x| < \varepsilon |x| \leq \varepsilon r \quad \text{for } x \in \overline{B}(0,r).$$

Thus by Lemma 26.4, we have $A\big(B(0,r)\big) \supset B\big(0,(1-2\varepsilon)r\big)$. From (3), we have also $|A(x)| \leq |x| + \varepsilon r < (1+2\varepsilon)r$ for $x \in B(0,r)$ so that $A\big(B(0,r)\big) \subset B\big(0,(1+2\varepsilon)r\big)$. Therefore

$$\mu_L^n\Big(B\big(0,(1-2\varepsilon)r\big)\Big) \leq \mu_L^n\Big(A\big(B(0,r)\big)\Big) \leq \mu_L^n\Big(B\big(0,(1+2\varepsilon)r\big)\Big).$$

Now by Proposition 24.34, we have $\mu_L^n\Big(B\big(0,(1-2\varepsilon)r\big)\Big) = (1-2\varepsilon)^n \mu_L^n\big(B(0,r)\big)$ and similarly $\mu_L^n\Big(B\big(0,(1+2\varepsilon)r\big)\Big) = (1+2\varepsilon)^n \mu_L^n\big(B(0,r)\big)$. Thus for every $r \in (0,\delta)$, we have

$$(1-2\varepsilon)^n \leq \frac{\mu_L^n\Big(A\big(B(0,r)\big)\Big)}{\mu_L^n\big(B(0,r)\big)} \leq (1+2\varepsilon)^n.$$

Then the limit inferior as well as the limit superior of the quotient as $r \to 0$ are bounded between $(1-2\varepsilon)^n$ and $(1+2\varepsilon)^n$. Since this holds for every $\varepsilon \in \left(0, \frac{1}{2}\right)$, the limit inferior and the limit superior are equal to 1. Thus we have

(4) $$\lim_{r \to 0} \frac{\mu_L^n\Big(A\big(B(0,r)\big)\Big)}{\mu_L^n\big(B(0,r)\big)} = 1.$$

Now $A = L^{-1} \circ T$ and $T = L \circ A$. Thus by Theorem 24.32 (Linear Transformation of the Lebesgue Integral), writing M_L for the matrix of the linear transformation L, we have

$$(5) \qquad \frac{\mu_L^n\big(T(B(0,r))\big)}{\mu_L^n\big(B(0,r)\big)} = \frac{\mu_L^n\big(L \circ A(B(0,r))\big)}{\mu_L^n\big(B(0,r)\big)} = |\det M_L| \frac{\mu_L^n\big(A(B(0,r))\big)}{\mu_L^n\big(B(0,r)\big)}.$$

Letting $r \to 0$ in (5), using (4), and recalling that $L = dT(0;\cdot)$ so that $M_L = J_T(0)$, we have (2).

2. Suppose $0 \in \Omega$, $T(0) = 0$ and $\det J_T(0) = 0$. Let us show that (2) holds for this case also by showing that the limit on the left side of (2) is equal to 0. Since $J_T(0)$ is the matrix of the linear transformation $dT(0;\cdot)$, $\det J_T(0) = 0$ implies that $dT(0;\cdot)$ is a singular linear transformation of $\mathbb{R}^n$ into $\mathbb{R}^n$. For brevity let us write L for $dT(0;\cdot)$. The singularity of L implies $\mu_L^n\big(L(\mathbb{R}^n)\big) = 0$ by (b) of Corollary 24.36 and then $\mu_L^n\big(L(\overline{B}(0,1))\big) = 0$. Since $\overline{B}(0,1)$ is a compact set in $\mathbb{R}^n$, the continuity of L implies that $L\big(\overline{B}(0,1)\big)$ is a compact set. For brevity let us write K for $L\big(\overline{B}(0,1)\big)$. For each $j \in \mathbb{N}$, let E_j be the set consisting of every point in $\mathbb{R}^n$ whose distance from K is less than $\frac{1}{j}$, that is,

$$E_j = \big\{x \in \mathbb{R}^n : \inf_{y \in K} |x - y| < \tfrac{1}{j}\big\}.$$

Clearly $K \subset E_j$ and $(E_j : j \in \mathbb{N})$ is a decreasing sequence so that $K \subset \bigcap_{j \in \mathbb{N}} E_j$. Now if $x \in \bigcap_{j \in \mathbb{N}} E_j$, then for every $j \in \mathbb{N}$ we have $x \in E_j$ so that there exists $y_j \in K$ such that $|x - y_j| \le \frac{2}{j}$. Thus x is a limit point of the compact set K and therefore $x \in K$. This shows that $K = \bigcap_{j \in \mathbb{N}} E_j$. Thus $\lim_{j \to \infty} \mu_L^n(E_j) = \mu_L^n(K) = 0$. Thus for an arbitrary $\varepsilon > 0$, there exists $j_0 \in \mathbb{N}$ such that $\mu_L^n(E_{j_0}) < \varepsilon$. Now since $L = dT(0;\cdot)$ and $T(0) = 0$, there exists $\delta > 0$ such that

$$|T(x) - L(x)| < \tfrac{1}{j_0}|x| \quad \text{for } x \in B(0,\delta).$$

This implies that for any $r \in (0,\delta)$, the set $T\big(B(0,r)\big)$ is contained in a set E that consists of every point in $\mathbb{R}^n$ whose distance from the set $L\big(\overline{B}(0,r)\big)$ is less than $\frac{r}{j_0}$. Now $\frac{1}{r}E$ consists of every point in $\mathbb{R}^n$ whose distance from $\frac{1}{r}L\big(\overline{B}(0,r)\big)$ is less than $\frac{1}{j_0}$. But we have $\frac{1}{r}L\big(\overline{B}(0,r)\big) = L\big(\overline{B}(0,1)\big) = K$. Thus $\frac{1}{r}E = E_{j_0}$. Then $T\big(B(0,r)\big) \subset E = rE_{j_0}$. This implies $\mu_L^n\big(T(B(0,r))\big) \le \mu_L^n(r^n E_{j_0}) < r^n \varepsilon$ by Proposition 24.34. Therefore we have

$$\frac{\mu_L^n\big(T(B(0,r))\big)}{\mu_L^n\big(B(0,r)\big)} = \frac{1}{r^n}\frac{\mu_L^n\big(T(B(0,r))\big)}{\mu_L^n\big(B(0,1)\big)} < \frac{\varepsilon}{\mu_L^n\big(B(0,1)\big)}.$$

Since this holds for all $r \in (0,\delta)$, we have

$$\limsup_{r \to 0} \frac{\mu_L^n\big(T(B(0,r))\big)}{\mu_L^n\big(B(0,r)\big)} \le \varepsilon.$$

By the arbitrariness of $\varepsilon > 0$, the limit superior is equal to 0. Thus the limit exists and is equal to 0. Since $\det J_T(0) = 0$, (2) holds.

3. Let us keep the assumption that $0 \in \Omega$ but drop the assumption that $T(0) = 0$. Let $q = T(0)$. Consider a mapping of Ω into $\mathbb{R}^n$ defined by $S(x) = T(x) - q$. Then $S(0) = 0$. The fact that T is of class $C^1(\Omega)$ implies that S is of class $C^1(\Omega)$. Furthermore $J_S(x) = J_T(x)$. Thus applying (2) to S we have

$$\lim_{r \to 0} \frac{\mu_L^n\big(S\big(B(0,r)\big)\big)}{\mu_L^n\big(B(0,r)\big)} = |\det J_S(0)| = |\det J_T(0)|.$$

Now $S\big(B(0,r)\big) = T\big(B(0,r)\big) - q$ so that by Theorem 24.27 (Translation Invariance of the Lebesgue Measure μ_L^n), we have $\mu_L^n\big(S\big(B(0,r)\big)\big) = \mu_L^n\big(T\big(B(0,r)\big)\big)$. Therefore (2) holds for T.

4. Let us consider the general case. We drop the assumption that $0 \in \Omega$ and let p be an arbitrary point in Ω. Let S be a translation on $\mathbb{R}^n$ defined by $S(x) = x + p$ for $x \in \mathbb{R}^n$. Its inverse mapping is given by $S^{-1}(x) = x - p$ for $x \in \mathbb{R}^n$. Let $W = S^{-1}(\Omega)$. Then W is an open set in $\mathbb{R}^n$, and since $S^{-1}(p) = 0$, we have $0 \in W$. Note that $S(W) = \Omega$. Let us define a mapping of W into $\mathbb{R}^n$ by setting $Z = T \circ S$ on W. If we let $(g_1, \dots, g_n)$ be the component functions of S and let $p = (a_1, \dots, a_n)$, then $g_i(x) = x_i + a_i$ for $x \in \mathbb{R}^n$ for $i = 1, \dots, n$. Thus $\frac{\partial g_i}{\partial x_j} = 1$ when $i = j$ and $\frac{\partial g_i}{\partial x_j} = 0$ when $i \neq j$. Then since $\frac{\partial g_i}{\partial x_j}$ for $i = 1, \dots, n$ and $j = 1, \dots, n$ are all continuous on $\mathbb{R}^n$, S is of class $C^1(\mathbb{R}^n)$. Thus $dS(x; \cdot)$ exists and $J_S(x) = I$, the $n \times n$ identity matrix, for every $x \in \mathbb{R}^n$. Since $J_S(x)$ is the matrix of $dS(x; \cdot)$, $dS(x; \cdot)$ is an identity mapping of $\mathbb{R}^n$ into $\mathbb{R}^n$.

Since T is differentiable at every point in Ω and since $S(W) = \Omega$, the Chain Rule for the differential implies that $Z = T \circ S$ is differentiable at every $x \in W$ and

$$dZ(x; \cdot) = dT\big(S(x); \cdot\big) \circ dS(x; \cdot) = dT\big(S(x); \cdot\big),$$

and

$$J_Z(x) = J_T\big(S(x)\big) J_S(x) = J_T\big(S(x)\big).$$

Thus Z is a mapping of class $C^1(W)$ into $\mathbb{R}^n$. Since $0 \in W$, (2) applies to Z and we have

$$\lim_{r \to 0} \frac{\mu_L^n\big(Z\big(B(0,r)\big)\big)}{\mu_L^n\big(B(0,r)\big)} = |\det J_Z(0)|.$$

Now $\det J_Z(0) = \det J_T\big(S(0)\big) = \det J_T(p)$. By the translation invariance of μ_L^n, we have $\mu_L^n\big(B(0,r)\big) = \mu_L^n\big(B(p,r)\big)$. We also have $Z\big(B(0,r)\big) = (T \circ S)\big(B(0,r)\big)$ and $S\big(B(0,r)\big) = B(0,r) + p = B(p,r)$. Thus $Z\big(B(0,r)\big) = T\big(B(p,r)\big)$. Therefore we have

$$\lim_{r \to 0} \frac{\mu_L^n\big(\big(B(p,r)\big)\big)}{\mu_L^n\big(B(p,r)\big)} = |\det J_T(p)|.$$

This proves (1). ∎

Proposition 26.6. *Let T be a mapping of class $C^1(\Omega)$ of an open set Ω in $\mathbb{R}^n$ into $\mathbb{R}^n$. Consider the set function ν on $\mathfrak{M}_L^n$ defined by $\nu(E) = (\mu_L^n \circ T)(E \cap \Omega)$ for $E \in \mathfrak{M}_L^n$.*

(a) *If T is a one-to-one mapping, then ν is a measure on $\mathfrak{M}^n_L$ and moreover $\nu \ll \mu^n_L$ on the measurable space $(\mathbb{R}^n, \mathfrak{M}^n_L)$.*
(b) *If in addition we assume that ν is finite on the compact sets in $\mathbb{R}^n$, then $\frac{d\nu}{d\mu^n_L} = \mathcal{D}_{\mathcal{S}}\nu$ a.e. on the measure space $(\mathbb{R}^n, \mathfrak{B}_{\mathbb{R}^n}, \mu^n_L)$.*

Proof. 1. By (b) of Proposition 26.3, $T(E \cap \Omega) \in \mathfrak{M}^n_L$ so that $\nu(E) = (\mu^n_L \circ T)(E \cap \Omega)$ is defined. Assume that T is a one-to-one mapping. Let us show that ν is a measure on the σ-algebra $\mathfrak{M}^n_L$. Clearly $\nu(\emptyset) = 0$. To show the countable additivity of ν on $\mathfrak{M}^n_L$, let $(E_j : j \in \mathbb{N})$ be a disjoint sequence in $\mathfrak{M}^n_L$. Since T is a one-to-one mapping, $(T(E_j \cap \Omega) : j \in \mathbb{N})$ is a disjoint sequence in $\mathfrak{M}^n_L$. Then by the countable additivity of μ^n_L on $\mathfrak{M}^n_L$, we have

$$\nu\Big(\bigcup_{j\in\mathbb{N}} E_j\Big) = (\mu^n_L \circ T)\Big(\bigcup_{j\in\mathbb{N}} E_j \cap \Omega\Big) = \mu^n_L\Big(\bigcup_{j\in\mathbb{N}} T(E_j \cap \Omega)\Big)$$
$$= \sum_{j\in\mathbb{N}} \mu^n_L(T(E_j \cap \Omega)) = \sum_{j\in\mathbb{N}} \nu(E_j).$$

This proves the countable additivity of ν on $\mathfrak{M}^n_L$ and thus ν is a measure on $\mathfrak{M}^n_L$.

2. Let us assume that ν is finite on the compact sets in $\mathbb{R}^n$. Now if E is a null set in $(\mathbb{R}^n, \mathfrak{M}^n_L, \mu^n_L)$, then so is $E \cap \Omega$. Then by (a) of Proposition 26.3, $T(E \cap \Omega)$ is a null set in $(\mathbb{R}^n, \mathfrak{M}^n_L, \mu^n_L)$ and thus $\nu(E) = (\mu^n_L \circ T)(E \cap \Omega) = 0$. This shows that $\nu \ll \mu^n_L$ on $(\mathbb{R}^n, \mathfrak{M}^n_L)$. In particular we have $\nu \ll \mu^n_L$ on $(\mathbb{R}^n, \mathfrak{B}_{\mathbb{R}^n})$. Then for the Radon-Nikodym derivative $\frac{d\nu}{d\mu^n_L}$ of ν with respect to μ^n_L on $(\mathbb{R}^n, \mathfrak{B}_{\mathbb{R}^n})$ and the symmetric derivative $\mathcal{D}_{\mathcal{S}}\nu$ of ν with respect to μ^n_L, we have $\frac{d\nu}{d\mu^n_L} = \mathcal{D}_{\mathcal{S}}\nu$ a.e. on $(\mathbb{R}^n, \mathfrak{B}_{\mathbb{R}^n}, \mu^n_L)$ by (3) of Theorem 25.36. ■

Remark 26.7. (a) For a one-to-one mapping T of $C^1(\Omega)$ of an open set Ω in $\mathbb{R}^n$ into $\mathbb{R}^n$, the measure ν on $\mathfrak{M}^n_L$ defined by $\nu(E) = (\mu^n_L \circ T)(E \cap \Omega)$ for $E \in \mathfrak{M}^n_L$ may not be finite on the compact sets in $\mathbb{R}^n$ and Theorem 25.36 may not be applicable. See the Example below.
(b) If we assume that the mapping T is bounded on Ω, that is, $T(\Omega) \subset B(0, a)$ for some $a \in (0, \infty)$, then $\nu(\mathbb{R}^n) = (\mu^n_L \circ T)(\mathbb{R}^n \cap \Omega) = (\mu^n_L \circ T)(\Omega) \le \mu^n_L(B(0, a)) < \infty$ so that ν is a finite measure on $(\mathbb{R}^n, \mathfrak{M}^n_L)$.

Example. Let T be a mapping of an open set $\Omega = \left(-\frac{\pi}{2}, \frac{\pi}{2}\right)$ in $\mathbb{R}$ into $\mathbb{R}$ given by $T(x) = \tan x$ for $x \in \Omega$. Then T is a mapping of class $C^1(\Omega)$ mapping Ω one-to-one onto $\mathbb{R}$. For the compact set $E = [0, 2]$ in $\mathbb{R}$, we have $T(E \cap \Omega) = T\left(\left[0, \frac{\pi}{2}\right)\right) = [0, \infty)$ so that $\nu(E) = (\mu_L \circ T)(E \cap \Omega) = \mu_L([0, \infty)) = \infty$.

Lemma 26.8. *Let T be a mapping of class $C^1(\Omega)$ of an open set Ω in $\mathbb{R}^n$ into $\mathbb{R}^n$. Assume* *further that T is a one-to-one mapping. Let*

(1) $$V_k = \{x \in \Omega : |T(x)| < k\} \quad \text{for } k \in \mathbb{N}.$$

Let us define a set function ν_k on $\mathfrak{M}_L^n$ by setting

(2) $$\nu_k(E) = (\mu_L^n \circ T)(E \cap V_k) \quad \text{for } E \in \mathfrak{M}_L^n.$$

Then ν_k is a finite measure on $\mathfrak{M}_L^n$, $\nu_k \ll \mu_L^n$ on $(\mathbb{R}^n, \mathfrak{M}_L^n)$, and for an arbitrary version $\frac{d\nu_k}{d\mu_L^n}$ of the Radon-Nikodym derivative of ν_k with respect to μ_L^n on $(\mathbb{R}^n, \mathfrak{B}_{\mathbb{R}^n})$, we have

(3) $$\frac{d\nu_k}{d\mu_L^n} = |\det J_T| \quad \text{a.e. on } (V_k, \mathfrak{B}_{\mathbb{R}^n} \cap V_k, \mu_L^n).$$

Proof. By the continuity of T, V_k is an open set in $\mathbb{R}^n$. Thus $T(E \cap V_k) \in \mathfrak{M}_L^n$ for every $E \in \mathfrak{M}_L^n$ by (b) of Proposition 26.3 (with Ω replaced by the open set V_k) so that $(\mu_L^n \circ T)(E \cap V_k)$ is defined.

Since T is a one-to-one mapping, ν_k is a measure on $\mathfrak{M}_L^n$ and moreover $\nu_k \ll \mu_L^n$ on $(\mathbb{R}^n, \mathfrak{M}_L^n)$ by (a) of Proposition 26.6. By (1) we have $T(V_k) \subset B(0, k)$ so that ν_k is a finite measure on $(\mathbb{R}^n, \mathfrak{M}_L^n)$ by (b) of Remark 26.7. Then by (b) of Proposition 26.6, we have

(4) $$\frac{d\nu_k}{d\mu_L^n} = \mathcal{D}_S \quad \text{a.e. on } (\mathbb{R}^n, \mathfrak{B}_{\mathbb{R}^n}, \mu_L^n).$$

On the other hand, by Theorem 26.5 for every $x \in V_k$ we have

(5) $$(\mathcal{D}_S\nu)(x) = |\det J_T(x)|.$$

By (4) and (5), we have (3). ∎

Theorem 26.9. *Let Ω be an open set in $\mathbb{R}^n$ and let T be a one-to-one mapping of Ω of class $C^1(\Omega)$ into $\mathbb{R}^n$. Then the following hold:*
(a) *For every $E \in \mathfrak{M}_L^n$, we have $T(E \cap \Omega) \in \mathfrak{M}_L^n$ and*

(1) $$\mu_L^n\big(T(E \cap \Omega)\big) = \int_\Omega \mathbf{1}_E |\det J_T| \, d\mu_L^n.$$

(b) *For every $E \in \mathfrak{M}_L^n$, we have*

(2) $$\int_{T(\Omega)} \mathbf{1}_E \, d\mu_L^n = \int_\Omega (\mathbf{1}_E \circ T) |\det J_T| \, d\mu_L^n.$$

(c) *If f is an extended real-valued $\mathfrak{M}_L^n$-measurable function on $T(\Omega)$, then*

(3) $$\int_{T(\Omega)} f \, d\mu_L^n = \int_\Omega (f \circ T) |\det J_T| \, d\mu_L^n,$$

in the sense that the existence of one side implies that of the other and the equality of the two. In particular, (3) holds for every nonnegative extended real-valued $\mathfrak{M}_L^n$-measurable function f on $T(\Omega)$.

Proof. 1. For every $E \in \mathfrak{M}_L^n$, we have $T(E \cap \Omega) \in \mathfrak{M}_L^n$ by (b) of Proposition 26.3. To prove (1), let us consider first $E \in \mathfrak{B}_{\mathbb{R}^n}$. As in Lemma 26.8, let $(V_k : k \in \mathbb{N})$ be an increasing sequence of open subsets of Ω defined by $V_k = \{x \in \Omega : |T(x)| < k\}$ and let $(\nu_k : k \in \mathbb{N})$ be a sequence of finite measures on $\mathfrak{M}_L^n$ defined by $\nu_k(E) = (\mu_L^n \circ T)(E \cap V_k)$ for $E \in \mathfrak{M}_L^n$. By Lemma 26.8, $\nu_k \ll \mu_L^n$ on $(\mathbb{R}^n, \mathfrak{M}_L^n)$, and for an arbitrary version $\frac{d\nu_k}{d\mu_L^n}$ of the Radon-Nikodym derivative of ν_k with respect to μ_L^n on $(\mathbb{R}^n, \mathfrak{B}_{\mathbb{R}^n})$, we have $\frac{d\nu_k}{d\mu_L^n} = |\det J_T|$ a.e. on $(V_k, \mathfrak{B}_{\mathbb{R}^n} \cap V_k, \mu_L^n)$. Thus if $E \in \mathfrak{B}_{\mathbb{R}^n}$, then

$$\text{(4)} \qquad \mu_L^n\big(T(E \cap V_k)\big) = \nu_k(E) = \nu_k(E \cap V_k)$$

$$= \int_{E \cap V_k} \frac{d\nu_k}{d\mu_L^n}\, d\mu_L^n = \int_{V_k} \mathbf{1}_E |\det J_T|\, d\mu_L^n.$$

Since $V_k \uparrow \Omega$ as $k \to \infty$, we have $E \cap V_k \uparrow E \cap \Omega$ and $T(E \cap V_k) \uparrow T(E \cap \Omega)$. Letting $k \to \infty$ in (4) and applying the Monotone Convergence Theorem (Theorem 8.5), we have (1) for the case $E \in \mathfrak{B}_{\mathbb{R}^n}$. If $E \in \mathfrak{M}_L^n$ then according to Proposition 24.13, E is the disjoint union of an F_σ-set F and a null set N in $(\mathbb{R}^n, \mathfrak{M}_L^n, \mu_L^n)$. Since $F \in \mathfrak{B}_{\mathbb{R}^n}$, we have the equality (i) $\mu_L^n\big(T(F \cap \Omega)\big) = \int_\Omega \mathbf{1}_F |\det J_T|\, d\mu_L^n$ by our result above. On the other hand, since $N \cap \Omega$ is a null set in $(\mathbb{R}^n, \mathfrak{M}_L^n, \mu_L^n)$, we have $\mu_L^n\big(T(N \cap \Omega)\big) = 0$ by (a) of Proposition 26.3. The fact that N is a null set in $(\mathbb{R}^n, \mathfrak{M}_L^n, \mu_L^n)$ implies also $\int_\Omega \mathbf{1}_N |\det J_T|\, d\mu_L^n = 0$. Thus we have the equality (ii) $\mu_L^n\big(T(N \cap \Omega)\big) = \int_\Omega \mathbf{1}_N |\det J_T|\, d\mu_L^n$. Adding the two equalities (i) and (ii) side by side, we have (1) for our $E \in \mathfrak{M}_L^n$. This completes the proof of (1).

2. Let us prove (2). First let $E \in \mathfrak{B}_{\mathbb{R}^n}$. Consider $T^{-1}(E) = \{x \in \Omega : T(x) \in E\}$. Since T is a continuous mapping of Ω into $\mathbb{R}^n$, it is a $\mathfrak{B}_{\mathbb{R}^n}/\mathfrak{B}_{\mathbb{R}^n}$-measurable mapping. This implies that $T^{-1}(E) \in \mathfrak{B}_{\mathbb{R}^n} \subset \mathfrak{M}_L^n$. Now

$$\int_{T(\Omega)} \mathbf{1}_E\, d\mu_L^n = \mu_L^n\big(E \cap T(\Omega)\big) = \mu_L^n\big(T\big(T^{-1}(E) \cap \Omega\big)\big)$$

$$= \int_\Omega \mathbf{1}_{T^{-1}(E)} |\det J_T|\, d\mu_L^n = \int_\Omega (\mathbf{1}_E \circ T) |\det J_T|\, d\mu_L^n,$$

where the third equality is by (1). This proves (2) for the case $E \in \mathfrak{B}_{\mathbb{R}^n}$. Now let $E \in \mathfrak{M}_L^n$. By Proposition 24.13, E is the disjoint union of an F_σ-set F and a subset N of a G_δ-set G with $\mu_L^n(G) = 0$. Then (i) $\int_{T(\Omega)} \mathbf{1}_F\, d\mu_L^n = \int_\Omega (\mathbf{1}_F \circ T)|\det J_T|\, d\mu_L^n$ and (ii) $0 = \int_{T(\Omega)} \mathbf{1}_G\, d\mu_L^n = \int_\Omega (\mathbf{1}_G \circ T)|\det J_T|\, d\mu_L^n$ by our result above. Now (ii) implies that $(\mathbf{1}_G \circ T)|\det J_T| = 0$ a.e. on $(\mathbb{R}^n, \mathfrak{B}_{\mathbb{R}^n}, \mu_L^n)$. Since $N \subset G$, we have $0 \le \mathbf{1}_N \le \mathbf{1}_G$. Thus we have $(\mathbf{1}_N \circ T)|\det J_T| = 0$ a.e. on $(\mathbb{R}^n, \mathfrak{M}_L^n, \mu_L^n)$ and then $\int_\Omega (\mathbf{1}_N \circ T)|\det J_T|\, d\mu_L^n = 0$. Since $\int_{T(\Omega)} \mathbf{1}_N\, d\mu_L^n = 0$ also, we have (iii) $\int_{T(\Omega)} \mathbf{1}_N\, d\mu_L^n = \int_\Omega (\mathbf{1}_N \circ T)|\det J_T|\, d\mu_L^n$. Adding (i) and (iii) side by side, we have (2) for our $E \in \mathfrak{M}_L^n$. This proves (2).

3. To prove (3), let us observe that if $f = \mathbf{1}_E$ where $E \in \mathfrak{M}_L^n$, then (3) reduces to (2). It follows then by the linearity of the integrals that (3) holds when f is a simple function on $(\mathbb{R}^n, \mathfrak{M}_L^n, \mu_L^n)$. If f is a nonnegative extended real-valued $\mathfrak{M}_L^n$-measurable function on $\mathbb{R}^n$, then there exists an increasing sequence of nonnegative simple functions $(\varphi_j : j \in \mathbb{N})$ such that $\varphi_j \uparrow f$ by Lemma 8.6. Then by the Monotone Convergence Theorem (Theorem 8.5), (3) holds for our f. If f is an extended real-valued $\mathfrak{M}_L^n$-measurable function on $\mathbb{R}^n$, then we write $f = f^+ - f^-$ and apply the result to f^+ and f^-. ∎

[II] Spherical Coordinates in $\mathbb{R}^n$

Proposition 26.10. (Polar Coordinates in $\mathbb{R}^2$) Consider a mapping $T = (g_1, g_2)$ of the open set $\Omega = (0, \infty) \times (-\pi, \pi)$ in $\mathbb{R}^2$ into $\mathbb{R}^2$ defined for $(r, \vartheta) \in (0, \infty) \times (-\pi, \pi)$ by

(1)
$$\begin{cases} x_1 = g_1(r, \vartheta) = r\cos\vartheta, \\ x_2 = g_2(r, \vartheta) = r\sin\vartheta. \end{cases}$$

T maps Ω one-to-one onto $\mathbb{R}^2 \setminus \Lambda$ where $\Lambda = (-\infty, 0] \times \{0\} \subset \mathbb{R} \times \mathbb{R}$. Let us note that $\mu_L^2(\Lambda) = 0$ by Corollary 24.36. The Jacobian matrix of T is given by

$$J_T(r, \vartheta) = \begin{pmatrix} \cos\vartheta & -r\sin\vartheta \\ \sin\vartheta & r\cos\vartheta \end{pmatrix}.$$

Thus T is of class $C^1(\Omega)$ with

(2)
$$\det J_T(r, \vartheta) = r \quad \text{for } (r, \vartheta) \in (0, \infty) \times (-\pi, \pi).$$

If f is a nonnegative extended real-valued $\mathfrak{M}_L^2$-measurable function f on $\mathbb{R}^2$, then since $\mu_L^2(\Lambda) = 0$ we have

(3)
$$\begin{aligned} \int_{\mathbb{R}^2} f(x_1, x_2)\,\mu_L^2(d(x_1, x_2)) &= \int_{\mathbb{R}^2 \setminus \Lambda} f(x_1, x_2)\,\mu_L^2(d(x_1, x_2)) \\ &= \int_{(0,\infty)\times(-\pi,\pi)} f(r\cos\vartheta, r\sin\vartheta) r\,\mu_L^2(d(r, \vartheta)) \end{aligned}$$

by Theorem 26.9. This equality holds for every extended real-valued $\mathfrak{M}_L^2$-measurable function f on $\mathbb{R}^2$ in the sense that if one of the two integrals exists then so does the other and the two are equal.

Example 1. For an open ball $B(0, a)$ in $\mathbb{R}^2$, we have

$$\begin{aligned} \mu_L^2(B(0, a)) &= \int_{B(0,a)} \mu_L^2(d(x_1, x_2)) = \int_{(0,a)\times(-\pi,\pi)} r\,\mu_L^2(d(r, \vartheta)) \\ &= \int_{(-\pi,\pi)} \Big[\int_{(0,a)} r\,\mu_L(dr)\Big] \mu_L(d\vartheta) = \pi a^2 \end{aligned}$$

by (3) of Proposition 26.10 and Theorem 23.17 (Tonelli). ∎

Example 2. Let $\alpha > 0$. By (3) of Proposition 26.10 and Theorem 23.17 (Tonelli), we have

$$\begin{aligned}\int_{\mathbb{R}^2} e^{-\alpha\{x_1^2+x_2^2\}}\,\mu_L^2(d(x_1,x_2)) &= \int_{(0,\infty)\times(-\pi,\pi)} e^{-\alpha r^2} r\,\mu_L^2(d(r,\vartheta))\\ &= \Big[\int_{(-\pi,\pi)}\mu_L(d\vartheta)\Big]\Big[\int_{(0,\infty)} re^{-\alpha r^2}\,\mu_L(dr)\Big].\end{aligned}$$

The integrand in the second factor is nonnegative. Thus computing the Lebesgue integral as an improper Riemann integral, we have

$$\int_{(0,\infty)} re^{-\alpha r^2}\,\mu_L(dr) = \int_0^\infty re^{-\alpha r^2}\,dr = \frac{1}{2\alpha}\lim_{R\to\infty}\Big[e^{-\alpha r^2}\Big]_R^0 = \frac{1}{2\alpha}.$$

Substituting this in the previous equation, we have

$$(1)\qquad \int_{\mathbb{R}^2} e^{-\alpha\{x_1^2+x_2^2\}}\,\mu_L^2(d(x_1,x_2)) = \frac{\pi}{\alpha}.$$

Now by Theorem 23.17 (Tonelli), we have

$$\int_{\mathbb{R}^2} e^{-\alpha\{x_1^2+x_2^2\}}\,\mu_L^2(d(x_1,x_2)) = \Big[\int_{\mathbb{R}} e^{-\alpha x_1^2}\,\mu_L(dx_1)\Big]\Big[\int_{\mathbb{R}} e^{-\alpha x_2^2}\,\mu_L(dx_2)\Big].$$

The last two factors are identical. Taking the square root and recalling (1), we have

$$(2)\qquad \int_{\mathbb{R}} e^{-\alpha\xi^2}\,\mu_L(d\xi) = \sqrt{\frac{\pi}{\alpha}}.$$

Then

$$(3)\qquad \int_{\mathbb{R}^n} e^{-\alpha|x|^2}\,\mu_L^n(dx) = \prod_{i=1}^n\Big[\int_{\mathbb{R}} e^{-\alpha x_i^2}\,\mu_L(dx_i)\Big] = \Big(\frac{\pi}{\alpha}\Big)^{n/2}$$

by Theorem 23.17 (Tonelli) and (2). ∎

In $\mathbb{R}^2$ the orthogonal coordinates (x, y) and the polar coordinates (r, ϑ) are related by

$$\begin{cases} x = r\cos\vartheta,\\ y = r\sin\vartheta,\end{cases}$$

for $(r,\vartheta)\in(0,\infty)\times(-\pi,\pi)$ and in $\mathbb{R}^3$ the orthogonal coordinates (x, y, z) and the spherical coordinates (r,φ,ϑ) are related by

$$\begin{cases} z = r\cos\varphi,\\ x = r\sin\varphi\cos\vartheta,\\ y = r\sin\varphi\sin\vartheta,\end{cases}$$

for $(r, \varphi, \vartheta) \in (0, \infty) \times (0, \pi) \times (-\pi, \pi)$. We generalize these to $\mathbb{R}^n$ for all $n \geq 2$ as follows.

Proposition 26.11. (Spherical Coordinates in $\mathbb{R}^n$) For $n \geq 2$, let Ω_n be an open set in $\mathbb{R}^n$ defined by

$$(1) \qquad \begin{cases} \Omega_2 = (0, \infty) \times (-\pi, \pi), \\ \Omega_3 = (0, \infty) \times (0, \pi) \times (-\pi, \pi), \\ \vdots \\ \Omega_n = (0, \infty) \times (0, \pi) \times \cdots \times (0, \pi) \times (-\pi, \pi). \end{cases}$$

Note that the first factor in Ω_n is $(0, \infty)$, the last factor is $(-\pi, \pi)$, and every factor in between is $(0, \pi)$. For the coordinates of a point in Ω_n, we write

$$(r, \varphi_1, \ldots, \varphi_{n-2}, \varphi_{n-1}) \in (0, \infty) \times (0, \pi) \times \cdots \times (0, \pi) \times (-\pi, \pi) = \Omega_n.$$

For $n \geq 2$, we define a mapping T_n of Ω_n into $\mathbb{R}^n$ inductively as follows. For $n = 2$, let

$$(2) \qquad T_2(r, \varphi_1) = (r \cos \varphi_1, r \sin \varphi_1),$$

for $(r, \varphi_1) \in \Omega_2$. For $n \geq 3$, let

$$(3) \qquad T_n(r, \varphi_1, \ldots, \varphi_{n-1}) = \big(r \cos \varphi_1, T_{n-1}(r \sin \varphi_1, \varphi_2, \ldots, \varphi_{n-1})\big),$$

for $(r, \varphi_1, \ldots, \varphi_{n-1}) \in \Omega_n$. We write $(x_1, \ldots, x_n)$ for the component functions of T_n.

Thus by (3) and (2), we have

$$(4) \qquad \begin{aligned} T_3(r, \varphi_1, \varphi_2) &= \big(r \cos \varphi_1, T_2(r \sin \varphi_1, \varphi_2)\big) \\ &= (r \cos \varphi_1, r \sin \varphi_1 \cos \varphi_2, r \sin \varphi_1 \sin \varphi_2) \end{aligned}$$

so that

$$\begin{cases} x_1 = r \cos \varphi_1, \\ x_2 = r \sin \varphi_1 \cos \varphi_2, \\ x_3 = r \sin \varphi_1 \sin \varphi_2. \end{cases}$$

Similarly by (3) and (4), we have

$$\begin{aligned} T_4(r, \varphi_1, \varphi_2, \varphi_3) &= \big(r \cos \varphi_1, T_3(r \sin \varphi_1, \varphi_2, \varphi_3)\big) \\ &= (r \cos \varphi_1, r \sin \varphi_1 \cos \varphi_2, r \sin \varphi_1 \sin \varphi_2 \cos \varphi_3, r \sin \varphi_1 \sin \varphi_2 \sin \varphi_3) \end{aligned}$$

so that

$$\begin{cases} x_1 = r \cos \varphi_1, \\ x_2 = r \sin \varphi_1 \cos \varphi_2, \\ x_3 = r \sin \varphi_1 \sin \varphi_2 \cos \varphi_3, \\ x_4 = r \sin \varphi_1 \sin \varphi_2 \sin \varphi_3. \end{cases}$$

In general for T_n, we have

$$(5)\qquad \begin{cases} x_1 = r\cos\varphi_1, \\ x_2 = r\sin\varphi_1\cos\varphi_2, \\ x_3 = r\sin\varphi_1\sin\varphi_2\cos\varphi_3, \\ \vdots \\ x_{n-1} = r\sin\varphi_1\cdots\sin\varphi_{n-2}\cos\varphi_{n-1}, \\ x_n = r\sin\varphi_1\cdots\sin\varphi_{n-2}\sin\varphi_{n-1}, \end{cases}$$

that is, with the understanding $\prod_{j=1}^{0}\sin\varphi_j = 1$, we have

$$(6)\qquad x_i = r\cos\varphi_i\prod_{j=1}^{i-1}\sin\varphi_j \quad \text{for } i = 1,\dots,n-1, \quad \text{and} \quad x_n = r\prod_{j=1}^{n-1}\sin\varphi_j.$$

From (5), adding $x_n^2 + x_{n-1}^2 + \cdots + x_1^2$ we have

$$(7)\qquad \sum_{i=1}^{n} x_i^2 = r^2.$$

Let us define

$$(8)\qquad \begin{cases} \Lambda_n = \{x \in \mathbb{R}^n : x_n = 0 \text{ and } x_{n-1} \in (-\infty, 0]\}, \\ \Delta_n = \mathbb{R}^n \setminus \Lambda_n = \{x \in \mathbb{R}^n : x_n \neq 0 \text{ or } x_{n-1} \in (0,\infty)\}, \end{cases}$$

that is, Λ_n is that half of the $n-1$ dimensional hyperplane in $\mathbb{R}^n$ determined by equation $x_n = 0$ which contains the negative x_{n-1}-axis but not the positive x_{n-1}-axis.

Let us show that T_n maps Ω_n one-to-one onto Δ_n. It is easily verified from (5) that if $y = (r, \varphi_1, \dots, \varphi_{n-1}) \in \Omega_n$, then $x = T_n(y) \in \mathbb{R}^n$. To show that $x \notin \Lambda_n$, let us note first that for $\sin\varphi_j$ and $\cos\varphi_j$ for $j = 1,\dots,n-1$ in (5), we have

$$(9)\qquad \begin{cases} \sin\varphi_j \in (0,1] & \text{for } j = 1,\dots,n-2, \\ \sin\varphi_{n-1} \in [-1,1], & \\ \cos\varphi_j \in (-1,1) & \text{for } j = 1,\dots,n-2, \\ \cos\varphi_{n-1} \in (-1,1]. & \end{cases}$$

If $x \in \Lambda_n$, then $x_n = 0$ which implies $\sin\varphi_{n-1} = 0$ by (5) and (9) and then $\varphi_{n-1} = 0$ so that $\cos\varphi_{n-1} = 1$ and thus $x_{n-1} > 0$ by (5) and (9), a contradiction. Thus $x \notin \Lambda_n$. This shows that $T_n(\Omega_n) \subset \Delta_n$.

It remains to show that for every $x \in \Delta_n$, there exists a unique $y \in \Omega_n$ such that $T_n(y) = x$. Let us prove this by induction on n. Now for $n = 2$, it is obvious that for every $x \in \Delta_2$, there exists a unique $y \in \Omega_2$ such that $T_2(x) = y$. Suppose the assertion is true for $n-1$ where $n \geq 3$. Let $x = (x_1,\dots,x_n) \in \Delta_n$ be arbitrarily chosen. Let $r = |x| > 0$. Then $\frac{x_1}{r} \in (-1,1)$ since $x_n \neq 0$. Thus there exists a unique $\vartheta_1 \in (0,\pi)$ such that $r\cos\vartheta_1 = x_1$. Now since $(x_2,\dots,x_n) \in \Delta_{n-1}$, by our induction hypothesis there exists a unique point

$(\rho, \vartheta_2, \ldots, \vartheta_{n-1}) \in \Omega_{n-1}$ such that $T_{n-1}(\rho, \vartheta_2, \ldots, \vartheta_{n-1}) = (x_2, \ldots, x_n)$. Then we have $\rho = \{x_2^2 + \cdots + x_n^2\}^{1/2} = \{r^2 - x_1^2\}^{1/2} = \{r^2 - r^2 \cos^2 \vartheta_1\}^{1/2} = r \sin \vartheta_1$. This shows that T_{n-1} maps the point $(r \sin \vartheta_1, \vartheta_2, \ldots, \vartheta_{n-1})$ to the point $(x_2, \ldots, x_n)$. Thus according to the definition of T_n by (3), T_n maps the point $(r, \vartheta_1, \ldots, \vartheta_{n-1}) \in \Omega_n$ to $(x_1, \ldots, x_n) \in \Delta_n$.

From (5), it is obvious that all the first order partial derivatives of $x_1, \ldots, x_n$ with respect to the variables $r, \varphi_1, \ldots, \varphi_{n-1}$ are continuous on the open set Ω_n. Thus T_n is a class $C^1(\Omega_n)$ mapping of Ω_n into $\mathbb{R}^n$. As we showed above T_n maps Ω_n one-to-one onto $\Delta_n = \mathbb{R}^n \setminus \Lambda_n$. Note that Λ_n, being a subset of an $n-1$ dimensional subspace of $\mathbb{R}^n$, has $\mu_L^n(\Lambda_n) = 0$. Expanding the Jacobian determinant $\det J_{T_n}$ along the first row, we obtain $\det J_{T_n}(r, \varphi_1, \ldots, \varphi_{n-1}) = r \sin^{n-2} \varphi_1 J_{T_{n-1}}(r, \varphi_2, \ldots, \varphi_{n-1})$. Using this recursion formula and recalling that $\det J_{T_2}(r, \varphi_1) = r$, we have

(10) $$\det J_{T_n}(r, \varphi_1, \ldots, \varphi_{n-1}) = r^{n-1} \prod_{j=1}^{n-2} \sin^{n-j-1} \varphi_j.$$

Let us note that $\det J_{T_n} > 0$ on Ω_n.

Observation 26.12. The volume of a ball in $\mathbb{R}^n$ can be expressed in terms of the gamma function. The gamma function is a real-valued function on $(0, \infty)$ defined by

(1) $$\Gamma(u) = \int_{(0,\infty)} x^{u-1} e^{-x} \, \mu_L(dx) \quad \text{for } u \in (0, \infty).$$

It has the properties that

(2) $$\Gamma(u+1) = u\Gamma(u) \quad \text{for } u \in (0, \infty),$$

(3) $$\Gamma(1) = 1,$$

(4) $$\Gamma\left(\tfrac{1}{2}\right) = \sqrt{\pi},$$

and in particular

(5) $$\Gamma(m+1) = m! \quad \text{for } m \in \mathbb{N}.$$

Proof. 1. Let us show that $\Gamma(u) \in (0, \infty)$ for $u \in (0, \infty)$. For fixed $u \in (0, \infty)$, the integrand in (1) is a strictly positive continuous function of $x \in (0, \infty)$ so that $\Gamma(u) > 0$. To show that $\Gamma(u) < \infty$, let us note that

$$\int_{(0.1]} x^{u-1} e^{-x} \, \mu_L(dx) \leq \int_{(0.1]} x^{u-1} \, \mu_L(dx) = \left[\frac{x}{u}\right]_0^1 < \infty.$$

Since $\lim_{x \to \infty} x^{u+1} e^{-x} = 0$, we have $0 \leq x^{u+1} e^{-x} \leq B$ for $x \in [1, \infty)$ for some $B > 0$. Then

$$\int_{[1,\infty)} x^{u-1} e^{-x} \, \mu_L(dx) = \int_{[1,\infty)} x^{u+1} e^{-x} x^{-2} \, \mu_L(dx) \leq B \int_{[1,\infty)} x^{-2} \, \mu_L(dx) < \infty.$$

Therefore $\Gamma(u) < \infty$.

2. To prove (2), note that since the integrand in (1) is nonnegative, the Lebesgue integral can be evaluated as an improper Riemann integral. Then by integration by parts, we have

$$\Gamma(u+1) = \int_0^\infty x^u e^{-x}\,dx = \Big[x\big(-e^{-x}\big)\Big]_0^\infty + \int_0^\infty u x^{u-1} e^{-x}\,dx$$
$$= u\int_0^\infty x^{u-1} e^{-x}\,dx = u\Gamma(u).$$

3. $\Gamma(1) = \int_0^\infty e^{-x}\,dx = 1$.

4. With a change of variable of integration $x = y^2$ for $y \in (0, \infty)$, we have

$$\Gamma\left(\tfrac{1}{2}\right) = \int_0^\infty x^{-1/2} e^{-x}\,dx = \int_0^\infty y^{-1} e^{-y^2} 2y\,dy = 2\int_0^\infty e^{-y^2}\,dy = \sqrt{\pi}.$$

5. (5) follows by iterated application of (2). ■

For $a > 0$ let us write B_a^n for $\{x \in \mathbb{R}^n : |x| < a\}$, an open ball with center at $0 \in \mathbb{R}^n$ and radius a. Let $\overline{B_a^n}$ be the closure of B_a^n, that is, $\overline{B_a^n} = \{x \in \mathbb{R}^n : |x| \le a\}$. For the volumes of these balls, we have $\mu_L^n\big(\overline{B_a^n}\big) = \mu_L^n\big(B_a^n\big)$ by Observation 25.5 and $\mu_L^n\big(B_a^n\big) = a^n \mu_L^n\big(B_1^n\big)$ by Proposition 24.34. Thus it remains to compute the volumes of the balls $\mu_L^n\big(B_1^n\big)$ for $n \in \mathbb{N}$.

Theorem 26.13. (Volume of a Ball in $\mathbb{R}^n$) *Let $B_a^n = \{x \in \mathbb{R}^n : |x| < a\}$ for $a > 0$. Then*

(1) $$\mu_L^n(B_1^n) = \frac{\pi^{n/2}}{\Gamma\left(\frac{n}{2}+1\right)} \quad \text{for } n \in \mathbb{N},$$

where Γ is the gamma function. Thus

(2) $$\mu_L^n(B_a^n) = \frac{\pi^{n/2}}{\Gamma\left(\frac{n}{2}+1\right)} a^n \quad \text{for } n \in \mathbb{N},$$

and in particular

$$\mu_L\big(B_a^1\big) = \tfrac{2}{1}a, \quad \mu_L^3\big(B_a^3\big) = \tfrac{2^2\pi}{1\cdot 3}a^3, \quad \mu_L^5\big(B_a^5\big) = \tfrac{2^3\pi^2}{1\cdot 3\cdot 5}a^5, \quad \mu_L^7\big(B_a^7\big) = \tfrac{2^4\pi^3}{1\cdot 3\cdot 5\cdot 7}a^7, \quad \ldots,$$

$$\mu_L^2\big(B_a^2\big) = \tfrac{\pi}{1!}a^2, \quad \mu_L^4\big(B_a^4\big) = \tfrac{\pi^2}{2!}a^4, \quad \mu_L^6\big(B_a^6\big) = \tfrac{\pi^3}{3!}a^6, \quad \mu_L^8\big(B_a^8\big) = \tfrac{\pi^4}{4!}a^8, \quad \ldots.$$

Proof. For brevity, let us write $V_n = \mu_L^n(B_1^n)$ for $n \in \mathbb{N}$. First let us note that we have $V_1 = \mu_L\big(B_1^1\big) = \mu_L\big((-1,1)\big) = 2$ and $V_2 = \mu_L^2\big(B_1^2\big) = \pi$ by Example 1 after Proposition 26.10. Let us prove the recursion formula

(3) $$V_n = \frac{2\pi}{n} V_{n-2} \quad \text{for } n \ge 3.$$

For an arbitrary $n \geq 3$, let us write $\mathbb{R}^n = \mathbb{R}^2 \times \mathbb{R}^{n-2}$. For an arbitrary point $(x_1, x_2) \in \mathbb{R}^2$, consider the $\mathbb{R}^2$-section of B_1^n at (x_1, x_2), that is,

$$(4) \qquad \begin{aligned} B_1^n(x_1, x_2, \cdot) &= \Big\{(x_3, \ldots, x_n) \in \mathbb{R}^{n-2} : (x_1, \ldots, x_n) \in B_1^n\Big\} \\ &= \Big\{(x_3, \ldots, x_n) \in \mathbb{R}^{n-2} : x_1^2 + \cdots + x_n^2 < 1\Big\} \\ &= \Big\{(x_3, \ldots, x_n) \in \mathbb{R}^{n-2} : x_3^2 + \cdots + x_n^2 < 1 - x_1^2 - x_2^2\Big\} \\ &= \begin{cases} B_\rho^{n-2} \quad \text{where } \rho = \{1 - x_1^2 - x_2^2\}^{1/2} & \text{if } x_1^2 + x_2^2 < 1, \\ \emptyset & \text{if } x_1^2 + x_2^2 \geq 1. \end{cases} \end{aligned}$$

Thus by Theorem 23.17 (Tonelli), (4), and the fact that $\mu_L^k\left(B_a^k\right) = a^k \mu_L^k\left(B_1^k\right)$, we have

$$\begin{aligned} V_n &= \mu_L^n(B_1^n) = \int_{\mathbb{R}^n} \mathbf{1}_{B_1^n}(x)\, \mu_L^n(dx) \\ &= \int_{\mathbb{R}^2} \mathbf{1}_{B_1^2}(x_1, x_2) \Big[\int_{\mathbb{R}^{n-2}} \mathbf{1}_{B_1^n}(x_1, x_2, \cdot)\, \mu_L^{n-2}(d(x_3, \ldots, x_n))\Big]\, \mu_L^2(d(x_1, x_2)) \\ &= \int_{\mathbb{R}^2} \mathbf{1}_{B_1^2}(x_1, x_2) \mu_L^{n-2}(B_\rho^{n-2})\, \mu_L^2(d(x_1, x_2)) \quad \text{where } \rho = \Big\{1 - x_1^2 - x_2^2\Big\}^{1/2} \\ &= \int_{\mathbb{R}^2} \mathbf{1}_{B_1^2}(x_1, x_2) \Big\{1 - x_1^2 - x_2^2\Big\}^{(n-2)/2} V_{n-2}\, \mu_L^2(d(x_1, x_2)) \\ &= V_{n-2} \int_{(0,1)\times(-\pi,\pi)} \{1 - r^2\}^{(n-2)/2} r\, \mu_L^2(d(r, \varphi)) \\ &= 2\pi\, V_{n-2} \int_{(0,1)} \{1 - r^2\}^{(n-2)/2} r\, \mu_L(dr) \\ &= \frac{2\pi}{n} V_{n-2}. \end{aligned}$$

This proves (3). Since $V_1 = 2$ and $V_2 = \pi$, by iterated application of (3), we have

$$(5) \qquad V_{2k-1} = \frac{2(2\pi)^{k-1}}{1 \cdot 3 \cdot 5 \cdots (2k-1)} \quad \text{and} \quad V_{2k} = \frac{\pi^k}{k!} \quad \text{for } k \in \mathbb{N}.$$

If $n \in \mathbb{N}$ is an even integer, say $n = 2k$ for some $k \in \mathbb{N}$, then by the second expression in (5) and by (5) in Observation 26.12 we have

$$(6) \qquad V_n = \frac{\pi^{n/2}}{\left(\frac{n}{2}\right)!} = \frac{\pi^{n/2}}{\Gamma\left(\frac{n}{2} + 1\right)}.$$

Suppose $n \in \mathbb{N}$ is an odd integer, say $n = 2k - 1$ for some $k \in \mathbb{N}$. Now by (2) and (4) in Observation 26.12, we have

$$\begin{aligned} \Gamma\Big(\frac{2k+1}{2}\Big) &= \Gamma\Big(k + \frac{1}{2}\Big) = \Big(k - \frac{1}{2}\Big)\Big(k - \frac{3}{2}\Big) \cdots \frac{3}{2} \cdot \frac{1}{2} \Gamma\Big(\frac{1}{2}\Big) \\ &= \Big(\frac{1}{2}\Big)^k (2k-1)(2k-3) \cdots 3 \cdot 1 \sqrt{\pi} \end{aligned}$$

so that by the first expression in (5) we have

$$(7) \qquad V_n = V_{2k-1} = \frac{\pi^{(2k-1)/2}}{\Gamma\left(\frac{2k+1}{2}\right)} = \frac{\pi^{n/2}}{\Gamma\left(\frac{n}{2}+1\right)}.$$

With (6) and (7), we have (1). Then (2) follows from (1) by Proposition 24.34. ∎

[III] Integration by Image Measure on Spherical Surfaces

Let us address the question of constructing a Borel measure on a spherical hypersurface in $\mathbb{R}^n$. For $a > 0$, consider the spherical hypersurface with center 0 and radius $a > 0$ in $\mathbb{R}^n$, that is, $S_a^{n-1} = \{x \in \mathbb{R}^n : |x| = a\}$. The collection of subsets of S_a^{n-1} defined by $\mathfrak{B}_{\mathbb{R}^n} \cap S_a^{n-1} = \{E \cap S_a^{n-1} : E \in \mathfrak{B}_{\mathbb{R}^n}\}$ is a σ-algebra of subsets of S_a^{n-1}. This is an immediate consequence of the fact that $\mathfrak{B}_{\mathbb{R}^n}$ is a σ-algebra of subsets of $\mathbb{R}^n$. Now since S_a^{n-1} is a closed set in $\mathbb{R}^n$, S_a^{n-1} is a member of $\mathfrak{B}_{\mathbb{R}^n}$ and thus $\mathfrak{B}_{\mathbb{R}^n} \cap S_a^{n-1} \subset \mathfrak{B}_{\mathbb{R}^n}$. If we restrict the Lebesgue measure μ_L^n on $\mathfrak{B}_{\mathbb{R}^n}$ to $\mathfrak{B}_{\mathbb{R}^n} \cap S_a^{n-1}$, then we have a measure μ_L^n on the measurable space $\left(S_a^{n-1}, \mathfrak{B}_{\mathbb{R}^n} \cap S_a^{n-1}\right)$. This measure is not a very useful measure on $\left(S_a^{n-1}, \mathfrak{B}_{\mathbb{R}^n} \cap S_a^{n-1}\right)$ since $\mu_L^n\left(S_a^{n-1}\right) = \mu_L^n\left(\overline{B_a^n} - B_a^n\right) = \mu_L^n\left(\overline{B_a^n}\right) - \mu_L^n\left(B_a^n\right) = 0$.

For $n \geq 2$, consider the set $\mathbb{R}^n \setminus \{0\}$. Every point in $\mathbb{R}^n \setminus \{0\}$ is uniquely determined by its distance from the origin $0 \in \mathbb{R}^n$ and its radial projection on the spherical hypersurface S_1^{n-1} with unit radius. Thus $\mathbb{R}^n \setminus \{0\}$ is represented by the Cartesian product $(0, \infty) \times S_1^{n-1}$. Let Φ be the mapping of $\mathbb{R}^n \setminus \{0\}$ into $(0, \infty) \times S_1^{n-1}$ determined by this representation. Let ν be the image measure on $\left((0, \infty) \times S_1^{n-1}, \sigma\left(\mathfrak{B}_{(0,\infty)} \times \mathfrak{B}_{S_1^{n-1}}\right)\right)$ of the Lebesgue measure μ_L^n on $\left(\mathbb{R}^n, \mathfrak{M}_L^n\right)$ by the mapping Φ. Let ρ^{n-1} be a measure on $\left((0, \infty), \mathfrak{B}_{(0,\infty)}\right)$ such that $\rho^{n-1}(E) = \int_E r^{n-1}\, \mu_L(dr)$ for $E \in \mathfrak{B}_{(0,\infty)}$. In what follows we construct a measure σ_1^{n-1} on $\left(S_1^{n-1}, \mathfrak{B}_{S_1^{n-1}}\right)$ such that $\nu = \rho^{n-1} \times \sigma_1^{n-1}$.

Observation 26.14. Let A be an arbitrary subset of $\mathbb{R}^n$. Let the topology on the set A be the topology induced by that of $\mathbb{R}^n$. Thus if $\mathfrak{O}$ is the collection of all open sets in $\mathbb{R}^n$, then we define $\mathfrak{O} \cap A = \{O \cap A : O \in \mathfrak{O}\}$ as the collection of all open sets in A. Consider the Borel σ-algebra of subsets of A, that is, $\mathfrak{B}_A = \sigma(\mathfrak{O} \cap A)$.
(a) We have $\mathfrak{B}_A = \mathfrak{B}_{\mathbb{R}^n} \cap A$.
(b) If $A \in \mathfrak{B}_{\mathbb{R}^n}$, then $\mathfrak{B}_A \subset \mathfrak{B}_{\mathbb{R}^n}$.

Proof. To prove (a), note that $\mathfrak{B}_A = \sigma(\mathfrak{O} \cap A) = \sigma(\mathfrak{O}) \cap A = \mathfrak{B}_{\mathbb{R}^n} \cap A$ by Theorem 1.15. Statement (b) is obvious. ∎

Definition 26.15. *Let $n \geq 2$. For $x \in \mathbb{R}^n \setminus \{0\}$, let $x^* = |x|^{-1}x \in S_1^{n-1}$. Let Φ be a mapping of $\mathbb{R}^n \setminus \{0\}$ into the product set $(0, \infty) \times S_1^{n-1}$ defined by $\Phi(x) = (|x|, x^*) \in (0, \infty) \times S_1^{n-1}$.*

It is evident that Φ is a one-to-one mapping of $\mathbb{R}^n \setminus \{0\}$ onto $(0, \infty) \times S_1^{n-1}$. Its inverse mapping is given by $\Phi^{-1}(r, z) = rz \in \mathbb{R}^n \setminus \{0\}$ for $(r, z) \in (0, \infty) \times S_1^{n-1}$. It will be shown that Φ is a $\mathfrak{B}_{\mathbb{R}^n \setminus \{0\}}/\sigma\left(\mathfrak{B}_{(0,\infty)} \times \mathfrak{B}_{S_1^{n-1}}\right)$-measurable mapping of $\mathbb{R}^n \setminus \{0\}$ into

$(0, \infty) \times S_1^{n-1}$ where $\mathfrak{B}_{\mathbb{R}^n \setminus \{0\}}$, $\mathfrak{B}_{(0,\infty)}$, and $\mathfrak{B}_{S_1^{n-1}}$ are the Borel σ-algebras of subsets of $\mathbb{R}^n \setminus \{0\}$, $(0, \infty)$, and S_1^{n-1} respectively. The set function ν on $\sigma\big(\mathfrak{B}_{(0,\infty)} \times \mathfrak{B}_{S_1^{n-1}}\big)$ defined by $\nu = \mu_L^n \circ \Phi^{-1}$ is then a measure on $\big((0, \infty) \times S_1^{n-1}, \sigma\big(\mathfrak{B}_{(0,\infty)} \times \mathfrak{B}_{S_1^{n-1}}\big)\big)$ by Theorem 1.44 (Image Measure). We shall show that this measure ν is the product measure of a measure ρ^{n-1} on $\big((0, \infty), \mathfrak{B}_{(0,\infty)}\big)$ and a measure σ_1^{n-1} on $\big(S_1^{n-1}, \mathfrak{B}_{S_1^{n-1}}\big)$.

Lemma 26.16. *The mapping Φ of $\mathbb{R}^n \setminus \{0\}$ into $(0, \infty) \times S_1^{n-1}$ in Definition 26.15 is $\mathfrak{B}_{\mathbb{R}^n \setminus \{0\}}/\sigma\big(\mathfrak{B}_{(0,\infty)} \times \mathfrak{B}_{S_1^{n-1}}\big)$-measurable.*

Proof. For $E \times F \in \mathfrak{B}_{(0,\infty)} \times \mathfrak{B}_{S_1^{n-1}}$, we have

$$\begin{aligned}\Phi^{-1}(E \times F) &= \big\{rz \in \mathbb{R}^n \setminus \{0\} : (r, z) \in E \times F\big\} \\ &= \big\{rz \in \mathbb{R}^n \setminus \{0\} : r \in E \text{ and } z \in S_1^{n-1}\big\} \cap \big\{rz \in \mathbb{R}^n \setminus \{0\} : r \in (0, \infty) \text{ and } z \in F\big\} \\ &= \big\{x \in \mathbb{R}^n \setminus \{0\} : |x| \in E\big\} \cap \big\{x \in \mathbb{R}^n \setminus \{0\} : x^* \in F\big\}.\end{aligned}$$

Now $x \mapsto |x|$ is a continuous mapping of $\mathbb{R}^n \setminus \{0\}$ into $(0, \infty)$ and is thus $\mathfrak{B}_{\mathbb{R}^n \setminus \{0\}}/\mathfrak{B}_{(0,\infty)}$-measurable. Since $E \in \mathfrak{B}_{(0,\infty)}$, we have $\{x \in \mathbb{R}^n \setminus \{0\} : |x| \in E\} \in \mathfrak{B}_{\mathbb{R}^n \setminus \{0\}}$. Also $x \mapsto x^*$ is a continuous mapping of $\mathbb{R}^n \setminus \{0\}$ into S_1^{n-1} and is thus $\mathfrak{B}_{\mathbb{R}^n \setminus \{0\}}/\mathfrak{B}_{S_1^{n-1}}$-measurable. Then since $F \in \mathfrak{B}_{S_1^{n-1}}$, we have $\{x \in \mathbb{R}^n \setminus \{0\} : x^* \in F\} \in \mathfrak{B}_{\mathbb{R}^n \setminus \{0\}}$. Thus $\Phi^{-1}(E \times F) \in \mathfrak{B}_{\mathbb{R}^n \setminus \{0\}}$. By the arbitrariness of $E \times F \in \mathfrak{B}_{(0,\infty)} \times \mathfrak{B}_{S_1^{n-1}}$, we have $\Phi^{-1}\big(\mathfrak{B}_{(0,\infty)} \times \mathfrak{B}_{S_1^{n-1}}\big) \subset \mathfrak{B}_{\mathbb{R}^n \setminus \{0\}}$. Then by Theorem 1.14, we have

$$\Phi^{-1}\big(\sigma\big(\mathfrak{B}_{(0,\infty)} \times \mathfrak{B}_{S_1^{n-1}}\big)\big) = \sigma\big(\Phi^{-1}\big(\mathfrak{B}_{(0,\infty)} \times \mathfrak{B}_{S_1^{n-1}}\big)\big) \subset \mathfrak{B}_{\mathbb{R}^n \setminus \{0\}}.$$

This proves the $\mathfrak{B}_{\mathbb{R}^n \setminus \{0\}}/\sigma\big(\mathfrak{B}_{(0,\infty)} \times \mathfrak{B}_{S_1^{n-1}}\big)$-measurability of Φ. ∎

Definition 26.17. *For $n \geq 2$, let us define a measure ρ^{n-1} on $\big((0, \infty), \mathfrak{B}_{(0,\infty)}\big)$ and a measure σ_1^{n-1} on $\big(S_1^{n-1}, \mathfrak{B}_{S_1^{n-1}}\big)$ by setting*

$$(1) \qquad \rho^{n-1}(E) = \int_E r^{n-1}\, \mu_L(dr) \quad \text{for } E \in \mathfrak{B}_{(0,\infty)},$$

$$(2) \qquad \sigma_1^{n-1}(F) = n\big(\mu_L^n \circ \Phi^{-1}\big)\big((0, 1] \times F\big) \quad \text{for } F \in \mathfrak{B}_{S_1^{n-1}}.$$

The set function ρ^{n-1} on $\mathfrak{B}_{(0,\infty)}$ defined above is a measure by (c) of Proposition 8.11. The fact that the set function σ_1^{n-1} on $\mathfrak{B}_{S_1^{n-1}}$ defined above is a measure is verified directly. In particular the countable additivity of σ_1^{n-1} on $\mathfrak{B}_{S_1^{n-1}}$ follows from the countable additivity of μ_L^n on $\mathfrak{B}_{\mathbb{R}^n \setminus \{0\}}$.

Theorem 26.18. *For $n \geq 2$, consider the measure space $(\mathbb{R}^n \setminus \{0\}, \mathfrak{B}_{\mathbb{R}^n\setminus\{0\}}, \mu_L^n)$ and the measurable space $\big((0,\infty) \times S_1^{n-1}, \sigma(\mathfrak{B}_{(0,\infty)} \times \mathfrak{B}_{S_1^{n-1}})\big)$. Let ν be the image measure of μ_L^n on the σ-algebra $\sigma(\mathfrak{B}_{(0,\infty)} \times \mathfrak{B}_{S_1^{n-1}})$ by the mapping Φ of $\mathbb{R}^n \setminus \{0\}$ into $(0,\infty) \times S_1^{n-1}$, that is, $\nu(G) = (\mu_L^n \circ \Phi^{-1})(G)$ for $G \in \sigma(\mathfrak{B}_{(0,\infty)} \times \mathfrak{B}_{S_1^{n-1}})$. Then $\nu = \rho^{n-1} \times \sigma_1^{n-1}$ on $\sigma(\mathfrak{B}_{(0,\infty)} \times \mathfrak{B}_{S_1^{n-1}})$.*

Proof. The set function ν on $\sigma(\mathfrak{B}_{(0,\infty)} \times \mathfrak{B}_{S_1^{n-1}})$ is a measure by Theorem 1.44 (Image Measure). Let us show that ν is identical with the product measure $\rho^{n-1} \times \sigma_1^{n-1}$ on $\sigma(\mathfrak{B}_{(0,\infty)} \times \mathfrak{B}_{S_1^{n-1}})$.

Let $\mathfrak{I}_{(0,\infty)}$ be the collection of $\emptyset$ and all intervals in $(0,\infty)$ of the type $(\alpha,\beta]$. It is easily verified that $\mathfrak{I}_{(0,\infty)}$ is a semialgebra of subsets of $(0,\infty)$ and $\sigma(\mathfrak{I}_{(0,\infty)}) = \mathfrak{B}_{(0,\infty)}$. By Lemma 23.2, $\mathfrak{I}_{(0,\infty)} \times \mathfrak{B}_{S_1^{n-1}}$ is a semialgebra of subsets of $(0,\infty) \times S_1^{n-1}$. Furthermore $\sigma(\mathfrak{I}_{(0,\infty)} \times \mathfrak{B}_{S_1^{n-1}}) = \sigma(\mathfrak{B}_{(0,\infty)} \times \mathfrak{B}_{S_1^{n-1}})$ by Lemma 24.23. Therefore to show that $\nu = \rho^{n-1} \times \sigma_1^{n-1}$ on the σ-algebra $\sigma(\mathfrak{B}_{(0,\infty)} \times \mathfrak{B}_{S_1^{n-1}})$, it suffices to show that $\nu = \rho^{n-1} \times \sigma_1^{n-1}$ on the semialgebra $\mathfrak{I}_{(0,\infty)} \times \mathfrak{B}_{S_1^{n-1}}$ according to Theorem 21.11.

With $F \in \mathfrak{B}_{S_1^{n-1}}$ and $\alpha > 0$, consider the two subsets of $\mathbb{R}^n \setminus \{0\}$:

$$\begin{cases} \Phi^{-1}\big((0,1] \times F\big) = \big\{x \in \mathbb{R}^n \setminus \{0\} : |x| \in (0,1] \text{ and } |x|^{-1}x \in F\big\}, \\ \Phi^{-1}\big((0,\alpha] \times F\big) = \big\{x \in \mathbb{R}^n \setminus \{0\} : |x| \in (0,\alpha] \text{ and } |x|^{-1}x \in F\big\}. \end{cases}$$

For any $x \in \mathbb{R}^n \setminus \{0\}$ with $|x| \in (0,1]$ we have $|\alpha x| \in (0,\alpha]$ and for any $y \in \mathbb{R}^n \setminus \{0\}$ with $|y| \in (0,\alpha]$ we have $|\frac{1}{\alpha}y| \in (0,1]$. Thus $\Phi^{-1}\big((0,\alpha] \times F\big) = \alpha\Phi^{-1}\big((0,1] \times F\big)$ and then $\mu_L^n\big(\Phi^{-1}((0,\alpha] \times F)\big) = \alpha^n \mu_L^n\big(\Phi^{-1}((0,1] \times F)\big)$ by Corollary 24.33. Thus for $0 < \alpha < \beta$, we have

$$\begin{aligned} \nu\big((\alpha,\beta] \times F\big) &= \mu_L^n\big(\Phi^{-1}((\alpha,\beta] \times F)\big) \\ &= \mu_L^n\big(\Phi^{-1}((0,\beta] \times F)\big) - \mu_L^n\big(\Phi^{-1}((0,\alpha] \times F)\big) \\ &= (\beta^n - \alpha^n)\mu_L^n\big(\Phi^{-1}((0,1] \times F)\big) = \tfrac{1}{n}(\beta^n - \alpha^n)\sigma_1^{n-1}(F) \\ &= \Big[\int_{(\alpha,\beta]} r^{n-1}\,\mu_L(dr)\Big]\sigma_1^{n-1}(F) = \rho^{n-1}\big((\alpha,\beta]\big)\sigma_1^{n-1}(F) \\ &= \big(\rho^{n-1} \times \sigma_1^{n-1}\big)\big((\alpha,\beta] \times F\big). \end{aligned}$$

This shows that $\nu = \rho^{n-1} \times \sigma_1^{n-1}$ on the semialgebra $\mathfrak{I}_{(0,\infty)} \times \mathfrak{B}_{S_1^{n-1}}$. ■

Theorem 26.19. *For $n \geq 2$, let F be an extended real-valued $\sigma(\mathfrak{B}_{(0,\infty)} \times \mathfrak{B}_{S_1^{n-1}})$-measurable function on a set $D \in \sigma(\mathfrak{B}_{(0,\infty)} \times \mathfrak{B}_{S_1^{n-1}})$. Let Φ be the mapping of $\mathbb{R}^n \setminus \{0\}$ into $(0,\infty) \times S_1^{n-1}$ as in Definition 26.15. Then*

$$\int_{\Phi^{-1}(D)} (F \circ \Phi)(x)\,\mu_L^n(dx) = \int_D F(r,z)\,\big(\rho^{n-1} \times \sigma_1^{n-1}\big)\big(d(r,z)\big),$$

in the sense that the existence of one side implies that of the other and the equality of the two. In particular if F is nonnegative, then the equality holds.

Proof. The mapping Φ of $\mathbb{R}^n \setminus \{0\}$ into $(0,\infty) \times S_1^{n-1}$ is $\mathfrak{B}_{\mathbb{R}^n\setminus\{0\}}/\sigma\big(\mathfrak{B}_{(0,\infty)} \times \mathfrak{B}_{S_1^{n-1}}\big)$-measurable by Lemma 26.16. Consider the image measure ν of μ_L^n on $\sigma\big(\mathfrak{B}_{(0,\infty)} \times \mathfrak{B}_{S_1^{n-1}}\big)$ by the mapping Φ, that is, $\nu(G) = \big(\mu_L^n \circ \Phi^{-1}\big)(G)$ for $G \in \sigma\big(\mathfrak{B}_{(0,\infty)} \times \mathfrak{B}_{S_1^{n-1}}\big)$. By Theorem 9.34 (Integration by Image Measure), we have

$$(1) \qquad \int_{\Phi^{-1}(D)} (F \circ \Phi)(x)\, \mu_L^n(dx) = \int_D F\, d\nu.$$

According to Theorem 26.18, $\nu = \rho^{n-1} \times \sigma_1^{n-1}$ on $\sigma\big(\mathfrak{B}_{(0,\infty)} \times \mathfrak{B}_{S_1^{n-1}}\big)$. Thus

$$(2) \qquad \int_D F\, d\nu = \int_D F(r,z)\, \big(\rho^{n-1} \times \sigma_1^{n-1}\big)(d(r,z)).$$

With (1) and (2), we have the proof. ∎

A function of $x \in \mathbb{R}^n$ is called a radial function if it depends only on $|x|$. In other words, a function g on $\mathbb{R}^n$ is a radial function if there exists a function h on $[0,\infty)$ such that $g(x) = h(|x|)$ for $x \in \mathbb{R}^n$. For integration of radial functions, we have the following specialization of Theorem 26.19:

Theorem 26.20. *For $n \geq 2$, let f be an extended real-valued $\mathfrak{B}_{(0,\infty)}$-measurable function on $(0,\infty)$. Then*

$$\int_{\mathbb{R}^n\setminus\{0\}} f(|x|)\, \mu_L^n(dx) = \sigma_1^{n-1}\big(S_1^{n-1}\big) \int_{(0,\infty)} f(r) r^{n-1}\, \mu_L(dr),$$

in the sense that the existence of one side implies that of the other and the equality of the two.

Proof. For $D = (0,\infty) \times S_1^{n-1}$, we have $\Phi^{-1}(D) = \mathbb{R}^n \setminus \{0\}$. Given an extended real-valued $\mathfrak{B}_{(0,\infty)}$-measurable function f on $(0,\infty)$, consider an extended real-valued $\sigma\big(\mathfrak{B}_{(0,\infty)} \times \mathfrak{B}_{S_1^{n-1}}\big)$-measurable function on D defined by $F(r,z) = f(r)$ for $(r,z) \in D$. Since $\Phi(x) = (|x|, x^*) \in D$ for $x \in \mathbb{R}^n \setminus \{0\}$, we have $(F \circ \Phi)(x) = F\big(\Phi(x)\big) = F\big(|x|, x^*\big) = f\big(|x|\big)$. Thus by Theorem 26.19, we have

$$\begin{aligned}
\int_{\mathbb{R}^n\setminus\{0\}} f(|x|)\, \mu_L^n(dx) &= \int_{(0,\infty)\times S_1^{n-1}} f(r)\, \big(\rho^{n-1} \times \sigma_1^{n-1}\big)\big(d(r,z)\big) \\
&= \Big[\int_{(0,\infty)} f(r) r^{n-1}\, \mu_L(dr)\Big]\Big[\int_{S_1^{n-1}} \sigma_1^{n-1}(dz)\Big] \\
&= \sigma_1^{n-1}\big(S_1^{n-1}\big) \int_{(0,\infty)} f(r) r^{n-1}\, \mu_L(dr). \ \blacksquare
\end{aligned}$$

We apply our integration formula for radial functions to compute the volume of an open sphere with radius $a > 0$ in terms of the surface area of a unit sphere in $\mathbb{R}^n$. Since the volume of a sphere is known already, we can find the surface area of the unit sphere from this relation.

Theorem 26.21. *For $n \geq 2$, the volume $\mu_L^n(B_a^n)$ of the open ball B_a^n with center 0 and radius $a > 0$ and the surface area $\sigma_1^{n-1}(S_1^{n-1})$ of the sphere S_1^{n-1} with center 0 and radius 1 are related by*

$$\mu_L^n(B_a^n) = \frac{1}{n} a^n \sigma_1^{n-1}(S_1^{n-1}), \tag{1}$$

and thus

$$\sigma_1^{n-1}\left(S_1^{n-1}\right) = n\mu_L^n(B_1^n) = \frac{n\pi^{n/2}}{\Gamma\left(\frac{n}{2}+1\right)}. \tag{2}$$

Proof. By Theorem 26.20, we have

$$\begin{aligned}
\mu_L^n(B_a^n) &= \mu_L^n(B_a^n \setminus \{0\}) = \int_{\mathbb{R}^n\setminus\{0\}} \mathbf{1}_{B_a^n\setminus\{0\}}(x)\, \mu_L^n(dx) \\
&= \int_{\mathbb{R}^n\setminus\{0\}} \mathbf{1}_{(0,a)}(|x|)\, \mu_L^n(dx) = \sigma_1^{n-1}(S_1^{n-1}) \int_{(0,\infty)} \mathbf{1}_{(0,a)}(r) r^{n-1}\, \mu_L(dr) \\
&= \sigma_1^{n-1}(S_1^{n-1}) \int_0^a r^{n-1}\, dr = \tfrac{1}{n} a^n \sigma_1^{n-1}(S_1^{n-1}).
\end{aligned}$$

This proves (1). Then since $\mu_L^n(B_a^n) = a^n \pi^{n/2} / \Gamma\left(\frac{n}{2}+1\right)$ by Theorem 26.13, (2) follows from (1). ∎

With the measure σ_1^{n-1} on the spherical hypersurface S_1^{n-1} with center 0 and radius 1 in $\mathbb{R}^n$, we can define a measure on the spherical hypersurface S_a^{n-1} with center 0 and radius $a > 0$ in $\mathbb{R}^n$ by projection as follows.

Lemma 26.22. *For $n \geq 2$, let P be a mapping of S_1^{n-1} into S_a^{n-1} defined by $P(x) = ax$ for $x \in S_1^{n-1}$.*
(a) *P maps S_1^{n-1} one-to-one onto S_a^{n-1} with inverse mapping $P^{-1}(y) = a^{-1}y$ for $y \in S_a^{n-1}$.*
(b) *P is $\mathfrak{B}_{S_1^{n-1}}/\mathfrak{B}_{S_a^{n-1}}$-measurable.*

Proof. (a) is obvious. To prove (b), let $E \in \mathfrak{B}_{S_a^{n-1}}$. Since $\mathfrak{B}_{S_a^{n-1}} = \mathfrak{B}_{\mathbb{R}^n} \cap S_a^{n-1}$ by Observation 26.14, we have $E = F \cap S_a^{n-1}$ for some $F \in \mathfrak{B}_{\mathbb{R}^n}$. Then

$$P^{-1}(E) = a^{-1}\left(F \cap S_a^{n-1}\right) = a^{-1}F \cap a^{-1}S_a^{n-1} = a^{-1}F \cap S_1^{n-1}.$$

Since $a^{-1}F \in \mathfrak{B}_{\mathbb{R}^n}$, we have $a^{-1}F \cap S_1^{n-1} \in \mathfrak{B}_{\mathbb{R}^n} \cap S_1^{n-1} = \mathfrak{B}_{S_1^{n-1}}$. Thus we have $P^{-1}(E) \in \mathfrak{B}_{S_1^{n-1}}$. This shows the $\mathfrak{B}_{S_1^{n-1}}/\mathfrak{B}_{S_a^{n-1}}$-measurability of P. ■

Definition 26.23. *For $n \geq 2$, consider the measure space $(S_1^{n-1}, \mathfrak{B}_{S_1^{n-1}}, \sigma_1^{n-1})$ and the measurable space $(S_a^{n-1}, \mathfrak{B}_{S_a^{n-1}})$. With the mapping $P(x) = ax$ for $x \in S_1^{n-1}$, let us define a measure σ_a^{n-1} on $(S_a^{n-1}, \mathfrak{B}_{S_a^{n-1}})$ by setting $\sigma_a^{n-1} = a^{n-1}(\sigma_1^{n-1} \circ P^{-1})$, that is, σ_a^{n-1} is a^{n-1} times the image measure of σ_1^{n-1} on $(S_a^{n-1}, \mathfrak{B}_{S_a^{n-1}})$.*

Observation 26.24. Needless to say, $\sigma_1^{n-1} \circ P^{-1}$ is a measure on $(S_a^{n-1}, \mathfrak{B}_{S_a^{n-1}})$ by Theorem 1.44 (Image Measure). Then $\sigma_a^{n-1} = a^{n-1}(\sigma_1^{n-1} \circ P^{-1})$ is a measure on $(S_a^{n-1}, \mathfrak{B}_{S_a^{n-1}})$. For the σ_a^{n-1} value of S_a^{n-1}, we have

$$\sigma_a^{n-1}(S_a^{n-1}) = a^{n-1}(\sigma_1^{n-1} \circ P^{-1})(S_a^{n-1}) = a^{n-1}\sigma_1^{n-1}(S_1^{n-1})$$

and thus by (2) of Proposition 26.21 and (2) of Proposition 26.13, we have

$$\sigma_a^{n-1}(S_a^{n-1}) = \frac{n\pi^{n/2}}{\Gamma\left(\frac{n}{2}+1\right)} a^{n-1} = \frac{n}{a}\mu_L^n(B_a^n) \quad \text{for } n \in \mathbb{N}.$$

In particular we have

$$\sigma_a^1(S_a^1) = \tfrac{2\pi}{1!}a, \quad \sigma_a^3(S_a^3) = \tfrac{2\pi^2}{1!}a^3, \quad \sigma_a^5(S_a^5) = \tfrac{2\pi^3}{2!}a^5, \quad \sigma_a^7(S_a^7) = \tfrac{2\pi^4}{3!}a^7, \quad \ldots,$$

$$\sigma_a^2(S_a^2) = \tfrac{2^2\pi}{1}a^2, \quad \sigma_a^4(S_a^4) = \tfrac{2^3\pi^2}{1\cdot 3}a^4, \quad \sigma_a^6(S_a^6) = \tfrac{2^4\pi^3}{1\cdot 3\cdot 5}a^6, \quad \sigma_a^8(S_a^8) = \tfrac{2^5\pi^4}{1\cdot 3\cdot 5\cdot 7}a^8, \quad \ldots.$$

Proposition 26.25. *For $n \geq 2$, the volume $\mu_L^n(B_a^n)$ of the open ball B_a^n with radius $a > 0$ in $\mathbb{R}^n$ and the surface area $\sigma_a^{n-1}(S_a^{n-1})$ of the spherical hypersurface S_a^{n-1} with radius a in $\mathbb{R}^n$ are related by*

(1) $$\frac{d}{da}\mu_L^n(B_a^n) = \sigma_a^{n-1}(S_a^{n-1}),$$

and

(2) $$\int_0^a \sigma_r^{n-1}(S_r^{n-1})\, dr = \mu_L^n(B_a^n).$$

Proof. Since $\mu_L^n(B_a^n) = \pi^{n/2}a^n / \Gamma\left(\frac{n}{2}+1\right)$ and $\sigma_a^{n-1}(S_a^{n-1}) = n\pi^{n/2}a^{n-1}/\Gamma\left(\frac{n}{2}+1\right)$, the equalities (1) and (2) are obvious. ■

Problems

Prob. 26.1. Calculate the volume of the n-dimensional ellipsoid $\frac{x_1^2}{a_1^2} + \cdots + \frac{x_n^2}{a_n^2} \leq 1$ where $a_1, \ldots, a_n > 0$.

Prob. 26.2. Let f be a real-valued $\mathfrak{M}_L^n$-measurable function on $\mathbb{R}^n$. For $a > 0$, let $B_a^n = \{x \in \mathbb{R}^n : |x| < a\}$. Prove the following:
(a) If $|f(x)| \leq C|x|^{-\alpha}$ on B_a^n for some $C > 0$ and $\alpha < n$, then f is μ_L^n-integrable on B_a^n.
(b) If $|f(x)| \geq C|x|^{-n}$ on B_a^n for some $C > 0$, then f is not μ_L^n-integrable on B_a^n.

Prob. 26.3. Let f and B_a^n be as in Prob. 26.2. Prove the following:
(a) If $|f(x)| \leq C|x|^{-\alpha}$ on $(B_a^n)^c$ for some $C > 0$ and $\alpha > n$, then f is μ_L^n-integrable on $(B_a^n)^c$.
(b) If $|f(x)| \geq C|x|^{-n}$ on $(B_a^n)^c$ for some $C > 0$, then f is not μ_L^n-integrable on $(B_a^n)^c$.

Chapter 7

Hausdorff Measures on the Euclidean Space

§27 Hausdorff Measures

[I] Hausdorff Measures on $\mathbb{R}^n$

Definition 27.1. *Consider $\mathfrak{P}(\mathbb{R}^n)$, the collection of all subsets of $\mathbb{R}^n$. The diameter $|E|$ of $E \in \mathfrak{P}(\mathbb{R}^n)$, $E \neq \emptyset$, is defined by $|E| = \sup\{|x - y| : x, y \in E\}$. We define $|\emptyset| = 0$.*

Note that $|E| \in [0, \infty]$ for $E \in \mathfrak{P}(\mathbb{R}^n)$ and $|E| < \infty$ if and only if E is a bounded set in $\mathbb{R}^n$. If $E \neq \emptyset$, then $|E| = 0$ if and only if E is a singleton. If E consists of two points x and y only, then $|E| = |x - y| > 0$.

Observation 27.2. (a) For every $E \in \mathfrak{P}(\mathbb{R}^n)$, we have $|\overline{E}| = |E|$.
(b) For every $E \in \mathfrak{P}(\mathbb{R}^n)$ and $\eta > 0$, there exists an open set G such that $G \supset E$ and $|G| \leq |E| + \eta$.

Proof. **1.** Since $E \subset \overline{E}$, it follows from Definition 27.1 that $|E| \leq |\overline{E}|$. Let $\varepsilon > 0$. If $x, y \in \overline{E}$ then there exist $x_0, y_0 \in E$ such that $|x - x_0| < \varepsilon$ and $|y - y_0| < \varepsilon$. Then $|x - y| \leq |x - x_0| + |x_0 - y_0| + |y_0 - y| \leq |E| + 2\varepsilon$. Thus $|\overline{E}| \leq |E| + 2\varepsilon$. Since this holds for every $\varepsilon > 0$, we have $|\overline{E}| \leq |E|$. Therefore $|\overline{E}| = |E|$.

2. Let $\eta > 0$. Let $G = \bigcup_{x \in E} B\big(x, \frac{\eta}{2}\big)$. Then G is an open set and $G \supset E$. Let $x, y \in G$. Then $x \in B\big(x_0, \frac{\eta}{2}\big)$ and $y \in B\big(y_0, \frac{\eta}{2}\big)$ for some $x_0, y_0 \in E$. Now we have $|x - y| \leq |x - x_0| + |x_0 - y_0| + |y_0 - y| \leq |E| + \eta$. Then $|G| \leq |E| + \eta$. ∎

Conventions. We adopt the convention that $|E|^0 = 1$ if $E \neq \emptyset$ and $|\emptyset|^0 = 0$.
(For $\xi \in \mathbb{R}$, $\xi \neq 0$, we have $\xi^0 = \xi^1 \cdot \xi^{-1} = 1$ but 0^0 is undefined. For a singleton $\{x\} \subset \mathbb{R}^n$ we have $|\{x\}| = 0$ and adopt the convention that $|\{x\}|^0 = 1$ but for $\emptyset \subset \mathbb{R}^n$ with $|\emptyset| = 0$ we adopt the convention that $|\emptyset|^0 = 0$.)

Definition 27.3. (a) *Let $\delta \in (0, \infty]$ and $E \in \mathfrak{P}(\mathbb{R}^n)$. By a δ-cover of E we mean a*

finite or infinite sequence $(V_i : i \in \mathbb{N})$ *in* $\mathfrak{P}(\mathbb{R}^n)$ *with* $|V_i| \in [0, \delta]$ *for* $i \in \mathbb{N}$ *such that* $\bigcup_{i \in \mathbb{N}} V_i \supset E$.
(b) *For* $s \in [0, \infty)$ *and* $\delta \in (0, \infty]$, *we define a set function* $\mathcal{H}_\delta^s$ *on* $\mathfrak{P}(\mathbb{R}^n)$ *be setting for every* $E \in \mathfrak{P}(\mathbb{R}^n)$

$$\mathcal{H}_\delta^s(E) = \inf \left\{ \sum_{i \in \mathbb{N}} |V_i|^s : (V_i : i \in \mathbb{N}) \text{ is a } \delta\text{-cover of } E \right\}.$$

We call the sum $\sum_{i \in \mathbb{N}} |V_i|^s$ *an estimate of* $\mathcal{H}_\delta^s(E)$ *by the* δ*-cover* $(V_i : i \in \mathbb{N})$ *of* E.

Note in particular that a ∞-cover of E is just a sequence $(V_i : i \in \mathbb{N})$ in $\mathfrak{P}(\mathbb{R}^n)$ such that $\bigcup_{i \in \mathbb{N}} V_i \supset E$ without any restriction on the diameters.

Observation 27.4. Let $(V_i : i \in \mathbb{N})$ be a δ-cover of a set $E \in \mathfrak{P}(\mathbb{R}^n)$. Suppose we replace a member V_i in the δ-cover by two disjoint subsets V' and V'' of V_i such that $E \cap V_i = E \cap (V' \cup V'')$, that is, $V' \cup V''$ still covers that part of E which is covered by V_i. Let $(W_i : i \in \mathbb{N})$ be the resulting sequence. Then $(W_i : i \in \mathbb{N})$ is a δ-cover of E with $\bigcup_{i \in \mathbb{N}} W_i \subset \bigcup_{i \in \mathbb{N}} V_i$, but the estimate of $\mathcal{H}_\delta^s(E)$ by the δ-cover $(W_i : i \in \mathbb{N})$ of E may not be smaller than that by the δ-cover $(V_i : i \in \mathbb{N})$, and in fact we may have $\sum_{i \in \mathbb{N}} |W_i|^s > \sum_{i \in \mathbb{N}} |V_i|^s$. This is shown by the following examples.

Example 1. Let $E = \left[0, \frac{1}{3}\right] \cup \left[\frac{2}{3}, 1\right]$. Consider a 1-cover $(V_i : i \in \mathbb{N})$ of E consisting of $V_1 = [0, 1]$ only. We have $\sum_{i \in \mathbb{N}} |V_i|^s = |V_1|^s = 1^s = 1$. If we replace V_1 with two disjoint subsets $W_1 = \left[0, \frac{1}{3}\right]$ and $W_2 = \left[\frac{2}{3}, 1\right]$, then $(W_i : i \in \mathbb{N}) = (W_1, W_2)$ is a 1-cover of E with $\sum_{i \in \mathbb{N}} |W_i|^s = \left(\frac{1}{3}\right)^s + \left(\frac{1}{3}\right)^s = 2\left(\frac{1}{3}\right)^s$. Since $\lim_{s \to 0} \left(\frac{1}{3}\right)^s = 1$, we have $2\left(\frac{1}{3}\right)^s > 1$ for sufficiently small $s > 0$ and for such $s > 0$ we have $\sum_{i \in \mathbb{N}} |W_i|^s > \sum_{i \in \mathbb{N}} |V_i|^s$.

Example 2. Consider a subset of $\mathbb{R}^n$ defined by

$$E = \left\{B\left(0, \tfrac{1}{2}\right) \setminus B\left(0, \tfrac{1}{4}\right)\right\} \cup \left\{B(0, 1) \setminus B\left(0, \tfrac{3}{4}\right)\right\}.$$

For a 2-cover $(V_i : i \in \mathbb{N})$ of E consisting of one member only given by $V_1 = B(0, 1)$, we have $\sum_{i \in \mathbb{N}} |V_i|^s = |V_1|^s = 2^s$. For a 2-cover $(W_i : i \in \mathbb{N})$ of E consisting of two members W_1 and W_2 only, given by $W_1 = \left\{B\left(0, \frac{1}{2}\right) \setminus B\left(0, \frac{1}{4}\right)\right\}$ and $W_2 = \left\{B(0, 1) \setminus B\left(0, \frac{3}{4}\right)\right\}$, we have $\sum_{i \in \mathbb{N}} |W_i|^s = |W_1|^s + |W_2|^s = 1^s + 2^s$. Thus $\sum_{i \in \mathbb{N}} |W_i|^s > \sum_{i \in \mathbb{N}} |V_i|^s$.

Proposition 27.5. *For every* $s \in [0, \infty)$ *and* $\delta \in (0, \infty]$, *the set function* $\mathcal{H}_\delta^s$ *on* $\mathfrak{P}(\mathbb{R}^n)$ *is an outer measure on* $\mathbb{R}^n$, *that is,* $\mathcal{H}_\delta^s$ *satisfies the following conditions:*

(1) $\mathcal{H}_\delta^s(E) \in [0, \infty]$ *for* $E \in \mathfrak{P}(\mathbb{R}^n)$.

(2) $\mathcal{H}_\delta^s(\emptyset) = 0$.

(3) $\mathcal{H}_\delta^s(E) \leq \mathcal{H}_\delta^s(F)$ *if* $E \subset F$.

(4) $\mathcal{H}_\delta^s\left(\bigcup_{i \in \mathbb{N}} E_i\right) \leq \sum_{i \in \mathbb{N}} \mathcal{H}_\delta^s(E_i)$.

Proof. (1) is immediate from Definition 27.3. To prove (2), note that by using a δ-cover of $\emptyset$ consisting of only one member $\emptyset$, we have $\mathcal{H}_\delta^s(\emptyset) \leq |\emptyset|^s = 0$ for every $s \in [0, \infty)$

including the case $s = 0$ for which we have $|\emptyset|^0 = 0$ by our convention. Thus $\mathcal{H}_\delta^s(\emptyset) = 0$. (When $s \neq 0$, by using a δ-cover of $\emptyset$ consisting of only one member $B(x, r)$ with $r < \frac{\delta}{2}$, we have $\mathcal{H}_\delta^s(\emptyset) \leq (2r)^s$. Since $\lim_{r \to 0}(2r)^s = 0$ when $s \neq 0$, we have $\mathcal{H}_\delta^s(\emptyset) = 0$. When $s = 0$, no matter what nonempty set $V \in \mathfrak{P}(\mathbb{R}^n)$ we may use as a δ-cover of $\emptyset$, we have $|V|^0 = 1$. Without using $\emptyset$ as a δ-cover of $\emptyset$ with the convention $|\emptyset|^0 = 0$, we would have $\mathcal{H}_\delta^0(\emptyset) = 1$.)

(3) follows from the fact that if $E \subset F$, then every δ-cover of F is also a δ-cover of E, that is, the collection of all δ-covers of F is a subcollection of the collection of all δ-covers of E.

To prove (4), note that if $\sum_{i \in \mathbb{N}} \mathcal{H}_\delta^s(E_i) = \infty$, then (4) holds trivially. Let us consider the case $\sum_{i \in \mathbb{N}} \mathcal{H}_\delta^s(E_i) < \infty$. In this case we have $\mathcal{H}_\delta^s(E_i) < \infty$ for every $i \in \mathbb{N}$. Let $\varepsilon > 0$ be arbitrarily given. For each $i \in \mathbb{N}$, by Definition 27.3 there exists a δ-cover $(V_{i,j} : j \in \mathbb{N})$ of E_i such that $\sum_{j \in \mathbb{N}} |V_{i,j}|^s < \mathcal{H}_\delta^s(E_i) + \frac{\varepsilon}{2^i}$. Now $\{V_{i,j} : j \in \mathbb{N}, i \in \mathbb{N}\}$ is a δ-cover of $\bigcup_{i \in \mathbb{N}} E_i$ so that

$$\mathcal{H}_\delta^s\big(\textstyle\bigcup_{i \in \mathbb{N}} E_i\big) \leq \displaystyle\sum_{i \in \mathbb{N}} \sum_{j \in \mathbb{N}} |V_{i,j}|^s \leq \sum_{i \in \mathbb{N}} \{\mathcal{H}_\delta^s(E_i) + \tfrac{\varepsilon}{2^i}\} = \sum_{i \in \mathbb{N}} \mathcal{H}_\delta^s(E_i) + \varepsilon.$$

Since this holds for every $\varepsilon > 0$, we have (4). ■

Theorem 27.6. *For fixed $s \in [0, \infty)$ and $E \in \mathfrak{P}(\mathbb{R}^n)$, $\mathcal{H}_\delta^s(E) \uparrow$ as $\delta \downarrow 0$. Thus $\mathcal{H}^s(E) := \lim_{\delta \downarrow 0} \mathcal{H}_\delta^s(E)$ exists in $[0, \infty]$. Furthermore $\mathcal{H}^s$ is an outer measure on $\mathbb{R}^n$.*

Proof. Let $0 < \delta_1 < \delta_2 \leq \infty$. Then every δ_1-cover of E is also a δ_2-cover of E. Thus the collection of all δ_1-covers of E is contained in the collection of all δ_2-covers of E. This implies that $\mathcal{H}_{\delta_2}^s(E) \leq \mathcal{H}_{\delta_1}^s(E)$. Thus $\mathcal{H}_\delta^s(E) \uparrow$ as $\delta \downarrow 0$.

The fact that $\mathcal{H}^s$ is an outer measure on $\mathbb{R}^n$ follows from the fact that $\mathcal{H}_\delta^s$ is an outer measure on $\mathbb{R}^n$ for every $\delta \in (0, \infty)$ and $\mathcal{H}_\delta^s \uparrow \mathcal{H}^s$ as $\delta \downarrow 0$. In particular the countable subadditivity of $\mathcal{H}^s$ can be verified as follows: By (4) of Proposition 27.5 and by $\mathcal{H}_\delta^s \uparrow \mathcal{H}^s$ as $\delta \downarrow 0$, we have $\mathcal{H}_\delta^s\left(\bigcup_{i \in \mathbb{N}} E_i\right) \leq \sum_{i \in \mathbb{N}} \mathcal{H}_\delta^s(E_i) \leq \sum_{i \in \mathbb{N}} \mathcal{H}^s(E_i)$. Then letting $\delta \downarrow 0$, we have $\mathcal{H}^s\left(\bigcup_{i \in \mathbb{N}} E_i\right) \leq \sum_{i \in \mathbb{N}} \mathcal{H}^s(E_i)$. ■

Definition 27.7. *For $s \in [0, \infty)$, we call the set function $\mathcal{H}^s$ on $\mathfrak{P}(\mathbb{R}^n)$ defined by*

$$\mathcal{H}^s(E) = \lim_{\delta \downarrow 0} \mathcal{H}_\delta^s(E) = \sup_{\delta > 0} \mathcal{H}_\delta^s(E)$$

for $E \in \mathfrak{P}(\mathbb{R}^n)$ the s-dimensional Hausdorff measure on $\mathbb{R}^n$.

Let us note that the s-dimensional Hausdorff measure $\mathcal{H}^s$ defined above is only an outer measure by Theorem 27.6. Since it is an outer measure, the collection $\mathfrak{M}(\mathcal{H}^s)$ of all $\mathcal{H}^s$-measurable subsets of $\mathbb{R}^n$ is a σ-algebra of subsets of $\mathbb{R}^n$ by Theorem 2.8 and if we restrict $\mathcal{H}^s$ to this σ-algebra, then $\left(\mathbb{R}^n, \mathfrak{M}(\mathcal{H}^s), \mathcal{H}^s\right)$ is a complete measure space by Theorem 2.9.

Remark 27.8. The s-dimensional Hausdorff measure $\mathcal{H}^s(E)$ of a subset E of a Euclidean space does not depend on the Euclidean space containing E on which $\mathcal{H}^s$ is defined. Suppose we regard $E \in \mathfrak{P}(\mathbb{R}^n)$ as a subset of $\mathbb{R}^{n+k}$ for some $k \in \mathbb{N}$. If we let $\alpha = \mathcal{H}^s(E)$ defined for E as a subset of $\mathbb{R}^n$ and $\beta = \mathcal{H}^s(E)$ defined for E as a subset of $\mathbb{R}^{n+k}$, then $\alpha = \beta$.

Proof. For $\delta \in (0, \infty]$, let $\alpha_\delta = \mathcal{H}^s_\delta(E)$ defined for E as a subset of $\mathbb{R}^n$ and let $\beta_\delta = \mathcal{H}^s_\delta(E)$ defined for E as a subset of $\mathbb{R}^{n+k}$. Then $\alpha = \lim_{\delta \downarrow 0} \alpha_\delta$ and $\beta = \lim_{\delta \downarrow 0} \beta_\delta$. Now every δ-cover of E by subsets of $\mathbb{R}^n$ is also a δ-cover of E by subsets of $\mathbb{R}^{n+k}$. This implies $\beta_\delta \leq \alpha_\delta$. On the other hand, if $(V_i : i \in \mathbb{N})$ is a δ-cover of E by subsets of $\mathbb{R}^{n+k}$, then since $|V_i \cap \mathbb{R}^n| \leq |V_i| \leq \delta$ for $i \in \mathbb{N}$, $(V_i \cap \mathbb{R}^n : i \in \mathbb{N})$ is a δ-cover of E by subsets of $\mathbb{R}^n$. Now $\sum_{i \in \mathbb{N}} |V_i \cap \mathbb{R}^n|^s \leq \sum_{i \in \mathbb{N}} |V_i|^s$. Thus corresponding to every δ-cover $(V_i : i \in \mathbb{N})$ of E by subsets of $\mathbb{R}^{n+k}$, there exists a δ-cover $(W_i : i \in \mathbb{N})$ of E by subsets of $\mathbb{R}^n$ with $\sum_{i \in \mathbb{N}} |W_i|^s \leq \sum_{i \in \mathbb{N}} |V_i|^s$. This implies $\alpha_\delta \leq \beta_\delta$. Thus $\alpha_\delta = \beta_\delta$ for every $\delta \in (0, \infty)$. Letting $\delta \downarrow 0$, we have $\alpha = \beta$. ∎

$\mathcal{H}^s_\delta(E)$ is a nonnegative extended real-valued function defined for a triple of variables $(s, \delta, E) \in [0, \infty) \times (0, \infty] \times \mathfrak{P}(\mathbb{R}^n)$. Let us fix one or two of the three variables s, δ, and E and study the behavior of $\mathcal{H}^s_\delta(E)$ as a function of the remaining one or two variables. We have shown above that for fixed s and δ, $\mathcal{H}^s_\delta(E)$ is an outer measure on $\mathbb{R}^n$. We have shown also that for fixed s and E, $\mathcal{H}^s_\delta(E)$ as a function of δ increases as $\delta \downarrow 0$ and $\mathcal{H}^s(E) := \lim_{\delta \downarrow 0} \mathcal{H}^s_\delta(E)$ is an outer measure on $\mathbb{R}^n$. We show next that for fixed s and E, $\mathcal{H}^s_\delta(E)$ as a function of $\delta \in (0, \infty]$ is either identically vanishing on $(0, \infty]$ or strictly positive on $(0, \infty]$.

Observation 27.9. Let $E \in \mathfrak{P}(\mathbb{R}^n)$ and $\delta \in (0, \infty]$.
(a) If $\mathcal{H}^0_\delta(E) = 0$ for some $\delta \in (0, \infty]$, then $E = \emptyset$.
(b) If $\mathcal{H}^0(E) = 0$, then $E = \emptyset$.

Proof. 1. Suppose $\mathcal{H}^0_\delta(E) = 0$. Then for an arbitrary $\varepsilon \in (0, 1)$, there exists a δ-cover $(V_i : i \in \mathbb{N})$ of E such that $\sum_{i \in \mathbb{N}} |V_i|^0 < \varepsilon$. Now if $E \neq \emptyset$, then $V_i \neq \emptyset$ so that $|V_i|^0 = 1$ for at least one $i \in \mathbb{N}$ and consequently $\sum_{i \in \mathbb{N}} |V_i|^0 \geq 1 > \varepsilon$, a contradiction. Thus $E = \emptyset$.

2. Suppose $\mathcal{H}^0(E) = 0$. Since $\mathcal{H}^0(E) \geq \mathcal{H}^0_\delta(E)$ for any $\delta \in (0, \infty]$ by Theorem 27.6, we have $\mathcal{H}^0_\delta(E) = 0$. Then by (a), we have $E = \emptyset$. ∎

Lemma 27.10. *Let $E \in \mathfrak{P}(\mathbb{R}^n)$ and $s \in [0, \infty)$ be fixed. Consider $\mathcal{H}^s_\delta(E)$ as a function of $\delta \in (0, \infty]$.*
(a) *If $\mathcal{H}^s_{\delta_0}(E) > 0$ for some $\delta_0 \in (0, \infty]$, then $\mathcal{H}^s_\delta(E) > 0$ for every $\delta \in [\delta_0, \infty]$.*
(b) *If $\mathcal{H}^s(E) > 0$, then $\mathcal{H}^s_\delta(E) > 0$ for every $\delta \in (0, \infty]$, or equivalently, if $\mathcal{H}^s_{\delta_0}(E) = 0$ for some $\delta_0 \in (0, \infty]$, then $\mathcal{H}^s(E) = 0$.*
(c) *$\mathcal{H}^s_\delta(E)$ as a function of $\delta \in (0, \infty]$ is either identically vanishing on $(0, \infty]$ or strictly positive on $(0, \infty]$. Indeed if $\mathcal{H}^s(E) = 0$ then $\mathcal{H}^s_\delta(E)$ is identically vanishing on $(0, \infty]$ and if $\mathcal{H}^s(E) > 0$ then $\mathcal{H}^s_\delta(E)$ is strictly positive on $(0, \infty]$.*

Proof. 1. To prove (a), suppose $\mathcal{H}^s_{\delta_0}(E) > 0$ for some $\delta_0 \in (0, \infty]$. Since $\mathcal{H}^s_\delta(E) \uparrow$ as $\delta \downarrow 0$, to show that $\mathcal{H}^s_\delta(E) > 0$ for every $\delta \in [\delta_0, \infty]$ it suffices to show that $\mathcal{H}^s_\infty(E) > 0$.

To show $\mathcal{H}^s_\infty(E) > 0$, assume the contrary, that is, $\mathcal{H}^s_\infty(E) = 0$.

Consider the case $s \in (0, \infty)$. Now $\mathcal{H}^s_\infty(E) = 0$ implies that for an arbitrary $\varepsilon > 0$, there exists a ∞-cover $(V_i : i \in \mathbb{N})$ of E such that $\sum_{i\in\mathbb{N}} |V_i|^s < \varepsilon^s$. Then $|V_i|^s < \varepsilon^s$, that is, $|V_i| < \varepsilon$, for every $i \in \mathbb{N}$ so that $(V_i : i \in \mathbb{N})$ is actually an ε-cover of E. Then $\mathcal{H}^s_\varepsilon(E) \leq \sum_{i\in\mathbb{N}} |V_i|^s < \varepsilon^s$. This implies $\mathcal{H}^s(E) = \lim_{\varepsilon\downarrow 0} \mathcal{H}^s_\varepsilon(E) \leq \lim_{\varepsilon\downarrow 0} \varepsilon^s = 0$ since $s > 0$. Since $\mathcal{H}^s_{\delta_0}(E) \leq \mathcal{H}^s(E)$, we have $\mathcal{H}^s_{\delta_0}(E) = 0$, contradicting the assumption $\mathcal{H}^s_{\delta_0}(E) > 0$.

Now if $s = 0$, then $\mathcal{H}^0_\infty(E) = 0$ implies that $E = \emptyset$ by (a) of Observation 27.9. Then $\mathcal{H}^s_{\delta_0}(E) = \mathcal{H}^s_{\delta_0}(\emptyset) = 0$, a contradiction.

2. To prove (b), suppose $\mathcal{H}^s(E) > 0$. Since $\mathcal{H}^s_\delta(E) \uparrow \mathcal{H}^s(E)$ as $\delta \downarrow 0$, there exists $\delta_0 \in (0, \infty]$ such that $\mathcal{H}^s_\delta(E) > 0$ for every $\delta \in (0, \delta_0]$. By (a), $\mathcal{H}^s_{\delta_0}(E) > 0$ implies that $\mathcal{H}^s_\delta(E) > 0$ for every $\delta \in [\delta_0, \infty]$. Thus $\mathcal{H}^s_\delta(E) > 0$ for every $\delta \in (0, \infty]$.

3. To prove (c), note that since $\mathcal{H}^s_\delta(E) \leq \mathcal{H}^s(E)$ for every $\delta \in (0, \infty]$, if $\mathcal{H}^s(E) = 0$ then $\mathcal{H}^s_\delta(E) = 0$ for every $\delta \in (0, \infty]$. On the other hand, if $\mathcal{H}^s(E) > 0$, then $\mathcal{H}^s_\delta(E) > 0$ for every $\delta \in (0, \infty]$ by (b). ∎

Remark 27.11. **(a)** If E is a bounded set in $\mathbb{R}^n$ then for every $s \in [0, \infty)$ and $\delta \in (0, \infty]$ we have $\mathcal{H}^s_\delta(E) < \infty$. Indeed if E is a bounded set then E is contained in a cube $Q = \times_{i=1}^n [-r, r]$ with sufficiently large $r > 0$. Now for any $\delta \in (0, \infty)$ there exist finitely many cubes, say $C_1, \ldots, C_N$, with $|C_i| \leq \delta$ for $i = 1, \ldots, N$ such that $\bigcup_{i=1}^N C_i \supset Q$. Then $\mathcal{H}^s_\delta(E) \leq \sum_{i=1}^N |C_i|^s \leq \sum_{i=1}^N \delta^s < \infty$.
(b) For a bounded set E in $\mathbb{R}^n$, $\mathcal{H}^s(E) = \lim_{\delta\downarrow 0} \mathcal{H}^s_\delta(E)$ may not be finite. In Proposition 27.32 below, we show that $\mathcal{H}^0$ is a counting measure on the σ-algebra $\mathfrak{P}(\mathbb{R}^n)$. This implies that if E has infinitely many elements then $\mathcal{H}^0(E) = \infty$ whether E is a bounded set or not. In Theorem 27.33 we show that if $\mathcal{H}^{s^*}(E) \in (0, \infty)$ for some $s^* \in [0, \infty)$ then $\mathcal{H}^s(E) = \infty$ for all $s \in [0, s^*)$ whether E is a bounded set or not.

Theorem 27.12. *For every $s \in [0, \infty)$, the s-dimensional Hausdorff measure $\mathcal{H}^s$ is a metric outer measure on $\mathbb{R}^n$. In particular we have $\mathfrak{B}_{\mathbb{R}^n} \subset \mathfrak{M}(\mathcal{H}^s)$, that is, every Borel set in $\mathbb{R}^n$ is $\mathcal{H}^s$-measurable.*

Proof. To show that $\mathcal{H}^s$ is a metric outer measure, we show that if $E, F \in \mathfrak{P}(\mathbb{R}^n)$ and $d(E, F) > 0$ then $\mathcal{H}^s(E \cup F) = \mathcal{H}^s(E) + \mathcal{H}^s(F)$. Let $\delta \in \big(0, d(E, F)\big)$ be arbitrarily fixed. Let $(W_i : i \in \mathbb{N})$ be an arbitrary δ-cover of $E \cup F$. Now if $x \in E$ and $y \in F$, then $|x - y| \geq d(E, F) > \delta$. Then for each $i \in \mathbb{N}$, since $|W_i| \leq \delta$, W_i cannot contain both x and y and thus W_i cannot intersect both E and F. Let $\{U_i : i \in \mathbb{N}\}$ be the subcollection of $\{W_i : i \in \mathbb{N}\}$ consisting of those W_i which intersect E and let $\{V_i : i \in \mathbb{N}\}$ be the subcollection of $\{W_i : i \in \mathbb{N}\}$ consisting of those W_i which intersect F. Then $(U_i : i \in \mathbb{N})$ is a δ-cover of E and $(V_i : i \in \mathbb{N})$ is a δ-cover of F. Thus

$$\sum_{i\in\mathbb{N}} |W_i|^s \geq \sum_{i\in\mathbb{N}} |U_i|^s + \sum_{i\in\mathbb{N}} |V_i|^s \geq \mathcal{H}^s_\delta(E) + \mathcal{H}^s_\delta(F).$$

Since this holds for an arbitrary δ-cover $(W_i : i \in \mathbb{N})$ of E, we have $\mathcal{H}^s_\delta(E \cup F) \geq \mathcal{H}^s_\delta(E) + \mathcal{H}^s_\delta(F)$. On the other hand since $\mathcal{H}^s_\delta$ is an outer measure, it is subadditive and thus the reverse inequality holds. Therefore $\mathcal{H}^s_\delta(E \cup F) = \mathcal{H}^s_\delta(E) + \mathcal{H}^s_\delta(F)$. Since this equality holds for every $\delta \in \big(0, d(E, F)\big)$, letting $\delta \downarrow 0$ in the last equality we have $\mathcal{H}^s(E \cup F) = \mathcal{H}^s(E \cup F) + \mathcal{H}^s(E \cup F)$. This shows that $\mathcal{H}^s$ is a metric outer measure. Then $\mathfrak{B}_{\mathbb{R}^n} \subset \mathfrak{M}(\mathcal{H}^s)$ by Theorem 2.19. ∎

[II] Equivalent Definitions of Hausdorff Measure

[II.1] Covering by Closed Sets and by Open Sets

The s-dimensional Hausdorff measure $\mathcal{H}^s$ has equivalent definitions using δ-covers by closed sets only or using δ-covers by open sets only. We show that if we use δ-covers by closed sets only, then $\mathcal{H}^s_\delta$ remains unchanged for every $\delta \in (0, \infty]$ and thus $\mathcal{H}^s$ is unchanged; if however we use δ-covers by open sets only, then $\mathcal{H}^s_\delta$ changes but $\mathcal{H}^s$ is unchanged.

Theorem 27.13. *With $s \in [0, \infty)$ and $\delta \in (0, \infty]$, let us define set functions $\mathcal{F}^s_\delta$ and $\mathcal{G}^s_\delta$ on $\mathfrak{P}(\mathbb{R}^n)$ by*

$$(1) \quad \begin{cases} \mathcal{F}^s_\delta(E) = \inf\left\{\sum_{i\in\mathbb{N}} |F_i|^s : (F_i : i \in \mathbb{N}) \text{ is a } \delta\text{-cover of } E \text{ by closed sets}\right\} \\ \mathcal{G}^s_\delta(E) = \inf\left\{\sum_{i\in\mathbb{N}} |G_i|^s : (G_i : i \in \mathbb{N}) \text{ is a } \delta\text{-cover of } E \text{ by open sets}\right\} \end{cases}$$

for $E \in \mathfrak{P}(\mathbb{R}^n)$. Then

$$(2) \qquad \mathcal{F}^s_\delta(E) \uparrow \quad \text{and} \quad \mathcal{G}^s_\delta(E) \uparrow \quad \text{as } \delta \downarrow 0.$$

Let us define set functions $\mathcal{F}^s$ and $\mathcal{G}^s$ on $\mathfrak{P}(\mathbb{R}^n)$ by setting

$$(3) \qquad \mathcal{F}^s(E) = \lim_{\delta\downarrow 0} \mathcal{F}^s_\delta(E) \quad \text{and} \quad \mathcal{G}^s(E) = \lim_{\delta\downarrow 0} \mathcal{G}^s_\delta(E).$$

Then $\mathcal{F}^s_\delta$, $\mathcal{G}^s_\delta$, $\mathcal{F}^s$ and $\mathcal{G}^s$ are all outer measures on $\mathbb{R}^n$. Furthermore for $0 < \delta < \delta'$, we have

$$(4) \qquad \mathcal{G}^s_{\delta'}(E) \le \mathcal{F}^s_\delta(E) = \mathcal{H}^s_\delta(E) \le \mathcal{G}^s_\delta(E),$$

and

$$(5) \qquad \mathcal{G}^s(E) = \mathcal{F}^s(E) = \mathcal{H}^s(E).$$

Proof. The fact that $\mathcal{F}^s_\delta$ and $\mathcal{G}^s_\delta$ are outer measures on $\mathbb{R}^n$, $\mathcal{F}^s_\delta \uparrow$ and $\mathcal{G}^s_\delta \uparrow$ as $\delta \downarrow 0$, and $\mathcal{F}^s$ and $\mathcal{G}^s$ are outer measures on $\mathbb{R}^n$ can be proved by the same argument as in Proposition 27.5 and Theorem 27.6 for $\mathcal{H}^s_\delta$ and $\mathcal{H}^s$.

Let us prove (4). Since the collection of all δ-covers of E by open sets is a subcollection of all δ-covers of E, we have $\mathcal{H}^s_\delta(E) \le \mathcal{G}^s_\delta(E)$. This proves the second inequality in (4).

By the same reason as above, we have $\mathcal{H}^s_\delta(E) \le \mathcal{F}^s_\delta(E)$. Let us show that the reverse inequality holds in this case. Now if $\mathcal{H}^s_\delta(E) = \infty$, then $\mathcal{F}^s_\delta(E) \le \mathcal{H}^s_\delta(E)$ holds trivially.

Suppose $\mathcal{H}^s_\delta(E) < \infty$. Then for an arbitrary $\eta > 0$, there exists a δ-cover $(V_i : i \in \mathbb{N})$ of E such that $\mathcal{H}^s_\delta(E) + \eta \geq \sum_{i\in\mathbb{N}} |V_i|^s$. Now $\overline{V_i}$ is a closed set and $|\overline{V_i}| = |V_i|$ for $i \in \mathbb{N}$. Then $(\overline{V_i} : i \in \mathbb{N})$ is a δ-cover of E by closed sets and thus we have $\sum_{i\in\mathbb{N}} |\overline{V_i}|^s \geq \mathcal{F}^s_\delta(E)$. Thus $\mathcal{H}^s_\delta(E) + \eta \geq \sum_{i\in\mathbb{N}} |V_i|^s = \sum_{i\in\mathbb{N}} |\overline{V_i}|^s \geq \mathcal{F}^s_\delta(E)$. By the arbitrariness of $\eta > 0$, we have $\mathcal{H}^s_\delta(E) \geq \mathcal{F}^s_\delta(E)$. Therefore $\mathcal{F}^s_\delta(E) = \mathcal{H}^s_\delta(E)$. This proves the equality in (4).

Let us prove the first inequality in (4). If $\mathcal{F}^s_\delta(E) = \infty$, then $\mathcal{G}^s_{\delta'}(E) \leq \mathcal{F}^s_\delta(E)$ holds trivially. So let us assume $\mathcal{F}^s_\delta(E) < \infty$. Then for an arbitrary $\eta > 0$, there exists a δ-cover $(F_i : i \in \mathbb{N})$ of E by closed sets such that

(6) $$\sum_{i\in\mathbb{N}} |F_i|^s \leq \mathcal{F}^s_\delta(E) + \eta.$$

Since $0 < \delta < \delta'$, for each $i \in \mathbb{N}$ there exists $\varepsilon_i > 0$ such that

(7) $$\begin{cases} \{|F_i| + 2\varepsilon_i\}^s < |F_i|^s + \frac{\eta}{2^i} \\ \delta + 2\varepsilon_i < \delta'. \end{cases}$$

Let $G_i = \bigcup_{x\in F_i} B(x, \varepsilon_i)$. Then G_i is an open set, $G_i \supset F_i$ and we have $|G_i| \leq |F_i| + 2\varepsilon_i \leq \delta + 2\varepsilon_i < \delta'$ so that $|G_i|^s \leq \{|F_i| + 2\varepsilon_i\}^s \leq |F_i|^s + \frac{\eta}{2^i}$ by (7). Thus $(G_i : i \in \mathbb{N})$ is a δ'-cover of E by open sets and by (6) we have

$$\sum_{i\in\mathbb{N}} |G_i|^s \leq \sum_{i\in\mathbb{N}} \{|F_i|^s + \tfrac{\eta}{2^i}\} \leq \mathcal{F}^s_\delta(E) + 2\eta.$$

Therefore by the definition of $\mathcal{G}^s_{\delta'}(E)$ as infimum on the collection of all δ'-covers of E by open sets, we have $\mathcal{G}^s_{\delta'}(E) \leq \mathcal{F}^s_\delta(E) \leq 2\eta$. Then by the arbitrariness of $\eta > 0$, we have $\sum_{i\in\mathbb{N}} |G_i|^s \leq \mathcal{F}^s_\delta(E)$. This proves the first inequality in (4).

Finally letting $\delta \downarrow 0$ in (4), we have $\mathcal{G}^s_{\delta'}(E) \leq \mathcal{F}^s(E) = \mathcal{H}^s(E) \leq \mathcal{G}^s(E)$. Then letting $\delta' \downarrow 0$, we have (5). ∎

If we use δ-covers by balls only then the resulting outer measure is no longer equivalent to $\mathcal{H}^s$ but still comparable.

Proposition 27.14. *With $s \in [0, \infty)$ and $\delta \in (0, \infty]$, let set functions $\mathcal{S}^s_\delta$ and $\mathcal{B}^s_\delta$ on $\mathfrak{P}(\mathbb{R}^n)$ be defined by*

(1) $$\begin{cases} \mathcal{S}^s_\delta(E) = \inf\left\{\sum_{i\in\mathbb{N}} |S_i|^s : (S_i : i \in \mathbb{N}) \text{ is a } \delta\text{-cover of } E \text{ by closed balls}\right\} \\ \mathcal{B}^s_\delta(E) = \inf\left\{\sum_{i\in\mathbb{N}} |B_i|^s : (B_i : i \in \mathbb{N}) \text{ is a } \delta\text{-cover of } E \text{ by open balls}\right\} \end{cases}$$

for $E \in \mathfrak{P}(\mathbb{R}^n)$. Then $\mathcal{S}^s_\delta$ and $\mathcal{B}^s_\delta$ are outer measures on $\mathbb{R}^n$ and $\mathcal{S}^s_\delta(E) \uparrow$ and $\mathcal{B}^s_\delta(E) \uparrow$ as $\delta \downarrow 0$. If we define set functions $\mathcal{S}^s$ and $\mathcal{B}^s$ on $\mathfrak{P}(\mathbb{R}^n)$ by setting

(2) $$\begin{cases} \mathcal{S}^s(E) = \lim_{\delta\downarrow 0} \mathcal{S}^s_\delta(E), \\ \mathcal{B}^s(E) = \lim_{\delta\downarrow 0} \mathcal{B}^s_\delta(E), \end{cases}$$

then $\mathcal{S}^s$ and $\mathcal{B}^s$ are outer measures on $\mathbb{R}^n$. Furthermore for $\delta \in (0, \infty)$, we have

(3) $$\mathcal{H}^s_\delta(E) \le \mathcal{S}^s_\delta(E) \quad \text{and} \quad \mathcal{S}^s_{2\delta}(E) \le 2^s \mathcal{H}^s_\delta(E),$$

(4) $$\mathcal{H}^s(E) \le \mathcal{S}^s(E) \le 2^s \mathcal{H}^s(E),$$

(5) $$\mathcal{S}^s_\delta(E) \le \mathcal{B}^s_\delta(E) \quad \text{and} \quad \mathcal{B}^s_{\alpha\delta}(E) \le \alpha^s \mathcal{S}^s_\delta(E) \quad \text{for every } \alpha \in (1, \infty),$$

(6) $$\mathcal{B}^s(E) = \mathcal{S}^s(E).$$

Proof. $\mathcal{S}^s_\delta$, $\mathcal{B}^s_\delta$, $\mathcal{S}^s$, and $\mathcal{B}^s$ are outer measures on $\mathbb{R}^n$ by the same argument as in Proposition 27.5 and Theorem 27.6.

Let $\delta \in (0, \infty)$. The first inequality in (3) follows from the fact that the collection of all δ-covers of E by closed balls is a subcollection of all δ-covers of E. To prove the second inequality in (3), let $(V_i : i \in \mathbb{N})$ be an arbitrary δ-cover of E. For each $i \in \mathbb{N}$, there exists a closed ball S_i with radius equal to $|V_i|$ containing V_i. Then $(S_i : i \in \mathbb{N})$ is a 2δ-cover of E by closed balls so that we have $\mathcal{S}^s_{2\delta}(E) \le \sum_{i\in\mathbb{N}} |S_i|^s = \sum_{i\in\mathbb{N}} \left(2|V_i|\right)^s = 2^s \sum_{i\in\mathbb{N}} |V_i|^s$. Since this holds for an arbitrary δ-cover $(V_i : i \in \mathbb{N})$ of E, the second inequality in (3) follows. Then letting $\delta \downarrow 0$ in the two inequalities in (3), we have (4).

To prove (5), let us note that corresponding to each δ-cover $(B_i : i \in \mathbb{N})$ of E by open balls, there exists a δ-cover $(S_i : i \in \mathbb{N})$ of E by closed balls with $\sum_{i\in\mathbb{N}} |S_i|^s = \sum_{i\in\mathbb{N}} |B_i|^s$. Indeed if we let $S_i = \overline{B}_i$, then S_i is a closed ball containing B_i with $|S_i| = |B_i|$. For this δ-cover $(S_i : i \in \mathbb{N})$ of E by closed balls, we have $\sum_{i\in\mathbb{N}} |S_i|^s = \sum_{i\in\mathbb{N}} |B_i|^s$. This implies the first inequality in (5). To prove the second inequality in (5), let $\alpha \in (1, \infty)$ be arbitrarily fixed. Let $(S_i : i \in \mathbb{N})$ be a δ-cover of E by closed balls. For each $i \in \mathbb{N}$, let B_i be an open ball concentric with S_i and with radius equal to α times that of S_i. Then $(B_i : i \in \mathbb{N})$ is an $\alpha\delta$-cover of E by open balls so that $\mathcal{B}^s_{\alpha\delta}(E) \le \sum_{i\in\mathbb{N}} |B_i|^s = \sum_{i\in\mathbb{N}} \left(\alpha|S_i|\right)^s = \alpha^s \sum_{i\in\mathbb{N}} |S_i|^s$. Since this holds for an arbitrary δ-cover $(S_i : i \in \mathbb{N})$ of E by closed balls, the second inequality in (5) holds. Now letting $\delta \downarrow 0$ in the first inequality of (5), we have $\mathcal{S}^s(E) \le \mathcal{B}^s(E)$. Letting $\delta \downarrow 0$ in the second inequality of (5), we have $\mathcal{B}^s(E) \le \alpha^s \mathcal{S}^s(E)$. Since this holds for every $\alpha \in (1, \infty)$, we have $\mathcal{B}^s(E) \le \mathcal{S}^s(E)$. Thus (6) holds. ∎

[II.2] Covering by Convex Sets

Definition 27.15. *For $x, y \in \mathbb{R}^n$ and $\lambda \in [0, 1]$, we call the point $\lambda x + (1 - \lambda)y$ in $\mathbb{R}^n$ a convex combination of x and y. We say that a nonempty set E in $\mathbb{R}^n$ is a convex set if E contains every convex combination of any two points in E.*

Note that a singleton in $\mathbb{R}^n$ and $\mathbb{R}^n$ itself are trivial examples of convex sets in $\mathbb{R}^n$. We exclude the empty set from the definition of a convex set. The set of all convex combinations of two distinct points in $\mathbb{R}^n$ is the line segment joining the two points. Recall also that a subset A of a line L in $\mathbb{R}^n$ is an interval of the line if and only if the line segment joining any two points in A is contained in A. The following is a geometric criterion for convexity of a set.

Proposition 27.16. *A nonempty set E in $\mathbb{R}^n$ is a convex set if and only if the intersection of E and an arbitrary line L in $\mathbb{R}^n$ is $\emptyset$ or a singleton or an interval.*

Proof. 1. Let E be a convex set and L be a line in $\mathbb{R}^n$. Suppose $E \cap L$ is neither $\emptyset$ nor a singleton. Let x and y be two distinct points in $E \cap L$. Since $x, y \in E$ and E is a convex set, the line segment joining x and y is contained in E. Also since $x, y \in L$, the line segment joining x and y is contained in L. Thus the line segment is contained in $E \cap L$. This shows that $E \cap L$ is an interval of L.

2. Conversely suppose E is such that its intersection with an arbitrary line L in $\mathbb{R}^n$ is $\emptyset$ or a singleton or an interval of L. Take any two distinct points x and y in E. Let L be the line through x and y. Then $E \cap L$ is neither $\emptyset$ nor a singleton and is therefore an interval. Then every convex combination of x and y is contained in this interval and hence contained in E. This shows that E is a convex set. ■

Proposition 27.17. *If E is a convex set in $\mathbb{R}^n$, then its closure $\overline{E}$ is a convex set.*

Proof. 1. Assume that E is a convex set. To show that $\overline{E}$ is a convex set we show that for any $x, y \in \overline{E}$ and $\lambda \in [0, 1]$ we have $\lambda x + (1-\lambda)y \in \overline{E}$. Now $\overline{E} = E \cup E'$ where E' is the derived set of E, that is, the set of all limit points of E. Thus if $x \in \overline{E}$ then x is a point in E or x is a limit point of E or both. In any case there exists a sequence $(x_i : i \in \mathbb{N}) \subset E$ such that $\lim_{i\to\infty} |x_i - x| = 0$. Similarly if $y \in \overline{E}$ then there exists a sequence $(y_i : i \in \mathbb{N}) \subset E$ such that $\lim_{i\to\infty} |y_i - y| = 0$. Now we have

$$
\begin{aligned}
&\big|\{\lambda x_i + (1-\lambda)y_i\} - \{\lambda x + (1-\lambda)y\}\big| \\
=&\big|\lambda(x_i - x) + (1-\lambda)(y_i - y)\big| \\
\leq&\lambda\big|x_i - x\big| + (1-\lambda)\big|y_i - y\big|.
\end{aligned} \tag{1}
$$

Thus we have

$$
\begin{aligned}
&\lim_{i\to\infty}\big|\{\lambda x_i + (1-\lambda)y_i\} - \{\lambda x + (1-\lambda)y\}\big| \\
\leq&\lambda \lim_{i\to\infty}|x_i - x| + (1-\lambda)\lim_{i\to\infty}|y_i - y| = 0.
\end{aligned} \tag{2}
$$

Now since E is a convex set and $x_i, y_i \in E$ and $\lambda \in [0, 1]$, we have $\lambda x_i + (1-\lambda)y_i \in E$. Thus (2) shows that $\lambda x + (1-\lambda)y$ is a limit point of E. Then $\lambda x + (1-\lambda)y \in E' \subset \overline{E}$. This shows that $\overline{E}$ is a convex set. ■

Proposition 27.18. *Let E be a convex set in $\mathbb{R}^n$.*
(a) *For every $x \in \mathbb{R}^n$, the translate of E by x, $E + x$, is a convex set in $\mathbb{R}^n$.*
(b) *For every linear transformation T of $\mathbb{R}^n$ into $\mathbb{R}^m$, $T(E)$ is a convex set in $\mathbb{R}^m$.*

Proof. These two statements are immediate consequences of the definition of a convex set. To prove (b) for instance, let $x, y \in T(E)$. To show that $T(E)$ is a convex set in $\mathbb{R}^m$, we show that for every $\lambda \in [0,1]$, the point $\lambda x + (1-\lambda)y$ in $\mathbb{R}^m$ is contained in $T(E)$. Now since $x, y \in T(E)$, there exist points p and q in E such that $T(p) = x$ and $T(q) = y$. Then by the linearity of T, we have $\lambda x + (1-\lambda)y = \lambda T(p) + (1-\lambda)T(q) = T\big(\lambda p + (1-\lambda)q\big)$. Since E is a convex set, $r := \lambda x + (1-\lambda)y$ is a point in E. Then $\lambda x + (1-\lambda)y = T(r) \in T(E)$. This shows that $T(E)$ is a convex set. ∎

Example. Every open ball $B(x,r)$ in $\mathbb{R}^n$ is a convex set.

Proof. By (a) of Proposition 27.18, it suffices to show that $B(0,r)$ is a convex set. For $x, y \in B(0,r)$ and $\lambda \in [0,1]$, let $z = \lambda x + (1-\lambda)y$. Then $|z| \le \lambda|x| + (1-\lambda)|y| < \lambda r + (1-\lambda)r = r$ so that $z \in B(0,r)$. This shows that $B(0,r)$ is a convex set.

Definition 27.19. *For $x_1, \ldots, x_k \in \mathbb{R}^n$ and $\lambda_1, \ldots, \lambda_k \in [0,1]$ such that $\lambda_1 + \cdots + \lambda_k = 1$, we call the point $x := \lambda_1 x_1 + \cdots + \lambda_k x_k$ in $\mathbb{R}^n$ a convex combination of $x_1, \ldots, x_k$.*

Lemma 27.20. *If E is a convex set in $\mathbb{R}^n$, then it contains every convex combination of any finite number of its points.*

Proof. We prove the lemma by induction on the number of points in E of which a convex combination is made. To start with, every convex combination of any two points in E is contained in E by the fact that E is a convex set. Now suppose for some $k \in \mathbb{N}$, every convex combination of any k points in E is contained in E. Take arbitrary $k+1$ points $x_1, \ldots, x_{k+1}$ in E, let $\lambda_1, \ldots, \lambda_{k+1} \in \mathbb{R}$ be such that $\lambda_1 + \cdots + \lambda_{k+1} = 1$ and consider the convex combination $x = \lambda_1 x_1 + \cdots + \lambda_{k+1} x_{k+1}$. If $\lambda_{k+1} = 1$, then we have $x = x_{k+1} \in E$. If $\lambda_{k+1} < 1$, then $c := \lambda_1 + \cdots + \lambda_k = 1 - \lambda_{k+1} > 0$ so that we can write $x = c\{c^{-1}\lambda_1 x_1 + \cdots + c^{-1}\lambda_k x_k\} + \lambda_{k+1}x_{k+1}$. Since $c^{-1}\lambda_1, \ldots, c^{-1}\lambda_k \in [0,1]$ and $c^{-1}\lambda_1 + \cdots + c^{-1}\lambda_k = 1$, we have $c^{-1}\lambda_1 x_1 + \cdots + c^{-1}\lambda_k x_k \in E$ by the induction hypothesis. Then since $c, \lambda_{k+1} \in [0,1]$ and $c + \lambda_{k+1} = 1$, we have $x \in E$ by the fact that E is a convex set. Therefore every convex combination of any $k+1$ points of E is contained in E. This proves the lemma by induction. ∎

Let $\mathfrak{C}$ be an arbitrary collection of convex sets in $\mathbb{R}^n$. If $\bigcap_{C \in \mathfrak{C}} C \neq \emptyset$, then $\bigcap_{C \in \mathfrak{C}} C$ is a convex set in $\mathbb{R}^n$. (Indeed if $x, y \in \bigcap_{C \in \mathfrak{C}} C$, then $x, y \in C$ for every $C \in \mathfrak{C}$. Then since C is a convex set we have $\lambda x + (1-\lambda)y \in C$ for every $\lambda \in [0,1]$. Since this holds for every $C \in \mathfrak{C}$, we have $\lambda x + (1-\lambda)y \in \bigcap_{C \in \mathfrak{C}} C$.)

Let E be a nonempty set in $\mathbb{R}^n$. $\mathbb{R}^n$ is a convex set containing E. The intersection of all convex sets in $\mathbb{R}^n$ containing E is nonempty and is therefore a convex. It is then the smallest convex set in $\mathbb{R}^n$ containing E.

Definition 27.21. *Let E be a nonempty set in $\mathbb{R}^n$. Let us call the smallest convex set in $\mathbb{R}^n$ containing E, that is, the intersection of all convex sets in $\mathbb{R}^n$ containing E, the convex hull of E and write $h(E)$ for it.*

Proposition 27.22. *Let E be a nonempty set in $\mathbb{R}^n$. Then the convex hull of E, $h(E)$, is* *equal to the set of all convex combinations of any finite number of points in E.*

Proof. Let F be the set of all convex combinations of any finite number of points in E. Clearly $E \subset F$. Let us show that F is a convex set in $\mathbb{R}^n$. Let x and y be two arbitrary points in F. Then $x = s_1x_1 + \cdots s_px_p$ and $y = t_1y_1 + \cdots t_qy_q$ where $x_1, \ldots, x_p, y_1, \ldots, y_q \in E$, $s_1, \ldots, s_p, t_1, \ldots, t_q \in [0, 1]$, $s_1 + \cdots + s_p = 1$, $t_1 + \cdots + t_q = 1$, and $p, q \in \mathbb{N}$. For $\lambda \in [0, 1]$, we have $\lambda x + (1-\lambda)y = \lambda s_1x_1 + \cdots + \lambda s_px_p + (1-\lambda)t_1y_1 + \cdots + (1-\lambda)t_qy_q$. The coefficients of the $p+q$ points in the last sum are all in $[0, 1]$ and their sum is equal to 1. Thus this sum is a convex combination of the points $x_1, \ldots, x_p, y_1, \ldots, y_q$ in E and is therefore in F. This shows that F is a convex set.

If G be an arbitrary convex set in $\mathbb{R}^n$ containing E then it contains every convex combination of any finite number of points in E by Lemma 27.20, and thus $G \supset F$. This shows that F is the smallest convex set in $\mathbb{R}^n$ containing E, that is, $F = h(E)$. ∎

Proposition 27.23. *For every nonempty set E in $\mathbb{R}^n$, we have $|h(E)| = |E|$.*

Proof. Since $E \subset h(E)$, we have $|E| \le |h(E)|$. It remains to prove the reverse inequality. Let c be an arbitrary positive number such that $c > |E|$. Let $x_0 \in E$ be arbitrarily chosen. Then for every $x \in E$, we have $|x - x_0| \le |E| < c$ so that $x \in B(x_0, c)$. Thus $E \subset B(x_0, c)$. Since $B(x_0, c)$ is a convex set, this implies that $h(E) \subset B(x_0, c)$.

Now let $y_1, y_2 \in h(E)$ be arbitrarily chosen. For an arbitrary $x_0 \in E$, we have $y_1 \in h(E) \subset B(x_0, c)$ by our result above. This implies that $x_0 \in B(y_1, c)$. By the arbitrariness of $x_0 \in E$, we have $E \subset B(y_1, c)$. Since $B(y_1, c)$ is a convex set, we have $h(E) \subset B(y_1, c)$. Then for our $y_2 \in h(E) \subset B(y_1, c)$, we have $|y_2 - y_1| < c$. Since this holds for arbitrary $y_1, y_2 \in h(E)$, we have $|h(E)| \le c$. Then since this holds for every $c > |E|$, we have $|h(E)| \le |E|$. ∎

Theorem 27.24. *For $s \in [0, \infty)$ and $\delta \in (0, \infty]$, let us define set functions $\mathfrak{C}_\delta^s$ and $\mathfrak{C}^s$ on $\mathfrak{P}(\mathbb{R}^n)$ by setting for every $E \in \mathfrak{P}(\mathbb{R}^n)$*

(1) $$\mathfrak{C}_\delta^s(E) = \inf\left\{ \textstyle\sum_{i\in\mathbb{N}} |C_i|^s : (C_i : i \in \mathbb{N}) \text{ is a } \delta\text{-cover of } E \text{ by convex sets}\right\}$$

(2) $$\mathfrak{C}^s(E) = \lim_{\delta \downarrow 0} \mathfrak{C}_\delta^s(E).$$

Then $\mathfrak{C}_\delta^s$ and $\mathfrak{C}^s$ are outer measures on $\mathbb{R}^n$ and furthermore $\mathfrak{C}_\delta^s = \mathcal{H}_\delta^s$ and $\mathfrak{C}^s = \mathcal{H}^s$.

Proof. The fact that $\mathfrak{C}_\delta^s(E) \uparrow$ as $\delta \downarrow 0$ so that $\lim_{\delta\downarrow 0} \mathfrak{C}_\delta^s(E)$ exists and the fact that $\mathfrak{C}_\delta^s$ and $\mathfrak{C}^s$ are outer measures follow by the same arguments as in Proposition 27.5 and Theorem 27.6.

Let us prove $\mathfrak{C}_\delta^s(E) = \mathcal{H}_\delta^s(E)$. Since the collection of all δ-covers of E by convex sets is a subcollection of the collection of all δ-covers of E, we have $\mathfrak{C}_\delta^s(E) \ge \mathcal{H}_\delta^s(E)$. On the other hand, corresponding to each δ-cover $(V_i : i \in \mathbb{N})$ of E there exists a δ-cover of E by convex sets, namely $\big(h(V_i) : i \in \mathbb{N}\big)$, from the fact that $|h(V_i)| = |V_i|$ by Proposition 27.23. Then $\sum_{i\in\mathbb{N}} |h(V_i)|^s = \sum_{i\in\mathbb{N}} |V_i|^s$. Thus $\mathfrak{C}_\delta^s(E) \le \mathcal{H}_\delta^s(E)$. This proves $\mathfrak{C}_\delta^s(E) = \mathcal{H}_\delta^s(E)$. Letting $\delta \downarrow 0$, we have $\mathfrak{C}^s(E) = \mathcal{H}^s(E)$. ∎

Theorem 27.25. *For $s \in [0, \infty)$ and $\delta \in (0, \infty]$, let us define set functions $\mathcal{K}_\delta^s$ and $\mathcal{K}^s$ on $\mathfrak{P}(\mathbb{R}^n)$ by setting for every $E \in \mathfrak{P}(\mathbb{R}^n)$*

(1) $\mathcal{K}_\delta^s(E) = \inf\left\{\sum_{i\in\mathbb{N}} |K_i|^s : (K_i : i \in \mathbb{N}) \text{ is a } \delta\text{-cover of } E \text{ by closed convex sets}\right\}$

(2) $\mathcal{K}^s(E) = \lim_{\delta\downarrow 0} \mathcal{K}_\delta^s(E)$.

Then $\mathcal{K}_\delta^s$ and $\mathcal{K}^s$ are outer measures on $\mathbb{R}^n$ and furthermore $\mathcal{K}_\delta^s = \mathcal{H}_\delta^s$ and $\mathcal{K}^s = \mathcal{H}^s$.

Proof. The fact that $\mathcal{K}_\delta^s(E) \uparrow$ as $\delta \downarrow 0$ and the fact that $\mathcal{K}_\delta^s$ and $\mathcal{K}^s$ are outer measures follow by the same arguments as in Proposition 27.5 and Theorem 27.6.

Let us show that $\mathcal{K}_\delta^s(E) = \mathcal{H}_\delta^s(E)$. Since the collection of all δ-covers of E by closed convex sets is a subcollection of the collection of all δ-covers of E, we have $\mathcal{K}_\delta^s(E) \geq \mathcal{H}_\delta^s(E)$. Let us prove the reverse inequality. For an arbitrary δ-cover $(V_i : i \in \mathbb{N})$ of E, consider $\left(\overline{h(V_i)} : i \in \mathbb{N}\right)$ where $h(V_i)$ is the convex hull of V_i. By Proposition 27.17, $\overline{h(V_i)}$ is a closed convex set. By Proposition 27.23, $\left|\overline{h(V_i)}\right| = |h(V_i)| = |V_i| \leq \delta$. Thus $\left(\overline{h(V_i)} : i \in \mathbb{N}\right)$ is a δ-cover of E by closed convex sets. This shows that corresponding to every δ-cover $(V_i : i \in \mathbb{N})$ of E there exists a δ-cover $\left(\overline{h(V_i)} : i \in \mathbb{N}\right)$ of E by closed convex sets such that $\sum_{i\in\mathbb{N}} \left|\overline{h(V_i)}\right|^s = \sum_{i\in\mathbb{N}} |V_i|^s$. This implies that $\mathcal{K}_\delta^s(E) \leq \mathcal{H}_\delta^s(E)$. Therefore we have $\mathcal{K}_\delta^s(E) = \mathcal{H}_\delta^s(E)$. Then letting $\delta \downarrow 0$, we have $\mathcal{K}^s(E) = \mathcal{H}^s(E)$. ■

[III] Regularity of Hausdorff Measure

Lemma 27.26. *Let $s \in [0, \infty)$ and $E \in \mathfrak{P}(\mathbb{R}^n)$.*

(a) *For every $\delta \in (0, \infty]$ and $\eta > 0$, there exists a δ-cover $(F_i : i \in \mathbb{N})$ of E by closed sets such that $\sum_{i\in\mathbb{N}} |F_i|^s \leq \mathcal{H}_\delta^s(E) + \eta$.*

(b) *For every $\delta, \delta' \in (0, \infty]$ and $\eta > 0$, there exists a $\delta + \delta'$-cover $(G_i : i \in \mathbb{N})$ of E by open sets such that $\sum_{i\in\mathbb{N}} |G_i|^s \leq \mathcal{H}_\delta^s(E) + \eta$.*

Proof. 1. By the definition of $\mathcal{F}_\delta^s(E)$ in Theorem 27.13, there exists a δ-cover $(F_i : i \in \mathbb{N})$ of E by closed sets such that $\sum_{i\in\mathbb{N}} |F_i|^s \leq \mathcal{F}_\delta^s(E) + \eta$. By Theorem 27.13, $\mathcal{H}_\delta^s(E) = \mathcal{F}_\delta^s(E)$. Thus $\sum_{i\in\mathbb{N}} |F_i|^s \leq \mathcal{H}_\delta^s(E) + \eta$.

2. For an arbitrary $\delta \in (0, \infty]$ and $\eta > 0$, by Definition 27.3 for $\mathcal{H}_\delta^s(E)$ there exists a δ-cover $(V_i : i \in \mathbb{N})$ of E such that $\sum_{i\in\mathbb{N}} |V_i|^s \leq \mathcal{H}_\delta^s(E) + \frac{\eta}{2}$. Let $\delta' \in (0, \infty]$ be arbitrarily given. For each $i \in \mathbb{N}$, by (b) of Observation 27.2 there exists an open set $G_i \supset V_i$ such that

$$\begin{cases} |G_i| \leq |V_i| + \delta' \leq \delta + \delta', \\ |G_i|^s \leq |V_i|^s + \frac{\eta}{2^{i+1}}. \end{cases}$$

Then $(G_i : i \in \mathbb{N})$ is a $\delta + \delta'$-cover of E by open sets and

$$\sum_{i\in\mathbb{N}} |G_i|^s \leq \sum_{i\in\mathbb{N}} \left\{|V_i|^s + \frac{\eta}{2^{i+1}}\right\} = \sum_{i\in\mathbb{N}} |V_i|^s + \frac{\eta}{2} \leq \mathcal{H}_\delta^s(E) + \eta. \quad ■$$

Theorem 27.27. *For $s \in [0, \infty)$, consider the s-dimensional Hausdorff measure $\mathcal{H}^s$ on $\mathbb{R}^n$.*

(a) *For every* $E \in \mathfrak{P}(\mathbb{R}^n)$, *there exists a* G_δ*-set* $G \supset E$ *such that* $\mathcal{H}^s(G) = \mathcal{H}^s(E)$. *Thus* $\mathcal{H}^s$ *is a Borel regular outer measure.*

(b) *For every* $E \in \mathfrak{M}(\mathcal{H}^s)$ *with* $\mathcal{H}^s(E) < \infty$, *there exists an* F_σ*-set* $F \subset E$ *such that* $\mathcal{H}^s(F) = \mathcal{H}^s(E)$.

(c) *If* $E \in \mathfrak{M}(\mathcal{H}^s)$ *and* $\mathcal{H}^s(E) < \infty$, *then for every* $\varepsilon > 0$ *there exists a closed set* $C \subset E$ *such that* $\mathcal{H}^s(E \setminus C) < \varepsilon$.

Proof. 1. If $E \in \mathfrak{P}(\mathbb{R}^n)$ and $\mathcal{H}^s(E) = \infty$, then by the monotonicity of the outer measure $\mathcal{H}^s$ we have $\mathcal{H}^s(\mathbb{R}^n) \geq \mathcal{H}^s(E) = \infty$. Thus $\mathbb{R}^n$ is an open set containing E with $\mathcal{H}^s(\mathbb{R}^n) = \mathcal{H}^s(E)$ and we are done.

Consider the case $\mathcal{H}^s(E) < \infty$. By (b) of Lemma 27.26, for every $i \in \mathbb{N}$, there exists a $\frac{2}{i}$-cover of E by open sets $(O_{i,j} : j \in \mathbb{N})$ such that $\sum_{j \in \mathbb{N}} |O_{i,j}|^s \leq \mathcal{H}^s_{1/i}(E) + \frac{1}{i}$. If we let $G = \bigcap_{i \in \mathbb{N}} \bigcup_{j \in \mathbb{N}} O_{i,j}$, then G is a G_δ-set containing E. For each $i \in \mathbb{N}$, $(O_{i,j} : j \in \mathbb{N})$ is a $\frac{2}{i}$-cover of G so that $\mathcal{H}^s_{2/i}(G) \leq \sum_{j \in \mathbb{N}} |O_{i,j}|^s \leq \mathcal{H}^s_{1/i}(E) + \frac{1}{i}$. Since $\mathcal{H}^s_\delta \uparrow \mathcal{H}^s$ as $\delta \to 0$, we have $\mathcal{H}^s(G) = \lim_{\delta \to 0} \mathcal{H}^s_\delta(G) = \lim_{i \to \infty} \mathcal{H}^s_{2/i}(G) \leq \lim_{i \to \infty} \{\mathcal{H}^s_{1/i}(E) + \frac{1}{i}\} = \mathcal{H}^s(E)$. On the other hand, since $G \supset E$, we have $\mathcal{H}^s(G) \geq \mathcal{H}^s(E)$ by the monotonicity of $\mathcal{H}^s$. Thus we have $\mathcal{H}^s(G) = \mathcal{H}^s(E)$. Since $G \in \mathfrak{B}_{\mathbb{R}^n} \subset \mathfrak{M}(\mathcal{H}^s)$ by Theorem 27.12, $\mathcal{H}^s$ is a Borel regular outer measure.

2. Let $E \in \mathfrak{M}(\mathcal{H}^s)$ and $\mathcal{H}^s(E) < \infty$. Let us show that for every $\varepsilon > 0$, there exists an F_σ-set $F_\varepsilon \subset E$ such that

(1) $$\mathcal{H}^s(F_\varepsilon) \geq \mathcal{H}^s(E) - \varepsilon.$$

Now by (a), there exists a G_δ-set $G \supset E$ with $\mathcal{H}^s(G) = \mathcal{H}^s(E)$. Let $G = \bigcap_{i \in \mathbb{N}} O_i$ where $(O_i : i \in \mathbb{N})$ is a sequence of open sets. We have $O_i \supset G \supset E$ for $i \in \mathbb{N}$. Since $\mathcal{H}^s$ is a measure on the σ-algebra $\mathfrak{M}(\mathcal{H}^s)$ and since $G \in \mathfrak{B}_{\mathbb{R}^n} \subset \mathfrak{M}(\mathcal{H}^s)$ and $E \in \mathfrak{M}(\mathcal{H}^s)$ with $\mathcal{H}^s(E) < \infty$, we have $\mathcal{H}^s(G \setminus E) = \mathcal{H}^s(G) - \mathcal{H}^s(E) = 0$, that is,

(2) $$\mathcal{H}^s\Big(\bigcap_{i \in \mathbb{N}} O_i \setminus E\Big) = \mathcal{H}^s\Big(\bigcap_{i \in \mathbb{N}} O_i\Big) - \mathcal{H}^s(E) = 0.$$

Now every open set in $\mathbb{R}^n$ is an F_σ-set. (Every point in an open set in $\mathbb{R}^n$ is contained in a closed cube with rational vertices contained in the open set. Thus every open set in $\mathbb{R}^n$ is a union of closed cubes with rational vertices. Since the collection of all closed cubes with rational vertices is a countable collection, an open set in $\mathbb{R}^n$ is an F_σ-set.) Thus for each $i \in \mathbb{N}$, O_i can be given as $O_i = \bigcup_{j \in \mathbb{N}} F_{i,j}$ where $(F_{i,j} : j \in \mathbb{N})$ is an increasing sequence of closed sets. Now $\mathcal{H}^s(E \setminus F_{i,j}) = \mathcal{H}^s\big(E \setminus (E \cap F_{i,j})\big) = \mathcal{H}^s(E) - \mathcal{H}^s(E \cap F_{i,j})$. Since $F_{i,j} \uparrow O_i$, $E \cap F_{i,j} \uparrow E \cap O_i = E$, and since $\mathcal{H}^s$ is a measure on the σ-algebra $\mathfrak{M}(\mathcal{H}^s)$, we have $\lim_{j \to \infty} \mathcal{H}^s(E \cap F_{i,j}) = \mathcal{H}^s(E)$. Thus $\lim_{j \to \infty} \mathcal{H}^s(E \setminus F_{i,j}) = 0$ for each $i \in \mathbb{N}$. Therefore for each $i \in \mathbb{N}$, there exists $j_i \in \mathbb{N}$ such that

(3) $$\mathcal{H}^s(E \setminus F_{i,j_i}) < \frac{\varepsilon}{2^i}.$$

Let

(4) $$F_* = \bigcap_{i \in \mathbb{N}} F_{i,j_i} \subset \bigcap_{i \in \mathbb{N}} O_i.$$

Now $E \setminus F_* = E \cap F_*^c = E \cap \big(\bigcup_{i \in \mathbb{N}} F_{i,j_i}^c\big) = \bigcup_{i \in \mathbb{N}} \big(E \cap F_{i,j_i}^c\big) = \bigcup_{i \in \mathbb{N}} \big(E \setminus F_{i,j_i}\big)$ so that $\mathcal{H}^s(E \setminus F_*) \le \sum_{i \in \mathbb{N}} \mathcal{H}^s(E \setminus F_{i,j_i}) < \varepsilon$ by (3) and then

$$\text{(5)} \qquad \begin{aligned} \mathcal{H}^s(F_*) &\ge \mathcal{H}^s(E \cap F_*) = \mathcal{H}^s\big(E \setminus (E \setminus F_*)\big) \\ &= \mathcal{H}^s(E) - \mathcal{H}^s(E \setminus F_*) > \mathcal{H}^s(E) - \varepsilon. \end{aligned}$$

By (4) and (2), we have $\mathcal{H}^s(F_* \setminus E) \le \mathcal{H}^s\big(\bigcap_{i \in \mathbb{N}} O_i \setminus E\big) = 0$. By (a), there exists a G_δ-set $G \supset F_* \setminus E$ such that $\mathcal{H}^s(G) = \mathcal{H}^s(F_* \setminus E) = 0$. Let $F_\varepsilon = F_* \setminus G = F_* \cap G^c$. Since G is a G_δ-set, G^c is an F_σ-set. Since F_* defined by (4) is a closed set, F_ε is an F_σ-set. Now

$$\begin{aligned} F_\varepsilon = F_* \cap G^c &\subset F_* \cap (F_* \setminus E)^c = F_* \cap (F_* \cap E^c)^c \\ &= F_* \cap (F_*^c \cup E) = F_* \cap E \subset E \end{aligned}$$

and $\mathcal{H}^s(F_\varepsilon) = \mathcal{H}^s(F_* \setminus G) \ge \mathcal{H}^s(F_*) - \mathcal{H}^s(G) = \mathcal{H}^s(F_*) > \mathcal{H}^s(E) - \varepsilon$ by (5). This completes the proof of (1).

Now for every $k \in \mathbb{N}$, there exists an F_σ-set $F_k \subset E$ such that $\mathcal{H}^s(F_k) > \mathcal{H}^s(E) - \frac{1}{k}$ by (1). Let $F = \bigcup_{k \in \mathbb{N}} F_k$. Then F is an F_σ-set and $\mathcal{H}^s(F) \ge \mathcal{H}^s(F_k) > \mathcal{H}^s(E) - \frac{1}{k}$ for every $k \in \mathbb{N}$ so that $\mathcal{H}^s(F) \ge \mathcal{H}^s(E)$. On the other hand since $F \subset E$, we have $\mathcal{H}^s(F) \le \mathcal{H}^s(E)$. Therefore $\mathcal{H}^s(F) = \mathcal{H}^s(E)$. This completes the proof of (b).

3. To prove (c), let $E \in \mathfrak{M}(\mathcal{H}^s)$ and $\mathcal{H}^s(E) < \infty$. By (b), there exists an F_σ-set $F \subset E$ with $\mathcal{H}^s(F) = \mathcal{H}^s(E)$. Since F is an F_σ-set, we have $F = \bigcup_{k \in \mathbb{N}} C_k$ where $(C_k : k \in \mathbb{N})$ is an increasing sequence of closed sets. Since $\mathcal{H}^s$ is a measure on the σ-algebra $\mathfrak{M}(\mathcal{H}^s)$ containing $\mathfrak{B}_{\mathbb{R}^n}$ and since $C_k \in \mathfrak{B}_{\mathbb{R}^n}$ and $C_k \uparrow F$ as $k \to \infty$, we have $\mathcal{H}^s(F) = \mathcal{H}^s\big(\lim_{k \to \infty} C_k\big) = \lim_{k \to \infty} \mathcal{H}^s(C_k)$. Since $\mathcal{H}^s(F) = \mathcal{H}^s(E) < \infty$, for every $\varepsilon > 0$, there exists $k_0 \in \mathbb{N}$ such that $\mathcal{H}^s(C_{k_0}) > \mathcal{H}^s(F) - \varepsilon$. Thus we have finally $\mathcal{H}^s(E \setminus C_{k_0}) = \mathcal{H}^s(E) - \mathcal{H}^s(C_{k_0}) = \mathcal{H}^s(F) - \mathcal{H}^s(C_{k_0}) < \varepsilon$. ∎

Next we have an estimate of $\mathcal{H}^s(E)$ by a δ-cover of E.

Proposition 27.28. *Let $s \in [0, \infty)$ and $E \in \mathfrak{M}(\mathcal{H}^s)$ with $\mathcal{H}^s(E) < \infty$.*
(a) *For every $\varepsilon > 0$ there exists $\delta_0 > 0$, depending on E and ε only, such that for any countable collection $\{W_i : i \in \mathbb{N}\} \subset \mathfrak{B}_{\mathbb{R}^n}$ with $|W_i| \in (0, \delta_0]$ for $i \in \mathbb{N}$, we have $\mathcal{H}^s\big(E \cap \bigcup_{i \in \mathbb{N}} W_i\big) < \sum_{i \in \mathbb{N}} |W_i|^s + \varepsilon$.*
(b) *In particular for every $\varepsilon > 0$ there exists $\delta_0 > 0$ such that for any δ_0-cover of E, $(W_i : i \in \mathbb{N}) \subset \mathfrak{B}_{\mathbb{R}^n}$, we have $\mathcal{H}^s(E) < \sum_{i \in \mathbb{N}} |W_i|^s + \varepsilon$.*

Proof. Since $\mathcal{H}^s_\delta(E) \uparrow \mathcal{H}^s(E) < \infty$ as $\delta \downarrow \infty$, for an arbitrary $\varepsilon > 0$ there exists $\delta_0 > 0$ such that

$$\text{(1)} \qquad \mathcal{H}^s(E) < \mathcal{H}^s_{\delta_0}(E) + \frac{\varepsilon}{2} \le \sum_{i \in \mathbb{N}} |V_i|^s + \frac{\varepsilon}{2},$$

for any δ_0-cover $(V_i : i \in \mathbb{N})$ of E. Let $\{W_i : i \in \mathbb{N}\}$ be an arbitrary countable collection in $\mathfrak{B}_{\mathbb{R}^n}$ with $|W_i| \in (0, \delta_0]$ for $i \in \mathbb{N}$. Consider the set $E \setminus \bigcup_{i \in \mathbb{N}} W_i$. There exists a δ_0-cover

$(U_i : i \in \mathbb{N})$ of $E \setminus \bigcup_{i\in\mathbb{N}} W_i$ such that

$$(2) \qquad \sum_{i\in\mathbb{N}} |U_i|^s < \mathcal{H}^s_{\delta_0}\big(E \setminus \textstyle\bigcup_{i\in\mathbb{N}} W_i\big) + \frac{\varepsilon}{2} \le \mathcal{H}^s\big(E \setminus \bigcup_{i\in\mathbb{N}} W_i\big) + \frac{\varepsilon}{2}.$$

Now $\{W_i : i \in \mathbb{N}\} \cup \{U_i : i \in \mathbb{N}\}$ is a δ_0-cover of E so that by (1) we have

$$(3) \qquad \mathcal{H}^s(E) < \sum_{i\in\mathbb{N}} |W_i|^s + \sum_{i\in\mathbb{N}} |U_i|^s + \frac{\varepsilon}{2}.$$

Then by (3) and (2), we have

$$\begin{aligned}\mathcal{H}^s\big(E \cap \textstyle\bigcup_{i\in\mathbb{N}} W_i\big) &= \mathcal{H}^s\big(E \setminus \big(E \setminus \textstyle\bigcup_{i\in\mathbb{N}} W_i\big)\big) = \mathcal{H}^s(E) - \mathcal{H}^s\big(E \setminus \textstyle\bigcup_{i\in\mathbb{N}} W_i\big)\\ &< \sum_{i\in\mathbb{N}} |W_i|^s + \sum_{i\in\mathbb{N}} |U_i|^s + \frac{\varepsilon}{2} - \sum_{i\in\mathbb{N}} |U_i|^s + \frac{\varepsilon}{2}\\ &= \sum_{i\in\mathbb{N}} |W_i|^s + \varepsilon. \quad \blacksquare\end{aligned}$$

[IV] Hausdorff Dimension

For $s \in [0, \infty)$, the s-dimensional Hausdorff measure $\mathcal{H}^s(E)$ of a set $E \in \mathfrak{P}(\mathbb{R}^n)$ is a nonnegative extended real number. If we keep $E \in \mathfrak{P}(\mathbb{R}^n)$ fixed and let s vary over $[0, \infty)$, then we have a nonnegative extended real-valued function on $[0, \infty)$. Let us denote this function by $\mathcal{H}^\bullet(E)$. The next lemma shows that for every $E \in \mathfrak{P}(\mathbb{R}^n)$, $\mathcal{H}^\bullet(E)$ is a decreasing function on $[0, \infty)$ and furthermore $\mathcal{H}^\bullet(E)$ either assumes no finite positive values at all or assumes some finite positive value at a unique point in $[0, \infty)$.

Lemma 27.29. *Let $E \in \mathfrak{P}(\mathbb{R}^n)$ and let $0 \le s < t < \infty$. For the function $\mathcal{H}^\bullet(E)$, we have*
(a) $\mathcal{H}^s(E) \ge \mathcal{H}^t(E)$.
(b) $\delta^{t-s}\mathcal{H}^s_\delta(E) \ge \mathcal{H}^t_\delta(E)$ *for every* $\delta \in (0, \infty]$.
(c) *If* $\mathcal{H}^s(E) < \infty$, *then* $\mathcal{H}^t(E) = 0$.
(d) *If* $\mathcal{H}^t(E) > 0$, *then* $\mathcal{H}^s(E) = \infty$.
(e) *There exists at most one point* $s^* \in [0, \infty)$ *such that* $\mathcal{H}^{s^*}(E) \in (0, \infty)$.

Proof. 1. Recall that we have $\mathcal{H}^s_\delta(E) = \inf\big\{\sum_{i\in\mathbb{N}} |V_i|^s : (V_i : i \in \mathbb{N}) \text{ is a } \delta\text{-cover of } E\big\}$ for $\delta \in (0, \infty]$ and $\mathcal{H}^s(E) = \lim_{\delta\downarrow 0} \mathcal{H}^s_\delta(E)$. Consider $\delta \in (0, 1)$ and let $(V_i : i \in \mathbb{N})$ be an arbitrary δ-cover of E. Then since $|V_i| \le \delta < 1$ and $0 \le s < t$, we have $|V_i|^s > |V_i|^t$ so that $\sum_{i\in\mathbb{N}} |V_i|^s > \sum_{i\in\mathbb{N}} |V_i|^t$. Therefore $\mathcal{H}^s_\delta(E) \ge \mathcal{H}^t_\delta(E)$ for $\delta \in (0, 1)$ and this implies $\lim_{\delta\downarrow 0} \mathcal{H}^s_\delta(E) \ge \lim_{\delta\downarrow 0} \mathcal{H}^t_\delta(E)$, that is, $\mathcal{H}^s(E) \ge \mathcal{H}^t(E)$. This proves (a).

2. To prove (b), let $\delta \in (0, \infty]$ and let $(V_i : i \in \mathbb{N})$ be an arbitrary δ-cover of E. If

$0 \le s < t < \infty$, then $t - s > 0$ so that $|V_i|^t = |V_i|^{t-s}|V_i|^s \le \delta^{t-s}|V_i|^s$. Thus we have

$$\begin{aligned}\mathcal{H}^t_\delta(E) &= \inf\left\{\textstyle\sum_{i\in\mathbb{N}}|V_i|^t : (V_i : i \in \mathbb{N}) \text{ is a } \delta\text{-cover of } E\right\}\\ &\le \delta^{t-s}\inf\left\{\textstyle\sum_{i\in\mathbb{N}}|V_i|^s : (V_i : i \in \mathbb{N}) \text{ is a } \delta\text{-cover of } E\right\}\\ &= \delta^{t-s}\mathcal{H}^s_\delta(E).\end{aligned}$$

3. To prove (c), note that by (b) we have $\mathcal{H}^t(E) = \lim_{\delta\downarrow 0}\mathcal{H}^t_\delta(E) \le \lim_{\delta\downarrow 0}\delta^{t-s}\mathcal{H}^s_\delta(E)$. Now since $t - s > 0$, we have $\lim_{\delta\downarrow 0}\delta^{t-s} = 0$. If $\mathcal{H}^s(E) < \infty$, that is, $\lim_{\delta\downarrow 0}\mathcal{H}^s_\delta(E) < \infty$, then $\mathcal{H}^t(E) = 0$.

4. To prove (d), let $\delta \in (0, \infty)$. Multiplying the inequality in (b) by δ^{s-t}, we have $\delta^{s-t}\mathcal{H}^t_\delta(E) \le \mathcal{H}^s_\delta(E) \le \mathcal{H}^s(E)$. Since $s - t < 0$, we have $\lim_{\delta\downarrow 0}\delta^{s-t} = \infty$. If $\mathcal{H}^t(E) > 0$, then $\lim_{\delta\downarrow 0}\delta^{s-t}\mathcal{H}^t_\delta(E) = \infty$ and thus $\mathcal{H}^s(E) = \infty$.

5. Take any two distinct points $s, t \in [0, \infty)$, say $s < t$. If $\mathcal{H}^s(E) \in (0, \infty)$, then $\mathcal{H}^t(E) = 0$ by (c). If $\mathcal{H}^t(E) \in (0, \infty)$, then $\mathcal{H}^s(E) = \infty$ by (d). This proves (e). ∎

We show next that for every $E \in \mathfrak{P}(\mathbb{R}^n)$ the function $\mathcal{H}^\bullet(E)$ vanishes on (n, ∞).

Lemma 27.30.
(a) $\mathcal{H}^n(Q) \in [2^{-n}, n^{n/2}]$ *for every unit cube* Q *in the class* $\mathfrak{I}^n = \mathfrak{I}^n_{oc} \cup \mathfrak{I}^n_{co} \cup \mathfrak{I}^n_o \cup \mathfrak{I}^n_c$.
(b) $\mathcal{H}^s(\mathbb{R}^n) = \infty$ *for* $s \in [0, n]$ *and* $\mathcal{H}^s(\mathbb{R}^n) = 0$ *for* $s \in (n, \infty)$.
(c) $\mathcal{H}^s(E) = 0$ *for* $s \in (n, \infty)$ *for every* $E \in \mathfrak{P}(\mathbb{R}^n)$.

Proof. 1. Let us show $2^{-n} \le \mathcal{H}^n(Q)$ first. For $\delta \in (0, \infty)$, let $(V_i : i \in \mathbb{N})$ be an arbitrary δ-cover of Q. For each $i \in \mathbb{N}$, let $x_i \in V_i$ be arbitrarily chosen and let C_i be a closed cube with center at x_i and with edge equal to $2|V_i|$. Then $V_i \subset C_i$ and $\mu^n_L(C_i) = (2|V_i|)^n$. Now $Q \subset \bigcup_{i\in\mathbb{N}} V_i \subset \bigcup_{i\in\mathbb{N}} C_i$ so that we have $\mu^n_L(Q) \le \sum_{i\in\mathbb{N}}\mu^n_L(C_i) = 2^n\sum_{i\in\mathbb{N}}|V_i|^n$. Since this holds for an arbitrary δ-cover $(V_i : i \in \mathbb{N})$ of Q, we have $\mu^n_L(Q) \le 2^n\mathcal{H}^n_\delta(Q)$. Since $\mathcal{H}^n(Q) = \lim_{\delta\downarrow 0}\mathcal{H}^n_\delta(Q)$, we have $\mu^n_L(Q) \le 2^n\mathcal{H}^n(Q)$. Since $\mu^n_L(Q) = 1$, we have $2^{-n} \le \mathcal{H}^n(Q)$.

Next let us show $\mathcal{H}^n(Q) \le n^{n/2}$. Let $\delta \in (0, \infty)$ be arbitrarily fixed. Let $k \in \mathbb{N}$ be so large that $\sqrt{n}/k < \delta$. Now Q is contained in the union of k^n closed cubes C_i, $i = 1, \dots, k^n$, with edge equal to $\frac{1}{k}$. The diameter of each of these k^n closed cubes is equal to $(n/k^2)^{1/2} = \sqrt{n}/k < \delta$. Thus $\{C_i : i = 1, \dots, k^n\}$ is a δ-cover of Q. This implies that $\mathcal{H}^n_\delta(Q) \le \sum_{i=1}^{k^n}|C_i|^n = k^n(\sqrt{n}/k)^n = n^{n/2}$. Since this holds for every $\delta \in (0, \infty)$, we have $\mathcal{H}^n(Q) = \lim_{\delta\downarrow 0}\mathcal{H}^n_\delta(Q) \le n^{n/2}$. This completes the proof of (a).

2. $\mathbb{R}^n$ is a countable disjoint union of unit cubes Q_i, $i \in \mathbb{N}$, in the class $\mathfrak{I}^n_{oc}$. Since $\mathcal{H}^n$ is a measure on the σ-algebra $\mathfrak{M}(\mathcal{H}^n)$ of $\mathcal{H}^n$-measurable subsets of $\mathbb{R}^n$ and since $Q_i \in \mathfrak{B}_{\mathbb{R}^n} \subset \mathfrak{M}(\mathcal{H}^n)$, we have $\mathcal{H}^n(\mathbb{R}^n) = \mathcal{H}^n\left(\bigcup_{n\in\mathbb{N}} Q_i\right) = \sum_{i\in\mathbb{N}}\mathcal{H}^n(Q_i) = \infty$ by the fact that $\mathcal{H}^n(Q_i) \ge 2^{-n} > 0$ for $i \in \mathbb{N}$. Thus $\mathcal{H}^n(\mathbb{R}^n) = \infty$. Then since $\mathcal{H}^\bullet(\mathbb{R}^n)$ is a decreasing function on $[0, \infty)$, we have $\mathcal{H}^s(\mathbb{R}^n) = \infty$ for $s \in [0, n]$.

To show that $\mathcal{H}^s(\mathbb{R}^n) = 0$ for $s \in (n, \infty)$, note that since $\mathcal{H}^s$ is a measure on the σ-algebra $\mathfrak{M}(\mathcal{H}^s)$ and since $Q_i \in \mathfrak{B}_{\mathbb{R}^n} \subset \mathfrak{M}(\mathcal{H}^s)$, we have $\mathcal{H}^s(\mathbb{R}^n) = \sum_{i \in \mathbb{N}} \mathcal{H}^s(Q_i)$. Now $\mathcal{H}^n(Q_i) \in (0, \infty)$ by (a). This implies that $\mathcal{H}^s(Q_i) = 0$ for $s \in (n, \infty)$ by (c) of Lemma 27.29. Thus $\mathcal{H}^s(\mathbb{R}^n) = 0$.

3. If $E \in \mathfrak{P}(\mathbb{R}^n)$, then $\mathcal{H}^s(E) \le \mathcal{H}^s(\mathbb{R}^n) = 0$ for $s \in (n, \infty)$. This proves (c). ∎

The next lemma shows that the 0-dimensional Hausdorff measure $\mathcal{H}^0$ counts the number of elements in a set $E \in \mathfrak{P}(\mathbb{R}^n)$. Thus the outer measure $\mathcal{H}^0$ is actually a counting measure on the σ-algebra $\mathfrak{P}(\mathbb{R}^n)$.

Lemma 27.31. *Let $\mathcal{N}(E)$ be the number of elements in a finite set E and let $\mathcal{N}(E) = \infty$ if E is an infinite set. Then for every $E \in \mathfrak{P}(\mathbb{R}^n)$, we have $\mathcal{H}^0(E) = \mathcal{N}(E)$. In particular $\mathcal{H}^0(E) = 0$ if and only if $E = \emptyset$.*

Proof. 1. Since $\mathcal{H}^0$ is an outer measure, we have $\mathcal{H}^0(\emptyset) = 0$.

2. Suppose E has k elements where $k \in \mathbb{N}$. Let $x_1, \dots, x_k$ be the distinct elements of E. Then $d(x_i, x_j) > 0$ for $i, j = 1, \dots, k$, $i \ne j$. Thus $\min_{i \ne j} d(x_i, x_j) > 0$. Let $\delta \in \left(0, \min_{i \ne j} d(x_i, x_j)\right)$. Let $(V_i : i \in \mathbb{N})$ be an arbitrary δ-cover of E. Since $|V_i| \le \delta$, each V_i can contain at most one element of E. Dropping those V_i which do not contain any element of E from the δ-cover of E and renumber the remaining ones as $(V_1, \dots, V_k)$. Then we have $E \subset \bigcup_{i=1}^k V_i$ and $\sum_{i=1}^k |V_i|^0 = k$. Thus the estimates of $\mathcal{H}^0_\delta(E)$ by δ-covers of E are bounded below by k and this lower bound is attained by the δ-cover $(V_1, \dots, V_k)$. Therefore we have $\mathcal{H}^0_\delta(E) = k$. Then $\mathcal{H}^0(E) = \lim_{\delta \downarrow 0} \mathcal{H}^0_\delta(E) = k$.

3. If E is an infinite set, for each $k \in \mathbb{N}$ let E_k be a subset of E consisting of k elements of E. By the monotonicity of the outer measure $\mathcal{H}^0$, we have $\mathcal{H}^0(E) \ge \mathcal{H}^0(E_k) = k$ for every $k \in \mathbb{N}$. Thus $\mathcal{H}^0(E) = \infty$. ∎

Proposition 27.32. *For the outer measure $\mathcal{H}^0$ on $\mathbb{R}^n$, we have the σ-algebra of all the $\mathcal{H}^0$-measurable sets $\mathfrak{M}(\mathcal{H}^0) = \mathfrak{P}(\mathbb{R}^n)$. Thus $\mathcal{H}^0$ is a measure on the σ-algebra $\mathfrak{P}(\mathbb{R}^n)$ and $\mathcal{H}^0(E) = \mathcal{N}(E)$, the number of elements in E, for every $E \in \mathfrak{P}(\mathbb{R}^n)$.*

Proof. According to Lemma 27.31, $\mathcal{H}^0(E) = \mathcal{N}(E)$ for every $E \in \mathfrak{P}(\mathbb{R}^n)$. Now for an arbitrary $A \in \mathfrak{P}(\mathbb{R}^n)$, we have $\mathcal{N}(A) = \mathcal{N}(A \cap E) + \mathcal{N}(A \cap E^c)$. Thus for every $A \in \mathfrak{P}(\mathbb{R}^n)$, we have $\mathcal{H}^0(A) = \mathcal{H}^0(A \cap E) + \mathcal{H}^0(A \cap E^c)$. This shows that every $E \in \mathfrak{P}(\mathbb{R}^n)$ satisfies the Carathéodory condition with respect to the outer measure $\mathcal{H}^0$ and hence $E \in \mathfrak{M}(\mathcal{H}^0)$. Thus we have $\mathfrak{P}(\mathbb{R}^n) \subset \mathfrak{M}(\mathcal{H}^0)$. On the other hand we have $\mathfrak{M}(\mathcal{H}^0) \subset \mathfrak{P}(\mathbb{R}^n)$. Therefore we have $\mathfrak{M}(\mathcal{H}^0) = \mathfrak{P}(\mathbb{R}^n)$. ∎

Summarizing (c), (d), and (e) of Lemma 27.29, (b) and (c) of Lemma 27.30 and Lemma 27.31, we have:

Theorem 27.33. *For every $E \in \mathfrak{P}(\mathbb{R}^n)$, consider the function $\mathcal{H}^s(E)$ of $s \in [0, \infty)$.*

(a) *$\mathcal{H}^\bullet(E)$ is a nonnegative extended real-valued decreasing function on $[0, \infty)$ vanishing on (n, ∞).*

(b) *There are three possibilities for the range of the function $\mathcal{H}^\bullet(E)$:*

(i) $\mathcal{H}^{\bullet}(E)$ *assumes the value* 0 *only.*
(ii) $\mathcal{H}^{\bullet}(E)$ *assumes only two values,* ∞ *and* 0.
(iii) $\mathcal{H}^{\bullet}(E)$ *assumes three values* ∞, 0, *and one finite positive value.*

(c) *If* $\mathcal{H}^{\bullet}(E)$ *assumes a finite positive value at some point* $s^* \in [0, \infty)$, *then we have* $\mathcal{H}^s(E) = \infty$ *for* $s \in [0, s^*)$, *with the understanding that* $[0, 0) = \emptyset$, *and* $\mathcal{H}^s(E) = 0$ *for* $s \in (s^*, \infty)$.

(d) $\mathcal{H}^{\bullet}(E)$ *is identically equal to* 0 *on* $[0, \infty)$ *if and only if* $E = \emptyset$. *If* $E \neq \emptyset$, *then* $\mathcal{H}^{\bullet}$ *has exactly one point of discontinuity in* $[0, \infty)$ *and the point of discontinuity is contained in* $[0, n]$.

Definition 27.34. *For* $E \in \mathfrak{P}(\mathbb{R}^n)$, $E \neq \emptyset$, *we define the Hausdorff dimension of* E, *denoted by* $\dim_H E$, *to be the unique number* $s^* \in [0, \infty)$ *at which the discontinuity of the function* $\mathcal{H}^{\bullet}(E)$ *occurs. For* $E = \emptyset$, *we set* $\dim_H \emptyset = 0$. *In other words, for* $E \neq \emptyset$, $\dim_H E$ *is defined as the unique number* $s^* \in [0, \infty)$ *such that* $\mathcal{H}^s(E) = \infty$ *for* $s \in [0, s^*)$ *and* $\mathcal{H}^s(E) = 0$ *for* $s \in (s^*, \infty)$.

Theorem 27.35. *For the Hausdorff dimension* $\dim_H E$ *of every* $E \in \mathfrak{P}(\mathbb{R}^n)$, *we have:*

(a) $\dim_H E \in [0, n]$.

(b) $\dim_H \mathbb{R}^n = n$.

(c) $\dim_H E = \sup\{s \in [0, \infty) : \mathcal{H}^s(E) > 0\} = \sup\{s \in [0, \infty) : \mathcal{H}^s(E) = \infty\}$.

(d) $\dim_H E = \inf\{s \in [0, \infty) : \mathcal{H}^s(E) = 0\} = \inf\{s \in [0, \infty) : \mathcal{H}^s(E) < \infty\}$.

(e) *If* $\mathcal{H}^s(E) \in (0, \infty)$, *then* $s = \dim_H E$.

(f) *The converse of (e) is false, that is,* $s = \dim_H E$ *does not imply* $\mathcal{H}^s(E) \in (0, \infty)$. *For instance* $\dim_H \emptyset = 0$ *but* $\mathcal{H}^0(\emptyset) = 0$ *and* $\dim_H \mathbb{R}^n = n$ *but* $\mathcal{H}^n(\mathbb{R}^n) = \infty$.

Proof. (a) and (b) are from Lemma 27.30. (c), (d), and (e) are from Theorem 27.33. ∎

Proposition 27.36. *Regarding the Hausdorff dimension, we have the following:*

(a) *If* $E, F \in \mathfrak{P}(\mathbb{R}^n)$ *and* $E \subset F$, *then* $\dim_H E \leq \dim_H F$.

(b) *For* $(E_i : i \in \mathbb{N}) \subset \mathfrak{P}(\mathbb{R}^n)$, *we have* $\dim_H \bigcup_{i \in \mathbb{N}} E_i = \sup\{\dim_H E_i : i \in \mathbb{N}\}$.
In particular, $\dim_H \bigcup_{i=1}^N E_i = \max\{\dim_H E_1, \ldots, \dim_H E_N\}$.

Proof. 1. Suppose $E \subset F$. For every $s \in [0, \infty)$, by the monotonicity of the outer measure $\mathcal{H}^s$ we have $\mathcal{H}^s(E) \leq \mathcal{H}^s(F)$. Thus we have $\mathcal{H}^{\bullet}(E) \leq \mathcal{H}^{\bullet}(F)$ on $[0, \infty)$. Since both $\mathcal{H}^s(E)$ and $\mathcal{H}^s(F)$ are nonnegative functions on $[0, \infty)$, this implies that we have $\{s \in [0, \infty) : \mathcal{H}^s(E) = 0\} \supset \{s \in [0, \infty) : \mathcal{H}^s(F) = 0\}$ and consequently we have $\inf\{s \in [0, \infty) : \mathcal{H}^s(E) = 0\} \leq \inf\{s \in [0, \infty) : \mathcal{H}^s(F) = 0\}$, that is, $\dim_H E \leq \dim_H F$ by (d) of Theorem 27.35.

2. To prove (b), note that for every $i \in \mathbb{N}$, we have $E_i \subset \bigcup_{j \in \mathbb{N}} E_j$ so that by (a) we have $\dim_H E_i \leq \dim_H \bigcup_{j \in \mathbb{N}} E_j$. Thus $\sup\{\dim_H E_i : i \in \mathbb{N}\} \leq \dim_H \bigcup_{i \in \mathbb{N}} E_i$. To prove the reverse inequality, let us write $t = \sup\{\dim_H E_i : i \in \mathbb{N}\}$ for brevity. Let $\varepsilon > 0$. Then for every $i \in \mathbb{N}$ we have $t + \varepsilon > \dim_H E_i$ so that $\mathcal{H}^{t+\varepsilon}(E_i) = 0$ by (d) of Theorem 27.35. Then we have $\mathcal{H}^{t+\varepsilon}\left(\bigcup_{i \in \mathbb{N}} E_i\right) \leq \sum_{i \in \mathbb{N}} \mathcal{H}^{t+\varepsilon}(E_i) = 0$ by the countable subadditivity of

$\mathcal{H}^{t+\varepsilon}$. Then by (d) of Theorem 27.35, we have $\dim_H \bigcup_{i\in\mathbb{N}} E_i \le t+\varepsilon$. Since this holds for every $\varepsilon > 0$, we have $\dim_H \bigcup_{i\in\mathbb{N}} E_i \le t$. ∎

Proposition 27.37. *If* $E \in \mathfrak{P}(\mathbb{R}^n)$ *is a countable set, then* $\dim_H E = 0$.

Proof. If $E = \emptyset$, then $\dim_H E = 0$ by Definition 27.34. If E is a nonempty finite set, say it has k elements where $k \in \mathbb{N}$, then $\mathcal{H}^0(E) = k \in (0,\infty)$ by Lemma 27.31. Then $\dim_H E = 0$ by (e) of Theorem 27.35.

Suppose E is a countable infinite set. If we show that $\mathcal{H}^s(E) = 0$ for every $s \in (0,1)$, then by (d) of Theorem 27.35 we have $\dim_H E = \inf\{s \in [0,\infty) : \mathcal{H}^s(E) = 0\} = 0$ and we are done. Let $s \in (0,1)$ be arbitrarily fixed. Let $\varepsilon > 0$ be arbitrarily given and let $\delta \in \big(0, [(2^s-1)\varepsilon]^{1/s}\big)$. Let $E = \{x_i : i \in \mathbb{N}\}$ and let $(V_i : i \in \mathbb{N})$ be a δ-cover of E such that $x_i \in V_i$ and $0 < |V_i| = \delta/2^i < \delta$ for $i \in \mathbb{N}$. Then $\sum_{i\in\mathbb{N}} |V_i|^s = \sum_{i\in\mathbb{N}} \big(\delta/2^i\big)^s = \delta^s \sum_{i\in\mathbb{N}} 1/2^{si}$. The series is a convergent geometric series with ratio $1/2^s < 1$ and first term $1/2^s$. Thus $\sum_{i\in\mathbb{N}} 1/2^{si} = 1/(2^s-1)$ and $\sum_{i\in\mathbb{N}} |V_i|^s = \delta^s/(2^s-1) < \varepsilon$. This implies that $\mathcal{H}^s_\delta(E) < \varepsilon$. Since this holds for every $\delta \in \big(0, [(2^s-1)\varepsilon]^{1/s}\big)$, we have $\mathcal{H}^s(E) = \lim_{\delta\downarrow 0} \mathcal{H}^s_\delta(E) \le \varepsilon$. By the arbitrariness of $\varepsilon > 0$, we have $\mathcal{H}^s(E) = 0$. ∎

Proposition 27.38. *Let* $E \in \mathfrak{P}(\mathbb{R}^n)$. *If* $E^\circ \neq \emptyset$, *then* $\dim_H E = n$.

Proof. Let O be a bounded and nonempty open set in $\mathbb{R}^n$. According to Theorem 28.1 to follow next, for any $\lambda > 0$, we have $\mathcal{H}^n(\lambda O) = \lambda^n \mathcal{H}^n(O)$. For a sufficiently large $\lambda > 0$, λO contains a unit cube Q in $\mathbb{R}^n$ so that $\mathcal{H}^n(\lambda O) \ge \mathcal{H}^n(Q) > 0$ by Lemma 27.30 and then $\mathcal{H}^n(O) = \lambda^{-n}\mathcal{H}^n(\lambda O) > 0$. Similarly for sufficiently small $\lambda > 0$, λO is contained in a unit cube Q in $\mathbb{R}^n$ so that $\mathcal{H}^n(\lambda O) \le \mathcal{H}^n(Q) < \infty$ by Lemma 27.30 and then $\mathcal{H}^n(O) = \lambda^{-n}\mathcal{H}^n(\lambda O) < \infty$. This shows that $\mathcal{H}^n(O) \in (0,\infty)$. Then we have $\dim_H O = n$ by (e) of Theorem 27.35.

Now let $E \in \mathfrak{P}(\mathbb{R}^n)$ with $E^\circ \neq \emptyset$. Let B be an arbitrary open ball contained in E°. Since B is a bounded and nonempty open set, we have $\dim_H B = n$ by our result above. Since $E \supset B$, we have $\dim_H E \ge \dim_H B = n$ by Proposition 27.36. On the other hand $E \subset \mathbb{R}^n$ implies $\dim_H E \le \dim_H \mathbb{R}^n = n$ by (b) of Theorem 27.35. Therefore we have $\dim_H E = n$. ∎

§28 Transformations of Hausdorff Measures

[I] Hausdorff Measure of Transformed Sets

Theorem 28.1. (Dilation) *For an arbitrary set $E \in \mathfrak{P}(\mathbb{R}^n)$, consider its dilation by $\lambda \in \mathbb{R}$, $\lambda \neq 0$, that is, $\lambda E = \{y \in \mathbb{R}^n : y = \lambda x$ for some $x \in E\}$. We have*
(a) $\mathcal{H}^s(\lambda E) = |\lambda|^s \mathcal{H}^s(E)$ *for every $s \in [0, \infty)$.*
(b) $\dim_H \lambda E = \dim_H E$.

Proof. 1. For any $x, y \in \mathbb{R}^n$, we have $d(\lambda x, \lambda y) = |\lambda x - \lambda y| = |\lambda||x - y| = |\lambda| d(x, y)$. It follows that for every $V \in \mathfrak{P}(\mathbb{R}^n)$, we have $|\lambda V| = |\lambda||V|$. This implies that if $(V_i : i \in \mathbb{N})$ is a δ-cover of E for some $\delta \in (0, \infty)$, then $(\lambda V_i : i \in \mathbb{N})$ is a $|\lambda|\delta$-cover of λE. Then for every $s \in [0, \infty)$, $\mathcal{H}^s_{|\lambda|\delta}(\lambda E) \leq \sum_{i \in \mathbb{N}} |\lambda V_i|^s = |\lambda|^s \sum_{i \in \mathbb{N}} |V_i|^s$. Since this holds for every δ-cover $(V_i : i \in \mathbb{N})$ of E, we have $\mathcal{H}^s_{|\lambda|\delta}(\lambda E) \leq |\lambda|^s \mathcal{H}^s_\delta(E)$. Letting $\delta \to 0$, we have $\mathcal{H}^s(\lambda E) \leq |\lambda|^s \mathcal{H}^s(E)$. Applying this result to the dilation of the set λE by the scalar λ^{-1}, we have $\mathcal{H}^s\left(\lambda^{-1}(\lambda E)\right) \leq |\lambda^{-1}|^s \mathcal{H}^s(\lambda E)$, that is, $\mathcal{H}^s(E) \leq |\lambda|^{-s} \mathcal{H}^s(\lambda E)$ and then $|\lambda|^s \mathcal{H}^s(E) \leq \mathcal{H}^s(\lambda E)$. Therefore $\mathcal{H}^s(\lambda E) = |\lambda|^s \mathcal{H}^s(E)$.

2. By (a) we have $\mathcal{H}^s(\lambda E) = |\lambda|^s \mathcal{H}^s(E)$ for every $s \in [0, \infty)$. Since $|\lambda|^s \neq 0$, this implies $\{s \in [0, \infty) : \mathcal{H}^s(\lambda E) = 0\} = \{s \in [0, \infty) : \mathcal{H}^s(E) = 0\}$. Then by (d) of Theorem 27.35, we have

$$\begin{aligned}\dim_H \lambda E &= \inf\left\{s \in [0, \infty) : \mathcal{H}^s(\lambda E) = 0\right\} \\ &= \inf\left\{s \in [0, \infty) : \mathcal{H}^s(E) = 0\right\} \\ &= \dim_H E. \quad \blacksquare\end{aligned}$$

Theorem 28.2. (Invariance of $\mathfrak{M}(\mathcal{H}^s)$ under Dilation) *Let $E \in \mathfrak{P}(\mathbb{R}^n)$ and $\lambda \in \mathbb{R}$, $\lambda \neq 0$. Then $\lambda E \in \mathfrak{M}(\mathcal{H}^s)$ if and only if $E \in \mathfrak{M}(\mathcal{H}^s)$.*

Proof. A subset $E \in \mathfrak{P}(\mathbb{R}^n)$ is in $\mathfrak{M}(\mathcal{H}^s)$, that is E is $\mathcal{H}^s$-measurable, if and only if for an arbitrary subset $A \in \mathfrak{P}(\mathbb{R}^n)$ we have $\mathcal{H}^s(A) = \mathcal{H}^s(A \cap E) + \mathcal{H}^s(A \cap E^c)$.

Let $E \in \mathfrak{M}(\mathcal{H}^s)$. To show that $\lambda E \in \mathfrak{M}(\mathcal{H}^s)$, note that for an arbitrary subset A of $\mathbb{R}^n$ we have by (a) of Theorem 28.1

$$\mathcal{H}^s(A \cap \lambda E) = \mathcal{H}^s\left(\lambda\left(\tfrac{1}{\lambda}A\right) \cap \lambda E\right) = \mathcal{H}^s\left(\lambda\left(\tfrac{1}{\lambda}A \cap E\right)\right) = |\lambda|^s \mathcal{H}^s\left(\tfrac{1}{\lambda}A \cap E\right)$$

and similarly

$$\mathcal{H}^s\left(A \cap (\lambda E)^c\right) = \mathcal{H}^s\left(\lambda\left(\tfrac{1}{\lambda}A\right) \cap \lambda E^c\right) = \mathcal{H}^s\left(\lambda\left(\tfrac{1}{\lambda}A \cap E^c\right)\right) = |\lambda|^s \mathcal{H}^s\left(\tfrac{1}{\lambda}A \cap E^c\right).$$

Thus

$$\begin{aligned}\mathcal{H}^s(A \cap \lambda E) + \mathcal{H}^s\left(A \cap (\lambda E)^c\right) &= |\lambda|^s\left\{\mathcal{H}^s\left(\tfrac{1}{\lambda}A \cap E\right) + \mathcal{H}^s\left(\tfrac{1}{\lambda}A \cap E^c\right)\right\} \\ &= |\lambda|^s \mathcal{H}^s\left(\tfrac{1}{\lambda}A\right) = \mathcal{H}^s(A)\end{aligned}$$

by the fact that $E \in \mathfrak{M}(\mathcal{H}^s)$ and by (a) of Theorem 28.1. This shows that $\lambda E \in \mathfrak{M}(\mathcal{H}^s)$.

Conversely if $\lambda E \in \mathfrak{M}(\mathcal{H}^s)$, then since $E = \lambda^{-1}(\lambda E)$, we have $E \in \mathfrak{M}(\mathcal{H}^s)$ by our result above. ∎

Definition 28.3. **(a)** *A mapping T of a set $E \in \mathfrak{P}(\mathbb{R}^n)$ into $\mathbb{R}^m$ is called a Hölder mapping of exponent $\alpha > 0$ with coefficient $c > 0$ if it satisfies the condition*

$$|T(x) - T(y)| \leq c|x - y|^\alpha \quad \text{for } x, y \in E.$$

(b) *A Hölder mapping of exponent 1 is called a Lipschitz mapping.*

Let us note that a Hölder mapping of a set $E \in \mathfrak{P}(\mathbb{R}^n)$ is uniformly continuous on E. Indeed for an arbitrary $\varepsilon > 0$, if we let $\delta = \left(\frac{\varepsilon}{c}\right)^{1/\alpha}$, then for any $x, y \in E$ such that $|x - y| < \delta$, we have $|T(x) - T(y)| \leq c|x - y|^\alpha < c\left(\frac{\varepsilon}{c}\right) = \varepsilon$.

Theorem 28.4. *Let T be a Hölder mapping of exponent $\alpha > 0$ with coefficient $c > 0$ of a set $E \in \mathfrak{P}(\mathbb{R}^n)$ into $\mathbb{R}^m$. Then*
(a) *$\mathcal{H}^{s/\alpha}(T(E)) \leq c^{s/\alpha}\mathcal{H}^s(E)$, that is, $\mathcal{H}^s(T(E)) \leq c^s\mathcal{H}^{\alpha s}(E)$, for every $s \in [0, \infty)$.*
(b) *$\dim_H T(E) \leq \frac{1}{\alpha}\dim_H E$.*
In particular if T is a Lipschitz mapping with coefficient $c > 0$, then
(c) *$\mathcal{H}^s(T(E)) \leq c^s\mathcal{H}^s(E)$ for every $s \in [0, \infty)$.*
(d) *$\dim_H T(E) \leq \dim_H E$.*

Proof. 1. For $\delta \in (0, \infty)$, let $(V_i : i \in \mathbb{N})$ be a δ-cover of E. Now for $x, y \in E \cap V_i$, we have $|T(x) - T(y)| \leq c|x - y|^\alpha \leq c|V_i|^\alpha$ so that $|T(E \cap V_i)| \leq c|V_i|^\alpha \leq c\delta^\alpha$. Since $E \subset \bigcup_{i \in \mathbb{N}} V_i$, we have $E \subset \bigcup_{i \in \mathbb{N}}(E \cap V_i)$. This implies $T(E) \subset \bigcup_{i \in \mathbb{N}} T(E \cap V_i)$. Thus $(T(E \cap V_i) : i \in \mathbb{N})$ is a $c\delta^\alpha$-cover of $T(E)$. Therefore for every $s \in [0, \infty)$, we have

$$\mathcal{H}^{s/\alpha}_{c\delta^\alpha}(T(E)) \leq \sum_{i \in \mathbb{N}} |T(E \cap V_i)|^{s/\alpha} \leq \sum_{i \in \mathbb{N}} \left(c|V_i|^\alpha\right)^{s/\alpha} = c^{s/\alpha} \sum_{i \in \mathbb{N}} |V_i|^s.$$

By the arbitrariness of the δ-cover $(V_i : i \in \mathbb{N})$ of E, we have $\mathcal{H}^{s/\alpha}_{c\delta^\alpha}(T(E)) \leq c^{s/\alpha}\mathcal{H}^s_\delta(E)$. Then letting $\delta \downarrow 0$, we have $\mathcal{H}^{s/\alpha}(T(E)) \leq c^{s/\alpha}\mathcal{H}^s(E)$. This proves (a).

2. By (a), we have $\mathcal{H}^{s/\alpha}(T(E)) \leq c^{s/\alpha}\mathcal{H}^s(E)$. Since $c^{s/\alpha} > 0$, the inequality implies that $\{s \in [0, \infty) : \mathcal{H}^s(E) = 0\} \subset \{s \in [0, \infty) : \mathcal{H}^{s/\alpha}(T(E)) = 0\}$. Then by (d) of Theorem 27.35, we have

$$\begin{aligned}
\dim_H T(E) &= \inf\left\{s \in [0, \infty) : \mathcal{H}^s(T(E)) = 0\right\} \\
&= \inf\left\{\tfrac{s}{\alpha} \in [0, \infty) : \mathcal{H}^{s/\alpha}(T(E)) = 0\right\} \\
&= \tfrac{1}{\alpha}\inf\left\{s \in [0, \infty) : \mathcal{H}^{s/\alpha}(T(E)) = 0\right\} \\
&\leq \tfrac{1}{\alpha}\inf\left\{s \in [0, \infty) : \mathcal{H}^s(E) = 0\right\} \\
&= \tfrac{1}{\alpha}\dim_H E. \;\blacksquare
\end{aligned}$$

The result above for Hölder mappings has the following extension.

Theorem 28.5. *Let T and S be two mappings of $\mathbb{R}^n$ into $\mathbb{R}^m$ satisfying the condition that there exist $c > 0$ and $\alpha > 0$ such that*

$$|T(x) - T(y)| \leq c|S(x) - S(y)|^\alpha \quad \text{for } x, y \in \mathbb{R}^n.$$

Then for every $E \in \mathfrak{P}(\mathbb{R}^n)$, we have

(a) $\mathcal{H}^{s/\alpha}\big(T(E)\big) \leq c^{s/\alpha}\mathcal{H}^s\big(S(E)\big)$ *for every $s \in [0, \infty)$.*

(b) $\dim_H T(E) \leq \frac{1}{\alpha}\dim_H S(E)$.

Proof Let $s \in [0, \infty)$. For arbitrary $\varepsilon > 0$ and $\delta \in (0, \infty]$, let $(V_i : i \in \mathbb{N})$ be a δ-cover of $S(E) \subset \mathbb{R}^m$ such that $\sum_{i\in\mathbb{N}} |V_i|^s \leq \mathcal{H}^s_\delta\big(S(E)\big) + \varepsilon \leq \mathcal{H}^s\big(S(E)\big) + \varepsilon$. Let $(W_i : i \in \mathbb{N})$ be a sequence of sets in $\mathbb{R}^m$ defined by $W_i = T\left(S^{-1}(V_i)\right)$ for $i \in \mathbb{N}$. Let us show that the sequence is a $c\delta^\alpha$-cover of $T(E)$. Now since $S(E) \subset \bigcup_{i\in\mathbb{N}} V_i$, we have

$$E \subset S^{-1}\big(S(E)\big) \subset S^{-1}\left(\bigcup_{i\in\mathbb{N}} V_i\right) = \bigcup_{i\in\mathbb{N}} S^{-1}(V_i).$$

Then $T(E) \subset T\big(\bigcup_{i\in\mathbb{N}} S^{-1}(V_i)\big) = \bigcup_{i\in\mathbb{N}} T\big(S^{-1}(V_i)\big) = \bigcup_{i\in\mathbb{N}} W_i$. To estimate $|W_i|$, let $w', w'' \in W_i$. Then there exist $z', z'' \in S^{-1}(V_i)$ such that $w' = T(z')$ and $w'' = T(z'')$. We have $|w' - w''| = |T(z') - T(z'')| \leq c|S(z') - S(z'')|^\alpha$. Since $z', z'' \in S^{-1}(V_i)$, we have $S(z'), S(z'') \in V_i$ so that $|S(z') - S(z'')| \leq \delta$. Thus $|w' - w''| \leq c\delta^\alpha$. This implies $|W_i| \leq c\delta^\alpha$. Therefore $(W_i : i \in \mathbb{N})$ is a $c\delta^\alpha$-cover of $T(E)$. We also have $|w' - w''| \leq c|S(z') - S(z'')|^\alpha \leq c|V_i|^\alpha$ and thus $|W_i| \leq c|V_i|^\alpha$. Therefore

$$\mathcal{H}^{s/\alpha}_{c\delta^\alpha}\big(T(E)\big) \leq \sum_{i\in\mathbb{N}} |W_i|^{s/\alpha} \leq \sum_{i\in\mathbb{N}} c^{s/\alpha}|V_i|^s \leq c^{s/\alpha}\left\{\mathcal{H}^s\big(S(E)\big) + \varepsilon\right\}.$$

From the arbitrariness of $\varepsilon > 0$, we have $\mathcal{H}^{s/\alpha}_{c\delta^\alpha}\big(T(E)\big) \leq c^{s/\alpha}\mathcal{H}^s\big(S(E)\big)$. Letting $\delta \downarrow 0$, we have $\mathcal{H}^{s/\alpha}\big(T(E)\big) \leq c^{s/\alpha}\mathcal{H}^s\big(S(E)\big)$. This proves (a). Then (b) follows by the same argument as in the proof of (b) of Theorem 28.4. ■

As an example of Lipschitz mapping we have the following:

Proposition 28.6. *Let T be a mapping of class $C^1(\Omega)$ of an open set Ω in $\mathbb{R}^n$ into $\mathbb{R}^m$, that is, the first order partial derivatives of the component functions $g_1, \ldots, g_m$ of T are all continuous on Ω. Let E be a convex set contained in Ω. If all the first order partial derivatives of $g_1, \ldots, g_m$ are bounded on E, then T is a Lipschitz mapping of E.*

Proof. By our assumption on the first order partial derivatives of $g_1, \ldots, g_m$, there exists a constant $B > 0$ such that $\left|\frac{\partial g_i}{\partial x_j}(p)\right| \leq B$ for every $p \in E$, $i = 1, \ldots, m$, and $j = 1, \ldots, n$. Let $p \in E$ and let $h = (h_1, \ldots, h_n) \in \mathbb{R}^n$ be such that $p + h \in E$. For each $i = 1, \ldots, m$, by the Mean Value Theorem for functions of n variables applied to g_i, there exists a point p_i^* on the line segment joining p to $p+h$ such that $g_i(p+h) - g_i(p) = \sum_{j=1}^n \frac{\partial g_i}{\partial x_j}(p_i^*)h_j$. Thus we have

$$\left|g_i(p+h) - g_i(p)\right|^2 \leq \sum_{j=1}^n \left[\frac{\partial g_i}{\partial x_j}(p_i^*)\right]^2 \cdot \sum_{j=1}^n h_j^2 \leq nB^2|h|^2.$$

Therefore we have

$$|T(p+h)-T(p)|^2 = \sum_{i=1}^{m} |g_i(p+h)-g_i(p)|^2 \le mnB^2|h|^2.$$

Then $|T(p+h)-T(p)| \le \sqrt{mn}B|h|$. Thus we have $|T(x)-T(y)| \le \sqrt{mn}B|x-y|$ for $x, y \in E$. ∎

Definition 28.7. (a) *We say that a mapping T of a set $E \in \mathfrak{P}(\mathbb{R}^n)$ into $\mathbb{R}^m$ is a bi-Lipschitz mapping with coefficients c_1 and c_2 where $0 < c_1 \le c_2$ if T satisfies the condition*

$$c_1|x-y| \le |T(x)-T(y)| \le c_2|x-y| \quad \textit{for } x, y \in E.$$

(b) *We say that T is an isometry if*

$$|T(x)-T(y)| = |x-y| \quad \textit{for } x, y \in E.$$

(Thus an isometry is a bi-Lipschitz mapping with coefficients $c_1 = c_2 = 1$.)

Observation 28.8. If T is a bi-Lipschitz mapping with coefficients c_1 and c_2 of a set $E \in \mathfrak{P}(\mathbb{R}^n)$ into $\mathbb{R}^m$, then T is a homeomorphism of E onto $T(E)$ and the inverse mapping T^{-1} of T is a bi-Lipschitz mapping with coefficients c_2^{-1} and c_1^{-1} of $T(E)$ into $\mathbb{R}^n$.

Proof. T is a Lipschitz mapping with coefficient c_2 of $E \in \mathfrak{P}(\mathbb{R}^n)$ into $\mathbb{R}^m$. Thus T is a continuous mapping of E into $\mathbb{R}^m$. The condition $c_1|x-y| \le |T(x)-T(y)|$ for $x, y \in E$ implies that if $T(x) = T(y)$ then $x = y$. Thus T is a one-to-one mapping. Let T^{-1} be the inverse mapping of T mapping $T(E) \subset \mathbb{R}^m$ one-to-one onto E. For every $x', y' \in T(E)$, we have unique $x, y \in E$ such that $x = T^{-1}(x')$ and $y = T^{-1}(y')$. The conditions $c_1|x-y| \le |T(x)-T(y)|$ and $|T(x)-T(y)| \le c_2|x-y|$ are rewritten as $|T^{-1}(x')-T^{-1}(y')| \le c_1^{-1}|x'-y'|$ and $c_2^{-1}|x'-y'| \le |T^{-1}(x')-T^{-1}(y')|$ respectively. Thus T^{-1} is a bi-Lipschitz mapping with coefficients c_2^{-1} and c_1^{-1} of $T(E)$ into $\mathbb{R}^n$. Since T and T^{-1} are Lipschitz mappings, they are continuous mappings. This shows that T is a homeomorphism. ∎

Remark 28.9. (a) A linear transformation T of $\mathbb{R}^n$ into $\mathbb{R}^m$ is a Lipschitz mapping. If T is non-singular, then it is a bi-Lipschitz mapping.
(b) Orthogonal transformations on $\mathbb{R}^n$, including rotations on $\mathbb{R}^n$ and reflexions on $\mathbb{R}^n$, are isometries.
(c) A translation on $\mathbb{R}^n$ is an isometry.

Proof. Let us prove (a). Let T be a linear transformation of $\mathbb{R}^n$ into $\mathbb{R}^m$. If T is identically vanishing on $\mathbb{R}^n$, then $|T(x)-T(y)| \le c|x-y|$ for $x, y \in \mathbb{R}^n$ holds trivially for every $c > 0$. If T is not identically vanishing on $\mathbb{R}^n$, then $\|T\| = \sup_{|x|\le 1}|T(x)| > 0$ and $|T(x)-T(y)| \le \|T\||x-y|$ for $x, y \in \mathbb{R}^n$. This shows that T is a Lipschitz mapping.

Suppose T is non-singular on $\mathbb{R}^n$. Then T maps $\mathbb{R}^n$ one-to-one onto the n-dimensional linear subspace $T(\mathbb{R}^n)$ of $\mathbb{R}^m$. The inverse mapping T^{-1} is defined on $T(\mathbb{R}^n)$ and is a linear transformation of $T(\mathbb{R}^n)$ to $\mathbb{R}^n$. Let $x, y \in \mathbb{R}^n$. Then $T(x), T(y) \in T(\mathbb{R}^n)$ and $\left|T^{-1}T(x) - T^{-1}T(y)\right| \leq \|T^{-1}\|\left|T(x) - T(y)\right|$, that is, $|x - y| \leq \|T^{-1}\|\left|T(x) - T(y)\right|$. Thus we have $\|T^{-1}\|^{-1}|x - y| \leq |T(x) - T(y)| \leq \|T\|\left|x - y\right|$. This shows that T is a bi-Lipschitz mapping. ∎

Theorem 28.10. *Let T be a bi-Lipschitz mapping with coefficients c_1 and c_2 of a set $E \in \mathfrak{P}(\mathbb{R}^n)$ into $\mathbb{R}^m$. Then*
(a) $c_1^s \mathcal{H}^s(E) \leq \mathcal{H}^s\big(T(E)\big) \leq c_2^s \mathcal{H}^s(E)$ *for every* $s \in [0, \infty)$.
(b) $\dim_H T(E) = \dim_H E$.
In particular if T is an isomery, then
(c) $\mathcal{H}^s\big(T(E)\big) = \mathcal{H}^s(E)$ *for every* $s \in [0, \infty)$.

Proof. If T be a bi-Lipschitz mapping with coefficients c_1 and c_2, then T is a Lipschitz mapping of E with coefficient c_2 so that by (c) of Theorem 28.4 we have the estimate $\mathcal{H}^s(T(E)) \leq c_2^s \mathcal{H}^s(E)$ for every $s \in [0, \infty)$. By Observation 28.8, T^{-1} is a Lipschitz mapping of $T(E)$ with coefficient c_1^{-1} so that by (c) of Theorem 28.4 again we have $\mathcal{H}^s\big(T^{-1}\big(T(E)\big)\big) \leq c_1^{-s}\mathcal{H}^s\big(T(E)\big)$, that is, $c_1^s \mathcal{H}^s(E) \leq \mathcal{H}^s\big(T(E)\big)$, for every $s \in [0, \infty)$. This proves (a). Also by (d) of Theorem 28.4, we have $\dim_H T(E) \leq \dim_H E$ as well as $\dim_H T^{-1}\big(T(E)\big) \leq \dim_H T(E)$, that is, $\dim_H E \leq \dim_H T(E)$. This proves (b). Since an isometry is a bi-Lipschitz mapping with coefficients $c_1 = c_2 = 1$, (c) follows from (a). ∎

Theorem 28.11. (Invariance of $\mathfrak{M}(\mathcal{H}^s)$ under Isometry) *Let $E \in \mathfrak{P}(\mathbb{R}^n)$ and let T be an isometry on $\mathbb{R}^n$. Then $T(E) \in \mathfrak{M}(\mathcal{H}^s)$ if and only if $E \in \mathfrak{M}(\mathcal{H}^s)$.*

Proof. By definition $\mathfrak{M}(\mathcal{H}^s)$ consists of all $E \in \mathfrak{P}(\mathbb{R}^n)$ such that for every $A \in \mathfrak{P}(\mathbb{R}^n)$ we have $\mathcal{H}^s(A) = \mathcal{H}^s(A \cap E) + \mathcal{H}^s(A \cap E^c)$. Suppose $E \in \mathfrak{M}(\mathcal{H}^s)$. Let us show that $T(E) \in \mathfrak{M}(\mathcal{H}^s)$. Now if T is an isometry on $\mathbb{R}^n$, then T is a one-to-one mapping of $\mathbb{R}^n$ onto $\mathbb{R}^n$ so that the inverse mapping T^{-1} exists and furthermore T^{-1} is an isometry. For an arbitrary $A \in \mathfrak{P}(\mathbb{R}^n)$, (c) of Theorem 28.10 implies

$$\begin{aligned}\mathcal{H}^s\big(A \cap T(E)\big) &= \mathcal{H}^s\big(TT^{-1}(A) \cap T(E)\big)\\ &= \mathcal{H}^s\big(T\big(T^{-1}(A) \cap E\big)\big) = \mathcal{H}^s\big(T^{-1}(A) \cap E\big),\end{aligned}$$

and similarly

$$\begin{aligned}\mathcal{H}^s\big(A \cap T(E)^c\big) &= \mathcal{H}^s\big(TT^{-1}(A) \cap T(E^c)\big)\\ &= \mathcal{H}^s\big(T\big(T^{-1}(A) \cap E^c\big)\big) = \mathcal{H}^s\big(T^{-1}(A) \cap E^c\big).\end{aligned}$$

Then by the fact that $E \in \mathfrak{M}(\mathcal{H}^s)$ and by (c) of Theorem 28.10 we have

$$\begin{aligned}\mathcal{H}^s\big(A \cap T(E)\big) + \mathcal{H}^s\big(A \cap T(E)^c\big) &= \mathcal{H}^s\big(T^{-1}(A) \cap E\big) + \mathcal{H}^s\big(T^{-1}(A) \cap E^c\big)\\ &= \mathcal{H}^s\big(T^{-1}(A)\big) = \mathcal{H}^s(A).\end{aligned}$$

This shows that $T(E) \in \mathfrak{P}(\mathbb{R}^n)$ if $E \in \mathfrak{M}(\mathcal{H}^s)$. Conversely if we assume that $T(E) \in \mathfrak{M}(\mathcal{H}^s)$, then since $E = T^{-1}\big(T(E)\big)$ and since T^{-1} is an isometry, we have $E \in \mathfrak{M}(\mathcal{H}^s)$ by our result above. ∎

[II] 1-dimensional Hausdorff Measure

We show that for every $E \in \mathfrak{P}(\mathbb{R})$, we have $\mathcal{H}^1(E) = \mu_L^*(E)$. The difference between the 1-dimensional Hausdorff (outer) measure $\mathcal{H}^1$ and the 1-dimensional Lebesgue outer measure μ_L^* is that while μ_L^* is an outer measure on $\mathbb{R}$, $\mathcal{H}^1$ is an outer measure on $\mathbb{R}^n$ for any $n \in \mathbb{N}$.

Theorem 28.12. *For every $E \in \mathfrak{P}(\mathbb{R})$, $\mathcal{H}_\delta^1(E)$ is independent of $\delta \in (0, \infty]$ and in fact $\mathcal{H}_\delta^1(E) = \mu_L^*(E)$ for every $\delta \in (0, \infty]$. Thus $\mathcal{H}^1(E) = \mu_L^*(E)$.*

Proof. A subset of $\mathbb{R}$ is a convex set if and only if it is an interval. Thus the collection of closed convex sets in $\mathbb{R}$ is the collection of closed intervals $\mathfrak{I}_c$. Then by Theorem 27.25, for every $E \in \mathfrak{P}(\mathbb{R})$ and $\delta \in (0, \infty]$, we have

$$\mathcal{H}_\delta^1(E) = \inf\left\{\sum_{i\in\mathbb{N}} |J_i| : J_i \in \mathfrak{I}_c, |J_i| \le \delta, i \in \mathbb{N} \text{ and } \bigcup_{i\in\mathbb{N}} J_i \supset E\right\}.$$

Recall that according to Observation 3.2, we have

$$\mu_L^*(E) = \inf\left\{\sum_{i\in\mathbb{N}} |J_i| : J_i \in \mathfrak{I}_c, i \in \mathbb{N} \text{ and } \bigcup_{i\in\mathbb{N}} J_i \supset E\right\}.$$

The collection on which we are taking infimum for $\mathcal{H}_\delta^1(E)$ is contained in the collection on which we are taking infimum for $\mu_L^*(E)$. Thus $\mu_L^*(E) \le \mathcal{H}_\delta^1(E)$. On the other hand every closed interval J in $\mathbb{R}$ is the union of at most countably many closed interval J_p, $p \in \mathbb{N}$, with disjoint interiors such that $|J_p| \le \delta$ for $p \in \mathbb{N}$ and $|J| = \sum_{p\in\mathbb{N}} |J_p|$. Thus for every sequence $(J_i : i \in \mathbb{N})$ in $\mathfrak{I}_c$ such that $\bigcup_{n\in\mathbb{N}} J_i \supset E$, there exists a sequence $(J_i' : i \in \mathbb{N})$ in $\mathfrak{I}_c$ such that $|J_i'| \le \delta$ for $i \in \mathbb{N}$, $\bigcup_{n\in\mathbb{N}} J_i' \supset E$, and $\sum_{i\in\mathbb{N}} |J_i'| = \sum_{i\in\mathbb{N}} |J_i|$. This implies $\mathcal{H}_\delta^1(E) \le \mu_L^*(E)$. Therefore $\mathcal{H}_\delta^1(E) = \mu_L^*(E)$. Then $\mathcal{H}^1(E) = \lim_{\delta\downarrow 0} \mathcal{H}_\delta^1(E) = \mu_L^*(E)$. ∎

Remark 28.13. We showed in Theorem 28.12 that $\mathcal{H}_\delta^1(E) = \mu_L^*(E)$ for every $\delta \in (0, \infty]$ and in particular $\mathcal{H}_\infty^1(E) = \mu_L^*(E)$ as long as $E \in \mathfrak{P}(\mathbb{R})$. For $E \in \mathfrak{P}(\mathbb{R}^n)$ where $n > 1$, $\mu_L^*(E)$ is undefined.

Example 1. Consider two line segments in $\mathbb{R}^2$ given by $E_1 = \{(x, 0) \in \mathbb{R}^2 : x \in (0, 1]\}$ and $E_2 = \{(0, y) \in \mathbb{R}^2 : y \in (0, 1]\}$. Let $E = E_1 \cup E_2$. If we regard E_1 as a subset of $\mathbb{R}$, then by Theorem 28.12 we have $\mathcal{H}^1(E_1) = \mu_L^*(E_1) = 1$. By the invariance of $\mathcal{H}^1$ under an isometry ((c) of Theorem 28.10), we have $\mathcal{H}^1(E_2) = \mathcal{H}^1(E_1) = 1$. Then since $\mathcal{H}^1$ is a measure on $\mathfrak{B}_{\mathbb{R}^2}$ and since $E_1 \cap E_2 = \emptyset$, we have $\mathcal{H}^1(E) = \mathcal{H}^1(E_1) + \mathcal{H}^1(E_2) = 2$. On the other hand the triangle in $\mathbb{R}^2$ defined by $V = \{(x, y) \in \mathbb{R}^2 : x, y \ge 0, x + y \le 1\}$ contains E so that $\mathcal{H}_\infty^1(E) \le |V| = \sqrt{2}$. Thus $\mathcal{H}_\infty^1(E) < \mathcal{H}^1(E)$. Note that since $\mathcal{H}^1(E) = 2 \in (0, \infty)$, we have $\dim_H E = 1$ by (e) of Theorem 27.35. Note also that $\mu_L^2(E) = 0$ while $\mu_L(E)$ is undefined since E is not a subset of $\mathbb{R}$.

Example 2. If E is a finite set in $\mathbb{R}^n$ with k elements, then $\mathcal{H}^0(E) = k$ by Proposition 27.31. On the other hand for any set $V \in \mathfrak{P}(\mathbb{R}^n)$ such that $V \supset E$ we have $|V|^0 = 1$. Thus $\mathcal{H}_\infty^s(E) \le 1$.

[III] Hausdorff Measure of Jordan Curves

Definition 28.14. *A Jordan curve G in $\mathbb{R}^n$ is the image of a one-to-one continuous mapping g of a finite closed interval $[a, b] \subset \mathbb{R}$ into $\mathbb{R}^n$. We call the mapping g a parametric representation of the Jordan curve G.*

Observation 28.15.

(a) A parametric representation of a Jordan curve G is not unique, that is, there exist distinct one-to-one continuous mappings g and h of $[a, b]$ into $\mathbb{R}^n$ such that we have

$$G = \{x \in \mathbb{R}^n : x = g(t), t \in [a, b]\} = \{x \in \mathbb{R}^n : x = h(t), t \in [a, b]\}.$$

(b) A Jordan curve has two endpoints and is not self-intersecting.

(c) A Jordan curve is a compact connected subset of $\mathbb{R}^n$.

(d) If $g : [a, b] \to \mathbb{R}^n$ is a parametric representation of a Jordan curve G in $\mathbb{R}^n$, then g is a homeomorphism of $[a, b]$ onto G.

Proof. Let φ be an arbitrary one-to-one continuous mapping of $[a, b]$ onto $[a, b]$. If $g : [a, b] \to \mathbb{R}^n$ is a parametric representation of a Jordan curve G in $\mathbb{R}^n$, then $h = g \circ \varphi$ is also a parametric representation of G. This proves (a). (b) is immediate from Definition 28.14. Let $g : [a, b] \to \mathbb{R}^n$ be a parametric representation of G. Since $[a, b]$ is a compact set in $\mathbb{R}$ and g is continuous on $[a, b]$, $G = g([a, b])$ is a compact set in $\mathbb{R}^n$. Since $[a, b]$ is a connected set in $\mathbb{R}$, the continuity of g implies that $G = g([a, b])$ is a connected set in $\mathbb{R}^n$. Finally since g is a one-to-one continuous mapping of the compact set $[a, b]$ in $\mathbb{R}$ into $\mathbb{R}^n$, g is a homeomorphism of $[a, b]$ to $G = g([a, b])$ by Proposition 4.35. ∎

Definition 28.16. *Let f be a mapping of $[a, b]$ into $\mathbb{R}^n$. For a partition of $[a, b]$ given by $\mathcal{P} : a = t_0 < \cdots < t_m = b$, let*

$$V_a^b(f, \mathcal{P}) = \sum_{k=1}^{m} |f(t_k) - f(t_{k-1})| \in [0, \infty).$$

Let $\mathfrak{P}_{a,b}$ be the collection of all partitions of $[a, b]$. Let

$$V_a^b(f) = \sup_{\mathcal{P} \in \mathfrak{P}_{a,b}} V_a^b(f, \mathcal{P}) \in [0, \infty].$$

We say that the mapping f is of bounded variation on $[a, b]$ and write $f \in BV([a, b])$ if $V_a^b(f) < \infty$. If $f \in BV([a, b])$, we call the function

$$v_f(t) = V_a^t(f) \quad \textit{for } t \in [a, b]$$

the total variation function of f.

Lemma 28.17. *Let f be a mapping of $[a, b]$ into $\mathbb{R}^n$ and let $f_1, \ldots, f_n$ be the component functions of f. Then for each $i = 1, \ldots, n$, we have*

$$V_a^b(f_i) \le V_a^b(f) \le \sum_{i=1}^{n} V_a^b(f_i). \tag{1}$$

In particular, f is a mapping of bounded variation on $[a, b]$ if and only if $f_1, \dots, f_n$ are all functions of bounded variation on $[a, b]$.

Proof. Let $\mathcal{P} \in \mathfrak{P}_{a,b}$ be given by $\mathcal{P} : a = t_0 < \cdots < t_m = b$. Now $|f(t_k) - f(t_{k-1})| = \left\{\sum_{i=1}^n |f_i(t_k) - f_i(t_{k-1})|^2\right\}^{1/2}$ so that for $i = 1, \dots, n$ we have

$$|f_i(t_k) - f_i(t_{k-1})| \le |f(t_k) - f(t_{k-1})| \le \sum_{i=1}^n |f_i(t_k) - f_i(t_{k-1})|.$$

Summing over $k = 1, \dots, m$, we have $V_a^b(f_i, \mathcal{P}) \le V_a^b(f, \mathcal{P}) \le \sum_{i=1}^n V_a^b(f_i, \mathcal{P})$ so that

$$\sup_{\mathcal{P} \in \mathfrak{P}_{a,b}} V_a^b(f_i, \mathcal{P}) \le \sup_{\mathcal{P} \in \mathfrak{P}_{a,b}} V_a^b(f, \mathcal{P}) \le \sum_{i=1}^n \sup_{\mathcal{P} \in \mathfrak{P}_{a,b}} V_a^b(f_i, \mathcal{P}).$$

This proves (1). Then by (1) we have $V_a^b(f) < \infty$ if and only if $V_a^b(f_i) < \infty$ for all $i = 1, \dots, n$. ∎

Lemma 28.18. *If f is a Lipschitz mapping of $[a, b]$ into $\mathbb{R}^n$, then f is a mapping of bounded variation on $[a, b]$.*

Proof. If f is a Lipschitz mapping, then there exists $c > 0$ such that $|f(t') - f(t'')| \le c|t' - t''|$ for $t', t'' \in [a, b]$. Let $\mathcal{P} \in \mathfrak{P}_{a,b}$ be given by $\mathcal{P} : a = t_0 < \cdots < t_m = b$. Then

$$V_a^b(f, \mathcal{P}) = \sum_{k=1}^m |f(t_k) - f(t_{k-1})| \le c \sum_{k=1}^m |t_k - t_{k-1}| = c(b - a)$$

and $V_a^b(f) = \sup_{\mathcal{P} \in \mathfrak{P}_{a,b}} V_a^b(f, \mathcal{P}) \le c(b - a) < \infty$. This shows that f is a mapping of bounded variation on $[a, b]$. ∎

Lemma 28.19. *Let f be a mapping of $[a, b]$ of bounded variation into $\mathbb{R}^n$. If f is continuous at some $t_0 \in [a, b]$, then its total variation function v_f is continuous at t_0.*

Proof. Let $t', t'' \in [a, b]$, $t' < t''$. By the same argument as in proving (1) in Lemma 28.17, we have

$$(1) \qquad V_{t'}^{t''}(f_i) \le V_{t'}^{t''}(f) \le \sum_{i=1}^n V_{t'}^{t''}(f_i).$$

Since $V_{t'}^{t''}(f_i) = v_{f_i}(t'') - v_{f_i}(t')$ by (b) of Lemma 12.15 and $V_{t'}^{t''}(f) = v_f(t'') - v_f(t')$ similarly, we have

$$(2) \qquad v_{f_i}(t'') - v_{f_i}(t') \le v_f(t'') - v_f(t') \le \sum_{i=1}^n \left\{v_{f_i}(t'') - v_{f_i}(t')\right\}.$$

Now if f is continuous at some $t_0 \in [a, b]$, then its component functions $f_1, \dots, f_n$ are all continuous at t_0. This implies that their total variation functions $v_{f_1}, \dots, v_{f_n}$ are all continuous at t_0 by Theorem 12.22. Then the second inequality in (2) implies that v_f is continuous at t_0. ∎

Definition 28.20. *Let G be a Jordan curve in $\mathbb{R}^n$ and let $g : [a, b] \to \mathbb{R}^n$ be a parametric representation of G. We define the length of G by*

$$\mathcal{L}(G) = V_a^b(g),$$

that is, $\mathcal{L}(G) = \sup_{\mathcal{P} \in \mathfrak{P}_{a,b}} V_a^b(g, \mathcal{P})$ where $V_a^b(g, \mathcal{P}) = \sum_{k=1}^m |g(t_k) - g(t_{k-1})|$ for the partition $\mathcal{P} : a = t_0 < \cdots < t_m = b$ of $[a, b]$. We say that G is rectifiable if $\mathcal{L}(G) < \infty$, that is, $g \in BV([a, b])$.

Remark 28.21. The length $\mathcal{L}(G)$ of a Jordan curve G in $\mathbb{R}^n$ does not depend on the parametric representation $g : [a, b] \to \mathbb{R}^n$ of G used in defining $\mathcal{L}(G)$.

Proof. Let $g : [a, b] \to \mathbb{R}^n$ and $h : [a, b] \to \mathbb{R}^n$ be two parametric representations of G. Let $\mathcal{P} : a = t_0 < \cdots < t_m = b$ be an arbitrary partition of $[a, b]$ and consider $V_a^b(g, \mathcal{P}) = \sum_{k=1}^m |g(t_k) - g(t_{k-1})|$. Since g and h are one-to-one mappings of $[a, b]$ onto G, there exists unique $s_k \in [a, b]$ such that $g(t_k) = h(s_k)$ for $k = 0, \ldots, m$. For the partition $\mathcal{P}' : a = s_0 < \cdots < s_m = b$ of $[a, b]$, we have $V_a^b(g, \mathcal{P}) = V_a^b(h, \mathcal{P}')$. This shows that $\{V_a^b(g, \mathcal{P}) : \mathcal{P} \in \mathfrak{P}_{a,b}\} \subset \{V_a^b(h, \mathcal{P}) : \mathcal{P} \in \mathfrak{P}_{a,b}\}$. Interchanging the roles of g and h, we have $\{V_a^b(h, \mathcal{P}) : \mathcal{P} \in \mathfrak{P}_{a,b}\} \subset \{V_a^b(g, \mathcal{P}) : \mathcal{P} \in \mathfrak{P}_{a,b}\}$. Thus we have $\{V_a^b(g, \mathcal{P}) : \mathcal{P} \in \mathfrak{P}_{a,b}\} = \{V_a^b(h, \mathcal{P}) : \mathcal{P} \in \mathfrak{P}_{a,b}\}$. Then $\sup_{\mathcal{P} \in \mathfrak{P}_{a,b}} V_a^b(g, \mathcal{P}) = \sup_{\mathcal{P} \in \mathfrak{P}_{a,b}} V_a^b(h, \mathcal{P})$. This shows the independence of $\mathcal{L}(G)$ from g. ∎

Observation 28.22. Let $g : [a, b] \to \mathbb{R}^n$ be a parametric representation of a Jordan curve G in $\mathbb{R}^n$. For $t', t'' \in [a, b]$, $t' < t''$, let

$$G_{g(t'),g(t'')} = g([t', t'']), \tag{1}$$

that is, the part of G that is the image of the subinterval $[t', t'']$ of $[a, b]$ by the mapping g. Then $G_{g(t'),g(t'')}$ is a Jordan curve in $\mathbb{R}^n$ with $g : [t', t''] \to \mathbb{R}^n$ as a parametric representation. Let us assume that G is rectifiable. Then $G_{g(t'),g(t'')}$ is rectifiable. Let us define a function φ on $[a, b]$ by setting

$$\varphi(t) = \mathcal{L}\left(G_{g(a),g(t)}\right) \quad \text{for } t \in [a, b], \tag{2}$$

that is, $\varphi = v_g$, the total variation function of g. Now φ is a real-valued increasing function on $[a, b]$ with $\varphi(a) = 0$ and $\varphi(b) = \mathcal{L}(G)$. Since g is a continuous mapping of $[a, b]$, $\varphi = v_g$ is a continuous function on $[a, b]$ by Lemma 28.19. Since g is a one-to-one mapping, if $t', t'' \in [a, b]$ and $t' < t''$ then $g(t') \neq g(t'')$. Then for at least one of the component functions $g_1, \ldots, g_n$, say g_i, we have $g_i(t') \neq g_i(t'')$. This implies that for the total variation function φ_i of g_i we have $\varphi_i(t') < \varphi_i(t'')$. Then by (2) in the Proof of Lemma 28.19, we have $\varphi(t'') - \varphi(t') \geq \varphi_i(t'') - \varphi_i(t') > 0$. This shows that φ is a strictly increasing function on $[a, b]$.

Now that φ is a strictly increasing continuous function on $[a,b]$ with $\varphi(a) = 0$ and $\varphi(b) = \mathcal{L}(G)$, its inverse function φ^{-1} is a strictly increasing continuous function on $[0, \mathcal{L}(G)]$ mapping $[0, \mathcal{L}(G)]$ onto $[a,b]$. Let us define a mapping γ of $[0, \mathcal{L}(G)]$ into $\mathbb{R}^n$ by

(3) $$\gamma(s) = \left(g \circ \varphi^{-1}\right)(s) \quad \text{for } s \in [0, \mathcal{L}(G)].$$

Then γ is a one-to-one continuous mapping of $[0, \mathcal{L}(G)]$ into $\mathbb{R}^n$ and in fact we have

$$\gamma\left([0, \mathcal{L}(G)]\right) = \left(g \circ \varphi^{-1}\right)\left([0, \mathcal{L}(G)]\right) = g\left([a,b]\right) = G.$$

Thus $\gamma : [0, \mathcal{L}(G)] \to \mathbb{R}^n$ is a parametric representation of G. Let us call the variable $s = \varphi(t)$ for $t \in [a,b]$ the arc length of G and call $\gamma(s)$ for $s \in [0, \mathcal{L}(G)]$ the parametric representation of G by the arc length.

For $s', s'' \in [0, \mathcal{L}(G)]$ and $s' < s''$, let t' and t'' be the unique points in $[a,b]$ such that $s' = \varphi(t')$ and $s'' = \varphi(t'')$. Then $\gamma(s') = g(t')$ and $\gamma(s'') = g(t'')$. Thus by (2) we have

(4) $$\begin{aligned} s'' - s' &= \varphi(t'') - \varphi(t') = \mathcal{L}\left(G_{g(t'),g(t'')}\right) \\ &= \mathcal{L}\left(G_{\gamma(s'),\gamma(s'')}\right) \geq |\gamma(s') - \gamma(s'')|. \end{aligned}$$

This shows that γ is a Lipschitz mapping with coefficient 1.

Theorem 28.23. *Let G be a Jordan curve in $\mathbb{R}^n$. Then $\mathcal{H}^1(G) = \mathcal{L}(G)$.*

Proof. 1. Let $g : [a,b] \to \mathbb{R}^n$ be a parametric representation of a Jordan curve G in $\mathbb{R}^n$. Let $t', t'' \in [a,b]$ and $t' < t''$ and consider $G_{g(t'),g(t'')} = g\left([t', t'']\right)$. Let $[g(t'), g(t'')]$ be the line segment in $\mathbb{R}^n$ joining $g(t')$ to $g(t'')$ and let T be the orthogonal projection of $\mathbb{R}^n$ onto the line containing $[g(t'), g(t'')]$. Then for any $x, y \in \mathbb{R}^n$ we have $|T(x) - T(y)| \leq |x - y|$ so that T is a Lipschitz mapping with coefficient 1. Note also that $T\left(G_{g(t'),g(t'')}\right) \supset [g(t'), g(t'')]$. Thus by (c) of Theorem 28.4, we have

(1) $$\begin{aligned} \mathcal{H}^1\left(G_{g(t'),g(t'')}\right) &\geq \mathcal{H}^1\left(T\left(G_{g(t'),g(t'')}\right)\right) \geq \mathcal{H}^1\left([g(t'), g(t'')]\right) \\ &= \mu_L\left([g(t'), g(t'')]\right) = |g(t') - g(t'')|, \end{aligned}$$

where the first equality is by the fact that $[g(t'), g(t'')] \subset \mathbb{R}$ and $\mathcal{H}^1(E) = \mu_L^*(E)$ for any $E \in \mathfrak{P}(\mathbb{R})$ according to Theorem 28.12. For $\mathcal{P} : a = t_0 < \cdots < t_m = b$, we have

$$V_a^b(g, \mathcal{P}) = \sum_{k=1}^m |g(t_k) - g(t_{k-1})| \leq \sum_{k=1}^m \mathcal{H}^1\left(G_{g(t_{k-1}),g(t_k)}\right) = \mathcal{H}^1(G),$$

where the inequality is by (1) and the last equality is from the fact that $G_{g(t_{k-1}),g(t_k)}$ is a compact set in $\mathbb{R}^n$ by Observation 28.15 so that $G_{g(t_{k-1}),g(t_k)} \in \mathfrak{B}_{\mathbb{R}^n} \subset \mathfrak{M}(\mathcal{H}^1)$ and by the fact that $\bigcup_{k=1}^m G_{g(t_{k-1}),g(t_k)} = G$ and moreover the intersection of any two of the sets $G_{g(t_{k-1}),g(t_k)}$, $k = 1, \ldots, m$, is either an empty set or a singleton for which the $\mathcal{H}^1$ measures are equal to 0. Then we have

(2) $$\mathcal{L}(G) = \sup_{\mathcal{P} \in \mathfrak{P}_{a,b}} V_a^b(g, \mathcal{P}) \leq \mathcal{H}^1(G).$$

2. If G is not rectifiable, then $\mathcal{L}(G) = \infty$ so that $\mathcal{H}^1(G) = \infty$ also by (2) and therefore $\mathcal{H}^1(G) = \mathcal{L}(G)$. Suppose G is rectifiable. Let $\gamma : [0, \mathcal{L}(G)] \to \mathbb{R}^n$ be the parametric representation of G by the arc length. As we noted in Observation 28.22, γ is a Lipschitz mapping with coefficient 1. Thus by (c) of Theorem 28.4 we have

$$(3) \qquad \mathcal{H}^1(G) = \mathcal{H}^1\big(\gamma\left([0, \mathcal{L}(G)]\right)\big) \leq \mathcal{H}^1\big([0, \mathcal{L}(G)]\big) = \mu_L\big([0, \mathcal{L}(G)]\big) = \mathcal{L}(G).$$

where the second equality is by Theorem 28.12. Then with (2) and (3), we have the equality $\mathcal{H}^1(G) = \mathcal{L}(G)$. ∎

§29 Hausdorff Measures of Integral and Fractional Dimensions

[I] Hausdorff Measure of Integral Dimension and Lebesgue Measure

In Theorem 28.12, we showed that for every $E \in \mathfrak{P}(\mathbb{R})$ we have $\mathcal{H}^1(E) = \mu_L^*(E)$. We show next that for every $n \in \mathbb{N}$ there exists a positive constant κ_n such that $\mathcal{H}^n(E) = \kappa_n \big(\mu_L^n\big)^*(E)$ for every $E \in \mathfrak{P}(\mathbb{R}^n)$, that is, the n-dimensional Hausdorff (outer) measure $\mathcal{H}^n$ is a constant multiple of the n-dimensional Lebesgue outer measure $\big(\mu_L^n\big)^*$ on $\mathbb{R}^n$. (One advantage of $\mathcal{H}^n$ over $\big(\mu_L^n\big)^*$ is that while $\big(\mu_L^n\big)^*$ is an outer measure on $\mathbb{R}^n$, $\mathcal{H}^n$ is an outer measure on $\mathbb{R}^m$ for every $m \in \mathbb{N}$.) We shall show that the constant κ_n is equal to the n-dimensional Hausdorff (outer) measure $\mathcal{H}^n(Q)$ of a unit cube Q in $\mathbb{R}^n$. Let us call a ball with diameter 1, that is, with radius $\frac{1}{2}$, a unit ball. We shall show also that the n-dimensional Lebesgue outer measure $\big(\mu_L^n\big)^*(S)$ of a unit ball S in $\mathbb{R}^n$ is equal to $\frac{1}{\kappa_n}$. From this it follows that for a ball B in $\mathbb{R}^n$, we have $\mathcal{H}^n(B) = |B|^n$.

Observation 29.1. In Lemma 27.30 we showed that if Q is a unit cube in the class $\mathfrak{I}^n = \mathfrak{I}^n_{oc} \cup \mathfrak{I}^n_{co} \cup \mathfrak{I}^n_o \cup \mathfrak{I}^n_c$, then $\mathcal{H}^n(Q)$ is a positive number in the range $\big[2^{-n}, n^{n/2}\big]$. We show now that $\mathcal{H}^n(Q)$ has the same value whether Q is in $\mathfrak{I}^n_{oc}$, $\mathfrak{I}^n_{co}$, $\mathfrak{I}^n_o$, or $\mathfrak{I}^n_c$.

Proof. By the translation invariance of the outer measure $\mathcal{H}^n$ implied by (c) of Theorem 28.10, we may assume without loss of generality that our cubes have center at $0 \in \mathbb{R}^n$. Let G_λ be a cube in the class $\mathfrak{I}^n_o$, that is, an open cube, with center at $0 \in \mathbb{R}^n$ and edge $\lambda > 0$ and let F_λ be a cube in the class $\mathfrak{I}^n_c$, that is, a closed cube, with center at $0 \in \mathbb{R}^n$ and edge $\lambda > 0$. Then $\big(F_{1-\frac{1}{k}} : k \in \mathbb{N}\big)$ is an increasing sequence of closed cubes with $\lim_{k\to\infty} F_{1-\frac{1}{k}} = \bigcup_{n\in\mathbb{N}} F_{1-\frac{1}{k}} = G_1$. Since $\mathcal{H}^n$ is a measure on $\mathfrak{B}_{\mathbb{R}^n}$, we have $\lim_{k\to\infty} \mathcal{H}^n\big(F_{1-\frac{1}{k}}\big) = \mathcal{H}^n(G_1)$. Now since $F_{1-\frac{1}{k}}$ is a dilation of F_1 by the scalar $1 - \frac{1}{k}$, we have $\mathcal{H}^n\big(F_{1-\frac{1}{k}}\big) = \big[1 - \frac{1}{k}\big]^n \mathcal{H}^n(F_1)$ by Theorem 28.1. Then $\lim_{k\to\infty} \mathcal{H}^n\big(F_{1-\frac{1}{k}}\big) = \lim_{k\to\infty} \big[1 - \frac{1}{k}\big]^n \mathcal{H}^n(F_1) = \mathcal{H}^n(F_1)$. This shows that $\mathcal{H}^n(G_1) = \mathcal{H}^n(F_1)$. If Q is a unit cube in the class $\mathfrak{I}^n_{oc}$ or in the class $\mathfrak{I}^n_{co}$, then $G_1 \subset Q \subset F_1$ so that $\mathcal{H}^n(G_1) \le \mathcal{H}^n(Q) \le \mathcal{H}^n(F_1)$. Then the equality $\mathcal{H}^n(G_1) = \mathcal{H}^n(F_1)$ implies $\mathcal{H}^n(G_1) = \mathcal{H}^n(Q) = \mathcal{H}^n(F_1)$. ∎

Theorem 29.2. *The n-dimensional Hausdorff (outer) measure $\mathcal{H}^n$ is a constant multiple of the n-dimensional Lebesgue outer measure $\big(\mu_L^n\big)^*$ on $\mathbb{R}^n$. Indeed we have*

(1) $$\kappa_n^{-1}\mathcal{H}^n(E) = \big(\mu_L^n\big)^*(E) \quad \text{for every } E \in \mathfrak{P}(\mathbb{R}^n)$$

where Q is a unit cube in $\mathbb{R}^n$ and

(2) $$\kappa_n = \mathcal{H}^n(Q).$$

Proof. 1. For $a = (a_1, \ldots, a_n) \in \mathbb{R}^n$ and $\lambda > 0$, consider a cube in $\mathbb{R}^n$ of the class $\mathfrak{I}^n_{oc}$ given by $Q(a, \lambda) = (a_1, a_1 + \lambda] \times \cdots \times (a_n, a_n + \lambda]$. Since a translation in $\mathbb{R}^n$ is an isometry, we have $\mathcal{H}^n\big(Q(a,\lambda)\big) = \mathcal{H}^n\big(Q(0,\lambda)\big)$ by Theorem 28.10. Since $Q(0,\lambda)$ is a dilation of $Q(0,1)$ by λ, we have $\mathcal{H}^n\big(Q(0,\lambda)\big) = \lambda^n \mathcal{H}^n\big(Q(0,1)\big) = \lambda^n \kappa_n$ by Theorem 28.1. Thus we have $\mathcal{H}^n\big(Q(a,\lambda)\big) = \lambda^n\kappa_n = \kappa_n\big(\mu_L^n\big)^*\big(Q(a,\lambda)\big)$.

2. Let O be a nonempty open set in $\mathbb{R}^n$. According to Proposition 24.10, there exists a disjoint sequence $(Q_j : j \in \mathbb{N})$ of cubes in the class $\mathfrak{I}_{oc}^n$ such that $O = \bigcup_{j\in\mathbb{N}} Q_j$. Since $\mathcal{H}^n$ and $\big(\mu_L^n\big)^*$ are measures on the σ-algebra $\mathfrak{B}_{\mathbb{R}^n}$, we have

$$\begin{aligned}\mathcal{H}^n(O) &= \mathcal{H}^n\Big(\bigcup_{j\in\mathbb{N}} Q_j\Big) = \sum_{j\in\mathbb{N}} \mathcal{H}^n(Q_j) = \kappa_n \sum_{j\in\mathbb{N}} \big(\mu_L^n\big)^*(Q_j)\\ &= \kappa_n\big(\mu_L^n\big)^*\Big(\bigcup_{j\in\mathbb{N}} Q_j\Big) = \kappa_n\big(\mu_L^n\big)^*(O).\end{aligned}$$

3. Let G be a G_δ-set in $\mathbb{R}^n$. If $\mathcal{H}^n(G) = \infty$ and $\big(\mu_L^n\big)^*(G) = \infty$ also, then the equality $\mathcal{H}^n(G) = \kappa_n\big(\mu_L^n\big)^*(G)$ holds trivially. Let us show that if at least one of $\mathcal{H}^n(G)$ and $\big(\mu_L^n\big)^*(G)$ is finite, then there exists an open set $O \supset G$ such that both $\mathcal{H}^n(O)$ and $\big(\mu_L^n\big)^*(O)$ are finite.

Consider first the case $\big(\mu_L^n\big)^*(G) < \infty$. By Proposition 24.6, there exists a sequence $(O_j : j \in \mathbb{N})$ in $\mathfrak{I}_o^n$ such that $\bigcup_{j\in\mathbb{N}} O_j \supset G$ and $\sum_{j\in\mathbb{N}} \big(\mu_L^n\big)^*(O_j) < \infty$. If we let $O = \bigcup_{j\in\mathbb{N}} O_j$, then O is an open set containing G and $\big(\mu_L^n\big)^*(O) \le \sum_{j\in\mathbb{N}} \big(\mu_L^n\big)^*(O_j) < \infty$. Since O is an open set, we have $\mathcal{H}^n(O) = \kappa_n\big(\mu_L^n\big)^*(O) < \infty$ by our result in **2**.

Next consider the case $\mathcal{H}^n(G) < \infty$. By Theorem 27.13 we have $\mathcal{G}_\delta^n(G) \uparrow \mathcal{H}^n(G)$ as $\delta \downarrow 0$ where $\mathcal{G}_\delta^n(G) = \inf\big\{\sum_{j\in\mathbb{N}} |O_j|^n : (O_j : j \in \mathbb{N}) \text{ is a } \delta\text{-cover of } G \text{ by open sets}\big\}$. Thus $\mathcal{H}^n(G) < \infty$ implies that $\mathcal{G}_\delta^n(G) < \infty$. Then there exists a sequence $(O_j : j \in \mathbb{N})$ of open sets with $|O_j| \le \delta$ such that $\bigcup_{n\in\mathbb{N}} O_j \supset G$ and $\sum_{j\in\mathbb{N}} |O_j|^n < \infty$. Let $x_j \in O_j$ be arbitrarily chosen. Then for every $x \in O_j$ we have $|x - x_j| \le |O_j|$ so that O_j is contained in an open cube C_j with center at x_j and with edge $2|O_j|$. If we let $O = \bigcup_{j\in\mathbb{N}} C_j$, then O is an open set containing G and

$$\big(\mu_L^n\big)^*(O) = \mu_L^n\Big(\bigcup_{j\in\mathbb{N}} C_j\Big) \le \sum_{j\in\mathbb{N}} \big(\mu_L^n\big)^*(C_j) = \sum_{j\in\mathbb{N}} \big(2|O_j|\big)^n = 2^n \sum_{j\in\mathbb{N}} |O_j|^n < \infty.$$

Since C_j is a translate of a dilation of a unit cube Q with center at $0 \in \mathbb{R}^n$ by $2|O_j|$, we have by Theorem 28.1 and Theorem 28.10

$$\begin{aligned}\mathcal{H}^n(O) &= \mathcal{H}^n\Big(\bigcup_{j\in\mathbb{N}} C_j\Big) \le \sum_{j\in\mathbb{N}} \mathcal{H}^n(C_j) = \sum_{j\in\mathbb{N}} \big(2|O_j|\big)^n \mathcal{H}^n(Q)\\ &= 2^n\mathcal{H}^n(Q) \sum_{j\in\mathbb{N}} |O_j|^n < \infty.\end{aligned}$$

Now let G be a G_δ-set in $\mathbb{R}^n$ for which at least one of $\mathcal{H}^n(G)$ and $\big(\mu_L^n\big)^*(G)$ is finite. We showed above that there exists an open set $O \supset G$ such that both $\mathcal{H}^n(O)$ and $\big(\mu_L^n\big)^*(O)$ are finite. Since G is a G_δ-set, there exists a decreasing sequence $(G_j : j \in \mathbb{N})$ of open sets containing G such that $G = \lim_{j\to\infty} G_j$. With the open sets $O \supset G$, we have $G = G \cap O = \lim_{j\to\infty} (G_j \cap O)$. Since $\mathcal{H}^n$ and $(\mu_L^n)^*$ are measures on the σ-algebra $\mathfrak{B}_{\mathbb{R}^n}$ and since

$$\mathcal{H}^n(G_j \cap O) < \infty \quad \text{and} \quad (\mu_L^n)^*(G_j \cap O) < \infty,$$

the convergence $G = \lim_{j\to\infty} G_j$ implies

$$\mathcal{H}^n(G) = \lim_{j\to\infty} \mathcal{H}^n(G_j \cap O) \quad \text{and} \quad (\mu_L^n)^*(G) = \lim_{j\to\infty} (\mu_L^n)^*(G_j \cap O).$$

Since $G_j \cap O$ is an open set, $\mathcal{H}^n(G_j \cap O) = \kappa_n (\mu_L^n)^*(G_j \cap O)$ by our result in **2**. Therefore we have $\mathcal{H}^n(G) = \kappa_n (\mu_L^n)^*(G)$.

4. Let $E \in \mathfrak{P}(\mathbb{R}^n)$. By Theorem 27.27, there exists a G_δ-set $G' \supset E$ such that $\mathcal{H}^n(G') = \mathcal{H}^n(E)$. By Lemma 24.11, there exists a G_δ-set $G'' \supset E$ such that $(\mu_L^n)^*(G'') = (\mu_L^n)^*(E)$. Then since $E \subset G' \cap G'' \subset G'$, we have $\mathcal{H}^n(E) = \mathcal{H}^n(G' \cap G'')$. Similarly by $E \subset G' \cap G'' \subset G''$, we have $(\mu_L^n)^*(E) = (\mu_L^n)^*(G' \cap G'')$. Since G' and G'' are G_δ-sets, so is $G' \cap G''$. Then by our result in **3**, we have $\mathcal{H}^n(G' \cap G'') = \kappa_n (\mu_L^n)^*(G' \cap G'')$. Thus $\mathcal{H}^n(E) = \kappa_n (\mu_L^n)^*(E)$. ∎

Corollary 29.3. *For every $n \in \mathbb{N}$, we have $\mathfrak{M}(\mathcal{H}^n) = \mathfrak{M}((\mu_L^n)^*) (= \mathfrak{M}_L^n)$, that is, the σ-algebra of $\mathcal{H}^n$-measurable subsets of $\mathbb{R}^n$ is equal to the σ-algebra of $(\mu_L^n)^*$-measurable subsets of $\mathbb{R}^n$.*

Proof. Let $E \in \mathfrak{P}(\mathbb{R}^n)$. Then $E \in \mathfrak{M}(\mathcal{H}^n)$ if and only if for every $A \in \mathfrak{P}(\mathbb{R}^n)$ we have

(1) $$\mathcal{H}^n(A) = \mathcal{H}^n(A \cap E) + \mathcal{H}^n(A \cap E^c),$$

and $E \in \mathfrak{M}((\mu_L^n)^*)$ if and only if for every $A \in \mathfrak{P}(\mathbb{R}^n)$ we have

(2) $$(\mu_L^n)^*(A) = (\mu_L^n)^*(A \cap E) + (\mu_L^n)^*(A \cap E^c).$$

According to Theorem 29.2, $\mathcal{H}^n(F) = \kappa_n (\mu_L^n)^*(F)$ for every $F \in \mathfrak{P}(\mathbb{R}^n)$. Thus if E satisfies (1), then it satisfies (2) so that $\mathfrak{M}(\mathcal{H}^n) \subset \mathfrak{M}((\mu_L^n)^*)$ and similarly if E satisfies (2), then it satisfies (1) so that $\mathfrak{M}((\mu_L^n)^*) \subset \mathfrak{M}(\mathcal{H}^n)$. Thus $\mathfrak{M}(\mathcal{H}^n) = \mathfrak{M}((\mu_L^n)^*)$. ∎

[II] Calculation of the n-dimensional Hausdorff Measure of a Unit Cube in $\mathbb{R}^n$

Let Q be a unit cube and S be a unit ball in $\mathbb{R}^n$. We show that $\kappa_n = \mathcal{H}^n(Q)$ and $\gamma_n = \mu_L^n(S)$ are related by $\kappa_n \gamma_n = 1$.

Definition 29.4. *For $x = (x_1, \dots, x_n) \in \mathbb{R}^n$ and $i = 1, \dots, n$, let $x^{[i]}$ be the point whose i-th coordinate is the negative of that of x. We say that a set E in $\mathbb{R}^n$ is symmetric with* *respect to the hyperplane* $H_i = \{(x_1, \dots, x_n) \in \mathbb{R}^n : x_i = 0\}$ *if for every* $x \in E$ *we have* $x^{[i]} \in E$.

Theorem 29.5. (Steiner Symmetrization) *Let* E *be a bounded set in* $\mathfrak{B}_{\mathbb{R}^n}$. *For every* $i = 1, \dots, n$, *there exists a set* $S_i E$ *in* $\mathbb{R}^n$ *with the following properties:*
(a) $S_i E \in \mathfrak{B}_{\mathbb{R}^n}$.
(b) $\mu_L^n(S_i E) = \mu_L^n(E)$.
(c) $|S_i E| \le |E|$.
(d) $S_i E$ *is symmetric with respect to the hyperplane* $H_i = \{(x_1, \dots, x_n) \in \mathbb{R}^n : x_i = 0\}$.
(e) *If* E *is symmetric with respect to a hyperplane* H_j *for some* $j = 1, \dots, n$, *then* $S_i E$ *is symmetric with respect to both* H_i *and* H_j.

Proof. For simplicity in notation, we prove the theorem for $i = 1$. Let π be the projection of $\mathbb{R}^n = \mathbb{R} \times \mathbb{R}^{n-1}$ onto $\mathbb{R}^{n-1}$, that is, $\pi(x_1, \dots, x_n) = (x_2, \dots, x_n)$ for $(x_1, \dots, x_n) \in \mathbb{R}^n$. Thus for every $E \in \mathfrak{P}(\mathbb{R}^n)$ we have

$$\pi(E) = \{y \in \mathbb{R}^{n-1} : \text{there exists } s \in \mathbb{R} \text{ such that } (s, y) \in E\}.$$

If E is a bounded set in $\mathfrak{B}_{\mathbb{R}^n}$, then $\pi(E)$ is a bounded set in $\mathbb{R}^{n-1}$. For every $y \in \mathbb{R}^{n-1}$, let $E(\cdot, y) = \{s \in \mathbb{R} : (s, y) \in E\}$ and call this subset of $\mathbb{R}$ the section of E at y. Note that if $y \notin \pi(E)$, then $E(\cdot, y) = \emptyset$. Now $\mathfrak{B}_{\mathbb{R}^n} = \sigma\big(\bigtimes_{i=1}^n \mathfrak{B}_{\mathbb{R}}\big) = \sigma\big(\mathfrak{B}_{\mathbb{R}} \times \mathfrak{B}_{\mathbb{R}^{n-1}}\big)$ by Theorem 24.24 and Proposition 23.5. Thus $E(\cdot, y) \in \mathfrak{B}_{\mathbb{R}}$ for every $y \in \mathbb{R}^{n-1}$ by Proposition 23.9 and $\mu_L(E(\cdot, y))$ is a real-valued $\mathfrak{B}_{\mathbb{R}^{n-1}}$-measurable function of $y \in \mathbb{R}^{n-1}$. For every $y \in \mathbb{R}^{n-1}$, let us define a set in $\mathbb{R}$ by

$$(1) \qquad I_y = \big(-\tfrac{1}{2}\mu_L(E(\cdot, y)), \tfrac{1}{2}\mu_L(E(\cdot, y))\big)$$

with the understanding that $I_y = \emptyset$ if $\mu_L\big(E(\cdot, y)\big) = 0$. Note that for $y \in \mathbb{R}^{n-1}$ such that $\mu_L\big(E(\cdot, y)\big) > 0$, I_y is an open interval of length $\mu_L\big(E(\cdot, y)\big)$ with center at the origin of $\mathbb{R}$. Let us define a set in $\mathbb{R}^n$ by setting

$$(2) \qquad S_1 E = \bigcup_{y \in \mathbb{R}^{n-1}} I_y \times \{y\}.$$

The section of this set at $y \in \mathbb{R}^{n-1}$ is equal to I_y, that is,

$$(3) \qquad (S_1 E)(\cdot, y) = I_y.$$

Consider the hyperplane $H_1 = \{(x_1, \dots, x_n) \in \mathbb{R}^n : x_1 = 0\}$. Let us show that $S_1 E$ is symmetric with respect to H_1. Let $x = (x_1, \dots, x_n) \in S_1 E$. By (2), $x_1 \in I_y$ where $y = (x_2, \dots, x_n) \in \mathbb{R}^{n-1}$ is such that $I_y \neq \emptyset$. Now $x_1 \in I_y$ implies $-x_1 \in I_y$ by (1). Thus $(-x_1, x_2, \dots, x_n) \in S_1 E$ by (2). This shows that $S_1 E$ is symmetric with respect to H_1.

To show that $S_1 E \in \mathfrak{B}_{\mathbb{R}^n}$, consider the nonnegative real-valued $\mathfrak{B}_{\mathbb{R}^{n-1}}$-measurable function $\frac{1}{2}\mu_L\big(E(\cdot, y)\big)$ of $y \in \mathbb{R}^{n-1}$. By Lemma 8.6, there exists an increasing sequence $(\varphi_j : j \in \mathbb{N})$ of nonnegative simple functions on $\big(\mathbb{R}^{n-1}, \mathfrak{B}_{\mathbb{R}^{n-1}}\big)$ such that we have $\varphi_j(y) \uparrow \frac{1}{2}\mu_L\big(E(\cdot, y)\big)$ for every $y \in \mathbb{R}^{n-1}$. Let $B_j = \{(s, y) \in \mathbb{R}^n : |s| \le \varphi_j(y)\}$ for $j \in \mathbb{N}$. Since $(x_1, \dots, x_n) \mapsto |x_1|$ is a real-valued $\mathfrak{B}_{\mathbb{R}^n}$-measurable function on $\mathbb{R}^n$ and $(s, y) \mapsto$

$\varphi_j(y)$ is also a real-valued $\mathfrak{B}_{\mathbb{R}^n}$-measurable function on $\mathbb{R}^n$, we have $B_j \in \mathfrak{B}_{\mathbb{R}^n}$ by (3) of Theorem 4.16. By the fact that $\varphi_j(y) \uparrow \frac{1}{2}\mu_L\big(E(\cdot, y)\big)$ for every $y \in \mathbb{R}^{n-1}$ and by (1) and (2), we have $S_1 E = \bigcup_{j\in\mathbb{N}} B_j$. Thus $S_1 E \in \mathfrak{B}_{\mathbb{R}^n}$.

The fact that $\mu_L^n(S_1 E) = \mu_L^n(E)$ follows from Proposition 23.16 for a measurable set in a product measure space. Indeed we have

$$\begin{aligned}\mu_L^n(S_1 E) &= \left(\mu_L \times \mu_L^{n-1}\right)(S_1 E) = \int_{\mathbb{R}^{n-1}} \mu_L\big((S_1 E)(\cdot, y)\big)\, \mu_L^{n-1}(dy)\\ &= \int_{\mathbb{R}^{n-1}} \mu_L(I_y)\, \mu_L^{n-1}(dy) = \int_{\mathbb{R}^{n-1}} \mu_L\big(E(\cdot, y)\big)\, \mu_L^{n-1}(dy)\\ &= \left(\mu_L \times \mu_L^{n-1}\right)(E) = \mu_L^n(E),\end{aligned}$$

where the second equality is by Proposition 23.16, the third equality is by (3), the fourth equality is by (1) and the fifth equality is by Proposition 23.16.

Let us show that $|S_1 E| \le |E|$. For every $y \in \mathbb{R}^{n-1}$, $E(\cdot, y)$ is a bounded set in $\mathbb{R}$. Let us define a finite closed interval by setting

$$(4) \qquad J_y = \big[\inf E(\cdot, y), \sup E(\cdot, y)\big] \subset \mathbb{R}.$$

Since $J_y \supset E(\cdot, y)$, we have by (1)

$$(5) \qquad \mu_L(J_y) = \sup E(\cdot, y) - \inf E(\cdot, y) \ge \mu_L\big(E(\cdot, y)\big) = \mu_L(I_y).$$

Let x' and x'' be two arbitrary points in $S_1 E$. By (2), we have $x' = (s', y')$ where $s' \in I_{y'}$ and $y' \in \pi(E)$ and similarly $x'' = (s'', y'')$ where $s'' \in I_{y''}$ and $y'' \in \pi(E)$. Let c' and c'' be respectively the midpoints of the closed intervals $J_{y'}$ and $J_{y''}$ defined by (4) with our y' and y''. Then by (1) and (5) we have

$$(6) \qquad \begin{aligned}|s' - s''| &\le |s'| + |s''| \le \tfrac{1}{2}\mu_L\big(E(\cdot, y')\big) + \tfrac{1}{2}\mu_L\big(E(\cdot, y'')\big)\\ &\le \tfrac{1}{2}\mu_L(J_{y'}) + \tfrac{1}{2}\mu_L(J_{y''}) \le |c' - c''| + \tfrac{1}{2}\mu_L(J_{y'}) + \tfrac{1}{2}\mu_L(J_{y''})\\ &= |t' - t''|,\end{aligned}$$

where t' is one of the two endpoints of $J_{y'}$ and t'' is one of the two endpoints of $J_{y''}$. Now $(t', y'), (t'', y'') \in \overline{E}$. By (6), we have $|x' - x''| = |(s', y') - (s'', y'')| \le |(t', y') - (t'', y'')|$. Since this holds for any two points x' and x'' of $S_1 E$, we have $|S_1 E| \le |\overline{E}|$. Then since the diameter of the closure of an arbitrary bounded set is equal to the diameter of the set, we have $|S_1 E| \le |E|$.

Consider the hyperplanes $H_j = \big\{(x_1, \dots, x_n) \in \mathbb{R}^n : x_j = 0\big\}$ for $j = 1, \dots, n$. Suppose E is symmetric with respect to a hyperplane H_j for some $j = 1, \dots, n$. Let us show that $S_1 E$ is symmetric with respect to H_j. Since $S_1 E$ is symmetric with respect to H_1 as we showed above, let us consider $j = 2, \dots, n$. For $x \in \mathbb{R}^n$, let $x^{[j]}$ be the point whose j-th coordinate is the negative of that of x and for $y \in \mathbb{R}^{n-1}$, $y = (x_2, \dots, x_n)$, let $y^{[j]}$ be the point whose $j-1$-th coordinate is the negative of that of y. Let x be an arbitrary point in $S_1 E$. By (2), $x = (s, y)$ where $s \in I_y$ and $y \in \pi(E)$. Then $x^{[j]} = \big(s, y^{[j]}\big)$. Since

$y \in \pi(E)$, there exists $s_0 \in \mathbb{R}$ such that $(s_0, y) \in E$. Then since E is symmetric with respect to H_j, we have $\left(s_0, y^{[j]}\right) \in E$. Thus $y^{[j]} \in \pi(E)$. This implies $x^{[j]} = \left(s, y^{[j]}\right) \in I_y \times \{y^{[j]}\} \subset S_1 E$ by (2). This shows that $S_1 E$ is symmetric with respect to H_j. ∎

The volume of a ball with radius $a > 0$ in $\mathbb{R}^n$ is equal to $\pi^{n/2} a^n / \Gamma\left(\frac{n}{2}+1\right)$ as we showed in Theorem 26.13. Thus $\gamma_n := \pi^{n/2}\left(\frac{1}{2}\right)^n / \Gamma\left(\frac{n}{2}+1\right)$ is the volume of a unit ball in $\mathbb{R}^n$. The volume of a ball with diameter $d > 0$ in $\mathbb{R}^n$ is then $\pi^{n/2}\left(\frac{d}{2}\right)^n / \Gamma\left(\frac{n}{2}+1\right) = \gamma_n d^n$. We have for instance $\gamma_1 = 1$, $\gamma_2 = \frac{\pi}{4}$, $\gamma_3 = \frac{\pi}{6}$, $\gamma_4 = \frac{\pi^2}{32}$, and so on.

Theorem 29.6. (Isodiametric Inequality) *If E is a bounded set in $\mathfrak{B}_{\mathbb{R}^n}$, then the volume of E does not exceed that of a ball with the same diameter as E, that is,*

$$(1) \qquad \mu_L^n(E) \le \gamma_n |E|^n,$$

where

$$(2) \qquad \gamma_n = \frac{\pi^{n/2}}{\Gamma\left(\frac{n}{2}+1\right)}\left(\frac{1}{2}\right)^n,$$

that is, the volume of a ball with diameter 1.

Proof. Let E be a bounded set in $\mathfrak{B}_{\mathbb{R}^n}$. Let S_i be the Steiner symmetrization with respect to the hyperplane $H_i = \left\{(x_1, \dots, x_n) \in \mathbb{R}^n : x_i = 0\right\}$ for $i = 1, \dots, n$ as defined in Theorem 29.5. Let $SE = S_n \cdots S_1 E$, the Steiner symmetrization of E by $S_1, \dots, S_n$ successively. Then by Theorem 29.5, we have $SE \in \mathfrak{B}_{\mathbb{R}^n}$, $\mu_L^n(SE) = \mu_L^n(E)$, $|SE| \le |E|$, and SE is symmetric with respect to H_i for $i = 1, \dots, n$. Thus for every $x \in SE$, we have $-x \in SE$. To show that SE is contained in a closed ball with center at $0 \in \mathbb{R}^n$ and radius $\frac{1}{2}|E|$, we show that SE cannot contain any point $x \in \mathbb{R}^n$ with $|x| > \frac{1}{2}|E|$. Suppose there exists $x \in SE$ with $|x| > \frac{1}{2}|E|$. Now $x \in SE$ implies that $-x \in SE$ also. Thus $|x - (-x)| \le |SE|$. But $|x - (-x)| = |x| + |-x| = 2|x| > |E| \ge |SE|$, a contradiction. This shows that SE is contained in a closed ball with center at $0 \in \mathbb{R}^n$ and radius $\frac{1}{2}|E|$. The volume of such a ball is given by $\pi^{n/2}\left(\frac{1}{2}|E|\right)^n / \Gamma\left(\frac{n}{2}+1\right) = \gamma_n |E|^n$ by Theorem 26.13. ∎

Proposition 24.10 is a covering theorem for open sets in $\mathbb{R}^n$ by binary cubes in $\mathbb{R}^n$. The next Proposition is a covering theorem for open sets in $\mathbb{R}^n$ by balls in $\mathbb{R}^n$.

Proposition 29.7. *Let G be a bounded open set in $\mathbb{R}^n$. The for every $\varepsilon > 0$ there exists a disjoint sequence $(S_k : k \in \mathbb{N})$ of closed balls with diameters not exceeding ε contained in G such that* $\mu_L^n\left(G \setminus \bigcup_{k \in \mathbb{N}} S_k\right) = 0$.

Proof. 1. Let $\varepsilon > 0$ be arbitrarily given. Let $(S_k : k \in \mathbb{N})$ be a disjoint sequence of closed balls contained in G with diameters not exceeding ε and satisfying the condition that if we let $\mathfrak{S}_k$ be the collection of all closed balls with diameters not exceeding ε and contained in the open set $G \setminus \bigcup_{i=1}^k S_i$ and if we let d_k be the supremum of the diameters of the members of $\mathfrak{S}_k$ then $|S_{k+1}| > \frac{1}{2} d_k$.

Let us show that a sequence $(S_k : k \in \mathbb{N})$ of the description above exists by mathematical induction. Let S_1 be an arbitrary closed ball with $|S_1| \le \varepsilon$ and contained in G. Then $G \setminus S_1$ is a non-empty open set so that $\mathfrak{S}_1 \neq \emptyset$. Thus we can select $S_2 \in \mathfrak{S}_1$ with $|S_2| > \frac{1}{2}d_1$. Suppose for some $k \in \mathbb{N}$ we have selected k disjoint closed balls $S_1, \cdots, S_k$ contained in G with diameters not exceeding ε and $|S_i| > \frac{1}{2}d_{i-1}$ for $i = 2, \dots, k$. Then since $\bigcup_{i=1}^k S_i$ is a closed set contained in the open set G, $G \setminus \bigcup_{i=1}^k S_i$ is a non-empty open set. Then $\mathfrak{S}_k \neq \emptyset$ and we can select $S_{k+1} \in \mathfrak{S}_k$ with $|S_{k+1}| > \frac{1}{2}d_k$. Thus by induction $(S_k : k \in \mathbb{N})$ exists.

2. Let us show that $\mu_L^n\left(G \setminus \bigcup_{k\in\mathbb{N}} S_k\right) = 0$. Suppose $\mu_L^n\left(G \setminus \bigcup_{k\in\mathbb{N}} S_k\right) > 0$. Let γ_n be the volume of a ball with diameter 1 in $\mathbb{R}^n$. Then $\mu_L^n(S_k) = |S_k|^n \gamma_n$. By the disjointness of the sequence $(S_k : k \in \mathbb{N})$ and by the fact that G is a bounded open set in $\mathbb{R}^n$, we have

$$\gamma_n \sum_{k\in\mathbb{N}} |S_k|^n = \sum_{k\in\mathbb{N}} \mu_L^n(S_k) = \mu_L^n\Big(\bigcup_{k\in\mathbb{N}} S_k\Big) \le \mu_L^n(G) < \infty.$$

Thus $\sum_{k\in\mathbb{N}} |S_k|^n < \infty$, which implies $\lim_{k\to\infty} |S_k|^n = 0$ and then $\lim_{k\to\infty} |S_k| = 0$. Let $\widetilde{S}_k$ be the closed ball concentric with S_k and having $|\widetilde{S}_k| = 4|S_k|$. Then $\mu_L^n(\widetilde{S}_k) = 4^n \mu_L^n(S_k)$ so that $\sum_{k\in\mathbb{N}} \mu_L^n(\widetilde{S}_k) = 4^n \sum_{k\in\mathbb{N}} \mu_L^n(S_k) < \infty$ and therefore $\lim_{N\to\infty} \sum_{k>N} \mu_L^n(\widetilde{S}_k) = 0$. Now since $\mu_L^n\left(G \setminus \bigcup_{k\in\mathbb{N}} S_k\right) > 0$, for sufficiently large $N \in \mathbb{N}$ we have

$$\mu_L^n\Big(\bigcup_{k>N} \widetilde{S}_k\Big) \le \sum_{k>N} \mu_L^n(\widetilde{S}_k) < \mu_L^n\Big(G \setminus \bigcup_{k\in\mathbb{N}} S_k\Big).$$

Thus there exists $x_0 \in \mathbb{R}^n$ such that

$$(1) \qquad x_0 \in G \setminus \bigcup_{k\in\mathbb{N}} S_k \subset G \setminus \bigcup_{k=1}^N S_k \quad \text{and} \quad x_0 \notin \bigcup_{k>N} \widetilde{S}_k.$$

Since $\bigcup_{k=1}^N S_k$ is a closed set contained in the open set G, there exists a closed ball S contained in G with center at x_0 and $|S| \le \varepsilon$ such that $S \subset G \setminus \bigcup_{k=1}^N S_k$ and then

$$(2) \qquad S \cap \bigcup_{k=1}^N S_k = \emptyset.$$

Now if $S \cap \bigcup_{i=1}^k S_i = \emptyset$ for some $k \in \mathbb{N}$, then $S \subset G \setminus \bigcup_{i=1}^k S_i$ so that $S \in \mathfrak{S}_k$ and then $|S| \le d_k < 2|S_{k+1}|$. Since $\lim_{k\to\infty} |S_k| = 0$ and $|S| > 0$, this cannot hold for every $k \in \mathbb{N}$. Therefore S intersects S_k for some $k \in \mathbb{N}$. Let S_{k_0} be the first in the sequence $(S_k : k \in \mathbb{N})$ to intersect S. Then

$$(3) \qquad S \cap S_{k_0} \neq \emptyset \quad \text{and} \quad S \cap (S_1 \cup \cdots \cup S_{k_0-1}) = \emptyset.$$

By (2), we have $k_0 > N$. Then by (1), we have $x_0 \notin \widetilde{S}_{k_0}$. Since $S \cap S_{k_0} \neq \emptyset$ and $x_0 \notin \widetilde{S}_{k_0}$, we have $\frac{1}{2}|S| + \frac{1}{2}|S_{k_0}| > 4\left(\frac{1}{2}|S_{k_0}|\right)$ that is, $|S| > 3|S_{k_0}|$. On the other hand, since

$S \cap (S_1 \cup \cdots \cup S_{k_0-1}) = \emptyset$ by (3), we have $S \in \mathfrak{S}_{k_0-1}$ so that $|S| \leq d_{k_0-1} < 2|S_{k_0}|$. Thus $3|S_{k_0}| < |S| < 2|S_{k_0}|$, a contradiction. Therefore $\mu_L^n(G \setminus \bigcup_{k \in \mathbb{N}} S_k) = 0$. ∎

Theorem 29.8. *Let $\kappa_n = \mathcal{H}^n(Q)$ and $\gamma_n = \mu_L^n(S)$ where Q is a unit cube and S is a unit ball in $\mathbb{R}^n$, that is, a ball with diameter 1.*
(a) *κ_n and γ_n are related by the equality $\kappa_n \gamma_n = 1$.*
(b) *The n-dimensional Lebesgue outer measure $(\mu_L^n)^*$ and the n-dimensional Hausdorff measure $\mathcal{H}^n$ are related by the equality $\gamma_n^{-1}(\mu_L^n)^*(E) = \mathcal{H}^n(E)$ for every $E \in \mathfrak{P}(\mathbb{R}^n)$.*
(c) *$\mathcal{H}^n(S) = 1$ for a unit ball S in $\mathbb{R}^n$. For a ball B in $\mathbb{R}^n$, we have $\mathcal{H}^n(B) = |B|^n$.*

Proof. 1. Consider $Q = (0,1) \times \cdots \times (0,1)$, an open unit cube in $\mathbb{R}^n$. By the equivalent definition of $\mathcal{H}^n$ by δ-covers of closed sets (Theorem 27.13), $\mathcal{H}_\delta^n(Q) = \mathcal{F}_\delta^n(Q)$ where $\mathcal{F}_\delta^n(Q) = \inf\left\{\sum_{i \in \mathbb{N}} |F_i|^n : (F_i : i \in \mathbb{N}) \text{ is a } \delta\text{-cover of } Q \text{ by closed sets}\right\}$ for every $\delta > 0$. Thus for an arbitrary $\varepsilon > 0$ and $\delta > 0$, there exists a δ-cover $(F_i : i \in \mathbb{N})$ of Q by closed sets such that

$$(1) \qquad \sum_{i \in \mathbb{N}} |F_i|^n \leq \mathcal{F}_\delta^n(Q) + \varepsilon = \mathcal{H}_\delta^n(Q) + \varepsilon \leq \mathcal{H}^n(Q) + \varepsilon.$$

Since $Q \subset \bigcup_{i \in \mathbb{N}} F_i$, the monotonicity and countable subadditivity of $(\mu_L^n)^*$ implies

$$(2) \qquad (\mu_L^n)^*(Q) \leq (\mu_L^n)^*\Big(\bigcup_{i \in \mathbb{N}} F_i\Big) \leq \sum_{i \in \mathbb{N}} (\mu_L^n)^*(F_i).$$

Now F_i is a closed set with $|F_i| \leq \delta$ so that F_i is a bounded set in $\mathfrak{B}_{\mathbb{R}^n}$. Thus by Theorem 29.6 (Isodiametric Inequality), we have $\mu_L^n(F_i) \leq \gamma_n |F_i|^n$. Then by (2) and (1), we have

$$(\mu_L^n)^*(Q) \leq \gamma_n \sum_{i \in \mathbb{N}} |F_i|^n \leq \gamma_n \left\{\mathcal{H}^n(Q) + \varepsilon\right\}.$$

Since this holds for an arbitrary $\varepsilon > 0$, we have $(\mu_L^n)^*(Q) \leq \gamma_n \mathcal{H}^n(Q)$. But $(\mu_L^n)^*(Q) = 1$ and $\mathcal{H}^n(Q) = \kappa_n$. Thus we have $1 \leq \gamma_n \kappa_n$.

Let us prove the reverse inequality $1 \geq \gamma_n \kappa_n$. Now since Q is a bounded open set in $\mathbb{R}^n$, according to Proposition 29.7 for every $\delta > 0$ there exists a disjoint sequence $(S_i : i \in \mathbb{N})$ of closed balls contained in Q with $|S_i| \leq \delta$ such that $\mu_L^n(Q \setminus \bigcup_{i \in \mathbb{N}} S_i) = 0$. Since μ_L^n is a measure on $\mathfrak{B}_{\mathbb{R}^n}$, we have

$$(3) \qquad \mu_L^n(Q) = \mu_L^n\Big(\bigcup_{i \in \mathbb{N}} S_i\Big) + \mu_L^n\Big(Q \setminus \bigcup_{i \in \mathbb{N}} S_i\Big) = \mu_L^n\Big(\bigcup_{i \in \mathbb{N}} S_i\Big) = \sum_{i \in \mathbb{N}} \mu_L^n(S_i).$$

By Theorem 29.2, we have

$$\mathcal{H}_\delta^n\Big(Q \setminus \bigcup_{i \in \mathbb{N}} S_i\Big) \leq \mathcal{H}^n\Big(Q \setminus \bigcup_{i \in \mathbb{N}} S_i\Big) = \kappa_n \mu_L^n\Big(Q \setminus \bigcup_{i \in \mathbb{N}} S_i\Big) = 0.$$

Then by the subadditivity of the outer measure $\mathcal{H}_\delta^n$, we have

$$(4) \qquad \mathcal{H}_\delta^n(Q) \leq \mathcal{H}_\delta^n\Big(\bigcup_{i \in \mathbb{N}} S_i\Big) + \mathcal{H}_\delta^n\Big(Q \setminus \bigcup_{i \in \mathbb{N}} S_i\Big) = \mathcal{H}_\delta^n\Big(\bigcup_{i \in \mathbb{N}} S_i\Big).$$

Since $(S_i : i \in \mathbb{N})$ is a δ-cover of $\bigcup_{i\in\mathbb{N}} S_i$, we have

$$(5) \qquad \mathcal{H}^n_\delta\Big(\bigcup_{i\in\mathbb{N}} S_i\Big) \le \sum_{i\in\mathbb{N}} |S_i|^n = \frac{1}{\gamma_n}\sum_{i\in\mathbb{N}} \mu^n_L(S_i),$$

where the equality is from the fact that S_i is a ball in $\mathbb{R}^n$ so that $\mu^n_L(S_i) = \gamma_n|S_i|^n$. By (4), (5), and (3), we have $\mathcal{H}^n_\delta(Q) \le \mu^n_L(Q)/\gamma_n$. Since this holds for every $\delta > 0$, we have $\mathcal{H}^n(Q) \le \mu^n_L(Q)/\gamma_n$. Since $\mathcal{H}^n(Q) = \kappa_n$ and $\mu^n_L(Q) = 1$, we have $\kappa_n\gamma_n \le 1$. This completes the proof of the equality $\kappa_n\gamma_n = 1$.

2. By Theorem 29.2, $\mathcal{H}^n(E) = \kappa_n\big(\mu^n_L\big)^*(E)$ for every $E \in \mathfrak{P}(\mathbb{R}^n)$. According to (a), $\kappa_n = \frac{1}{\gamma_n}$. Thus we have $\big(\mu^n_L\big)^*(E) = \gamma_n\mathcal{H}^n(E)$. This proves (b).

3. For every $E \in \mathfrak{P}(\mathbb{R}^n)$, we have $\mathcal{H}^n(E) = \kappa_n\big(\mu^n_L\big)^*(E)$ by Theorem 29.2. If S is a unit ball in $\mathbb{R}^n$, then $\big(\mu^n_L\big)^*(S) = \gamma_n$ by (2) of Theorem 29.6. Thus we have $\mathcal{H}^n(S) = \kappa_n\big(\mu^n_L\big)^*(S) = \kappa_n\gamma_n$. Since $\kappa_n\gamma_n = 1$ by (a), we have $\mathcal{H}^n(S) = 1$. Let $B = B\left(x, \frac{\lambda}{2}\right)$, an open ball with center at x and diameter λ. Now $B\left(x, \frac{\lambda}{2}\right) = \lambda B\left(\lambda^{-1}x, \frac{1}{2}\right)$ by Proposition 24.34. Then $\mathcal{H}^n(B) = \mathcal{H}^n\left(\lambda B\left(\lambda^{-1}x, \frac{1}{2}\right)\right) = \lambda^n\mathcal{H}^n\left(B\left(\lambda^{-1}x, \frac{1}{2}\right)\right) = \lambda^n = |B|^n$ by Theorem 28.1 and by the fact that $B\left(\lambda^{-1}x, \frac{1}{2}\right)$ is a unit ball in $\mathbb{R}^n$ for which the n-dimensional Hausdorff measure is equal to 1 as we showed above. ∎

[III] Transformation of Hausdorff Measure of Integral Dimension

Let T be a non-singular linear transformation of $\mathbb{R}^n$ into $\mathbb{R}^n$. By Theorem 29.2 and Theorem 24.31, we have

$$\mathcal{H}^n\big(T(E)\big) = \kappa_n\big(\mu^n_L\big)^*\big(T(E)\big) = \kappa_n|\det M_T|\big(\mu^n_L\big)^*(E) = |\det M_T|\mathcal{H}^n(E)$$

for every $E \in \mathfrak{P}(\mathbb{R}^n)$. Let us consider a non-singular linear transformation of $\mathbb{R}^k$ into $\mathbb{R}^n$.

Observation 29.9. Let T be a linear transformation of $\mathbb{R}^k$ into $\mathbb{R}^n$. It is a well-known fact in linear algebra that there exists a linear transformation T^* of $\mathbb{R}^n$ into $\mathbb{R}^k$, called the adjoint of T, such that $\langle Tx, y\rangle = \langle x, T^*y\rangle$ for every $x \in \mathbb{R}^k$ and $y \in \mathbb{R}^n$ where $\langle\cdot,\cdot\rangle$ is the inner product. The matrix of T, M_T, is an $n \times k$ matrix and the matrix of T^*, M_{T^*}, is a $k \times n$ matrix and the two matrices are related by $M_{T^*} = M^t_T$. The composition T^*T is a linear transformation of $\mathbb{R}^k$ into $\mathbb{R}^k$ and its matrix $M_{TT^*} = M_{T^*}M_T$ is a $k \times k$ matrix. The transformation T^*T is positive semidefinite, that is, $\langle T^*Tx, x\rangle \ge 0$ for every $x \in \mathbb{R}^k$, and $\det M_{TT^*} \ge 0$.

Now let T be a non-singular linear transformation of $\mathbb{R}^k$ into $\mathbb{R}^n$ where $k \le n$. The image of $\mathbb{R}^k$ by T, $T(\mathbb{R}^k)$, is a k-dimensional linear subspace of $\mathbb{R}^n$. Let R be a rotation on $\mathbb{R}^n$ mapping $T(\mathbb{R}^k)$ onto the k-dimensional linear subspace $\mathbb{R}^k \times \{0\} = \{y \in \mathbb{R}^n : y_j = 0 \text{ for } j > k\}$ of $\mathbb{R}^n$. Let $S = RT$, a non-singular linear transformation of $\mathbb{R}^k$ onto $\mathbb{R}^k \times \{0\}$. We have $S^*S = T^*R^*RT = T^*T$. Thus $M_{S^*S} = M_{T^*T}$ and $\det M_{S^*S} = \det M_{T^*T}$. If we identify $\mathbb{R}^k \times \{0\}$ with $\mathbb{R}^k$, then S is a non-singular linear transformation of $\mathbb{R}^k$ into $\mathbb{R}^k$. Now $M_{S^*S} = M_{S^*}M_S$. Since M_S is a square matrix and $M_{S^*} = M^t_S$ with

$\det M_{S^*} = \det M_S^t = \det M_S$, we have $\det M_{S^*S} = \det M_{S^*} \det M_S = (\det M_S)^2$. Therefore $|\det M_S| = \sqrt{\det M_{S^*S}} = \sqrt{\det M_{T^*T}}$.

Definition 29.10. *Let $k \le n$. For a non-singular linear transformation T of $\mathbb{R}^k$ into $\mathbb{R}^n$, we define $J(T) = \sqrt{\det M_{T^*T}}$ where T^* is the adjoint linear transformation of T.*

Note that if T is a non-singular linear transformation of $\mathbb{R}^n$ into $\mathbb{R}^n$ then $\det M_{T^*T} = \det M_{T^*} \det M_T = (\det M_T)^2$ so that $J(T) = \sqrt{\det M_{T^*T}} = |\det M_T|$.

Theorem 29.11. *Let T be a non-singular linear transformation of $\mathbb{R}^k$ into $\mathbb{R}^n$ where $k \le n$. Then $\mathcal{H}^k(T(E)) = J(T)\mathcal{H}^k(E)$ for every $E \in \mathfrak{P}(\mathbb{R}^n)$.*

Proof. Consider the case $k = n$. We have by Theorem 29.2 and Theorem 24.31 the equalities $\mathcal{H}^n(T(E)) = \kappa_n (\mu_L^n)^*(T(E)) = \kappa_n |\det M_T| (\mu_L^n)^*(E) = |\det M_T| \mathcal{H}^n(E)$. Then since $J(T) = |\det M_T|$ for a non-singular linear transformation T of $\mathbb{R}^n$ into $\mathbb{R}^n$, we have $\mathcal{H}^n(T(E)) = J(T)\mathcal{H}^n(E)$.

Next consider the case $k < n$. Let $S = RT$ where R is a rotation on $\mathbb{R}^n$ mapping the k-dimensional linear subspace $T(\mathbb{R}^k)$ to $\mathbb{R}^k \times \{0\}$. If we identify $\mathbb{R}^k \times \{0\}$ with $\mathbb{R}^k$, then S is a non-singular linear transformation of $\mathbb{R}^k$ into $\mathbb{R}^k$ with $|\det M_S| = \sqrt{\det M_{T^*T}} = J(T)$ by Observation 29.9. Thus by Theorem 24.31, we have $(\mu_L^k)^*(S(E)) = J(T)(\mu_L^k)^*(E)$. Now $\mathcal{H}^k(S(E)) = \kappa_k (\mu_L^k)^*(S(E))$ by Theorem 29.2. A rotation on $\mathbb{R}^n$ is an isometry so that we have $\mathcal{H}^k(S(E)) = \mathcal{H}^k(RT(E)) = \mathcal{H}^k(T(E))$ by Theorem 28.10. Thus we have $\mathcal{H}^k(T(E)) = \mathcal{H}^k(S(E)) = \kappa_k (\mu_L^k)^*(S(E)) = \kappa_k J(T)(\mu_L^k)^*(E) = J(T)\mathcal{H}^k(E)$. ∎

Let $1 \le k \le n$. Let Ω be an open set in $\mathbb{R}^k$. Let us consider a one-to-one mapping T of Ω into $\mathbb{R}^n$ of class $C^1(\Omega)$. The differential $dT(x, \cdot)$ exists at every $x \in \Omega$ and its matrix is an $n \times k$ matrix $\left[\frac{\partial T_i}{\partial x_j}(x)\right]$ where $T_1, \dots, T_n$ are the component functions of T.

Lemma 29.12. *Let $1 \le k \le n$. Let T be a continuous one-to-one mapping of an open set Ω in $\mathbb{R}^k$ into $\mathbb{R}^n$. Then for every $E \in \mathfrak{B}_{\mathbb{R}^k}$ such that $E \subset \Omega$, we have $T(E) \in \mathfrak{B}_{\mathbb{R}^n}$.*

Proof. An open set Ω in $\mathbb{R}^k$ is a countable union of closed balls in $\mathbb{R}^k$ and is thus a σ-compact set. If T is a continuous mapping of Ω into $\mathbb{R}^n$, then $T(K)$ is a compact set in $\mathbb{R}^n$ for every compact set K contained in Ω. Thus $T(\Omega)$ is a σ-compact set in $\mathbb{R}^n$ so that $T(\Omega) \in \mathfrak{B}_{\mathbb{R}^n}$.

Now let T be a continuous one-to-one mapping of an open set Ω in $\mathbb{R}^k$ into $\mathbb{R}^n$. Let $\mathfrak{A} = \{E \subset \Omega : T(E) \in \mathfrak{B}_{\mathbb{R}^n}\}$. Let us show that $\mathfrak{A}$ is a σ-algebra of subsets of Ω. Now since Ω is an open set in $\mathbb{R}^k$, we have $T(\Omega) \in \mathfrak{B}_{\mathbb{R}^n}$ as we showed above. Thus $\Omega \in \mathfrak{A}$. Let E be an arbitrary member of $\mathfrak{A}$. Consider $\Omega \setminus E$. Since T is a one-to-one mapping, we have $T(\Omega \setminus E) = T(\Omega) \setminus T(E)$. The fact that both $T(\Omega)$ and $T(E)$ are members of $\mathfrak{B}_{\mathbb{R}^n}$ implies that $T(\Omega \setminus E) \in \mathfrak{A}$. Thus $\mathfrak{A}$ is closed under complementations. To show that $\mathfrak{A}$ is closed under countable unions, let $(E_j : j \in \mathbb{N})$ be a sequence in $\mathfrak{A}$. Then $T\left(\bigcup_{j\in\mathbb{N}} E_j\right) = \bigcup_{j\in\mathbb{N}} T(E_j) \in \mathfrak{A}$ since $T(E_j) \in \mathfrak{A}$ for $j \in \mathbb{N}$. This shows that $\mathfrak{A}$ is a σ-algebra of subsets of Ω.

If E is an open set in Ω in its relative topology from $\mathbb{R}^k$, then E is an open set in $\mathbb{R}^k$ and thus $T(E) \in \mathfrak{B}_{\mathbb{R}^n}$ by our opening remark and therefore $E \in \mathfrak{A}$. Thus the σ-algebra $\mathfrak{A}$ contains all open sets in Ω and therefore it contains the Borel σ-algebra $\mathfrak{B}_\Omega$ of subsets of Ω, that is, $\mathfrak{A} \supset \mathfrak{B}_\Omega$. Let $\mathfrak{O}$ be the collection of all open sets in $\mathbb{R}^k$. Then $\mathfrak{O} \cap \Omega$ is the collection of all open sets in Ω in its relative topology. Thus $\mathfrak{B}_\Omega = \sigma_\Omega(\mathfrak{O} \cap \Omega) = \sigma(\mathfrak{O}) \cap \Omega = \mathfrak{B}_{\mathbb{R}^k} \cap \Omega$ by Theorem 1.15. Therefore we have $\mathfrak{A} \supset \mathfrak{B}_{\mathbb{R}^k} \cap \Omega$. This shows that if $E \in \mathfrak{B}_{\mathbb{R}^k} \cap \Omega$, then $T(E) \in \mathfrak{B}_{\mathbb{R}^n}$. In particular if $E \in \mathfrak{B}_{\mathbb{R}^k}$ and $E \subset \Omega$, then $T(E) \in \mathfrak{B}_{\mathbb{R}^n}$. ∎

Lemma 29.13. *Let $1 \le k \le n$. Let T be a mapping of class $C^1(\Omega)$ of an open set Ω in $\mathbb{R}^k$ into $\mathbb{R}^n$ such that T and the differential $dT(x, \cdot)$ for each $x \in \Omega$ are one-to-one mappings. Then for an arbitrary $c > 1$, there exist a disjoint sequence $(B_j : j \in \mathbb{N})$ in $\mathfrak{B}_{\mathbb{R}^k}$ such that $\bigcup_{j\in\mathbb{N}} B_j = \Omega$ and a sequence $(L_j : j \in \mathbb{N})$ of non-singular linear transformations of $\mathbb{R}^k$ into $\mathbb{R}^n$ such that for each $j \in \mathbb{N}$ we have*

(1) $$c^{-1}|L_j(z)| \le |dT(x,z)| \le c|L_j(z)| \quad \text{for } x \in B_j \text{ and } z \in \mathbb{R}^k$$

and

(2) $$c^{-1}|L_j(x) - L_j(y)| \le |T(x) - T(y)| \le c|L_j(x) - L_j(y)| \quad \text{for } x, y \in B_j.$$

Proof. Let us observe that if a linear transformation L of $\mathbb{R}^k$ into $\mathbb{R}^n$ is such that $|dT(x,z)| \le c|L(z)|$ for all $z \in \mathbb{R}^k$ for some $x \in \Omega$ and $c > 0$, then L is non-singular. In fact if $L(z) = 0$ for some $z \in \mathbb{R}^k$, then $dT(x,z) = 0$. Since $dT(x, \cdot)$ is a non-singular linear transformation of $\mathbb{R}^k$ into $\mathbb{R}^n$, we have $z = 0$. This shows that L is non-singular.

The normed linear space $\mathbf{L}$ of the linear transformations L of $\mathbb{R}^k$ into $\mathbb{R}^n$ with the norm $\|L\| = \sup_{|x|\le 1} |L(x)|$ has a countable dense subset $\mathbf{L}_0 = \{L_r : r \in \mathbb{N}\}$ consisting of non-singular linear transformations. (For instance we may take the collection of all linear transformations of $\mathbb{R}^k$ into $\mathbb{R}^n$ the entries of whose matrices are rational numbers and then drop those which are singular.) Let $c > 1$ be arbitrarily given. Let $\varepsilon > 0$ be so small that $c^{-1} + \varepsilon < 1 < c - \varepsilon$. Then let $c_0 > 1$ be such that

(3) $$c^{-1} + \varepsilon < c_0^{-1} < 1 < c_0 < c - \varepsilon.$$

For each $r \in \mathbb{N}$ and $p \in \mathbb{N}$, let $E_{r,p}$ be the set consisting of all $x \in \Omega$ such that

(4) $$c_0^{-1}|L_r(z)| \le |dT(x,z)| \le c_0|L_r(z)| \quad \text{for every } z \in \mathbb{R}^k$$

and

(5) $$c^{-1}|L_r(x) - L_r(y)| \le |T(x) - T(y)| \le c|L_r(x) - L_r(y)|$$
$$\text{for every } y \in \Omega \text{ such that } |y - x| \le \tfrac{1}{p}.$$

Let us show that $E_{r,p} \in \mathfrak{B}_{\mathbb{R}^k}$. For each $z \in \mathbb{R}^k$, let

(6) $$F_z = \left\{x \in \Omega : c_0^{-1}|L_r(z)| \le |dT(x,z)| \le c_0|L_r(z)|\right\}.$$

Let $\{z_i : i \in \mathbb{N}\}$ be a countable dense subset of $\mathbb{R}^k$. For a fixed $x \in \Omega$, $|dT(x,z)|$ and $|L_r(z)|$ are continuous real-valued functions of $z \in \mathbb{R}^k$. Thus

$$\bigcap_{z\in\mathbb{R}^k} F_z = \bigcap_{i\in\mathbb{N}} F_{z_i}. \tag{7}$$

Since T is a mapping of class $C^1(\Omega)$, the entries of the matrix of T are the first order partial derivatives of the component functions of T which are continuous functions of $x \in \Omega$. Thus for each $z \in \mathbb{R}^k$, $|dT(x,z)|$ is a continuous function of $x \in \Omega$ and therefore $F_z \in \mathfrak{B}_{\mathbb{R}^k}$ and in particular $F_{z_i} \in \mathfrak{B}_{\mathbb{R}^k}$ for every $i \in \mathbb{N}$. Then by (7) we have $\bigcap_{z\in\mathbb{R}^k} F_z \in \mathfrak{B}_{\mathbb{R}^k}$. This shows that the set of $x \in \Omega$ satisfying (4) is a member of $\mathfrak{B}_{\mathbb{R}^k}$. By similar argument, the set of all $x \in \Omega$ satisfying (5) is a member of $\mathfrak{B}_{\mathbb{R}^k}$. Then as the intersection of two members of $\mathfrak{B}_{\mathbb{R}^k}$, $E_{r,p}$ is a member of $\mathfrak{B}_{\mathbb{R}^k}$.

By the definition of $E_{r,p}$ as a subset of Ω, we have $\bigcup_{r\in\mathbb{N}}\bigcup_{p\in\mathbb{N}} E_{r,p} \subset \Omega$. Let us show that the union is actually equal to Ω. Let $x \in \Omega$ be arbitrarily chosen. Let us show that $x \in E_{r,p}$ for some $r \in \mathbb{N}$ and $p \in \mathbb{N}$. Now since $dT(x,\cdot)$ is a one-to-one mapping, we have $\delta_0 = \inf_{|z|=1} |dT(x,z)| > 0$. Choose $\delta > 0$ satisfying $\delta \le (c_0 - 1)\delta_0$ and $\delta \le \left(1 - c_0^{-1}\right)\delta_0$. Then select $L_r \in \mathbf{L}_0$ so that $\|L_r - dT(x,\cdot)\| < \delta$. Then for every $z \in \mathbb{R}^k$, we have by the triangle inequality of the norm in $\mathbb{R}^n$ and the definitions of δ and δ_0

$$\begin{aligned}|L_r(z)| &\le |dT(x,z)| + |L_r(z) - dT(x,z)| \le |dT(x,z)| + \delta|z| \\ &\le |dT(x,z)| + (c_0 - 1)\delta_0|z| \le |dT(x,z)| + (c_0 - 1)|dT(x,z)| \\ &= c_0|dT(x,z)|.\end{aligned}$$

Similarly we have

$$\begin{aligned}|L_r(z)| &\ge |dT(x,z)| - |L_r(z) - dT(x,z)| \ge |dT(x,z)| - \delta|z| \\ &\ge |dT(x,z)| - \left(1 - c_0^{-1}\right)\delta_0|z| \ge |dT(x,z)| - \left(1 - c_0^{-1}\right)|dT(x,z)| \\ &= c_0^{-1}|dT(x,z)|.\end{aligned}$$

Thus we have $c_0^{-1}|L_r(z)| \le |dT(x,z)| \le c_0|L_r(z)|$, that is, x satisfies (4). Since L_r is a non-singular linear transformation, we have $\eta = \inf_{|z|=1} |L_r(z)| > 0$. The differentiability of T at x implies that there exists $p \in \mathbb{N}$ such that

$$\begin{aligned}&|T(y) - T(x) - dT(x, y-x)| \le \varepsilon\eta|y-x| \le \varepsilon|L_r(y-x)| \\ &\text{for } x \in \Omega \text{ such that } |y-x| \le \tfrac{1}{p}.\end{aligned} \tag{8}$$

Then we have

$$\begin{aligned}|T(y) - T(x)| &\le |T(y) - T(x) - dT(x, y-x)| + |dT(x, y-x)| \\ &\le \varepsilon|L_r(y-x)| + c_0|L_r(y-x)| \le c|L_r(y-x)|,\end{aligned}$$

where the second inequality is by (8) and (4) and the last inequality is by (3). Similarly we have

$$|T(y)-T(x)| \geq |dT(x,y-x)| - |T(y)-T(x)-dT(x,y-x)|$$
$$\geq c_0^{-1}|L_r(y-x)| - \varepsilon|L_r(y-x)| \geq c^{-1}|L_r(y-x)|.$$

This shows that x satisfies (5). Thus $x \in E_{r,p}$. Therefore $\Omega = \bigcup_{r\in\mathbb{N}}\bigcup_{p\in\mathbb{N}} E_{r,p}$.

For our $E_{r,p} \in \mathfrak{B}_{\mathbb{R}^k}$, let $\{E_{r,p,q} : q \in \mathbb{N}\}$ be an arbitrary countable collection in $\mathfrak{B}_{\mathbb{R}^k}$ such that the diameter $|E_{r,p,q}| < \frac{1}{p}$ for $q \in \mathbb{N}$ and $\bigcup_{q\in\mathbb{N}} E_{r,p,q} = E_{r,p}$. Now $\Omega = \bigcup_{r\in\mathbb{N}}\bigcup_{p\in\mathbb{N}}\bigcup_{q\in\mathbb{N}} E_{r,p,q}$. Let $(A_j : j \in \mathbb{N})$ be an arbitrary numbering of the countable collection $\{E_{r,p,q} : q \in \mathbb{N}, p \in \mathbb{N}, r \in \mathbb{N}\}$. If $A_j = E_{r,p,q}$ in this renumbering, let $L_j = L_r$. If we let $B_1 = A_1$ and $B_j = A_j \setminus \bigcup_{i=1}^{j-1} B_i$ for $j \geq 2$, then $(B_j : j \in \mathbb{N})$ is a disjoint sequence in $\mathfrak{B}_{\mathbb{R}^k}$ and the sequence $(L_j : j \in \mathbb{N})$ of the corresponding non-singular linear transformations of $\mathbb{R}^k$ into $\mathbb{R}^n$ satisfy conditions (1) and (2). ■

Theorem 29.14. *Let $1 \leq k \leq n$. Let T be a mapping of class $C^1(\Omega)$ of an open set Ω in $\mathbb{R}^k$ into $\mathbb{R}^n$ such that T and the differential $dT(x,\cdot)$ for each $x \in \Omega$ are one-to-one mappings. Then for every $E \in \mathfrak{B}_{\mathbb{R}^k}$ such that $E \subset \Omega$, we have $T(E) \in \mathfrak{B}_{\mathbb{R}^n}$ and*

$$\text{(1)} \qquad \mathcal{H}^k\big(T(E)\big) = \int_E J\big(dT(x,\cdot)\big)\,\mathcal{H}^k(dx).$$

If f is a real-valued $\mathfrak{B}_{\mathbb{R}^n}$-measurable function on $T(\Omega)$, then

$$\text{(2)} \qquad \int_{T(\Omega)} f(y)\,\mathcal{H}^k(dy) = \int_\Omega (f\circ T)(x)J\big(dT(x,\cdot)\big)\,\mathcal{H}^k(dx),$$

in the sense that the existence of one side implies that of the other and the equality of the two. In particular the equality holds if f is nonnegative valued.

Proof. 1. If $E \in \mathfrak{B}_{\mathbb{R}^k}$ and $E \subset \Omega$, then $T(E) \in \mathfrak{B}_{\mathbb{R}^n}$ by Lemma 29.12. For $c > 1$ arbitrarily given, let $(B_j : j \in \mathbb{N})$ be a disjoint sequence in $\mathfrak{B}_{\mathbb{R}^k}$ such that $\bigcup_{n\in\mathbb{N}} B_j = \Omega$ and let $(L_j : j \in \mathbb{N})$ a sequence of non-singular linear transformations of $\mathbb{R}^k$ into $\mathbb{R}^n$ satisfying (1) and (2) of Lemma 29.13. Let $E_j = E \cap B_j$ for $j \in \mathbb{N}$. By (1) of Lemma 29.13, for every $x \in E_j$ we have $c^{-1}|L_j(z'-z'')| \leq |dT(x,z'-z'')| \leq c|L_j(z'-z'')|$ for $z', z'' \in \mathbb{R}^k$. Thus by (a) of Theorem 28.5, we have

$$\text{(3)} \qquad c^{-k}\mathcal{H}^k\big(L_j(A)\big) \leq \mathcal{H}^k\big(dT(x,A)\big) \leq c^k\mathcal{H}^k\big(L_j(A)\big)$$

for $x \in E_j$ and $A \in \mathfrak{P}(\mathbb{R}^k)$. Applying Theorem 29.11 to the two non-singular linear transformations $dT(x,\cdot)$ and L_j of $\mathbb{R}^k$ into $\mathbb{R}^n$, we have

$$\text{(4)} \qquad \begin{cases} \mathcal{H}^k\big(dT(x,A)\big) = J\big(dT(x,\cdot)\big)\mathcal{H}^k(A), \\ \mathcal{H}^k\big(L_j(A)\big) = J(L_j)\mathcal{H}^k(A) \end{cases}$$

for $x \in E_j$ and $A \in \mathfrak{P}(\mathbb{R}^k)$. By (3) and (4), we have

$$\text{(5)} \qquad c^{-k}J(L_j) \leq J\big(dT(x,\cdot)\big) \leq c^kJ(L_j) \quad \text{for } x \in E_j.$$

By (2) of Lemma 29.13, we have $c^{-1}|L_j(x-y)| \le |T(x)-T(y)| \le c|L_j(x-y)|$ for $x, y \in E_j$. Thus by (a) of Theorem 28.5, we have

$$(6) \qquad c^{-k}\mathcal{H}^k\big(L_j(E_j)\big) \le \mathcal{H}^k\big(T(E_j)\big) \le c^k\mathcal{H}^k\big(L_j(E_j)\big).$$

By the second expression in (4), we have $\mathcal{H}^k\big(L_j(E_j)\big) = J(L_j)\mathcal{H}^k(E_j)$. Using this in (6), we have

$$(7) \qquad c^{-k}J(L_j)\mathcal{H}^k(E_j) \le \mathcal{H}^k\big(T(E_j)\big) \le c^k J(L_j)\mathcal{H}^k(E_j).$$

Applying (7) and (5), we have

$$c^{-2k}\mathcal{H}^k\big(T(E_j)\big) \le c^{-k}J(L_j)\mathcal{H}^k(E_j) = \int_{E_j} c^{-k}J(L_j)\,\mathcal{H}^k(dx)$$
$$\le \int_{E_j} J\big(dT(x,\cdot)\big)\,\mathcal{H}^k(dx) \le c^k J(L_j)\mathcal{H}^k(E_j) \le c^{2k}\mathcal{H}^k\big(T(E_j)\big).$$

Letting $c \downarrow 1$, we have $\mathcal{H}^k\big(T(E_j)\big) = \int_{E_j} J\big(dT(x,\cdot)\big)\,\mathcal{H}^k(dx)$ for every $j \in \mathbb{N}$. Summing these equalities over $j \in \mathbb{N}$, we have (1).

2. To prove (2), note first that $T(\Omega) \in \mathfrak{B}_{\mathbb{R}^n}$ by Lemma 29.12. Now since T is a continuous mapping of $\Omega \in \mathfrak{B}_{\mathbb{R}^k}$ into $\mathbb{R}^n$, T is a $\mathfrak{B}_{\mathbb{R}^k}/\mathfrak{B}_{\mathbb{R}^n}$-measurable mapping by Theorem 1.43. Thus for every $F \in \mathfrak{B}_{\mathbb{R}^n}$ such that $F \subset T(\Omega)$, we have $T^{-1}(F) \in \mathfrak{B}_{\mathbb{R}^k}$ so that there exists $E \in \mathfrak{B}_{\mathbb{R}^k}$, $E \subset \Omega$, such that $F = T(E)$. Then by (1), we have the equality $\mathcal{H}^k\big(T(E)\big) = \int_E J\big(dT(x,\cdot)\big)\,\mathcal{H}^k(dx)$, that is, $\mathcal{H}^k(F) = \int_{T^{-1}(F)} J\big(dT(x,\cdot)\big)\,\mathcal{H}^k(dx)$. Thus

$$\int_{T(\Omega)} \mathbf{1}_F(y)\,\mathcal{H}^k(dy) = \int_\Omega \mathbf{1}_{T^{-1}(F)} J\big(dT(x,\cdot)\big)\,\mathcal{H}^k(dx)$$
$$= \int_\Omega (\mathbf{1}_F \circ T)(x) J\big(dT(x,\cdot)\big)\,\mathcal{H}^k(dx),$$

proving (2) for the particular case $f = \mathbf{1}_F$ where $F \in \mathfrak{B}_{\mathbb{R}^n}$ and $F \subset T(\Omega)$. Then by the linearity of the integrals with respect to the integrands, (2) holds when f is a nonnegative simple function on the measure space $\big(T(\Omega), \mathfrak{B}_{\mathbb{R}^n} \cap T(\Omega), \mathcal{H}^k\big)$. If f is a nonnegative real-valued $\mathfrak{B}_{\mathbb{R}^n}$-measurable function on $T(\Omega)$, then there exists an increasing sequence $(\varphi_j : j \in \mathbb{N})$ nonnegative simple functions on $\big(T(\Omega), \mathfrak{B}_{\mathbb{R}^n} \cap T(\Omega), \mathcal{H}^k\big)$ such that $\varphi_j \uparrow f$ on $T(\Omega)$ by Lemma 8.6. Applying the Monotone Convergence Theorem (Theorem 8.5), we have (2) for our function f. If f is an arbitrary real-valued $\mathfrak{B}_{\mathbb{R}^n}$-measurable function on $T(\Omega)$, then we write $f = f^+ - f^-$ and apply the result above to f^+ and f^-. ∎

[IV] Hausdorff Measure of Fractional Dimension

Definition 29.15. *A set E in a topological space X is said to be totally disconnected if for every pair of distinct points x_1 and x_2 in E there exists a pair of disjoint open sets O_1 and O_2 in X such that $x_1 \in O_1$, $x_2 \in O_2$, and $E \subset O_1 \cup O_2$.*

Proposition 29.16. *Let $E \in \mathfrak{P}(\mathbb{R}^n)$. If $\dim_H E < 1$, then E is totally disconnected.*

Proof. Let $x_1 \in E$ be arbitrarily fixed and define a mapping T of $\mathbb{R}^n$ into $\mathbb{R}$ by setting $T(y) = |y - x_1|$ for $y \in \mathbb{R}^n$. Then we have $|T(y) - T(z)| = \big||y - x_1| - |z - x_1|\big| \le |y - z|$ for any $y, z \in \mathbb{R}^n$, so that T is a Lipschitz mapping. Thus by (d) of Theorem 28.4, we have $\dim_H T(E) \le \dim_H E < 1$. This implies that $\mathcal{H}^1\big(T(E)\big) = 0$ by the definition of the Hausdorff dimension. By Theorem 29.2, we have $\mu_L^*\big(T(E)\big) = \kappa_1^{-1}\mathcal{H}^1\big(T(E)\big) = 0$. This implies that $T(E)$ cannot contain any open interval in $\mathbb{R}$.

Let $x_2 \in E$, $x_2 \ne x_1$, be arbitrarily chosen. We have $T(x_2) = |x_2 - x_1| > 0$. Consider the open interval $\big(0, T(x_2)\big)$ in $\mathbb{R}$. Since $T(E)$ cannot contain any open interval in $\mathbb{R}$, there exists $\xi \in \big(0, T(x_2)\big)$ such that $\xi \notin T(E)$. Since T is a Lipschitz mapping of $\mathbb{R}^n$ into $\mathbb{R}$, it is a continuous mapping. Thus $O_1 := \{y \in \mathbb{R}^n : T(y) < \xi\}$ and $O_2 := \{y \in \mathbb{R}^n : T(y) > \xi\}$ are two open sets in $\mathbb{R}^n$. Clearly they are disjoint. Since $T(x_1) = 0 < \xi$, we have $x_1 \in O_1$, and since $T(x_2) > \xi$ we have $x_2 \in O_2$. For any $x \in E$, since $\xi \notin T(E)$ we have $T(x) \ne \xi$ so that $x \in O_1 \cup O_2$. This shows that $E \subset O_1 \cup O_2$. Thus we have shown that for any pair of distinct points x_1 and x_2 in E, there exists a pair of disjoint open sets O_1 and O_2 in X such that $x_1 \in O_1$, $x_2 \in O_2$, and $E \subset O_1 \cup O_2$. This shows that E is a totally disconnected set. ∎

For examples of sets with Hausdorff dimensions in the range $(0, 1)$, we define Cantor sets C_λ with ratio $\lambda \in \big(0, \frac{1}{2}\big)$ next. The Cantor ternary set defined in §4 is a particular case of C_λ with $\lambda = \frac{1}{3}$.

Definition 29.17. *Let λ be a positive number in the range $\big(0, \frac{1}{2}\big)$. We define a sequence $(\mathfrak{J}_k : k \in \mathbb{Z}_+)$ of classes of closed intervals in $\mathbb{R}$ inductively as follows. Let $\mathfrak{J}_0 = \{J_{0,1}\}$ where $J_{0,1} = [0, 1]$. Let $\mathfrak{J}_1 = \{J_{1,1}, J_{1,2}\}$ where $J_{1,1}$ and $J_{1,2}$ are the two closed intervals of length λ obtained by removing an open interval of length $1 - 2\lambda$ from the center of $J_{0,1}$. Suppose for some $k \in \mathbb{N}$, the collection $\mathfrak{J}_{k-1} = \{J_{k-1,i} : i = 1, \dots, 2^{k-1}\}$ of 2^{k-1} closed intervals of length λ^{k-1} has been defined. We define $\mathfrak{J}_k = \{J_{k,i} : i = 1, \dots, 2^k\}$ as the collection of 2^k closed intervals of length λ^k obtained by removing an open interval of length $(1 - 2\lambda)\lambda^{k-1}$ from the center of each of the members of $\mathfrak{J}_{k-1}$. We continue the process indefinitely. Let $\mathfrak{J} = \bigcup_{k \in \mathbb{Z}_+} \mathfrak{J}_k$. For each $k \in \mathbb{Z}_+$, let $E_k = \bigcup_{i=1}^{2^k} J_{k,i}$. We define the Cantor set C_λ with ratio $\lambda \in \big(0, \frac{1}{2}\big)$ by setting $C_\lambda = \bigcap_{k \in \mathbb{Z}_+} E_k$.*

Note that $E_k \in \mathfrak{B}_{\mathbb{R}}$ for every $k \in \mathbb{Z}_+$ so that $C_\lambda \in \mathfrak{B}_{\mathbb{R}}$. Also $\mu_L(E_k) = 2^k\lambda^k = (2\lambda)^k$ and $\mu_L(C_\lambda) = \lim_{k\to\infty} \mu_L(E_k) = 0$ since $2\lambda < 1$.

Proposition 29.18. *The Cantor set C_λ has the following properties:*
(a) *C_λ is a null set in $\big(\mathbb{R}, \mathfrak{B}_{\mathbb{R}}, \mu_L\big)$.*
(b) *C_λ is an uncountable set.*
(c) *C_λ is a compact set in $\mathbb{R}$.*
(d) *C_λ is a perfect set in $\mathbb{R}$.*
(e) *C_λ is nowhere dense in $\mathbb{R}$, that is, $\big(\overline{C_\lambda}\big)^\circ = \emptyset$.*

Proof. These statements can be proved in the same way as Theorem 4.33 for the Cantor ternary set. ■

Lemma 29.19. *Regarding the classes of closed intervals $\mathfrak{J}_k$ for $k \in \mathbb{Z}_+$ and $\mathfrak{J} = \bigcup_{k \in \mathbb{Z}_+} \mathfrak{J}_k$ in Definition 29.17, we have the following:*
(a) *If x is an endpoint of a member of $\mathfrak{J}$, then $x \in C_\lambda$.*
(b) *If $x \in C_\lambda$ and x is not the left endpoint of any member of $\mathfrak{J}$, then there exists a sequence $(x_m : m \in \mathbb{N})$ such that x_m is the left endpoint of a member of $\mathfrak{J}$, $x_m < x$, and $x_m \uparrow x$. Similarly if $x \in C_\lambda$ and x is not the right endpoint of any member of $\mathfrak{J}$, then there exists a sequence $(x_m : m \in \mathbb{N})$ such that x_m is the right endpoint of a member of $\mathfrak{J}$, $x_m > x$, and $x_m \downarrow x$.*
(c) *If $x \notin C_\lambda$ and there exists a point $\xi \in C_\lambda$ such that $x < \xi$, then there exists $\alpha \in C_\lambda$ such that $x < \alpha$, α is the left endpoint of a member of $\mathfrak{J}$ and $[x, \alpha) \cap C_\lambda = \emptyset$. Similarly if $y \notin C_\lambda$ and there exists a point $\eta \in C_\lambda$ such that $\eta < y$, then there exists $\beta \in C_\lambda$ such that $\beta < y$, β is the right endpoint of a member of $\mathfrak{J}$ and $(\beta, y] \cap C_\lambda = \emptyset$.*

Proof. 1. If x is a left endpoint of a member of $\mathfrak{J}$, then x is the left endpoint of a member of $\mathfrak{J}_k$ for some $k \in \mathbb{Z}_+$. Then x is the left endpoint of a member of $\mathfrak{J}_m$ for every $m \geq k$ so that $x \in E_m$ for every $m \geq k$. Then $x \in C_\lambda = \bigcap_{k \in \mathbb{Z}_+} E_k$. Similarly for the case that x is a right endpoint of a member of $\mathfrak{J}$.

2. Suppose $x \in C_\lambda$ and x is not the left endpoint of any member of $\mathfrak{J}$. Now since $x \in C_\lambda = \bigcap_{k \in \mathbb{Z}_+} E_k$, we have $x \in E_k$ for every $k \in \mathbb{Z}_+$. For every $n \in \mathbb{N}$, there exists $k \in \mathbb{Z}_+$ such that $\lambda^k < \frac{1}{n}$. Since $x \in E_k = \bigcup_{i=1}^{2^k} J_{k,i}$, we have $x \in J_{k,i}$ for some $i = 1, \ldots, 2^K$. Since $|J_{k,i}| = \lambda^k < \frac{1}{n}$, the distance between the left endpoint ξ_n of $J_{k,i}$ and x is less than $\frac{1}{n}$. Thus $\xi_n < x$ and $x - \xi_n < \frac{1}{n}$. Then $\lim_{n \to \infty} \xi_n = x$. Let $x_m = \max\{\xi_1, \ldots, \xi_m\}$ for $m \in \mathbb{N}$. Then $(x_m : m \in \mathbb{N})$ is a sequence of left endpoints of members of $\mathfrak{J}$, $x_m < x$, and $x_m \uparrow x$. Similarly for the case that $x \in C_\lambda$ and x is not the right endpoint of any member of $\mathfrak{J}$.

3. Suppose $x \notin C_\lambda$ and there exists a point $\xi \in C_\lambda$ such that $x < \xi$. Then $[x, \infty) \cap C_\lambda$ is nonempty and is a compact set. Let α be the nearest point in the compact set $[x, \infty) \cap C_\lambda$ from x. Then $[x, \alpha) \cap C_\lambda = \emptyset$ so that α must be a left endpoint of a member of $\mathfrak{J}$ by (b). Similarly for the case that $y \notin C_\lambda$ and there exists a point $\eta \in C_\lambda$ such that $\eta < y$. ■

With the notations introduced in Definition 29.17, let $C' = C_\lambda \cap J_{1,1}$ and $C'' = C_\lambda \cap J_{1,2}$. Then $C' \cap C'' = \emptyset$, $C' \cup C'' = C_\lambda$. Moreover it is evident from the construction of C_λ that $C' = \lambda C_\lambda$, a dilation of C_λ by λ, and $C'' = C' + (1 - \lambda)$, a translate of C' by $1 - \lambda \in \mathbb{R}$, as can be seen by the construction of C_λ. Since $C_\lambda \in \mathfrak{B}_\mathbb{R} \subset \mathfrak{M}(\mathcal{H}^s)$ for an arbitrary $s \in [0, \infty)$, we have $C' \in \mathfrak{M}(\mathcal{H}^s)$ by Theorem 28.2 and, since a translation is an isometry, $C'' \in \mathfrak{M}(\mathcal{H}^s)$ by Theorem 28.11. Then since $\mathcal{H}^s$ is a measure on the σ-algebra $\mathfrak{M}(\mathcal{H}^s)$, we have $\mathcal{H}^s(C_\lambda) = \mathcal{H}^s(C') + \mathcal{H}^s(C'')$. Now $\mathcal{H}^s(C') = \lambda^s \mathcal{H}^s(C_\lambda)$ by Theorem 28.1 and $\mathcal{H}^s(C'') = \mathcal{H}^s(C')$ by (c) of Theorem 28.10 so that we have

$$(1) \qquad \mathcal{H}^s(C_\lambda) = 2\lambda^s \mathcal{H}^s(C_\lambda) \quad \text{for every } s \in [0, \infty).$$

Now consider the nonnegative extended real-valued function $\mathcal{H}^s(C_\lambda)$ of $s \in [0, \infty)$. The fact that C_λ is an infinite set implies that $\mathcal{H}^0(C_\lambda) = \infty$ by Lemma 27.31 and the fact that

$C_\lambda \subset \mathbb{R}$ implies that $\mathcal{H}^s(C_\lambda) = 0$ for $s > 1$ by (c) of Lemma 27.30. Since $C_\lambda \neq \emptyset$, $\mathcal{H}^\bullet(C_\lambda)$ has exactly one point of discontinuity and the discontinuity occurs at a point in $[0, 1]$ according to (d) of Theorem 27.33. Now ∞, a finite positive number, and 0 are the three possible values of $\mathcal{H}^\bullet(C_\lambda)$ at its point of discontinuity $s \in [0, 1]$. If we assume that $\mathcal{H}^s(C_\lambda) \in (0, \infty)$, then dividing (1) by this positive number $\mathcal{H}^s(C_\lambda)$ we have $1 = 2\lambda^s$. Thus $\log 2 + s \log \lambda = 0$ and therefore

$$s = -\frac{\log 2}{\log \lambda} = \frac{\log 2}{\log \lambda^{-1}}. \tag{2}$$

The calculation above for the point of discontinuity $s = \log 2/\log \lambda^{-1}$ of the function $\mathcal{H}^\bullet(C_\lambda)$ was based on the assumption that the value of $\mathcal{H}^\bullet(C_\lambda)$ at the point of discontinuity is a finite positive number. In what follows we define $s^* = \log 2/\log \lambda^{-1}$ and show that $\mathcal{H}^{s^*}(C_\lambda)$ is indeed a finite positive number. By (e) of Theorem 27.35, this implies that $\dim_H C_\lambda = s^* = \log 2/\log \lambda^{-1}$.

Lemma 29.20. *For* $\lambda \in \left(0, \frac{1}{2}\right)$, *let* $s^* = \log 2/\log \lambda^{-1} = -\log 2/\log \lambda \in (0, 1)$.
(a) *For every* $x \in \mathbb{R}$, *we have* $2^x = \lambda^{-xs^*}$.
(b) *If* $J \in \mathfrak{J}_{k_0}$ *for some* $k_0 \in \mathbb{Z}_+$, *then* $|J|^{s^*} = \sum_{J_{k,i} \subset J} |J_{k,i}|^{s^*}$ *for any* $k \geq k_0$.
(c) *For every* $k \in \mathbb{Z}_+$, *we have* $\sum_{i=1}^{2^k} |J_{k,i}|^{s^*} = 1$.
(d) *For an arbitrary open interval in* $\mathbb{R}$ *and* $k \in \mathbb{Z}_+$, *we have* $\sum_{J_{k,i} \subset I} |J_{k,i}|^{s^*} \leq 4|I|^{s^*}$.

Proof. 1. To prove (a), note that for $x \in \mathbb{R}$, we have

$$\begin{aligned} 2^x &= e^{x \log 2} = \exp\left\{x \log 2 (\log \lambda^{-1})^{-1} \log \lambda^{-1}\right\} \\ &= \exp\left\{x s^* \log \lambda^{-1}\right\} = \lambda^{-xs^*}. \end{aligned}$$

2. If $J \in \mathfrak{J}_{k_0}$ for some $k_0 \in \mathbb{Z}_+$ then for any $k \in \mathbb{Z}_+$ such that $k \geq k_0$, the interval J contains 2^{k-k_0} members of $\mathfrak{J}_k = \left\{J_{k,i} : i = 1, \ldots, 2^k\right\}$. Since $|J_{k,i}| = \lambda^k$ for every $i = 1, \ldots, 2^k$, we have $\sum_{J_{k,i} \subset J} |J_{k,i}|^{s^*} = 2^{k-k_0}\lambda^{ks^*}$. Now $2^{k-k_0} = \lambda^{(k_0-k)s^*}$ by (a). Thus

$$\sum_{J_{k,i} \subset J} |J_{k,i}|^{s^*} = \lambda^{(k_0-k)s^*}\lambda^{ks^*} = \lambda^{k_0 s^*} = |J|^{s^*}.$$

This proves (b).

3. (c) is a particular case of (b). Indeed $J_{0,1} = [0, 1] \in \mathfrak{J}_0$ contains all members of $\mathfrak{J}_k = \left\{J_{k,i} : i = 1, \ldots, 2^k\right\}$ for an arbitrary $k \in \mathbb{Z}_+$ so that by (b) we have

$$1 = |J_{0,1}|^{s*} = \sum_{J_{k,i} \subset J_{0,1}} |J_{k,i}|^{s^*} = \sum_{i=1}^{2^k} |J_{k,i}|^{s^*}.$$

4. Now suppose that an open interval I in $\mathbb{R}$ contains no members of the class $\mathfrak{J}_k = \left\{J_{k,i} : i = 1, \ldots, 2^k\right\}$ for some $k \in \mathbb{Z}_+$. Then the inequality in (d) holds trivially. Suppose

I contains some member of $\mathfrak{J}_k$. Let k_0 be the smallest nonnegative integer such that I contains some member of $\mathfrak{J}_{k_0}$. Then $k_0 \leq k$. Let $J_{k_0,j_1}, \ldots, J_{k_0,j_p}$ be all the members of $\mathfrak{J}_{k_0}$ that intersect I. Then $p \leq 4$ for otherwise I would contain some member of $\mathfrak{J}_{k_0-1}$. Now since $p \leq 4$ and $J_{k_0,j_n} \subset I$ for $n = 1, \ldots, p$, we have

$$4|I|^{s^*} \geq \sum_{n=1}^{p} |J_{k_0,j_n}|^{s^*} = \sum_{n=1}^{p} \sum_{J_{k,i} \subset J_{k_0,j_n}} |J_{k,i}|^{s^*} \geq \sum_{J_{k,i} \subset I} |J_{k,i}|^{s^*},$$

where the equality is by (b). ∎

Lemma 29.21. *Let $p \in [0, 1]$ and $\lambda \in [0, 1]$. Then for any $a_1, a_2 > 0$, we have*

$$\{\lambda a_1 + (1-\lambda)a_2\}^p \geq \lambda a_1^p + (1-\lambda)a_2^p, \tag{1}$$

$$a_1^p + a_2^p \geq \{a_1 + a_2\}^p. \tag{2}$$

Moreover if $a_n > 0$ for $n \in \mathbb{N}$, then

$$a_1^p + \cdots + a_N^p \geq \{a_1 + \cdots + a_N\}^p \quad \textit{for every } N \in \mathbb{N}, \tag{3}$$

$$\sum_{n \in \mathbb{N}} a_n^p \geq \Big(\sum_{n \in \mathbb{N}} a_n\Big)^p. \tag{4}$$

Proof. Consider the function $f(x) = -x^p$ for $x \in (0, \infty)$. Since $f'(x) = -px^{p-1}$ is an increasing function of $x \in (0, \infty)$, f is a convex function on $(0, \infty)$ by (b) of Theorem 14.10. The convexity of f implies $f\big(\lambda a_1 + (1-\lambda)a_2\big) \leq \lambda f(a_1) + (1-\lambda) f(a_2)$. Multiplying both sides by -1 we obtain (1).

To prove (2), we may assume without loss of generality that $a_1 \leq a_2$. Assume further that $a_1 < a_2$. Consider $0 < a_0 < a_1 < a_2 < a_1 + a_2$. Applying Proposition 14.4, we have

$$\frac{f(a_1) - f(a_0)}{a_1 - a_0} \leq \frac{f(a_2) - f(a_0)}{a_2 - a_0} \leq \frac{f(a_2) - f(a_1)}{a_2 - a_1}$$
$$\leq \frac{f(a_1 + a_2) - f(a_1)}{(a_1 + a_2) - a_1} \leq \frac{f(a_1 + a_2) - f(a_2)}{(a_1 + a_2) - a_2}.$$

Multiplying the first and the last members in the chain of inequalities above by -1, we have

$$-\frac{f(a_1) - f(a_0)}{a_1 - a_0} \geq -\frac{f(a_1 + a_2) - f(a_2)}{a_1},$$

that is,

$$\frac{a_1^p - a_0^p}{a_1 - a_0} \geq \frac{\{a_1 + a_2\}^p - a_2^p}{a_1}.$$

Letting $a_0 \downarrow 0$, we obtain

$$\frac{a_1^p}{a_1} \geq \frac{\{a_1 + a_2\}^p - a_2^p}{a_1},$$

that is, $a_1^p + a_2^p \geq \{a_1 + a_2\}^p$. For the case $a_1 = a_2$, by letting $a_2 \downarrow a_1$, we have $a_1^p + a_1^p \geq \{a_1 + a_1\}^p$. This proves (2). By repeated application of (2), we obtain (3). Letting $N \to \infty$ in (3), we obtain (4) by the continuity of the power function. ∎

Theorem 29.22. *For the Cantor set C_λ with $\lambda \in \left(0, \frac{1}{2}\right)$, if we let $s^* = \log 2/\log \lambda^{-1}$, then*

$$\tfrac{1}{4} \leq \mathcal{H}^{s^*}(C_\lambda) \leq 1,$$

and therefore we have $\dim_H C_\lambda = s^*$.

Proof. 1. A set in $\mathbb{R}$ is a convex set if and only if it is an interval. Now for every $k \in \mathbb{Z}_+$, the class $\mathfrak{J}_k = \{J_{k,j} : i = 1, \dots, 2^k\}$ is a λ^k-cover of C_λ by closed convex sets. Thus by Theorem 27.25 and by (c) of Lemma 29.20, we have $\mathcal{H}^{s^*}_{\lambda^k}(C_\lambda) \leq \sum_{i=1}^{2^k} |J_{k,i}|^{s^*} = 1$. Since $\mathcal{H}^{s^*}_\delta(C_\lambda) \uparrow \mathcal{H}^{s^*}(C_\lambda)$ as $\delta \downarrow 0$, we have $\mathcal{H}^{s^*}(C_\lambda) = \lim_{\delta\downarrow 0} \mathcal{H}^{s^*}_\delta(C_\lambda) = \lim_{k\to\infty} \mathcal{H}^{s^*}_{\lambda^k}(C_\lambda) \leq 1$.

2. We show that if $(I_j : j \in \mathbb{N})$ is a sequence of open intervals and $\bigcup_{j\in\mathbb{N}} I_j \supset C_\lambda$, then

$$(1) \qquad \sum_{j\in\mathbb{N}} |I_j|^{s^*} \geq \frac{1}{4}.$$

Now since $s^* \in (0, 1)$, (3) and (4) of Lemma 29.21 imply that if we replace intersecting members of $(I_j : j \in \mathbb{N})$ by their union we do not increase the sum $\sum_{j\in\mathbb{N}} |I_j|^{s^*}$. Therefore in proving (1), we may assume without loss of generality that $(I_j : j \in \mathbb{N})$ is a disjoint sequence. Now since C_λ is a compact set, $\bigcup_{j\in\mathbb{N}} I_j \supset C_\lambda$ implies that there exists $N \in \mathbb{N}$ such that $\bigcup_{j=1}^N I_j \supset C_\lambda$. Since C_λ has no interior points, we can enlarge I_j slightly so that the endpoints of the interval are in C_λ^c. Thus for an arbitrary $\varepsilon > 0$, let I'_j be an open interval such that $I'_j \supset I_j$, the endpoints of I'_j are in C_λ^c, and

$$(2) \qquad \sum_{j\in\mathbb{N}} |I'_j|^{s^*} \leq \sum_{j\in\mathbb{N}} |I_j|^{s^*} + \varepsilon.$$

Let δ be the minimum of the distances from the endpoints of $I'_1, \dots, I'_N$ to C_λ. Since the endpoints are outside of the compact set C_λ, we have $\delta > 0$. Let $k \in \mathbb{Z}_+$ be so large that $\lambda^k < \delta$.

Consider the class $\mathfrak{J} = \{J_{k,i} : i = 1, \dots, 2^k\}$ of closed intervals each with length λ^k. Let us show that every member $J_{k,i}$ of $\mathfrak{J}_k$ is contained in one of the open intervals $I'_1, \dots, I'_N$. Let $a < b$ be the endpoints of $J_{k,i}$. By (a) of Lemma 29.19, we have $a, b \in C_\lambda \subset \bigcup_{j=1}^N I'_j$. Suppose $a \in I'_n$ and $b \in I'_m$ where $n \neq m$. If we let α be the right endpoint of i'_n and β be the left endpoint of I'_m, then $a < \alpha < \beta < b$. This contradicts the fact that $b - a = \lambda^k < \delta$ while $\alpha - a \geq \delta$ and $b - \beta \geq \delta$. Thus a and b must be contained in the same member of $\{I'_1, \dots, I'_N\}$. This implies that $J_{k,i}$ is contained in one member of $\{I'_1, \dots, I'_N\}$. Therefore every member of $\mathfrak{J}_k$ is contained in one member of $\{I'_1, \dots, I'_N\}$. Then by (d) and (c) of Lemma 29.20, we have

$$4\sum_{j=1}^N |I'_j|^{s^*} \geq \sum_{j=1}^N \sum_{J_{k,i}\subset I'_j} |J_{k,i}|^{s^*} = \sum_{j=1}^{2^k} |J_{k,i}|^{s^*} = 1.$$

Then by (2), we have

$$\sum_{j=1}^{N} |I_j|^{s^*} + \varepsilon \geq \sum_{j=1}^{N} |I_j'|^{s^*} \geq \frac{1}{4}.$$

Thus by the arbitrariness of $\varepsilon > 0$, we have (1).

3. Let $\delta > 0$. Let $(K_j : j \in \mathbb{N})$ be a δ-cover of C_λ by closed convex sets in $\mathbb{R}$. Now a closed convex set in $\mathbb{R}$ is a closed interval. Thus for an arbitrary $\varepsilon > 0$, we can select an open interval $I_j \supset K_j$ such that $\sum_{j\in\mathbb{N}} |I_j|^{s^*} - \varepsilon \leq \sum_{j\in\mathbb{N}} |K_j|^{s^*}$. Since $(I_j : j \in \mathbb{N})$ is a sequence of open intervals such that $\bigcup_{j\in\mathbb{N}} I_j \supset C_\lambda$, we have $\sum_{j\in\mathbb{N}} |I_j|^{s^*} \geq \frac{1}{4}$ by (1). Then $\sum_{j\in\mathbb{N}} |K_j|^{s^*} \geq \frac{1}{4} - \varepsilon$. Now by Theorem 27.25, we have

$$\begin{aligned}\mathcal{H}_\delta^{s^*}(C_\lambda) &= \inf\left\{\textstyle\sum_{j\in\mathbb{N}} |K_j|^{s^*} : (K_j : j \in \mathbb{N}) \text{ is a } \delta\text{-cover of } C_\lambda \text{ by closed convex sets}\right\} \\ &\geq \frac{1}{4} - \varepsilon.\end{aligned}$$

By the arbitrariness of $\varepsilon > 0$, we have $\mathcal{H}_\delta^{s^*}(C_\lambda) \geq \frac{1}{4}$ and $\mathcal{H}^{s^*}(C_\lambda) = \lim_{\delta\downarrow 0} \mathcal{H}_\delta^{s^*}(C_\lambda) \geq \frac{1}{4}$. Thus $\frac{1}{4} \leq \mathcal{H}^{s^*}(C_\lambda) \leq 1$. Then $\dim_H C_\lambda = s^*$ by (e) of Theorem 27.35. ■

Corollary 29.23. *For every $s \in (0, 1)$ there exists a subset E of $\mathbb{R}$ with $\dim_H E = s$. Indeed for the Cantor set C_λ with $\lambda = \exp\{-\frac{1}{s}\log 2\}$ we have $\dim_H(C_\lambda) = s$.*

Proof. According to Theorem 29.22, for a Cantor set C_λ with $\lambda \in (0, \frac{1}{2})$, we have $\dim_H(E) = \log 2/\log \lambda^{-1}$. Now the function $\varphi(\lambda) = \log 2/\log \lambda^{-1}$ for $\lambda \in (0, \frac{1}{2})$ is a strictly increasing continuous function mapping $(0, \frac{1}{2})$ one-to-one onto $(0, 1)$. Its inverse function φ^{-1} is then a strictly increasing continuous function mapping $(0, 1)$ one-to-one onto $(0, \frac{1}{2})$. Solving $s = \log 2/\log \lambda^{-1}$ for λ, we obtain the inverse function $\lambda = \varphi^{-1}(s) = \exp\{-\frac{1}{s}\log 2\}$ for $s \in (0, 1)$. Then for $s \in (0, 1)$, for the Cantor set $C_{\varphi^{-1}(s)}$ we have $\dim_H C_{\varphi^{-1}(s)} = \varphi(\varphi^{-1}(s)) = s$ by Theorem 29.22. ■

We show next that actually $\mathcal{H}^{s^*}(C_\lambda) \geq 1$. This is a shaper estimate than the estimate $\mathcal{H}^{s^*}(C_\lambda) \geq \frac{1}{4}$ in Theorem 29.22. However the simple argument in obtaining $\mathcal{H}^{s^*}(C_\lambda) \geq \frac{1}{4}$ can be generalized in other situations where precise value of the Hausdorff measure is not necessary.

Theorem 29.24. *For the Cantor set C_λ with $\lambda \in (0, \frac{1}{2})$, we have $\mathcal{H}^{s^*}(C_\lambda) = 1$ where $s^* = \log 2/\log \lambda^{-1}$.*

Proof. 1. For an arbitrary $\delta > 0$, let $I = (a, b)$ be an open interval such that $a, b \in C_\lambda^c$ and $I \cap C_\lambda \neq \emptyset$. Then there exists $\xi \in C_\lambda$ such that $a < \xi$. Thus by (c) of Lemma 29.19, there exists $\alpha \in C_\lambda$ such that $a < \alpha$, α is the left endpoint of a member of $\mathfrak{J} = \bigcup_{k\in\mathbb{Z}_+} \mathfrak{J}_k$ and $(a, \alpha) \cap C_\lambda = \emptyset$. Similarly there exists $\beta \in C_\lambda$ such that $\beta < b$, β is the right endpoint of a member of $\mathfrak{J}$, and $(\beta, b) \cap C_\lambda = \emptyset$. Thus we have $I = (a, b) = (a, \alpha) \cup [\alpha, \beta] \cup (\beta, b)$ and $(a, \alpha), (\beta, b) \subset C_\lambda^c$.

Now if a point is the left (resp. right) endpoint of a member of $\mathfrak{J}$, then it is the left (resp. right) endpoint of a member of $\mathfrak{J}_k$ for some $k \in \mathbb{Z}_+$ and then it is the left (resp. right) endpoint of a member of $\mathfrak{J}_\ell$ for every $\ell \geq k$. Therefore there exists $k \in \mathbb{Z}_+$ such that α is the left endpoint of a member of $\mathfrak{J}_k$ and β is the right endpoint of a member of the same $\mathfrak{J}_k$. Let $\{J_{k,i_p} : p = 1, \ldots, q\}$ be the collection of all members of $\mathfrak{J}_k$ contained in $[\alpha, \beta]$ and enumerated from left to right. We have

$$(1) \qquad I \cap C_\lambda \subset J_{k,i_1} \cup \cdots \cup J_{k,i_q}.$$

Let us show that

$$(2) \qquad |I|^{s^*} \geq |J_{k,i_1}|^{s^*} + \cdots + |J_{k,i_q}|^{s^*}.$$

Now if $q = 1$, then since $I \supset J_{k,i_1}$, we have $|I|^{s^*} \geq |J_{k,i_1}|^{s^*}$ so that (2) is valid in this case. Suppose $q \geq 2$. Consider the collection $\mathfrak{G}$ of $q-1$ open intervals, each between two consecutive members of $\{J_{k,i_p} : p = 1, \ldots, q\}$. If two members of $\mathfrak{G}$ have the same length then there must be a member of $\mathfrak{G}$ with a greater length. Since the member of $\mathfrak{G}$ are all contained in $[\alpha, \beta]$, there exists a unique member with a maximal length. Let $G = (\alpha', \beta')$ be the member of $\mathfrak{G}$ with the maximal length. Let $F_1 = [\alpha, \alpha']$ and $F_2 = [\beta', \beta]$ so that $\{F_1, G, F_2\}$ is a disjoint collection and $[\alpha, \beta] = F_1 \cup G \cup F_2$. Now G is contained in a closed interval J in the class $\mathfrak{J}_\ell$ for some $\ell \in \mathbb{Z}_+$ with $|G|/|J| = 1 - 2\lambda$. Let J' and J'' be the two closed intervals in the class $\mathfrak{J}_{\ell+1}$ which constitute $J \setminus G$. Then $|J'|/|J| = |J''|/|J| = \lambda$. The fact that G is the unique open interval with maximal length in the collection $\mathfrak{G}$ implies that $F_1 \subset J'$ and $F_2 \subset J''$. Thus $|F_1|/|J| \leq |J'|/|J| = \lambda$. Consequently

$$\frac{|F_1|}{|G|} = \frac{|F_1|/|J|}{|G|/|J|} \leq \frac{\lambda}{1-\lambda}$$

so that $|F_1| \leq \frac{\lambda}{1-\lambda}|G|$. Similarly we have $|F_2| \leq \frac{\lambda}{1-\lambda}|G|$. Thus we have

$$\begin{aligned}
|I| \geq |[\alpha, \beta]| &= |F_1| + |G| + |F_2| \\
&= \left(\frac{\lambda}{1-2\lambda}\right)^{-1}\left\{\frac{\lambda}{1-2\lambda}|F_1| + \frac{\lambda}{1-2\lambda}|G| + \frac{\lambda}{1-2\lambda}|F_2|\right\} \\
&\geq \left(\frac{\lambda}{1-2\lambda}\right)^{-1}\left\{\frac{\lambda}{1-2\lambda}|F_1| + \frac{|F_1| + |F_2|}{2} + \frac{\lambda}{1-2\lambda}|F_2|\right\} \\
&= \left(\frac{\lambda}{1-2\lambda}\right)^{-1}\left\{\frac{\lambda}{1-2\lambda} + \frac{1}{2}\right\}\{|F_1| + |F_2|\} \\
&= \frac{1}{2\lambda}\{|F_1| + |F_2|\}.
\end{aligned}$$

Since $s^* \in (0, 1)$, we have $\left(\frac{1}{2}\{a_1 + a_2\}\right)^{s^*} \geq \frac{1}{2}a_1^{s^*} + \frac{1}{2}a_2^{s^*}$ for $a_1, a_2 \in (0, \infty)$ by (1) of Lemma 29.21. Thus

$$|I|^{s^*} \geq \left\{\frac{|F_1| + |F_2|}{2\lambda}\right\}^{s^*} \geq \frac{|F_1|^{s^*}}{2\lambda^{s^*}} + \frac{|F_2|^{s^*}}{2\lambda^{s^*}}.$$

By (a) of Lemma 29.20, we have $2\lambda^{s^*} = 1$. Thus

$$|I|^{s^*} \geq |F_1|^{s^*} + |F_2|^{s^*}.$$

Applying this process of reduction to F_1 and F_2 in the place of $[\alpha, \beta]$ and repeating, we finally have (2).

2. Let $I_1, \dots, I_N$ be open intervals with endpoints in C_λ^c such that $C_\lambda \subset \bigcup_{j=1}^N I_j$. By dropping those intervals in the collection that are disjoint from C_λ we do not increase the sum $\sum_{j=1}^N |I_j|^{s^*}$. Thus let us assume that none of the open intervals is disjoint from C_λ. By our result in **1**, for each $j = 1, \dots, N$ there exists $k_j \in \mathbb{Z}_+$ such that for the collection $\{J_{k_j,i_p} : p = 1, \dots, q_j\}$ of all members of the class $\mathfrak{J}_{k_j}$ contained in I_j we have according to (2)

$$|I_j|^{s^*} \geq \sum_{p=1}^{q_j} |J_{k_j,i_p}|^{s^*}.$$

Let $k = \max_{j=1,\dots,N} k_j$. By (b) of Lemma 29.20, for each $j = 1, \dots, N$ and $p = 1, \dots, q_j$ we have $|J_{k_j,i_p}|^{s^*} = \sum_{J_{k,i} \subset J_{k_j,i_p}} |J_{k,i}|^{s^*}$. Thus for each $j = 1, \dots, N$, we have

$$|I_j|^{s^*} \geq \sum_{p=1}^{q_j} \sum_{J_{k,i} \subset J_{k_j,i_p}} |J_{k,i}|^{s^*} = \sum_{J_{k,i} \subset I_j} |J_{k,i}|^{s^*}.$$

Then we have

$$(3) \qquad \sum_{j=1}^N |I_j|^{s^*} \geq \sum_{J_{k,i} \subset \bigcup_{j=1}^N I_j} |J_{k,i}|^{s^*} \geq \sum_{i=1}^{2^k} |J_{k,i}|^{s^*} = 1$$

where the equality is by (c) of Lemma 29.20.

3. Let $\delta > 0$. Let $(K_j : j \in \mathbb{N})$ be a δ-cover of C_λ by closed convex sets in $\mathbb{R}$. Recall that a closed convex set in $\mathbb{R}$ is a closed interval. Let $\varepsilon > 0$ be arbitrarily given. Now since C_λ has no interior points, we can select an open interval I_j such that $I_j \supset K_j$, the endpoints of I_j are in C_λ^c, and

$$(4) \qquad \sum_{j \in \mathbb{N}} |I_j|^{s^*} \leq \sum_{j \in \mathbb{N}} |K_j|^{s^*} + \varepsilon.$$

Then $\bigcup_{j \in \mathbb{N}} I_j \supset C_\lambda$. Since C_λ is a compact set, there exists $N \in \mathbb{N}$ such that $\bigcup_{j=1}^N I_j \supset C_\lambda$. Then by (4) and (3), we have

$$\sum_{j \in \mathbb{N}} |K_j|^{s^*} + \varepsilon \geq \sum_{j \in \mathbb{N}} |I_j|^{s^*} \geq \sum_{j=1}^N |I_j|^{s^*} \geq 1.$$

By the arbitrariness of $\varepsilon > 0$, we have $\sum_{j \in \mathbb{N}} |K_j|^{s^*} \geq 1$. Then

$$\mathcal{K}_\delta^{s^*}(C_\lambda) = \inf\left\{\sum_{j \in \mathbb{N}} |K_j|^{s^*} : (K_j : j \in \mathbb{N}) \text{ is a } \delta\text{-cover of } C_\lambda \text{ by closed convex sets}\right\} \geq 1.$$

Since $\mathcal{H}_\delta^{s^*}(C_\lambda) = \mathcal{K}_\delta^{s^*}(C_\lambda)$ by Theorem 27.25, we have $\mathcal{H}_\delta^{s^*}(C_\lambda) \geq 1$. Thus we have $\mathcal{H}^{s^*}(C_\lambda) = \lim_{\delta \downarrow 0} \mathcal{H}_\delta^{s^*}(C_\lambda) \geq 1$. Then since $\mathcal{H}^{s^*}(C_\lambda) \leq 1$ by Theorem 29.22, we have $\mathcal{H}^{s^*}(C_\lambda) = 1$. ∎

Bibliography

[1] L. Ambrosio and P. Tilli, *Selected Topics on "Analysis in Metric Spaces"*, Scuola Normale Superiore, Pisa, 2000.

[2] H. Bauer, *Wahrscheinlichkeitstheorie und Grundzüge der Maßtheorie*, 3rd ed., Walter de Gruyter, Berlin, 1978.

[3] P. Billingsley, *Probability and Measure*, 2nd ed., John Wiley and Sons, New York, 1979.

[4] A. M. Bruckner, J. B. Bruckner and B. S. Thomson, *Real Analysis*, Prentice-Hall, Upper Saddle River, New Jersey, 1997.

[5] R. C. Buck, *Advanced Calculus*, 3rd ed., McGraw-Hill, New York, 1978.

[6] R. M. Dudley, *Real Analysis and Probability*, Wadsworth & Brooks/Cole, 1989.

[7] L. C. Evans and R. F. Gariepy, *Measure Theory and Fine Property of Functions*, CRC Press, Boca Raton, 1992.

[8] K. J. Falconer, *The Geometry of Fractal Sets*, Cambridge University Press, UK, 1985.

[9] G. B. Folland, *Real Analysis*, John Wiley and Sons, New York, 1985.

[10] P. Halmos, *Measure Theory*, Van Nostrand, New York, 1950.

[11] E. Hewitt and K. Stromberg, *Real and Abstract Analysis*, Springer Verlag, New York, 1969.

[12] P. Mattila, *Geometry of Sets and Measures in Euclidean Spaces*, Cambridge University Press, UK, 1995.

[13] I. P. Natanson, *Theorie der Funktionen einer reellen Veränderlichen*, Akademie Verlag, Berlin, 1961.

[14] C. A. Rogers, *Hausdorff Measures*, Cambridge University Press, UK, 1970, 1998.

[15] H. L. Royden, *Real Analysis*, 3rd ed., Macmillan, New York, 1988.

[16] W. Rudin, *Principles of Mathematical Analysis*, 2nd ed., McGraw-Hill, New York, 1964.

[17] W. Rudin, *Real and Complex Analysis,* 3rd ed., McGraw-Hill, New York, 1987.

[18] E. C. Titchmarsh, *The Theory of Functions,* 2nd ed., Oxford University Press, UK, 1939.

Index

C, D, L, O, P, R, and T are abbreviations for Corollary, Definition, Lemma, Observation, Proposition, Remark, and Theorem respectively.

D

E

I

J

L

T